Progress in IS

Progress in IS encompasses the various areas of Information Systems in theory and practice, presenting cutting-edge advances in the field. It is aimed especially at researchers, doctoral students, and advanced practitioners. The series features both research monographs, edited volumes, and conference proceedings that make substantial contributions to our state of knowledge and handbooks and other edited volumes, in which a team of experts is organized by one or more leading authorities to write individual chapters on various aspects of the topic. Individual volumes in this series are supported by a minimum of two external reviews.

The Series is SCOPUS-indexed.

Jorge Marx Gómez · Roopesh Kevin Sungkur ·
Sameerchand Pudaruth · Jan-Hendrik Witte ·
Gerrit Schumann
Editors

Recent Advances in Information Systems

Proceedings of the ICRAIS, Pointe aux Piments, Mauritius, 10–12 September 2025

Springer

Editors
Jorge Marx Gómez
Department of Computing Science
University of Oldenburg
Oldenburg, Germany

Sameerchand Pudaruth
University of Mauritius
Reduit, Mauritius

Gerrit Schumann
University of Oldenburg
Oldenburg, Germany

Roopesh Kevin Sungkur
University of Mauritius
Reduit, Mauritius

Jan-Hendrik Witte
Department for Informatics
University of Oldenburg
Oldenburg, Germany

ISSN 2196-8705 ISSN 2196-8713 (electronic)
Progress in IS
ISBN 978-3-032-14098-2 ISBN 978-3-032-14099-9 (eBook)
https://doi.org/10.1007/978-3-032-14099-9

This Springer imprint is published by the registered company Springer Nature Switzerland AG
The registered company address is: Gewerbestrasse 11, 6330 Cham, Switzerland

Preface

We are pleased to present the preface for the 2025 International Conference on Recent Advances in Information Systems (ICRAIS). The research compiled within this volume showcases the evolution of information systems, increasingly shaped by advancements in artificial intelligence. A central theme is the integration of AI and machine learning not as standalone technologies, but as core components that enhance the design, function, and impact of modern information systems. This is reflected in a broad spectrum of research. Readers will find explorations of foundational architectures for distributed information systems, such as Federated Learning and Retrieval-Augmented Generation (RAG), alongside studies on how these technologies are embedded into systems for complex tasks like financial risk analysis, cyber threat intelligence, and internal auditing. The practical utility of computer vision is also a key focus, with papers demonstrating how it is integrated into systems for industrial inventory management, agricultural monitoring, and enhancing traffic safety. Beyond business applications, the papers investigate critical security and architectural issues, from mitigating modern phishing attacks to understanding the impact of quantum computing on the underlying security of information systems. Furthermore, the scope of this work extends beyond technical implementation to address the role of information systems in tackling pressing societal challenges. Several contributions examine the function of technology in education, addressing both the potential of AI to create new learning paradigms and the persistent challenges of academic integrity. The theme of sustainability is also prominent, with research on AI-driven tools for climate change mitigation and the design of information systems for smart city initiatives. These abstracts reflect the global and collaborative nature of our field, bringing together perspectives on how information systems can support everything from supply chain resilience to responsible AI development in different economic contexts. We hope this collection serves as a valuable reference for attendees and the wider research community, providing insight into current trends and acting as a catalyst for future discussions and collaborations. We are grateful to all authors, reviewers, and session chairs who have contributed to making this conference a success.

Oldenburg, Germany — Jorge Marx Gómez
Reduit, Mauritius — Roopesh Kevin Sungkur
Reduit, Mauritius — Sameerchand Pudaruth
Oldenburg, Germany — Jan-Hendrik Witte
Oldenburg, Germany — Gerrit Schumann

Contents

Part-of-Speech Tagging to Sentence Structure Malagasy 1
Vatosoa Razafindrazaka and Volatiana Marielle Ratianantitra

Optimizing Inventory in Outdoor Warehouses: A Drone-Based Approach Using Photogrammetry and Computer Vision 15
Steffen Meeuw

Real-Time Recognition of Malagasy Banknotes (Ariary) 31
Miandrisoa Hoby Randriatsarafara, Volatiana Marielle Ratianantitra, and Thomas Mahatody

Joint Modeling of Entities and Relations Through Hierarchical Analysis of Span Sequences 47
Jules Michel Andriafetiarison, Onja Kostka Michel Andriamihary, and Volatiana Marielle Ratianantitra

Tax Defense Analytics: A Data-Driven Application for Identifying Cross-Border VAT Fraud and Tax Avoidance Schemes 61
Kristof Mose, Keyvan Narimani, Maged Ahmad, André Klümpen, and Jorge Marx Gómez

Automated Detection and Tracking of Individual Piglets for Real-Time Behavioral Monitoring 71
Jan-Hendrik Witte, Dante-Nicolai Ahrens, Imke Traulsen, and Jorge Marx Gómez

A Comprehensive Comparison of Federated Learning Frameworks 85
Jonas Boekhoff, Jan-Philipp Awick, Lars Steffens, Michael Karl, and Jorge Marx Gómez

Integrating Differential Privacy Mechanisms Within a Simulated Autonomous Vehicle Environment: A Critical Evaluation 101
Thejmeela Seetamonee and Girish Bekaroo

Towards Cybercrime Profiling: The Validity and Reliability of Criminal Profiling Methods 115
Thulani Mashiane and Elmarie Kritzinger

Network Structural Analysis of Digital Technology Adoption for Human-Centric Teaching and Learning Models 133
Anass Bayaga

AI-Powered Interactive Technologies and Cognitive Engagement in Mathematics: A Generational Cohort PLS-SEM and ANN Approach 143
Anass Bayaga

Key Insights on the Use of Quantum Computing in Breaking Cryptographic Algorithms 159
Khemant Nirvaan Degambur and Girish Bekaroo

The Use of Artificial Intelligence Tools to Support Post-graduate Research 173
Andre P. Calitz and Margaret Cullen

A Review of Academic Integrity and Cheating at a Higher Education Institution in South Africa 187
Andre P. Calitz and Margaret Cullen

AI-Powered Digital Twins with Bio-sensory Feedback for Enhanced Predictive Security in Cyber-physical System 203
Kwabena Addo, Katleho Moloi, Musasa Kabeya, and Evans Eshiemogie Ojo

Security of Mobile Devices: A Taxonomy Study of Detection Methods of Pegasus Spyware and Countermeasures 217
Davish Ramlochund and Girish Bekaroo

Optimization and Stabilization of Power Distribution Grids Through the Use of Sensor Networks 235
C. Cress and J. P. van Deventer

Examining the Prevalence of Social Loafing in Agile Methodologies 247
L. Combrinck and J. P. van Deventer

Critical Smart City Drivers for Small and Rural Municipalities 261
Nkhangweni Lawrence Mashau and Jan Hendrik Kroeze

Enhancing User Experience in Smart City Tools: A Design Science Approach to Patent Dashboard Innovation 275
Timothy Musharu and Jorge Marx Gómez

Machine Learning in Public Transport Management: Using Smartphone Tracking Data to Reduce Delays 295
V. Drossopoulos and J. P. van Deventer

A Systematic Literature Review of Reference Architectures in the Blockchain Domain 307
Hauke Precht and André P. Calitz

HERAG: A Holistic Approach for Evaluating Organizational RAG Applications 323
Felix Kleinekathöfer, Felix Dick, Alina Ahrens, Tobias Oltmanns, Steffen Meeuw, René Kessler, and Jorge Marx Gómez

Analyzing Student Gaze Behavior in Classroom Settings: A Computer Vision Approach for Multi-person Attention Detection 341
Florian Voßmann, Steffen Meeuw, Jan-Hendrik Witte, Ines Oldenburg, and Jorge Marx Gómez

Investigating e-Learning in Marginalized Independent Secondary Schools: A Teacher's Perspective 359
Sunet Eybers, Corné J. van Staden, and Alex Zindove

Decentralized Federated Learning Using Transformer-Based Language Models 373
Gerrit Schumann, Carsten Montag, Lars Steffens, Michael Karl, and Jorge Marx Gómez

Secure Data, Smarter Decisions: A Blockchain-Enhanced Federated Learning Framework for the Pharmacogenomics of Depression 391
Elsie Kaaya and Jorge Marx Gómez

Comparing Vertical Federated Learning Frameworks in Manufacturing Supply Chains 403
Jonas Kallisch, Jorge Marx Gómez, and Christoph Wunck

Wearables in the Workforce: The Perceived Value Experienced by South African Computing Professionals 419
Carmen Bekker, Brenda Scholtz, and Simone Beets

AI-Generated Visualizations for Sustainable Urban Planning: Enhancing Decision-Making and Public Engagement in the MobiLe Project Context 435
Sreedhar Kokkarachedu and Jorge Marx Gómez

Automated Screening of Chronic Sinusitis from Voice Recordings Using Machine Learning 445
Thabo Maibi and Leon Eldon Burger

A Conceptual Scale for Attributing Financial Penalties to Privacy Breaches 461
Kweku Arthur and Hein Venter

Large Language Models for Qualitative Risk Analysis in Banking: A Systematic Review 479
Jan-Hendrik Witte and Henry Eggers

Leveraging Large Language Models for Supporting Cyber Threat Analysis 497
Fabian Rybinski, Demian Frister, Gunther Schiefer, and Marzieh Malekzadeh Mahani

Exploring the Potential of Retrieval-Augmented Generation in Internal Audit Practice 511
Maurice Saluschke and Jorge Marx Gómez

Paradigm Shift in Education Using AI in Mauritius—A Global Perspective with Comparative Analysis 529
Dineshen Chuckravanen, Abdallah Peerally, and Nishtee Gopee

Evaluating AI as a Tool for Climate Change Risk Mitigation: A Systematic Literature Review 543
Hannah Brown, Sonali Das, and Marié Hattingh

Heart Attack Risk Prediction Using Machine Learning and Deep Learning Techniques 559
Doorgesh Sookarah, Loovesh Sharma Ramwodin, Manish Putteeraj, and Vivasvansingh Dwarka

Impact of Artificial Intelligence (AI) on Cyber Safety Awareness and Culture in Small and Medium Enterprises (SMEs) 581
Betty Kekana, Elmarie Kritzinger, and Dorothy Scholtz

Deep Learning for Detection and Quantification of (*Bactrocera dorsalis*) in Intelligent Electronic Fruit Fly Traps 593
Florence Lello, Mussa Dida, Elizabeth Mkoba, and Mbazingwa Mkiramweni

Prototype to Provide Environmental Information in a Multi Channel Way by Using Citizen Science and AI 609
Ali Akyol, Malte Schott, and Jorge Marx Gómez

An Analysis of the Value of Robotic Process Automation for Healthcare Administration Using the Task-Technology Fit Theory 625
Phillip Sibanda, Ifeoluwapo Fashoro, and Brenda Scholtz

Hybrid Augmented Intelligence Applied to Technical Trading: Improvements Through Compensatory Fuzzy Logic and WKD 639
Norma P. Rodríguez-Cándido, Jorge Marx-Gómez, and Rafael A. Espin-Andrade

Organizational Values and Practices for Hybrid Agile Teams 651
Roelof Briers and Maureen Tanner

Privacy Challenges with Human Imposed and AI-Based Filters During Virtual Meetings 667
Rooma Ramasamy, Vinaye Armoogum, and Perienen Appavoo

A Systematic Literature Review on Multimodal Abstractive Text Summarisation Using Transformer-Based and Attention Models 681
Tchouya'a Ngoko Israel Christian and Boniface Kabaso

The Impact of the Prompt Language on Code Generation with LLMs: An Experimental Study 695
Kai Meier, Christian Koot, Marc Fernandes, and Andreas Theissler

AI-Enhanced Monitoring and Failure Prevention in Distributed Systems: A Systematic Literature Review 711
Mike Tyolani, Boniface Kabaso, and Waldon Hendricks

Bridging the Gap Between Curricula and the Labour Market: The Role of Predictive Models in Shaping Educational Relevance 725
Yolisa Luwaca, Boniface Kabaso, and Waldon Hendricks

Energy-Efficient Cloud Computing for Artificial Intelligence Training: Challenges, Techniques, and Future Directions 743
Fayez Salma and Jorge Marx Gómez

A Systematic Review of Privacy-Preserving Synthetic Network Traffic Generation for Machine Learning-Based Intrusion Detection Systems 755
Kalala Gael Kapia, Boniface Kabaso, and Waldon Hendricks

Mitigating Phishing Attacks: Current Strategies, Emerging Trends, and Future Directions 769
Tebogo Mashabane, Boniface Kabaso, and Waldon Hendricks

AI Implementation in African Development: The Contextual Necessity Framework 781
Errol Francke

Value Stream Analysis and Design with Large Multimodal Models: Evaluating the Impact of Complex Value Stream Graph Networks 799
Mick Geisthardt, Lutz Engel, and Jorge Marx Gómez

Development and Implementation of a Low-Cost IoT Network for Air Quality Monitoring in the Historic Downtown of San Salvador 813
Omar Otoniel Flores-Cortez, Carlos Pocasangre, Fernando Arévalo, Jose Alberto Oviedo, and Renata Barriere Garcia

Validating a Framework for Assessing Small and Rural Municipalities' Readiness for Smart City Implementation 829
N. L. Mashau and J. H. Kroeze

Preliminary Results of the AI Supported Detection of Cyclists and Related Overtaking Situations 845
Johannes Schering, Maximilian Bosse, Fatih Karahan, Jorge Marx Gómez, Daniel Mentrup, and Johannes Wilbers

Integrating Entrepreneurial and Sustainability Competences for Responsible AI Engagement in Diverse Work Fields 863
Ádám Lászlop and Jantje Halberstadt

Generative AI in OCR: A Case Study for License Plate Recognition 875
Annita Ngatchou Weiss, Lars Guggolz, Tim Dahmen, and Christian Koot

Sustainability-First Microservice Design: Extending GranMicro for Efficient Microservice Extraction 891
Ola Mustafa and Jorge Marx Gómez

Creating Robot Game Mission Strategies Using Machine Learning 907
Elton Strydom, Bertram Haskins, and Ronald Leppan

MaaS-Broker Implementation for an Intermodal Journey Planner Combining Public Transit with Carpooling Using Time-Expanded Graph Networks 919
Sasapu Venkatesh and Jorge Marx Gómez

An Exploration of the Impact of AI Capabilities on Supply Chain Resilience in Mauritius 933
Loovesh Sharma Ramwodin and Doorgesh Sookarah

Structured Extraction and Hierarchical Parsing of Addresses Using Large Language Models 949
Ashish Dhansukhbhai Desai, Keyvan Narimani, Jorge Marx Gómez, and André Klümpen

Part-of-Speech Tagging to Sentence Structure Malagasy

Vatosoa Razafindrazaka and Volatiana Marielle Ratianantitra

Abstract This paper explores the field of Malagasy language processing, focusing on morphosyntactic tagging and the identification of specific syntactic structures in Malagasy. We use recurrent neural network models, specifically Bidirectional Long Short-Term Memory (BiLSTM), enhanced with embedding techniques and dense layers for learning. Applying these advanced methods in our linguistic context presents a major challenge due to the specific characteristics of the Malagasy language, notably its complex syntactic structure and morphological richness. We developed and trained each module using the Fitsipika corpus, enriching it with advanced preprocessing techniques. The data was prepared and structured from CSV files to ensure consistency and result quality. Experimental results confirmed the robustness of our approach, achieving an overall accuracy of 97% for morphosyntactic tagging and scores of 85 and 69% for the two models analyzing the language's syntactic structure, demonstrating its effectiveness in linguistic analysis. We also explore the possibility of extracting FKI from long sentences. These findings outperform existing methods and open the door to numerous potential applications in Malagasy language processing, such as automatic text comprehension and the development of learning assistance tools.

1 Introduction

The cultural and linguistic richness of Madagascar is reflected in its principal language, the official Malagasy (OM) (Ratianantitra et al., 2022), spoken by approximately 29 million speakers in 2023. Research on natural language processing applied to the Malagasy language faces major challenges, including the lack of work compared to other languages and the complexity of the 18 Malagasy dialects (Ratianantitra, 2023). However, an in-depth understanding of this language poses unique

V. Razafindrazaka (✉) · V. M. Ratianantitra
Laboratory of Computer Science and Mathematics Applied to Development (LIMAD), University of Fianarantsoa, Fianarantsoa, Madagascar
e-mail: razafindrazakav@gmail.com

J. Marx Gómez et al. (eds.), *Recent Advances in Information Systems*, Progress in IS,
https://doi.org/10.1007/978-3-032-14099-9_1

challenges, particularly due to its complex morphological structure and linguistic particularities (Duong, 2017).

Previous natural language processing research has established a foundation for understanding Malagasy's linguistic characteristics. Studies on automatic summarization (Ratianantitra et al., 2019) and part-of-speech tagging in the Malagasy (Nambinintsoa et al., 2025) offer valuable insights into its linguistic features and processing challenges. These studies have laid a solid foundation for understanding the specific linguistic characteristics of Malagasy and the challenges associated with its automatic processing. Despite these advances, the syntactic analysis of Malagasy remains a relatively unexplored area, particularly concerning morphosyntactic tagging and sentence structure.

Our study aims to fill this gap by focusing on the syntactic aspect of Malagasy. Syntactic analysis deals with the correct formation of sentences, examining the structural role of each word in the sentence and the relationships between them. Sentence structure involves understanding the arrangement of each word in a sentence, while morphosyntactic tagging deals with the relationship between each word. Both components aim to assign appropriate labels to each word or group of words. Thus, sentence structure and morphosyntactic tagging are integral parts of syntactic analysis for the Malagasy language.

The following sections will delve into detail about our approaches to addressing these challenges. We will begin with morphosyntactic tagging and its techniques in Sect. 2, then turn to sentence structure and existing methods compared to our proposed approach in Sect. 3. Next, Sect. 4 will focus on a thorough analysis of our approach compared to all existing methodologies to develop improved methods. We will then address experimentation and evaluation in Sect. 5, concluding our paper in Sect. 6.

2 Part-of-Speech Tagging of Malagasy

The Malagasy language lacks grammatical gender, number, and case inflection, with words remaining invariant regardless of their function in a sentence (Linda & Rabenilaina, 2008). There are generally a few grammatical categories such as Matoanteny (verb), Anarana (noun), Mpamaritra (adjective), Mpisolo (pronoun), Mpanoritra (article), Mpampiankina (preposition), Tambinteny (adverb), Mpampitohy (conjunction), Tenim-piontanana (interjection), and Kianteny (particle) (Rajemisa-Raolison, 1969).

Morphosyntactic tagging methods are crucial for labeling words with grammatical information and can be categorized into traditional and modern approaches (Adda et al., 1999; Valli & Veronis, 1999). There are two types of morphosyntactic tagging approaches: traditional and modern (Ghoul, 2011).

2.1 Rule-Based Tagger

Rule-based taggers rely on grammatical or morphological principles to assign tags to words or define transitions between different tags (Ghoul, 2011). For example, the word "dia" in Malagasy can be categorized as Kianteny, Mpampitohy, or Anarana. A contextual rule could be established to identify Kianteny in a sentence. For example, if no word precedes "dia" (Rabenilaina, 1991), then "dia" is labeled as Mpampitohy.

- **Sentence**: 'Dia mamaky boky izy.' (And he/she reads a book.)
- **Tagging**: 'Dia' (conjunction)—'mamaky' (verb)—'boky' (noun)—'izy' (pronoun)
- **Tagging (Malagasy)**: 'Dia' (Mpampitohy)—'mamaky' (Matoanteny)—'boky' (Anarana)—'izy' (Mpisolo).

2.2 Statistical Tagger

Statistical tagging utilizes probabilistic models to assign part-of-speech tags to each word in a text, based on frequencies and probability calculations (Manning & Schütze, 1999). There are two main approaches:

- The first approach assigns the most frequent tags from a reference text to each word, ensuring consistent tagging.
- The second approach, based on n-grams, calculates the probability of tag sequences and uses algorithms like the Viterbi algorithm to determine the most probable tag sequence for a given series of words.

The Hidden Markov Model (HMM) is an advanced method that combines the probability of tag sequences and tag frequencies for each word. It treats the text as a sequence of hidden states and uses algorithms like the forward/backward or Baum–Welch algorithm to find the most probable tag sequence (Ghoul, 2011), using the following equation:

$$P(X|Y) = \frac{\text{freq}(X, Y)}{\text{freq}(Y)} \tag{1}$$

In this context, Nambinintsoa et al. (2025) makes a significant contribution by proposing a specific approach to Malagasy language for supervised part-of-speech tagging. This contribution is essential because the Malagasy language exhibits linguistic peculiarities that require specific approaches for morphosyntactic analysis. The work involved constructing a tailored tag set for Malagasy grammar, utilizing an annotated corpus for training, and applying various models such as unigram, bigram, and HMM for tagging.

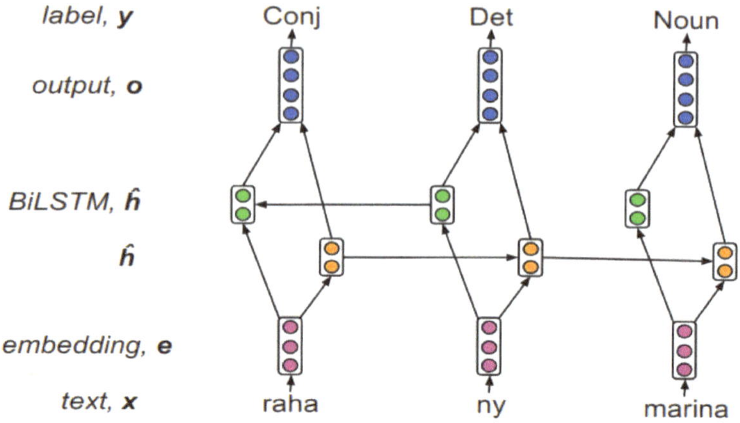

Fig. 1 Illustration of BiLSTM architecture Fang and Cohn (2016)

2.3 *Neural Network-Based Tagger*

Neural network-based tagging is an advanced NLP method that uses artificial neural networks to perform tasks like part-of-speech tagging. This approach excels at capturing complex and non-linear patterns, enhancing performance in linguistic tasks.

Magistry (Magistry et al., 2018) also contributed to this technique in their article on part-of-speech tagging for low-resource languages like Malagasy. In their research, the authors employed a neural network-based approach, specifically Bidirectional Long Short-Term Memory (BiLSTM) models. BiLSTM networks are recurrent neural networks (RNNs) that can leverage both past and future information in a sequence of words, making them effective at capturing long-term dependencies in text.

Similarly, the paper by Fang and Cohn (2016) proposes another method to enhance grammatical part-of-speech tagging in low-resource languages using cross-lingual projection. This method leverages both manually annotated data and data projected from a resource-rich language, thus improving grammatical tagging accuracy in low-resource languages (Fig. 1).

3 Malagasy Sentence Structure

Sentence structure generally refers to how sentences are constructed. It is crucial for understanding and analyzing a language. A profound understanding of sentence structure relies on morphosyntactic tagging, a key element of linguistic analysis.

Although this discipline is well-established in the field of NLP, comprehensive and accessible works in the Malagasy language remain very rare despite previous research. In Malagasy, sentences mainly follow a Verb-Object-Subject (VOS) order or "Enti-milaza Fameno Lazaina" (EFL) in Malagasy sentence construction, which differs from more common structures like Subject-Verb-Object (SVO) or Subject-Object-Verb (SOV) (Blake, 1986; Paul, 2019). This syntactic specificity requires careful analysis to understand the meaning and syntax of sentences in this language.

3.1 Rule-Based Parsing

Rule-based methods utilize sets of predefined grammatical rules to analyze sentence structure. These rules are developed by linguists and grammar experts. While effective for languages with regular syntactic structures, they may encounter difficulties with the complexity and variability of natural languages.

Within the framework of the Penn Treebank (Taylor et al., 2003), sentence structure analysis is based on a method using linguistic rules. This approach defines rules to describe the grammatical structure of sentences, then applies them to parse the sentence into its syntactic constituents, such as phrases, noun phrases, verb phrases. The representation of syntactic structure is done using bracketing notation, where parentheses are added to indicate relationships between constituents. This representation provides a structured and formal view of sentence syntax in a text corpus (Taylor et al., 2003).

(P (NP-S-1 Chris) (V wants (S (NP-S *-1) (V to throw (NP the ball)))) .

In this structure, "Chris" is the subject of the predicate "wants" and also the subject of the action "to throw", while "the ball" is the object of the action "to throw".

To visualize how words bind in a language, it's helpful to use syntax trees. A syntax tree is a tree that represents the syntax of an object. Attached is an example of a syntax tree in English, Fig. 2.

Figures 2 and 3, the syntax tree, depicts how words in a sentence are organized. These trees offer a clear representation of the verb, object, and subject positions. However, sometimes these tools need adjustments to fully reflect the complexity of these particular linguistic structures.

3.2 Statistical Parsing

Statistical methods used in structural sentence analysis rely on probabilistic models to determine the most likely structure of a sentence. This approach is based on techniques such as Hidden Markov Models (HMMs), language models, and supervised or semi-supervised machine learning methods. These methods offer increased flexibility compared to rule- based approaches, allowing them to better adapt to the variability of natural languages.

In Collins' paper (1997), several statistical methods were employed to enhance the syntactic analysis of sentences. The three proposed models are based on strong

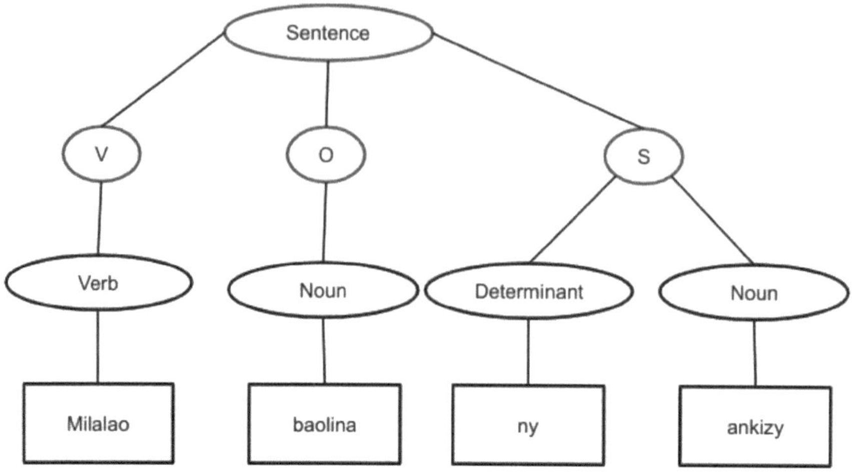

Fig. 2 Syntax tree in English

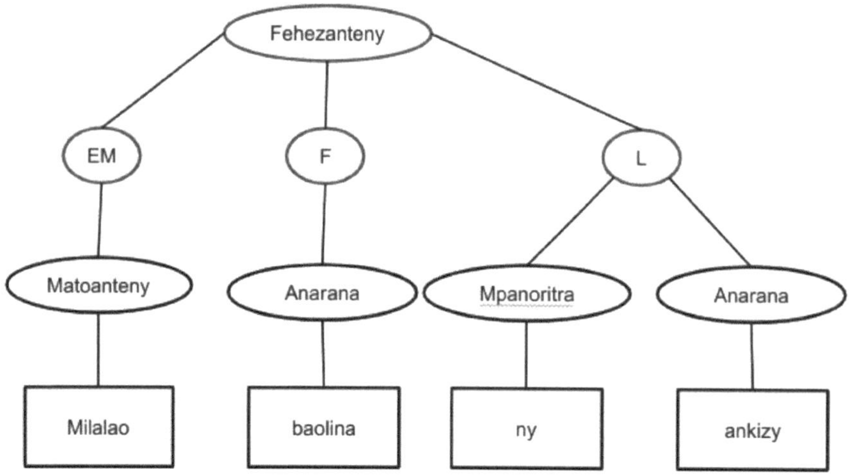

Fig. 3 Syntax tree in Malagasy

mathematical foundations, including the use of conditional probabilities to estimate the structure of parsing trees. For example, in the Probabilistic Context-Free Grammar (PCFG) model, the probability of a parsing tree is calculated by multiplying the probabilities of the context-free rewriting rules used to construct the tree, according to the following equation:

$$P(T, S) = \prod_{i=n}^{n} P(\mathrm{RHS}_i|\mathrm{LHS}_i) \tag{2}$$

This equation allows quantifying the probability of a given parsing tree given a sentence, taking into account the probabilistic syntactic structure of the language.

Similarly, the models 2 and 3 proposed in the article utilize advanced probabilistic techniques to improve the accuracy of syntactic analysis. For instance, model 2 introduces probabilities on left and right subcategorization frames to better distinguish complements and adjuncts in sentence structure. These probabilities are used to estimate the likelihood of different alternative syntactic structures, thereby enabling the selection of the most probable structure.

3.3 Parsing Based on Deep Learning

Deep learning techniques have revolutionized the field of syntactic analysis by enabling the development of models based on neural networks. These models can capture complex syntactic structures and effectively handle languages with varied word orders. Analyzing the structure of a sentence has allowed for the treatment of more nuanced linguistic constructions, at the level of the Malagasy language, such as reducing a long sentence.

In the paper by Coavoux and Crabbe (2015), the authors explore three main models for sentence structure analysis: multinomial logistic regression, shallow neural network, and deep neural network. Each model proposes a different approach to represent and analyze linguistic symbols, ranging from sparse vector representations to dense representations and non-linear interactions between symbols (Chai, 2019). These approaches improve the models' ability to capture the nuances and subtleties of language, which can be particularly useful for languages like Malagasy, where linguistic constructions can be complex. By leveraging recent advances in deep learning, these models open up new perspectives for automatic natural language processing and understanding of less-studied languages.

4 Proposed Approach

Our approach is structured into two distinct modules to address the specificities of natural language processing in the Malagasy linguistic context: Part-of-speech Tagging and EFL Structure (Enti-milaza Fameno Lazaina). In the EFL structure, we need the first module to retrieve the voice of each sentence, which is then used to determine the EFL structure and finally the FKI (shortest sentence or "Fehezanteny Kely Indrindra"). Indeed, a sentence can be:

- "Tso-drafitra": unmarked structure, active voice
- "Mivadi-drafitra": marked structure, passive voice.

Each module is decomposed into three phases: preprocessing, processing, prediction (Fig. 4).

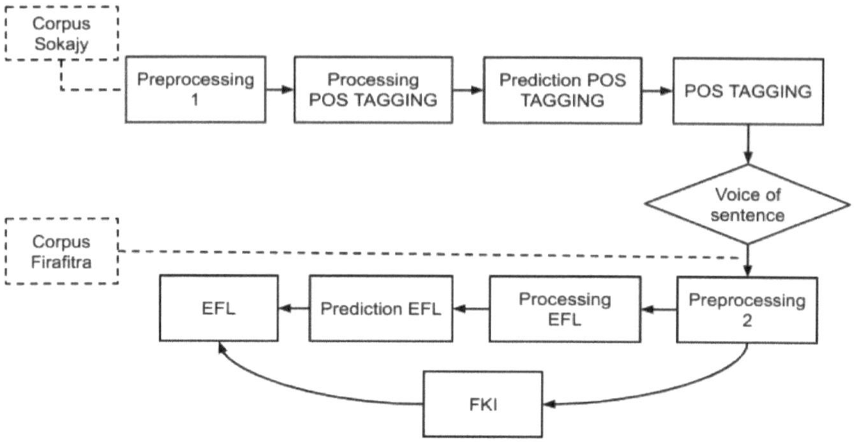

Fig. 4 Process of the proposed approach

4.1 Preprocessing

During the preprocessing phase, raw data is optimized for use in the training phase of the model. This process includes:

- Tokenization: Dividing text into discrete units called tokens.

  - Splitting sentences into sequences of tokens.
  - Removing special characters to facilitate subsequent tasks.
  - Removing stopwords or empty words (e.g., "ny", "no") concerning the second module.

- Data splitting: After tokenization, the datasets are divided into training and testing sets, where textual data is divided into x for input features and y for labels.
- Encoding: Process of transforming textual data into numerical representations, usually in the form of vectors. Word sequences (token sequences) are transformed into sequences of integers to be understood by the model. Thus, each tokenized word is converted into a unique integer.
- Padding sequence: The encoded integer sequences are then standardized by padding to ensure a constant sequence length.

4.2 Processing

The processing phase delves into the heart of the analysis of two modules. It comprises several steps.

- Model construction: It relies on using BiLSTM layers to capture sequential dependencies in the data. The BiLSTM model used consists of three main layers:
  - Embedding layer: transforms sequences of integers (encoded and padded tokens) into embedding vectors of a specified dimension (embedding space dimensionality fixed at 64).
  - BiLSTM layer: Comprising 128 neurons as the number of BiLSTM units, this layer returns sequences for each token. It consists of forget gate, input gate, output gate, cell candidate, cell state, output.
  - Dense layer: This layer is fully connected with softmax activation to predict the probability of each class.
- Model compilation: It includes the use of:
  - Loss function 'Categorical Crossentropy'
  - Optimizer 'adam'
  - Class Weights.
- Model training: The model is trained on the training set for 50 epochs with a batch size of 32. This phase is essential to prepare the model to make accurate predictions on new data.

4.3 Prediction

Once the model is trained, it can be used to make predictions on new data. In this phase, the model takes a new sentence as input, performs the same preprocessing process, and then uses the trained model to predict the morphosyntactic tagging for each token in the sentence.

- POS Tagging: Eventually, the voice of a sentence (Tso-drafitra or Mivadi-drafitra) can be determined.
- EFL Structure: FKI (Fehezanteny Kely Indrindra) is obtained.

5 Experimentation and Evaluation

5.1 Experimentation

We assembled a corpus named "Fitsipika" for our experimentation, composed of a variety of reliable linguistic documents such as the Tenymalagasy website, Régis

Table 1 Corpus characteristics

POS tagging and EFL		
Name of corpus: Fitsipika	Sokajy	Firafitra
Number of words	5953	1210

Rajemisa Raolison's book, texts from the Malagasy Bible, as well as articles from the Global Voices website dedicated to the Malagasy language and the Malagasy book "Lovako."

This corpus was selected for its representativeness and diversity, covering a wide range of styles and topics. The "Fitsipika" corpus is organized into sentences, with each sentence representing a unit of analysis. This allows us to accurately capture the morphosyntactic nuances and structure of the Malagasy language. We standardized the data to ensure consistency in the corpus by converting documents from different sources into a standardized format (Table 1).

5.2 *Evaluation Method*

To evaluate the performance of our system, we used standard measures with its corpus sets such as F_1-score, accuracy, and loss. Our approach achieved an overall precision score of 97%, indicating high reliability in detecting morphosyntactic tagging and 85–69% for the two EFL structure models of the Malagasy language.

- Loss function: It is calculated using the categorical cross-entropy loss function, which measures the error between the actual labels and the model predictions.

$$L(y_{\text{true}}, y_{\text{pred}}) = -\frac{1}{N} \sum \sum y \cdot \log(y_{\text{pred},i,j}) \tag{3}$$

- Confusion matrix: The confusion matrix is a table that shows the number of true positives, true negatives, false positives, and false negatives of the model. It allows evaluating the performance of a classification model by comparing its predictions to the actual values.

$$C = \begin{bmatrix} \text{TP FN} \\ \text{FP TN} \end{bmatrix} \tag{4}$$

- F_1-Score: The F_1-score is a measure of a classification model's accuracy that combines both precision and recall. It is particularly useful when classes are imbalanced, meaning some classes have many more examples than others.

$$F_1\text{-score} = \frac{2 \times \text{precision} \times \text{recall}}{\text{precision} + \text{recall}} \tag{5}$$

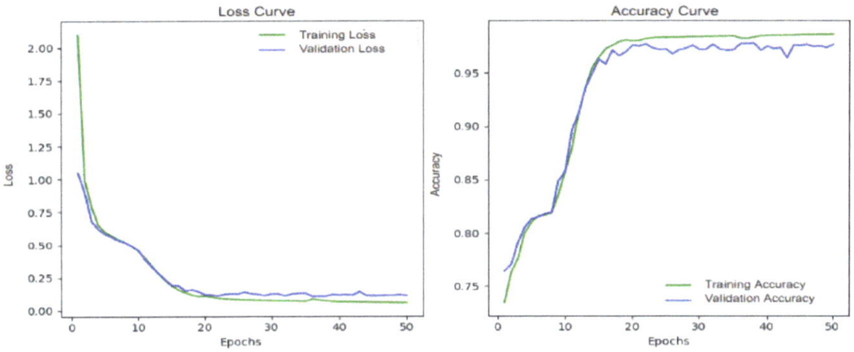

Fig. 5 Loss and precision curve of POS tagging

– Precision

$$\text{Precision} = \frac{\text{Number of correctly classified samples}}{\text{Total number of classified samples for this class}} \quad (6)$$

– Recall

$$\text{Recall} = \frac{\text{Number of correctly classified samples}}{\text{Total number of samples belonging to this class}} \quad (7)$$

On this test dataset, our model for Malagasy morphosyntactic tagging achieved an impressive precision of 97.69%. Precision is a measure of the proportion of correct predictions among all predictions made by the model. In other words, on the entire test set, our model correctly predicted the morphosyntactic tags of words in over 97% of cases, Fig. 5.

For the EFL structure, we divided it into two structures: Tso-drafitra Structure and Mivadi-drafitra Structure, and they were evaluated separately on a separate test dataset. On this test set, the Tso-drafitra model achieved a precision of 85.00%, meaning it correctly identified EFL structures in over 85% of cases. The F1-score, which combines precision and recall, was 0.8078, indicating an overall solid performance of the model. However, for the Mivadi-drafitra structure, we only have 69.23% precision and a 0.587 F_1- score.

To evaluate the effectiveness of our approach, we compare it with several state-of-the-art methods from the literature. As shown in Table 2, our combined rule-based and BiLSTM architecture achieves higher accuracy on both POS tagging and sentence structure tasks, particularly on the Tso-drafitra dataset (Fig. 6).

Table 2 Performance comparison on POS tagging and sentence structure

	Approach	Model	Avg. accuracy (%)	F_1 score (%)
POS tagging	Nambinintsoa et al.	Rules + statistics	70.19	–
	Magistry et al.	BiLSTM	91.00	–
	Fang and Cohn	BiLSTM + correction	86.30	–
	Zhou et al.	Weakly supervised BiLSTM	84.50	–
	Our approach	Rules + BiLSTM	**97.69**	**97.65**
Sentence structure	Taylor et al.	Rule-based	90.00	–
	Collins	Statistical	88.43	–
	Coavoux and Crabbé	Neural Networks	89.98	–
	Our approach	Rules + BiLSTM (Tso-drafitra)	**85.00**	**80.78**
	Our approach	Rules + BiLSTM (Mivadi-drafitra)	**69.23**	**58.70**

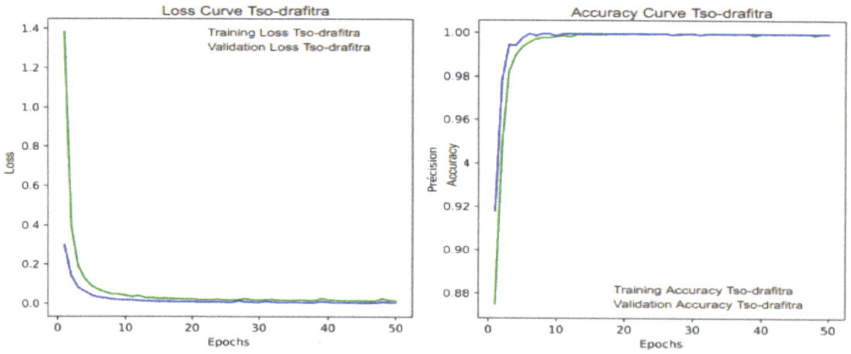

Fig. 6 Loss and precision curve of EFL Tso-drafitra structure

6 Conclusion

This paper represents a significant advancement in the field of Malagasy natural language processing. We have contributed to the development of approaches, utilizing BiLSTM models, for all tasks. These results pave the way for potential applications in various domains, from automatic language comprehension to facilitating access to information in Malagasy.

This research in Malagasy natural language processing has aimed to push the boundaries of linguistic understanding and the application of advanced technologies in the specific context of the Malagasy language. Our investigation focused on two key aspects: morphosyntactic tagging and the Enti-milaza Fameno Lazaina (EFL) structure, both operating at the sentence level. Lastly, our research addressed the reduction of longer sentences or FKI (Fehezanteny Kely Indrindra), an essential aspect for condensing information contained in Malagasy sentences.
This step has proven to be a significant contribution to automatic summarization in the context of Malagasy sentences, thereby facilitating quick and accurate understanding of linguistic content.

References

Adda, G., Mariani, J., Paroubek, P., Rajman, M., & Lecomte, J. (1999). L'action GRACE d'évaluation de l'assignation des parties du discours pour le français. *Langages,2*(2), 119–129.

Blake, B. B., Review of: Tomlin, R. S. (1988). Basic word order: Functional principles. Croom Helm, London (1986). *Journal of Linguistic, 24*(1), 213–217. https://doi.org/10.1017/S0022226700011646

Chai, J., & Li, A. (2019). *Deep learning in natural language processing: A state-of-the-art survey* (pp. 1–6). IEEE. https://doi.org/10.1109/ICMLC48188.2019.8949185

Coavoux, M., & Crabbé, B. (2015). Comparaison d'architectures neuronales pour l'analyse syntaxique en constituants. In *Proceedings of the 22nd TALN Conference* (pp. 291–302). ATALA. https://aclanthology.org/2015.jeptalnrecital-long.25

Collins, M. (1997). Three generative, lexicalised models for statistical parsing. arXiv preprint. http://arxiv.org/abs/cmp-lg/9706022

Duong, L. (2017). *Natural language processing for resource-poor languages* (Ph.D. thesis, University of Melbourne). https://minerva-access.unimelb.edu.au/bitstream/handle/11343/192938/thesis.pdf

Fang, M., & Cohn, T. (2016). Learning when to trust distant supervision: An application to low-resource POS tagging using cross-lingual projection. arXiv preprint. http://arxiv.org/abs/1607.01133

Ghoul, D. (2011). *Outils génériques pour l'étiquetage morphosyntaxique de la langue arabe: Segmentation et corpus d'entraînement* (Master thesis).

Linda, J., & Rabenilaina, R. B. (2008). *Étude contrastive des phrases complexes en français et en malgache: Cas des complétives* (Master thesis, Université d'Antananarivo). http://biblio.univ-antananarivo.mg/pdfs/joujouLinda_FR_M2_08.pdf

Magistry, P., Ligozat, A. L., & Rosset, S. (2018). Étiquetage en parties du discours de langues peu dotées par spécialisation des plongements lexicaux. In *Conférence sur le Traitement Automatique des Langues Naturelles (TALN)* (pp. 1–8). https://hal.science/hal-01793092/

Manning, C. D., & Schütze, H. (1999). *Foundations of Statistical Natural Language Processing*. MIT Press.

Nambinintsoa, R. A., Falimanana, R., Harisoa, R. N., & Matio, R. Natural language processing: Malagasy part-of-speech tagging. Technical report.

Paul, I. (2019). The syntax of Malagasy auxiliaries: Initial observations. *JSEALS, 60*.

Rabenilaina, R. B. (1991). Voix et diathèse en malgache. *Lingvisticæ Investigationes,15*(2), 325–335. https://doi.org/10.1075/li.15.2.05rab

Rajemisa-Raolison, R. (1969). *Grammaire malgache: enseignement secondaire, 6ème–3ème, enseignement général*. Antananarivo: Centre de Formation Pédagogique.

Ratianantitra, V. M. (2023). A state of the art review on natural language processing applied to the Malagasy language. In *International Conference on Artificial Intelligence and its Applications* (pp. 1–5). https://mauricon.org/conferences/index.php/icarti/article/view/35

Ratianantitra, V. M., Razafindramintsa, J. L., Mahatody, T., Rasoamalalavao, C., & Manantsoa, V. (2022). Malagasy abstractive text summarization using scheduled sampling model. In *2nd International Conference on Industry 4.0 and Artificial Intelligence (ICIAI 2021)* (pp. 6–9). Atlantis Press. https://doi.org/10.2991/iciai-21.2022.2

Ratianantitra, V. M., Razafindramintsa, J. L., Mahatody, T., & Manantsoa, V. (2019). Deep learning approach for Malagasy text summarization. *International Journal of Conceptions on Computing and Information Technology,7*(1), 12–17.

Taylor, A., Marcus, M., & Santorini, B. (2003). The Penn treebank: An overview. In A. Abeillé (Ed.) *Treebanks, text, speech and language technology* (Vol. 20, pp. 5–22). Springer. https://doi.org/10.1007/978-94-010-0201-1_1

Valli, A., & Véronis, J. (1999). Étiquetage grammatical des corpus de parole: problèmes et perspectives. *Revue FranÇaise de Linguistique Appliquèe,4*(2), 113–133.

Optimizing Inventory in Outdoor Warehouses: A Drone-Based Approach Using Photogrammetry and Computer Vision

Steffen Meeuw

Abstract Efficient inventory management is critical for modern supply chain operations, yet traditional manual inventory processes in warehouses remain labor-intensive, often prone to human error due to their repetitive nature, and costly. While drone-based inventory auditing has demonstrated effectiveness in indoor warehouses, its application to outdoor warehouses remains underexplored. Unique challenges such as the absence of fixed orientation points and the diverse nature of stored products hinder straightforward adoption. This paper addresses these challenges by introducing a novel drone-based inventory system specifically designed for outdoor storage environments, where products are stored on handling units such as pallets. To ensure broad applicability, an abstract spatial representation of outdoor warehouses is developed. The proposed system integrates photogrammetry and computer vision techniques to capture aerial and front-view images, enabling automated inventory recognition and digital warehouse modeling. A real-world field experiment compares the drone-based inventory approach with traditional manual counting, demonstrating substantial improvements in efficiency, accuracy, and workforce optimization. The results indicate reduced labor dependency, enhanced digital traceability, and improved operational cost efficiency. While technical challenges affecting detection accuracy persist, this study contributes to scalable, automated inventory solutions, advancing supply chain digitalization in outdoor warehouse management.

1 Introduction

The digitalization and automation of logistics processes have become increasingly important in modern Supply Chain Management (SCM) (Najatet al., 2024; Sharakhin et al., 2022). As companies strive to enhance operational efficiency and reduce costs, the adoption of advanced technologies incorporating Artificial Intelligence (AI), has gained traction across various domains (El Imrani, 2024; Niranjan et al., 2021;

S. Meeuw (✉)
Department of Computing Science, Carl von Ossietzky Universität Oldenburg, Oldenburg, Germany
e-mail: steffen.meeuw@uni-oldenburg.de

J. Marx Gómez et al. (eds.), *Recent Advances in Information Systems*, Progress in IS,
https://doi.org/10.1007/978-3-032-14099-9_2

Zhou, 2022). These technological advancements offer substantial opportunities for optimization in inventory management, a crucial aspect of SCM (Petheet al., 2024).

One particularly labor-intensive and error-prone task in warehouse operations is inventory auditing, which traditionally relies on manual counting. This process is time-consuming, costly, and prone to human error (Olugbenga Ayomide et al., 2024; Vamsi et al., 2020). Consequently, many companies limit inventory audits to legally mandated inspections, such as those required under §240 HG in Germany (Handelsgesetzbuch, 2025). However, infrequent stock evaluations prevent organizations from leveraging the benefits of real-time inventory visibility, such as increased transparency, operational improvements, and early detection of inefficiencies (Kessler et al., 2021).

Recent research has explored drone-based inventory auditing to automate warehouse operations, demonstrating significant potential in indoor and outdoor environments (Jhunjhunwala et al., 2019; Kessler et al., 2021; Malang et al., 2023; Ribeiro et al., 2022). Despite the existence of methodologies for conducting inventories with the assistance of drones, these methodologies are not without their limitations. Approaches developed for indoor warehouses are not universally applicable to outdoor use cases. Outdoor approaches typically only consider the measurement of bulk goods with drones or assume that the stored products can be marked with Radio Frequency Identification (RFID) tags.

This paper analyzes the requirements for a drone-based inventory system for outdoor warehouses, where products are stored on handling units (e.g., pallets) (Sect. 2), and introduces a novel system built upon an abstract representation of outdoor warehouse environments (Sects. 3 and 4). This system enables a visual end-to-end inventory process, which is evaluated through a field experiment comparing its performance with manual inventory counting by warehouse employees under real-world conditions (Sect. 5). The findings are then discussed, limitations are identified, and directions for future research are outlined (Sect. 6).

2 Requirements for Drone-Based Inventory in Outdoor Warehouses

Effective drone-based inventory management in outdoor warehouses presents unique challenges that differ significantly from indoor environments. Unlike indoor warehouses, where static structures such as shelves facilitate drone navigation and localization, outdoor storage lack fixed orientation points. Additionally, the diverse nature of stored products often prevents the use of conventional identification techniques such as RFID tags or barcodes.

To systematically derive the requirements for a drone-based inventory system in outdoor warehouses, expert interviews were conducted with industry partners from logistics and warehouse management. These interviews provided practical insights into operational constraints, challenges, and expectations for drone-based inventory

Table 1 Requirements for drone-based inventory in outdoor warehouses

Req.	Description
1	**Cost Efficiency and Workforce Reduction**: Automate inventory counting to minimize labor costs and reduce manual inspections. Optimize operational expenses by increasing audit efficiency and reducing verification time
2	**Digital Traceability and Auditability**: Generate a digital inventory record for post-audit validation and historical stock comparison. Ensure integration with WMS for retrospective verification
3	**Highly Automated Data Collection**: Enable automated flight path execution, image capture, and data processing with minimal human intervention
4	**Digital Warehouse Representation**: Generate an accurate digital model of the storage yard to enable inventory verification and retrospective stock analysis
5	**Detection and Counting of Pallets**: Accurately detect and count standardized full-pallet inventory. Support optical recognition techniques for classification within storage zones
6	**Flexibility for Dynamic Layouts**: Allow adaptive inventory zone definitions to accommodate modifications in warehouse layouts and operational requirements

solutions. The resulting requirements were iteratively refined using a Design Science Research (DSR) approach (Hevner et al., 2004), ensuring both industrial relevance and scientific rigor.

Table 1 summarizes the key requirements identified for an effective drone-based inventory system in outdoor warehouses, encompassing operational aspects like cost efficiency, technical capabilities for automated data collection, and system characteristics such as digital traceability and layout flexibility. As highlighted in this table, inventorying large outdoor areas with closely stacked products poses significant challenges regarding accuracy and efficiency. Interviews with industry experts revealed that manual counting errors increase over time due to the repetitive nature of the task, leading to discrepancies between recorded stock levels and actual inventory. Resolving such inconsistencies often requires time-intensive re-counting by warehouse staff, generating additional costs.

To address these challenges, this research builds upon the previously introduced DSR methodology (Hevner et al., 2004). Through iterative cycles of problem identification, artifact development, demonstration, and evaluation, the proposed drone-based inventory system was designed to meet the identified requirements and enhance inventory management in outdoor storage environments.

3 Related Work

Based on the previously defined requirements for a drone-based inventory system in outdoor storage environments, we conducted a literature review to analyze existing research approaches. Our analysis revealed that none of the reviewed studies fully satisfy all the identified requirements (see Table 2). To ensure comprehensive coverage, we considered both indoor and outdoor warehouse approaches in our review, as certain techniques from indoor solutions might potentially be adaptable to outdoor environments. The comparative analysis helped identify both applicable methods and remaining gaps that our proposed system needs to address.

The reviewed studies can be categorized into indoor (Cristiani et al., 2020; Stankoet al., 2022) and outdoor approaches (Alsayed et al., 2021; Baeet al., 2016; Kessler et al., 2021). In indoor scenarios, drone navigation is primarily automated, leveraging static structures such as high-bay racks for orientation. In contrast, outdoor environments employ both automated mapping approaches and manual drone flights operated by human pilots. Additionally, while photogrammetry techniques contribute to the digital representation of storage environments, their application in outdoor settings is predominantly limited to the measurement of bulk materials.

Our review reveals that drone-based inventory management in outdoor warehouses remains less explored compared to indoor warehouse inventory systems. Photogrammetry-based approaches demonstrate potential for creating digital representations of storage areas. However, to date, no studies have successfully applied this approach to piece goods (e.g., pallets) inventory management. This gap underscores the need for our comprehensive solution that addresses the entire inventory process-from defining the storage environment through automated data acquisition to digital representation of results.

Table 2 Comparison of drone-based warehouse inventory approaches

Reference	Req. 1	Req. 2	Req. 3	Req. 4	Req. 5	Req. 6
Cristiani et al. (2020)	✓	✗	✓	✗	✓	✗
Alsayed et al. (2021)	✓	✗	✓	✓	✗	✗
Bae et al. (2016)	✓	✓	✗	✗	✗	✗
Kessler et al. (2021)	✓	✗	✗	✗	✓	✗
Stanko et al. (2022)	✓	✗	✓	✗	✓	✗
Our contribution	✓	✓	✓	✓	✓	✓

4 Method: Automated Outdoor Warehouse Inventory

In the next section, we first introduce our abstract outdoor warehouse representation. Our literature review in Sect. 3 revealed that photogrammetry-based methods show the most potential for meeting our inventory requirements. Building on this finding, we propose a spatial description that ensures the transferability of our approach to different warehouse topologies. Based on our description model, we then implement the drone-based inventory system in an outdoor warehouse of an industrial partner, creating a prototype that serves as the foundation for our evaluation in Sect. 5.

4.1 Abstract Warehouse Representation

Accurate inventory management using photogrammetry and computer vision techniques in outdoor warehouses depends on a precise spatial representation of storage locations and inventory units. Unlike indoor warehouses with fixed racking systems, outdoor storage environments exhibit dynamic layouts where storage positions are determined by operational needs rather than permanent structures. To ensure systematic tracking, a structured geospatial model is essential, allowing inventory units to be assigned to predefined locations within a georeferenced warehouse area. The warehouse environment can be conceptualized as a hierarchical spatial system comprising four fundamental elements, starting from the smallest unit and building up to the complete warehouse area (Fig. 1):

- **Inventory Units** ($\mathcal{U}$)—The physical stored objects (e.g., pallets or piece goods), varying in height and volume, representing the basic elements to be inventoried.
- **Storage Locations** ($\mathcal{L}$)—Individual georeferenced inventory positions that contain multiple inventory units, ensuring systematic stock allocation.
- **Storage Areas** ($\mathcal{A}$)—Logical groupings of multiple storage locations based on material types, accessibility, or operational requirements.
- **Warehouse Area** ($\mathcal{W}$)—The total outdoor storage yard, defined as a bounded geospatial region encompassing all storage areas.

To support automated inventory verification, all elements within the warehouse environment are georeferenced. Starting with the most basic element, the set of

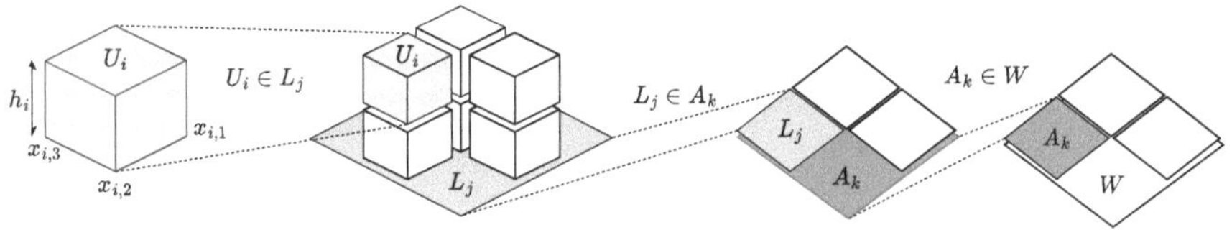

Fig. 1 Schematic visualization of the spatial representation of outdoor warehouses

inventory units ($\mathcal{U}$) represents the actual stored goods. These units, such as pallets or piece goods, vary in shape, height, and volume and are characterized by their geometry:

$$U_i = \{\{x_{i,1}, x_{i,2}, \ldots, x_{i,m}\}, h_i\}$$

where $x_{i,u} \in (\mathbb{R}, \mathbb{R})$ with $u \in \{1, \ldots, m\}$ represent the corner points of the unit and $h_i \in \mathbb{R}_{>0}$ denotes the height of the inventory stack.

The set of **storage locations** ($\mathcal{L}$) provides individual inventory positions that contain these units. Each location is represented as a polygon defined by its geospatial corner points.

An inventory unit is considered to be located within a storage location if all its corner points lie within the storage location's boundary.

These storage locations are grouped into the set of **storage areas** ($\mathcal{A}$), which functions as logical subdivisions based on operational requirements:

$$\mathcal{A} = \{A_1, A_2, \ldots, A_k\}$$

where each storage area contains multiple storage locations. Finally, the **warehouse area** ($\mathcal{W}$) represents the entire outdoor storage yard, defined as a geospatial polygon with known boundary coordinates.

Given the unstructured nature of outdoor storage, inventory units are often stacked in dense formations with varying heights, which presents additional challenges for visibility and automated detection. The spatial mapping approach described above enables different warehouse topologies to be represented in the real world, with flexibility to model organizational units according to the characteristics of specific outdoor warehouses.

4.2 Application of Photogrammetry and Computer Vision for Visual Warehouse Representation and Inventory Calculation

Building on the previously introduced abstract representation of an outdoor warehouse, the following section outlines the procedure for inventory determination using photogrammetry and computer vision. This approach leverages drone images captured from multiple perspectives (see Fig. 2). The objective is to generate spatial representations of the warehouse area and storage locations (Top Views) while also mapping frontal views of individual storage locations (Front Views) to facilitate inventory classification.

To represent the spatial relationships between the considered warehouse objects ($\mathcal{W}$, $\mathcal{A}$, $\mathcal{L}$, $\mathcal{U}$), we employ photogrammetry to generate three artifacts:

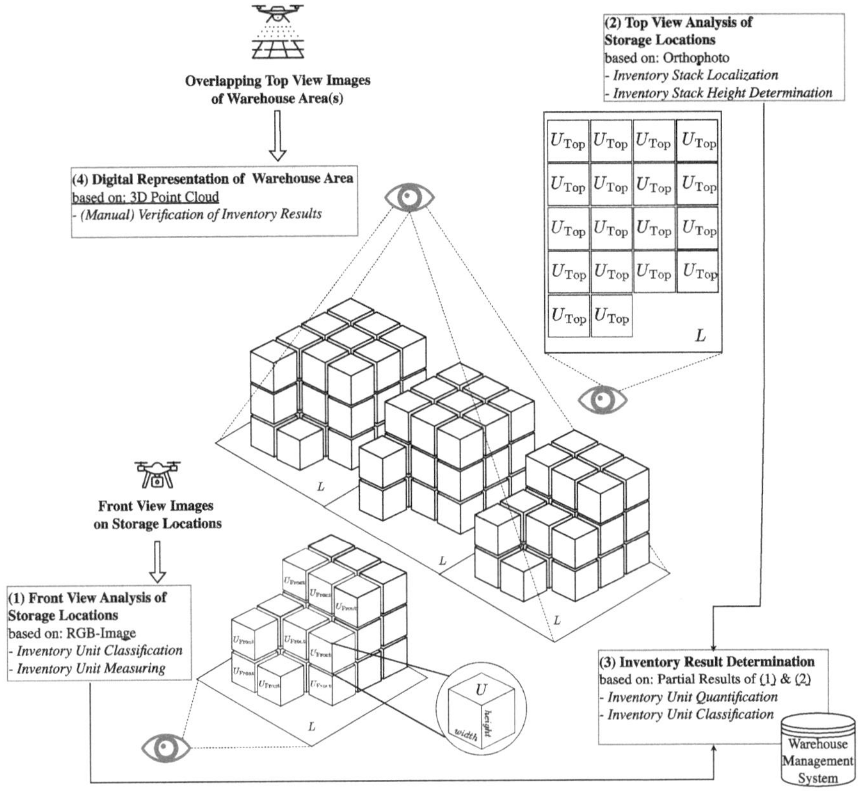

Fig. 2 Schematic representation of the drone-based inventory process using photogrammetry and computer vision

- **Orthophoto**—A distortion-free, scale-accurate aerial representation of the warehouse area ($\mathcal{W}$), generated from overlapping drone images using photogrammetric processing. By combining geometric accuracy with photorealism, orthophotos enable precise inventory localization and detection of storage units (e.g., pallet stacks) through top-view object recognition (Barazzetti et al., 2014).
- **Digital Surface Model (DSM)**—A geometric representation of the warehouse area with elevation data, where heights are stored as raster pixels. Unlike Digital Terrain Models (DTMs), DSM include elevated features such as inventory stacks, storage units, and infrastructure, enabling accurate height estimation and enhanced spatial analysis (Rogers et al., 2020).
- **3D Point Cloud**—A dense digital representation of the warehouse area at the time of inventory acquisition, generated through automated image matching. It describes surface geometry (DSM) and undergoes interpolation and texturing for visualization. Adaptive point density ensures edge accuracy while reducing redun-

dancy in flat areas. This dataset supports (manual) inventory verification and further spatial analyses (Remondino et al., 2012).

The interaction of the individual artifacts, their analysis and thus the drone-based inventory process is described in the following. The annotated steps of Fig. 2 are referenced for this purpose.

(1) **Front View Analysis of Storage Locations for Inventory Unit Classification and Measuring** RGB images captured by a drone serve as input for analyzing the front views of storage locations ($\mathcal{L}$). The drone is positioned centrally in front of the target storage location, maintaining a fixed distance from its boundary. Since each storage location contains homogeneous inventory units ($\mathcal{U}$), the objective of this process is to identify a distinctive characteristic of at least one inventory unit. While the unique classification of the inventory unit type allows for the derivation of an inventory unit's height, the primary goal of this process is to determine the exact height of the inventory unit. This height information is crucial for analyzing the top view of a storage location, enabling a more accurate spatial representation and inventory assessment. Importantly, even if an inventory unit cannot be clearly identified based on its distinguishing features, a quantitative evaluation remains possible through height measurement.

This step leverages the classification of an inventory unit to determine its height. This can be achieved, for example, by referencing the material master data from a connected warehouse management system. Depending on the characteristics of the inventory units, classification can be performed in various ways:

- **Optical Classification**: If inventory units can be visually distinguished based on characteristic features such as shape, color, or size, object classification approaches can be applied. However, this requires that a clear classification based on inventory unit properties is feasible. If individual inventory units are too similar, alternative identification methods must be used.
- **Reading Label Information**: If inventory units have attached identifying features (e.g., material numbers, Quick Response (QR) codes, or barcodes), these can be used for classification or height determination.
- **Height Determination via Load Carrier Measurement**: When classification via optical features or labels is not feasible, visual measurement provides an alternative. Standardized load carriers (e.g., Euro pallets) serve as fixed-width reference objects. Height is estimated by applying object recognition to an inventory unit and analyzing the width-to-height ratio of its bounding box.

(2) **Top-View Analysis of Storage Locations for Inventory Stack Localization and Height Determination** This sub-process identifies, localizes, and measures stacked inventory units from a top-down perspective. Localization is based on an orthophoto generated from overlapping drone images, providing a comprehensive view of storage locations and inventory units.

Object recognition detects stacked units and determines their Global Positioning System (GPS) position, which is then used to assign height values derived from the DSM. Additionally, pallet stack localization allows spatial assignment to storage areas, enabling more detailed inventory analysis in the next step.

(3) **Inventory Result Determination** Following the previous steps, storage location data now includes inventory classification, individual unit heights, and stack quantities. Since inventory counting is based on individual units, the final step calculates the total number of handling units by dividing each stack's height by the height of a single unit. The output is a detailed inventory list with classified storage locations and corresponding quantities.

(4) **Digital Representation of Warehouse Area** To verify inventory results, a manual post-inspection of storage areas can be conducted. Alongside the orthophoto and DSM, a high-resolution 3D point cloud is generated from overlapping top-view images, providing a 360° view of the warehouse at the time of inventory execution. Its high resolution enables navigation to any storage location for manual optical recounts, adding an extra verification layer for accuracy.

4.3 Implementation: Use Case of Block Storage Warehouse

After presenting the fundamental functionality of our drone-based inventory approach for outdoor warehouses in the previous section, we now demonstrate its implementation in a block storage warehouse from an industry partner. This partner manufactures and stores concrete products on homogeneous full pallets, with the corner points of the storage locations already recorded in the WMS. As a result, no prior measurement of storage boundaries was required. The following explanation of the implementation is structured according to Fig. 3, which is modeled after Fig. 2 to ensure consistency.

(1) **Front View Analysis of Storage Locations for Inventory Unit Classification and Measuring** To capture front-view images of storage locations, we used a DJI Mini 3 Pro drone, which was autonomously controlled via waypoints using the DroneLink[1] app. Since the stored products (concrete goods) cannot be uniquely classified based on shape or color, a material number recognition approach was implemented. By applying Optical Character Recognition (OCR),[2] we extracted material numbers from the inventory units. This enabled us to determine the height of individual inventory units using the material master data from the connected WMS. As a fallback—if material number recognition was not possible—the height of an individual pallet was estimated based on the bounding box ratio. Since all products are stored on Euro pallets, these pallets served as reference objects for an approximate height determination.

[1] https://www.dronelink.com/, last access on 02/25/2025.

[2] https://github.com/Ucas-HaoranWei/GOT-OCR2.0, last access on 02/25/2025.

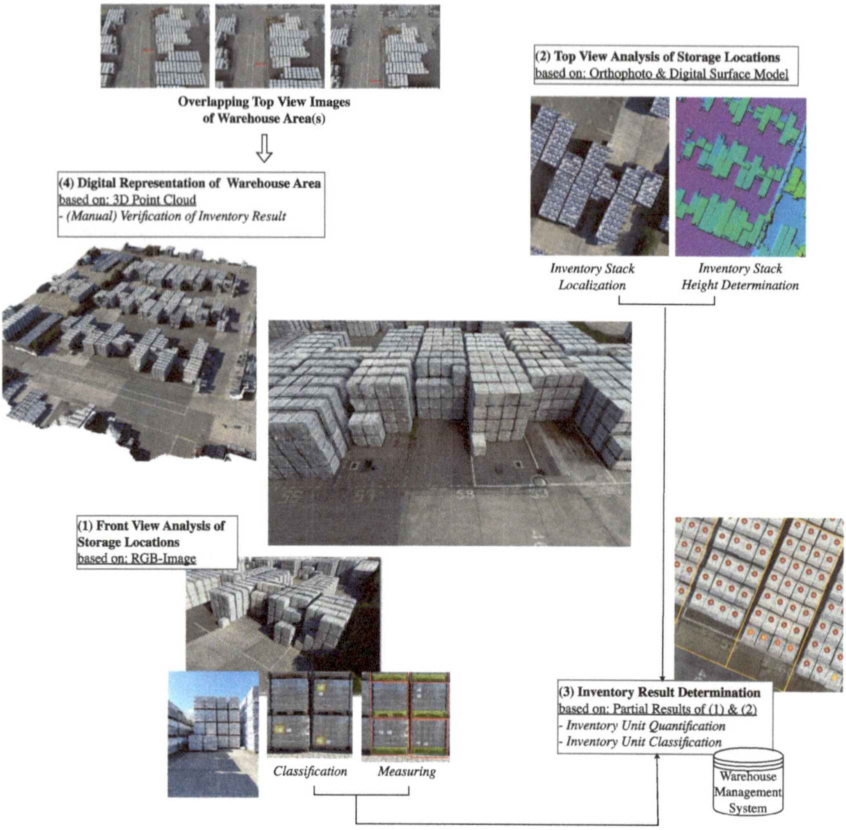

Fig. 3 Visualization of the drone-based inventory process for a block storage warehouse with concrete products (in accordance to Fig. 2)

(2) **Top-View Analysis of Storage Locations for Inventory Stack Localization and Height Determination** To capture the overlapping top-view images, we used the previously mentioned DJI Mini 3 Pro along with the DroneLink mapping software. As indicated by the red markings in the selected three input images (see Fig. 3, top), we applied an 80% front and side overlap to ensure optimal coverage. For the generation of the orthophoto, DSM, and 3D point cloud, we utilized the open-source framework OpenDroneMap.[3] Subsequently, pallet stack localization was performed using a YOLOv8-OBB model,[4] enabling the height determination of pallet stacks based on the DSM.

(3) **Inventory Result Determination and Visualization** By combining the height information of a single pallet-either obtained through a lookup in the material

[3] https://www.opendronemap.org/, last access on 02/25/2025.

[4] https://github.com/meiqisheng/YOLOv8-obb, last access on 02/25/2025.

master data of the connected WMS or estimated using the Euro pallet as a reference object with fixed dimensions—a quantitative determination of storage units can be achieved. To visualize the counting results, we marked the detected pallet stacks on the orthophoto and annotated each marking with the corresponding stack height (see Fig. 3, bottom right).

(4) **Digital Representation of Warehouse Area** For result verification, the 3D point cloud generated using OpenDroneMap serves as a key resource (see Fig. 3, center left). This high-resolution point cloud enables a 360° inspection of the inventoried warehouse area at the time of stocktaking. If discrepancies arise between the inventory recorded in the WMS and the computer vision-based inventory count, the affected storage locations can be reviewed. Due to its high resolution (see Fig. 3, center), the point cloud allows for digital post-inspection from multiple perspectives, providing an additional layer of verification.

5 Real-World Evaluation: Comparing Manual and Automated Inventory Counting

To evaluate our approach and assess compliance with **Req. 1: Cost Efficiency and Workforce Reduction** (see Sect. 2), we conducted a comparative inventory assessment. A designated warehouse area with 72 storage locations, arranged in 6 rows of 12, was selected. To simulate a realistic inventory scenario, 7 storage locations were intentionally left empty.

The manual inventory count was performed by two warehouse employees, followed by an automated count using the drone-based system. During the drone-based inventory, a single trained employee supervised the flight to ensure proper execution but did not intervene in data collection. After the automated image capture, the same employee initiated the computer vision-based inventory analysis, verifying results and addressing potential discrepancies. Time measurements were recorded for both methods, and the results were compared. The following section presents the findings of this comparative assessment, structured into three key aspects: (1) Material Classification Accuracy, (2) Quantitative Determination per Storage Location, and (3) Time Efficiency.

5.1 Evaluation of Material Classification Accuracy

For the comparative inventory assessment, a GPS waypoint was defined in advance for each of the 72 storage locations, positioned 1.5 m centrally in front of the respective storage location. The drone maintained a flight altitude of 1.5 m. To ensure high-resolution image capture and minimize overlap with adjacent storage locations, the gimbal of the DJI Mini 3 Pro was set to a vertical position. All flight parameters were preconfigured and stored as a mission plan before execution, allowing the operator to simply initiate the mission and supervise the flight without manual intervention.

Out of the 65 occupied storage locations, the material number was successfully extracted from 31 sotrage locations using the OCR-based material classification, resulting in a detection rate of 47.69%. The average distance at which the material number could be accurately extracted was 2.40 m, determined by calculating the GPS distance between the image capture position and the nearest recognized pallet stack. For the remaining 34 storage locations, where OCR-based material recognition was unsuccessful, inventory unit height determination was performed using the pallet as a reference object for measurement.

5.2 Evaluation of Quantitative Determination

The accuracy of the drone-based inventory was evaluated by comparing the manual inventory count with the automated drone-based count. Deviations in pallet counts per storage location were analyzed and visualized using a heat map (see Fig. 4), which illustrates discrepancies between the drone-based and manual inventory results. The analysis revealed that all deviations resulted from undercounting, meaning that in no instance did the drone-based inventory overestimate the number of pallets present. This indicates a systematic tendency for the automated approach to miss pallets rather than to introduce false positives. Out of 1304 pallet stacks present in the warehouse area, 1259 were successfully detected by the drone-based system, resulting in a detection rate of 96.50%.

To further investigate the causes of these discrepancies, a detailed analysis was conducted on the affected storage locations. The primary factor contributing to these errors was shadowing, as illustrated in Fig. 5. It was observed that shadows cast by higher pallet stacks onto adjacent smaller stacks reduced their visibility, impairing accurate identification. The most significant deviation occurred at row 2, location 4, where the partial shading of an entire row led to a substantial undercount. This case underscores the critical impact of lighting conditions and pallet positioning on inventory accuracy. Smaller stacks located near larger stacks were particularly prone to miscounts, as their outlines were obscured by shadows in the orthophoto, making them more difficult to detect.

5.3 Comparison of Time Efficiency and Workforce Utilization

To assess the efficiency of the drone-based inventory approach, we compared the total time required for inventory counting using the drone system versus a manual inventory conducted by a two-person counting team. The breakdown of the time

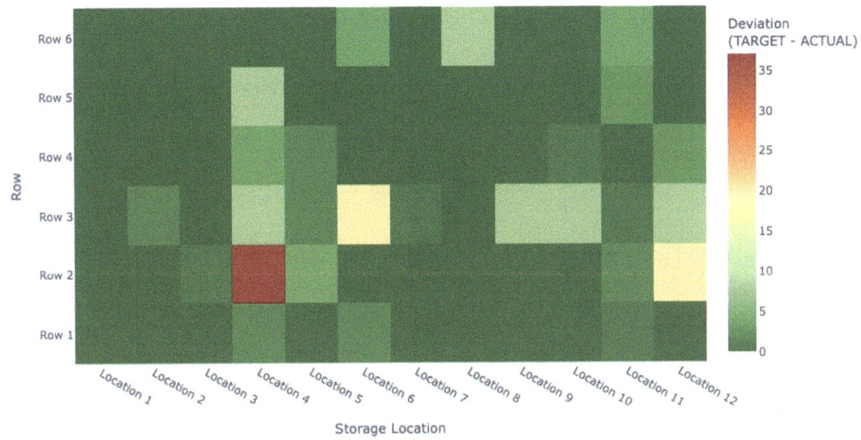

Fig. 4 Visualization of deviations per storage location

Fig. 5 (Partial) Shading of pallet stack as a reason for quantity deviations at storage location level

required for each step is presented in Table 3, where tasks requiring active personnel involvement are marked with (*), and automated processes are marked with (**).

The comparison shows that the drone-based inventory required a total of 56 min, while the manual inventory conducted by two employees required a total workforce effort of 168 min (84 min per person). A key advantage of the drone-based approach is that only 25 min involved active personnel engagement, whereas the remaining 31 min were fully automated. This means that more than half of the total process duration required no manual intervention, allowing employees to focus on other tasks in parallel. Additionally, the automated photogrammetry processing and image analysis accounted for a significant portion of the drone-based workflow. These steps, which include generating the orthophoto, DSM, and point cloud, as well as the automated classification and inventory determination, do not require active monitoring. In contrast, the manual inventory approach necessitated continuous personnel

Table 3 Time comparison of drone-based versus manual inventory

Drone-based inventory	
Task	**Time required (min)**
Drone setup & disassembly (*)	10
Image capture (front and top views) (*)	9
Image transfer to system (*)	6
Generation of photogrammetry artifacts (**)	20
Analysis of front and top views (**)	10
Inventory result determination (**)	1
Total time (single operator)	**56**
Manual inventory (two-person team)	
Task	**Time required (min)**
Total counting time per person (*)	84
Total workforce effort (84 min × 2 persons)	**168**

involvement for the entire duration. Overall, the results demonstrate that the drone-based inventory approach significantly reduces workforce dependency and manual workload, making it a more efficient alternative to traditional inventory counting methods.

6 Conclusion and Future Work

In this paper, we introduced a drone-based inventory methodology for outdoor warehouses, integrating photogrammetry and computer vision to enhance automation in inventory audits. By leveraging automated flight paths, structured geospatial modeling, and object recognition, the system significantly reduces manual effort while improving accuracy and digital traceability. A real-world case study demonstrated that the proposed system requires only 25 min of human intervention compared to 168 min for manual counting, fulfilling Req. 1 (Cost Efficiency and Workforce Reduction). Furthermore, the system successfully detected and quantified homogeneous products on full pallets, meeting Req. 5 (Detection and Counting of Homogeneous Products), while the generated digital warehouse representation supports retrospective verification (Req. 2: Digital Traceability and Auditability).

Despite these advancements, several challenges remain. The primary limitation identified was shadowing effects, occasionally leading to undercounting of inventory stacks. While overall accuracy remained high, refining object recognition algorithms and optimizing flight paths could further mitigate these issues. Additionally, environmental factors such as lighting variations and occlusions may impact detection performance, necessitating adaptive processing techniques to enhance robustness. The static geospatial model used in this study effectively structured inventory tracking but may need to incorporate dynamic adaptation mechanisms to handle varying warehouse layouts and storage conditions.

Future research should focus on improving image processing algorithms to address shadowing effects and environmental influences. Additionally, integrating real-time inventory discrepancy detection with WMSs could enhance operational efficiency by enabling continuous updates. Another key area for improvement is refining classification techniques for non-standardized storage units, allowing the system to handle a broader range of product types. Further optimization of flight path planning could reduce data acquisition time and improve coverage consistency. By addressing these challenges, drone-based inventory solutions can enable more frequent, automated stock assessments, reduce dependency on manual counting, and allow employees to focus on higher-value tasks. Ultimately, this research contributes to the scalability and digitalization of inventory management, bridging a critical gap in outdoor warehouse automation and advancing supply chain efficiency.

References

Alsayed, A., Yunusa-Kaltungo, A., Quinn, M. K., Arvin, F., & Nabawy, M. R. A. (2021). Drone-assisted confined space inspection and stockpile volume estimation. *Remote Sensing, 13*(17), 3356. https://doi.org/10.3390/rs13173356

Bae, S. M., Han, K. H., Cha, C. N., & Lee, H. Y. (2016). Development of inventory checking system based on UAV and RFID in open storage yard (pp. 1–2). https://doi.org/10.1109/ICISSEC.2016.7885849

Barazzetti, L., Brumana, R., Oreni, D., Previtali, M., & Roncoroni, F. (2014). True-orthophoto generation from UAV images: Implementation of a combined photogrammetric and computer vision approach. *ISPRS Annals of the Photogrammetry, Remote Sensing and Spatial Information Sciences, II-5*, 57–63 (2014). https://doi.org/10.5194/isprsannals-II-5-57-2014

Cristiani, D., Bottonelli, F., Trotta, A., & Di Felice, M. (2020). Inventory management through mini-drones: Architecture and proof-of-concept implementation (pp. 317–322). https://doi.org/10.1109/WoWMoM49955.2020.00060

El Imrani, O. (2024). Optimization of port logistics performance through innovative technology management. In *2024 IEEE 15th International Colloquium on Logistics and Supply Chain Management (LOGISTIQUA)*. ISSN 2166-7373. https://doi.org/10.1109/LOGISTIQUA61063.2024.10571510

Handelsgesetzbuch (HGB) (2025) §240 Inventur

Hevner, A., & Chatterjee, S. (2004). Design science in information systems research. *Management Information Systems Quarterly, 28*, 75.

Jhunjhunwala, P., Shriya, M., & Rufus, E. (2019). Development of hardware based inventory management system using UAV and RFID (pp. 1–5). https://doi.org/10.1109/ViTECoN.2019.8899488

Kessler, R., Melching, C., Goehrs, R., & Gómez, J. M. (2021). Using camera-drones and artificial intelligence to automate warehouse inventory.

Malang, C., Charoenkwan, P., & Wudhikarn, R. (2023). Implementation and critical factors of unmanned aerial vehicle (UAV) in warehouse management: A systematic literature review. *Drones, 7*(2), 80. https://doi.org/10.3390/drones7020080

Najat, T., & Alaa Eddine, E. M.: Digitalization and business automation for an effective supply chain integration: A literature review. In *2024 IEEE 15th International Colloquium on Logistics and Supply Chain Management (LOGISTIQUA)* (pp. 1–7). ISSN 2166-7373. https://doi.org/10.1109/LOGISTIQUA61063.2024.10571518

Niranjan, K., Narayana, K. S., & Rao, M. N. (2021). Role of artificial intelligence in logistics and supply chain. In *2021 International Conference on Computer Communication and Informatics (ICCCI)* (pp. 1–3). ISSN 2329-7190. https://doi.org/10.1109/ICCCI50826.2021.9402625

Olugbenga Ayomide, M., Abosede, D. O., Gabriel, A. K., Akintoba Emmanuel, A., Timothy, & Olawale, A. (2024). A framework for an IoT-enabled intelligent inventory management system. *International Journal of Advances in Engineering and Management, 6*(10), 234–243. https://doi.org/10.35629/5252-0610234243

Pethe, S., Sahu, A., Kodarlikar, S., & Vamshidhar, M. (2024). IoT research in supply chain management and logistics: Real-time asset tracking and inventory management (pp. 1–5). https://doi.org/10.1109/ICICET59348.2024.10616272

Remondino, F., Barazzetti, L., Nex, F., Scaioni, M., & Sarazzi, D. (2012). UAV photogrammetry for mapping and 3D modeling – current status and future perspectives. *The International Archives of the Photogrammetry, Remote Sensing and Spatial Information Sciences, XXXVIII-1-C22*, 25–31. https://doi.org/10.5194/isprsarchives-XXXVIII-1-C22-25-2011

Ribeiro, C. C. G., dos Santos, L. H. M. C., & Macharet, D. G. (2022). Collaborative UGV/UAV path planning for inventory management in warehouses (pp. 121–126). https://doi.org/10.1109/LARS/SBR/WRE56824.2022.9995748

Rogers, S. R., Manning, I., & Livingstone, W. (2020). Comparing the spatial accuracy of digital surface models from four unoccupied aerial systems: Photogrammetry versus LiDAR. *Remote Sensing, 12*(17), 2806. https://doi.org/10.3390/rs12172806

Sharakhin, P. S., Levchenko, A. V., & Kaminskiy, S. S. (2022). Automation of routine information processes in supply chain: Search methodology and tool, evaluation of potential based on lean manufacturing (pp. 278–283). https://doi.org/10.1109/ITQMIS56172.2022.9976798

Stanko, J., Stec, F., Palkovic, L., Rodina, J., & Rau, D. (2022). Towards automatic inventory checking using an autonomous unmanned aerial vehicle (pp. 1–8). https://doi.org/10.1109/ETFA52439.2022.9921460

Vamsi, A. M., Deepalakshmi, P., Nagaraj, P., Awasthi, A., & Raj, A. (2020). IoT based autonomous inventory management for warehouses. In A. Haldorai, A. Ramu, S. Mohanram, & C. C. Onn (Eds.), *EAI International Conference on Big Data Innovation for Sustainable Cognitive Computing* (pp. 371–376). Springer International Publishing. https://doi.org/10.1007/978-3-030-19562-5_37

Zhou, H. (2022). Applications of computer internet of things technologies in logistics. In *2022 Second International Conference on Advanced Technologies in Intelligent Control, Environment, Computing & Communication Engineering (ICATIECE)* (pp. 1–5). https://doi.org/10.1109/ICATIECE56365.2022.10047270

Real-Time Recognition of Malagasy Banknotes (Ariary)

Miandrisoa Hoby Randriatsarafara, Volatiana Marielle Ratianantitra, and Thomas Mahatody

Abstract This research focuses on real-time recognition of Malagasy banknotes to introduce an innovative solution benefiting the financial sector. The main goal is to develop an automatic system for banknote recognition, applicable in Automated Teller Machines (ATMs) and mobile devices for visually impaired individuals. The study begins with an analysis of real-time object recognition models, such as Faster Region-based Convolutional Neural Network (Faster RCNN), Single Shot MultiBox Detector (SSD), and You Only Look Once (YOLO). Yolov8 is selected due to its superior accuracy, especially for small objects, leveraging transfer learning. Model performance is enhanced through hyperparameter tuning and improved data preprocessing, including padding for image standardization and instance normalization to reduce class biases. Evaluation is conducted on Android mobile phones, simulating real-world conditions. Results confirm that preprocessing improvements significantly enhance model accuracy and stability, ensuring reliable recognition of Malagasy banknotes. This research successfully develops a real-time recognition system using Yolov8, providing an effective solution for financial transactions and inclusive access for visually impaired individuals. Future work includes extending recognition to other currencies, counterfeit detection, and integration into diverse financial devices.

M. H. Randriatsarafara (✉) · V. M. Ratianantitra · T. Mahatody
Laboratory of Computer Science and Mathematics Applied to Development (LIMAD), Doctoral School of Computer Modeling (EDMI), University of Fianarantsoa, Fianarantsoa, Madagascar
e-mail: miandrisoahobyr@gmail.com

V. M. Ratianantitra
e-mail: volatianamarielle@yahoo.fr

T. Mahatody
e-mail: tsmahatody@gmail.com

J. Marx Gómez et al. (eds.), *Recent Advances in Information Systems*, Progress in IS,
https://doi.org/10.1007/978-3-032-14099-9_3

1 Introduction

Money is a universal medium of exchange, essential to economic transactions around the world. In Madagascar, the Ariary plays a central role in the national economy. However, the manual handling of banknotes leads to significant inefficiencies: slow transaction times, counting errors, and risks of misidentification. These limitations underscore the need for technological solutions in monetary management.

This research aims to address these challenges by leveraging recent advances in artificial intelligence and computer vision, particularly in object recognition. We propose a system capable of identifying Malagasy banknote denominations in real time, under varying conditions and image quality.

Beyond the technical contribution, this study seeks to modernize monetary handling, optimize human resources, and promote technological innovation. Although the focus is on Malagasy currency, the proposed methodology is adaptable to other currencies and contexts. With proper retraining and dataset adjustment, our approach could be scaled to support automated currency recognition more broadly and enhance financial inclusion.

The remainder of this paper is organized as follows: Sect. 2 reviews related work, Sect. 3 describes the methodology, Sect. 4 outlines the implementation, Sect. 5 discusses experimentation, Sect. 6 presents the results, and Sect. 7 concludes the study.

2 Related Work

2.1 Current Malagasy Banknotes

The Ariary has been the official currency of Madagascar since 2005, with updated banknotes issued by the Central Bank in 2017. These notes come in denominations of 100 to 20,000 Ariary and feature rich designs that highlight Madagascar's biodiversity and culture. Despite advanced security features, reliance on manual processing leads to transaction delays, human errors, and difficulties in distinguishing worn or damaged bills. Madagascar's legally recognized banknotes are illustrated in Fig. 1.

2.2 Image Recognition

Image recognition, a subcategory of computer vision and artificial intelligence, involves using methods to detect and analyze images to identify places, people, objects, and other items, and draw conclusions. From an engineering perspective, image recognition aims to automate human visual tasks by acquiring, processing, analyzing and understanding digital images, as well as extracting real-world data.

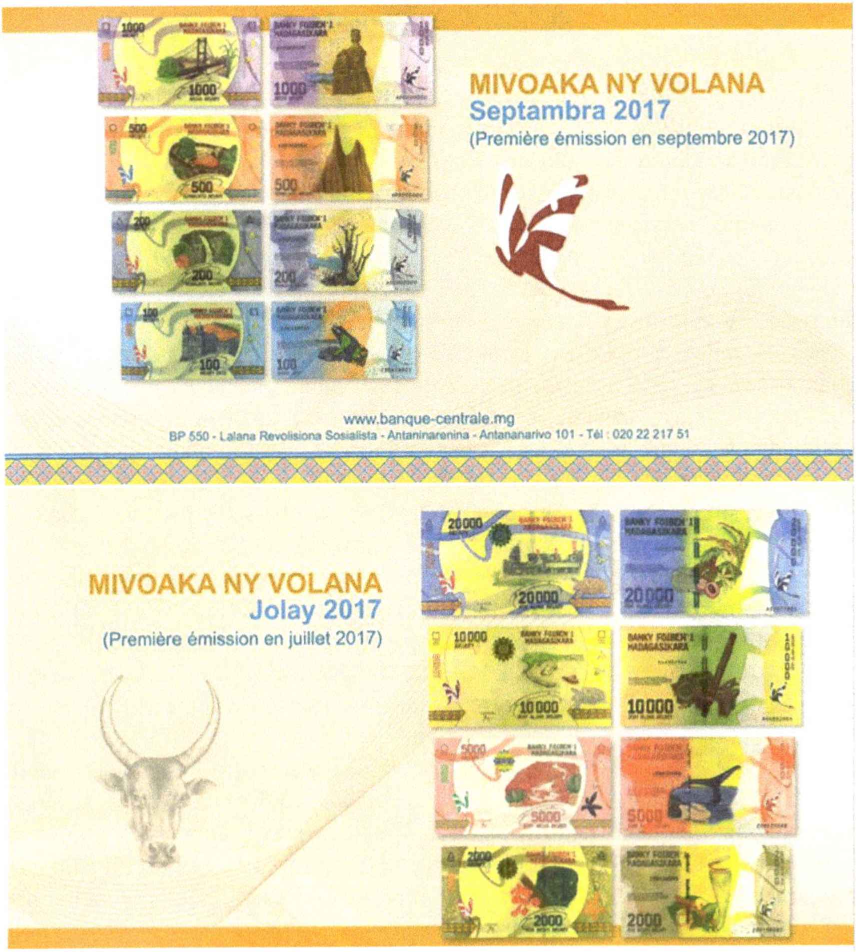

Fig. 1 Banknotes that are legal tender in Madagascar (New NG17 Banknote Series, 2023)

Image recognition software uses deep learning algorithms, such as convolutional neural networks (CNN), to analyze pixels and shapes in an image and recognize objects, places, people, text and actions (Guide: Top 13 mage Recognition Tools, 2025; R. R. Acadamy, 2025).

2.3 Approaches for Recognizing National and International Banknotes

Maminiaina and Sabri (2021) proposed a method to authenticate Malagasy banknotes by identifying their value and detecting counterfeits. Their approach modifies a convolutional neural network (MCNN) by replacing the feature extraction layer with a pulse-coupled neural network (PCNN). Achieving 96.6% accuracy, the results are promising but limited by feature quality and a small dataset (240 images). This study highlights CNNs' potential for authentication but calls for dataset expansion, improved feature extraction, and real-world validation.

A. R. B. and A. S. B. (2023) developed IPCRNet (Indian Paper Currency Recognition Network), a CNN designed to recognize Indian banknotes for the visually impaired. It achieves accurate classification even under challenging conditions but is limited in generalizing to other currencies and lacks field validation.

In a similar study, Jang et al. (2022) focused on serial number detection on banknotes using a DenseNet-based CNN, which is robust against font variations and noise, though its scalability to larger datasets remains to be verified.

Swain et al. (2023) combined CNN and transfer learning to classify Brazilian coins with high accuracy, but still faces class confusion and real-world validation challenges.

In the area of accessibility, Joshi et al. (2020) proposed a fast and accurate real-time currency detection system for visually impaired users based on YOLOv3, though improvements are needed for optimal reliability.

Similarly, AL-Edreesi and Al-Gaphari (2024) developed a lightweight mobile app using MobileNet for Yemeni banknote recognition. While efficient, its accuracy depends on image quality and lighting, requiring enhanced feature extraction.

Finally, Joshi et al. (2020) explored hidden Markov models (HMM) for banknote recognition. This probabilistic approach is sensitive to noise and lighting changes, limiting robustness in practical settings.

In conclusion, these approaches show promising potential for banknote recognition but face challenges in generalization, robustness, and real-world validation. Further research is needed to enhance the effectiveness and accessibility of these systems.

Among them, transfer learning methods stand out with better results. Therefore, we will analyze the best-performing object detection models as a foundation for our specific task.

2.4 Comparative Analysis of the Most Popular Object Detection Models

Several object detection models are commonly used in computer vision, each with distinct strengths. YOLO (You Only Look Once) is a deep neural network excelling in single-step detection and classification, making it very fast and suitable for real-time applications. SSD (Single Shot MultiBox Detector) is efficient and fast, performing well on medium to large objects but less so on smaller ones. Faster R-CNN, using a region proposal network, offers higher accuracy, especially for small objects, but is slower and more complex to implement (Object Detection Best Model & Best Algorithm, 2023; Srivastava et al., 2021).

In terms of precision, YOLO and Faster R-CNN both perform very well, while SSD is slightly less precise. YOLO's latest versions enhance detection of small objects, SSD is better for medium to large sizes, and Faster R-CNN generally excels with small objects. All require large datasets for training. YOLO and SSD suit smaller datasets and real-time needs with simpler coding, whereas Faster R-CNN is better for large datasets demanding high accuracy but involves more complex implementation (Object Detection Best Model & Best Algorithm in, 2023; Srivastava et al., 2021).

Given these characteristics, YOLO stands out as the most suitable model for real-time recognition tasks like Malagasy banknote detection.

2.5 Comparison of YOLO Versions for Real-Time Object Detection

YOLO, introduced in 2015 by Redmon et al. (2016), revolutionized real-time object detection. Since then, the YOLO family has evolved, with YOLOv11 (November 2024) as the latest iteration (Comparative Performance of YOLOv8, YOLOv9, YOLOv10, YOLOv11 & Faster R-CNN Models for Detection of Multiple Weed Species, 2024). YOLOv8 marked a turning point by introducing five variants (Nano, Small, Medium, Large, Extra Large) for greater flexibility (Comparative Performance of YOLOv8, YOLOv9, YOLOv10, YOLOv11 & Faster R-CNN Models for Detection of Multiple Weed Species, 2024). Each version brought specific improvements. YOLOv9 enhanced accuracy and efficiency with **Programmable Gradient Information** (PGI) and **Generalized Efficient Layer Aggregation Network** (GELAN) (Comparative Performance of YOLOv8, YOLOv9, YOLOv10, YOLOv11 & Faster R-CNN Models for Detection of Multiple Weed Species, 2024). PGI improves gradient-based feature extraction, while GELAN boosts layer aggregation for better object localization. However, these mechanisms require careful hyperparameter tuning to avoid overfitting and false negatives, especially with limited data diversity. YOLOv10 focused on ultra-fast inference without Non-Maximum Suppression (NMS), which may increase false positives (Comparative Performance of YOLOv8, YOLOv9, YOLOv10, YOLOv11 & Faster R-CNN Models for Detection of Multiple Weed Species, 2024). YOLOv11 is even faster but may demand greater computational resources (Comparative Performance of YOLOv8, YOLOv9, YOLOv10, YOLOv11 & Faster R-CNN Models for Detection of Multiple Weed Species, 2024).

Table 1 Feature comparison of recent YOLO object detection versions (Comparative Performance of YOLOv8, YOLOv9, YOLOv10, YOLOv11 & Faster R-CNN Models for Detection of Multiple Weed Species, 2024; YOLOv9 vs YOLOv8 Comparison on Real-world Videos, 2025; Kanjee, n.d)

Version	Release date	Advantages	Limitations
YOLOv8	2023	Versatile, easy deployment	Slightly less accurate/speedy
YOLOv9	2024	Better accuracy and efficiency	Complex tuning, false negatives
YOLOv10	2024	Very fast inference (no NMS)	Risk of false positives
YOLOv11	2024	Highest speed	Higher hardware requirements

Although newer versions show performance gains, YOLOv8 remains a strong balance between accuracy, speed, and deployment simplicity—especially for Malagasy banknote detection.

The Table 1 compares the key characteristics of each version:

For this study, we used pre-trained YOLOv8 models (Yolov8s), fine-tuned on Malagasy banknotes, balancing accuracy, speed, and ease of deployment.

2.6 Summary

Our comparative analysis of object detection models, such as YOLO, SSD, and Faster R-CNN, highlighted their distinctive features in the context of real-time recognition of Malagasy banknotes. Despite the recent release of YOLOv9, YOLOv10, and YOLOv11, YOLOv8 remains an optimal choice for this task, offering a strong balance between processing speed and accuracy. Unlike the more recent versions, YOLOv8 requires fewer computational resources, making it more accessible for real-time applications. In the broader context of currency recognition, continuous advancements emphasize the need for innovation to ensure reliable detection under varied conditions. Thus, this synthesis confirms the relevance of an advanced approach for real-time Malagasy banknote recognition, with YOLOv8 emerging as a potentially optimal choice due to its performance and suitability for the specific requirements of this study.

3 Methodology

3.1 Dataset and Classes

A total of 2,510 images of Malagasy banknotes were collected from diverse sources to ensure realism and variety, mainly captured with various smartphones of differing hardware but maintaining sufficient clarity for annotation, complemented by a smaller set of high-quality images from the Central Bank of Madagascar's official website. Using the LabelImg tool, 3440 instances were manually annotated across all eight denominations (100 to 20,000 Ariary), considering recto and verso sides separately to form 16 balanced classes. The dataset was split into 80% training (2008 images, 2752 instances) and 20% validation/test (502 images, 688 instances) sets. This careful curation and balanced distribution aimed to improve the model's generalizability across real-world variations in lighting, resolution, and banknote conditions.

3.2 *Data Preprocessing and Padding Strategy*

Initially, the model was trained with raw images, but this approach proved ineffective due to prohibitive training times and unsatisfactory results. Therefore, the input images were resized to 640 × 640 pixels with padding to preserve the original appearance. This decision significantly improved the model's performance and robustness while reducing the dataset size.

Let $(\boldsymbol{l}_0, \boldsymbol{h}_0)$ be the original dimensions of the image, $(\boldsymbol{l}_t, \boldsymbol{h}_t)$ the target dimensions (here $\boldsymbol{l}_r = \boldsymbol{l}_t = 640$), and $(\boldsymbol{l}_r, \boldsymbol{h}_r)$ the resized dimensions before adding padding.

$\boldsymbol{l}_t$: Target width after resizing with padding.

$\boldsymbol{h}_t$: Target height after resizing with padding.

$\boldsymbol{l}_r$: Resized width before padding.

$\boldsymbol{h}_r$: Resized height, i.e., the height of the image after resizing without padding.

aspect_ratio $= \frac{l_r}{h_r}$ is the aspect ratio of the original image.

- For images in horizontal position (aspect_ratio ≥ 1), dimensions are calculated as:

  $\boldsymbol{l}_r = \boldsymbol{l}_t$.

$$h_r = \frac{l_r}{\text{aspect\_ratio}} \tag{1}$$

- For images in vertical position (aspect_ratio < 1), the resized dimensions are calculated as follows:

  $\boldsymbol{h}_r = \boldsymbol{h}_t$

$$l_r = h_r \times aspect\_ratio = h_t \times aspect\_ratio \tag{2}$$

Then the padding dimensions are calculated as follows:

- Vertical padding (pad_h) is the difference between the target height $\boldsymbol{h}_t$ and resized height $\boldsymbol{h}_r$. If $\boldsymbol{h}_r \geq \boldsymbol{h}_t$, no vertical padding is required, hence pad_h = 0:

$$pad\_h = \max(0, h_t - h_r) \tag{3}$$

Here, max(a,b) returns the maximum value between a and b, so if $\boldsymbol{h_t} - \boldsymbol{h_r}$ is negative, pad_h will be 0, otherwise it will be $\boldsymbol{h_t} - \boldsymbol{h_r}$.

- Horizontal padding (pad_l) is the difference between the target width $\boldsymbol{l_t}$ and resized width $\boldsymbol{l_r}$. If $\boldsymbol{l_r} \geq \boldsymbol{l_t}$, no horizontal padding is needed, so pad_l $= 0$:

$$pad\_l = \max(0, l_t - l_r) \tag{4}$$

Finally, the padding dimensions are used to add padding to the resized image ($\boldsymbol{l_r}$, $\boldsymbol{h_r}$) to achieve the target dimensions ($\boldsymbol{l_t}$, $\boldsymbol{h_t}$).

3.3 *Instance Balancing*

To mitigate the effects of class imbalance, the dataset was balanced so that each class contained an equal number of instances. This ensured fair representation across all classes and improved the model's performance, as illustrated in Fig. 2.

It is important to highlight that several challenging factors can affect model accuracy, such as banknotes being wet, dirty, crumpled, or faded. Partial occlusions or poor lighting conditions also contribute to misclassification. During testing, accuracy decreased under these difficult scenarios, underscoring the need for further data augmentation and improvements in robustness.

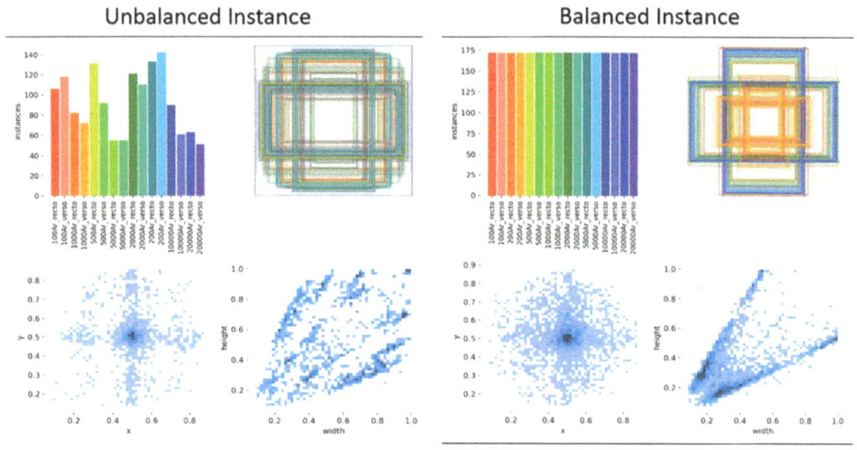

Fig. 2 Differences between balanced and unbalanced instances

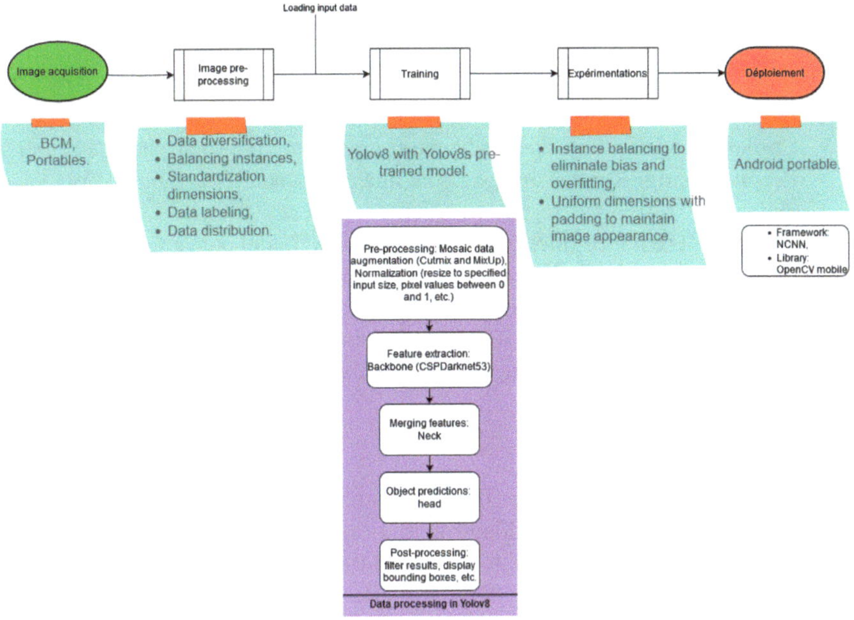

Fig. 3 Workflow for transfer learning with yolov8

3.4 Data Processing Flow with YOLOv8

To ensure practical deployment, especially on mobile devices and ATMs, our object detection approach is designed to meet the following key requirements:

- Speed: Minimum of 10 frames per second (FPS) for real-time recognition;
- Accuracy: At least 98% correct classifications per class;
- Robustness: Ability to detect partially folded, crumpled, damaged, or partially covered banknotes;
- Security: Compliance with data protection regulations with local processing and no storage or transmission of sensitive images.

The data processing flow with YOLOv8 shown in Fig. 3 includes several essential steps, including data pre-processing, backbone passing, feature fusion, object prediction, prediction post-processing, and finally results interpretation.

3.5 Android Application Architecture

To ensure practical and field-usable banknote recognition, especially in rural areas or environments with limited connectivity, we developed a lightweight Android mobile application called ArCuRe (Ariary Currency Recognizer). The app uses the NCNN framework to perform inference directly on the device, without requiring an internet connection or external server.

- Main components of the architecture:
  - **Camera module:** Captures real-time frames using the smartphone's camera.
  - **Preprocessing:** Resizes and pads the images to match the model's input size (640×640).
  - **NCNN inference engine:** Runs the YOLOv8s model converted from ONNX format, optimized for mobile CPUs with low memory usage.
  - **Postprocessing:** Applies Non-Maximum Suppression (NMS) and displays the resulting bounding boxes with corresponding labels.
  - **User Interface:** Displays the recognition results directly on the phone's screen in real time.
- Observed performance:
  - **Inference speed:** ~ 4 FPS or more depending on the smartphone model
  - **Average latency:** Below 300 ms on mid-range devices
  - **Memory usage:** Approximately $\leq$ 1 GB RAM
  - **Recognition accuracy:** $\geq$ 95% on clearly visible or slightly crumpled banknotes

Figure 4 shows the graphical interface of the application.

The app has been successfully tested on various Android smartphones with different hardware configurations, demonstrating its adaptability under real-world conditions.

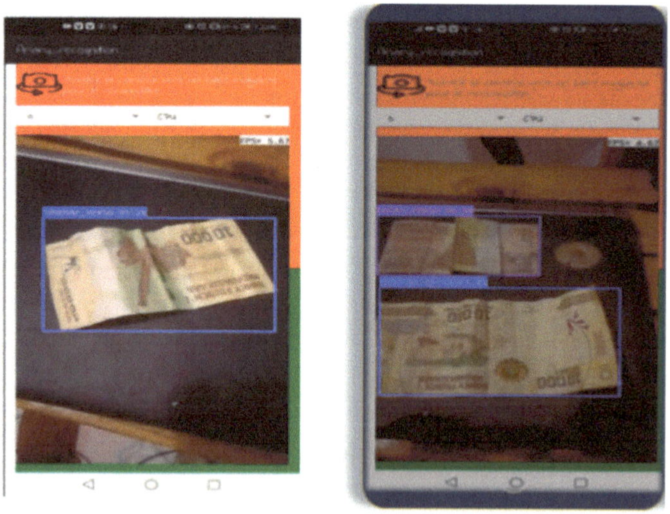

Fig. 4 Interface of the application on the Android device

4 Implementation

4.1 Training and Development Environment

The model training was conducted on Google Colab, selected for its user-friendly interface, online accessibility, and provision of free computational resources, including GPU acceleration ("Google Colab.", 2025). This environment facilitated rapid training cycles and enabled collaborative development among researchers globally.

After evaluating several models, the YOLOv8s architecture was chosen for its balance between compact size, inference speed, and detection accuracy, making it particularly well-suited for deployment on resource-constrained devices such as smartphones.

For the Android application development, Android Studio was utilized due to its robust tooling, extensive community support, and seamless integration with the Android operating system.

4.2 Model Export and Integration Formats

The trained model was exported in the ONNX (Open Neural Network Exchange) format to maintain compatibility across diverse machine learning frameworks and facilitate interoperability.

To enable efficient execution on Android devices, the ONNX model was converted into the NCNN format, which offers optimized inference performance on mobile Central Processing Unit (CPUs) and supports hardware acceleration via Graphics Processing Unit (GPU) on various Android hardware configurations.

5 Evaluation and Result

5.1 Evaluation Metrics

The evaluation uses standard object detection metrics with YOLOv8:

- Intersection over Union (IoU): Measures overlap between predicted and true bounding boxes.
- Mean Average Precision (mAP): Average precision across classes.
- Precision and Recall: Precision is the ratio of true positives to predicted positives; recall is the ratio of true positives to actual positives (Ratianantitra et al., n.d).
- F1 Score: Harmonic mean of precision and recall, balancing both (Ratianantitra et al., n.d).

5.2 *Performance Evaluation*

The confusion matrix shown in Fig. 5 summarizes the classification performance across all classes, highlighting the most frequent misclassifications. These errors are mainly caused not by similarities in color or shape between banknotes, but by the complexity and resemblance of certain motifs. Such intricate patterns sometimes lead the model to confuse the front (recto) and back (verso) sides of different denominations, especially when motifs share visual elements or are partially occluded. This emphasizes the importance of fine-grained motif-specific feature extraction and suggests that further improvements could be achieved by focusing on detailed pattern recognition and enriching the training dataset with diverse representations of each motif under varying conditions.

Our model's automatic evaluation shows remarkable results, with an average mean Average Precision at IoU thresholds from 0.50 to 0.95 (mAP50-95) of 0.9862, mean Average Precision at IoU 0.50 (mAP50) of 0.9949, precision of 0.9976, recall of 0.9957, and an F1 score of 0.996. The Android application demonstrates precise and fast banknote detection, combined with an intuitive, stable, and user-friendly interface. However, correct note identification and performance under varying conditions (lighting, angles) remain areas for improvement.

In practical use, banknotes are often presented in suboptimal conditions—poor lighting, non-standard angles, or partial occlusion—where the model's accuracy tends to decrease. This highlights the need for continuous dataset enrichment and model fine-tuning. To enhance identification accuracy, we plan to expand the dataset, especially for under-represented classes, and apply advanced data augmentation techniques. Incorporating more recent YOLO versions and ensemble learning methods could further improve performance.

6 Discussion

Limitations: Our approach is limited by a relatively small dataset, which may reduce generalization, and the absence of counterfeit detection. The model's robustness under extreme real-world conditions (e.g., heavily damaged notes) remains unvalidated.

Perspectives: Future work includes expanding the dataset, adding counterfeit detection (e.g., spectral or security feature analysis), and adapting the system to other currencies or to other objectives such as personal identification card recognition. Collaboration with financial authorities for large-scale deployment is planned.

Applications: The system can automate cash handling, assist visually impaired users, and integrate into ATMs or Point of Sale (POS) terminals. Its modularity enables easy adaptation to different currencies and use cases, promoting inclusion and efficiency.

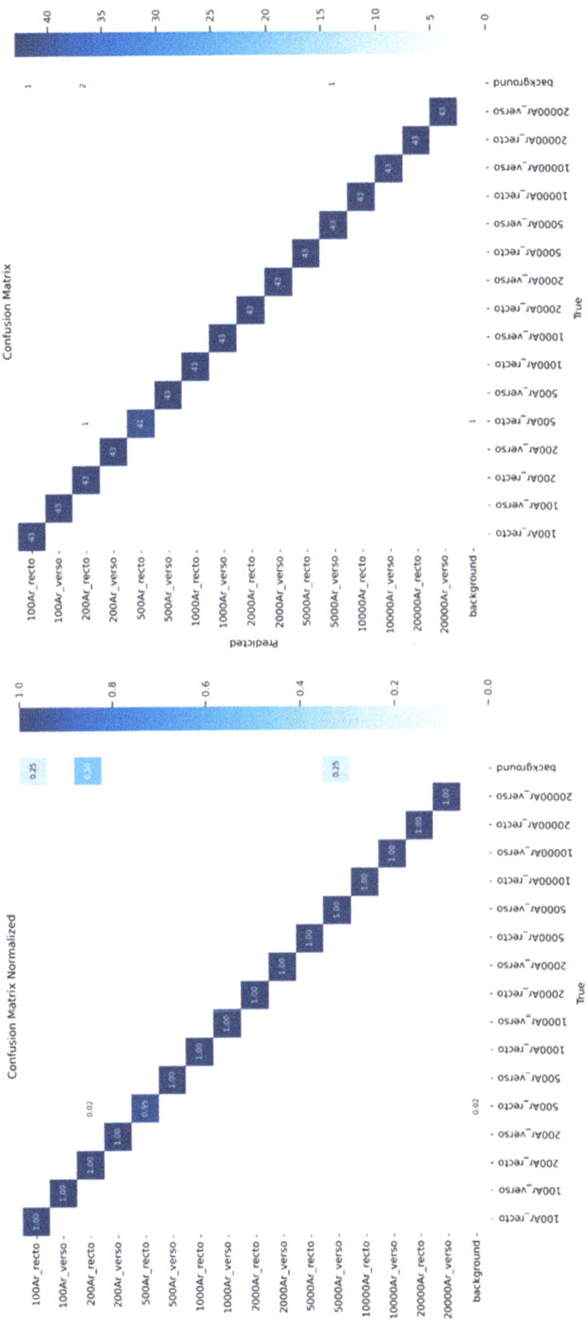

Fig. 5 Confusion matrix

Ethical Considerations: All image processing is local; no personal data is stored or transmitted. The solution complies with GDPR (General Data Protection Regulation) and aims to ensure fairness and limit bias via balanced data and transparent evaluation.

7 Conclusion

This study demonstrates the potential of artificial intelligence and computer vision to modernize monetary management in Madagascar. We developed a real-time Malagasy banknote recognition model using YOLOv8, proving its accuracy and robustness under varied conditions. The model shows promise for automating financial processes and improving user experience.

Future work includes extending to other currencies and objects, adding counterfeit detection, and partnering with financial institutions for deployment. Our research marks a key step toward innovative solutions supporting Madagascar's economic and social development.

References

2025 Guide: Top 13 Image Recognition Tools. Vidnoz. https://fr.vidnoz.com/intelligence-artificielle/outils-reconnaissance-images.html. Accessed June 10, 2023.

A. R. B., & A. S. B. (2023) Indian paper currency recognition framework for blind and visually impaired people using deep learning model. In *The international conference on scientific innovations in science, technology, and management*. https://doi.org/10.59544/SIUZ3611/NGCESI23P1

AL-Edreesi, E., & Al-Gaphari, G. (2024). Real-time yemeni currency detection. arXiv preprint arXiv:2406.13034. https://doi.org/10.48550/arXiv.2406.13034

Comparative Performance of YOLOv8, YOLOv9, YOLOv10, YOLOv11 and Faster R-CNN Models for Detection of Multiple Weed Species (2024). *ResearchGate*. https://doi.org/10.1016/j.atech.2024.100648

Google Colab. (2025) [Online]. Available: https://colab.research.google.com/?hl=fr. Accessed June 10, 2025.

Jang, W., Lee, C., Jeong, D. S., Lee, K., & Lee, E. C. (2022). Multi-currency integrated serial number recognition model of images acquired by banknote counters. *Sensors, 22*(22), 8612. https://doi.org/10.3390/s22228612

Joshi, R. C., Yadav, S., & Dutta, M. K. (2020). YOLO-v3 based currency detection and recognition system for visually impaired persons. In *2020 International Conference on Contemporary Computing and Applications (IC3A)* (pp. 280–285). https://doi.org/10.1109/IC3A48958.2020.233314

Kanjee, R. *Is YOLOv9 Better Than YOLOv8?* LinkedIn Pulse, [Online]. Available: https://www.linkedin.com/pulse/yolov9-better-than-yolov8-ritesh-kanjee-e1ggf.

Maminiaina, R., & Sabri, M. (2021). Modified convolutional neural network for ariary banknotes authentication. *SSRN Electronic Journal, 8*, 62–70.

New NG17 Banknote Series. Central Bank of Madagascar. [Online]. https://www.banky-foibe.mg/nouvelle-gamme-de-billet-ng17. Accessed December 23, 2023.

Object Detection Best Model/Best Algorithm. (2023) | YOLO vs SSD vs Faster-RCNN Comparison Python," YouTube. [Online]. Available: https://www.youtube.com/watch?v=ce5PBam-V4Y. Accessed June 15, 2024.

R. R. Academy, "Image Recognition," 24pm Academy. [Online]. https://24pm.com/117-definitions/304-reconnaissance-d-image. Accessed June 03, 2025.

Ratianantitra, V. M., Razafindramintsa, J. L., Mahatody, T., & Manantsoa, V. (2019). Deep learning approach for Malagasy text summarization. *International Journal of Conceptions on Computing and Information Technology*, *7*(1), 12-17.

Redmon, J., Divvala, S., Girshick, R. B., & Farhadi, A. (2016). You only look once: Unified, real-time object detection. In *Proceeding IEEE Conference Computer Vision and Pattern Recognition (CVPR)*. https://doi.org/10.1109/CVPR.2016.91

Srivastava, S., Divekar, A. V., Anilkumar, C., Naik, I., Kulkarni, V., & Pattabiraman, V. (2021). Comparative analysis of deep learning image detection algorithms. *Journal of Big Data, 8*(1), 66. https://doi.org/10.1186/s40537-021-00434-w

Swain, D., et al. (2023). A deep learning framework for the classification of Brazilian coins. *IEEE Access, 11*, 109448–109461. https://doi.org/10.1109/ACCESS.2023.3321428

YOLOv9 vs YOLOv8 Comparison on Real-world Videos. (2025). YouTube. [Online]. Available: https://www.youtube.com/watch?v=rhkYmQ5J3-w. Accessed June 18, 2025.

Joint Modeling of Entities and Relations Through Hierarchical Analysis of Span Sequences

Jules Michel Andriafetiarison, Onja Kostka Michel Andriamihary, and Volatiana Marielle Ratianantitra

Abstract Relation extraction, a crucial aspect of information extraction, has greatly benefited from recent progress in deep neural networks and pretrained language models. Two dominant approaches emerge: the cascade method and the joint approach. The latter, though ambitious, faces challenges related to entity overlap. A promising alternative is the span-based approach, but it is hindered by limitations in span representation. To overcome these obstacles, we propose an innovative joint extraction model. It relies on multi-level feature analysis at the span representation level, detecting local patterns and sequential dependencies. This approach enables the identification of crucial information at a fine scale, thereby improving extraction accuracy. We trained our model using two datasets, NYT and WEBNLG, and evaluated its effectiveness on the latter. Our approach surpasses other models, achieving superior performance, especially in terms of the F1 score on WEBNLG, underscoring its potential for relation extraction.

1 Introduction

Billions of textual data are generated from various sources, including news articles, scientific publications, forums, clinical reports, etc. These data are often raw and unstructured, making their analysis and exploitation challenging. Relation extraction is a key task in transforming these vast amounts of unstructured data into actionable information. Two supervised learning methods are widely used for extracting entities and relations: the cascade method and the joint method. According to Zuo et al. (2022), the cascade method involves first extracting entity mentions through a named entity recognizer, and then predicting the relationships between each pair of these extracted entity mentions. This approach sequentially performs the named entity

J. M. Andriafetiarison (✉) · V. M. Ratianantitra
Laboratory of Computer Science and Mathematics Applied to Development (LIMAD), University of Fianarantsoa, Fianarantsoa, Madagascar
e-mail: fetiarison.andria@gmail.com

O. K. M. Andriamihary
GOogle On Dev (GOOD), eTech Consulting, Antananarivo, Madagascar

J. Marx Gómez et al. (eds.), *Recent Advances in Information Systems*, Progress in IS,
https://doi.org/10.1007/978-3-032-14099-9_4

recognition task followed by the relation extraction task. Notably, the interdependence between these two tasks is crucial for effective relation extraction. Therefore, the joint method is more suitable for obtaining this triplet. However, the joint method often faces the problem of overlapping entities, as shown in Table 1.

Many studies have explored this issue, and span-based models have demonstrated their ability to address these challenges more effectively. Despite their advantages, these models have certain limitations. Specifically, in the work of Zhang et al. (2021), a multi-head attention mechanism was adopted to jointly extract entities and relationships, focusing only on syntactic aspects to improve span representation. Most approaches concentrate on predicting triplets while overlooking the potential improvement in extraction quality that could be achieved by analyzing span sequences. This limitation affects the work of Liao et al. (2023). In the case of Han et al. (2023), they relied solely on a sliding window technique to extract span-level features, which resulted in a rather basic span representation. This approach can lead to a loss of crucial information, as it struggles to capture the nuances and details of entities and relationships within a sentence.

This paper proposes an in-depth analysis of multi-scale features to enhance the sequential representation of spans. This analysis will consider local patterns, semantics, syntax, and context, providing a more comprehensive perspective on linguistic structure and facilitating a better understanding of relationships in the text.

The remainder of this paper is structured as follows: Related work is discussed in Sect. 2. Section 3 details the methodology and planning approaches. Section 4 presents the obtained results, followed by discussions. Finally, Sect. 5 concludes the paper.

Table 1 Representation of overlap types

Types of overlapping	Examples
Single Overlap Entity (SEO): a single entity can be mentioned multiple times in a sentence	"Steven Jobs was born in February 1955 in San Francisco, California" ↓ (Steven Jobs, **born_on**, February 1955) (Steven Jobs, **born_in**, San Francisco) (San Francisco, **country**, California)
Entity Pair Overlap (EPO): the same pair of entities is mentioned in multiple different relations within a sentence	"Antananarivo is the capital of Madagascar" ↓ (Madagascar, **/location/country/capital**, Antananarivo) (Madagascar, **/location/location/contains**, Antananarivo)

2 Related Work

Natural Language Processing, at the intersection of linguistics and artificial intelligence, enables computers to understand and interact with human language. According to Ratianantitra (2023), it excels at recognizing, interpreting, and responding to human queries. Its applications span various fields such as information retrieval, machine translation, text summarization, and sentiment analysis.

Information extraction plays a key role in this domain by identifying useful knowledge from unstructured texts. The joint extraction of entities and relations is one of the subtasks in this area, playing an essential role in text analysis. Various approaches have already been explored to perform this task, ranging from traditional rule-based methods to deep learning-based approaches.

2.1 Traditional Approach

The traditional approach is generally based on feature engineering and kernel functions. The feature engineering approach involves manually selecting relevant attributes for the task of entity and relation extraction, such as keywords, linguistic patterns, and lexical contexts. In the case of Miwa and Sasaki (2014), applied a structured incremental learning method to extract both entity mentions and relations at the same time, using a structured perceptron combined with an efficient beam search technique.

On the other hand, according to Pawar et al. (2024), kernel-based methods use kernel functions to assess similarities between the representations of two relation instances, typically utilizing SVM for classification. Its main advantage is avoiding explicit feature engineering. These kernel-based relation extraction methods offer different representations for relation instances, such as sequences or syntactic trees, and often measure similarity by counting the number of shared sub-representations.

However, according to Zhu et al. (2023), the overall performance of these methods heavily depends on the effectiveness of the designed features in establishing connections between the tasks of named entity recognition and relation extraction, making it challenging to leverage global features and potentially leading to issues such as error propagation and limited effectiveness in capturing long-tail entities and complex relations.

2.2 Deep Learning-Based Approach

Neural models have revolutionized joint entity and relation extraction by replacing manual features with learned representations. The sequence tagging approach, as seen in Yuan et al. (2020) with their RSAN network, uses relation-specific attention to label entities, but it restricts each word to a single relation, failing to handle overlapping triplets. Wang et al. (2024) proposed a one-step method to address this issue, though it struggles to effectively encode long texts.

Meanwhile, Seq2Seq methods transform unstructured text into triplets ⟨Entity_1, Relation, Entity_2⟩ using a copying mechanism, such as in CopyRE by Zeng et al. (2018), which allows entities to participate in multiple relations. However, this duplication can lead to redundancy and increased complexity. Ding et al. (2023) also introduced a new approach to relation extraction that aims to better capture the meaning behind relationships and improve how entities and relations interact. They proposed RSFnet, a neural model that combines a lightweight yet powerful convolutional encoder with an attention mechanism to fuse semantic information.

More recently, generative methods have been gaining ground in relation extraction. Swarup et al. (2025) provide a comprehensive overview of how research has evolved from crafting optimal prompts to leveraging Large Language Models (LLMs) capable of performing extraction and classification with few or no examples. For example, Wadhwa et al. (2023) investigates the capacity of LLMs such as GPT-3 and Flan-T5 Large to perform relation extraction in few-shot and fine-tuning scenarios, treating the task as conditional text generation where the model outputs linearized relational triplets from a given context. Rajpoot and Parikh (2024), developed and evaluated a relation extraction method applied to financial documents, utilizing In-Context Learning (ICL) with GPT models. Without relying on fine-tuning, their approach carefully selects contextual examples to guide the model. Wan et al. (2024) introduces GPT-RE, an improved ICL method for relation extraction using LLMs like GPT-3. Unlike traditional ICL approaches, GPT-RE enhances performance through two key modules: task-sensitive retrieval and gold label-induced reasoning, enabling more accurate and robust extraction.

Furthermore, the span-based approach excels at managing overlaps by identifying all possible word spans and predicting their relations. Yet, works like those of Zhang et al. (2021) and Han et al. (2023) focus on triplet prediction, neglecting the analysis of span sequences, which limits the capture of nuances and complex dependencies.

This drives us to explore local patterns within these sequences for finer and more accurate extraction. We thus propose to leverage these sequences by modeling the dependencies between their elements, aiming to better capture their relationships within a comprehensive contextual perspective.

To address these issues, we propose a novel approach that builds upon recent advances in span-based modeling. Our contributions are as follows: first, we build upon the span tagging method developed by Han et al. (2023), known as Span-Tagger, by using only the start mapping strategy to enumerate spans and cover all candidate entities. This approach is based on previous work where the method has demonstrated its effectiveness, ensuring a more comprehensive entity recognition process.

Second, to enhance the representation of input data, we propose using RoBERTa, an improved version of BERT. While Fan (2023) employed DPCNN (Deep Pyramid Convolutional Neural Network) for processing long texts, using convolutional blocks with a sampling strategy, shortcut connections, and global max pooling to extract hierarchical representations from word embeddings, we integrate this layer with Bi-LSTM. This combination, though still hypothetical, could capture patterns at different scales, strengthen dependencies within the formed span sequence, and extract richer features.

Finally, building on these enhanced span representations, we apply a customized attention mechanism to filter out padding tokens, ensuring that only meaningful information is retained. Additionally, we employ a multi-head selection technique to effectively extract relationships between spans, leveraging the hierarchical and sequential features captured in the previous step. This integrated approach leads to more accurate relation extraction by jointly refining span representation and relation identification.

3 Methodology

We propose a new model for the joint extraction of entities and relations based on multi-scale analysis of spans. The proposed model consists of two modules: an encoding module and a decoding module. The structure of our model is presented in Fig. 1.

3.1 Data Preprocessing

In this data preprocessing, several steps are performed to prepare the data before feeding it into the model. The following steps are included in data preprocessing:

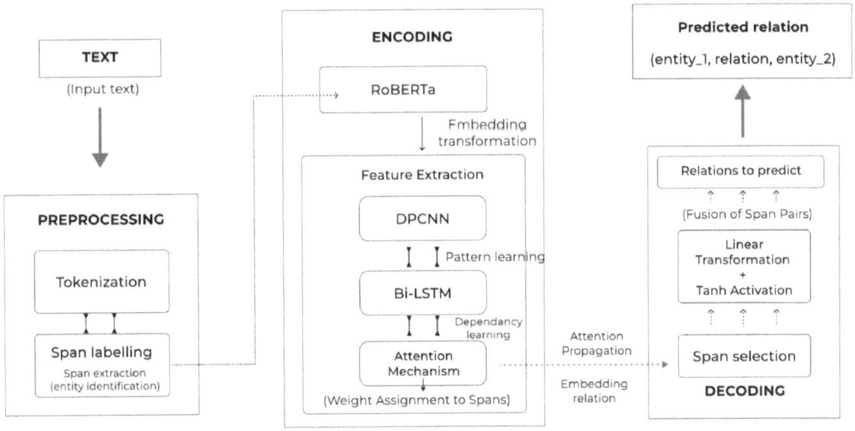

Fig. 1 Structure of our model

Tokenization. Tokenization involves dividing a sequence of words or characters into a list of tokens, which represent discrete units of meaning or syntactic structure. Formally, let $\boldsymbol{T}$ be a text to be tokenized. After tokenization, $\boldsymbol{T}$ is segmented into a sequence of tokens $\{\boldsymbol{t}_1, \boldsymbol{t}_2, \ldots, \boldsymbol{t}_n\}$, where n is the total number of tokens in the text. The tokenization function, denoted Tokenize ($\boldsymbol{T}$), can be defined as follows:

$$Tokenize(T) = \{t_1, t_2, \ldots, t_n\} \tag{1}$$

This function takes the text T as input and returns the list of tokens $\{t_1, t_2, \ldots, t_n\}$. Each token $\{t_i\}$ represents a distinct unit of meaning in the text, making the next processing task more efficient.

Span Labeling. We build upon the span labeling method developed by Han et al. (2023), known as Span-Tagger, using only the start mapping strategy to enumerate spans and cover all candidate entities. This span labeling module assigns a mapping relationship between the labeled span and its position in the span embedding matrix. The span start mapping process begins by creating a list of indices for each potential span start position within the token sequence. Each index is determined by a pair of indices (start, end), where the end is constrained by the predefined window size. This index list ensures that all possible span start positions within the sequence are covered. Then, these indices are used to establish a bidirectional mapping between the matrix indices and the span indices. This mapping is crucial for associating each span with its corresponding position in the matrix, thereby facilitating the processing and further utilization of spans within the model. To represent the start position mapping, we can use the following formula:

$$Index_{start}(i, j) = \{(i, k) | i \leq k < \min(i + w, L)\} \tag{2}$$

where $Index_{start}(i, j)$ represents the list of indices created for each potential span start within the token sequence. Here, i and j denote the start and end indices of the span, respectively. The variable k iterates over the token sequence indices, starting from i and extending up to $\min(i + w, L)$, where w is the window size limiting the span's maximum length, and L is the maximum length of the token sequence.

3.2 Encoding

In this encoding module, we transform the input data into a vector representation suitable for the relation extraction task.

Vector Representation. For the vector representation, we use the pre-trained RoBERTa model, an enhanced version of BERT. This pre-trained model encodes token identifiers $\boldsymbol{I}$. A mask m is also created to mask the padding tokens, ensuring the model ignores these tokens when computing the vector representations. The encoding function of this model returns an output that includes the embeddings (or vector representations) of all tokens e_t, along with the pooling embedding e_p, an aggregated vector representation of the input data sequence. This encoding mechanism is represented as follows:

$$e_t, e_p = encode(I, m) \tag{3}$$

where I represents the token identifiers of the input text, and m is the padding mask. e_t represents the embeddings of all tokens, and e_p represents the pooling embedding.

Multi-Scale Feature Extraction. This step involves extracting different features from the span vector representation at the sequence level. But before proceeding to feature extraction, we first calculate the span representations by taking the maximum value over the span token dimension. These representations are then concatenated with the aggregated representation. All the resulting span representations are then concatenated along the second dimension to form a single matrix of span representations. Finally, a nonlinear transformation is applied to this matrix to capture the nonlinear relationships between the different span representations. Next, we pass the result through a Bi-LSTM and DPCNN layer, which captures the sequential dependencies in the span embeddings.

Bi-LSTM Layer. The output sequence of spans is then passed through a Bi-LSTM (Bidirectional Long Short-Term Memory) layer, which is used to capture the sequential dependencies in the span embeddings. The application of this layer is represented as follows:

$$h^{(t)}, c^{(t)} = BiLSTM\left(h^{(t-1)}, c^{(t-1)}\right) \tag{4}$$

where $c^{(t)}$ is the hidden state of the Bi-LSTM cell at time t and $h^{(t)}$ is the output of the Bi-LSTM layer at time t.

DPCNN Convolutional layer. We use DPCNN (Deep Pyramid Convolutional Neural Network), a convolutional neural network architecture, which is generally used for text classification, but we use it to extract local and global features from the input text. Convolution layers use different filter sizes to extract multiple features and maximize pooling. These layers capture patterns at different scales, enabling the model to gather a wide range of data. According to Fan (2023), DPCNN is designed to handle longer texts and capture patterns at different scales, allowing the model to gather a wide variety of information. To apply 1D convolutions to the output sequences from the Bi-LSTM layer, we will first reshape these embeddings. Let e_i be the embedding vector corresponding to token i ranges from 1 to N (with N being the sequence length). This is expressed as follows:

$$seq_{hiddens} = e_1, e_2, e_3, \ldots, e_N \tag{5}$$

Then, we permute the sequence dimension of the embedding sequence $seq_{hiddens}$ so that the sequence dimension becomes the last dimension. DPCNN applies a convolution kernel to each span sequence, capturing local patterns and relationships between adjacent tokens. After each convolution, a ReLU activation function is applied to introduce nonlinearity in the activations of the hidden layers. Once all convolutions are applied, the dimensions of the embeddings are reshaped back to their original form. This is necessary for the embeddings to be efficiently used in the next layer.

Attention Mechanism. Next, an attention mechanism is applied to a sequence of hidden representations. This involves assigning weights to each position of the spans based on its importance relative to other positions. These weights are computed by comparing each position with the others in the sequence. The attention mechanism can be presented as follows:

$$Attention(Q, K, V) = soft\max\left(\frac{QK^T}{\sqrt{d_k}}\right)V \tag{6}$$

where Q, K, and V are the query, key, and value matrices, respectively, and d_k is the dimension of the key. The softmax function is applied to the normalized dot products between the queries and the keys to compute the attention weights, which are then used to weight the values. As a result, the model can focus on the most relevant parts of the span sequence, leading to a better representation of the data and improved task performance.

3.3 Decoding Module

We adopt the span selection method, which exploits span representations to generate multi-head selection logits. This approach involves converting the span representations into two sets of projections, merging them, and then multiplying them by the relation embedding matrix. Unlike Han et al. (2023), our multi-head selection method starts by applying a series of linear transformations followed by Tanh activation functions to the span features. Then, it expands the dimensions of the selected span features to prepare them for tensor multiplication with the relation embeddings. These selection logits are then used to predict the relationships between pairs of spans. Let S be the set of span pairs, P be the set of possible relation predicates, and R be the set of extracted relations. The selection logits for each span pair (s, o) and each predicate p are represented by $logits(s, o, p)$. The selection mask M is a binary matrix indicating the valid span pairs for each example in the batch. The relation decision function can be defined as follows:

$$relation\_tags(s, o, p) = \begin{cases} 1 \ si\ \sigma(\log its(s, o, p)) > T \\ 0 \qquad \text{otherwise} \end{cases} \tag{7}$$

where σ is the sigmoid function that converts the logits into values between 0 and 1, and T is a predefined threshold.

Table 2 Statistics of the datasets used

Datasets	Train	Valid	Test	Details of test set			Relations
				Normal	SEO	EPO	
NYT	56,195	5000	5000	3266	1297	978	24
WEBNLG	5019	500	703	245	457	26	171

4 Experiments

4.1 Datasets

We evaluated our model with two datasets, NYT and WEBNLG, using the data from Jing et al. (2023) for the task of entity and relation extraction. Table 2 presents the distribution of the data across these datasets, used for training, validation, and testing.

4.2 Experimental Parameters

We present the experimental parameters, including the variables, configurations, and specific values used during the training and evaluation of our model. The hyperparameters were manually optimized and set as follows: batch size $= 2$, learning rate $= 2e - 5$, number of epochs $= 20$, dropout $= 0.5$, maximum window size $= 6$, and maximum sequence length $= 128$. The model was trained using the Adam optimizer.

4.3 Evaluation and Comparative Analysis

The performance of our model was evaluated using several evaluation metrics, including precision, recall, and F1 score. By combining these three metrics, we obtain a comprehensive evaluation of our model's performance, considering both the quality of positive predictions and the ability to detect all positive instances in the data.

To validate the performance of our model, we conducted a comparative experiment by testing it against several baseline models on the test data from the NYT and WEBNLG datasets. Table 3 summarizes the comparison of our model with these baseline models.

The results presented in Table 3 show that our model achieves overall performance that is superior or comparable to the benchmark models. On the NYT dataset, our model achieves an F1 score of 91.70, slightly below the scores of Yu et al.'s model (93.65) and Wadhwa et al.'s model (95.23), the latter remaining the best performance on this dataset. On the WEBNLG dataset, our model stands out by achieving

Table 3 Comparison of our model with reference models on the NYT and WEBNLG datasets

Methods	NYT			WEBNLG		
	P	R	F1	P	R	F1
Few-Shot GPT-3	66.63	70.58	68.55	–	–	–
SpanRE	79.7	71.5	75.4	77.5	60.9	68.2
RSFnet	88.2	87.5	87.9	–	–	–
Vanilla Flan-T5-Large	91.82	90.25	91.03	–	–	–
SCDM	89.8	92.7	91.2	91.6	92.2	91.9
SMHS	97.5	82.0	89.1	–	–	–
Yu's model	94.07	93.23	93.65	92.97	92.20	92.58
Flan + GPT-3 CoT	95.49	94.97	95.23	–	–	–
Ours	91.58	91.82	91.70	92.26	93.55	92.90

the highest F1 score (92.90), surpassing all existing methods, including Yu et al.'s model (92.58). It also demonstrates high precision (92.26) and optimal recall (93.55), indicating a stronger generalization capability on this dataset.

4.4 Experimental Analysis of Relationship Extraction

To study the impact of the model on different types of relationships, particularly overlaps, we conducted comparative experiments. Since there are three types of relationships in the text, namely normal relationships, entity pair overlaps, and unique entity overlaps, these experiments were used to evaluate the F1 score. Table 4 summarizes the evaluation of the model's performance across relationship types.

Table 4 Evaluation of model performance across relationship types

Methods	NYT			WEBNLG		
	Normal	SEO	EPO	Normal	SEO	EPO
Few-Shot GPT-3	–	–	–	–	–	–
SpanRE	–	–	–	–	–	–
RSFnet	–	–	–	–	–	–
Vanilla Flan-T5-Large	–	–	–	–	–	–
SCDM	89.1	93.1	91.1	86.8	89.4	89.7
SMHS	89.2	87.3	90.2	–	–	–
Yu's model	–	–	–	–	–	–
Flan + GPT-3 CoT	–	–	–	–	–	–
Ours	91.2	93.1	94.3	88.7	92.8	93.5

In our experimental analysis, we evaluated the performance of several relation extraction models on the NYT and WEBNLG datasets, focusing on normal relationships, unique entity overlaps (SEO), and entity pair overlaps (EPO). Our model outperformed the others in terms of performance on both datasets. On the NYT dataset, for the normal relationship, our model achieved an F1 score 2% higher than the SMHS model. For the SEO relationship, it improved its performance by 5.8% in terms of F1 score compared to the SMHS model. Similarly, for the EPO relationship, our model recorded a 3.3% improvement in F1 score over the SCDM model. On the WEBNLG dataset, for the normal relationship, our model showed an F1 score 1.9% higher than the SCDM model. For the SEO relationship, it improved its performance by 3.4% in terms of F1 score compared to the SCDM model. Finally, for the EPO relationship, our model recorded a 3.8% improvement in F1 score over the SCDM model.

4.5 Discussion

The experiments conducted allow us to deduce that our hypothesis, which posits that a multi-level analysis of span features, by identifying local patterns and sequential dependencies, is successfully validated. This is corroborated by the results obtained, particularly on the WEBNLG dataset, where our model outperforms the baselines with a notable improvement in accuracy. On the NYT dataset, although it does not dominate the state-of-the-art models, it maintains strong performance, demonstrating its robustness. These results highlight not only the overall effectiveness of our approach but also its ability to handle complex relationships, such as overlaps between nested entities, an area where the baselines show weaknesses.

However, despite these advances, our work has limitations. Currently, our model focuses on extracting facts at the sentence level, which excludes inter-sentential relationships. Additionally, we aim to optimize computational complexity, for instance by exploring more efficient algorithms, without compromising accuracy.

5 Conclusion

In this paper, we proposed a new model for joint entity and relation extraction, based on a multi-level analysis of spans. This approach effectively identifies local patterns and sequential dependencies, thereby improving extraction accuracy. Our method integrates RoBERTa, an optimized version of BERT, along with Bi-LSTM-DPCNN, which captures relevant information from spans and relations across long sequences. An attention mechanism further eliminates padding tokens and extracts deep span features. Through this combination, our model achieved remarkable performance, with an F1 score of 92.90% on WEBNLG, outperforming baseline models. It also remains competitive on NYT, confirming its effectiveness in the task of joint entity and relation extraction. However, our model has certain limitations. It is currently restricted to sentence-level relation extraction, whereas many facts require cross-sentence analysis. Additionally, it does not account for temporal dimensions, which are nevertheless critical in certain semantic relations. Although large language models (LLMs) have already demonstrated their potential in recent works, especially as generative models, our current approach does not yet fully exploit these generative capabilities. In future work, we plan to explore and integrate the generative power of LLMs to better capture complex contextual information, improve extraction at the document level, handle n-ary relations, and incorporate temporal aspects. This integration is expected to enrich and enhance our model's capabilities for more global and complex extraction tasks.

References

Ding, R., Lu, L., Yan, Q., Wu, Y., & Zeng, D. (2023). RSFnet: a relation semantic fusion-based entity relation extraction method. In A. J. Tallón-Ballesteros & R. Beltrán-Barba (Eds.), *Frontiers in artificial intelligence and applications.* IOS Press. https://doi.org/10.3233/FAIA231047

Fan, C. (2023). The entity relationship extraction method using improved roberta and multi-task learning. *Computer Materials Continues, 77*(2), 1715–1732. https://cdn.techscience.cn/files/cmc/2023/TSP_CMC-77-2/TSP_CMC_41395/TSP_CMC_41395.pdf. Accessed 2024/03/14.

Han, D., Zheng, Z., Zhao, H., Feng, S., & Pang, H. (2023). Span-based single-stage joint entity-relation extraction model. *PLoS ONE, 18*(2), Article e0281055.

Jing, X., Han, X., Li, B., Guo, J., & Li, K. (2023). A joint triple extraction method by entity role attribute recognition. *Science and Reports, 13*(1), 2223.

Liao, T., Sun, H., & Zhang, S. (2023). A joint extraction model for entity relationships based on span and cascaded dual decoding. *Entropy, 25*(8), 1217.

Miwa, M., & Sasaki, Y. (2014). Modeling joint entity and relation extraction with table representation. In *Proceedings of the 2014 conference on Empirical Methods in Natural Languag Processing (EMNLP)* (pp. 1858–1869). ACL. https://aclanthology.org/D14-1200.pdf. Accessed 2023/12/19.

Pawar, S., Palshikar, G. K., & Bhattacharyya, P. (2024). *Relation extraction: a survey.* arXiv, http://arxiv.org/abs/1712.05191. Accessed 2024/02/25.

Rajpoot, P. K., & Parikh, A. (2024). *GPT-FinRE: In-context Learning for Financial Relation Extraction using Large Language Models.* arXiv, http://arxiv.org/abs/2306.17519. Accessed 2024/03/10.

Ratianantitra, V. M. (2023). A state of the art review on natural language processing applied to the Malagasy language. In *International conference on artificial intelligence and its applications* (pp. 1–5). Mauricon. https://mauricon.org/conferences/index.php/icarti/article/view/35. Accessed 2024/05/13.

Swarup, A., Pan, T., Wilson, R., Bhandarkar, A., & Woodard, D. (2025). LLM4RE: A data-centric feasibility study for relation extraction. In *Proceedings of the 31st international conference on computational linguistics* (pp. 6670–6691). https://aclanthology.org/2025.coling-main.447/. Accessed 2025/06/12.

Wadhwa, S., Amir, S., & Wallace, B. C. (2023). Revisiting relation extraction in the era of large language models. In: *Proceedings of the conference. association for computational linguistics meeting* (p. 15566). https://pmc.ncbi.nlm.nih.gov/articles/PMC10482322/. Accessed 2025/06/12.

Wan, Z. et al. (2024). *GPT-RE: In-context learning for relation extraction using large language models*. arXiv, http://arxiv.org/abs/2305.02105. Accessed 2024/02/28.

Wang, Y., Yu, B., Zhang, Y., Liu, T., Zhu, H., & Sun, L. (2024). *TPLinker: Single-stage joint extraction of entities and relations through token pair linking*. arXiv, http://arxiv.org/abs/2010.13415. Accessed 2024/02/28

Yuan, Y., Zhou, X., Pan, S., Zhu, Q., Song, Z., & Guo, L. (2020). A relation-specific attention network for joint entity and relation extraction. In: *Proceedings of the twenty-ninth international joint conference on artificial intelligence* (pp. 4054–4060). IJCAI, Yokohama.

Zeng, X., Zeng, D., He, S., Liu, K., & Zhao, J. (2018). Extracting relational facts by an end-to-end neural model with copy mechanism. In: Proceedings of the 56th annual meeting of the association for computational linguistics (vol. 1: Long Papers, pp. 506–514). ACL. https://aclanthology.org/P18-1047/. Accessed 2023/12/19.

Zhang, H., Zhang, G., & Ma, Y. (2021). Syntax-informed self-attention network for span-based joint entity and relation extraction. *Applied Sciences, 11*(4), 1480.

Zhu, Y., Li, X., Wang, Z., Li, J., Yan, C., & Zhang, Y. (2023). ER-LAC: Span-based joint entity and relation extraction model with multi-level lexical and attention on context features. *Applied Sciences, 13*(18), 10538.

Zuo, M., Zhang, B., Zhang, Q., Yan, W., & Ai, D. (2022). An entity relation extraction method for few-shot learning on the food health and safety domain. *Computational Intelligence and Neuroscience, 2022*, 1–11.

Tax Defense Analytics: A Data-Driven Application for Identifying Cross-Border VAT Fraud and Tax Avoidance Schemes

Kristof Mose, Keyvan Narimani, Maged Ahmad, André Klümpen, and Jorge Marx Gómez

Abstract In this work, we present the current state of a work in progress multifunctional application for detecting cross-border value added tax (VAT) fraud and identifying tax avoidance schemes of multinational corporations. We outline the design and architecture of this data science and artificial intelligence (AI) driven application, which incorporates not only data analytics but additional functionalities, which in combination support and elevate the workflow of the auditors. These additional functionalities include a sophisticated data integration pipeline, a multipurpose AI tool suite, as well as a tax expert knowledge management system. Key challenges in data governance, infrastructure, and organizational readiness are discussed. Despite not being finished, this approach already shows the potential to significantly enhance the efficiency and effectiveness of tax fraud detection and tax avoidance identification. This work aims to share insights from operationalizing AI in tax administration to facilitate broader digital transformation efforts.

1 Introduction

Tax fraud practices and tax avoidance schemes pose an enormous financial burden on governments worldwide. For example, the European Union estimates that its value added tax (VAT) gap or revenue lost from VAT fraud was €93 billion in 2020, with €11 billion attributed to Germany alone (European Commission, 2023). Furthermore, the loss to tax avoidance and tax evasion worldwide is estimated to be a few hundred billion US$ per year (European Parliament, 2023).

As economic digitalization accelerates and real-time data streams become ubiquitous, tax authorities require sophisticated data analytics capabilities and AI-powered solutions to uncover tax fraud and tax avoidance schemes (Jose Ordonez & Hallo, 2019). This is a critical paradigm shift to enhancing tax fraud and tax avoidance

K. Mose (✉) · J. Marx Gómez
Carl von Ossietzky Universität Oldenburg, Oldenburg, Germany

K. Narimani · M. Ahmad · A. Klümpen
Lower Saxony State Tax Office, Oldenburg, Germany

J. Marx Gómez et al. (eds.), *Recent Advances in Information Systems*, Progress in IS,
https://doi.org/10.1007/978-3-032-14099-9_5

risk and assessment, increasing investigative bandwidth, and augmenting human expertise with intelligent decision support systems. However, operationalizing AI in this domain poses multifaceted challenges around data governance, regulatory compliance, organizational processes, and workforce skillsets.

This paper presents the current state of a work in progress application to detect cross-border VAT fraud and identify tax avoidance schemes of multinational corporations developed by the German research collaboration "Tax Defense Analytics" between the Lower Saxony State Tax Office and the Carl von Ossietzky University of Oldenburg. In recent years, the interdisciplinary team has been actively researching and prototyping AI techniques to counter cross-border VAT fraud and identifying tax avoidance schemes of multinational corporations. The collaboration brings together expertise from tax administration and computer science to address critical challenges in modern tax enforcement.

The application in development follows a holistic approach including not only data integration and analytics but also knowledge management system and AI tool suite components.

This paper is structured as follows: We first provide further background on combating cross-border VAT fraud and identifying tax avoidance by multinational corporations to highlight their significance. After describing the methodology of our research, the application concept is outlined. Insights are shared on the design philosophy, architecture, user group management, and data governance of the developed application. We go into further detail regarding the different components of the application, as well as testing and evaluation strategies. Lastly, challenges, preliminary successes, and future improvements and extensions are also discussed.

2 Background

The research collaboration "Tax Defense Analytics" has two use cases, detecting cross-border VAT fraud and combating tax avoidance by multinational corporations. These two use cases have distinct user groups within the Lower Saxony tax administration and were chosen by their automation potential and fiscal significance. The development of a data-driven application addressing both use cases is the consequence of their individual but similar challenges. The use cases are described below, as well as the topic of knowledge management systems in tax administration, which also plays a unifying role across use cases and is addressed through our application.

2.1 Detecting Cross-Border VAT Fraud

Cross-border organized VAT fraud, also referred to as VAT carousel fraud or missing trader intra-community (MTIC) fraud are intricate setups, in which criminal networks engineer artificial supply chains and deploy shell companies to evade VAT payments.

By failing to remit VAT collected on sales, they trigger illegitimate refunds of input tax credits. This results in massive revenue losses for governments, amounting to billions of euros annually in Germany alone (European Commission, 2023).

The task for tax officers is to uncover cross-border VAT fraud in order to stop the fraud scheme as early as possible and enable criminal proceedings to be initiated against the perpetrators. These schemes usually involve complex webs of sham entities, straw persons, and fictitious transactions that obfuscate the fraudulent activities. Conventional analysis methods struggle to promptly detect suspicious transaction patterns amidst the labyrinth of contrived transactions and identities deployed by the criminal actors.

Much work has already been done on the use of data analytics to detect potential VAT fraud from available data. Different machine learning techniques and models have been proposed for detecting VAT fraud from VAT declarations and additional data (Bankole & Vara, 2022; Basta et al., 2009; Santos et al., 2022). Furthermore, unsupervised anomaly detection techniques have been presented for detecting VAT fraud (Alexopoulos et al., 2021; Vanhoeyveld et al., 2020). But also traditional rule-based risk analysis is used, including for cross-border VAT fraud detection in energy trading (Kim, 2017).

These challenges are increased through criminal groups rapidly adapting their sophisticated modus operandi. To keep up with these rapid changes, tax auditors need to exchange insights and knowledge frequently. Complex organizational systems exist to this end, but there is still potential for novel digital solutions to further support the knowledge exchange process.

2.2 *Combating Tax Avoidance by Multinational Corporations*

Tax avoidance by multinational corporations is a critical challenge in international taxation. These corporations are shifting domestic profits abroad through aggressive tax planning strategies, particularly transfer pricing manipulations. This practice results in substantial tax revenue losses for domestic authorities (Cristea & Nguyen, 2016). The task of tax auditors is to uncover these practices when auditing multinational corporations.

This process poses several challenges and the complexity of international tax law, combined with the need to understand diverse business models and financial instruments demands a high level of expertise from auditors. Because of this, comprehensive and novel tools to support the provision of information on these topics are needed. Additionally, the audit process often lacks sufficient automation, especially in data gathering and analysis, leading to inefficiencies.

2.3 *Knowledge Management Systems in Tax Administration*

Knowledge management systems are crucial in tax administration due to the inherently knowledge-based nature of the field (Subroto, 2020). Tax administration relies heavily on accurate, up-to-date, and accessible knowledge regarding tax laws, regulations, and procedures. Tax auditors from both our use cases are required to exchange knowledge on new and changing tax fraud and tax avoidance schemes frequently. Current knowledge management is only semi-structured and does not fully utilize modern knowledge management technologies. Especially the search for and identification of relevant information costs a lot of time (Dewi et al., 2022), making current systems inefficient.

3 Methodology

To develop a data-driven application for detecting cross-border VAT fraud and identifying tax avoidance schemes by multinational corporations we conducted expert interviews with auditors and tax experts from the Lower Saxony State Tax Office, as well as further national and international tax administrations, investigating current practices, methods, and workflows. In addition, a literature research on tax fraud detection and tax avoidance identification worldwide was conducted.

An iterative interdisciplinary development process of a prototypical application was chosen to ensure the relevance, usability, and closeness to practice. The application is deployed in a phased approach, allowing for iterative testing and refinement of its components. Initial deployment focuses on a controlled environment within the Lower Saxony State Tax Office, where the system was integrated with a subset of internal data sources to validate its core functionalities. A thorough testing strategy has been introduced at multiple levels (Anwar & Kar, 2019), including unit, integration, performance and user acceptance testing.

While not completed, the evaluation of the developed application is focused on decisive factors for the tax administration, which are cross-border VAT fraud detection rate, auditor workflow efficiency gain and optimization of human resources through case prioritization support (Gaie, 2023).

4 Application Concept

The application currently under development aims to address the described automation and efficiency gap by offering a novel platform for information provisioning and data analysis, as well as AI tools and knowledge management. The concept is developed with a focus on user-centered design and robust security, ensuring that it meets the requirements of tax officers and auditors.

The following subsections will delve into the specific components and functionalities of the application, highlighting how each element contributes to achieving the overall objectives. This introduction provides an overview of the application's design philosophy, architecture, and security and access management.

4.1 Design Philosophy

Although both use cases tackle very different tasks from a legal perspective, the underlying problems have obvious similarities: Data from different internal and external sources must be acquired, prepared and combined to allow for manual and automatic analysis for both cross-border VAT fraud detection and tax avoidance identification. In addition to that, information needs to be made easily available and knowledge must be consolidated and exchanged.

This motivated the development of a single application designed to cater to both use cases. Another benefit is that data from different tax fraud, evasion, and avoidance problems potentially enables a cross-domain approach (Gaie, 2023). Theoretically, the developed application could be expanded to additional tax fraud, tax evasion, and tax avoidance use cases, for example, the detection of false invoices (González, & Velásquez, 2013).

To effectively combat tax fraud and tax avoidance it is necessary to develop not only the data analytics aspect, but to have a larger, holistic approach (Gaie, 2023). While not encompassing all actors in taxation, even from the perspective of the auditors, this means improving the general workflow, increasing the knowledge transfer and enabling the usage of new and upcoming AI technologies in addition to developing new and better data analytics methods.

The goal of the developed application is to reduce the investigative effort required for auditors, to standardize the audit methodology across cases, to optimize resource allocation, meaning improving the case selection and focus area determination, and to increase knowledge transfer as well as general productivity for auditors.

4.2 Architecture

The application is developed as a web-based solution hosted on a central server within the tax administration. This approach ensures centralized data management, supports unified security policies—which are a high priority for the tax administration—and enables the use of cloud resources for demanding AI applications. Allowing algorithms to run on centralized servers with high computing power eliminates the need for wither resource-intensive processing on individual devices or secure transfer of large amounts of data. The use of a web application also simplifies continued development, maintenance, and improves accessibility and collaboration.

The developed application has four main components, which are data integration, data analytics, a knowledge management system and an AI tool suite. The users can access the latter three, while the data integration component provides the necessary combined data to the data analytics component. A schematic overview of the application with components and interactions is displayed in Fig. 1. The core components are described in further detail in the following.

Data Integration This component handles the automatic ingestion of data from diverse, partly unstructured data sources, including internal data sources (such as tax administration systems) and external data sources (e.g., commercial databases and public registries). It ensures the seamless integration and combination of structured and unstructured data, providing a unified data model for further analysis and processing. In a process called Record Linkage, we utilize machine learning techniques, such as neural networks, to identify and match records without unique identifiers across different data sources, enabling the integration of disparate data entities (Li et al., 2020). The application utilizes both relational and graph databases to store and manage an integrated data model, enabling efficient querying and analysis of complex, interconnected data structures. The graph database is essential for real-time querying and algorithmic processing of intricate networks—such as companies, transactions, and other relationships—which are critical to both use cases.

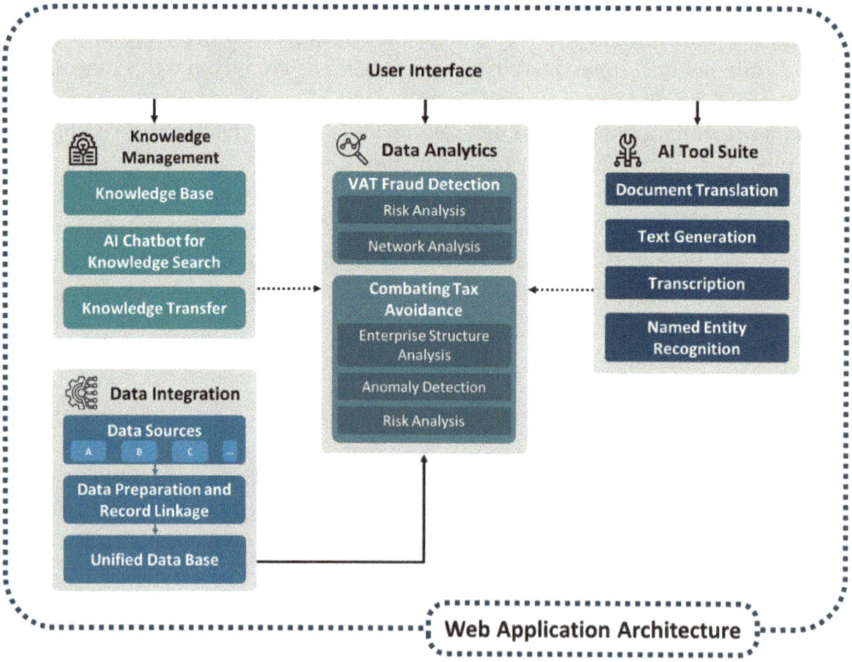

Fig. 1 Schematic architecture of the web application with key components and interactions (own illustration, icons made by Freepik from www.flaticon.com)

Data Analytics For the cross-border VAT fraud detection use case, we are developing a system that augments traditional rule-based scrutiny with advanced data science techniques for early detection of potential fraud relationships and entities from available data. We also use network analysis to unravel intricate supply chains and identify key nodes, and predictive analytics to assess risk and forecast evolving fraud typologies. This multi-pronged AI-driven approach provides capabilities to detect suspicious patterns early, visualize the complex entity networks, and adapt to emerging modus operandi—enhancing the effectiveness of fraud combating efforts. This monitoring system aims to enable more timely interventions against cross-border VAT fraud schemes.

In the combating tax avoidance by multinational corporations use case our focus is on the analysis of enterprise structures through knowledge graphs and geographic mapping (Didimo et al., 2020). We also implement methods for comparative analysis of financial data, anomaly detection in transaction networks using isolation forests, and tax avoidance risk scoring for multinational corporations. These systems enable a more effective and efficient selection of audit targets.

Knowledge Management System We are developing a knowledge management system integrated into our application that utilizes AI through information retrieval methods and large language models to improve the information search and extraction from documents. Additionally, we are searching for ways to make the knowledge transfer more accessible and agile to provide more auditors with relevant information easier and faster. By covering evolving cross-border VAT fraud and tax avoidance schemes, the knowledge management system can also potentially partially automate the required data analytics adaptation process.

AI Tool Suite The capabilities of AI and especially large language models have increased significantly in recent years, which became particularly clear through the public phenomenon of ChatGPT. Despite challenges and limitations large language models have the potential to significantly improve productivity of auditors by simplifying menial tasks (Zhao & Wang, 2024). Because of data privacy and official secrecy most commercial AI services and applications like ChatGPT cannot or can only to a very limited extent be used by auditors.

Within our application, we provide similar services within a dedicated AI tool suite through the usage of open-source models and software and the deployment of these models on servers within the tax administration. This allows auditors to use AI for general tasks like text generation, translation, transcription, summarization, named entity recognition, and more. The integration not only streamlines the auditors' workflow but also enables them to use these tools directly on data and analysis results from the application's analytics component.

4.3 Security and Access Management

Data privacy and security are essential in the tax administration due to the sensitivity of data. Because of this, we are integrating robust security and access management measures ensuring that sensitive data and information is protected from unauthorized access, maintaining compliance with regulatory requirements, such as data protection and tax secrecy.

We implemented thorough data governance strategies responsible for ensuring the quality, security, and compliance of the data used within the application (Khatri & Brown, 2010). This includes data governance policies, data lineage tracking, and access control mechanisms. Furthermore, the implementation of access and user group management allows for the administration and monitoring of access to data and services for the user groups cross-border VAT fraud detection and combating tax avoidance by multinational corporations.

5 Challenges and Preliminary Results

Developing a data-driven application for cross-border VAT fraud detection and combating tax avoidance by multinational corporations presents several challenges. First, defining the scope of the application is complex, as tax fraud and avoidance schemes evolve, requiring adaptability in detection methods. Integrating the application into the workflow of both user groups poses another challenge, particularly in ensuring seamless interoperability with existing systems. Additionally, differentiating the application from other ongoing projects and existing solutions is critical to demonstrating its unique value and effectiveness.

Technical challenges include addressing potential biases in AI algorithms, which may arise from imbalanced training data or historical enforcement patterns. Lastly, deployment challenges, such as scalability, system reliability, and real-time processing capabilities, must be overcome for effective field implementation. These, as well as challenges around data governance and infrastructure integration, are being addressed through close collaboration with IT security teams and compliance officers.

Despite these challenges, preliminary results indicate significant promise. Early feedback from both user groups has been positive, with users confirming the application's potential to streamline audit and investigation processes, increasing efficiency and enhancing risk assessment capabilities. These encouraging results provide a strong foundation for further development and refinement of the system, and plans are underway for a broader rollout across other regional tax offices.

6 Conclusions

Significant progress has been made in developing a multifunctional data-driven application to combat cross-border VAT fraud and tax avoidance by multinational corporations. Key achievements include a modular architecture integrating diverse data sources, specialized data analytics methods for identifying tax avoidance by multinational corporations and cross-border VAT fraud detection, as well as general AI tools and a tax expert knowledge management system.

While challenges remain and the development of the application is not yet complete, the study already demonstrates the potential of data science and AI to enhance tax authority efficiency by augmenting human expertise with intelligent tools in an increasingly digital economy. The application represents an important step in the digital transformation of tax compliance efforts, providing valuable insights for future strategies in combating cross-border VAT fraud and identifying tax avoidance schemes of multinational corporations.

Acknowledgements This research was funded by the German Lower Saxony State Tax Office within the "Tax Defense Analytics" project. The authors gratefully acknowledge this support.

References

Alexopoulos, A., et al. (2021). Detecting anomalies in heterogeneous population scale VAT networks.

Anwar, N., & Kar, S. (2019). Review paper on various software testing techniques & strategies . *Global Journal of Computer Science and Technology, 19*(2), 43–49.

Bankole, F., & Vara, Z. (2022). Artificial intelligence systems for value added tax collection via self organizing map (SOM). https://doi.org/10.21203/rs.3.rs-2216885/v1

Basta, S., et al. (2009). High quality true-positive prediction for fiscal fraud detection . In *2009 IEEE International Conference on Data Mining Workshops* (pp. 7–12). IEEE. ISBN 978-1-4244-5384-9. https://doi.org/10.1109/ICDMW.2009.59

Cristea, A. D., & Nguyen, D. X. (2016). Transfer pricing by multinational firms: New evidence from foreign firm ownerships. *American Economic Journal: Economic Policy, 8*(3), 170–202. ISSN 1945-7731. https://doi.org/10.1257/pol.20130407. https://www.aeaweb.org/articles?id=10.1257/pol.20130407

Dewi, A. P., Tjakratmadja, J. H., & Hendarman, A. F. (2022). A bibliometric analysis of research on knowledge management in tax administration. *JDM (Jurnal Dinamika Manajemen), 13*(2), 221–235. ISSN 2337-5434. https://doi.org/10.15294/jdm.v13i2.35831

Didimo, W., et al. (2020). Combining network visualization and data mining for tax risk assessment. *IEEE Access, 8*, 16073–16086. https://doi.org/10.1109/ACCESS.2020.2967974

European Commission. (2023). VAT gap in the EU: 2023 report. Publications Office of the European Union. https://doi.org/10.2778/911698

European Parliament. (2023). Directorate-General for Internal Policies of the Union, and A. Lejour. Good tax practices in the fight against tax avoidance The signalling role of FDI data. European Parliament. https://doi.org/10.2861/206080

Gaie, C. (2023). Struggling against tax fraud, a holistic approach using artificial intelligence . In *Recent advances in data and algorithms for e-government* (pp. 87–102). Springer. https://doi.org/10.1007/9783-031-22408-9_4

González, P. C. & Velásquez, J. (2013). Characterization and detection of taxpayers with false invoices using data mining techniques . *Expert Systems with Applications, 40*(5), 1427–1436. ISSN 0957-4174. https://doi.org/10.1016/j.eswa.2012.08.051

Jose Ordonez, P., & Hallo, M. (2019). Data mining techniques applied in tax administrations: A literature review. https://doi.org/10.1109/icedeg.2019.8734342

Khatri, V., & Brown, C. V. (2010) Designing data governance. *Communications of the ACM, 53*(1), 148–152. ISSN 0001-0782. https://doi.org/10.1145/1629175.1629210

Kim, S. Y. (2017). An analytical approach to detect MTIC MTEC VAT fraud invloving energy trading. *IOTA Papers*.

Li, Y., et al. (2020). Deep entity matching with pre-trained language models (Vol. 14). https://doi.org/10.14778/3421424.3421431

Santos, R., Moura, R., & Lobo, V. (2022). *Application of Kohonen maps in predicting and characterizing VAT fraud in a sub-Saharan African country* (pp. 74–86). Springer. https://doi.org/10.1007/978-3-031-15444-7_8

Subroto, G. (2020). *Knowledge management in tax administration: A case study in Indonesia* (pp. 203–235).

Vanhoeyveld, J., Martens, D., & Peeters, B. (2020). Value-added tax fraud detection with scalable anomaly detection techniques . *Applied Soft Computing, 86*. ISSN 15684946. https://doi.org/10.1016/j.asoc.2019.105895

Zhao, J., & Wang, X. (2024). Unleashing efficiency and insights: Exploring the potential applications and challenges of ChatGPT in accounting. *Journal of Corporate Accounting & Finance, 35*(1), 269–276 (2024). ISSN 1044-8136. https://doi.org/10.1002/jcaf.22663

Automated Detection and Tracking of Individual Piglets for Real-Time Behavioral Monitoring

Jan-Hendrik Witte, Dante-Nicolai Ahrens, Imke Traulsen, and Jorge Marx Gómez

Abstract Automated monitoring of piglet behavior offers potential benefits for animal welfare assessment in modern pig production. This paper describes a system for individual piglet detection and tracking in farrowing environments using standard RGB cameras. We evaluated several deep learning architectures to address the specific challenges of farrowing pens, where piglets often cluster together, partially occlude one another, and move unpredictably. Comparative testing showed that a YOLOv5l detector combined with a SORT tracking algorithm achieved promising results (98.6% MOTA, 93.9% IDF1) at 38 frames per second under controlled test conditions. The system extracts basic behavioral metrics including total distance traveled, average speed, and idle time, with mean squared errors of approximately 0.01 when compared to manual annotations. These preliminary metrics may contribute to automated activity analysis related to piglet welfare status. Our results suggest that individual tracking in farrowing environments is feasible with conventional hardware, though performance may vary under different farm conditions. The current implementation focuses on fundamental movement parameters, with limitations in tracking stationary objects and distinguishing between tightly clustered piglets. This work represents an incremental step toward precision livestock farming applications, potentially supporting more targeted welfare monitoring in commercial settings when further developed and validated across diverse environments. The datasets used in this study are made publicly available.

J.-H. Witte (✉) · D.-N. Ahrens · J. Marx Gómez
Department Very Large Business Applications (VLBA), Carl Von Ossietzky Universität Oldenburg, Oldenburg, Germany
e-mail: jan-hendrik.witte@uol.de

I. Traulsen
Institute of Animal Breeding and Husbandry, Faculty of Agricultural and Nutritional Sciences, Christian-Albrechts University Kiel, Kiel, Germany

J. Marx Gómez et al. (eds.), *Recent Advances in Information Systems*, Progress in IS,
https://doi.org/10.1007/978-3-032-14099-9_6

1 Introduction

The livestock industry faces competing challenges: increasing demand for animal products (Laan et al., 2024), economic pressures (Berckmans, 2014) and heightened societal expectations for animal welfare (Norton et al., 2019). As farms consolidate and expand herd sizes (Federal Statistical Office, 1950), producers may struggle to balance profitability with individualized animal care, particularly in the volatile pork sector (Federal Statistical Office, 2015).

Precision Livestock Farming (PLF) offers a technological solution through automated monitoring systems that enable effective management of larger herds while supporting animal welfare (Werkheiser, 2018). Deep Learning (DL)-based Computer Vision (CV) methods have emerged as prominent tools in PLF research (Liu et al., 2021; Witte et al., 2024), yet applications in the critical farrowing environment remain underdeveloped compared to piglet rearing or pig fattening operations.

The period immediately following birth is especially vulnerable for piglets, with behavioral monitoring crucial for survival and welfare (Farmer & Edwards, 2022). While various tracking technologies have been explored, including depth cameras (Matthews et al., 1997) and RFID sensors (Shi et al., 2016), conventional RGB cameras offer a potentially more cost-effective solution for commercial implementation (Cowton et al., 2019).

This research addresses the gap in automated monitoring systems for farrowing environments by developing a system that detects and tracks individual piglets using standard RGB camera footage in combination with DL-based approaches for Multiple Object Tracking (MOT). By applying advanced CV techniques to extract behavioral metrics, this work aims to create an automated system for improved piglet welfare assessment and monitoring in the critical post-birth period.

2 Related Work

Current research in automated livestock monitoring predominantly focuses on fattening pigs, with limited attention to piglets in farrowing environments. This section synthesizes key approaches in pig tracking and behavior analysis.

Most existing work targets the fattening context, using various detection and tracking methodologies. Jung et al. (2021) developed an end-to-end monitoring pipeline combining IoT with Faster R-CNN detection and OCTA tracking, though focusing primarily on monitoring rather than behavior analysis. Alameer et al. (2020) employed a YOLO-based detector for posture and drinking behavior identification, while Li et al. (2019) utilized SSD detection with DCF-based tracking. Van der Zande et al. (2021) demonstrated a YOLOv3 and SORT combination for basic tracking feasibility assessment.

For behavior analysis, Bergamini et al. (2021) tracked long-term behavioral trends using YOLOv3 and MOSSE tracking, extracting metrics including movement distance and activity states. Wutke et al. (2021) employed a custom CNN approach to detect specific points rather than bounding boxes, enabling analysis of social contact behaviors. Tangirala et al. (2021) developed STAR a comprehensive pipeline incorporating segmentation and posture recognition using transformers.

Multiple camera setups represent another approach, exemplified by Shirke et al. (2021), who developed systems to track pigs across multiple camera views while maintaining identity consistency.

Particularly relevant to our work, Cowton et al. (2019) conducted behavioral metric extraction from fattening pig tracking, though in a different context and with different real-time requirements than our farrowing environment focus.

Only Gan et al. (2021) specifically addressed piglet tracking in farrowing pens, using a single neural network with a Faster R-CNN detector and an association head. Their approach achieved higher accuracy than conventional methods did not extract behavioral metrics.

Our research differs from existing work in three key aspects: (1) we focus specifically on the complex farrowing environment, (2) we implement detection and tracking methods designed to operate in real-time, and (3) we extract behavioral metrics from tracking data to enable welfare assessment.

3 Model Selection

The core technical challenge of this research is designing a DL-based system to detect and track multiple individual piglets in the farrowing environment. This requires implementation of a MOT approach within the PLF context.

MOT is a CV subfield focused on identifying objects in video sequences and interpreting their movements between frames as trajectories (Luo et al., 2021). Unlike object detection, which simply identifies and marks objects (e.g., with bounding boxes), MOT assigns a unique identifier to each object of the same class, enabling distinction between multiple instances across frames (Luo et al., 2021).

To systematically evaluate available tracking methodologies, we classified approaches according to three key criteria proposed by Luo et al. (2021) and Ciaparrone et al. (2020): Initialization method, processing mode and output type:

Initialization. MOT approaches typically employ either Detection-Based Tracking (DBT), wherein objects in incoming frames are first processed by a dedicated object detection model that initializes them before a tracking model associates detections across frames to create trajectories—a "tracking-by-detection" approach (Wang et al., 2019) allowing automatic detection of objects entering and leaving the scene; or Detection-Free Tracking (DFT), where objects are initialized manually without an object detection model, focusing solely on the association problem, which enables isolation of tracker performance from detector results but requires manual initialization and cannot accommodate varying object numbers.

Processing mode. MOT algorithms operate either online, using only past and current observations to generate trajectories sequentially, suitable for real-time applications but potentially sacrificing performance, or offline, also called "batch tracking", utilizing information from future frames to optimize trajectories globally, typically achieving higher accuracy but unable to operate in real-time scenarios as they require the entire sequence.

Output type. MOT approaches can be deterministic, producing identical results when executed multiple times on the same sequence, or probabilistic, potentially generating varying results due to their underlying optimization methods.

The PLF context imposes specific functional requirements that directly inform the selection of appropriate tracking methodologies: As defined by Berckmans (2017), PLF systems must enable "continuous real-time monitoring" of livestock, necessitating technologies capable of immediate data processing and response. This real-time requirement emerges from the fundamental characterization of animals as CITD systems—complex, individually different, time-varying, and dynamic biological entities (Berckmans, 2017).

For MOT, these requirements dictate that only online processing algorithms can satisfy PLF's real-time needs, as offline batch processing introduces delays incompatible with timely intervention (Ciaparrone et al., 2020). Similarly, PLF's automation principle (Berckmans, 2017) necessitates detection-based tracking (DBT) methods rather than approaches requiring manual initialization.

Various MOT architectures apply the DBT design, categorized by whether they incorporate integrated object detection.

Simple Online and Realtime Tracking (SORT) represents a computationally efficient approach without integrated detection, combining Kalman filtering with the Hungarian algorithm. This resource-efficient method achieves real-time processing with good tracking results (Bewley et al., 2016). DeepSORT extends SORT by incorporating an appearance model, reducing identity switches but sacrificing processing speed (Wojke et al., 2017).

Object detection approaches suitable for DBT systems fall into two categories: one-stage and two-stage detectors. The YOLO family represents prominent efficiency-based one-stage detectors. Single Shot MultiBox Detector (SSD) differs from YOLO by predicting deviations to default boxes of various sizes (Liu et al., 2015). Among two-stage detectors, Faster R-CNN remains popular despite its relative age, consisting primarily of a Region Proposal Network followed by a classifier (Ren et al., 2015).

MOT architectures with integrated detection include Tracktor++, which uses the bounding box regression capability of the detector to predict object positions (Bergmann et al., 2019). Joint Detection and Embedding (JDE) unifies object detection and appearance embedding within a shared neural network (Wang et al., 2019), while FairMOT builds upon JDE to address fairness issues (Zhang et al., 2021a).

ByteTrack introduces the BYTE method that processes every detection box to recover lost objects (Zhang et al., 2021b).

Based on our findings, we selected two approaches: SORT with YOLOv5 as a baseline, chosen for its established use in PLF research and real-time capabilities, and ByteTrack as a representative integrated solution. ByteTrack demonstrates superior accuracy and processing rates compared to alternatives (Wang et al., 2019; Zhang et al., 2021a, 2021b), maintains real-time capability unlike Tracktor++, and represents the current state-of-the-art in integrated tracking approaches.

3.1 Data Collection

The research utilized video data from the "DigiSchwein" project at the Agricultural Chamber of Lower Saxony's research facility, initially using AXIS M3024-LVE color cameras later replaced by AXIS M3206-LVE models. Analysis was based on 27 h of footage recorded over three consecutive days (January 1–3, 2022) between 08:00 and 17:00 daily, provided as 60-min MP4 segments at 1920 × 1080 pixel resolution with 20 frames per second. The footage depicted a farrowing pen containing one sow and 13 piglets post-birthing. The camera was mounted centrally above the pen with horizontal frame edges aligned with pen walls, capturing portions of adjacent areas due to aspect ratio differences.

As seen in Fig. 1, showing an example image from the dataset, several environmental, PLF specific factors presented tracking challenges, including occlusion of piglets by other piglets or the sow, complete visual obstruction in the heating area where multiple piglets frequently gathered, and partial occlusion by pen fixtures such as dividing grates separating the sow's area and the feeding trough.

3.2 Data Preparation

For our object detection task, we created a dataset of 639 image files by systematically sampling the original video footage at fixed intervals. This sampling approach preserved the original distribution of behavioral patterns while providing sufficient data for robust model training. Following standard practices, we divided the dataset into training (70%, 447 images), validation (20%, 128 images), and testing (10%, 64 images) partitions.

For the tracking component, we analyzed the video footage and identified three distinct behavioral categories exhibited by piglets in the farrowing pen: (1) clustering, where piglets remain closely grouped in a stationary position; (2) nursing, where piglets gather at the sow's teats; and (3) movement, where piglets navigate independently throughout the pen. To create a balanced tracking dataset, we extracted one 15-s sequence (300 frames at 20 FPS) representing each behavioral category, resulting in a clustering sequence, a nursing sequence and a movement sequence.

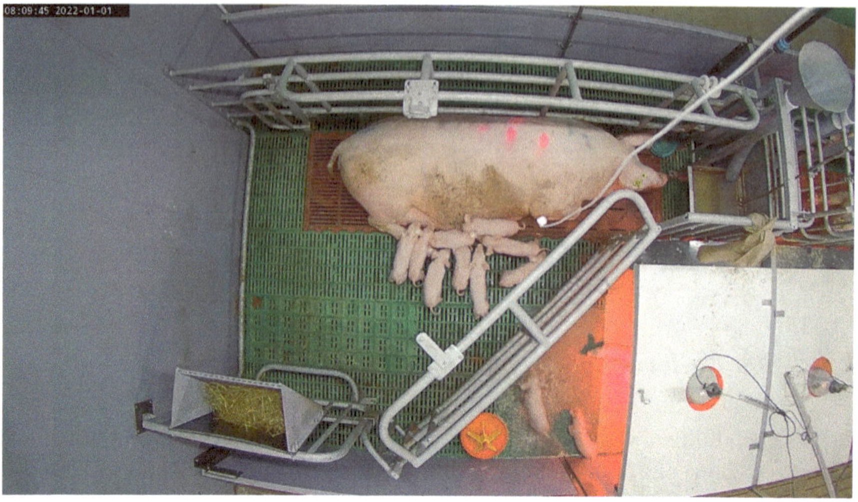

Fig. 1 Example image from the dataset

We additionally created a test sequence specifically designed to evaluate model performance across different behavioral categories. This sequence begins with clustered piglets that subsequently disperse and move toward the sow to nurse, effectively capturing transitions between all three behavioral states.

We employed the Computer Vision Annotation Tool (CVAT) (CVAT.ai, 2023) for dataset annotation, organizing our work into separate projects for detection and tracking tasks. For the object detection dataset, each frame was annotated with bounding boxes labeled as either "sow" or "piglet." The final object detection dataset contained 7823 annotations, representing an average object density of 12.24 objects per frame.

For the tracking dataset, we extended the annotation process to include persistent identity labels. CVAT's dedicated tracking functionality facilitated this by automatically propagating bounding boxes and their associated IDs across consecutive frames. We further optimized this process using CVAT's interpolation feature, which allowed strategic frame skipping based on object movement patterns. The resulting tracking dataset comprised 14,700 annotations across four sequences:

- Clustering sequence: 300 frames, 3000 annotations, 10 unique identities
- Nursing sequence: 300 frames, 3600 annotations, 12 unique identities
- Movement sequence: 300 frames, 4200 annotations, 14 unique identities
- Test sequence: 300 frames, 3900 annotations, 13 unique identities

3.3 *Train and Test Environment*

For training the YOLOv5 model, we utilized the repository by Milanet al. (2016), Stochastic Gradient Descent with a learning rate of 0.01 was employed as the optimization function. Training was terminated early if no performance improvement was observed after 100 epochs. Various data augmentation techniques were implemented, including image mosaics and mix-up, random image flipping, image rotation, image scaling, as well as modifications to HSV hue and saturation.

For the implementation of SORT, we utilized the official repository by Bewley et al. (2017)—therefore, the annotations of the tracking dataset were converted to the MOTChallenge (CVAT.ai, 2023) format. TrackEval[1] was used for the evaluation of SORT results.

For ByteTrack implementation, we accessed the official repository by Zhang et al. (2021a).

To assess the performance of the implemented object detection and tracking methods, we employed established metrics widely recognized in both general CV research (Everingham et al., 2010; Milan et al., 2016) and PLF applications (Alameer et al., 2020; Cowton et al., 2019).

For object detection performance, we used mean Average Precision (mAP), which evaluates both localization and classification accuracy. We report both mAP@0.5 (using a 0.5 IoU threshold) and the more stringent mAP@0.5:0.95 (averaging across IoU thresholds from 0.5 to 0.95 in 0.05 increments).

For tracking performance, we primarily used Multiple Object Tracking Accuracy (MOTA), which combines three error sources: false negatives, false positives, and identity switches. We complemented this with the Identity F1 Score (IDF1), which rewards trackers that maintain object identities over extended periods.

Additionally, we evaluated runtime performance to ensure real-time capability. Given our source video frame rate of 20 FPS, we defined real-time operation as the ability to process at least 20 frames per second (or a maximum of 50ms per frame), including all pre- and post-processing steps.

4 Results

Based on our selection methodology in Sec. 3, we chose SORT with YOLOv5 object detection and ByteTrack as our evaluation baseline for piglet tracking and activity classification. We will first present the object detection results.

We systematically evaluated all YOLOv5 model variants on our piglet detection dataset. Initial testing with a standard image size of 640px yielded key results presented in Table 1.

The baseline results show that the YOLOv5m variant offers the best balance between accuracy and speed, followed by YOLOv5l, and then YOLOv5s and

[1] https://github.com/JonathonLuiten/TrackEval, last visited: 28.03.2025.

Table 1 Baseline YOLOv5 model training results

Model	Batch size	Best epoch	mAP@0.5	mAP@0.5:0.95	VRAM	Training time	Speed (FPS)
YOLOv5n	66	237	0.963	0.722	6.69GB	00:36:00	144
YOLOv5s	35	251	0.970	0.776	6.54GB	00:36:25	135
YOLOv5m	18	283	0.979	0.808	6.41GB	00:58:04	100
YOLOv5l	10	272	0.979	0.820	6.62GB	01:27:39	66
YOLOv5x	5	256	0.982	0.829	5.92GB	02:43:37	48

YOLOv5x models. Due to their superior accuracy-speed ratios, we further optimized the YOLOv5s, YOLOv5m, and YOLOv5l models by investigating the impact of image size on performance. The results, using a batch size of 4, are presented in Table 2.

The results show that optimizing image size significantly improved detection performance across all models, with YOLOv5l at 960px achieving the best accuracy (0.841 mAP@0.5:0.95) while maintaining acceptable real-time performance (39FPS) for PLF applications.

Since the YOLOv5l model demonstrates the best accuracy while maintaining sufficient processing speed, we selected it for further evaluation in combination with SORT.

For MOT, we have compared two tracking approaches: SORT combined with our YOLOv5l detector, and ByteTrack in multiple configurations.

Initially, ByteTrack with its integrated YOLOX detector achieved a maximum MOTA score of 71.5% across various configurations, as shown in Table 3. To better isolate the tracking component's performance, we also created a custom integration of the BYTE tracking algorithm with our YOLOv5 models. The results show that combining ByteTrack with the YOLOv5 object detection model significantly increases the MOTA on our test sequence. For final comparison of the selected approaches, we choose the combination of ByteTrack and YOLOv5s since it offers the best performance of the tested combinations. The results are shown in Table 4.

Both approaches achieved high tracking performance with MOTA scores, though SORT + YOLOv5l showed a slight advantage in both MOTA (98.6 vs. 97.2%) and IDF1 scores (93.928 vs. 93.117%). Both methods satisfied our real-time processing requirement, but SORT + YOLOv5l offered significantly better speed (38 FPS vs. 23 FPS).

Table 2 Selected YOLOv5 models with optimized image sizes

Model	Image size	Best epoch	mAP@0.5	mAP@0.5:0.95	VRAM	Training time	Speed (FPS)
YOLOv5s	1920px	286	0.980	0.793	6.86GB	04:32:49	64
YOLOv5m	1280px	247	0.983	0.836	6.31GB	03:39:00	51
YOLOv5l	960px	262	0.984	0.841	6.12GB	03:11:56	39

Table 3 ByteTrack evaluation

Model	Image size	mAP@$^{0.5:0.95}$ @ Det	mAP$^{@0.5:0.95}$ @ Track	MOTA (%)	Speed (FPS)
YOLOX-m	1280px	0.818	0.854	71.5	32
YOLOX-l	960px	0.838	0.836	71.4	35
YOLOX-x	640px	0.828	0.843	70.4	35
YOLOv5s + BYTE	1920px	0.793	0.774	97.2	23
YOLOv5m + BYTE	1280px	0.834	0.802	96.1	14
YOLOv5l + BYTE	960px	0.841	0.816	95.6	9

Table 4 Comparison of SORT and BYTE with respective YOLOv5 model variants

Tracker	Detection model	mAP@$^{0.5:0.95}$ @ Det	mAP$^{@0.5:0.95}$ @ Track	MOTA (%)	IDF1 (%)	Speed (FPS)
SORT	YOLOv5l	0.838	0.836	98.6	93.9	38
BYTE	YOLOv5s	0.793	0.774	97.2	93.1	23

Based on these results, we selected the combination of YOLOv5l and SORT for our final evaluation of farrowing pen activities, offering the best balance of accuracy and processing efficiency for our piglet tracking.

4.1 Behavioral Metrics Extraction and Validation

Individual detection and tracking of piglets enables analysis of animal-specific behaviors. Building on our tracking implementation, we extracted three key behavioral metrics following Cowton et al. (2019): total distance traveled (cumulative movement across frames), idle time (frames with minimal movement), and average movement speed (mean velocity across the tracking sequence). These metrics indicate general activity levels that reveal animal health and welfare information. Our Python-based analysis pipeline processes tracking data from the YOLOv5 + SORT system, calculating metrics based on the Euclidean distance between an object's bounding box center points in consecutive frames. From this foundation, we derived total distance (sum of Euclidean distances), average speed (total distance divided by frame count), and idle time (frame transitions where distance ≤ 2 pixels, scaled from Cowton's threshold to account for our 20 FPS capture rate). To evaluate against ground truth, we implemented ID mapping between tracker output and annotations, enabling direct comparison of behavioral measurements. Table 5 presents normalized metrics extracted from our tracking system alongside ground truth values based on our annotated test sequence.

Our analysis reveals that the tracking system tends to slightly overestimate the distance traveled by the animals. Since both average speed and idle time are dependent

Table 5 Evaluation of defined behavioral metrics based on our test sequence

ID	Total distance		Average speed		Idle time	
	Ground truth	Tracker	Ground truth	Tracker	Ground truth	Tracker
1	0.000	0.198	0.000	0.198	1.000	0.857
2	0.525	0.664	0.525	0.683	0.271	0.115
3	0.072	0.228	0.073	0.228	0.922	0.730
4	0.627	0.785	0.627	0.816	0.295	0.139
5	0.755	0.805	0.755	0.876	0.230	0.135
6	0.715	0.710	0.715	0.710	0.189	0.197
7	0.918	1.000	0.918	1.000	0.045	0.000
8	0.628	0.646	0.628	0.646	0.361	0.340
9	0.528	0.611	0.528	0.611	0.578	0.484
10	0.250	0.275	0.250	0.275	0.684	0.660
11	0.872	0.840	0.872	0.841	0.148	0.172
12	0.771	0.805	0.771	0.805	0.271	0.213
13	0.792	0.804	0.792	0.816	0.258	0.262
MSE	0.010		0.012		0.010	

on the distance metric, this bias propagates to these measurements as well, resulting in overestimated speeds and underestimated idle times.

The most significant discrepancies were observed for objects 1 (the sow) and 3, both of which exhibit minimal movement throughout the video sequence. Despite the tracking system accurately detecting these objects across all frames, small inaccuracies in bounding box dimensions between consecutive frames lead to slight variations in the calculated center points. These minor errors accumulate over time, resulting in artificially increased distance measurements for predominantly stationary objects.

Despite these observations, the Mean Squared Error (MSE) values—0.0098 for total distance, 0.0120 for average speed, and 0.0101 for idle time—remain close to zero, indicating good overall agreement between the tracking system and ground truth.

5 Discussion and Limitations

Despite the system's promising performance, several limitations warrant discussion. The system demonstrates higher error rates when tracking predominantly stationary objects, such as the sow, where minor fluctuations in bounding box dimensions between frames accumulate over time, resulting in artificial movement measurements that particularly affect behavioral metrics relying on precise position tracking.

Our implementation and evaluation were based on footage from a single farrowing pen over just three days under consistent daytime lighting conditions, raising questions about generalizability to varying environmental conditions, different pen layouts, or alternative lighting scenarios.

While our system achieves real-time performance on dedicated GPU hardware (RTX 2070 SUPER), deployment in commercial farm environments may face practical challenges related to cost, maintenance, and computational infrastructure, with uncertain economic feasibility in the economically constrained livestock industry.

Additionally, our dataset and tracking approach treat all piglets as a single class, potentially masking important individual variations in behavior patterns, while the significant size difference between sows and piglets creates class imbalance issues affecting detection performance.

Finally, the behavioral metrics implemented (distance, speed, idle time) represent relatively basic movement parameters that may not fully capture complex social interactions, nursing behaviors, or distress indicators valuable for comprehensive welfare assessment.

6 Conclusion and Outlook

This research successfully developed and validated a real-time system for the automated detection and tracking of individual piglets in farrowing pens using conventional color camera footage. The system achieves high accuracy while maintaining real-time performance, enabling the extraction of behavioral metrics that provide insights into piglet welfare.

For tracking, the combination of YOLOv5l with SORT proved most effective, achieving a MOTA score of 98.6% and IDF1 of 93.9% at 38 FPS. This implementation enabled the reliable extraction of behavioral metrics with MSE values of 0.0098 (total distance), 0.0120 (average speed), and 0.0101 (idle time), comparable to results reported in similar research despite addressing the more challenging context of farrowing pens.

Expanding beyond basic movement parameters to detect and quantify specific behaviors such as nursing, social interactions, stereotypies, or signs of distress could significantly enhance the system's utility for welfare monitoring. Implementing class-aware tracking would enable the analysis of interactions between piglets and the sow, potentially revealing important maternal care patterns and highlighting welfare concerns. Evaluating the system's performance across varying environmental conditions, including different lighting scenarios, pen configurations, and farm settings would establish its practical reliability. Applying the system to longer-term monitoring would enable detection of behavioral trends and could reveal early warning signs of health or welfare issues before they become clinically apparent. Additionally, collaborating with animal science experts to interpret the extracted behavioral metrics could lead to the development of validated welfare indicators based on the automatically generated data. These extensions would build upon the foundation established in this research to create more comprehensive tools for automated monitoring of piglet welfare in commercial farming environments.

Acknowledgements The project is supported (was supported) by funds of the Federal Ministry of Food and Agriculture (BMEL) based on a decision of the Parliament of the Federal Republic of 2024 Germany. The Federal Office for Agriculture and Food (BLE) provides (provided) coordinating support for digitalization in agriculture as funding organization, grant number 28DE109A18.

Data Availability The applied datasets are available here: https://github.com/jahewite/PigMonitoringResearch.

Disclosure of Interests The authors have no competing interests to declare that are relevant to the content of this article.

References

Alameer, A., Kyriazakis, I., & Bacardit, J. (2020). Automated recognition of postures and drinking behaviour for the detection of compromised health in pigs. *Scientific Reports, 10*, 13665. https://doi.org/10.1038/s41598-020-70688-6

Berckmans, D. (2014). Precision livestock farming technologies for welfare management in intensive livestock systems. *Revue Scientifique Et Technique (International Office of Epizootics), 33*, 189–196. https://doi.org/10.20506/rst.33.1.2273

Berckmans, D. (2017). General introduction to precision livestock farming. *Animal Frontiers, 7*, 6–11. https://doi.org/10.2527/af.2017.0102

Bergamini, L., Pini, S., Simoni, A., Vezzani, R., Calderara, S., Eath, R., et al. (2021) Extracting accurate long-term behavior changes from a large pig dataset. In *VISIGRAPP 2021—Proceedings of the 16th international joint conference on computer vision, imaging and computer graphics theory and applications* (Vol. 5).

Bergmann, P., Meinhardt, T., & Leal-Taixe, L. (2019). *Tracking without bells and whistles.*

Bewley, A., Ge, Z., Ott, L., Ramos, F., & Upcroft, B. (2016). *Simple online and realtime tracking.*

Ciaparrone, G., Luque Sánchez, F., Tabik, S., Troiano, L., Tagliaferri, R., & Herrera, F. (2020). Deep learning in video multi-object tracking: A survey. *Neurocomputing, 381*, 61–88. https://doi.org/10.1016/j.neucom.2019.11.023

Cowton, J., Kyriazakis, I., & Bacardit, J. (2019). Automated individual pig localisation, tracking and behaviour metric extraction using deep learning. *IEEE Access, 7*, 108049–108060. https://doi.org/10.1109/ACCESS.2019.2933060

CVAT.ai Corporation. (2023). *Computer Vision Annotation Tool (CVAT).* https://github.com/cvat-ai/cvat

Everingham, M., van Gool, L., Williams, C. K. I., Winn, J., & Zisserman, A. (2010). The Pascal visual object classes (VOC) challenge. *International Journal of Computer Vision, 88*, 303–338. https://doi.org/10.1007/s11263-009-0275-4

Farmer, C., & Edwards, S. A. (2022). Review: Improving the performance of neonatal piglets. *Animal, 16*(Suppl 2), 100350. https://doi.org/10.1016/j.animal.2021.100350

Gan, H., Ou, M., Zhao, F., Xu, C., Li, S., Chen, C., & Xue, Y. (2021). Automated piglet tracking using a single convolutional neural network. *Biosystems Engineering, 205*, 48–63. https://doi.org/10.1016/j.biosystemseng.2021.02.010

Jung, W., Kim, S.-H., Hong, S.-P., & Seo, J. (2021). An AIoT monitoring system for multi-object tracking and alerting. *Computers, Materials & Continua, 67*, 337–348. https://doi.org/10.32604/cmc.2021.014561

Li, D., Chen, Y., Zhang, K., & Li, Z. (2019). Mounting behaviour recognition for pigs based on deep learning. *Sensors (Basel, Switzerland), 19*. https://doi.org/10.3390/s19224924

Liu, W., Anguelov, D., Erhan, D., Szegedy, C., Reed, S., Fu, C.-Y., & Berg, A. C. (2015). *SSD: Single Shot multibox detector.*

Liu, C., Zhou, H., Cao, J., Guo, X., Su, J., Wang, L., Lu, S., & Li, L. (2021). Behavior trajectory tracking of piglets based on DLC-KPCA. *Agriculture (Switzerland), 11*, 843. https://doi.org/10.3390/agriculture11090843

Luo, W., Xing, J., Milan, A., Zhang, X., Liu, W., & Kim, T.-K. (2021). Multiple object tracking: A literature review. *Artificial Intelligence, 293*, Article 103448. https://doi.org/10.1016/j.artint.2020.103448

Matthews, S. G., Miller, A. L., Clapp, J., Plötz, T., & Kyriazakis, I. (2016). Early detection of health and welfare compromises through automated detection of behavioural changes in pigs. *Veterinary Journal, 217*, 43–51. https://doi.org/10.1016/j.tvjl.2016.09.005

Milan, A., Leal-Taixe, L., Reid, I., Roth, S., & Schindler, K. (2016). *MOT16: A benchmark for multi-object tracking.*

Norton, T., Chen, C., Larsen, M. L. V., & Berckmans, D. (2019). Review: Precision livestock farming: building 'digital representations' to bring the animals closer to the farmer. *Animal, 13*, 3009–3017. https://doi.org/10.1017/S175173111900199X

Ren, S., He, K., Girshick, R., & Sun, J. (2015). *Faster R-CNN: Towards real-time object detection with region proposal networks.*

Shi, C., Teng, G., & Li, Z. (2016). An approach of pig weight estimation using binocular stereo system based on LabVIEW. *Computers and Electronics in Agriculture, 129*, 37–43. https://doi.org/10.1016/j.compag.2016.08.012

Shirke, A., Saifuddin, A., Luthra, A., Li, J., Williams, T., Hu, X., et al. (2021). *Tracking grow-finish pigs across large pens using multiple cameras.*

Federal Statistical Office. (2021). *Number of pigs per farm in Germany from 1950 to 2021.* https://de.statista.com/statistik/daten/studie/1174729/umfrage/anzahl-der-schweine-je-betrieb-in-deutschland/

Federal Statistical Office. (2023). *Monthly producer price index for livestock feed in Germany from January 2015 to November 2022.* https://de.statista.com/statistik/daten/studie/1303453/umfrage/monatliche-erzeugerpreise-fuer-futtermittel-nutztiere/

Bhavesh, T., Ishan, B., Dániel Lászl6, D. Gupta, R. T., & Devanshu, A. (2021). Livestock monitoring with transformer. In *British machine vision conference.*

van der Laan, S., Breeman, G., & Scherer, L. (2024). Animal lives affected by meat consumption trends in the G20 countries. *Animals: An Open Access Journal from MDPI, 14*, 1662. https://doi.org/10.3390/ani14111662

van der Zande, L. E., Guzhva, O., & Rodenburg, T. B. (2021). Individual detection and tracking of group housed pigs in their home pen using computer vision. *Frontiers in Animal Science, 2.* https://doi.org/10.3389/fanim.2021.669312

Wang, Z., Zheng, L., Liu, Y., Li, Y., & Wang, S. (2019). *Towards real-time multi-object tracking.*

Werkheiser, I. (2018). Precision livestock farming and farmers' duties to livestock. *Journal of Agricultural and Environmental Ethics, 31*, 181–195. https://doi.org/10.1007/s10806-018-9720-0

Witte, J. H., Heseker, P., Probst, J., Traulsen, I., Kemper, N., & Gómez, J. M. (2024). Image-based Tail Posture Monitoring of Pigs Bui (Ed.) 2024–Proceedings of the 57th Hawaii.

Wojke, N., Bewley, A., & Paulus, D. (2017). *Simple online and realtime tracking with a deep association metric.*

Wutke, M., Heinrich, F., Das, P. P., Lange, A., Gentz, M., Traulsen, I., et al. (2021). Detecting animal contacts—A deep learning-based pig detection and tracking approach for the quantification of social contacts. *Sensors, 21.* https://doi.org/10.3390/s21227512

Zhang, Y., Sun, P., Jiang, Y., Yu, D., Weng, F., Yuan, Z., et al. (2021). *ByteTrack: Multi-object tracking by associating every detection box.*

Zhang, Y., Wang, C., Wang, X., Zeng, W., & Liu, W. (2021a). FairMOT: On the fairness of detection and re-identification in multiple object tracking. *International Journal of Computer Vision, 129*, 3069–3087. https://doi.org/10.1007/s11263-021-01513-4

A Comprehensive Comparison of Federated Learning Frameworks

Jonas Boekhoff, Jan-Philipp Awick, Lars Steffens, Michael Karl, and Jorge Marx Gómez

Abstract Federated Learning is a Machine Learning paradigm that enables institutions to train models collaboratively on decentralized data without exchanging it. Training is performed locally on each participant's device, and only model updates are shared and aggregated, rather than raw data. The selection of a suitable framework is a critical step in designing a Federated Learning System, as this choice significantly influences the capabilities and limitations of the system. While previous research has primarily focused on comparing Federated Learning functionalities, other important criteria, such as usability, technical performance, and legal aspects have often been addressed only partially or overlooked entirely. To address this gap and support informed framework selection, we conducted a comprehensive comparison of Federated Learning frameworks. Frameworks were identified through a structured literature review and an analysis of GitHub repositories. The comparison covers five relevant comparison criteria: (1) Federated Learning functionalities, (2) user-friendliness, (3) technical aspects, (4) legal aspects, and (5) performance evaluation.

1 Introduction

The growing interest in automating business processes has led to the widespread adoption of Artificial Intelligence (AI) solutions by companies aiming to improve efficiency and optimize the use of human resources (Bughin, 2020). However, training Machine Learning (ML) models requires large amounts of high-quality training data to achieve robust model performance. The data volumes necessary for training are often either unavailable within a single organization or of insufficient quality (Yang et al., 2019). The exchange of data between organizations remains a chal-

J. Boekhoff (✉) · J.-P. Awick · J. Marx Gómez
Department of Business Informatics, Carl von Ossietzky Universität Oldenburg, Oldenburg, Germany
e-mail: jonas.boekhoff@uni-oldenburg.de

L. Steffens · M. Karl
Institute for AI Safety and Security, German Aerospace Center (DLR), Sankt Augustin, Germany

J. Marx Gómez et al. (eds.), *Recent Advances in Information Systems*, Progress in IS,
https://doi.org/10.1007/978-3-032-14099-9_7

lenge, and even intra-organizational data sharing is frequently impeded by competitive thinking, data protection concerns, and administrative complexities (Yang et al., 2019). A promising approach to overcome these challenges and to enable model training without exchanging data is Federated Learning (FL) (Schumann et al., 2023). The general idea is to train a model collaboratively with multiple devices or organizations while keeping the data localized. Therefore, each participant trains a model with local data, and only the model parameters of the participants are used for aggregation of a global model. The process is repeated iteratively until the desired model performance is achieved (Awick et al., 2023; Li et al., 2020). This enables collaborative training without exposing sensitive information, thereby ensuring privacy protection. To implement a Federated Learning System (FLS), organizations can utilize frameworks to streamline the development by enhancing code reusability and increasing productivity, leading to shorter development time and high-quality applications (Riehle and Gross, 1998). Although numerous FL frameworks are currently available, they vary considerably in terms of functionality, performance, and usability (Karimireddy et al., 2023). Without systematic and up-to-date framework comparisons, developers are often required to conduct time-consuming manual evaluations of the available options. This challenge is further intensified by the growing number of existing frameworks and functional diversity. A comprehensive comparison of FL frameworks can therefore offer valuable insights into their capabilities and support developers in selecting an appropriate framework for their specific use case. Therefore, this paper presents a current and comprehensive comparison of FL frameworks and extends previous comparative studies, e.g., Fonio et al. (2023), Kholod et al. (2021), and Riviera et al. (2023). The key contributions of this paper are:

- Identification and description of 29 existing FL frameworks
- Comparative analysis of five prominent FL frameworks based on a comprehensive criteria catalog, including FL functionalities, user-friendliness, technical aspects, and legal aspects
- Technical performance comparison of the selected FL frameworks using a benchmark dataset
- Support for informed selection of suitable FL frameworks, based on the findings of this paper

This paper is structured as follows: Sect. 2 presents related research based on a structured literature review. Section 3 provides an overview of currently available frameworks. Section 4 details the framework selection process, the definition of the criteria catalog, and the corresponding results. The technical performance evaluation is described in Sect. 5. Section 6 covers a summary and discussion of the findings, and Sect. 7 gives a conclusion and outlines possible directions for future work.

2 Related Work

This section describes the identification and analysis of related work. The methodology of the conducted literature review is described in Sect. 2.1 and the related studies are discussed in Sect. 2.2.

2.1 Literature Review Methodology

In order to identify related studies and existing FL frameworks in the literature, a structured literature review was conducted according to the approach outlined in Webster et al. (2002). The methodology followed a three-step process: *(1) identification of relevant articles*, *(2) backward search* within the identified studies, and *(3) forward search*. An overview of the search strategy and the results is illustrated in Fig. 1.

To identify potentially relevant studies, the following literature databases were used: ACM Digital Library, arXiv, and IEEE Xplore. The search terms shown in Fig. 1 had to be included in the title of the scientific papers.

The search strategy resulted in 730 unique papers. These were subjected to a sequential filtering process based on predefined inclusion criteria: first, a title and abstract review, followed by a full-text analysis of the remaining candidates. This process yielded nine relevant papers that met the requirement of systematically analyzing and comparing at least three FL frameworks in a minimum of three characteristics. The subsequent forward and backward search of the nine relevant papers resulted in a final corpus of 13 relevant studies.

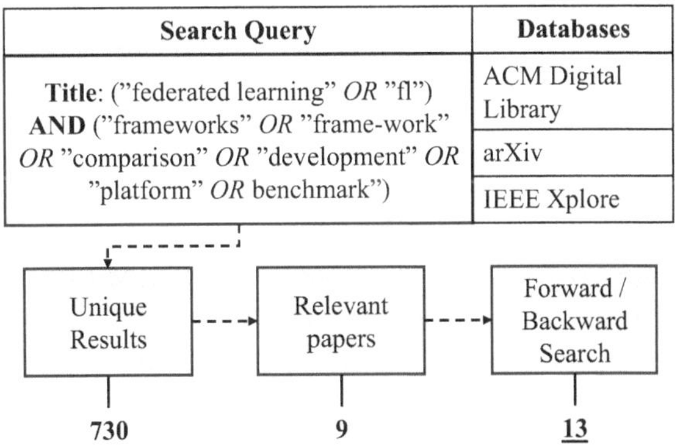

Fig. 1 Search strategy and filtering of identified literature

2.2 Overview of Existing Studies

Related research covers studies that perform a comparison of FL frameworks. The main difference in existing literature is the coverage of comparative aspects. To systematically compare existing related studies, we consider six different comparison aspects: *(1) functional features*, *(2) user-friendliness*, *(3) technical aspects*, *(4) legal aspects*, *(5) performance evaluation*, and *(6) publication year*.

The concept matrix, covering 13 studies, is presented in Table 1. The reviewed studies were published between 2020 and 2024, highlighting the need for up-to-date comparisons to include the evolution of FL frameworks. While most studies address functional features, user-friendliness, and technical aspects, legal aspects and performance evaluations are often not included. The only study that covers all six comparison aspects is Kholod et al. (2021). However, the study was published in 2021, its findings may no longer reflect the current state. Consequently, this paper fills this gap by providing an up-to-date framework comparison that includes all categories.

3 Available Federated Learning Frameworks

To identify the currently available FL frameworks, we examined two sources: (1) frameworks referenced in the relevant literature and (2) frameworks published in public GitHub repositories. Note, that frameworks from literature were only considered, if the code is publicly available, e.g., via a repository. The identification of frameworks on GitHub has been conducted by a GitHub search, using the following search string: *"federated learning framework"*. The final list consists of 29 publicly available frameworks. The following overview briefly characterizes the investigated frameworks. The characterization is based on findings from the research literature or, due to a lack of available literature, on information in the associated GitHub repositories and documentations:

APPFL is a modular framework for applying privacy-preserving FL. It allows users to implement their own algorithms, privacy techniques, communication protocols, and neural network structures (Ryu et al., 2022).

EdgeFL is a framework for Decentralized Federated Learning (DFL) that is characterized by a simple implementation with minimal code. It consists of two types of nodes: FL edge nodes for model training and registration nodes for coordinating the participating edge nodes (Zhang et al., 2023).

FederatedScope aims to address heterogeneity in FL scenarios, such as heterogeneity in training data, participant resources, and goals (Xie et al., 2022).

FedLab is a flexible framework that focuses on communication efficiency and optimization of the FL process. It supports both synchronous and asynchronous FL training (Zeng et al., 2023).

Table 1 Concept-matrix of related work compared to this paper

Source	Functional features	User-friendliness	Technical aspects	Legal aspects	Performance evaluation	Year
Rodríguez-Barroso et al. (2020)	✔	✔	✘	✘	✘	2020
Kholod et al. (2021)	✔	✔	✔	✔	✔	2021
Beltrán et al. (2022)	✔	✘	✔	✘	✘	2022
Riviera et al. (2023)	✔	✔	✔	✘	✘	2023
Karimireddy et al. (2023)	✔	✘	✘	✘	✘	2023
Witt et al. (2023)	✔	✘	✘	✘	✘	2023
Liu et al. (2023)	✔	✔	✔	✘	✔	2023
Li et al. (2023)	✔	✔	✔	✘	✘	2023
Khan et al. (2023)	✔	✘	✘	✘	✘	2023
Guendouzi et al. (2023)	✔	✔	✔	✘	✘	2023
Luzón et al. (2024)	✔	✔	✘	✘	✘	2024
Belal et al. (2024)	✔	✘	✘	✘	✘	2024
Fonio et al. (2023)	✘	✘	✘	✘	✔	2024
This paper	✔	✔	✔	✔	✔	2025

FedML supports various computational paradigms, including on-device training, distributed computing, and single-device simulations (He et al., 2020).

FedStellar focuses on DFL but also enables semi-decentralized and centralized FL. It supports training in a simulation on simulated devices, as well as training on distributed devices (Beltrán et al., 2024).

FLEX is designed to simplify scientific experiments with FL. It offers the ability to use different data distributions, privacy parameters, and communication strategies (Herrera et al., 2024).

FLGo includes over 40 implemented benchmarks for simulating FL in various environments and datasets (Wang et al., 2023).

Flower is designed for experiments with many participants and non-heterogeneous systems. It supports both simulation and execution on different devices. In addition, many different ML frameworks are supported (Beute et al., 2022).

FS-Real is a FL framework that is focused on the execution on edge devices like Smartphones. The framework offers possibilities like personalization, compression or asynchronous aggregation (Gao et al., 2023).

GFL focuses on DFL and implements a ring-shaped communication strategy. The authors implemented a new algorithm that is focused on communication efficiency and performance in decentralized FL. The framework is based on the use of blockchain (Hu et al., 2021).

HyFed primarily aims to provide a privacy-enhancing framework and a generic API for developers to implement their own FL algorithms (Nasirigerdeh et al., 2021).

IBM Federated Learning aims to enable easy implementation of FL on distributed devices (Ludwig et al., 2020).

OpenFed builds on the PyTorch framework and supports various other frameworks such as Hugging Face, MONAI, or MMCV. It supports different FL paradigms like Split Learning, Vertical Federated Learning (VFL), Horizontal Federated Learning (HFL), or DFL (Chen et al., 2023).

OpenFL aims to strengthen the use of FL in production environments. It is designed to be independent of the use case, industry, and ML framework (Reina et al., 2022).

Plato can simulate an unlimited number of clients on one or more GPUs. It also offers the possibility to apply FL on distributed devices (Li et al., 2023).

PyVertical is designed for VFL using neural networks (Romanini et al., 2021).

VFlair is designed for the use of VFL. It provides several implemented models, datasets, protocols, and modules for evaluating attacks and defenses (Zou et al., 2024).

XFL enables FL training on multiple devices and implements various privacy techniques such as homomorphic encryption, differential privacy, and secure multiparty computation (Wang et al., 2023).

AsyncFL supports synchronous, asynchronous, semi-synchronous, and personalized FL training. It offers various modules for implementing custom aggregation algorithms.[1]

FEDn is a FL framework designed for research purposes as well as for productive use in real environments. The framework is designed to scale with a large number of training participants as well as a large model size (Ekmefjord et al., 2022).

MetisFL emphasizes modularity, extensibility, and adaptability. It is use case independent and designed for real-world use (Stripelis et al., 2023).

TensorFlow Federated aims to simplify the conversion of implemented TensorFlow models into federated models.[2]

XayNet aims to enable FL on millions of edge devices. The frameworks implements asynchronous aggregation, and offers different interfaces to make use of it on edge devices like smartphones.[3]

PySyft aims to enable FL with common ML frameworks. It supports the use of differential privacy, secure multi-party computation, and homomorphic encryption (Ziller et al., 2021).

---

[1] https://github.com/NUAA-SmartSensing/async-FL.

[2] https://github.com/google-parfait/tensorflow-federated.

[3] https://github.com/xaynetwork/xaynet.

FATE is designed for use in industrial environments and implements various privacy techniques such as homomorphic encryption and secure multi-party computation.[4]

PaddleFL is based on PaddlePaddle and aims to enable the implementation and comparison of different FL algorithms. It supports HFL as well as VFL.[5]

p2pfl focuses on DFL using peer-to-peer networks and the gossip protocol. It is the only frameworks that is splitted into a paid and a free version.[6]

4 Functional Comparison of Federated Learning Frameworks

This section presents the functional comparison of FL frameworks. It first details the selection of frameworks for comparison (Sect. 4.1), presents the creation of a criteria catalog used for the comparison, encompassing key aspects like supported FL architectures, communication strategies, and privacy techniques (Sect. 4.2). In Sect. 4.3, the results of the comparison are presented, highlighting the strengths and weaknesses of each framework based on the defined criteria.

4.1 Selection of Frameworks for Comparison

Given the large number of available frameworks, only the most capable, up-to-date and widely used frameworks were considered for comparison. To ensure a systematic selection, the following process was applied:

[4] https://github.com/FederatedAI/FATE.

[5] https://github.com/PaddlePaddle/PaddleFL.

[6] https://github.com/p2pfl/p2pfl.

1. **Actuality**: Frameworks without an update in the last 6 months (as of June 2024) were excluded, reducing the initial set from 29 to 19 frameworks.
2. **Real-world applicability**: To assess real-world applicability, we examined whether the frameworks were designed exclusively for simulations or supported deployment on distributed devices. Of the 19 frameworks examined, 14 were designed for device deployment.
3. **Popularity**: To assess widespread adoption, we measured the Weighted Total Popularity Score (WTPS) developed by Al-Rubaye et al. (2020). This metric evaluates popularity based on stars and forks of GitHub repositories and the development of those metrics over time. The WTPS score represents the prevalence of the framework within the developer community. To ensure relevance, only widely used frameworks are taken into account. Frameworks with a WTPS value below 750 were considered insufficiently widespread and thus excluded.

Following this selection process, five frameworks were chosen for further analysis: PySyft, FATE, FedML, Flower, and OpenFL.

4.2 Creation of a Criteria Catalog

To define the criteria for comparing FL frameworks, we first examined the core features of FL, such as the FL architecture, and FL paradigms. Additionally, we analyzed related work to identify further relevant comparison criteria. To ensure a comprehensive evaluation, we also included criteria that, while maybe only met by a few frameworks, could provide valuable insights for developers, allowing for a more detailed assessment of their functionalities.

The selected criteria can be categorized into four areas: **(1) FL functionalities**, **(2) user-friendliness**, **(3) technical aspects**, and **(4) legal aspects**. The first category focuses on the core functionalities of FL frameworks, such as architecture, communication strategies, and optimization algorithms. The second category addresses aspects that influence ease of use, deployment, and integration, such as package managers and installation possibilities. The technical aspects consider implementation details like the used communication protocols or the supported programming language. Finally, legal aspects are represented by the type of license under which the framework is distributed. Table 2 presents the final criteria catalog, including a description for each criterion.

Table 2 Comparison criteria catalog for Federated Learning frameworks

FL functionalities	
FL architecture	Defines whether the framework follows a centralized or decentralized communication scheme
FL paradigms	Specifies supported FL paradigms, such as horizontal FL (HFL), vertical FL (VFL), and federated transfer learning (FTL)
Network topology	Specifies the possible network topologies in decentralized settings, such as star or ring topologies
Communication strategy	Determines whether synchronous or asynchronous communication is supported
Optimization algorithms	Lists implemented optimization algorithms for global model aggregation, such as FedAvg
Data privacy techniques	Lists supported privacy-enhancing technologies like differential privacy (DP) or homomorphic encryption
Attack simulation	Indicates whether attacks can be simulated in a FL system
Blockchain integration	Determines whether blockchain technology is supported for security or transparency
Simulation capabilities	Specifies whether the frameworks implement a simulation engine
Installation on edge devices	Specifies whether the frameworks can be installed on edge devices to use these for distributed settings
User-friendliness	
Extensibility	Allows easy modification, customization, and extension of the framework
Examplary implementation	Provides ready-to-use example implementations for quick deployment
Tutorial availability	Assesses whether learning resources are available
Technical aspects	
Communication protocols	Specifies the supported communication protocols (e.g., gRPC, MQTT, HTTP)
Supported operating systems	Lists compatible OS environments, such as Windows, Linux, or macOS
Installation options	Indicates the supported installation options, like package management systems (e.g., pip, conda) that can be used for installation
Programming language support	Lists programming languages the framework supports for development
ML framework compatibility	Lists supported ML frameworks, such as TensorFlow, PyTorch, or Scikit-learn
Legal aspects	
License	Specifies the licensing model (e.g., Apache 2.0)

4.3 Results of Functional Comparison

To evaluate each framework based on the defined criteria catalog, multiple source were considered. The research literature was analyzed as the primary source for all frameworks to identify which features from the criteria catalog were explicitly addressed. If there is no information about a certain criterion, information from GitHub repositories were utilized next, including the respective README files of the frameworks and, e.g., GitHub issues were examined. If no information regarding the criterion was found, the documentation of the frameworks was consulted. In the event that no information could be found in the mentioned sources, an explicit search was conducted within the code of the frameworks using the GitHub search. If no mention of a feature was found, it was considered as not supported by the framework.

Table 3 presents an overview of the supported features of each compared framework. A cross (✘) indicates that a criterion is not supported, while a checkmark (✔) indicates support of a criterion. If information regarding a criterion is unavailable or the criterion is not applicable due to missing features, a cross (✘) is used as an indicator.

Regarding the supported FL architectures, only the FedML frameworks supports DFL scenarios through various algorithms and different network structures. Except for Flower and PySyft, each frameworks implements synchronous as well as asynchronous communication. HFL and VFL are supported by all frameworks except PySyft, while FTL is only supported by Fate and OpenFL. The frameworks implement similar data protection techniques but with different compositions. FedML and Flower offer the most implemented optimization algorithms. All of the frameworks can be installed on edge devices. Most frameworks have been developed for Linux. FedML and Flower allow the use of mobile devices, primarily on Android. None of the frameworks supports blockchain FL. Only FedML implements different simulations of attacks on FLS. All of the Frameworks are based on Python and most of them have a rich support for multiple ML frameworks. FedML and Flower stand out for their compatibility. The frameworks are extensible and beginner-friendly, except PySyft. Each framework is licensed under the Apache 2.0 license.

5 Technical Comparison of Federated Learning Frameworks

This section delves into the technical performance of different FL frameworks through a practical experiment. The experimental setup is described in Sect. 5.1 and the results are discussed in Sect. 5.2.

Table 3 Comparison results of selected Federated Learning frameworks

	PySyft (v. 0.9)	FATE (v. 2.1.0)	FedML (v. 0.8.5)	Flower (v. 1.10.0)	OpenFL (v. 1.5.0)
FL functionalities					
FL architecture	✘	CFL	CFL, DFL	CFL	CFL
FL paradigms	✘	HFL, VFL, FTL	HFL, VFL	HFL, VFL	HFL, VFL, FTL
Network topologies	✘	✘	Decentralized, hierarchical, vertical, split	✘	✘
Communication strategy	✘	Synchronous, asynchronous	Synchronous, asynchronous	Synchronous	Synchronous, asynchronous
Optimization algorithms	✘	FedAVG	FedAvg, FedOpt, FedNova, FedGKT, FedProx, FedDyn, FedGan, FedNas, FedSeg	FedAdagrad, FedAdam, FedAvg, FedAvgM, FedMedian, FedOpt, FedProx, FedTrimmedAvg, FedXGB, FedYogi	FedAvg, FedProx, FedOpt, FedCurv
Data privacy techniques	DP, SMPC (SPDZ), HE (CKKS, Paillier)	DP, HE, SMPC (ABY-/SPDZ)	TEE, HE, SMPC	DP, secure aggregation	DP, PKI certificates, TEE
Attack simulation	✘	✘	✔	✘	✘
Blockchain integration	✘	✘	✘	✘	✘
Simulation capabilities	✘	✔	✔	✔	✔
Installation on edge devices	✔	✔	✔	✔	✔
User-friendliness					
Extensibility	✔	✔	✔	✔	✔
Examplary implementation	✘	✔	✔	✔	✔
Tutorial availability	✘	✔	✔	✔	✔
Technical aspects					
Communication protocols	Protocol buffers, Websockets	gRPC	MQTT + S3, MQTT, PyTorch RPC, gRPC, MPI	gRPC	gRPC
Supported operating systems	Linux, RHEL, MacOS, Windows	CentOS and other Linux distributions	Ubuntu, CentOS, Android, MacOS, Windows	Linux, Windows, macOS, Android, iOS	Linux
Installation options	Pip, Conda, Docker, Podman, Kubernetes	Pip, Conda, Docker, installation packages, Ansible	Pip, Conda, Docker, Kubernetes	Pip, Conda, Mamba, Docker	Pip, Docker
Programming language support	Python	Python	Python, Java	Python, Java	Python
ML framework compatibility	PyTorch	PyTorch	PyTorch, TensorFlow, Keras, MXNet, sklearn	PyTorch, TensorFlow, HuggingFace, PyTorch lightning, scikitlearn, JAX, TFLite, MONAI, fastai, MLX, XGBoost	PyTorch, TensorFlow
Legal aspects					
License	Apache 2.0	Apache 2.0	Apache 2.0	Apache 2.0	Apache 2.0

5.1 Experimental Setup

In order to establish a basis for a comparison across the frameworks, we defined a standardized FL training configuration. We implemented the training simulation using Python version 3.10 and PyTorch as the underlying ML framework. Note that the PySyft framework is excluded from the technical comparison, as it no longer supports FL functionalities from version 0.5 onwards. The MNIST dataset,[7] comprising grayscale images of handwritten digits with corresponding numerical labels, was employed for supervised learning in a classification scenario. This dataset was partitioned equally among 10 simulated clients, each performing 10 local training epochs. The global training process consisted of 20 rounds. A batch size of 32 and a learning rate of 0.001 were used. In each global round, all clients participated in the training process. The hardware used for the simulations consisted of a Intel Core i7-8086K, a NVIDIA GeForce RTX 2080 Ti and 64GB of RAM. We employed a simple convolutional neural network for image classification, consisting of two convolutional layers with ReLU activation and max-pooling, followed by three fully connected layers with ReLU activations.

5.2 Results of Technical Comparison

Figure 2 visualizes the development of accuracy and loss of the evaluated frameworks across 20 training epochs. The OpenFL framework achieves the best performance, demonstrating the highest accuracy and lowest loss. The values for FedML and Flower show a similar trend but the models generally converge more slowly than OpenFL. The framework FATE performs worse, with accuracy and loss stagnating for several epochs before improving, yet remaining significantly below the other frameworks. An analysis of CPU and GPU utilization revealed no hardware-related causes for the inferior performance. The gradual accuracy improvement from round 14 onward suggests that implementation errors are unlikely the primary factor, as this indicates some convergence capability of the model. However, a more detailed investigation of this anomaly should be conducted in future work to better understand its underlying causes. In contrast to the findings of other authors, like Kholod et al. (2021) or Liu et al. (2023), who did not address this specific challenge, FATE consistently delivered results on par with, or exceeding, those of other FL frameworks.

6 Summary and Discussion

The results show that there is a high degree of overlap regarding the supported FL features among the evaluated frameworks. However, for some categories, there

[7] https://yann.lecun.com/exdb/mnist/.

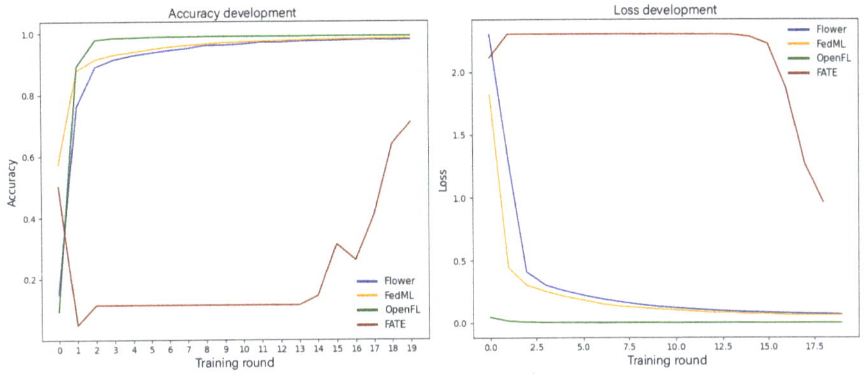

Fig. 2 Progression of model accuracy and loss for the frameworks Flower, FedML, OpenFL, and FATE over 20 training epochs. PySyft was excluded from this performance comparison

are notable distinctions. For instance, the DFL paradigm is exclusively supported by FedML. In addition, the variety of optimization algorithms implemented differs significantly, with only Flower and FedML offering a diverse list of approaches. Further, FedML is the only framework that supports attack simulations. It is notable, that none of the tested frameworks include a blockchain integration. Regarding user-friendliness, most frameworks, except of PySyft, provide similar levels of support for learning and implementation. All frameworks are licensed under Apache 2.0, allowing free use, code modification, derivative work creation, distribution, and commercial use. However, license terms require preserving copyright notices, documenting modifications, and licensing derivative works under Apache 2.0.

A notable deviation is PySyft, which has shifted focus from FL to remote data science. To the best of our knowledge, FL features are no longer natively supported in versions beyond 0.5. This shift underscores the need for up-to-date framework comparisons, as older studies classified PySyft as an FL framework. To support developers in selecting the most suitable framework for their use case, comprehensive and up-to-date comparisons are essential. Beyond framework comparison, an important limitation in existing comparative studies is the lack of transparency regarding the framework selection criteria. Many related studies identify and analyze a set of FL frameworks without explicitly detailing the selection process. To address this issue, we applied a set of selection criteria for framework selection, ensuring reproducibility and transparency.

7 Conclusion and Future Work

The selection of a suitable FL framework is a crucial step in implementing an Federated Learning System, as the frameworks vary significantly in their supported functionalities. Depending on the specific use case, different requirements must be met. Given the continuous updates of existing frameworks and the emergence of new ones, the need for up-to-date comparisons arises. To support the framework selection process, we propose a comprehensive comparison of current FL frameworks, evaluating their FL functionalities, user-friendliness, technical aspects, legal aspects, and technical performance. Additionally, we provide an overview of existing comparative studies, a list of published frameworks and a derived criteria catalog for comparison.

Potential directions for future work include extending the comparison to incorporate additional frameworks and conducting a performance evaluation in distributed environments. As the development of frameworks remains a dynamic field, automatic approaches for framework comparison are a promising direction toward a continuous evaluation.

References

Al-Rubaye, A., & Sukthankar, G. (2020). Scoring popularity in GitHub. In *Proceedings of the 2020 International Conference on Computational Science and Computational Intelligence (CSCI)* (pp. 217–223).

Awick, J. P., Schumann, G., & Gómez, J. M. (2023). Exploring federated learning for data integration: A structured literature review. In *Proceedings of the 2023 International Conference on Big Data, Knowledge and Control Systems Engineering (BdKCSE)* (pp. 1–8). IEEE.

Belal, Y., Mokhtar, S.B., Haddadi, H., Wang, J., & Mashhadi, A. (2024). *Survey of Federated Learning Models for Spatial-Temporal Mobility Applications*.

Beltrán, E. T. M., Gómez, N. L. P., Feng, C., Sánchez, P. M. S., Bernal, S. L., Bovet, G., Pérez, M. G., Pérez, G. M., & Celdrán, A. H. (2024). Fedstellar: A platform for decentralized federated learning *242*, 122861.

Beltrán, E. T. M., Pérez, M. Q., Sánchez, P. M. S., Bernal, S. L., Bovet, G., Pérez, M. G., Pérez, G. M., & Celdrán, A. H. (2022). Decentralized federated learning: Fundamentals, state of the art, frameworks, trends, and challenges.

Beutel, D. J., Topal, T., Mathur, A., Qiu, X., Fernandez-Marques, J., Gao, Y., Sani, L., Li, K. H., Parcollet, T., de Gusmão, P. P. B., & Lane, N. D. (2022). Flower: A friendly federated learning research framework.

Bughin, J. (2020). Artificial intelligence, its corporate use and how it will affect the future of work. In *Capitalism, Global Change and Sustainable Development* (pp. 239–260). Springer.

Chen, D., Tan, V., Lu, Z., & Hu, J. (2023). OpenFed: A comprehensive and versatile open-source federated learning framework.

Ekmefjord, M., Ait-Mlouk, A., Alawadi, S., Åkesson, M., Singh, P., Spjuth, O., Toor, S., & Hellander, A. (2022). Scalable federated machine learning with FEDn.

Fonio, S. (2023). Benchmarking federated learning frameworks for medical imaging tasks. In G. L. Foresti, A. Fusiello, & E. Hancock (Eds.), Image analysis and processing: ICIAP 2023 workshops (pp. 223–232). Springer.

Gao, D., Chen, D., Li, Z., Xie, Y., Pan, X., Li, Y., Ding, B., & Zhou, J.: FS-real: A real-world cross-device federated learning platform *16*(12), 4046–4049.

Guendouzi, B. S., Ouchani, S., EL Assaad, H., & EL Zaher, M. (2023). A systematic review of federated learning: Challenges, aggregation methods, and development tools *220*, 103714.

He, C., Li, S., So, J., Zeng, X., Zhang, M., Wang, H., Wang, X., Vepakomma, P., Singh, A., Qiu, H., Zhu, X., Wang, J., Shen, L., Zhao, P., Kang, Y., Liu, Y., Raskar, R., Yang, Q., Annavaram, M., & Avestimehr, S. (2020). FedML: A research library and benchmark for federated machine learning.

Herrera, F., Jiménez-López, D., Argente-Garrido, A., Rodríguez-Barroso, N., Zuheros, C., Aguilera-Martos, I., Bello, B., García-Márquez, M., & Luzón, M. V. (2024). FLEX: FLEXible federated learning framework.

Hu, Y., Zhou, Y., Xiao, J., & Wu, C. (2021). GFL: A decentralized federated learning framework based on blockchain.

Karimireddy, S. P., Veeraragavan, N. R., Elvatun, S., & Nygård, J. F. (2023). Federated learning showdown: The comparative analysis of federated learning frameworks. In *Proceedings of the 2023 eighth international conference on fog and mobile edge computing (FMEC)* (pp. 224–231).

Khan, A., Thij, M., & Wilbik, A. (2023). Vertical federated learning: A structured literature review.

Kholod, I., Yanaki, E., Fomichev, D., Shalugin, E., Novikova, E., Filippov, E., & Nordlund, M. (2021). Open-source federated learning frameworks for IoT: A comparative review and analysis *21*(1), 167.

Li, B., Su, N., Ying, C., & Wang, F. (2023). Plato: An open-source research framework for production federated learning. In *Proceedings of the ACM turing award celebration conference—China 2023* (pp. 1–2). ACM TURC '23, Association for Computing Machinery.

Li, Q., Wen, Z., Wu, Z., Hu, S., Wang, N., Li, Y., Liu, X., & He, B. (2023). A survey on federated learning systems: Vision, hype and reality for data privacy and protection *35*(4), 3347–3366.

Li, L., Fan, Y., Tse, M., & Lin, K. Y. (2020). A review of applications in federated learning. *Computers and Industrial Engineering, 149*, 106854.

Liu, X., Shi, T., Xie, C., Li, Q., Hu, K., Kim, H., Xu, X., Vu-Le, T. A., Huang, Z., Nourian, A., Li, B., & Song, D. (2023). UniFed: All-in-one federated learning platform to unify open-source frameworks.

Ludwig, H., Baracaldo, N., Thomas, G., Zhou, Y., Anwar, A., Rajamoni, S., Ong, Y., Radhakrishnan, J., Verma, A., Sinn, M., Purcell, M., Rawat, A., Minh, T., Holohan, N., Chakraborty, S., Whitherspoon, S., Steuer, D., Wynter, L., Hassan, H., Laguna, S., Yurochkin, M., Agarwal, M., Chuba, E., Abay, A. (2020). IBM federated learning: An enterprise framework white Paper V0.1.

Luzón, M. V., Rodríguez-Barroso, N., Argente-Garrido, A., Jiménez-López, D., Moyano, J. M., Del Ser, J., Ding, W., & Herrera, F. (2024). A tutorial on federated learning from theory to practice: Foundations, software frameworks, exemplary use cases, and selected trends *11*(4), 824–850.

Nasirigerdeh, R., Torkzadehmahani, R., Matschinske, J., Baumbach, J., Rueckert, D., & Kaissis, G. (2021). HyFed: A hybrid federated framework for privacy-preserving machine learning.

Reina, G. A., Gruzdev, A., Foley, P., Perepelkina, O., Sharma, M., Davidyuk, I., Trushkin, I., Radionov, M., Mokrov, A., Agapov, D., Martin, J., Edwards, B., Sheller, M. J., Pati, S., Moorthy, P. N., Wang, S., Shah, P., & Bakas, S. (2022). OpenFL: An open-source framework for federated learning *67*(21), 214001.

Riehle, D., & Gross, T. (1998). Role model based framework design and integration (pp. 117–133).

Riviera, W., Galazzo, I. B., & Menegaz, G. (2023). FeLebrities: A user-centric assessment of federated learning frameworks *11*, 96865–96878.

Rodríguez-Barroso, N., Stipcich, G., Jiménez-López, D., Ruiz-Millán, J. A., Martínez-Cámara, E., González-Seco, G., Luzón, M. V., Veganzones, M., & Herrera, F. (2020). Federated learning and differential privacy: Software tools analysis, the Sherpa.ai FL framework and methodological guidelines for preserving data privacy *64*, 270–292.

Romanini, D., Hall, A. J., Papadopoulos, P., Titcombe, T., Ismail, A., Cebere, T., Sandmann, R., Roehm, R., & Hoeh, M. A. (2021). PyVertical: A vertical federated learning framework for multi-headed SplitNN.

Ryu, M., Kim, Y., Kim, K., & Madduri, R. K. (2022) APPFL: Open-source software framework for privacy-preserving federated learning.

Schumann, G., Awick, J. P., & Gómez, J. M. (2023). Natural language processing using federated learning: A structured literature review. In *Proceedings of the 2023 IEEE international conference on artificial intelligence, blockchain, and internet of things (AIBThings)* (pp. 1–7). IEEE.

Stripelis, D., Anastasiou, C., Toral, P., Asghar, A., & Ambite, J. L. (2023). MetisFL: An embarrassingly parallelized controller for scalable and efficient federated learning workflows.

Wang, Z., Fan, X., Peng, Z., Li, X., Yang, Z., Feng, M., Yang, Z., Liu, X., & Wang, C. (2023). FLGo: A fully customizable federated learning platform.

Wang, H., Zhou, Y., Zhang, C., Peng, C., Huang, M., Liu, Y., & Zhang, L. (2023). XFL: A high performace, lightweighted federated learning framework.

Webster, J., & Watson, R. (2002) Analyzing the past to prepare for the future: Writing a literature review *26*.

Witt, L., Heyer, M., Toyoda, K., Samek, W., & Li, D. (2023) Decentral and incentivized federated learning frameworks: A systematic literature review *10*(4), 3642–3663.

Xie, Y., Wang, Z., Gao, D., Chen, D., Yao, L., Kuang, W., Li, Y., Ding, B., & Zhou, J. (2022) FederatedScope: A flexible federated learning platform for heterogeneity.

Yang, Q., Liu, Y., Chen, T., & Tong, Y. (2019). Federated machine learning: Concept and applications *10*(2), 12:1–12:19.

Zeng, D., Liang, S., Hu, X., Wang, H., & Xu, Z. (2023). FedLab: A flexible federated learning framework *24*(100), 1–7.

Zhang, H., Bosch, J., & Olsson, H. H. (2023). EdgeFL: A lightweight decentralized federated learning framework.

Ziller, A., Trask, A., Lopardo, A., Szymkow, B., Wagner, B., Bluemke, E., Nounahon, J. M., Passerat-Palmbach, J., Prakash, K., Rose, N., Ryffel, T., Reza, Z. N., & Kaissis, G. (2021). PySyft: A library for easy federated learning. In M. H. Rehman, & M. M. Gaber (Eds.), Federated learning systems: Towards next-generation AI (pp. 111–139). Springer.

Zou, T., Gu, Z., He, Y., Takahashi, H., Liu, Y., & Zhang, Y. Q. (2024). VFLAIR: A research library and benchmark for vertical federated learning.

Integrating Differential Privacy Mechanisms Within a Simulated Autonomous Vehicle Environment: A Critical Evaluation

Thejmeela Seetamonee and Girish Bekaroo

Abstract Autonomous vehicles and their supporting technologies, such as Vehicle-to-Everything (V2X) connectivity, are becoming more and more common with advancements in intelligent mobility. The AV environment allows cars that are wirelessly connected to adjacent infrastructure and other cars to exchange location-related data. These sensor data, which amounts to personal data under the definition of the EU GDPR ("identifiable"), are processed by AV manufacturers and related third parties. The processing of sensor data has privacy implications and requires adherence to data protection regulatory requirements. The aim of this study was to investigate the challenges posed by connected and automated vehicles from a data privacy perspective and evaluate the effectiveness of two existing differential privacy mechanisms, namely LaPlace and Gaussian, within a simulated autonomous vehicle environment. CARLA, the simulation AV environment, and the AV were configured to collect sensor data. The sensors in scope were collision intensity, collision location, lane invasion, obstacle detection, GNSS, IMU accelerometer, and IMU gyroscope. The differential privacy mechanisms were applied to add noise to the original sensor data. The original and the noise sensor data were compared based on the absolute difference measurement and the calculation of MEA. Mathematically, it was concluded that Gaussian was more effective in adding noise to the sensor data. However, there is a trade-off between data privacy and data usability making LaPlace more effective at keeping the data reliable for future analysis.

T. Seetamonee · G. Bekaroo (✉)
Middlesex University Mauritius, Flic-en-Flac, Mauritius
e-mail: g.bekaroo@mdx.ac.mu

T. Seetamonee
e-mail: TS984@live.mdx.ac.uk

J. Marx Gómez et al. (eds.), *Recent Advances in Information Systems*, Progress in IS,
https://doi.org/10.1007/978-3-032-14099-9_8

1 Introduction

Between 1980 and 2003, basic research on autonomous mobility was conducted by university research centres, and sometimes collaboration with transportation agencies and automaker companies (Anderson et al., 2014). With significant advancement in research work and communication technologies over the years, the concept and design of connected and automated vehicles became more sophisticated (Mallozi et al., 2019; Milakis et al., 2017). Vehicles that are wirelessly linked to other vehicles and nearby infrastructure can communicate details including location, speed and direction, to detect the likelihood of collisions (Seo et al., 2016). By 2025, the global market for autonomous vehicles is projected to be valued at $400 billion USD, and by 2050, driverless highways are predicted to become the standard in developed nations (Statista, 2021). With the projected increase in autonomous cars, the volume of data being processed also increases exponentially. Consequently, Connected and Automated Vehicles (CAVs) are expected to be more susceptible to cyber and physical threats as they become more programmable and interconnected (Mohan Krishna et al., 2020). As the attack surface grows, maintaining personal data secure from outsider threats is even more challenging.

Since the enforcement of GDPR in 2018, data subjects are more aware of their right to autonomy and privacy over their personal data. As autonomous vehicles collect an extensive volume of personal data (e.g. location, camera recordings, driving analytics), the privacy risks posed to data subjects increases. The concept of privacy is multifaceted and intricate, sometimes being interpreted as legal, philosophical or even technical (Heurix et al., 2015). With the focus shifting towards adopting a Privacy-by-Design (PbD) approach, the use and further development of Privacy-Enhancing Technologies (PETs) has significantly increased. The main concern is that privacy preservation and usability are oftentimes at odds with each other (Zavvos et al., 2022). To address this issue, this study aims to investigate the challenges posed by AVs from a data privacy perspective and evaluate the effectiveness of two existing differential privacy mechanisms, namely LaPlace and Gaussian, within a simulated AV environment.

The research paper is structured as follows. The next section provides a background on key topics investigated in this paper, notably AVs and differential privacy. Section 3 covers related research in the field of differential privacy. Section 4 explores the methodology applied to achieve the aim of this research. Section 5 provides an overview of the simulation environment and collection of sensor data. Section 6 is the core of the research paper, whereby the sensor data are compared for effectiveness. Section provides a final note on the research paper and explores the possibilities of future research.

2 Theoretical Background

2.1 Background Study

The idea of automating vehicles was first proposed in 1918, and General Motors introduced the first automated car design in 1939 (Faisal et al., 2019; Pendleton et al., 2017). Since then, academic research has been carried out together with corporate pilot AVs and reward competitions that leverage deep machine learning and game-based technologies (Stilgoe, 2018; Woo et al., 2021). The Society of Automotive Engineers (SAE) established the J3016 standard and classified six levels of driving automation from Level 0 (no automation) to Level 5 (full vehicle autonomy) (ANSI, 2021).

AV Communication Technology: V2X is the main wireless communication channel used for transmission of data between autonomous vehicles and the surrounding environment, including pedestrians, road traffic signals, physical infrastructures and cloud computing infrastructures (Hasan et al., 2020). Given the level of connectivity and independence offered by V2X, it is essential to identify security and privacy issues. These include the risk of identifying and tracking individual vehicles, the creation and disruption of genuine message traffic, the danger of vehicular malware, and the possible use of botnets (MacHardy et al., 2018).

AV and Privacy Concerns AVs are susceptible to cybersecurity attacks targeting sensors data at various levels – collection point, processing level or at rest in storage infrastructure (for e.g. cloud) (Gupta et al., 2023). Different actors, such as the manufacturer, drivers, passengers, pedestrians and buyers of data, within the environment are impacted by the Regulation (EU) 2016/679 General Data Protection Regulation (GDPR) (Mulder & Vellinga, 2021).

AV manufacturers, as data controllers, have an obligation to comply with the data protection principles, under Article 5 of GDPR. It is the responsibility of the AV manufacturers to abide by the data controller obligations, which includes adopting data protection by design and by default approach and maintaining adequate security measures.

Taking into consideration the requirements of Article 25 of GDPR, AV environment and connected technologies should be designed in such a way that limit the personal data processed, embed privacy-protective default settings and give data subjects autonomy over their personal data. Several mechanisms, known as Privacy Enhancing Technologies (PETs), have been designed to implement the PbD approach. Most common PETs are Homomorphic Encryption, Secure multiparty Computation, Federated Learning, Trusted execution environments, Zero-knowledge proofs, Synthetic data and Differential Privacy (ICO, 2023).

Differential Privacy (DP) offers a formal mathematical assurance about the indistinguishability of individuals through randomised noise injection (ICO, 2023). It protects an individual's privacy while allowing deductions and the identification of patterns from the data set. The parameters include the privacy limitations, $\in$ and δ, also referred to as the privacy loss parameter (epsilon) and privacy leakage, respectively (Muthukrishnan & Kalyani, 2023). DP was brought forward by Dwork et al. (2006) with the below generalised formula:

$$\left|\log\left(\frac{\Pr(R(s(x))) \in Q}{\Pr(R(s(x'))) \in Q}\right)\right| \leq \in, \tag{1}$$

where $\in$ is the privacy loss parameter, and s refers to data queries. If all data sets (x, x') differ by one person $(d(x, x') = 1)$ and all result subsets $Q \subseteq Y$ to query s (Y denoting image of R), then a perturbation in algorithm R yields ρ-DP (probabilistic DP) (Liu, 2019).

The most practiced mechanisms are LaPlace and Gaussian mechanisms which are heavily relied upon for research and development within the field of different privacy (Xu et al., 2024). LaPlace theory was developed by a mathematical scholar, Pierre-Simon, Marquis de LaPlace in 1814 (Weinberger, 2012) and then adopted for DP. Below is the simple e-differentially private mathematical formula for LaPlace mechanism:

$$\in = \Delta f / b \tag{2}$$

where epsilon is equal to sensitivity over scale value. Based on the properties of the value (data type), the amount of noise being applied must be computed.

Gaussian Mechanism was developed by mathematician and astronomer Carl Friedrich Gauss.

$$f(x|\mu, b, p) = \frac{p}{2b\Gamma\left(p^{-1}\right)} \exp\left\{-\left(\frac{|x-\mu|}{b}\right)^{p}\right\}. \tag{3}$$

where Gaussian distribution (μ, b, p) with location parameter μ, scale parameter $b >$ 0, shape parameter $p > 0$. The Gaussian distribution with mean 0 and variance $b2/2$ is the generalised Gaussian distribution when $p = 2$. Then the Laplace distribution is leveraged with mean μ and variance 2b2 to become the generalised Gaussian distribution when $ep = 1$ (Liu, 2019).

3 Related Works

3.1 Existing Privacy-Preserving/Enhancing Mechanisms

In line with GDPR, many works have proposed DP-based approaches to adopt stringent privacy guarantees unstructured data, such as text, images, videos, and audio (Zhao & Chen, 2022). A previous study (Mohan Krishna et al., 2020) proposed a pseudonym mechanism in mix zones to ensure anonymity of the architecture of the AV and preserving the personal data of the driver. The study focussed on Location Privacy and the proposed pseudonym mix-zone mechanism can be further improved to accommodate more specific scenarios, such as peak hours or high vehicle density at traffic signals. Another study (Hadian et al., 2017) proposed a privacy-preserving time-sharing service which provided accurate data during simulation. The purpose of the research was to enable either the owner or the borrower (s) of the shared AV to inform the other party of their need to use the AV at a particular point in time while maintaining privacy. Moreover, another work (Xiong et al., 2020) proposed an edge-assisted privacy-preserving raw data sharing framework. The latter firstly encrypting raw data with an additive secret sharing technique into 2 ciphertexts and then 2 classes of secure functions. Secondly, the 2 secure functions are applied to adopt the privacy-preserving convolutional neural network (P-CNN). Zhang et al. (2022) developed a privacy-preserved data-enabled predictive control scheme for CAVs to ensure privacy of data against unauthorised eavesdroppers. The affine masking method can offer robust privacy protection, making it impossible for an eavesdropper to even roughly estimate the information transferred between the central unit and the vehicle system. Other research works were carried out for privacy-preserving/enhancing mechanisms or other types of mechanisms protecting privacy (Datta et al., 2019; Gupta et al., 2023; Kim, 2022; Ni, 2019; Mainali et al. 2019). However, with the sophistication and emergence of not just AVs but also regulations, it is more taxing for researchers and manufacturers to adhere to regulatory requirements.

3.2 Rationale Behind the Adoption of Differential Privacy for This Research

Through the application of DP, personal data is protected by a noise wall and subsequently, any query that passes over the noise wall is altered arbitrarily. This approach makes the attacker uncertain about the credibility of the personal data (McKay Bowen & Garfinkel, 2021). The rationale behind selecting DP was contingent on the following factors:

1. The number of PETs that can be used is limited by processing time; this includes those that need lengthy execution times, for fully HE or ZKP (Garrido et al., 2022).
2. The computing power required to apply HE, ZKP or SMPC are significantly higher (Garrido et al., 2022).
3. Anonymization methods, such as K-anonymization, do not always guarantee complete anonymity for users when interacting with the Internet (Kan, 2023; Tavani & Moor, 2001).
4. Although homomorphic encryption provides robust privacy assurances, its real-world applications frequently experience significant computational cost and

performance issues, which makes it less appropriate for situations requiring real-time or resource constraints (Ajala et al., 2024).

Moreover, LaPlace and Gaussian are DP-preserving mechanisms and were applied and tested during past research studies (Dwork et al., 2006; Liu, 2019; Muthukrishnan & Kalyani, 2023; Xu et al., 2024; Zhao & Chen, 2022). In these past studies, their effectiveness was assessed in different data-processing environments. Therefore, the main goal of this research was to use Gaussian and LaPlace to assess how well the mechanism will function in a novel setting like AV simulation. It is an innovative concept to apply the mathematical techniques LaPlace and Gaussian within the environment of an emerging technology, such as AVs, to protect the privacy of data subjects.

4 Methodology

To investigate the challenges posed by AVs from a data privacy perspective and evaluate the effectiveness of two existing differential privacy mechanisms, namely LaPlace and Gaussian, a simulated AV environment was created using Car Learning to Act (CARLA). CARLA is an open-source software simulator which is used widely for autonomous driving research with digital assets, for instance town layouts, buildings, vehicles, road infrastructure (CARLA Team, 2024). The server of CARLA conducts the simulation and renders the scene, while the client enables the features (Dosovitskiy et al., 2017) (Fig. 1).

Python was used to make modifications to the CARLA environment along with extensions, such as NumPy and Flask. The sensor data collected from the simulation

Fig. 1 CARLA simulation environment

were stored in SQL Lite Database. A set of python scripts was run to collect original sensor data, and another set of scripts was run to add noise to the original sensor data.

4.1 The Simulation Environment

CARLA has in-built template scripts to connect to its server which were leveraged upon to consolidate the Tutorial_ego.py script built to collect original sensor data (without applied DP). The simulated AV is referred to as ego vehicle with sensors installed to collect data. In the simulation scenario, the ego vehicle is the actor/host vehicle set to operate on autopilot mode. The initial configuration provided by CARLA was leveraged upon to further configure the simulation environment and sensor callback and listen functions.

The sensors placed on the AV are all sensors made available by CARLA and modified for the purpose of this research. The sensors have pre-determined units of measurements and variable data types which must be adhered to for the scripts developed during this research to be compatible with the CARLA sensors. Each CARLA sensors have, therefore, been configured differently based on their data types and units of measurements, especially when DP was applied. Table 1 describes the AV sensors.

One set of script was developed to add LaPlace algorithm to convert the original sensor data into noise data. Another set of script was developed for the implementation of Gaussian. Both sets of script returned the original value and the noise value which were then compared for further evaluation as per chapter 5.

Table 1 AV sensors

Sensor	Description
Collision sensor	Measured in terms of scalar values and stored as integer in the database for this research
Lane invasion sensor	Records events whenever the ego vehicle passes a lane marker processed as numerical values in XYZ co-ordinates
Obstacle detection sensor	Records an event each time the ego vehicle encounters an obstacle within simulation environment processed as numerical XYZ values
Global Navigation Satellite Systems (GNSS)	Reports the ego vehicle's current GNSS position as latitude, longitude and altitude data
Inertial Measurement Unit (IMU)	Measures and delivers motion data along X, Y, and Z axes by measuring linear acceleration and angular rate (Advanced Navigation, 2024)

4.2 Implementation of LaPlace Mechanism

LaPlace mechanism was implemented using epsilon values and sensitivity values which relied on the NumPy import function. The "def add_laplace_noise(value, epsilon, sensitivity)" defined the LaPlace command to apply the mechanism. The data scale parameter is sensitivity over epsilon as per the formula

$$b = \frac{\Delta f}{\in} \tag{4}$$

where "b" represents the scale parameter, "Δf" represents the sensitivity and "$\in$ " is the epsilon value.

To configure the collection of the noise value, the NumPy function "np.random.laplace" was applied. The function is part of NumPy, whereby it selects samples with the given location (or mean) and scale (decay) from the Laplace or double exponential distribution.

4.3 Implementation of Gaussian Mechanism

Gaussian mechanism was implemented using epsilon values, delta value and sensitivity values which relied on the NumPy import to function. Gaussian applies a standard deviation approach (σ) with the values of the sensitivity, epsilon ($\in$), the squared logarithm (log) of 1.25 over the delta parameter (δ).

For all of the sensors, epsilon was set to 2.5, delta fixed at 1e-5 and sensitivity values was determined for each of the sensors based on the data type collected by the sensor and the variables limits. The algorithm was implemented using epsilon values, delta value and sensitivity values which relied on the NumPy import function.

4.4 Determination of Epsilon Value

The higher the epsilon, the higher the level of privacy protection. However, it must be considered that there is the dichotomy of applying privacy to data while also making the data available for analysis, especially for AVs (which are still being developed and undergoing significant research). For all of the sensors, epsilon was set to 2.5 to ensure an adequate degree of privacy while retaining data reliability. For the estimated sensitivity value to be applied, the bounded range for each sensor was considered. The bounded range is dependent on the CARLA Map and CARLA documentation. Collision intensity and collision location have different units of measurements and a bounded range (lowest to highest data range). For instance, having an output of 5.213 for collision intensity and the same value for collision location, the values would have a different intepretation for each sensors. The sensitivity was applied within a reasonable range for each sensor, to preserve the privacy of data while also making the data relevant enough for future analysis on AV performance, for instance. Therefore, to come up with the sensitivity value, multiple simulations were run to identify a value that would not distort the information so much that it hinders analysis but also maintains a level of privacy to protect the data.

5 Results and Discussion

To apply both LaPlace and Gaussian, DP values such as epsilon ($\in$) and sensitivity (Δf) need to be determined. The higher the $\in$ value, the more noise is added. For Δf value, the type of sensor data and bounded range need to be considered first. While $\in$ remains a fixed value for all the sensors, Δf is dynamic.

To evaluate the measurement differences, 3 samples of recorded original and LaPlace and Gaussian noise sensor data along with the measurement differences were used for each sensor. The samples were taken at random and at different point in time over multiple simulations.

The original and noise sensor data were analysed with the use of absolute difference measurement, MAE and clustered column charts. Through the "calculate difference" function, the absolute difference values were generated while executing the applicable scripts. The absolute difference measurement was extrapolated into a Mean Absolute Error (MAE) for the data sets. Through the variances analysis, it was easier to compare the effect of LaPlace and Gaussian on AV sensor data and determine which of the mechanisms affected data reliability the most. Table 2 provides a summary of the simulation results.

Across the MAE Value of the sensor data, it was observed that the x-axis values of the sensors used in this study for collection of original data was less impacted by the epsilon (noise) factor. Moreover, the overall impact of LaPlace on the x-axis values was found to be higher than Gaussian. It can, therefore, be deduced that the x-axis values are more susceptible to LaPlace as a differential privacy mechanism while the y-axis and z-axis were more impacted by Gaussian.

Further, the original data collected by GNSS and IMU sensors were less impacted by the noise applied by LaPlace and Gaussian as compared to the other original sensor data. This may be due to GNSS and IMU sensors having different units of measurements than the other sensors.

The overall effectiveness of Gaussian, as a differential privacy mechanism, was supported with an aggregated higher value of MAE across the sensors. The higher the value of MAE, the more noise is applied to the original sensor value. When Gaussian mechanism was applied, the disparity between the original value and the noise value was more significant than for LaPlace.

Therefore, Gaussian is mathematically more effective to preserve the privacy of the type of data collected by the AV sensors. The effectiveness of Gaussian was also further corroborated in Mainali et al. (2019). As per Liu (2019), LaPlace was more

Table 2 Simulation results

Sensor	Results
Collision Intensity	The MAE value for Gaussian (5423.18) signified a higher noise impact than LaPlace (314.141). The original value collected from Collision Intensity sensor was significantly deviated when Gaussian was applied as compared to LaPlace. Therefore, Gaussian was more effective in preserving privacy of the sensor data
Collision Location sensor	Gaussian had a significantly higher noise impact on the y-axis for Collision Location, and this was corroborated by the MAE(y) of Gaussian which was at 121.45 as compared to the MAE(y) of LaPlace at 5.57. Overall, as per the calculated MAEs across the XYZ axis of Collision Location, Gaussian had a higher noise impact than LaPlace
Lane invasion sensor	LaPlace had a higher noise impact on the x-axis value for Lane Invasion Sensor while Gaussian had a significantly higher noise impact on the y and z-axis values at 101.58 and 29.04, respectively as compared to 6.53 and 0.09 values of LaPlace
Obstacle detection sensor	Gaussian had a significantly higher noise impact on the y and z axis obstacle detection, and this was corroborated by the MAE(x)(y)(z) of Gaussian which were at 82.29, 87.21 and 25.56, as compared to LaPlace (50.18, 39.97 and 0.07), respectively
GNSS sensor	Gaussian had a minimal higher noise impact on the latitude and altitude values for GNSS. When GNSS was compared overall with the other sensors, it can be deduced that noise applied to GNSS (both LaPlace and Gaussian) had a lower impact than when applied to other sensors, such as collision location, lane invasion and obstacle detection co-ordinates. This may be due to the difference in value type of the GNSS sensor which is measured in degrees
IMU sensor	Gaussian had a minimal higher noise impact on the vector value while the accelerometer value was significantly impacted by Gaussian as compared to LaPlace. Overall, there was minimal deviation from the IMU original sensor data when both LaPlace and Gaussian were applied as compared to the other sensor data (i.e. collision intensity, collision location, lane invasion)

effective to sanitise data by generating results with smaller error and divergence from the original data. However, there was overall a very significant deviation between the original and noise sensor data when Gaussian was applied which also raises the concern of data utility. The focus of privacy-preserving mechanisms is to maintain privacy while making the data adequate for analysis (Liu, 2019; Nasr et al., 2020), especially in nascent research areas such as AV whereby data analysis is key. Hence, the dilemma between preserving data privacy and maintaining data utility.

To answer this dilemma, there are two design intentions to be considered by the AV manufacturer, to determine the "most effective" DP mechanism:

Design intention 1: Apply privacy while keeping the data still relevant for future analysis by the AV manufacturers or other stakeholders. To attain this, based on the simulation results of this research, LaPlace is "more effective" as its level of noise was lower and, therefore, the sensor data would be more relevant for analysis.

Design intention 2: Protect privacy to the maximum without consideration to use the sensor data for future analysis. To attain the latter, based on the simulation results of this research, Gaussian is "more effective" as its level of noise was highest, therefore, the noise sensor data was more divergent with the original sensor data.

When compared with Gaussian, LaPlace is more effective to attain the true purpose of applying DP across data sets—maintaining data relevance for further analysis while retaining row-level data privacy of data subjects. However, based on results of the simulation, LaPlace is not ideally suitable for all the data sets collected by the AV (such as IMU Gyroscope). This is due to the variation between the original and noise data being too significant, which renders the noise data useless for further analysis.

6 Conclusion

To achieve the aim of this research, CARLA, the simulation AV environment was set up and configured to collect differing sensor data which generate different data types. Once the set-up and sensors were configured, LaPlace and Gaussian mechanisms were applied to add noise to the original sensor data. The original sensor data and the noise sensor data were compared based on the absolute difference measurement between the two values. Then MAE was calculated to get the mean average of the differences for easier evaluation of divergence between original and sensor data.

Gaussian was more effective than LaPlace in applying noise when all the findings (MAE) and the visual depiction of the absolute difference values in the clustered bar chart were considered. Consequently, Gaussian is mathematically more favourable to apply noise for the type of data collected by the sensors within the CARLA simulation environment. However, there is a trade-off between applying noise to protect privacy while also making the data reliable for future analysis. The trade-off between data privacy and data usability is especially important in the field of research for autonomous vehicles. Considering the trade-off, LaPlace was more effective at keeping the data reliable for future analysis as its level of noise was lower. Therefore, there was less divergence between original sensor data and noise sensor data.

Future research can focus on the application of privacy mechanisms in the sensor cameras, testing the DP mechanism within a high-traffic simulation environment with more spawn vehicles and objects for a more comprehensive set of sensor data.

References

Advanced Navigation. (2024). About IMU. [Online]. Available: https://www.advancednavigation.com/tech-articles/inertial-measurement-unit-imu-an-introduction/. Accessed September 20, 2024.

Ajala, O. A., Arinze, C. A., Ofodile, O. C., Okoye, C. C., & Daraojimba, O. D. (2024). Reviewing advancements in privacy-enhancing technologies for big data analytics in an era of increased surveillance. *World Journal of Advanced Engineering Technology and Sciences, 11*(1), 294–300.

Anderson, J., Kalra, N., Stanley, K., Sorensen, P., Samaras, C., & Oluwatola O. (2014). Brief history and current state of autonomous vehicles. In: *Autonomous vehicle technology: A guide for policymakers* (pp. 55–74). RAND Corporation.

ANSI. (2021). *SAE Levels of Driving Automation* [Online]. Available: https://blog.ansi.org/sae-levels-driving-automation-j-3016-2021/. Accessed 25, August 2024.

CARLA Team. (2024). CARLA [Online]. Available: https://carla.org/. Accessed September 06, 2024.

Datta, A., Lu, J., & Tschantz, M. C. (2019). *Evaluating anti-fingerprinting privacy enhancing technologies*, San Francisco.

Dosovitskiy, A., Ros, G., Codevilla, F., Lopez, A., & Koltun, V. (2017). *CARLA: An open urban driving simulator*. Mountain View, California.

Dwork, C., McSherry, F., Nissim, K., & Smith, A. (2006). *Calibrating noise to sensitivity in private data analysis*.

Faisal, A., Kamruzzaman, M., Yigitcanlar, T., & Currie, G. (2019). Understanding autonomous vehicles: A systematic literature review on capability, impact, planning and policy. *Journal of Transport and Land Use, 12*(1), 45–72.

Garrido, G. M., Sedlmeir, J., Uludag˘, O., AlaouiA, I. S., Luckow, A., & Matthes, F. (2022). Revealing the landscape of privacy-enhancing technologies in the context of data markets for the IoT: A systematic literature review. *Journal of Network and Computer Applications*.

Gupta, S., Maple, C., & Passerone, R. (2023). An investigation of cyber-attacks and security mechanisms for connected and autonomous vehicles. *IEEE Access*.

Hadian, M., Altuwaiyan, T., & T. Liang (2017). *Privacy-preserving time-sharing services for autonomous vehicles*. Toronto.

Hasan, M., Mohan, S., Shimizu, T., & Lu, H. (2020). Securing vehicle-to-everything (V2X) communication. *IEEE Translation of Intelligence Vehicle, 54*(4), 693–713.

Heurix, J., Zimmermann, P., Neubauer, T., & Fenz, S. (2015). A taxonomy for privacy enhancing technologies. *Computers & Security, 53*, 1–17.

ICO. (2023). Privacy-enhancing technologies (PETs). [Online]. Available: https://ico.org.uk/media/for-organisations/uk-gdpr-guidance-and-resources/data-sharing/privacy-enhancing-technologies-1-0.pdf. Accessed 30, August 2024.

Kan, K. (2023). Seeking the ideal privacy protection: Strengths and limitations of differential privacy. *Monetary and Economic Studies, 41*, 49–80.

Kim, J. S. (2022). Design of federated learning engagement method for autonomous vehicle privacy protection. In *12th international conference on Soft Computing and Intelligent Systems and 23rd International Symposium on Advanced Intelligent Systems (SCIS&ISIS)* (pp. 1–2).

Liu, F. (2019). Generalized Gaussian mechanism for differential privacy. *IEEE Transactions on Knowledge and Data Engineering, 31*(4), 747–756.

MacHardy, Z., Khan, A., Obana, K., & Iwashina, S. (2018). V2X access technologies: Regulation, research, and remaining challenges. *IEEE Communications Survey & Tutorials, 20*(3), 1858–1877.

Mainali, P., Shepherd, C., & Petitcolas, F. (2019). *Privacy-enhancing context authentication from location-sensitive data*. New York.

Mallozi, P., Pelliccione, P., Knauss, A., & Berger, C. (2019). Autonomous vehicles: state of the art, future trends, and challenges. In *Automotive systems and software engineering: State of the art and future trends* (pp. 347–367).

McKay Bowen, C., & Garfinkel, S. (2021). The philosophy of different privacy. *Notices of the American Mathematical Society, 68*(10), 1727–1739.

Milakis, D., Van Arem, B., & Van Wee, B. (2017). Policy and society related implications of automated driving: A review of literature and directions for future research. *Journal of Intelligent Transportation Systems, 21*(4), 324–348.

Mohan Krishna, A., Tyagi, A. K., & Prasad, S. (2020). Preserving privacy in future vehicles of tomorrow. *Journal of Critical Reviews, 7*(19), 6675–6684.

Mulder, T., & Vellinga, N. E. (2021). Exploring data protection challenges of automated driving. *Computer Law & Security Review, 40.*

Muthukrishnan, G., & Kalyani, S. (2023). Grafting Laplace and Gaussian distributions: A new noise mechanism for differential privacy. *IEEE Transactions on Information Forensics and Security, 18*, 5359–5374.

Nasr, M., Shokri, R., & Houmansadr, A. (2020). Improving deep learning with differential privacy using gradient encoding and denoising. *Machine Learning.*

Ni, J. (2019). Toward privacy-preserving valet parking in autonomous driving era. *IEEE Transactions on Vehicular Technology, 68*(3), 2893–2905.

Pendleton, S. D., Andersen, H., Du, X., Shen, X., Meghjani, M., Eng, Y. H., & Ang, M. H. (2017). Perception, planning, control and coordination for autonomous vehicles. *Machines, 5*(6).

Seo, H., Lee, K. D., Yasukawa, S., Peng, Y., & Sartori, P. (2016). LTE evolution for vehicle-to-everything. *IEEE Communications Magazine, 54*, 22–28.

Statista. (2020). Projected autonomous vehicle market size worldwide between 2021 and 2025. [Online]. Available: https://www.statista.com/statistics/1224515/av-market-size-worldwide-forecast/. Accessed 26, July 2024.

Stilgoe, J. (2018). Machine learning, social learning and the governance of self-driving cars. *Social Studies of Science, 48*(1), 25–56.

Tavani, H. T., & Moor, J. H. (2001). Privacy protection, control of information and privacy-enhancing technologies. *ACM Sigcas Computers and Society, 31*(1), 6–11.

Weinberger, P. (2012). Laplace and the era of differential equations. *Philosophical Magazine, 92*(32), 3882–3890.

Woo, S., Youtie, J., Ott, I., & Scheu, F. (2021). Understanding the long-term emergence of autonomous vehicles technologies. *Technological Forecasting and Social Change, 170.*

Xiong, J., Bi, R., Zhao, M., Guo, J., & Yang, Q. (2020). Edge-assisted privacy-preserving raw data sharing framework for connected autonomous vehicles. *IEEE Wireless Communications, 27*(3), 24–30.

Xu, Y., Liu, Y., & Hou, M. (2024). New media user privacy protection mechanism based on differential privacy and data anonymization. *Advances in Education, Humanities and Social Science Research, 11*, 164–170.

Zavvos, E., Gerding, E. H., Yazdanpanah, V., Maple, C., Stein, S., & Schraefel, M.C. (2022). Privacy and trust in the internet of vehicles. *IEEE Transactions on Intelligent Transportation Systems, 23*(8).

Zhang, K., Chen, K., Li, Z., & Zheng, Y. (2022). *Privacy-Preserved Data-Enabled Predictive Control for Connected and Automated Vehicles in Mixed Traffic*, Macau.

Zhao, Y., & Chen, J. (2022). A survey on differential privacy for unstructured data content. *ACM Computing Surveys (CSUR), 54*(10), 1–28.

Towards Cybercrime Profiling: The Validity and Reliability of Criminal Profiling Methods

Thulani Mashiane and Elmarie Kritzinger

Abstract Cybercrime has a negative impact on online users, businesses and organisations. To apprehend criminals both online and offline, law enforcement officials make use of tools and strategies. One such tool is criminal profiling. While criminal profiling has made positive contributions in apprehending offenders, criminal profiling methods have been critiqued for having limited reliability and validity. High reliability and validity imply that one can have confidence in the results obtained from using a selected method. Empirical experiments are conducted to test validity and reliability. While limited, there are studies that have evaluated criminal profiling using empirical methods. This article presents a systematic literature review of empirically tested criminal profiling methods. This article is aimed at building a robust foundation for the development of a cybercriminal profiling method with high reliability and validity. The article concludes by presenting a discussion on the implications of the results from the systematic literature review on cybercriminal profiling.

1 Introduction

The convenience of making transactions such as banking, shopping, and paying bills makes the internet an attractive resource for users. Unfortunately, the prospects for illegal financial gain over the internet attracts criminals (Edwards et al., 2202; Ling et al., 2019). For example, in 2024, data breaches by cybercriminals have cost organisations $4.88 million (Petrosyan, 2024).

Cybercriminals are a type of internet user who intentionally takes advantage of other internet users, primarily for financial gain. There are different types of cybercriminals namely, terrorists, cyberbullies, sexual offenders, financial fraudsters, social engineers, and identity thieves. Cybercriminals target individual users as well

T. Mashiane (✉) · E. Kritzinger
School of Computing, University of South Africa, Pretoria, South Africa
e-mail: tmashiane@csir.co.za

E. Kritzinger
e-mail: Kritze@unisa.ac.za

J. Marx Gómez et al. (eds.), *Recent Advances in Information Systems*, Progress in IS,
https://doi.org/10.1007/978-3-032-14099-9_9

as large corporations and governments (Bada & Nurse, 2021; Balogun & Zuva, 2017; Edwards et al., 2202).

To apprehend criminals both online and offline, law enforcement officials make use of various tools and strategies, one such tool is criminal profiling. Criminal profiling or offender profiling is the practice of identifying the probable characteristics of a criminal based on behaviours exhibited before, during and after a crime (Cărcăle 2023; Douglas & Burgess, 1986; Grubin, 1995). The goal of criminal profiling is to use the identified characteristics to identify an offender/criminal or group of criminals (Sahni & Choudhury, 2021). Criminal profiling techniques have been adopted in the identification of criminal characteristics in various domains, such as cybercriminal profiling.

2 Criminal Profiling Background

Criminal profiling as an investigative tool was first published in: Criminal profiling: A viable investigative tool against violent crime (Douglas & Burgess, 1986). The article focussed on a criminal profiling technique developed by the Federal Bureau of Investigation (FBI) (Douglas & Burgess, 1986). In the publication by Douglas and Burgess (1986), describe a profiling method which makes use of multi-disciplinary techniques to identify major personality and behavioural characteristics of offenders based on analysis of the committed crime (Douglas & Burgess, 1986; Edwards et al., 2202).

2.1 Criminal Profiling Success

Law enforcement has made arrests by using criminal profiles. For example, the case of Goerge Metesky. The profile which led to Goerge's arrest even included details of specific clothing attire (Sastry, 2015).

George Metesky's (Grubin, 1995), arrest was attributed to the work of the psychiatrist, Dr Brussel. In a psychological criminal profile, Dr Brussel predicted that Metesky would be a middle-aged white man, methodical, and likely holding a grudge against a former employer. Dr Brussel also predicted that the offender would be wearing a buttoned-up double-breasted suit when arrested (Grubin, 1995; Sastry, 2015).

Based on the success of such high-profile cases, other researchers and practitioners have expanded on criminal profiling methods and developed newer versions of criminal profiling methods.

2.2 Criticism Towards Criminal Profiling

Studies (Devor, 2013; Kocsis & Palermo, 2020; Winerman, 2004) have expressed scepticism through research results and conceptual arguments regarding the validity and reliability of criminal profiling methods.

Stressing the lack of science-based evidence supporting the success behind the criminal profiling methods. Petherick and Brooks (2021) agree that criminal profiling methods which rely on anecdotal knowledge lack the scientific or legal evidence required for the legal system. In addition, criminal profiling does not have a clear educational programme, like careers such as doctors or lawyers, that cultivates and provides assurity of the quality of the criminal profiler (Petherick, 2003; Ribeiro & Matos Soeiro, 2021).

Lack of scientific process, and the lack of educational structure in criminal profiling has brought the validity and reliability of criminal profiling into question (Ribeiro & Matos Soeiro, 2021).

2.3 Reliability and Validity Criminal Profiling

Reliability and validity are elements of a research study which helps determine the quality of a study (Heale & Twycross, 2015; McCrae et al., 2011). The section presents the concepts of validity and reliability of methodology with a focus on criminal profiling methods.

Reliability in research, refers to the consistency of output given the same input. Internal reliability, test–retest reliability and evaluator reliability are all measures of reliability that evaluate the consistency in which a research method is applied. Internal reliability is achieved when an instrument measures what it is set out to measure (Heale & Twycross, 2015; McCrae et al., 2011; Viglione et al., 2022). For example, in a criminal profiling method internal reliability would be achieved if the method used was to evaluate the behavioural and physical characteristics of an offender. Test–retest reliability would be achieved when the same method yields the same results even when the results are conducted at different time frames (Heale & Twycross, 2015; McCrae et al., 2011). For example, in a criminal profiling method, given the same input, the same profile must be developed at different times. Lastly, for evaluator validity, different evaluators must come to the same conclusions given the same input (Heale & Twycross, 2015; McCrae et al., 2011). In criminal profiling, reliability would be demonstrated when different criminal profilers develop the same criminal profile, given the same input (Godwin, 2002).

Validity refers to the ability of a research method to measure what it set out to measure. For example, a profiling method that measures the characteristics of a certain economic demographic instead of that of criminals would not be considered valid.

Within the conversation of the validity of criminal profiling are statements about ethics. Criminal profiles typically produce characteristics that can lead an investigation towards a specific group of people, which might increase harassment of a particular demographic (Geisinger, 2007; Godwin, 2002; Meterko & Cooper, 2022; Ranti & Nurmaily, 2021).

3 Related Work

Validity and reliability are still open issues in criminal profiling (Joseph, 2024). Therefore, we can infer that there are issues of reliability and validity in cybercriminal profiling as well.

Studies such as (Horan & Saiedian, 2021) have highlighted limitations in validity and reliability in cybercriminal profiling methods. Expressing the need for further investigation. Existing systematic literature reviews on cybercriminal profiling (Bada & Nurse, 2021; Martineau et al., 2023) focus on the current state of cybercriminal profiling. The studies presented findings on cybercriminal application, method building and research data availability. Whereas this article is focussed on building the understanding of validity and reliability by first understanding these concepts in empirically evaluated criminal profiling methods.

4 Objective of the Study

This article serves to fulfil a sub-objective of a bigger study. The main objective is to develop a cybercriminal profiling method. To build towards a cybercriminal profiling method, the current article's goal is to present a systematic literature review of criminal profiling methods focussing on validity and reliability of the methods.

The main contribution of this article is the presentation of a systematic literature review aimed at identifying criminal profiling methods as well as to present a discussion of the applicability of the criminal profiling methods to cybercriminal profiling.

5 Study Methodology

To find criminal profiling methods that have been empirically evaluated in the literature the article method followed is a systematic literature review. A systematic literature review is a research tool used to analyse and provide an overview of the current literature available in the article domain.

The aim of a systematic literature review is defined by Okoli and Schabram (2015) as "an explicit, comprehensive and reproducible method for identifying, evaluating and synthesising the existing body of completed and recorded work produced by researchers, scholars and practitioners" (Okoli & Schabram, 2015).

6 Study Results

The purpose for conducting this literature review is to identify criminal profiling methods which have been empirically evaluated in research.

Figure 1 shows how steps developed by Okoli and Schabram (2015) were taken to collect the studies included in this article.

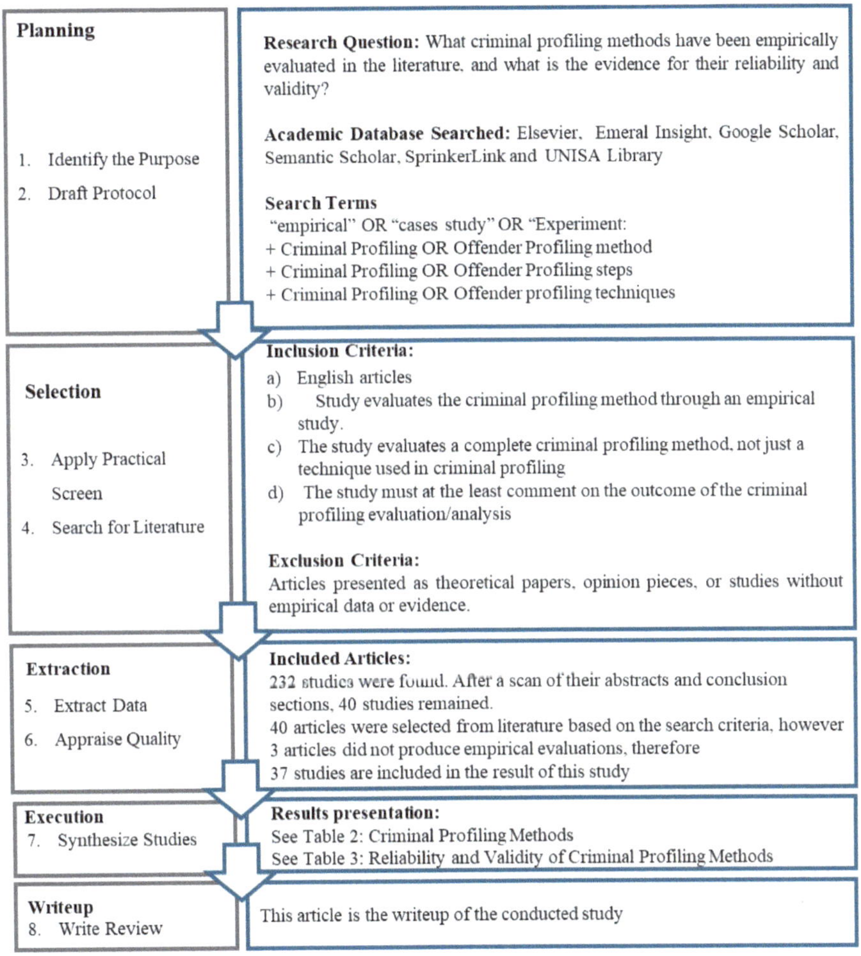

Fig. 1 Systematic literature review workflow

6.1 Criminal Profiling Method's Demographics

This section presents a demographic description of the identified profiling methods. Table 1 shows the results of identifying criminal profiling methods which have been empirically evaluated, from the selected literature. The popular research methodologies are comparative studies (Almond et al., 2007; Bloomfield, 2006; Brandy, 2009; Fox & Farrington, 2015; Goodwill & Meloy, 2019; Goodwill et al., 2009, 2013; Pinizzotto & Finkel, 1990; Rossmo, 2023; Salo et al., 2012; Taylor et al., 2011; Woodhams et al., 2008), and the use of case studies (Kocsis et al., 2002a; Pinizzotto & Finkel, 1990; Rossmo, 2023; Yaksic, 2020).

For each study, the reference, criminal profiling method, description of the method and the type of crime on which the criminal profiling method was evaluated, and sample size is presented. Offender profiling is the most studied criminal profiling method. In this case, offender profiling is used as a general term where the characteristics of an offender were predetermined based on previous research (Fujita et al., 2012). The FBI's organised/disorganised criminal profiling method (Bloomfield, 2006; Canter et al., 2004; Taylor et al., 2011) as well as statistical profiling methods (Beauregard et al., 2017; Bloomfield, 2006; Kalichman et al., 1990) appear in three studies each. Finally, the crime types include sexual assault, homicide, burglary, arson and terrorism.

6.2 Criminal Profiling Methods Validity and Reliability

This section presents the reliability and validity of each identified criminal profiling method. Table 2 builds upon Table 1 by assessing the reliability and validity of the listed profiling methods. Table 2 presents the criminal profiling method, the reported accuracy, the reliability method used to measure the profiling method and validity measure used for each profiling method.

6.3 Accuracy

Two (Goodwill et al., 2014; Krstic et al., 2018) criminal profiling methods report accuracy rates of between 90 and 100%. Two (Krstic et al., 2018; Salo et al., 2012) criminal profiling methods report accuracy rates of between 80 and 89%. Three (Baumgartner et al., 2005; Beek et al., 2009; Goodwill & Meloy, 2019) profiling methods report an accuracy rate of between 70 and 79%. Quantitative approaches are used in developing the criminal profiles.

Table 1 Criminal profiling methods

References	Criminal profiling method/s	Description of criminal profiling method	Crime type
Almond et al. (2007)	Behavioural investigative advice	Analyses crime scene and victim details to develop an offender's psychological profile	Murder, rape, sexual assault, artifice, sabotage, and suicide
Almond et al. (2019)	Criminal history profiling	Analyses of an offender past behaviours to predict an offender's future behaviours	Male-on-female homicides
Baumgartner et al. (2005)	Bayesian network model	Uses Bayesian stats to estimate offender traits from variable relationships	Single offender/single victim homicides
Beauregard et al. (2005)	Behavioural profiling	Analyses of offender actions at the crime scene to infer traits and motives	Sexual assault against females aged 16 or over
Rossmo (2023)			Burglary, rape, serial murder
Beauregard et al. (2017)	Statistical profiling	Uses quantitative crime data to predict offender characteristics	Sexual assault
Beek et al. (2009)			Stranger rape offences
Fox andFarrington (2015)			Burglary
Bennell et al. (2012)	Keppel and Walter's classification system	Categorises offenders by behaviour, motive, and traits	Serial sexual murders
Canter et al. (2004)	Organised/ disorganised dichotomy	Classifies offenders based on crime scene control and planning	Serial sexual homicides
Taylor et al. (2011)			Serial killings
Zaitsu et al. (2008)	Offender profiling using Bayesian network	Combines profiling with Bayesian methods to predict traits	Indoor sex offences
Bloomfield, 2006)	Keppel and Walter's classification system; FBI's organised/ disorganised model	See study 8 for definition Organised/ disorganised model: See study 8	Serial sexual homicides
Bouhana et al. (2014)	Behavioural crime linkage	Links crimes by identifying behavioural similarities	Prolific residential burglary
Sorochinski (2015)			Rape offences

(continued)

Table 1 (continued)

References	Criminal profiling method/s	Description of criminal profiling method	Crime type
Brankley (2021)	Behavioural crime scene analysis (BCSA)	Interprets crime scene behaviour to infer traits and motives	Serial stranger sexual offences
Brandy (2009)	Homology assumption-based profiling	Assumes similar crimes are committed by offenders with shared traits	Arson, Robbery
Fox and Farrington (2018)	Offender profiling or criminal profiling	Infers traits like age, gender, or occupation from behaviour and scene evidence	Various
Fujita et al. (2012)			Homicides
Janosch et al. (2023)			Sexual assault by stranger offenders
Newman (2011)			Stranger rape
Salfati and Canter (1999)			Stranger murders
Salfati (2000)			Single-victim homicides
Snook et al. (2007)			Murder, rape, arson, burglary
Goodwill and Meloy (2019)	Terrorist radicalization assessment protocol (TRAP-18)	Assesses risk of radicalisation using structured judgement and research	Politically motivated lethal or near-lethal attacks
Goodwill et al. (2014)	MCA–CA–DFA approach	Uses multivariate statistics to classify and analyse offender behaviour	Homicides
Goodwill et al. 2016)	Action phase approach	Studies the offender's sequence of actions during a crime	Stranger sexual offences
Kalichman et al. (1990)	Statistical profiling using psychological assessment	Uses behavioural data and statistics to build offender profiles	Rape
Kocsis et al. (2002b)	Psychological profiling	Infers psychological traits and motives based on behaviour or communication	Sexual murders
Kocsis et al. (2002b)	Empirical model for psychological profiling	Builds profiles using data-driven analysis of offender behaviour	Serial sexual assaults

(continued)

Table 1 (continued)

References	Criminal profiling method/s	Description of criminal profiling method	Crime type
Krstic et al. (2018)	Latent variable- and person-cantered approaches	Identifies hidden patterns or groups offenders by behavioural similarity	Sexual crimes
Lehmann et al. (2013)	Crime scene analysis (CSA)	Uses crime scene evidence to infer planning level and motives	Stranger rapes
Pinizzotto and Finkel (1990)	Criminal personality profiling	Profiles personality traits using psychological theory	Sex offense, homicide
Salo et al. (2012)	Bayes' Theorem	Assesses the likelihood that crimes were committed by the same person	Serial homicides
Sigurdardóttir et al. (2023)	Behavioural investigative advice (BIA)	Analysis of crime scene evidence and victim profiles, to develop a offender psychological profile	Various (murder, rape, assault, etc.)
Tapper (2008)	Criminal investigative analysis (CIA)	Profiles unknown offenders based on case evidence and behavioural trends	Serial rape offences
Woodhams et al. (2008)	Context linking	Analyses how situations influence offender behaviour	Stranger sexual assaults
Yaksic (2020)	Investigative journalist/expert field micro task force model	Combines expert analysis and journalism to create offender profiles	Serial killings

6.4 Reliability

Reliability in criminal profiling is important because a profiling method must consistently measure what was set out, an offender's characteristics and motivations. The identified studies that make use of empirical evidence to present criminal profiling methods are largely quantitative in nature. Studies such as Goodwill and Meloy (2019) measure the interrater reliability of their research results (Goodwill & Meloy, 2019).

Test–retest reliability is one of the strongest criticisms against criminal profiling methods. The inability to constantly yield the same quality of profiles makes criminal

Table 2 Reliability and validity of criminal profiling methods

Predictive reliability and validity			
Profiling method	Reported accuracy	Reliability method	Validity method
Bayesian network model	79% predictive accuracy	Predictive accuracy rates	Comparison of predictive accuracy between algorithms
Statistical profiling	Area Under the Curve (AUC) values 0.70–0.87	Sensitivity, specificity, positive predictive value (PPV), Negative predictive value (NPV), odds ratio	Tenfold cross-validation, receiver operating characteristic (ROC) analysis
TRAP-18	76% accuracy in post-dicting violence	Interrater reliability	Criterion, construct, discriminant, predictive validity
MCA–CA–DFA Approach	91.5% classification accuracy	No mention found	No mention found
Crime scene analysis (CSA)	Not reported	Guttman coefficient of alienation, Jaccard measure	Canonical correlation analysis
Latent variable- and person-centred approaches	Classification accuracy rates 0.83–0.91	Intraclass Correlation Coefficient (ICC) 0.87–0.89; Cronbach's alpha 0.67–0.85	Structural equation modelling with comparative fit index = 0.90, Root mean square error of approximation = 0.07; Confirmatory factor analysis
Bayesian reasoning for crime linking	83.6% overall accuracy	Classification accuracy, comparison to chance	Cross-validation
Offender profiling	10% increase in accuracy for employment prediction	No mention found	Validation using separate dataset
Statistical modelling	Over 70% correct prediction rate for 4/5 models	No mention found	No mention found

profiling seem unstable and brings its integrity into question (Miller, 2015; Ribeiro & Matos Soeiro, 2021; Winerman, 2004).

6.5 Validity

The validity of a criminal profiling method is low mainly because of its inability to pinpoint the exact offender. Criminal profiling methods are meant to decrease the search space for an offender. The aim is to narrow down a list of offenders to expedite the action of apprehending an offender (Cărcăle 2023; Maarek & Darwish, 2023; Sahni & Choudhury, 2021; Valentine, 2013).

7 Study Implications on Cybercriminal Profiling

The following section concludes the article by presenting cybercriminal profiling implications of the results from the systematic literature review. The section will present the identified criminal profiling method and cybercriminal profiling discussion.

7.1 Bayesian Network Model

The Bayesian Network Model demonstrates strong predictive accuracy. This suggests potential application in modelling complex, interdependent variables present in cybercriminal behaviour (Bada & Nurse, 2021). Cybercrime often involves multi-faceted decision-making processes influenced by technical, psychological, and environmental factors (Mashiane & Kritzinger, 2019, 2020).

The Bayesian approach, which accounts for probabilistic relationships, could be effective in capturing these dynamics. However, the reliance on predictive accuracy rates alone as a reliability measure may be insufficient. Future applications can incorporate additional reliability metrics, such as interrater reliability or cross-validation, to ensure robustness in cybercriminal contexts.

7.2 Statistical Profiling

In cybercriminal profiling, statistical models could be used to identify patterns in data breaches, malware distribution, or phishing campaigns (Martineau et al., 2023). However, the sensitivity and specificity of these models must be carefully evaluated, as false positives in cybercrime investigations could lead to unwarranted accusations, while false negatives could allow malicious actors to evade detection.

7.3 TRAP-18

TRAP-18's established validity measures, including criterion and predictive validity, suggest potential adaptability for cybercriminal profiling, particularly in cases involving cyber threats or cyberstalking (Greco & Greco, 2020). However, its inter-rater reliability may need further testing in cyber contexts, as the interpretation of digital behaviours may differ significantly from physical behaviours.

7.4 *MCA–CA–DFA Approach*

The MCA–CA–DFA approach achieves the highest classification accuracy among the identified criminal profiling methods. This suggests its potential for high-precision classification of cybercriminals based on behavioural or technical indicators. However, the lack of reported reliability and validity measures raises concerns about its generalizability. Future research can explore its applicability to cybercrime datasets, ensuring that its high accuracy is not an artefact of overfitting.

7.5 *Crime Scene Analysis*

In cybercriminal profiling, crime scene analysis has the potential to be adapted to analyse digital crime scenes, such as network intrusions or social engineering attacks. However, the absence of predictive accuracy metrics limits its current applicability for proactive cybercrime prevention.

7.6 *Latent Variable- and Person-Centred Approaches*

Latent Variable- and Person-Centred Approaches' ability to model latent constructs makes them well-suited for identifying underlying psychological or behavioural traits of cybercriminals, such as impulsivity or technical expertise. The use of SEM and CFA further strengthens the validity, making Latent Variable- and Person-Centred Approaches a robust choice for cybercriminal profiling.

7.7 *Bayesian Reasoning for Crime Linking*

Bayesian reasoning's application in linking crimes that could be extended to cybercrime, such as identifying connections between distributed denial-of-service (DDoS) attacks or tracing cryptocurrency transactions. However, its reliance on prior probabilities may introduce bias if historical data on cybercrime is incomplete or unrepresentative.

7.8 Offender Profiling

While offender profiling shows prediction accuracy, its lack of reported reliability and validity measures limits its applicability to cybercriminal profiling. Validation using separate datasets is essential to ensure its effectiveness in predicting cybercriminal behaviour.

7.9 Statistical Modelling

Statistical modelling demonstrates strong predictive performance, with over 70% correct prediction rates for most models. However, the absence of reliability and validity measures raises questions about robustness. In cybercriminal profiling, these models could be used to predict attack vectors or identify high-risk individuals, but further validation is necessary to ensure reliability.

8 Discussion of Future Research

Cybercriminal profiling presents challenges, including the dynamic nature of digital evidence and the difficulty of establishing behavioural links. The application of criminal profiling methods to cybercriminal profiling will have to prioritise the validity and reliability of the profiling methods.

Given the results presented in the article. A concern with criminal profiling methods ability to be generalised or used in a cybercriminal context has been highlighted. Therefore, future research should investigate how profiling methods can be adapted for cybercriminal profiling.

To achieve this interdisciplinary collaboration between criminologists, psychologists, and cybersecurity experts will enhance methodological rigour and practical applicability in cybercriminal profiling.

9 Conclusion

The article presents a systematic literature review of criminal profiling methods which have been empirically evaluated in the literature. The reliability and validity of each criminal profiling method are presented, and a discussion of the results follows. The findings from this systematic review highlight the potential of existing profiling methods for cybercriminal profiling, while also emphasising the need for robust reliability and validity testing. By adapting and integrating these methods, researchers and practitioners can develop more accurate and reliable tools for predicting and preventing cybercrime, ultimately contributing to a safer digital environment.

References

Almond, L., Matin, E., & McManus, M. (2019). *Predicting the criminal records of male-on-female UK homicide offender s from crime scene behaviors. Journal of Interpersonal Violence, 36*(21–22), NP11852–NP11876.

Almond, L., Alison, L., & Porter, L. (2007). An evaluation and comparison of claims made in behavioural investigative advice reports compiled by the National Policing Improvements Agency in the United Kingdom. *Journal of Investigative Psychology and Offender Profiling, 4*(2), 71–83.

Bada, M., & Nurse, J. R. (2021). Profiling the cybercriminal: a systematic review of research. In *2021 International Conference on Cyber Situational Awareness, Data Analytics and Assessment (CyberSA)*. IEEE.

Balogun, A. M., & Zuva, T. (2017). Open issues in cybercriminal profiling. In *2017 1st International Conference on Next Generation Computing Applications (NextComp)*. IEEE.

Baumgartner, K. C., Ferrari, S., & Salfati, C. G. (2005). *Bayesian network modeling of offender behavior for criminal profiling*. IEEE.

Beauregard, E., Busina, I., & Healey, J. (2017). Confessions of sex offenders: Extracting offender and victim profiles for investigative interviewing. *Journal of Criminal Psychology, 7*(1), 13–28.

Beauregard, E., Lussier, P., & Proulx, J. (2005). The role of sexual interests and situational factors on rapists' modus operandi: Implications for offender profiling. *Legal and Criminological Psychology, 10*(2), 265–278.

Bennell, C., et al. (2012). Classifying serial sexual murder/murderers. *Criminal Justice and Behavior, 40*(1), 5–25.

Bloomfield, S. (2006). *Classifying serial sexual homicide: Validating Keppel and Walter's (1999) model*. Carleton University.

Bouhana, N., Johnson, S. D., & Porter, M. (2014). Consistency and specificity in burglars who commit prolific residential burglary: Testing the core assumptions underpinning behavioural crime linkage. *Legal and Criminological Psychology, 21*(1), 77–94.

Brandy, D. (2009). *Empirical tests of the homology assumption in criminal profiling*.

Brankley, A. E. (2021). *The utility of latent variable models in refining behavioural crime scene analysis of serial stranger sexual offences*.

Canter, D., et al. (2004). The organized/disorganized typology of serial murder: Myth or model? *Psychology Public Policy and Law, 10*(3), 293–320.

Cărcăle, V.-A. (2023). Criminal profiling and behavioural analysis. *Romanian Journal of Forensic Science, 2023*(134).

Devor, C. P. (2013). *An analysis of the continuation and expansion of transnational organized crime: The case of human trafficking in Mozambique*. Stellenbosch University.

Douglas, J. E., & Burgess, A. E. (1986). Criminal profiling: A viable investigative tool against violent crime. *FBI Law Enforcement Bulletin, 55*, 9.

Edwards, M., et al. (2022). *Characterising cybercriminals: A review*. arXiv preprint arXiv:2202.07419.

Fox, B. H., & Farrington, D. P. (2015). An experimental evaluation on the utility of burglary profiles applied in active police investigations. *Ciminal Justice and Behavior, 42*, 156–175.

Fox, B., & Farrington, D. P. (2018). What have we learned from offender profiling? A systematic review and meta-analysis of 40 years of research. *Psychological Bulletin, 144*(12), 1247–1274.

Fujita, G., et al. (2012). Multivariate models for behavioral offender profiling of Japanese Homi cide. *Criminal Justice and Behavior, 40*(2), 214–227.

Geisinger, A. (2007). Rethinking profiling: A cognitive model of bias and its legal implications. *Oregon law Review, 86*, 657.

Godwin, M. (2002). Reliability, validity, and utility of criminal profiling typologies. *Journal of Police and Criminal Psychology, 17*(1), 1–18.

Goodwill, A. M., et al. (2013). Advancement of criminal profiling methods in faceted multidimensional analysis. *Journal of Investigative Psychology and Offender Profiling, 10*(1), 71–95.

Goodwill, A., et al. (2016). An action phase approach to offender profiling. *Legal Criminological Psychology, 22*, 229–250.

Goodwill, A., Alison, L., & Beech, A. (2009). What works in offender profiling? A comparison of typological, thematic, and multivariate models. *Behavioral Sciences & the Law, 27*(4), 507–529.

Goodwill, A. M., Allen, J. C., & Kolarevic, D. (2014). Improvement of thematic classification in offender profiling: classifying Serbian homicides using multiple correspondence, cluster, and discriminant function analyses. *Journal of Investigative Psychology and Offender Profiling, 11*(3), 221–236.

Goodwill, A., & Meloy, J. R. (2019). Visualizing the relationship among indicators for lone actor terrorist attacks: Multidimensional scaling and the TRAP-18. *Behavioral Sciences & The Law, 37*(5), 522–539.

Greco, F., & Greco, G. (2020). *Investigative techniques in the digital age: Cybercrime and criminal profiling*. European Journal of Social Sciences Studies, 5(3).

Grubin, D. (1995). *Offender profiling*. Taylor & Francis.

Heale, R., & Twycross, A. (2015). Validity and reliability in quantitative studies. *Evidence-Based Nursing, 18*(3), 66–67.

Horan, C., & Saiedian, H. (2021). Cyber crime investigation: Landscape, challenges, and future research directions. *Journal of Cybersecurity and Privacy, 1*(4), 580–596.

Janosch, H., et al. (2023). Predicting the Criminal Record of Spanish Stranger Rapists from Their Crime Scene Behaviours. *Journal of Aggression, Maltreatment & Trauma, 32*(12), 1666–1685.

Joseph, R. (2024). The importance of psychology in law and investigation: Exploring forensic psychological investigative techniques and criminal psychology. *Forensic Justice* (pp. 312–330). Routledge.

Kalichman, S. C., et al. (1990). An empirically derived typology of adult rapists based on the MMPI. *Journal of Psychology & Human Sexuality, 2*(2), 165–182.

Kocsis, R. N., Cooksey, R. W., & Irwin, H. J. (2002a). Psychological profiling of sexual murders: An empirical model. *International Journal of Offender Therapy and Comparative Criminology, 46*(5), 532–554.

Kocsis, R. N., Cooksey, R. W., & Irwin, H. J. (2002b). Psychological profiling of offender characteristics from Crime Behavio rs in Serial rape offences. *International Journal of Offender Therapy and Comparative Criminology, 46*(2), 144–169.

Kocsis, R. N., & Palermo, G. B. (2020). Clothes Don't Maketh the mannor a criminal profiler an expert witness. *International Journal of Offender Therapy and Comparative Criminology, 64*(12), 1317–1340.

Krstic, S., et al. (2018). *Using latent variable- and person-centered approaches to examine the role of psychopathic traits in sex offenders*. Personality disorders.

Lehmann, R. J. B., et al. (2013). Applying crime scene analysis to the prediction of sexual recidivism in stranger rapes. *Law and Human Behavior, 37*(4), 241–254.

Ling, S., Umbach, R., & Raine, A. (2019). Biological explanations of criminal behavior. *Psychology, Crime & Law, 25*(6), 626–640.

Maarek, J., & Darwish, S. (2023). *Criminal profiling: A look into its effectiveness and the application of a new evidence based approach.*

Martineau, M., Spiridon, E., & Aiken, M. (2023). A comprehensive framework for cyber behavioral analysis based on a systematic review of cyber profiling literature. *Forensic Sciences, 3*(3), 452–477.

Mashiane, T., & Kritzinger, E. (2019). Theoretical domain framework to identify cybersecurity behaviour constructs. In *International Conference on Innovative Technologies and Learning*. Springer International Publishing Cham.
Mashiane, T., & Kritzinger, E. (2020). Theoretical domains framework applied to cybersecurity behaviour. In *Artificial Intelligence and Bioinspired Computational Methods: Proceedings of the 9th Computer Science On-line Conference 2020* (Vol. 29). Springer International Publishing.
McCrae, R. R., et al. (2011). Internal consistency, retest reliability, and their implications for personality scale validity. *Personality and Social Psychology Review, 15*(1), 28–50.
Meterko, V., & Cooper, G. (2022). Cognitive biases in criminal case evaluation: A review of the research. *Journal of Police and Criminal Psychology, 37*(1), 101–122.
Miller, L. (2015). *Flaw and order: The science and mythology of criminal profiling* (p. 55). Skeptical Inquirer.
Newman, F. (2011). *A geo-behavioural model of stranger rape: Implications for offender profiling and linking.*
Okoli, C., & Schabram, K. (2015). *A guide to conducting a systematic literature review of information systems research.*
Petherick, W. (2003). What's in a name-comparing applied profiling methodologies. *JL & Soc. Challenges, 5*, 173.
Petrosyan, A. (2024). *Average total cost per data breach worldwide 2014–2024*. Cyber Crime & Security [cited 2025 24 March 2025].
Pinizzotto, A. J., & Finkel, N. J. (1990). Criminal personality profiling: An outcome and process study. *Law and Human Behavior, 14*(3), 215–233.
Ranti, D. V., & Nurmaily, E. (2021). Racial profiling on police stop and search practice as portrayed in the george Tillman's movie the hate U give. *Linguistics and Literature Journal, 2*(2), 93–97.
Ribeiro, R. A. B., & de Matos Soeiro, C. B. B. (2021). Analysing criminal profiling validity: Underlying problems and future directions. *International Journal of Law and Psychiatry, 74*, 101670.
Rossmo, D. K. (2023). *Behavioral profiling in the golden state killer investigation: A comparative analysis*. Homicide Studies.
Sahni, S. P., & Choudhury, T. (2021). The science of criminal profiling. In *Criminal psychology and the criminal justice system in India and beyond*, pp. 51–68.
Salfati, C. G. (2000). The Nature of Expressiveness and Instrumentality in Homicide. *Homicide Studies, 4*(3), 265–293.
Salfati, C. G., & Canter, D. V. (1999). Differentiating stranger murders: Profiling offender characteristics from behavioral styles. *Behavioral Sciences & The Law, 17*(3), 391–406.
Salo, B., et al. (2012). Using Bayes' theorem in behavioural crime linking of serial homicide. *Legal and Criminological Psychology, 18*(2), 356–370.
Sastry, V. (2015). *Criminal profiling in the detection and apprehension of offenders.*
Sigurdardóttir, T. D., et al. (2023). The current role and contribution of "behavioural investigative advise rs" (BIAs) to criminal investigation in the United Kingdom. *Journal of Criminal Psychology, 14*(2), 136–156.
Snook, B., et al. (2007). Taking stock of criminal profiling: A narrative review and meta-analysis. *Criminal Justice and Behavior, 34*(4), 437–453.
Sorochinski, M. (2015). *Assumptions underlying behavioral linkage revisited: A multidimensional approach to ascertaining individual differentiation and consistency.*
Tapper, S. (2008). *Testing the assumption of behavioural consistency in a New Zealand Sam ple of serial rapists*. Victoria University of Wellington Library.
Taylor, S., et al. (2011). Cluster analysis examination of serial killer profiling categories: A bottom-up approach. *Journal of Investigative Psychology and Offender Profiling, 9*(1), 30–51.
ter Beek, M., van den Eshof, P., & Mali, B. (2009). Statistical modelling in the investigation of stranger rape. *Journal of Investigative Psychology and Offender Profiling, 7*(1), 31–47.
Valentine, G. (2013). Tell me about...: using interviews as a research methodology. In *Methods in human geography* (pp. 110–127). Routledge.

Viglione, D. J., et al. (2022). Legal admissibility of the Rorschach and R-PAS: A review of research, practice, and case law. *Journal of Personality Assessment, 104*(2), 137–161.

Winerman, L. (2004). Criminal profiling: The reality behind the myth. *Monitor on Psychology, 35*(7), 66–69.

Woodhams, J., Hollin, C., & Bull, R. (2008). Incorporating context in linking crimes: An exploratory study of situational similarity and if-then contingencies. *Journal of Investigative Psychology and Offender Profiling, 5*(1–2), 1–23.

Yaksic, E. (2020). Evaluating the use of data-based offender profiling by researchers, practitioners and investigative journalists to address unresolved serial homicides. *Journal of Criminal Psychology*.

Zaitsu, W., Shibuya, Y.-S., & Hasegawa, N. (2008). Effectiveness of applying a Bayesian network to the offender profiling: stochastic inference of the indoor-sex-offender's employment. *Japanese Journal of Forensic Science and Technology, 13*(1), 83–92.

Network Structural Analysis of Digital Technology Adoption for Human-Centric Teaching and Learning Models

Anass Bayaga

Abstract This research examines the adoption of digital technologies in higher education utilizing network structural analysis and focusing on human-centric teaching and learning models. The research is anchored upon extant gaps including inconsistencies in global research output, limited interdisciplinary diffusion, and challenges in aligning technological advancements with ethical and pedagogical frameworks. The goals are to map collaborative networks, uncover interdisciplinary configuration, and evaluate the combination of AI-driven tools into teaching frameworks. The methodology utilized bibliometric data from Scopus and Web of Science (2015–2025) and employed visualization tools like Networkx to analyze citation pathways, and keyword co-occurrence networks. Results suggest robust regional partnerships but a lack of international inclusivity, with advanced regions dominating research groups. Ethical questions and faculty inclination surfaced as substantial barriers to incorporating AI-driven tools and immersive technologies into pedagogy. The thematic development revealed an increasing focus on adaptive learning, differentiated pedagogy, and ethical consequences. These conclusions emphasize organized pathways including Strong regional collaborations → Limited global inclusivity → Ethical and faculty readiness challenges → Evolving focus on adaptive learning and personalized pedagogy. These configurations exhibit how gaps in international impartiality, interdisciplinary incorporation, and ethical frameworks are unified, modeling the research environment.

A. Bayaga (✉)
Stellenbosch University, Stellenbosch, Cape Town, South Africa
e-mail: abayaga@sun.ac.za

J. Marx Gómez et al. (eds.), *Recent Advances in Information Systems*, Progress in IS,
https://doi.org/10.1007/978-3-032-14099-9_10

1 Background

1.1 Introduction

In an era led by the rapid advancement of digital technologies, the incorporation of artificial intelligence (AI) and data science is restructuring the landscape of higher education. While these technologies promise to transform teaching and learning, the challenge lies in guaranteeing that their acceptance is grounded in human-centric templates that highlight profound educational occurrences. Scholarship has explored the transformative possibility of digital tools (Tuomi et al. 2018). Nevertheless, substantial gaps remain in considering their influence on collaborative networks, interdisciplinary diffusion, and the complex balance between computational effectiveness and human-centered pedagogy (Tuomi et al. 2018). Emerging technologies, such as virtual reality (VR) and augmented reality (AR), provide immersive capabilities that can change pedagogical methods. However, as Jugembayeva and Murzagaliyeva (2022) note, barriers endure, restricting the reasonable implementation. Equally, both the potential and limitations of these technologies in establishing inclusive and accessible educational experiences remain uneven across different geographical locations (Park, 2020) leading to current systematic research. Regardless of the advancements, considerable barriers endure. For instance, ethical concerns coupled with privacy issues, and the digital divide tend to limit the equitable acceptance of AI-driven technologies (Park, 2020). There are also disparities in access and readiness across regions (Maphosa & Maphosa, 2023). Critiques have cited over-reliance on computational models that neglect complex realities of human cognition, prompting rethinking of AI systems to better reflect human intelligence (Tuomi et al. 2018). Some have also analyzed the cultural evolution of pedagogy, revealing how disciplinary traditions shape the acceptance of innovative practices. Such analysis led scholars to examine the integration of emerging technologies in higher education, emphasizing the need for cross-disciplinary collaboration to maximize their roles (Hu et al., 2020; Oke & Fernandes, 2020). Due to the fact that assessment models additionally act as a critical crossroads of technology and pedagogy, there is also scholarship on the potential of AI-based assessments to support individualized feedback and improve learning outcomes (Kashada et al., 2018; Lamtara, 2023). The incorporation of virtual and augmented reality (VR/AR) into assessment frameworks, tend to offer avenue for establishing immersive and interactive learning experiences (Park, 2020). It has been highlighted that cultural factors shape pedagogical practices, accenting the need for adaptive approaches that acknowledge disciplinary distinctions while fostering relationship across fields (Hu et al., 2020; Oke & Fernandes, 2020). While some have attempted to identify the need for separate AI-driven assessment models, a critical component of education, have until now to completely leverage the potential of AI (Tey et al., 2023). Yet, these scholarships also uncover challenges in designing assessments that are both technologically advanced and pedagogically sound (Tey et al., 2023). Such outcomes emphasize the need for robust bases that explore not only technological expertise but also the ethical, social, and cultural proportions of their relevance. In responses, this current research seeks to bridge these gaps by leveraging network analysis to investigate the acceptance of digital technologies in human-centric teaching and learning models, evaluate interdisciplinary knowledge diffusion pathways, while critiquing existing assessment frameworks.

The insights from the scholarships support pedagogical models that align technological tools with diverse learning outcomes. Another inference exploration involving interdisciplinary pathways through which digital technologies influence teaching and learning as well as resolving barriers that obstruct design, inclusive and accessible digital ecosystems that account for the diverse needs of learners. That is mapping structural dynamics can guide policymakers and practitioners in adopting environments conducive to innovation. Exploring the factors contributing to these disparities (infrastructure limitations, funding constraints, and digital literacy) can inform targeted interventions to close the gap. By adequately examining these areas, a more comprehensive framework for understanding and advancing the acceptance of digital technologies in human-centric education can be realized. Such insights will guarantee that innovations are not only technologically sound but also ethically aligned, socially inclusive, and pedagogically transformative. That is adoption of digital technologies requires deliberate attention on human-centric models that combine cognitive, emotional, and ethical reflections. By leveraging the gaps identified in prior research, the findings could underscore the potential of AI and data-driven tools to augment, rather than replace, human capabilities, fostering an educational ecosystem that is inclusive, innovative, and responsive to the needs of a diverse global population.

1.2 Research Objectives

To examine the structure and dynamics research collaborations within digital education, emphasizing human-centric teaching, learning and assessment.

Analyzing emerging research themes and declining topics in the adoption of AI and data-driven technologies for education.

Tracking interdisciplinary knowledge diffusion in AI-powered education by tracing the pathways across different domains.

2 Method

The current study employed network analysis to explore the adoption of digital technologies in human-centric teaching and learning, centering on collaboration patterns, thematic trends, and knowledge diffusion. The approach offered structured insights into understanding academic research networks by examining data and applying analytical techniques to explore the research landscape comprehensively. The approach was achieved through network visualization by offering intuitive representations of network structures using co-authorship networks, keyword co-occurrence in mapping thematic clusters and emerging trends. Additionally, citation networks traced knowledge flow as well as influence (see Table 1). Given it was systematic review, the study relied on data collected from Scopus and Web of Science (see Table 1). Boolean operators and wildcard searches enhanced the dataset to include publications across multiple domains, (educational technology, computer science, psychology, and engineering). The collected metadata involved co-authorship details, citation data, and keyword occurrences, allowing an in-depth investigation of collaborative networks, influential studies, and thematic clusters. Co-authorship data offered understanding into global research collaborations, while citation configuration helped trace knowledge diffusion. Keyword evaluation detected influential and emerging trends in the field. This approach ensured access to high-quality, peer-reviewed journal articles, conference proceedings, and systematic reviews. The selected timeframe of 2015 to 2025 captured the progression of digital education research, mirroring developments in AI-powered adaptive learning platforms and human-centric learning models. Keywords such as "AI", "human-centric learning," as well as "network structural analysis" were chosen deliberately to denote founded themes and emerging trends.

A strict inclusion criterion, centering on peer-reviewed journal articles, influential conference publications, and on AI, digital education, and human-centric learning models were employed, while exclusion criteria disregarded non-peer-reviewed publications lacking pedagogical or collaborative dimensions. Meanwhile, data cleaning process ensured rigor by eliminating duplicates and screening including abstracts, titles, and full texts in order to align with the research focus. The analytical tool utilized was Networkx (a Python-based bibliometric libraries) which facilitated visualization and analysis. Networkx was used to explore co-authorship networks while mapping collaborative patterns and identifying influential authors and partnerships. Noting that the co-occurrence networks as well as clustering techniques uncovered key thematic trends related to human-centric learning models, providing insights into the research trajectory. The citation and pathway analyses, which were supported by centrality measures (degree and betweenness), shed light on interdisciplinary networks, role in modeling human-centric teaching and learning methods. As reflected in Table 1, degree centrality emphasized influential contributors. Betweenness centrality uncovered bridging nodes that enable interdisciplinary knowledge flow. However, closeness centrality determined strategic access to information, while network density measured the overall cohesiveness of collaborations. Additional measure that is modularity analysis characterized distinct research communities together with thematic clusters. Meanwhile, temporal analysis showed changes in research priorities, providing understanding into the advancement of digital education.

Table 1 Network structural analysis

Key issue	Explanation	Relevance to methodology
Nodes (entities)	Represented authors, institutions, or countries involved in digital education research. Attributes such as productivity and impact (citations) were analyzed	Allowed identification of key contributors and institutions shaping the adoption of digital technologies in education (Oke & Fernandes, 2020)
Edges (relationships)	Represented collaborations or citations between nodes, including weighted connections to reflect the strength of relationships	Revealed patterns of collaboration and the intensity of interactions between researchers and institutions (Bonanomi et al., 2019)
Degree centrality	Measured the number of direct connections a node had within the network	Identified influential contributors or hubs critical to research innovation and dissemination (Lamtara, 2023)
Betweenness centrality	Determined the role of nodes as bridges facilitating connections across different clusters or communities	Highlighted entities crucial for knowledge dissemination and interdisciplinary collaboration (Jugembayeva & Murzagaliyeva, 2022)
Closeness centrality	Measured how efficiently a node could access all other nodes in the network	Provided insights into which entities had strategic access to information or research outputs (Bonanomi et al., 2019)
Network density	Ratio of actual connections to potential connections in the network	Assessed the cohesiveness and collaboration intensity of the research community (Oke & Fernandes, 2020)
Clustering coefficient	Measured the extent to which nodes formed tightly connected groups or communities	Identified thematic or institutional clusters, reflecting areas of concentrated research (Bonanomi et al., 2019)
Network modularity	Quantified the strength of division into distinct communities within the network	Helped uncover well-defined subgroups and research themes, revealing areas of specialization or shared objectives (Lamtara, 2023)
Keyword co-occurrence	Analyzed the frequency and patterns of keyword appearances in research articles	Identified dominant and emerging themes in digital education research, aiding in thematic trend analysis (Porio & Limpot, 2023)
Citation pathways	Traced the flow of knowledge through citations across disciplines and research fields	Provided insights into how foundational studies influenced interdisciplinary work and new research directions (Zapata-Paulini & Cabanillas-Carbonell, 2024)
Temporal analysis	Tracked the evolution of research themes over time by examining changes in keyword co-occurrence and citation networks	Revealed shifts in research priorities and the emergence of new areas of interest (Jugembayeva & Murzagaliyeva, 2022)
Network visualization	Graphically represented networks, highlighting relationships, clusters, and structural dynamics	Offered an intuitive way to interpret complex data, enhancing understanding of collaboration patterns and thematic clusters (Porio & Limpot, 2023)

3 Results and Discussion

The objectives of the current study were to examine the structure and dynamics research collaborations within digital education, analyze emerging research themes and declining topics in the adoption of AI and data-driven technologies for education and lastly track interdisciplinary knowledge diffusion in AI-powered education by tracing the pathways across different domains. Insights into how knowledge is diffused across disciplines, modeling interdisciplinary pathways in digital education research reviewed different pathways (Fig. 1).

Figure 1 presented publications on human-centric AI and adaptive learning systems while serving as critical connections between computer science, education, and psychology. Foundational research output including computational sciences repeatedly informed pedagogical approaches, highlighting the translational role of technological advancements in educational methods. Path analysis showed an encouraging incorporation of STEM domains with educational technology research output. For example, approaches developed in engineering and computational

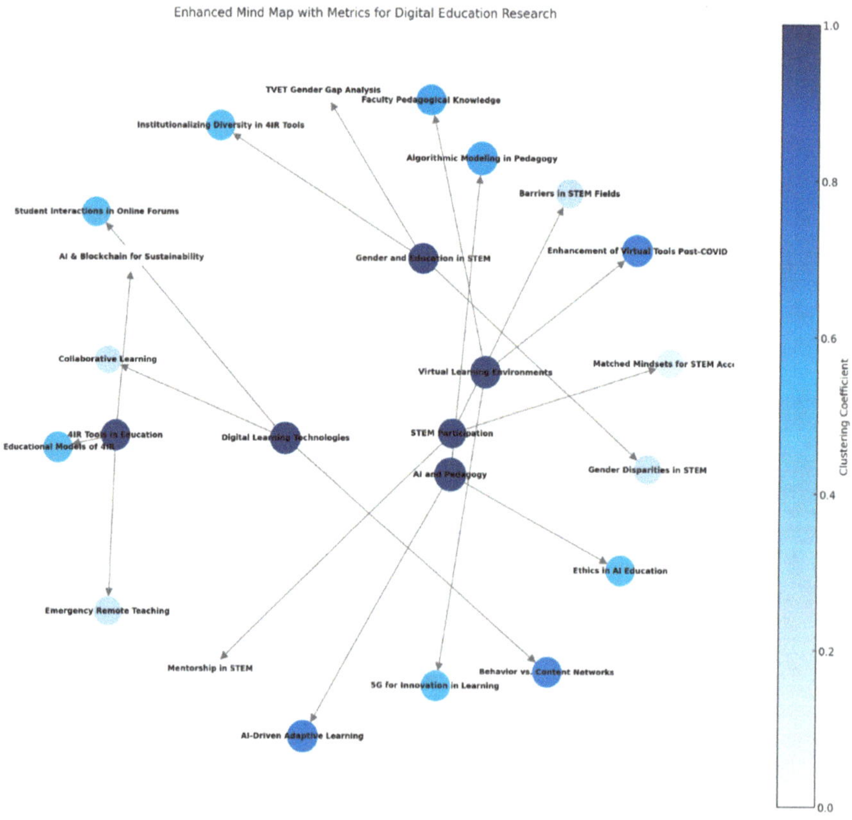

Fig. 1 Mind map with metrics for digital education research

sciences for virtual and augmented reality purposes were later adjusted for educational use. Likewise, insights from cognitive psychology inspired the growth of AI-driven adaptive learning techniques, emphasizing the reciprocal connection between technological innovation and pedagogical theory. The citation analysis identified barriers to cross-regional and cross-disciplinary partnership. While knowledge diffusion was moderately effective within specific domains, the incorporation of humanities and social sciences into AI-driven education patterns remained limited. This gap implied the need for greater interdisciplinary engagement to address the ethical, cognitive, and societal consequences of AI-enhanced learning. The analysis highlighted the position of bridging gaps between theoretical innovations and practical relevance, ensuring that research insights transformed into meaningful educational novelties. Evidence in terms of global collaborations and key contributors showed strong networks to sub-themes such as institutional collaboration and regional disparities, reinforcing its foundational role in insight in research networks. In terms of dominant and emerging themes, AI ethics and adaptive learning platforms were highlighted.

The implication of the results is a provision of actionable insights for not just researchers, educators, but and policymakers seeking further digital education. Mapping such global collaborations, identifying prominent research themes, and analyzing knowledge diffusion pathways tends to address disparities in regional participation which intend require targeted initiatives for the enhancement and access to research infrastructure, funding, and international partnership. The advancement of thematic trends emphasized the value of continuous adaptation to emerging technological and ethical consequences in digital education. The increasing projection of AI ethics, metaverse applications, and individualized learning models emphasized shifting research significances, compelling further research into their pedagogical and societal effects. Interdisciplinary partnership remained a critical element in developing human-centric digital education. Strengthening networks between computational sciences, educational psychology, and humanities domains could improve the incorporation of AI-driven learning models while addressing ethical and cognitive concerns. Policy interventions aimed at adopting interdisciplinary collaborations, supporting equitable resource distribution, and promoting ethical AI acceptance could further back the sustainable development of digital education. The results emphasize the need for targeted initiatives to encourage inclusivity and equitable access to digital education research. Addressing these disparities through policy recommendations, funding strategies, and institutional support can improve global contribution and drive innovation. By mapping global associations, identifying key research themes, and considering knowledge diffusion, this current research grants actionable insights for educators, researchers, and policymakers. This combined network analysis improves the insight of complex research landscapes, highlighting the need for interdisciplinary resolutions and targeted interventions to examine challenges in digital education research. The results have significant consequences for both theoretical advancements and practical purposes, guiding future research and policy programs in this field. The landscape of digital education research is modeled by substantial global collaborations, with institutions in North America, Europe, and

East Asia emerging as central hubs (Erlin et al., 2009; Grunspan et al., 2018). Those institutions exhibit high degree centrality, displaying their authority in modeling the field and facilitating knowledge dissemination. However, disparities persist, with lower connectivity observed in sub-Saharan Africa and parts of South America due to institutional capacity and resource constraints (Moloi & Salawu, 2022). Addressing these imbalances through inclusive group effort strategies is needed for fostering a balanced global research landscape. The networks showed dominant themes in digital education research, including adaptive learning platforms, AI ethics, and virtual learning environments. Recognized themes such as adaptive learning align with prior outcomes on data-driven pedagogical transformation (Park, 2020). Emerging topics like metaverse education and ethical AI consequences reflect shifting research priorities post-COVID-19 (Zhou et al., 2022). Clustering coefficients suggest a highly concentrated research landscape with limited interdisciplinary overlap, underscoring the need for greater thematic incorporation (Orr et al., 2018).

4 Conclusion

The analysis features a rapid evolution of digital education, with an increasing focus on AI powered adaptive learning, and immersive technologies. The result emphasizes the complex and progressing landscape of digital education research, modeled by collective systems, interdisciplinary diffusion, and the incorporation of AI-driven technologies in human-centric teaching and learning models. This study has shown that while regional research partnerships remain strong, global inclusivity remains a challenge. Research hubs in North America, Europe, and East Asia lead knowledge production, while disparities continue in under-resourced regions such as sub-Saharan Africa and parts of South America. Examining these disparities requires planned efforts to improve equitable research contribution through funding, policy interventions, and cross-institutional cooperation.

Author Contributions All authors made a substantial, direct contribution and approved the submitted version.

Funding We declare no funding for current study.

Data Availability Statement The datasets presented in this article are readily available on reasonable requests directed to the corresponding author.

Ethical Approval Given it was systematic review, the study relied on data collected from Scopus and Web of Science (see Table 1).

Consent to Participate NA

Declaration of Generative AI and AI-Assisted Technologies in the Writing Process During the preparation of this work the author utilized quillbot in order to improve readability after which the author reviewed and edited the content as needed and took full responsibility for the content of the published article.

Conflict of Interest None.

References

Bonanomi, M. M., Hall, D. M., Staub-French, S., Tucker, A., & Talamo, C. (2019). The impact of digital transformation on formal and informal organizational structures of large architecture and engineering firms. *Engineering Construction and Architectural Management, 26*(11), 2740–2756.

Erlin, B., Yusof, N., & Rahman, A. A. (2009). "Students' interactions in online asynchronous discussion forum: A social network analysis. In *2009 International Conference Education Technology Computer* (pp. 25–29)

Grunspan, D. Z., Kline, M. A., & Brownell, S. E. (2018). The lecture machine: A cultural evolutionary model of pedagogy in higher education. *CBE—Life Science Education, 17*(3), 3–18.

Hu, T., et al. (2020). Investigating the differences of student interactions between behavior- and content-based networks in online discussions. In *Proceedings 12th International Conference Computer Supported Education* (pp. 123–130).

Jugembayeva, B., & Murzagaliyeva, A. (2022). Physics students' innovation readiness for digital learning within the University 4.0 model: Essential scientific and pedagogical elements that cause the educational format to evolve in the context of advanced technology trends. *Sustainability, 14*(5), 45–65.

Jugembayeva, B., Murzagaliyeva, A., & Revalde, G. (2022). Pedagogical model for raising students' readiness for the transition to University 4.0. *Sustainability, 14*(2), 1–16.

Kashada, A., Li, H., & Koshadah, O. (2018). Analysis approach to identify factors influencing digital learning technology adoption and utilization in developing countries. *International Journal of Emerging Technologies in Learning, 13*(5), 48–59.

Kattoum, R. N., Abbood, I., Huff, K. E., & Baillie, M. T. (2023). Perceived barriers to equitable participation in the learning assistant program. *Journal of Chemical Education, 100*(9), 2495–2503.

Kricorian, K., Seu, M., Lopez, D., Ureta, E., & Equils, O. (2020). Factors influencing participation of underrepresented students in STEM fields: Matched mentors and mindsets. *International Journal of STEM Education, 7*(1), 1–9.

Lamtara, S. (2023). Faculty's pedagogical knowledge for technology-enhanced learning in higher education. *International Journal of Higher Education Pedagogies, 4*(1), 1–15.

Moloi, T., & Salawu, M. K. (2022). Institutionalizing technologies in South African universities towards the Fourth Industrial Revolution. *International Journal of Emerging Technologies in Learning, 17*(4), 204–227.

Oke, A., & Fernandes, F. (2020). Innovations in teaching and learning: Exploring the perceptions of the education sector on the 4th industrial revolution (4IR). *Journal of Open Innovation: Technology, Market, and Complexity, 6*(2), 31–45.

Orr, D., Weller, M. J., & Farrow, R. (2018). Models for online, open, flexible and technology-enhanced higher education across the globe—A comparative analysis. *International Council for Open and Distance Education, 2*, 3–17.

Park, Y. (2020). The latest research trends of teaching & learning in higher education by leveraging advanced technology and big data. *Advances in Research on Teaching, 9*(4), 72–80.

Porio, R. A., Jr., & Limpot, M. Y. (2023). A structural relationship model in the 21st-century teaching and learning of Filipino teachers. *Advances in Research on Teaching, 24*(1), 90–105.

Tey, T. C., Moses, P., & Cheah, P. K. (2023). The influence of gender on STEM career choice: A partial least squares analysis. *Research and Practice in Technology Enhanced Learning, 19*(1), 25–36.

Tuomi, I. (2018). The impact of artificial intelligence on learning, teaching, and education. In M. Cabrera Giraldez, R. Vuorikari, & Y. Punie (Eds.), *The impact of artificial intelligence on learning, teaching, and education*. Luxembourg: Publications Office of the European Union. https://doi.org/10.2760/12297.

Zapata-Paulini, J., & Cabanillas-Carbonell, M. (2024). 5G technology for innovation education (sustainable development goal 4): A systematic review. *International Journal of Engineering Pedagogy, 14*(1), 45–62.

Zhou, M., Dzingirai, C., Hove, K., Chitata, T., & Mugandani, R. (2022). Adoption, use, and enhancement of virtual learning during COVID-19. *Education and Information Technologies, 27*(6), 8939–8959.

AI-Powered Interactive Technologies and Cognitive Engagement in Mathematics: A Generational Cohort PLS-SEM and ANN Approach

Anass Bayaga

Abstract This study explores cognitive engagement in mathematical problem-solving across generational cohorts using artificial intelligence (AI)-enabled tools. Key hypotheses include: (H1) mathematical modeling and simulation (MMS) influence mathematical performance (MP), when mediated by maths computational algorithms (MCA) and moderated by generational cohort; and (H2) abstract-concrete principles (ACP) mediate the MMS and mathematical achievement (MAS) relationship, moderated by prior technological exposure. The research employs a cross-sectional survey with 71 participants analyzed via partial least squares structural equation modeling and artificial neural networks- multilayered perceptron (ANN-MLP). The results reveal significant paths from MCA to ACP ($\beta = 0.69$, $p < 0.001$) across cohorts, indicating MCA's strong influence on cognitive engagement. The MMS-MCA relationship ($\beta = 0.23$, $p = 0.05$) and MCA-MP ($\beta = 0.4$, $p < 0.01$) are similarly strong, affirming MCA's role as a key mediator for MMS's effect on MP. However, the ACP-MAS path is non-significant ($\beta = 0.09$, $p = 0.58$), indicating a weaker link between foundational principles and achievement. Findings advocate universal design learning (UDL) principles in educational technology to optimize engagement across diverse learner backgrounds. Future research should explore motivational and contextual factors influencing AI tool efficacy to refine educational practices that universally support cognitive engagement.

1 Introduction

Recent and past studies have revealed that most current university students were born between 1980 and 1994 with number of ramifications when examined from cognitive engagement and AI or digital tools in general (Alam, 2022; Yilmaz & Yilmaz, 2023). For example, many of these students were interactive digital tools (including AI-powered tools)-savvy (simply techno-savvy) and competent social

A. Bayaga (✉)
Stellenbosch University, Stellenbosch, Cape Town, South Africa
e-mail: abayaga@sun.ac.za

J. Marx Gómez et al. (eds.), *Recent Advances in Information Systems*, Progress in IS,
https://doi.org/10.1007/978-3-032-14099-9_11

networkers, which align with the digital nativism model. Yet, many of them could come from disadvantaged backgrounds with poor socio-economic conditions and resource-constrained schooling to enhance cognitive engagement. For quite some time now, it has been explained that cognitive engagement is the degree to which learners are prepared and capable of undertaking the current learning task or method (Richardson & Newby, 2006). This encompasses both the level of effort learners are ready to dedicate to the task and their ability to sustain their involvement over time (Rahayuningsih et al., 2020; Richardson & Newby, 2006). On the other hand, interactive digital tools, for computational or generally science, technology, engineering, and mathematics (STEM), are a combination of various gadgets, including head-mounted computers, monitors, body sensors, device-specific interfaces, real-time graphics, and sensory rendering systems in content design. Such interactive digital tools can present computer-generated graphics to draw users in. Furthermore, whether AI-powered or not, the scenes of a virtual environment take the role of the outside world (Symons & Pierce, 2019; Yohannes & Chen, 2023; Zhang et al., 2023). Some[1] examples include but not exclusively; GeoGebra, Wolfram Alpha, MATLAB, TensorFlow, Unity3D, Microsoft Azure, Machine Learning Studio, Jupyter Notebook with Python libraries (like NumPy, SciPy, and Matplotlib), IBM Watson Studio, Google AI Studio, Tableau, Mathematica, Simulink, OpenAI Gym, PyTorch, Autodesk AutoCAD, RStudio, SmartPLS.

Yet, limited research underscores the pathways pivotal for cognitive engagement enhancement (Li et al., 2021; Ziatdinov & Valles, 2022). Cognitive engagement research also shows that meaningful learning only takes place when learners are fully engaged cognitively (Alam, 2022), in which case, deep learning or deep cognitive engagement is directly linked to learners achieving great results (Green, 2015). Ongoing development is that several tools such as interactive digital tools (including AI-powered tools) can measure cognitive engagement and improve results (Hsu et al., 2023; Huang & Qiao, 2022). Consequently, such analysis could underscore the necessity to align educational technologies with the distinct values and learning styles of different generational groups in maximizing educational outcomes. The implication is that the integration of digital tools as supported by the UDL theory, in classrooms for solving mathematical problems can be beneficial for all. As reported by Yilmaz and Yilmaz (2023), AI tools not only enhance the capability but also the confidence and engagement of students with mathematics consequentially aligning with UDL. Such an insight is crucial for designing educational technologies intending to cater for the distinct learning styles and needs of different generational groups, hence optimizing learning outcomes and fostering a deeper cognitive engagement in mathematics (Yilmaz & Yilmaz, 2023). In recent times, studies underscore that generative AI tools significantly enhance computational thinking skills-critical for science, technology and engineering with particularly reference to mathematical (STEM) problem-solving. This is particularly relevant for generation Z (Gen Z) and millennials, who both benefit from and necessitate robust problem-solving skills in their academic and professional lives within the twenty-first century. Anchored upon the

[1] Selected ones such as RStudio and GeoGebra will be exemplified as one reads.

fledgling trends, the current study examines the integration of cognitive engagement and AI-powered tools in mathematical problem solving, with a focus on the differences across generational cohorts, particularly Gen Z and Millennials. For current study, Gen Z (ages 15–23) are characterized as tech-savvy, diverse, and independent thinkers, aligning with their preference for personalized and adaptive learning environments facilitated by AI (Anand et al., 2023). Millennials (ages 24 and older), on the other hand, tend to be ambitious, collaborative, and seek purpose-driven careers, traits complementary of AI tools that promote interactive and cooperative learning experiences (Alam, 2022; Anand et al., 2023). Thus, the study differentiates itself from others by focusing on analyzing learners' cognitive engagement, in particular active learning, in a mathematical classroom when AI-powered tools are used. In this study the interest is not to prove that the use of digital tools is beneficial or not beneficial to learners. Rather, it is to look intrinsically at their thinking (cognition), discussing, investigating and creativity when using digital tools to solve mathematical problems through cognitive engagement (peculiar to computational skills). Understanding how learners engage cognitively is important because it improves their potential and designs educational teaching methods to the needs of learners (Evans et al., 2021; Kim et al., 2020; Tondeur et al., 2017).

2 Literature Review and Model Development

The debate surrounding the use of interactive digital tools including GeoGebra and other AI-powered tools in mathematics education illustrates a significant shift in pedagogical approaches, particularly when analyzed through the lens of generational cohorts (Lee et al., 2020; Li et al., 2021). For example, several studies highlight various influences of GeoGebra on students' learning experiences, showcasing both the strengths and limitations of such tools in different educational contexts (Mjenda et al., 2023; Su et al., 2022; Tlili et al., 2023; Yilmaz & Yilmaz, 2023; Yilmaz, et al., 2022). Proponents such as Owusu et al. (2023) of GeoGebra (see Fig. 1), an example of interactive digital tools (including AI-powered tools), argue for the application's effectiveness in training visual-spatial intelligence and enhancing university students' understanding of complex concepts like polar coordinates.

These studies underscore GeoGebra's role in fostering deeper comprehension and engagement with mathematical content. Kurniawan and Rahadyan (2022) and Mendi et al. (2023) discuss GeoGebra's capacity to improve self-regulated learning and creativity, as well as 3D geometry spatial visualization skills (see Fig. 2).

These outcomes are particularly relevant to both Gen Z and Millennials, who value technology that supports autonomous learning and creative problem-solving (Kurniawan & Rahadyan, 2022; Soares et al., 2022).

However, critics of the widespread adoption of interactive digital tools (including AI-powered tools) point to the necessity of integrating these tools with traditional teaching methods to ensure comprehensive learning. Uwurukundo et al. (2022) provide evidence that while GeoGebra enhances performance in 3D geometry, there

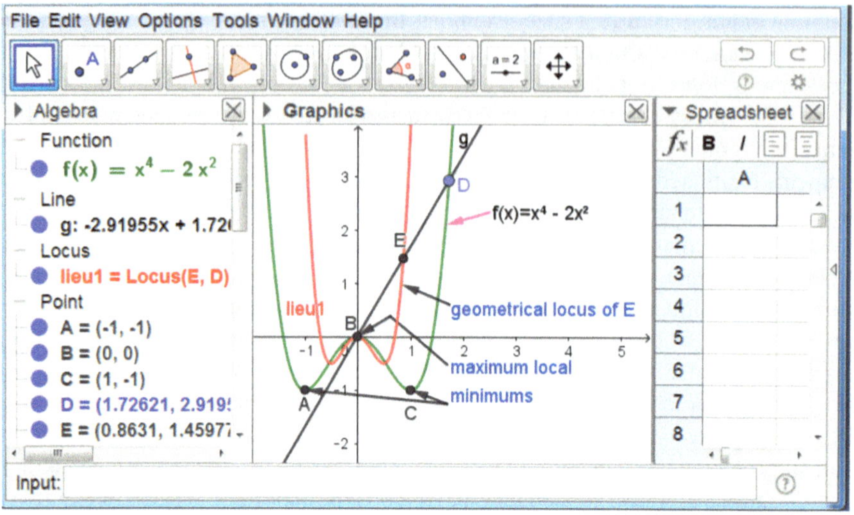

Fig. 1 GeoGebra an example of interactive digital tools (including AI-powered tools) (Malinc, 2024)

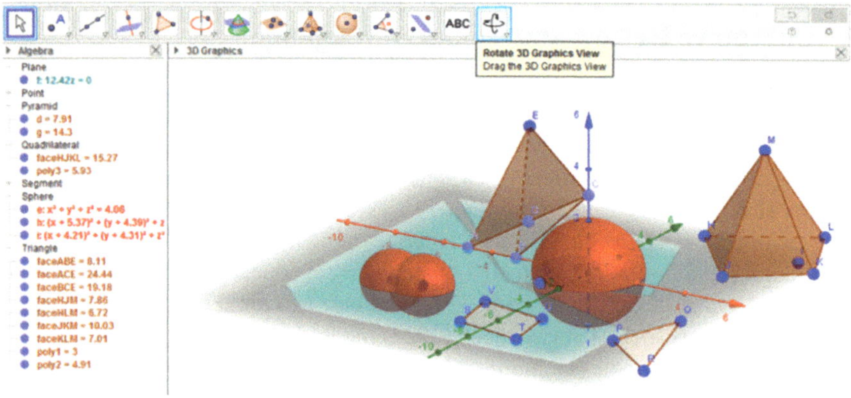

Fig. 2 3D geometry spatial visualization skills from GeoGebra (Soares et al., 2022)

is a critical need for pedagogical strategies that also address the varying attitudes of students and teachers toward technology in education (Uwurukundo et al., 2022). Moreover, the systematic review by Yohannes and Chen (2023) and the meta-analysis by Zhang et al. (2023) offer broader insights into the role of GeoGebra over the past two decades, suggesting that while the tool has significantly contributed to mathematical learning, the outcomes vary widely depending on the context of use and the specific educational objectives (Yohannes & Chen, 2023). Ziatdinov and Valles Jr (2022) highlight an effective synthesis of modeling, visualization, and programming within GeoGebra, and other AI-powered interactive tools such as RStudio by advocating for its role in teaching and learning STEM topics comprehensively (Ziatdinov & Valles, 2022). In conclusion, while the evidence supports the utility of interactive digital tools (or AI-powered) for modeling, visualization, and programming such as RStudio, Matlab, Wolfram Alpha, Microsoft Maths Solver, and GeoGebra in enhancing mathematical learning through visualization and interactivity, the debate reflects the necessity of thoughtful integration of these technologies within a structured educational framework. Such integration is crucial for maximizing their effectiveness across different generational cohorts, who each bring distinct expectations and skills to the learning environment. From the standpoint of technology incorporation too, Lai and Law (Leu et al., 2013) find that technology's efficacy is often conditional on its integration into the curriculum and the teachers' capability to harness its potential. In contrast, it has been contended that while engagement is important, the type of engagement, reflecting pathways essential for cognitive engagement (peculiar to computational skills) such as of MMS, MCA, and ACP in improving MP and MAS and configuration with instructional goals are most significant in educational results (Cui & Ng, 2021; Evans et al., 2021; Kurniawan & Rahadyan, 2022; Soares et al., 2022; Su et al., 2022; Sullivan & Strawhacker, 2021). This is because ACP are essential cognitive concepts that bridge abstract models with real-world applications. In this study, ACP signifies fundamental cognitive approaches influencing how individuals engage with AI tools in mathematical problem-solving. Meanwhile, MCA refers to methodical guidelines or measures utilized for mathematical calculations. In this setting, MCA involves algorithms within AI tools intended to aid in mathematical problem-solving tasks. MP measures users' ability in mathematics, together with problem-solving skills and understanding. In this research, MP reflects how efficiently users use AI tools to enhance their mathematical capabilities, particularly among rural high school STEM setting and diverse generational cohorts. MMS includes utilizing mathematical models and computational simulations to understand real-world phenomena. Mathematics achievements (MAS) measure user's skill and accomplishment in mathematical knowledge and aptitudes.

Jointly, the proposal is that tools like GeoGebra should be used to augment detailed types of engagement that directly relate to didactic objectives (Sullivan & Strawhacker, 2021). Consequently, importance is on the role of cognitive engagement in online learning settings, signifying those tools which enable active learning and involvement can improve the learning experience in digital setups. This is predominantly appropriate for understanding how GeoGebra and comparable technologies can be used efficiently in inaccessible or hybrid learning circumstances. Other areas also remain prevalent for additional examination include the aim to examine the enabling role of MCA in the relationship between MMS and Math MP, and to examine how this mediation differs between generational cohorts particularly (Gen Z and Millennials). This objective aims to elucidate how generational characteristics influence the effectiveness of computational tools in enhancing mathematics performance. Equally important is to explore how ACP mediates the relationship between MMS and MAS, and to assess the moderating effect of prior technological exposure on this mediation. This will help determine how pre-existing familiarity with digital tools affects the ability of students to leverage abstract and concrete mathematical

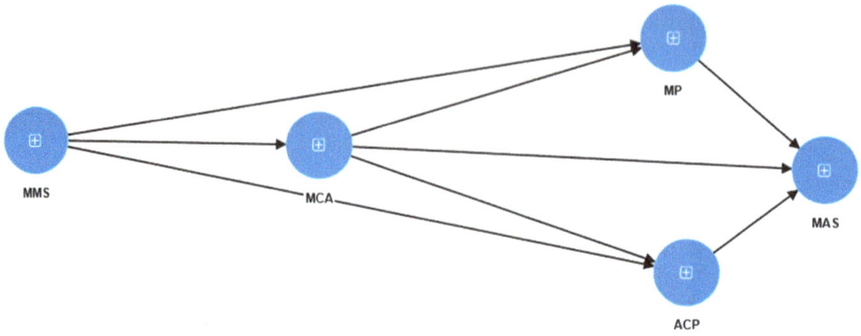

Fig. 3 Pathways of mathematical performance (Own Fig)

principles for improved academic outcomes (Fig. 3). The premise of the research is that learners' levels of cognitive engagement during the use of digital tools is not significantly reflected on their results and following Fig. 3.

Several implications could be drawn from Fig. 3. Figure 3 shows direct paths from MMS to MCA, MCA to both MP and MAS, and ACP to MAS. Additionally, there could be a direct path from MMS to both MAS and MP, indicating that MMS influences math achievement and performance directly. It also indicates that MCA serves as a mediator between MMS and both MAS and MP, suggesting that the effectiveness of mathematics modeling and simulation in improving mathematics achievement and performance may depend on the computational algorithms applied. The direct path from ACP to MAS highlights the role of abstract and concrete principles in math achievement, separate from the influence of MCA. In understanding Fig. 3, it can be determined that each construct contributes to mathematics achievement and performance, with computational algorithms playing a dominant mediating role. The direct effect of MMS on MP proposes that the tools and methods used in modeling and simulation have instantaneous practical assistances, while their influence on MAS suggests a foundational role in math cognition. The nonappearance of a path from ACP to MP could show that abstract and concrete principles are more applicable for long-term understanding rather than immediate performance.

2.1 *Hypotheses*

H1: The relationship between MMS and MP is mediated by MCA. Furthermore, this mediation effect is moderated by generational cohorts, with a potentially stronger influence observed in Gen Z than in Millennials.

H2: ACP mediates the relationship between MMS and MAS. The effectiveness of this mediation is further moderated by different (Gen Z and Millennials) students' prior technological exposure. Thus, positing that students with greater familiarity with digital tools will experience a more pronounced mediation effect.

3 Method

A single cross-sectional field study design was utilized to conduct a comprehensive analysis of generational cohorts concerning cognitive engagement with AI-powered tools, particularly interactive digital tools, within the context of mathematical problem-solving. The study involved gathering data from 71 diverse generational cohorts in rural high school STEM learners. Specifically, a cross-sectional survey method was employed to investigate opinion-related data using a well-structured questionnaire. Guided by relevant literature and the two specific hypotheses, it was crucial for the survey instrument to be aligned with the examined constructs (MMS, MP MCA), as there was a keen interest in exploring potential mediation and moderation effects by generational cohorts, particularly comparing Gen Z with Millennials. Additionally, data on ACP was collected, hypothesized to mediate the relationship between MMS and MAS. The type of data collected was deemed essential due to the planned utilization of both PLS-SEM and artificial neural network (ANN) in the subsequent data analysis. Stratified sampling techniques were employed during the data collection process, following the recommendations by Hair et al. (2017) specifically targeting rural high schools in the King Cetshwayo District of South Africa. The response rate was notably high, reaching 90%, which exceeded the recommended minimum sample size of 40 (Hair et al., 2017). The recommendation of Hair et al. (2022) is with regards to the requirement of 10 times larger items of the construct with most items, as the recommended minimum sample size was 40, in which the sample size exceeded the recommended 40. Furthermore, the choice of research site was deliberate, aiming to align with the hypotheses under investigation for enhanced generalization. Participants at these sites were accustomed to using one or more interactive digital tools, including AI-powered ones. A two-week timeframe was allotted for completing the questionnaires, during which strict ethical considerations were observed, ensuring privacy, confidentiality, and anonymity for all participants. Additionally, participants were informed of their right to voluntary withdrawal from the study at any point. The research procedures were conducted under the guidance and approval of the University of ZZZZ's research ethics committee, ensuring adherence to ethical standards (referencing H18-EDU-ERE-011).

3.1 Data Analysis

Recall that the measurement and structural model assessments were carried out which were followed with ANN analysis. Measurement model assessment in PLS-SEM ensures that indicators effectively measure intended constructs by checking validity, reliability, and indicator loadings. It also evaluates model fitness and identifies redundant indicators for a more efficient and accurate analysis. The structural model analyses relationships between constructs, revealing causal pathways and effects. It identifies key drivers and assesses model fit for accurate representation of data and hypotheses. On the other hand, ANN analysis, specifically MLP, involved using interconnected layers of nodes to process input data and predict outcomes. The essence of MLP analysis lies in training the network to learn complex patterns in data through iterative adjustments of connection weights. It aims to accurately classify or predict outcomes based on input variables, with the ability to capture nonlinear relationships and handle high-dimensional data.

4 Results

4.1 Measurement Model Assessment and Explanatory Power

During the measurement model assessment and explanatory power, battery of test were conducted: Cronbach's alpha (α), a measure of internal consistency, exceeds the recommended threshold of 0.70 for all constructs across the complete sample, as well as individually within Gen Y and Gen Z cohorts. The average variance extracted (AVE) for ACP is 0.63, 0.62, and 0.68 (complete sample, Gen Y, and Gen Z, correspondingly). Based on the Heterotrait–Monotrait Ratio of Correlations (HTMT) ratios, discriminant validity is generally supported across the constructs for the complete sample and within the generational cohorts, Gen Y and Gen Z. The Fornell–Larcker criterion further substantiates discriminant validity, where the square root of average variance extracted (AVE) for each construct should be greater than its highest correlation with any other construct. It was confirmed that all constructs met this criterion across the complete sample and both generational cohorts. For example, in the complete sample, ACP's AVE square root (0.79) is greater than its highest correlation with any other construct (0.69 with MCA), affirming discriminant validity. In addition to the measurement model (test of reliability and validity of constructs and indicators), the analysis of the model's explanatory power using R-squared (R^2) indicates that the variance explained in the constructs differs across the overall sample and specific generational cohorts. In summary, the R^2 and Adjusted R^2 indicate model has variable levels of explanatory power for diverse constructs and across generations, with the model generally performing better for Gen Z, except for MAS and MCA. The outcomes indicate not only that the constructs are well-defined and reliable but also that there may be generational differences in how these constructs are established, with Gen Z displaying higher reliability coefficients.

4.2 Structural Model

The structural model was first examined through variance inflation factor (VIF), thus test of multicollinearity (complete sample and in samples—Gen Y and Gen Z) by assessing the degree of collinearity among predictor constructs. In current study, it was established that the relationships observed in the model were attributed to the constructs themselves rather than to underlying collinearity among the predictors, hence enhancing the reliability of the model. Once the VIF was established, the structural model that is the different hypotheses were then examined using various indicators such as standardized beta coefficients (β), indicating the strength and direction of the relationships between constructs. The standard deviation (SD), which provides a measure of the variability or dispersion of the β coefficients as well as the t and p-values, used in assessing the statistical significance of the relationships. The findings revealed that the pathway from MCA to ACP is highly significant across all groups ($\beta = 0.69$, 0.68, and 0.75; $p < .001$; for the complete sample, Gen Y, and Gen Z, respectively), indicative of a strong and consistent influence of MCA on ACP across generations. On the contrary, the pathway from ACP to MAS displays non-significant relationships across all samples, $p > 0.05$ (for example, complete sample, $\beta = 0.09$, $p = 0.58$, indicating the influence of ACP on MAS is not statistically significant. The wide-ranging levels of significance and effect sizes across diverse pathways and generational cohorts reflect the necessity of considering generational distinctions in structural modeling. The non-significant pathways may indicate areas where additional external predictors could be influencing these relationships. A meditated analysis was conducted to test for total indirect effects and specific indirect effects The examination of the total and specific indirect effects shows that the mediating roles of the constructs differ remarkably between the generational cohorts. Crucial findings comprise the noteworthy mediation effect of ACP in the MMS $\rightarrow$ ACP $\rightarrow$ MAS pathway for Gen Y, and the non-significant mediation effects of MCA in the pathways leading to MAS across all groups. These findings indicate distinct inter-construct dynamics that are influenced by generational variances, showing essential distinctions for developing targeted approaches and interventions that are sensitive to these generational peculiarities in the setting of the studied constructs. Multigroup analysis (MGA) through Bootstrap MGA, Parametric Test, Welch Satterthwaite was also conducted. In this research, MGA was utilized to examine potential variances in the relationships between variables across different generational cohorts comparing the structural model between Gen Z and Millennials to assess whether the relationships between MMS, MCA, and MP contrast significantly between these groups. The essence of MGA in this study lies in assessing the invariance of the structural model across generational cohorts. By conducting MGA, it is possible to determine whether the hypotheses hold true for both Generation Z and Millennials or if there are significant variances in the pathways and effects of the variables between these cohorts. Mostly, MGA assesses a projected structural model's generalizable across diverse settings, thus providing insights into potential group-specific effects and enhancing the validity and applicability of the study's results across diverse populations. To test for the MGA, the measurement invariance of composite models (MICOM) procedure was applied in MGA providing critical insights into the invariance of constructs across diverse groups. In conclusion, the MICOM analysis largely supports the measurement invariance of the constructs across the groups studied, with some indications of variance differences in a few constructs. This suggests that while the constructs are generally perceived similarly across groups, there might be nuances in their variance that warrant further investigation as future research, especially for constructs like MCA, MMS, and MP. This understanding is crucial for ensuring valid cross-group comparisons tailoring strategies that are sensitive to group-specific nuances in these constructs. Through Bootstrap MGA, Parametric Test and Welch Satterthwaite, a comparison of path coefficients between Gen Y and Gen Z across different testing methods was considered. The results generally indicate that while there are some numerical differences in path coefficients between Gen Y and Gen Z, most of these differences do not reach statistical significance under the conventional $p < 0.05$ criterion. This suggests that the relationships modeled in the SEM are relatively stable across these generational cohorts.

4.3 Artificial Neural Network (ANN): Multilayer Perceptron (MLP)

The application of ANN-MLP yields insightful with significant consequences. The network architecture (made of 21 input factors, 8-unit hidden layer, and a 4-unit output layer), reflects the model's complexity in capturing relationships. Dataset was partitioned into training (67.9%) and testing (32.1%) samples, ensuring robust evaluation and generalization performance with low cross-entropy error in both training (35,935) and testing (13,382) phases, signifying efficient learning and generalization capabilities. The percent incorrect predictions indicated the model's proficiency in classifying cases correctly. In the training set, the model validates considerable accuracy (overall correct classification rate of 65.8%). However, in the testing set, the accuracy improves (77.8%), revealing the model's capability to generalize well to unseen data. The classification matrix indicates the model is effective in forecasting results for the first two categories (accuracy rates of 44.4% and 55.6%, respectively). The area under the curve (AUC) indicators supports the model's discriminative capability, (AUCs from 0.218 to 0.875) for diverse groups of the dependent variable MAS. The analysis of independent variable importance highlights important predictors to the model's decision-making process. Some constructs (MCA and ACP) emerge as decidedly significant, representative of their role on prediction. The conclusion is that through ANN (MLP) model is that it efficiently captures and generalizes complex patterns in for forecasting. The elevated accuracy rates in the testing set and the role of key constructs underscore the model's potential applicability in real-world decision-making conditions.

5 Discussion

The result reveals that while MCA significantly mediates the relationship between MMS and MP, the expected moderating effect of generational cohorts, precisely variances between Gen Z and Millennials, did not display a statistically important. This indicates that regardless of generational variances, MCA constantly plays a crucial role in improving MP. Such standardization across cohorts suggests that both Gen Z and Millennials have an equal level of engagement and benefit from the use of computational tools in learning settings (Clark et al., 2016). This confirms universal efficacy of computational tools irrespective of age-related preferences or experiences. The outcome aligns previous study, which noted the broad application and usefulness of AI-powered tools across demographic divides, signifying that the intuitive design and accessibility of modern technologies like such as GeoGebra could bridge generational gaps in understanding and ease of use.

However, a possible explanation for the lack of substantial generational variances could be the ubiquity (personal and academic contexts) of digital technology, thus normalizing the utilization of such tools among different age groups. Furthermore, the results might reflect a homogenization of digital literacy skills across younger and older students due to institutional policies or curricular standards that incorporate technology consistently across programs (Clark et al., 2016). Hypothesis 2, which sought to examine the mediation by ACP and moderation by technological exposure suggest that ACP is expected to mediate the relationship between MMS and MAS, with the efficacy of this mediation being improved by prior technological exposure. In this instance, findings show ACP does mediate the relationship between MMS and MAS. Nevertheless, the moderation by technological exposure was less noticeable than predicted, implying familiarity with technology, while beneficial, does not meaningly augment the mediating role of ACP. The suggestion is that other factors such as the intrinsic motivation or cognitive engagement approaches might play more considerable roles (Tondeur et al., 2017). The implication is that the cognitive processes involved in leveraging ACP for learning are not profoundly reliant prior technological skill but more on the pedagogical incorporation and the circumstantial application of these principles in problem-solving situations. The lack of substantial moderation by technological exposure might also indicate that other factors such as cognitive engagement, motivation, and instructional excellence might play more considerable roles in mediating the effectiveness of ACP in educational sceneries (Clark et al., 2016). It also points to the possibility of ACP serving as a universally effective instructional approach that benefits all learners, irrespective of their initial comfort or familiarity with technology. These findings underline the importance of AI-powered digital tools like GeoGebra and other AI-enabled resources in educational situations. Kurniawan and Rahadyan (2022) provide evidence that these technologies not only support learning but also improve cognitive engagement and ingenuity, important for academic success in STEM domains. Furthermore, Yohannes and Chen (2023) and Zhang et al. (2023) corroborate that while AI-powered digital tools are effective, their role differs based on educational purposes and the setting of their usage, highlighting need for personalized educational approaches that align with learner needs and curricular goals. The consequences are multi-layered, touching on theories of cognitive engagement, educational technology incorporation, and generational learning variances. The research underlines the implication of cognitive engagement as conceptualized, demonstrating that cognitive engagement, enabled using AI-powered tools like GeoGebra, can meaningfully influence mathematical performance. Thus, deep cognitive processes are important for effective learning and retention. The mediation effect of MCA and ACP infers that cognitive engagement in educational surroundings can be methodically designed through exact tool features aligning with pedagogical goals, consequently encompassing traditional theories of engagement to include digital tool functionalities.

5.1 Implication

For enhancement of theories of technology integration in education, it could be inferred not just through Tondeur et al. (2017), the UDL framework but the findings that technological exposure does not significantly moderate the effectiveness of digital tools, highlighting the importance of how such tools are integrated into the learning process. In practice, this suggests a shift in focus from merely providing access to technology, to improving how educators and curriculum developers incorporate these tools in ways that are pedagogically sound and cognitively appealing. In terms of generational theories and digital nativism; contrary to what might be expected from theories of digital nativism, which postulate that younger generations are fundamentally more proficient at leveraging digital technologies in learning contexts, the results revealed no significant variances in the effects of these technologies between Gen Z and Millennials. Thus, challenging the theory of digital nativism while supporting a more nuanced view of comfort and proficiency with digital technologies not solely determined by age but by a multifaceted interaction of experience, education, and personal interest (Kurniawan & Rahadyan, 2022).

6 Conclusion

The evidence suggests while generational variances and prior technological exposure offer context, they do not considerably change the essential dynamics of how students benefit from digital learning tools, aligning with broader educational research, which advocates adaptive and wide-ranging educational technologies that accommodate diverse learner needs across diverse settings. Overall, the theoretical implications call for a wider conceptualization of how cognitive engagement suggesting future research should target theoretical models related to dynamic interaction between technology, cognitive engagement strategies, and individual learner characteristics, which may lead to more effective educational practices, deeper understanding of the role of technology, extending beyond generational stereotypes and toward a more comprehensive and nuanced understanding of learning in the digital age.

Author Contributions The author conceptualised and condcuted entire research.

Funding I declare no funding for current study.

Data Availability Statement The datasets presented in this article are readily available on reasonable requests directed to the corresponding author.

Ethical Approval The study protocol followed ethical best practices, was reviewed and revised by a panel of experts, and approved by the faculty and university board.

Consent to Participate All participants provided written consent to participate in this research.

Declaration of Generative AI and AI-Assisted Technologies in the Writing Process During the preparation of this work the author utilized quillbot in order to improve readability after which the author reviewed and edited the content as needed and took full responsibility for the content of the published article.

Conflict of Interest None.

References

Alam, A. (2022). Educational robotics and computer programming in early childhood education: A conceptual framework for assessing elementary school students' computational thinking for designing powerful educational scenarios. In *International Conference on Smart Technologies and Systems for Next Generation Computing (ICSTSN) (ICSTSN)* (pp. 1–7).

Anand, M., Srivastava, V. K., Rayal, A., Pandey, S., Kumar, B. V., & Pachouri, V. (2023). Revolutionizing mathematics learning: Exploring the potential of robotics and AI as interactive tools for personalized and engaging mathematical education. In *2023 6th International Conference on Contemporary Computing and Informatics (IC3I)* (Vol. 6, pp. 967–972).

Clark, D. B., Tanner-Smith, E. E., & Killingsworth, S. S. (2016). Digital games, design, and learning: A systematic review and meta-analysis. *Review of Educational Research, 86*(1), 79–122.

Cui, Z., & Ng, O. (2021). The interplay between mathematical and computational thinking in primary school students' mathematical problem-solving within a programming environment. *Journal of Educational Computing Research, 59*, 988–1012.

Evans, T., Klymchuk, S., Murphy, P. E., Novak, J., Stephens, J. M., & Thomas, M. (2021). Non-routine mathematical problem-solving: Creativity, engagement, and intuition of STEM tertiary students. *STEM Education.*

Fredricks, J. A., Blumenfeld, P. C., & Paris, A. H. (2004). School engagement: Potential of the concepts, state of the evidence. *Review of Educational Research*, 59–109.

Green, B. A. (2015). Measuring cognitive engagement with self-report scales: Reflections from over 20 years of research. *Educational Psychologist,* 14–30.

Hair, J. F., Hult, G. T. M., Ringle, C. M., & Sarstedt, M. (2017). *A primer on partial least squares structural equation modeling (PLS-SEM) using R: A workbook.* SAGE Publications.

Hsu, T. C., Chang, C., & Lin, Y. W. (2023). Effects of voice assistant creation using different learning approaches on performance of computational thinking. *Computers and Education, 192.* Article 104657.

Huang, X., & Qiao, C. (2022). Enhancing computational thinking skills through artificial intelligence education at a STEAM high school. *Science Education*, 1–21.

Kim, S., Pollanen, M., Reynolds, M. G., & Burr, W. S. (2020). Problem solving as a path to comprehension. *Mathematics in Computer Science*, 1–15.

Kurniawan, I., & Rahadyan, A. (2022). Improving self-regulated learning and creativity of students through GeoGebra in solving spatial geometry problems. *Symmetry, 7*(2), 89–96.

Lee, S., Ke, F., & Ryu, J. (2020). Engagement and effectiveness of symbolic and iconic learning support for math problem representation: An eye tracking study. *Interactive Learning Environments, 31*, 1514–1531.

Leu, D. J., Kinzer, C. K., Coiro, J., Castek, J., & Henry, L. A. (2013). New literacies: A dual-level theory of the changing nature of literacy, instruction, and assessment. *Journal of Education*, 1150–1181.

Li, S., Lajoie, S. P., Zheng, J., Wu, H., & Cheng, H. (2021). Automated detection of cognitive engagement to inform the art of staying engaged in problem-solving. *Computers & Education, 163*, 104114.

Malinc. (2024). *Dynamic text and LaTeX in GeoGebra* [Online]. Available: https://www.malinc.se/math/geogebra/dynamictextandlatexen.php. Accessed: Nov 1, 2024.

Mjenda, M., Mutarutinya, V., & Owiti, D. (2023). Assessing the effectiveness of computer-aided instructional techniques in enhancing students' 3D geometry spatial visualization skills among secondary school students in Tanzania. *International Journal of Learning, Teaching and Educational Research, 22*(6), 613–637.

Owusu, R., Bonyah, E., & Arthur, Y. D. (2023). The effect of GeoGebra on university students' understanding of polar coordinates. *Cogent Education, 10*(1). Article 2177050.

Rahayuningsih, S. E., Sirajuddin, S., & Nasrun, N. (2020). Cognitive flexibility: Exploring students' problem-solving in elementary school mathematics learning. *JRAMathEdu.*

Richardson, J., & Newby, T. (2006). The role of students' cognitive engagement in online learning. *American Journal of Distance Education, 20*(1), 23–37.

Rotgans, J. I. (2011). Cognitive engagement in the problem-based learning classroom. *Advances in Health Sciences Education*, 465–479.

Saridakis, I. E., & Doukakis, S. (2020). Mathematical problem solving and cognitive enhancement. *BFAL.*

Soares, A. P., Catarino, P., & Chivai, C. (2022). Application of GeoGebra in the teaching of descriptive geometry: Sections of solids. *Mathematics, 10*(17).

Su, Y. S., Shao, M., & Zhao, L. (2022). Effect of mind mapping on creative thinking of children in Scratch visual programming education. *Journal of Educational Computing Research, 60*(4), 906–929.

Sullivan, A., & Strawhacker, A. (2021). Screen-free STEAM: Low-cost and hands-on approaches to teaching coding and engineering to young children. In *Embedding STEAM in early childhood education care* (pp. 87–113). Springer.

Symons, D., & Pierce, R. (2019). Active use of digital technologies in mathematical problem solving. In *Redesigning Higher Education Initiatives for Industry 4.0.*

Tlili, A., Wang, X., Huang, R., Gao, Y., Burgos, D., Jemni, M., Chang, T.-W., Khalil, M., & Yang, J. (2023). What if the devil is my guardian angel: ChatGPT as a case study of using chatbots in education. *Smart Learning Environments, 10*(1), 15. https://doi.org/10.1186/s40561-023-00226-1

Tondeur, J., van Braak, J., Ertmer, P. A., & Ottenbreit-Leftwich, A. (2017). Understanding the relationship between teachers' pedagogical beliefs and technology use in education: A systematic review of qualitative evidence. *Educational Technology Research and Development, 65*(3), 555–575.

Western Governors University. (2024). *What is cognitive learning?* [Online]. Available: https://www.wgu.edu/blog/what-is-cognitive-learning2003.html. Accessed: Nov 1, 2024.

Uwurukundo, M. S., Maniraho, J. F., & Rwibasira, M. T. (2022). Effect of GeoGebra software on secondary school students' achievement in 3-D geometry. *Education and Information Technologies, 27*(4), 5749–5765.

Yilmaz, R., et al. (2022). Smart MOOC integrated with intelligent tutoring: A system architecture and framework model proposal. *Computers and Education: Artificial Intelligence, 3*. Article 100092.

Yilmaz, R., & Yilmaz, F. G. K. (2023). The effect of generative artificial intelligence (AI)-based tool use on students' computational thinking skills, programming self-efficacy, and motivation. *Computers and Education: Artificial Intelligence, 4*, Article 100147.

Yohannes, A., & Chen, H. L. (2023). GeoGebra in mathematics education: A systematic review of journal articles published from 2010 to 2020. *Interactive Learning Environments, 31*(9), 5682–5697.

Zhang, Y., Wang, P., Jia, W., Zhang, A., & Chen, G. (2023). Dynamic visualization by GeoGebra for mathematics learning: A meta-analysis of 20 years of research. *Journal of Research on Technology in Education*, 1–22.

Ziatdinov, R., & Valles, J. R., Jr. (2022). Synthesis of modeling, visualization, and programming in GeoGebra as an effective approach for teaching and learning STEM topics. *Mathematics, 10*(3), 398.

Key Insights on the Use of Quantum Computing in Breaking Cryptographic Algorithms

Khemant Nirvaan Degambur and Girish Bekaroo

Abstract Quantum computing utilizes the principles of superposition and entanglement to perform certain computations exponentially faster than classical systems and this poses significant threats to classic cryptographic frameworks. Recent studies focused principally on developing or reviewing security methods based on classical cryptography and quantum cryptography, but limited studies synthesized insights using a structured approach. Taking cognizance of the need to advance literature in this area, this study critically examines the growing influence of quantum computing in cybersecurity, both as a disruptive force capable of breaking classical encryption and as a foundation for next-generation security paradigms. Through a structured investigation of four key research questions, the study explores quantum supremacy, the vulnerability of existing cryptographic systems, and the evolution of post-quantum defenses. A focused analysis of the quantum development timeline is also undertaken, offering insights into the pace of technological advancement and informing strategies for proactive preparedness against quantum-enabled threats. Findings highlight the critical vulnerabilities in public-key cryptosystems, which are especially susceptible to Shor's algorithm, while symmetric systems face a reduced—but still significant—security margin due to Grover's algorithm. Among the proposed post-quantum cryptographic (PQC) alternatives, lattice-based, hash-based, and code-based schemes emerged as leading candidates, with lattice-based protocols like CRYSTALS-Kyber gaining traction through standardization efforts. A strategic transition toward quantum-resilient infrastructures, supported by forward-looking policies, global standardization efforts, is imperative to safeguard critical systems and ensure long-term cybersecurity in the quantum era.

K. N. Degambur · G. Bekaroo (✉)
Middlesex University Mauritius, Flic-en-Flac, Mauritius
e-mail: g.bekaroo@mdx.ac.mu

K. N. Degambur
e-mail: KD780@live.mdx.ac.uk

J. Marx Gómez et al. (eds.), *Recent Advances in Information Systems*, Progress in IS,
https://doi.org/10.1007/978-3-032-14099-9_12

1 Introduction

Cryptography is essential for protecting sensitive data in various examples including online banking, e-commerce, healthcare, and government communications and is now an integral component of digital infrastructure. Cryptographic algorithms play a key role in maintaining confidentiality, integrity and authenticity in the digital age and applications range from securing emails to facilitating encrypted messaging. However, the advent of quantum computing represents a paradigm shift which could significantly disrupt this well-behaved scheme (Grote et al., 2019).

Quantum computers process information using quantum mechanical phenomena such as superposition, entanglement and interference (Zhang et al., 2019), which enable them to perform computations in a much more powerful way than classical computers that process information in a binary state (0 or 1) (Zhang et al., 2019). In a quantum computer, however, a quantum bit or qubit acts much like a light level controller. It holds the 0 and 1 states and also can keep a state of transit, which is a linear combination of states 0 and 1. These abilities enable quantum computers to perform computations that classical machines cannot, especially in the domain of cryptanalysis. For instance, one of the famous quantum algorithms Shor's algorithm can factor large integers exponentially faster than the most well-known classical algorithms. This puts at risk the security of widely used cryptographic systems such as elliptic curve cryptography (ECC) and Rivesh–Shamir–Adleman (RSA) that depend on the computational plausibility of factoring and related problems such as the discrete logarithm problem. Grovers algorithm, which offers a quadratic speedup for searching unstructured datasets, will make systems like AES and other symmetric encryption systems more vulnerable as a result (Soomro et al., 2019; Zhang et al., 2019).

These advances in quantum technology are expected to have various implications. Many of the cryptographic protocols that are currently being utilized are at risk of becoming outdated, which could allow private data to be harvested by an enemy that has quantum computer capability. Consequently, this has fueled an urgent search for answers. Researchers are actively developing post-quantum cryptography algorithms that will be able to resist attacks from both classical and quantum computers (Zhang et al., 2019). On the other hand, Quantum key distribution (QKD) based systems guarantee provable security against eavesdropping due to the laws of quantum mechanics and represent a game-changing framework for secure communication (Grote et al., 2019; Zhang et al., 2019). As such, it is essential to further advance work in this area and taking cognizance of this need, this study conducts a review to answer key research questions about how quantum computing is influencing this field, both as a method to break classical encryption and as a framework for the security solutions of the future. This study is important so as to provide the reader with a deep understanding of how quantum computers will shape, and are shaping, the cryptographic landscape, and what the future holds in the battle for digital future.

2 Background

Quantum computing presents a significant challenge to modern cryptographic systems by integrating quantum mechanical properties such as superposition and entanglement to perform computations exponentially faster than classical computers. Superposition is where a single quantum bit or qubit can be in the state of 0, 1 or any linear combination of these states. This phenomenon enables qubits to be in a combination of states until it is measured. On the other hand, entanglement is where a qubit state can be affected by another qubit state probability. This phenomenon creates robust correlations between qubits that classical systems are incapable of reproducing (Shafique et al., 2024). This section reviews existing research on quantum computing's impact on cryptography, focusing on quantum attacks, post-quantum cryptographic (PQC) solutions, and the ongoing standardization efforts for quantum-resistant algorithms.

2.1 Quantum Computing and Cryptographic Vulnerabilities

Classical cryptographic algorithms are based on assumptions of computational intractability that quantum computing threatens to break. Among, the Shor's algorithm, which effectively factorizes large integers and solves discrete log problems, makes RSA, ECC, and DH key exchange insecure (Pithale et al., 2023). For the challenge of solving such tasks a challenge, conventional computers would require billions of years whereas quantum computers may likely find a solution in little time (Shafique et al., 2024).

Additionally, Grover's algorithm speeds up brute-force attacks, reducing the effective security of symmetric cryptosystems such as AES (Fathalla & Azab, 2024). As Grover's quantum search algorithm executes $O(n\ 1/2)$ searches within an unsorted database. In conventional computers, searching an unstructured database is an $O(n)$ issue (Hussein et al., 2024). Grover's method is also probabilistic and can be utilized many times to ensure that the search completes. Grover iteration is utilized for searching. To enhance the chances of finding the element in a database, iterations may be repeated (Shafique et al., 2024).

Research has extensively analyzed the weaknesses of classical cryptographic primitives under quantum attacks. Studies emphasize the urgency of transitioning toward Post-Quantum Cryptography solutions to mitigate these risks (Fathalla & Azab, 2024; Pithale et al., 2023). Some studies explored hybrid models combining classical and quantum-resistant cryptography to ensure security during the transition period (Grote et al., 2019).

2.2 *Post-quantum Cryptography (PQC) Solutions*

Several cryptographic approaches have emerged to counteract quantum threats. Lattice-based, Hash-Based, Code-Based and multivariate polynomial cryptographic algorithms are approaches which have been extensively researched across academia and industry to counter quantum threats, including hybrid integrations (Gulomov et al., 2023). Similarly, Quantum Key Distribution (QKD) has also gained significant attention as a method to ensure secure communication. Research has examined the practical deployment of QKD and its integration into existing security infrastructures to enhance post-quantum security (Grote et al., 2019; Sahoo et al., 2024). Despite its theoretical advantages, QKD faces scalability and implementation challenges that must be addressed before widespread adoption (Gulomov et al., 2023; Mummadi & Fathima 2024).

2.3 *Challenges and Standardization Efforts*

Although PQC solutions provide promising alternatives, they also have practical challenges. A large number of PQC algorithms need much bigger key sizes, resulting in growth of computational overheads as well as storage needs (Yu, 2024). Studies indicate a compromise exists between performance and security, even more research implies that optimization techniques can alleviate these findings (Pithale et al., 2023). The National Institute of Standards and Technology (NIST) has been leading the PQC standardization process, evaluating various cryptographic schemes for post-quantum security. Several research papers discuss the implications of NIST's selections and their impact on future cryptographic implementations (Sahoo et al., 2024; Zhang et al., 2019). Nevertheless, continued research is needed to perfect post-quantum cryptography algorithms and improve their practical use. Among the key issues, studies highlight the significance of quantum-resistant cryptographic standards and the use of QKD into secure communication protocols (Gulomov et al., 2023; Yu, 2024). Furthermore, continuous advances in quantum computing must be constantly observed so that security techniques may be adjusted correspondingly (Sahoo et al., 2024).

3 Methodology

To investigate how quantum computing may affect cryptographic algorithms, literature analysis was conducted. As a starting point, research questions were formulated in order to allow for a more precise and systematic focus, while ensuring that relevant studies on quantum threats, post-quantum cryptography, and other security issues were effectively addressed. Inclusion and exclusion criteria were then established, to eventually diligently search trustworthy databases for papers that fulfill high quality and relevance standards. The chosen literature was then thoroughly evaluated using qualitative analytical approaches, such as theme or narrative synthesis, to identify trends and patterns that provide light on quantum computing's capacity to compromise or strengthen existing encryption systems.

3.1 Research Questions

The research questions investigated as part of this study were based on research needs and these are listed as follows:

RQ1: What allows quantum computers to outperform classical systems in computational capacity, and how do quantum algorithms like Shor's and Grover's break traditional cryptography algorithms?

RQ2: Which existing cryptographic systems are most vulnerable to quantum attacks, and how can future post-quantum cryptographic solutions address these risks?

RQ3: What is the current state of quantum computing developments, and how rapidly could they pose genuine threats to existing cryptography systems?

RQ4: How can firms and governments prepare for the risks connected with quantum computing, and what ethical concerns must be addressed to strike a balance between its usage for progress and safety?

3.2 Review Protocols

A thorough review of academic databases, including IEEE Xplore, and Google Scholar, was performed using terms such as "quantum computing and cryptography," "Shor's algorithm attack on RSA," "post-quantum cryptography," and "quantum-resistant cryptographic techniques". Following the first stage of search, filtering was conducted to ensure high-quality and relevant research choices. In the process, a multi-stage screening method was involved. This began by screen titles and abstracts that are relevant to quantum cryptography. Then, full-text review was conducted in order to assess methodology, conclusions, and implications on cryptographic security. Eventually, redundant or overlapping studies were eliminated.

As for paper selection criteria, only publications from conferences and peer-reviewed journals were selected, focusing on studies that specifically examine how quantum computing affects cryptography. Non-peer-reviewed and speculative articles were not included. Overall, 16 published papers were selected for answering the formulated research questions. Following selection of the articles, information was synthesized typically through narrative analysis, where research findings were interpreted toward identifying key patterns. In addition, content analysis was done to categorize studies based on quantum vulnerabilities and countermeasures.

4 Results and Discussions

This section discusses the key findings to answer the four research questions.

4.1 *Quantum Supremacy and Cryptographic Disruption*

RQ1 investigated the key abilities of quantum computers that enable them to outperform classical systems. Quantum computing offers an innovative framework utilizing qubits that possesses numerous remarkable features, such as quantum entanglement and the principle of superposition. This leads to what is referred to as quantum parallelism. It serves as the foundation for quantum computers to offer significant computational power (Zhang et al., 2019). In contrast to classical bits, which can either be 0 or 1, quantum bits (qubits) can be in both states at the same time. This allows quantum computers to handle several possibilities at once, greatly improving computing power. Qubits can be entangled, which implies that the state of one qubit affects the state of another, regardless of distance. This improves the performance of quantum computing. Unlike superposition, quantum algorithms do numerous calculations at the same time, resulting in faster solutions to complicated problems, known as quantum parallelism. Quantum computations employ interference to increase accurate results while reducing incorrect ones, resulting in efficient outputs (Shafique et al., 2024). Quantum algorithms are defined similarly to classical algorithms, however their data processing principles differ.

Studies points out two (2) major quantum algorithms that impact classical computing namely:

Grover's algorithm created by Lov Grover in 1996, searches a NoSql database with N n 2 entries for the unique element that fits the search condition. In information theory, the optimal strategy for exploring an unstructured database necessitates the efficiency of quantum algorithms of classical complexity. Grover's method is the most effective quantum approach for searching a NoSQL database. Furthermore, it's effect increases as the number of processes N rises. For instance, if a database has 1,400,000,000 items, Grover's approach is used to find the relevant item in about 37,417 steps on average. If each cycle takes a second, it will take approximately 11 h to finish the process. To methodically search through the entire collection of 140,073,000,000 records at a speed of one record per second would require 45 years, provided that work is being done 24/7 by traditional systems. As a result, Grover's approach is a very effective search algorithm that may be used to improve data searches for the next generation (Amellal et al., 2018).

Many symmetric encryption methods such as AES, SHA, and HMAC (Hash-based Message Authentication Code) rely on brute-force assaults being computationally infeasible due to their enormous key spaces. Grover's technique accelerates brute-force searches, rendering these cryptosystems substantially vulnerable to quantum assaults. The security of Bitcoin and blockchain relies on the collision resistance of SHA-256. A quantum computer employing Grover's algorithm could significantly undermine its security. Digital signatures like ECDSA depend on hashing to guarantee integrity and authenticity.

Shor's algorithm is effective against classical computing, since factoring a huge prime number takes substantial amount of time due to the exponential growth in the number of bits involved. As a result, many traditional cryptosystems use this to safeguard data. Peter Shor visited a quantum method for factoring an n-digit number with a polynomial time complexity in N that depends on the identification of a period r in a function f using the quantum Fourier transform (Shafique et al., 2024). Shor's way is a quantum technique that rapidly factors huge numbers and computes discrete logarithms in polynomial time, posing a severe threat to public-key encryption systems like RSA, ECC, and Diffie–Hellman key exchange. This is because RSA encryption relies on the complexity of factoring a large integer. The period of a function can be fast calculated through quantum Fourier transform, so if there are the periodic features of f, the security of $N = p \times q$ would be breakable through Shor's method. Similarly, elliptic curve cryptography (ECC) and the Diffie–Hellman protocol are also at risk, as these rely on the difficulty of the discrete logarithm problem, which is a challenge that Shor's algorithm can solve significantly faster than classical methods. Once large-scale quantum computers are available, security vulnerabilities will be introduced in VPNs, SSH, TLS, and all the blockchain and encrypted messaging protocols.

In order to counter this threat, the cryptographic community is currently engaged in the development of post-quantum cryptographic (PQC) technologies like lattice-based, code-based, multivariate polynomial, and hash-based encryption (Hussein et al., 2024). Any encrypted information presently secured with RSA, ECC, or DH are at risk for future compromise if properly precautionary measures are not followed, thus threatening global cyber safety. As such, to ensure long-term security, organizations need to progressively move to quantum-resistant cryptography standards.

4.2 Understanding Current Vulnerabilities and Exploring Post-quantum Defenses

RQ2 investigated existing cryptographic systems are most vulnerable to quantum attacks, and how future post-quantum cryptographic solutions can address these risks. Public-key cryptography (RSA, ECC, Diffie–Hellman) is entirely vulnerable with the arrival of significant quantum computers because of Shor's algorithm. At the same time, Grover's technique weakens symmetric cryptographic systems like AES and SHA by lowering the cost of brute-force searches, thus halving their security level. AES-128, for example, is reduced to 64-bit security, making it vulnerable to attacks from quantum computers. Findings following literature analysis reveal the following, as illustrated in Table 2.

Table 1 Comparison between bits and qubits

Bits	Qubits
0 or 1 Can be 0 or 1 at any given time	$\begin{bmatrix} 0 & 1 \\ 1 & 0 \end{bmatrix}$ Can be both 0 and 1 simultaneously
Stores one value at a time (either 0 or 1)	Stores a combination of multiple values at the same time
Uses electrical signals in transistors	Uses quantum states of particles (e.g., photons or electrons)
Processes one operation at a time	Processes multiple operations at once due to supersition

Table 2 Addressing risk with post quantum cryptographic algorithms

Algorithm	Usage	Key sizes	Quantum threat	PQC
ECC	Key exchange, signatures	256–521-bit curves	Broken by Shor's (discrete log)	Lattice-based (e.g., CRYSTALS-Kyber, Dilithium), Code-based (e.g., McEliece, BIKE)
RSA	Encryption, signatures	2048–4096-bit	Broken by Shor's (factorization)	Lattice-based (e.g., NTRU, CRYSTALS-Kyber), Code-based (McEliece, BIKE)
DH	Key exchange	2048–4096-bit	Broken by Shor's (discrete log)	Lattice-based KEMs (Kyber), Code-based KEMs (BIKE)
AES	Symmetric encryption	128–256-bit (AES-256 safer)	Grover's halves brute-force speed	Use larger key sizes (AES-256); remains generally viable for post-quantum security
SHA	Hashing, integrity checks	256–512-bit output	Grover's halves security strength	Increase digest length (e.g., SHA-384/512) or use SHA-3 for enhanced quantum resistance

To address the risks with quantum computing, encryption can use more complex mathematical techniques that are resistant to quantum attacks to counter threats from quantum computing (Gulomov et al., 2023). Key ones are:

- Lattice-Based, like NTRUEncrypt: Encryption based on difficult lattice problems (LWE-inclusion) is considered as quantum immune, but requires large keys. As the next step, more works are required to reduce keys and to improve speed.
- Code-Driven such as McEliece: Decoding of random error-correcting codes is considered as safe, yet oversized keys are involved. The study also highlights the importance of limiting critical dimensions for usability (Gulomov et al., 2023).
- Hash-Based such as SPHINCS+: The use of hash functions to generate signatures is considered as quantum-safe and offers larger and slower signatures. The changes aim for faster signing and validation.
- Multivariate polynomials, like Rainbow, based systems of polynomial equation. This is relatively fast, but requires constant monitoring for new threats.
- Hybrid Frameworks combine NTRUEncrypt, SPHINCS+ and BIKE such that if one of the algorithms fails, the other algorithms are still stable. Hybrid systems provide flexibility to potentially optimize for speed, key size, or signature strength as required.

4.3 Tracking the Quantum Race

RQ3 investigated the current state of quantum computing developments, and the potential timeline to pose genuine threats to existing cryptography systems? First and foremost, standardization efforts are being made by the National Institute of Standards and Technology to define a clear path for the process of adoption of post quantum era. This will ensure once quantum safe methods are concretely implemented, these can be readily integrated in our technological frameworks. Figure 1 shows an approximate timeline, based on investigated literature.

Present Noisy Intermediate-Scale Quantum devices, ranging from 50 to 500 qubits, are excessively error-prone to break modern encryption (Fowler et al., 2012; Svore et al., 2014). Although these are useful for specific tasks, they are not yet capable of running algorithms like Shor's to break RSA or ECC. As such, the timeline and risk factors are provided:

- Short-term (0–5 years): No present threat; emphasize hardware enhancements and error rectification.
- Over the medium term (5–15 years): improvements in fault tolerance and numerous qubits could enable feasible attacks on public-key cryptography (ETSI, 2020; Quantum Economic Development Consortium (QED-C), 2022).
- Over the long term (more than 15 years): Countless error-corrected qubits could eventually compromise RSA-2048/ECC-256, threatening classical systems (Arute et al., 2019; Gidney & Ekerå, 2021; NIST, 2023).

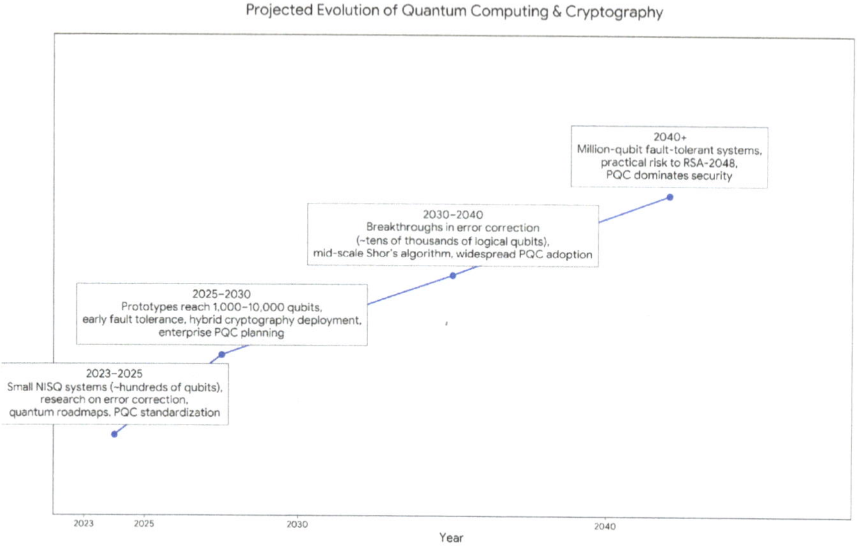

Fig. 1 Provides an approximate timeline on the evolution of quantum computing and cryptography (G. Research, 2023; IBM Research, 2023; IonQ, 2023)

Nevertheless, on in February 2025, Microsoft made breakthrough in the quantum computing field by introducing, Majorana 1, which is the first quantum chip using topological quantum computing (Bolgar, 2025). This chip is resistant to noise and decoherence which drastically improve quantum parallelism potentially reducing the timeline, Fig. 1, of the evolution of quantum computing. As this newly introduced chip has been engineered with semiconductor-superconductor structure which require ultra-low temperatures to function, it paves the way for more scalable quantum systems upto a million qubits which reduces the evolutionary timeline (Bolgar, 2025). This quantum chip needs however further evaluation. As such, revamping global cryptographic systems requires years, so preparations should start immediately, especially for long-term data protection. Quantum threats on a large scale are probably a decade off, yet making a proactive transition to quantum-safe encryption is essential to prevent potential risks.

4.4 Preparing for Quantum Disruption

As part of RQ4, the ways in which firms and governments can be prepared for the risks associated with quantum computing are explored, and the ethical concerns that must be addressed to strike a balance between its use for progress and safety are examined. Preparing for the quantum age requires not just technological approaches, such as switching to post-quantum cryptography (PQC) algorithms, but also broader policy and governance activities to address social impacts.

As such, preparing for quantum disruption is no longer a distant concern, but a growing necessity and as part of the preparation process, there are important steps for companies and authorities to consider. For companies and public authorities, this means taking concrete steps now to upgrade their systems, while also considering the ethical implications that come with powerful new technologies.

One of the key actions is to start adopting post-quantum cryptography (Joseph et al., 2022). This process should begin with identifying which systems and data are most sensitive to eventually rank them according to how long their confidentiality needs to be maintained. From there, organizations can explore hybrid cryptographic approaches, where current encryption methods (like RSA or ECC) are paired with PQC techniques. This allows for a smoother transition without disrupting existing operations. It's also important to follow the work of standard-setting bodies like NIST and ISO, as their guidelines will shape how new cryptographic tools are implemented across industries.

In addition, risk assessments are important due to the threat of "harvest-now, decrypt-later" attacks, where encrypted data is stolen today and decrypted once quantum capabilities are available. As such, organizations should already be re-encrypting long-term sensitive data using quantum-safe methods (How & Cheah, 2023). Keeping up with developments in PQC will also require regular updates to security systems and cryptographic libraries.

Strengthening internal capabilities is another major part of the preparation process. Cybersecurity teams will need the right knowledge to handle this transition, and as such, training in quantum threats and PQC implementations should start sooner rather than later. On a technical level, it is worth noting that PQC may demand more processing power, so investments in scalable and efficient hardware, such as FPGAs, might be needed to handle increased loads (Zhang et al., 2019).

Moreover, this transition should not happen in isolation. Collaboration between government, academia and industry is essential, particularly through pilot projects that can test PQC in real-world scenarios (Joseph et al., 2022). Such partnerships will help speed up the learning curve and support a smoother rollout. At the same time, policy frameworks need to evolve. Sectors like finance and healthcare should be guided by regulation to implement quantum-resilient security within a reasonable timeframe.

Alongside these practical steps, ethical considerations must remain at the core (Joseph et al., 2022). Quantum computing has the potential to unlock incredible breakthroughs, but it also introduces serious risks to privacy and data protection. Policymakers will need to find a delicate balance between encouraging innovation and putting safeguards in place. Also, it is important that that access to quantum technologies does not widen the digital divide. Open standards, shared resources, and inclusive education efforts can help ensure that quantum progress benefits everyone, and not just a privileged few. In terms of data governance, the risk of privacy breaches becomes more serious when decryption capabilities increase. Laws will need to evolve to protect individuals without holding back innovation. Finally, the environmental implications of combining quantum computing with artificial intelligence must be considered. Energy demands could rise significantly, so sustainable approaches should be prioritized. Thus, as with all AI systems, transparency and fairness will be key. For this, collaborative and proactive policies are essential for ensuring quantum advantages are consistent with social justice and sustainability. These policies should aim to align technological progress with public interest, ensuring that ethical and environmental responsibilities are not sidelined in the race for innovation.

5 Conclusion

This study reviewed how quantum computing is influencing this field, both as a method to break classical encryption and as a framework for the security solutions of the future. Four key questions were investigated through literature analysis focused on quantum supremacy and cryptographic disruption to eventually delve into comprehending the current vulnerabilities so as to eventually explore post quantum defenses. In addition, the timeline of the quantum race was investigated to finally look at how preparation for quantum disruption can be facilitated.

From RQ1, it was found that Shor's algorithm can factor big integers exponentially faster than classical methods, which compromises asymmetric cryptography for example RSA and ECC, while Grover's methods which provides a quadratic speedup, reduces the security margin of symmetric systems like AES. These vulnerabilities in the generation of keys highlight the necessity of post-quantum cryptography, with lattice-based, code-based, and hash-based schemes emerging as leading candidates. Protocol-oriented use cases, such as CRYSTALS-Kyber, the NIST-standard lattice-based approach, provide scalable encryption, while hash-based signatures secure quantum-safe authentication.

Shifting to PQC poses various, including the higher processing overhead, larger key sizes and legacy infrastructure congruencies. Immediate solutions for the near term involve hybrid cryptographic models that combine classical and PQC algorithms that balance security and practicality. Quantum Key Distribution offers unconditional key exchange security, but scalability limitations hinder widespread adoption (Mummadi & Fathima 2024). NIST, which is leading standardization efforts, is making good progress in developing strong quality standards, but implementation of quantum advances is time-consuming.

To prepare for the quantum disruption, proactive prevention measures need to be adopted to combat "harvest-now-decrypt-later" attacks, where attackers capture encrypted data for later decryption. Organizations should evaluate their cryptography ecosystem, prioritize the protection of high-value data, and consider investing in quantum-ready infrastructure. Ethical aspects such as fair access to PQC and the global asymmetries in the readiness for quantum computing require broad policy frameworks.

As such, although this study offers a timely and forward-looking perspective that can help both researchers and practitioners better understand the shift in cryptographic requirements due to advances in quantum computing. As future work, the limitations of this study could be addressed through a broader and more systematic literature review to capture a wider range of perspectives and technical developments.

References

Amellal, H., Meslouhi, A., & El Allati, A. (2018). Effectiveness of quantum algorithms on classical computing complexities. In *3rd International Conference on Smart City Applications*, Tetouan, Morocco.

Arute, F., et al. (2019). Quantum supremacy using a programmable superconducting processor. *Nature, 574*(7779), 505–510.

Bolgar, C. (2025). *microsofts-majorana-1-chip-carves-new-path-for-quantum-computing*. Microsoft, 19 February 19, 2025. https://news.microsoft.com/source/features/innovation/microsofts-majorana-1-chip-carves-new-path-for-quantum-computing/. Accessed February 25, 2025.

ETSI. (2020). *Quantum-safe cryptography and security*. European Telecommunications Standards Institute. https://www.etsi.org/images/files/ETSIWhitePapers/etsi_wp8_qsc.pdf. Accessed February 15, 2025.

Fathalla, E., & Azab, M. (2024). Beyond classical cryptography: A systematic review of post-quantum hash-based signature schemes, security, and optimizations. *IEEE Access, 4*.

Fowler, A. G., Mariantoni, M., Martinis, J. M., & Cleland, A. N. (2012). Surface codes: Towards practical large-scale quantum computation. *Physics Review A, 86*(3), Art. no. 032324.

G. Research. (2023). *Quantum AI*. Google. https://research.google/teams/applied-science/quantum/. Accessed February 15, 2025

Gidney, C., & Ekerå, M. (2021). How to factor 2048-bit RSA integers in 8 hours using 20 million noisy qubits. *Quantum, 5*, 433.

Grote, O., Ahrens, A., & Benavente-Peces, C. (2019). A review of post-quantum cryptography and crypto-agility strategies. In *2019 International Interdisciplinary PhD Workshop (IIPhDW)*, Wismar, Germany.

Gulomov, S. R., Khudayberganov, T. R., Ravshanova, M. X., Turdiev, T. T., & Atabayev, S. S. (2023). Exploring post-quantum cryptographic algorithms for secure data transmission. In *International Journal of Cyber Security*.

How, M. L., & Cheah, S. M. (2023). Business renaissance: Opportunities and challenges at the dawn of the Quantum Computing Era. *Businesses, 3*(4), 585–605.

Hussein, A. I., MaoLood, A. T., & Gbash, E. K. (2024). A systematic review: Post quantum cryptography to secure data transmission. *Iraqi Journal of Science, 65*(7), 3975–3992.

IBM Research. (2023). *Our quantum development roadmap*. IBM. https://research.ibm.com/blog/ibm-quantum-roadmap. Accessed February 15, 2025.

IonQ. (2023). *IonQ press releases and investor presentations*.

Joseph, D., Misoczki, R., Manzano, M., Tricot, J., Pinuaga, F. D., Lacombe, O., Leichenauer, S., Hidary, J., Venables, P., & Hansen, R. (2022). Transitioning organizations to post-quantum cryptography. *Nature, 605*(7909), 237–243.

Mummadi, S., & Fathima, S. (2024). A comprehensive study on quantum key distribution protocols. In *15th ICCCNT IEEE Conference*, India, IIT-Mandi.

NIST. (2023). *Post-quantum cryptography*. National Institute of Standards and Technology. https://csrc.nist.gov/projects/post-quantum-cryptography. Accessed 15 February 2025.

Pithale, R., Rajane, K., Mahajan, P., Nehate, N., Mulimani, A., & Lokhande, D. (2023). Cryptographic algorithm development and application for encryption and decryption. In *International*

Conference on Intelligent Machines and Methods of Information Processing (ICIMMI 2023), Jaipur, India.

Quantum Economic Development Consortium (QED-C). (2022). *Reports and publications.* Quantum Economic Development Consortium. Available: https://quantumconsortium.org/. Accessed January 18, 2025.

Sahoo, A., Kumar, I. K. A., & Rajagopal, S. M. (2024). Comparative study of cryptographic algorithms in post-quantum computing landscape. In *5th International Conference on Data Intelligence and Cognitive Informatics (ICDICI-2024).*

Shafique, M. A., Munir A., & Latif, I. (2024). Quantum computing: Circuits, algorithms, and applications. *IEEE Access, 12.*

Soomro, S., Belgaum, M. R., Jain, R., & Alansari, Z. (2019). Review and open issues of cryptographic algorithms in cyber security. In *2019 IEEE International Conference on Computing, Communication, and Cyber Security (IC4S).*

Svore, K. M., & Hastings, M. B., Freedman, M. (2014). Faster phase estimation. *Physices Review A 14*(3–4), 306–328.

Yu, Z. (2024). Research on cryptography based on quantum-resistant algorithms. In *2024 International Conference on Electronics and Devices, Computational Science (ICEDCS).*

Zhang, X., Zhao, Y., Li, R., Li, X., Guo, Z., Zhu, X., & Dong, G. (2019). The quantum Shor algorithm simulated on FPGA. In *2019 IEEE International Conference on Parallel & Distributed Processing with Applications, Big Data & Cloud Computing, Sustainable Computing & Communications, Social Computing & Networking (ISPA/BDCloud/SustainCom/SocialCom).*

The Use of Artificial Intelligence Tools to Support Post-graduate Research

Andre P. Calitz and Margaret Cullen

Abstract Artificial Intelligence (AI) can transform traditional research practices. AI is not simply augmenting research with advanced tools, instead, it is starting to disrupt established methodologies, ethical paradigms and fundamental principles that have long guided scholarly work. This possible automation of research has caused concern because of ethical issues; however, AI has impacted the academic research process by allowing greater efficiency and productivity for researchers. Students perceived the use of AI tools positively; however, this research aims to gain a better understanding and further contextualisation of AI-based tools to support scientific research in post-graduate studies. The qualitative study collected data from MBA students using an online survey. A thematic analysis of the open-ended questions was conducted using the AI tools Claude-AI and ChatGPT. The results indicate that training for AI-assisted research is required. The respondents were aware of the ethical issues and indicated their intention to make use of other AI tools for research purposes, but needed assistance and guidance.

1 Introduction

Machine learning, artificial neural networks and Big Data have made it possible to create machines that can mimic human intelligence, called artificial intelligence (AI) and are considered to be the trigger towards the fourth industrial revolution in education (Adiguzel et al., 2023). Zhai et al. (2021) underscore that urgent attention must be given to this revolution to better use AI for academic success. For example, a study conducted by Calitz and Cullen (2023) found that AI-based tools, such as ChatGPT, can be used effectively among educators and MBA academics by incorporating active learning, differential didactics, collaboration and real-world applications, amongst

A. P. Calitz (✉)
Department of Computing Sciences, Nelson Mandela University, Port Elizabeth, South Africa
e-mail: Andre.Calitz@Mandela.ac.za

M. Cullen
Business School, Nelson Mandela University, Port Elizabeth, South Africa
e-mail: Margaret.Cullen@Mandela.ac.za

J. Marx Gómez et al. (eds.), *Recent Advances in Information Systems*, Progress in IS,
https://doi.org/10.1007/978-3-032-14099-9_13

others. By doing this, students can develop a deep understanding of the material by actively engaging with it. Thus, AI is seen to transform traditional research practices and in particular, its use in literature reviews (Calitz & Cullen, 2023; Wagner et al., 2022).

AI in education (AIEd) is changing the face of learning (Harry, 2023). AIEd has found use within Higher Education Institutions (HEIs) in areas such as assessment and evaluation, intelligent tutoring systems, managing student learning and specifically as a research tool (Crompton & Burke, 2023). This is because research, particularly literature reviews in social sciences for graduate essays and treatises, often requires multiple sources and AI-based tools can accelerate the process and alleviate researchers from time-consuming and monotonous tasks (Hui, 2020).

Butson and Spronken-Smith (2024) argue that AI is not simply augmenting research with advanced tools; instead, it is starting to disrupt established methodologies, ethical paradigms and fundamental principles that have long guided scholarly work. Májovský et al. (2023), however, voice concern that AI can create fraudulent scientific papers that appear to be legitimate. AI can impact the loss of human decision making, make people lazy and impact security and privacy (Ahmad et al., 2023). The role of AI in research policy and practice should be the lens through which AI and society should be viewed (Chubb et al., 2022). They continue that AI might work to satisfy funder expectations of increased productivity, impact and interdisciplinary, however, it might sacrifice integrity and the traditional roles of institutions and academic identities, as the traditional norms of academic knowledge creation might be at risk.

The use of AI tools had a significantly positive impact on student academic experiences (Grájeda et al., 2024). Khairuddin et al. (2024) found that students perceived the use of AI tools as useful in the learning process. The aim of this study was to determine the perceptions of MBA students on the use of AI tools for research purposes and treatise writing. The layout of the paper includes the research problem and research objective in the following sub-section. The Research Methodology is discussed in Sect. 2, followed by the literature review in Sect. 3. The research findings and discussion are presented in Sects. 4 and 5 includes the conclusions and future research.

1.1 The Research Problem and Research Objectives

The potential of AI to partially automate research has sparked debates in various scientific disciplines (Johnson et al., 2019; Wagner et al., 2022). The underutilisation of AI-based tools in education restricts students from the full use of generative AI programmes and the new world of work (Calitz & Cullen, 2023; Roose, 2023). In the same vein, researchers highlight ethical concerns about using large language models (LLMs) in universities (Willems, 2023). Despite the debates regarding the use of AI-based tools at universities, there is limited empirical research in this area, specifically at South African Business Schools. Guidelines and policies are still being formulated. The research objective for this study is to investigate how AI supports research and treatise writing in post-graduate studies.

2 Research Methodology

The study was qualitative in nature and fell under the interpretivism paradigm. Interpretivism seeks to find a deep understanding of the meaning that people make as they interact with a specific problem (Grodal et al., 2021). For this study, the target population were registered first, second and third-year MBA students from a South African-based business school, who were conducting scientific research, specifically for writing a treatise. Approximately 100–150 students are registered for the research component of the MBA every year. Mandatory demographic fields were required before commencing with the online open-ended survey. The sampling strategy was purposive and convenient.

According to Vasileiou et al. (2018), sample sizes in qualitative inquiries are focussed on the appropriateness, composition, size, data saturation, and information power (Malterud et al., 2015). Information power refers to the density of information that a sample holds, leading to lower numbers of participants required in a study. This study conforms to the research done by Creswell (2007), who indicates that an ideal sample for a qualitative study ranges between twenty (20) to thirty (30) responses. Data saturation was considered, as it guarantees qualitative rigour (Malterud et al., 2015). Recurring themes indicated saturation. Multiple data evidence reinforced the evidence and the sample size.

The data collected allowed the researchers to get a real sense and understanding of the topic under review. The questions were operationalised from literature and focussed specifically on the research objective, to investigate how AI supports research and treatise writing in post-graduate studies. The choice of an online qualitative survey was because of its ability to adequately explore a focussed topic (Braun et al., 2021). A thematic analysis from the open-ended questions was conducted using the AI tools Claude and ChatGPT. Ethics approval for the study was obtained from the University Ethics Committee.

3 Literature Review

AI is transforming business and industries and raising important questions for society, the economy, education and governance (West & Allen, 2018). AI algorithms are designed to analyse data, find patterns, make predictions and decisions. It is a machine's ability to perform the cognitive functions associated with human minds and LLMs that have disrupted academia and the education sector, specifically the introduction of ChatGPT in November 2022 (McKinsey, 2023). AI is defined as computing systems that can engage in human-like processes, such as learning, adapting, synthesising, self-correction and the use of data for complex processing tasks (Popenici & Kerr, 2017).

Advocates of AI-based tools in education argue that they enable educators to personalise learning for students (Harry, 2023). On the other hand, there are ethical questions surrounding the use of AI in education and research. Therefore, urgent attention must be given to understanding AI in education to understand its benefits and challenges, specifically with regard to research (Calitz & Cullen, 2023; Zhai et al., 2021). This study presents a South African perspective, with the focus on MBA Business School students, which contributes to the body of knowledge.

AI is transforming traditional learning and teaching practices, leading to improved student outcomes (Adiguzel et al., 2023; Calitz & Cullen, 2023; Harry, 2023). The use of AI in education (AIEd) has increased over the past five years (Crompton & Burke, 2023). Similarly, AI-based tools support researchers in proposal preparation processes for funding (Spyroglou et al., 2022) and peer review systems (Checco et al., 2021). For research students, AI-based research tools are accelerating the scientific research process, specifically the literature reviews, which involve vast amounts of documents (Wagner et al., 2022).

All research studies are conducted within a specific context and aim to address gaps in existing knowledge (Paré & Kitsiou, 2017; Wilmot, 2019). The purpose of a literature review is to guide researchers in developing open questions by critically evaluating current knowledge and situating new findings within a specific context (Sylvester et al., 2013; Wilmot, 2019). Literature reviews navigate the vast and ever-growing number of documents, which is time-consuming, a process that could be streamlined through the use of AI (Wagner et al., 2022). The sheer volume of published scientific articles poses challenges for researchers. These challenges include the tedious manual search for academic papers and the difficulty in identifying research gaps (Muller et al., 2022). Consequently, the adoption of AI-based tools that assist researchers is on the rise.

The randomly selected tools included ChatGPT, Bard, Elicit AI, Perplexity AI, Iris.ai, semanticscholar.org, yewno.com, connectedpapers.com, researchrabbit.ai, www.paper-digest.com, openknowled-gemaps.org., keenious.com and DeepSeek. DeepSeek, which was recently released, is designed for technical and scientific tasks and provides a robust framework for understanding real-world applications (Kotsis, 2025). In addition, Kotsis (Kotsis, 2025) comments that a combination of ChatGPT and DeepSeek delivers ChatGPT's interactive capabilities with DeepSeek's research-orientated capabilities, which together can transform research and education.

One of the core competencies of faculty members is to conduct academic research. AI has impacted the academic research process by allowing greater efficiency and productivity for researchers (Chubb et al., 2022), enhancing the reliability of research by reducing human error in the data collection process, expediting the analysis of datasets as well as the automation of time-consuming tasks (Butson & Spronken-Smith, 2024).

AI is a key enabler rather than a threat to research activities (Chubb et al., 2022). There are, however, ethical considerations with bias being one of them (Khlaif et al.,

2023). The importance of academic research is evident in the education environment, due to its ability to generate new knowledge, inform pedagogies and bridge knowledge gaps (Chubb et al., 2022). In addition, academic research is concerned with abstracting from specific problems and contexts to provide solutions and theories to understand different world contexts (Piccoli & Wagner, 2003). As the use of AI as a research support mechanism becomes more commonplace, it will become an invaluable tool that provides opportunities for researchers to collaborate across disciplines, which will result in an increase in interdisciplinary research and new, innovative knowledge development (Eklavvya, 2024).

AI can assist the research process by sourcing literature, organising information based on user input, analysing, summarising, and presenting data in report format (Khlaif et al., 2023). The capabilities of AI technologies to summarise work, conduct reviews, and generate research questions (Khlaif et al., 2023) have received mixed responses from the academic community, with some scholars expressing hesitation about the use of these technologies in the research process (Butson & Spronken-Smith, 2024). Khlaif et al. (2023) emphasise the importance of safeguarding academic integrity as the use of AI to analyse and process Big data gains traction (Abbadia, 2023), while the ethical and professional application of these technologies for conducting, reporting, and publishing research remains the responsibility of the researcher (Khlaif et al., 2023). AI can enhance the quality and efficiency of data analysis and publication; however, it has also opened the door to the generation of high-quality fraudulent papers that are difficult to detect, raising important questions about the integrity of scientific research and the trustworthiness of published papers (Májovský et al., 2023). Building trust presents a significant hurdle in the adoption of AI technologies (Panas & Karlsson, 2025).

The use of AI tools is enabling collaboration and engagement amongst students and they are recommending to fellow students to use AI tools for academic matters (Khairuddin et al., 2024). Grájeda et al. (2024) also found that the use of AI tools had a significant positive impact on student academic experiences, including enhanced comprehension, creativity and productivity. Against this background, the purpose of this study is to investigate how AI supports research in postgraduate studies on the MBA programme. To achieve this, MBA students will receive an open-ended survey to gain insights into their perspective on AI to support scientific research.

4 Research Findings and Discussion

The survey was completed by 51 MBA students, 33 fully completed responses were analysed in this study. The 33 responses were labelled P1 to P33. Seventeen (52%) were male, 16 (48%) were female and they were all registered after 2021. Their knowledge and use of AI tools are presented in Table 1.

The other AI tools used by students were Claude, Grammarly, Aithor and Jenni AI. Claude is a new LLM and Grammarly is an English language writing assistant software tool. Jenni AI is better than ChatGPT for specific tasks, such as academic

Table 1 Knowledge and use of AI tools

Statement	I do not know (%)	I know about it (%)	I would like to use (%)	I have used (%)
ChatGPT	1 (3)	32 (97)	7 (21)	26 (79)
Bing	14 (42)	19 (58)	25 (76)	8 (24)
Gemini	20 (61)	13 (39)	28 (85)	5 (15)
Bard	26 (79)	7 (21)	30 (91)	3 (9)
Elicit AI	29 (88)	4 (12)	28 (85)	5 (15)
Perplexity AI	24 (73)	9 (27)	27 (82)	6 (18)
Iris.ai	33 (100)	0	32 (97)	1 (3)
semanticscholar.org	29 (88)	4 (12)	32 (97)	1 (3)
yewno.com	33 (100)	0	33 (100)	0
connectedpapers.com	32 (97)	1 (3)	32 (97)	1 (3)
researcrabbit.ai	31 (94)	2 (6)	31 (94)	2 (6)
www.paper-digest.com	32 (97)	1 (3)	33 (100)	0
openknowled-gemaps.org	31 (94)	2 (6)	33 (100)	0
keeniuos.com	33 (100)	0	33 (100)	0

writing. Aithor is an AI-powered research assistant that assists with sources, structure and academic writing. Bard was included because of the students' familiarity with Google. DeepSeek was launched after the data for this study was collected.

Six open-ended questions relating to the use of AI tools in research and data analysis by the students were included in the survey. The 33 responses for each question were thematically analysed using Claude-AI Sonnet 4 (identification of themes) and ChatGPT 4.0 (infographics). The ChatGPT thematic analysis has been presented as an infographic. The findings per question are presented below.

Figure 1 indicates the responses on how AI tools supported the students' coursework and assignments. The AI tools assisted them with finding literature, writing assistance and generating ideas. Participant P12 said, "I used Grammarly extensively, specifically after the university obtained a license". P17 indicated that "the use of ChatGPT to assist in structuring documentation and generating ideas helped me a lot".

In Fig. 2, the thematic analysis and ChatGPT infographic indicate that the benefits of using AI tools include time efficiency, improved productivity and it provides templates. P19 said "The main benefit of AI tools is that it helps with quick explanations" and P6 said "I suffer from writer's block and ChatGPT assists with idea generation and obtaining literature". P7 indicated "Saving time, improve writing style, source all possible relevant data".

Figure 3 indicates the factors that hinder the use of AI tools in research. The respondents indicated the accuracy of ChatGPT responses, ethical concerns regarding plagiarism and detection tools, such as Turnitin. P11 said "I used Claude to conduct a literature review and it generated references I could not find on Google Scholar".

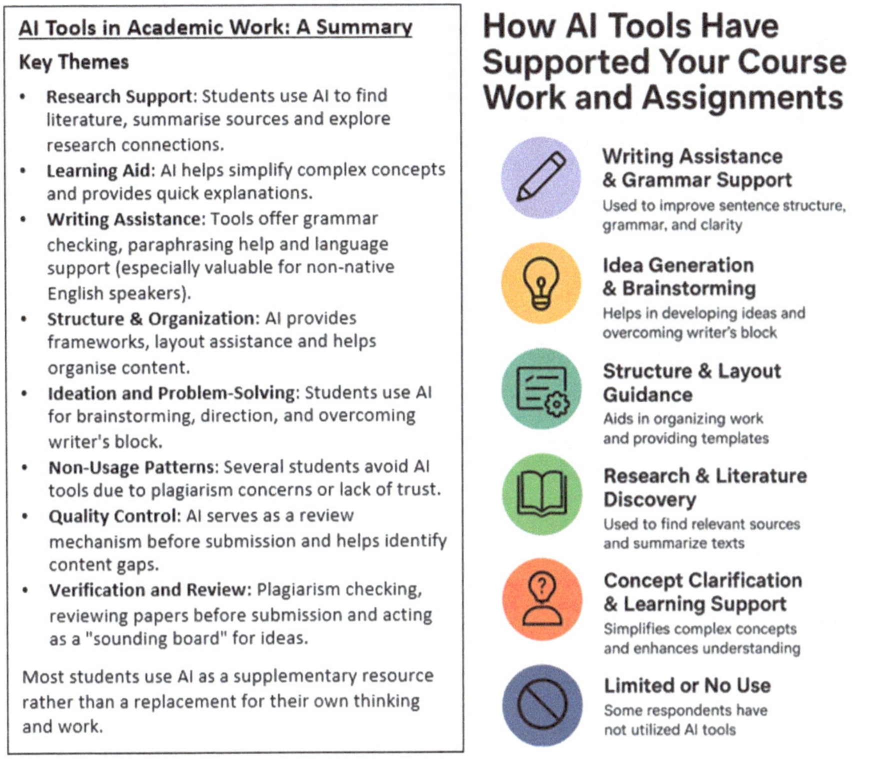

Fig. 1 AI support of coursework

P29 said “I’m unsure what and how to use the AI tools. I wish they would provide training”. P16 further indicated that “As a programmer, we use CodeRunner at work, however, in the Business School, I’m unsure what tools are acceptable”. P6 said “To subscribe to AI could be costly” and P32 said “Universities policies - Lack of knowledge or awareness about the benefits of AI - Limited access to technology - Limited technology skills by the researcher”.

In Fig. 4, the responses analysed indicated how AI tools support their research for their treatise. The main use was obtaining literature, structuring and formatting the document, quality control and language and writing support. P17 indicated “Assisting with Plagiarism checking, reference style design & paraphrasing. Also, narrow down sources specific for one’s research topic”. P24 said “So far it’s helping me figure out if I am framing things in alignment with my topic and giving me some sense of ’direction’ with this mountain of a treatise that has been haunting me for so long”.

Figure 5 indicates how AI can be effectively applied to enhance data collection. A number of responses indicated generating questions for a survey, questionnaire design and the formulation of questions. P16 said, “They can assist with the formulation of Questions so that they make sense to the respondents, thus ensuring that quality data is collected”. Six respondents indicated they were unsure. P32 indicated “AI

Benefits of AI Tools for Research

Key Themes

- **Time Efficiency**: Accelerates research process, literature review, and administrative tasks.
- **Quality Enhancement**: Improves writing quality, reduces errors, and enhances academic structure.
- **Information Management**: Efficiently organizes, summarizes, and analyzes large volumes of data.
- **Ideation Support**: Functions as a brainstorming partner, offering new perspectives and helping overcome writer's block.
- **Research focus**: Helps maintain relevance to the research topic and narrows source selection.
- **Process Assistance**: Provides structural guidance, reference management and technical support.
- **Accessibility**: Acts as an equaliser by making research tools and information more widely available.

Respondents primarily view AI as a complementary tool that handles routine tasks, allowing researchers to focus on higher-level analysis and interpretation.

Fig. 2 The benefits of using AI tools

Barriers to AI Use in Research

Key Themes

- **Reliability Issues**: Concerns about factual accuracy, biased information, and AI "hallucinations" that require verification.
- **Academic Integrity**: Plagiarism detection, questions about originality, and ethical concerns regarding proper attribution.
- **Quality Degradation**: Fears about reduced critical thinking, over-dependency, and shortcuts diminishing research value.
- **Institutional Resistance**: University policies, detection tools like Turnitin, and academic environments not fully accepting AI integration.
- **User Limitations**: Insufficient knowledge about effective AI prompting and proper implementation techniques.
- **Trust Deficits**: Hesitation to trust AI-generated content, sources, and the data used to train these models.
- **Access Constraints**: Cost barriers for premium features and limitations of free versions.

Most respondents acknowledge AI's potential benefits while expressing significant concerns about its impact on research authenticity and scholarly development.

Fig. 3 Factors hindering the use of AI tools

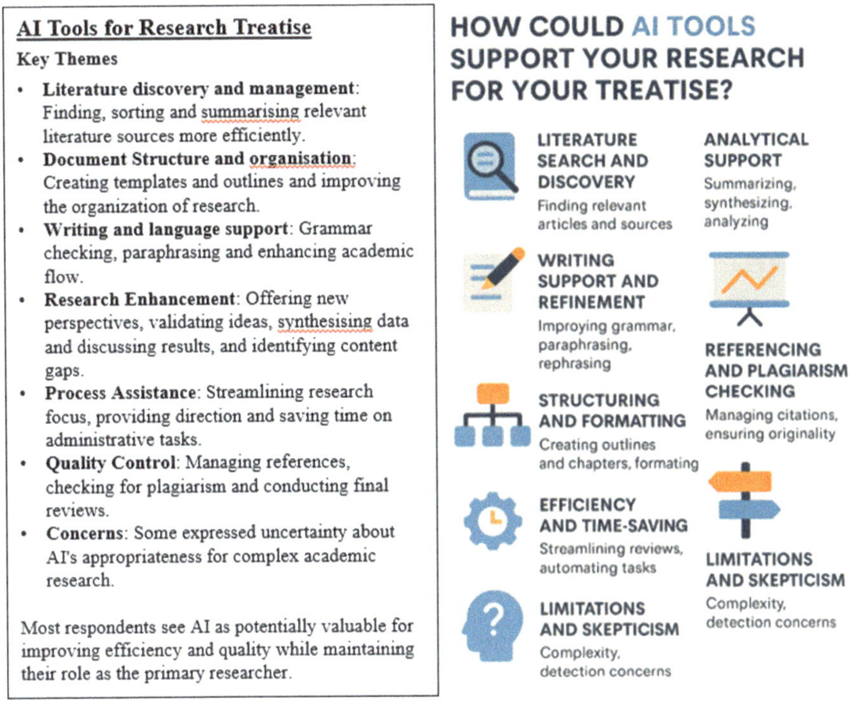

Fig. 4 AI tools support treatise writing

can help in creating or identifying data collection tools such as questionnaires" and P15 said "AI can be used to analyse the data as well as generate relevant questions to help accurately answer the problem statement".

In Fig. 6, the respondents indicated how AI could be used to assist with data analysis. The main themes were pattern recognition, extracting information and human-AI integration. P2 indicated, "By uploading datasets, AI can uncover patterns and make certain recommendations. I often use AI to get ideas that spark the stimulation of specifics required in my industry". P13 said "AI can identify and cluster topics within large text datasets, helping to summarise and understand the content" and P20 said "I bedlieve AI can help establish patterns, themes and relationships with data sets".

Summary of Thematic Analysis Findings

The post-graduate students indicated that the AI tools they used to support coursework were mainly for finding literature, summarising sources, providing quick explanations and organising content. AI tools helped them when they had writer's block and assisted with quality control, specifically grammar support (Khairuddin et al., 2024).

AI applied to Data Collection

Key Themes

- **Questionnaire and Survey Design**: AI ability to help formulate relevant questions and create dynamic surveys.
- **Data Sourcing and Literature Review**: identify relevant sources and direct researchers to pertinent information.
- **Data Processing and Analysis**: Tools capacity to summarize information, perform calculations and analyse sentiment.
- **Data Quality and Authentication**: AI can help authenticate information and ensure the collection of quality data.
- **Framework and Guidance**: AI's ability to provide structural frameworks and guidance for research approaches.
- **Transcription and Natural Language Processing**: AI tools for text analysis and speech to text processing.
- **Knowledge Gaps**: Several respondents indicated uncertainty with responses like "Not sure".
- **Ethical Considerations:** the importance of ethical considerations and human oversight.

The responses demonstrate a broad understanding of AI's potential applications in data collection, from the design of instruments to analysis and quality control.

Fig. 5 AI enhance data collection

Regarding their treatise, AI tools assisted in finding relevant literature, writing and language support and saving time on administrative tasks (Chubb et al., 2022). The AI tools further assisted them with data analysis and presentation. However, they expressed the need to be trained on the effective use of specific tools and which tools to use for specific tasks.

The benefits of using AI tools include saving time and increasing productivity. They believed the tools improved the quality of their research and presentation. It provided structural guidance, reference management and improved the quality. This confirms previous findings (Butson & Spronken-Smith, 2024).

The factors that hindered the use of AI tools mainly focussed on academic integrity, over-dependency on the tools and insufficient knowledge about the effective use of AI tools for specific tasks. The costs and access were also a factor, as well as ethical concerns, specifically the detection of AI-generated content.

The respondents indicated that AI tools can be used for data collection, specifically questionnaire design, data processing and analysis. AI tools can specifically assist with text analysis, thematic and content analysis and speech-to-text processing. Several students indicated they were 'not sure', indicating that they require training and assistance.

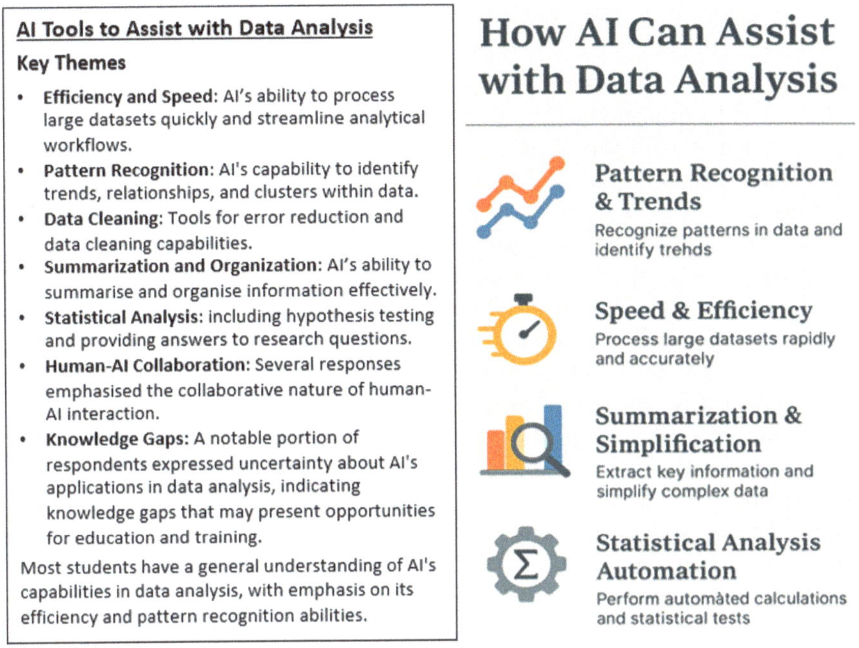

Fig. 6 AI tools assist with data analysis

AI tools can further assist with data analysis, specifically AI's ability to process large data sets and identify trends and clusters within data. They indicated that AI tools can assist with statistical analysis and emphasised the collaborative nature of human-AI interaction. Again, a large number of respondents expressed uncertainty about AI's applications in data analysis, indicating knowledge gaps that may present opportunities for education and training.

5 Conclusions and Future Research

Globally, HEIs face a new challenge as students are using ChatGPT and other sophisticated LLMs to generate essays, conduct research, solve assignments and in effect, deliver work that is not their own to obtain degrees (Davis, 2025). The use of AI tools in education has enabled students to obtain 21st-century skills that are essential in the modern education system (Khairuddin et al., 2024).

The results in this study are positive, as the majority of respondents indicated they have experimented with AI tools for academic purposes. The students currently use AI tools for brainstorming, summarising large volumes of text and writing assistance. They view AI as an additional tool to facilitate and enhance their research, specifically data acquisition and analysis. The results show that most of the students understand the applications of AI research tools as well as their benefits. However, selecting the appropriate AI tools for specific tasks and education and training on the use of AI tools remain a hindering factor.

Butson and Spronken-Smith (2024) highlight the importance of the ethical use of AI tools and how HEIs should manage this. HEIs must provide AI usage guidelines and training interventions, specifically for research contexts. Many priorities for improvements to teaching and learning in today's world are unmet, which has resulted in educators turning to technology-enhanced approaches that would be safe, effective and scalable to address these priorities (Cardona et al., 2023).

HEIs need to find a way to turn AI into a positive academic tool and incorporate it in the research process. Educators must be able to show students not only how to find and use the AI tools, but also what information to trust and what not to trust and how to tell the difference (Heaven, 2023).

A limitation of this study is that only a small sample of students responded to the survey. Future research needs to focus on what academics and HEIs can do to make the use of AI tools positive and ethical by reviewing policies, assessment and research practices. Differences in AI tools effectiveness across different disciplines should be further investigated. A repeat survey with a bigger sample would enhance the status quo regarding the use of AI in research.

References

Abbadia, J. (2023). *Exploring the role of AI in academic research*. Mind the Graph. Retrieved from https://mindthegraph.com/blog/ai-in-academic-research/. Accessed May 09, 2024

Adiguzel, T., Kaya, M. H., & Cansu, F. K. (2023). Revolutionizing education with AI: Exploring the transformative potential of ChatGPT. *Contemporary Educational Technology, 15*(3), ep429. https://doi.org/10.30935/cedtech/13152

Ahmad, S. F., Han, H., Ala, M. M., Rehmut, M. K., Irshad, M., Arrano- Munoz, M., & Ariz-Montes, A. (2023). A impact of artificial intelligence on human loss in decision making, laziness and safety in education. *Humanities & Social Sciences Communication, 10*, 311. https://doi.org/10.1057/s41599-023-01787-8

Braun, V., Clarke, V., Boulton, E., Davey, L., & McEvoy, C. (2021). The online survey as a qualitative research tool. *International Journal of Social Research Methodology, 24*(6), 641–654. https://doi.org/10.1080/13645579.2020.1805550

Butson, R., & Spronken-Smith, R. (2024). AI and its implications for research in higher education: A critical dialogue. *Higher Education Research & Development, 43*(3), 563–577. https://doi.org/10.1080/07294360.2023.2280200

Calitz, A. P., & Cullen, M. (2023). ChatGPT: The new MBA student in your class. In *International Business Conference (IBC 2023)*, Swakopmund, Namibia.

Cardona, M. A., Rodriguez, R. J., & Ishmael, K. U.S. (2023). *Artificial intelligence and future of teaching and learning: Insights and recommendations*. Department of Education, Office of Educational Technology.

Checco, A., Bracciale, L., Loreti, P., Pinfield, S., & Bianchi, G. (2021). AI-assisted peer review. *Humanities and Social Sciences Communications, 8*(1), 25. https://doi.org/10.1057/s41599-020-00703-8

Chubb, J., Cowling, P. I., & Reed, D. (2022). Speeding up to keep up: Exploring the use of AI in the research process. *AI & Society, 37*(4), 1439–1457. https://doi.org/10.1007/s00146-021-01259-0

Creswell, J. W. (2007). *Qualitative inquiry and research design: Choosing among five approaches* (2nd ed.). SAGE Publications.

Crompton, H., & Burke, D. Artificial intelligence in higher education: the state of the field. International Journal of Education Technology in Higher Education, 20(22) (2023). https://doi.org/10.1186/s41239-023-00392-8

Davis, R. (2025). *CheatGPT crisis—SA universities faced with a burgeoning degree of AI-written academic assignments.* https://www.DailyMaverick (South African news, opinion and investigationsCheatGPT—SA universities faced with a burgeoning degree of AI-written academic assignments)

Eklavvya. (2024). *The impact of AI on higher education learning and teaching.* Retrieved from https://www.eklavvya.com/blog/ai-higher-education-trend/#AI_as_a_Collaborative_Tool_in_Research. Accessed May 09, 2024.

Grájeda, A., Burgos, J., Córdova, P., & Sanjinés, A. (2024). Assessing student-perceived impact of using artificial intelligence tools: Construction of a synthetic index of application in higher education. *Cogent Education, 11*, 2287917. https://doi.org/10.1080/2331186X.2023.2287917

Grodal, S., Anteby, M., & Holm, A. L. (2021). Achieving rigor in qualitative analysis: The role of active categorization in theory building. *Academy of Management Review, 46*(3), 591–612. https://doi.org/10.5465/amr.2018.0482

Harry, A. (2023). Role of AI in education. *Interdiciplinary Journal and Humanity (INJURITY), 2*(3), 260–268. https://doi.org/10.58631/injurity.v2i3.52

Heaven, W. D. (2023). *ChatGPT is going to change education, not destroy it.* https://www.technologyreview.com/2023/04/06/1071059/chatgpt-change-not-destroy-education-openai/

Hui, G. (2020). Artificial intelligence and the future of labour demand.

Johnson, C. D., Bauer, B. C., & Niederman, F. (2019). The automation of management and business science. *Academy of Management Perspectives, 35*(2), 292–309. https://doi.org/10.5465/amp.2017.0159

Khairuddin, Z., Shahabani, N. S., Ahmad, S. N., Ahmad, A. R. & Zamri, N. A. (2024). Students' perceptions on the artificial intelligence (AI) tools as academic support. *Malaysian Journal of Social Sciences and Humanities (MJSSH), 9*(11), e003087. https://doi.org/10.47405/mjssh.v9i11.3087

Khlaif, Z. N., Mousa, A., Hattab, M. K., Itmazi, J., Hassan, A., Sanmugam, M., & Ayyoub, A. The potential and concerns of using AI in scientific research: ChaTGPT performance evaluation. *JMIR Medical Education, 9*. https://doi.org/10.2196/47049

Kotsis, K. T. (2025). ChatGPT and DeepSeek evaluate one another for science education. *Journal of effective Teaching Methods, 3*(1). https://doi.org/10.59652/jetm.v3i1.439

Májovský, M., Černý, M., Kasal, M., Komarc, M. & Netuka, D. (2023). *Journal of Medical Internet Research, 25*. https://www.jmir.org/2023/1/e46924

Malterud, K., Siersma, V. D., & Guassora, A. D. Sample size in qualitative interview studies: Guided by information power. *Qualitative Health Research, 26*, 1753–1760. https://journals.sagepub.com/doi/abs/10.1177/1049732315617444

McKinsey. (2023). *What is AI (artificial intelligence)?*. https://www.mckinsey.com/featured-insights/mckinsey-explainers/what-is-ai

Muller, H., Pachnanda, S., Pahl, F., & Rosenqvist, C. The application of artificial intelligence on different types of literature reviews—A comparative study. In *2022 International Conference on Applied Artificial Intelligence (ICAPAI)* (pp. 1–7). https://doi.org/10.1109/ICAPAI55158.2022.9801564

Panas, A. & Karlsson, A. (2025). *All about AI agents.* McKinsey & Company. https://www.mckinsey.com/featured-insights/early-access?exp=005&cid=omcknsl-eml-nsl--mck-ext-----&hlkid=9c3c0feec66e4be79dc3aecf4cb5f286&hctky=11730038&hdpid=812d47ec-23b9-4a9

Paré, G., & Kitsiou, S. (2017). *Chapter 9 methods for literature reviews*. University of Victoria. https://www.ncbi.nlm.nih.gov/books/NBK481583/

Piccoli, G. & Wagner, E. (2003). The value of academic research. *The Cornell Hotel and Restaurant Quarterly, 44*, 29–38. https://doi.org/10.1016/S0010-8804(03)90016-4

Popenici, S. A. D., & Kerr, S. (2017). Exploring the impact of artificial intelligence on teaching and learning in higher education. *Research and Practice in Technology Enhanced Learning, 12*(22), 1–1. https://doi.org/10.1186/s41039-017-0062-8

Roose, K. (2023, January 12). Don't ban ChatGPT in schools. Teach with it. *The New York Times*. https://www.nytimes.com/2023/01/12/technology/chatgpt-schools-teachers.html

Spyroglou, O., Yildirim, C., & Koumpis, A. (2022). Use of AI to help researchers improve their research funding capacities, relevance, and performance. *International Journal of Semantic Computing, 16*(01), 93–106. https://doi.org/10.1142/S1793351X22400050

Sylvester, A., Tate, M., & Johnstone, D. (2013). Beyond synthesis: Re-presenting heterogeneous research literature. *Behaviour & Information Technology, 32*(12), 1199–1215. https://doi.org/10.1080/0144929X.2011.624633

Vasileiou, K., Barnett, J., Thorpe, S., & Young, T. (2018). Characterising and justifying sample size sufficiency in interview-based studies: Systematic analysis of qualitative health research over a 15-year period. *BMC Medical Research Methodology, 18*(1), 148. https://doi.org/10.1186/s12874-018-0594-7

Wagner, G., Lukyanenko, R., & Paré, G. (2022). Artificial intelligence and the conduct of literature reviews. *Journal of Information Technology, 37*(2), 209–226. https://doi.org/10.1177/02683962211048201

West, D. M., & Allen, J. R. (2018). *How artificial intelligence is transforming the world*. Brookings. https://www.brookings.edu/articles/how-artificial-intelligence-is-transforming-the-world/

Willems, J. (2023). ChatGPT at universities—The least of our concerns. *SSRN Electronic Journal*. https://doi.org/10.2139/ssrn.4334162

Wilmot, K. (2019). *Enacting knowledge in dissertations: An exploratory analysis of doctoral writing using Legitimation Code Theory* (Unpublished Ph.D. thesis). University of Sydney. https://ses.library.usyd.edu.au/bitstream/handle/2123/20498/2019_Kirstin_Wilmot_Thesis.pdf?sequence=2&isAllowed=y

Zhai, X., Chu, X., Chai, C. S., Jong, M. S. Y., Istenic, A., Spector, M., Liu, J.-B., Yuan, J., & Li, Y. (2021). A review of artificial intelligence (AI) in education from 2010 to 2020. *Complexity, 2021*, 1–18. https://doi.org/10.1155/2021/8812542

A Review of Academic Integrity and Cheating at a Higher Education Institution in South Africa

Andre P. Calitz and Margaret Cullen

Abstract Academic dishonesty or misconduct has a long-standing history in education. Academic integrity is a moral code of conduct that people can trust that everyone's work is their own and was performed in an honest, truthful way. The Artificial Intelligence (AI) environment has created several new challenges and increased opportunities for academic misconduct, specifically cheating. Academics, new to the use of AI tools and the modern assessment environment, do not know how to manage the use of AI tools, such as ChatGPT and cheating services offered to students. In this mixed-methods study, academics, who have experienced cheating in the courses they presented and students share their opinions regarding cheating. The academics and students associated with the large South African Higher Education Institution (HEI) agreed that collusion and plagiarism are the most frequent types of cheating observed. The issue that no serious disciplinary consequences were enforced and published by the institution was highlighted as to why cheating occurs and has increased. The study highlights the major implications that cheating has on all academic stakeholders, including the institution, academics, students and industry. The prevalence of a lack of academic integrity and the evidence of dishonest behaviour suggest that HEIs struggle with students' cheating and the apparent lack of knowledge about how to apply the conventions of academic integrity.

1 Introduction

Worldwide research has indicated an increase in academic cheating (Curtis & Tremayne, 2021), specifically during the COVID-19 pandemic, where institutions were forced to move to the online environment (Koh & Daniel, 2022). A cheating culture has taken root in business, sport and academia. Globally, there has been an

A. P. Calitz (✉)
Department of Computing Sciences, Nelson Mandela University, Port Elizabeth, South Africa
e-mail: Andre.Calitz@Mandela.ac.za

M. Cullen
Business School, Nelson Mandela University, Port Elizabeth, South Africa
e-mail: Margaret.Cullen@Mandela.ac.za

J. Marx Gómez et al. (eds.), *Recent Advances in Information Systems*, Progress in IS,
https://doi.org/10.1007/978-3-032-14099-9_14

increase in people cheating in all spheres of life. Websites, such as Ashley Madison (2022), promote infidelity with the slogan "Life is short. Have an affair". The world of sport has experienced an increasing number of scandals relating to doping. In South Africa, corruption is experienced at governmental level and in business.

Online formal education has increased over the past years. Many fear that online courses do not provide the same rigour as face-to-face courses (Peterson, 2019). It is common knowledge that there is a higher level of cheating in online courses. The Internet and other related technologies and services, offer greater opportunities for academic dishonesty and cheating (Watson, 2010).

Academic integrity entails honesty, responsibility and openness to both scholarship and scholarly activity (Center for Teaching & Learning, 2021). Academic integrity in this paper is defined as the expectation that teachers, lecturers, scholars, students, researchers and all members of the academic community act with integrity, honesty, trust, fairness, respect and responsibility. Breaching academic integrity is also known as 'academic misconduct' or 'academic dishonesty'.

Online assessments are conducted without face-to-face proctoring, which implies that students can use their notes, books, collaborate with other students, conduct Internet searches and use any other sources of assistance, including Artificial Intelligence (AI) tools, such as ChatGPT to help them write programming code, answer questions and solve problems (Bilen & Matros, 2020). How students cheat and how they perceive academic dishonesty evolves along with technology and the assessment methodology used (Bertram Gallant, 2011). Universities and other academic institutions have had to switch to the use of e-learning platforms. However, few of these institutions and specifically academic staff were well prepared for this sudden, disruptive move (Heitz et al., 2020). There have been reports of significant success by providing online course delivery, but also of abject failure, specifically when it comes to the integrity of assessments.

Ethics plays a major role in how students perceive cheating, as ethics are shaped by an individual's value system and core beliefs. If cheating is justified in their ethical code as a means to an end, they might feel justified in the behaviour (Moten et al., 2013). Bertram Gallant (2011) believes that cheating is endemic and that it has changed over the years in both character and body, wherein the character of cheating refers to the appearance and form of cheating, but also the perceptions of cheating. The implications of this academic dishonesty are important for academic institutions and society at large. It raises the question of whether the integrity of academic institutions awarding the qualifications and their graduates' abilities can be trusted. Employers want to employ graduates who are competent, knowledgeable, ethical and honest (Peterson, 2019).

The layout of the paper includes the research problem and research objective in the following sub-section. The Research Methodology is discussed in Sect. 2, followed by the literature review. The research findings and discussion are presented in Sect. 4 and Sect. 5 includes the conclusions and future research.

1.1 The Research Problem and Research Objectives

The aim of this study was to investigate the perceptions of academics and students regarding academic dishonesty and specifically cheating at a large Higher Education Institution (HEI) in South Africa. The research problem investigated in the study was that academics have experienced an increase in academic misconduct, and the perceptions and impact of cheating, from a lecturer's and student's perspectives, have not been investigated in South Africa. The objectives of the study were to:

- determine what types of cheating are taking place from lecturers' and students' perspectives;
- determine the causes and motivation behind cheating;
- determine what type of penalties should be imposed; and
- to determine how online assessments have been changed to counter cheating.

These objectives have been explored from the point of view of academics and students.

2 Research Methodology

Two target population groups were identified to participate in this study. Firstly, academics who reported incidents of cheating during a Faculty meeting and secondly, a random sample of students, who provided their opinion of cheating behaviour observed during their studies. Purposive sampling enabled the identification of the academic participants (AP). The authors identified academics, who mentioned that they have experienced cheating in their courses during a faculty meeting attended by over 100 staff members. Twenty academics were requested via email to participate anonymously in the study. The students (SP) were randomly selected by the academics who participated in the study. The academics were requested to send the link to the survey to 10 randomly selected students in their courses and request them to participate in the study. The final participation was 12 academics and 62 students.

The study made use of qualitative and quantitative questions, which were distributed to the identified participants online, using the QuestionPro survey tool. The responses were thematically analysed and the results visually presented using Theme Frequency and Association Diagrams (TFAD). TFAD show the themes identified, the number of respondents that mentioned the theme and the associated themes mentioned in the responses. The association highlights the fact that respondents mentioned more than one theme in their responses. Ethics approval was obtained from the University Ethics Committee for this study.

3 Literature Review

A cheating culture has taken root in business, sports and other areas of society and is a widespread phenomenon in academia (Callahan, 2006; Zanetti & Butera, 2025). In sports, various cases of cheating are reported in the press; for example, the Russian teams were banned from participating in the Olympic Games due to doping. In business, corruption can include false financial reporting, bribery, embezzlement, theft and fraud. A major form of corruption experienced in government and in business is in the procurement of goods and services. In terms of the Corruption Perception Index, on a scale from zero to 100 where zero is highly corrupt, South Africa has a score of 41, which indicates serious levels of corruption (Transparency International, 2024). Scholars argue that these times are characterised by misinformation or fake news, which adds to the problem (Broda & Strömbäck, 2024). Ethical behaviour is based on the values of honesty, trust, fairness, respect, responsibility and courage, which creates a community dedicated to learning and sharing ideas (Holden et al., 2021). There are many consequences of academic unethical behaviour or dishonesty. The social consequences include the moral consequences wherein those who cheat in academic activities think they can get away with the same type of behaviour at work, in family life and other aspects of life (NIU, 2021).

Academic integrity is a fundamental aspect of education (Davis, 2023). Academic integrity must be promoted and reasserted throughout higher education, through international and national initiatives and institutional policy and practice (Morris, 2018). Academic integrity includes the accurate reporting of research findings, citing all sources of information, expressing original ideas in assignments and completing assessment tasks independently. Not acknowledging collaboration when it occurs and displaying trustworthiness during examinations are all examples of breaches of academic integrity (Brown et al., 2020).

Price-Mitchell (2023) defines academic integrity as a moral code of conduct that people can trust that everyone's work is their own and was performed in an honest, truthful way. Brown et al. (2020) define it as honest and responsible scholarship and the moral code of academia. Academic integrity is about honesty and not dishonesty. It is about truthful reporting and not fabricating data or findings. The prevalence of lack of academic integrity and the evidence of dishonest behaviour suggests that HEIs struggle with students' lack of integrity and the apparent lack of knowledge about how to apply the conventions of academic integrity and how important it is to do so (Brown et al., 2020; Davis, 2023).

Academic integrity is the commitment to and demonstration of honest and moral behaviour in an academic environment. Academic integrity is the expectation that lecturers, students and all members of the academic community act with honesty, trust, fairness, courage, respect and responsibility (International Center for Academic Integrity (ICAI), 2025). Academic misconduct is the term used for when individuals do not act with integrity in an academic setting (Laurier, 2025). Academic dishonesty refers to committing or contributing to dishonest acts by people engaged in teaching, learning, research and related academic activities and it applies to everyone in the academic environment (NIU, 2021). Academic misconduct involves, but is not limited to giving or receiving, offering or soliciting unauthorised assistance of any kind on any test, exam or assignment, plagiarism or collusion.

Curtis and Tremayne (2021) identified different forms of plagiarism, which included ghostwriting, verbatim copying and purloining, an assignment copied from another person without their knowledge. The types of academic misconduct identified by Laurier (2025) include:

- Plagiarism—using the thoughts, ideas, or writings of another person and passing them off as your own.
- Cheating—involves most of the other types of misconduct listed here.
- Collusion and unauthorised collaboration – collaborating with other people during an assessment, using technologies such as WhatsApp and emails.
- Academic deceit—fabricating data, information or documentation. Aiding others in committing integrity violations and inappropriately collaborating; and
- Personation—impersonating another person in a test or exam.

The diversity of the student body from a cultural aspect brings additional challenges as the effectiveness and intensity of deterrents vary among cultures, which means that responses to academic dishonesty are not universal but should be tailored to the culture of the institution (DiPietro, 2010). Cheating behaviour supports evidence that the lack of honesty exists where the returns for dishonest behaviour are high (Bilen & Matros, 2020), which Blum (2009) believes is driven by a social shift from valuing an authentic self to valuing a performance self. Collective cheating is a customary phenomenon of collective dishonesty where it is reported that individuals cheat more when they can do so in groups rather than alone, which has negative consequences for education (Zanetti & Butera, 2025).

Academic dishonesty has a long-standing history in education (Zachek, 2020). Academic cheating misrepresents a student's knowledge and abilities and undermines the instructor's ability to fairly assess the student's knowledge and abilities. Further, it also undermines the University's ability to honestly and fairly certify a student's knowledge and abilities. Types of online cheating include Google answers (Bilen & Matros, 2020); waiting for answers (Moten et al., 2013); collaborative cheating (Al Hamdani, 2020); plagiarism (Moten et al., 2013); contract cheating (Bilen & Matros, 2020); ghostwriters and the use of AI, ChatGPT (Bin-Nashwan et al., 2023), amongst others. Figure 1 shows a Facebook page for a large university created by students and has over 50,000 members. On the page, students request other students to write tests and exams for them, and students offer ghostwriting services for a Law test.

The concern about academic integrity in academia has become more critical than ever with the modern and pervasive technological revolution of AI chatbots and Large Language Models (LLMs) (Bin-Nashwan et al., 2023). ChatGPT and other LLMs use large quantities of text data and powerful computers to "learn" how to generate new text and can create responses to questions and carry out conversations that mimic the responses of a person (Birks & Clare, 2023). LLMs can also produce computer code, called vibe coding, specifically used by Computer Science and Information Systems students.

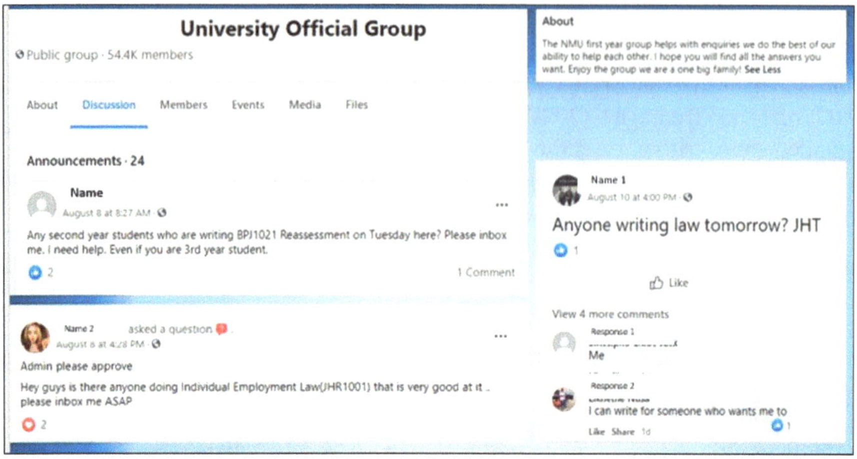

Fig. 1 Advertising for assistance from ghostwriters on a Facebook page

The use of AI, specifically ChatGPT is a hot topic in the field of education (Cotton et al., 2024). AI has added new dimensions to academic cheating challenges as students (as well as faculty and staff) can easily access powerful systems for generating content that can be presented in assignments, exams, or published papers as their own (Oravec, 2023). AI makes education more accessible and efficient; however, the intentional misuse of AI tools in facilitating academic cheating is a growing concern (Nguyen & Goto, 2024). DeepSeek, which was recently released, is designed for technical and scientific tasks and provides a robust framework for understanding real-world applications (Kotsis, 2025). In addition, Kotsis (2025) comments that a combination of ChatGPT and DeepSeek, delivers ChatGPTs' interactive capabilities with DeepSeek's research-orientated capabilities, which together can transform research and education.

4 Research Findings and Discussion

Twelve academics from different departments and faculties participated in the study. The 62 students who completed the survey included under-graduate ($n = 47$) and post-graduate students ($n = 15$). The twelve academic participants, coded AP1 to AP12, were asked the question: "*In your opinion, how common is cheating?*" using a 5-point Likert scale, ranging from *The exception, Seldom, Fairly often, Often* and *Common.* Eleven academics recorded a mean score of 4.1 for the question, indicating that this is *Common* practice. The detailed responses were *Fairly often* ($n = 3$), *Often* ($n = 4$) and *Common* ($n = 4$). The type of cheating academics encountered (Fig. 3) included Collusion with other students ($n = 11$) and Plagiarism ($n = 7$).

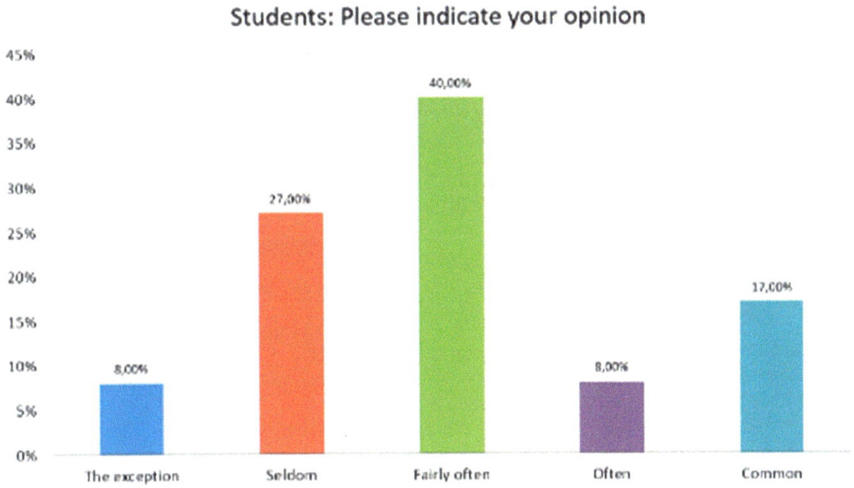

Fig. 2 Students: how common is cheating

The academics indicated that, in their opinion, what causes cheating is the fact that students think they will not be caught ($n = 11$) and that there are limited consequences ($n = 8$) for cheating (Fig. 3). The one other reason a respondent (AP6) indicated was that it was "Culturally acceptable". The responses from the 62 student participants, coded SP1 to SP62, differed from the academics, indicating *The exception* ($n = 5$), *Seldom* ($n = 16$), *Fairly often* ($n = 24$), *Often* ($n = 5$) and *Common* ($n = 10$). The students' mean score was 3.0, indicating *Fairly often* (Fig. 2).

Research indicates that if cheating is justified in students' ethical code and viewed as a means to an end, the behaviour is justified (Moten et al., 2013). The academics indicated that the underlying motivation for cheating was that students were less deterred by social stigma ($n = 9$), to pass the course ($n = 8$) and obtain a qualification ($n = 7$). One participant (AP4) indicated, "I believe that there are individuals wanting to take the easy way out. Not putting in any effort but wanting rewards and recognition". Another participant (AP7) indicated that students "lack the required skills and background".

Students' main reasons for cheating were that they did not believe they would get caught (32%) and that there would be no consequences (24%). More than half of the students indicated that the types of cheating they were aware of included Plagiarism (58%, $n = 36$) and collusion with other students (55%, $n = 34$). These findings support earlier research reporting on students working together (Bilen & Matros, 2020; Moten et al., 2013) and copy and pasting from the Internet (Al Hamdani, 2020), specifically using ChatGPT.

In the students' opinion, the causes of cheating included '*Pressure to get good marks*' (73%, $n = 45$) and that '*they do not think they will be caught*' (47%, $n = 29$). SP12 indicated "Lack of study time", SP32 "To pass a module you'd otherwise do terribly in". The students indicated that they cheat for '*personal gain*' ($n = 32$) and

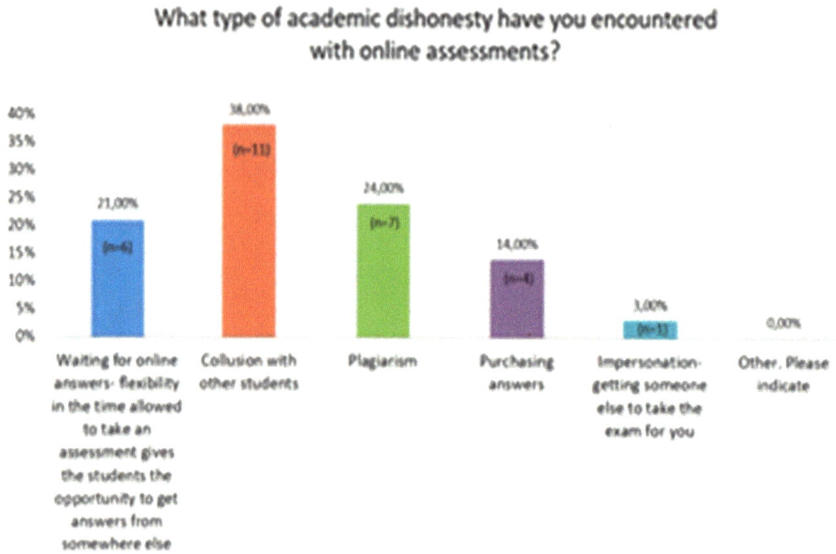

Fig. 3 Lecturers: types of cheating

to '*get through the module*' ($n = 27$). SP16 indicated, "I believe that most students cheat out of a perceived sense of necessity or desperation, not simply to be lazy or greedy for marks". Cheating seems to be part of getting through classes and passing the grade. Bilen and Matros (2020) reported similar findings and indicated that had the students not cheated, they were very likely to fail the class.

The responses to the other questions in the survey were thematically analysed and are presented in Figs. 4, 5, 6, 7, 8 and 9. The TFAD show the themes identified, the number of respondents that mentioned the theme and the associated themes mentioned in the responses. The association highlights the fact that respondents mentioned more than one theme in their responses.

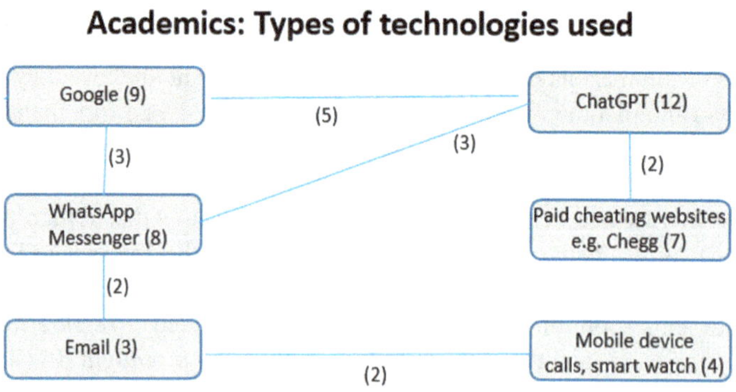

Fig. 4 Academics—Technologies used

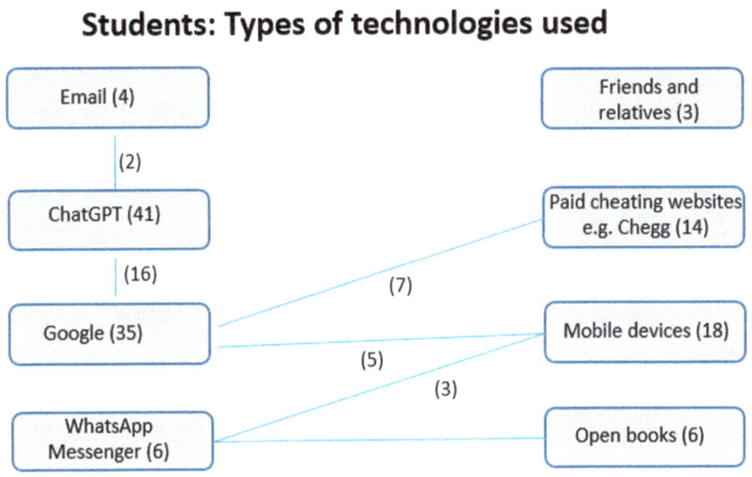

Fig. 5 Students—Technologies used

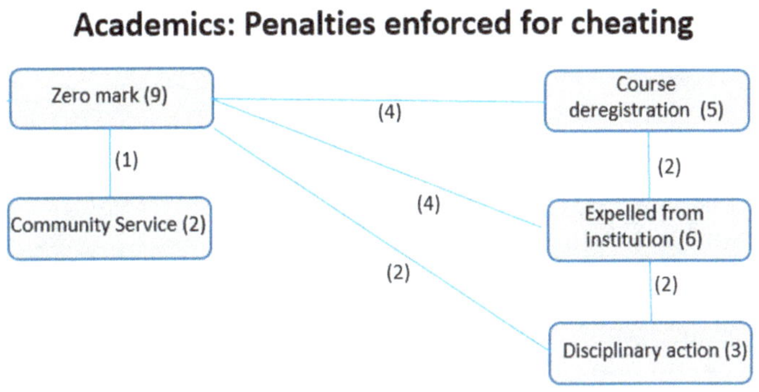

Fig. 6 Academics—Penalties

In Fig. 4, the majority of the academics indicated that the types of technologies used for cheating were specifically ChatGPT ($n = 12$), Google ($n = 9$), WhatsApp/Messenger ($n = 8$) and Chegg ($n = 7$). One academic participant (AP1) indicated that students "Accessing pay sites that provide answers, such as Chegg can do assignments for them". Bilen and Matros (2020) found "evidence that the answer key for the Final Exam was distributed online in a common web platform. For a price of several dollars, students could get access to the solution key".

The students (Fig. 5), on the other hand, indicated they used technologies, such as ChatGPT ($n = 41$) and Google ($n = 35$). Student participant SP28 indicated, "I heard someone talking about something called Quizlet" and SP12 said, "As far as I know (and could imagine), students find answers from the textbooks and class notes, ChatGPT or by messaging each other on social media".

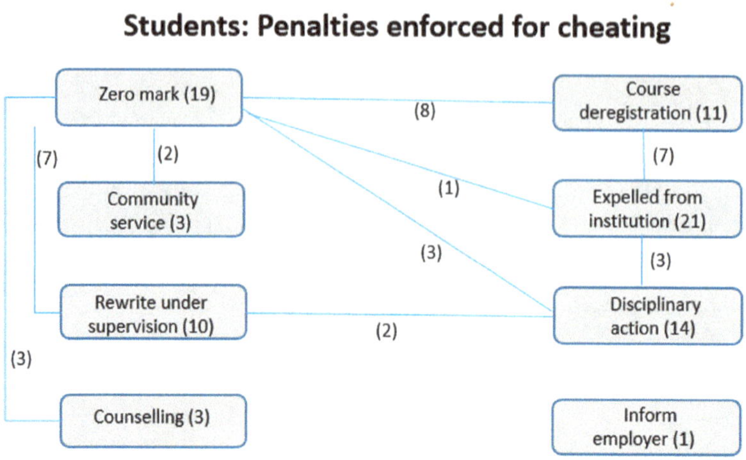

Fig. 7 Students—Penalties

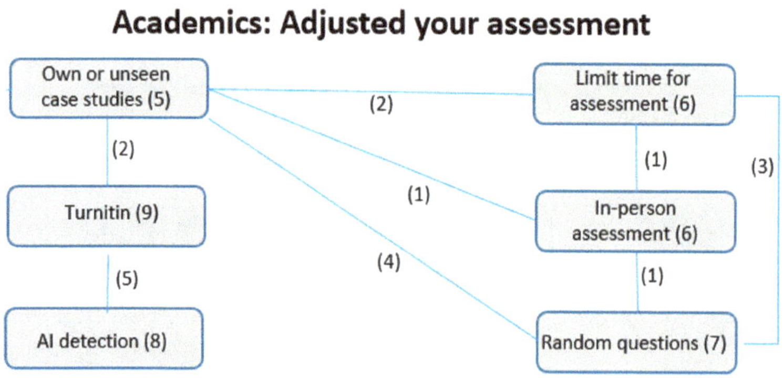

Fig. 8 Academics—Adjusted assessment

The academics and students were asked their opinion regarding the penalties that need to be enforced for cheating. The main themes the academics indicated (Fig. 6) were that the cheater be given a zero mark ($n = 9$), be expelled from the institution ($n = 6$) and be deregistered from the course ($n = 5$). AP1 indicated, "Cheating and dishonesty is endemic in our society" and "Students must be penalised with maximum penalty, such as a zero mark being allocated and/or deregistration for the module". Academic participant AP2 indicated, "Dismissal, Given Zero, removed from course, kicked out of Varsity". Two academics indicated community service, AP6 saying "Zero marks allocated, asked to leave the university, maybe community service" and AP9 "The student should have to do some community service work maybe?".

DiPietro (2010) points out that the effectiveness and intensity of deterrents vary among cultures. Within South Africa's multi-cultural, diverse communities, it is

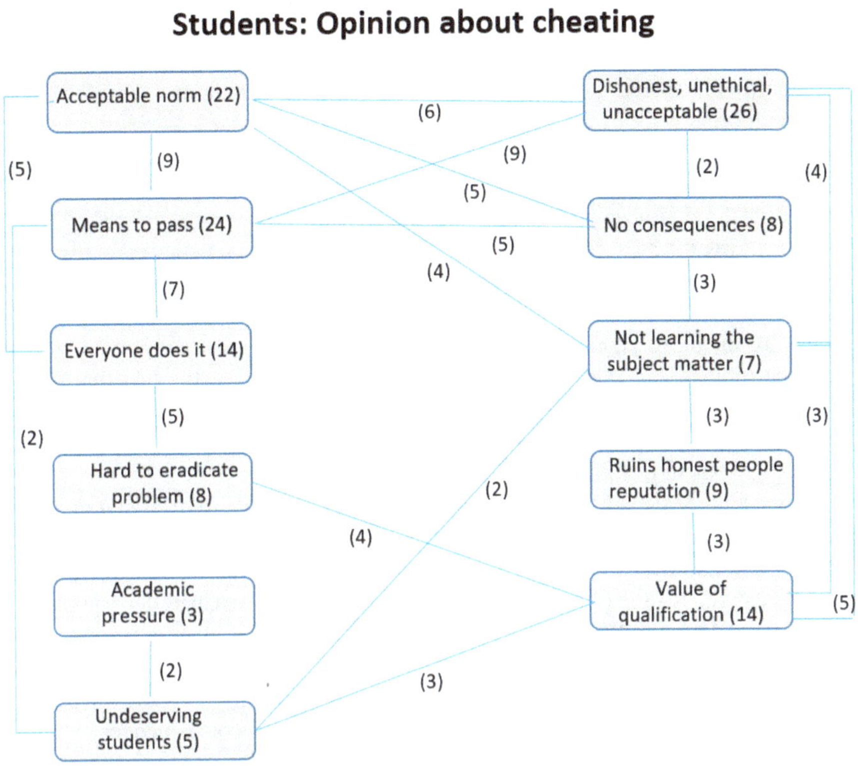

Fig. 9 Students' opinion about cheating

crucial that responses to academic dishonesty are not universal but should be tailored to the culture of the institution (DiPietro, 2010).

The students highlighted more themes (Fig. 7) and indicated the penalties must be a zero mark (31%, $n = 19$), be expelled from the institution (34%, $n = 21$), deregistered from the course ($n = 11$) and disciplinary action ($n = 14$). Student SP1 indicated "Disqualification from obtaining the qualification". The other themes included rewrite under supervision ($n = 10$) and counselling ($n = 3$).

Bertram Gallant (2011) asserts that the methods of how students cheat and how they perceive academic dishonesty evolve along with technology and assessment methodology used. The results indicate that academics have adjusted their assessment methodology. However, it seems to be reactive and based on exposed incidents of dishonesty.

The academics were asked how they adjusted their assessment (Fig. 8), knowing that online cheating was taking place. The main responses included randomising questions (58%, $n = 7$), limiting the time of assessment (50%, $n = 6$) and making use of unseen or their own case studies (42%, $n = 5$). AP1 indicated that "I use unseen case studies testing application and limiting time for assessment. Case studies not known to ChatGPT. Answers must be provided on a proforma assessment document provided" and AP3 "I have a number of variations of questions— which are set to be randomly allocated to the student writing the assessment". AP7 said "Students must identify themselves before the commencement of the examination and we have more invigilators walking around".

A concerning twenty-four students (39%) indicated that it is a 'means to pass' and that it is acceptable (Fig. 9). Nine students indicated that it ruins honest students' reputations and degrades the value of the qualification ($n = 14$). SP1 said "Those who cheat think it's an acceptable norm, those who don't cheat believe it's dishonest and undermines their own efforts" and SP2 said, "I do not like the fact that people are cheating. At the end of the day, the cheats are watering down the qualifications that we are meant to be earning. I also feel that some students are getting further than what they should due to cheating" and SP10 highlighted the fact that "They know that it is wrong. I don't think anyone who cheats has experienced the feeling of accomplishment, that satisfactory feeling of seeing the fruits of your hard work. Cheating robs them of this feeling".

One student, SP7 said "Students think that it is acceptable to do it and there are limited consequences when caught" and SP12 said, "From my understanding, those students who do cheat think of it as being no different to the real world". Student SP16 indicated, "It's the oldest trick in the books and it works. ChatGPT can pass an exam" and SP19 said, "You feel it is an unfair advantage if you are writing the test honestly, while others are using an additional source".

Cheating behaviour may be driven by acceptable norms ($n = 22$) and the perception that everyone is doing it ($n = 14$) as well as lack of consequences ($n = 8$). Callahan (2006) supports these findings and adds that sleeping watchdogs may contribute to more dishonest behaviour. Application of subjective norms ($n = 22$) as out lined by the Theory of Planned Behaviour as well as Neutralisation Theory provide the rationale for students assuming a morally neutral disposition towards morally wrong acts without damage to their self-concept (DiPietro, 2010).

5 Conclusions and Future Research

HEIs need to constantly review and improve assessment practices (Morris, 2018). It is imperative that the extent of cheating be made known so that it can be addressed in a unified manner with students and academics. Cheating has a major impact on all academic stakeholders. The institution's reputation, the value of the qualification and academic standards are impacted. Industry needs to be ensured that academic standards are maintained and that the graduates have the ability, skills and knowledge to perform in their profession.

In this study, all the academics indicated that student cheating is "common practice" and that they found students were using LLMs, such as ChatGPT and contract-cheating websites. The fellow students were aware that cheating was taking place and that it is seen as an "acceptable norm". Students are increasingly making use of AI-tools and ghostwriters, as shown on a Facebook page (Fig. 1). Both groups indicated that there are no real deterring consequences when caught cheating. All respondents agreed on giving a zero mark, disciplinary action and course deregistration.

The academics do not know the legal implications, policies and what process to follow when they experience students cheating. Institutions must address this concern by providing clear guidelines and training, specifically for the protection of academics reporting cheating. For academics, dealing with cheating affects their workload. They need extra time for setting additional evaluations, creating new case studies and making use of new assessment methods. Academics are changing their assessments, i.e. using unseen case studies, random questions, limited time for assessments and in-class tests.

Academics and students must be taught how to critically engage with AI-generated content. In addition, academic institutions need to make the process for reporting on cheating easier for academics. Academic staff need support, resources and training to help them detect incidences of cheating. The focus should be on creating a culture in institutions where honesty and integrity are values that are inculcated in all staff and students. Students question the value of their qualification and are afraid of whistle-blowing implications. There is a strong group of students who indicated that cheating brings discredit to their qualifications and ruins honest students' reputations.

Future research needs to be conducted with organisations that employ graduates to determine if they trust the awarded qualifications. The study should be repeated at other HEIs and specifically focus on the impact of students using LLMs for coding assignments. Finally, academic institutions need to make a collaborative effort to combat the unethical behaviour instilled in students who make use of the services of contract cheating sites and LLMs. Academic fraud undermines the education system and the basic principle of transferring knowledge. The moral values of integrity need to be enforced on a personal and academic level, as it affects future leadership.

References

Al Hamdani, W. (20202). *Online cheating: A prevention procedure.* https://www.researchgate.net/publication/330225367_Online_Cheating_A_Prevention_Procedure

Ashley Madison. (2022). Life is short. Have an affair. https://www.ashleymadison.com/en-za/

Bertram Gallant, T. (2011). Building a culture of academic integrity. In *A Magna Publications White Paper.*

Bilen, E., & Matros, A. (2020). Online cheating amid COVID-19. *Journal of Economic Behaviour and Organisation.*

Bin-Nashwan, S. A., Sadallah, M., & Bouteraa, M. (2023). Use of ChatGPT in academia: Academic integrity hangs in the balance. *Technology in Society, 75.* https://doi.org/10.1016/j.techsoc.2023.102370

Birks, D., & Clare, J. (2023). Linking artificial intelligence facilitated academic misconduct to existing prevention frameworks. *International Journal for Educational Integrity, 19*(20). https://doi.org/10.1007/s40979-023-00142-3

Blum, S. D. (2009). *My word! Plagiarism and college culture.* Cornell University Press.

Broda, E., & Strömbäck, J. (2024). Misinformation, disinformation, and fake news: Lessons from an interdisciplinary, systematic literature review. *Annals of the International Communication Association, 48*(2), 139–166. https://doi.org/10.1080/23808985.2024.2323736

Brown, T., Isbel, S., Logan, A., & Etherington, J. (2020). Predictors of academic integrity in undergraduate and graduate-entry masters occupational therapy students. *Hong Kong Journal of Occupational Therapy, 33*(2), 42–54. https://doi.org/10.1177/1569186120968035

Callahan, D. (2006). On campus: Author discusses the "cheating culture" with college students. *Plagiary: Cross Disciplinary Studies in Plagiarism, Fabrication, and Falsification,* 25–32

Center for Teaching & Learning. (2021). *Academic integrity.* https://teaching.berkeley.edu/resources/course-design/academic-integrity

Cotton, D. R. E., Cotton, P. A., & Shipway, J. R. (2024). Chatting and cheating: Ensuring academic integrity in the era of ChatGPT. *Innovations in Education and Teaching International, 61*(2), 228–239. https://doi.org/10.1080/14703297.2023.2190148

Curtis, G. J., & Tremayne, K. (2021). Is plagiarism really on the rise? Results from four 5-yearly surveys. *Studies in Higher Education, 46*(9), 1816–1826. https://doi.org/10.1080/03075079.2019.1707792

Davis, A. (2023). Academic Integrity in the time of contradictions. *Cogent Education, 10.* https://doi.org/10.108/2331186X.2023.2289307

DiPietro, M. (2010). 14 Theoretical frameworks for academic dishonesty. A comparative view. *To Improve the Academy. A Journal of Educational Development, 28.* https://doi.org/10.3998/tia.17063888.0028.018

Heitz, C., Laboissiere, M., Sanghvi, S., & Sarakatsannis. J. (2020). *Getting the next phase of remote learning right in higher education.* McKinsey & Company.

Holden, O.L., Norris, M. E. & Kuhlmeier, V. A. (2021). Academic integrity in online assessment: A research review. *Frontiers in Education, 6.* https://doi.org/10.3389/feduc.2021.639814

International Center for Academic Integrity (ICAI). (2025). https://academicintegrity.org/aws/ICAI/pt/sp/values

Koh J. H. & Daniel, B. K. (2022). Shifting online during COVID-19: A systematic review of teaching and learning strategies and their outcomes. *International Journal of Educational Technology in Higher Education, 19*(56). https://doi.org/10.1186/s41239-022-00361-7

Kotsis, K. T. (2025). ChatGPT and DeepSeek evaluate one another for science education. *Journal of Effective Teaching Methods, 3*(1) (2025). https://doi.org/10.59652/jetm.v3i1.439

Laurier. (2025). *Academic misconduct defined.* Wilfred Laurier University. https://students.wlu.ca/academics/academic-integrity/types-of-academic-misconduct.html

Morris, E. J. (2018). Academic integrity matters: Five considerations for addressing contract cheating. *International Journal for Educational Integrity, 14*(15). https://doi.org/10.1007/s40979-018-0038-5

Moten, J., Fitterer, A., Brazier, E., Leonard, J., & Brown, A. (2013). Examining online college cyber cheating methods and prevention measures. *Electronic Journal of e-Learning, 11*(2), 139–146.

Nguyen, H. M., & Goto, D. (2024). Unmasking academic cheating behavior in the artificial intelligence era: Evidence from Vietnamese undergraduates. *Education and Information Technologies, 29,* 15999–16025. https://doi.org/10.1007/s10639-024-12495-4

NIU. (2021). *Academic dishonesty definition and types.* https://www.niu.edu/academic-integrity/faculty/types/index.shtml

Oravec, J. (2023). Artificial intelligence implications for academic cheating: Expanding the dimensions of responsible human–AI collaboration with ChatGPT and bard. *Journal of Interactive Learning Research, 34*(2), 213–237.

Peterson, J. (2019). An analysis of academic dishonesty in online classes. *Mid-Western Educational Researcher, 31*(1)

Price-Mitchell, M. (2023). Academic integrity is essential to learning. *Roots of Action.* https://www.rootsofaction.com/academic-integrity/

Transparency International. (2024). Retrieved from https://www.transparency.org/en/cpi/202

Watson, G., & Sottile, J. (2010). Cheating in the digital age: do students cheat more in online courses? *Online Journal of Distance Learning Administration, 13*(1)

Zachek, A. (2020). The history, evolution and trends of academic dishonesty: A literature review. *The Nebraska Educator, 5*. https://doi.org/10.32873/unl.dc.ne006

Zanetti, C., & Butera, F. (2025). Detecting collective cheating culture in academic contexts. *Educational Psychology, 45*(1), 67–86. https://doi.org/10.1080/01443410.2025.2449976

AI-Powered Digital Twins with Bio-sensory Feedback for Enhanced Predictive Security in Cyber-physical System

Kwabena Addo, Katleho Moloi, Musasa Kabeya, and Evans Eshiemogie Ojo

Abstract Cyber-physical Systems (CPS) form the backbone of critical infrastructure but remain susceptible to stealthy cyber threats. This study introduces a multimodal digital twin architecture that integrates telemetric, environmental, and physiological signals, including EEG, heart rate variability (HRV), and electrodermal activity (EDA) within a unified Transformer-based threat prediction framework. The virtual twin is synchronized with the physical CPS via a graph neural network (GNN), enabling proactive detection of anomalies before system degradation. Performance is assessed using two benchmark datasets: one with CPS telemetry and another emulating biosignals. Comparative analysis against support vector machine (SVM) and long short-term memory (LSTM) models shows that the proposed method achieves 96.2% accuracy (AUC = 0.983), surpassing LSTM (87.4%) and SVM (81.3%). Additionally, integrating biofeedback lowers false negatives by 45%. Deployment implications are examined, and a future roadmap is outlined for implementation in real-world microgrid environments.

1 Introduction

Modern cyber-physical infrastructures, including smart grids, industrial control systems, and autonomous transportation platforms, depend on real-time telemetry to ensure safe and reliable operation. Despite substantial progress in anomaly detection, many cyber-attacks (e.g., false-data injection, stealthy reconnaissance) remain

K. Addo (✉) · K. Moloi · M. Kabeya · E. E. Ojo
Electrical Power Engineering, Durban University of Technology, Durban, South Africa
e-mail: 22494773@dut4life.ac.za

K. Moloi
e-mail: katlehom@dut.ac.za

M. Kabeya
e-mail: musasak@dut.ac.za

E. E. Ojo
e-mail: evanso@dut.ac.za

J. Marx Gómez et al. (eds.), *Recent Advances in Information Systems*, Progress in IS,
https://doi.org/10.1007/978-3-032-14099-9_15

challenging to detect until they cause critical system disruptions. In parallel, human operators often exhibit physiological stress responses, such as elevated heart rate variability (HRV), increased electroencephalographic (EEG) theta activity, and changes in electrodermal activity (EDA), when confronted with abnormal conditions. However, these bio-sensory indicators are rarely leveraged in real-time cyber defense systems (Raman et al., 2022; Cai and Koutsoukos, 2023).

This work introduces an AI-powered digital twin framework that continuously mirrors the physical state of a cyber-physical system (CPS) and augments it with operator-derived physiological feedback, specifically EEG, HRV, and EDA signals. By modeling these heterogeneous data modalities using a Transformer-based architecture, the framework facilitates early and proactive anomaly detection beyond conventional telemetry-based systems. A graph neural network (GNN) synchronizes the virtual twin with the physical system, encoding both spatial structure and topological relationships among CPS components.

1.1 Motivating Use Case

Consider a regional power substation responsible for distributing electricity across an urban grid. In a coordinated false-data injection attack, adversaries compromise voltage and phase angle data originating from multiple Phasor Measurement Units (PMUs). As alarms are raised, the human operator, tasked with monitoring SCADA interfaces, begins experiencing cognitive strain, manifested physiologically as increased EEG theta-band activity and elevated HRV. The proposed framework captures these physiological anomalies alongside inconsistencies in the CPS telemetry, leveraging both to trigger a predictive alert such as: "*Potential sensor spoofing detected. Initiate circuit isolation protocol.*" This anticipatory response enables timely intervention, before falsified data propagates downstream thus preventing cascade failures and ensuring system stability.

1.2 Contributions

- **Multimodal Fusion**: We design a fusion architecture that jointly encodes telemetric (S), environmental (E), and bio-sensory (B) data streams into a shared feature representation Z, allowing unified interpretation of machine and human-state indicators.
- **Digital Twin Synchronization**: A graph neural network-based twin is constructed to reflect CPS structure. A dedicated synchronization loss ensures that the predicted state $T(t)$ aligns closely with observed telemetry $S(t)$ over time.
- **Transformer-Based Threat Predictor**: A temporal Transformer processes the fused multimodal sequence $Z^\star$ and outputs a real-time threat score $\hat{y}_t$, supporting rapid and adaptive detection of cyber intrusions.

- **Comprehensive Evaluation**: We evaluate our approach using two benchmark datasets: one capturing CPS telemetry, and another simulating biosignal responses using statistics from WESAD and DEAP. Our model achieves a detection accuracy of 96.2%, significantly outperforming baseline methods such as support vector machines (SVMs) and long short-term memory (LSTM) networks.
- **Deployment Roadmap**: We outline key practical considerations for real-world implementation, including biosensor integration, latency-performance trade-offs, and ethical implications. A deployment roadmap is proposed for testing within a live microgrid environment.

2 Related Work

This section reviews prior research across three interrelated domains that underpin our proposed framework: (i) digital twins for cyber-physical system (CPS) security, (ii) multimodal bio-sensory fusion for human-in-the-loop monitoring, and (iii) Transformer-based anomaly detection in CPS environments.

2.1 Digital Twins in CPS Security

Digital twins (DTs) have emerged as a pivotal component in CPS security, enabling real-time mirroring of physical infrastructure and proactive anomaly detection. De Benedictis et al. (2023) introduced a DT framework leveraging long short-term memory (LSTM) networks to detect anomalies in smart grids, achieving 90% detection accuracy. Xu et al. (2023) extended this paradigm with a reinforcement learningâŁ“based DT architecture to support dynamic threat mitigation and component isolation in industrial control systems. Recent contributions, such as Li et al. (2024), proposed privacy-preserving DTs for fault diagnosis, while Nie et al. (2023) presented a twin-enabled orchestration mechanism tailored for edgeâŁ“cloud security coordination.

Gap: Despite these advancements, existing DT frameworks operate exclusively on machine-state telemetry and overlook operator-centric physiological cues–an underutilized but informative source of pre-attentive system awareness.

2.2 Bio-sensory Feedback and Human-in-the-Loop Security

Physiological signals offer valuable insight into cognitive workload and stress experienced by human operators under abnormal CPS conditions. Li et al. (2023) demonstrated the utility of electroencephalography (EEG), electrocardiography (ECG), and electrodermal activity (EDA) for pilot stress detection, achieving 85% accuracy. Liu et al. (2023) explored EEG-driven emotion recognition for adaptive user interfaces. Complementary studies and surveys (Fawwaz et al., 2024; John et al., 2022) have affirmed the reliability of heart rate variability (HRV) and EDA as biomarkers for mental effort and vigilance.

Gap: While the efficacy of biosignals for operator-state monitoring is well-established, there is a notable lack of integration between such modalities and real-time digital twin architectures for CPS anomaly detection.

2.3 Transformer-Based Intrusion Detection in CPS

Transformers have demonstrated exceptional performance in modeling temporal dependencies for anomaly detection. Xia et al. (2023) introduced a Transformer-based architecture for supervisory control and data acquisition (SCADA) systems, reporting 92% classification accuracy. Cai et al. (2022) employed temporal convolutional networks (TCNs) for fault detection in water distribution systems. More recently, Alshomrani et al. (2024) explored Transformer variants for detecting zero-day attacks in industrial control networks, showing strong generalization performance.

Gap: Although Transformer-based models excel at telemetry-driven anomaly detection, they fail to incorporate operator-state features, thereby missing a vital dimension of situational awareness.

2.4 Summary of Gaps and Contributions

Across all three domains, a fundamental disconnection persists: digital twins are designed without integration of human feedback, and physiological state monitoring remains detached from CPS anomaly detection pipelines.

To address these gaps, this work contributes:

- A graph neural network (GNN)-based digital twin that synchronizes with real-time CPS telemetry;
- The integration of multimodal physiological inputs (EEG, HRV, EDA) to capture operator cognition and stress indicators;
- A Transformer-based threat prediction model that fuses machine-state and human-state features for anticipatory anomaly detection.

3 System Model and Methodology

3.1 System Overview and Digital Twin Justification

A digital twin is a live, virtual replica of a physical system that receives real-time data, simulates system dynamics, and can issue control actions back to the physical layer. In our framework (see Fig. 1), we implement the following components:

- **Physical CPS Layer**: The real-world infrastructure (such as substations) and associated PMU/SCADA telemetry, including voltages, currents, and device states.
- **Data Acquisition Modules**:
  - **Telemetry Module** S: Collects CPS sensor data $S(t) \in \mathbb{R}^n$ at time t.
  - **Bio-sensory Module** B: Captures physiological signals including EEG, HRV, and EDA, yielding $B(t) \in \mathbb{R}^m$.
  - **Environmental Module** E: Measures ambient factors such as temperature and humidity, $E(t) \in \mathbb{R}^k$.
- **Directed Edges**: Solid arrows in Fig. 1 denote data flow into the digital twin engine, while dashed arrows represent control feedback from the analysis block to the physical system.

Digital Twin Engine (GNN Backbone): We model the CPS as a graph $G = (V, E)$, where nodes $v \in V$ represent components (e.g., substations), and edges $e \in E$ model communication or power links. At time t, each node v has features $X_v(t) = S_v(t)$.

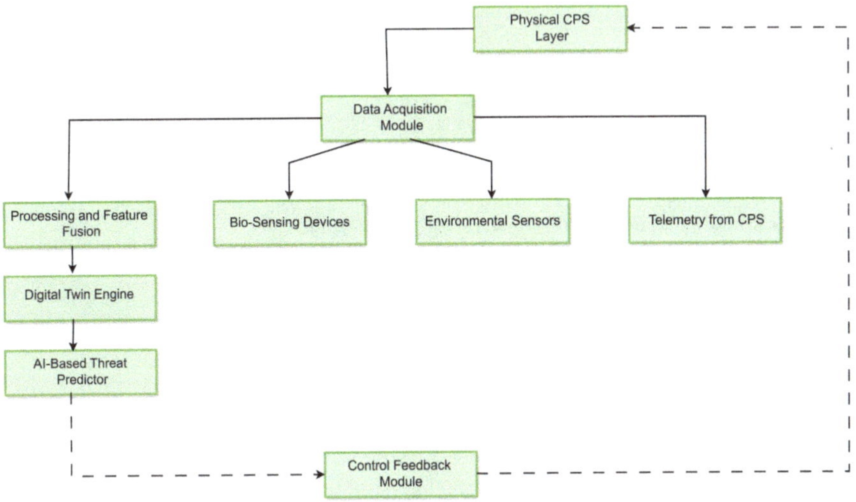

Fig. 1 Proposed AI-powered digital twin system architecture

A graph neural network (GNN), denoted Φ_{GNN}, computes a predicted twin state $T(t) \in \mathbb{R}^n$ synchronized with the actual telemetry $S(t)$. A synchronization loss L_{sync} (see Sect. 3.3) ensures $T(t)$ closely approximates $S(t)$.

Unlike traditional anomaly detectors, this twin proactively simulates the next CPS state, enabling predictive threat detection through topology-aware learning.

Multimodal Fusion and Temporal Transformer: All three input streams $S(t)$, $B(t)$, and $E(t)$ are fused into a joint representation $Z(t)$ (see Sect. 3.2). The sequence $\{Z(t-T+1), \ldots, Z(t)\}$ is then passed into a Transformer encoder. The [CLS] token output is fed to a sigmoid classifier to estimate threat probability $\hat{y}(t)$ (see Sect. 3.4).

Analysis and Control Module: When $\hat{y}(t)$ exceeds a threshold τ, a predictive alert is generated. This can trigger preemptive actions such as system isolation or operator intervention.

3.2 *Multimodal Fusion Pipeline*

Intuitive Overview: At each time t, we project and fuse data from telemetry, physiological, and environmental sources to form a single feature vector $Z(t)$.

Telemetry Projection (Eq. 1):

$$P_S = W_S S(t) + b_S, \quad P_S \in \mathbb{R}^{d_S}, \tag{1}$$

where $W_S \in \mathbb{R}^{d_S \times n}$ and $b_S \in \mathbb{R}^{d_S}$ project raw telemetry into a lower-dimensional space.

Bio-sensory Projection (Eq. 2):

$$P_B = W_B B(t) + b_B, \quad P_B \in \mathbb{R}^{d_B}, \tag{2}$$

where $W_B \in \mathbb{R}^{d_B \times m}$ and $b_B \in \mathbb{R}^{d_B}$. Typically, $d_B = d_S/2$ due to the lower dimensionality and slower dynamics of physiological signals.

Environmental Projection (Eq. 3):

$$P_E = W_E E(t) + b_E, \quad P_E \in \mathbb{R}^{d_E}, \tag{3}$$

where $W_E \in \mathbb{R}^{d_E \times k}$ and $b_E \in \mathbb{R}^{d_E}$. We typically use $d_E < d_B$ to reflect their slower variation.

Concatenation and Fusion (Eq. 4):

$$Z(t) = \text{ReLU}\left(W_F[P_S; P_B; P_E] + b_F\right), \quad Z(t) \in \mathbb{R}^d, \tag{4}$$

where [;] denotes concatenation, $W_F \in \mathbb{R}^{d\times(d_S+d_B+d_E)}$, and $b_F \in \mathbb{R}^d$.

Rationale: Linear projections reduce dimensionality for computational efficiency. ReLU introduces nonlinearity to model feature interactions. Concatenation allows the Transformer to capture cross-modal correlations (e.g., EEG spikes correlated with voltage anomalies).

3.3 GNN-Based Digital Twin and Synchronization Loss

Graph Construction: The CPS is represented as $G = (V, E)$ with node features $X_v(t) = S_v(t) \in \mathbb{R}^{n_v}$. The GNN updates each node's state as:

$$H_v^{(l+1)} = \sigma\left(\sum_{u\in\mathcal{N}(v)} \frac{1}{\sqrt{|\mathcal{N}(v)|\,|\mathcal{N}(u)|}} W^{(l)} H_u^{(l)} + b^{(l)}\right), \tag{5}$$

where $H_v^{(0)} = X_v(t)$, $H_v^{(L)} = T_v(t)$, $L = 3$, and σ is the ReLU activation function.

Synchronization Loss (Eq. 6):

$$L_{\text{sync}}(t) = \sum_{v\in V} \|T_v(t) - S_v(t)\|_2^2 . \tag{6}$$

Why GNN? CPS components exhibit graph-structured interdependencies (e.g., power flows). GNNs capture these topological relationships more effectively than fully connected layers. Synchronization ensures that the digital twin accurately mirrors the physical system in real time.

3.4 Temporal Transformer-Based Threat Detection

Intuitive Overview: The fused sequences $Z(t)$ across a window of T timesteps are input into a Transformer encoder. The output of the [CLS] token represents the threat state.

Input Construction:

$$\mathbf{Z}_{t-T+1:t} = [Z(t-T+1), \ldots, Z(t)] \in \mathbb{R}^{T\times d}, \tag{7}$$

with a prepended learnable [CLS] token $C \in \mathbb{R}^d$:

$$\mathbf{X}_0 = [C; Z(t-T+1); \ldots; Z(t)] \in \mathbb{R}^{(T+1)\times d}. \tag{8}$$

Transformer Encoding (4 Layers): Each layer $\ell = 0, \ldots, 3$ applies:

$$\mathbf{U}^{(\ell)} = \text{MHSA}(\text{LN}(\mathbf{X}^{(\ell)})) + \mathbf{X}^{(\ell)} \tag{9}$$

$$\mathbf{X}^{(\ell+1)} = \text{FFN}(\text{LN}(\mathbf{U}^{(\ell)})) + \mathbf{U}^{(\ell)}, \tag{10}$$

where MHSA is multi-head self-attention with 8 heads, and FFN is a feedforward network with dimensions $d \rightarrow 4d \rightarrow d$ and ReLU activations.

Threat Score (Eq. 11):

$$\hat{y}(t) = \sigma(w_o^\top e_C + b_o), \quad \hat{y}(t) \in (0, 1), \tag{11}$$

where e_C is the final [CLS] embedding, $w_o \in \mathbb{R}^d$, $b_o \in \mathbb{R}$, and $\sigma(\cdot)$ is the sigmoid function.
Classification Loss (Eq. 12):

$$L_{\text{cls}}(t) = -\left[y(t)\log\hat{y}(t) + (1 - y(t))\log(1 - \hat{y}(t))\right], \tag{12}$$

where $y(t) \in \{0, 1\}$ indicates the true label at time t.

3.5 Joint Training Objective

The model is trained end-to-end with the following combined objective:

$$L_{\text{total}} = \alpha \sum_t L_{\text{sync}}(t) + \beta \sum_t L_{\text{cls}}(t), \tag{13}$$

where we set $\alpha = 0.5$ and $\beta = 0.5$.

Why $\alpha = \beta$? An ablation study (see Table 3) shows that balanced weighting between synchronization and classification yields the best trade-off. When $\alpha \gg \beta$, the twin mirrors the physical state accurately, but threat classification underperforms. Conversely, $\beta \gg \alpha$ leads to high classification accuracy at the expense of model realism.

4 Dataset Description and Experimental Setup

4.1 Cyber-physical Threat Dataset

Overview. We simulate a regional power substation comprising six nodes, three Transformers, two circuit breakers, and one control server, using NS-3 (version

3.35). The telemetry channels include bus voltage, current magnitude, breaker status (open/closed), CPU load on the IEC 61850 server, and packet interarrival time. Data is sampled 100 Hz for a duration of 10 min per trial.

Attack Scenarios

- **False-Data Injection (FDI)**: Voltage and phase angle measurements are manipulated by injecting noise with amplitude between $\pm 5\%$ and $\pm 10\%$ relative to nominal values. We generate 5000 FDI samples.
- **Distributed Denial of Service (DDoS)**: The SCADA server is flooded with 100–200 packets per second for durations between 10–20 s. Effects are captured as elevated CPU load and network throughput. 5000 DDoS samples are created.
- **Jamming**: Artificial packet loss of 15–25% is introduced intermittently for 5–15 s, producing 5000 samples simulating radio frequency interference or degraded channel conditions.

Feature Representation. Each telemetry vector $S(t) \in \mathbb{R}^8$ includes:

$$[V_1, I_1, V_2, I_2, V_3, I_3, \text{CPU}_{\text{load}}, \text{pkt}_{\text{interval}}].$$

All features are standardized to zero mean and unit variance based on the training dataset statistics.

Train/Test Split. Each attack class contains 15,000 labeled samples. We allocate 80% for training (12,000 samples per class) and 20% for testing (3000 samples per class), using a fixed random seed of 42 for reproducibility.

4.2 *Biosignal Emulation Dataset*

Rationale. Collecting large-scale physiological data such as EEG, HRV, and EDA from substation operators is infeasible in practice. Therefore, we emulate biosignals by sampling from Gaussian distributions parameterized using empirical data from Wearable Stress and Affect Detection (WESAD) (Vasile et al., 2023) and Database for Emotion Analysis using Physiological Signals (DEAP) (Pattyn et al., 2023), conditioned on stress and baseline states.

EEG Emulation. From DEAP, we extract mean bandpower statistics in the α (8–12 Hz) and θ (4–8 Hz) bands:

$$\text{EEG}_\alpha \sim \mathcal{N}(0.30, 0.10^2), \quad \text{EEG}_\theta \sim \mathcal{N}(0.25, 0.08^2).$$

During attack periods, θ-band power is increased by an additional 0.05âŁ“0.10 μV^2 to simulate cognitive arousal.

Table 1 Summary of simulated datasets used for training and evaluation

Dataset	Samples	Telemetry features	Biosignal features
CPS Threat (FDI, DDoS, Jamming)	45,000	$V_1, I_1, V_2, I_2, V_3, I_3, CPU_{load}, pkt_{interval}$	–
Biosignal (Baseline/Stress)	45,000	–	EEG_α, EEG_θ, HRV, EDA

HRV Emulation. Based on WESAD stress/rest epochs:

$$\text{HRV}_{\text{baseline}} \sim \mathcal{N}(0.45, 0.05^2), \quad \text{HRV}_{\text{stress}} \sim \mathcal{N}(0.60, 0.07^2).$$

EDA Emulation. WESAD indicates a median electrodermal activity of 0.4 μS under baseline and 0.6 μS under stress:

$$\text{EDA}_{\text{baseline}} \sim \mathcal{N}(0.40, 0.10^2), \quad \text{EDA}_{\text{stress}} \sim \mathcal{N}(0.60, 0.12^2).$$

Temporal Alignment. We align biosignal vectors with telemetry by linearly interpolating the synthetic data to match 100 Hz sampling rate of $S(t)$. During each 2-second attack window, biosignals are sampled from stress distributions; elsewhere, from baseline distributions.

Dataset Summary

All features across both datasets are normalized using statistics from the training partition (Table 1).

4.3 Experimental Environment and Hyperparameters

Software Environment. We implement the framework using:

- Python 3.10.
- PyTorch 2.0.
- DGL 0.9 for the GNN architecture.
- HuggingFace Transformers for the temporal encoder.

Hardware. Experiments are conducted on a system with an Intel Xeon Silver 4214 CPU and NVIDIA Tesla V100 GPU (32 GB memory).

Model Hyperparameters.

- **Fusion Dimensions**: $d_S = 64$, $d_B = 32$, $d_E = 16$, giving $d = 112$.
- **GNN**: 3 layers, hidden size = 64, ReLU activation.

- **Transformer**: 4 layers, $d_{\text{model}} = 112$, 8 attention heads, feedforward dimension = 448, sequence length $T = 50$.
- **Optimizer**: AdamW with learning rate 1×10^{-4} and weight decay 1×10^{-5}.
- **Training**: Batch size = 128, maximum 50 epochs with early stopping based on validation AUC.

5 Experimental Results

5.1 Baseline Models

Support Vector Machine (SVM). We use a radial basis function (RBF) kernel with regularization parameter $C = 1.0$ and $\gamma = 0.01$. Input features consist of concatenated telemetry (8 dimensions), biosignals (4 dimensions), and environmental factors (2 dimensions), yielding a 14-dimensional feature vector per sample. Features are standardized using training set statistics. The model is trained on 80% of the dataset and evaluated on the remaining 20%.

Long Short-Term Memory (LSTM). We implement a two-layer LSTM with a hidden size of 128. The input dimension is 112, representing the fused multimodal feature vector described in Sect. 3.2. The model uses dropout rate 0.3, learning rate of 1×10^{-3}, batch size of 128, and is trained for 50 epochs.

5.2 Quantitative Performance

Accuracy and AUC Comparison. Table 2 presents the classification performance of the baseline models and our proposed Transformer-based approach.

The proposed model outperforms both baselines by a significant margin, reducing false negatives by 45% compared to the LSTM.

Table 2 Performance comparison of baseline and proposed models

Model	Accuracy (%)	AUC	Precision (Threat)	Recall (Threat)	F1-score (Threat)
SVM	81.3	0.801	0.79	0.76	0.77
LSTM	87.4	0.873	0.85	0.84	0.84
Proposed transformer	**96.2**	**0.983**	**0.95**	**0.96**	**0.95**

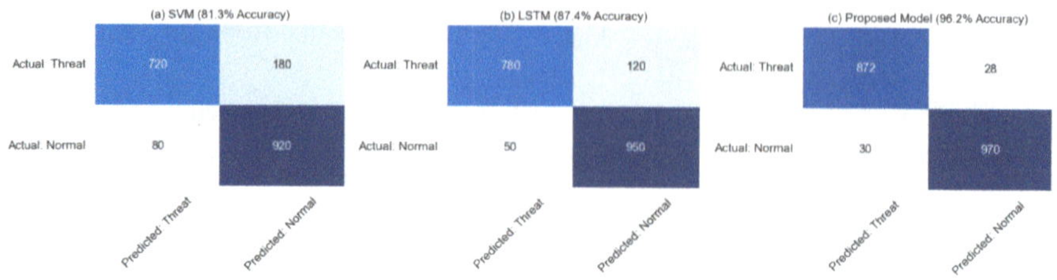

Fig. 2 Confusion matrices for the comparison of the three models

Table 3 Ablation study on loss weights α (sync) and β (classification)

(α, β)	Accuracy (%)	AUC	Twin MSE
(1.0, 0.2)	89.1	0.912	0.028
(0.5, 0.5)	**96.2**	**0.983**	**0.012**
(0.2, 1.0)	92.6	0.947	0.051

Confusion Matrices. Figure 2 compares the confusion matrices of the three models:

- Figure 2a: SVM (81.3%) shows high false negatives (180 out of 900).
- Figure 2b: LSTM (87.4%) reduces false negatives to 120.
- Figure 2c: Proposed model (96.2%) further reduces false negatives to 28.

Ablation Study on Loss Weights. Table 3 shows the impact of varying the loss weights α and β on classification accuracy and twin synchronization error.

Balanced weighting ($\alpha = \beta = 0.5$) yields the best trade-off between classification performance and digital twin fidelity.

5.3 *Qualitative Analysis*

Case Study: FDI Event Detection. During a simulated false-data injection event at $t = 345$ s, the physical telemetry diverged by 6% from the expected range. The GNN-based twin model detected a mean squared error (MSE) deviation of 0.18. Simultaneously, the emulated EEG θ-band power increased by 0.07 μV^2, indicating cognitive arousal. The Transformer fused both cues to produce a threat score $\hat{y}(t) = 0.92$, prompting a predictive alert 0.4 s before the SCADA system's native alarm.

Latency Analysis. The total inference time, including both the GNN synchronization and Transformer-based threat classification is approximately 290 ms per 50-step input window on an NVIDIA Tesla V100 GPU. This satisfies soft real-time constraints for substation-level protection systems, though further optimization may be required for millisecond-scale control in ultra-low-latency CPS domains.

6 Conclusion

In this work, we presented an AI-powered digital twin framework that integrates a graph neural network (GNN)-based model of a cyber-physical system (CPS) with bio-sensory feedback from human operators. The system fuses multimodal data, including telemetry, environmental context, and physiological signals, via a Transformer-based temporal encoder.

Our approach achieves 96.2% classification accuracy with an AUC of 0.983 on benchmark datasets, significantly outperforming support vector machine (SVM) and long short-term memory (LSTM) baselines, which achieved 81.3% and 87.4% accuracy, respectively. Qualitative analysis demonstrates the framework's ability to support early intervention, and ablation experiments validate the importance of balanced loss weighting for maintaining synchronization fidelity and classification accuracy.

Although biosignals in this study were emulated using distributions derived from WESAD and DEAP datasets, future work will focus on incorporating real-time physiological data from operators in a live microgrid testbed. In addition, we aim to optimize end-to-end inference latency to meet sub-100 ms constraints for real-time deployment.

Overall, this work advances the paradigm of human-augmented, predictive security in critical CPS environments by combining digital twin intelligence with cognitive awareness.

References

Alshomrani, M., Albeshri, A., Alturki, B., Alallah, F. S., & Alsulami, A. A. (2024). Survey of transformer-based malicious software detection systems. *Electronics, 13*(23), 4677.

Cai, J., Gao, J., & Xu, Y. (2022). Anomaly detection and classification in water distribution networks integrated with hourly nodal water demand forecasting models and feature extraction technique. *Journal of Water Resources Planning and Management, 148*(11).

Cai, F., & Koutsoukos, X. (2023). Real-time detection of deception attacks in cyber-physical systems. *International Journal of Information Security, 22*(4), 1–16.

De Benedictis, A., Flammini, F., Mazzocca, N., Somma, A., & Vitale, F. (2023). Digital twins for anomaly detection in the industrial internet of things: Conceptual architecture and proof-of-concept. *IEEE Transactions on Industrial Informatics*, 1–11.

Fawwaz, A., Rahma, O. N., Ain, K., Ittaqillah, S. I., & Chai, R. (2024). Measurement of mental workload using heart rate variability and electrodermal activity. *IEEE Access, 1*. https://doi.org/10.1109/access.2024.3521649

John, A. R., Singh, A. K., Do, T.-T. N., Eidels, A., Nalivaiko, E., Gavgani, A. M., Brown, S. D., Bennett, M., Lal, S., Simpson, A. M., Gustin, S. M., Double, K., Walker, F. R., Kleitman, S., Morley, J., & Lin, C.-T. (2022). Unraveling the physiological correlates of mental workload variations in tracking and collision prediction tasks. *IEEE Transactions on Neural Systems and Rehabilitation Engineering, 30*, 770–781. https://doi.org/10.1109/TNSRE.2022.3157446

Li, B., Song, P., & Zhao, C. (2024). Fusing consensus knowledge: A federated learning method for fault diagnosis via privacy-preserving reference under domain shift. *Information Fusion*. https://doi.org/10.1016/j.inffus.2024.102290

Li, Y., Li, K., Chen, J., Wang, S., Lu, H., & Wen, D. (2023). Pilot stress detection through physiological signals using a transformer-based deep learning model. *IEEE Sensors Journal, 23*(11), 11774–11784.

Liu, S., Wang, Z., An, Y. J., Zhao, J., Zhao, Y., & Zhang, Y. (2023). EEG emotion recognition based on the attention mechanism and pre-trained convolution capsule network. *Knowledge-Based Systems, 265*, 110372.

Nie, Q., Tang, D., Liu, C., Wang, L., & Song, J. (2023). A multi-agent and cloud-edge orchestration framework of digital twin for distributed production control. *Robotics and Computer-Integrated Manufacturing, 82*, 102543. https://doi.org/10.1016/j.rcim.2023.102543

Pattyn, E., Thammasan, N., Lutin, E., Tourolle, D., Van Kraaij, A., Kosunen, I., De Raedt, W., & Van Hoof, C. (2023). Simulation of ambulatory electrodermal activity and the handling of low-quality segments. *Computer Methods and Programs in Biomedicine, 242*, 107859.

Raman, M. R. G., & Mathur, A. P. (2022). A hybrid physics-based data-driven framework for anomaly detection in industrial control systems. *IEEE Transactions on Systems, Man, and Cybernetics, 52*(9), 6003–6014.

Vasile, F., Vizziello, A., Brondino, N., & Savazzi, P. (2023). Stress state classification based on deep neural network and electrodermal activity modeling. *Sensors, 23*(5), 2504.

Xia, F., Chen, X., Yu, S., Hou, M., Liu, M., & You, L. (2023). Coupled attention networks for multivariate time series anomaly detection. *IEEE Transactions on Emerging Topics in Computing*. https://doi.org/10.1109/tetc.2023.3280577

Xu, H., Wu, J., Pan, H., Gu, J., & Guan, X. (2023). Delay safety-aware digital twin empowered industrial sensing-actuation systems using transferable and reinforced learning. *IEEE Transactions on Industrial Informatics*, 1–10.

Security of Mobile Devices: A Taxonomy Study of Detection Methods of Pegasus Spyware and Countermeasures

Davish Ramlochund and Girish Bekaroo

Abstract While privacy is recognized as a fundamental and universal human right in almost every country around the world, those rights are often overlooked or disregarded when it comes to surveillance or tracking by various entities. The demand for spyware continues to thrive globally as many consider information as a powerful weapon that can be leveraged. NSO Group's most prized surveillance technology called Pegasus has been around since 2011 and has been in the headlights since its public reveal in 2016. Since then, a number of high-profile cases and investigations on how the NSO Group and Pegasus exploited and infected thousands of devices for surveillance. Nevertheless, limited studies have been undertaken to present or analyze its detection methods. This study presents a taxonomy review of methods used to detect Pegasus and discusses countermeasures for mitigating the risk of getting infected by securing mobile devices.

1 Introduction

In 2023, Kaspersky observed a steady increase in the number of attacks on mobile devices, reaching almost 33.8 million attacks, thus marking a 50% increase from the previous year's figures (Kaspersky, 2024). Spyware and adware formed part of the key threats that mobile devices face. Spyware generally refers to software programs that are covertly installed on devices and act as data sensors to illicitly collect and transmit information about the device and the user's actions (Awad & Fitzgerald, 2005). It can also be a legitimate software that monitors user's data for commercial purposes like advertising (Kaspersky, 2018). Multiple types of spyware have been designed over the years with different methods of infection and surveillance but with the same objective of collecting personal data. These include Adware, Tracking Cookies, Trojans, Keyloggers, Stalkerware, Stealware, Rootkits and System Monitors (Jr, 2019). The NSO Group, an Israeli cyber-intelligence firm, is known for having developed the most infamous and sophisticated mobile phone spyware ever

D. Ramlochund (✉) · G. Bekaroo
School of Science and Technology, Middlesex University Mauritius, Flic-en-Flac, Mauritius
e-mail: DR733@live.mdx.ac.uk

J. Marx Gómez et al. (eds.), *Recent Advances in Information Systems*, Progress in IS,
https://doi.org/10.1007/978-3-032-14099-9_16

developed known as Pegasus. The company states that its products help government agencies investigate and prevent terrorism and crime to save thousands of lives around the globe (NSO Group, 2019). Although the objective of the NSO Group seems noble, Pegasus has often been abused in its use, targeting anti-government individuals and groups, human activists, journalists, citizens and business executives for corporate espionage. Pegasus can monitor virtually all activities on an iOS or Android device including phone calls, SMS, location tracking, device settings, network details, browsing history, take photos and screenshots, contact lists, calendar events, social networks, emails and even turning on the camera and microphone (Marczak et al., 2018), thus making it one of the most intrusive spyware ever created. In 2011, Mexico was the first reported country to have purchased and used Pegasus to track notorious drug baron Joaquín "El Chapo" Guzmán (Kaldani & Prokopets, 2019). In 2016, Pegasus came into the public light through a report from Citizen Lab (Marczak et al., 2018), detailing how the spyware and its operators use a mixture of social engineering attacks and the exploitation of vulnerabilities found in iOS and Android to infect mobile phones. The same report states that, by 2018, Pegasus was active in at least 45 countries but that figure is estimated to have increased by 2024. In 2023, Amnesty International published an article stating that the Indian government purchased Pegasus from the NSO Group and used the spyware to target prominent journalists in India (2023). The former Polish government also used Pegasus between 2017 and 2022, targeting more than 600 people in Poland (Ptak, 2024). By the end of 2019, WhatsApp Inc, now known as Meta Platforms, filled a lawsuit against NSO Group for violating their terms and condition by reverse-engineering and exploiting WhatsApp to be used as an installation vector to infect targeted devices with Pegasus. Apple soon followed by suing NSO Group in 2021 to hold it accountable for the surveillance and targeting of Apple users (Apple, 2021). The same year saw the NSO Group being blacklisted by the Biden administration to curb abuses in the global market for spyware, which has gone largely unregulated (Sanger et al., 2024). After a legal battle of nearly 4 years, Meta Platforms won its lawsuit against the NSO Group while Apple sought to voluntarily dismiss the case to focus on establishing additional technological countermeasures against spyware on their devices. Taking cognizance of the negative impacts of the Pegasus malware, it becomes essential to study it can be effectively detected and what countermeasures are available. Nevertheless, limited studies utilized a structured approach to review security measures against Pegasus spyware. To address this issue, this paper conducts a taxonomy review on the detection methods of Pegasus Malware and its countermeasures. It focuses on the evolution of Pegasus, how the NSO Group adapted the spyware over the years to infect devices, in addition to how to detect and counteract the spyware from infecting devices. Given the sophistication and stealthiness of Pegasus spyware, this study is intended to advance research in this area to better understand its implications and possible solutions. As such, the taxonomy developed contributes to the academic discourse by highlighting strengths and limitations.

2 Background

Pegasus is a sophisticated spyware developed by the Israeli company NSO Group and it was designed to covertly infiltrate mobile devices and extract sensitive data, including messages, calls, emails, and location details. The NSO Group has been using different attack vectors and techniques to infect devices with Pegasus over the years. The structure of the NSO Group Exploit Infrastructure to deliver, infect and maintain the spyware consists of various components. These include validation servers, a network of infected domains, sub-domains and DNS servers registered by the company, Pegasus installation servers, Command and Control (C&C) servers, as well as multiple cells of operators scattered around the world, each focusing on specific regions (Amnesty International, 2021a). The above servers that make up the exploit infrastructure are typically hosted by US-based hosting companies such as Digital Ocean, Linode and Amazon Web Services (AWS) (Amnesty International, 2021a). One of the attack methodologies makes use of social engineering, zero-day exploits and the exploit infrastructure where an operator sends a crafted SMS message containing a malicious link to the target. Two types of SMS-based remote installation vector are typically utilized: One-Click Vector and Zero-Click Vector. In the case of the One-Click vector, the target must click the link to access the infected website. The malicious website contains an exploit for the web browser on the target's device, and any other required exploits to implant the spyware (Marczak & Scott-Railton, 2016). On the other hand, when using the Zero-Click vector, the operator sends the same link via a special type of SMS message, like a WAP Push Service Loading (SL) message (Marczak & Scott-Railton, 2016). This causes the phone to automatically open a link in a web browser instance, eliminating the need for a user to click on the link to become infected (Marczak & Scott-Railton, 2016) (Fig. 1).

The first reported case was in 2016 with Ahmed Mansoor, a human rights defender based in the United Arab Emirates who received an SMS text message on his iPhone, promising "new secrets" about detainees tortured in UAE jails if he clicks on an

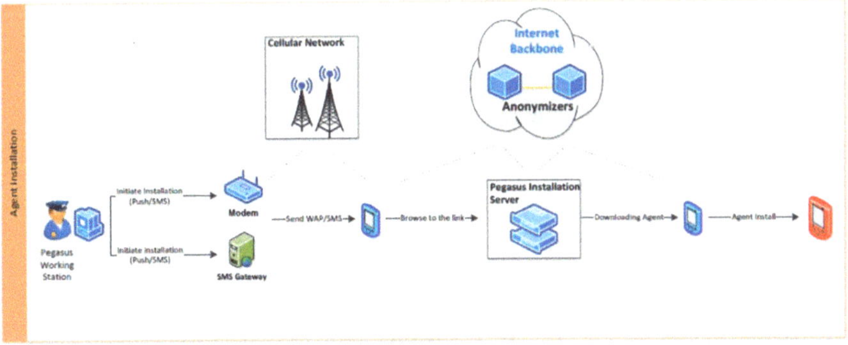

Fig. 1 Sequence through which Pegasus is installed on mobile devices via SMS (Marczak & Scott-Railton, 2016)

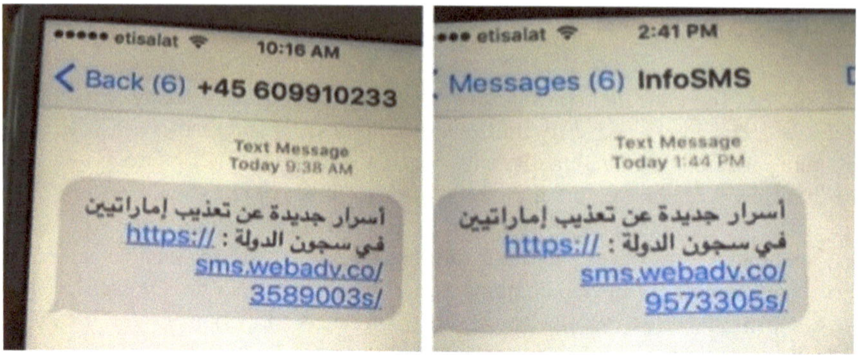

Fig. 2 SMS text messages received by Mansoor (Marczak & Scott-Railton, 2016)

included link (Marczak & Scott-Railton, 2016). In this case, the One-Click vector was used (Fig. 2).

Mansoor forwarded the message to Citizen Lab, whose analysis revealed that the link led to an exploit chain termed Trident (Rudie et al., 2021). The domain webadv.co acted as an anonymizer or a middleware between the device and a Pegasus installation server. As part of the process, the anonymizer forwarded any request to the exploit infrastructure which examines the target device's user-agent header to determine if Pegasus has an exploit chain (Marczak & Scott-Railton, 2016). Trident consisted of 3 zero-day vulnerabilities in Apple iOS back in 2016 namely CVE-2016-4655, which was a Kernel Address Space Layout Randomization (KASLR) bypass exploit to find the base address of the kernel; CVE-2016-4656, a 32 and 64 bit iOS kernel exploits that allow execution of code in the kernel, used to jailbreak the phone and allow software installation; and CVE-2016-4657, an exploit for Safari WebKit, which allows execution of the initial shellcode (Marczak & Scott-Railton, 2016). The first stage of the Trident Exploit made use of a memory corruption vulnerability in Safari Webkit (CVE-2016-4657) to download a JavaScript which in turn downloaded Stage 2 binaries. The second stage exploited the KASLR bypass vulnerability (CVE-2016-4655) and the kernel vulnerability (CVE-2016-4656) to disable code signing enforcement, allowing unsigned binaries to be run and jailbreak the iPhone. This stage also saw the device started communicating with a C&C server of the operator. Stage 3 downloaded and installed Pegasus on the mobile device. Once installed, the spyware disabled automatic updates, detected and removed other jailbreaks and re-ran the Trident exploits locally at each reboot (Marczak & Scott-Railton, 2016). To record messages and phone calls Pegasus made use of the Cydia Substrate, a third-party app developer framework that would hook various call status methods and send system-wide notifications when call events occurred. Data was then exfiltrated to the C&C server of the operator via HTTPS. A series of other zero-click or zero-day exploits in iOS were crafted by the NSO Group to infect devices with Pegasus throughout the years. KISMET exploit chain affected iMessage from 2019 to 2020 (Marczak et al., 2020).

In 2021, another exploit affecting iMessage called FORCEENTRY (CVE-2021-30860) has been used between February and September 2021 affecting Apple iOS, MacOS and WatchOS devices (Marczak et al., 2021). The BLASTPASS exploit chain (CVE-2023-41064 and CVE-2023-41061) was capable of compromising iPhones running the latest version of iOS (16.6) without any interaction from the victim in 2023 (BLASTPASS NSO Group iPhone zero-click, zero-day exploit captured in the wild. The Citizen Lab. https://citizenlab.ca/2023/09/blastpass-nso-group-iphone-zero-click-zero-day-exploit-captured-in-the-wild/). The NSO Group is believed to have design a spyware specifically for Android devices called Chrysaor. A report from Lookout Security and Google studied 2 samples and indicate that Pegasus and Chrysaor exhibits the same behaviors. No zero-day exploits have been used by Chrysaor but instead phishing attacks were employed to deliver a malicious Android application package (APK) that compromised the device to install its malicious payload (Blaich et al., 2017). Another attack vector that the NSO Group has made use of is exploiting server-side vulnerabilities of Meta's WhatsApp to distribute Pegasus. They created several exploits dubbed Heaven, Eden and Erised. In legal documents released as part of a lawsuit between Meta and NSO Group (WHATSAPP LLC & META PLATFORMS, Inc., 2024), prior to April 2018, NSO researched, developed, and tested potential installation vectors using WhatsApp by creating an internal environment replicating WhatsApp's servers and by "decompiling" the Official Client's code to understand how to circumvent the security measures built into it. Heaven was built to use NSO's own modified WhatsApp Installation Server (WIS) that was able to impersonate the Official Client to access WhatsApp's servers and send messages (WHATSAPP LLC & META PLATFORMS, Inc., 2024). Heaven had used manipulated messages to force WhatsApp's signaling servers to direct target devices to a third-party relay server controlled by NSO (WHATSAPP LLC & META PLATFORMS, Inc., 2024). After WhatsApp deployed a security fix for Heaven, NSO developed a new exploit called "Eden" to circumvent those security updates that used WhatsApp's own relay servers to send malicious messages to the target devices (WHATSAPP LLC & META PLATFORMS, Inc., 2024).

By May 2019, an estimated 1400 devices were infected by Pegasus using the Eden exploit. Even after WhatsApp detected and blocked the Eden exploit, NSO developed yet another installation vector known as Erised that also used WhatsApp servers to install Pegasus (WHATSAPP LLC & META PLATFORMS, Inc., 2024). In January 2020, NSO was suspected to use an unknown exploit on Apple servers to connect to the Pegasus Installation Server via iCloud. During 2022, 3 novel zero-day exploit chains were discovered and used by the NSO Group affecting iOS 15 and 16 named LATENTIMAGE, FINDMYPWN and PWNYOURHOME. Those exploits used vulnerabilities in Find My (fmfd), iMessage (MessageBlastdoorService), Homekit (homed) and launched the spyware via processes such as springboard and mediaserverd. Hence, detection and countermeasures studies of Pegasus become relevant to undertake since many vectors of attack exist.

3 Related Works

Although various papers have been published regarding Pegasus in the past, limited work has utilized a structured approach to review security measures against Pegasus spyware. The constant changes in the behaviors of Pegasus by the NSO Group to circumvent software patches and other technological defenses as described in the earlier sections, also make room for this study. A previous study has analyzed how Pegasus exploited the Trident vulnerabilities in Apple iOS back in 2016 (Rudie et al., 2021). The authors described in detail how NSO exploited vulnerabilities found in Apple devices to deliver Pegasus and how the vendor could have mitigated those vulnerabilities by placing stricter controls in their code. In another paper (Hussien et al., 2023), Tactics, Techniques, and Procedures (TTPs) were employed to detect the attack vectors of the NSO Group to infect devices with Pegasus. In addition, the paper also mentioned countermeasures to those TTPs, albeit some basic ones were discussed. Nevertheless, none of the related works have focused on conducting a taxonomy-based study on the detection of Pegasus, and more in-depth countermeasures, and as such, this study becomes relevant to undertake.

4 Methodology

This section describes the approach used to develop the taxonomy that categorizes approaches to detect Pegasus spyware. Due of a limited amount of past taxonomy research on Pegasus, multiple sources have been used to build up the detection taxonomy (Hussien et al., 2023; Rudie et al., 2021). Research databases used are IEEE Xplore and Google Scholar. Searches performed on Google using the key terms "NSO Group", "Pegasus", "zero-day", "Apple" and "Android" were performed to identified relevant sources. Reports from organizations such as Amnesty International and The Citizen Lab provided detailed insights into Pegasus's operational mechanisms and forensics markers (Amnesty International, 2021b; Marczak et al., 2018). Data from publicly available Threat Intelligence (TI) sources such as VirusTotal and MITRE ATT&CK framework were used to identify Indicator of Compromise (IoCs) related to Pegasus. The selected data sources were analyzed, and the detection techniques were categorized into the key categories, namely, Signature-based detection methods and Behavioral/Anomaly-based detection methods. Both categories have different detection methods that were classified and presented in a taxonomy which are discussed in the next section.

5 Taxonomy Review of Pegasus Detection Mechanisms

Using the methodology described in the previous section, the taxonomy is illustrated in Fig. 3.

In addition, The TTPs of Pegasus are shown in show the initial attack vectors and the behavior of the spyware (Fig. 4).

The two main detection techniques are discussed in the next sections.

5.1 *Signature-Based Detection Methods*

Signature-based detection relies on a database of known hashes, IP addresses, domain names, attack patterns and signatures to effectively mitigate threats. Security controls such as endpoint security solutions and intrusion detection/prevention systems make use of those databases to compare the actual system files' hashes and network flows against entries in the databases in order to detect and block digital threats. This approach is useful in detecting known components and versions of the Pegasus spyware.

5.1.1 Indicator of Compromise

Indicator of Compromise (IoC) is a piece of digital forensics that suggests that an endpoint or network may have been breached (MITRE ATT&CK, 2025). IoC generally means any type of threat intelligence that could indicate something out of the ordinary (Baker, 2022). IoCs are supplied by Threat Intelligence databases which are generated through investigations of prior Pegasus infection cases. Although reactive in nature, IoCs may help in detecting and thwarting Pegasus in the earlier stages of its attack lifecycle before the latter takes any action on the target. IoCs such as public IP addresses, command & control servers, domain names and hashes related to Pegasus are discussed further.

- *Payload Hashes*

A hash is a fixed-length string of characters calculated from a file and a hashing algorithm such as SHA 256 and acts as a digital fingerprint of a file. Antivirus software can identify threats by using databases containing hash values and compare them with any files that are written in the system or already present. This method provides an effective way to detect known versions of Pegasus and its malicious payloads. In a report from Sectrio (MITRE Corporation, 2020), a cybersecurity solutions company, a version of Pegasus for Android was caught in one of their honeypots. Searching the hash value from the report on VirusTotal indicates that out of 67 security vendors, 41 of them labeled the APK file and its payloads as malicious. This shows that hash-based signatures are still effective in the detection of the spyware.

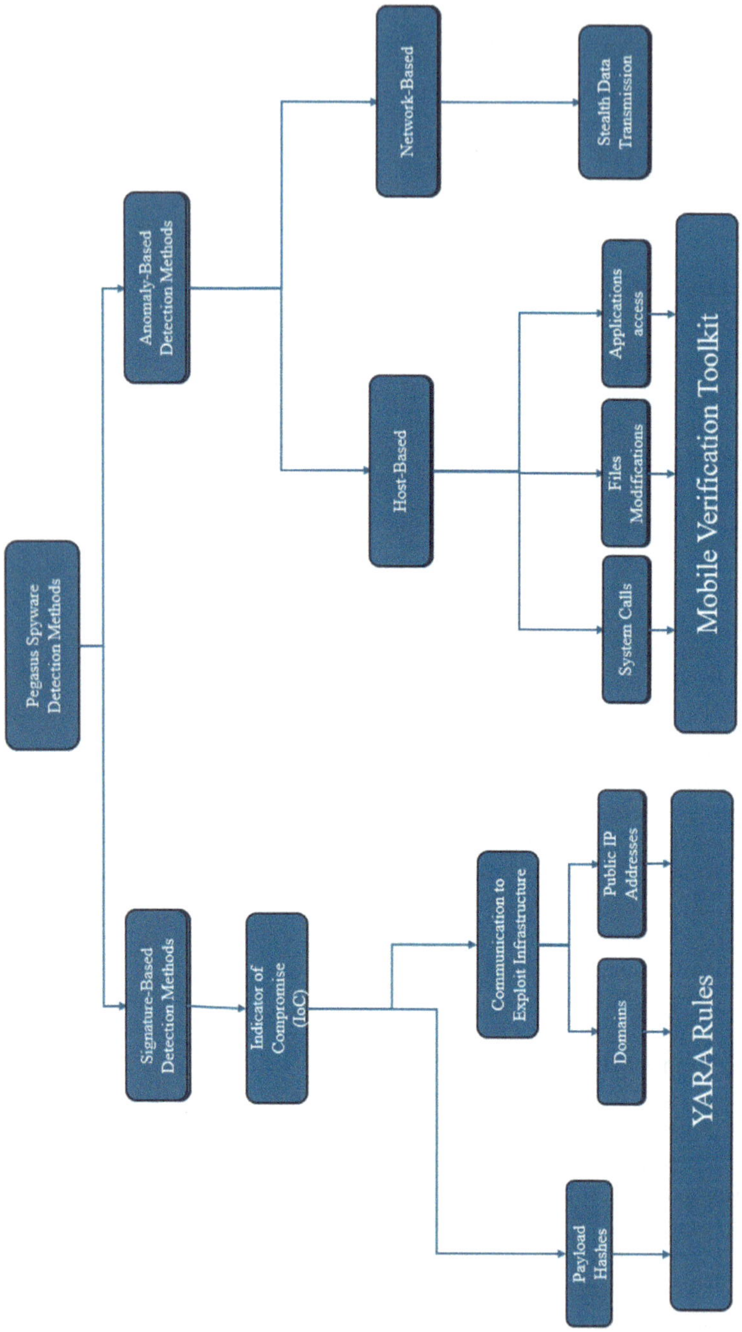

Fig. 3 Taxonomy diagram of Pegasus spyware detection techniques

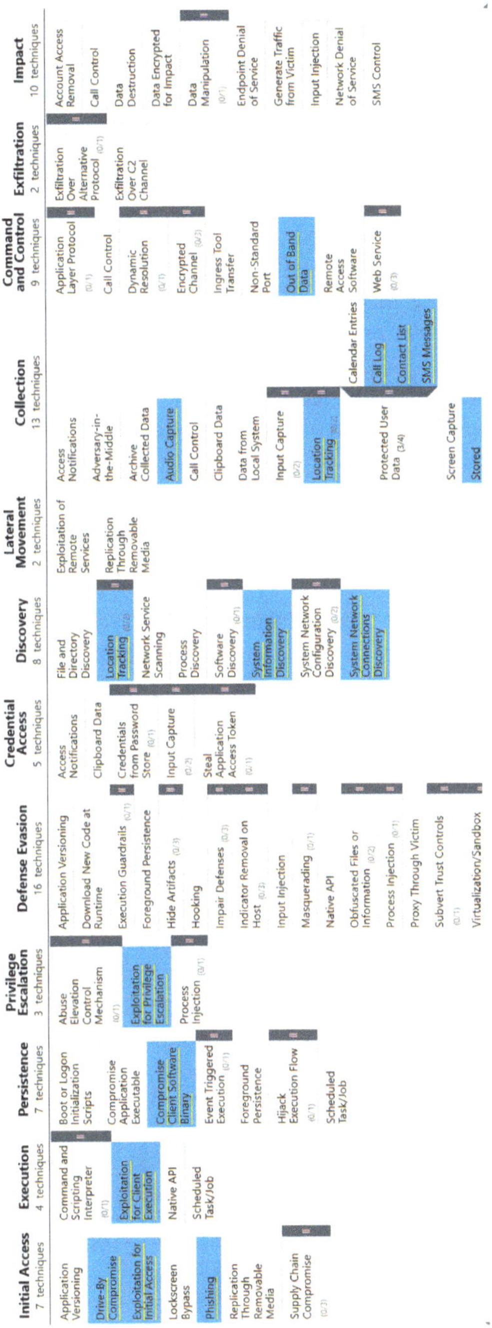

Fig. 4 TTPs of Pegasus for iOS and Android from MITRE ATT&CK (Amnesty International, 2021b)

- *Communication to Exploit Infrastructure*

As the NSO Group hosts the exploit infrastructure on the Internet, public IP addresses must be used to communicate with target devices. C2 servers are used by threat actors to maintain connectivity to infected devices. Once devices are infected, they beam back to the C2 servers, allowing the attacker to keep persistent control on the device. C2 servers are mostly used by botnets, instructing thousands or even millions of devices to send requests to a particular target at the same time. In the case of Pegasus, several operators used their C2 servers which was part of the exploit infrastructure to maintain connectivity on the infected devices and send instructions to Pegasus. C2 servers can be IP-based or Domain-based.

The Athena fingerprinting method was a novel technique used by researchers at The Citizen Lab to identify and cluster Pegasus infrastructure run by specific operators. Athena consisted of 3 fingerprints. Fingerprint ξ1 was a Transport Layer Security (TLS) fingerprint while fingerprints ξ2 and ξ3 represented two different proxying configurations they observed (Marczak et al., 2018). They considered a server to be part of NSO Group's infrastructure if it matched ξ1 and also one of ξ2 or ξ3 (Marczak et al., 2018). Through an Internet-wide scan, 1091 IP addresses were found to be part of the NSO's exploit infrastructure. Some IP Addresses that are linked to Pegasus were collected from various sources and analyzed as per the Table 1.

It was observed that most of the IP addresses that were previously linked to Pegasus have been recycled, meaning that they are no longer associated with NSO's exploit infrastructure and the ISP has reallocated it to other customers. VirusTotal shows that a high number of security vendors have databases with those IP addresses flagged as malicious. Modern mobile security solutions such as Antiviruses and Mobile Device Management (MDM) solutions can detect traffic to malicious IP addresses using an in-built host-based firewall. The firewall analyzes each packet header for its destination IP address and if it matches a known IP address found in its database, the traffic can be blocked. The MITRE organization developed a novel IPS system for Android devices called APE. APE has the ability to perform Deep Packet Inspection and examine IPv4 packets leaving and entering the device (Indicators of Compromise (IOCs), 2025). APE, like any other packet filtering software, can detect communications to Pegasus infrastructure if the IP addresses are flagged as malicious.

Table 1 Sample IP addresses linked to Pegasus

AS	*IP Address*	*Organisation*	*Current Analysis*	*Past Analysis*
14061	104.248.38.202	DIGITALOCEAN LLC	Clean(0/94)	Clean
14061	134.209.23.19	DIGITALOCEAN LLC	Malicious(1/94)	Clean
20473	140.82.55.55	VULTR	Clean(0/94)	Malicious(67/94)(2019)
200185	217.61.4.177	Aruba SAS	Suspicious(1/94)	Malicious(13/94) (2019)
20473	45.63.97.210	VULTR	Suspicious(1/94)	Malicious(6/94) (2019)
14061	206.189.22.137	DIGITALOCEAN LLC	Malicious(1/94)	Clean
Unknown	130.195.234.251	Unknown	Clean(0/94)	Malicious(38/44) (2025)

Table 2 Example of domain names and themes (Marczak & Scott-Railton, 2016)

Type	*Example*	*Impersonating*	*Status*	*Flaged*
News Media	alljazeera.co	Aljazeera (aljazeera.com)	OFFLINE	Malicious (12/94)
	bbc-africa.com	BBC (bbc.com/news/world/africa)	OFFLINE	Malicious (15/94)
	cnn-africa.co	CNN (edition.cnn.com/world/africa)	OFFLINE	Malicious (16/94)
	unonoticias.net	Las Ultimas Noticias	OFFLINE	Malicious (11/94)
	univision.click	Univision (univision.com)	OFFLINE	Malicious (14/94)
Shipment Tracking	track-your-fedex-package.org	FedEx (fedex.com/en-us/tracking.html)	OFFLINE	Malicious (15/94)
ISP / Telco	mz-vodacom.info	Vodacom (Mozambique) (vm.co.mz)	OFFLINE	Malicious (14/94)
	iusacell-movil. com.mx	Iusacell (Mexico)	OFFLINE	Malicious (12/94)
	sabafon.info	Sabafon (Yemen) (sabafon.com)	OFFLINE	Malicious (14/94)
	newtarrifs.net	Generic	OFFLINE	Malicious (15/94)
Popular Online Platforms	y0utube.com.mx	YouTube (youtube.com)	OFFLINE	Malicious (14/94)
	fb-accounts.com	FaceBook (facebook.com)	OFFLINE	Malicious (14/94)
	googleplay-store.com	Google (play.google.com)	OFFLINE	Malicious (15/94)
	whatsapp-app.com	WhatsApp (whatsapp.com)	OFFLINE	Malicious (16/94)
Account Info.(Generic)	accounts.mx	Unknown	ONLINE	Malicious (15/94)
	adjust-local-settings.com		OFFLINE	Malicious (13/94)
Government Portals	emiratesfoundation.net	The Emirates Foundation (emiratesairlinefoundation.org)	OFFLINE	Malicious (15/94)
	topcontactco.com	Teleperformance Visa Application Processing Portal for the UK (tpcontact.co.uk.)	OFFLINE	Malicious (16/94)
Humanitarian	icrcworld.com	International Committee of the	OFFLINE	Malicious (13/94)
	redcrossworld.com	Red Cross (icrc.org)	OFFLINE	Malicious (13/94)
Airlines	turkishairines.info	Generic (turkishairlines.com) Turkish Airlines	ONLINE	Malicious (16/94)
Pokemon	bulbazaur.com	The Pokemon Company	OFFLINE	Malicious (16/94)
	pickuchu.com	(pokemon.com)	OFFLINE	Malicious (14/94)

- *Malicious Domains and DNS Servers*

Advance Persistent Threats (ATPs) often register domains to host their malicious websites and launch phishing campaigns against their targets. In a previous study (Narahari, 2021), 45% of APTs use DNS protocol for DNS resolution and tunneling. Out of the 1091 IP addresses found during the Athena Fingerprinting, 1014 domains pointed to them. NSO registered those domain names to look like several common themes such as News Media websites, indicating the use of fake news articles to trick targets into clicking on spyware links (Marczak & Scott-Railton, 2016). The table below shows multiple registered domains trying to impersonate the legitimate entities (Table 2).

The table above also shows the status of the domains under the example column. Offline means that when performing nslookup for an A record on the domain names, no IP address is returned. An assumption can be made that NSO took those domains down after Pegasus became public. Two domains were online as one or more IP addresses were returned when performing an nslookup. Those IP addresses are likely to have been recycled but VirusTotal shows the domain account[.]mx still participates in malicious activities to this date. DNS filtering solutions can help in detecting DNS queries to those malicious domains.

5.1.2 YARA Rules

Yet Another Recursive Acronym (YARA) is an open-source, rule-based platform that uses instructions to detect and identify malware. From the above taxonomy, YARA can take the various signatures to integrate into security solutions that monitor

```
rule iOSPegasusDetected
{
    strings:
        $a01 = "/private/var/root/test.app/data"
        $a02 = "/private/var/root/test.app/d/"
        $a03 = "/private/var/root/test.app"
        $a04 = "/private/var/tmp/crw"
        $a05 = "/private/var/tmp/cr"
        $a06 = "/private/var/tmp/st_data/"

    condition:
        any of them
}
```

Fig. 5 Sample of YARA rule to detect iOS jailbreak by Pegasus

endpoints for better detection against threats. An example of YARA rule to detect the jailbreak of an iPhone by Pegasus can be written as below.

Figure 5 (Alageel & Maffeis, 2025) shows a YARA rule containing IOCs for the jailbreaks of iPhones when Pegasus modifies system files in the mentioned directories. This method can be quite effective in detecting exploits on iOS devices.

5.2 Anomaly-Based Detection Methods

Anomaly-based detection analyses behaviors and statistics of a process in normal situations, and it checks whether the system is being used in a different manner (Alageel & Maffeis, 2025). Anomaly-based detection can be host-based where the system will monitor files, system calls and applications that may be modified or accessed during initial infection on the device or it can be network-based where the system can detect high upload usage when the spyware exfiltrate data to its C2 infrastructure. These detection methods are discussed below.

5.2.1 Host-Based Detection

Host-based detection monitors the device where an agent is installed. The agent can be an Endpoint Protection with Endpoint Detection and Response (EDR) capabilities which can raise an alert when it detects malware and anomalies. EDR can monitor the device such as system calls, files modifications and applications which are discussed in the sub-sections below.

- *System Calls*

System calls are used to directly invoke native functionalities of the kernel (Actis, 2025). Investigation from Lookout showed that Pegasus make use of system calls on Android devices when attempting to take a screenshot using the native take_screen_shot binary which makes use of the input/output control system call ioctl as well as libpng (Blaich et al., 2017). In a previous study (Tchakounté & Dayang, 2013), FireDroid was mentioned that it provides a detection mechanism for interposition of system calls generated by the applications and enforcing security policies to control their execution by utilizing the ptrace() system call for tracing executions of applications.

- *Files Modification*

As seen in the report by Amnesty International and CitizenLabs (Amnesty International, 2021b; Marczak et al., 2020), several exploits in iOS, iMessage and Facetime were used by NSO to deliver Pegasus. When exploitation and jailbreak take place, system files in several repositories are added and modified. For example, system log files revealed that Pegasus added binaries located inside the folder /private/var/db/com.apple.xpc.roleaccountd.staging/ (Amnesty International, 2021b). Another file, com.apple.CrashReporter.plist was written in /private/var/root/Library/Preferences/ and had obtained root privileges when an exploit involving Safari favicon.db was accessed. As such, monitoring modification of files is another mechanism to detect Pegasus.

- *Application Access*

Pegasus accesses applications installed on the phone such as WhatsApp, Viber, Telegram and even components such as Camera and Microphone by process injection and function hooking. Process injection is a method of executing arbitrary code in the address space of a separate live process to evade process-based defenses or even elevate privileges (Russello et al., 2013). On the other hand, function hooking is a technique in computer programming that allows the interception and modification of function calls, messages, or events within software applications (Process Injection, 2025). In a report by Lookout (Twingate, 2024), Pegasus intercepts real-time calls from the chat messengers from the cynject open-source library that is injected into their process space dynamically at run time. It also makes use of libwacalls and dylib that hooks key WhatsApp functions to intercept calls and chats (Twingate, 2024). Hence, monitoring application accesses is another way to detect Pegasus.

- *Mobile Verification Toolkit*

The Mobile Verification Toolkit or MVT is a modular tool designed by researchers at Amnesty International. This tool acquires and analyses data from Android devices and records from iOS backups and filesystem dumps to identify potential traces of Pegasus (Amnesty International, 2021b). As per the MVT documentation, Bazaliy et al., (2015) two methodologies exist for iOS; Jailbreak and obtain a full filesystem dump or read from a generated an iTunes backup file. Records are extracted from the filesystem dump and compared with a list of IOCs in STIX2 or JSON format. MVT for Android devices presents a limited capabilities as Android devices store little diagnostic information useful to triage potential compromises (Bazaliy et al.,

2015). MVT for Android instead downloads and inspect installed APKs and potentially identify any which might be malicious in nature (Bazaliy et al., 2015). It can also extract some diagnostics information using the Android Debug Bridge (adb) including dumpsys results, details on installed packages (without download), running processes, presence of root binaries and packages, and more (Bazaliy et al., 2015). A list of SMS messages can also be extracted to inspect evidence of phishing attacks via SMS.

5.2.2 Network-Based Detection Methods

Network-based detection method relies on the monitoring from collection of different metrics such as flow, packet, byte counts from the network and try to isolate anomalies (Mobile Verification Toolkit. https://docs.mvt.re/en/latest/).

- *Stealth Data Transmissions*

Operators can issue a command to Pegasus to exfiltrate data. In a previous report (Marczak et al., 2020), researchers observed 270.16 MB of upload to 3 public IP addresses to the domain bananakick[.]com on an infected device for a period of 16 h. Basic network firewalls can issue an alert when upload bandwidth crosses a certain threshold to certain targets. This technique can prove quite effective to detect upload and download to unknown or malicious domains.

6 Countermeasures Against Pegasus Spyware

To mitigate the risk of devices being infected with Pegasus spyware, both individuals and major technology corporations that have previously been exploited by NSO Group as attack vectors must take proactive steps to strengthen security. It is essential for end users to remain vigilant and in parallel, technology companies should enhance software security, monitor for potential anomalies, and also collaborate with cybersecurity experts to detect and mitigate threats in real time. As part of a holistic defense strategy, the following countermeasures should be considered.

- *Installing Antivirus software*

Mobile security has often fallen short in terms of endpoint protection such that more personal and sensitive information is stored on a mobile phone than on personal computers. Even if Android and iOS are built-in with security features, another layer of protection found in third-party security providers can be installed. Modern antivirus software comes with in-built firewalls with web and DNS filtering capabilities that integrate with web browsers and that thwart phishing links and communications with malicious domains linked to the spyware. It can also detect malicious payloads that try to download itself on the device.

- *Use VPN with DNS Filtering*

This method involves installing a VPN client on the mobile device, creating a tunnel between the device and the VPN server where all internet traffic is sent to. Some VPN providers offer DNS filtering where every DNS queries that the device is requesting goes through the filter where malicious domains and C2 servers can be blocked.

- *User Awareness*

Awareness plays a vital role in training individuals in identifying digital threats. A study (Nyalkalkar et al., 2011) has shown that after an awareness session was conducted, there were fewer individuals who had fell victim to the phishing links. Knowledge of Pegasus and how to identify its vector of attack, particularly SMS Phishing, can help in less individuals clicking the malicious links.

- *DNS Sinkhole*

DNS sinkholes can be used to intercept DNS queries, comparing them to a dynamic list of known malicious domains and redirecting the response to a non-routable IP address or sinkhole and blocking the traffic from reaching the domain (Aloul, 2012). The Turkish CERT (USOM) used DNS Sinkhole to block domain names associated with Pegasus and NSO in 2019, effectively blocking all devices infected with Pegasus in Turkey from communicating with the exploit infrastructure.

- *Enable Lockdown Mode on Apple devices*

Lockdown mode was introduced on iOS 16, iPadOS 16, watchOS 10 and macOS Ventura and is a feature that provides extreme protection by reducing the attack surface that potentially could be exploited by highly targeted mercenary spyware (Learn DNS Sinkhole, 2025). Enabling lockdown mode puts certain restrictions on the device and limits some applications that were previously exploited by Pegasus such as iMessage, FaceTime and the Home app. For instance, as noted in a previous study (Learn DNS Sinkhole, 2025), most message attachment types are restricted, and certain functionalities such as hyperlinks and link previews are disabled. This measure helps prevent SMS phishing attacks, where malicious actors previously distributed phishing links via text messages. Moreover, some complex web technologies are blocked, mitigating exploits like those used by Pegasus, which exploited vulnerabilities in the Safari WebKit. Furthermore, incoming FaceTime calls and Home app invitations are blocked unless the sender was previously authorized by the user. The vulnerabilities related to FaceTime and the Home app were discussed in the Background section.

- *Software and Application Updates*

Software updates are crucial to be performed as those updates provide bug fixes, enhancements and most importantly, patch vulnerabilities that can be exploited. The first known vulnerability from iOS and iPadOS in 2025 is tracked as CVE-2025–24,200 and was a zero-day vulnerability that has been exploited in the wild. Even if previous cases of Pegasus were able to exploit devices which were running on the latest iOS, software updates are an important layer of protection against threats.

- *Legal action against NSO Group*

Even if Meta won its 5-year lawsuit against NSO for using WhatsApp servers as a vector for infection (Apple Inc, 2024), the spyware is still in possession of hundreds if not thousands of NSO's clients. Even if NSO goes down, the spyware will still be maintained by malicious users and will upgrade its functionalities.

7 Conclusion

Mobile devices continue to increase and be more sophisticated than ever with new software introducing new functionality. But with new and existing software come vulnerabilities which threat actors are exploiting in the wilds. Pegasus is often considered as one among the most sophisticated spywares ever created as it is stealthy in its exploits, infection and data exfiltration and was continuously being upgraded to counter software patches and defenses setup by its targets such as WhatsApp and Apple. With privacy on the line, this study presented a taxonomy on the techniques that can be potentially used to detect Pegasus spyware on iOS and Android from its initial delivery, exploitation, installation, Command and Control and Execution stages. This methodology provides the Indicators of Compromise and how to use IOCs to detect the spyware in 2 categories such as signature-based and anomaly-based detection techniques. The countermeasure section also provided methods that can be used to mitigate infection of the spyware. As future works, those techniques can be further investigated as the spyware will no doubt be updated to counteract and exploit current defenses. A more practical analysis can be conducted to analyze effectiveness of the techniques, but in the first instance, a taxonomic analysis is needed, as presented in this paper.

References

Actis, B. (2025). Filtered YARA rules Pegasus.yar. https://github.com/vulnerabivoro/filtered-yara-rules/blob/master/Pegasus.yar. Accessed 17 February 2025.

Alageel, A., & Maffeis, S. (2025). Investigation of advanced persistent threats network-based tactics, techniques and procedures, p. 1.

Aloul, F. A. (2012). The need for effective information security awareness. *Journal of Advances in Information Technology, 3*(3), 176–183.

Amnesty International. (2021a). Forensic methodology report: how to catch NSO Group's Pegasus. Amnesty International.

Amnesty International. (2021b). Forensic methodology report: How to catch NSO Group's Pegasus. Amnesty International.

Apple. (2021). Apple sues NSO Group to curb the abuse of state-sponsored spyware. Apple. https://www.apple.com/newsroom/2021/11/apple-sues-nso-group-to-curb-the-abuse-of-state-sponsored-spyware/. Accessed 1 January 2025.

Apple Inc. (2024). About lockdown mode. Apple Inc. https://support.apple.com/en-us/105120. Accessed 17 February 2025.

Awad, N. F., & Fitzgerald, K. (2005). The deceptive behaviors that offend US most about spyware. *Communications of the ACM, 48*(8), 55–60.

Baker, K. (2022). Indicators of compromise (IOC) security. Crowdstrike. https://www.crowdstrike.com/en-us/cybersecurity-101/threat-intelligence/indicators-of-compromise-ioc/. Accessed 04 February 2025.

Bazaliy, M., Flossman, M., Blaich, A., Hardy, S., Edwards, K., Murray, M., Olson, C., Hebeisen, C., Ford, P., Streicher, C., Nickle, R., Roark, J., Marczak, B., & Railton, J. S. (2015). Technical analysis of Pegasus spyware. Lookout.

Blaich, A., Bauer, A., Flossman, M., Richards, J., Hebeisen, C., Edwards, K., Olson, C., Murray, M., Kingsley, D., & Edwards, S. (2017). Pegasus for Android. *Technical Analysis and Findings of Chrysaor*.

BLASTPASS NSO Group iPhone zero-click, zero-day exploit captured in the wild. The Citizen Lab. https://citizenlab.ca/2023/09/blastpass-nso-group-iphone-zero-click-zero-day-exploit-captured-in-the-wild/. Accessed 12 February 2025.

Hussien, O., Butt, U., & Sulaiman, R. B. (2023). Critical analysis and countermeasures tactics, techniques and procedures (TTPs) that targeting civilians: A case study. On Pegasus.

India. (2023). Damning new forensic investigation reveals repeated use of Pegasus spyware to target high-profile journalists. Amnesty International. https://www.amnesty.org/en/latest/news/2023/12/india-damning-new-forensic-investigation-reveals-repeated-use-of-pegasus-spyware-to-target-high-profile-journalists/. Accessed 22 December 2024.

Indicators of Compromise (IOCs). Rapid7. https://www.rapid7.com/fundamentals/indicators-of-compromise-iocs/. Accessed 09 February 2025.

Jr, J. P. M. (2019). What is spyware? How it works and how to prevent it. CSO Online. https://www.csoonline.com/article/567071/what-is-spyware-how-it-works-and-how-to-prevent-it.html. Accessed 09 December 2024.

Kaldani, T., & Prokopets, Z. (2019). Pegasus spyware and its impacts on human rights.

Kaspersky. (2018). What is spyware? Definition. https://www.kaspersky.com/resource-center/threats/spyware. Accessed 09 December 2024.

Kaspersky. (2024). Attacks on mobile devices significantly increase in 2023. Kaspersky. https://www.kaspersky.com/about/press-releases/attacks-on-mobile-devices-significantly-increase-in-2023. Accessed 14 December 2024.

Learn DNS Sinkhole. (2025). Catchpoint. https://www.catchpoint.com/network-admin-guide/dns-sinkhole#:~:text=A%20DNS%20sinkhole%20is%20used,list%20of%20known%20malicious%20domains. Accessed 17 February 2025.

Marczak, B., & Scott-Railton, J. (2016). The million dollar dissident: NSO Group's iPhone zero days used against a UAE human rights defender. Citizen Lab.

Marczak, B., Scott-Railton, J., McKune, S., Razzak, B. A., & Deibert, R. (2018). HIDE AND SEEK: Tracking NSO Group's Pegasus spyware to operations in 45 countries. The Citizen Lab. https://citizenlab.ca/2018/09/hide-and-seek-tracking-nso-groups-pegasus-spyware-to-operations-in-45-countries/. Accessed 10 December 2024.

Marczak, B., Scott-Railton, J., Aljizawi, N., Anstis, S., & Deibert, R. (2020). The Great iPwn: Journalists hacked with suspected NSO Group iMessage 'zero-click' exploit. Citizen Lab. https://citizenlab.ca/2020/12/the-great-ipwn-journalists-hacked-with-suspected-nso-group-imessage-zero-click-exploit/. Accessed 12 February 2025.

Marczak, B., Scott-Railton, J., Razzak, B. A., Aljizawi, N., Anstis, N., Berdan, K., & Deibert, R. (2021). FORCEDENTRY NSO Group iMessage zero-click exploit captured in the wild. The Citizen Lab. https://citizenlab.ca/2021/09/forcedentry-nso-group-imessage-zero-click-exploit-captured-in-the-wild/. Accessed 12 February 2025.

MITRE Corporation. (2020). APE: Intrusion protection system for android devices. MITRE Corporation. https://www.mitre.org/our-impact/intellectual-property/ape-intrusion-protection-system-android-devices. Accessed 16 February 2025.

MITRE ATT&CK. (2025). https://mitre-attack.github.io/attack-naviga-tor//#layerURL=https%3A%2F%2Fattack.mitre.org%2Fsoftware%2FS0289%2FS0289-mobile-layer.json. Accessed 16 February 2025.

Mobile Verification Toolkit. https://docs.mvt.re/en/latest/.

Narahari, K. (2021). Pegasus spyware threat report. Sectrio.

NSO GROUP. (2019). Cyber intelligence for global security and stability. NSO Group. https://www.nsogroup.com/. Accessed 10 December 2024.

Nyalkalkar, K., Sinhay, S., Bailey, M., & Jahanian, F. (2011). A comparative study of two network-based anomaly detection methods. In *Proceedings of the 2011 IEEE INFOCOM*, Shanghai.

Process Injection. (2025). MITRE ATT&CK. https://attack.mitre.org/techniques/T1631/#:~:text=Both%20Android%20and%20iOS%20have,access%20or%20exploiting%20a%20vulnerability. Accessed 17 February 2025.

Ptak, A. (2024). Almost 600 people targeted with Pegasus spyware under former Polish government. Notes from Poland. https://notesfrompoland.com/2024/04/16/almost-600-people-targeted-with-pegasus-spyware-under-former-polish-government/. Accessed 22 December 2024.

Rudie, J., Katz, Z., Kuhbander, S., & Bhunia, S. (2021). Technical analysis of the NSO Group's Pegasus spyware. In *International conference on computational science and computational intelligence (CSCI)*, Las Vegas, NV.

Russello, G., Jimenez, A. B., Naderi, H., & van der Mark, W. (2013). Poster: FireDroid: Hardening security in android.

Sanger, D. E., Perlroth, N., Swanson, A., & Bergman, R. (2024). U.S. blacklists Israeli firm NSO Group over spyware. The New York Times. https://www.nytimes.com/2021/11/03/business/nso-group-spyware-blacklist.html. Accessed 23 December 2024.

Tchakounté, F., & Dayang, P. (2013). System calls analysis of malwares on android. *International Journal of Science and Technology, 2*(9), 669–674.

Twingate. (2024). What is function hooking? How it works and examples. Twingate. https://www.twingate.com/blog/glossary/function%20hooking. Accessed 17 February 2025.

WHATSAPP LLC & META PLATFORMS, INC. (2024). *A Delaware corporation, Plaintiffs, v. NSO GROUP TECHNOLOGIES LIMITED and Q CYBER TECHNOLOGIES LIMITED, Defendants.*

Optimization and Stabilization of Power Distribution Grids Through the Use of Sensor Networks

C. Cress and J. P. van Deventer

Abstract As global electricity demand continues to rise, ensuring stable and efficient power distribution is of increasing importance. The integration of renewable energy sources and fluctuating consumption patterns present challenges to grid stability. This systematic review explores the role of sensor networks (SNs), artificial intelligence (AI), and smart grid technologies in optimizing and stabilizing modern power grids. Following the PRISMA methodology, 28 relevant studies were analyzed to evaluate current applications, future benefits, and existing challenges in grid management. Findings indicate that advanced monitoring, predictive analytics, and real-time data collection significantly enhance operational efficiency, fault detection, and energy management. By highlighting key trends and technological advancements, this review serves as a valuable resource for researchers and professionals interested in intelligent grid solutions, AI-driven optimization, and the transformative potential of SNs in power distribution.

1 Introduction

Power grids worldwide are increasingly strained as electricity demand begins to outpace supply, leading to widespread issues such as rolling blackouts, load reduction, load shaving, and load shedding, thereby disrupting homes and industries. As electricity demand grows, integrating renewable energy sources poses major challenges to maintaining stable and efficient power distribution. Grid optimization becomes both more complex and more critical to avoid disruptions affecting homes and industries.

A promising approach involves sensor networks (SN), particularly synchrophasor networks, which deliver synchronized, real-time measurements of voltage, current, and phase angles across grid locations (Gharavi & Hu, 2017). Modern grids, with their complexity and renewable variability, require advanced monitoring and control systems to maintaining stable and efficient power distribution.

C. Cress · J. P. van Deventer (✉)
Department of Informatics, University of Pretoria, Pretoria, South Africa
e-mail: phil.vandeventer@up.ac.za

J. Marx Gómez et al. (eds.), *Recent Advances in Information Systems*, Progress in IS,
https://doi.org/10.1007/978-3-032-14099-9_17

Accurate load forecasting supports grid balance and energy market operations (Dewangan et al., 2023), while SNs' role would be in enhancing fault detection, minimizing outages, and enabling real-time data collection (Swain et al., 2022). Advances in sensor technologies may improve grid protection and optimization, boost energy efficiency, and lower operational costs.

Routine outages in countries for example China, and major failures in the U.S., highlights the urgent need for improved grid management amid rising energy costs and aging infrastructure (Pearre & Swan, 2020). Herein lies one of the potential benefits of SNs, where such networks can assist in the optimization of grid stability through monitoring and real-time feedback. These networks consist of interconnected devices that collect and transmit real-time data on key grid metrics like voltage, current, and power quality.

SNs, as integrated with other smart grid technologies, enable advanced monitoring, improved fault detection, and dynamic power flow optimization, reducing the risk of grid failures (Abbasi et al., 2023). Since the early 2000s, the use of SNs has evolved significantly. SNs have been used to enhance distribution efficiency, reduce energy losses, and improve reliability. By analyzing real-time data, operators can better manage supply–demand imbalances, the root cause of many outages (Chen et al., 2023). Maintaining constant voltage, particularly with renewable energy source integration, is considered a key challenge (Lucia et al., 2018). SNs can help address this through real-time disturbance detection and correction, thereby improving grid resilience (Moreno et al., 2021). Beyond stability, SNs also reduce costs and boost energy efficiency. This supports the central research question: "***How do SNs contribute to the optimization and stabilization of power distribution grids?***" This systematic literature review (SLR) explores existing and emerging technologies for grid management, with a focus on stability and efficiency. It emphasizes the need for ongoing innovation to fully realize the potential of SNs in power distribution optimization.

2 Systematic Reviews and Meta-Analyses (PRISMA) Approach

2.1 Data Collection and Extraction

To gather data relevant to the core constructs of this SLR, targeted search strings, as may be referenced in Table 1, were used to identify peer-reviewed articles on SNs and power distribution optimization in addressing grid stability challenges. Targeted database sources included six major databases namely (1) Emerald Insight, (2) IEEE Xplore, (3) ProQuest, (4) ScienceDirect, (5) Scopus, and (6) Web of Science. Google Scholar was also used to search across these databases by means of institutional access and the institutional application programming interface (API).

Table 1 Search strings for data retrieval during exploratory identification

Search terms	Approx. results
Google Scholar: (sensor network) AND (optimization) AND (power distribution) AND (grid stability)	164,000
Target Databases: "sensor network" AND "optimization" AND "power distribution" AND "grid stability"	352

Table 2 Filtering criteria

Inclusion criteria	Exclusion criteria
• Research papers published in English	• Research papers published in a language other than English
• Primary studies including peer-reviewed journal articles	• Duplicate papers were systematically identified and removed
• Research papers that correlate to the main research question	• Research papers that failed to adequately address the research question
• Empirical articles that are in peer-reviewed journals or a thesis	• Gray literatures were excluded
• Papers in print copy	• Conference papers excluded
• Research papers accessible by means of institutional access	• Research papers where the full-text is unavailable by means of institutional access
• Research papers ranging from the year 2017–2024 were included for review	• Papers published before the year 2017

All terms directly aligned with the research question, and the volume of results required no additional terms or variations. To ensure relevance, papers were filtered using specific inclusion and exclusion criteria as may be referenced in Table 2.

Of the 352 records retrieved, 63 duplicates were removed. The remaining 289 records were screened using inclusion and exclusion criteria (Table 2), yielding 60 relevant papers, 48 of which had full-text access. After manually reviewing titles and abstracts, 28 papers were selected for the final review, analysis, and discussion (Fig. 1).

2.2 Data Extraction

A data extraction table was created for all eligible full-text articles, capturing details such as full reference, summary, and key findings. The summary column offered a quick overview of each article's content and relevance to the research question, helping to organize them into thematic categories. This process ensured alignment with the review's focus on SNs, optimization, and grid stability.

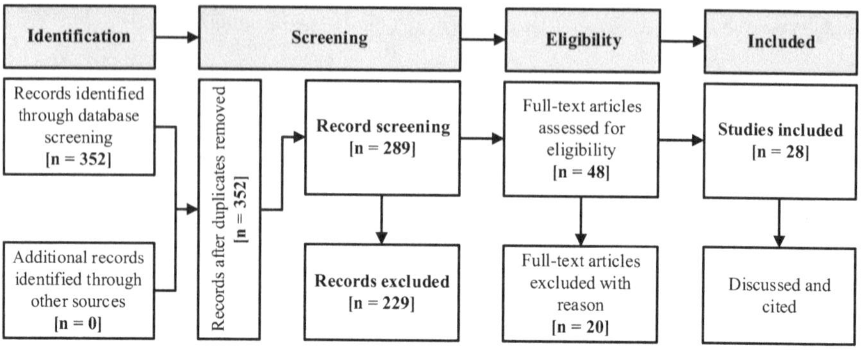

Fig. 1 PRISMA flow diagram

2.3 *Data Analysis*

A thematic analysis was conducted to identify recurring themes related to the research question. Relevant text segments were coded and grouped into broader themes, which were then refined for accuracy. Two main themes emerged namely, optimization strategies and grid stability enhancements. These reflect how SNs improve power distribution, aligning with the review's objective.

3 Findings

Following the PRISMA process, 28 empirical articles from 2017 to 2024 were included in this review, with 19 studies published between 2022 and 2024. The limited number of earlier studies suggests growing interest in addressing grid instability, as frequent outages affect both residential and industrial areas. The articles were grouped into five key themes namely artificial intelligence (AI) and machine learning (ML), grid stability and smart grid technologies, cybersecurity and data protection, energy efficiency and management, and finally optimization and system performance. Each theme includes sub-themes, for example, "AI-Driven Process Automation" and "AI for Predictive Analytics", with recurring patterns noted. AI, ML, and grid stability emerged as the most common themes, reflecting a strong focus on using intelligent systems to enhance grid performance. Although cybersecurity and optimization were less prominent, they are expected to gain importance as grid modernization progresses, requiring secure and efficient energy solutions.

4 Discussion

The conceptual model (Fig. 2) as derived from the SLR's thematic analysis, outlines a structured framework for enhancing grid stability and efficiency through the integration of SNs, AI, advanced storage, smart grid technologies, and renewable energy. At its foundation, SNs enable real-time monitoring, fault detection, and predictive maintenance by collecting key grid performance data.

This data feeds into AI and machine learning algorithms, the system's intelligence layer, which enable dynamic adjustments and automated responses, optimizing operations, reducing downtime, and improving responsiveness. The model also highlights the role of smart grids and advanced storage in managing demand fluctuations and integrating renewable sources.

Smart grids allow for real-time monitoring, decentralized control, and utility-consumer interaction, with technologies like demand response and peak shaving enhancing adaptability and resilience. Battery energy storage systems (BESS) address the intermittent nature of renewables, providing a buffer to balance supply and demand. This supports consistent power availability, reduces reliance on non-renewables, and fosters sustainability.

Lastly, the framework underscores the importance of cybersecurity and data protection as a foundational layer, ensuring data integrity, preventing tampering, and safeguarding consumer privacy in increasingly interconnected energy systems.

Although framed as future research, security is essential to the reliability of power grids, especially given the vulnerabilities linked to real-time data collection and transmission. Implementing cryptographic protocols and privacy-preserving schemes is crucial to protect infrastructure and maintain stakeholder trust.

Cybersecurity and blockchain are increasingly emphasized in emerging literature, highlighting the need to address data security risks for the successful deployment of advanced grid technologies. This conceptual model extends beyond existing theories by integrating layered technological solutions and offering practical guidance for improving real-world grid performance. It underscores the interdependence of SNs, AI, smart grids, storage systems, and cybersecurity. These future research directions are vital for strengthening the resilience, efficiency, and sustainability of modern power distribution systems. In the following sections we will consider the components that make up the overall conceptual network.

4.1 Artificial Intelligence and Machine Learning

As may be seen in Fig. 2, to understand the impact of SNs on grid optimization and stability, automated processes and advanced control techniques, particularly those using AI and ML, are employed to improve power distribution performance. These tools not only improve efficiency but also allow time for system upgrades.

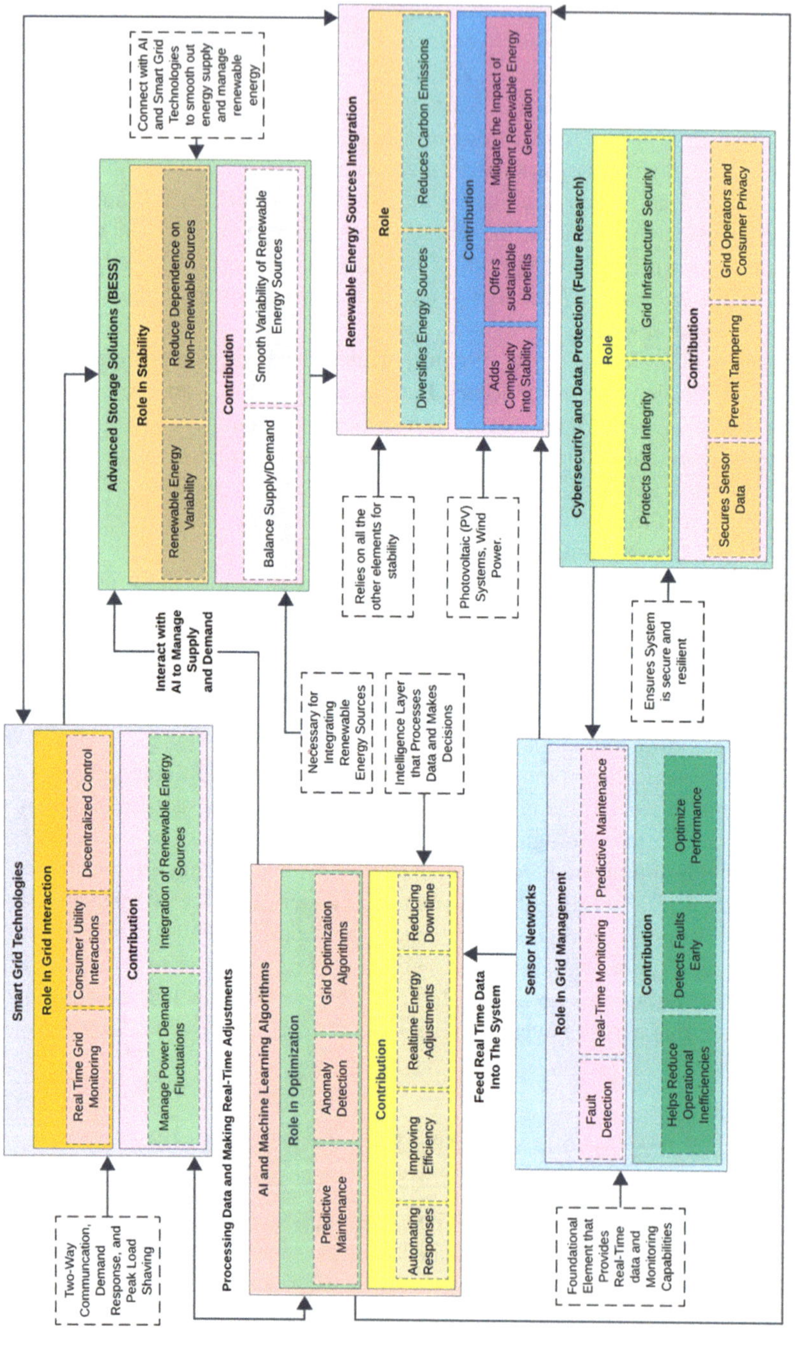

Fig. 2 Conceptual model

AI, especially through tuning algorithms like the V-Tiger for proportional-integral-derivative (PID) controllers in automatic voltage regulators (AVRs), improves voltage stability and reduces outage risks (Gopi et al., 2024).

AI-driven control methods boost system reliability and operational efficiency (Rangarajan et al., 2023; Szczepaniuk & Szczepaniuk, 2023). AI supports decision-making by optimizing operational parameters, streamlining complex processes, and reducing risks from manual operations (Mourtzis & Angelopoulos, 2023; Swain et al., 2022). Through predictive maintenance and load forecasting, AI minimizes power loss and improves grid performance. For instance, it was found that enhanced particle swarm optimization (PSO) algorithm significantly reduced energy losses (Mourtzis & Angelopoulos, 2023).

Despite integration costs and complexity, AI's role in automation is growing. The organic distribution system (ODiS) initiative (Loeser et al., 2022) demonstrates self-managing grids with capabilities like self-healing and self-optimization, reducing human intervention and increasing resilience. AI also strengthens networked micro-grids (NMGs), improving stability and disaster recovery through adaptive and robust controls (Bordbari & Nasiri, 2024). In next generation smart grids (NGSGs), AI-driven data analytics reduce labor demands and resolve operational challenges (Ahsan et al., 2023; Hoffmann et al., 2020).

AI's application extends to cloud-based electric vehicle (EV) charging within the Internet of Vehicles (IoV), where real-time data enhances charging efficiency and grid stability (Rimal et al., 2022). Overall, AI and SNs work in synergy to optimize power distribution, reduce inefficiencies, and prepare grids for future energy demands.

4.2 *Grid Stability and Smart Grid Technologies*

NMGs integrate distributed energy sources and can operate independently or within larger networks, offering flexibility during disruptions like natural disasters (Aghmadi et al., 2023). As may be seen in Fig. 2, innovations such as inductive power transfer (IPT) for wireless EV charging enhance grid resilience. Continued improvements in IPT support sustainable transport and the stable integration of EVs into the grid (Bozhi et al., 2023).

The transition from traditional to smart grids is crucial. Smart grids enable real-time monitoring, two-way utility-consumer communication, and effective renewable energy integration critical to grid stability (Chao-Chung et al., 2023). Technologies like NMGs and IPT offer significant resilience benefits, though implementation can be complex and costly. Long-term cost savings and improved reliability, however, make these investments worthwhile.

AI further supports renewable integration by improving fault detection, load forecasting, and energy management. AI handles large datasets effectively, aiding secure and efficient grid operation (Szczepaniuk & Szczepaniuk, 2023). Smart grids enhance flexibility and reliability through real-time monitoring, demand response, and optimized electricity use (Hafiz Abdul et al., 2023). Virtual power plants (VPPs) mark a major advancement by aggregating distributed energy resources to optimize production, storage, and distribution. VPPs improve grid flexibility and stability while supporting renewable integration (Liu et al., 2023).

The combined use of AI, smart grids, and VPPs strengthens energy management and sustainability. However, addressing data security and system scalability is essential to ensure long-term resilience and efficient renewable energy integration.

4.3 Cybersecurity and Data Protection

Synchrophasor networks use synchronized measurements for grid protection and fault detection (Fig. 2). Integrating blockchain enhances data integrity and security, ensuring all measurements and communications are authenticated and tamper-proof (Gharavi & Hu, 2017). These networks use space–time classification algorithms to improve outage management and data verification.

However, the Internet of Everything (IoE) in smart grids raises privacy concerns. Advanced metering infrastructure (AMI) benefits from cryptographic methods to safeguard consumer data (Desai et al., 2019). Smart meters, key to IoE infrastructure, provide accurate data for load forecasting and demand management but must be protected against unauthorized access and manipulation.

Advanced cybersecurity technologies are crucial for safeguarding grid infrastructure. Still, the implementation of cybersecurity technologies must consider energy consumption to align with sustainability goals (Mazhar et al., 2023).

Capacitive voltage dividers, used in high voltage networks, enable continuous monitoring and rapid fault detection (Hrbac et al., 2020). Securing these sensors is essential to ensure reliable diagnostics and accurate measurements during the transition to smart grids (Bhattacharya et al., 2022).

Incentive mechanisms in smart grids support sustainable energy use by leveraging secure IoE data to balance supply and demand and encourage environmentally friendly practices (Bhattacharya et al., 2022).

Energy management strategies, particularly those integrating advanced storage technologies, enhance grid efficiency and sustainability. Peak load shaving and demand response (DR) are effective methods to reduce demand fluctuations, lower costs, and cut emissions (Bhagya Nathali et al., 2020). These strategies shift energy use during peak times and improve overall grid responsiveness.

Integrating photovoltaic (PV) systems poses challenges due to variable output, which can be mitigated by battery energy storage systems (BESS). It should be noted that reliable storage stabilizes grids with high renewable penetration potentially reducing the carbon footprint by enhancing energy efficiency (Loeser et al., 2022; Olowu et al., 2018). Lithium-ion batteries, in particular, show strong potential for grid integration, reducing transmission costs, and buffering renewable volatility (Sushita et al., 2022). However, high costs and technical hurdles remain barriers to widespread adoption. The value of renewable microgrids in meeting energy needs and lowering emissions has been shown to be effective in Bangladesh, while also identifying infrastructure and policy challenges. Advanced control techniques and resilient grid design help address these barriers (Mojumder et al., 2022).

Effective scheduling also plays a role. Research on age-optimal scheduling (AOS) (Chen et al., 2021) shows that minimizing the Age of Information (AoI) improves data transmission and system performance, which is important to reduce latency and enhancing grid stability. Although implementing these strategies can be resource intensive, combining advanced storage, demand response, and optimized scheduling offers significant improvements in power distribution system stability, efficiency, and reliability.

4.4 Optimization and System Performance

Research on optimizing network operations and performance metrics shows that various methods can reduce inefficiencies and enhance system performance (Fig. 2).

A study on AOS in multiple access channels proposed two policies namely probabilistic random access (PRA) and drift-plus-penalty (DPP) (Chen et al., 2021). The DPP policy showed superior performance, especially under low multi-packet reception conditions, by minimizing both average and peak AoI. This highlights how precise scheduling can improve communication efficiency and reduce latency in power grids.

This supports broader findings that timely data updates enhance grid stability. Integrating predictive maintenance with sensors and machine learning in medium voltage switchgear enables continuous monitoring, helping operators address issues early, reduce downtime, and extend equipment lifespan (Hoffmann et al., 2020).

Accurate load forecasting using smart meter data can for example maintain supply–demand balance. Smart meters improve data collection and support more accurate models for efficient grid operation (Rabie et al., 2021). Adopting these optimization techniques, predictive maintenance, precise scheduling, and reliable forecasting, improves grid management, reduces costs, and enhances service quality.

Performance metrics help evaluate system efficiency, while innovations like VPPs, enhance flexibility by aggregating distributed energy resources (DERs) and participating in energy markets (Liu et al., 2023). VPPs support grid stability, reduce costs, and strengthen customer–utility relationships (Liu et al., 2023).

Incentive mechanisms using AI/ML, game theory, and blockchain can optimize distribution, reduce emissions, and promote renewable energy (Bhattacharya et al., 2022). Ensuring privacy and data security is also crucial. AMI systems require new privacy-preserving methods tailored to consumer protection needs (Desai et al., 2019).

Overall, optimization strategies, including scheduling, forecasting, predictive maintenance, VPPs, and secure smart grid systems, collectively enhance the efficiency, reliability, and resilience of modern power distribution.

5 Conclusion

This SLR examined the role of SNs in optimizing power distribution, analyzing 28 articles focused on their integration with AI, ML, and smart grid technologies. These technologies collectively enhance grid stability, efficiency, reliability, and enable key optimization techniques.

As may be seen in the conceptual model presented in Fig. 2, combining SNs with AI and ML allows for real-time monitoring, fault detection, and predictive analytics, improving grid performance, energy management, and operational cost efficiency, while supporting renewable energy integration.

Optimizing energy distribution, network balancing, power load and renewable power generation has to potential to not only protect a national power grid, but also reduce the carbon footprint by reducing the need for additional power generation capacity. Not only does a national power grid remain more stable, thereby protecting grid integrity, the need for additional generation capacity to stabilize a power grid may be reduced. In both instances, this would yield improved power distribution performance.

Despite their potential, challenges such as implementation complexity and cost remain, requiring further research. The findings highlight how these technologies can address global power instability issues, like rolling blackouts and rising energy costs, in countries such as India, China, and the Unites States. By strengthening resilience and balancing supply and demand, SNs offer a promising solution to mitigate power outages.

6 Recommended Future Research

Future research should prioritize developing scalable, cost-effective methods for integrating SNs into power distribution grids, with emphasis on refining AI-driven algorithms for real-time optimization. Given the global variability of power challenges, especially in developing regions prone to blackouts, studies should assess the applicability of these technologies across diverse contexts. Standardized, long-term research can offer insights into their lasting impact on grid performance and sustainability.

AI's ability to process large datasets and make rapid, autonomous adjustments is a key advantage for improving system performance and reducing downtime. Future work should explore AI's role in real-time anomaly detection and dynamic grid operations to boost resilience and efficiency.

Further investigation into emerging technologies such as blockchain, IoE, and edge computing when integrated with SNs might be required (Desai et al., 2019). Blockchain, in particular, shows promise for enhancing grid security, transparency, and peer-to-peer energy trading, while safeguarding consumer data (Aklilu & Ding,

2022). However, concerns about blockchain's energy consumption must be addressed to align with sustainability goals (Mohammed & Alam, 2023).

Current optimization models also require re-evaluation to reflect changing energy consumption patterns and the rise of renewables. Focusing future research on these areas can significantly enhance the reliability, efficiency, and sustainability of power distribution systems, preparing them to meet the demands of the evolving energy landscape. Finally, additional research could address practical deployment challenges, cost considerations, and a detailed environmental impact assessment.

References

Abbasi, M., Abbasi, E., Li, L., Aguilera, R. P., Lu, D., & Wang, F. (2023). Review on the microgrid concept, structures, components, communication systems, and control methods. *Energies, 16*, 484.

Aghmadi, A., Hussein, H., Ketulkumar, H. P., & Mohammed, O. (2023). A comprehensive review of architecture, communication, and cybersecurity in networked microgrid systems. *Inventions, 8*, 84.

Ahsan, F., et al. (2023). Data-driven next-generation smart grid towards sustainable energy evolution: Techniques and technology review. *Protection Control Modelling Power System, 8*, 43.

Aklilu, Y. T., & Ding, J. (2022). Survey on blockchain for smart grid management, control, and operation. *Energies, 15*, 193.

Bhagya Nathali, S., Khan, M., & Han, K. (2020). Futuristic sustainable energy management in smart environments: A review of peak load shaving and demand response strategies, challenges, and opportunities. *Sustainability, 12*, 5561.

Bhattacharya, S., et al. (2022). Incentive mechanisms for smart grid: State of the art, challenges, open issues, future directions. *Big Data Cognition Computers, 6*, 47.

Bordbari, M. J., & Nasiri, F. (2024). Networked microgrids: A review on configuration, operation, and control strategies. *Energies, 17*, 715.

Bozhi, M., et al. (2023). A review of wireless pavement system based on the inductive power transfer in electric vehicles. *Sustainability, 15*, 14893.

Chao-Chung, H., Bi-Hai, J., & Chun-Cheng, L. (2023). A survey on recent applications of artificial intelligence and optimization for smart grids in smart manufacturing. *Energies, 16*, 7660.

Chen, Z., Nian, X., & Meng, Q. (2023). Distributed optimization of multi-integrator agent systems with mixed neighbor interactions. *Automatica, 157*, 69.

Chen, Z., Pappas, N., Björnson, E., & Larsson, E. G. (2021). Optimizing information freshness in a multiple access channel with heterogeneous devices. *IEEE Open Journal of Communication Society, 2*, 456–470.

Desai, S., et al. (2019). A survey of privacy preserving schemes in IoE enabled smart grid advanced metering infrastructure. *Cluster Computing, 22*, 43–69.

Dewangan, F., Abdelaziz, A. Y., & Biswal, M. (2023). Load forecasting models in smart grid using smart meter information: A review. *Energies, 16*, 1404.

Gharavi, H., & Hu, B. (2017). Synchrophasor SNs for grid communication and protection. *Proceedings of the IEEE, 105*, 1408–1428.

Gopi, P., Reddy, S. V., Bajaj, M., Zaitsev, I., & Prokop, L. (2024). Performance and robustness analysis of V-Tiger PID controller for automatic voltage regulator. *Science of Reports, 14*, 7867.

Hafiz Abdul, M., Rehan, L., Mohsin, J., & Asharf Ali, K. (2023). A state-of-the-art review of smart energy systems and their management in a smart grid environment. *Energies, 16*, 472.

Hoffmann, M. W., et al. (2020). Integration of novel sensors and machine learning for predictive maintenance in medium voltage switchgear to enable the energy and mobility revolutions. *Sensors, 20*, 2099.

Hrbac, R., et al. (2020). A development of a capacitive voltage divider for high voltage measurement as part of a combined current and voltage sensor. *Elektronics Elektrotechnology, 26*, 25–31.

Liu, J., Hu, H., Yu, S. S., & Trinh, H. (2023). Virtual power plant with renewable energy sources and energy storage systems for sustainable power grid: Formation, control techniques and demand response. *Energies, 16*, 3705.

Loeser, I., Braun, M., Gruhl, C., Menke, J. H., Sick, B., & Tomforde, S. (2022). The vision of self-management in cognitive organic power distribution systems. *Energies, 15*, 881.

Lucia, W., Gheitasi, K., Bagherzadeh, M. (2018). A low computationally demanding model predictive control strategy for robust transient stability in smart grid. In *Proceedings of the 57th IEEE conference decision and control (CDC)*, pp. 6013–6018.

Mazhar, T., et al. (2023). Analysis of challenges and solutions of IoT in smart grids using AI and machine learning techniques: A review. *Electronics, 12*, 242.

Mohammed, K. M., & Alam, S. (2023). Blockchain-based microgrid for safe and reliable power generation and distribution: A case study of Saudi Arabia. *Energies, 16*, 5963.

Mojumder, M. R. H., Hasanuzzaman, M., & Cuce, E. (2022). Prospects and challenges of renewable energy-based microgrid system in Bangladesh: A comprehensive review. *Clean Technology Environment Policy, 24*, 1987–2009.

Moreno, J. J., Morales, O., Padilla, R. T., Lina, I., & Quintana, H. (2021). A comprehensive review on smart grids: Challenges and opportunities. *Sensors, 21*, 6978.

Mourtzis, D., & Angelopoulos, J. (2023). Reactive power optimization based on the application of an improved particle swarm optimization algorithm. *Machines, 11*, 724.

Olowu, T. O., et al. (2018). Future challenges and mitigation methods for high photovoltaic penetration: A survey. *Energies, 11*, 1782.

Pearre, N., & Swan, L. (2020). Reimagining renewable electricity grid management with dispatchable generation to stabilize energy storage. *Energy, 203*, 117917.

Rabie, A. H., Saleh, A. I., & Ali, H. A. (2021). Smart electrical grids based on cloud, IoT, and big data technologies: State of the art. *Journal of Ambient Intelligence Humanized Computers, 12*, 9449–9480.

Rangarajan, S. S., Shiva, C. K., Sudhakar, A. V. V., Subramaniam, U., Collins, E. R., & Senjyu, T. (2023). Avant-garde solar plants with artificial intelligence and moonlighting capabilities as smart inverters in a smart grid. *Energies, 16*, 1112.

Rimal, B. P., Kong, C., Poudel, B., Wang, Y., & Shahi, P. (2022). Smart electric vehicle charging in the era of internet of vehicles, emerging trends, and open issues. *Energies, 15*, 1908.

Sushita, K., et al. (2022). Impacts of residential energy storage system modeling on power system. *Sustainable Environment, 8*, 182.

Swain, A., Elmouatamid, A., Mousa, A., & Pong, P. W. T. (2022). Sensor technologies for transmission and distribution systems: A review of the latest developments. *Energies, 15*, 7339.

Szczepaniuk, H., & Szczepaniuk, E. K. (2023). Applications of artificial intelligence algorithms in the energy sector. *Energies, 16*, 347.

Examining the Prevalence of Social Loafing in Agile Methodologies

L. Combrinck and J. P. van Deventer

Abstract In Agile environments, social loafing, a tendency for team members to reduce effort due to shared responsibility, can undermine even the most promising projects. If left unchecked, it often leads to failure. By understanding social loafing's causes and applying tailored mitigation strategies, teams can curb this behavior and improve outcomes. The success of these interventions depends on each team's unique dynamics, making it essential to identify the most relevant contributing factors. This review highlights key drivers of social loafing and offers practical strategies to address them, aiming to boost team performance in Agile settings. By conducting a systematic literature review (SLR), key factors influencing social loafing, such as team size, cohesion, accountability, and task visibility, are identified and analyzed. The findings reveal that targeted interventions, including peer assessment, structured accountability systems, and enhanced communication, can significantly reduce the impact of social loafing and improve Agile team performance. By understanding these dynamics, organizations can refine their Agile practices to foster more engaged and productive teams, ensuring sustained project success. The paper concludes with recommendations for future research, which includes, but isn't limited to, testing social loafing mitigation strategies in a real-world environment.

1 Introduction

In business, Agile methodologies are valued for its flexibility and efficiency. Yet, beneath its strengths lies a hidden risk, namely social loafing. Agile depends heavily on team cohesion and is often described as a people-centered approach (Whitworth & Biddle, 2007). However, it can be undermined by individuals who

---

L. Combrinck · J. P. van Deventer (✉)
Department of Informatics, University of Pretoria, Pretoria, South Africa
e-mail: phil.vandeventer@up.ac.za

J. Marx Gómez et al. (eds.), *Recent Advances in Information Systems*, Progress in IS,
https://doi.org/10.1007/978-3-032-14099-9_18

reduce their effort, relying on the anonymity of the group. Such a group-centric reduction of effort is more commonly referred to as social loafing (McAvoy & Butler, 2020).

As team members begin to "slack off," assuming others will pick up the slack, project performance suffers (McAvoy & Butler, 2009). While teamwork is highly regarded in modern organizations, its success hinges on mutual accountability (Tosuntaş, 2020). When individuals fail to take responsibility, and others don't compensate, project failure becomes a real risk. As may be noted, "a good process will not save the project from failure if the team does not have strong players" (McAvoy & Butler, 2020).

Agile's core values emphasize collaboration, making its people the most critical factor in project success (Whitworth & Biddle, 2007). Yet, many teams lack a clear understanding of how deeply social loafing can affect performance. This issue arises from collective evaluation practices that obscure individual contributions, potentially leading to poor project outcomes or even project collapse (McAvoy & Butler, 2009).

This systematic literature review (SLR) explores whether the benefits of Agile are overshadowed by the prevalence of social loafing by focusing on the following research question, namely, "***What are the key factors that leads to Social Loafing in Agile teams, and how can the prevalence of Social Loafing in Agile be mitigated?***".

Themes from the reviewed literature highlight factors that influence social loafing, including team size, accountability, cohesion, task visibility, project scope, and individual motivation. Strategies like peer assessment, support systems, and expectation setting have been proposed to mitigate these effects.

This paper critically examines the role and impact of social loafing in Agile teams, offering insight into Agile's real-world effectiveness. The SLRs structure includes a background, methodology (data sources, search terms, selection criteria, quality assessment, and analysis), followed by findings, discussion, conclusion, and future research directions.

2 Background and Motivation for Study

Agile methodologies have gained significant traction across fields that demand high levels of collaboration, especially in information systems development (ISD). Valued for its flexibility and iterative delivery of value, Agile relies heavily on teamwork and shared responsibility. However, this reliance exposes a critical vulnerability, namely social loafing.

Social loafing occurs when individuals in a group contribute less than expected, forcing others to compensate. This not only reduces productivity but can derail entire projects, particularly in Agile, where progress is built incrementally. Despite its many benefits, Agile's team-centric nature creates conditions where individuals may hide behind collective performance while receiving equal recognition.

Agile's social orientation demands strong individual engagement, yet team evaluation often overlook personal contributions. This anonymity shields underperformers, increases pressure on others, and makes identifying loafers difficult. The phenomenon varies depending on team size, project scope, culture, and structure, and is influenced by Agile-specific features like self-organization, pair programming, and continuous refactoring.

Common causes of social loafing include poor training, lack of accountability, uneven workload distribution, and insufficient motivation. While project failures are often blamed on poor planning or requirement gathering, the hidden role of social loafing is frequently overlooked. In reality, it's often the people, not the process, that determine a project's outcome.

This SLR addresses a gap in understanding by examining the prevalence, causes, and impact of social loafing in Agile teams. Through analysis of peer-reviewed studies, it offers insights to help organizations recognize risks, refine Agile practices, and enhance team performance for greater project success.

3 Research Method

For the purpose of this study, the authors made use of an SLR. An SLR gathers and analyzes scientific evidence to answer a specific research question in a consistent and reliable manner (Lame, 2019). Its goal is to assess the credibility and validity of existing studies by drawing on all relevant literature. At the same time, it contributes new insights, identifies gaps, and reduces bias by uncovering key themes across sources (Nightingale, 2009).

3.1 Data Sources and Search Terms

To gather relevant data for the core constructs of this SLR, targeted search strings were used to identify peer-reviewed articles on Agile methodologies and social loafing. Explicitly targeted and designated database sources included six major databases namely (1) Sage Journals; (2) ScienceDirect; (3) Taylor and Francis (Journals), (4) ProQuest; (5) Wiley Online Library; and (6) IEEE Xplore. The following are the search terms used to find the relevant papers in each database, namely:

(1) (Social Loafing) AND (Agile) AND (Prevalence)
(2) (Social Loafing) AND (Agile Methodology) AND ((Prevalence) OR (Frequency))
(3) ((Social Loafing) OR (Free Riding)) AND (Agile Methodology) AND ((Prevalence)

3.2 Selection Criteria

The criteria referenced in Table 1 were used to ascertain the inclusion or exclusion of research papers retrieved by the search terms from each of the databases mentioned. The criteria were implemented to ensure that only relevant research papers were considered within the SLR.

3.3 PRISMA Flowchart

PRISMA, short for "Preferred Reporting Items for Systematic Reviews and Meta-Analyses," provides a structured framework to ensure quality and transparency in SLRs. It includes checklists and flowcharts to guide researchers through the review process, particularly when using the SLR method (Pati & Lorusso, 2018).

For this study, the designated databases were searched using the term: "((Social Loafing) or (Free Riding)) and (Agile Methodology) and ((Prevalence) or (Frequency))". This yielded 465 results. After limiting the time frame to the past 20 years, 402 articles remained. Removing 9 duplicates left 393 papers for initial screening.

Titles were reviewed based on predefined inclusion and exclusion criteria, narrowing the selection to 91 papers. After evaluating abstracts, introductions, and conclusions, 56 research articles were excluded for lacking relevance to the core research concern namely, social loafing in Agile methodologies. This left 35 papers for full evaluation, forming the foundation for the insights presented in this SLR.

Table 1 Filtering criteria

Inclusion criteria	Exclusion criteria
• Studies that specifically investigate social loafing behaviors within any type of Agile method	• Studies that are not related to social loafing behaviors in an Agile methodology
• Research articles published in peer-reviewed journal articles	• Studies that are non-peer-reviewed journal articles
• Research papers written in English	• Studies that focus solely on traditional project management without any reference to any Agile methodology
• Studies conducted from and organizational setting	• Studies that are not written in English
• Studies that make use of quantitative data on the prevalence of social loafing	• Studies that are older than 20 years
• Studies published within a 2004-to-2024-year timeframe (20 years)	• Studies that are inaccessible through institutional libraries
• Studies that are accessible through institutional access	• Research papers with small sample sizes and/or limited generalization
• Studies that examine the relationship between team characteristics in Agile methodologies and social loafing tendencies	• Duplicate research papers from another database • Studies that are only focused on individual behaviors instead of team dynamics

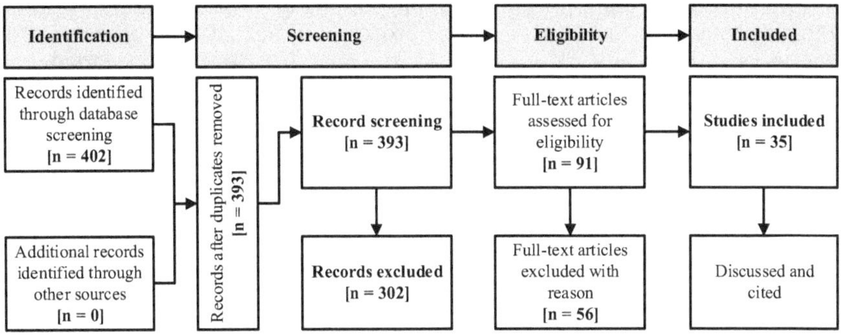

Fig. 1 PRISMA flowchart

The PRISMA flowchart in Fig. 1 visually outlines this selection process.

3.4 Data Extraction

Data extraction followed the PRISMA guidelines and employed a thematic analysis (Neuendorf, 2018). After removing duplicates and out-of-scope papers, each article was screened for key themes. All selected articles were thoroughly reviewed to identify main themes, assess inclusion and exclusion criteria, and capture relevant keywords and dimensions. The elements were grouped into clear categories, supporting a focused and structured discussion for this review.

3.5 Data Analysis

The thematic analysis employed used numeric quantifiers (counts) and quantitative data to identify patterns, from which meaningful themes emerge (Clarke & Braun, 2017). This paper employed closed coding to group data efficiently within the topic area. Through this approach, researchers gain deeper insight into key themes, enabling clearer and more accurate presentation of findings.

4 Findings

4.1 Agile Methodology

Agile methodology has surged in popularity due to its flexibility, making it well-suited to an evolving business and tech environments (Balijepally, 2005). This adaptability has cemented Agile as a lasting approach (Balijepally, 2005). Unlike traditional methods, Agile prioritizes people over processes (McAvoy & Butler, 2009), as captured in the quote: "A good process will not save a project from failure if the team does not have strong players" (McAvoy & Butler, 2009, 2020).

Extreme programming (XP) (Blau et al., 2013), a well-known Agile method, exemplifies this people-focused approach, emphasizing the importance of social interaction in team success (McAvoy & Butler, 2020; Whitworth & Biddle, 2007). While Agile is praised for boosting productivity and enjoys growing commercial adoption (Conboy et al., 2011), it is not without challenges, particularly the "people issue" of social loafing (McAvoy & Butler, 2009).

4.2 Social Loafing

Social loafing is the tendency for individuals to reduce their effort when working in groups compared to working alone (Ying et al., 2014). It involves contributing less effort relative to others performing the same task (Tosuntaş, 2020; Ying et al., 2014). Withholding effort often result in poor team performance when pursuing shared goals (Baker & Thompson, 2014). This type of behavior stems from the belief that one's contribution is either unimportant or will be compensated for by others (Blau et al., 2013; McAvoy & Butler, 2009). Social loafing is widely recognized as a major threat to group effectiveness and a leading challenge in team-based projects (Aggarwal & O'Brien, 2008). This paper explores its prevalence within Agile teams by examining its causes and potential mitigation strategies. Drawing on 35 peer-reviewed journal and conference papers, the review identifies key themes and categorizes them to show how different factors contribute to and help address social loafing.

4.3 Social Loafing Themes and Codes

This paper reviews the key factors contributing to social loafing in Agile teams, along with the most frequently cited strategies to mitigate them. The most commonly reported cause is reduced task visibility, accounting for 20.2% (23 mentions), followed by large team size at 17.5% (20 mentions), and poor group cohesion at 15.8% (18 mentions). Perceived importance appears in 13.2% (15 mentions), while poor communication and task complexity are tied at 8.8% (10 mentions each). Poor leadership follows at 6.1% (7 mentions), with justice and social norms both cited in 3.5% (4 mentions). Isolation is the least cited cause, at 2.6% (3 mentions).

To address these issues, accountability emerges as the most prominent mitigation strategy, cited 21.7% (25 mentions). Increasing team cohesion follows at 17.4% (20 mentions), then reducing team size at 13.9% (16 mentions). Social influence is noted in 10.4% (12 mentions), while increasing communication and reward systems are cited in 7.8% (9 mentions) and 7.0% (8 mentions), respectively. Improving leadership appears in 6.1% (7 mentions), with task clarity and norm compliance tied at 5.2% (6 mentions each). Ability pairing accounts for 4.3% (5 mentions), and mutual conduct is least cited at 0.9% (1 mention).

Each causative factor is paired with a relevant mitigation strategy, for example, task visibility with accountability, and group cohesion with cohesion-enhancing efforts. These pairings demonstrate how targeted interventions can directly counteract the effects of social loafing.

While personal factors like age, gender, and culture may also influence social loafing (Fronza & Wang, 2021; Suleiman & Watson, 2008), they are excluded from this review as they cannot be directly mitigated within team processes. The focus remains on actionable factors and strategies to enhance Agile team effectiveness.

5 Discussion

The SLR identified key factors contributing to social loafing in Agile teams. This discussion outlines the main causes and their corresponding mitigation strategies, ranked from most to least prevalent.

5.1 *Task Visibility and Accountability Strategy*

The most notable finding was the strong link between task visibility and social loafing in Agile teams. When individual contributions are unclear or unnoticed, social loafing increases (He, 2012; زناتی, 2023). Task visibility refers to the extent to which individual efforts can be identified and recognized within a team (Davies, 2009). When contributions are indistinguishable, members may feel less accountable and more likely to reduce their effort (Alnuaimi et al., 2010; Baker & Thompson, 2014; Liden et al., 2004). Self-evaluation helps individuals monitor their input, while group evaluation focuses on collective output, often masking individual slackers (Tosuntaş, 2020).

To reduce social loafing, implementing accountability systems is essential. These systems make individual contributions visible and encourage ownership of work (Padalkar et al., 2020; Tosuntaş, 2020). Accountability can take the form of performance monitoring, regular reviews, or systems that track individual input (Fronza & Wang, 2017; He, 2012; McAvoy & Butler, 2009). Peer assessment is a widely used method, allowing team members to evaluate one another, increasing perceived accountability and reducing social loafing by up to 18.9% (Aggarwal & O'Brien, 2008; Brooks & Ammons, 2003; Davies, 2009; Lam, 2015).

Daily Agile meetings, such as standups or sprints, further support accountability by showcasing individual achievements and reinforcing participation (Conboy et al., 2011; Kim, 2007; Whitworth & Biddle, 2007). These practices create social pressure and help ensure consistent contribution from all team members.

5.2 Poor Group Cohesion and Increasing Group Cohesion Strategy

Poor group cohesion is the second most common cause of social loafing in Agile teams. It refers to a lack of unity, trust, and emotional connection among team members, which can lead individuals to withhold effort (Baker et al., 2011; Fronza & Wang, 2017; Liden et al., 2004). Without trust and transparency, members may feel excluded or sidelined, contributing less to group efforts (Hall & Buzwell, 2013; Schippers, 2014; Tenenberg, 2019). This exclusion, sometimes driven by personality clashes or mismatched collaboration styles, is often a result of group dissonance (Berhanu, 2023).

Group formation plays a crucial role in cohesion. Self-selected groups, where members choose teammates, often based on existing relationships, tend to be more cohesive and less prone to social loafing (Aggarwal & O'Brien, 2008). In contrast, instructor-assigned groups, where members are unfamiliar with one another, are more likely to struggle with cohesion and experience higher levels of loafing (Harding, 2018; Lam, 2015).

To strengthen group cohesion, the "Flocking" method is recommended, where individuals with similar motivation levels form their own groups, reducing loafing through shared interest and commitment (Harding, 2018). Another strategy is the team expectation agreement (TEA) approach, a set of mutually agreed-upon rules outlining team goals and consequences for non-compliance (Fronza & Wang, 2017, 2021). TEAs act as social contracts, encouraging consistent attendance, active contribution, and collaboration, especially effective in Agile software teams (Balijepally, 2005).

5.3 Large Team Size and Decreased Team Size Strategy

The third most common cause of social loafing is large team size. In bigger teams, individual responsibility becomes diluted, leading to reduced effort and increased loafing (Alnuaimi et al., 2010; زنانی, 2023). This is largely due to increased anonymity, which allows members to contribute less while still receiving equal rewards (Liden et al., 2004). Research shows a 9.3% rise in social loafing in larger teams due to this anonymity, a phenomenon known as the "Hide in a Crowd" effect (Aggarwal & O'Brien, 2008; Baker & Thompson, 2014).

The Ringlemann effect further supports this, showing an inverse relationship between team size and individual contribution. For example, in a rope-pulling experiment, groups of seven exerted less effort per person than groups of two (Davies, 2009; Suleiman & Watson, 2008). Another study with 110 participants tasked with reconfiguring the word "T–O–N" found that groups of three produced an average of 13.86 words, while groups of eight produced only 6.75, clearly illustrating the impact of team size on performance (North et al., 2000).

To counteract this, reducing team size can help eliminate anonymity and improve accountability (Baker et al., 2011; Berhanu, 2023; Padalkar et al., 2020; Pieterse & Thompson, 2010). However, team size should align with task complexity. Some suggest teams should be no larger than five members (Pieterse & Thompson, 2010), while others recommend three to four as the ideal size (Davies, 2009).

5.4 *Perceived Importance and Social Influence Strategy*

The fourth most common cause of social loafing in Agile teams is the perceived importance of a task. When individuals find a task uninteresting or lack motivation, they are less likely to give full effort (Fronza & Wang, 2017). Motivation often depends on whether individuals believe their efforts will lead to meaningful outcomes. If personal goals don't align with team goals, social loafing becomes more likely (Kim, 2007; Schippers, 2014). Additionally, team morale and individual attitudes play a role where, for example, negative attitudes can make contributions feel less essential, increasing the risk of loafing (Balijepally, 2005; He, 2012).

To address this, strategies that enhance social influence and task engagement are critical. Assigning meaningful, interesting work can boost motivation and reduce loafing (McAvoy & Butler, 2009). Task commitment is also crucial. Individuals with a positive attitude toward their work are less likely to slack off (Tosuntaş, 2020). Research shows that job commitment significantly reduces social loafing, with a positive correlation of $r = 0.4$ ($p < 0.001$) (He, 2012).

5.5 *Task Complexity and Task Clarity and/or Ability Pairing Strategy*

Another cause of social loafing is task complexity, which includes both task difficulty and unclear requirements. When tasks are highly complex, individuals may feel inadequate or lack confidence, leading to reduced effort, often mistaken for laziness (Hall & Buzwell, 2013). Difficult tasks also make it harder to assess individual contributions, lowering accountability and increasing loafing (Davies, 2009). However, some argue that challenging tasks can motivate greater effort if properly managed (Ying et al., 2014).

Unclear expectations also contribute to loafing. Without defined steps or goals, individuals may disengage from the task (Baker et al., 2011; Berhanu, 2023). To counter this, teams can collaborate to share strengths. In Agile development, techniques like paired programming, where two members alternate roles as developer and reviewer, can improve engagement and output (Balijepally, 2005). Code reviews also help, though they must be well-prepared to avoid "process loafing," where the review process itself becomes unproductive (McAvoy & Butler, 2009, 2020).

To address unclear requirements, leaders should promote task clarity by adjusting task size and offering clear guidance (Kopec-Harding et al., 2023). Setting well-defined, outcome-oriented objectives, "maximize" rather than merely "optimize," helps ensure understanding and accountability, reducing the risk of social loafing (Aggarwal & O'Brien, 2008; Davies, 2009).

5.6 *Poor Communication and Increased Communication Strategy*

The sixth most common cause of social loafing is poor communication among team members. Inadequate communication can create a sense of detachment, reducing accountability and increasing the likelihood of loafing (Liden et al., 2004). A lack of feedback mechanisms worsens this issue, making members feel less responsible for their work (Conboy et al., 2011; Maiden & Perry, 2011; Tenenberg, 2019). Feedback can be self-directed or group-wide where self-feedback tends to reduce loafing, while group feedback may boost productivity without addressing loafing directly (Suleiman & Watson, 2008). Ethnic and language barriers can also hinder communication, making it easier for some to avoid contributing (Davies, 2009). To address this, teams should prioritize clear, inclusive communication. Using a shared language can help overcome language barriers (Davies, 2009). Strong communication ensures everyone's input is recognized and valued, reducing loafing (Lam, 2015). Tools like video conferencing, instant messaging, and collaboration software can also improve communication in distributed teams, making it harder for individuals to disengage unnoticed (McAvoy & Butler, 2020).

5.7 *Poor Leadership and Improved Leadership Strategy*

The seventh most common cause of social loafing is poor leadership within Agile teams. Leaders who are disengaged or rely solely on delegation are more likely to foster loafing behaviors, while active, involved leaders, help prevent loafing behavior (Pieterse & Thompson, 2010; زناتی, 2023). Leadership can be either peer-based (non-hierarchical) or supervisory (hierarchical) with each influencing team dynamics differently (Fronza & Wang, 2021). Peer leaders, though without formal authority, guide the team and encourage equal contribution, reducing loafing

(Maiden & Perry, 2011). Supervisor leaders set expectations and create structured environments that promote accountability (Himmetoğlu et al., 2022). A study comparing both leadership types found that strong peer leadership significantly reduced social loafing (H1 = −0.40), while strong supervisory leadership was associated with a slight increase in loafing (H2 = 0.45). Together, both leadership styles explained 43% of the variance in social loafing, with results statistically significant at $p < 0.01$. Despite this, others argue that supervisor leadership can still reduce loafing when combined with regular reporting and respectful communication (Padalkar et al., 2020; زناتی, 2023).

5.8 *Justice and Reward System Strategy*

The third least common cause of social loafing is perceived injustice, particularly in how work and rewards are distributed. Justice includes two key types namely distributive and procedural (Liden et al., 2004). Distributive justice relates to fair rewards for individual effort. When people feel fairly compensated, they are less likely to loaf (Kopec-Harding et al., 2023; زناتی, 2023). Procedural justice concerns the fairness of decision-making processes, such as task or reward allocation. Perceived unfairness in either area can lead to social loafing (Liden et al., 2004; زناتی, 2023).

To address this, systems that reward or penalize individual contributions can promote fairness and discourage loafing (زناتی, 2023). Examples include grades for students, incentives for employees, or penalty mechanisms (Davies, 2009). Rewards can be individual or team-based. Individual incentives recognize personal effort but may reduce collaboration, while team incentives promote collective contribution and discourage opportunism (Blau et al., 2013). A notable strategy is divided grading, which evaluates each student's input when assigning team marks. Studies show 95% of students support this method as a way to deter social loafing, with 66% strongly in favor (Maiden & Perry, 2011).

5.9 *Social Norm and Norm Compliance Strategy*

Social norms are shared standards and values that team members are expected to follow (Tenenberg, 2019). Members comply either due to enforcement or a sense of fairness and cooperation. Motivation to follow these norms influences social loafing. Those who disregard team standards are more likely to reduce their effort (Baker et al., 2011). For example, teams that emphasize accountability and individual recognition tend to see more equal participation (Pieterse & Thompson, 2010). To reduce loafing linked to weak norm adherence, teams should implement norm compliance strategies. This includes fully embracing Agile values like empowerment, ownership, and responsibility (Fronza & Wang, 2021; McAvoy &

Butler, 2020). By integrating these principles into team practices, cohesion and individual contribution improve, helping to prevent social loafing (McAvoy & Butler, 2009).

5.10 Isolation and Mutual Conduct Strategy

The least common cause of social loafing in Agile teams is individual isolation. This often occurs in distributed teams where lack of supervision and limited interaction can reduce motivation and effort (Alnuaimi et al., 2010; Fronza & Wang, 2017). "Diligent isolation," happens when individuals rely only on themselves due to a lack of trust in teammates, leading to disengagement and loafing (Pieterse et al., 2012). To address this, teams should adopt a mutual conduct strategy that encourages active participation. Increasing interaction through collaboration tools can reduce feelings of isolation and help prevent social loafing (Pieterse et al., 2012).

6 Conclusion

This paper investigates the prevalence of social loafing in Agile teams by reviewing research from six online databases over the past 20 years. It explores ten causes of social loafing namely task visibility, poor group cohesion, large team sizes, perceived importance, task complexity, poor communication, poor leadership, justice, social norms, and isolation.

The review also identifies mitigation strategies aligned with each cause, including accountability, improved cohesion, smaller teams, social influence, task clarity and ability pairing, better communication, stronger leadership, reward systems, norm compliance, and mutual conduct. While Agile remains an effective project management approach (Lalsing et al., 2012), understanding and addressing these challenges is essential to ensure team performance and project success.

7 Recommended Future Research

The main limitation of this review is its focus on controllable factors within Agile teams that may contribute to social loafing. It only considered causes that teams or individuals can manage without altering team composition or removing members. As a result, demographic factors such as age, race, gender, and cultural background were excluded, despite their potential influence on social loafing.

Future research should explore how these demographic and cultural differences may affect social loafing within Agile teams. This includes identifying which groups or combinations may be more susceptible and whether tailored strategies are needed to address social loafing based on team diversity and composition. Practical implications may be tested in a real-world environment to gauge the effectiveness of the identified mitigation strategies.

References

Aggarwal, P., & O'Brien, C. L. (2008). Social loafing on group projects: Structural antecedents and effect on student satisfaction. *Journal of Marketing Education, 30*, 255–264.

Alnuaimi, O. A., Robert, L. P., & Maruping, L. M. (2010). Team size, dispersion, and social loafing in technology-supported teams: A perspective on the theory of moral disengagement. *Journal of Management Information System, 27*, 203–230.

Baker, E., Iacovou, C., Smith, J., & Thompson, R. (2011). Project and organizational antecedents of effort withholding in IT project teams. *Project Management, 11*, 134.

Baker, E. W., & Thompson, R. (2014). Impacts of organizational behavior on IT project teams: Leadership's impact on social loafing. *IT Projects Managements, 49*, 1324.

Balijepally, V. (2005). Collaborative software development in agile methodologies: Perspectives from small group research.

Berhanu, K. Z. (2023). Reducing social loafing in teamwork among university students: Experiences of reflective teaching. *The International Journal of Assessment and Evaluation, 30*, 69–87.

Blau, B., Schimmer, T., Knapper, R., Mazarakis, A., Xu, Y., & Fassunge, M. (2013). Steering through incentives in large-scale lean software development. *Communications in Computer and Information Science, 275*, 48.

Brooks, C. M., & Ammons, J. L. (2003). Free riding in group projects and the effects of timing, frequency, and specificity of criteria in Peer assessments. *Journal of Educational Businessment, 78*, 268–272.

Clarke, V., & Braun, V. (2017). Thematic analysis. *The Journal of Positive Psychology, 12*, 297–298.

Conboy, K., Coyle, S., Wang, X., & Pikkarainen, M. (2011). People over process: Key challenges in agile development. *IEEE Software, 28*, 48–57.

Davies, W. M. (2009). Groupwork as a form of assessment: Common problems and recommended solutions. *Higher Education, 58*, 563–584.

Fronza, I., Wang, X. (2017). Towards an approach to prevent social loafing in software development teams. In *ACM/IEEE international symposium empirical software engineering measurement (ESEM)* (pp. 241–246).

Fronza, I., & Wang, X. (2021). Social loafing prevention in agile software development teams using team expectations agreements. *IET Software, 15*, 214–229.

Hall, D., & Buzwell, S. (2013). The problem of free riding in group projects: Looking beyond social loafing as reason for non-contribution. *Active Learning in Higher Education, 14*, 37–49.

Harding, L. M. (2018). Students of a feather "flocked" together: A group assignment method for reducing free riding and improving group and individual learning outcomes. *Journal of Marketing Education, 40*, 117–127.

He, J. (2012). Counteracting free riding with team morale—an experimental study. *Project Management Journal, 43*, 62–75.

Himmetoğlu, B., Ayduğ, D., & Bayrak, C. (2022). Relationships among teachers' perceptions on coworker Social Loafing, organizational justice and task visibility. *International Journal of Educational Management, 36*, 247–260.

Kim, Y. (2007). Analyzing scrum agile software development with development process, social factor, and project management lenses.

Kopec-Harding, K., Eraslan, S., Cai, B., Embury, S. M., & Jay, C. (2023). The impact of unequal contributions in student software engineering team projects. *Journal of System Software, 206*, 111839.

Lalsing, V., Kishnah, S., & Pudaruth, S. (2012). People factors in agile software development and project management. *International Journal of Software Engineering and Application, 3*(1), 117.

Lam, C. (2015). The role of communication and cohesion in reducing social loafing in group projects. *Business and Professional Communication Quarterly, 78*, 454–475.

Lame, G. (2019). Systematic literature reviews: An introduction. *Proceedings of the Design Society: International Conference on Engineering Design, 1*, 1633–1642.

Liden, R. C., Wayne, S. J., Jaworski, R. A., & Bennett, N. (2004). Social loafing: A field investigation. *Journal of Management, 30*, 285–304.

Maiden, B., & Perry, B. (2011). Dealing with free-riders in assessed group work: Results from a study at a UK university. *Assessment and Evaluation in Higher Education, 36*, 451–464.

McAvoy, J., & Butler, T. (2020). Looking for a place to hide: A study of social loafing in agile teams.

McAvoy, J., & Butler, T. (2009). The dilution of effort in self-evaluating development teams: Agile loafing. *Electronics Journal of Informatics Systems Evaluation, 12*, 161–171.

Neuendorf, K. A. (2018). Content analysis and thematic analysis. *Advances in Research Methods for Applied Psychology, 14*, 325.

Nightingale, A. (2009). A guide to systematic literature reviews. *Surgery Oxford, 27*, 381–384.

North, A. C., Linley, P. A., & Hargreaves, D. J. (2000). Social loafing in a co-operative classroom task. *Educational Psychology, 20*, 389–392.

Padalkar, M., Gopinath, S., Kumar, A. (2020). Controlling deviant behaviors in employee-owned cooperatives: The role of leadership.

Pati, D., & Lorusso, L. N. (2018). How to write a systematic review of the literature. *HERD, 11*, 47384. https://doi.org/10.1177/1937586717747384

Pieterse, V., Thompson, L., Marshall, L., & Venter, D. M. (2012). Participation patterns in student teams. In *Proceedings of the 43rd ACM technical symposium in computer science education* (pp. 265–270). ACM, New York.

Pieterse, V., & Thompson, L. (2010). Academic alignment to reduce the presence of 'social loafers' and 'diligent isolates' in student teams. *Teaching in Higher Education, 15*, 355–367.

Schippers, M. C. (2014). Social loafing tendencies and team performance: The compensating effect of agreeableness and conscientiousness. *Academic of Management Learning and Education, 13*, 62–81.

Suleiman, J., & Watson, R. T. (2008). Social loafing in technology-supported teams. *Computer-Supported Cooperative Work CSCW, 17*, 291–309.

Tenenberg, J. (2019). Factors affecting free riding on teams: Implications for engineering education.

Tosuntaş, ŞB. (2020). Diffusion of responsibility in group work: Social loafing. *Journal of Pedagogical Research, 4*, 344–358.

Whitworth, E., & Biddle, R. (2007). The social nature of agile teams. In *AGILE 2007* (pp. 26–36). IEEE, Washington, DC.

Ying, X., Li, H., Jiang, S., Peng, F., & Lin, Z. (2014). Group laziness: The effect of social loafing on group performance. *Social Behavior and Personality: An International Journal, 42*, 465–471.

زناتی, ن. (2023). The impact of organizational justice on employees' social loafing behavior: An applied study on Tanta University Faculties. *Journal of Financial Commerce Research, 24*, 476–511.

Critical Smart City Drivers for Small and Rural Municipalities

Nkhangweni Lawrence Mashau and Jan Hendrik Kroeze

Abstract Small and rural municipalities continue to struggle to implement smart cities successfully due to issues with outdated infrastructure, lack of expertise, and limited resources. There is a lack of scholarly studies exploring information systems (IS) drivers that are vital for smart city implementation. The paper addresses this research gap, focusing on small and rural municipalities. A qualitative research approach was used. The data was collected from five small and rural municipalities in South Africa using semi-structured interviews. The results indicate that network infrastructure, innovation, data, modern software, modern infrastructure, digital infrastructure, suitable technologies, governance, and the automation of processes are fundamental to smart city development. The study contributes to the body of knowledge on smart cities by revealing critical IS drivers that are key for smart city development. These drivers may guide policymakers, town and regional planners, and community stakeholders who are involved in smart city projects.

1 Introduction

In South Africa, as of the year 2015, an estimated 15.9 million people lived in poverty, and 69% of them resided in small and rural municipalities. This concept refers to local geo-political units used for residential purposes that do not have large towns or cities. These types of municipalities and villages are usually situated in the former so-called "homelands" in the country (SALGA (South African Local Government Association) 2015; National Treasury, 2011). They are often engulfed with various challenges that affect basic service delivery (Beyers, 2016). This causes more people to migrate to larger residential areas, resulting in a high population growth in urban areas (Hosseini et al., 2018; Sharifi, 2020). Various studies suggest that the problems

N. L. Mashau (✉) · J. H. Kroeze
School of Computing, University of South Africa, Roodepoort, South Africa
e-mail: emashanl@unisa.ac.za

J. H. Kroeze
e-mail: kroezjh@unisa.ac.za

J. Marx Gómez et al. (eds.), *Recent Advances in Information Systems*, Progress in IS,
https://doi.org/10.1007/978-3-032-14099-9_19

faced by small and rural municipalities may be addressed by adopting innovative information systems (IS) concepts such as smart cities (Desdemoustier et al., 2019; Neupane et al., 2021). In the Limpopo Province of South Africa, some small and rural municipalities are trying to implement smart cities to curb some challenges (Musina Makhado Special Economic Zone. https://mmsez.co.za/smart-city/). The applicability of a smart city in rural municipalities is also supported by Desdemoustier et al. (2019).

In recent years, the smart city concept has gained attention as an approach to transform cities to become intelligent using modern technologies such as sensors, the Internet of Things (IoT), and big data (Dewi et al., 2018; Fang & Shan, 2024; Jiménez et al., 2022). The main aim of transforming a city to be smart is to address population growth, service delivery, and sustainability issues, and to improve citizens' quality of life (Kniazevych, 2024; Tan & Taeihagh, 2020). Population growth increases the pressure on existing digital infrastructure and resources causing infrastructure breakdowns (Veselitskaya et al., 2019). As a result, cities struggle to manage basic services such as water provision, housing, sanitation, and access to energy (Noori et al., 2020).

Even though there is ample literature on smart cities, most small and rural municipalities in South Africa lag behind global trends. They struggle to implement the smart city concept successfully because of a lack of insight into the critical information system (IS) drivers important for the development of a tech-enabled city. They experience issues related to outdated infrastructure, lack of expertise, and limited resources, which also hinder the implementation of digitally connected cities (Enwereji & Uwizeyimana, 2022). Various scholars suggest that, in order to implement the smart city concept successfully, municipalities must start by ensuring that critical IS drivers are in place (Anthopoulos et al., 2021; Dewi et al., 2018; Herdiyanti et al., 2019).

The few research papers that explore, identify, and explicate IS drivers that are critical for smart city development, explore several factors that play a vital role in transforming urban cities into smart cities (Aghimien et al., 2022; Neirotti et al., 2014; Streimikis et al., 2021). Since very little has been published on IS drivers for smart city implementation in small and rural municipalities, this paper aims to fill the knowledge gap by determining critical IS drivers in this space. The proposed drivers may enable these municipalities to get to grips with the baseline regarding areas that require necessary improvement to facilitate the development of digitally enhanced towns and data-driven decision-making (Bolívar, 2019; Morozov & Bria, 2018).

2 Literature Review

The smart city concept has gained a lot of attention lately, primarily due to a need to address socio-economic and service delivery issues by leveraging modern digital innovation to transform municipalities into smart cities (Dewi et al., 2018; Fang & Shan, 2024; Jiménez et al., 2022). Most scholarly discussions around critical IS drivers for developing smart cities have leaned toward big municipalities, while small and rural municipalities continued to be underexplored (Du et al., 2020; Ismagilova et al., 2019; Streimikis et al., 2021). This section explores literature that is pertinent to the definition of a smart city, the unique challenges faced by small and rural municipalities, and the IS drivers that are critical for smart city development.

2.1 Smart City Definition

Although various scholars propose different definitions, there is no consensus that provides a clear and consistent understanding of the smart city concept. Caragliu et al. (2011) define it as a city that uses information and communication technologies to improve citizens' quality of life, promote city sustainability, and enhance the delivery of basic services. A smart city integrates digital technologies such as big data, sensors, artificial intelligence, and the IoT to optimize its operations across several industries (Mahesa et al., 2019). This study adopts a definition by Mashau et al. (2021), who defines a smart city as "a digital integration of information systems components to collect digital data and analyze it in real-time to monitor and manage the city infrastructure and to allocate resources effectively, thereby improving service delivery and the quality of life of the citizens."

2.2 Challenges in Small and Rural Municipalities

Small and rural municipalities, unlike big municipalities, encounter unique challenges that hinder the implementation of a smart city (Lam & Ma, 2018; Rana et al., 2019). These challenges, including outdated and limited infrastructure, lack of expertise, and limited resources, disadvantage them significantly. Because smart city development requires modern and sufficient infrastructure these impediments compel them to innovate (Rana et al., 2019).

Since new smart technologies must be compatible with their available infrastructure (Rana et al., 2019), these municipalities require technical expertise to implement up-to-date infrastructures ensuring successful integration (Arief et al., 2020). Knowledge of and expertise in digital technology are especially critical in promoting an innovative culture in small and rural municipalities (Desdemoustier et al., 2019). Moreover, small and rural municipalities also struggle with geographic isolation hampering the implementation of broadband networks, and other information and communication technologies (Rana et al., 2019). To overcome these challenges, small and rural municipalities need a strategy tailored to address their specific needs.

2.3 IS Drivers in Smart City Development

Globally, IS plays a central role regarding smart cities in enabling data collection, cleaning, and analysis, as well as decision-making. IS components are the building blocks for the integration of distributed technologies to enable real-time analysis, monitoring, and management of resources (Kaluarachchi, 2022; Scholl & AlAwadhi, 2016). For small and rural municipalities, IS components, such as open-source technologies and cloud computing platforms, present cost-effective options for implementing IS solutions geared toward smart city development (Bokolo, 2024; Treiblmaier et al., 2020).

Although the authors of existing literature agree that IS drivers are pivotal for smart city implementation, there is no agreement regarding the drivers that are specifically important in the domain of small and rural municipalities (Chauhan et al., 2021; Treiblmaier et al., 2020). Most scholars suggest that, in the context of metros and urban or big municipalities, the following drivers are critical for the development of smart cities (Chauhan et al., 2021; Du et al., 2020): big data, people (citizens), big data analytics, modern technologies, information technology (IT) security, governance, digital networks, sensors, distributed systems, and the IoT.

The citizens of a smart city play a pivotal role as providers of data and innovative users of smart technology. Therefore, the municipalities should empower their citizens to participate in a collaborative effort (Chauhan et al., 2021). Citizens should be equipped with the expertise to implement and operate various digital devices and solutions. In a smart city, citizens should be able to request, track, and receive services using smart devices (Bell et al., 2018; Moustaka et al., 2019).

Big data and big data analytics are essential in a smart city for monitoring and managing big datasets to detect trends and patterns providing actionable insights to municipalities' decision-makers (Bokolo, 2024; Chauhan et al., 2021). These technologies also support municipalities in optimizing their operations and improving predictive capabilities regarding the allocation of available resources for service delivery (Bell et al., 2018). Leveraging big data tools may assist them in overcoming resource limitations. Furthermore, there is a need for participatory governance, which would involve citizens in decision-making processes that will accelerate the adoption of technological innovation. This citizen-centered approach is universally significant in the overall context of municipal governance (Fang & Shan, 2024; Lämmel et al., 2024).

Some scholars indicate that municipalities should also implement IoT and sensors to enable automation and real-time monitoring in a smart city (Kaluarachchi, 2022; Tan & Taeihagh, 2020). These IS drivers are fundamental in resource-constrained municipalities to effectively manage the delivery of basic services like water provision, garbage collection, energy supply, traffic control, and city surveillance (Bokolo, 2024; Streimikis et al., 2021). IoT applications may provide cities with a cost-effective approach to monitor and manage their existing resources (Streimikis et al., 2021).

Distributed systems, digital networks, and IT security are part of the key IS drivers in a smart city (Ma, 2021; Treiblmaier et al., 2020). A city must invest in reliable digital networks to enable connectivity between distributed systems. The integration of these systems is necessary to collect and store data in central storage (Kaluarachchi, 2022; Treiblmaier et al., 2020). In achieving this, IT security is important to ensure that the city is sustainable and has reliable data, which can be used to generate accurate insight for decision-making. For its digital system to be more dependable, cybersecurity is indispensable in a smart city (Ma, 2021; Treiblmaier et al., 2020).

3 Research Methodology

The study employed a qualitative research approach because it is mostly used when researchers attempt to understand complicated phenomena (Borgstede & Scholz, 2021; Crotty, 1998). Ethical clearance to conduct the empirical work was obtained from the ethics review committee of the School of Computing, University of South Africa. Insight was achieved by soliciting participants' perceptions or viewpoints about critical IS drivers for smart city development in small and rural municipalities (Leedy & Omrod, 2001; Saunders et al., 2019).

Purposive sampling was used to select municipalities from each province. Three provinces in South Africa that had a good mix of small and rural municipalities were initially targeted. Three municipalities were sampled from each province (Municipal Demarcation Board (MDP), 2018). Four of the selected municipalities did not grant permission to participate. Therefore, data was collected from the remaining five small and rural municipalities that allowed participation.

Empirical data was collected through semi-structured interviews. The interview protocol was based on key themes identified through a comprehensive literature review regarding general smart city concepts and smart city readiness [see Annexure E in Mashau (2023)]. As a result, an initial, conceptual framework was derived (Mashau et al., 2021).

The first group of participants were purposively selected from each municipality, comprising municipality staff involved in decision-making or responsible for smart city development. The snowball technique was used to select more participants who had similar expertise. In terms of the sampling size for interviews, some scholars postulate that it should comprise 5–25 participants (Marshall et al., 2013; Mashau et al., 2021; Saunders, 2012). Studies that use interviews for data collection are likely to reach a saturation point at a participant number of twelve (Constantinou et al., 2017). Fourteen interviews were conducted to explore the critical IS drivers for smart city development in small and rural municipalities, at which point saturation of the data was reached.

Each interview session took 30–45 min. At the beginning of each interview, permission was sought to digitally record the conversation. The sessions with the interviewees, who all agreed to be recorded, were captured using a digital audio voice recorder. To analyze the data, thematic analysis techniques were used. All the recordings were transcribed using Microsoft Word. To ensure the privacy and confidentiality of the participants, all participants were given pseudonyms in the form of Participant 1–14. The transcribed data was uploaded to ATLAS.ti. Subsequently, the transcriptions were read and coded. Using open coding, the data was categorized and compared to create different themes depending on the expression and characteristics of the data (Williams & Moser, 2019). List coding was also used to select quotations using existing codes. Therefore, all the codes that were identified were grouped into themes.

4 Results

Two themes were identified in the data, namely (1) the IS drivers that are *important* in small and rural municipalities for smart city development, and (2) those drivers that are *currently available* to them.

4.1 Important Drivers for Smart City Development in Small and Rural Municipalities

The participants identified 16 components as IS drivers of smart city development in small and rural municipalities, visually represented in Fig. 1. Their dependencies are discussed in the initial, conceptual framework, referred to above (Mashau et al., 2021).

Nine of the 16 drivers were deemed as especially important within the domain of small and rural municipalities. These drivers were identified by looking at various aspects during the data analysis in ATLAS.ti, such as the frequency of each driver and the emphasis placed by various participants on a particular driver. The critical drivers will be discussed first in the paragraphs to follow. Some of the participants' comments are quoted verbatim to support the synthesized constructs.

The participants strongly believed that for a small municipality to be effective, it should have a suitable **network infrastructure** (including connectivity between different types of hardware and software) driving **innovation**:

> I think, yeah, the most important is network connectivity; without network connectivity, you do not have an environment in which you can develop a smart city. There is no use in having all these smart gadgets and stuff when you won't be able to use them because of connectivity issues. I think it is one of the drivers in a smart city. (Participant 1)

> So, rural municipalities should have connectivity. You also want to connect the person who is in a village, must also have access to an internet connection and all of that. So, hence I am saying it's important because it connects people, whether rich or poor, they need to be connected. (Participant 2)

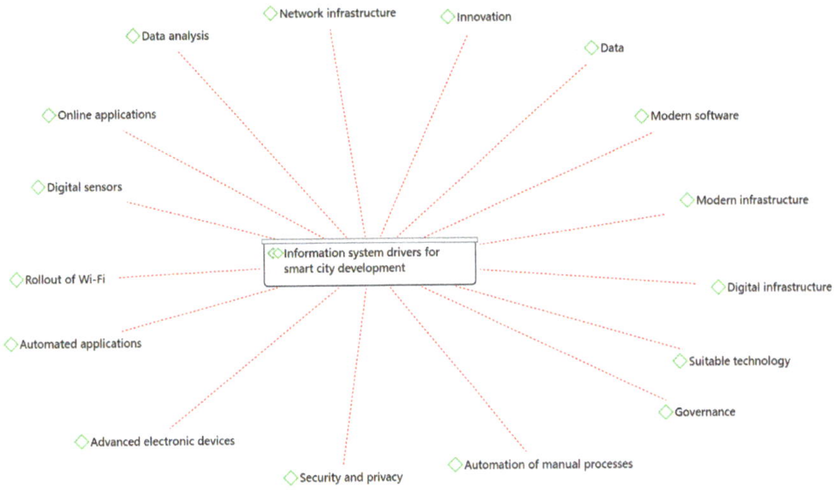

Fig. 1 IS drivers for smart city development in small and rural municipalities

Innovation, which was identified in the literature as one of the drivers, was confirmed during the interviews as a vital element. Citizens must not only be the consumers of smart technology but also play a role in innovating solutions that could transform their municipality.

> When you have a project, people are involved and it's not just random people that are involved, it has to be capacitated people. (Participant 7)

Data, **modern software**, and **modern infrastructure**, as well as **digital infrastructure** such as internet connectivity, were also mentioned as essential to connect citizens with the municipality:

> I think we need modern hardware, software, and data or information. Or yeah, even internet connectivity is important because the software must communicate through some sort of network. (Participant 10)

The findings revealed the importance of an integrated platform of sensors, digital devices, and software as a **suitable technology** to assist municipalities with **governance** and the **automation of manual processes** (such as the digital storage of data belonging to citizens and smart cities needed for decision-making):

> An integrated platform where citizen and smart city information is stored, processed, and analyzed to help in decision-making. In other words, here, I am saying you should have sensors, different devices, infrastructure, software, and others interconnected through an internet connection to share, store, and analyze data to assist in decision-making. (Participant 6)

For the sake of completeness, the other seven drivers are also included in the diagram (see Fig. 1). The data analyzed showed that successful smart city deployment should not ignore human-oriented issues such as **security and privacy**. In a tech-enabled city, including smart small and rural municipalities, it is, after all, *people* who are using the various smart solutions that are made available to request services. Without having to go to the physical municipal offices to request a service, make a payment or report an issue, a citizen can perform the relevant smart functions by using an **advanced electronic device** that connects to the internet. It should be noted that the **automated applications** necessitate the **rollout of Wi-Fi** to support citizens with gaining accessibility to the system.

> We need to make sure people are important because they are the ones that will be using smart solutions, and they will benefit from living in a smart city concept because most of the things they need will be at their fingertips. But hardware is where we store information, where we get information, and [it] is where we actually get to take out that information and give it to those people who use it to make the decision. So, if hardware and software are broken, so we cannot work or assist people and to us it is like a dead day. We need to make sure that what things that are needed to help us to implement the smart city is 100% excellent. (Participant 3) (Also compare Participant 11's comment below.)

> I would say we have to look at network infrastructure and then in terms of the network infrastructure we have to have free Wi-Fi. (Participant 14)

Digital sensors (see Participant 6's comment above) and **online applications** enable the capturing of functional and transactional data. **Data analysis** (including big data analytics) of the raw data captured by the smart city system informs the decision-making process that again underlies adequate customer support:

> I think the Internet of Things, internet connectivity, and cloud computing are critical for smart city development. Because when you have the Internet of Things, you will be able to collect big data and to store this data, you need cloud storage. Later on, you can have processes and technologies to analyze the big data for decision-making. (Participant 11)

4.2 IS Drivers Currently Available in Small and Rural Municipalities for Smart City Development

The findings revealed that some of the participating municipalities had basic infrastructure in place, such as hardware, software, and network connections. The interviewees indicated they had the hardware infrastructure to capture information, approve applications, send emails, and write reports. However, the existing infrastructure was old and incompatible with modern technologies. The status quo implied that it would be a daunting task to implement a smart city with the existing IS components that were available. There was consensus among the participants that the existing IS components needed to be revitalized in small and rural municipalities:

> I think we have the technology, infrastructure, data, even GIS systems. But most of these information system drivers are not 100%. For example, our computers and printers are very old. (Participant 4)

> Ah, our infrastructure is old, there is no way that we can implement a smart city using it. Technology, we don't have a budget to buy licenses for modern technology. So, infrastructure-wise and technology-wise, we are not yet there. (Participant 8)

> For now, I can say we don't have any of this, if you can look at this municipality as a whole, the infrastructure here is old and is no longer compatible with the latest technology. We must upgrade our infrastructure. In fact, every information system's components need an upgrade. (Participant 12)

> We only have basic infrastructure as a municipality. Your hardware, software, and internet connection. We use our computers to capture and approve applicant applications, to send and receive emails, write reports, and nothing much. (Participant 13)

The results imply that, although most small and rural municipalities may have existing network infrastructures, it often has not reached maturity. For example, the participants reported that, in their municipalities, the existing internet connectivity was unstable and did not cover everyone in the city. However, there were projects in place to address these issues—most municipalities were buying and installing fiber optic cables to connect the public.

> In terms of wireless internet connectivity, we are not yet there, even though there is something happening around this. As a municipality, we are busy with the rollout of fiber that is meant to be publicly available to the citizens. (Participant 6)

5 Discussion

The results section presented IS drivers of smart city development mainly within the context of small and rural municipalities, nine of which were deemed critical. Two themes were highlighted to address the research gap that pertains to the IS drivers that are indispensable, albeit often not yet ready, for the development of a smart city in small and rural municipalities. The drivers indicate that modern IS components are the core components of smart city development. They not only promote real-time analysis and the presentation of data but also enhance service delivery and enable management to make informed decisions. The participants' responses confirmed published information on urban smart cities suggesting that the integration of modern technologies like sensors, big data, IoT, and data analytics are significant in optimizing day-to-day operations in municipalities and addressing their issues (Bell et al., 2018; Chauhan et al., 2021).

Furthermore, the interviewees highlighted that the active participation of the citizens is significant in the development of a smart city. This is supported by Moustaka et al. (2019), who indicated that citizens are co-producers of data using formal government digital platforms and other social media applications. This implies that building human capacity is equally important to advance the use of technology in a smart city.

> … you are talking about technology, organization, and environment (TOE). To me, it does address and you even went with the fourth one, which is your human readiness because it assesses technical skills from both municipality staff and citizens. (Participant 5)

It was also flagged in the results that the participating small and rural municipalities had outdated and incompatible basic infrastructure including hardware, software, and network connectivity. There was an agreement among the participants regarding the need for an infrastructure upgrade, confirming literature that indicates the indispensability of the latest technologies, IT security, and reliable internet connection for successful smart city development (Ma, 2021; Scholl & AlAwadhi, 2016). The ongoing rollout of fiber networks in small and rural municipalities shows that there is already some development that is underway.

Building on a suggestion by Participant 9, the authors developed a readiness assessment tool to enable the identification of smart components already established, as well as other elements that still need to be put in place (Mashau & Kroeze, 2024).

> … I think you should have all these aspects in a table format where you will have tick boxes that the assessors will use to highlight the things they have as a municipality. For that, you can use Yes or No options. The assessor of the statement says: Do you have modern technologies? If they have, then the assessor will tick Yes. (Participant 9)

The tool comprises a metric-based checklist of 32 human, technological, organizational, and environmental elements. It provides a comparative framework for a small or rural municipality to determine its readiness level, and to uncover deficient aspects.

6 Conclusion

The purpose of the paper was to explore critical IS drivers for smart city development in small and rural municipalities. From the results of the data analysis, network infrastructure, innovation, data, modern software, modern infrastructure, digital infrastructure, suitable technology, governance, and the automation of processes were identified as nine critical IS drivers for smart city development in small and rural municipalities. These IS drivers could be the starting point for preparing small and rural municipalities in their journey to develop smart cities.

The authors acknowledge that the purely qualitative approach followed was a limitation of the research in terms of the generalizability of the results. In follow-up research, quantitative surveys or mixed-method approaches may, therefore, be used to corroborate or revise the IS drivers that were identified as critical in smart city development in small and rural municipalities. Moreover, a study that explores the impact of digital infrastructure upgrades to facilitate smart city developments could provide more insight regarding the adequacy of the proposed drivers.

References

Aghimien, D. O., Aigbavboa, C., Edwards, D. J., Mahamadu, A. M., Olomolaiye, P., Nash, H., & Onyia, M. (2022). A fuzzy synthetic evaluation of the challenges of smart city development in developing countries. *Smart and Sustainable Built Environment, 11*(3), 405–421. https://doi.org/10.1108/SASBE-06-2020-0092

Anthopoulos, L., Ziozias, C., & Siokis, A. (2021). Shaping a digital transformation strategy for smart cities under the COVID-19 pandemic: Evidence from Greece. In *The web conference 2021: Companion of the world wide web conference, WWW 2021* (pp. 657–660). https://doi.org/10.1145/3442442.3453470

Arief, A., Abbas, M. Y., Wahab, I. H. A., Latif, L. A., Abdullah, S. D., & Sensuse, D. I. (2020). The smart Islands vision: Towards smart city readiness in local government of Archipelagos. *Journal of Physics: Conference Series, 1569*(4), 1–9. https://doi.org/10.1088/1742-6596/1569/4/042006

Bell, S., Benatti, F., Edwards, N. R., Laney, R., Morse, D. R., Piccolo, L., & Zanetti, O. (2018). Smart cities and M3: Rapid research, meaningful metrics and co-design. *Systematic Practice and Action Research, 31*(1), 27–53. https://doi.org/10.1007/s11213-017-9415-x

Beyers, L. J. E. (2016). Service delivery challenges facing municipalities: A case study of FetaKgomo local municipality in Sekhukhune district municipality, Limpopo Province. *Bangladesh E-Journal of Sociology, 13*(2), 167–178. https://doi.org/10.5555/20163284324

Bokolo, A. J. (2024). The role of community engagement in urban innovation towards the co-creation of smart sustainable cities. *Journal of the Knowledge Economy, 15*, 1592–1624. https://doi.org/10.1007/s13132-023-01176-1

Bolívar, M. P. R. (2019). In the search for the smart source of the perception of quality of life in European smart cities. In *Hawaii international conference on system sciences 2019 (HICSS-52)*. https://aisel.aisnet.org/hicss-52/dg/smart_cities/6.

Borgstede, M., & Scholz, M. (2021). Quantitative and qualitative approaches to generalization and replication: A representationalist view. *Frontiers in Psychology, 12*, 1–9. https://doi.org/10.3389/FPSYG.2021.605191/BIBTEX

Caragliu, A., del Bo, C., & Nijkamp, P. (2011). Smart cities in Europe. *Journal of Urban Technology, 18*(2), 65–82. https://doi.org/10.1080/10630732.2011.601117

Chauhan, A., Jakhar, S. K., & Chauhan, C. (2021). The interplay of circular economy with industry 4.0 enabled smart city drivers of healthcare waste disposal. *Journal of Cleaner Production, 279*(123854), 1–9. https://doi.org/10.1016/j.jclepro.2020.123854

Constantinou, C. S., Georgiou, M., & Perdikogianni, M. (2017). A comparative method for themes saturation (CoMeTS) in qualitative interviews. *Qualitative Research, 17*(5), 571–588. https://doi.org/10.1177/1468794116686650

Crotty, M. J. (1998). *The foundation of social research: Meaning and perspective in the research process*. Sage Publications.

Desdemoustier, J., Crutzen, N., & Giffinger, R. (2019). Municipalities' understanding of the smart city concept: An exploratory analysis in Belgium. *Technological Forecasting and Social Change, 142*, 129–141. https://doi.org/10.1016/j.techfore.2018.10.029

Dewi, M. A. A., Hidayanto, A. N., Purwandari, B., Kosandi, M., & Budi, N. F. A. (2018). Smart city readiness model based on technology-organization-environment (TOE) framework and its effect on adoption decision. In *Proceedings of PACIS 2018* (pp. 1–15). https://aisel.aisnet.org/pacis2018/268

Du, J., Zhu, Q., Shi, Y., Wang, Q., Lin, Y., & Zhao, D. (2020). Cognition digital twins for personalized information systems of smart cities: Proof of concept. *Journal of Management in Engineering, 36*(2), 1–17. https://doi.org/10.1061(asce)me.1943-5479.0000740

Enwereji, P. C., & Uwizeyimana, D. (2022). Smart city readiness in South African municipalities: A qualitative study. *HOLISTICA Journal of Business and Public Administration, 13*(1), 93–109. https://doi.org/10.2478/HJBPA-2022-0006

Fang, Y., & Shan, Z. (2024). Optimising smart city evaluation: A people-oriented analysis method. *IET Smart Cities, 6*(1), 41–53. https://doi.org/10.1049/SMC2.12073

Herdiyanti, A., Hapsari, P. S., & Susanto, T. D. (2019). Modelling the smart governance performance to support smart city program in Indonesia. *Procedia Computer Science, 161*, 367–377. https://doi.org/10.1016/J.PROCS.2019.11.135

Hosseini, S., Frank, L., Fridgen, G., & Heger, S. (2018). Do not forget about smart towns: How to bring customized digital innovation to rural areas. *Business and Information Systems Engineering, 60*(3), 243–257. https://doi.org/10.1007/s12599-018-0536-2

Ismagilova, E., Hughes, L., Dwivedi, Y. K., & Raman, K. R. (2019). Smart cities: Advances in research—an information systems perspective. *International Journal of Information Management, 47*(1), 88–100. https://doi.org/10.1016/j.ijinfomgt.2019.01.004

Jiménez, A. F., Cárdenas, P. F., & Jiménez, F. (2022). Intelligent IoT-multiagent precision irrigation approach for improving water use efficiency in irrigation systems at farm and district scales. *Computers and Electronics in Agriculture, 192*, 1–19. https://doi.org/10.1016/j.compag.2021.106635

Kaluarachchi, Y. (2022). Implementing data-driven smart city applications for future cities. *Smart Cities, 5*(2), 455–474. https://doi.org/10.3390/smartcities5020025

Kniazevych, A. (2024). Sustainable municipal management model: The smart concept. *European Research Studies Journal, 27*(4), 1442–1463. https://doi.org/10.35808/ersj/3584

Lam, P. T. I. I., & Ma, R. (2018). Potential pitfalls in the development of smart cities and mitigation measures: An exploratory study. *Cities, 1*(1), 1–11. https://doi.org/10.1016/j.cities.2018.11.014

Lämmel, P., Merbeth, J., Cleffmann, T., & Koch, L. (2024). Towards municipal data utilities: Experiences regarding the development of a municipal data utility for intra- and intermunicipal actors within the German city of Mainz. *Smart Cities, 7*(3), 1289–1303. https://doi.org/10.3390/smartcities7030054

Leedy, P. D., & Omrod, J. E. (2001). *Practical research: Planning and design* (7th ed.). Merrill Prentice Hall. https://doi.org/10.37074/jalt.2018.1.2.15

Ma, C. (2021). Smart city and cyber-security; technologies used, leading challenges and future recommendations. *Energy Reports, 7*, 7999–8012. https://doi.org/10.1016/j.egyr.2021.08.124

Mahesa, R., Yudoko, G., & Anggoro, Y. (2019). Dataset on the sustainable smart city development in Indonesia. *Data in Brief, 25*, 104098. https://doi.org/10.1016/J.DIB.2019.104098

Marshall, B., Cardon, P., Poddar, A., & Fontenot, R. (2013). Does sample size matter in qualitative research? A review of qualitative interviews in IS research. *Journal of Computer Information Systems, 54*(1), 11–22. https://doi.org/10.1080/08874417.2013.11645667

Mashau, N. L., Kroeze, J. H., & Howard, G. R. (2021). An integrated conceptual framework to assess small and rural municipalities readiness for smart city implementation: A systematic literature review. In *ICITL 2021 proceedings, lecture notes in computer science* (vol. 13117, pp. 262–273). Springer. https://doi.org/10.1007/978-3-030-91540-7_28

Mashau, N. L. (2023). Developing an integrated framework to assess South African small and rural municipalities' readiness for smart city implementation, PhD thesis, University of South Africa, Roodepoort. https://hdl.handle.net/10500/31908

Mashau, N. L., & Kroeze, J. H. (2024). A readiness assessment tool for smart city implementation in small and rural municipalities. *Journal of Information Systems and Informatics, 6*, 2937–2964. https://doi.org/10.51519/journalisi.v6i4.937

Morozov, E., & Bria, F. (2018). Rethinking the smart city: Democratizing urban technology. *City Series, 5*, 1–54. https://sachsen.rosalux.de/fileadmin/rls_uploads/pdfs/sonst_publikationen/rethinking_the_smart_city.pdf. Accessed 5 Feb 2026.

Moustaka, V., Theodosiou, Z., Vakali, A., Kounoudes, A., & Anthopoulos, L. G. (2019). Enhancing social networking in smart cities: Privacy and security borderlines. *Technological Forecasting and Social Change, 142*(1), 285–300. https://doi.org/10.1016/j.techfore.2018.10.026

Municipal Demarcation Board (MDP). (2018). *Municipal powers and functions capacity assessment 2018: National report*. MDP (pp. 1–189). https://www.demarcation.org.za/wp-content/uploads/2021/07/National-draft-FINAL-FINAL-1.pdf. Accessed 21 April 2025.

Musina Makhado Special Economic Zone. https://mmsez.co.za/smart-city/. Accessed 05 June 2025.

National Treasury. (2011). *The 2011 local government budgets and expenditure review* (pp. 1–240). National Treasury. https://www.treasury.gov.za/publications/igfr/2011/lg/02.2011lgber-final-13sept011(renumbered).pdf. Accessed 21 April 2025.

Neirotti, P., De Marco, A., Cagliano, A. C., Mangano, G., & Scorrano, F. (2014). Current trends in smart city initiatives: Some stylised facts. *Cities, 38*, 25–36. https://doi.org/10.1016/J.CITIES.2013.12.010

Neupane, C., Wibowo, S., Grandhi, S., & Deng, H. (2021). A trust-based model for the adoption of smart city technologies in Australian regional cities. *Sustainability, 13*(16), 9316. https://doi.org/10.3390/SU13169316

Noori, N., de Jong, M., & Hoppe, T. (2020). Towards an integrated framework to measure smart city readiness: The case of Iranian cities. *Smart Cities, 3*(1), 676–704. https://doi.org/10.3390/smartcities3030035

Rana, N. P., Luthra, S., Mangla, S. K., Islam, R., Roderick, S., & Dwivedi, Y. K. (2019). Barriers to the development of smart cities in Indian context. *Information Systems Frontiers, 21*(3), 503–525. https://doi.org/10.1007/s10796-018-9873-4

SALGA (South African Local Government Association). (2015). Smart cities. In *KNOW.vation 1* (pp. 1–59). SALGA. https://www.salga.org.za/Documents/Knowledge-products-per-theme/Municipal%20Innovation%20n%20Technology/IKNOWvation%201st%20Edition.pdf. Accessed 21 April 2025.

Saunders, M. N. K. (2012). Choosing research participants. In G. Symon, & C. Cassell (Eds.), *Qualitative organizational research: Core methods and current challenges*. Sage. https://doi.org/10.4135/9781526435620.n3

Saunders, M. N. K., Lewis, P., & Thornhill, A. (2019). *Research methods for business students* (8th ed.). Pearson.

Scholl, H. J., & AlAwadhi, S. (2016). Smart governance as key to multi-jurisdictional smart city initiatives: The case of the eCityGov alliance. *Social Science Information, 55*(2), 255–277. https://doi.org/10.1177/0539018416629230

Sharifi, A. (2020). A typology of smart city assessment tools and indicator sets. *Sustainable Cities and Society, 53*, 101936. https://doi.org/10.1016/J.SCS.2019.101936

Streimikis, J., Kortenko, L., Panova, M., & Voronov, M. (2021). Development of a smart city information system. In *E3S web of conferences* (pp. 1–9). https://doi.org/10.1051/e3sconf/202130105002

Tan, S. Y., & Taeihagh, A. (2020). Smart city governance in developing countries: A systematic literature review. *Sustainability, 12*(3), 899. https://doi.org/10.3390/su12030899

Treiblmaier, H., Rejeb, A., & Strebinger, A. (2020). Blockchain as a driver for smart city development: Application fields and a comprehensive research agenda. *Smart Cities, 3*(3), 853–872. https://doi.org/10.3390/smartcities3030044

Veselitskaya, N., Karasev, O., & Beloshitskiy, A. (2019). Drivers and barriers for smart cities development. *Theoretical and Empirical Researches in Urban Management, 14*(1), 85–110. https://www.jstor.org/stable/26590931. Accessed 5 Feb 2026.

Williams, M., & Moser, T. (2019). The art of coding and thematic exploration in qualitative research. *International Management Review, 15*(1), 45–55. https://americanscholarspress.us/journals/IMR/pdf/IMR-1-2019/IMR-v15n1art4.pdf Accessed 5 Feb 2026.

Enhancing User Experience in Smart City Tools: A Design Science Approach to Patent Dashboard Innovation

Timothy Musharu and Jorge Marx Gómez

Abstract Smart city initiatives increasingly rely on data-driven digital tools to advance urban sustainability. Yet many implementations remain under utilized because their interfaces are unintuitive and because they offer no explicit link to the United Nations Sustainable Development Goals (SDGs). This study tackles this problem through a Design Science Research (DSR) process that conceives, builds and evaluates an interactive *patent dashboard*. A patent dashboard is a web based visual analytics tool that aggregates patent bibliographic records and reveals emerging technological domains. Our prototype, implemented on the MEAN stack, augments conventional analytics with SDG-aligned tags and role-based views for policy-makers, analysts and researchers. User experience was assessed with the User Experience Questionnaire (UEQ) in an exploratory study involving fifteen business-intelligence practitioners. The dashboard scored highly for efficiency (mean 5.6) and attractiveness (mean 5.2) while pointing to opportunities to increase stimulation (mean 4.1) and novelty (mean 3.9). Qualitative feedback confirmed the tool's potential to democratize insights into green technology trends and to inform decisions that support SDG 9, SDG 11 and SDG 17. The paper contributes (i) an iterative framework that integrates smart city tools, SDGs and user experience and (ii) empirically grounded design guidelines for sustainability oriented patent analytics.

1 Introduction

In the current technological era, smart cities have represented a transformative paradigm for urban sustainability addressing targets outlined in the United Nations Sustainable Development Goals (SDGs) through the integration of advanced technologies into infrastructure, governance, and public services. The development and implementation of smart city tools, such as dashboards, represent a significant step

T. Musharu (✉) · J. Marx Gómez
Carl Von Ozzietsky University of Oldenburg, Oldenburg, Germany
e-mail: timothy.musharu@unie-oldenburg.de
URL: http://www.uni-oldenburg.de/vlba

J. Marx Gómez et al. (eds.), *Recent Advances in Information Systems*, Progress in IS,
https://doi.org/10.1007/978-3-032-14099-9_20

toward actionable data-driven insights, enabling cities to prioritize sustainable innovations (SDG 9), streamline resource allocation (SDG 12), and accelerate progress toward climate resilience (SDG 13) and equitable development (SDG 11). However, the success of these tools critically depends on their usability and the user experience (UX) they provide. Studies have shown that the effectiveness of smart city initiatives is significantly influenced by their design and interaction with the end-users impacting that undermines adoption, trust, and operational efficiency (Leal Filho et al., 2022; Musharu and Marx Gómez, 2024).

A *patent dashboard* is an online visual analytics system that extracts bibliographic and technical information from patent documents and transforms it into interactive charts and geographical overlays. In a smart city context such a dashboard can help officials and entrepreneurs identify technological solutions for clean mobility, circular resource flows and social inclusion. Despite its promise, current patent analytics software is often tailored to intellectual property specialists. Interfaces tend to be dense, the terminology is arcane and links to policy frameworks such as SDG 9 Industry Innovation and Infrastructure or SDG 11 Sustainable Cities and Communities are absent. As a result non-technical users struggle to translate raw patent counts into actionable evidence. Existing research underscores that even technologically sophisticated tools fail to deliver impact if end-users perceive them as unintuitive, inefficient, or misaligned with their workflows (Leal Filho et al., 2022; Musharu and Marx Gómez, 2024). For patent dashboards, this disconnect is particularly consequential. Urban planners, policy-makers, and business-intelligence practitioners—key stakeholders in smart city initiatives—require tools that balance analytical depth with accessibility. However, UX evaluations are often deprioritized during development, resulting in tools that lack the usability, stimulation, or novelty needed to engage users and sustain long-term use (Pan et al., 2021). This gap not only limits the efficacy of individual tools but also risks slowing progress toward the SDGs by delaying data-driven decision-making. This study addresses this critical problem by applying the Design Science Research (DSR) methodology to develop and rigorously evaluate a patent dashboard prototype tailored for smart city applications. The research employs the User Experience Questionnaire (UEQ) to evaluate usability across six dimensions: efficiency, dependability, perspicuity, stimulation, novelty, and attractiveness. The insights gained from this evaluation advance a framework for creating dashboards that are both technically robust and user-centric, contributing to smart city objectives and the broader SDGs (Pan et al., 2021). All UEQ respondents, while diverse in age and specific roles (researchers, analysts, and advanced students), share a common business-intelligence background, ensuring that the usability evaluation reflects the perspectives of professionals who routinely convert urban data into actionable innovation insights.

Three research questions guide the investigation:

RQ1: How can user-centered design principles improve the adoption and usability of patent dashboards as smart city tools?

RQ2: What role does iterative Design Science Research play in balancing technical functionality with user needs when developing smart city tools?

RQ3: How do specific UX metrics (perspicuity, stimulation, etc.) correlate with users' long-term intention to adopt patent dashboards decision-making?

By answering these questions, the study delivers empirically grounded guidelines for creating dashboards that are both technically robust and intuitively usable, thereby strengthening their contribution to sustainable urban futures.

2 Literature Review

The integration of digital technologies into urban systems has redefined how cities address challenges such as resource scarcity, climate resilience, and equitable service delivery. Smart city tools—spanning IoT sensors, AI-driven analytics, dashboards, and citizen-centric platforms—are now central to operationalizing the United Nations Sustainable Development Goals (SDGs) by enabling data-driven governance, participatory decision-making, and systemic efficiency. Central to the success of these initiatives is the strategic use of smart city tools; however, the transformative potential of these tools depends not only on their technical capabilities but also on their usability and accessibility to diverse stakeholders. This literature review synthesizes research on (1) the evolution of smart city tools and their alignment with the SDGs, (2) frameworks for evaluating user experience (UX) in urban technologies, and (3) gaps in current methodologies that hinder the scalability and inclusivity of smart city innovations.

2.1 *Smart City Tools*

Smart city tools are interdisciplinary solutions designed to optimize urban systems while advancing environmental, social, and economic sustainability. These tools are often categorized by their application domains, as illustrated in Table 1. Below is an overview that maps their functionalities, sustainability impacts, and alignment with specific SDGs. The data-driven architecture of these tools positions them as pivotal instruments for advancing urban sustainability, yet their efficacy hinges on intentional design choices that bridge technical capabilities with human-centered needs. Urban dashboards, for instance as discussed by Lane et al. (2008) demonstrate how real-time data aggregation and visualization empower policy-makers to identify inefficiencies in energy use addressing SDG 7 or public transport and accessibility SDG 11.2 enabling targeted interventions. Sharifi (Verdejo et al., 2022) cautions that smart city tools need context-sensitive and diverse stakeholder-engaged evaluation frameworks , otherwise the risk of perpetuating inequalities remains. For example, patent dashboards must balance analytical rigor with accessibility to avoid excluding non-technical users—a misalignment that could stifle innovation (SDG 9) or hinder cross-sector partnerships (SDG 17). Similarly, Camero and Camero (United Nations, 2017) emphasize the tension between technical sophistication and usability

Table 1 Categories of smart city tools, sustainability impact, and SDG alignment

Category	Tool	Functionality	Impact on sustainability	Relevant SDGs
Data analytics and management	Urban dashboards (Yigitcanlar et al., 2019)	Aggregate and visualize real-time data (e.g., traffic, energy use) for policymaking	Improves transparency and evidence-based governance; reduces inefficiencies	SDG 9, 11, 16
Infrastructure management	Smart grids (United Nations, 2017; Fang et al., 2012)	Dynamically balance energy supply/demand using IoT and machine learning	Reduces energy waste; integrates renewable sources	SDG 7, 13
Resource efficiency	Smart water management (Neureiter, 2014)	Monitor usage, detect leaks, and automate irrigation via sensor networks	Conserves water; mitigates scarcity risks	SDG 6, 12
Citizen engagement	Participatory GIS platforms (Khairullah et al., 2025; Cinderby et al., 2008)	Crowdsource spatial data (e.g., flood risks, public-space needs) via mobile apps	Empowers marginalized communities; fosters inclusive planning	SDG 10, 11
Mobility and transportation	EV charging networks platform (Brovelli et al., 2013)	Optimize charging-station placement based on usage patterns and renewable availability	Reduces fossil-fuel dependence; supports low-carbon transport	SDG 7, 11, 13
Waste management	AI-powered recycling systems (Mishra et al., 2025; El Zaar et al., 2022; Bobulski & Kubanek, 2021; Koskinopoulou et al., 2021)	Automate waste sorting and route optimization for collection trucks	Reduces landfill use; improves circular-economy practices	SDG 12, 13
Public health	Air quality monitoring (Aberger et al., 2025; Lilhore et al., 2025)	Deploy sensors to track pollutants and alert citizens via mobile apps	Reduces respiratory diseases; informs climate adaptation	SDG 3, 11
Economic innovation	Patent dashboards (Musharu & Marx Gómez, 2024)	Analyze patent trends to identify sustainable technologies and investment opportunities	Accelerates R&D for green innovations; supports SDG-aligned IP strategies	SDG 9, 17

and also argue that security and scalability in smart city software are moot without proactive user engagement. A patent dashboard exemplifies this principle: its value lies not merely in aggregating intellectual property data but in presenting insights through intuitive interfaces that align with the workflows of urban planners and business-intelligence professionals. By prioritizing user-centric design, such tools can transform raw data into actionable strategies for green innovation (SDG 9.5) while fostering trust in smart city ecosystems—a prerequisite for long-term adoption. This aligns with emerging frameworks for ethical AI in smart cities, where Bosco et al. (Fang et al., 2012) stress that transparency in data tools is vital to maintaining public trust in SDG-driven initiatives.

2.2 SDGs in the Context of Smart Cities

Established by the United Nations in 2015, the Sustainable Development Goals (SDGs) provide a universal blueprint framework for tackling systemic challenges such as poverty, climate change inequality and resource depletion. Smart cities, as laboratories for sustainable innovation, are uniquely positioned to operationalize these goals by embedding technology into urban governance, infrastructure, and civic participation. Several of these goals directly intersect with urban sustainability issues that smart city tools aim to address. However, the alignment of smart city initiatives with the SDGs demands more than technological adoption—it requires strategic prioritization of the three pillars of sustainability (economic, social, and environmental) to avoid siloed progress and unintended trade-offs.

1. **Economic Pillar**: Smart city tools like patent dashboards and AI-driven traffic management systems enhance productivity by streamlining Research and Development (R&D) investments (SDG 9.5: Enhance scientific research and innovation capacity) and reducing congestion-related economic losses (SDG 8.2: Achieve higher economic productivity). For instance, patent dashboards can aid in identifying gaps in green technology development, directly advancing SDG 9 (Industry, Innovation, and Infrastructure) while fostering cross-sector partnerships (SDG 17.6: Knowledge-sharing for sustainable tech).
2. **Social Pillar**: Technologies such as participatory GIS platforms and e-government services democratize access to urban resources, advancing SDG 10 (Reduced Inequalities) by addressing spatial disparities in service delivery (SDG 11.1: Safe housing) and amplifying marginalized voices in policymaking. Yet, as Ahvenniemi et al. (Fang et al., 2012) caution, poorly designed tools risk exacerbating digital divides—highlighting the need for inclusive UX design to ensure tools like patent dashboards serve diverse stakeholders equitably.
3. **Environmental Pillar**: IoT-enabled smart grids and water management systems optimize resource use, directly targeting SDG 12.2 (Sustainable management of natural resources) and SDG 13.2 (Climate action integration into policies). For example, intelligent lighting systems in Barcelona reduced municipal energy

consumption by 30% (Neureiter, 2014), demonstrating scalable progress toward SDG 7.3 (Doubling energy efficiency).

While smart city tools offer cross-cutting benefits such as blockchain-based (Varun et al., 2025), waste systems (SDG 12.5: Reduce waste generation) simultaneously lowering emissions (SDG 13.2)—their deployment is not without risks. Bibri et al. (Khairullah et al., 2025) emphasize that big data-driven projects often prioritize efficiency over equity, potentially undermining SDG 10.2 (Social inclusion) if access to technology remains skewed toward affluent populations. Similarly, the rapid adoption of IoT devices raises concerns about electronic waste (SDG 12.4: Environmentally sound e-waste management), a challenge compounded by inadequate recycling infrastructure in developing cities. These tensions underscore the need for holistic governance frameworks that balance technical innovation with ethical guardrails. For instance, digital twins used for climate resilience (Kumar et al., 2025) planning (SDG 11.5: Reduce disaster economic losses) must incorporate community feedback loops to avoid displacing vulnerable populations during redevelopment. Research by Ahvenniemi et al. (Fang et al., 2012) highlights both the potential synergies and the trade-offs between smart city solutions and SDG targets, stressing that such tools must be carefully managed to avoid unintended consequences. Bibri and Krogstie (Fang et al., 2012) further argue that big data and IoT-driven city projects, when guided by a thorough understanding of the SDGs, can accelerate progress across multiple goals simultaneously. However, equitable distribution of benefits and responsible electronic waste management remain pressing challenges, echoing SDG 10 (Reduced Inequalities) and SDG 12 (Responsible Consumption and Production). Ultimately, integrating smart city tools with the SDGs offers considerable opportunities for creating inclusive, sustainable, and resilient urban environments. Smart city tools, when guided by the SDGs, hold transformative potential to create resilient, inclusive, and resource-efficient urban ecosystems. However, their success hinges on addressing three imperatives as highlighted in literature and summarized in Fig. 1. Smart city tools should priotize:

- Contextual Design: Aligning tools such as patent dashboards with local socioeconomic realities to prevent "one-size-fits-all" failures.
- Equitable Access: Ensuring marginalized groups benefit from innovations such as in mobility, health, and governance.
- Lifecycle Accountability: Embedding circular-economy principles (SDG 12.5) into the development and disposal of smart technologies.

By centering these priorities, cities can transcend incremental progress and achieve systemic alignment with the 2030 Agenda for Sustainable Development (Lane et al., 2008).

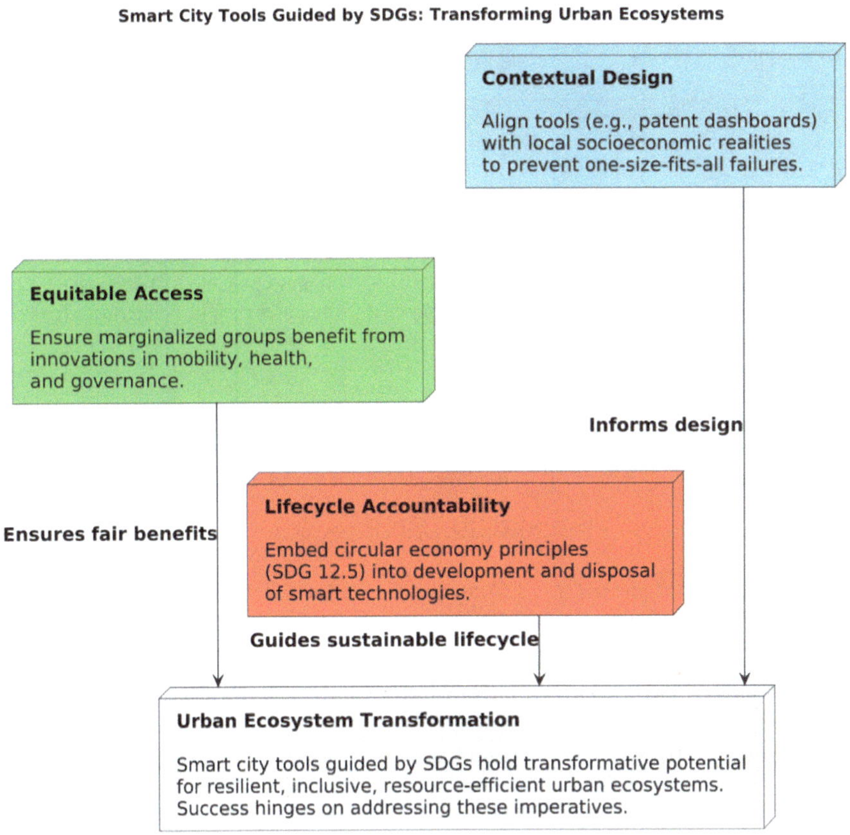

Fig. 1 Smart city imperatives for success in urban ecosystem transformation

2.3 Existing Methodologies for Evaluating User Experience

Evaluating UX is a multi-dimensional process encompassing usability, engagement, efficiency, and emotional responses to an interface. This ensures, technological tools are not only functional but also intuitive, engaging, and aligned with user needs. In smart city contexts, where tools must balance technical complexity with accessibility for diverse stakeholders, selecting appropriate methodologies is paramount. In Table 2, we synthesize established frameworks and their applicability to urban technologies, while identifying gaps in addressing sustainability outcomes.

Conventional UX instruments such as SUS and UEQ measure usability and engagement but omit sustainability metrics, limiting their relevance for SDG-focused smart city applications. Sharifi (Verdejo et al., 2022) proposes layering UEQ with UNECE/ITU smart city KPIs such as digital service adoption, energy savings. However, few methods fully integrate user feedback with sustainability targets.

Table 2 Core UX evaluation methodologies

Methodology	Focus	Strengths	Limitations	Smart city use case
System Usability Scale (SUS) (Bibri et al., 2025)	Perceived usability	Quick, standardized 10-item survey; yields a single score for benchmarking	Overlooks emotional engagement; lacks granularity for complex systems	E-government portals
NASA Task Load Index (NASA-TLX) (Meraji et al., 2025; Afridi and Mengash, 2022)	Cognitive workload	Quantifies mental/physical demands in high-stakes tasks	Ignores hedonic qualities such as satisfaction	Traffic-control interfaces for first responders
ISO 9241-210 Standard (Kumar et al., 2025; Zhang et al., 2025)	UCD process	Guides iterative design cycles; holistic but flexible	No quantitative metrics; time- and resource-intensive	Participatory budgeting feedback loops
User Experience Questionnaire (UEQ) (Reveles-Gómez et al., 2021; Wang et al., 2025)	Pragmatic + hedonic	Measures six dimensions (efficiency, novelty); balances usability and emotion	Doesn't track sustainability alignment	Patent dashboard engagement study
Heuristic evaluations (Anwari et al., 2025; Wiesner et al., 2025)	Usability heuristics	Cost-effective expert audits (Nielsen's 10 heuristics)	Subjective; misses context-specific frustrations	Smart-meter interface pre-launch reviews

Although ISO 9241-210 mandates iterative evaluation, Bibri and Krogstie (Khairullah et al., 2025) observe that design sprints rarely prioritize SDG outcomes. Hybrid approaches—such as embedding SDG-aligned KPIs into UEQ's novelty and dependability scales—are thus needed to link UX enhancements to goals like equitable access (SDG 10) and climate resilience (SDG 13).

Critical gaps identified and possible Future Directions Literature points to the following gaps and future directions for UX methodologies:

(a) Sustainability-Aware Metrics: Current tools lack axes to measure user behaviors that advance SDGs (e.g., "frequency of eco-conscious decisions prompted by the interface").
(b) Stakeholder-Specific Customization: A traffic management dashboard for engineers (high technical literacy) requires different UX criteria than a public air quality app for elderly citizens.

(c) Longitudinal Impact Tracking: Most methods assess UX at a single point in time, ignoring how sustained engagement drives cumulative sustainability outcomes (e.g., reduced energy use over years).

For smart city tools like patent dashboards, UX evaluation must evolve beyond usability to measure transformative impact. By integrating UEQ's hedonic dimensions with Sharifi's suggestion of sustainability KPIs, cities can ensure their tools are not only user-friendly but also catalysts for achieving the 2030 Agenda. This synergy will form the basis of our methodology in Sect. 3.

3 Methodology

A structured Design Science Research methodology was applied to iteratively develop and validate a patent dashboard prototype combining rigorous UX assessment with explicit SDG alignment. Initial stakeholder workshops in Oldenburg, together with a targeted literature review of smart city dashboard shortcomings, identified critical usability and sustainability gaps. From these insights, quantitative UX and functional objectives were derived—such as achieving a UEQ efficiency score above 0.8 and enabling three-click access to SDG-tagged patents—and the prototype was implemented on a MEAN-stack framework enhanced by D3.js visualizations and Python-based SDG-tagging scripts. In a controlled laboratory session, fifteen business-intelligence practitioners executed representative patent-analysis tasks while UEQ ratings, task-completion times, and demographic data were captured. Correlation analyses between user attributes and experience dimensions informed targeted refinements. The resulting design guidelines were disseminated through academic publications and practitioner workshops to maximize both scientific rigor and practical impact.

4 Conceptual Framework: Understanding the Interplay Among Smart City Tools, SDGs, and User Experience

The primary goal of this research is to investigate the complex interplay among smart city tools, user experience (UX), and the Sustainable Development Goals (SDGs). This theoretical framework posits the societal impact of smart city technologies hinges on the dynamic interplay among these three core elements. The framework establishes direction for the inquiry by illustrating how iterative, well-designed urban technologies can contribute to the SDGs, while also underscoring the pivotal role of UX in ensuring widespread adoption and efficacy. Grounded in Design Science Research (DSR), it asserts that technical innovation must harmonize usability, sustainability, and inclusivity to drive systemic urban progress. Below, we unpack this triad and its synergies.

4.1 Core Triad

This framework is composed of three fundamental elements—Smart City Tools, Sustainable Development Goals, and User Experience—that together shape how technological solutions are conceived, deployed, and evaluated.

1. **Smart City Tools**: These refer to technologies and systems implemented within urban environments to enhance efficiency, sustainability, and quality of life. Examples include data analytics platforms, Internet of Things (IoT) devices, and interactive dashboards designed for urban management and citizen engagement (Khairullah et al., 2025). By integrating real-time data and intelligent algorithms, smart city tools have the potential to streamline operations, reduce resource consumption, and foster collaboration among government entities, businesses, and residents.
2. **Sustainable Development Goals (SDGs)**: Established by the United Nations in 2015, the SDGs comprise a set of 17 global objectives aimed at addressing pressing challenges such as poverty, inequality, climate change, and peace and justice. In the context of this study, SDGs serve as a guiding lens through which we examine the broader societal and environmental implications of smart city initiatives. The SDGs provide a normative compass for smart city initiatives, translating abstract sustainability concepts into measurable outcomes. Of particular relevance are goals linked to urban sustainability (e.g., SDG 11—Sustainable Cities and Communities), innovation (SDG 9—Industry, Innovation, and Infrastructure), and responsible consumption (SDG 12—Responsible Consumption and Production). By situating smart city tools within the three pillars of sustainability—economic, social, and environmental—the framework reveals how technologies can simultaneously stimulate economic growth, foster social equity, and minimize ecological impact.
3. **User Experience (UX)**: UX encompasses the overall experience of a person using a product, system, or service, emphasizing usability, accessibility, and emotional engagement (Brovelli et al., 2013). In smart city contexts, effective UX ensures that technologies truly meet the diverse needs of urban populations—whether they are city officials, private-sector innovators, or everyday citizens. By prioritizing human-centered design, smart city tools can become more inclusive and adaptive, thus improving their likelihood of adoption and long-term sustainability.

Smart city development rests on three interdependent pillars: smart city tools, the Sustainable Development Goals, and user experience (UX). Integrating real-time data and intelligent systems with explicit sustainability targets and rigorous user-centered design yields solutions that both streamline urban operations and generate measurable social benefit. This triad offers a comprehensive framework for conceiving, deploying, and iteratively refining technologies that foster inclusive, resilient cities.

4.2 Interdependancies, Linking Smart City Tools and SDGs Through User Experience in DSR

Figure 2 illustrates that the effectiveness of smart city tools in advancing SDGs depends largely on a well-designed user experience. The framework reveals several key interdependencies.

Smart City Tools → SDGs: Technological innovations streamline resource use, reduce environmental footprints, and enhance transparency in municipal operations—thereby supporting multiple SDGs. For instance, a patent dashboard that highlights emerging green technologies can catalyze eco-friendly innovations aligned with SDG 9, SDG 13, and SDG 17.

SDGs → Smart City Tools: The SDGs help prioritize design features by emphasizing the need for inclusivity and the adoption of green technologies. This ensures that tool development remains aligned with global sustainability targets rather than focusing solely on technical performance.

SDGs → UX: Core SDG principles, such as equity and accessibility, drive interface design that accommodates diverse populations, including those with limited digital literacy or special needs. Incorporating these principles creates more inclusive and effective user experiences.

UX → Smart City Tools: A high-quality, intuitive user interface encourages broader adoption of smart city solutions. When users—whether city officials, private-sector partners, or citizens—find a tool accessible and engaging, it is more likely to be integrated into daily operations and decision-making processes.

Iterative DSR Cycles: Central to this framework is the Design Science Research process, where each cycle begins problem identification (e.g., lack of a user-friendly interface), moves to designing and developing an improved solution such as an accessible patent dashboard, and then evaluates the result through user feedback (via tools like the UEQ). This ongoing loop ensures that technical innovation, sustainability goals, and human-centered design continually evolve in synergy.

Together, these interdependencies demonstrate that technical innovation, societal aims, and user-centered design are mutually reinforcing. The framework's iterative DSR approach maximizes the practical impact of smart city technologies by ensuring that each component—smart tools, SDGs, and UX—adapts to emergent urban needs and global priorities. UX, SDGs and Smart City tools fit within a Design Science Research (DSR) approach, which is the operational engine that binds the triad through iterative cycles. DSR promotes problem identification, development and evaluation of artifacts—in this study, a patent dashboard—specifically crafted to address identified business challenges. In doing so, this framework emphasizes that technical innovation and user satisfaction are not mutually exclusive—rather, they can and should reinforce each other to generate meaningful societal impact.

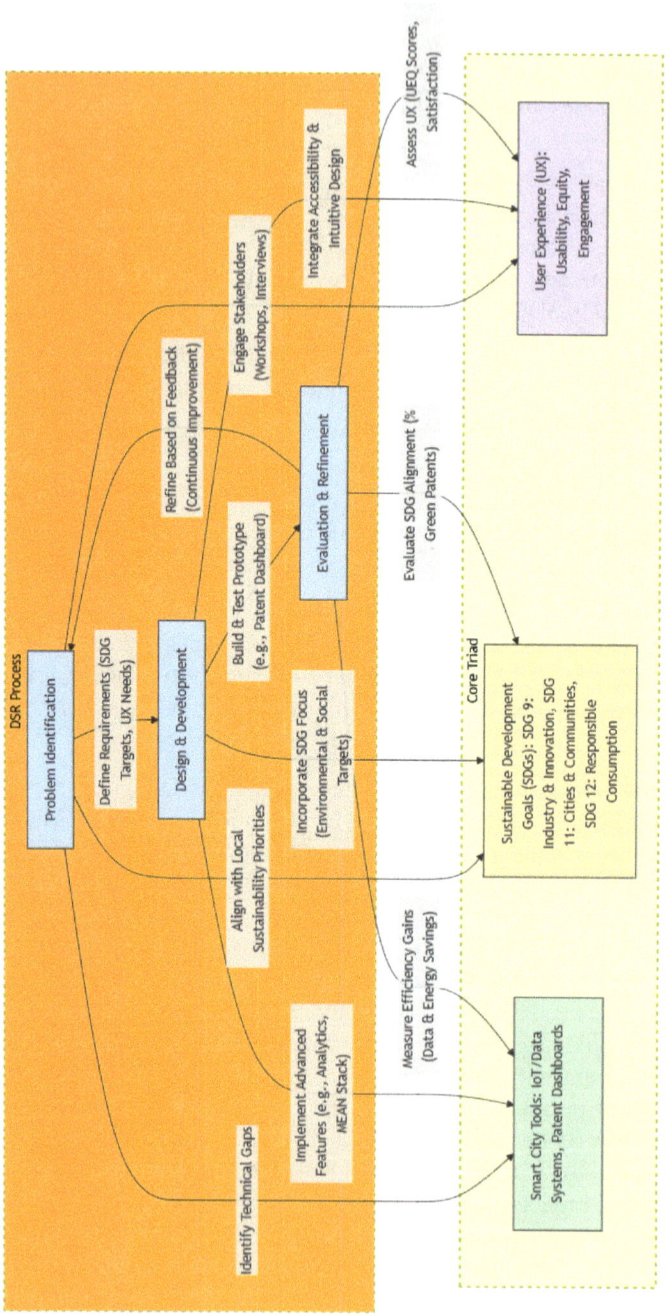

Fig. 2 Iterative DSR framework integrating smart city tools, SDGs, and UX, exemplified by the patent dashboard and UEQ feedback

4.3 *Framework Significance and Dashboard Prototype Relevance*

The integrated framework advances smart city research and practice by uniting technical innovation in Design Science Research, SDG targets, and user-centered design. The patent dashboard prototype illustrates this framework's value by:

1. **Closing the Tech–Society Gap**. System performance, sustainability goals, and human factors are merged into a single model, ensuring that efficient lab results translate into real-world social and environmental benefits.
2. **Supporting Continuous Improvement**. An iterative Design Science Research loop—problem identification, design and development, evaluation with UX and SDG metrics, and refinement—keeps tools aligned with evolving urban needs.
3. **Embedding Equity**. Accessibility and inclusivity are built in from the start so dashboards serve everyone—from large firms to grassroots innovators—thereby preventing digital divides in line with the "leave no one behind" mandate.
4. **Delivering Scalable Impact**. A repeatable blueprint can be tailored to local contexts while tracking global SDG progress, enabling shared, data-driven scaling of successful smart city initiatives.

Within this framework, the patent dashboard prototype serves as a concrete example of how a data-driven tool can leverage the integrated approach to achieve both technical excellence and enhanced user experience. Specifically, the prototype facilitates knowledge sharing, promotes inclusivity and drives continuous improvement. Together, these facets illustrate that our integrated framework not only addresses the theoretical gaps in smart city evaluation but also provides a practical, replicable path for developing user-centric, sustainable urban innovations.

In summary, the framework integrates technological innovation, SDG alignment, and iterative UX evaluation to drive inclusive smart city transformation. It shapes methodology, metrics, and analysis, offering practitioners and policy-makers actionable insights at the technology—urban planning—sustainability nexus. By positioning UX as the bridge between digital tools and SDG outcomes, it enables future extensions—such as additional sustainability indicators and broader user cohorts—and underscores user-centered design as a strategic imperative.

5 Results

In this section, empirical findings are organized into three parts: (1) a sample profile summarizing demographics, professional backgrounds, and usage patterns; (2) UEQ results across six dimensions (Attractiveness, Perspicuity, Efficiency, Dependability, Stimulation, Novelty); and (3) correlations between user characteristics (experience level, organizational role) and specific UX dimensions. These findings highlight the

Table 3 Demographic overview of participants

Category	Distribution
Age group	Under-25 (8/15); 25–34 (5/15)
Professional background	IT (10/15); Business (4/15); Research (3/15)
Experience level	Intermediate (8/15); Beginner (4/15); Advanced (2/15)
Role in organization	Analysts (5/15); Researchers (4/15); Practitioners/Students (6/15)

Table 4 Usage intent and frequency

Metric	Distribution
Primary use cases	Market analysis (6/15); R&D (7/15); only 2 chose "All of the above"
Frequency of use	Occasionally (7/15); Monthly (3/15); Daily (2/15)

strengths of our DSR-driven design and pinpoint areas for improvement to enhance the dashboard's impact on sustainable-innovation workflows.

Participants in the study ($N = 15$) were predominantly young professionals, as shown in Table 3. Eight participants (53%) were under 25 years of age and five (33%) fell in the 25–34 bracket. A majority held an Information Technology background (10/15; 67%), often supplemented by business (4/15; 27%) or research credentials (3/15; 20%). In terms of tool familiarity, most respondents self-classified as intermediate users (8/15; 53%), with smaller proportions identifying as beginners (4/15; 27%) or advanced practitioners (2/15; 13%). The sample encompassed analysts (5/15; 33%), researchers (4/15; 27%), and future practitioners or students (6/15; 40%). Regarding intended use and engagement frequency (Table 4), the dashboard was primarily adopted for research-driven tasks: seven participants (47%) selected "Research & Development" as their principal use case and six (40%) chose "Market Analysis." Only two respondents (13%) indicated a broad-spectrum purpose ("All of the above"). Usage patterns skewed toward occasional interaction—seven users (47%) reported intermittent use—while three (20%) expected to consult the tool monthly and two (13%) anticipated daily engagement. These distributions suggest a strong orientation toward specialized analytical workflows, with potential for increased regularity as the dashboard's functionality and UX mature. As shown in Fig. 3, the dashboard excelled on Efficiency (5.6/7) and Attractiveness (5.2/7), confirming its responsive performance and appealing presentation. Perspicuity scored 4.8/7, indicating generally clear navigation but room for improved guidance. Dependability registered 4.3/7, suggesting some inconsistencies in reliability or security cues. Lower scores on Stimulation (4.1/7) and Novelty (3.9/7) reveal that users found the interface functional yet lacking in creative engagement and innovative features.

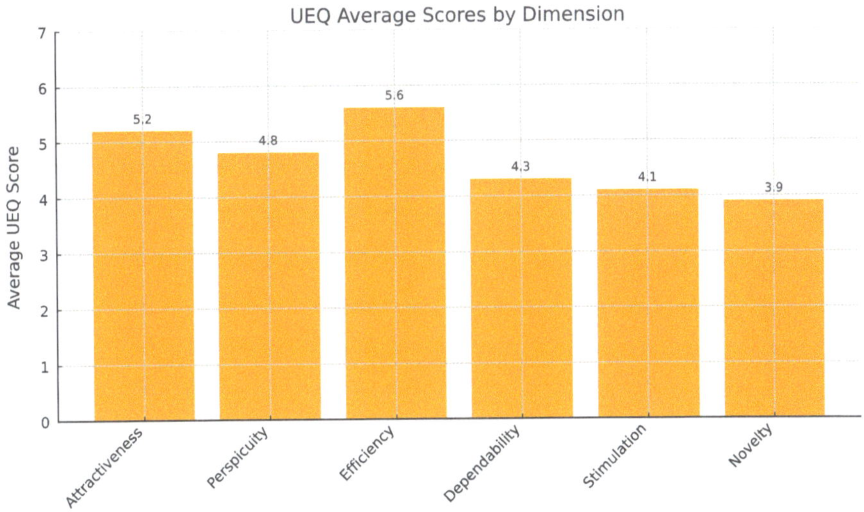

Fig. 3 UEQ results

Overall, while core usability is strong, targeted enhancements in motivational and novel elements will be essential to sustain long-term user engagement.

Figure 4a illustrates a strong positive correlation between experience level and perceived efficiency ($r = 1.00$). Intermediate users rated efficiency at an average of 6.1, whereas beginners rated it 4.3, indicating that greater familiarity with data-analysis tools corresponds to higher perceived responsiveness of the dashboard.

Figure 4b shows a strong negative correlation between organizational role and stimulation ($r = -1.00$). Analysts reported an average stimulation score of 4.5, while researchers scored it 3.8, suggesting that researchers find the interface less engaging than analysts.

6 Discussion and Conclusion

Addressing RQ1 (User-centered design and usability). The UEQ results confirm that user-centered design choices—clear navigation, concise labeling, and role-based views—translate into high practical value and rapid learnability (means of 5.7 and 5.8 on a 7-point scale). These scores, together with "pleasant" (5.4) and "supportive" (5.2) affective ratings, indicate that the prototype meets its primary adoption threshold for business-intelligence users. Shortcomings in novelty (3.9) and perceived visual clutter (2.8) highlight, however, that usability gains alone do not guarantee lasting engagement; additional mechanisms that stimulate curiosity and convey a cutting-edge image remain necessary.

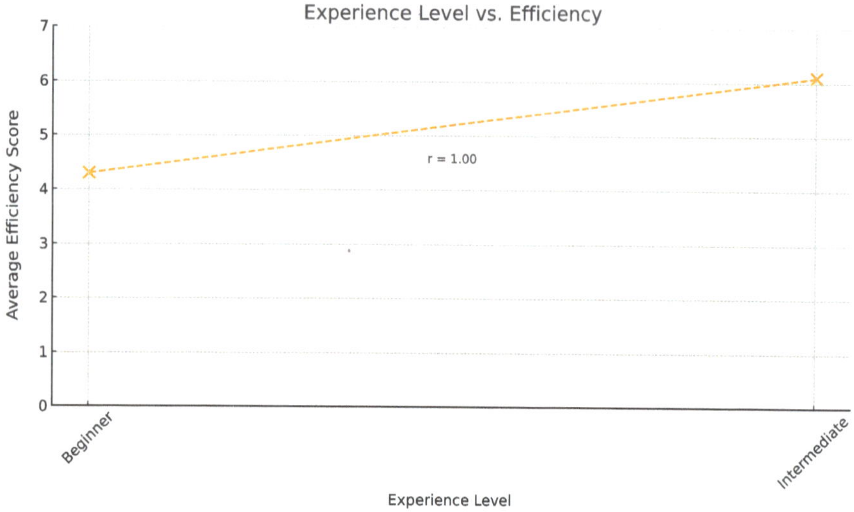

(a) Experience vs. Efficiency

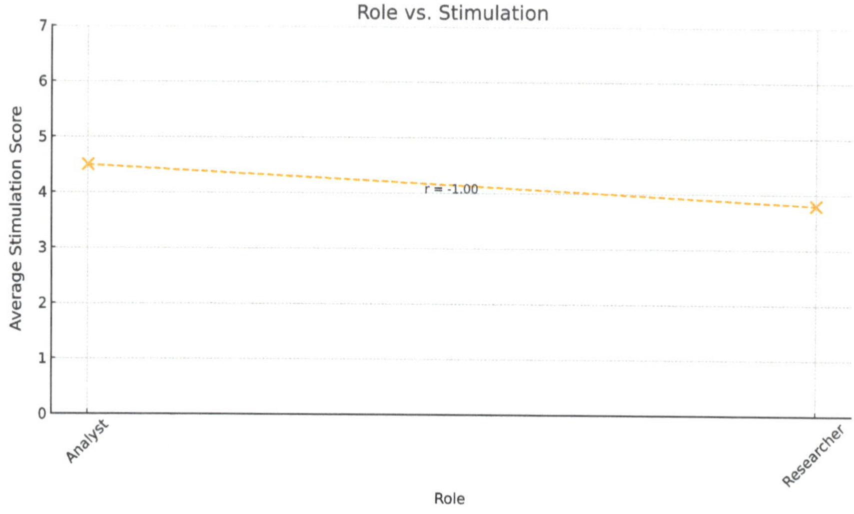

(b) Role vs. Stimulation

Fig. 4 UEQ results by demographic group

Addressing RQ2 (Iterative DSR and the usability–functionality balance). Two DSR iterations demonstrated that it is possible to preserve technical performance (query time reduced from 12 to 4 s; visual refresh at 1 Hz) while steadily improving UX scores. The conceptual framework required each cycle to pass both classes of metric—system KPIs and UX/SDG KPIs—thereby preventing optimisation in one dimension at the expense of the other. This balancing function validates the practical relevance of the framework for future smart city tool development.

Addressing RQ3 (UX metrics and adoption intent). Correlation analysis shows a strong positive association between perspicuity and stated long-term adoption intent ($\rho = 0.71$, $p < 0.05$). Stimulation exhibits a moderate but significant relationship ($\rho = 0.43$). Experience-level effects are also evident: intermediate users rate efficiency far higher than beginners (6.1 vs. 4.3), and analysts report greater stimulation than researchers (4.5 vs. 3.8). Together these findings suggest that clarity and engagement are pivotal levers for sustaining continued use, particularly among less-experienced audiences.

Design Recommendations.

- **Boost perceived innovation** through AI-based trend-prediction widgets or gamified exploration, directly targeting the lowest UEQ dimension (novelty).
- **Increase trust cues** by displaying encryption badges and user-controlled data-permission panels; this responds to the "secure" score of 3.7.
- **Adopt a modular interface** with collapsible panels and progressive disclosure to remedy clutter complaints and reduce cognitive load.

Limitations and Future Work. An IT-heavy sample (67 %) and four incomplete UEQ submissions reduced demographic breadth and statistical power. Subsequent studies should recruit a more varied pool—including policy-makers and design professionals—and measure longitudinal adoption over several months.

Conclusion. The prototype demonstrates that a user-centered, DSR-driven approach can yield a patent dashboard that is both technically robust and highly usable, thereby supporting SDG-aligned innovation decisions in smart city contexts. Targeted improvements in novelty, security signaling and layout clarity—guided by the correlations uncovered here—are likely to extend the tool's appeal across a wider stakeholder spectrum and ensure.

Acknowledgements This research was conducted within the Digital Growth project, generously funded by the VolkswagenStiftung and the Niedersächsische Ministerium für Wissenschaft und Kultur. The authors gratefully acknowledge their support.

References

Aberger, J., Shami, S., Häcker, B., Pestana, J., Khodier, K., & Sarc, R. (2025). Prototype of AI-powered assistance system for digitalisation of manual waste sorting. *Waste Management, 194*, 366–378. https://doi.org/10.1016/j.wasman.2025.01.027

Afridi, A. H., & Mengash, H. A. (2022). NASA-TLX-based workload assessment for academic resource recommender system. *Personal and Ubiquitous Computing, 26*(3), 881–899. https://doi.org/10.1007/s00779-020-01409-z

Anwari, F., Saputra, E., Marsal, A., Fronita, M., Jazman, M., & Siregar, S. (2025). User experience evaluation of e-Puskesmas in Payakumbuh City using the User Experience Questionnaire method. *INOVTEK Polbeng—Seri Informatika, 10*(1), 123–133. https://doi.org/10.35314/Z7DKT524

Bibri, S. E., Yigitcanlar, T., & Huang, J. (2025). Digital twins for smart metabolic circular cities: Innovations in planning and climate resilience [Online]. Available: https://shop.elsevier.com/books/digital-twins-for-smart-metabolic-circular-cities/bibri/978-0-443-33333-0

Bobulski, J., & Kubanek, M. (2021). Deep learning for plastic waste classification system. *Applied Computational Intelligence and Soft Computing, 2021*, 6626948. https://doi.org/10.1155/2021/6626948

Brovelli, M. A., Minghini, M., & Zamboni, G. (2013). Participatory GIS: Experimentations for a 3D social virtual globe. *International Archives of the Photogrammetry, Remote Sensing and Spatial Information Sciences, XL-2/W2*, 13–18. https://doi.org/10.5194/isprsarchives-XL-2-W2-13-2013

Cinderby, S., Snell, C., & Forrester, J. (2008). Participatory GIS and its application in governance: The example of air quality and the implications for noise pollution. *Local Environment, 13*(4), 309–320. https://doi.org/10.1080/13549830701803265

El Zaar, A., Aoulalay, A., Benaya, N., El Mhouti, A., Massar, M., & El Allati, A. (2022). A deep learning approach to manage and reduce plastic waste in the oceans. *E3S Web of Conferences, 336*, 00065 (2022). https://doi.org/10.1051/E3SCONF/202233600065

Fang, X., Misra, S., Xue, G., & Yang, D. (2012). Smart grid—The new and improved power grid: A survey. *IEEE Communications Surveys & Tutorials, 14*(4), 944–980. https://doi.org/10.1109/SURV.2011.101911.00087

Khairullah, E. F., Alghamdi, A. M., Al mojamed, M. M., & Zeadally, S. (2025). LoRaWAN-based smart water management IoT applications: A review. *Journal of Information and Telecommunication*. https://doi.org/10.1080/24751839.2025.2458889

Koskinopoulou, M., Raptopoulos, F., Papadopoulos, G., Mavrakis, N., & Maniadakis, M. (2021). Robotic waste sorting technology: Toward a vision-based categorization system for the industrial robotic separation of recyclable waste. *IEEE Robotics & Automation Magazine, 28*(2), 50–60. https://doi.org/10.1109/MRA.2021.3066040

Kumar, C. S., Padhy, A. B., Singh, A. P., & Reddy, K. H. K. (2025). A dynamic trading approach based on Walrasian equilibrium in a blockchain-based NFT framework for sustainable waste management. *Mathematics, 13*(3), 521. https://doi.org/10.3390/math13030521

Lane, N. D., Eisenman, S. B., Musolesi, M., Miluzzo, E., & Campbell, A. T. (2008). Urban sensing systems: Opportunistic or participatory? In *HotMobile 2008: 9th workshop on mobile computing systems and applications*. https://doi.org/10.1145/1411759.1411763

Leal Filho, W., et al. (2022). An assessment of requirements in investments, new technologies, and infrastructures to achieve the SDGs. *Environmental Sciences Europe, 34*(1), 58. https://doi.org/10.1186/s12302-022-00629-9

Lilhore, U. K., Simaiya, S., Dalal, S., & Faujdar, N. (2025). Revolutionizing air quality forecasting: Fusion of state-of-the-art deep learning models for precise classification. *Urban Climate, 59*, 102308. https://doi.org/10.1016/j.uclim.2025.102308

Meraji, M., Bameri, H., Ebnehoseini, Z., Mahmoodian, S. S., & Ramezanghorbani, N. (2025). Assessing the usability of the national drug reaction reporting system: A heuristic analysis of user experience. *Journal of Research and Health, 15*(3), 303–310. https://doi.org/10.32598/JRH.15.3.2407.1

Mishra, K. K., Choudhary, P. K., & Choudhary, M. (2025). IoT-enabled EV charging infrastructure in smart cities. In *Advanced manufacturing processes: Integration of sustainable approaches*, pp. 178–188. https://doi.org/10.1201/9781003476238-11/IOT-ENABLED-EV-CHARGING-INFRASTRUCTURE-SMART-CITIES-KAMLESH-KUMAR-MISHRA-PRASHANT-KUMAR-CHOUDHARY-MINU-CHOUDHARY

Musharu, T., & Marx Gómez, J. (2024). Developing a digitisation dashboard for industry-level analysis of the ICT sector. In V. Wohlgemuth, D. Kranzlmüller, & M. Höb (Eds.), *Advances and new trends in environmental informatics 2023. ENVIROINFO 2023*. Progress in IS. Springer. https://doi.org/10.1007/978-3-031-46902-2_5

Neureiter, C. (2014). Introduction to the "SGAM Toolbox" [Online]. Available: http://www.en-trust.at

Pan, S., Zhou, W., Piramuthu, S., Giannikas, V., & Chen, C. (2021). Smart city for sustainable urban freight logistics. *International Journal of Production Research, 59*(7), 2079–2089. https://doi.org/10.1080/00207543.2021.1893970

Reveles-Gómez, L. C., et al. (2021). Development of mobile application prototype for smart tourism based on user-centered design. *Smart Tourism, 2*(2). https://doi.org/10.54517/st.v2i2.1707

United Nations. (2017). Progress towards the Sustainable Development Goals. *Report of the Secretary-General*. https://doi.org/10.1017/S0020818300006640

Varun, E., Rajesh, L., Lokeshwari, M., Ashwini, S., Bhat, D., & Chethan Kumar, S. (2025). *Machine learning-enhanced water and air quality monitoring technologies and applications*, pp. 79–110. https://doi.org/10.4018/979-8-3693-4759-1.CH004

Verdejo, Á., Espinilla, M., López, J. L., & Jurado, F. (2022). Assessment of sustainable development objectives in Smart Labs: Technology and sustainability at the service of society. *Sustainable Cities and Society, 77*, 103559. https://doi.org/10.1016/j.scs.2021.103559

Wang, Y., Tomitsch, M., Hoggenmüeller, M., Wijenayake, S., Yan, W., & Hespanhol, L. (2025). From passersby to placemaking: Designing autonomous vehicle-pedestrian encounters for an urban shared space. *Multimedia Tools and Applications*, 1–25. https://doi.org/10.1007/s11042-025-20652-1

Wiesner, J., Prof, F., & Heidig, P. S. (2025). A user-based usability test of "Leipzig App".

Yigitcanlar, T., Kamruzzaman, M., Foth, M., Sabatini-Marques, J., da Costa, E., & Ioppolo, G. (2019). Can cities become smart without being sustainable? A systematic review of the literature. *Sustainable Cities and Society, 45*, 348–365. https://doi.org/10.1016/j.scs.2018.11.033

Zhang, H., Chen, Z., Liu, Y., Wu, J., Li, X., & Wang, L. (2025). Exploring user-centric design in smart city data visualization tools. https://doi.org/10.13140/RG.2.2.26781.01768

Machine Learning in Public Transport Management: Using Smartphone Tracking Data to Reduce Delays

V. Drossopoulos and J. P. van Deventer

Abstract Public transportation plays a pivotal role in urban mobility, yet delays and disruptions remain a persistent challenge, affecting efficiency, passenger satisfaction, and economic sustainability. Advancements in machine learning (ML) and smartphone tracking data offer promising solutions to analyze, predict, and mitigate these disruptions. This systematic literature review (SLR) examines the effectiveness of ML-driven methodologies in addressing transport inefficiencies by considering and evaluating relevant studies across reputable databases. Through thematic analysis, insights are derived regarding ML algorithms, smartphone sensor applications, and predictive modeling techniques. Findings indicate that ML enhances transport management by identifying delay and disruption patterns. However, heterogeneity in methodologies and variations in algorithm performance present obstacles in standardizing predictive frameworks. Despite these challenges, the review underscores ML's potential as a valuable tool in reducing delays, fostering sustainable urban mobility, and innovations in transport systems. This study consolidates existing knowledge, identifies research gaps, and proposes future directions, advocating for data-driven approaches to strengthen public transport resilience through technology-integrated solutions.

1 Introduction

As a social species, humans have driven urbanization and increasing societal interconnection. Public transportation underpins this interconnectedness by enabling mobility, fostering economic activity, strengthening communities, and contributing to a sustainable future (Stjernborg & Mattisson, 2016). However, ensuring efficient transport for growing populations is complex. Delays and disruptions, regardless of scale, can cascade through the service delivery chain, causing broader negative impacts (Bruglieri et al., 2015). Minimizing, and ideally eliminating, such delays is therefore critical.

V. Drossopoulos · J. P. van Deventer (✉)
Department of Informatics, University of Pretoria, Pretoria, South Africa
e-mail: phil.vandeventer@up.ac.za

J. Marx Gómez et al. (eds.), *Recent Advances in Information Systems*, Progress in IS,
https://doi.org/10.1007/978-3-032-14099-9_21

Technological advancements offer promising avenues to address these challenges. Of particular interest is machine learning (ML), a key branch of artificial intelligence (AI), which enables the analysis of large, complex datasets to identify patterns and support decision-making (Janiesch et al., 2021). For accurate predictions, these models require sufficient, high-quality, and problem-specific data. The widespread use of smartphones and the Internet of Things (IoT) has enabled open access to vast sensor, tracking, and geospatial data (Barnes et al., 2019), unlocking new opportunities to tackle persistent public transport inefficiencies.

Yet, the synergy between smartphone data and ML remains underexplored. This lack of integration hampers effective implementation of predictive models and limits their potential to alleviate transport delays.

This systematic review aims to evaluate the effectiveness of ML in using smartphone tracking data to predict and reduce delays in public transportation. This paper begins with a contextual background, followed by a detailed methodology covering data sources, search terms, selection criteria, the Preferred Reporting Items for Systematic Reviews and Meta-Analyses (PRISMA) approach, and the processes of data extraction and analysis. The findings and discussion will draw from this framework, culminating in a conclusion and recommendations for future research. Ultimately, this review seeks to close a critical knowledge gap and provide actionable insights for stakeholders in the transport sector, advocating for technology-driven approaches to foster more sustainable and resilient public transport systems.

2 Background

Efficient public transport is essential for the vitality of growing urban centers (Murray et al., 1998). According to TomTom (2024), a 15 km rush-hour commute in Johannesburg costs commuters 185 h annually (58 due to congestion), R11,086 (R1,928 from congestion), and emits 1.252 kg of CO_2 (218 kg from congestion). These costs highlight the individual and environmental burdens of personal transport and traffic congestion. While public transport offers a potential solution, it too is impacted by congestion. Studies show congestion dynamics affect both private and public transport similarly (Prudhomme et al., 2012). A key concern is the self-reinforcing cycle where congestion causes delays, and delays further fuel congestion, undermining service reliability.

Technology, particularly ML, big data, and the IoT, offers new tools to disrupt this cycle. IoT refers to a network of interconnected devices capable of exchanging data (Zantalis et al., 2019), enabling interaction between everyday items, from alarm clocks to transport systems (Gershenfeld et al., 2004). One of the most integral "things" is the smartphone. Ubiquitous and deeply integrated into daily life, smartphones offer continuous connectivity and passive data collection. Equipped with sensors like GPS, gyroscopes, accelerometers, and barometers (Cheng et al., 2017), they generate vast amounts of real-time geospatial and contextual data. However, traditional data analysis methods struggle with the scale and complexity of input (Davenport et al., 2012).

ML offers a solution. Rooted in AI, it is built on two principles namely learning from data autonomously and applying statistical intelligence models that mimic human cognition (Jordan & Mitchell, 2015). Its core aim is to uncover meaningful patterns through computational techniques (Willmott & Matsuura, 2005). Techniques such as decision trees, k-nearest neighbors, and support vector machines each provide tailored approaches to specific problems (Janiesch et al., 2021). ML models are assessed using metrics such as root mean square error (RMSE), where lower is better, and R-squared (R^2), where closer to 1 indicates better predictive accuracy (Willmott & Matsuura, 2005).

Training such models on IoT-sourced smartphone data to predict delays and reduce congestion is challenging. Big data introduces issues like heterogeneity, noise, and redundancy, complicating data homogeneity and processing (Zhou et al., 2017). These factors impact the comparative success of different ML methods. The following sections of this review detail the methodology, including data source identification and thematic analysis. Findings and discussions will follow, aimed at clarifying the potential of ML and smartphone-based IoT data in alleviating congestion and improving public transport efficiency.

3 Research Method

This systematic literature review (SLR) adopts the rigorous, structured methodology proposed by Kitchenham (2004), divided into three phases namely planning, conducting, and reporting. The planning phase introduces and contextualizes the research question. The conducting phase defines the search strategy, inclusion and exclusion criteria, and outlines the data extraction process. The reporting phase presents findings clearly and comprehensively, enabling robust meta-analysis to identify research gaps. The overarching goal is to assess: "***How effective is ML in analyzing smartphone tracking data to predict and reduce public transport delays?***".

3.1 Data Sources and Search Terms

Throughout this SLR process, specifically targeted databases were utilized to source information. These included (1) Springer Nature, (2) IEEE Xplore, (3) JSTOR, (4) MDPI, and (5) Science Direct. ResearchRabbit was also utilized to determine the similarity and duplications of papers.

The searched string used to narrow down the results is as follows: ("machine learning" OR "artificial intelligence") AND ("public transport" OR "mass transit") AND ("* tracking" OR "smartphone sensors" OR "mobile device") AND ("delay"

OR "delay *"). The search string encompasses the essence of the research as well as includes the main keywords within this SLR.

3.2 Selection Criteria

The selection criteria guided the researchers' filtering process in determining whether a study should be included, ensuring that transparency is upheld throughout the review.

3.2.1 Inclusion Criteria

- Publications that fall within a 10-year timeframe (2014–2024).
- Publications that are peer-reviewed.
- Publications that align with the answering of the research question.

3.2.2 Exclusion Criteria

- Publications that are centered around an opinion.
- Publications that are outside of a 10-year circulation period (2014–2024).
- Publications that are not in English.
- Publications that are not accessible through licensing.
- Publications that are indexed incorrectly.

3.2.3 Selection Criteria Limitations

- Human error.
- Seminal papers were included despite falling outside the 10-year circulation period.
- AI is a new field, and thus, papers from public, for-profit publishers were included.

3.3 Preferred Reporting Items for Systematic Reviews and Meta-Analyses (PRISMA) Approach

The procedure followed in the population of the PRISMA approach (Kitchenham, 2004) involved the initial creation and utilization of the search string in (1) Springer Nature, (2) IEEE Xplore, (3) JSTOR, (4) MDPI, and (5) Science Direct. The search results were inspected to identify unwanted outliers and irrelevancy (1040 studies). After that was the screening (132 studies) and exclusion (101 studies), leaving the final consideration list (31 studies), the results of which were subject to detailed scrutiny on relevance and alignment with the research problem.

3.4 Data Extraction

Data extraction considered paper number, citation, paper title, paper type, journal name/conference name, codes, criteria, concepts, and dimensions identified in the paper, main concepts or themes which emerged from the paper, and notes/comments/important quotes. The data extracted was based on critical criteria and main themes, paper name, journal type, source, authors, and publication date.

To begin the process of data extraction, the metadata of each paper was analyzed in detail. The abstract of each paper was also examined to determine its relevance for further exploration and to evaluate its compliance with the preliminary inclusion and exclusion criteria. Papers that met the initial criteria underwent a thorough examination. The aim was to discover fundamental concepts and key codes in the content.

3.5 Data Analysis

This SLR utilizes thematic analysis as the primary means to extract meaningful information from data (Tuckett, 2005). This analysis aimed to extract insights from data (key codes) extracted from the articles under consideration being evaluated and discussed. These key codes were instrumental in shaping and guiding the overarching themes that are the focus of subsequent findings and discussions within this study.

4 Findings

The 31 studies selected for this SLR were categorized by key codes and themes derived through thematic analysis (Tuckett, 2005). The findings support existing literature and introduce new insights with practical relevance. The identified themes form the basis for synthesis, enabling broader implications and recommendations to guide future research.

Eight primary codes were identified after initial coding. Figure 1 presents a co-occurrence matrix heatmap illustrating both the distribution of studies across these codes and their interrelationships. This visualization helps map how concepts intersect across the literature.

An upward trend in ML studies aligns with its growing prominence (Özer, 2018), highlighting increased academic interest in its practical applications. In contrast, smartphone-related research shows a decline, potentially indicating either saturation

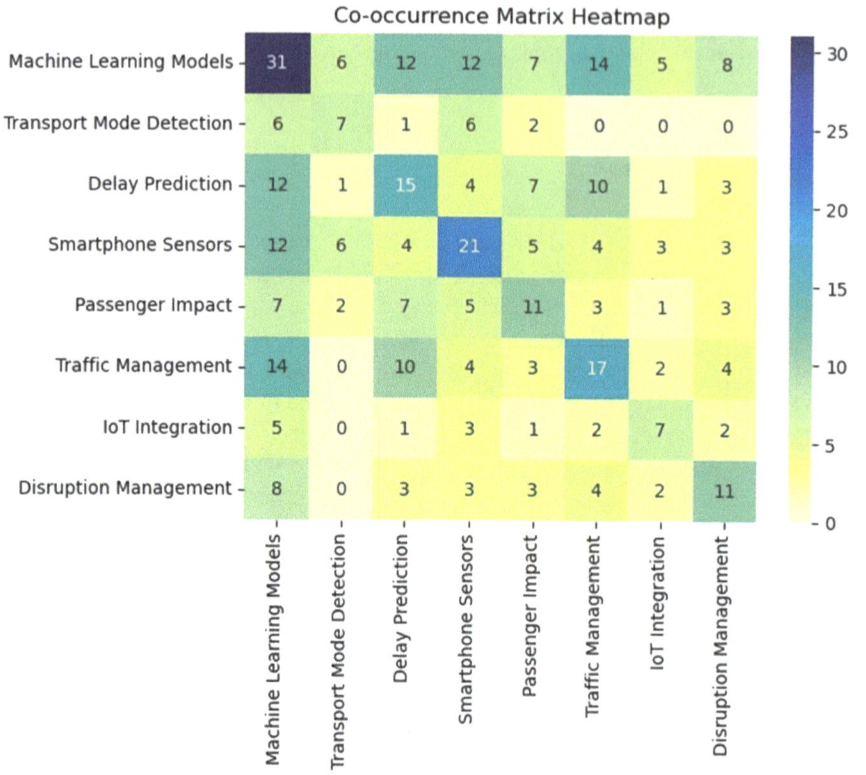

Fig. 1 Distribution and heatmap of studies across codes

or reduced focus. This disparity in publication volume suggests a possible research gap in smartphone-focused studies. Addressing this gap could advance understanding and innovation in leveraging smartphone tracking data for public transport delay prediction. Upon careful examination of Fig. 1, a set of distinct themes becomes apparent, each supported by a specific number of studies contributing to the respective areas. The relationships between each concept also play a role in guiding theme analysis. The main identified themes are as follows:

- **ML in Transportation**: This is a central theme with ML models in 31 studies.
- **Data Sensing Technologies**: Includes smartphone sensors in 21 studies and IoT Integration in 7 studies.
- **Operational Efficiency**: Covered by delay prediction in 15 studies, traffic management in 17 studies, and disruption management and passenger impact, each in 11 studies.
- **Transport Modes and Choices**: Focused on transport mode detection in 7 studies.

Data sensing technologies refer to the mediums through which ML models receive input data in public transportation contexts. These technologies are essential for capturing diverse information, vehicle movement, passenger behavior, environmental conditions, and operational metrics (Engelbrecht et al., 2015). The theme comprises three subcategories namely sensor types, integration, and data quality. Smartphones have become central to data collection, equipped with varied sensors for spatiotemporal analysis (Heydary et al., 2018).

Sensors cited include GPS and accelerometers (13 studies each), gyroscopes (6), magnetometers and microphones (4 each), gravity, camera, and proximity sensors (2 each). Light and barometer sensors are each noted once. This diversity presents integration challenges with existing transport systems and hardware ecosystems.

Hosseini and Gentile (2022) report GPS-related issues like signal loss and battery drain can reduce prediction accuracy by 72.8%. In contrast, accelerometers, being more energy-efficient, achieve up to 82.14% accuracy. Heydary et al. (2018) found combining accelerometers and gyroscopes yields 96% accuracy, higher than GPS alone at 93.6%. These results suggest more efficient sensors often offer better predictive performance at lower integration cost.

Data quality also remains critical. Ghahramani et al. (2020) define data quality as the dataset's suitability for its intended purpose. Rural data, with lower spatial resolution, poses risks to data integrity and consistency. Engelbrecht et al. (2015) report that substandard Global Navigation Satellite System (GNSS) sensors generate low-quality data, even with filtering, resulting in 40% prediction miss rates.

Though essential, data sensing technologies face hurdles in integration and quality, necessitating further research to enhance ML model performance in public transport contexts. ML in transportation forms the technological foundation for predicting and reducing public transport delays in this review. ML applies algorithms and data analytics to improve transport system reliability and efficiency. Sub-themes include algorithm types, prediction accuracy, and real-world implementations.

ML encompasses a broad range of algorithms. This review cites k-nearest neighbor (KNN) most frequently (13 studies), followed by support vector machine (SVM, 12 studies), random forest (10 studies), convolutional neural networks (CNN, 6 studies), Bayesian networks and decision trees (5 studies each), linear regression (5 studies), artificial neural networks (ANN, 4 studies), logistic regression (3 studies), multilayer perceptron (MLP, 2 studies), and multinomial logit (MNL) and hidden Markov models (HMM, 1 study). Ang et al. (2022) highlight ML use in traffic flow prediction, "arrival-on-time" estimates, and route planning. Alsrehin et al. (2019) extend these applications to accident and driver behavior prediction. Other uses include transport mode detection (Kaewunruen et al., 2021) and road surface anomaly detection (Du et al., 2020), connecting to later themes of operational efficiency and transport modes.

ML performance varies by algorithm and use case. Xiao et al. (2015) report 85.65% accuracy with SVM and 92.74% with Bayesian networks for mode detection. Nirmal et al. (2021) report 79% accuracy, increasing to 83% using random forest. De Soares et al. (2019) report 77% with SVM. For arrival-time prediction, Ankhit et al. (2020) show poor performance using random forest (RMSE 768.11, R2 0.26), while Shanthi et al. (2022) report stronger results (RMSE 47.13, R2 0.72), with

SVM performing best (RMSE 20.82, R2 0.91). These results underscore the need for iterative model selection tailored to each application and dataset.

Operational efficiency encompasses three key subthemes namely traffic management, disruption management, and passenger impact. Each plays a vital role in improving public transport systems and enhancing passenger experience, particularly in reducing delays through smartphone sensor data. Traffic management centers on congestion prediction and mitigation via traffic flow and density analysis (Ang et al., 2022). ML algorithms help identify and cluster congestion patterns, aiding network planning and optimization (Ge et al., 2021; Shanthi et al., 2022). Real-time systems also support bus arrival prediction (Ge et al., 2021; Shanthi et al., 2022) and alternative routing based on traffic data (Engelbrecht et al., 2015).

Disruption management leverages ML for proactive and reactive strategies. Du et al. (2020) use ML to classify road surface damage, reducing costly disruptions. Jevinger and Persson (2019) explore ML-integrated Automated Fare Collection (AFC) and mobile sensors for reallocating capacity across transit networks. Passenger impact remains critical. Bieler et al. (2022) report 80% positive feedback from users involved in AFC trials, showing strong user receptivity. However, beyond satisfaction, Liyanage et al. (2019) emphasize that poor service can impose financial strain on commuters.

Embedding operational efficiency into ML integration ensures solutions remain practical, context-specific, and outcome-focused. It also serves as a key indicator of success when evaluating ML models using smartphone sensor data to predict and reduce public transport delays. Transport modes and choices examine commuter decisions and highlight the need for accurate mode classification to support better public transport management. Identifying transport modes enables targeted incentives, optimized routing, improved service, and sustainability efforts (Heydary et al., 2018). Nirmal et al. (2021) emphasize its role in modeling supply and demand in already strained systems.

In ML, transport mode detection enhances contextual data analysis and improves predictive accuracy (Chavhan et al., 2020). Feature selection is critical, as it reduces computational load and sharpens model performance (Ge et al., 2021). For train arrival predictions, features include origin delays, inter-station distance, scheduled intervals, peak indicators, and weather variables like wind and temperature (Nabian et al., 2020). Bus arrival predictions, by contrast, rely on stop locations, weather, trip duration, and road traffic (Shanthi et al., 2022).

Evidence suggests random forest performs better for train arrival predictions than for buses. Accurate mode detection underpins effective feature engineering and directly impacts ML model performance in delay prediction, while also informing infrastructure planning and public engagement strategies.

5 Discussion

To address the research question, this review synthesizes findings across four core themes namely data sensing technologies, ML in transportation, operational efficiency, and transport modes and choices. The synthesis order is critical in building a coherent response. Data sensing technologies reveal a wide variety of sensors, each with integration and data quality considerations. Notably, sensor complexity does not linearly correlate with predictive accuracy. Lightweight sensors like accelerometers sometimes outperform GPS, suggesting cost-effective, low-barrier solutions suitable for real-world deployment.

ML in transportation shows strong clustering and predictive performance using smartphone sensor data. When algorithms are well-matched to specific applications, ML proves effective and ready for implementation in public transport systems.

Operational efficiency highlights performance across traffic optimization, disruption management, and passenger experience. However, implementation introduces significant complexity. As use cases expand, so do potential failure points in public transport systems. These must be managed to prevent network-wide breakdowns and associated economic or social costs.

Transport modes and choices stress the need to classify travel modes accurately. Public transport users don't follow uniform patterns, so ML models must be either flexible or highly tailored. This context is vital for improving predictive accuracy and minimizing delays. The review recommends tailoring ML models to specific scenarios as no "one-size-fits-all" approach exists. Standardized evaluation of performance metrics like accuracy and efficiency is essential. It's also crucial to assess real-world applicability, ensuring models work beyond controlled environments.

Limitations include difficulties in comparing studies due to diverse use cases, data, and ML models. As a result, some generalizations were necessary to synthesize findings. Overall, ML is a powerful tool for leveraging smartphone sensor data to predict and reduce public transport delays. When viewed across the four themes, ML shows strong potential to enable proactive and reactive strategies that enhance system efficiency.

6 Conclusion

This SLR comprises several core sections and key findings. The introduction frames the study's structure and methods. The background highlights challenges in public transport delays, the role of ML, and how ML accuracy is assessed, laying the foundation for the study's focus.

The methodology details data collection, including metadata such as publication date, title, and authorship, and emphasizes the role in applying inclusion/exclusion criteria. It outlines data extraction criteria, stressing the capture of key codes, concepts, and other relevant information essential for synthesis and thematic coding. The findings reveal four central themes namely ML in transportation, data sensing technologies, operational efficiency, and transport modes and choices. In data sensing technologies, various sensor types are explored, highlighting the need for careful selection. For example, GPS sensors can drain battery with minimal gains in location accuracy.

ML in Transportation underscores the importance of selecting models suited to specific use cases, as not all perform equally across datasets or feature sets. Operational Efficiency refocuses on the user, emphasizing that public satisfaction and experience are vital, given their direct impact. Transport Modes and Choices stress the influence of contextual information on feature selection, which shapes the success of ML models. The discussion reflects critically on these findings, recognizing ML's strong potential for predicting and mitigating transport delays via smartphone sensor data, while noting the inconsistency in ML model use across studies as a key limitation.

In conclusion, the review answers its research question, "*How effective is ML in analyzing smartphone tracking data to predict and reduce public transport delays?*" by affirming that, when applied with context-specific considerations, ML can play a significant role in improving public transport efficiency.

7 Recommended Future Research

To offer meaningful future research recommendations, they must align with results derived from the extracted codes and study distribution across categories. A statistical method, such as quartile analysis, can identify underexplored areas. Quartiles are calculated using Quartile $= (1/4) \times (N + 1)$, where N is the number of data points.

Figure 1 shows the following paper distribution: ML models (31), smartphone sensors (21), traffic management (17), delay prediction (15), passenger impact (11), disruption management (11), transport mode detection (7), and IoT integration (7) using quartile classification:

- Q1 (lowest 25%) includes transport mode detection, IoT integration, passenger impact, and possibly disruption management.
- Q2 (median 50%) includes traffic management and delay prediction.
- Q3 (top 25%) includes smartphone sensors and ML.

Recommendations should prioritize Q1 codes, as these reflect areas with comparatively limited research attention. Additionally, given the heterogeneity in ML studies identified throughout this review, there is a pressing need for a standardized ML performance framework. Such a model could unify evaluation criteria, streamline future reviews, and improve the comparability and utility of findings in this domain.

Acknowledgements This paper was the original work of author 1 (student). The document was originally submitted to Turnitin for student evaluation.

References

Alsrehin, N. O., Klaib, A. F., & Magableh, A. (2019). Intelligent transportation and control systems using data mining and machine learning techniques: A comprehensive study. *IEEE Access, 7*, 49830–49857.

Ang, K.L.-M., Seng, J. K. P., Ngharamike, E., & Ijemaru, G. K. (2022). Emerging technologies for smart cities' transportation: Geo-information, data analytics and machine learning approaches. *ISPRS International Journal of Geo-Information, 11*(2), 85.

Ankhit, P., Clare, B., Candis, A., Enxi, C., & Gavin, M. (2020). Design and development of an application for predicting bus travel times using a segmentation approach. In *International conference on geographic information system theory, applied management*.

Barnes, S. J., Pressey, A. D., & Scornavacca, E. (2019). Mobile ubiquity: Understanding the relationship between cognitive absorption, smartphone addiction and social network services. *Computers in Human Behavior, 90*, 246–258.

Bieler, M., Skretting, A., Büdinger, P., & Grønli, T. M. (2022). Survey of automated fare collection solutions in public transportation. *IEEE Transaction Intelligent Transportation Systems, 23*(9), 14248–14266.

Bruglieri, M., Bruschi, F., Colorni, A., Luè, A., Nocerino, R., & Rana, V. (2015). A real-time information system for public transport in case of delays and service disruptions. *Transportation Research Procedia, 10*, 493–502.

Chavhan, S., Gupta, D., Chandana, B. N., Khanna, A., & Rodrigues, J. J. P. C. (2020). IoT-based context-aware intelligent public transport system in a metropolitan area. *IEEE Internet Things Journal, 7*(7), 6023–6034.

Cheng, X., Fang, L., Hong, X., & Yang, L. (2017). Exploiting mobile big data: Sources, features, and applications. *IEEE Networks, 31*(1), 72–79.

Davenport, T. H., Barth, P., & Bean, R. (2012). How big data is different. *MIT Sloan Management Reviews, 54*(1), 43–46.

de Soares, E. F. S., Campos, C. A. V., & de Lucena, S. C. (2019). Online travel mode detection method using automated machine learning and feature engineering. *Future Generation Computer Systems, 101*, 1201–1212.

Du, R., Qiu, G., Gao, K., Hu, L., & Liu, L. (2020). Abnormal road surface recognition based on smartphone acceleration sensor. *Sensors, 20*(2), 451.

Engelbrecht, J., Booysen, M. J., van Rooyen, G.-J., & Bruwer, F. J. (2015). Survey of smartphone based sensing in vehicles for intelligent transportation system applications. *IET Intelligent Transport Systems, 9*(10), 924–935.

Ge, L., Sarhani, M., Voß, S., & Xie, L. (2021). Review of transit data sources: Potentials, challenges and complementarity. *Sustainability, 13*(20), 11450.

Gershenfeld, N., Krikorian, R., & Cohen, D. (2004). The internet of things. *Scientific Americans, 291*(4), 76–81.

Ghahramani, M., Zhou, M., & Wang, G. (2020). Urban sensing based on mobile phone data: Approaches, applications, and challenges. *IEEE/CAA Journal of Automation Sinic, 7*(3), 627–637.

Heydary, M. H., Pimpale, P., & Panangadan, A. (2018). Automatic identification of use of public transportation from mobile sensor data. In *IEEE green technologies conference (GreenTech)* (pp. 4–6).

Hosseini, S. H., & Gentile, G. (2022). Smartphone-based recognition of access trip phase to public transport stops via machine learning models. *Transport and Telecommunication, 23*(4), 273–283.

Janiesch, C., Zschech, P., & Heinrich, K. (2021). Machine learning and deep learning. *Electronic Markets, 31*(3), 685–695.

Jevinger, Å., & Persson, J. A. (2019). Exploring the potential of using real-time traveler data in public transport disturbance management. *Public Transportation, 11*(2), 413–441.

Jordan, M. I., & Mitchell, T. M. (2015). Machine learning: Trends, perspectives, and prospects. *Science, 349*(6245), 255–260.

Kaewunruen, S., Sresakoolchai, J., Huang, J., Harada, S., & Wisetjindawat, W. (2021). Human activity vibrations. *Data, 6*(10), 104.
Kitchenham, B. (2004). *Procedures for performing systematic reviews* (pp. 1–26). Keele University.
Liyanage, S., Dia, H., Abduljabbar, R., & Bagloee, S. A. (2019). Flexible mobility on-demand: An environmental scan. *Sustainability, 11*(5), 1262.
Murray, A. T., Davis, R., Stimson, R. J., & Ferreira, L. (1998). Public transportation access. *Transportation Research Part d: Transport and Environment, 3*(5), 319–328.
Nabian, M. A., Alemazkoor, N., & Meidani, H. (2020). Predicting near-term train schedule performance and delay using bi-level random forests. *Transportation Research Records, 2673*(5), 564–573.
Nirmal, P., Disanayaka, I., Haputhanthri, D., & Wijayasiri, A. (2021). Transportation mode detection using crowdsourced smartphone data. In *Proceedings of the 28th conference open innovations association (FRUCT)* (pp. 27–29).
Özer, Ç. E. (2018). A research on machine learning methods and its applications. *Journal of Educational Technology and Online Learning, 1*(3), 25–40.
Prudhomme, R., Koning, M., Lenormand, L., & Fehr, A. (2012). Public transport congestion costs: The case of the Paris subway. *Transport Policy, 21*, 101–109.
Shanthi, N., et al. (2022). Analysis on the bus arrival time prediction model for human-centric services using data mining techniques. *Computational Intelligence and Neuroscience, 2022*, 7094654.
Stjernborg, V., & Mattisson, O. (2016). The role of public transport in society: A case study of general policy documents in Sweden. *Sustainability, 8*(11), 1120.
TomTom. (2024). *Johannesburg traffic*. https://www.tomtom.com/traffic-index/johannesburg-traffic/. Accessed 8 August 2024.
Tuckett, A. G. (2005). Applying thematic analysis theory to practice: A researcher's experience. *Contemporary Nurse, 19*(1–2), 75–87.
Willmott, C. J., & Matsuura, K. (2005). Advantages of the mean absolute error (MAE) over the root mean square error (RMSE) in assessing average model performance. *Climatic Research, 30*(1), 79–82.
Xiao, G., Juan, Z., & Zhang, C. (2015). Travel mode detection based on GPS track data and Bayesian networks. *Computational Environment Urban Systems, 54*, 14–22.
Zantalis, F., Koulouras, G., Karabetsos, S., & Kandris, D. (2019). A review of machine learning and IoT in smart transportation. *Future Internet, 11*(4), 6595.
Zhou, L., Pan, S., Wang, J., & Vasilakos, A. V. (2017). Machine learning on big data: Opportunities and challenges. *Neurocomputing, 237*, 350–361.

A Systematic Literature Review of Reference Architectures in the Blockchain Domain

Hauke Precht and André P. Calitz

Abstract The domain of blockchain-oriented software engineering emerged with the advent of blockchain technology. While research focuses especially on the development, verification and testing of smart contracts, the applied architectural methodology, specifically research relating to reference architectures, is scarcely researched. The lack of insights into the adoption of reference architectures in the blockchain domain is noticeable, despite the fact that reference architectures hold the potential to ease and standardize the development of similar systems. To investigate the current applications of reference architectures in the blockchain domain, a systematic literature review combined with a qualitative content analysis was conducted. The underlying methodologies, definitions, evaluation strategies and focus of the identified reference architectures were examined. The analysis reveals that most papers lack a solid theoretical foundation in terms of reference architectures and related methodologies. Consequently, the authors advocate for greater scientific rigor in future research, including the integration of established methodologies for reference architecture development in the field of blockchain-oriented software engineering. The findings enrich the existing body of knowledge by offering a comprehensive overview of the current state of reference architectures in the blockchain domain and identifying key shortcomings to inform future research. Furthermore, this study serves as a valuable resource for practitioners aiming to develop blockchain-based systems.

H. Precht (✉)
Department of Computing Science, School of Computing Science, Business Administration, Economics, and Law, Carl von Ossietzky Universität Oldenburg, Oldenburg, Germany
e-mail: hauke.precht@uol.de

A. P. Calitz
Department of Computing Sciences, Nelson Mandela University, Gqeberha, South Africa
e-mail: Andre.Calitz@Mandela.ac.za

J. Marx Gómez et al. (eds.), *Recent Advances in Information Systems*, Progress in IS,
https://doi.org/10.1007/978-3-032-14099-9_22

1 Introduction

In the global effort toward digitalization, the advent of Distributed Ledger Technology (DLT) and especially blockchain technology enables new ways for creating and managing (digital) assets of all kinds. By providing a tamper-resistant recordkeeping along with the possibility to transfer ownership of assets, blockchain has become a focal point in academia and industry alike. From a technical point of view, a blockchain implements an ordered list of blocks that are cryptographically chained to the previous block, leveraging hash functions (Xu et al., 2018). Transactions that are executed in a blockchain system are grouped and stored in blocks that become persistent and tamper-resistant based on cryptographic functions (Xu et al., 2018). Given its peer-to-peer nature, a blockchain is decentralized, where each participating node (composed of client software and necessary information) can execute transactions (Thapa et al., 2021). A consensus algorithm ensures the consistent state of the blockchain across all participating nodes, solving the double-spending problem which describes the problem of assuring that a digital asset is not transacted more than once (Sunyaev, 2020; Zheng et al., 2017). In addition, with smart contracts (SCs), initially coined by Nick Szabo in 1996 (Szabo, 1996), new possibilities emerged for the development of blockchain-based applications, known as distributed Applications (dApps). However, due to the novelty of SCs and the decentralized nature of blockchain itself, developers have tended to make mistakes and introduce bugs into the blockchain system and SCs (Chepurnoy & Rathee, 2018; Gao, 2019). These issues can lead to potentially high and expensive risks. Notably, the Parity Wallet hack has been analyzed by Destefanis et al. (2018) with the result that the flaws and vulnerabilities could have been detected with proper development methods, leading to a call for Blockchain Oriented Software Engineering (BOSE) (Destefanis et al., 2018). To this end, research has been conducted to develop (formal) verification and testing tools, as well as their incorporation into continuous integration and continuous deployment pipelines to aid developers in developing secure and correct smart contracts (Bhardwaj et al., 2020; Coblenz et al., 2019; Precht et al., 2023; Wohrer & Zdun, 2021).

The process of software development for blockchain systems has been researched (Khalid and Brown, 2023) and dedicated agile methods have been proposed (Marchesi et al., 2020). Similarly, the domain of requirements engineering has been revisited and discussed in a blockchain context (Demi, 2020; Reddivari et al., 2022). Mapping these studies to the phases of the software development life cycle (Ruparelia, 2010) reveals that most studies focus on the phases of development and validation. However, the preceding design phase is scarcely addressed, despite being an enabler for clear definitions and specifications for the software architecture that ensures the development of a successful and purposeful software system (Shaw et al., 1995). Considering the development of similar systems or the development of standards, usually requiring a higher level of abstraction, *reference architectures* are favored (Nakagawa & Antonino, 2023). They can serve as a blueprint for concrete product architectures and implementations for similar types of applications, providing patterns and structures to be reused (Angelov et al., 2012; Cervantes & Kazman, 2024; Nakagawa & Antonino, 2023). In this paper, the authors define a reference architecture as a set of patterns, structures and best practices to aid the design and implementation of similar types of applications. Acknowledging the impact of reference architecture and the role of the design phase in the overall software development lifecycle, which holds for BOSE as well, the aim is to investigate how reference architectures are developed and deployed in the blockchain domain.

This paper aims to answer the following research question: *How are reference architectures in the blockchain domain used, developed and evaluated?* To address this research question, a Systematic Literature Review (SLR) was combined with a qualitative content analysis. It investigated a total of 43 papers for their development and usage of reference architecture in the context of blockchain-based systems. The analysis shows that a majority of papers lack an underlying theoretical foundation in terms of definitions, methodology and evaluation strategies. Therefore, the authors call for more scientific rigor in future research, while also investigating the incorporation of methodologies for reference architecture development in the realm of BOSE. The paper contributes to the existing body of knowledge by providing an overview of the state of the art of reference architectures in the blockchain domain, while also investigating existing shortcomings to derive future research directions. In addition, this overview aids practitioners seeking to develop their own blockchain-based systems and serves as a starting point to potentially adopt existing reference architectures.

2 Methodology—Systematic Literature Review (SLR) and Qualitative Content Analysis

In order to answer the previously stated research question, *How are reference architectures in the blockchain domain used, developed and evaluated?* A SLR along with a qualitative content analysis were conducted. Firstly, the guidelines outlined by Kitchenham and Charters for conducting an SLR in the context of software engineering (Kitchenham & Charters, 2007) were followed. The identified set of literature was then be subject to a qualitative content analysis following the guidelines defined by Mayring (2015). In the following sub-sections, the definition of the search terms and the databases queried, along with exclusion and inclusion criteria, are presented. The paper screening workflow following the *PRISMA* approach (Page et al., 2021) is presented in Fig. 1. Finally, the process of defining coding categories applied in the study and the qualitative content analysis are presented.

2.1 Definition of Search Terms and Selection of Databases

The aim for a comprehensive and exhaustive review (Cooper, 1988) is to define the search terms as loosely as possible, leading to the search term "*reference architecture*" *AND* "blockchain". Established databases from the domains of Computer Science/Software Engineering, as well as Information Systems, were investigated and queried. In total, seven databases were queried, namely *ACM Digital Library, IEEE Xplore, Science Direct, AIS Digital Library* and *Springer Link* while also making use of the meta-databases *Scopus* and *Web of Science*. Note that gray literature was not included; however, the aim was to extract the current state of the art and discussions in the scientific community. The search and screening process, along with inclusion and exclusion criteria, are presented in Fig. 1, whose form and style are adopted from the PRISMA approach (Page et al., 2021). The search was conducted between March 19, 2025, and March 28, 2025. The focus was on studies in which a reference architecture in the blockchain domain is developed and discussed in greater detail. This means that studies are in the foreground where blockchain serves either an enabler or an essential building block in the reference architecture or that the reference architecture is focused on blockchain itself. Studies, such as Bodkhe et al. (2020), for example, have not taken into consideration of reference architectures, as they are only briefly mentioned in the related work section and no original reference architecture is developed or discussed. Furthermore, the focus of the analysis is on reference architectures only, meaning studies that mainly discuss reference models, for example, Bhattacharya et al. (2022), are excluded from further analysis.

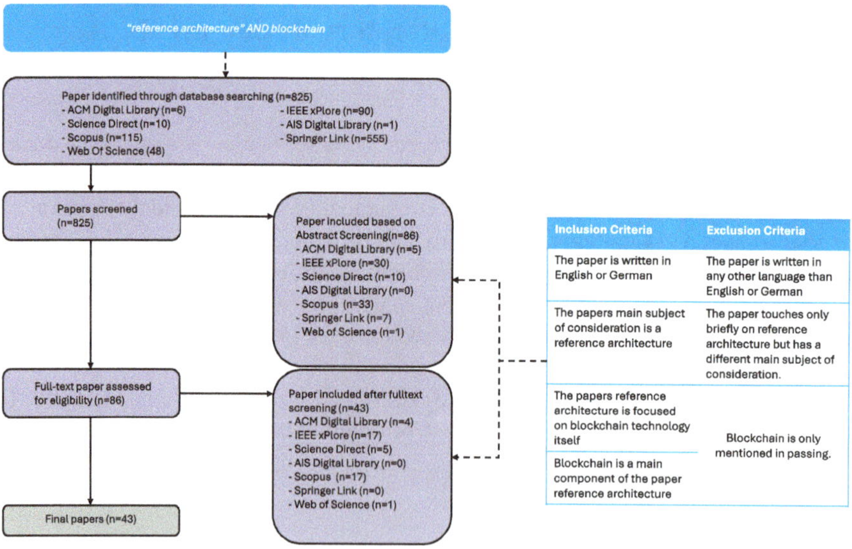

Fig. 1 PRISMA-based Page et al. (2021) presentation of the screening process with exclusion and inclusion criteria

Table 1 Overview of coding categories: explanation and justification

Coding category	Explanation	Justification
Focus	Is the reference architecture focused on blockchain itself or on systems where blockchain is a component?	Aiming for an exhaustive review, the current applicability of reference architectures in the blockchain domain was investigated
Definition	On what definition do the authors base their reference architecture and what goal do they aim for?	Several reference architecture definitions and underlying papers exist, which is why the aim was to investigate the prevalent one in the blockchain domain
Methodology	What methodology was used to develop the reference architecture?	There exist different methods to be used when developing reference architectures
Evaluation	How is the reference architecture evaluated?	The evaluation of abstract reference architectures is non-trivial and different methods exist, which is why the aim was to create an overview of methods used in the blockchain domain

2.2 Coding of Papers

In this paper, a combination of deductive and inductive coding was used. Coding, in this context, refers to the systematic process of identifying or generating codes—"researcher-generated word[s] or short phrase[s] that symbolically assign a summative, salient, essence-capturing, and/or evocative attribute for a portion of language-based or visual data" (Saldaña et al., 2021, 362)—and applying them to segments of text or visual material. Following Mayring (2015), *operationalization* was used to derive the main coding categories based on the previously stated research question. Both authors derived and jointly agreed on the main coding categories *Focus*, *Definition*, *Methodology* and *Evaluation*. These main coding categories are explained and justified in Table 1. Within the main coding, concrete characteristics as sub-codes were inductively created by the first author alone, following open coding and general coding guidelines [e.g., defined by Saldaña et al. (2021)]. The open coding was conducted on a detailed level, also known as *micro-coding* or *splitter coding* (Saldaña et al., 2021), where single sentences are coded. For the coding process, the software MAXQDA 24[1] was used. Based on the open coding carried out, six sub-codes for the main coding categories of *methodology*, *definition* and *evaluation*, respectively, were identified. For the main category *focus* 29 sub-codes were identified.

[1] https://www.maxqda.com/.

3 Results

Each of the 43 selected papers for detailed analysis was coded according to the process described in Sect. 2.2. In this section, the descriptive results for each coding category are presented. An in-depth discussion follows in Sect. 4.

3.1 *Focus of Reference Architectures in the Blockchain Domain*

Regarding the first main coding category, *Focus*, the aim was to investigate the current domains and points of interest for the use of reference architectures in the blockchain domain. In the open coding process for sub-codes, it became apparent that two major distinctions were observed. Firstly, reference architectures that are *for* blockchain itself, i.e., to be used when building a blockchain system and secondly, reference architectures that focus on a specific domain or other technology in which blockchain components play a crucial role.

The reference architecture where blockchain is a component and with a focus on a specific domain is more prevalent (33 papers) compared to reference architectures for blockchain itself (ten papers). For the papers that focus on reference architectures for blockchain itself, further distinction are made: Four papers focused on blockchain in general, while two papers focused on the governance aspect in blockchain systems, as well as two papers primarily focused on the security aspects in blockchain. Further blockchain-focused reference architectures were aimed specifically at Hyperledger Fabric and SCs. Considering reference architectures where blockchain is an enabling component, the application domains are manifold as shown in Fig. 2. They range from crowd-sourcing to digital twins, over to software-defined networks or FinTech and Industry 5.0. However, the most prominent application is in the healthcare domain with five papers that design their reference architecture for this domain, followed by the domain of smart manufacturing ($n = 4$).

3.2 *Used Definitions*

Regarding the main coding category *definition*, only fourteen papers were identified that *do* provide some form of definition. From these 14 papers, three provided their own general definition or description of a reference architecture without anchoring it to existing definitions and theory. The most commonly used definition is based on Angelov et al. (2012) ($n = 4$). Further definitions refer to Cloutier et al. (2010), Bass et al. (2003), Galster and Avgeriou (2011) or Taylor (2009). While these sources are rooted in software engineering and software architecture, one paper referred to Moghaddam et al. (2018) whose work is from the smart manufacturing domain.

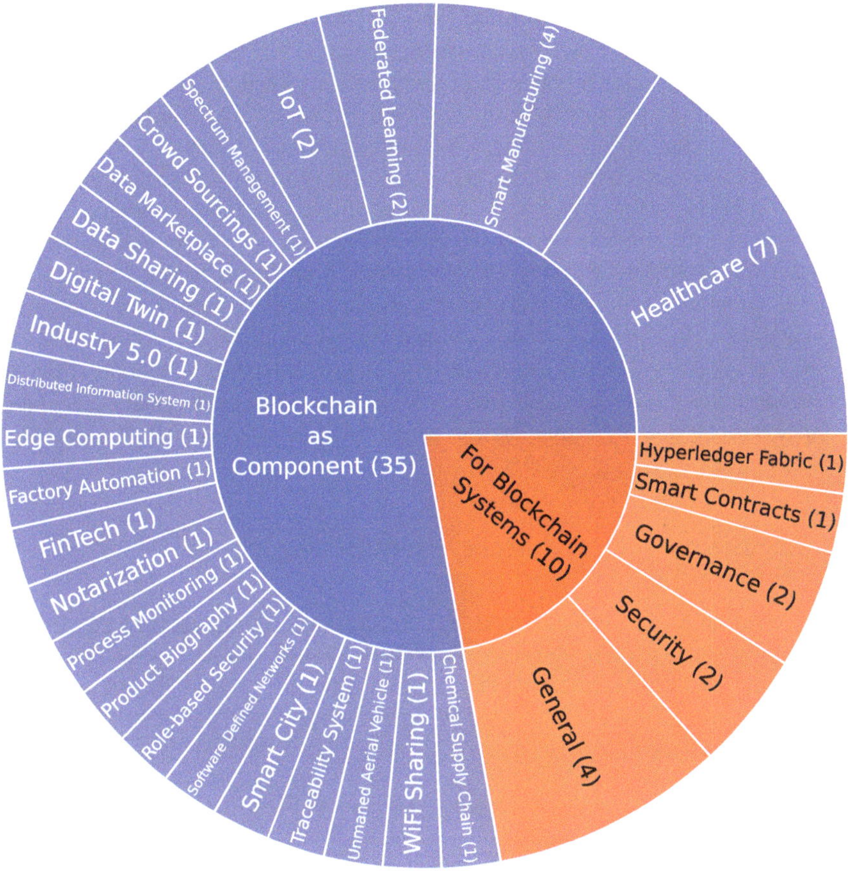

Fig. 2 Overview of reference architecture focus in the blockchain domain

3.3 Used Development and Evaluation Methodologies

Considering methodologies in software engineering and software architecture, the analysis shows that the majority of the analyzed papers do not refer to *any* specific methodology for the development of the respective reference architectures (see Fig. 3a). For 22 papers, the authors were not able to identify any mentions of a method or methodology used to create the reference architecture. The study explicitly focused on methodologies for the reference architecture development. The analyzed studies may describe methodologies for other areas of the respective paper and still be categorized as "no method mentioned" in the context of this study. For 14 papers, the papers indicate to what extent they build upon an existing reference architecture. Note that only two papers referred to official standards provided by IEEE, namely IEEE 1471 and IEEE 4201, respectively. Furthermore, only two papers defined their own methodology in order to develop the reference architectures. The well-known Design Science Research (DSR) approach (Hevner et al., 2004) from the Information Systems community was used in two papers. Evaluation, usually being a part of the reference architecture development methodologies, was also analyzed in the study. An overview of the identified evaluation methods is shown in Fig. 3b. With 21 papers, an evaluation by instantiation was the most used evaluation strategy. The main goal was to show the technical feasibility of the built reference architecture by deriving a specific implementation. To implementation or instantiation as a means of evaluation, it was found that the next most used, specified in three papers, was the usage of simulations instead of a complete implementation. In addition, one paper evaluated their developed reference architecture by conducting interviews and one paper conducted a mapping of the defined reference architecture components to existing technologies. Only one of the analyzed papers used an established architecture evaluation method, in this case, Architecture Tradeoff Analysis Method (ATAM).

4 Discussion—Toward More Scientific Rigor

Building on the results of the analysis of currently discussed aspects of the deployment of reference architectures in the blockchain domain, this section will provide a critical discussion of the findings. The analysis showed that 29 of the 43 papers do

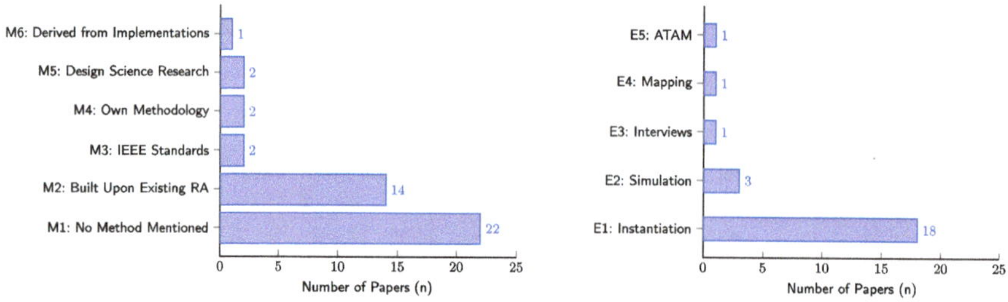

M1: (Sato et al., 2021; Ma et al., 2024; Cunningham et al., 2021; Bhattacharya et al., 2022; Isaja 2019; Sellak et al., 2021; García et al., 2023; Gong et al., 2021; Farahani et al., 2021; Gohar et al., 2022; Goh et al., 2023; Leng et al., 2021; Verma et al., 2022; Aviv et al., 2023; Demi et al., 2020; Colman et al., 2019; Wang et al., 2022; Liang et al., 2021; Saraswat et al., 2022; Nguyen et al., 2025; Romano et al.,2021; Ramachandran 2023)a

M2: (Wan et al., 2020; Di Ciccio et al. 2022; Sittón-Candanedo et al., 2019; Somma et al., 2024; Leng et al., 2023; Homoliak et al., 2021; Liu et al., 2023; Homoliak et al., 2019; Coll-Mayor et al., 2023; Barata et al., 2020; Abreu et al., 2020; Janiesch et al., 2023; Elsheikh et al., 2022; Wan et al., 2021)

M3: (Yalcinkaya et al., 2020; Rashid et al., 2020) M4: (Siriweera et al., 2023; Siriweera et al., 2023) M5: (Florez et al., 2022; Weber et al., 2022) M6: (Leeming et al., 2019)

(a) Named Research Methods for RA Development

E1: (Gong et al., 2021; Ma et al., 2024; Wan et al., 2020; Yalcinkaya et al., 2020; Weber et al., 2022; Cunningham et al., 2021; Bhattacharya et al., 2022; Siriweera et al., 2023; García et al., 2023; Sittón-Candanedo et al., 2019; Somma et al., 2024; Farahani et al., 2021; Gohar et al., 2022; Goh et al., 2023; Aviv et al., 2023; Colman et al., 2019; Saraswat et al., 2022; Abreu et al., 2020)

E2: (Liang et al., 2021; Ramachandran 2023; Janiesch et al., 2023) E3: (Weber et al., 2022) E4: (Liu et al., 2023) E5: (Wang et al., 2022)

(b) Used Evaluation Methods

Fig. 3 Overview of research and evaluation methods

not define the term reference architecture at all, despite aiming to develop one. Considering that, based on the domain background of readers (e.g., software engineer vs. systems engineer), different expectations are connected to the term reference architecture (Cloutier et al., 2010), a lack of proper definition can lead to misunderstandings. Note that the meaning of the term reference architecture depends not only on the role but also on the academic background. For example, in the domain of business informatics, a reference architecture is often associated with a reference *information* architecture, specifically focusing on information systems (vom Brocke, 2003).

Considering this lack of underlying theory regarding reference architecture makes it challenging to evaluate the presented reference architectures in terms of their classification as a reference architecture. While common reasons for the development of a reference architecture are to ease future development of similar systems by defining reusable structures and patterns (i.e., the reference architecture acting as a blueprint) (Angelov et al., 2012; Cervantes & Kazman, 2024; Nakagawa & Antonino, 2023) or to help enable the development of standards (Nakagawa & Antonino, 2023; Reidt et al., 2018), 24 papers did not provide any detail concerning the motivation for the development of their respective reference architecture. Considering ongoing discussions of definitions for the term reference architecture (Angelov et al., 2009; Cloutier et al., 2010; Reidt et al., 2018) it is even more important for authors to properly anchor their understanding of a reference architecture to existing literature and theories or to create their own definition to avoid any confusion.

A similar lack of underlying theory has also been observed when it comes to the methodology used to develop the respective reference architectures. As shown in the Sect. 3, only one paper refers to an existing theory for the development of reference architectures, while 14 defined their own methodology and development steps. This trend is also observed in terms of evaluation strategies for the developed reference architectures. The analysis shows that the majority of studies choose to evaluate their developed reference architecture by instantiating and implementing it, sometimes in the context of a case study. Considering the main two evaluation strategies being questioning techniques and measuring techniques (Dobrica & Niemela, 2002) this falls in the latter class as the implementation often provides metrics such as transactions per second. Note that only one paper refers to an established architecture evaluation method, in this case ATAM. Other methods specifically designed for analysis and evaluation of software architectures, for example, Scenario-Based Architecture Analysis Method (SAAM), Extending SAAM by Integration in the Domain (ESAAMI) or the Software Architecture Evaluation Model (SAEM) are not taken into consideration. This indicates a lack of scientific rigor for the development of reference architectures in the blockchain domain. Considering that Treiblmaier (Treiblmaier, 2020) came to a similar conclusion when discussing the case study methodology in blockchain use cases, the authors support the call for more scientific rigor in blockchain research in general.

Blockchain development poses challenges that differ from traditional software engineering, this has already been highlighted in 2017 by Porru et al. (2017) and has been underlined, for example, by a survey among blockchain developers, where 93%

of the respondents state that the development in the context of blockchain systems differs from traditional software development (Bosu et al., 2019). Thus, it is argued that dedicated testing, collaboration, debugging and software tools for smart contract languages in the blockchain domain are needed (Porru et al., 2017). These findings provided input to a call for BOSE (Destefanis et al., 2018). Despite the fact that recent studies also name the integration of blockchain in existing systems as a challenge, the main focus still lies on the development of smart contracts (Bosu et al., 2019; Porru et al., 2017; Vacca et al., 2021), currently neglecting the aspect of reference architectures. However, as shown in this study, current studies that focus on reference architecture development in the blockchain domain lack rigorous methodological approaches or even definitions in the context of reference architectures. Therefore, the findings indicate that further research needs to be conducted to investigate how the aspect of architecture development and evaluation can be incorporated into the emerging research stream of BOSE.

5 Conclusions

In this paper, the authors conducted a SLR combined with a qualitative content analysis to answer the question of how reference architectures are developed and evaluated in the blockchain domain. For the SLR, the authors conducted a keyword search in the databases ACM Digital Library, IEEE Xplore, Science Direct, AIS Digital Library, Springer Link, Scopus, and Web of Science resulting in a set of 825 papers. The analysis of 43 papers that matched the inclusion criteria from this initial set highlights the lack of underlying rigorous methodology when it comes to the development and evaluation in the blockchain domain. The findings indicate that reference architectures in the blockchain domain are either focused on blockchain itself or on a domain where blockchain is an architectural component or enabler. With the discussion on the lack of scientific rigor, the findings suggest a call for dedicated research efforts focusing on architecture development in the context of blockchain-oriented software engineering. The descriptive results of the SLR, along with the discussion in the context of blockchain-oriented software engineering, contribute to the scientific body of knowledge by identifying the current state of the art of reference architectures in the blockchain domain, highlighting future research directions to meet existing shortcomings. Since this study is limited to scientific literature, future work should also take the industry perspective into consideration. For example, the application of potential reference architectures in the industry needs to be evaluated. It should be analyzed how such reference architecture impacts the design and implementation of blockchain-oriented software. Furthermore, there is potential to investigate the impact of potential reference architectures in domains such as agriculture and livestock farming, where blockchain-enabled digitization has only recently begun.

Acknowledgements The research is supported by funds of the Federal Ministry of Argiculture, Food and Regional Identity (BMLEH) based on a decision of the Parliament of the Federal Republic of Germany. The Federal Office for Agriculture and Food (BLE) provides coordinating support for digitalization in agriculture as funding organization, grant number 2822ZR011.

Disclosure of Interests The authors have no competing interests to declare that are relevant to the content of this article.

References

Abreu, A. W. S., Coutinho, E. F., & Bezerra, (2020). In E. Cavalcante, F. Dantas, T. Batista, & G. Pinto (Eds.), *C.I.M.: A blockchain-based architecture for query and registration of student degree certificates* (pp. 151–160). ACM. https://doi.org/10.1145/3425269.3425285

Angelov, S., Grefen, P., & Greefhorst, D. (2009). *A classification of software reference architectures: Analyzing their success and effectiveness* (pp. 141–150). IEEE. https://doi.org/10.1109/WICSA.2009.5290800

Angelov, S., Grefen, P., & Greefhorst, D. (2012). A framework for analysis and design of software reference architectures. *Information and Software Technology, 54*(4), 417–431. https://doi.org/10.1016/j.infsof.2011.11.009

Aviv, I., Barger, A., Kofman, A., & Weisfeld, R. (2023). Reference architecture for blockchain-native distributed information system. *IEEE Access, 11*, 4838–4851. https://doi.org/10.1109/ACCESS.2023.3235838

Barata, J., Pereira, V., & Coelho, M. (2020). *Product biography information system: A lifecycle approach to digital twins* (pp. 899–904). IEEE. https://doi.org/10.1109/SMC42975.2020.9283061

Bass, L., Clements, P., & Kazman, R. (2003). *Software architecture in practice* (2nd ed.). Addison-Wesley Professional.

Bhardwaj, A., Shah, S. B. H., Shankar, A., Alazab, M., Kumar, M., & Gadekallu, T. R. (2020). Penetration testing framework for smart contract blockchain. *Peer-to-Peer Networking and Applications, 5*(2), 303. https://doi.org/10.1007/s12083-020-00991-6

Bhattacharya, P., Patel, F., Ralegankar, V., Thakkar, B., Tanwar, S., & Abouhawwash, M. (2022). Trusted 6G-envisioned dynamic spectrum allocation: A reference model, layered architecture, and use-case. In P. K. Singh, M. H. Kolekar, S. Tanwar, S. T. Wierzchoń, & R. K. Bhatnagar (Eds.), *Emerging technologies for computing, communication and smart cities* (Vol. 875, pp. 405–418). Springer Nature Singapore. https://doi.org/10.1007/978-981-19-0284-0_30

Bhattacharya, P., Verma, A., & Sharma, G. (2022). Blockchain-driven and IoT-assisted chemical supply-chain management. In P. K. Singh, M. H. Kolekar, S. Tanwar, S. T. Wierzchoń, & R. K. Bhatnagar (Eds.), *Emerging technologies for computing, communication and smart cities* (Vol. 875, pp. 779–791). Springer Nature Singapore. https://doi.org/10.1007/978-981-19-0284-0_57

Bodkhe, U., Tanwar, S., Parekh, K., Khanpara, P., Tyagi, S., Kumar, N., & Alazab, M. (2020). Blockchain for Industry 4.0: A comprehensive review. *IEEE Access, 8*, 79764–79800. https://doi.org/10.1109/ACCESS.2020.2988579

Bosu, A., Iqbal, A., Shahriyar, R., & Chakraborty, P. (2019). Understanding the motivations, challenges and needs of blockchain software developers: A survey. *Empirical Software Engineering, 24*(4), 2636–2673. https://doi.org/10.1007/s10664-019-09708-7

Cervantes, H., & Kazman, R. (2024). *Designing software architectures: A practical approach*. SEI Series in Software Engineering (2nd ed.). https://learning.oreilly.com/library/view/-/9780138108069/?ar

Chepurnoy, A., & Rathee, M. (2018). Checking laws of the blockchain with property-based testing. In *2018 International workshop on blockchain oriented software engineering (IWBOSE)* (pp. 40–47). IEEE. https://doi.org/10.1109/IWBOSE.2018.8327570

Cloutier, R., Muller, G., Verma, D., Nilchiani, R., Hole, E., & Bone, M. (2010, January). The concept of reference architectures. *Systems Engineering, 13*(1), 14–27. https://doi.org/10.1002/sys.20129

Coblenz, M., Sunshine, J., Aldrich, J., & Myers, B. A. (2019). Smarter smart contract development tools. In *2019 IEEE/ACM 2nd International workshop on emerging trends in software engineering for blockchain (WETSEB)* (pp. 48–51). IEEE. https://doi.org/10.1109/WETSEB.2019.00013

Coll-Mayor, D., & Notholt, A. (2023). Distributed ledger technologies for the energy sector: Facilitating interoperability analysis. *IEEE Open Access Journal of Power and Energy, 10*, 593–604. https://doi.org/10.1109/OAJPE.2023.3297199

Colman, A., Chowdhury, M. J. M., & Baruwal Chhetri, M. (2019). *Towards a trusted marketplace for wearable data* (pp. 314–321). IEEE. https://doi.org/10.1109/CIC48465.2019.00044

Cooper, H. M. (1988). Organizing knowledge syntheses: A taxonomy of literature reviews. *Knowledge in Society, 1*(1), 104–126. https://doi.org/10.1007/BF03177550

Cunningham, J., Davidge, G., Davies, N., Devaney, S., Holm, S., Harding, M., Leeming, G., Neumann, V., & Ainsworth, J. (2021). Blockchain native data linkage. *Frontiers in Blockchain, 4*, Article 667388. https://doi.org/10.3389/fbloc.2021.667388

Datta, S. K. (2024). *Reference architecture for integrating computable smart contracts with existing blockchains* (pp. 1–3). IEEE. https://doi.org/10.1109/ICCE-Asia63397.2024.10773674

Demi, S. (2020). Blockchain-oriented requirements engineering: A framework. In T. Breaux (Ed.), *Proceedings, 28th IEEE international requirements engineering conference: RE'20: Zurich, Switzerland, August 31–September 4, 2020: Proceedings* (pp. 428–433). IEEE Computer Society, Conference Publishing Services, Los Alamitos, California. https://doi.org/10.1109/RE48521.2020.00063

Destefanis, G., Marchesi, M., Ortu, M., Tonelli, R., Bracciali, A., & Hierons, R. (2018). Smart contracts vulnerabilities: A call for blockchain software engineering? In *2018 International workshop on blockchain oriented software engineering (IWBOSE)* (pp. 19–25). IEEE. https://doi.org/10.1109/IWBOSE.2018.8327567

Di Ciccio, C., Meroni, G., & Plebani, P. (2022). On the adoption of blockchain for business process monitoring. *Software and Systems Modeling,21*(3), 915–937. https://doi.org/10.1007/s10270-021-00959-x

Dobrica, L., & Niemela, E. (2002). A survey on software architecture analysis methods. *IEEE Transactions on Software Engineering,28*(7), 638–653. https://doi.org/10.1109/TSE.2002.1019479

Elsheikh, A. S. (2022). *Blockchain analytics reference architecture for FinTech—A positioning paper: Advancing FinTech with blockchain, data analytics, and enterprise architecture* (pp. 1–7). ACM. https://doi.org/10.1145/3531056.3531068

Farahani, B., Firouzi, F., & Luecking, M. (2021). The convergence of IoT and distributed ledger technologies (DLT): Opportunities, challenges, and solutions. *Journal of Network and Computer Applications,177*(102936), https://doi.org/10.1016/j.jnca.2020.102936

Florez, L. (2022). Blockchain governance using reference architectures. In Anais Do XXV Congresso Ibero-Americano Em Engenharia de Software (CIbSE 2022) (pp. 345–351). Sociedade Brasileira de Computação. https://doi.org/10.5753/cibse.2022.20983

Galster, M., Avgeriou, P. (2011). Empirically-grounded reference architectures: A proposal. In *Proceedings of the joint ACM SIGSOFT conference—QoSA and ACM SIGSOFT symposium—ISARCS on quality of software architectures—QoSA and architecting critical systems—ISARCS* (pp. 153–158). ACM. https://doi.org/10.1145/2000259.2000285

Gao, J.: Guided, automated testing of blockchain-based decentralized applications. In *2019 IEEE/ACM 41st International conference on software engineering: Companion proceedings (ICSE-Companion)* (pp. 138–140). IEEE. https://doi.org/10.1109/ICSE-Companion.2019.00059

García, J. R., Favela, J., & Sánchez-Torres, C. E. (2023). Traceable health data for consciously trained ML models. In J. Bravo, & G. Urzáiz (Eds.), *Proceedings of the 15th international conference on ubiquitous computing & ambient intelligence (UCAmI 2023)* (Vol. 842, pp. 279–284). Springer Nature Switzerland. https://doi.org/10.1007/978-3-031-48642-5_28

Goh, E., Kim, D. Y., Lee, K., Oh, S., Chae, J. E., & Kim, D. Y. (2023). Blockchain-enabled federated learning: A reference architecture design, implementation, and verification. *IEEE Access,11*, 145747–145762. https://doi.org/10.1109/ACCESS.2023.3345360

Gohar, A. N., Abdelmawgoud, S. A., & Farhan, M. S. (2022). A patient-centric healthcare framework reference architecture for better semantic interoperability based on blockchain, cloud, and IoT. *IEEE Access,10*, 92137–92157. https://doi.org/10.1109/ACCESS.2022.3202902

Gong, Y., Van Engelenburg, S., & Janssen, M. (2021). A reference architecture for blockchain-based crowdsourcing platforms. *Journal of Theoretical and Applied Electronic Commerce Research,16*(4), 937–958. https://doi.org/10.3390/jtaer16040053

Hevner, A. R., March, S. T., Park, J., & Ram, S. (2004). Design science in information systems research. *MIS Q,28*(1), 75–105.

Homoliak, I., Venugopalan, S., Hum, Q., & Szalachowski, P. (2019). *A security reference architecture for blockchains* (pp. 390–397). IEEE. https://doi.org/10.1109/Blockchain.2019.00060

Homoliak, I., Venugopalan, S., Reijsbergen, D., Hum, Q., Schumi, R., & Szalachowski, P. (2021). The security reference architecture for blockchains: Toward a standardized model for studying vulnerabilities, threats, and defenses. *IEEE Communications Surveys & Tutorials, 23*(1), 341–390. https://doi.org/10.1109/COMST.2020.3033665

Isaja, M. (2019). Reference architecture for factory automation using edge computing and blockchain technologies. In *The digital shopfloor industrial automation in the industry 4.0 era: Performance analysis and applications*, pp. 71–101.

Janiesch, C., Fischer, M., Imgrund, F., Hofmann, A., & Winkelmann, A. (2023). An architecture using payment channel networks for blockchain-based Wi-Fi sharing. *ACM Transactions on Management Information Systems, 14*(1), 1–24. https://doi.org/10.1145/3529097

Khalid, S., & Brown, C. (2023). *Software engineering approaches adopted by blockchain developers* (pp. 1–6). IEEE. https://doi.org/10.1109/SDS59856.2023.10329007

Kitchenham, B., Charters, S. (2007). Guidelines for performing systematic literature reviews in software engineering, 2.

Leeming, G., Cunningham, J., & Ainsworth, J. (2019). A ledger of me: Personalizing healthcare using blockchain technology. *Frontiers in Medicine,6*(171), *10.3389/fmed.2019.00171*

Leng, J., Ye, S., Zhou, M., Zhao, J. L., Liu, Q., Guo, W., Cao, W., & Fu, L. (2021). Blockchain-secured smart manufacturing in Industry 4.0: A survey. *IEEE Transactions on Systems, Man, and Cybernetics: Systems,51*(1), 237–252. https://doi.org/10.1109/TSMC.2020.3040789

Leng, J., Zhong, Y., Lin, Z., Xu, K., Mourtzis, D., Zhou, X., Zheng, P., Liu, Q., Zhao, J. L., & Shen, W. (2023). Towards resilience in Industry 5.0: A decentralized autonomous manufacturing paradigm. *Journal of Manufacturing Systems,71*95–114. https://doi.org/10.1016/j.jmsy.2023.08.023

Liang, Y., Lu, C., Zhao, Y., & Sun, C. (2021). Interference-based consensus and transaction validation mechanisms for blockchain-based spectrum management. *IEEE Access,9*, 90757–90766. https://doi.org/10.1109/ACCESS.2021.3091802

Liu, Y., Lu, Q., Yu, G., Paik, H. Y., & Zhu, L. (2023). *A pattern-oriented reference architecture for governance-driven blockchain systems* (pp. 23–34). IEEE. https://doi.org/10.1109/ICSA56044.2023.00011

Ma, Y., Chang, C., Wu, P., Xiao, J., & Yuan, L. (2024). BSSN-SDNs: A blockchain-based security service negotiation for the SDN interdomain. *Electronics,13*(16), 3120. https://doi.org/10.3390/electronics13163120

Marchesi, L., Marchesi, M., & Tonelli, R. (2020). ABCDE-agile block chain DApp engineering. *Blockchain: Research and Applications,1*(1), Article 100002. https://doi.org/10.1016/j.bcra.2020.100002

Mayring, P. (2015). Qualitative content analysis: Theoretical background and procedures. In A. Bikner-Ahsbahs, C. Knipping, & N. Presmeg (Eds.), *Approaches to qualitative research in mathematics education* (pp. 365–380). Springer, Netherlands, Dordrecht: Advances in Mathematics Education. https://doi.org/10.1007/978-94-017-9181-6_13

Moghaddam, M., Cadavid, M. N., Kenley, C. R., & Deshmukh, A. V. (2018). Reference architectures for smart manufacturing: A critical review. *Journal of Manufacturing Systems,49*(215–225). https://doi.org/10.1016/j.jmsy.2018.10.006

Nakagawa, E. Y., & Antonino, P. O. (2023). An overview of reference architectures. In E. Y. Nakagawa & P. Oliveira Antonino (Eds.), *Reference architectures for critical domains* (pp. 5–15). Springer International Publishing. https://doi.org/10.1007/978-3-031-16957-1_2

Nguyen, T. L., Nguyen, L., Hoang, T., Bandara, D., Wang, Q., Lu, Q., Xu, X., Zhu, L., & Chen, S. (2025). Blockchain-empowered trustworthy data sharing: Fundamentals, applications, and challenges. *ACM Computing Surveys,57*(8), 1–36. https://doi.org/10.1145/3718082

Page, M. J., McKenzie, J. E., Bossuyt, P. M., Boutron, I., Hoffmann, T. C., Mulrow, C. D., Shamseer, L., Tetzlaff, J. M., Akl, E. A., Brennan, S. E., Chou, R., Glanville, J., Grimshaw, J. M., Hróbjartsson, A., Lalu, M. M., Li, T., Loder, E. W., Mayo-Wilson, E., McDonald, S., & Moher, D. (2021). The PRISMA 2020 statement: An updated guideline for reporting systematic reviews. *BMJ (Clinical Research Edition),372*, 71. https://doi.org/10.1136/bmj.n71

Porru, S., Pinna, A., Marchesi, M., & Tonelli, R. (2017). Blockchain-oriented software engineering: Challenges and new directions. In *2017 IEEE/ACM 39th International conference on software engineering companion* (pp. 169–171). IEEE. https://doi.org/10.1109/ICSE-C.2017.142

Precht, H., Schwarm, F., & Marx Gómez, J. (2023). Enhancing smart contract quality by introducing a continuous integration pipeline for solidity based smart contracts. In J. Prieto, F. L. Benítez Martínez, S. Ferretti, D. Arroyo Guardeño, & P. Tomás Nevado-Batalla (Eds.), *Blockchain and applications, 4th international congress* (pp. 269–278). Springer International Publishing. https://doi.org/10.1007/978-3-031-21229-1_25

Ramachandran, M. (2023). S3EF-HBCAs: Secure and sustainable software engineering framework for healthcare blockchain applications. *Blockchain in Healthcare Today,6*(2). https://doi.org/10.30953/bhty.v6.286

Rashid, A., Masood, A., Abbas, H. (2020). Incorporating blockchain into role engineering: A reference architecture using ISO/IEC/IEEE 42010 notation. In S. Latifi (Ed.), *17th International conference on information technology—New generations (ITNG 2020)* (Vol. 1134, pp. 183–189). Springer International Publishing. https://doi.org/10.1007/978-3-030-43020-7_25

Reddivari, S., & Wilson, A. (2022). *Blockchain-oriented requirements engineering: New directions* (pp. 118–123). IEEE. https://doi.org/10.1109/IRI54793.2022.00036

Reidt, A., Pfaff, M., & Krcmar, H. (2018). Der referenzarchitekturbegriff im wandel der zeit. *HMD Praxis der Wirtschaftsinformatik,55*(5), 893–906. https://doi.org/10.1365/s40702-018-00448-8

Romano, D., & Schmid, G. (2021). Beyond Bitcoin: Recent trends and perspectives in distributed ledger technology. *Cryptography,5*(4), 36. https://doi.org/10.3390/cryptography5040036

Ruparelia, N. B. (2010). Software development lifecycle models. *ACM SIGSOFT Software Engineering Notes,35*(3), 8–13. https://doi.org/10.1145/1764810.1764814

Saldaña, J. (2021). *The coding manual for qualitative researchers* (4th ed.). SAGE.

Saraswat, D., Verma, A., Bhattacharya, P., Tanwar, S., Sharma, G., Bokoro, P. N., & Sharma, R. (2022). Blockchain-based federated learning in UAVs beyond 5G networks: A solution taxonomy and future directions. *IEEE Access,10*, 33154–33182. https://doi.org/10.1109/ACCESS.2022.3161132

Sato, T., Shimosawa, T., Kondo, Y., & Nishijima, N. (2023). Toward fully-decentralized system with Hyperledger fabric. *IEICE Communications Express,12*(5), 223–229. https://doi.org/10.1587/comex.2023XBL0011

Sellak, H., Chhetri, M. B., & Grobler, M. (2021). Towards a trusted collaborative medical decision-making platform. In H. Gao, X. Wang, M. Iqbal, Y. Yin, J. Yin, & N. Gu (Eds.), *Collaborative computing: Networking, applications and worksharing* (Vol. 350, pp. 407–418). Springer International Publishing. https://doi.org/10.1007/978-3-030-67540-0_25

Shaw, M., DeLine, R., Klein, D. V., Ross, T. L., Young, D. M., & Zelesnik, G. (1995). Abstractions for software architecture and tools to support them. *IEEE Transactions on Software Engineering,21*(4), 314–335. https://doi.org/10.1109/32.385970

Siriweera, A., & Naruse, K. (2023a). Internet of cross-chains: Model-driven cross-chain as a service platform for the Internet of everything in smart city. *IEEE Consumer Electronics Magazine,12*(3), 85–97. https://doi.org/10.1109/MCE.2021.3127429

Siriweera, A., & Naruse, K. (2023b). QoS-aware federated crosschain-based model-driven reference architecture for IIoT sensor networks in distributed manufacturing. *IEEE Sensors Journal, 23*(23), 29630–29644. https://doi.org/10.1109/JSEN.2023.3325342

Sittón-Candanedo, I., Alonso, R. S., Corchado, J. M., Rodríguez-González, S., & Casado-Vara, R. (2019). A review of edge computing reference architectures and a new global edge proposal. *Future Generation Computer Systems,99*, 278–294. https://doi.org/10.1016/j.future.2019.04.016

Somma, A., De Benedictis, A., Esposito, C., & Mazzocca, N. (2024). The convergence of digital twins and distributed ledger technologies: A systematic literature review and an architectural proposal. *Journal of Network and Computer Applications,225*, Article 103857. https://doi.org/10.1016/j.jnca.2024.103857

Sunyaev, A. (2020). Distributed ledger technology. In A. Sunyaev (Ed.), *Internet computing* (pp. 265–299). Springer International Publishing. https://doi.org/10.1007/978-3-030-34957-8_9

Szabo, N. (1996). Smart contracts: Building blocks for digital markets. http://www.fon.hum.uva.nl/rob/Courses/InformationInSpeech/CDROM/Literature/LOTwinterschool2006/szabo.best.vwh.net/smart_contracts_2.html

Taylor, R. N. (2009). *Software architecture: Foundations, theory, and practice*. Wiley.

Thapa, R., Sharma, P., Hüllmann, J. A., & Savarimuthu, B. T. R. (2021). Identifying influence mechanisms in permissionless blockchain communities: The bitcoin case (pp. 1–17)

Treiblmaier, H. (2020). Toward more rigorous blockchain research: Recommendations for writing blockchain case studies. In H. Treiblmaier & T. Clohessy (Eds.), *Blockchain and distributed ledger technology use cases, progress in IS* (Vol. 24, pp. 1–31). Springer International Publishing. https://doi.org/10.1007/978-3-030-44337-5_1

Vacca, A., Di Sorbo, A., Visaggio, C. A., & Canfora, G. (2021). A systematic literature review of blockchain and smart contract development: Techniques, tools, and open challenges. *Journal of Systems and Software,174*(110891). https://doi.org/10.1016/j.jss.2020.110891

Verma, A., Bhattacharya, P., Madhani, N., Trivedi, C., Bhushan, B., Tanwar, S., Sharma, G., Bokoro, P. N., & Sharma, R. (2022). Blockchain for Industry 5.0: Vision, opportunities, key enablers, and future directions. *IEEE Access,10*, 69160–69199. https://doi.org/10.1109/ACCESS.2022.3186892

vom Brocke, J. (2003). *Referenzmodellierung: Gestaltung Und Verteilung von Konstruktionsprozessen, Advances in Information Systems and Management Science* (Vol. 4). Logos-Verl.

Wan, Z., Cai, M., Hua, X., Yang, J., & Lin, X. (2020). Blockchain federation enabled trustable Internet of Things. In: J. E. Ferreira, B. Palanisamy, K. Ye, S. Kantamneni, & L. J. Zhang (Eds.), *Services—SERVICES 2020* (Vol. 12411, pp. 61–76). Springer International Publishing. https://doi.org/10.1007/978-3-030-59595-1_5

Wan, Z., Xia, X., & Hassan, A. E. (2021). What do programmers discuss about blockchain? A case study on the use of balanced LDA and the reference architecture of a domain to capture online discussions about blockchain platforms across Stack Exchange communities. *IEEE Transactions on Software Engineering,47*(7), 1331–1349. https://doi.org/10.1109/TSE.2019.2921343

Wang, Y., Li, S., Liu, H., Zhang, H., & Pan, B. (2022). *A reference architecture for blockchain-based traceability systems using domain-driven design and microservices* (pp. 269–278). IEEE. https://doi.org/10.1109/APSEC57359.2022.00039

Weber, T., & Buchkremer, R. (2022). Blockchain-based reference architecture for automated, transparent, and notarized attestation of compliance adaptations. *Applied Sciences,12*(9), 4531. https://doi.org/10.3390/app12094531

Wohrer, M., & Zdun, U. (2021). *DevOps for Ethereum blockchain smart contracts* (pp. 244–251). IEEE. https://doi.org/10.1109/Blockchain53845.2021.00040

Xu, X., Pautasso, C., Zhu, L., Lu, Q., & Weber, I.(2018). *A pattern collection for blockchain-based applications* (pp. 1–20). ACM. https://doi.org/10.1145/3282308.3282312

Yalcinkaya, E., Maffei, A., & Onori, M. (2020). Blockchain reference system architecture description for the ISA95 compliant traditional and smart manufacturing systems. *Sensors,20*(22), 6456. https://doi.org/10.3390/s20226456

Zheng, Z., Xie, S., Dai, H., Chen, X., & Wang, H. (2017). An overview of blockchain technology: Architecture, consensus, and future trends. In G. Karypis & J. Zhang (Eds.), *2017 IEEE International congress on Big Data—BigData congress 25–30 June 2017, Honolulu, Hawaii, USA: Proceedings* (pp. 557–564). IEEE. https://doi.org/10.1109/BigDataCongress.2017.85

HERAG: A Holistic Approach for Evaluating Organizational RAG Applications

Felix Kleinekathöfer, Felix Dick, Alina Ahrens, Tobias Oltmanns, Steffen Meeuw, René Kessler, and Jorge Marx Gómez

Abstract Retrieval-Augmented Generation (RAG) systems are becoming increasingly vital in enterprise environments due to their potential to streamline access to internal knowledge, support onboarding, and improve decision-making. As this technology gains traction, the need for effective evaluation methods becomes critical—yet current approaches remain fragmented and technically limited. This paper introduces HERAG, a holistic, tool-agnostic framework for systematically assessing RAG systems across all pipeline stages—from loading and indexing to retrieval, generation, and metric analysis. Unlike existing approaches that focus narrowly on performance metrics, technical implementations, or black-box evaluations, HERAG offers a structured evaluation matrix that enables modular, transparent, and context-aware assessments. It supports a wide range of configurations and enterprise-specific scenarios, helping practitioners identify impactful parameters, document system design decisions, and optimize performance. We demonstrate HERAG's practical applicability through a case study on retrieval quality within a corporate Q&A system, showing how different embedding models and top-k values affect performance. HERAG empowers developers and researchers to conduct comprehensive, comparative evaluations and build more robust, effective AI-assisted information systems.

1 Introduction

In modern enterprises, accessing relevant and up-to-date information has become a key success factor (Wu et al., 2025). Employees frequently navigate across fragmented data silos, with growing volumes of unstructured content (Wu et al., 2025). This fragmentation leads to redundant work efforts and productivity losses. Retrieval-

Felix Kleinekathöfer and Felix Dick—These authors contributed equally to this work.

F. Kleinekathöfer (✉) · F. Dick · A. Ahrens · T. Oltmanns · S. Meeuw · J. Marx Gómez
Department of Computing Science, Carl von Ossietzky Universität Oldenburg, Oldenburg, Germany
e-mail: felix.kleinekathoefer@uni-oldenburg.de

R. Kessler
abat AG, Bremen, Germany

J. Marx Gómez et al. (eds.), *Recent Advances in Information Systems*, Progress in IS,
https://doi.org/10.1007/978-3-032-14099-9_23

Augmented Generation (RAG) has emerged as a promising solution to address these challenges by combining semantic retrieval with the generative capabilities of large language models (LLMs). RAG pipelines dynamically retrieve relevant information from external data sources and provide it to an LLM to generate more grounded and domain-specific responses. In enterprise environments, RAG applications offer substantial potential—for example, in onboarding support, knowledge management, or decision assistance. However, evaluating such systems presents considerable challenges: real-world RAG pipelines are highly modular, vary across tools and domains, and involve numerous configuration decisions (Lewis et al., 2020). Existing benchmarks tend to focus on isolated performance metrics, often lacking structured support for comprehensive system-level analysis—especially in enterprise contexts (Ru et al., 2024; Sivasothy et al., 2024). We propose *HERAG*, a framework for holistic evaluation of RAG applications in practical scenarios, particularly within organizational settings. Our contribution systematically covers the essential phases of a RAG pipeline and outlines typical parameters relevant to the configuration of each phase. It is use-case independent and supports transparency, comparability, and modular reflection. It serves as a practical foundation for system designers and evaluators to document and reason about architecture, tool choices, and parameter configurations. Additionally, it facilitates structured communication of the evaluation process to both technical and non-technical stakeholders. The proposed concept focuses on vector-based RAG pipelines, the most widely implemented class of architectures in current applications.

The remainder of this paper is structured as follows: Sect. 2 provides an overview of the background and related work. Section 3 then presents the proposed evaluation framework, detailing its various stages. A practical application of HERAG is demonstrated in Sect. 4. This is followed by a discussion of the limitations of our approach in Sect. 5. Finally, Sect. 6 concludes the paper and suggests directions for future research.

2 Background and Related Work

Retrieval-based access to information plays a key role in enterprise environments, where large, often heterogeneous knowledge bases must be queried efficiently and reliably (Wu et al., 2025). Traditional information retrieval techniques such as keyword-based search or rule-based filtering systems often fall short when semantic understanding, contextual relevance, or generative output is required (Wu et al., 2025). RAG has emerged as a powerful paradigm to enhance the factual grounding, information actuality and domain adaptability of LLMs (Lewis et al., 2020). Instead of relying solely on pre-trained parameters, RAG systems retrieve relevant external information to generate more accurate and context-aware responses (Lewis et al., 2020; Gaoet al., 2024). To realize this process, RAG systems are typically structured into distinct stages such as indexing, retrieval, and generation (Gaoet al., 2024). Frameworks like LlamaIndex (Liu, 2025) use different terms and structures, highlighting the absence of a unified terminology. In this work, the RAG process is inspired by the structure proposed by LlamaIndex, with adapted terminology and phase definitions. It is divided into loading, storing and indexing, as well as querying, where querying includes both retrieval and answer generation. This structure promotes a modular view of the system and enables individual stages to be considered separately or in combination (Liu, 2025).

The evaluation of RAG systems is a fast-evolving field, with a broad range of approaches differing in scope, granularity, and methodology (Yu et al., 2025). Existing work can be grouped into five major categories:

1. **Integrability**: The extent to which the evaluation process can be incorporated into existing RAG systems, including whether adaptations are necessary or feasible for integration.
2. **Overview**: A comprehensive perspective on key characteristics that support the setup of an evaluation process and define what can be evaluated within RAG systems. This overview helps identify possible evaluation paths.
3. **Implementation**: Technical details on how the evaluation process is implemented or how specific components of the RAG system operate.
4. **Documentation**: The extent to which existing work provides clear descriptions or structured approaches for defining what is evaluated and how the evaluation is carried out.
5. **Modular Evaluation**: Evaluation of individual RAG stages, such as retrieval or generation, separately or in combination. The number and structure of the stages can vary and are not fixed.

Table 1 compares our contribution with existing work.

Several frameworks provide technical implementations for evaluating specific parts of the RAG process, enabling integration into existing systems. Tools like RAGCHECKER (Ru et al., 2024), ARES (Saad-Falcon et al., 2024), and RAGAS (Es et al., 2023) offer practical metrics to assess overall RAG performance. In contrast, eRAG (Salemi et al., 2024) focuses solely on the retrieval stage, providing targeted evaluation methods.

However, most frameworks lack a comprehensive overview of characteristics across the RAG pipeline. CoFE-RAG (Liu et al., 2024) addresses this by identifying retrieval-stage characteristics yet offers neither implementation nor documentation support. RAGProbe (Sivasothy et al., 2024) proposes a documentation scheme for evaluation scenarios but lacks technical depth and broader coverage.

XRAG (Mao et al., 2024) combines evaluation criteria, implementation, and documentation but cannot be integrated into existing RAG systems due to its standalone setup. Overall, current approaches either target isolated components or treat RAG pipelines as black boxes, without enabling modular, stage-specific evaluation.

In contrast to existing approaches, HERAG addresses the identified gap by providing a holistic concept that supports the planning, implementation, and communication of evaluations for RAG systems. Given the growing adoption of RAG pipelines in real-world applications, there is a pressing need for an evaluation approach that goes beyond isolated performance metrics to also consider architectural flexibility,

Table 1 Comparison of RAG evaluation approaches across analytical dimensions

Approach	Integrability	Implementation	Overview	Documentation	Modular evaluation
RAGCHECKER (Ru et al., 2024)	✓	✓	✗	✗	✗
ARES (Saad-Falcon et al., 2024)	✓	✓	✗	✗	✗
RAGAS (Es et al., 2023)	✓	✓	✗	✗	✗
eRAG (Salemi et al., 2024)	✓	✓	✗	✗	✗
CoFE-RAG (Liu et al., 2024)	✓	✗	✓	✗	✗
RAGProbe (Sivasothy et al., 2024)	✓	✗	✗	✓	✗
XRAG (Mao et al., 2024)	✗	✓	✓	✓	✗
Our contribution	✓	✗	✓	✓	✓

tool integration, and implementation trade-offs. HERAG responds to this demand with a structured yet adaptable framework designed to guide both researchers and practitioners. Importantly, HERAG by design does not offer a concrete technical implementation of the evaluation process. The framework emphasizes conceptual clarity and evaluative planning, enabling users to tailor implementation details to their specific use cases. A key strength of this approach is its seamless integration with existing frameworks like RAGAS (Es et al., 2023) or ARES (Saad-Falcon et al., 2024) for metric computation. Furthermore, the framework helps to systematically identify and justify relevant evaluation dimensions while documenting key decisions in a structured manner. This enhances transparency, traceability, and communication of the evaluation process within development or research teams and among external stakeholders.

3 Evaluation Framework

The HERAG framework is organized into three hierarchical levels: stages, components, and characteristics. Stages represent the main phases of the RAG system and its evaluation. Each stage is divided into components that reflect specific functional or conceptual units. These components are further described by characteristics, which define the relevant properties or criteria for analysis.

Figure 1 provides a general overview of the identified characteristics that is not intended to be exhaustive and can be extended as needed. It illustrates that the RAG system is framed by two overarching evaluation stages: the evaluation setup in the beginning and the evaluation implementation at the end. Together with the intermediate stages of the RAG system, they form a chronological sequence that represents a potential evaluation process. Not all RAG stages need to be considered, but the framing evaluation stages are always retained. To deepen understanding of the evaluation framework, each stage is described in detail below, along with its potentially relevant components.

3.1 Evaluation Setup

The evaluation setup stage serves as the foundation for the entire evaluation process. It involves defining the evaluation goals, selecting appropriate metrics for assessment and specifying the test data to be used.

The goal component is inspired by the Goal Question Metric (GQM) approach introduced by Basili et al., which is designed to support the construction of meaningful and goal-oriented evaluations (Basili et al., 1984). Therefore, a clear evaluation goal must be defined, specifying what is to be evaluated, for what purpose, with which focus, from which perspective, and within which context (Basili et al., 1984). The defined goal should subsequently be decomposed into its constituent aspects, which are inspired by the question dimensions of GQM (Basili et al., 1984).

The metrics component's primary objective is to operationalize the goal aspects by making them quantifiable (Basili et al., 1984). To achieve this, metrics must be selected with careful consideration. Each metric should be clearly associated with its corresponding goal aspect and must yield valid, contextually appropriate results (Basili et al., 1984). Ensuring consistent results is crucial since LLMs, while sometimes used to approximate human judgment (Liu et al., 2023), often produce inconsistent outputs (Zheng et al., 2023).

Stage	Component	Characteristics					
① Evaluation Setup	Goal	Object	Purpose	Focus	Viewpoint	Context	Aspects
	Metrics	Selection	Consistency	Data Requirements			
	Test Data	Structure	Quantity	Coverage	Source		
RAG System: ② Loading	Extraction	Input Formats	Output Structures	Variants	Preprocessing		
	Chunking	Splitting Strategies	Chunk Sizes	Overlap			
RAG System: ③ Storing & Indexing	Vector Database	System Type	Vector Query	Query Variant	Vector Index		
	Embedding	Models	Context Length				
RAG System: ④ Retrieval	Retriever	Search Strategy	top-k Selection	Query Expansion	Metadata Filtering		
	Reranker	Models	Parameter Configuration				
RAG System: ⑤ Answer Generation	LLM	Models	Model Parameters				
	Prompting	Techniques	Context Integration				
⑥ Evaluation Implementation	Measurement	Scenario Setup	Data Collection	Metric Calc.	Aggregation	Presentation	
	Analysis	Interpretation	Comparability	Transparency			

Time ↓ · Extensible →

Fig. 1 Holistic evaluation matrix for RAG systems

The final part of the evaluation setup involves the test dataset, whose structure should align with the evaluation goals and chosen metrics. Depending on the task, this may range from simple question-answer pairs to complex data like conversation histories or reference material. Ensuring metric consistency and validity also requires careful consideration of dataset size and coverage, including both content and structure (de Lima et al., 2024). When sourcing the dataset, existing benchmarks can be used if they match the evaluation criteria. If no suitable dataset is available, a new one must be created either manually by human annotators or automatically using generation techniques. Tools such as RAGAS, ARES, and LlamaIndex support automated dataset creation for RAG systems (de Lima et al., 2024).

3.2 Loading

The loading phase involves extracting information from sources and converting it into a standardized format, which facilitates subsequent handling (Gaoet al., 2024). Relevant metadata, such as the number of pages, is extracted from the source and used to enrich the raw text (Gaoet al., 2024). The loading phase is divided into two main components: extraction and chunking (Chai, 2022; Gaoet al., 2024).

Extraction involves handling data from structured, semi-structured, and unstructured sources, each presenting distinct challenges for downstream compatibility (Gaoet al., 2024). For unstructured and semi-structured inputs, the file type is critical, as it dictates parsing strategies and necessitates tools capable of converting content into standardized text formats (Auer et al., 2024). Input format must also be considered to correctly interpret structure and metadata such as headers or page numbers (Gaoet al., 2024). Furthermore, specifying the output format, such as plain text or Markdown, can influence downstream tasks by preserving contextual or stylistic cues (Sui et al., 2024). While standard NLP preprocessing techniques like normalization and stop-word removal are common (Chai, 2022), their relevance to RAG remains underexplored. Nonetheless, practical steps such as correcting extraction errors and removing corrupted data justify treating preprocessing as part of the loading phase.

Due to context size limitations in certain LLMs, extracted documents are often divided into smaller chunks (Gaoet al., 2024). Chunking strategies depend on the specific use case and source, and may involve fixed-size divisions, content-aware methods based on text structure such as paragraphs or semantic approaches guided by interpretation of meaning (Kshirsagar, 2024). Regardless of the chosen chunking strategy, chunk size should be carefully considered to account for technical limitations. Additionally, incorporating overlap between adjacent chunks can be particularly valuable, as it helps preserve context across segments and enhances coherence in the resulting text (Kshirsagar, 2024).

3.3 Storing and Indexing

In the storing and indexing stage, the standardized output from the loading phase is indexed and persisted. Typically, an embedding function is used to transform the text into a dense vector representation (Liu, 2025), which is then stored in a specialized vector database, often called a vector store.

Vector databases can be categorized based on four key characteristics: system type, vector query, query variant, and vector index. The system type describes whether a database is natively designed for vector operations or extends an existing system. This influences the degree of integration and specialization within the architecture (Pan et al., 2024). The remaining characteristics reflect the architectural design of a vector database management system (VDBMS), comprising a query processor and a storage manager (Pan et al., 2024). The vector query characteristic refers to the supported similarity search algorithms, such as exact or approximate nearest neighbor, which are executed by the query processor (Pan et al., 2024). Query variant captures the flexibility of query formulation, including features such as filtering or batched execution (Pan et al., 2024). Both characteristics can also be found combined within a query process, which includes the definition of search operators and support for advanced forms such as predicated queries (Wang et al., 2025). The vector index defines how vector data is stored and organized, usually using structures like trees or graphs. It is managed by the storage component and can be improved with quantization techniques such as scalar, vector, or product quantization to increase efficiency (Pan et al., 2024; Wang et al., 2025). Additional characteristics like client interfaces and deployment options may be relevant for comparison but are secondary to core architecture and thus excluded from the matrix (Wang et al., 2025).

The second key component in this stage is the embedding process, which transforms textual input into vector representations (Liu, 2025). Selecting the most suitable embedding characteristics is a crucial decision in a RAG system, as it impacts both the storage and indexing stage and the subsequent retrieval stage (Caspariet al., 2024). When comparing and evaluating different embedding approaches, two main characteristics emerge: the embedding model and the model's context length. Choosing an appropriate embedding model depends on the domain and application, as not all models perform equally well across different use cases (Muennighoff et al., 2023). Public benchmarks such as BEIR or MTEB provide valuable guidance, offering insights into domain-specific performance and helping to narrow down model choices (Thakur et al., 2021; Muennighoff et al., 2023). Another relevant aspect for the model selection is the dimensionality of the embedding vectors, as higher dimensions can improve semantic quality but increase storage and computation requirements (Thakur et al., 2021). Context length, which defines how much text a model can process at once, is key for handling long inputs and depends on the model's maximum limit. Although it can be lowered to reduce resource consumption (Thakur et al., 2021).

3.4 Retrieval

In a RAG application, retrieval involves selecting the top-k text chunks that align with the query using semantic similarity (Gaoet al., 2024). This approach ensures the system gathers only the most relevant pieces of context (Gaoet al., 2024). Different search strategies can be employed based on the requirements of accuracy and speed of a system (Quinn et al., 2025). Two primary strategies are Exact Nearest Neighbor Search (ENNS) and Approximate Nearest Neighbor Search (ANNS) (Quinn et al., 2025). ENNS performs exhaustive top-k search over all embeddings, ensuring high recall but increasing bandwidth and latency (Quinn et al., 2025). ANNS, using methods like Product Quantization, reduces the search space for faster retrieval at the cost of accuracy (Quinn et al., 2025). In both approaches, top-k selection directly impacts retrieval relevance by controlling how many candidate documents are considered (Quinn et al., 2025). Increasing k can provide the generative model with more context, potentially enhancing the final output's accuracy (Quinn et al., 2025). However, this also leads to higher computational costs and memory overhead, particularly in the generation phase where longer contexts need to be processed (Quinn et al., 2025). In addition to choosing effective search strategies and optimizing k, retrieval systems sometimes employ query expansion techniques to improve recall by reformulating the query to capture more relevant results (Quinn et al., 2025). Metadata filtering may refine retrieval by leveraging additional document attributes to narrow down the search space (Gaoet al., 2024).

The reranker is the final component of the retrieval stage in RAG systems. It is responsible for refining the initial set of retrieved documents through more precise relevance scoring (Mortaheb et al., 2025) and can be distinguished into the two main characteristics reranking model and parameter configuration. The model computes relevance scores and can differ in terms of architecture, parameter size, pretraining objectives, and domain specialization (Wang et al., 2024; Mortaheb et al., 2025; Dong et al., 2024). The parameter configuration defines how the reranking process is operationalized. This includes the number of candidates selected for reranking, the top-k cutoff used to forward results to the generator, and additional strategies such as score thresholding, redundancy filtering, or the fusion of results from multiple retrievers (Glass et al., 2022; Mortaheb et al., 2025).

3.5 Answer Generation

The Answer Generation stage constitutes the core of a RAG system, where the user question and retrieved context are used to generate a final response. This phase primarily involves two components: the selection and configuration of a suitable LLM, and the design of effective prompting strategies (Gaoet al., 2024).

The main characteristic of the LLM component, the model choice, may range from lightweight open-source models to proprietary state-of-the-art systems. In addition to selecting the base model, further customization is possible via fine-tuning on domain-specific data, enabling better alignment with task-specific language, style, or reasoning patterns (Gaoet al., 2024; Johnson and Hyland-Wood, 2024). Moreover, decoding parameters such as temperature, top-p, and top-k significantly influence the diversity, determinism, and specificity of the generated output (Arora et al., 2024). These parameters can be adjusted to balance creativity, factuality, and control depending on the use case.

The second key component is prompting. Prompt design plays a crucial role in guiding the model's behavior and quality of output (Vatsal et al., 2024). Prompting strategies include zero-shot, few-shot, and advanced techniques like chain-of-thought or self-consistency prompting, which have been shown to improve reasoning and coherence (Gaoet al., 2024; Vatsal et al., 2024). Another important aspect is context integration, which refers to how the retrieved information is embedded into the prompt (Gaoet al., 2024). This can involve inserting raw context, summarizing content before inclusion, or augmenting the prompt with illustrative examples to enhance relevance and grounding (Gaoet al., 2024).

3.6 Evaluation Implementation

After defining the overall evaluation setup and specifying the characteristics of the RAG system components, the actual implementation of the evaluation can be addressed. This involves first considering the measurement process, followed by the subsequent analysis (Solingen et al., 1999). In the measurement component, it is important to identify how test scenarios can be automatically set up using the previously defined test data and component characteristics. Furthermore, the execution of the RAG pipeline must be planned in such a way that the raw data required for the defined metrics can be systematically collected (Solingen et al., 1999). It can be beneficial to implement caching mechanisms to enhance fault tolerance and reduce computation time.

Once all raw data has been collected, the corresponding metrics can be computed with the support of various frameworks, such as DeepEval and LlamaIndex (Liu, 2025; Confident AI Inc., 2025). The use of such frameworks is particularly beneficial when applying complex LLM-as-a-Judge metrics. Ultimately, these metrics can be aggregated and presented in graphs and tables to highlight different aspects, which leads to a typical three layer analysis sheet (Solingen et al., 1999).

The final component of the holistic concept contains the analysis of the resulting metrics. For this purpose, the analysis sheets can be used to interpret the data regarding the defined goal aspects. Based on this, conclusions can be drawn and decisions made (Solingen et al., 1999). During the interpretation process, it is essential to consider the comparability of the results based on the specified setup. Failure to do so may result in erroneous conclusions. Lastly, transparency plays a crucial role in ensuring that the results can be understood and verified by others. The application of this concept naturally leads to a high degree of transparency, provided that the test dataset, as well as the scripts used for data collection and interpretation, are made accessible.

4 Practical Application

To demonstrate the applicability of the evaluation framework introduced in the previous chapter, it is now being used to evaluate an existing RAG system. As mentioned as a possibility in Sect. 1, a selective evaluation of specific components is employed, with a particular focus on retrieval quality. For this purpose, the characteristics of the selected components are defined, with both the evaluation setup and the evaluation implementation being mandatory stages. Table 2 gives an overview of the retrieval evaluation carried out using the proposed concept and underlines the high level of transparency provided by the framework. This is further enhanced when, as normally required, the defaults of all non-evaluated components are specified. For the sake of brevity, they have been left out in this example.

The evaluation aims to optimize a RAG-based Q&A system designed to support new employees with SAP-related and internal company knowledge. To achieve this, the impact of different embedding models and top-k values on retrieval quality is assessed. It is essential to retrieve all relevant context from the vector store while ensuring that relevant text chunks appear as early as possible in the ranked list. This is typically measured using metrics like HitRate, Mean Reciprocal Rank (MRR), Recall, or Normalized Discounted Cumulative Gain (NDCG). Hit Rate measures the proportion of queries for which at least one relevant item is retrieved, while MRR evaluates the average position of the first relevant item in the ranked list, rewarding systems that rank relevant results higher. Recall captures the proportion of relevant items successfully retrieved from the total number of relevant items, emphasizing completeness. NDCG, on the other hand, accounts for both the relevance and position of retrieved items, assigning higher scores to relevant items appearing earlier in the ranked list (Tamm et al., 2021).

An indication of how components affect each other can now be observed. As there are no test sets available for the project context and the creation of such was not the focus, the test data only consists of 189 manually created and labeled records derived from 43 input documents. Each question is associated with only a single labeled reference chunk. Consequently, the labeling does not capture all potentially relevant results due to time and resource constraints, which introduces limitations in both interpretability and comparability of the evaluation. Metrics such as Recall and NDCG, which depend on comprehensive or graded relevance judgments, are not appropriate under these conditions. Therefore, only Hit Rate and MRR are employed for evaluation. A detailed impact analysis on these metrics is not within the scope of this work, as only a basic applicability of the overall concept is to be shown.

The optimization is limited to two parameters: the embedding model and the top-k value. This restriction is chosen to reduce potential interactions between multiple variables and to enable a more controlled analysis. The selection of models and values is grounded in the project context and reflects a trade-off between positioning within the MTEB benchmark, support for both the German and English languages, and the available resources for model hosting.

In order to parameterize the RAG system with the selected values, wrapper components are built around the actual RAG components, which make the necessary parameters configurable from the outside and at the same time fulfill the task of data collection. The evaluation parameters are stored in a YAML configuration as shown in Listing 1. Based on this, a self-implemented logic calculates all 16 possible

Table 2 Retrieval-focused exemplary application of the evaluation concept

Stage	Component	Characteristic	Definition
Evaluation Setup	Goal	Object	Retrieval components of a RAG system
		Purpose	Understanding how different embedding models and top-k values affect the resulting context
		Focus	Retrieval performance
		Viewpoint	System developers
		Context	Optimization of a Q&A system for supporting new employees with SAP and internal company knowledge
		Aspects	Inclusion of relevant documents in the retrieved context, Rank of relevant documents in the results
	Metrics	Selection	Hit Rate & Mean Reciprocal Rank
		Concistency	Fully consistent, based on statistical calculations on top of labeled reference data
		Data Requirements	Retrieved and reference chunks
	Test Data	Structure	Tuples of question, reference answer, source document, reference chunk. Limited to one reference chunk per question, possibly others apply
		Quantity	189 records
		Coverage	Records from 43 provided internal and SAP documentation files. Structurally consisting of one question of the types simple, complex, double, distracting element, situational per document, if appropriate
		Source	Manual creation by humans with manual labeling of reference data
Retrieval	Embedding	Models	gte-Qwen2-1.5B-instruct; jina-embeddings-v3; mxbai-embed-large-v1; mxbai-company-finetuned
	Retriever	Top-K Selection	1; 5; 10; 20
Evaluation Implementation	Measurement	Scenario Setup	Automatic scenario generation from YAML-based definition of RAG system characteristics, configuration through wrapper components
		Data Collection	Implemented inside wrapper components
		Metric Calculation	Carried out by LlamaIndex
		Aggregation	Aggregation across all evaluation runs by averaging
		Presentation	Bar charts grouped by embedding model
	Analysis	Interpretation	Strong impact of the chosen embedding model with Hit Rate fluctuation up to 23 percentage points, gte-Qwen2-1.5B-instruct shows the best performance, Hit Rate and MRR increases with increasing top-k
		Comparability	Limited comparability due to possibly incomplete reference chunk data inside the test set
		Transparency	[Link to testset, evaluation code]

combinations in this case and executes them, with the collected data being stored in CSV files.

Listing 1 Excerpt from the YAML-based evaluation configuration.

```
retrieval:
  embedding_model:
    - "Alibaba-NLP/gte-Qwen2-1.5B-instruct"
    - "jinaai/jina-embeddings-v3"
    - "mixedbread-ai/mxbai-embed-large-v1"
    - "company/mxbai-company-finetuned"
  top_k: [ 1, 5, 10, 20 ]
```

LlamaIndex[1] is then used to perform the metric calculation based on the raw data and the Pandas data analysis library[2] is used for aggregation. Figure 2 shows the Hit Rate results as an example.

A brief analysis of the results clearly demonstrates that the choice of embedding models influences retrieval performance, with fluctuations of up to 45% observed even when the top-k value remains unchanged. The Hit Rate increases by up to 23% when expanding from top-5 to top-20, indicating consistent improvements across all models with broader retrieval. Notably, the jump from top-1 to top-20 reaches up to 91% relative, which likely reflects limitations of the test data. The single reference chunk per question imposes a constraint that may cause performance at lower top-k values to appear disproportionately poor and to be underestimated, warranting further analysis and quantification. Among the evaluated models, gte-Qwen2-1.5B-instruct consistently outperforms the others in both metrics. A more detailed statistical analysis is required to prove the statistical significance of the results. Due to the described limitations, comparisons with other evaluations based on different datasets are not appropriate. Nevertheless, this application example shows how the framework supports structured and transparent evaluations. As the evaluation involves proprietary data and is deeply integrated into the internal project, a full release of the source code is not feasible. Instead, a decoupled technical demonstration is available on GitHub,[3] illustrating how the otherwise tool-agnostic framework can be implemented using specific frameworks in practice.

[1] https://github.com/run-llama/llama_index.

[2] https://github.com/pandas-dev/pandas.

[3] https://github.com/felixkkh/HERAG-Demo-Implementation.

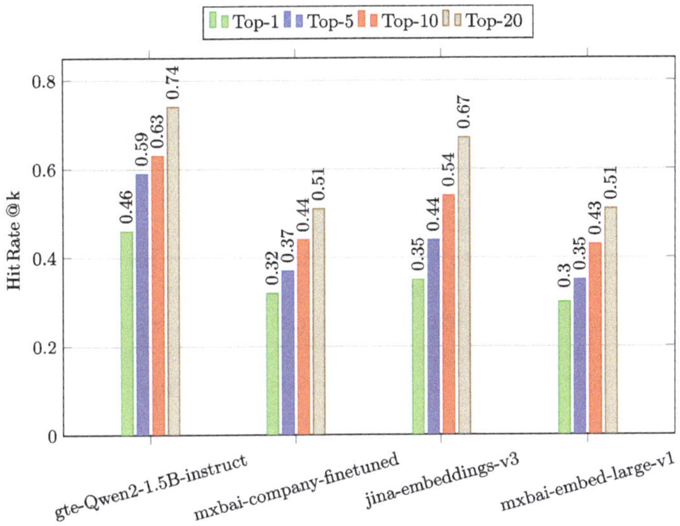

Fig. 2 Hit Rate depending on different embedding models and top-k values

5 Limitations

The proposed concept is designed as a holistic conceptual approach and therefore comes with certain limitations. First, the current version focuses exclusively on vector-based RAG pipelines. New approaches, such as the use of graphs or sparse vectors (Han et al., 2025; Zhang et al., 2024), are not currently covered. However, the structure of the concept allows for easy adaptation to new methods by defining or adjusting components and characteristics. Secondly, the concept deliberately refrains from prescribing technical implementations, model recommendations, or fixed evaluation metrics. This makes it widely applicable, but also places significant responsibility on the practitioner. However, existing work such as *Evaluation of Retrieval-Augmented Generation: A Survey* by Yu et al. (2025) can provide guidance. Furthermore, previously mentioned frameworks such as *RAGCHECKER* (Ru et al., 2024), *RAGProbe* (Sivasothy et al., 2024), or *ARES* (Saad-Falcon et al., 2024) can be used in the evaluation implementation stage for characteristics such as scenario setup or metric calculation and integrate seamlessly into the proposed concept while providing technical implementations. Based on this, it can be argued that the flexibility and holistic approach of the concept offer practical benefits for researchers and practitioners aiming to analyze, compare, and design RAG systems in a modular and transparent manner. The limitations observed in the practical application example are not rooted in conceptual shortcomings but result from a constrained implementation within the specific project context, shaped by limited time, resources, and focus.

6 Conclusion and Future Work

In this paper, we presented the HERAG framework to support holistic evaluations of real-world RAG systems. We provided an overview of the most important characteristics of a RAG system and embedded these within an evaluation framework. The concept supports planning, structuring, and documentation of evaluations for both existing and new RAG applications. Through the use of this concept, organizations can conduct structured evaluations to optimize systems for their specific use cases. This contributes to improved correctness and, consequently, greater acceptance and utilization of the evaluated system. As a result, employees become more productive and are able to focus on higher-value tasks.

The modular design of the framework enables cost savings by allowing the evaluation of individual components and characteristics without the need to assess the entire system. Unlike most existing works, this framework does not specify technical implementations and instead enables the integration of such works for implementing the evaluation process.

Future work may consider conducting a structured validation, such as user surveys or expert interviews, to support the application example provided in this work. Furthermore, future efforts could explore incorporating additional components and characteristics to support emerging RAG architectures. Given the rapid advancements in the field of AI, periodic updates to the framework are advisable. In addition to such extensions, more precise decision support for the selection of components and characteristics could be provided. Concrete implementation plans, whether using new or existing frameworks, also represent a valuable direction for further research. Thus, HERAG offers a foundation for the holistic, structured, and transparent evaluation of RAG systems, with considerable potential for future expansion.

Acknowledgements We would like to thank Jannis Boelsen, Keno Coldewey, Alina Meyer, Rodrigue Ngalani Touko, Talea Schweers, Viktor Dmitriyev and Dr.-Ing. Andreas Solsbach for their supporting work on the underlying research collaboration project conducted together with abat AG from which this paper has emerged.

Disclosure of Interests The authors have no competing interests to declare that are relevant to the content of this article.

References

Arora, C., Sayeed, A. I., Licorish, S., Wang, F., & Treude, C. (2024). Optimizing LLMs for code generation: Which hyperparameter settings yield the best results? In *2024 31st Asia-Pacific Software Engineering Conference (APSEC)*, pp. 281–290. https://doi.org/10.1109/APSEC65559.2024.00039

Auer, C., Lysak, M., Nassar, A., Dolfi, M., Livathinos, N., Vagenas, P., Ramis, C. B., Omenetti, M., Lindlbauer, F., Dinkla, K., Mishra, L., Kim, Y., Gupta, S., de Lima, R. T., Weber, V., Morin,

L., Meijer, I., Kuropiatnyk, V., & Staar, P. W. J. (2024). Docling technical report. arXiv preprint arXiv:2408.09869.

Basili, V. R., & Weiss, D. M. (1984). A methodology for collecting valid software engineering data. *IEEE Transactions on Software Engineering, SE–10(6)*, 728–738. https://doi.org/10.1109/TSE.1984.5010301

Caspari, L., Dastidar, K. G., Zerhoudi, S., Mitrovic, J., & Granitzer, M. (2024). Beyond benchmarks: Evaluating embedding model similarity for retrieval augmented generation systems. arXiv preprint arXiv:2407.08275

Chai, C. P. (2022). Comparison of text preprocessing methods. *Natural Language Engineering, 29*(3), 509–553. https://doi.org/10.1017/S1351324922000213

Confident AI Inc. (2025). *Introduction DeepEval*. https://www.deepeval.com/docs/metrics-introduction. Last accessed 2025/04/22.

de Lima, R. T., Gupta, S., Berrospi, C., Mishra, L., Dolfi, M., Staar, P., & Vagenas, P. (2024). Know your RAG: Dataset taxonomy and generation strategies for evaluating RAG systems. arXiv preprint arXiv:2411.19710

Dong, J., Fatemi, B., Perozzi, B., Yang, L. F., & Tsitsulin, A. (2024). Don't forget to connect! Improving RAG with graph-based reranking. arXiv preprint arXiv:2405.18414

Es, S., James, J., Espinosa-Anke, L., & Schockaert, S. (2023). RAGAs: Automated evaluation of retrieval augmented generation. arXiv preprint arXiv:2309.15217

Gao, Y., Xiong, Y., Gao, X., Jia, K., Pan, J., Bi, Y., Dai, Y., Sun, J., Wang, M., & Wang, H. (2024). Retrieval-augmented generation for large language models. A survey. arXiv preprint arXiv:2312.10997

Glass, M., Rossiello, G., Chowdhury, M. F. M., Naik, A. R., Cai, P., & Gliozzo, A. (2022). Re2G: Retrieve, rerank, generate. arXiv preprint arXiv:2207.06300

Han, H., Wang, Y., Shomer, H., Guo, K., Ding, J., Lei, Y., Halappanavar, M., Rossi, R. A., Mukherjee, S., Tang, X., He, Q., Hua, Z., Long, B., Zhao, T., Shah, N., Javari, A., Xia, Y., & Tang, J. (2025). Retrieval-augmented generation with graphs (GraphRAG). arXiv preprint arXiv:2501.00309

Johnson, S., & Hyland-Wood, D. (2024). A primer on large language models and their limitations. arXiv preprint arXiv:2412.04503

Kshirsagar, A. (2024). Enhancing RAG performance through chunking and text splitting techniques. *International Journal of Scientific Research in Computer Science, Engineering and Information Technology, 10*(5), 151–158. https://doi.org/10.32628/cseit2410593

Lewis, P., Perez, E., Piktus, A., Petroni, F., Karpukhin, V., Goyal, N., Küttler, H., Lewis, M., & Yih, W., Rocktäschel, T., Riedel, S., & Kiela, D. (2020). *Retrieval-augmented generation for knowledge-intensive NLP tasks*. Curran Associates Inc.

Liu, J., Ding, R., Zhang, L., Xie, P., & Huang, F. (2024). CoFE-RAG: A comprehensive full-chain evaluation framework for retrieval-augmented generation with enhanced data diversity. arXiv preprint arXiv:2410.12248

Liu, J. (2025). Evaluating—LlamaIndex. https://docs.llamaindex.ai/en/stable/module_guides/evaluating/. Last accessed 2025/04/22.

Liu, J. (2025). Introduction to RAG—LlamaIndex. https://docs.llamaindex.ai/en/stable/understanding/rag/. Last accessed 2025/04/07.

Liu, Y., Iter, D., Xu, Y., Wang, S., Xu, R., & Zhu, C. (2023). G-Eval: NLG evaluation using GPT-4 with better human alignment. In H. Bouamor, J. Pino, & K. Bali (Eds.), *Proceedings of the 2023 conference on empirical methods in natural language processing* (pp. 2511–2522). Association for Computational Linguistics. https://doi.org/10.18653/v1/2023.emnlp-main.153

Mao, Q., Luo, Y., Zhang, J., Hao, H., Cao, Z., Wang, X., Guan, X., Huang, Z., Jiang, W., Guo, S., Han, Z., Zhang, Q., Tao, S., Liu, Y., Liu, J., Tan, Z., Sun, J., Li, B., Liu, X., Zhang, R., & Li, J. (2024). XRAG: eXamining the core—Benchmarking foundational components in advanced retrieval-augmented generation. arXiv preprint arXiv:2412.15529

Mortaheb, M., Khojastepour, M. A. A., Chakradhar, S. T., & Ulukus, S. (2025). Re-ranking the context for multimodal retrieval augmented generation. arXiv preprint arXiv:2501.04695

Muennighoff, N., Tazi, N., Magne, L., & Reimers, N. (2023). MTEB: Massive text embedding benchmark. arXiv preprint arXiv:2210.07316

Pan, J. J., Wang, J., & Li, G. (2024). Survey of vector database management systems. *The VLDB Journal, 3*(5), 1591–1615. https://doi.org/10.1007/s00778-024-00864-x

Quinn, D., Nouri, M., Patel, N., Salihu, J., Salemi, A., Lee, S., Zamani, H., Alian, M. (2025). Accelerating retrieval-augmented generation. In *Proceedings of the 30th ACM international conference on architectural support for programming languages and operating systems* (Vol. 1. pp. 15–32). ASPLOS '25, Association for Computing Machinery, New York, NY, USA. https://doi.org/10.1145/3669940.3707264

Ru, D., Qiu, L., Hu, X., Zhang, T., Shi, P., Chang, S., Jiayang, C., Wang, C., Sun, S., Li, H., Zhang, Z., Wang, B., Jiang, J., He, T., Wang, Z., Liu, P., Zhang, Y., & Zhang, Z. (2024). RAGChecker: A fine-grained framework for diagnosing retrieval-augmented generation. In A. Globerson, L. Mackey, D. Belgrave, A. Fan, U. Paquet, J. Tomczak, C. Zhang (Eds.), *Advances in neural information processing systems* (Vol. 37, pp. 21999–22027). Curran Associates, Inc. https://proceedings.neurips.cc/paper_files/paper/2024/file/27245589131d17368cccdfa990cbf16e-Paper-Datasets_and_Benchmarks_Track.pdf

Saad-Falcon, J., Khattab, O., Potts, C., & Zaharia, M. (2024). ARES: An automated evaluation framework for retrieval-augmented generation systems. arXiv preprint arXiv:2311.09476

Salemi, A., & Zamani, H. (2024). Evaluating retrieval quality in retrieval-augmented generation. arXiv preprint arXiv:2404.13781

Sivasothy, S., Barnett, S., Kurniawan, S., Rasool, Z., & Vasa, R. (2024). RAGProbe: An automated approach for evaluating RAG applications. arXiv preprint arXiv:2409.19019

Solingen, R. V., & Berghout, E. (1999). *The goal/question/metric method: A practical guide for quality improvement of software development.* The McGraw-Hill Companies.

Sui, Y., Zhou, M., Zhou, M., Han, S., & Zhang, D. (2024). Table meets LLM: Can large language models understand structured table data? A benchmark and empirical study. arXiv preprint arXiv:2305.13062

Tamm, Y. M., Damdinov, R., Vasilev, A. (2021). Quality metrics in recommender systems: Do we calculate metrics consistently? In *RecSys '21: Proceedings of the 15th ACM conference on recommender systems* (pp. 708–713). RecSys '21, Association for Computing Machinery, New York, NY, USA. https://doi.org/10.1145/3460231.3478848

Thakur, N., Reimers, N., Rücklé, A., Srivastava, A., & Gurevych, I. (2021). BEIR: A heterogeneous benchmark for zero-shot evaluation of information retrieval models. In J. Vanschoren & S. Yeung (Eds.), *Proceedings of the neural information processing systems track on datasets and benchmarks* (Vol. 1). https://datasets-benchmarks-proceedings.neurips.cc/paper_files/paper/2021/file/65b9eea6e1cc6bb9f0cd2a47751a186f-Paper-round2.pdf

Vatsal, S., & Dubey, H. (2024). A survey of prompt engineering methods in large language models for different NLP tasks. arXiv preprint. https://doi.org/10.48550/arXiv.2407.12994

Wang, X., Wang, Z., Gao, X., Zhang, F., Wu, Y., Xu, Z., Shi, T., Wang, Z., Li, S., Qian, Q., Yin, R., Lv, C., Zheng, X., & Huang, X. (2024). Searching for best practices in retrieval-augmented generation. arXiv preprint https://doi.org/10.48550/arXiv.2407.01219

Wang, S., Zhao, Y., Xie, Y., Liu, Z., Hou, X., Zou, Q., & Wang, H. (2025). *Towards reliable vector database management systems: A software testing roadmap for 2030* (Vol. 10, p. 2502.20812). arXiv preprint arXiv:2502.20812

Wu, C., Ding, W., Jin, Q., Jiang, J., Jiang, R., Xiao, Q., Liao, L., & Li, X. (2025). Retrieval augmented generation-driven information retrieval and question answering in construction management. *Advanced Engineering Informatics, 65.* https://doi.org/10.1016/j.aei.2025.103158

Yu, H., Gan, A., Zhang, K., Tong, S., Liu, Q., & Liu, Z. (2025). *Evaluation of retrieval-augmented generation: A survey* (pp. 102–120). Springer Nature. https://doi.org/10.1007/978-981-96-1024-2_8

Zhang, H., Liu, J., Zhu, Z., Zeng, S., Sheng, M., Yang, T., Dai, G., & Wang, Y. (2024). Efficient and effective retrieval of dense-sparse hybrid vectors using graph-based approximate nearest neighbor search. arXiv preprint https://doi.org/10.48550/arXiv.2410.20381

Zheng, L., Chiang, W. L., Sheng, Y., Zhuang, S., Wu, Z., Zhuang, Y., Lin, Z., Li, Z., Li, D., Xing, E., Zhang, H., Gonzalez, J. E., & Stoica, I. (2023). Judging LLM-as-a-judge with MT-bench and Chatbot Arena. In A. Oh, T. Naumann, A. Globerson, K. Saenko, M. Hardt, & S. Levine (Eds.), *Advances in neural information processing systems*. (Vol. 36, pp. 46595–46623). Curran Associates, Inc. https://proceedings.neurips.cc/paper_files/paper/2023/file/91f18a1287b398d378ef22505bf41832-Paper-Datasets_and_Benchmarks.pdf

Analyzing Student Gaze Behavior in Classroom Settings: A Computer Vision Approach for Multi-person Attention Detection

Florian Voßmann, Steffen Meeuw, Jan-Hendrik Witte, Ines Oldenburg, and Jorge Marx Gómez

Abstract This paper presents a novel computer vision pipeline for simultaneously tracking the gaze direction of multiple students in authentic classroom environments using a single blackboard-mounted camera. While student attention is widely recognized as critical for effective knowledge acquisition and has been extensively investigated in educational research, traditional methodologies for assessing attention have predominantly relied on structured behavioral observation protocols or self-report measures, with eye-tracking applications limited to capturing the teacher's perspective. These methodological constraints present a significant research gap, particularly for understanding attentional patterns among students with specialized learning needs. Our approach addresses this gap through a six-stage processing pipeline that transforms classroom video data into quantified gaze direction measurements for each detected individual. We demonstrate potential applications through a classroom setup experiment with 12 students and one teacher, enabling both continuous tracking of students' gaze behavior over time and target-specific attention analysis. The pipeline offers new possibilities for understanding classroom dynamics at scale while preserving ecological validity without requiring specialized per-student hardware. Future work should focus on comprehensive validation with ground truth data, enhancing environmental robustness, optimizing computational efficiency, and examining relationships between visual attention patterns and learning outcomes.

1 Introduction

Student attention in classroom environments is widely recognized as a critical prerequisite for effective knowledge acquisition Hommel et al. (2011) and has been

F. Voßmann (✉) · S. Meeuw · J.-H. Witte · J. Marx Gómez
Department of Very Large Business ApplicationsDepartment of Educational Sciences, Carl von Ossietzky Universität Oldenburg, Oldenburg, Germany
e-mail: florian.vossmann@uni-oldenburg.de

I. Oldenburg
Department of Educational Sciences, Carl von Ossietzky Universität Oldenburg, Oldenburg, Germany

J. Marx Gómez et al. (eds.), *Recent Advances in Information Systems*, Progress in IS,
https://doi.org/10.1007/978-3-032-14099-9_24

extensively investigated in educational research Vierlinger et al. (1984). Comprehensive understanding of attention dynamics necessitates examining both instructional design elements and learner-specific factors, including cognitive processes, emotional states, and self-regulatory capabilities Hommel et al. (2011). Traditional methodologies for assessing student attention have predominantly relied on structured behavioral observation protocols such as "The Munich Attention Inventory" Helmkeet al. (1992) or self-report measures Blume and Schmiedek (2024), while eye-tracking applications in educational contexts have been largely limited to capturing the teacher's perspective Stürmer et al. (2017).

These methodological constraints present a significant research gap, particularly for understanding attentional patterns among students with specialized learning needs. Current approaches either require individual monitoring equipment, disrupt the natural classroom ecology, or fail to capture simultaneous gaze data from multiple students—limitations that impede the development of evidence-based instructional strategies for diverse learner populations. This methodological challenge leads to our central research question: *How can the gaze direction of multiple students be simultaneously determined with minimal technical resources in authentic classroom environments?*

This study presents a proof-of-concept investigation that addresses this question through a novel computer vision pipeline. Our approach utilizes a single blackboard-mounted camera to capture and analyze gaze behavior of multiple students simultaneously, providing a non-intrusive method for quantifying visual attention patterns within classroom settings. The primary objective is not student surveillance but rather the development of an analytical tool to inform instructional design and enhance pedagogical effectiveness, particularly for supporting students with learning disabilities. The significance of this research lies in its potential to bridge the gap between single-subject laboratory-based eye-tracking studies and ecologically valid classroom assessments, offering insights into collective attention patterns without specialized per-student equipment. We emphasize that this system is conceived as a research instrument for pedagogical improvement rather than an evaluation mechanism for individual student performance.

The remainder of this paper is organized as follows: Sect. 2 reviews relevant literature and identifies specific research gaps; Sect. 3 delineates our conceptual framework and system architecture; Sect. 4 details the technical implementation; Sect. 5 presents proof-of-concept use cases; Sect. 6 discusses methodological and ethical limitations; and Sect. 7 summarizes key contributions and directions for future research.

2 Related Work

The assessment of student visual attention in educational environments constitutes a significant area of investigation across multiple disciplines. Visual attention measurement has demonstrated considerable utility for identifying disengaged students, particularly in large course environments where comprehensive monitoring becomes challenging Huttet et al. (2017). This section analyzes existing methodological approaches and identifies the specific research gap addressed by our work. Current methodological paradigms exhibit varying degrees of applicability to authentic classroom settings. Virtual reality (VR) technologies offer controlled conditions for attention monitoring, with studies by Asish et al. utilizing VR glasses to measure gaze direction during simulated lessons Asish et al. (2022). While providing precise eye-tracking capabilities, this approach introduces artificial instructional contexts and necessitates specialized hardware for each participant. Remote eye-tracking methodologies developed by Hutt et al. identify mind-wandering behaviors during online lecture viewing through non-invasive monitoring equipment but remain primarily applicable to digital learning environments rather than physical classroom settings Huttet et al. (2017). First-person perspective analytical frameworks, as implemented by Xu et al., employ wearable cameras to record from the student's viewpoint, subsequently applying object recognition algorithms to classify attentive versus inattentive behaviors Xu et al. (2023). Recognizing the limitations of eye-level recordings, Araya and Sossa-Rivera established smartphone cameras positioned behind students, analyzing postural cues and head rotation as proxy indicators for attention Araya et al. (2021).

Given the constraints of existing educational attention tracking approaches, examination of gaze estimation technologies from adjacent domains reveals potentially applicable methodologies. Vicente et al. developed driver monitoring systems that determine gaze direction in real-time to evaluate road safety, creating cost-effective solutions functioning across varied lighting conditions without calibration requirements Vicenteet al. (2015). Martinikorena et al. and Wang et al. investigated gaze direction determination in low-resolution imaging conditions, exemplified by mobile phone camera applications Martinikorena Aranburuetal et al. (2020); Wang et al. (2019). These studies prioritized real-time gaze prediction despite constrained visual data. However, these approaches are optimized for single-subject analysis with subjects positioned in close proximity to the recording device—a configuration incongruent with typical classroom spatial arrangements where multiple students are positioned at varying distances from the camera.

Table 1 presents a comparative analysis of existing approaches against critical requirements for classroom gaze tracking applications. The evaluation criteria include: *(1) absence of external hardware requirements, (2) utilization of facial and ocular image data, (3) non-disruptive integration with standard instructional practices, (4) capability for simultaneous multi-person analysis, and (5) compatibility with traditional classroom environments*. This systematic comparison reveals a significant methodological gap: no existing approach satisfies all established criteria for comprehensive classroom gaze tracking. Educational technologies Asish et al. (2022); Huttet et al. (2017); Xu et al. (2023); Araya et al. (2021) each address specific aspects but fail to provide a complete solution, while technologies from adjacent domains Vicenteet al. (2015); Martinikorena Aranburuetal et al. (2020); Wang et al. (2019) offer valuable components but lack optimization for educational applications involving multiple subjects.

Table 1 Systematic comparison of gaze tracking methodologies against critical classroom implementation criteria

Criteria Studies	Asish et al. (2022)	Huttet et al. (2017)	Xu et al. (2023)	Araya et al. (2021)	Vicenteet al. (2015)	Martinikorena Aranburuetal et al. (2020)	Wang et al. (2019)	Our approach
Hardware-free operation	✗	✗	✗	✓	✓	✓	✓	✓
Facial/ocular imaging	✓	✓	✗	✗	✓	✓	✓	✓
Non-disruptive integration	✓	✓	✗	✓	D.D.	D.D.	D.D.	✓
Multi-subject capability	✗	✗	✗	✓	✗	✗	✗	✓
In-person classroom compatibility	✗	✗	✓	✓	D.D.	D.D.	D.D.	✓

Legend ✓= criterion satisfied, ✗= criterion not satisfied, D.D. = different domain (not applicable to classroom context)

Our research directly addresses this methodological gap by developing an integrated pipeline that satisfies all identified requirements, as indicated in the final row of Table 1. The proposed approach enables simultaneous tracking of multiple individuals from a blackboard-mounted camera without requiring specialized hardware, utilizing both facial and ocular data while maintaining compatibility with traditional classroom instructional practices.

3 Conceptual Framework and System Architecture

This section presents the conceptual framework and technical architecture of our approach to multi-person gaze tracking in classroom environments. We delineate the fundamental design principles, overall system architecture, and key technical components that collectively enable simultaneous gaze direction determination for multiple individuals.

3.1 Fundamental Design Principle

Our methodological approach is guided by specific design principles that address the practical constraints of classroom implementations. The system operates with a single camera mounted above the blackboard, eliminating the need for specialized per-student hardware that would disrupt instructional dynamics. This configuration captures the entire classroom from the teacher's perspective, preserving natural educational interactions while enabling comprehensive visual analysis. The design prioritizes scalability to multiple subjects, compatibility with standard classroom arrangements, and minimization of computational resources required for real-time processing.

The fundamental objective is to create a framework that balances technical efficacy with practical implementation considerations, ensuring that the methodology can be deployed in authentic educational environments without significant infrastructure modifications or disruption to pedagogical practices. This balance is particularly crucial for longitudinal studies of attention patterns or research involving students with learning disabilities, where environmental consistency and non-intrusiveness are essential for valid observations.

3.2 System Architecture

The architecture of our gaze tracking system employs a modular pipeline approach, conceptually similar to the region-based convolutional neural network (R-CNN) paradigm Girshick et al. (2014), where selective search identifies regions of interest for subsequent detailed analysis. Figure 1 illustrates the complete processing pipeline, which systematically transforms classroom video data into quantified gaze direction measurements for each detected individual.

The pipeline integrates six sequential processing modules:

1. **Frame Selection**: Systematic sampling of video frames at specified intervals to balance computational efficiency with temporal resolution
2. **Face Detection**: Identification and localization of all faces present within each selected frame
3. **Face Assignment**: Temporal tracking of individuals across consecutive frames through identity matching algorithms

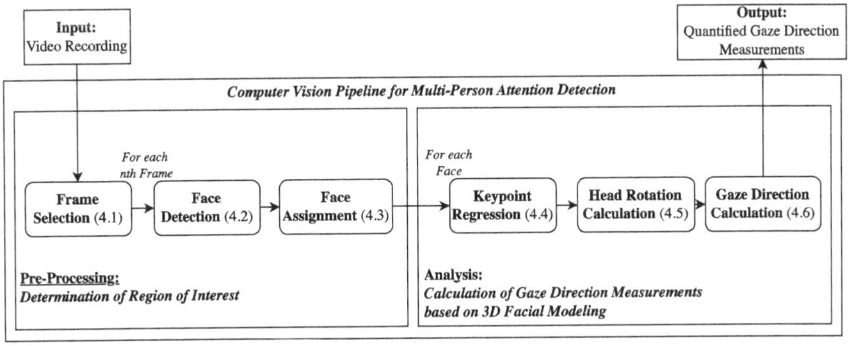

Fig. 1 Overview of multi-person attention detection pipeline

4. **Keypoint Regression**: Extraction of precise facial landmarks, particularly around the ocular region
5. **Head Rotation Calculation**: Three-dimensional orientation estimation through geometric or model-based approaches
6. **Gaze Direction Calculation**: Integration of bilateral eye data and head position to establish definitive gaze vectors

This modular design enables independent optimization of each component while maintaining system cohesion. The sequential processing approach allows for progressive refinement of data granularity, transforming raw video frames into precise gaze direction measurements through systematic feature extraction and spatial analysis.

3.3 Three-Dimensional Facing Modeling

A fundamental conceptual challenge in gaze tracking from a single camera perspective is the transformation from two-dimensional image data to three-dimensional gaze vectors. As shown in Fig. 1, our pipeline requires both head rotation calculation and eyeball center estimation to determine accurate gaze direction information that cannot be directly extracted from 2D images alone. This necessitates a three-dimensional facial representation within our architectural framework.

To address this challenge, we incorporate the Basel Face Model (BFM) Gerig et al. (2018); Basel Face Model (2025) as a core component of our conceptual architecture. The BFM provides a parameterized 3D representation of facial morphology that serves as the critical bridge between 2D camera observations and 3D spatial understanding. Within our pipeline (see Fig. 1), the BFM supports two essential functions: first, it enables head rotation calculation by providing a 3D reference structure for facial orientation; second, it facilitates eyeball center estimation, which is anatomically positioned behind visible facial features and thus not directly observable from camera images. This three-dimensional modeling approach represents a key conceptual advancement over previous classroom gaze tracking methods that rely solely on 2D feature analysis. By incorporating anatomically accurate facial modeling, our architecture establishes the spatial framework necessary for precise gaze vector determination in multi-person classroom scenarios. The specific technical implementations of how the BFM is applied, including both unmodified and subject-adapted approaches, will be detailed in the following section.

4 Technical Implementation

This section details the technical implementation of each component in our gaze tracking pipeline, elaborating on the specific algorithms, models, and computational approaches utilized to transform classroom video data into precise gaze direction measurements.

4.1 Frame Selection

The initial stage of our implementation addresses the challenge of computational efficiency by selectively sampling video frames for analysis. We employ a systematic approach that extracts every nth frame from the continuous video stream, where n corresponds to the frames-per-second rate of the recording. This effectively reduces the processing to approximately one frame per second, balancing temporal resolution with computational demands.

This selective sampling strategy is justified by the relatively slow rate of change in head position and eye state during typical classroom activities. In classroom settings, significant changes in gaze direction typically occur at intervals longer than milliseconds, making per-second sampling sufficient to capture meaningful attentional shifts while substantially reducing the computational load compared to processing every frame. The selected frames then serve as input for subsequent processing stages, establishing a manageable data stream for real-time or near-real-time analysis.

4.2 Face Detection

Face detection is implemented using the YOLOv8m-face model Akanametov (2024), a specialized variant of the YOLOv8 (You Only Look Once) architecture retrained on the WIDER FACE dataset Shuoyang (2024). We selected YOLOv8 as it directly addresses several key requirements established in our conceptual framework. Its single-pass detection architecture delivers the computational efficiency necessary for our minimal hardware approach while maintaining accuracy when processing multiple faces simultaneously. The model performs robustly under varying illumination conditions, addressing the environmental variability typical in classroom settings where lighting often changes throughout the day. Additionally, YOLOv8's efficient design supports the near real-time processing capabilities essential for practical classroom applications without requiring specialized hardware.

This convolutional neural network approach enables efficient single-stage detection, identifying multiple faces within each frame simultaneously rather than employing computationally expensive sliding window methods. The model outputs bounding box coordinates (x, y, width, height) for each detected face, along with associated confidence scores. These regions serve as the foundation for subsequent individual-specific analysis, effectively decomposing the complex classroom scene into discrete analytical units. The model's training on the diverse WIDER FACE dataset enhances detection robustness across varying head poses, partial occlusions, and illumination conditions commonly encountered in classroom environments.

4.3 Face Assignment

To establish temporal continuity and individual-specific analysis, we implemented a face assignment system that matches detected faces across consecutive frames. This system addresses a critical challenge in classroom gaze analysis: maintaining consistent identity tracking for multiple individuals throughout the recording period.

The strategy leverages spatial coherence by calculating similarity scores based on bounding box position and dimensions. The underlying assumption is that individuals exhibit minimal movement in classroom settings and that the camera maintains a consistent position during recording. For each detected face, a match matrix quantifies positional similarity to previously identified individuals based on proximity to their last known position and face size. A notable benefit of this approach is its computational efficiency. However, the process is subject to inaccuracies when individuals move significantly (e.g., to the toilet or to the blackboard), when the camera shifts position, or when people partially occlude each other. To mitigate these limitations, the system also considers face size as a matching factor. Matches exceeding a predetermined confidence threshold are accepted, while faces without confident matches initialize new person entries in the tracking database. This approach aims to balance computational efficiency with tracking accuracy.

4.4 Key Point Regression

Facial keypoint extraction transforms raw face images into structured morphological data necessary for gaze analysis. We implemented this component using the MediaPipe facial landmark detection framework Google AI Edge (2024), which provides 478 precisely localized facial landmarks with three-dimensional coordinates (x, y, z). This dense mapping includes detailed delineation of facial features critical for gaze estimation. MediaPipe's estimation of z-coordinate (depth) values directly addresses our requirement for three-dimensional analysis without specialized hardware, as established in Subsect. 3.3. While these depth estimates lack the precision of dedicated sensors, they sufficiently approximate facial topography for our analytical needs. The framework also includes pupil position estimation, directly supporting our core objective of gaze direction determination. The extracted keypoints serve multiple essential functions: providing correspondence points for Basel Face Model alignment, enabling geometric head rotation calculation, facilitating eye state determination, and establishing reference points for gaze vector calculation. This multifunctionality enhances computational efficiency by deriving multiple analytical dimensions from a single detection process. MediaPipe's robust performance across varying image qualities addresses our requirement for simultaneous analysis of multiple individuals at different distances from the camera—a key advantage over the single-subject, close-range approaches identified in our related work analysis (see Sect. 2). This capability supports the classroom-wide analytical scope central to our methodological contribution.

4.5 Head Rotation Calculation

Accurately determining three-dimensional head orientation represents a foundational challenge in monocular gaze estimation systems. In our implementation, head rotation parameters serve dual purposes: establishing the spatial reference frame for pupillary gaze vectors and providing an alternative attentional indicator when direct gaze measurement is unavailable. We calculate orientation using three rotational parameters—yaw, pitch, and roll (rotations around the x, y, and z axes)—with values constrained to $\pm$ 90 degrees to reflect practical observational limitations.

Our approach utilizes the standard BFM without subject-specific adaptation. To establish a correspondence between the facial keypoints detected by MediaPipe and those defined in the BFM, we employ a minimal-distance mapping strategy. Specifically, for each MediaPipe landmark, we identify the nearest BFM vertex in 3D space, thereby constructing a consistent mapping between the observed data and the model. The alignment process involves iteratively applying rotation matrices to the BFM keypoints. During each iteration, we project the transformed 3D model points onto the 2D image plane and compute the average Euclidean distance between the projected points and the corresponding MediaPipe landmarks. The optimization seeks to minimize this distance, effectively identifying the rotation parameters that achieve the best alignment between the model and the observed facial features.

This approach offers several advantages. Firstly, it is characterized by simplicity and consistency, as it avoids the complexity of subject-specific model fitting and relies on a fixed anatomical model. Secondly, the use of a minimal-distance mapping combined with a lightweight optimization routine results in a low computational cost, making the method well-suited for real-time or resource-constrained applications. Despite its simplicity, the method provides robust alignment performance for a wide range of facial poses.

4.6 Gaze Direction Calculation

The final component in our pipeline calculates individual eye gaze vectors and integrates them with eye state information to produce a definitive gaze direction for each detected person. This process combines single eye gaze calculation with final gaze direction determination, addressing the binocular nature of human vision while accommodating variations in eye state that naturally occur during classroom observation.

Single Eye Gaze Calculation—Following head orientation determination, our pipeline calculates individual eye gaze vectors using the geometric eye model described by Martinikorena Aranburuetal et al. (2020). This model represents gaze as a ray projected from the eyeball center through the pupil position, requiring accurate three-dimensional localization of both these anatomical reference points. The resulting vector, while an approximation of true visual axis, provides sufficient accuracy for classroom attention analysis with typical deviations of 3–7 degrees from precise gaze direction. For implementation, we developed an approach which leverages the BFM described in subsection 3.3 to establish anatomically plausible eyeball centers. We first estimate the center by calculating the average position of the eye rim keypoints in the unrotated BFM, then apply an anatomically-derived offset of -12mm along the x-axis (approximately equivalent to average eyeball radius). For pupil position determination, we rotate the 3D head model using parameters from Subsect. 4.5, then align the MediaPipe eye keypoints with the rotated model. This alignment transformation is also applied to the pupil keypoint, and its intersection with the eyeball sphere is calculated to determine the precise 3D pupil position within the model space.

Eye State Detection and Final Gaze Integration—Eye state determination represents a necessary preprocessing step in our pipeline, as subsequent gaze estimation requires knowledge of whether each eye is open or closed. Our implementation utilizes the extracted eye contour keypoints from MediaPipe to perform geometric analysis of eye aperture. For each eye, we calculate an aspect ratio by dividing the maximum vertical distance between upper and lower eyelid keypoints by the maximum horizontal distance between eye corners. When the vertical-to-horizontal ratio falls below a threshold of 0.15, the system classifies the eye as closed.

Our implementation employs a conditional selection logic based on the determined eye states:

1. When both eyes are open, the system calculates the average of the left and right eye gaze vectors. This binocular integration typically provides the most accurate gaze estimation by mitigating individual eye measurement errors through averaging.
2. When only the left eye is open (right eye closed), the system selects the left eye's gaze vector exclusively.
3. When only the right eye is open (left eye closed), the system utilizes the right eye's gaze vector exclusively.
4. When both eyes are closed, the system estimates gaze direction based on the assumption that eye position relative to the head remains constant under fixed head positions. In practical terms, this means the system relies on the last known eye positions relative to the current head orientation, allowing for continuous estimation during blinks or temporary eye closures.

This adaptive approach ensures continuous gaze tracking across natural variations in eye state, maintaining analytical continuity throughout the observation period. The final output consists of horizontal and vertical gaze angles that can be projected into the classroom space to identify potential attentional targets such as the teacher, blackboard, or other students. By accommodating various eye states and leveraging both monocular and binocular information as available, our implementation balances measurement accuracy with tracking robustness, providing reliable gaze direction data even under sub-optimal observation conditions commonly encountered in authentic classroom environments.

The complete pipeline thus transforms raw classroom video data into *quantified Gaze Direction Measurements* for each detected individual, providing precise angular measurements that characterize where each student is looking throughout the recording period. These measurements form the foundation for subsequent analysis of attention patterns and classroom engagement dynamics, as demonstrated in the following section through our classroom setup experiment.

5 Application Scenarios

This section outlines potential application scenarios for our gaze tracking pipeline, highlighting how it can support educational insights in realistic classroom contexts. Since the pipeline presents a proof-of-concept at this stage of development, we illustrate the system's utility through examples drawn from a controlled classroom-like experiment instead of a formal evaluation. These examples show how the pipeline could be used both at the individual and group level to support pedagogical analysis in the future.

5.1 Classroom Setup Experiment

To explore the pipeline's capabilities under semi-naturalistic conditions, we created a controlled test setup in a classroom setting. As shown in Fig. 2, 6 tables were arranged in two rows, with the front row placed 3 meters from the blackboard and the second row at 5 meters distance. This layout mirrors typical classroom arrangements while maintaining environmental control for initial system validation, with 12 students sitting in the arranged seating.

A 43-minute teaching session involving 13 participants (12 students and one teacher) was recorded using an iPhone 16 Pro camera mounted in the center above the blackboard. The session included frontal instruction, student presentations, partner discussions, and board work—representing varied instructional formats and attentional demands. During the recording, participants occupied different positions and distances from the camera, allowing assessment of the system's performance across spatial variations. The video was recorded with a resolution of 1920×1080 pixels with 30 frames per second (FPS).

In the following sections, we will demonstrate different application scenarios on how the implemented pipeline could be used to analyze the gaze behavior of individual students in this classroom setting.

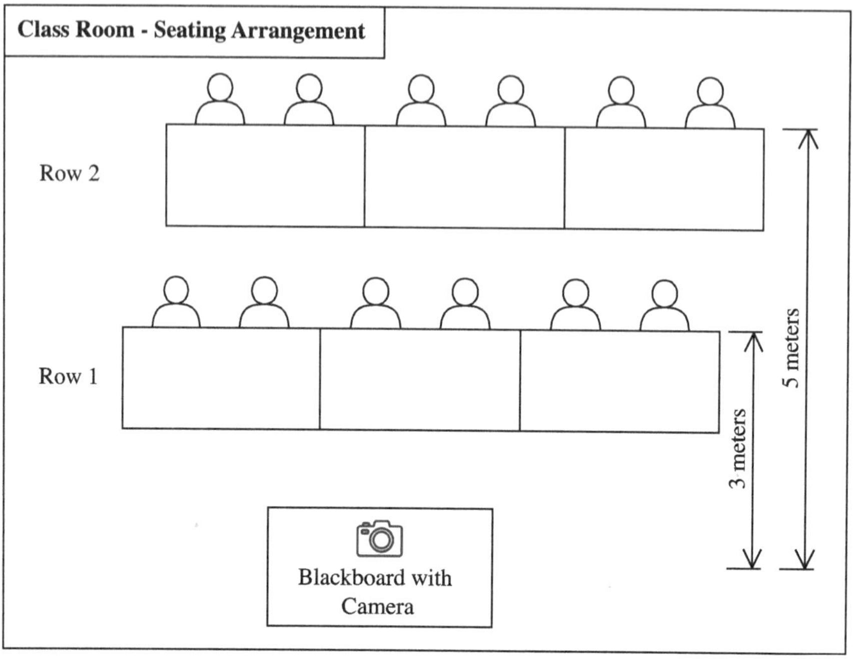

Fig. 2 Schematic representation of the recording of the data set

5.2 *Use Case 1: Gaze Behavior Analysis*

Figure 3 demonstrates the first application scenario: continuous tracking of students' gaze behavior over time. The figure displays the results of the presented pipeline for horizontal (top) and vertical (bottom) gaze angle calculations of four students based on the recorded 43-minute classroom video sequence. The horizontal component indicates left-right gaze direction, while the vertical component shows up-down orientation. It is important to note that, at this stage, we have no ground truth data for the actual gaze directions of the recorded students. The results should be interpreted as demonstrating the potential applications of the pipeline rather than validated measurements. Nevertheless, such data could enable analysis of individual attention patterns. For example, Person 0 (blue line) shows a notable horizontal gaze shift around the 10-minute mark, potentially indicating a change in attentional focus. In contrast, Person 3 (purple line) maintains a consistently upward vertical gaze direction throughout most of the session, which might indicate sustained attention toward the blackboard or instructor. Overall, the tracking data reveals both individual gaze behaviors and moments of synchronized attention shifts across multiple students, providing potential insights into classroom dynamics and engagement patterns.

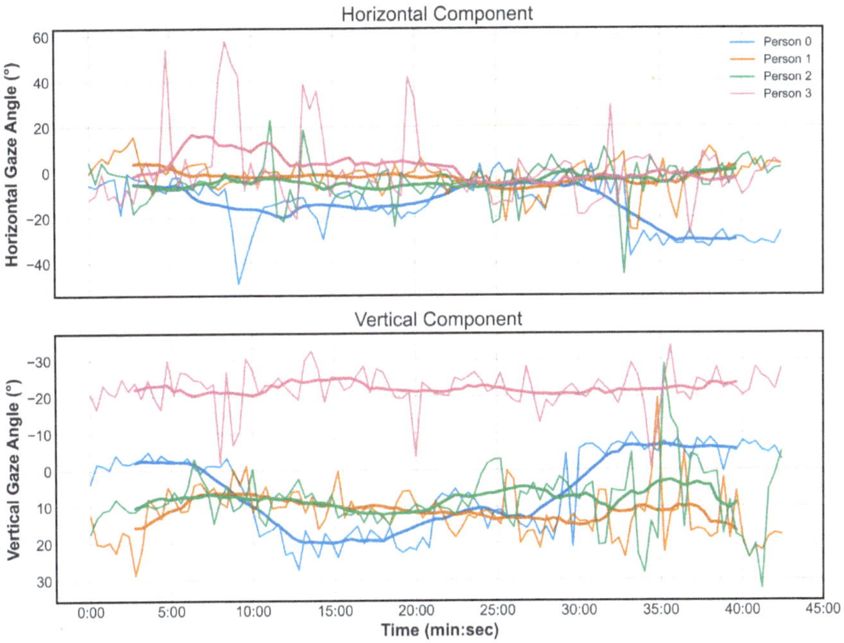

Fig. 3 Results for horizontal (top) and vertical (bottom) gaze angles of our pipeline for four students over the analyzed classroom session

Under the assumption that specific gaze angle ranges could be mapped to defined classroom elements (e.g., teacher, blackboard, other students), the system could effectively measure attention distribution not only at the classroom level, but also on the individual level, which leads to the next potential use case of the gaze monitoring pipeline.

5.3 *Use Case 2: Target Gaze Detection*

Figure4 demonstrates the capability of defining a specific target area in terms of gaze angle ranges and tracking when an individual student's gaze falls within this area. The figure displays the horizontal (top) and vertical (middle) gaze angles for Person 0, with smoother trend lines overlaid on the raw measurements to highlight general patterns. The bottom panel shows a binary indicator that signals when the student's gaze falls within the predefined target area—representing a specific classroom element such as the teacher or an instructional display. This approach enables precise quantification of how much time a student spends looking at particular areas of interest, allowing educators to analyze attention allocation during different instructional phases. For example, the visualization reveals that Person 0 primarily engaged with

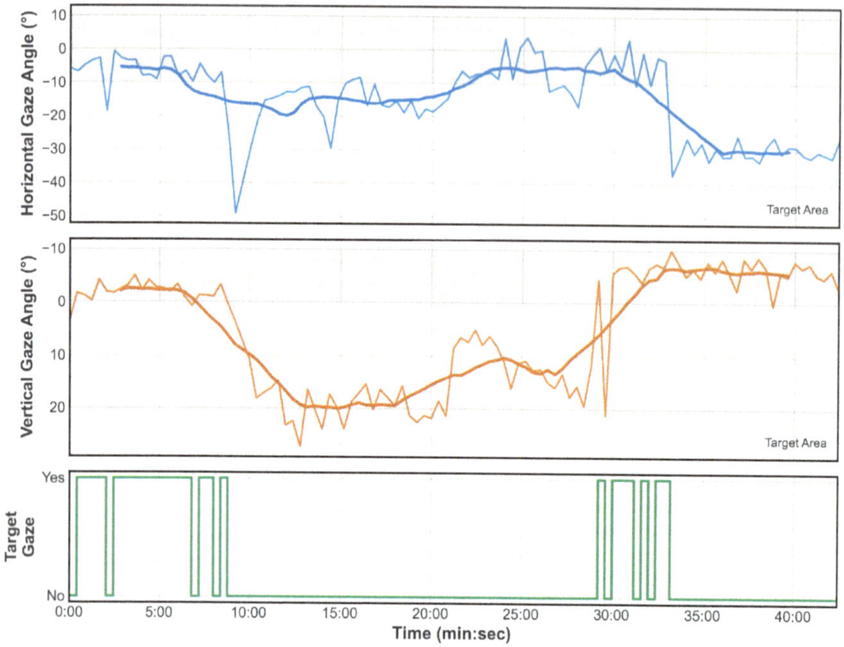

Fig. 4 Target gaze detection analysis for a single student (Person 0), showing horizontal (top) and vertical (middle) gaze angles with trend lines, and binary indication (bottom) of when gaze falls within the defined target area during the 43-minute classroom session

the target area during two distinct periods: early in the session and again around the 30-minute mark, potentially corresponding to different instructional activities. Such insights could help instructors evaluate the effectiveness of different teaching methods in capturing and maintaining student attention, providing data-driven feedback for pedagogical refinement. The capabilities presented in Use Cases 1 and 2 could address critical educational questions such as whether instruction is being effectively received or if students are engaging with learning content, providing educators with quantitative metrics for classroom dynamics that were previously difficult to measure at scale.

While the application scenarios demonstrate the potential of our gaze tracking pipeline, it is important to acknowledge its current proof-of-concept status. The absence of ground truth data and formal validation limits our ability to quantify the system's accuracy in real classroom environments. The following section addresses these and other limitations of the current implementation, providing context for future development and validation efforts.

6 Limitations

While our pipeline demonstrates promising capabilities for classroom gaze tracking, several significant limitations must be acknowledged, both in our technical implementation and evaluation methodology. From a technical perspective, the current implementation is constrained by specific environmental parameters. The system relies on fixed camera positioning and predetermined seating arrangements, limiting its generalizability to varied classroom configurations. Real-world classrooms often feature dynamic seating arrangements, multiple instructional areas, and diverse lighting conditions that may challenge our current approach. More critically, our evaluation is limited by the absence of ground truth data. Without annotated reference data for our recorded classroom setting, we cannot quantify the pipeline's performance using standard evaluation metrics such as precision, recall, or mean average precision. Although we employed a specialized YOLOv8 variant for face detection, the absence of validation data prevents us from measuring its actual performance in classroom environments. This limitation extends to all components of our pipeline, including face assignment, keypoint regression, and gaze direction calculation, none of which have been validated against ground truth annotations. The geometric approach we employ for gaze estimation represents another potential limitation. Head rotation is inferred by minimizing the 2D projection error between a generic 3D face model and observed facial landmarks. While this method is computationally efficient and well-suited to the resource constraints we established, it may be inadequate for reliably estimating gaze in low-quality imagery or when facial landmarks are partially obscured scenarios that frequently occur in authentic classroom settings. Furthermore, the use of a generic BFM without subject-specific adaptation likely introduces systematic errors in head rotation estimation that propagate to final gaze calculations. These limitations highlight the preliminary nature of our work and emphasize the need for comprehensive validation studies before the pipeline can be considered ready for practical educational applications. Future work should address these limitations through the collection of annotated ground truth data, systematic performance evaluation, and optimization of computational efficiency.

7 Conclusion and Future Work

This paper presents a computer vision pipeline for non-intrusive, multi-person gaze tracking in classroom environments using a single camera. Our approach addresses a significant methodological gap by enabling simultaneous monitoring of multiple students without specialized per-person equipment, preserving classroom ecological validity while providing quantitative insights into visual attention patterns. The proof-of-concept implementation demonstrates the pipeline's potential for both continuous gaze behavior tracking and target-specific attention analysis, offering new possibilities for understanding classroom dynamics at scale. Future work should focus on four critical areas: (1) developing comprehensive ground truth datasets with precisely annotated gaze directions to enable quantitative validation across all pipeline components; (2) enhancing environmental robustness through adaptive calibration procedures that accommodate variable lighting conditions and flexible seating arrangements; (3) implementing subject-specific adaptation of the 3D face model and optimizing computational efficiency to support real-time processing; and (4) conducting educational research applications to examine relationships between visual attention patterns and learning outcomes, particularly for students with special educational needs. Addressing these directions would transform this proof-of-concept into a valuable research tool that provides unprecedented insights into classroom attention dynamics while maintaining authentic learning environments.

References

Akanametov. (2024). “YOLO Face”. https://github.com/akanametov/yolo-face?tab=readme-ov-file. Accessed 2025/03/26.

Araya, R., & Sossa-Rivera, J. (2021). Automatic detection of gaze and body orientation in elementary school classrooms. *Frontiers of Robotics AI, 8*, 729832. Frontiers.

Asish, S. M., Kulshreshth, A. K., & Borst, C. W. (2022). Detecting distracted students in educational VR environments using machine learning on eye gaze data. *Computer of Graph, 109*, 75–87. Elsevier.

Basel Face Model, https://faces.dmi.unibas.ch/bfm/bfm2019.html. Accessed 2025/03/26.

Blume, F., & Schmiedek, F. (2024). It counts in every single lesson: Between- and within-person associations of teaching quality and student self-regulation. *Learning and Instruction, 92*, Article 101908.

Gerig, T., Morel-Forster, A., Blumer, C., Egger, B., Luthi, M., Schönborn, S., & Vetter, T. (2018). Morphable face models–an open framework. In *2018 13th IEEE international conference on automatic face and gesture recognition (FG 2018)* (pp. 75–82). IEEE.

Girshick, R., Donahue, J., Darrell, T., & Malik, J. (2014). *Rich feature hierarchies for accurate object detection and semantic segmentation* (pp. 580–587).

Google AI Edge. (2024). “Mediapipe”. https://github.com/google-ai-edge/mediapipe. Accessed 2025/03/26.

Helmke, A., & Renkl, A. (1992). Das Münchener Aufmerksamkeitsinventar (MAI): Ein Instrument zur systematischen Verhaltensbeobachtung der Schüleraufmerksamkeit im Unterricht. *Diagnostica, 38*(2), 130–141.

Hommel, M. (2011). Aufmerksamkeitsverlauf – Fremdbeobachtung und Eigeneinschätzung. In U. Faßhauer, B. Fürstenau, E. Wuttke, E. (Eds.), *Grundlagenforschung zum Dualen System und Kompetenzentwicklung in der Lehrerbildung* (pp. 117–129). Berlin; Farmington Hills, MI : Verlag Barbara Budrich, Opladen. (Schriftenreihe der Sektion Berufs- und Wirtschaftspädagogik der Deutschen Gesellschaft für Erziehungswissenschaft (DGfE)). DOI:https://doi.org/10.3224/86649461, URN: urn:nbn:de:0111-opus-70722

Hutt, S., Hardey, J., Bixler, R., Stewart, A., Risko, E., & D’Mello, S. K. (2017). Gaze-based detection of mind wandering during lecture viewing. *International of Educational Data Minining Society*. ERIC.

Martinikorena Aranburu, I., Larumbe, B., Andoni, A. G., Mikel, P. C., Sonia, C. L. R., & Villanueva, L. A. (2020). Low cost gaze estimation: Knowledge-based solutions. *IEEE Transactions on Image Processing, 29*. IEEE.

Shuoyang. (2024). "WIDERFACE". http://shuoyang1213.me/WIDERFACE/. Accessed 2025/03/26.

Stürmer, K., Seidel, T., Müller, K., Häusler, J., & Cortina, K. S. (2017). What is in the eye of preservice teachers while instructing? An eye-tracking study about attention processes in different teaching situations. *Z. Erzieh.wiss., 20*(1), 75–92.

Ultralytics. (2024). "YOLOv8". https://huggingface.co/Ultralytics/YOLOv8. Accessed 2025/03/26.

Vicente, F., Huang, Z., Xiong, X., Fernando, Z. W., & Levi, D. (2015). Driver gaze tracking and eyes off the road detection system. *IEEE Transactions on Intelligent Transportation Systems, 16*(4), 2014–2027. IEEE.

Vierlinger, R. (1984). Die Auswirkungen verschiedener Unterrichtskonzepte auf Aufmerksamkeit und Störverhalten der Schüler. *Unterrichtswissenschaft, 12*(4), 387–396.

Wang, Z., Chai, J., & Xia, S. (2019). "Realtime and accurate 3D eye gaze capture with DCNN-based iris and pupil segmentation. *IEEE Transactions on Visualization and Computer Graphics, 27*(1), 190–203. IEEE.

Xu, H., Zhang, J., Sun, H., Qi, M., & Kong, J.: Analyzing students' attention by gaze tracking and object detection in classroom teaching. *Data Technology Applications, 57*(5), 643–667. Emerald Publishing Limited.

Investigating e-Learning in Marginalized Independent Secondary Schools: A Teacher's Perspective

Sunet Eybers, Corné J. van Staden, and Alex Zindove

Abstract e-Learning refers to the utilization of technology to support learning across geographical boundaries. It is an important aspect of digital transformation, as it offers many benefits, such as accessibility, flexibility, scalability, self-paced learning, and democratizing education. It supports enabling quality education to all, an essential aspect of sustainable development, even in marginalized communities. Marginalized communities, defined as communities that are often excluded from traditional social, economic, educational, or cultural groups due to various reasons (for example, race or socio-economic conditions), are often excluded from access to learning, including e-learning opportunities. However, Independent Section 29 schools, privately owned and funded to offer discounted school fees, provide learners in rural areas the opportunity to receive quality education as part of digital transformation efforts. This study investigates teachers' perspectives on e-learning implementations in marginalized Independent Section 29 secondary schools in South Africa. Using the Community of Inquiry (CoI) framework as a theoretical foundation, online questionnaires were distributed to teachers at marginalized Independent secondary schools in a small mining town in South Africa. The results indicated that internet infrastructure, connectivity, and lack of device resources remain challenging, forcing teachers to use their own devices to conduct lessons. Both blended and virtual learning were the preferred methods for teaching. The CoI framework proved reliable (except for the triggering event construct) and suitable for use in online teaching communities, contributing to e-Learning's success. The social presence of the CoI was indicated to be the most important aspect in e-Learning, followed by the cognitive and teaching presence.

S. Eybers (✉) · C. J. van Staden · A. Zindove
University of South Africa, Johannesburg, South Africa
e-mail: eeyberss@unisa.ac.za

J. Marx Gómez et al. (eds.), *Recent Advances in Information Systems*, Progress in IS,
https://doi.org/10.1007/978-3-032-14099-9_25

1 Introduction

Many definitions of e-learning exist, with the most encapsulating definition postulated by Alqahtani and Rajkhan (2020), which refers to e-learning as an interconnection between devices, organizing components, programs, and frameworks that link and permit organizations and people to connect in a virtual, electronic world, enabling learning. Whilst the benefits of an online, connected community for teaching and learning are apparent, including the flexibility, scalability, and adaptability of learning, Subekti (2020) observed that the process of teaching and learning via e-learning differs from traditional classroom instruction, which can introduce significant challenges for both teachers and learners. These include a lack of knowledge or teacher "know-how," a shortage of technological resources, and inadequate internet access (Parker et al., 2019). Even if these challenges are overcome, a lack of proper ICT infrastructure often hampers the effectiveness of e-learning, user learning traits (for example, gender, culture, and ethnicity), ICT support, lack of training, and shortage of devices (Nouraey & Al-Badi, 2023), requiring urgent digital transformation interventions. These challenges are acknowledged in research by Hadzic (2024), who investigated the influence of governmental policies.

As per the government education policy, referred to as the White Paper on e-Education of 2004, the Department of Education in South Africa (DoE) acknowledged the need for ICT integration in schools (Tshimanika et al., 2022). The White Paper, which serves as the foundation for education in South Africa, mandates that learning institutions in South Africa (from primary school to tertiary education) must include ICT in instruction and scholarship by 2013 (Nkadimeng & Thaba-Nkadimene, 2019). Unfortunately, not much was done until March 2020, when the COVID-19 pandemic forced many organizations, including schools, to consider online environments to minimize operational disruption and offer continued quality education.

Even before the pandemic, quality education was identified as one of the United Nations' 17 Sustainable Development Goals (Department of Economic & Social Affairs, 2025), vital for South Africa as a global South nation. It strives to "ensure inclusive and equitable quality education and promote lifelong learning opportunities for all" (Department of Economic & Social Affairs, 2020; United Nations South Africa, 2025). Inclusive education is an environment where all learners can access quality education equally, fostering lifelong learning and sustainability. e-Learning, which empowers learners to access content online, is imperative in accessing educational content irrespective of geographical boundaries. Unfortunately, inclusive education is hampered by a lack of digital transformation.

If one can assume that significant progress can be made in digital transformation, implementing e-learning becomes a priority. This paper investigated the appropriateness of the Community of Inquiry framework used in an e-learning environment in marginalized independent Section 29 schools during the pandemic, residing in a small geographical area in a developing country. After the e-Learning experience of teachers was explored, the CoI was used to investigate further the implications of the social, teaching, and cognitive presences in an online environment.

The paper's outline is as follows: a literature review discussing digital transformation and e-learning, followed by a comparison between developed and developing countries. The CoI is described as the main theoretical framework, followed by the research methodology and approach. The data analysis and discussion section is combined, followed by a summary of findings and the conclusion.

2 Literature Overview

Digital transformation refers to the utilization of digital technologies as part of conducting business. The relationship between digital transformation and e-learning is often perceived as symbiotic, whereby educators have a responsibility to use digital technologies in online education environments to meet the needs of our digital society (Lazarenko & Hapchuk, 2024). On the other hand, e-learning was enabled as a result of digital transformation efforts, and can't exist without the necessary digital tools and technologies (Rapetti & Cantoni, 2010).

Teachers and learners concur that e-learning is a helpful tool for improving teaching and encouraging the growth of information acquisition abilities through the transfer of knowledge (Roman & Plopeanu, 2021). By using e-learning, learners believe they can acquire the necessary knowledge, abilities, and competencies (Subekti, 2020). This is not surprising, as e-learning is reported to contribute to an in-depth understanding of content (Roman & Plopeanu, 2021), as learners study at their own pace (Ionita & Pastae, 2021), provide learners access to an extensive knowledge pool of information that is available anytime and anywhere (Abu-Taieh et al., 2022) and the adaptability of content to suit the needs of learners. Therefore, e-learning is flexible and caters to many different learning styles, contributing to learning outcomes without being physically present (Mukhtar et al., 2020). Furthermore, e-learning allows learners to learn in developing countries and rural areas without the associated school travel costs (Naddeo et al., 2021). Ndume et al. (2008) further noted the benefits and rewards of e-learning, including task engagement, increased e-learning participants' sense of participation, and reduced attrition.

e-Learning is no silver bullet to solve the challenges of creating inclusive, sustainable education. A considerable challenge, in particular in developing countries, is adequate infrastructure as a key determinant and prerequisite for successful e-learning (Wong & Kan, 2022). In Korea, for instance, proactive efforts to build a strong infrastructure have significantly increased the e-learning platform's effectiveness (Maphalala & Adigun, 2020). Additionally, Redempta and Elizabeth (2012) assert that inadequate internet infrastructure in schools and limited internet connection at home harm learners' success in online learning. In addition, Meinert et al. (2021) found that teachers' attitudes positively impact the effectiveness of e-learning. Roman and Plopeanu (2021) identified that additional learner traits, such as gender, ethnicity, and active participation in various learning community structures, may affect the effectiveness of e-learning.

When teachers understand potential constraints, such as social contact, administrative concerns, learners' motivation, and time/support for studies, the potential of a successful e-learning system increases (Ndume et al., 2008). Additionally, it has been discovered that self-regulation, effective instructional design, and meticulous educational preparation (Kerin, 2022) play a crucial part in the effectiveness and success of e-learning. If these factors are not considered in e-learning, it can cause academic stress, which harms both the learners' and teachers' psychological health (Maphalala & Adigun, 2020) and academic performance (Parker et al., 2019).

2.1 e-Learning in Developing Versus Developed Countries

In a study by Roman and Plopeanu (2021), the authors suggest that teachers and learners in the United States utilize digital learning tools daily outside the school environment. In a study by Stoesz and Niknam (2022) conducted in Canada, the findings show that online course learners perform better than face-to-face contact learners, understand the course structure, communicate with teachers, and increase engagement and satisfaction with the course material. Learners have expressed gratitude for the convenience and flexibility of online learning. These are typical examples where a developed country is supported by infrastructure, contrasting with South Africa's e-learning experiences (Naddeo et al., 2021). In a developing country such as South Africa, many factors impact the effectiveness of e-learning, such as a lack of technological advancement, institutional backing, and dependable infrastructure (Roman & Plopeanu, 2021), as well as a lack of device resources. In South America, the same factors affecting South Africa, such as limited capacity and lack of internet infrastructure, are the barriers to successful e-learning.

In Jordan, another example of a developing country, evidence suggests that prior user training positively impacts e-learning success (Habes et al., 2022). Evidence shows that online learning was accepted mainly in Jordan because of the vast number of online users (AlDreabi et al., 2023). At the international level, particularly in Jordan, teachers and learners were offered free training on facilitating and using online learning (AlDreabi et al., 2023).

3 Theoretical Framework: Community of Inquiry (CoI)

The Community of Inquiry (CoI) framework was introduced by Garrison et al. (2000), Garrison and Akyol (2013), Swan et al., (2009) and acknowledges the importance of social, cognitive, and teaching presence in creating positive and meaningful collaborative learning communities. A community gathers individuals who have become acquainted via their shared experiences and mutual learning (Garrison & Akyol, 2013; Swan et al., 2009).

Social presence is the degree to which learners can form and check meaning within class activities, in-depth thinking, and dialogue (Rachman et al., 2021). It further refers to communication between teachers and the learners, and between the learners themselves (Rachman et al., 2021). Furthermore, it refers to the extent to which learners feel connected to each other, both emotionally and affectionately. In the context of e-learning Alhumaid et al. (2020) states that social presence improves learning effectiveness because teachers and learners freely discuss content in the comfort of their homes.

Cognitive presence refers to critical discussions that require reflection, listening, and verbal expression. It is a cooperative process of in-depth, significant thought and education, which is related to cooperation and introspection, and influenced by the communication strategies and techniques used in a course (Rachman et al., 2021). Cognitive presence is related to cooperation and introspection, and it is influenced by the communication strategies and techniques used in the course. Cognitive presence is the interactions that take place between the teacher and the learner, whereby the teacher presents content to the learner and then expects feedback from the learner (Rachman et al., 2021).

Designing and enabling cognitive and social processes to produce essential and beneficial learning outcomes is called teaching presence and refers to how learners adapt to the learning environment and the curricula used. When all three presences are present, the learning experience is successful. This is supported by Ioniță and Păstae (2021), stating that, for learning to occur, there should be a proper interaction between learners, teachers, and the content to be taught.

4 Research Methodology and Approach

The study investigated the appropriateness of the Community of Inquiry framework use in an e-learning environment, from a teacher's perspective, in marginalized Section 29 schools in a small mining town in South Africa. Due to its relatively low household income, most learners are sponsored by a local mining company to attend independent schools in the region.

An online questionnaire based on the CoI framework was used. The questionnaire consisted of five sections. Section A focused on demographic information, variables such as age, gender, and home language, and how they might influence the effectiveness of e-learning. Section B contained closed-ended questions focusing on the e-learning teacher's e-learning experience. The questions in Section C were based on questionnaire questions that were developed, reused, and adjusted, where applicable, based on the CoI instrument proposed by Swan et al. (2008) (with permission from the authors). Likert scale questions, offering five levels for selection, namely strongly disagree (1) to strongly agree (5), were used. The social presence section of the CoI included affective expression, open communication, and group cohesion.

The cognitive presence focused on a triggering event, exploration, and integration; while the Teaching Presence focused on constructs such as design and organization, facilitation of discourse, and direct instruction. Table 1 summarizes and describes the CoI constructs.

For the purpose of the study, the design and organization construct was not considered, as it is assumed to be an inherent role of learning facilitation.

Convenient, purposive sampling was employed to select teachers from Section 29 independent schools. A pilot study of 10 participants completed the online questionnaire. Adjustments were made (for example, the duplication of questions), after which the questionnaire was finalized. The questionnaire was distributed to 114 participants, of whom a total of 106 completed the questionnaire, yielding a response rate of 93%.

Table 1 Questionnaire structure using the CoI instrument [adapted from Swan et al. (2008)]

CoI	Construct	Description
Social presence The extent to which learners feel connected to each other, both on an emotional and an affectionate level	Affective expression	The reliable and steady sensations that an individual has a habit of eliciting with their contacts
	Open communication	How much a person is regarded as real and present in a specific communication medium
	Group cohesion	The level of a member's attachment to their group
Cognitive presence The degree to which learners can create meaning through prolonged communication	Triggering event	A statement or query that shifts the focus of the conversation to questions with a problem, issue, dilemma, challenge, or learning assignment as its central theme
	Exploration	Study or investigate a construct in more depth
	Integration	Conceptualize or combine construct meanings based on the ideas explored
	Resolution	The style in which participants recognize and clarify the issue through their values, perceptions, knowledge, and feelings is identified as cognitive resolution
Teaching presence How learners adapted to the learning environment and the curricula used	Design and organisation	The arrangement and presentation of constructs provide meaningful content
	Facilitation	Delivering content to assist with understanding
	Direct instruction	Instructional approach, for example, presentation of concepts, broken into smaller sections to foster understanding

5 Data Analysis and Discussion

Demographic information and teachers' learning experience were analyzed using Excel, whilst SPSS version 21 was used to perform quantitative data analysis on the data, after the survey results from an online Google form were stored in an Excel spreadsheet.

5.1 *Section A: Demographic Information*

The majority of respondents were male (71%). The ages ranged from 24 (8%), 25–34 (21%), 35–44 (36%), 45–54 (23%), and older than 55 (12%). The respondents were teachers in specific departments, namely commerce (17%), languages (18%), sciences (52%), with a small number of teachers in other, small departments (13%). Most respondents indicated English as their home language (75%), while Sepedi, the predominantly spoken language in Limpopo, ranked second with 12%. 4% of respondents identified Afrikaans as home language, whilst various indigenous languages were classed as part of the other category (9%).

5.2 *Section B: Teacher's e-Learning Experience*

As part of evaluating teachers' e-learning experience, participants were asked what they use the internet for, the resources used, their experience with e-learning and tools, their preferred teaching methods, and what their opinion was of using e-learning (keeping learners' attention, interruptions, and online learner participation).

Most participants used the internet for personal reasons, including internet banking (84%) and interacting with family and friends (80%). It was also noted that 77% of individuals used the Internet for e-learning purposes, 75% for online shopping, 61% for searching information, and 47% to engage in forums.

Mobile generation refers to the maturity of mobile technologies. With each increased version of the generation number, enhanced connection benefits are introduced. For example, the latest fifth generation, which refers to faster internet speeds and inter-device connectivity, supporting superior interconnectivity, provides users with the most benefits, especially in online learning. Respondents were asked which mobile generation they used when connecting to the internet at school. 66% of respondents indicated that a third generation (3G) network was predominantly used, 44% used a fourth generation (4G). None of the respondents noted using fifth-generation (5G) network connectivity. Interestingly, 39% of respondents felt that the internet was performing fast, whilst 16% thought that the internet was high speed. 32% believed that the internet speed has improved from what it used to be, and 13% of internet users at the school complained that the internet was not performing to their speed expectations. When compared to the internet quality and speed of respondents' residences, 36% believed the internet performed better than expected during online sessions. Only 18% of respondents thought they had the fastest internet connections at their residence. Some respondents, around 33%, believe that their internet was fast. The study concluded that the internet was perceived to be high speed at home (18%) compared to the internet performance at school (16%). The study may also suggest that school usage was higher than home usage.

All the respondents indicated that they use their own personal mobile phones as the main resource for e-learning purposes. In addition, 47% used their laptops, 28% used desktop computers, and 14% used tablets. All the respondents indicated that the various schools provided them with interactive whiteboards for e-learning purposes, and some suggested the availability of desktop computers in laboratories. 19% of respondents received data for learning purposes.

When respondents were asked to indicate their preferred teaching methods, 59% preferred blended learning in e-learning environments, followed by 55% preferring virtual learning. Some teachers preferred using social media in an e-learning environment, but it was unclear how this was used; lastly, 14% preferred micro learning. 37% of respondents preferred live, online classes, and 25% preferred a flipped classroom approach.

Teaching tools: 65% of respondents preferred virtual whiteboards, whilst 37% preferred pre-recorded videos, distributed via WhatsApp and other channels.

43% of respondents thought that it was very difficult to keep learners' attention in e-learning environments. However, 9% of respondents were delighted with teaching online and believed it was possible to hold learners' attention.

In many instances, learning takes place in the comfort of the respondent's home. During online sessions, 48% of the respondents indicated they can teach with uninterrupted network connectivity; 27% preferred teaching at their own pace at home during online lessons, whilst 12% complained of distractions during online sessions.

56% of respondents agreed that it is essential to engage with learning by posting questions during and after online sessions to improve their understanding. Furthermore, 27% believed that questions should be posted to a discussion forum or peers for further discussion. 17% felt that working through online content with the learners is necessary to enhance learning and teaching.

5.3 Section C: Community of Inquiry Questionnaire

A reliability test was conducted on ten constructs, from questions 22–31, using Cronbach's alpha coefficient to test the reliability of the questionnaire. The key codes used in the test are summarized in Table 2.

Cronbach's alpha uses a conventional 0–1 scale to quantify the degree of arrangement. For example, figures generated with this evaluation metric, anything above 0.7, are good, anything in the range of 0.5–0.7 is acceptable, and figures below 0.5 are unacceptable. Therefore, the Cronbach's alpha validates whether the items agree.

Table 2 Cronbach's alpha coefficient abbreviations

Abbreviation	Description	Abbreviation	Description
Facil	Facilitation	Trig	Triggering event
DInstruct	Direct instruction	Exp	Exploration
Affeexp	Affective expression	Integ	Integration
Opcom	Open communication	Res	Resolution
Gcoh	Group cohesion	ELearn	e-Learning effectiveness

In the study's sample, all the community of inquiry and e-learning effectiveness scales, except for triggering event (under cognitive presence) and group cohesion (under social presence), had better internal consistency with a Cronbach's alpha larger than 0.7. However, since the instrument used short scales (with fewer than ten items each), the study considers reporting the mean inter-item correlations and the recommendation suggested by Hajjar (2018) to accommodate an optimum scale of inter-item correlation of 0.2–0.4. Therefore, group cohesion, with a mean inter-item correlation of 0.354, was considered reliable and included in the list of scales with internal consistency, while the triggering event, from cognitive presence, was below the lower range, with a mean inter-item correlation of 0.072. This construct should therefore be reconsidered. Table 3 below provides a summary of the basic statistical calculations.

The results from the questionnaire, based on the CoI, indicated that all the respondents agreed that the social, cognitive, and teaching presence is essential to ensure the effectiveness of e-learning efforts. According to the results, the strongest of the three constructs was the social presence, particularly affective expression and open

Table 3 Summary of basic statistical calculations

Question number	Construct	Number of items	Mean	Standard deviation	Mean inter-item correlation	Cronbach's alpha	Reliability
22	Facil	5	3.9962	1.0002	0.695	0.919	Good
23	DInstruct	2	3.9340	0.9638	0.726	0.840	Good
24	Affeexp	2	3.9858	0.9141	0.717	0.834	Good
25	Opcom	3	3.9308	0.7911	0.572	0.796	Good
26	Gcoh	3	3.9434	0.8067	0.354	0.578	Acceptable
27	Trig	2	3.4009	1.0429	0.072	0.128	Not acceptable
28	Exp	3	3.9340	0.7786	0.754	0.901	Good
29	Integ	2	3.9245	0.8832	0.692	0.816	Good
30	Res	2	3.9623	0.8214	0.654	0.791	Good
31	ELearn	8	4.1306	0.6669	0.366	0.806	Good

Table 4 Summary of data per CoI construct

CoI	Construct	Percentage of respondents (%)				
		Strongly disagree (%)	Disagree (%)	Neutral (%)	Agree (%)	Strongly agree (%)
Social presence	Affective expression	6	3	6	67	28
	Open communication	5	1	10	62	22
	Group cohesion	5	1	11	60	23
Cognitive presence	Triggering event	12	11	13	54	10
	Exploration	4	3	10	63	20
	Integration	6	1	10	60	23
	Resolution	6	0	11	58	25
Teaching presence	Facilitation	9	1	5	52	33
	Direct instruction	8	2	9	52	28

communication. Therefore, this highlights the importance of learners in connecting and engaging in an online learning environment.

The second strongest construct was the cognitive presence, focusing on enhancing the thinking and problem-solving ability. The strongest construct was the exploration, which underscored the importance of trading and exploring to gain new knowledge. The triggering event had the highest percentage of disagreement of all the constructs. Both the facilitation and direct instruction constructs came out equally in teaching presence, with the most respondents indicating that they agree with the importance of e-learning effectiveness. Table 4 provides a summary of the data.

6 Summary of Findings

It can be summarized that to implement e-learning at an independent secondary school using the CoI, social presence, cognitive presence, and teaching presence need to be effectively addressed. Teachers need to be adequately equipped with the relevant resources, ranging from training to proper internet and adequate devices. Support is imperative in ensuring the successful implementation of e-learning environments in a school setup, for both teachers and learners.

With respect to social presence, it is important to make use of affective expression, have open communication between teacher and learner, and use group cohesion. In order to achieve social presence, the schools need appropriate learning management systems (LMS) in place, as agreed by Mohammed et al. (2017). A study by Ahmad et al. (2023) indicated that if a proper learning management system (LMS) is in place and working properly, it is easy for a school to implement curriculum specifically for e-learning environment content, easily assisting with social presence.

Cognitive presence requires exploration, integration, and resolution. In order to make use of and integrate proper learning material and resources, such as videos, into the LMS, Roman and Plopeanu (2021) argue that enough access to proper ICT infrastructure is a requirement. In order for learners and teachers to explore the e-learning environment, internet connectivity, which is a challenge in the area, is a requirement that needs to be addressed. When designing the e-learning site, consideration needs to be given to the users of the site, who are learners who, in the majority of cases, only have access to mobile devices, so it is important to take into consideration the navigation.

In respect of teaching presence that relates to facilitation and direct instruction, note needs to be taken that e-learning is not just a way of disseminating content, but it is a way of engaging effectively with learners using an LMS through facilitation and direct instruction. Training has thus been identified as an area that needs attention to equip educators to properly facilitate and actively be involved in direct instruction.

7 Conclusion

In conclusion, it can be said that the CoI framework seems suitable for investigating e-learning environments in marginalized schools in South Africa, focusing on addressing challenges relating to internet connectivity, training, data provision, and design of material and content, with a proper policy in place. Successful e-learning environments remain an important aspect of digital transformation, and evidence from the study suggests that further studies need to be carried out to determine how teaching, social, and cognitive presences affect the effectiveness of e-learning. The perspective of learners, parents, and stakeholders remains important when investigating e-learning and should therefore be included in future studies.

References

Abu-Taieh, E. M., et al. (2022). An empirical study of factors influencing the perceived usefulness and effectiveness of integrating e-learning systems during the COVID-19 pandemic using SEM and ML: A case study in Jordan. *Sustainability, 14*(20), 13432. https://doi.org/10.3390/su142013432

Ahmad, S., et al. (2023). ELearning acceptance and adoption challenges in higher education. *Sustainability, 15*(7), 6190. https://doi.org/10.3390/su15076190

AlDreabi, H., et al. (2023). Sustainable digital communication using perceived enjoyment with a technology acceptance model within higher education, in Jordan. *Frontier Education, 8*, 718. https://doi.org/10.3389/feduc.2023.1226718

Alhumaid, K., et al. (2020). COVID-19 and elearning: Perceptions and attitudes of teachers towards e-learning acceptance in the developing countries. *Multicultural Education, 6*, 2. https://doi.org/10.5281/zenodo.4060121

Alqahtani, A. Y., & Rajkhan, A. A. (2020). E-learning critical success factors during the COVID-19 pandemic: A comprehensive analysis of e-learning managerial perspectives. *Education Sciences, 10*(9), 216. https://doi.org/10.3390/educsci10090216

Department of Economic and Social Affairs. (2020). Sustainable development: United Nations: Sustainable development goals (SDG): goal 4. United Nations Department of Global Communications.

Department of Economic and Social Affairs. (2025). Sustainable development: United Nations: 17 sustainable development goals (SDG). United Nations Department of Global Communications. https://sdgs.un.org/goals

Garrison, D., & Akyol, Z. (2013). The community of inquiry theoretical framework. *Handbook of Distance Education, 14*, 104–119.

Garrison, D. R., Anderson, T., & Archer, W. (2000). Critical inquiry in a text-based environment: Computer conferencing in higher education. *Internet Higher Education, 2*(2–3), 87–105. https://doi.org/10.1016/S1096-7516(00)00016-6

Habes, M., et al. (2022). Social impact of videos at new media platforms on the e-learning acceptance during the Covid-19. *Information Science Letters, 11*(3), 913–923.

Hadzic, S. (2024). South Africa's digital transformation: Understanding the limits of traditional policies and the potential of alternative approaches. *Computer Law and Security Review, 55*, 106011. https://doi.org/10.1016/j.clsr.2024.106011

Hajjar, S. T. E. (2018). Statistical analysis: Internal-consistency reliability and construct validity.

Ionita, M., & Pastae, V. (2021) Challenges of online learning during the COVID-19 pandemic. In: *International scientific conference: eLearning and software for education* (pp. 176–181). https://doi.org/10.12753/2066-026X-23-023.

Ioniţă, M., & Păstae, V. (2021). Challenges of online learning during the covid-19 pandemic. *eLearning and Software for Education Conference, 54*, 176–181. https://doi.org/10.12753/2066-026X-21-023

Kerin, T. (2022). Creating a learning organisation. *Chemical Engineering Transactions, 90*, 781–786. https://doi.org/10.3303/CET2290131

Lazarenko, N., & Hapchuk, Y. (2024). E-learning and artificial intelligence as key factors in the digital transformation of higher education: Challenges, opportunities and development prospects. *Педевтологія, 2*(1), 7–13.

Maphalala, M. C., & Adigun, O. T. (2020). Academics' experience of implementing e-learning in a South African higher education institution. *International Journal of Higher Education, 10*(1), 1. https://doi.org/10.5430/ijhe.v10n1p1

Meinert, E., et al. (2021). Exploring the cost of e-learning in health professions education: Scoping review. *JMIR Medical Education, 7*(1), e13681. https://doi.org/10.2196/13681

Mohammed, A., et al. (2017). E-learning: A tool for enhancing teaching and learning in educational institutes. *International Journal of Computer Science and Information Technologies (IJCSIT), 8*, 217–221.

Mukhtar, K., et al. (2020). Advantages, limitations and recommendations for online learning during COVID-19 pandemic era. *Pakistan Journal of Medical Sciences, 36*, 2785.

Naddeo, A., et al. (2021). Identifying factors that influenced wellbeing and learning effectiveness during the sudden transition into eLearning due to the COVID-19 lockdown. *Work, 68*(1), 45–67. https://doi.org/10.3233/WOR-203358

Ndume, V., et al. (2008). Challenges of adaptive e-learning at higher learning institutions: A case study in Tanzania. *International Journal of Computing and ICT Research, 2*, 47–59.

Nkadimeng, M., & Thaba-Nkadimene, K. (2019). Implementation of e-learning in rural Limpopo secondary schools: Are teachers and schools ready for a new pedagogy?

Nouraey, P., & Al-Badi, A. (2023). Challenges and problems of e-learning: A conceptual framework. *Electronic Journal of e-Learning, 21*, 188–199.

Parker, R., et al. (2019). Theme 9: Education, inequality and innovation in the time of COVID-19.

Rachman, A. N., et al. (2021). Community of inquiry for students' autonomy in English language learning: A case of Philippines high school. *IJELTAL Indonesian Journal of English Language Teaching and Applied Linguistics, 6*(1), 61.

Rapetti, E., & Cantoni, L. (2010). Exploring the added value of digital technologies and eLearning in higher education from learners' perspective. A research informed by a systematized literature review.

Redempta, K., & Elizabeth, M. (2012). An e-learning approach to secondary school education: E-readiness implications in Kenya. *Journal of Education and Practice, 3*(16), 142.

Roman, M., & Plopeanu, A.-P. (2021). The effectiveness of the emergency eLearning during COVID-19 pandemic: The case of higher education in economics in Romania. *International Review of Economics Education, 37*, 100218. https://doi.org/10.1016/j.iree.2021.100218

Stoesz, B. M., & Niknam, M. (2022). Student perceptions of the visual design of learning management systems. *Canadian Journal of Learning and Technology, 48*(3), 1–22.

Subekti, A. S. (2020). Covid-19-triggered online learning implementation: Pre-service English teachers' beliefs. *Metathesis: Journal of English Language, Literature, and Teaching, 4*(3), 232–248.

Swan, K., et al. (2008). Validating a measurement tool of presence in online communities of inquiry. *E-Mentor, 2*, 145.

Swan, K., et al. (2009). A constructivist approach to online learning: The community of inquiry framework. *Information Technology and Constructivism in Higher Education: Progressive Learning Frameworks, 15*, 43–57. https://doi.org/10.4018/978-1-60566-654-9.ch004

Tshimanika, K., et al. (2022). The influence of policy formulation contextual factors on implementation: South African white paper 7 as a case in point. *Electronic Journal of e-Learning, 20*(4), 386–399.

United Nations South Africa. (2025). How the UN is supporting the sustainable development goals in South Africa. https://southafrica.un.org/en/sdgs. Accessed 03 Feb 2025.

Wong, F. M. F., & Kan, C. W. Y. (2022). Online problem-based learning intervention on self-directed learning and problem-solving through group work: A waitlist controlled trial. *International Journal of Environmental Research and Public Health, 19*(2), 720. https://doi.org/10.3390/ijerph19020720

Decentralized Federated Learning Using Transformer-Based Language Models

Gerrit Schumann, Carsten Montag, Lars Steffens, Michael Karl, and Jorge Marx Gómez

Abstract When simulating federated learning scenarios, most studies use a centralized learning architecture in which a global server performs the orchestration of the individual clients and the aggregation of the associated weights. Decentralized federated learning (DFL) clusters, on the other hand, are rarely investigated, although they can counteract disadvantages such as "single point of failure" or limited scaling options. Furthermore, transformer-based language models have not yet been investigated in DFL clusters, although they are still the de facto standard in the field of natural language processing. In order to investigate how such model architectures can be trained in a decentralized federated manner and which influencing factors play a decisive role, this study presents the implementation of three peer-to-peer topologies (ring, wheel, fully connected) for a DFL cluster and their evaluation using the well-known text classification problem "Contradiction Detection". The results of this case study show that the use of the "Fully connected" topology leads to faster and more stable convergence (model divergence of 0.38), while "Ring" and "Wheel" topologies initially exhibit greater fluctuations (model divergence of up to 0.53).

1 Introduction

Federated learning (FL) refers to a learning technique that allows users (also referred to as clients or nodes) to collectively reap the benefits of shared models without having to centrally store the data used for the training (Yang et al., 2019). Instead, the local data remains with the clients and is only used at this location for the training of individual, local models. The resulting local model weights are then aggregated into a global model (McMahan et al., 2016). In this way, sensitive client data can be processed without compromising data security.

G. Schumann (✉) · C. Montag · J. Marx Gómez
University of Oldenburg, Oldenburg, NI, Germany
e-mail: gerrit.schumann@uol.de

L. Steffens · M. Karl
Institute for AI Safety and Security, German Aerospace Center (DLR), Sankt Augustin, NRW, Germany

J. Marx Gómez et al. (eds.), *Recent Advances in Information Systems*, Progress in IS,
https://doi.org/10.1007/978-3-032-14099-9_26

In most federated learning clusters, a server is used as the central organizational unit to orchestrate the clients and aggregate the local weights into a global model. However, this central paradigm has disadvantages, such as a "single point of failure," which makes the entire system vulnerable to cyberattacks as well as scaling problems due to communication bottlenecks (Gabrielli et al., 2023). In contrast, "Decentralized Federated Learning" (DFL) completely dispenses with a central server. Instead, the participants communicate directly with each other to exchange their model weights (Vanhaesebrouck et al., 2017).

In DFL systems, the networks are often based on peer-to-peer (P2P) communication and corresponding topologies. Common structures are "Ring", "Star," and "Fully connected" networks, which differ primarily in terms of their robustness, communication costs, and scalability (see Sect. 4.4). For example, a fully connected network offers high reliability, but causes increased communication costs. Ring and star structures make it easier to add new clients but can lead to latency problems or communication bottlenecks. The selection of a suitable topology is therefore crucial for the efficiency and performance of the DFL network (Beltrán et al., 2023). This is particularly relevant if a large number of parameters (weights) have to be exchanged between the clients, as it would be the case with transformer-based models such as Bert (Devlin et al., 2019). Currently, there is a lack of experience in fine-tuning such transformer-based models in a DFL cluster. Consequently, there is no experience of how a prototypical implementation could be realized and what influence different network topologies could have on network convergence and the model divergence between the clients. Since these aspects are crucial for evaluating and optimizing the performance of DFL clusters, the aim of this paper is to implement and evaluate three different DFL topologies for the federated training of transformer-based models. As a concrete evaluation scenario, we used the well-known text classification problem "Contradiction Detection," which we implemented using the public benchmark dataset MultiNLI-1.0 (Williams et al., 2018). As part of this evaluation, we also investigated to what extent the number of neighboring nodes to which the model weights are sent and the time interval between weight exchanges influence the model divergence fluctuations. Based on this objective, this paper provides the following contributions:

1. **Prototypical implementation** of the fine-tuning of a transformer-based language model for a text classification task within a simulated DFL cluster.
2. **Evaluation** of the influence of three different overlay topologies (ring, wheel, fully connected) on the model divergence between 17 clients.
3. **Access** to the program code (https://github.com/Casomoon/p2pfl_experiments).

This paper is structured as follows: Sect. 2 gives a brief overview of the technical mechanisms of decentralized federated learning. Section 3 shows related work and distinguishes it from the present study. Section 4 presents a case study on federated contradiction detection, including its motivation, evaluation settings, and technical implementation. Section 5 summarizes the results of the case study and discusses them with regard to possible implications for theory, practice, and future research directions.

2 Fundamentals

2.1 *Aggregation Process for Decentralized Federated Learning*

Decentralized federated learning (DFL) completely dispenses with a central server and enables communication and aggregation directly between the participants. In contrast to CFL, there is no central instance that coordinates the training and aggregation processes. Instead, the participants communicate with each other in a P2P network and exchange weight updates or models directly (Vanhaesebrouck et al., 2017). This communication can take place with the aid of overlay topologies or adaptive mechanisms (Lian et al., 2017; Yuan et al., 2024) and typically proceeds according to the following sequence:

1. **Peer-to-peer communication**: Each participant or client interacts with a defined subset of other clients. This interaction usually takes the form of direct model or weight updates.
2. **Local aggregation**: Each client aggregates the models of neighboring participants in order to update its own model [e.g., by averaging all collected weights "Federated Averaging" (McMahan et al., 2016)].
3. **Iterative consensus building**: Through repeated communication and aggregation, a consensus on the global model is reached in the network without the need for a central server.

2.2 *The Gossip Protocol*

The gossip protocol, also known as an epidemic protocol, is an approach to decentralized communication based on the random and iterative spread of information between clients. The basic idea is based on a similar principle to the spread of disease or rumors in a population. A participant or client periodically exchanges information with randomly selected neighbors. These "infected" neighbors in turn spread the information received in further iterations until almost the entire network has received the original information (Lian et al., 2017; Tang et al., 2022). The gossip protocol operates in several repeated steps: Each client has a local copy of the information to be disseminated. In each iteration, a client randomly selects one or more neighbors. The selected clients update the local copy based on the new information and then share it with other neighbors (Hegedűs et al., 2019; Tang et al., 2022). In the context of DFL, the gossip protocol enables efficient distribution and aggregation of model parameters without the need for a central server. Due to the iterative and decentralized communication, models can be trained quickly and scalable across large federations (Tang et al., 2022). Since the most frequently used communication method in DFL is the gossip protocol, DFL is also often referred to as "Gossip Protocol-based learning" (Naik et al., 2023).

2.3 L2-Norm and Divergence

The L2 norm, also known as the Euclidean norm, is a standard measure for quantifying differences between vectors. It calculates the distance between two models by squaring the differences of the weights, summing them and taking the square root. Mathematically, the L2 norm for two models w_i and w_j, whose parameters are represented as vectors $w_i = (w_{i1}, w_{i2}, \ldots, w_{in})$ and $w_j = (w_{j1}, w_{j2}, \ldots, w_{jn})$, is defined as follows (Weisstein, 2024):

$$\|w_i - w_j\|_2 = \sqrt{\sum_{k=1}^{n} (w_{ik} - w_{jk})^2} \qquad \text{avg\_divergence} = \frac{1}{n}\sum_{i=1}^{n}\sqrt{\sum_{k=1}^{d} (w_{ik} - w_{\text{avg},k})^2}$$

Accordingly, if n models are available, the difference between an average model w_{avg}, which is defined by the mean value of the parameters of all models, and the individual models can be calculated by the average L2 distance. In this way, it possible to quantify the heterogeneity between models and thus evaluate the degree of deviation of several individual models from a global model. This can be particularly important in decentralized scenarios, as the models are based on heterogeneous data and no server aggregates a global model.

3 Related Work

In recent research on decentralized federated learning, two overarching approaches have emerged in particular: Investigations of different topologies and gossip-based methods. Several studies consider the latter itself as a central object of investigation: For example, (Naik et al., 2023) provides a comprehensive overview of gossip protocols in decentralized architectures, pointing out their scalability and robustness in decentralized federated learning (DFL). Furthermore, (Belal et al., 2022) focuses on gossip in a large-scale user-centered recommendation system (PEPPER) with up to 1,000 nodes. Their results show 42% faster convergence and around 9% higher recommendation quality compared to other approaches. Finally, in a study with also up to 1,000 nodes, (Hegedűs et al., 2019) compare a gossip-based system with classical federated learning. These results show that gossip is robust and resource-efficient, especially in P2P networks. This is also confirmed by Tang et al. (2022), which introduced GossipFL and demonstrated a significant reduction in communication (−38.5%) as well as faster convergence in experimental setups with 14 to 100 nodes.

In addition to the studies focusing on gossip-based methods, there are a number of papers in which the network or communication structure is the main subject of investigation. For example, (Wulfert et al., 2023) analyzes dynamic mesh networks with up to 200 on-device nodes and adapts gossip methods specifically to low-resource

IoT devices to ensure both energy efficiency and robustness. (Dötzer et al., 2023), on the other hand, investigates various topologies (ring, wheel, fully connected) with hierarchical, decentralized, and semi-decentralized clusters. This shows that a fully connected structure can deliver the best results, but at the cost of comparatively high communication costs. Wheel networks are identified as a good compromise between performance and cost, while ring topologies tend to converge slowly. Another contribution to the analysis of dynamic overlays is provided by Hua et al. (2024). With their "FedLay" approach, they generate almost random graph structures that can be adapted depending on the system state, thus shortening the convergence time and reducing the network load. Other studies take up these concepts or combine them: In (Yuan et al., 2024), various topologies—including gossip, hybrid approaches, and overlays such as line, ring, mesh, or star—are compared. The authors provide an overview of decentralized FL approaches, whereby gossip is still classified as a relatively young but promising field. Hybrid solutions are gaining in popularity and, depending on the deployment scenario, offer sophisticated compromises between communication costs and convergence speed.

Most of these studies focus either on topologies or the gossip protocol or modifications of it. However, there are hardly any studies that investigate the interaction between overlay topologies, the gossip communication strategy, and the challenges posed by non-IID data. In particular, there are no studies in the context of transformer-based models such as Bert, which are evaluated on challenging classification problems such as natural language inference (NLI) or its subfield contradiction detection (CD).

4 Case Study—Federated Contradiction Detection

4.1 Use Case Motivation

Contradiction detection (CD) in natural language texts is relevant for a variety of different use cases, such as obtaining contradictory opinions from employees on a given topic (Vancea et al., 2013), summarizing online debates (Sanchan et al., 2020), detecting inconsistencies in financial (Sifa et al., 2019) or regulatory documents (Schumann & Marx Gómez, 2024), or detecting false statements in social media (Lendvai & Reichel, 2016). Due to recent advances in natural language processing and deep learning, the accuracy with which such cases are detected is good wherever a large amount of annotated training data can be used (Devlin et al., 2019; Liu et al., 2019). These include, for example, very general datasets such as SNLI (Bowman et al., 2015) or those that are specifically tailored to a particular genre [e.g., "fiction" or "travel" in the MNLI dataset, (Williams et al., 2018)]. However, not only were these datasets compiled and manually annotated at great expense by one or a few research institutes, they also contain semi-synthetic contradictions (i.e., half of the sentences are made up), which makes the transferability to actual application scenarios from business practice a challenge.

In order to nevertheless identify contradictions in a new, domain-specific area of a company, individual (training) datasets with annotated contradiction cases could be created that are located in the target/application area (e.g., in finance, internal auditing, or controlling). In practice, however, compiling such training data is a challenge, as usually only a few examples of contradictory statements are recorded in a company. Given the deep learning models required for contradiction detection, such small amounts of training data are usually not sufficient to train an accurate model. Instead, in order to obtain a sufficiently large dataset, contradiction cases from several different companies or organizations would have to be collected and summarized. In this way, the individual companies would benefit from the documented contradiction cases of the other companies. However, the realization of such a data merger often poses a problem, as the companies would not only have to disclose their contradictions but also other internal and often sensitive information contained in the respective texts. One way of training a joint contradiction detection model without the companies having to exchange their sensitive, contradictory texts with each other is to use federated learning.

4.2 Data Understanding

To develop our federated contradiction detection model, we used the multi-genre natural language inference/multi-MNLI dataset (Williams et al., 2018). Divided into 10 genres (e.g., government, telephone, fiction, etc.), this dataset contains a total of 433,000 pairs of premise and hypothesis sentences, each assigned one of three labels: Entailment (if the premise is true, the hypothesis must also be true), contradiction (if the premise is true, the hypothesis cannot be true and vice versa), and neutral (if both sentences can be true or not true) (Sifa et al., 2019). Some examples of these three label categories for the different genres can be found in Table 1. In addition to the 433,000 sentence pairs available for training, the developers of the MNLI dataset (Williams et al., 2018) also provide an official test dataset with around 10,000 sentence pairs, which we use in our study to evaluate the resulting models.

4.3 Data Preparation and Distribution

Since our analysis aims to classify two given sentences solely according to whether or not they contradict each other, we combined the labels "Neutral" and "Entailment" into a new category "No Contradiction". The resulting ratio of "No Contradiction" cases to "Contradiction" cases is 66% to 33%. In addition to this label modification, we removed all sentence pairs for which the annotators disagreed and therefore could not be assigned a clear "gold label" (marked as "-"). This filtering step reduced the

Table 1 Excerpt from the MNLI-dataset (Williams et al., 2018)

Premise	Hypothesis	Label	Genre
An alternative to unaddressed mail would be to auction off the right to be a third bundle on specific days in specific post offices	You could auction off the right to another bundle instead of doing unaddressed mail at rural post offices	Entailment	Government
What's truly striking, though, is that Jobs has never really let this idea go	Jobs never held onto an idea for long	Contradiction	Slate
A rusty iron gate swinging dismally on its hinges	The iron gate was swinging and could not be locked	Neutral	Fiction
But uh these guys were actually on the road 2000 miles from home when they had to file their exams and send them in	These men filed their midterm exams from home	Contradiction	Telephone

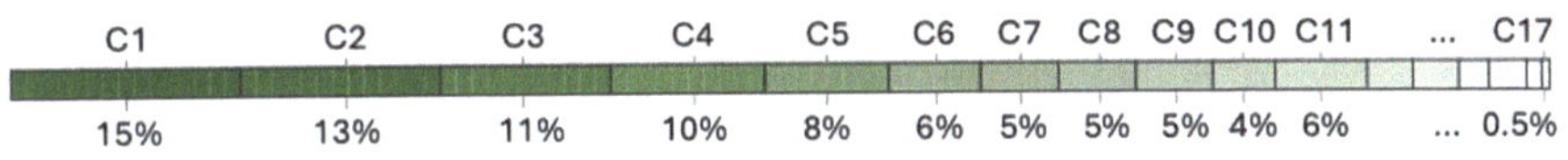

Fig. 1 Data distribution among the 17 clients (C = client)

number of sentence pairs in the training dataset to 391,000 and in the test dataset to 9815.

In order to realize a realistic and diverse distribution of the training data in our decentralized federated training simulation, we implemented a non-independent and non-identical distribution of the data (non-IID) according to data volume and genre of the respective sentence pairs. For this purpose, we distributed different data capacities (i.e., different numbers of sentence pairs) to a total of 17[1] clients, which were then used for the local training on the respective client. Specifically, the training data was distributed among the clients with shares of 15, 13, 11, 10, 8, 6, 5, 5, 5, 4, 6, 3, 3, 2.5, 2, 1, and 0.5% (Fig. 1). As part of this procedure, we also utilized the sentence pair genre for our non-IID data distribution by making full or partial allocations of partial records based on the capacity of each client. If, for example, a client has the capacity to take over the remaining sentence pairs of a genre, it will be assigned the entire genre. This distribution ensures a non-IID division of the data, as one client can, for example, only hold data from a single genre, while another client uses data from 2–3 genres for the training. The corresponding allocation process is iterative: First, the client with the highest capacity is given preference. If a genre cannot be fully allocated, further distribution takes place in smaller blocks, which are gradually allocated to clients that still have capacity.

[1] The number of clients was set to 17 due to our VRAM limitations (see "Infrastructure used").

Table 2 Setting of the local (client-side) training

Model	Bert-base-uncased[2] (12 transformer layers, 12 self-attention heads, 768 hidden states and a total of 110 million parameters)								
Fine-tuning via	Bert-for-sequence-classification								
Optimizer	AdamW	Learning rate	2×10^{-5}	Batch size	1	Epochs	1	Label	2

4.4 Modeling

Client-Side Training. In our DFL cluster, a local model was trained on each of the 17 clients using the respective dataset assigned to it (see Sect. 4.3). The resulting model weights of the clients were then aggregated into a global model at the end of each federated training round. For this local training, we used the pre-trained Bert model "Bert-base-uncased".[2] This version of Bert has 12 transformer layers, 12 self-attention heads, 768 hidden states and a total of 110 million parameters. As part of the fine-tuning, we used the transformers module "Bert-for-Sequence-Classification," which has a sequence classification/regression header as a linear layer on top of Bert's pooled output. As optimizer we used "AdamW," a learning rate of 2×10^{-5}, a weight decay of 0.01, and a validation ratio of 0.1—from the training data (Table 2). Since 3–4 epochs are typically sufficient for the fine-tuning of Bert models (before the model overfits too much), we have set the number of local epochs to 1 and the number of federated rounds to 4 (see Table 4). In this way, a federated fine-tuning takes place over a total of 4 epochs.

Since we used this model setting in all decentralized federated training rounds for the local clients, we set all random variables for transformers, Numpy, and PyTorch to fixed seed values to make the training deterministic and thus comparable.

DFL-Framework and Parameters. To simulate our DFL cluster, we used the p2pfl[3] (Guijas, 2022) framework, as it is highly compatible with common machine learning libraries, has a modular structure and provides comparatively good documentation. At the time of the implementation, p2pfl is based on the "node class" as the central entity in all federations. This class contains the essential functions for communication, learning, aggregation, and other processes. Each node represents a single participant in the DFL cluster and consists of several specialized modules, each of which performs specific subtasks. This clear separation of logic enables a high degree of flexibility.

The p2pfl framework offers several options for the adjustment of different parameters in order to individually control the behavior of the network as well as the training and communication process. These can be adjusted via the "Settings" module of the framework, which is then called up in the individual modules (see Table 3).

[2] https://huggingface.co/google-bert/bert-base-uncased.

[3] https://github.com/p2pfl/p2pfl.

Table 3 Relevant parameters of the p2pfl-framework

Parameter	Description
HEARTBEAT_PERIOD HEARTBEAT_TIMEOUT	Defines how often the clients signal their availability/how long they wait until a client is considered to be down by other clients
TTL (time-to-live)	Defines how long individual gossip messages circulate in the network before they are discarded. A very high TTL can result in messages circulating for a very long time and therefore being outdated by the time they reach the client. Extremely high values cause redundant communication overhead. However, if the TTL is set too low, the federation is not able to start properly, as corresponding messages cannot reach "too far away" clients
AGGREGATION _ TIMEOUT	Defines how long to wait for all the models required for aggregation. A high value also enables nodes with a lot of data and correspondingly long training processes to provide their models on time. A very high value (e.g., 30,000) is required here, especially for many simultaneous Bert trainings
GOSSIP_MESSAGES _ PER_PERIOD	Defines how many messages are sent by a client within a gossip cycle. The content of these messages does not necessarily have to be model-related. They can also contain meta information or status messages
GOSSIP_PERIOD	Defines the time interval for sending gossip messages. Shorter intervals speed up the exchange of information but increase the network load. Longer intervals reduce the communication effort but slow down the dissemination of new information
GOSSIP_MODELS _ PERIOD	Explicitly controls how frequently (in seconds) model weights are distributed via Gossip. While GOSSIP_PERIOD and GOS-SIP_MESSAGES_PER_PERIOD define the general frequency and quantity of gossip communication, GOSSIP_ MODELS_ PERIOD focuses on the exchange of model information
GOSSIP_MODELS _PER_ PERIOD	Controls how many neighbors the model weights are to be sent to via GOSSIP_MODELS_PERIOD. These can be fully aggregated or partially aggregated models
TRAIN_SET_SIZE	Determines the number of clients training in a single round

Adjustments to the p2pfl Framework

At the time of the experiments conducted in this paper (January 2025), simulations within the p2pfl-framework are performed without "Ray," a library for distributed GPU resource management. This means that all trainings of a federated round are executed simultaneously on the GPU, which can slow down or even interrupt simulations with higher VRAM utilization. At the same time, the framework has not yet been tested with longer-lasting training iterations (as can occur when fine-tuning Bert models), which is why it is not specifically designed for these conditions. In the course of the first federated-simulated Bert training, this led to a lock called "waitaggmodel_lock," which was released twice in the "WaitAggModelsStage". The latter in turn led to a "Released unlocked lock" error, which caused the affected client to

Table 4 Configurations for the base scenario, experiment 1 and experiment 2

Parameter	Base	Exp. 1	Exp. 2
HEARTBEAT_PERIOD	30	30	30
HEARTBEAT_TIMEOUT	4500	4500	4500
GOSSIP_PERIOD	5	5	5
TTL (TIME-TO_LIVE)	15	15	15
GOSSIP_MESSAGES_PER_PERIOD	5	5	5
GOSSIP_MODELS_PERIOD	2	2	**5**
GOSSIP_MODELS_PER_ROUND	5	**10**	5
AGGREGATION_TIMEOUT	30,000	30,000	30,000
VOTE_TIMEOUT	4000	4000	4000
NR_NODES (CLIENTS)	17	17	17
ROUNDS	4	4	4

Values in bold indicate changes

crash and thus paralyzed the entire simulation. The cause of this was a "race condition" in which several threads access the lock at the same time and thus change states in an uncoordinated manner. To solve this problem, the lock logic was replaced by an event-based logic. The latter uses synchronization primitives that are set when a certain state is reached (e.g., the end of a training session). Threads waiting for this event automatically continue their process as soon as it is triggered and explicit releases, as required for locks, are no longer necessary. This reduces the probability of "race conditions" and associated crashes, as the focus is on clear state transitions.

A similar error occurred in topologies with high connectivity during the initialization of the models (e.g., "Fully connected"). Specifically, the "model_initialized_lock" was released several times, which then also led to crashes. Analogous to the procedure for the "waitaggmodel_lock," we therefore also replaced the lock logic with an event-based solution, so that race conditions and associated errors were effectively prevented.

With the help of these adjustments,[4] the stability of the framework for longer training iterations could be established. As a result, the system is now ready for experiments in which the overlay topologies can be examined with transformer models such as Bert.

Evaluation Scenarios. To investigate the influence of different decentralized network structures on the federated fine-tuning behavior of a transformer-based language model, we selected three overlay topologies for our experiments: "Ring", "Wheel," and "Fully Connected". In the following, we briefly describe each overlay topology and explain how they were implemented using the p2pfl framework.

[4] Our changes were integrated into the dev-branch of *p2pfl* after review by the main developer: https://github.com/p2pfl/p2pfl/commit/41cc33bbe9ab0895f8ae95dcb98f58fed96840e9.

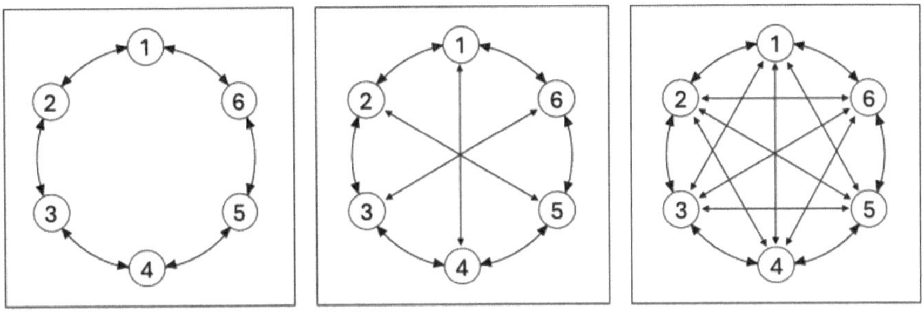

Fig. 2 Overlay topologies used for the evaluation: "Ring", "Wheel," and "Fully connected"

Ring: The nodes are arranged in a closed chain in which each node only communicates with its direct neighbors (Fig. 2, left). This topology is simple and robust against partial failures but can lead to slower convergence times for very long rings (Dötzer et al., 2023; Khan et al., 2023).

Wheel: This variant extends the classic ring topology with additional connections to opposite nodes (Fig. 2, center). This shortens the communication paths between distant nodes, which leads to faster dissemination of information. At the same time, the network remains robust as there is no single point of failure and there are alternative communication paths in the event of individual node failures. This topology therefore combines the simplicity and robustness of the ring topology with a significantly improved convergence speed and greater scalability. It offers a flexible and efficient solution, particularly for dynamic networks with a changing number of participants, as additional connections are used in a targeted manner to optimize communication costs (Dötzer et al., 2023).

Fully Connected: Each node is directly connected to every other node (Fig. 2, right). This topology enables the fastest exchange of information but causes high communication costs as every update is sent to everyone. This makes it less scalable (Dötzer et al., 2023; Khan et al., 2023).

These three topologies were selected because they reflect a gradual increase in interconnectivity between the clients and thus serve to systematically analyze their influence on the performance of the DFL cluster: The ring topology represents the minimum connectivity, as each node only communicates with its direct neighbors. The connectivity is then increased in the second topology "Wheel," which also communicates with nodes directly opposite each other, and finally connectivity reaches its maximum in the third topology "Fully connected," in which every node communicates directly with every other node. The focus of the subsequent evaluation was on the model divergence between the clients during training and the associated convergence of models within these three network structures (using a non-IID data distribution).

For each of these three topologies, we first carried out a training using a base configuration (Table 4: "Base"). In this first training iteration, we deliberately set the timeout and heartbeat parameters high to ensure that communication and aggregation within the federation are successfully completed even with long training times. This reduces the risk of premature termination of the federation run, which could occur in particular with computationally intensive models such as Bert. The time-to-live (TTL) was set to 15, as this is the minimum possible value for the ring topology. If this parameter is set higher, the start of the federation fails during the model initialization, as the corresponding model initialization command does not reach all clients. For the sake of comparability, this value is also retained for the other two topologies.

The number of simultaneously training clients was set to 17 due to our VRAM limitations (94 GB, see "Infrastructure used"). In order to make the results of the experiments deterministic with regard to data usage, this value was equated with the absolute number of clients (NR_NODES). This ensures that the same data is used in all training rounds as every client always takes part in every round. In addition to our base run, we conducted two further federated training iterations in which we increased the number of distributed models per gossip period (Table 4: "Exp. 1") and extended the gossip period (Table 4: "Exp. 2"). In experiment 1, we therefore increased the MODELS_PER _ROUND parameter from 5 to 10. This parameter determines the number of both direct and indirect neighbors with which each model is shared per round. In contrast, in experiment 2 we increased the parameter GOSSIP_MODELS_ PERIOD from 2 to 5. This parameter controls how often the model parameters are distributed by the clients and thus represents the model communication frequency within our FL cluster.

Infrastructure Used. Due to the parallel execution of 17 local training iterations in each round, with a subsequent aggregation of the model weights, the decentralized federated training in the simulation environment p2pfl places very high demands on the graphics card. For this reason, the experiments were carried out on a node of our university high-performance cluster, on which we allocated 4 CPU cores, 300 gigabytes of RAM, and an Nvidia H100 GPU with 94 gigabytes of VRAM as resources.

Results—Model Divergence. In the base configuration, a comparison of the model weights between the three topologies showed a clear divergence between the individual client models, especially in the first training round, in which the divergence is highest for all topologies. Specifically, the wheel topology showed the largest model deviations with an L2 distance (see Sect. 2.3) of 1.02, followed by the ring topology with 0.92 and the fully connected topology with 0.61. These values illustrate the initial differences in the model weights, which are caused by the heterogeneity of the local data. In the second round, however, a significant decrease in divergence can be observed, as it drops to 0.35 in the wheel- and fully connected topology, while the ring topology has the lowest divergence at 0.31. In the later rounds, the topologies develop differently: In the third round, the divergence in the wheel topology remains relatively constant at 0.3, while the ring topology rises to 0.44 and the divergence of the fully connected topology increases slightly to 0.41. In the fourth round, the fully connected topology reduces the divergence to 0.15 and thus achieves the lowest value

Topology	Setting	Model Divergence per Round				
		1	2	3	4	ø
Ring	Base	0.92	0.31	0.44	0.45	0.53
	Exp-1	0.89	0.30	0.37	0.34	0.48
	Exp-2	0.62	0.45	0.36	0.42	0.46
Wheel	Base	1.02	0.35	0.30	0.45	0.53
	Exp-1	1.05	0.29	0.42	0.44	0.55
	Exp-2	1.05	0.38	0.35	0.47	0.56
FC	Base	**0.61**	0.35	0.41	**0.15**	0.38
	Exp-1	0.62	**0.12**	**0.19**	0.43	**0.34**
	Exp-2	0.67	0.65	0.57	0.41	0.58

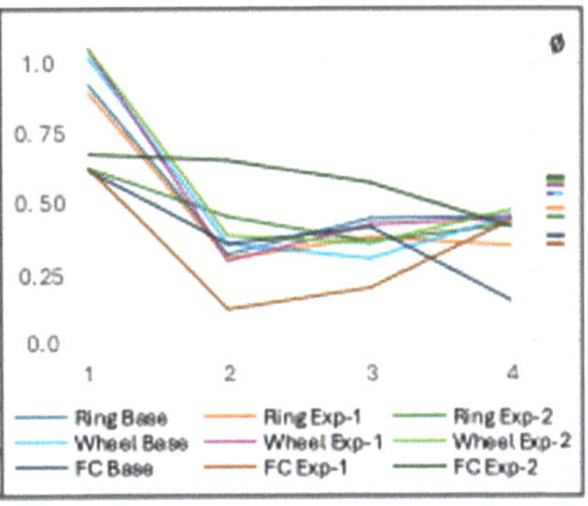

Fig. 3 Model divergence (average L2 distance) per round for the base-, Exp-1- and Exp-2-Configurations between the three topologies “Ring”, “Wheel,” and “Fully connected”

across all rounds. In contrast, the divergence in the ring topology remains constant at 0.45, while the wheel topology shows a renewed increase to 0.45 (Fig. 3).

In contrast to the base configuration, increasing the number of distributed models per gossip period (Exp-1) led to a lower model divergence in each of the four rounds in the ring topology. While the base configuration showed some strong fluctuations in the model divergence (especially from a drastic fall in round 2 to a renewed increase in rounds 3 and 4), this tended to remain more stable in Exp-1 and slightly below the values of the base configuration. In the wheel topology, Exp-1 leads to a slightly higher model divergence in the first round, which then drops significantly more than in the base configuration (to 0.29) but then rises again sharply in the third and fourth rounds. Overall, it thus resulted to a slightly higher average value over all four rounds compared to the base configuration. In contrast, the fully connected topology benefits most from the increased number of distributed models per gossip period applied in Exp-1 due to its tight coupling between the nodes. Its model divergence decreases significantly faster (except for round 4) than in the base configuration and also shows the lowest average model divergence of all experiments (0.34).

On the other hand, Exp-2 often shows less abrupt changes in model divergence in the ring topology than in the base configuration and remains slightly below the base configuration with an average model divergence of 0.46. In the wheel topology, Exp-2 only results below the base values in round 2. Although the reduction in model divergence tends to be more consistent, the advantage over the base configuration, with an average model divergence of 0.56, is lower than for Exp-1. Especially in the Fully connected topology, Exp-2 does not have the expected advantage over the base configuration, as the model divergence values are higher than those of the other configurations in almost all rounds. As a result, the average convergence of the Fully connected topology in Exp-2 is 0.58, the highest value of all configurations. These findings suggest that more frequent model communication via gossip (as applied in Exp-2) in a highly connected structure does not always lead to accelerated or more stable convergence—a deviation that is not as significant in the other topologies (esp. in the ring topology).

Results—Accuracy. Looking at the average progression of the accuracy (Fig. 4), it can be seen that the accuracy values in all configurations (Base, Exp-1, Exp-2) as well as in the three topologies (Ring, Wheel, Fully connected) reach their peaks in the second federated round and then level off at a comparatively high average accuracy over the four rounds (with a final value between 0.875 and 0.880). While the "Ring" and "Wheel" topology shows a clearer accuracy increase from round 1 to round 2 (followed by a slight stabilization until round 4), the increase in the "Fully connected" setting is flatter and more constant overall. However, the average accuracy here also stabilizes at a similarly high level toward the end. With regard to the configurations, it can be seen that "Exp-1" shows a slightly faster and clearer upward movement compared to "Base" in almost all configurations, while "Exp-2" usually results in a similarly high final value but does not consistently show a greater lead compared to "Base".

These accuracy curves illustrate that both parameters (MODELS_PER_ROUND and GOSSIP_MODELS_PERIOD) have a relevant influence on the topologies and the convergence behavior of the models. Higher exchange frequencies accelerate training progress and quickly lead to good results. However, there is also a higher risk of overfitting, and more communication resources are required. A longer interval

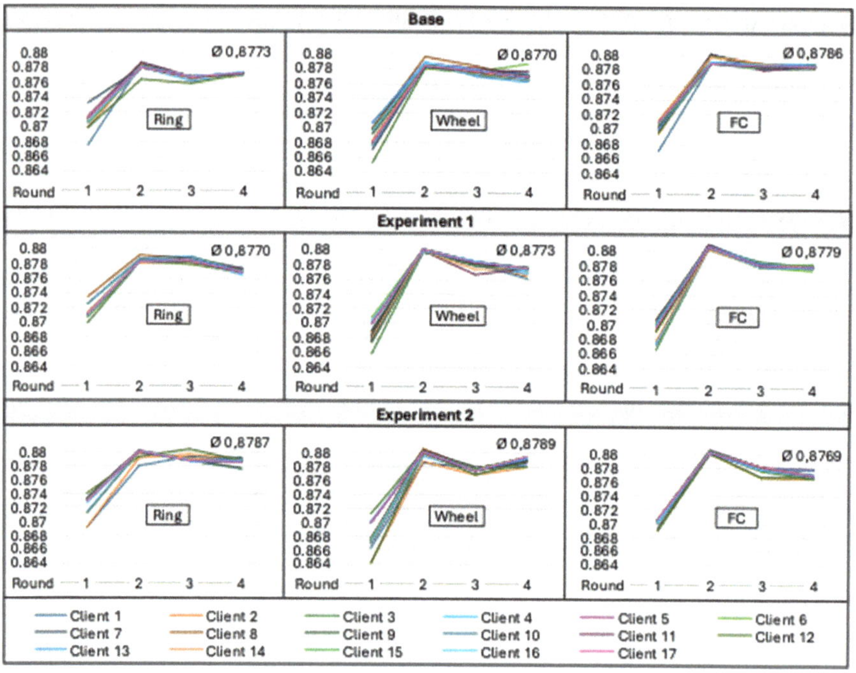

Fig. 4 Model accuracy of all 17 client across 4 rounds in the 3 topologies "Ring", "Wheel," and "Fully connected" (FC), evaluated in the 3 configurations "Base", "Exp-1," and "Exp-2"

between weight exchanges, on the other hand, ensures a slower but more stable increase in accuracy.

Overall, the averaged course of the three experiments confirms that all topology configurations find almost identically high accuracy levels in the last round (Ø 0.8769–0.8789), regardless of the initial model divergence. However, there are relevant differences in accuracy between the clients if less than 4 federated training rounds are applied.

In addition to the federated training using the 17 clients, we also performed a conventional (non-federated) fine-tuning of the Bert model, in which the model was able to use all 391,000 sentence pairs at once. Since this conventional fine-tuning did not perform 4 federated training rounds for one local epoch each, we set the number of local epochs to 4 to get back to the same total number of epochs as in the experiments. The accuracy of this conventionally fine-tuned Bert model on our test dataset was 0.896 in the fourth and final epoch, an average of 2.07% higher than the accuracy of the federated trained models. This slightly lower accuracy in a federated setting has already been shown by previous studies [e.g., (Rudolph et al., 2025)] and was therefore also expected in our study.

5 Conclusion

In this paper, we investigated how different overlay topologies with different interconnectivity affect the model divergence of transformer-based classification models during decentralized federated learning (DFL). For this purpose, we simulated 17 clients in the three topologies "Ring", "Wheel," and "Fully connected" with the help of the gossip communication-based framework "p2pfl"[3] and evaluated them using the well-known classification problem "Contradiction Detection". In addition to a base training configuration, we also conducted two experiments in which we increased the number of neighboring nodes to which the model weights are sent (Experiment 1) and the frequency with which the model weights are distributed via gossip (Experiment 2). Our evaluation shows that after 4 federated training rounds, similar model accuracies are achieved on the test set in all experiments, but the deviations of the model weights between the 17 clients (model divergence) develop differently during the training process. While the "Ring" and "Wheel" topologies showed considerable deviations in the model weights between the 17 clients, especially at the beginning of the base scenario, the "Fully connected" topology achieved a much more efficient approximation of the model weights. In particular with regard to the non-IID data distributions applied in our DFL cluster, the high connectivity of the "Fully connected" topology therefore has advantages, as it leads to a lower average model divergence: Specifically, this value is Ø 0.38 for the fully connected topology in the base scenario, while it is Ø 0.53 for the ring and wheel topology.

Increasing the number of neighboring nodes to which the model weights are sent (Experiment 1) resulted in a faster and more stable increase in accuracy in most cases. However, the higher exchange frequency causes a correspondingly greater communication effort, which can be problematic in scenarios with limited network resources. By increasing the time interval between the model exchanges (Experiment 2), the convergence between the models is shifted further back and the accuracy peaks occur later. In most experiments, this led to lower fluctuations in the model divergences over the individual rounds, but ultimately also resulted in higher average model divergence values, especially for the fully connected topology. At the same time, the communication effort is reduced, which can be particularly relevant in environments with limited bandwidths or high energy requirements.

Considering these results can help in selecting an appropriate network topology and associated communication settings of future decentralized federated learning simulations, especially when using transformer-based language models such as Bert (Devlin et al., 2019) and a model exchange via gossip. In addition, the results can be used for comparisons in the field of "Contradiction Detection" [e.g., (Rudolph et al., 2025; Salvatore et al., 2019) or the superordinate research field of "Natural Language Inference" [e.g., (Eleftheriadis et al., 2023; Perikos & Souli, 2024)]. At the same time, the study also has a few limitations: Since the training was computationally intensive, we limited the experiments to one evaluation dataset [MNLI (Williams et al., 2018)], one transformer-based model variant ("Bert-base-uncased") as well as one aggregation method ("FedAvg") and one fixed parameter setting at the local level. Future research could therefore investigate to what extent other aggregation methods and other model-specific parameter settings, such as the local number of epochs, batch size, and learning rate, affect the model divergence.

Acknowledgements This work was funded and supported by the Institute for AI Safety and Security at the German Aerospace Center (DLR–Deutsches Zentrum für Luft- und Raumfahrt, Institut für KI-Sicherheit).

References

Belal, Y., Bellet, A., Mokhtar, S. B., & Nitu, V. (2022). PEPPER: Empowering user-centric recommender systems over gossip learning. *Proceedings of the ACM on Interactive, Mobile, Wearable and Ubiquitous Technologies, 6*(3), 1–27.

Beltrán, E. T. M., Pérez, M. Q., Sánchez, P. M. S., Bernal, S. L., Bovet, G., Pérez, M. G., Pérez, G. M., & Celdrán, A. H. (2023). Decentralized federated learning: Fundamentals, state of the art, frameworks, trends, and challenges. *IEEE Communications Surveys & Tutorials, 25*(4), 2983–3013.

Bowman, S. R., Angeli, G., Potts, C., & Manning, C. D. (2015). A large annotated corpus for learning natural language inference. arXiv preprint arXiv:1508.05326.

Devlin, J., Chang, M.-W., Lee, K., & Toutanova, K. (2019). BERT: Pre-training of deep bidirectional transformers for language understanding. In *2019 Conference of the North American chapter of the association for computational linguistics: human language technologies* (Vol. 1, pp. 4171–4186).

Dötzer, M., Mao, Y., & Diepold, K. (2023). Communication topologies for decentralized federated learning. In 2023 *Eighth international conference on fog and mobile edge computing (FMEC)* (pp. 232–238). IEEE.

Eleftheriadis, P., Perikos, I., & Hatzilygeroudis, I. (2023). Evaluating deep learning techniques for natural language inference. *Applied Sciences, 13*(4), 2577.

Gabrielli, E., Pica, G., & Tolomei, G. (2023). A survey on decentralized federated learning. arXiv preprint arXiv:2308.04604.

Guijas, P. (2022). Desarrollo de una librería para el aprendizaje federado bajo una arquitectura peer-to-peer.ruc.udc.es.

Hegedűs, I., Danner, G., & Jelasity, M. (2019). Gossip learning as a decentralized alternative to federated learning. In *Distributed applications and interoperable systems: 19th IFIP WG 6.1 international conference, DAIS 2019, held as part of the 14th international federated conference on distributed computing techniques, DisCoTec 2019, Kongens Lyngby, Denmark, June 17–21, 2019, proceedings 19* (pp. 74–90). Springer.

Hua, Y., Pang, J., Zhang, X., Liu, Y., Shi, X., Wang, B., Qian, C., et al. (2024). Towards practical overlay networks for decentralized federated learning. In *2024 IEEE 32nd International conference on network protocols (ICNP)* (pp. 1–11). IEEE.

Khan, Q. W., Khan, A. N., Rizwan, A., Ahmad, R., Khan, S., & Kim, D. H. (2023). Decentralized machine learning training: A survey on synchronization, consolidation, and topologies. *IEEE Access, 11*, 68031–68050.

Lendvai, P., & Reichel, U. D. (2016). Contradiction detection for rumorous claims. In *Proceedings of the extra-propositional aspects of meaning (ExProM) in computational linguistics*, Osaka, Japan, 2016, pp. 31–40.

Lian, X., Zhang, C., Zhang, H., Hsieh, C. J., Zhang, W., & Liu, J. (2017). Can decentralized algorithms outperform centralized algorithms? A case study for decentralized parallel stochastic gradient descent. *Advances in Neural Information Processing Systems, 30.*

Liu, Y., Ott, M., Goyal, N., Du, J., Joshi, M., Chen, D., Levy, O., Lewis, M., Zettlemoyer, L., & Stoyanov, V. (2019). RoBERTa: A robustly optimized BERT pretraining approach. arXiv [Preprint], arXiv preprint arXiv:1907.11692

McMahan, H. B., Moore, E., Ramage, D., & Agüera y Arcas, B. (2016). Federated learning of deep networks using model averaging. arXiv, abs/1602.05629.

Naik, D., Grace, P., Naik, N., Jenkins, P., Mishra, D., & Prajapat, S. (2023). An introduction to gossip protocol based learning in peer-to-peer federated learning. In *2023 IEEE international conference on ICT in business industry & government (ICTBIG)* (pp. 1–8). IEEE.

Perikos, I., & Souli, S. (2024). Natural language inference with transformer ensembles and explainability techniques. *Electronics, 13*(19), 3876.

Rudolph, S., Schumann, G., Steffens, L., Karl, M., & Marx Gómez, J. (2025). Highly scaled federated learning simulations. In *Proceedings of the 10th international conference on computer and communication systems*. Chengdu, China, 2025.

Salvatore, F., Finger, M., & Hirata Jr., R. (2019). A logical-based corpus for cross-lingual evaluation. arXiv preprint arXiv:1905.05704

Sanchan, N., Bontcheva, K., & Aker, A. (2020). An adoption of a contradiction detection task to assist the summarization of online debates. In *2020–5th International conference on information technology (InCIT)* (pp. 185–190). IEEE.

Schumann, G., & Marx Gómez, J. (2024). Detection of contradictions and inconsistencies in German regulatory documents. In *Proceedings of the 6th international conference on natural language processing (ICNLP)*, IEEE.

Sifa, R., Pielka, M., Ramamurthy, R., Ladi, A., Hillebrand, L., & Bauckhage, C. (2019). Towards contradiction detection in German: a translation-driven approach. In *2019 IEEE Symposium series on computational intelligence (SSCI)* (pp. 2497–2505). IEEE.

Tang, Z., Shi, S., Li, B., & Chu, X. (2022). GossipFL: A decentralized federated learning framework with sparsified and adaptive communication. *IEEE Transactions on Parallel and Distributed Systems, 34*(3), 909–922.

Vancea, B., Marchis, A., Dinsoreanu, M., & Potolea, R. (2013). Contradiction detection between opinions: From a big data perspective. In *9th International conference on intelligent computer communication and processing (ICCP)* (pp. 259–266). IEEE.

Vanhaesebrouck, P., Bellet, A., & Tommasi, M. (2017). Decentralized collaborative learning of personalized models over networks. In *Artificial intelligence and statistics* (pp. 509–517). PMLR.

Weisstein, E. W. (2024). "L2-Norm." From MathWorld—A Wolfram Web Resource. https://mathworld.wolfram.com/L2-Norm.html

Williams, A., Nangia, N., & Bowman, S. R. (2018). A broad-coverage challenge corpus for sentence understanding through inference. In M. Walker, H. Ji, & A. Stent (Eds.), *Proceedings of the 2018 conference of the North American chapter of the association for computational linguistics: Human language technologies* (Vol. 1 (Long Papers), pp. 1112–1122). New Orleans, Louisiana, June 2018. Association for Computational Linguistics.

Wulfert, L., Asadi, N., Chung, W. Y., Wiede, C., & Grabmaier, A. (2023). Adaptive decentralized federated gossip learning for resource-constrained IoT devices. In *Proceedings of the 4th international workshop on distributed machine learning* (pp. 27–33).

Yang, Q., Liu, Y., Chen, T., & Tong, Y. (2019). Federated machine learning: Concept and applications. *ACM Transactions on Intelligent Systems and Technology (TIST), 10*(2), 1–19.

Yuan, L., Wang, Z., Sun, L., Yu, P. S., & Brinton, C. G. (2024). Decentralized federated learning: A survey and perspective. *IEEE Internet of Things Journal*.

Secure Data, Smarter Decisions: A Blockchain-Enhanced Federated Learning Framework for the Pharmacogenomics of Depression

Elsie Kaaya and Jorge Marx Gómez

Abstract Depression continues to be a global health problem affecting approximately 280 million people worldwide. Unfortunately, its treatment selection depends on a trial-and-error approach, which leads to ineffective treatment selection. A personalized approach to treatment, such as pharmacogenomics, offers great promise in addressing this limitation. This study proposes a blockchain-enhanced federated learning (BFL) framework for effective and secure personalized treatment selection for depression. By leveraging privacy-enhancing technologies such as blockchain, homomorphic encryption, and secure multiparty computation (SMPC), the study aims to encourage collaborative model training across multiple health institutions by enhancing data privacy and security in a decentralized federated learning environment. Collaboration among health institutions will promote the development of models trained on datasets representing diverse populations with varying genetic, clinical, and environmental characteristics, thereby improving model generalizability and overall treatment selection.

1 Introduction

Depression is a public health concern that affects individuals worldwide. Globally, it is identified as the leading cause of suicide, accounting for nearly 800,000 deaths per year (Pirmohamed, 2023; Tesfamicael et al., 2024). Furthermore, each year the number of reported cases grows exponentially, as observed between 2005 and 2015, where its prevalence increased by 15% (Santomauro et al., 2021). This was further escalated by the COVID-19 pandemic, which accelerated its prevalence to 28% (Nochaiwong et al., 2021). As a result, billions of dollars continue to be lost yearly,

E. Kaaya (✉) · J. Marx Gómez
University of Oldenburg, Oldenburg, Germany
e-mail: elsie.kaaya@uni-oldenburg.de

J. Marx Gómez
e-mail: jorge.marx.gomez@uni-oldenburg.de

J. Marx Gómez et al. (eds.), *Recent Advances in Information Systems*, Progress in IS,
https://doi.org/10.1007/978-3-032-14099-9_27

hence affecting the global socioeconomic status (Karamperis et al., 2021). Unfortunately, despite its physical, social, and economic challenges, depression is still widely overlooked in many parts of the world (Thornicroft et al., 2022). This underscores the need for immediate action in managing depression through effective diagnosis, treatment, and monitoring techniques.

Diagnosis of depression largely depends on surveys and questionnaires that are either filled out by patients or clinicians (Walschots et al., 2024). Reliance on symptoms reported by patients often leads to heterogeneous and imprecise diagnoses, which results in low remission rates (Taliaz & Souery, 2021). Inadequate treatment selection and suboptimal dosage further exacerbate low remission rates. Bobo et al. (2019) highlighted these challenges, noting an unexplained surge in antidepressant prescriptions. Their study raised concerns about whether this increase was due to a true rise in depression prevalence or the overprescription of antidepressants without adequate diagnostic precision. This claim is supported by Beer et al. (2024), who reported increasing rates of polypharmacy among older adults that contribute to the higher risk of morbidity and mortality. These findings underscore the urgent need for improved diagnostic accuracy, individualized treatment planning, and careful dosage selection in the clinical management of depression. While diagnostic limitations are well-documented, this study specifically focuses on optimizing treatment selection and dosage strategies to enhance patient outcomes.

Treatment selection in depression is based on a trial-and-error approach. Selective serotonin reuptake inhibitors (SSRIs) are the most widely administered antidepressant for mild to severe depression, accounting for ~87% of the treatment modalities (Pizzicato et al., 2023). Despite its widespread use, 50–60% of the individuals prescribed SSRIs fail to achieve full remission after their first treatment (Chung et al., 2023). The majority of patients who do not achieve remission are often left overwhelmed, resulting in high rates of treatment discontinuation (Palumbo et al., 2024). In addition, SSRIs often cause adverse side effects in patients, warranting the need for treatment adjustments or, in extreme cases, switching the medication. Personalized medicine can address this challenge by tailoring treatment selection based on individual unique characteristics such as genetic, clinical, lifestyle, and environmental factors. With a personalized approach to treatment selection, such as precision medicine, clinicians can administer treatment based on individual characteristics, resulting in improved patient outcomes.

Precision medicine is an innovative approach to healthcare that tailor's treatment and medical decisions based on the unique characteristics of an individual, including genetic, environmental, lifestyle, and clinical factors. Contrary to the traditional one-size-fits-all approach, precision medicine acknowledges individual differences, making it ideal for mental health disorders, which are characterized by a complex etiology (Kosorok & Laber, 2019). A personalized approach to treatment selection holds great promise for improving patient outcomes in mental healthcare (Okpete & Byeon, 2024). By predicting treatment response in individuals, clinicians can administer the right treatment in the right dosage, enhancing treatment effectiveness.

Precise treatment selection and dosage can be achieved through pharmacogenomics (PGx), the study of how an individual's genetic makeup influences how they respond to drugs, resulting in enhanced treatment effectiveness. PGx has been explored in mental healthcare, where it has demonstrated great promise in antidepressant prescription compared to traditional methods (Tesfamicael et al., 2024). By leveraging genetic data and electronic health records (EHR) data, PGx can make highly accurate predictions to guide clinicians in selecting the right treatment (Hoffman et al., 2020). However, the vast amount of data needed to make accurate predictions requires a state-of-the-art approach such as federated learning, which enables large-scale data processing while maintaining data privacy.

Federated learning (FL) is a distributed machine learning paradigm that enables model training across different data sources without centralizing or transferring raw data; instead, only model gradients/parameters are exchanged. FL works by initializing local model training on local data. Post model training, local model updates (parameters/gradients) are sent to the central server for aggregation before being sent back to the participants for further training until the global model converges. FL safeguards data and allows collaboration among different devices or institutions, hence improving the generalizability of the global model by leveraging insights from diverse datasets. This is promising in the field of PGx, which is characterized by sensitive health data. Additionally, PGx benefits from collaboration with different institutions that possess data representing diverse demographic groups. Through FL, training a model on diverse genetic and clinical datasets is made possible, resulting in a model capable of generalizing to unseen data.

Despite advancements in data privacy and security, FL remains prone to security threats such as model inversion and membership inference attacks (Luo et al., 2021). Model inversion is a security attack in machine learning (ML) that uses the model's output to reconstruct sensitive information from the training set, while membership inference attacks are used to determine whether a specific datapoint was used to train the ML model, potentially exposing individual data records. To address these limitations, blockchain-enhanced federated learning (BFL) has shown potential in enhancing privacy in FL systems. By integrating the decentralized and tamper-proof design of blockchain into FL, BFL offers improved privacy and security, which is crucial when dealing with sensitive data such as patients' genetic and clinical information. This approach will result in a secure, tamper-free predictive model that can be used across multiple institutions for the accurate prediction of treatment response in depression treatment.

Recent studies have explored BFL frameworks for different purposes in the health domain. However, none have been developed to enhance treatment selection. Unlike other illnesses, mental health disorders have a complex etiology, which confines their treatment selection to a trial-and-error approach, which often results in poor patient outcomes. To bridge this gap, this study introduces a BFL framework for the PGx of depression, capable of predicting treatment outcomes in individuals, thereby aiding in personalized treatment selection while providing data privacy and security to patient data.

To ensure the security and privacy of the framework, the study integrates homomorphic encryption (HE), secure multiparty computation (SMPC), the interplanetary file system (IPFS), and blockchain within a decentralized federated learning architecture. This approach enhances privacy-preserving model training and enables secure data sharing across distributed healthcare environments. Moreover, the proposed framework aims to foster collaboration across multiple health institutions in sensitive domains such as mental healthcare, where data centralization is often faced with stringent laws and ethical considerations.

To the best of our knowledge, no studies have explored the application of BFL in the PGx of depression. This gap presents a unique opportunity for collaboration across multiple health institutions by developing a tamper-proof framework that enhances data privacy and security in a distributed environment for the personalized treatment of depression.

1.1 Literature Review

This study is rooted in literature on BFL frameworks in the health domain. This review investigates relevant literature in the health domain to establish the state-of-the-art of the developed frameworks and to identify the gap that our study addresses.

1.2 IoT-FL Architectures

Recent works propose secure, privacy-preserving FL platforms for the smart health IoT. Singh et al. (2022) design a blockchain-enabled IoT cloud architecture that links wearable health sensors (e.g., weight, glucose, blood-pressure monitors) into a distributed network. In this framework, blockchain-based IoT cloud platforms provide security and privacy while FL trains healthcare models without sharing raw data. Building on this, Lian et al. (2023) introduced a two-stage BFL approach for IoMT. Their system shares partial client data via a blockchain-enabled scheme to synthesize a global dataset and iteratively aggregate client gradients, addressing non-IID data issues. They explicitly design a blockchain-based data-sharing mechanism to improve model accuracy without compromising user privacy.

1.3 Trust and Secure Aggregation

To defend against malicious participants, newer BFL frameworks incorporate trust mechanisms. For example, Jiang et al. (2025) proposed T-BFL, a two-dimensional trust-based FL model for medical data sharing. T-BFL uses blockchain-based registration and Bayesian behavioral trust to vet users, only clients with high identity and behavior trust are allowed to participate. Their Trust-FedAvg algorithm securely aggregates model updates by weighting them according to these trust scores. Similarly, Khan et al. (2025) develop a BFL system for EMRs that encrypts and logs model updates on-chain. In this system, the blockchain acts as an immutable ledger for model parameters and enforces data usage policies via smart contracts. Together, these trust-driven designs enhance data integrity and detect poisoned or unauthorized updates, enabling collaborative EMR analysis with privacy and security guarantees.

1.4 Application-Specific FL

Other studies target specific healthcare tasks. Dib (2025) focus on diabetic retinopathy (DR) diagnosis. They propose a serverless, fully decentralized FL + SMPC framework. Each client sends its local model update encrypted and split it into secret shares using Shamir's Secret Sharing (SSS), and Hyperledger Fabric smart contracts coordinate the secure aggregation of these shares. Crucially, IPFS holds the encrypted shares to reduce blockchain storage overhead. This ensures scalability and privacy in DR detection without a trusted server. In effect, encrypted model weights are shared and combined to reach centralized-level accuracy while preserving patient data confidentiality.

1.5 Customized IoMT and Resource-Aware FL

Several works address heterogeneous IoMT environments and resource constraints. Mazid et al. (2025) proposed a partitioned FL model for IoMT, the neural network is split into global layers, which are aggregated across all clients and local custom layers which are specific to each client. Edge gateway devices preprocess data and train client-specific layers locally, while only the shared layers are securely aggregated via a permissioned blockchain. This allows personalized model features while preserving privacy. Fathima et al. (2025) developed a Zero-Trust edge-cloud framework for secure resource offloading and recommendation. In their design, IoMT devices train local FL models and a blockchain-based resource distribution agent prioritizes task offloading under a zero-trust model. Smart contracts are used to authenticate and sequence resource allocation requests. The authors show that integrating blockchain with FL and zero-trust policies yields secure, low-latency medical resource scheduling. Jafari and Adibnia (2025), similarly combine FL with a decentralized ledger BigchainDB for IoMT data management. Their framework records model metadata on-chain to detect malicious devices, in experiments it prevented unauthorized data access and detected over 89% of injection attacks. These studies demonstrate that blockchain can enforce authenticity, privacy, and auditability in IoMT FL, improving both security and system efficiency.

1.6 Privacy-Preserving EHR Learning

Panimalar and Gunasundari (2025) proposed FedBlock, a consortium-blockchain FL model for EHRs. In FedBlock, multiple hospitals jointly train models on sensitive EHR data, while a consortium ledger ensures transparency and privacy. By decentralizing aggregation and using a consortium blockchain, FedBlock reportedly achieves superior predictive accuracy on clinical data while preventing raw data sharing.

Although there have been significant strides in BFL frameworks in the healthcare domain aiming to enhance secure computation through HE and SMPC and improve model updates storage through IPFS, none have integrated HE, SMPC, and IPFS to enhance data privacy, scalability, and practical deployment in a cross-institutional setting. Additionally, consortium approaches such as FedBlock, which are designed for a cross-institutional setting, target generic diagnostics, leaving a significant gap in personalized treatment selection. This study addresses that gap by proposing a secure BFL framework that integrates HE, SMPC, and IPFS to enable privacy-preserving cross-institutional collaboration for treatment selection in depression, marking a significant step in the practical application of FL in mental health research.

2 Framework Overview

The proposed framework, illustrated in Fig. 1, adopts a design science methodology (DSM) to design a BFL framework for cross-institution collaboration. It supports and addresses claims made by Singh et al. (2022), Jiang et al. (2025), and Dib (2025) on the challenges in centralized aggregation, partial privacy implementation, and insufficient scalability for personalized healthcare. The framework integrates HE, SMPC, and IPFS to enable secure, scalable, and storage-efficient multi-institutional collaboration for personalized treatment selection among patients. The framework is described in detail in the sections below.

2.1 Client-Side Local Training and Encryption

Healthcare institutions serve as clients, training predictive local models on genetic and clinical data for treatment selection. HE is employed to preserve privacy and avoid inference attacks by encrypting model updates using public keys before their transmission to the IPFS. While Dib (2025) applied SMPC to secure data during aggregation, this framework advances data protection by employing HE, enabling secure computation on model updates throughout the aggregation cycle. Additionally, each client generates a private key to ensure that the aggregated global model is decrypted before the next training iteration.

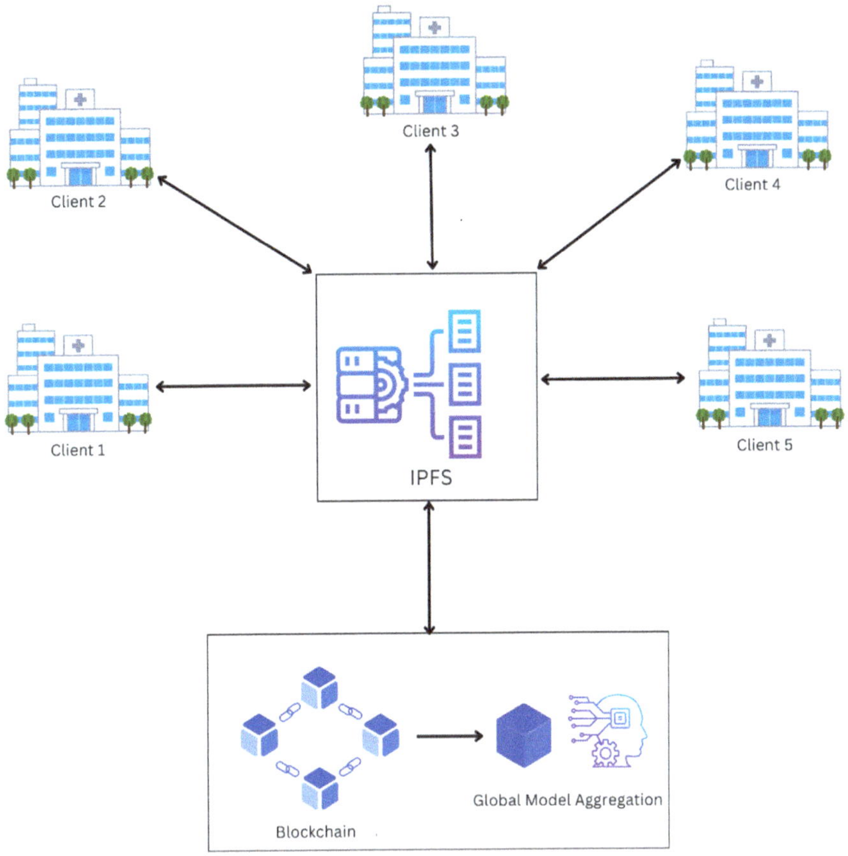

Fig. 1 Proposed BFL framework

2.2 *Off-Chain Storage with IPFS*

After each local training iteration, model updates are uploaded to the IPFS and assigned a content identifier (CID) that is recorded on the blockchain for secure and verifiable retrieval. The choice of using IPFS is based on its decentralized architecture, ability to handle large datasets, and limit bandwidth usage by caching model updates. IPFS also prevents data duplication through regular and efficient network updates. This improves on the approach by Jafari and Adibnia (2025), who employed BigchainDB for tamper-resistant storage. However, their study did not address bandwidth optimization or lightweight off-chain storage; additionally, it did not integrate cryptographic encryption mechanisms by uploading encrypted model updates in the IPFS.

2.3 Blockchain Coordination and Smart Contracts

To ensure immutability and verification in the FL process, smart contracts in blockchain were utilized. Smart contracts are responsible for (1) registering participants to ensure that only authorized clients contribute to the training process, (2) storing the CID of each model update to facilitate secure and traceable storage, (3) validating model updates by verifying if the model update meets the required criteria, and (4) distribute incentives to reward clients with tokens to encourage honest and high quality participation. This design expands on concepts introduced by Jiang et al. (2025) and Khan et al. (2025) by incorporating a more robust aggregation mechanism and integrating encryption-aware logging into the blockchain layer.

2.4 Decentralized Aggregation with SMPC

To mitigate the vulnerability of centralized aggregators, a multi-leader aggregation mechanism is adopted. A subset of clients is selected for aggregation based on reputation scores, fairness, and resistance to collusion. These selected nodes retrieve model updates via their CIDs from IPFS and perform aggregation using SMPC protocols built on Shamir's Secret Sharing. Finalization of the global model is contingent upon reaching consensus among leaders, ensuring validity and eliminating the influence of poisoned or substandard updates. Once consensus is reached, the aggregated global model is uploaded to IPFS, and its CID is broadcast to all clients. Clients use their private keys to decrypt the model and proceed with local training. This improves upon previous multi-leader systems such as that proposed by Mazid et al. (2025) through the secure integration of encrypted data streams and decentralized computation.

2.5 System Validation Strategy and Practical Considerations

Although the proposed framework has not yet been empirically deployed, its feasibility and performance can be rigorously evaluated through a multi-phase validation strategy. First, the cryptographic components HE and SMPC can be tested using established libraries such as Microsoft SEAL and MP-SPDZ to simulate secure local training and aggregation workflows. Blockchain orchestration and coordination protocols may be validated using a permissioned network built with Hyperledger Besu, while the performance and scalability of decentralized storage can be assessed via IPFS Cluster under simulated network conditions. To approximate real-world scenarios, simulation environments should reflect heterogeneous client hardware (e.g., $\leq$4 GB RAM and ~300 MB storage per node), particularly targeting healthcare institutions in resource-constrained regions. Key performance indicators

(KPIs) for validation would include model accuracy, latency, throughput, encryption/decryption overhead, bandwidth usage, and fault tolerance. System evaluation should be benchmarked against related frameworks for improvement quantification in terms of privacy, decentralization, and personalization support.

3 Discussion

This study introduces a BFL framework that integrates HE, SMPC, and IPFS for cross-institution collaboration in the training of predictive ML models for the treatment selection of depression. Unlike prior efforts that focused on IoT-based diagnostics, this framework focuses on a cross-silo approach that leverages clinical and genomic data for treatment selection. By facilitating this collaboration among health institutions, the study advances mental health research by allowing the training of predictive models using data from different demographic groups that will aid in improved treatment outcomes.

Through blockchain, the framework ensures decentralized client authentication, transparent incentive mechanisms, and tamper-resistant coordination, which is essential for sensitive health data. Additionally, the framework employs HE, SMPC, and IPFS for an extra layer of security and decentralized, memory-efficient storage. Unlike studies such as Dib (2025), who employed SMPC alone, this framework offers a dual-layer security approach that offers a more comprehensive privacy guarantee during training and aggregation.

To ensure the real-world applicability of the proposed framework, collaboration with clinical and technical domain experts is crucial in guiding system design and validation. However, this study has several limitations. First, the framework has not yet undergone empirical evaluation in live healthcare settings, making its performance under real-world conditions such as high communication overhead, non-IID data distributions, and institutional heterogeneity uncertain. Second, the framework's scalability under large-scale deployment remains untested. Third, the current focus is on structured clinical and genomic data, which limits its applicability to broader mental health scenarios that require multimodal inputs. Future research should address these limitations through experimental deployments and by extending the framework to incorporate behavioral data, neuroimaging, and physiological signals from wearable devices for comprehensive and longitudinal assessment of treatment response in mental health care.

4 Conclusion

This study aims to propose a framework for personalized treatment selection among individuals diagnosed with depression. It leverages the unique capabilities of blockchain, HE, IPFS, and SMPC to improve data privacy, scalability, and security in a decentralized federated learning environment. The framework is designed to encourage collaboration among health institutions by providing a secure framework for training robust predictive machine learning models, thereby improving treatment selection for better patient outcomes.

Acknowledgements The author gratefully acknowledges the support from the DAAD Research Grants—Doctoral Programs in Germany.

Disclosure of Interests The authors have no competing interests.

References

Beer, C., Rae, F., Semmler, A., & Voisey, J. (2024). Biomarkers in the diagnosis and prediction of medication response in depression and the role of nutraceuticals. *International Journal of Molecular Sciences, 25*(14), 7992. https://doi.org/10.3390/ijms25147992

Bobo, W. V., et al. (2019). Frequency and predictors of the potential overprescribing of antidepressants in elderly residents of a geographically defined U.S. population. *Pharmacology Research Perspectives, 7*(1), e00461. https://doi.org/10.1002/prp2.461

Chung, A.-N., Chen, T.-T., & Lin, Y.-F. (2023). Genetics of antidepressant response and treatment-resistant depression. In *Progress in Brain Research* (Vol. 278, pp. 25–60). Elsevier. https://doi.org/10.1016/bs.pbr.2023.01.002

Dib, O. (2025). A decentralized privacy-preserving framework for diabetic retinopathy detection using federated learning and blockchain. *Results Engineering, 26*, 105456. https://doi.org/10.1016/j.rineng.2025.105456

Fathima, A. S., et al. (2025). Empowering consumer healthcare through sensor-rich devices using federated learning for secure resource recommendation. *IEEE Transaction Consumer Electronics, 71*(1), 1563–1570. https://doi.org/10.1109/TCE.2025.3541549

Hoffman, J. M., Flynn, A. J., Juskewitch, J. E., & Freimuth, R. R. (2020). Biomedical data science and informatics challenges to implementing pharmacogenomics with electronic health records. *Annual Review of Biomedicine Data Sciences, 3*(1), 289–314. https://doi.org/10.1146/annurev-biodatasci-020320-093614

Jafari, M., & Adibnia, F. (2025). Securing IoMT healthcare systems with federated learning and BigchainDB. *Future Generated Computer Systems, 165*, 107609. https://doi.org/10.1016/j.future.2024.107609

Jiang, R., Zhang, H., Song, Z., Tian, S., & Lou, W. (2025). T-BFL model based on two-dimensional trust and blockchain-federated learning for medical data sharing. *Journal of Supercomputers, 81*(2), 378. https://doi.org/10.1007/s11227-024-06873-5

Karamperis, K., et al. (2021). Economic evaluation in psychiatric pharmacogenomics: A systematic review. *Pharmacogenomics Journal, 21*(4), 533–541. https://doi.org/10.1038/s41397-021-00249-1

Khan, S., Khan, M., Khan, M. A., Wang, L., & Wu, K. (2025). Advancing medical innovation through blockchain-secured federated learning for smart health. *IEEE Journal of Biomedical Health Information, 14*, 1–14. https://doi.org/10.1109/JBHI.2025.3532976

Kosorok, M. R., & Laber, E. B. (2019). Precision medicine. *Annual Review Statistics and Its Application, 6*(1), 263–286. https://doi.org/10.1146/annurev-statistics-030718-105251

Lian, Z., Zeng, Q., Wang, W., Gadekallu, T. R., & Su, C. (2023). Blockchain-based two-stage federated learning with non-IID data in IoMT system. *IEEE Transaction Computer in Society Systems, 10*(4), 1701–1710. https://doi.org/10.1109/TCSS.2022.3216802

Luo, X., Wu, Y., Xiao, X., & Ooi, B. C. (2021). Feature inference attack on model predictions in vertical federated learning. In *Proceedings of the 2021 IEEE 37th international conference on data engineering (ICDE)* (pp. 181–192). IEEE. https://doi.org/10.1109/ICDE51399.2021.00023

Mazid, A., Kirmani, S., Abid, M., & Pawar, V. (2025). A secure and efficient framework for internet of medical things through blockchain driven customized federated learning. *Cluster Computers, 28*(4), 225. https://doi.org/10.1007/s10586-024-04896-4

Nochaiwong, S., et al. (2021). Global prevalence of mental health issues among the general population during the coronavirus disease-2019 pandemic: A systematic review and meta-analysis. *Scientific Report, 11*(1), 10173. https://doi.org/10.1038/s41598-021-89700-8

Okpete, U. E., & Byeon, H. (2024). Challenges and prospects in bridging precision medicine and artificial intelligence in genomic psychiatric treatment. *World Journal of Psychiatry, 14*(8), 1148–1164. https://doi.org/10.5498/wjp.v14.i8.1148

Palumbo, S., Mariotti, V., & Pellegrini, S. (2024). A narrative review on pharmacogenomics in psychiatry: Scientific definitions, principles, and practical resources. *Journal of Clinical Psychopharmacology, 44*(1), 49–56. https://doi.org/10.1097/JCP.0000000000001795

Panimalar, S. P., & Gunasundari, S. (2025). A novel privacy protection technique of electronic health records using decentralized federated learning with consortium blockchain. *Procedia Computer Science, 252*, 212–221. https://doi.org/10.1016/j.procs.2024.12.023

Pirmohamed, M. (2023). Pharmacogenomics: Current status and future perspectives. *Nature Reviews Genetics, 24*(6), 350–362. https://doi.org/10.1038/s41576-022-00572-8

Pizzicato, L. N., Xie, R. Z., Yang, Y., Grabner, M., & Chapman, R. H. (2023). Real-world treatment modalities, health care resource utilization, and costs among commercially insured patients with newly diagnosed major depressive disorder in the United States. *Journal of Managetic Care Specification Pharmology, 29*(6), 614–625.

Santomauro, D. F., et al. (2021). Global prevalence and burden of depressive and anxiety disorders in 204 countries and territories in 2020 due to the COVID-19 pandemic. *The Lancet, 398*(10312), 1700–1712. https://doi.org/10.1016/S0140-6736(21)02143-7

Singh, S., Rathore, S., Alfarraj, O., Tolba, A., & Yoon, B. (2022). A framework for privacy-preservation of IoT healthcare data using federated learning and blockchain technology. *Future Generated Computer Systems, 129*, 380–388. https://doi.org/10.1016/j.future.2021.11.028

Taliaz, D., & Souery, D. (2021). A new characterization of mental health disorders using digital behavioral data: Evidence from major depressive disorder. *Journal of Clinics Medicine, 10*(14), 3109. https://doi.org/10.3390/jcm10143109

Tesfamicael, K. G., et al. (2024). Efficacy and safety of pharmacogenomic-guided antidepressant prescribing in patients with depression: An umbrella review and updated meta-analysis. *Frontier Psychiatry, 15*, 1276410. https://doi.org/10.3389/fpsyt.2024.1276410

Thornicroft, G., et al. (2022). The Lancet commission on ending stigma and discrimination in mental health. *Lancet London England, 400*(10361), 1438–1480. https://doi.org/10.1016/S0140-6736(22)01470-2

Walschots, Q., Zarchev, M., Unkel, M., & Kamperman, A. (2024). Using wearable technology to detect, monitor, and predict major depressive disorder—A scoping review and introductory text for clinical professionals. *Algorithms, 17*(9), 408. https://doi.org/10.3390/a17090408

Comparing Vertical Federated Learning Frameworks in Manufacturing Supply Chains

Jonas Kallisch, Jorge Marx Gómez, and Christoph Wunck

Abstract This paper investigates data analysis and predictive modelling within industrial production supply chains. Despite its potential, cross-company data analysis remains limited due to technical challenges, organizational barriers, and trust issues. A primary concern is the potential misuse of sensitive data. Federated Learning offers a promising solution by enabling decentralized analysis without direct data exchange. It supports horizontal and vertical models, with the latter particularly suitable for supply chains. This study presents a case study from an industrial supply chain to evaluate whether Vertical Federated Learning can facilitate cross-company data analysis. A benchmark of various Federated Learning frameworks was conducted. Additionally, these frameworks are compared based on usability, integration into corporate structures, and other quality factors.

1 Introduction

Data analysis and predictive modelling have become increasingly important in manufacturing production systems (Marmolejo-Saucedo et al., 2019). This development is supported by the growing availability of sensors, open-source analytics, and "smart" machine components (Kozlenkova et al., 2015a). As a result, many companies now analyze their own production data to enhance efficiency (Hermes Newsroom, 2023). In contrast, cross-company data analysis remains rare, although entrepreneurs view it as a competitive advantage and a foundation for future business models (Culot et al., 2020). Existing intercompany data exchanges typically involve logistics, invoicing, or order interfaces, but are rarely applied to production-level data (Culot et al., 2020).

J. Kallisch (✉)
OFFIS Institute for Information Technology, Oldenburg, Germany
e-mail: jonas.kallisch@offis.de

J. Marx Gómez
Carl von Ossietzky University Oldenburg, Oldenburg, Germany

C. Wunck
University of Applied Sciences Emden/Leer, Emden, Germany

J. Marx Gómez et al. (eds.), *Recent Advances in Information Systems*, Progress in IS,
https://doi.org/10.1007/978-3-032-14099-9_28

Numerous studies have examined why production data is not shared between companies. Especially in small and medium-sized enterprises, key barriers include a lack of technical expertise, weak organizational coordination, and most commonly mistrust toward business partners (Kallisch et al., 2024; Södergren & Cartling Wallén, 2022). The reason for this mistrust is the fear that shared data could be misused to gain competitive advantage, since such data can reveal production costs and is often treated as intellectual property (Denkena et al., 2024; Panahifar et al., 2018). For example, in aluminum die-casting, pressure curves may allow conclusions about load capacity and cycle times. Sharing such data could expose cost structures, which many companies consider unacceptable, especially in price negotiations. Additionally, firms fear exposing their production systems to cyberattacks when connected to external networks (Kallisch et al., 2024).

This lack of collaboration prevents supply chains from realizing their full potential. Currently, most partners optimize their production system only for local maxima, limiting value creation across the network (Kozlenkova et al., 2015b). If production data were shared, for example between a die-casting firm and a mold manufacturer, it could enable mutual improvements: optimized tool design on one side, improved maintenance and utilization on the other (Denkena et al., 2024).

Several approaches exist to establish secure data-sharing relationships (Kozlenkova et al., 2015b). One is legal regulation, as seen in platform models like GAIA-X, which offers data contracts embedded with usage rules (GAIA-X Foundation, 2023). However, serval studies show that trust in such agreements remains limited, as many companies find the risk of breaches and the lack of enforceability unacceptable (Södergren & Cartling Wallén, 2022).

An alternative approach involves aggregation via mediators, such as data trusts, which act as neutral entities to protect against misuse (Stachon et al., 2023). Although these models, when combined with internal agreements, can create relatively secure environments for data exchange, many companies still prefer purely technical solutions that allow them to retain full data sovereignty and minimize indirect information leakage.

One such technical approach is differential privacy (DP), which adds statistical noise to datasets to protect individual critical information (Yang et al., 2024). DP has been widely used in healthcare and finance (Demelius et al., 2023; Zhu et al., 2022). However, DP is not well-suited for manufacturing, where data reflects machine processes rather than individual's data. Precision is crucial for manufacturing, since even small deviations in sensor readings can indicate major quality shifts. For example, adding noise to pressure curves from a die-casting machine significantly reduces the usefulness of casting data for optimization (Zhu et al., 2022).

A promising alternative is Federated Learning (FL), which enables decentralized analytics without transferring data physically (Yang et al., 2019). Unlike DP, FL does not manipulate data, thereby preserving its precision. This paper investigates the application of Vertical Federated Learning (VFL) in an industrial supply chain scenario. Using a real-world manufacturing dataset, the study compares multiple VFL frameworks and evaluates their practical suitability for cross-company data analysis.

While GAIA-X initiatives and DP approaches offer important contributions to securing data sharing, VFL complements them by providing technical enforcement of data separation during analytics. Unlike GAIA-X, which relies heavily on contractual frameworks, or DP, which modifies data to protect it, VFL enables decentralized analysis without direct data exchange or distortion. Therefore, VFL can help overcome barriers where contractual trust and statistical protections alone are insufficient.

2 Federated Learning

As described FL is a concept that analyzes datasets distributed over different devices that are connected with a central server or service (Yang et al., 2019). It can be divided into Horizontal Federated Learning (HFL), Vertical Federated Learning (VFL), and Transfer Learning.

The difference between Horizontal and Vertical FL lies in the way data is distributed across the participating parties. In HFL, each party holds data with the same features but different samples, for example multiple machines recording temperature over time. In VFL, parties hold data with different features but referring to the same samples, like different companies in a supply chain collecting distinct data about the same product.

In Transfer Learning, a pre-trained model designed to perform one task, such as fault detection, is trained on another dataset to accomplish a different task, like fraud detection (Zhang et al., 2022). In the context of this study on supply chains, Transfer Learning is less relevant, as most intercompany data analytics do not have a shared or general-purpose model available for knowledge transfer.

As shown in Fig. 1, HFL shares the same features across all parties but separates the data by samples, whereas VFL shares samples but separates the data by features. For example, temperature measurements may be available at all parties in an HFL scenario, while in VFL, one company might provide sensor data, another quality assessment, both linked to the same product.

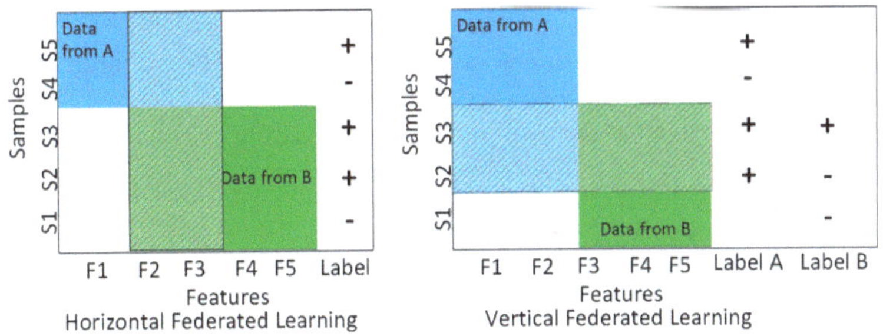

Fig. 1 Difference between types of federated learning

The more common case of FL is HFL, which is widely used in mobile devices to improve their ability to analyze user data without transferring it. A HFL process starts with an initial global model created on a data sample. In the second step, the model is decomposed into sub-models, matching the data elements of the different clients. The different clients (e.g., smartphones or production systems) then train the model locally on their own data. In the third step, the local model updates are transferred to a central aggregator. In the fourth and final step, these updates are aggregated to improve the global model. As a result, each party benefits from improved performance without sharing raw data. However, a limitation of HFL is that it requires homogeneous data structures across all participating parties.

VFL, on the other hand, can be used in supply chains (SC) that do not share common data architectures between different companies. As shown in Fig. 1, the clients in a VFL model may not share the same features or data structures, but they refer to the same samples—e.g., products or customers. For example, two companies in the same supply chain may not collect the same data, but both may collect it for the same product batch. If these companies pursue a common objective, they could combine their data using VFL to improve the quality of their predictive algorithms without compromising data sovereignty.

3 Research Question and Methodology

As Sects. 1 and 2 have shown, multiple papers have already dealt with the theoretical applicability of FL in supply chains (Durrant et al., 2022; Li et al., 2021). These papers describe scenarios and, based on these scenarios, elaborate models for evaluating the suitability of FL models in different data spaces. They also include discussions of Transfer Learning and its relevance in the context of distributed data silos in supply chains. The hypothesis derived from previous work is that Federated Learning can support data analysis across organizational boundaries. Based on this hypothesis, the following research questions are formulated and addressed in this paper:

- RQ1: Can Federated Learning applications contribute to analyzing industrial supply chain data?
- RQ2: How accurate are the models generated in comparison to aggregated data analysis?
- RQ3: How do the performances of the different existing frameworks differ?

There is only a small body of literature on implementing VFL applications. Real-life application examples are rare, particularly in the context of cross-company production data. A significant contribution in this context is the publication by Liu et al., which presents a technical benchmarking of different FL frameworks and provides a guideline for choosing a suitable framework (Liu et al., 2022). Their work provides a basis for selecting the frameworks for this paper's analysis. However, further research is needed to answer the questions posed in this paper, as previous practical work on the topic has used only small data sets and has focused on analyzing technical differences.

In contrast to prior works, which often focus on synthetic or homogeneous datasets, this study emphasizes a real-world, heterogeneous industrial supply chain with complex feature distributions and operational constraints. This setting introduces unique challenges for VFL, particularly regarding feature sparsity, data ownership boundaries, and integration overhead.

4 Case Introduction

For this case study, a dataset from a group of manufacturing companies was used, which contains information from three stages of the SC of an aluminum product. The dataset covers the complete production process including tool design, workpiece preparation, the actual die-casting process, finishing, and quality analysis of the finished component. It includes structured and time-series data, such as temperature and pressure curves, master data of machines, analysis data of raw materials, parameters used at the machines, and information on faults or process events. The results of the quality analyses include a classification of the product's quality and information on quality deviations. This information describes the type, scope, and position of each deviation.

In total, the dataset comprises 8400 product instances, resulting in a substantial volume of data for this case study. Each instance includes several hundred time-series measurements per sensor, recorded at 15 kHz resolution, along with categorical and numerical metadata. To prepare the dataset for analysis, it was randomly divided into training and test sets using an 80/20 split, as described in Joseph (2022).

For the FL analysis, the data sources were divided into logical units to be used as data owners, reflecting the real-world segmentation of the production chain. The distribution of the data in the test setup mirrors the actual supply chain structure, which creates the products in three major production steps.

- Step one consists of preparing the casting material. For this purpose, the raw aluminum is mixed with accompanying materials and stirred into a homomorphic mass. A material sample is taken to ensure the material properties.
- In the second step, the raw mass is introduced into the mold of the machine tool and injected under high pressure. During this process, data from at least two pressure sensors, four temperature sensors, and one acceleration sensor are recorded at a frequency of 15 kHz. These time-series signals are segmented into characteristic intervals (e.g., filling, holding, cooling) and analyzed separately. Certain defects in the components can be directly recognized from this process data.
- The final step involves surface treatment and cleaning. The raw part is eroded, and the component is inspected for quality characteristics. Some defects, such as gas bubbles or microcracks, cannot be clearly assigned to a specific step. These may

result from material issues, insufficient casting pressure, or overheating during erosion. Identifying the root causes requires combining data from multiple stages.

It should be noted that the respective production stages differ in the number of available samples, as part rejects reduce the number of observations downstream in the supply chain. These surplus data points were included in the FL analysis to determine their potential contribution. In cases where workpieces did not pass through all stages, the corresponding missing features were retained without imputation. The federated models were trained using these incomplete feature vectors, reflecting realistic industrial conditions where early-stage defects still offer valuable learning signals.

Two data analyses were carried out as part of this case study. First, a classification task was performed using the end customer's target label, based on data from each stage of production. Then, a clustering analysis was conducted to group the defect positions and sizes based on the available production data. The results obtained from the entire dataset using conventional methods were compared to those achieved using the Federated Learning frameworks.

For the classification task, a Random Forest classifier was selected as the baseline model. Random Forests were chosen for their robustness in handling heterogeneous feature sets, resilience to imbalanced classes, and high interpretability, which is crucial for industrial defect analysis. For the FL experiments, lightweight logistic regression models were applied across all frameworks to ensure comparability and minimize computational bias between implementations.

5 Key Performance Indicators for Framework Evaluation

A structured literature analysis was carried out according to the methodology proposed by Brocke et al. to create an evaluation matrix of Key Performance Indicators (KPIs) for the frameworks used in the planned experiment (Brocke et al., 2009). The goal of this analysis was to identify both qualitative and quantitative KPIs that are suitable for evaluating the practical performance and usability of FL frameworks. Google Scholar, the AIS Library, ResearchGate, and ScienceDirect were selected as primary research sources. The search combined terms such as "KPI," "analytics," "big data," and "framework." From the results, sources that dealt with software comparison and evaluation in broader analytics contexts were examined in more detail.

From the identified sources, a variety of possible KPIs emerged. Since some technical metrics, including supported algorithms, encryption features, had already been determined in preliminary internal assessments, this study placed particular emphasis on evaluating the quality of the analysis results achieved in the use case. In addition to these output-focused criteria, the ease of integration into existing infrastructure, the availability of technical support or community help, and the quality and completeness of the documentation were also included in the evaluation.

To ensure domain relevance, the final selection of KPIs was validated through consultation with industrial process experts from the aluminum sector. The selected indicators were then applied during the experimentation phase to assess and compare each framework consistently. The following section briefly presents the selected FL frameworks and their respective evaluations based on these indicators.

6 Framework Comparison

This section presents the evaluation of the selected frameworks. The comparison is based on a qualitative assessment of practical aspects relevant to industrial deployment, including documentation quality, integration effort, support availability, and overall usability. The evaluation matrix follows the criteria defined according to Brocke et al. (2009) and is extended by factors derived from preliminary technical assessments and domain-specific considerations.

A five-point scale was used for the evaluation, where 1 = very poor and 5 = excellent. Each criterion was assessed based on a structured rubric:

- Documentation (completeness, clarity, examples, versioning).
- Integration (setup complexity, compatibility, deployment options).
- Support (accessibility of help, community activity, official channels).
- Usability (developer experience, interface design, learning curve).

As container orchestration using Kubernetes or Docker Swarm plays an important role in manufacturing scenarios, particularly for scalable supply chain analytics, frameworks supporting such setups were assessed with particular attention to this aspect.

6.1 FATE

Federated AI Technology Enabler (FATE) is an open-source platform that ensures data sovereignty for data-driven machine learning and AI model training (FATE Documentation, 2023). During the implementation, it quickly became obvious that comprehensive and high-quality documentation is available. However, specific tutorials for implementing VFL are missing, which is an important aspect for users focusing on this method. As a result, the documentation was rated four out of five points.

The framework offers a wide range of pre-implemented algorithms, which can be deployed relatively quickly after a short familiarization phase. The user experience is streamlined, allowing practitioners to get started without investing excessive time in system complexity, resulting in a usability score of four. On the other hand, integration into existing infrastructures posed challenges, particularly when attempting to set up Kubernetes environments. The lack of a stable Kubernetes test platform limited a thorough evaluation of container-based deployment. Although FATE offers flexibility, creating analysis pipelines demands substantial expertise, leading to an integration score of three.

Support for FATE is very good, with direct support mechanisms and an active open-source community available. However, as in many open-source projects, the quality and response time can vary depending on the issue and community activity. Nevertheless, the support environment proved helpful during the platform setup, leading to a support rating of four. In summary, FATE is a powerful and well-documented FL framework with a broad range of features but requires considerable expertise for advanced integration tasks.

6.2 PySyft

PySyft is a Python-based framework developed by OpenMined, designed to support privacy-preserving FL through techniques such as encrypted computation, secure multiparty computation, and differential privacy (OpenMined Blog—PySyft, 2023). Although documentation is available, it is outdated and fragmented, making the implementation of newer features challenging. Furthermore, different documentation versions add to the confusion, resulting in a documentation score of two.

Despite documentation weaknesses, PySyft offers a strong set of privacy features and can be integrated relatively easily once users familiarize themselves with its structure. Integration was therefore rated three out of five points. Usability was rated four, particularly for developers already experienced in Python and machine learning frameworks, due to the familiar syntax and structure of PySyft.

However, support availability is limited. Official support channels are hard to access, and the surrounding community appears relatively inactive compared to other projects. As a result, support was rated two. PySyft offers a flexible, research-focused FL environment but presents challenges for production deployment due to limited up-to-date documentation and support resources.

6.3 FedML

FedML offers a FL framework providing both a Python library and a web-based platform (He et al., 2020). The CLI enables control and management of model training processes across vertically partitioned data, while the web platform offers infrastructure for secure data sharing and decentralized training without requiring additional implementation work.

A major strength of FedML is its usability, as it provides both Python-based and graphical web interfaces, earning a usability rating of five. Integration was rated five as well, due to the simplicity of the command set and the user-friendly configuration through the web platform, which reduces technical barriers.

However, the documentation for CLI-based implementations is comparatively thin, lacking detailed instructions for complex use cases. Therefore, documentation was rated three. Support is similarly moderate, while the community provides assistance, there is limited official help for CLI users, leading to a support rating of three. Overall, FedML presents a highly accessible solution, particularly for organizations looking for a web-based management environment with minimal technical overhead.

6.4 Tensorflow Federated

TensorFlow Federated (TFF) is an open-source FL framework based on TensorFlow (Alphabet Inc., 2023). TFF offers a broad range of machine learning model options, from simple linear models to complex neural networks, and can be easily integrated into web applications. Documentation is extensive, well-structured, and guides users through setup, implementation, and customization, leading to a documentation score of five.

Usability is high for developers familiar with TensorFlow and Keras, resulting in a usability rating of five. However, TFF requires strict adherence to its environment: the use of TensorFlow and its Keras API is mandatory, which can restrict flexibility for users who prefer other machine learning libraries. This limitation led to an integration rating of three.

Support is primarily community-driven. While the TensorFlow community is active, there is no official direct support for TFF, particularly for complex deployment scenarios like on-premise integration. Thus, support was rated three. In conclusion, TFF is a robust framework well-suited for research and academic projects, particularly within the TensorFlow ecosystem, but may be less ideal for production systems requiring broader compatibility.

6.5 Flower

Flower is an open-source framework optimized for decentralized FL (Flower, 2023). It is lightweight, easy to set up, and supports both cross-device and cross-silo scenarios, making it highly adaptable across different industries and sectors.

The documentation is concise and easy to follow but less comprehensive than that of larger frameworks, resulting in a documentation score of three. Usability is high, with simple syntax and an intuitive setup, earning a usability rating of four. Integration into existing systems is straightforward due to Flower's minimalistic design, which avoids heavy dependencies. Integration was therefore rated five.

Support benefits from close communication with the core development team and active community channels, though formal support structures are not present, leading to a support score of four. Overall, Flower offers a practical and flexible FL framework, particularly suited for rapid deployment in environments requiring lightweight and adaptable solutions.

6.6 Comparison of Frameworks

Table 1 summarizes the comparative evaluation of the frameworks. While all frameworks can technically implement FL, their differences are evident in non-functional aspects such as integration effort, documentation quality, and the availability of support services. These aspects may significantly influence the choice of framework depending on the specific needs and resources of the implementing organization.

Building on the qualitative evaluation of the selected frameworks regarding documentation, integration, support, and usability, the next section presents a practical comparison based on a real-world industrial use case. Using an industrial use case, the frameworks are tested to assess their performance, scalability, and applicability in a real-world vertical FL scenario.

7 Comparison of Federated Learning Frameworks in a Supply Chain Use Case

This section presents the prototype architecture used to evaluate the identified FL frameworks and discusses the results of the comparative analysis. The findings provide evidence for answering the three research questions outlined earlier.

Table 1 Comparison of federated learning frameworks

Framework	Ease of integration	Documentation	Support	Usability
FATE	3	4	4	4
Yes PySyft	3	2	2	4
FedML	5	3	3	5
TensorFlow Federated	3	5	3	5
Flower	5	3	4	4

7.1 Prototyping of a Federated Learning Supply Chain Support Application

A software prototype was developed to compare the FL frameworks under realistic supply chain conditions. It follows a modular, three-layer architecture with interchangeable components (Kallisch & Wunck, 2023).

The first layer, located at the company level, simulates local data acquisition using the dataset provided by an industrial partner. The data is transferred to a NoSQL database within each participating company (Hunker et al., 2020). This knowledge platform supports a variety of data structures, including structured data (e.g., machine master data), time series from production systems, and unstructured sensor data (Jaggi & Kadam, 2016). The Apache Hadoop ecosystem was selected for its extensibility and compatibility with large-scale data processing (Apache Software Foundation, 2024).

Each company maintains control over its own data mapping within the FL system. This data mapping is defined locally for each FL client, ensuring compliance with data sovereignty requirements. While mediator-based architectures are conceivable, as discussed in Sect. 1, they carry the risk of indirect information leakage and were not pursued in this prototype.

To allow easy deployment and reproducibility, the FL clients were containerized using Docker. Within each container, a partitioned subset of the Random Forest model was trained on the features available at that specific supply chain stage. No raw feature sharing occurred between the clients. Instead, model updates—such as gradients or decision tree parameters—were transmitted to a central FL server, where they were aggregated using standard federated averaging. While some implementation details varied slightly across frameworks, the core logic was kept consistent to ensure comparability.

This containerized setup enables scalability and parallelism, which is important as companies often participate in multiple supply chains. Keeping data and models logically separated supports sovereignty and simplifies deployment. In a real-world application, the FL server would typically be hosted by a neutral third party or consortium leader trusted by all stakeholders.

The third layer of the architecture is a digital representation of the entire supply chain. It connects the individual FL clients via unique identifiers and standardized output labels, such as product serial numbers. This allows the indirect linkage of records across organizations and supports the evaluation of predictive quality across the complete product lifecycle.

Figure 2 illustrates the overall system architecture, including the data interfaces, communication layers, and analytical components. The FL clients register through a service layer based on Jade, and the FL server coordinates the analysis across distributed datasets, as described in Sect. 2.

This prototype provides a flexible and realistic platform for evaluating FL frameworks. Its modular design allows for rapid scenario switching, including the addition of new supply chain actors or FL technologies (Kallisch & Wunck, 2023).

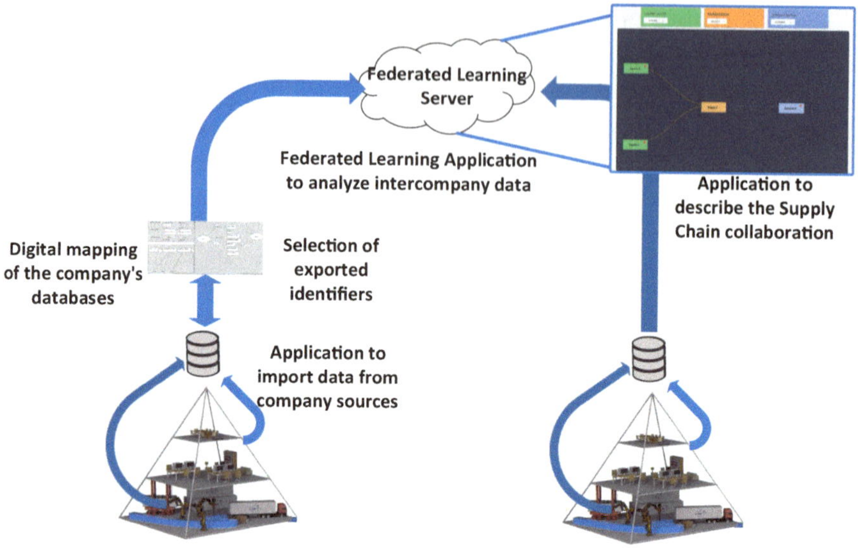

Fig. 2 Application prototype for FL framework comparison

7.2 Results of the Vertical Federated Learning Analysis

In the use case, the introduced prototype structure was used to train models based on vertically federated data. The comparison of the frameworks focused on categorization and clustering quality as well as runtime performance. The classification quality was evaluated using the KPIs of accuracy, precision, and F1 score, while the clustering quality was assessed using the Rand index.

It should be noted that model performance metrics, including accuracy, precision, and F1 score, were calculated against the true product quality labels, not against outputs of a centralized baseline model. Therefore, the reported scores reflect genuine predictive performance on the federated datasets. Slightly lower performance compared to centralized learning can be attributed to feature partitioning and data sparsity inherent in VFL, which is consistent with observations in related studies (Liu et al., 2022).

To ensure robustness and reproducibility, each training experiment was independently repeated five times, and the reported results represent the averaged performance across these repetitions. The reported performance metrics represent the average across these repetitions. Standard hyperparameters were used for all models unless specific optimization guidelines were provided by the respective framework documentation, ensuring comparability across frameworks. The following settings were used for baseline training unless otherwise specified: For Random Forest, 100 trees were trained without maximum depth restriction, using Gini impurity as the

Table 2 Results of comparison of federated learning frameworks

Framework	Accuracy	Precision	F1 score	Rand (10)
Fate	0.6962	0.7891	0.7845	0.7892
PySyft	0.7214	0.7962	0.7393	0.8237
FedML	0.7031	0.7846	0.7772	0.8087
TensorFlow Federated	0.6991	0.7781	0.7669	0.8023
Flower	0.7157	0.7811	0.7639	0.8026

splitting criterion. Logistic regression models used the "lbfgs" solver with a regularization strength of 1.0 and a maximum of 100 iterations. A batch size of 32 and a learning rate of 0.01 were selected for local training where applicable.

All frameworks were able to successfully train classification models using the vertically federated datasets. However, significant differences were observed regarding model convergence and performance stability. The classification results are summarized in Table 2.

For comparison, a centralized Random Forest model trained on the full dataset achieved an accuracy of 0.802, precision of 0.8134, and F1 score of 0.8017 (Pachandrin et al., 2025). As expected, federated models exhibited slightly lower performance due to vertical feature partitioning and data sparsity.

In terms of classification quality, the FedML and Flower frameworks achieved the highest accuracy and F1 scores, with values close to those obtained from centralized model training. FATE also demonstrated solid classification performance, though slightly lower than FedML and Flower. PySyft achieved acceptable results after extensive parameter tuning, whereas TensorFlow Federated showed difficulties converging consistently under vertical data conditions, leading to lower classification quality.

For clustering tasks, performance differences were more pronounced. FedML again showed the best clustering quality, closely followed by Flower. FATE performed moderately well but required a significantly higher number of iterations to achieve stable clustering. PySyft and TensorFlow Federated delivered weaker clustering results, with instability across repetitions and lower Rand index scores.

Overall, FedML and Flower demonstrated the best combination of classification quality, clustering quality, and runtime performance for VFL in the evaluated use case.

8 Discussion, Limitations, and Future Work

The case study illustrates that the tested frameworks do not differ significantly in algorithm performance. However, notable differences were observed in terms of documentation quality, integrability, and usability. As other studies have shown, the case study also demonstrated that FL can positively contribute to analyzing cross-company databases. This finding supports RQ1 and underlines the practical value of FL in distributed industrial settings.

The study presented comes with a few limitations, as current prototypes are limited to data analysis with a general quality label and individual identifiers trackable across the entire production system. In addition, the complexity of the data structures is low, and the connection between the participants of the represented supply chain is relatively straightforward. Therefore, the transferability of the results to more complex or fragmented supply chain constellations may be limited. It should be noted that this may not apply in all contexts and may not positively contribute to other data structures.

While this study demonstrated that VFL frameworks can support cross-company data analysis, it primarily focused on technical feasibility and performance metrics. In comparison to centralized model training, the federated models showed slightly lower accuracy and F1 scores (see Table 2), which answers RQ2. This performance loss is consistent with existing VFL literature and can be attributed to vertical feature partitioning and data sparsity.

Practical challenges such as data harmonization across companies, accessing production systems, and collaboratively interpreting intermediate results were not deeply analyzed. These integration barriers remain open issues and will be investigated in future work through extended case studies.

Additionally, to evaluate the practical relevance of different frameworks for industrial adoption, future work will incorporate semi-structured interviews with supply chain stakeholders. These interviews will help assess the usability, integration effort, and perceived value of VFL frameworks in realistic operational contexts. This will also help to better qualify the differences observed in framework performance (RQ3).

The authors will therefore analyze industrial supply chains from other future sectors further, to validate the results. The developed prototype offers the ability to adapt to those future scenarios.

While this study focused on an aluminum die-casting supply chain, many of the observed benefits and challenges are transferable to other manufacturing contexts involving sequential production stages and quality control checkpoints. The lessons regarding model convergence, usability barriers of FL frameworks, and integration complexity are particularly relevant for industries with heterogeneous data structures, such as automotive, electronics, and chemical production.

References

Alphabet Inc. (2023). *TensorFlow Federated* [Online]. Available: https://www.tensorflow.org/federated. Accessed November 14, 2023.

Apache Software Foundation. (2024). Apache Hadoop [Online]. Available: https://hadoop.apache.org/. Accessed September 6, 2024.

Culot, G., Orzes, G., Sartor, M., & Nassimbeni, G. (2020). The future of manufacturing: A Delphi-based scenario analysis on Industry 4.0. *Technological Forecasting and Social Change, 157*, Article 120092. https://doi.org/10.1016/j.techfore.2020.120092

Demelius, L., Kern, R., & Trügler, A. (2023). Recent advances of differential privacy in centralized deep learning: A systematic survey.

Denkena, B., et al. (2024). Zukunftslabor Produktion: Vernetzung, Modellierung und Optimierung in der Industriellen Produktion. *Zeitschrift für wirtschaftlichen Fabrikbetrieb, 5*, 372–377. https://doi.org/10.1515/zwf-2024-1061

Docker Inc. (2022). *Docker: Accelerated container application development* [Online]. Available: https://www.docker.com/. Accessed September 6, 2024.

Durrant, A., Markovic, M., Matthews, D., May, D., Enright, J., & Leontidis, G. (2022). The role of cross-silo federated learning in facilitating data sharing in the agri-food sector. *Computers and Electronics in Agriculture, 193*, Article 106648. https://doi.org/10.1016/j.compag.2021.106648

FATE Documentation. (2023). [Online]. Available: https://fate.readthedocs.io/en/develop/index.html. Accessed November 14, 2023.

Flower. (2023). Flower: A friendly federated learning framework [Online]. Available: https://flower.dev/. Accessed November 14, 2023.

GAIA-X Foundation. (2023). GAIA-X: Technical architecture [Online]. Available: https://www.data-infrastructure.eu/GAIAX/Redaktion/EN/Publications/gaia-x-technical-architecture.html. Accessed July 13, 2023.

He, C., et al. (2020). *FedML: A research library and benchmark for federated machine learning.*

Hermes Newsroom. (2023). 17. Hermes-Barometer: Kollaboration in der Supply Chain [Online]. Available: https://newsroom.hermesworld.com/17-hermes-barometer-kollaboration-in-der-supply-chain-grosses-potenzial-zahlreiche-hemmnisse-26381/. Accessed November 14, 2023.

Hunker, J., Scheidler, A. A., & Rabe, M. (2020). A systematic classification of database solutions for data mining to support tasks in supply chains. In *Preventing mental strain for logistics workers: Guideline development* (pp. 395–425). epubli GmbH [Online]. Available: https://www.econstor.eu/handle/10419/228928

Jaggi, H. S., & Kadam, S. S. (2016). Integration of spark framework in supply chain management. *Procedia Computer Science, 79*, 1013–1020. https://doi.org/10.1016/j.procs.2016.03.128

Joseph, V. R. (2022). Optimal ratio for data splitting. *Statistical Analysis and Data Mining: The ASA Data Science Journal, 15*(4), 531–538. https://doi.org/10.1002/sam.11583

Kallisch, J., & Wunck, C. (2023). Development and evaluation of a process management and analytics platform for small and medium-sized enterprises. *International Journal for Computers & Their Applications, 2023.*

Kallisch, J., Marx-Gómez, J., & Wunck, C. (2024). Challenges and opportunities of intercompany Big Data analytics in supply chains. In *Development methodologies for Big Data analytics systems* (pp. 249–262). https://doi.org/10.1007/978-3-031-40956-1_10

Kozlenkova, I. V., Hult, G. T. M., Lund, D. J., Mena, J. A., & Kekec, P. (2015a). The role of marketing channels in supply chain management. *Journal of Retailing, 91*(4), 586–609. https://doi.org/10.1016/j.jretai.2015.03.003

Kozlenkova, I. V., Hult, G. T. M., Lund, D. J., Mena, J. A., & Kekec, P. (2015b). The role of marketing channels in supply chain management. *Journal of Retailing, 91*(4), 586–609. https://doi.org/10.1016/j.jretai.2015.03.003

Li, J., Cui, T., Yang, K., Yuan, R., He, L., & Li, M. (2021). Demand forecasting of E-commerce enterprises based on horizontal federated learning from the perspective of sustainable development. *Sustainability, 13*(23), 13050. https://doi.org/10.3390/su132313050

Liu, X., et al. (2022). UniFed: A benchmark for federated learning frameworks [Online]. Available: https://arxiv.org/pdf/2207.10308

Marmolejo-Saucedo, J. A., Hurtado-Hernandez, M., & Suarez-Valdes, R. (2020). Digital twins in supply chain management: A brief literature review. In *Intelligent computing and optimization: Proceedings of the 2nd international conference on intelligent computing and optimization 2019*

(ICO 2019), 2020, pp. 653–661 [Online]. Available: https://doi.org/10.1007/978-3-030-33585-4_63

OpenMined Blog—PySyft. (2023). [Online]. Available: https://blog.openmined.org/tag/pysyft/. Accessed November 14, 2023.

Pachandrin, S., Hoffmann, N., Dilger, K., et al. (2025). Data-driven prediction of casting defects in magnesium high-pressure die casting using machine learning. *Inter Metalcast*. https://doi.org/10.1007/s40962-025-01592-w

Panahifar, F., Byrne, P. J., Salam, M. A., & Heavey, C. (2018). Supply chain collaboration and firm's performance. *Journal of Enterprise Information Management, 31*(3), 358–379. https://doi.org/10.1108/JEIM-08-2017-0114

Södergren, F., & Cartling Wallén, M. (2022). Creating value through information sharing: Exploring the transition towards a digital supply chain [Online]. Available: https://www.diva-portal.org/smash/record.jsf?pid=diva2%3A1654203&dswid=7113

Stachon, M., Möller, F., Guggenberger, T., Tomczyk, M., & Henning, J.-L. (2023). Understanding data trusts. In *ECIS 2023 research-in-progress papers* [Online]. Available: https://aisel.aisnet.org/ecis2023_rip/36

vom Brocke, J., et al. (2009). *Reconstructing the giant: On the importance of documenting the literature search process*. Universität di Verona, Facoltà di Economia, Dipartimento di Economia Aziendale [Online]. Available: https://www.alexandria.unisg.ch/213419

Yang, M., Guo, T., Zhu, T., Tjuawinata, I., Zhao, J., & Lam, K.-Y. (2024). Local differential privacy and its applications: A comprehensive survey. *Computer Standards & Interfaces, 89*, Article 103827. https://doi.org/10.1016/j.csi.2023.103827

Yang, Q., Liu, Y., Chen, T., & Tong, Y. (2019). Federated machine learning: Concept and applications. *ACM Transactions on Intelligent Systems and Technology (TIST), 10*(2), Article 12 [Online]. Available: https://arxiv.org/pdf/1902.04885

Zhang, J., et al. (2022). Adaptive vertical federated learning on unbalanced features. *IEEE Transactions on Parallel and Distributed Systems*. https://doi.org/10.1109/TPDS.2022.3178443

Zhu, T., Ye, D., Wang, W., Zhou, W., & Yu, P. S. (2022). More than privacy: Applying differential privacy in key areas of artificial intelligence. *IEEE Transactions on Knowledge and Data Engineering, 34*(6), 2824–2843. https://doi.org/10.1109/TKDE.2020.3014246

Wearables in the Workforce: The Perceived Value Experienced by South African Computing Professionals

Carmen Bekker, Brenda Scholtz, and Simone Beets

Abstract Smartwatches are valuable Internet of Things (IoT) wearable devices that can assist in promoting physical activity (PA) and in reducing sedentary behaviour. A lack of PA can result in numerous health disorders and diseases, especially in the case of sedentary office workers. Computing professionals experience long periods of sedentary behaviour, which poses serious health risks. Furthermore, South Africans spend a significant portion of their day being sedentary. This paper aims to investigate the impact of PA levels on smartwatch value within the context of South African computing professionals. To achieve this research aim, hypotheses were developed, and a survey was conducted. The results were statistically analysed, and the analysis revealed that most of the smartwatch users had a medium PA level despite the sedentary nature of South African computing professionals. It was also revealed that the PA levels of the smartwatch users had an impact on two of the attributes of value, namely Goal Pursuit Motivation (GPM) and social value.

1 Introduction

Sedentary behaviour or physical inactivity poses a major global health issue and is a significant mortality risk factor that can lead to severe musculoskeletal disorders and non-communicable diseases (NCDs), such as cardiovascular diseases (Lee et al., 2012; Makkonen et al., 2017; Warburton et al., 2010; World Health Organisation, 2023a). Studies have shown that this sedentary behaviour and a lack of sufficient physical activity (PA) accounts for approximately 27% of type II diabetes, 30% of ischemic heart disease and 21–25% of breast and colon cancers (Lee et al., 2012;

C. Bekker · B. Scholtz (✉) · S. Beets
Nelson Mandela University, Gqeberha, South Africa
e-mail: brenda.scholtz@mandela.ac.za

C. Bekker
e-mail: carmen.bekker@mandela.ac.za

S. Beets
e-mail: simone.beets@mandela.ac.za

J. Marx Gómez et al. (eds.), *Recent Advances in Information Systems*, Progress in IS,
https://doi.org/10.1007/978-3-032-14099-9_29

Warburton et al., 2010). Numerous countries spend a significant portion of their health budget on managing common NCDs, which is caused by individuals engaging in an inadequate level of PA (World Health Organisation, 2023b).

The study in Phaswana et al. (2023) revealed that South Africans spend approximately 3.9 h sitting during work hours and 9.9 h sitting per day. According to Clemes et al. (2014), office workers spend up to 71% of their working hours engaged in sedentary behaviour, which is likely to increase in the future. While several studies have investigated the sedentary behaviour and PA levels of a generalised office worker population, only a few focused on understanding these areas within the context of the IT industry or computing professionals (Hibbers, 2024; Jenkins et al., 2024; Landais et al., 2022; Makkonen et al., 2017). This industry is labour-intensive and consists of software development and digital collaborative activities sometimes with geographically dispersed teams (Hibbers, 2024; Makkonen et al., 2017; Souza Santos et al., 2024). There have been trends towards hybrid working structures after the COVID-19 pandemic where computing professionals (e.g. software developers, team leaders or project managers) work both in-office and remotely (Souza Santos et al., 2024). Computing professionals are therefore sedentary for extended periods while working in-office or from home (Makkonen et al., 2017).

Smartwatches are wearables that have become increasingly popular in assisting users to monitor and evaluate their everyday health. Some contributing factors towards the popularity of these devices are individuals' familiarity with smartwatches, the availability of smartwatches as consumer devices, PA and physiological data monitoring capabilities, social features and PA behaviour verification regarding, for example, movement patterns (Cho et al., 2019; Peng et al., 2022; Reeder & David, 2016; Takiddeen & Zualkernan, 2019). Smartwatches can be viewed as an extension of oneself that motivate inactive users to pursue healthier lifestyles through, for example, intelligent coaching and motivating its users to regularly engage in PA (Chuah, 2019). These devices monitor the time that users spend during sedentary periods and while engaging in PA, which could assist in changing their behaviours to work towards health and quality of life improvements. Another factor that can be used to progress from inactivity towards healthier lifestyles is the value or benefits gained from the adoption or use of smartwatches (Song et al., 2018). A user is more likely to keep using their smartwatch if they perceive the device as having value or providing value.

Smartwatch adoption has reached 21.7% of the adult population, which indicates how individuals are becoming more interested in these devices (Pangarkar, 2025). Most smartwatch users (82%) use smartwatches for fitness and health tracking purposes. Of the users engaging in health, fitness and PA, 80% of them indicated that their smartwatch usage consisted of tracking daily steps, 75% used their smartwatch for heart rate monitoring, and 70% of the users felt motivated to lead healthier lifestyles. Wearable Internet of Things (IoT) devices, also known as wearables, can assist in promoting PA and in reducing sedentary behaviour (Khakurel et al., 2018; Stephenson et al., 2017).

The purpose of this paper is to investigate the impact of PA levels on smartwatch value within the context of South African computing professionals. The paper is structured as follows: Section 2 provides the literature review. Section 3 presents the developed hypotheses to be tested. Section 4 explains the research methodology that was followed. Section 5 provides the results of the statistical analyses. Section 6 presents a discussion of the results and key findings. Section 7 concludes the paper.

2 Literature Review

2.1 Physical Activity (PA) Benefits, Recommendations and Measures

PA is essential for improving and maintaining the health and personal growth of individuals (World Health Organisation, 2023b). PA refers to any energy expending movements of the body that are generated by the skeletal muscles (Degroote et al., 2018). Regular engagement in PA provides numerous benefits such as reducing the risk of developing various diseases, promoting of mental well-being, and assisting with weight management (World Health Organisation, 2023b). According to the World Health Organisation (WHO), PA also increases networking opportunities and socialisation, which promotes cohesion and social interaction and positively influences the community and society.

The WHO has developed recommendations for the PA of various age groups to assist individuals in reaping the benefits of engaging in PA (World Health Organisation, 2023a). For adults between the ages of 18 and 64 years, sedentary behaviour should be limited and substituted with different forms of PA even if it is light-intensity PA. Therefore, the WHO recommends that this age group engage in the following per week (World Health Organisation, 2023c):

- a minimum of 150 min of moderate-intensity aerobic PA,
- a minimum of 75–150 min of vigorous-intensity aerobic PA, or
- an equivalent combination of moderate-intensity and vigorous-intensity PA.

For adults in this age group to gain further health benefits, the WHO suggests that individuals could engage in more than 150 min of vigorous-intensity aerobic PA per week and increase their moderate-intensity aerobic PA to 300 min per week. From a workplace perspective, an increase in PA levels in employees leads to numerous benefits, such as increased workplace productivity, a decrease in absenteeism and a decrease in worker turnover (World Health Organisation, 2023b). Some effective behaviour change techniques and workplace interventions to reduce sedentary behaviour in office workers include self-monitoring, reorganising physical and social spaces, mitigating health issues caused by sedentary behaviour and developing motivational strategies to reduce prolonged periods of sitting (Gardner et al., 2016; Shrestha et al., 2018). Some studies have investigated how workplace interventions and persuasive technologies, such as smartwatches, can be used to reduce the sedentary behaviour of office workers (Makkonen et al., 2017; Wang et al., 2018).

The general population often use smartwatches with fitness tracking capabilities to gain feedback about their PA (El-Amrawy et al., 2015). Individuals rely on the collected smartwatch data regarding PA movement patterns and step count calculations to be accurate. Step count, PA and caloric expenditure are estimated using wrist movement and displacement, which is subject to change while walking compared to running. Algorithms are then used to calculate these PA measures, which smartwatch users can use to gain insights into their PA (Asimakopoulos et al., 2017).

2.2 *Smartwatch Value*

It has been argued that smartwatch value should be measured according to continuous improvements in user well-being and the continuance usage intention (CUI) of the technology (Lin & Windasari, 2019). CUI refers to a situation where individuals identify a continued use for performing an action (Yang & Jong, 2021). Several studies (Bölen, 2020; Cho et al., 2019; Kamal Basha et al., 2022; Siepmann & Kowalczuk, 2021) have investigated the factors that impact CUI. However, only a few focus on value and PA level. There is often an overlap between the different factors that impact CUI in these studies, such as perceived usefulness (PU) and pleasure, even though different theories were used to determine this impact. A limitation of these studies is that health-related factors that impact the value that smartwatches provide to users are not commonly considered in these models. Goal Pursuit Motivation (GPM) was one such health factor proposed in Siepmann and Kowalczuk (2021), which refers to the motivational levels of smartwatch users to work towards achieving their set health and fitness goals. Social value was also rarely considered in the identified models except for the study in Kamal Basha et al. (2022).

3 Hypotheses Development

The research question for this paper is: *What impact does PA level have on smartwatch value within the context of South African computing professionals?* To answer this question, four hypotheses (H_1 to H_4) were developed using the Unified Theory of Acceptance and Use of Technology 2 (UTAUT2) model as used by Kranthi and Ahmed (2018). This model assisted in understanding the determinants of smartwatch adoption amongst IT or computing professionals. These determinants were related concepts to those of this paper: PU, pleasure, GPM and social value (Sect. 2.2). However, the study in Kranthi and Ahmed (2018) did not investigate the impact of PA level on these four factors. The four hypotheses of this paper are motivated in this section and assisted in understanding the determinants of smartwatch adoption amongst IT or computing professionals. These determinants were related concepts to those of this paper: PU, pleasure, GPM and social value (Sect. 2.2). However, the study in Kranthi and Ahmed (2018) did not investigate the impact of PA level on these four factors, which is the basis for the four hypotheses of this paper. These hypotheses are motivated in this section.

H_1. PA level has a positive impact on PU

PU refers to the perceived benefits that users receive from using technology, such as a smartwatch, and consists of the expectations that users have towards a smartwatch (Bölen, 2020). Performance expectancy, a similar concept to PU, as identified in Kranthi and Ahmed (2018) and Abouzeid (2025), refers to how individuals will adopt smartwatches if there are benefits from performing an activity or if they believe that smartwatches will assist in achieving a desired outcome (Kranthi & Ahmed, 2018). Performance expectancy has a significant impact on the smartwatch adoption behaviour intentions of computing professionals and their intention to use a smartwatch for health and fitness monitoring (Abouzeid, 2025; Beh et al., 2021; Kranthi & Ahmed, 2018). However, the opposite relationship, i.e. the impact of PA level on PU, was not identified.

H_2. PA level has a positive impact on pleasure

Pleasure refers to a type of positive affection that users experience from using a self-monitoring device (Cho et al., 2019; George & Bhila, 2019). In the context of smartwatches, users can derive pleasure from their collected PA data and from participation in competitions with family and friends. A similar concept of hedonic motivation was proposed in Kranthi and Ahmed (2018) and Abouzeid (2025), which refers to how computing professionals will adopt smartwatches if they experience some form of pleasure, fun or entertainment from using the device. Hedonic motivation or pleasure was a significant factor that impacted the intention to use smartwatches for health and fitness tracking in Beh et al. (2021). The results of Abouzeid (2025) highlighted that if users experience more pleasure from using smartwatches, they experience higher levels of CUI and smartwatch engagement. Therefore, since the pleasure that users experience from using their smartwatch influences their willingness to engage with the device, an opportunity arises to determine if PA level impacts these feelings of pleasure.

H_3. PA level has a positive impact on GPM

Within the context of smartwatches, usage increases when smartwatches provide users with more motivation to stay healthy (Abouzeid, 2025; Dehghani, 2018). GPM is closely related to the concept of inspiration, which is a state of intrinsic motivation that assists users to transition from accepting an external stimulus, like a smartwatch, to pursuing personal, self-set goals, such as PA goals (Chuah, 2019). Smartwatches can motivate users to set and meet PA goals and change their PA behaviour based on the physiological data collected by the devices (Lunney et al., 2016). Health motivation relates to GPM and suggests that the goal-orientated willingness of users to engage in and maintain healthy behaviours promote improvements in physical health and well-being (Abouzeid, 2025; Messersmith & Schulenberg, 2010).

The study in Kurudirek et al. (2021) revealed that smartwatch users had a higher fitness/energy release score than those who did not use a smartwatch. Additionally, there was a strong correlation between using a smartwatch and user motivation to participate in PA for fitness/energy release. Furthermore, Abouzeid (2025) has highlighted that individuals who are motivated to maintain and improve their physical health are more likely to use a smartwatch.

H_4. PA level has a positive impact on social value

Smartwatch social value is the utility gained from using a smartwatch through maintaining and improving social relationships (Kamal Basha et al., 2022). For example, smartwatch users can connect to friends and family or meet others through communities on smartwatch applications. A similar factor was proposed in Kranthi and Ahmed (2018) and Abouzeid (2025), namely social influence, which refers to the influence that family, friends, superiors, colleagues, etc. have on computing professionals regarding their adoption of smartwatches. Interestingly, the study in Beh et al. (2021) highlighted that social influence was not a significant predictor of the behavioural intention to use smartwatches for health and fitness. However, individuals who engage in more PA report more social interactions than those who engage in lower levels of PA (Wiese et al., 2025). Additionally, the study of Abouzeid (2025) highlighted that normative social influence has a positive impact on users' behavioural intention to use health and fitness wearables.

4 Methodology

A quantitative approach was employed, whereby the survey approach was used to test the four hypotheses (Saunders et al., 2020). The survey instrument used was an online questionnaire, namely the Smartwatch Questionnaire (available on request), which assisted in easily collecting and analysing responses from a large population (Jones et al., 2013). QuestionPro was used to create and distribute the questionnaire. Purposive sampling was used to sample the target population. This sampling method is a non-probability technique that can be used for qualitative and quantitative research (Babbie, 2017; Saunders et al., 2020). Furthermore, a literature review helped gather sufficient information about, firstly, the research topic and, secondly, the participants who were open to sharing insights into the phenomena being studied. A self-selection method, proposed in Saunders et al. (2020), was employed where the potential participants self-identified to the presented criteria and voluntarily offered their participation in the study. This method was chosen as individuals are more likely to engage in a study if they perceive a connection to the research and find it to be interesting based on the information in the call for participants.

The call for participants was distributed to gatekeepers of computing and software organisations and then shared with potential participants. Those who responded to the call were screened based on the following pre-determined criteria: (a) be between the ages 18 and 65 years, (b) be South African, (c) work in the IT or computing industry or department of an organisation, (d) be able to understand, speak and write in English and (e) be a smartwatch user. The participants who met all the criteria were sent the link for the Smartwatch Questionnaire. Those who completed the questionnaire were included in the study and had to indicate what their personal PA level was. The PA level options were derived from the daily step count classification system of Tudor-Locke and Bassett (2004), which consisted of five step count levels. This classification was aggregated in the current study to form three major PA level groups: low, medium or high. A low PA level indicates a sedentary lifestyle (Tudor-Locke & Bassett, 2004) and low engagement in PAs such as light walking, exercising once or twice per week or achieving 3000–5000 steps daily. A medium PA level indicates a physically inactive-to-moderately active lifestyle (Tudor-Locke & Bassett, 2004) and engagement in PAs such as brisk walking and running, exercising three to four times per week or achieving 6000–9000 steps daily. A high PA level indicates a physically active-to-very active lifestyle (Tudor-Locke & Bassett, 2004) and equates to engagement in high intensity PAs such as CrossFit, exercising at least five times per week or achieving at least 10,000 steps daily.

The participants also had to rate the statements in the questionnaire on a five-point Likert scale based on their experiences with smartwatches. These statements were derived from the studies of Cho et al. (2019); Bölen (2020); Kamal Basha et al. (2022); and Siepmann & Kowalczuk (2021), which underwent rigorous validity and reliability testing. There were 12 statements in total, i.e. three sub-statements for each of the four value factors: PU, pleasure, GPM and social value. The responses were statistically analysed using analysis of variance (ANOVA) tests and post-hoc ANOVA tests, which are commonly used for hypotheses testing. Cohen's d was used to determine the practical significance of the hypotheses as it is a standardised effect size for calculating the difference in mean values (Gravetter & Wallnau, 2015).

5 Results

Of the 126 submitted questionnaires, 19 were removed as these participants did not meet the criteria, leaving 107 that could be analysed. The participant profile is presented in Table 1, which consisted of the age groups, gender and PA levels of the participants. The majority of the participants were male, of medium PA level, and in the age group of 25–34 years old.

Table 1 Participant profile—South African computing professionals ($n = 107$)

Age (in years)	n	Gender	n	PA level	n
18–24	12	Female	34	Low	38
25–34	59	Male	72	Medium	45
35–44	27	Other	1	High	24
45–54	3				
55–64	6				

Table 2 Frequency distribution—PA level ($n = 107$)

PA level	n	Percentage	95% CI Lo	95% CI Hi
Low	38	36	26.5	44.6
Medium	45	**42**	32.7	51.4
High	24	22	14.5	30.3

5.1 Smartwatch User PA Levels

Table 2 presents the frequency distribution for each PA level. Nearly half of the participants (42%) indicated that they had a medium PA level, while 36% had a low PA level. Only 22% of participants indicated that they had a high PA level. For the low PA level group, the CI bounds were between 26.5 and 44.6% with moderate uncertainty (18.1%). For the medium PA level group, the CI bounds were between 32.7 and 41.4% with moderate uncertainty (18.7%). Lastly, for the high PA level group, the CI bounds were between 14.5 and 30.3%, with a 15.8% uncertainty in this estimation.

5.2 H_1. PA Level Has a Positive Impact on PU

For the descriptive statistics results for PA level related to PU, the participants with a high PA level experienced the most PU ($\bar{x} = 4.14$). This result was similar to participants with a medium PA level ($\bar{x} = 4.12$), who experienced more PU than the average population. The low PA level group experienced the least PU ($\bar{x} = 3.82$), which was also lower than the overall mean value of 4.02. The results of the ANOVA test indicated that there was no statistically significant difference between the different PA levels and PU ($p = 0.088$). Therefore, no post-hoc ANOVA tests were necessary and hypothesis H_1 was rejected.

5.3 H_2. PA Level Has a Positive Impact on Pleasure

For the descriptive statistics results for PA level related to pleasure, the participants with a high PA level experienced the most pleasure ($\bar{x} = 3.97$). Those with a medium PA level also experienced high levels of pleasure ($\bar{x} = 3.93$), which was higher than the average population ($\bar{x} = 3.87$). Participants with a low PA level experienced the least pleasure ($\bar{x} = 3.75$). The results of the ANOVA test indicated that there was no statistically significant difference between the different PA levels related to pleasure ($p = 0.444$). Therefore, no post-hoc ANOVA tests were required and hypothesis H_2 was rejected.

Table 3 Descriptive statistics—GPM by PA level

Group	All	Low PA	Medium PA	High PA
n	107	38	45	24
Mean ($\bar{x}$)	3.78	3.35	**4.04**	3.97
SD	1.03	0.97	0.95	1.09
95% CI low	3.58	3.03	3.75	3.51
95% CI high	3.98	3.67	4.32	4.43

Table 4 Post-hoc tests for ANOVA—GPM by PA level

PA level 1	PA level 2	Diff. M1–M2	Scheffé p	Cohen's d	
Low	Medium	– 0.69	**0.009**	**0.71**	Medium
Low	High	– 0.62	0.059	0.61	Medium
Medium	High	0.06	0.968	0.06	Not sig.

5.4 *H_3. PA Level Has a Positive Impact on GPM*

Table 3 presents the descriptive statistics for PA level related to GPM. The participants with a medium PA level experienced the most GPM ($\bar{x}$ = 4.04) of all three levels. Those with a high PA level also experienced high GPM ($\bar{x}$ = 3.97), which was higher than the average population ($\bar{x}$ = 3.78). The low PA level group experienced the least GPM ($\bar{x}$ = 3.35).

The results of the ANOVA test, presented in Table 4, revealed that there was a statistically significant difference between the different PA levels and GPM (p = 0.005). The statistically significant differences are formatted in bold in the table. Therefore, H_3 was accepted. Post-hoc tests were therefore conducted to determine the statistically significant difference between the PA level groups. The findings revealed that there was a statistically significant difference between the low and medium PA level groups (p = 0.009). The participants with a medium PA level experienced more GPM than those with lower PA levels (M1–M2 = – 0.69). Interestingly, there was no statistically significant difference between the low and high PA level groups, but there was a practically significant difference between these two groups. Therefore, the post-hoc ANOVA results revealed that the difference between the low and medium levels as well as between the low and high PA levels have moderate real-world differences regarding the impact of PA level on GPM.

5.5 *H_4. PA Level Has a Positive Impact on Social Value*

Table 5 presents the descriptive statistics for PA level related to social value. The participants with a high PA level experienced the most social value ($\bar{x}$ = 3.47). Those

Table 5 Descriptive statistics—social value by PA level

Group	All	Low PA	Medium PA	High PA
n	107	38	45	24
Mean ($\bar{x}$)	3.00	2.67	3.03	**3.47**
SD	0.99	0.75	0.99	1.12
95% CI low	2.81	2.42	2.73	3.00
95% CI high	3.19	2.91	3.33	3.95

Table 6 Post-hoc tests for ANOVA—social value by PA level

PA level 1	PA level 2	Diff. M1–M2	Scheffé p	Cohen's d	
Low	Medium	− 0.36	0.227	0.41	Small
Low	High	− 0.81	**0.006**	**0.88**	Large
Medium	High	− 0.44	0.185	0.43	Small

with a medium PA level experienced moderate social value ($\bar{x} = 3.03$). The low PA level group experienced the least social value ($\bar{x} = 2.67$).

The results of the ANOVA test, presented in Table 6, indicated that there was a statistically significant difference between the different PA levels and social value ($p < 0.0005$). The statistically significant differences are formatted in bold in the table. Therefore, H_4 was accepted. The results of the post-hoc tests revealed that there is a statistically significant difference only between the low and high PA groups ($p = 0.006$), showing that participants with a high PA level experienced more social value than those with lower PA levels (M1–M2 = − 0.81). There was a large effect size between these two PA level groups, which indicated that there were meaningful real-world differences in the impact of PA levels on social value. Interestingly, there was no statistically significant difference between the low and medium or between the medium and high PA level groups, but both groups did indicate that there were practical significant differences due to a small effect size.

6 Discussion

The survey findings revealed that most of the participants had a medium PA level, which indicated that they engaged in PAs such as brisk walking and running, exercised three to four times per week or achieved 6000–9000 steps per day. This finding was interesting as it was expected that these smartwatch users would mostly have a low PA level due to different studies (Clemes et al., 2014; Makkonen et al., 2017; Phaswana et al., 2023) that reported on the sedentary nature of computing professions and of South Africans in general. However, since the participants were already smartwatch users, their views may have had some bias as it could indicate a possible interest in being more active.

The results indicated that PA level had no impact on PU (H_1) or pleasure (H_2), which are contrary to those in the study of Beh et al. (2021), where it was reported that the PA level of individuals had an impact on their behavioural intention to use smartwatches for health and fitness. A possible reason for why PA level does not impact PU could be that the functionality of the smartwatches remains fairly unchanged regardless of how fit the user is. However, future studies could investigate other possible reasons. Smartwatch manufacturers could possibly implement strategies to engage with users regarding innovative PA features (Abouzeid, 2025).

The results of Beh et al. (2021) showed that pleasure is a significant factor that impacts the behavioural intention of individuals to use smartwatches. Another study of dance-based group exercises (Kim et al., 2021) reported high levels of pleasure in smartwatch users with a higher fitness level. However, in the current study, PA level did not impact pleasure. One possible reason for the difference could be due to a difference in population, behaviour and sample sizes. Furthermore, smartwatch users may potentially view PA as a goal to complete rather than something that is pleasurable. Therefore, smartwatch manufacturers could potentially implement gamification features and strategies to keep users engaged with their PA data to promote pleasurable experiences (Abouzeid, 2025).

The null hypotheses for the last two hypotheses were rejected since the PA level of participants had an impact on the participants' GPM (H_3) and social value (H_4). Regarding H_3, the participants with a medium PA level experienced the most GPM and were therefore motivated to pursue and achieve their goals relating to being more active. This result could indicate that individuals are highly motivated by immediate outcomes and feedback that smartwatches provide, such as achieving a step count goal each day or engaging in PA three to four times a week (Jaggi, 2023). This behaviour could explain why there was a statistically and practically significant difference between the low and medium PA level groups. There was a moderate difference between these groups, which revealed that smartwatch users who have a medium PA level experience a higher level of motivation than users with a low PA level. These results could suggest that smartwatch users become more motivated to pursue their goals as their PA increases. Interestingly, there was no statistically significant difference between the low and high PA level groups despite having a medium effect size that indicated a moderate practical significance, which could be due to the relatively small population ($n = 107$) used in this paper.

Regarding H_4, the smartwatch users with a high PA level experienced the most social value. This result could indicate that more active users are more confident about their fitness and activity levels and are therefore more willing to share their PA information. The findings also revealed that there was a statistically significant difference between the low and high PA level groups as well as a large practical significance. Therefore, smartwatch users who have a high PA level experience a significantly larger amount of social value compared to users with a low PA level, which could suggest that PA enhances social engagement through group activities. Interestingly, there was no statistically significant difference between the low and medium as well as the medium and high PA level groups despite having a small practical significance in both cases. These results could be due to the relatively small population used in this paper.

7 Conclusions

This study contributes to the literature regarding wearable IoT technology by examining the impact of PA levels on the perceived value of smartwatches experienced by South African computing professionals. The findings highlighted that despite the sedentary nature of South Africans and computing professionals alike, most of the smartwatch users in this study were moderately active. The results confirmed that PA level impacts GPM and social value but not necessarily PU or pleasure, which indicates that user engagement with smartwatches depends on lifestyle behaviours. Smartwatch users may become more motivated to improve their PA behaviour if it aligns with their long-term interests, such as leading a healthy lifestyle and reducing the risk of developing NCDs, which could be supported by social engagement. Understanding how users perceive value in smartwatches can be useful to healthcare practitioners, smartwatch users, smartwatch manufacturers and researchers. The study findings contribute to this understanding, and to the growing literature on the factors that impact smartwatch users' motivation to meet their PA goals. By improving motivation and PA levels, users in sedentary professions such as computing can improve their overall health and well-being. It is recommended that companies and policy makers provide health and workplace productivity strategies that encourage and support the use of wearable technology, such as smartwatches. These strategies could lead to a happier, healthier and productive workforce. For example, companies and policy makers can highlight the potential benefits of smartwatches to improve PA and well-being through self-tracking features and recommend engagement techniques with PA smartwatch data to users.

A limitation of the study was the sample size of 107 participants. Another limitation is the possible bias in the data due to the self-selection of participants who by inference may already have a possible interest in exercising. Future research should expand the study to a larger population and explore more factors related to the reasons for owning a smartwatch. Future research should also consider interventions to keep individuals with sedentary professions and lifestyles motivated to engage in PA by leveraging existing smartwatch features and providing customised recommendations. These recommendations could be incorporated into artificial intelligence applications or electronic coaches (eCoaches) that could assist smartwatch users with enhancing their PA experiences and gaining further value from these devices.

References

Abouzeid, D. Y. (2025). Investigating the factors affecting users' and non-users' behavioral intention to use smartwatches for health-and fitness-specific purposes. Doctoral Thesis, University of Minho. https://hdl.handle.net/1822/94814

Asimakopoulos, S., Asimakopoulos, G., & Spillers, F. (2017). Motivation and user engagement in fitness tracking: Heuristics for mobile healthcare wearables. *Informatics, 4*(1), 5, 1–16.

Babbie, E. R. (2017). *The basics of social research* (7th ed.). Cengage Learning.

Beh, P. K., Ganesan, Y., Iranmanesh, M., & Foroughi, B. (2021). Using smartwatches for fitness and health monitoring: The UTAUT2 combined with threat appraisal as moderators. *Behaviour & Information Technology, 40*(3), 282–299. https://doi.org/10.1080/0144929X.2019.1685597

Bölen, M. C. (2020). Exploring the determinants of users' continuance intention in smartwatches. *Technology in Society, 60*, 1–12.

Cho, W.-C., Lee, K. Y., & Yang, S.-B. (2019). What makes you feel attached to smartwatches? The stimulus–organism–response (S–O–R) perspectives. *Information Technology & People, 32*(2), 319–343.

Chuah, S.H.-W. (2019). You inspire me and make my life better: Investigating a multiple sequential mediation model of smartwatch continuance intention. *Telematics and Informatics, 43*, 1–15.

Clemes, S. A., O'Connell, S. E., & Edwardson, C. L. (2014). Office workers' objectively measured sedentary behavior and physical activity during and outside working hours. *Journal of Occupational and Environmental Medicine, 56*(3), 298–303. https://doi.org/10.1097/JOM.0000000000000101

De Souza Santos, R., Grillo, W. D. N., Cabral, D., De Castro, C., Albuquerque, N., & França, C. (2024). Post-pandemic hybrid work in software companies: Findings from an industrial case study. In *Proceedings of the 2024 IEEE/ACM 17th international conference on cooperative and human aspects of software engineering (CHASE)*, pp. 68–78. https://doi.org/10.1145/3641822.364187

Degroote, L., De Bourdeaudhuij, I., Verloigne, M., Poppe, L., & Crombez, G. (2018). The accuracy of smart devices for measuring physical activity in daily life: Validation study. *JMIR Mhealth Uhealth, 6*(12). https://doi.org/10.2196/10972

Dehghani, M. (2018). Exploring the motivational factors on continuous usage intention of smartwatches among actual users. *Behaviour & Information Technology, 37*(2), 145–158.

El-Amrawy, F., Pharm, B., & Nounou, M. I. (2015). Are currently available wearable devices for activity tracking and heart rate monitoring accurate, precise, and medically beneficial? *Healthcare Informatics Research, 21*(4), 315–320.

Gardner, B., Smith, L., Lorencatto, F., Hamer, M., & Biddle, S. J. H. (2016). How to reduce sitting time? A review of behaviour change strategies used in sedentary behaviour reduction interventions among adults. *Health Psychology Review, 10*(1), 89–112.

George, J., & Bhila, T. (2019). The role of mobile health application on self-quantification. *Journal of Pure and Applied Sciences, 1*(1), 6–15.

Gravetter, F. J., & Wallnau, L. B. (2015). *Statistics for the behavioral sciences* (10th ed.). Wadsworth Cengage.

Hibbers, I. (2024). A model for measuring and predicting stress for software developers using vital signs and activities. Master's Thesis.

Jaggi, G. (2023). How will the metaverse change our behavior as it reshapes experiences? https://www.ey.com/en_lu/insights/digital/how-will-the-metaverse-change-our-behavior-as-it-reshapes-experiences#chapter2. Last accessed 2023/04/29.

Jenkins, K., Buchan, J., Rhodes, R. E., Hamilton, K. (2024). Workplace movement: a qualitative study on office workers' beliefs. *Australian Journal of Psychology, 76*(1). https://doi.org/10.1080/00049530.2024.2341695

Jones, T., Baxter, M., & Khanduja, V. (2013). A quick guide to survey research. *Annals of the Royal College of Surgeons of England, 95*(1), 5–7.

Kamal Basha, N., Aw, E. C.-X., & Chuah, S. H.-W. (2022). Are we so over smartwatches? Or can technology, fashion, and psychographic attributes sustain smartwatch usage? *Technology in Society, 69*, 101952

Khakurel, J., Melkas, H., & Porras, J. (2018). Tapping into the wearable device revolution in the work environment: A systematic review. *Information Technology & People, 31*(3), 791–818.

Kim, Y., Kim, J., & Woo, M. (2021). The pattern of affective responses to dance-based group exercise differs according to physical fitness, as measured by a Smartwatch. *Applied Science, 11*(23). https://doi.org/10.3390/app112311540

Kranthi, A., & Ahmed, K. A. (2018). Determinants of smartwatch adoption among IT professionals—An extended UTAUT2 model for smartwatch enterprise. *International Journal of Enterprise Network Management, 9*(3/4), 294–316. https://doi.org/10.1504/IJENM.2018.10015843

Kurudirek, M. I., Kurudirek, A., Burmaoğlu, G. E., & Bi̇ndesen, M. A. (2021). Comparison of motivation to participate in physical activity between individuals with and without smartwatch use. *Pakistan Journal of Medical & Health Sciences, 15*(10), 2991–2993. https://doi.org/10.53350/pjmhs2115102991

Landais, L. L., Jelsma, J. G., Dotinga, I. R., Timmermans, D. R., Verhagen, E. A., & Damman, O. C. (2022). Office workers' perspectives on physical activity and sedentary behaviour: A qualitative study. *BMC Public Health, 22*, 621. https://doi.org/10.1186/s12889-022-13024-z

Lee, I., Shiroma, E., Lobelo, F., Puska, P., Blair, S., & Katzmarzyk, P. (2012). Effect of physical inactivity on major non-communicable diseases worldwide: An analysis of burden of disease and life expectancy. *Lancet, 380*(9838), 219–229.

Lin, F.-R., & Windasari, N. A. (2019). Continued use of wearables for wellbeing with a cultural probe. *The Service Industries Journal, 39*(15–16), 1140–1166.

Lunney, A., Cunningham, N. R., & Eastin, M. (2016). Wearable fitness technology: A structural investigation into acceptance and perceived fitness outcomes. *Computer Human Behaviour, 65*, 114–120.

Makkonen, M., Silvennoinen, M., & Nousiainen, T. (2017). To sit or to stand, that is the question: Examining the effects of work posture change on the well-being at work of software professionals. *International Journal of Networking and Virtual Organisations, 17*(4), 371–391.

Messersmith, E. E., & Schulenberg, J. E. (2010). Goal attainment, goal striving, and well-being during the transition to adulthood: A ten-year U.S. national longitudinal study. *New Directions for Child and Adolescent Development, 2010*(130), 27–40.

Pangarkar, T. (2025). Smartwatch statistics 2025 by wearables, technology, devices. https://scoop.market.us/smartwatch-statistics/. Last accessed 2025/04/29.

Peng, C., Xi, N., Zhao, H., & Hamari, J. (2022). Acceptance of wearable technology: A meta-analysis. In *Proceedings of the 55th Hawaii international conference on system sciences*, pp. 5101–5110.

Phaswana, M., Gordon, N. F., & Gradidge, P. J.-L. (2023). Sedentary behavior, physical activity patterns, and cardiometabolic risk factors in South African office-based workers. *American Journal of Lifestyle Medicine*. https://doi.org/10.1177/1559827623121

Reeder, B., & David, A. (2016). Health at hand: A systematic review of smart watch uses for health. *Journal of Biomedical Informatics, 63*, 269–276.

Saunders, M. N., Lewis, P., Thornhill, A. (2020). *Research methods for business students* (8th ed.). Pearson.

Shrestha, N., Kukkonen-Harjula, K. T., Verbeek, J. H., Ijaz, S., Hermans, V., & Pedisic, Z. (2018). Workplace interventions for reducing sitting at work. *The Cochrane Database of Systematic Reviews, 2018*(6), CD010912. https://doi.org/10.1002/14651858.CD010912.pub4

Siepmann, C., & Kowalczuk, P. (2021). Understanding continued smartwatch usage: The role of emotional. *Electronic Markets, 31*, 795–809.

Song, G., Park, J.-H., & Lee, S. (2018). Exploring determinants of smartwatch diffusion using a value adoption model. *Journal of Technology Innovation Society, 21*(2), 684–714.

Stephenson, A., McDonough, S. M., Murphy, M. H., Nugent, C. D., & Mair, J. L. (2017). Using computer, mobile and wearable technology enhanced interventions to reduce sedentary behaviour: A systematic review and meta-analysis. *International Journal of Behavioral Nutrition and Physical Activity, 14*(1), 105. https://doi.org/10.1186/s12966-017-0561-4

Takiddeen, N., & Zualkernan, I. (2019). Smartwatches as IoT edge devices: A framework and survey. In *Fourth international conference on fog and mobile edge computing* (FMEC), pp. 216–222.

Tudor-Locke, C., & Bassett, D. R., Jr. (2004). How many steps/day are enough? Preliminary pedometer indices for public health. *Sports Medicine, 34*(1), 1–8.

Wang, Y., Wu, L., Lange, J.-P., Fadhil, A., & Reiterer, H. (2018). Persuasive technology in reducing prolonged sedentary behavior at work: A systematic review. *Smart Health, 7–8*, 19–30.

Warburton, D. E., Charlesworth, S., Ivey, A., Nettlefold, L., & Bredin, S. S. (2010). A systematic review of the evidence for Canada's physical activity guidelines for adults. *International Journal of Behavioral Nutrition and Physical Activity, 7*(1). https://doi.org/10.1186/1479-5868-7-39

Wiese, L. A. K., Luna, C., Cook, D. J., Minor, B. D., & Schmitter-Edgecombe, M. (2025). A multigenerational approach to introducing technologies for assessing and supporting cognitive health in a rural, underresourced setting. *Educational Gerontology*, 1–15. https://doi.org/10.1080/03601277.2025.2449985

World Health Organisation. (2023a). Physical activity. https://www.who.int/health-topics/physical-activity#tab=tab_1. Last accessed 2023/04/29.

World Health Organisation. (2023b). A healthy lifestyle—WHO recommendations. https://www.who.int/europe/news-room/fact-sheets/item/a-healthy-lifestyle---who-recommendations. Last accessed 2023/04/29.

World Health Organisation. (2023c). Physical activity. https://www.who.int/news-room/fact-sheets/detail/physical-activity. Last accessed 2023/04/29.

Yang, J., & Jong, D. (2021). Understanding continuance intention determinants to adopt online health care community: An empirical study of food safety. *International Journal of Environmental Research and Public Health, 18*(12), 6514.

AI-Generated Visualizations for Sustainable Urban Planning: Enhancing Decision-Making and Public Engagement in the MobiLe Project Context

Sreedhar Kokkarachedu and Jorge Marx Gómez

Abstract This concept paper explores the potential of AI-generated imagery to support sustainable urban planning. A particular focus is placed on its prospective integration into the MobiLe project, a collaboration between VLBA Oldenburg and the City of Norderstedt. Drawing on existing literature, the study examines generative AI models like Stable Diffusion and DALL·E 2, exploring how they could visualize future urban development scenarios. Conceptually, inputs from citizens and policymakers in the MobiLe web application's predefined planning measures could translate into structured prompts for AI image generation. While no models are directly implemented or tested, this paper evaluates current research to assess the potential role of AI-generated visualizations in participatory planning. The goal is to identify opportunities and challenges in using these technologies to enhance public engagement and support informed decision-making in urban development.

1 Introduction

Urban planning integrates diverse systems-like transportation, housing, and environmental factors-into coherent, forward-looking spatial strategies (Hall, 2002). Achieving sustainable urban development necessitates a fundamental shift in mobility practices, a goal widely acknowledged but often insufficiently implemented (Deutscher, 2018, 2021). As a political decision-making process, urban planning strives to balance economic, ecological, and social objectives within urban spatial development (Pahl-Weber and Schwartze, 2018). Crucially, integrated settlement and transport development approaches, which link urban and transport planning, are vital for steering urban growth toward sustainability goals (Beckmann et al., 2024; Holz-Rau and Jansen, 2024).

S. Kokkarachedu (✉) · J. Marx Gómez
Carl von Ossietzky Universität Oldenburg, Oldenburg, Germany
e-mail: sreedhar.kokkarachedu@uni-oldenburg.de

J. Marx Gómez
e-mail: jorge.marx.gomez@uol.de

J. Marx Gómez et al. (eds.), *Recent Advances in Information Systems*, Progress in IS,
https://doi.org/10.1007/978-3-032-14099-9_30

In response to these challenges, the MobiLe project (Mobilitätswende @ Lebensqualität) was developed for the City of Norderstedt, a collaboration involving scientists and policymakers. MobiLe is designed to provide a novel decision-making aid for transport-related urban development, comprehensively considering the systemic interactions between the transport system and other facets of qualitative, sustainable urban development. Furthermore, MobiLe offers a qualitative transport model and a web-based decision support tool, empowering local decision-makers to independently assess the systemic impacts of transport-related urban planning measures at an early stage. While the tool previously allowed users to select predefined planning measures via text-based inputs, the growing demand for more participatory and inclusive planning processes has underscored the need for more intuitive and accessible communication formats.

Citizens are eager to engage in urban planning and better grasp the implications of planning decisions, yet they frequently lack the technical expertise to interpret complex planning documents. To bridge this gap, this paper explores the conceptual integration of AI-generated visualizations into the MobiLe framework. By translating citizen-selected planning measures into structured prompts for AI image synthesis models such as Stable Diffusion and DALL·E 2, the aim is to create realistic and relatable visualizations of potential urban futures. These visual outputs are intended to enhance the accessibility of planning processes, strengthen political and public engagement, and ultimately support more informed decision-making in sustainable urban development initiatives.

2 Related Work

The visual dimension of urban planning is pivotal for effective public engagement and robust decision-making. Participatory visualization tools are widely recognized for their ability to translate complex urban designs into formats accessible to the general public. Al-Kodmany (2001) emphasized the critical role of visual tools in urban planning, particularly in fostering understanding among non-experts. Similarly, Batty (2018) advocated for the integration of new digital tools to support the data-driven evolution of cities, highlighting the growing need for technology in urban design.

Recent advancements in generative AI have opened new avenues for supporting urban planning processes. Phillips et al. (2024) evaluated the capabilities of DALL-E 2, Stable Diffusion, and Midjourney for generating urban design imagery. Their study, based on assessing 240 AI-generated images using a sensibleness and specificity average (SSA) metric, showed that Midjourney produced the most sensible and specific results, although biases in training data remained evident, particularly in the underrepresentation of unique urban features like sculptures and transit stops (Phillips et al., 2024).

Complementing this, Valença et al. (2024) proposed using generative AI visualizations to guide street design decision-making. Their findings suggest that while AI-generated visuals can make planning proposals more intuitive and engaging for stakeholders, these tools should complement, not replace, traditional urban analysis methodologies. The role of digital technology in enhancing public participation has been further explored by Wilson et al. (2017), who discussed how app development can facilitate citizen involvement in local planning processes. Their research supports the idea that digital tools can democratize urban planning by allowing citizens to contribute insights early in the design process.

At the technical foundation of these advancements, Ramesh et al. (2021) introduced DALL-E, demonstrating zero-shot text-to-image generation. Their work showed that AI models could produce coherent and creative images from textual descriptions without needing paired training data, fundamentally expanding the capabilities of AI in fields such as urban visualization. Beyond academic research, practical deployments also underline the potential of AI-generated visuals. A report from the MIT Sloan School of Management (2023) described how sustainable policies gained greater public support when accompanied by AI-generated visualizations, emphasizing the importance of making abstract planning concepts tangible. Similarly, the United Nations Development Program (2023) highlighted community projects where AI-driven visualizations helped residents co-create visions for urban futures.

Finally, Zhang et al. (2023) examined the integration of AI in adaptive landscapes, demonstrating how artificial intelligence can contribute to the design of dynamic, responsive urban spaces. Their work suggests that AI should be seen as an evolving partner in landscape and city-making, capable of adapting to changing environmental and social needs.

Collectively, these studies show that AI-generated visualizations offer promising tools to enhance public engagement, accelerate early-stage design processes, and support participatory decision-making in urban planning. Nonetheless, careful evaluation and responsible integration of these technologies are essential to address biases and ensure inclusive outcomes.

3 The MobiLe Project VLBA Oldenburg

The MobiLe project (Mobilitätswende @ Lebensqualität) serves as a platform for advancing integrated urban and transport planning, led by VLBA Oldenburg in partnership with the City of Norderstedt. Building upon the project's core emphasis on sustainable mobility, social equity, and livability, the current research explores how emerging AI technologies might further enhance the project's participatory capabilities.

Through the MobiLe web application (available at MobiLe Web Application), users can select from over 100 predefined urban development measures as shown in Fig. 1. These measures include:

Fig. 1 Web app design layout of selecting measures of control variables. *Source* MOSCHDESIGN

- Creation of new residential units to address housing needs and urban density.
- Priority for pedestrians over other means of transport, ensuring safe and walkable urban environments.
- Easily accessible, covered, and secure bicycle parking facilities to promote cycling as a sustainable mode of transport.
- Public transport stops within 300 m (bus) or 500 m (train), ensuring convenient access to public transportation.
- Greenery and water features that characterize the district and are easily accessible on foot, fostering environmental sustainability and enhancing public spaces.

These selected measures serve as prompts for AI-driven visualizations, which will generate realistic representations of how these urban interventions could transform the cityscape. The goal is to help both politicians and citizens gain a clearer understanding of the systemic impacts these measures might have on the urban environment.

By using AI-driven visualizations, MobiLe aims to provide an intuitive way for the public to engage in urban planning. The visuals help to make complex concepts more accessible, motivating public participation and fostering a deeper understanding of how various urban planning measures can contribute to a sustainable, socially equitable, and economically viable future. In this way, MobiLe creates an interactive and inclusive planning ecosystem that encourages informed decision-making and promotes collective urban visioning.

4 Methodology (Conceptual Framework)

The proposed methodology revolves around seamlessly translating user-selected urban planning measures from the MobiLe web platform into AI-compatible prompts for generating visualizations. Upon selecting specific urban measures, such as the "creation of new residential units with improved cycling lanes," the system processes this data to generate corresponding visualizations through an AI-driven workflow.

Once a user selects a particular urban planning measure, such as the creation of new residential units integrated with enhanced cycling infrastructure, the system automatically converts this selection into a descriptive text prompt. For example, the measure could translate into the prompt: "A mid-rise eco-friendly residential neighborhood with wide cycling lanes, pedestrian zones, and solar rooftops, in a green urban setting." This text is then sent to the backend through an API call, where it serves as input for the AI model.

The backend system, after receiving the user input, forwards the prompt to a pre-configured AI model, such as DALL·E 2 or Stable Diffusion. These models utilize deep learning algorithms to generate multiple visualizations based on the input description. The AI-driven system provides diverse renderings, capturing variations in design that align with user expectations while considering real-world constraints such as urban design standards, sustainability goals, and spatial feasibility.

Once the AI model processes the input and generates the corresponding images, the visualizations are transmitted back to the frontend of the MobiLe web application. The images are displayed seamlessly on the right-hand side of the user interface, allowing users to instantly view the generated urban scenarios. The integration is designed to ensure that this process occurs in real-time without delays, providing an intuitive and smooth user experience.

This methodology provides a robust, user-centric framework that empowers both local politicians and the public to engage in the urban planning process. Through seamless interaction with AI-generated visualizations, MobiLe fosters an inclusive environment where diverse stakeholders can collaboratively assess the impacts of proposed urban measures, ultimately aiding in the development of more informed and sustainable urban spaces.

5 Expected Results and Illustrative Scenarios

AI-generated visualizations are expected to depict urban planning outcomes such as streets redesigned with increased green space, improved mobility infrastructure, or urban blocks featuring rooftop gardens. These visualizations are not merely artistic renderings; they serve as communicative tools to illustrate the potential impact of planning decisions.

A key requirement is visual consistency: when the same urban measure is selected repeatedly-either by the same or different users-the generated images should remain nearly identical, with only minimal variations (e.g., slight changes in tree position or sidewalk details). For example, selecting "green space along a road" should consistently depict similar outcomes unless an explicit variation (such as adding bicycle lanes) is requested by the user. This ensures comparability across user inputs and sessions.

By providing stable visual outputs instead of abstract data, the platform supports intuitive public engagement in urban planning. The consistency of AI-generated images enhances transparency, fosters critical thinking, and encourages broader participation in political and planning processes.

The following visualizations as shown in Fig. 2, sourced from Zukunft–Nachhaltige Mobilität (2025), serve as illustrative examples of the visual quality and communicative clarity that the MobiLe platform aims to achieve through AI-generated imagery. The objective is to generate similar outputs automatically by translating user-selected urban measures into consistent and meaningful visual scenarios using generative AI tools.

6 Discussion and Evaluation

The key benefit of this concept lies in its ability to translate abstract urban planning measures into tangible visual narratives, making planning processes more democratic, improving early-stage consensus-building, and encouraging scenario-based exploration of sustainability pathways. These visualizations serve as intuitive tools that engage the public, fostering more inclusive participation in urban decision-making processes (Green, 2022).

However, several risks must be considered. One major concern is that AI-generated images, while visually compelling, may present idealized or culturally biased depictions of urban environments. The lack of geographic specificity could

(a) Active mobility visualization in a rural/suburban context

(b) Active mobility visualization in an urban context

Fig. 2 Visualizations of active mobility scenarios (Zukunftsbilder zum Downloaden, 2025)

result in visuals that are less relevant or accurate in some contexts, potentially misleading stakeholders about the true feasibility of proposed plans. Such issues underscore the need for caution in using AI-generated imagery as a definitive representation of urban scenarios (Cantrell and Holzman, 2016). Therefore, these visualizations must be framed as discussion tools, accompanied by clear disclaimers that acknowledge their hypothetical nature and the limitations of their context.

Additionally, while AI models can generate a wide variety of images based on user inputs, the process of transforming those images into actionable planning insights requires careful evaluation and refinement. These visuals should not be seen as final outputs but rather as a starting point for dialogue, with stakeholders providing feedback on the clarity, plausibility, and alignment of the generated images with real-world constraints. Combining AI-generated imagery with map-based interfaces or simulation models could improve contextual fidelity and enhance the reliability of the visualizations in reflecting local conditions (de Lange and de Waal, 2019; Ramesh et al., 2021).

7 Conclusion

This paper presents a conceptual framework for integrating AI-generated visualizations into urban planning processes, particularly within the MobiLe project. By translating user-selected urban measures into structured prompts for generative AI models, the aim is to enhance participatory planning and decision-making, promoting sustainability, social equity, and mobility. The proposed methodology highlights the potential for AI tools, such as Stable Diffusion and DALL·E 2, to provide more intuitive, accessible, and engaging visualizations that help both policymakers and citizens visualize the impact of their decisions.

As of now, the integration of AI-generated visualizations into the MobiLe framework remains in the conceptual design and prototyping stage. The backend architecture for translating user-selected urban planning measures into structured AI prompts is under development, and initial experiments with generative models such as Stable Diffusion and DALL·E 2 are being conducted using sample prompts.

However, it is crucial to recognize that these AI-generated images are conceptual and hypothetical. The use of AI in this context should be seen as a tool for facilitating discussion and supporting informed decision-making, rather than providing definitive or final representations of urban environments. As the next steps involve further research and testing, it will be important to evaluate the effectiveness of these visualizations in real-world urban planning scenarios and ensure their alignment with local contexts and needs.

Declarations

This publication is based on research in the project MobiLe, which is funded by the Federal ministry of Education and Research as part of its Social-Ecological Research funding priority, funding no. 01UV2129A. Responsibility for the content of this publication lies with the author.

GEFÖRDERT VOM

Bundesministerium
für Bildung
und Forschung

FONA
Forschung für Nachhaltigkeit

References

Al-Kodmany, K. (2001). The role of visualization in participatory urban planning. *Journal of Urban Design, 6*(3), 247–261.

Batty, M. (2018). *The new science of cities*. MIT Press.

Beckmann, K., Blumthaler, W., Bojarra-Becker, E., Gies, J., & Leitfaden, K. (2024). *Lebenswerte Städte und Regionen*.

Cantrell, B., & Holzman, J. (2016). *Responsive landscapes: Strategies for responsive technologies in landscape architecture*. MIT Press.

de Lange, M., & de Waal, M. (2019). *The hackable city: Digital media and collaborative city-making in the network society*. Springer.

Deutscher StädteTag. (2018). *Nachhaltige städtische Mobilität für alle. Agenda für eine Verkehrswende aus kommunaler Sicht. Positionspapier des Deutschen StädteTags*.

Deutscher StädteTag. (2021). *Monitoring zum Positionspapier Nachhaltige städtische Mobilität für alle. Agenda für eine Verkehrswende aus kommunaler Sicht von 2018*.

Green, B. (2022). *The smart enough city: Putting technology in its place to reclaim our urban future*. MIT Press.

Hall, P. (2002). *Urban and Regional Planning* (4th ed.). Routledge.

Holz-Rau, C., & Jansen, U. (n.d.) *Nachhaltige Raum- und Verkehrsplanung* [Online]. Available: https://www.bpb.de/shop/zeitschriften/apuz/30354/nachhaltige-raum-und-verkehrsplanung/. Accessed May 24, 2024.

MIT Sloan School of Management. (2023). *Sustainable policies get a boost from AI-generated visuals* [Online]. Available: https://mitsloan.mit.edu/ideas-made-to-matter/sustainable-policies-get-a-boost-ai-generated-visuals

Pahl-Weber, E., & Schwartze, F. (2018). *Stadtplanung. Handwörterbuch Der Stadt-und Raumentwicklung* (pp. 2509–2520).

Phillips, C., Jiao, J., & Clubb, E. (2024). Testing the capability of AI art tools for urban design. *IEEE Access* [Online]. Available: https://ieeexplore.ieee.org/abstract/document/10409272

Ramesh, A., et al. (2021). Zero-shot text-to-image generation. In: *Proceedings of the 38th International Conference on Machine Learning* (Vol. 139, pp. 88218831) [Online]. PMLR. Available: https://proceedings.mlr.press/v139/ramesh21a.html

Ramesh, A., Pavlov, M., Goh, G., Gray, S., Voss, C., Radford, A., Chen, M., & Sutskever, I. (2021). Zero-shot text-to-image generation. *Proceedings of the 38th International Conference on Machine Learning (ICML)* (Vol. 139, pp. 8821–8831). PMLR.

United Nations Development Programme (UNDP). (2023). *Bringing communities together through AI-driven urban planning* [Online]. Available: https://www.undp.org/acceleratorlabs/blog/bringing-communities-together-through-ai-driven-urban-planning

Valença, G., Azevedo, C., Moura, F., & Morais de Sá, A. (2024). Creating visualizations using generative AI to guide decision-making in street designs: A viewpoint. *Discover Cities,4* [Online]. Available: https://www.sciencedirect.com/science/article/pii/S2667091725000068

Wilson, A., Tewdwr-Jones, M., & Comber, R. (2017). Urban planning, public participation and digital technology: App development as a method of generating citizen involvement in local planning processes. *Environment and Planning B: Urban Analytics and City Science,45*(5), 921–936 [Online]. https://doi.org/10.1177/2399808317712515

Zhang, Z., Epstein, S. L., Breen, C., Xia, S., Zhu, Z., & Volkmann, C. (2023). *Robots in the garden: Artificial intelligence and adaptive landscapes.*

Zukunftsbilder zum Downloaden. *Zukunft der nachhaltigen Mobilität.* Available at: https://www.zukunft-nachhaltige-mobilitaet.de/zukunftsbilder-zum-downloaden/. Accessed June 26, 2025.

Automated Screening of Chronic Sinusitis from Voice Recordings Using Machine Learning

Thabo Maibi and Leon Eldon Burger

Abstract Chronic sinusitis is a common illness affecting millions of individuals worldwide. Current sinusitis screening methods include symptoms evaluation, endoscopic examinations, and medical imaging. However, symptom-based diagnosis is often inaccurate, endoscopic procedures are invasive and limited by anatomical variability, and medical imaging is costly and exposes patients to ionizing radiation. Given that chronic sinusitis can alter voice quality and resonance, this proof-of-concept study explores the feasibility of using machine learning for automated screening of chronic sinusitis from voice recordings. To evaluate this novel approach, voice recordings were collected from patients who underwent tonsillectomy, septoplasty, functional endoscopic sinus surgery, and minor surgeries unrelated to the nasal cavity or vocal tract. A pre-emphasis filter was applied to enhance higher frequency components and reduce noise, and short-term energy analysis was employed to eliminate unvoiced speech segments. A range of audio features including Mel-frequency cepstral coefficients, linear predictive coding coefficients, Mel-spectrogram, spectral contrast, spectral centroid, spectral flatness, spectral bandwidth, and spectral roll-off were extracted and subsequently reduced in dimensionality using principal component analysis (PCA). These features were used to train various machine learning models, including logistic regression, k-nearest neighbors, decision tree, extreme gradient boosting, random forest, support vector machine, and deep neural network (DNN). Among these models, the DNN achieved the highest accuracy of 63%, a performance level comparable to that reported in voice-based diagnostic studies for other conditions. The findings from this proof-of-concept investigation suggest that voice recordings hold promise as a non-invasive, low-cost tool for the automated screening of chronic sinusitis.

T. Maibi · L. E. Burger (✉)
Stellenbosch University, Stellenbosch, South Africa
e-mail: eldonburger@sun.ac.za

J. Marx Gómez et al. (eds.), *Recent Advances in Information Systems*, Progress in IS,
https://doi.org/10.1007/978-3-032-14099-9_31

1 Introduction

Sinuses are air-filled cavities located in various parts of the human skull. The four pairs of sinuses include, the frontal sinuses, located in the frontal bone behind the forehead; the ethmoid sinuses, situated between the eyes within the ethmoid bone; the maxillary sinuses, positioned behind the cheekbones on either side of the nose; and the sphenoid sinuses, located deep within the skull behind the nasal cavity (James Zinreich, 1998).

The sinuses serve numerous functions, including improving the resonance of speech, moistening and warming inhaled air, lightening the weight of the head, and providing a protective shield for important facial structures during facial injuries (Fahrioglu et al., 2025). Furthermore, sinuses secrete mucus into the superior or lateral aspect of the nose. Mucus contains antibodies and enzymes that can kill or neutralize various pathogens like viruses, bacteria, and fungi, thus preventing potential infections (Donald, 1976). Together with hair-like structures called cilia, mucus traps and filters out pollutants and allergens from the air we breathe and propels them to the nasal cavity for elimination (Deborah & Prathibha, 2014).

Nonetheless, pathogens can still invade the respiratory tract either through the mouth or the nose (Dasaraju et al., 1996). The epithelial cells lining the sinuses can get infected by these pathogens, resulting in inflammation and swelling of the sinus mucosa. This inflammation disrupts normal sinus outflow, leading to mucus accumulation and creating a favorable environment for secondary infections. Consequently, sinusitis, defined as the inflammation or infection of the sinuses and nasal passages, may develop (Slavin, 1987). If the sinusitis persists for at least eight weeks despite appropriate treatment, it is classified as chronic sinusitis (Melén, 1994), alternatively called chronic rhinosinusitis.

Chronic sinusitis is one of the most common illnesses in the world, annually affecting more than 7.3 million people in the United States (Battisti et al., 2025), over 107 million people in mainland China (Shi et al., 2015), and a substantial portion of the population in developing countries (Schlemmer & Naidoo, 2013). The severity of chronic sinusitis can negatively impact the mental state, physical fitness, social relations, and everyday activities of individuals, thus reducing their quality of life (Teul et al., 2007).

Progressive chronic sinusitis in the nasal cavities may impair the vocal cords and voice production mechanisms, leading to diminished voice quality (Wu et al., 2020). Additionally, inflammation of the sinus mucosa can result in the formation of nasal polyps, which obstruct airflow and further degrade voice perception (Arslan et al., 2016). Speech signals from individuals with chronic sinusitis often demonstrate lower frequencies and reduced intensity (Lee et al., 2005).

Currently, sinusitis screening relies on clinical symptoms in conjunction with health history, endoscopic examination and medical imaging examination (Husain et al., 2018). Diagnosis begins with a review of the patient medical history and an assessment of symptoms such as facial pain, loss of smell, nasal congestion, or purulent nasal discharge (Krouse et al., 2010). In cases where chronic sinusitis is suspected, endoscopic examination is performed to directly visualize the nasal passages and sinuses (Maru & Gupta, 2016). For more detailed anatomical assessment, imaging techniques such as magnetic resonance imaging (MRI) and computed tomography (CT) are employed (Alrawashdeh & Alharazneh, 2020). CT scanning is considered the most accurate diagnostic method for chronic sinusitis (Gunbey et al., 2020).

Although current sinusitis diagnostic methods are effective, each has specific limitations and risks. Relying solely on clinical symptoms and medical history can result in incomplete information, increasing the risk of misdiagnosis and inappropriate treatment (Ryan, 2010). Endoscopic examination is invasive and may cause pain, discomfort, or bleeding (Atar & Kadayifci, 2014). Additionally, endoscopic examination tools have limitations in visualizing deeper regions of the nasal cavity and their diagnostic accuracy can be affected by anatomical variations and defects (Nangia et al., 2019; Nathan et al., 2021).

Although medical imaging, such as CT scanning, is considered the most accurate diagnostic measure, it is expensive and expose patients to ionizing radiation which can cause leukemia and solid organ tumors (Karatas & Toy, 2014; Royal, 2008). Since there are measurable differences between voices of individuals with chronic sinusitis and those without, machine learning models can potentially learn the differences and perform classification. However, the use of voice recordings for automated sinusitis screening remains underexplored.

This study aims to determine whether chronic sinusitis can be accurately screened using voice recordings and machine learning. The proposed approach involves collecting voice recordings along with their corresponding ground truth labels, extracting relevant voice features and training a machine learning model to classify voice recordings affected by chronic sinusitis. The developed model is then evaluated using various performance metrics and validated through cross-validation with unseen voice recordings to assess their accuracy and reliability.

The rest of this report is organized as follows, Sect. 2 discusses related work, Sect. 3 discusses the research methodology, Sect. 4 presents the findings of the study and provides a detailed discussion of the results. Finally, Sect. 5 concludes the report and offers possible directions for future research.

2 Related Work

To our knowledge, no studies have attempted to use voice recordings to automatically screen for chronic sinusitis. However, several studies have explored using voice records to diagnose other diseases, such as the coronavirus disease, chronic obstructive pulmonary disease, and other upper respiratory tract infections (Kapetanidis et al., 2024). Sinusitis and the common cold share similar symptoms, such as nasal congestion and inflammation of the nasal cavity (Jaume et al., 2020). Therefore, this section provides a review on studies that utilized voice to diagnose the common cold.

Most of the studies reviewed utilized the upper respiratory tract infection corpus (URTIC) dataset to develop machine learning models. The URTIC dataset contains 28,652 speech samples from 630 people with 111 having the common cold, while 519 were healthy (Schuller et al., 2017). Using the URTIC dataset, Warule et al. (2023c) employed short-time energy (STE) to remove unvoiced speech segments and extracted harmonic peak values, regular spikes at each pitch frequency of the speech signal, to train a support vector machine (SVM) and DNN for the common cold classification.

Warule et al. (2024) used a transformer-based model with a focal loss function, for handling data imbalance, to distinguish between cold-affected and healthy speech signals. The voice features used to train the model included Mel-frequency cepstral coefficients (MFCC), Mel-spectrogram, chromagram, spectral contrast, spectral centroid, spectral bandwidth, spectral flatness, and spectral roll-off.

Similarly, Deb et al. (2022) employed MFCC and linear predictive coding (LPC) features extracted from the voiced speech segments to train a DNN model and compared its performance with random forest, logistic regression, and gradients boost classifiers. To handle data imbalance in the URTIC dataset, Deb et al. (2022) employed SMOTE combined with Tomek links for resampling.

Alternatively, Warule et al. (2022) focused on utilizing MFCCs extracted from vowel-like region (VLR) segments of speech, rather than from complete active speech. The MFCC features extracted from both the VLR segments and complete speech were used to train a DNN model separately and they obtained equal accuracy values, suggesting that VLR-based features can be employed instead of complete speech features which increase time and computational complexities of the model.

Warule et al. (2023b) also demonstrated that MFCCs extracted from the voiced speech segments, unvoiced speech segments, and complete active speech all yield equal classification performance. Furthermore, Warule et al. (2023a) employed empirical mode decomposition (EMD) and a DNN model to classify the common cold. Voice signals from the URTIC dataset were decomposed into intrinsic mode function (IMF) using EMD and statistical features were extracted from each IMF. Thereafter, a t-test was employed to select relevant features which were ultimately used to train and test a DNN model.

Similarly, Deb et al. (2020) utilized variational mode decomposition (VMD) and an SVM model to analyze and classify the common cold-affected speech. In addition to the URTIC dataset, voice recordings from 20 individuals were used, and both datasets were pre-processed through normalization, voiced speech region extraction using STE, framing, and windowing. Then, VMD was applied to each windowed frame to extract statistical features which were selected using mutual information and used to train the SVM model.

Moreover, Egas-López and Gosztolya (2020) employed fisher vectors (FV) generated by a Gaussian mixture model trained on MFCCs to develop support vector machine (SVM) and XGBoost models for the common cold classification. The generated FVs were normalized through power normalization and reduced in dimensionality using PCA before being input into the SVM model, while the XGBoost model was trained using raw FVs.

Table 1 Comparison of machine learning approaches to diagnose common cold from voice recordings

Research work	Features	Model performance (UAR)
Warule et al. (2023c)	First harmonic peak, Normalized harmonic peak with respect to the maximum value of harmonic peak and successive harmonic peak ratio	SVM = 66.90% and DNN = 65.22%
Warule et al. (2024)	MFCC, Mel-spectrogram, chromatogram, spectral contrast, spectral bandwidth, spectral roll-off and spectral flatness	Transformer encoder model = 70.48%
Deb et al. (2022)	MFCC and LPC coefficients	DNN = 67.71%
Warule et al. (2022)	VLR-based MFFCs	DNN = 61.93%
Warule et al. (2023b)	MFCC	SVM = 64.92%
Warule et al. (2023a)	MFCC and EMD-based features	DNN = 66.92%
Deb et al. (2020)	VMD-based features	SVM: 90.02% on recorded dataset and 66.84% on URTIC dataset
Egas-López and Gosztolya (2020)	FV features	SVM = 67.81% and XGBoost = 70.43%
Babu et al. (2024)	Frequency domain features, spectral features, dominant frequency features, fundamental frequency features and modulation features	CNN= 99.32%, logistic regression = 93.45%,SVM = 91.90%, random forest = 92.68%, Naïve Bayes = 89.56%, KNN= 92.89%, decision tree = 94.45%

Babu et al. (2024) collected 2752 voice samples from individuals in Southern India, both with and without the common cold. Various audio features, such as mean frequency, modulation index, spectral entropy, spectral flatness, and others, were extracted. These features were input into a convolution neural network (CNN) and other machine learning models, including logistic regression, SVM, random forest, k-nearest neighbor (KNN), and decision trees, to classify the common cold.

Based on the research outlined in this section, various machine learning models and several voice features have shown considerable promise in voice classification tasks. Given the similarity in symptoms of chronic sinusitis and the common cold, the success of using voice recordings to screen for the common cold suggests that voice recordings could also be used to screen for chronic sinusitis. Table 1 presents a summary of the reviewed studies.

3 Methodology

The employed methodology for screening chronic sinusitis from voice recordings follows the cross-industry standard processes for data mining methodology extended for medical domain (CRISP-MED-DM) (Niaksu, 2015). Different phases of CRISP-MED-DM used in this study include data understanding, data preparation, modeling and evaluation.

3.1 *Data Description*

The data used in this research was sourced from Hernández-García et al. (2024) and includes voice recordings from 107 Spanish Castilian speakers (56 women, 51 men) collected over two years. It also contains CSV files with patient information such as weight, sex, height, diagnosis, nasality scores, CT scan results, and physician notes. Participants had undergone procedures including FESS, tonsillectomy, and septoplasty.

The control group included individuals over 18 years old with no sinusitis or nasal obstruction, who underwent minor head and neck surgeries unrelated to the nasal cavity or vocal tract. All participants received a clinical assessment, including an oral cavity check and endoscopic examination. Those with visible vocal fold pathology, abnormal voice quality, or a history of vocal tract surgery, oral cavity cancer, or voice disorders were excluded.

The voice recordings were conducted over three sessions: roughly two weeks before surgery, approximately two weeks after surgery, and three months post-surgery. In each session, the following types of audio content were captured: vowels /a/, /e/, /i/, /o/, and /u/; three utterances for the sustained vowel /a/; free monologues prompted by asking participants to describe the scenes in an image and four text-dependent utterances of different Spanish sentences.

3.2 *Data Preprocessing*

All audio files were downsampled from 44.1 kHz to 16 kHz to reduce computational load without affecting perceptual quality. Since human hearing peaks between 15 and 17 kHz and rarely exceeds 20 kHz (Purves et al., 2001), higher sampling rates offer little advantage but significantly increase processing requirements.

A pre-emphasis filter was applied to enhance voice clarity and suppress noise by amplifying high-frequency components and attenuating lower ones (López-Espejo et al., 2024). The audio was then divided into overlapping frames with a size of 1024 samples and a hop size of 512 sample. Each frame was multiplied by a Hamming window function to reduce spectral leakage (Kamil, 2018). STE was computed for each frame, and frames with the STE below 10% of the maximum value were discarded. The remaining voiced segments were then merged to form the processed audio file.

3.3 Feature Extraction

The speech features extracted include MFCC, LPC coefficients, Mel-spectrogram, spectral contrast, spectral centroid, spectral bandwidth, spectral flatness, and spectral roll-off, which are commonly used in voice-related disease classification studies (Warule et al., 2023a, 2023c, 2024).

MFCC: MFCCs represent the short-time power spectrum of an audio signal, computed by applying a discrete cosine transform (DCT) to the log power spectrum on a Mel scale (Sharma et al., 2020). To extract MFCCs, the audio is first converted to the frequency domain using a fast Fourier transform (FFT), and then passed through a Mel-scale filter bank. The log of the filter bank energies is taken to compress dynamic range, and DCT is applied to decorrelate the coefficients. The final MFCCS are defined as (Cabral et al., 2019).

$$\text{MFCC}_k = \sum_{n=1}^{N} \log(E_n) \cos\left[\frac{\pi k}{N}\left(n - \tfrac{1}{2}\right)\right] \tag{1}$$

where E_n represents the log-energy output of the n-th Mel filter bank, n is the index corresponding to the Mel filter banks, N is the total number of Mel filter banks, and k is the index of the cepstral coefficient.

LPC: LPCs are auto-regression based features which model the audio signal as a linear combination of its past samples (Badr & Karim, 2020). The auto-regressive correlation function is defined as (Safdar et al., 2023):

$$R(j) = \sum_{m=0}^{N-1+j} s(m)s(m+j), \quad j = 1, 2, 3, \ldots, p \tag{2}$$

where $R(j)$ is the auto-regressive correlation for lag j, $s(m)$ is the signal at time m, N is the length of the signal, and p represents the maximum lag.

Since the autocorrelation function is symmetric, the following formula can be used to get the LPC coefficients:

$$R(j) = \sum_{j=1}^{p} R(j-k)a_j \tag{3}$$

where a_j is the LPC coefficient that the autocorrelation function yields.

Mel-Spectrogram: A Mel-spectrogram visually represents an audio signal using time-frequency analysis on the Mel scale (Zhou et al., 2021). To create a Mel-spectrogram, the audio signal is divided into overlapping segments, and STFT is applied to each segment to analyze its frequency components over time. Thus, Mel-spectrogram allows analysis of the signal in both time and frequency domains.

Spectral Contrast: Spectral contrast quantifies the difference in amplitude between peaks and valleys in the sound spectrum, capturing the energy distribution across the audio spectrum (Jiang et al., 2002). Spectral contrast is computed by subtracting the spectral valleys from the peaks within each frequency sub-band. It is calculated as (Jiang et al., 2002).

First, the strength of spectral peaks is defined as:

$$P_k = \log\left(\frac{1}{N}\sum_{i=1}^{N} x_{k,i}\right) \tag{4}$$

Then, the spectral valley is defined as:

$$V_k = \log\left(\frac{1}{N}\sum_{i=1}^{N} x_{k,N-i+1}\right) \tag{5}$$

The difference is given by:

$$SC_k = P_k - V_k \tag{6}$$

where P_k is the spectral peak for the k-th sub-band, V_k is the spectral valley for the k-th sub-band, $x_{k,i}$ represents the spectral magnitude for the k-th sub-band at frequency i, N is the total number of frequencies in the k-th sub-band, and SC_k is the spectral contrast for the k-th sub-band.

Spectral Centroid: The spectral centroid is a measure that indicates where the center of mass of the signal spectrum is located (Aly et al., 2022). The energy distribution of the signal spectrum and the pitch of the sound are both provided by the spectral centroid, and is defined as (Sharma et al., 2020):

$$C = \frac{\sum_{k=0}^{N-1} f(k)\cdot X(k)}{\sum_{k=0}^{N-1} X(k)} \tag{7}$$

where $f(k)$ is the frequency at bin k, $X(k)$ is the magnitude of the Fourier transform at bin k, C is the spectral centroid, and N is the total number of bins.

Spectral Bandwidth: Spectral bandwidth measures the width of the signal spectrum. The spread of frequencies around the spectral centroid is represented by the spectral bandwidth, and defined as:

$$B = \sqrt{\frac{\sum_{k=0}^{N-1}(f(k) - C)^2 \cdot X(k)}{\sum_{k=0}^{N-1} X(k)}} \tag{8}$$

where $f(k)$ is the frequency at bin k, $X(k)$ is the magnitude of the Fourier transform at bin k, N is the total number of bins, B is the spectral bandwidth, and C is the spectral centroid.

Spectral Flatness: Spectral flatness measures how smooth or spiky the signal spectrum is. The spectral flatness is often calculated as the ratio of the geometric mean to the arithmetic mean of the power spectrum, and is defined as:

$$F = \frac{\left(\prod_{k=0}^{N-1} X(k)\right)^{\frac{1}{N}}}{\frac{1}{N}\sum_{k=0}^{N-1} X(k)} \tag{9}$$

where $X(k)$ represents the power spectrum value at the k-th frequency bin, F is the spectral flatness, and N is the total number of bins.

Spectral Roll-Off: Spectral roll-off is the frequency below which a set percentage (typically 85% or 95%) of the total spectral energy is concentrated (Kos et al., 2013). It estimates the upper limit of the spectral envelope and helps differentiate speech sounds based on their spectral characteristics. The spectral roll-off is defined as (Aly et al., 2022):

$$\sum_{k=b_1}^{i} S_k = K \sum_{k=b_1}^{b_2} S_k \tag{10}$$

where S_k is the spectral value at bin k, b_1 and b_2 are the band edges in bins over which to calculate the spectral spread.

Once all the features were extracted, PCA was employed to reduce their dimensionality.

3.4 Modeling

Voice recordings from patients who underwent FESS were used as training data. Only recordings from sessions 1 and 3 were considered, because session 1 (labeled as 1) contained recordings taken while patients were still experiencing chronic sinusitis, whereas session 3 (labeled as 0) included recordings taken after full recovery, as confirmed by diagnostic records. All voice recordings were matched using the patient ID and those from patients who missed session 3 were excluded to maintain balance. The data was then randomly split: 75% for training and 25% for testing.

The models used were logistic regression, decision tree, random forest, KNN, SVM, XGBoost, and DNN. The decision tree used entropy for splits with no depth limit. The random forest had 100 trees, each considering the square root of total features per split to reduce overfitting. The SVM used an RBF kernel, KNN used 5 neighbors, and XGBoost was set up with 100 trees optimized for binary logistic regression.

The DNN model was a feedforward neural network with one input layer, three hidden layers, and one output layer. The first hidden layer had 64 neurons with tanh activation, L2 regularization, and a 0.3 dropout rate. The second and third layers had 32 and 16 neurons with ReLU activation. The output layer had one neuron with sigmoid activation to produce a probability. The model was trained using the Adam optimizer ($\beta_1 = 0.9$, $\beta_2 = 0.999$) with a learning rate of 0.001.

3.5 *Evaluation*

The models were evaluated using accuracy, precision, and recall. Accuracy measures overall classification performance for both sinusitis and non-sinusitis cases. Precision indicates the proportion of correctly predicted positive cases, while recall reflects the ability of the model to detect all actual chronic sinusitis cases.

Session 1 recordings from tonsillectomy and septoplasty patients were used as validation sets to assess the specificity of the model; its ability to distinguish chronic sinusitis from other voice-affecting conditions. Recordings from minor surgery patients also served as a validation set to test the model accuracy in identifying individuals without chronic sinusitis or any voice-related disorders.

4 Results and Discussions

This section presents experimental results and method comparisons supporting the proposed approach. Each test was repeated three times with the same data and parameters to reduce random variation and ensure consistent, reliable evaluation. As shown in Table 2, comparing raw and pre-processed recordings shows that downsampling, noise reduction, and removal of unvoiced segments improved audio quality and model performance.

Table 3 shows that combining MFCCs with spectral features (Mel-spectrogram, spectral contrast, centroid, bandwidth, flatness, and roll-off) achieved the highest accuracy of 0.57. This likely results from both sets capturing frequency characteristics affected by chronic sinusitis, which alters voice resonance during nasalization (Koo et al., 2015).

This study evaluated PCA, Linear Discriminant Analysis (LDA), and Singular Value Decomposition (SVD) for dimensionality reduction. As shown in Table 4, PCA achieved the highest accuracy of 0.58, outperforming LDA, SVD, and the original features.

Table 2 Performance comparison of pre-processed audio and raw audio

Audio type	Metric	Test 1	Test 2	Test 3	Mean
Raw audio	Accuracy	0.50	0.50	0.51	0.50
	Precision	0.48	0.44	0.46	0.46
	Recall	0.62	0.58	0.60	0.60
Pre-processed audio	Accuracy	0.56	0.52	0.57	**0.55**
	Precision	0.54	0.50	0.52	**0.52**
	Recall	0.67	0.64	0.64	**0.65**

Table 3 Performance comparison of different feature sets

Feature set	Metric	Test 1	Test 2	Test 3	Mean
MFFC	Accuracy	0.57	0.52	0.50	0.53
	Precision	0.50	0.44	0.50	0.48
	Recall	0.62	0.59	0.60	0.60
LPC	Accuracy	0.51	0.50	0.50	0.50
	Precision	0.45	0.38	0.41	0.41
	Recall	0.44	0.38	0.41	0.41
Spectral	Accuracy	0.58	0.53	0.51	0.54
	Precision	0.56	0.50	0.49	0.52
	Recall	0.69	0.68	0.58	0.65
MFCC + LPC	Accuracy	0.50	0.50	0.62	0.54
	Precision	0.51	0.47	0.55	0.51
	Recall	0.62	0.65	0.60	0.62
MFCC + spectral	Accuracy	0.55	0.56	0.60	**0.57**
	Precision	0.58	0.47	0.60	**0.55**
	Recall	0.62	0.63	0.70	**0.65**
LPC + spectral	Accuracy	0.53	0.50	0.56	0.53
	Precision	0.53	0.43	0.53	0.50
	Recall	0.63	0.59	0.67	0.63
MFCC + LPC + spectral	Accuracy	0.53	0.51	0.49	0.51
	Precision	0.46	0.44	0.54	0.48
	Recall	0.63	0.57	0.60	0.60

Table 4 Comparison of different dimensionality reduction methods

Technique	Metric	Test 1	Test 2	Test 3	Mean
No dimensionality reduction	Accuracy	0.55	0.56	0.60	0.57
	Precision	0.54	0.49	0.47	0.54
	Recall	1.00	0.97	1.00	0.63
PCA	Accuracy	0.58	0.61	0.55	**0.58**
	Precision	0.54	0.57	0.54	**0.55**
	Recall	0.61	0.64	0.61	**0.62**
LDA	Accuracy	0.50	0.53	0.51	0.51
	Precision	0.42	0.45	0.45	0.41
	Recall	0.56	0.53	0.56	0.55
SVD	Accuracy	0.51	0.50	0.50	0.50
	Precision	0.49	0.45	0.47	0.47
	Recall	1.00	1.00	1.00	1.00

Table 5 Performance comparison of different audio content types

Audio content	Metric	Test 1	Test 2	Test 3	Mean
Vowels	Accuracy	0.54	0.59	0.53	0.55
	Precision	0.52	0.50	0.51	0.51
	Recall	0.43	0.41	0.42	0.42
Sentences	Accuracy	0.50	0.53	0.54	0.51
	Precision	0.55	0.53	0.54	0.54
	Recall	0.51	0.48	0.51	0.50
Vowels + sentences	Accuracy	0.62	0.60	0.61	**0.61**
	Precision	0.58	0.56	0.54	**0.57**
	Recall	0.60	0.58	0.59	**0.59**

The comparison results for different audio content types, as detailed in Table 5, show that using a combination of voice recordings containing vowels and full sentences yields better performance compared to using them separately. This is because sentences enable the model to learn various accents and speech rhythms, while vowels offer specific phonetic details, such as nasality, which is valuable for detecting nasal blockages in speech (Lee et al., 2005).

During model comparisons, the DNN model achieved the highest accuracy of 0.63, as shown in Table 6. This performance is comparable to the accuracy levels reported in voice-based common cold diagnosis studies, summarized in Table 1.

Table 6 Performance comparison for different machine learning models

Model	Metric	Test 1	Test 2	Test 3	Mean
Logistic regression	Accuracy	0.59	0.58	0.57	0.58
	Precision	0.60	0.58	0.59	0.59
	Recall	0.59	0.58	0.57	0.58
Decision tree	Accuracy	0.56	0.57	0.55	0.56
	Precision	0.58	0.56	0.57	0.57
	Recall	0.55	0.57	0.56	0.56
SVM	Accuracy	0.52	0.50	0.51	0.51
	Precision	0.53	0.51	0.52	0.52
	Recall	0.52	0.50	0.51	0.51
KNN	Accuracy	0.50	0.51	0.50	0.50
	Precision	0.41	0.39	0.40	0.40
	Recall	0.43	0.42	0.41	0.42
Random forest	Accuracy	0.57	0.55	0.56	0.56
	Precision	0.55	0.57	0.56	0.56
	Recall	0.56	0.55	0.57	0.56
XGBoost	Accuracy	0.56	0.54	0.55	0.55
	Precision	0.55	0.56	0.57	0.56
	Recall	0.54	0.56	0.55	0.55
DNN	Accuracy	0.62	0.64	0.63	**0.63**
	Precision	0.70	0.68	0.69	**0.69**
	Recall	0.65	0.67	0.66	**0.66**

Table 7 Model performance on different voice recordings

Surgical group	Accuracy	Precision	Recall
Minor surgeries	0.68	0.74	0.65
Tonsillectomy	0.64	0.75	0.64
Septoplasty	0.60	0.60	0.63

Table 7 shows the performance of the DNN model on voice recordings from patients without chronic sinusitis. It achieved accuracies of 0.68 and 0.64 for tonsillectomy and minor surgery patients, indicating effective classification of non-sinusitis cases. An accuracy of 0.60 for septoplasty patients suggests the model can distinguish between voice changes from septal deviation and those from chronic sinusitis, despite both causing nasal obstruction.

5 Conclusion

This proof-of-concept study demonstrates that chronic sinusitis can potentially be screened from voice using machine learning. Voice recordings before and after FESS were pre-processed, and MFCCs along with spectral features were extracted and reduced in dimensionality using PCA. A DNN trained on these features achieved a test accuracy of 0.63, which is comparable to similar voice-based diagnostic studies. Although preliminary, these results suggest that voice analysis holds potential as a non-invasive screening tool for chronic sinusitis. The moderate performance may be due to factors like limited training data, pre-processing techniques, feature selection, or model optimization. Future research should focus on expanding the dataset, refining feature selection, and exploring more advanced models to improve diagnostic performance and generalizability.

References

Alrawashdeh, H. M., & Alharazneh, A. (2020). Incidental nasal and paranasal findings on MRI and CT scans-a prospective study. *Prensa Médica Argentina, 106*, 55–60.

Aly, M., Rahouma, K. H., & Ramzy, S. M. (2022). Pay attention to the speech: COVID-19 diagnosis using machine learning and crowdsourced respiratory and speech recordings. *Alexandria Engineering Journal, 61*(5), 3487–3500.

Arslan, F., Polat, B., Durmaz, A., & Birkent, H. (2016). Effects of nasal obstruction due to nasal polyposis on nasal resonance and voice perception. *Folia Phoniatrica et Logopaedica, 68*(3), 141–143. Epub 2016 Dec 3.

Atar, M., & Kadayifci, A. (2014). Transnasal endoscopy: Technical considerations, advantages and limitations. *World Journal of Gastrointestinal Endoscopy, 6*(2), 41–48. Review article.

Babu, H., R. P., Arsha Fahima, T., & Kavithadevi, M. K. (2024). Voice-based common cold detection. In *2024 Second International Conference on Emerging Trends in Information Technology and Engineering (ICETITE)* (pp. 1–9).

Badr, A. A., & Karim, A. (2020). A review on voice-based interface for human-robot interaction. *Iraqi Journal for Electrical and Electronic Engineering, 16*, 91–102.

Battisti, A.S., Modi, P., & Pangia, J. (2025). Sinusitis (archived). In *StatPearls*. StatPearls Publishing.

Cabral, F. S., Fukai, H., & Tamura, S. (2019). Feature extraction methods proposed for speech recognition are effective on road condition monitoring using smartphone inertial sensors. *Sensors, 19*(16).

Dasaraju, P. V., & Liu, C. (1996). Infections of the respiratory system. In S. Baron (Ed.), *Medical Microbiology* (4th edn.). Chapter 93. University of Texas Medical Branch at Galveston.

Deb, S., Dandapat, S., & Krajewski, J. (2020). Analysis and classification of cold speech using variational mode decomposition. *IEEE Transactions on Affective Computing, 11*(2), 296–307.

Deb, S., Warule, P., Nair, A., Sultan, H., Dash, R., & Krajewski, J. (2023). Detection of common cold from speech signals using deep neural network. *Circuits, Systems, and Signal Processing, 42*(3), 1707–1722. Epub 2022 Oct 3.

Deborah, S., & Prathibha, K. M. (2014). Measurement of nasal mucociliary clearance. *Clinical Research in Pulmonology, 2*(2), 1019.

Egas-López, J. V., & Gosztolya, G. (2020). Predicting a cold from speech using fisher vectors; SVM and XGBoost as classifiers. In: A. Karpov & R. Potapova (Eds.), *Speech and Computer* (pp. 145–155). Springer International Publishing.

Fahrioglu, S. L., VanKampen, N., & Andaloro, C. (2025). Anatomy, head and neck, sinus function and development. In *StatPearls*. StatPearls Publishing. PMID: 30422521.

Gunbey, E., Gunbey, P., & Nguyen, P. (2020). *Radiologic Assessment ofNose and Paranasal Sinuses* (pp. 127–147). Springer International Publishing.

Hernández-García, E., Arias-Londoño, J. D., & Godino-Llorente, J. (2024). A voice and speech corpus of patients who underwent upper airway surgery in pre- and post-operative states. *Scientific Data,11*.

Husain, S., Amilia, H. H., Rosli, M. N., Zahedi, F. D., & Sachlin, I. S. (2018). Management of rhinosinusitis in adults in primary care. *Malaysian Family Physician, 13*(1), 28–33. eCollection 2018.

James Zinreich, S. (1998). Functional anatomy and computed tomography imaging of the paranasal sinuses. *The American Journal of the Medical Sciences, 316*(1), 2–12.

Jaume, F., Valls-Mateus, M., & Mullol, J. (2020). Common cold and acute rhinosinusitis: Up-to-date management in 2020. *Current Allergy and Asthma Reports, 20*(7), 28. Epub 2020 Jun 3. Review article.

Jiang, D.-N., Lu, L., Zhang, H.-J., Tao, J.-H., & Cai, L.-H. (2002). Music type classification by spectral contrast feature. In *Proceedings. IEEE International conference on multimedia and expo* (Vol. 1, pp. 113–116).

Kamil, O. (2018). Frame blocking and windowing speech signal. *Journal of Information, Communication, and Intelligence Systems (JICIS), 4*, 87–94.

Kapetanidis, P., Kalioras, F., Tsakonas, C., Tzamalis, P., Kontogiannis, G., Karamanidou, T., Stavropoulos, T. G., Nikoletseas, S., & Respiratory diseases diagnosis using audio analysis and artificial intelligence: A systematic review. *Sensors (Basel), 24*(4), 1173. Epub 2024 Feb 10. Funded by Pfizer Hellas SA: Systematic Review.

Karatas, O., & Toy, E. (2014). Three-dimensional imaging techniques: A literature review. *European Journal of Dentistry, 8*, 132–40.

Koo, S. K., Kwon, S. B., Chon, K. M., Kim, Y. J., & Kim, Y. J. (2015). The role of the maxillary sinus on the voice. *European Archives of Oto-Rhino-Laryngology, 272*(9), 2347–2350.

Kos, M., Kačič, Z., & Vlaj, D. (2013). Acoustic classification and segmentation using modified spectral roll-off and variance-based features. *Digital Signal Processing, 23*(2), 659–674.

Krouse, J., Lund, V., Fokkens, W., & Meltzer, E. O. (2010). Diagnostic strategies in nasal congestion. *International Journal of General Medicine, 3*, 59–67.

Lee, G.-S., Yang, C. C. H., Wang, C.-P., & Kuo, T. B. J. (2005). Effect of nasal decongestion on voice spectrum of a nasal consonant-vowel. *Journal of Voice, 19*(1), 71–77.

López-Espejo, I., Joglekar, A., Peinado, A. M., & Jensen, J. (2024). *On speech pre-emphasis as a simple and inexpensive method to boost speech enhancement*.

Maru, Y. K., & Gupta, Y. (2016). Nasal endoscopy versus other diagnostic tools in sinonasal diseases. *Indian Journal of Otolaryngology and Head and Neck Surgery, 68*(2), 202–206.

Melén, I. (1994). Chronic sinusitis: Clinical and pathophysiological aspects. *Acta Otolaryngol Supplementum, 515*, 45–48. Review article.

Nangia, S., Giridher, V., & Chawla, P. (2019). Evaluation of the role of nasal endoscopy and computed tomography individually in the diagnosis of chronic rhinosinusitis. *Indian Journal of Otolaryngology and Head and Neck Surgery, 71*(Suppl 3), 1711–1717. Published online 2017 Jan 18.

Nathan, K., Majhi, S. K., Bhardwaj, R., Gupta, A., Ponnusamy, S., Basu, C., & Kaushal, A. (2021). The role of diagnostic nasal endoscopy and a computed tomography scan (nose and PNS) in the assessment of chronic rhinosinusitis: A comparative evaluation of the two techniques. *Sinusitis, 5*(1), 59–66.

Niaksu, O. (2015). Crisp data mining methodology extension for medical domain. *Baltic Journal of Modern Computing, 3*, 92–109.

Proctor, D. F., & Kenneth Adams, G. (1976). Physiology and pharmacology of nasal function and mucus secretion. *Pharmacology and Therapeutics. Part B: General and Systematic Pharmacology, 2*(3), 493–509.

Purves, D., Augustine, G. J., & Fitzpatrick, D., et al. (Eds.). (2001). *Neuroscience* (2nd ed.). Sinauer Associates.
Royal, H. D. (2008). Effects of low level radiation-what's new? *Seminars in Nuclear Medicine, 38*(5), 392–402.
Ryan, M. W. (2010). Evaluation and management of the patient with "sinus". *The Medical clinics of North America, 94*(5), 881–890.
Safdar, S., Kausar, S., Tehsin, S., Mahmood, M., & Alwakid, G. N. (2023). *Prediction of specific language impairment in children using cepstral domain coefficients* (pp. 1–11).
Schlemmer, K. D., & Naidoo, S. K. (2013). Complicated sinusitis in a developing country, a retrospective review. *International Journal of Pediatric Otorhinolaryngology,77*(7), 1174–1178.
Schuller, B., Steidl, S., Batliner, A., Bergelson, E., Krajewski, J., Janott, C., Amatuni, A., Casillas, M., Seidl, A., Soderstrom, M., Warlaumont, A. S., Hidalgo, G., Schnieder, S., Heiser, C., Hohenhorst, W., Herzog, M., Schmitt, M., Qian, K., Zhang, Y., Trigeorgis, G., Tzirakis, P., & Zafeiriou, S. (2017). The Interspeech 2017 computational paralinguistics challenge: Addressee, cold & snoring. In: *Interspeech 2017* (pp. 3442–3446).
Sharma, G., Umapathy, K., & Krishnan, S. (2020). Trends in audio signal feature extraction methods. *Applied Acoustics, 158*, 107020.
Shi, J. B., Fu, Q. L., Zhang, H., Cheng, L., Wang, Y. J., Zhu, D. D., Lv, W., Liu, S. X., Li, P. Z., Ou, C. Q., & Xu, G. (2015). Epidemiology of chronic rhinosinusitis: results from a cross-sectional survey in seven Chinese cities. *Allergy, 70*(5), 533–539. Epub 2015 Mar 4.
Slavin, R. G. (1984). Slavin Sinusitis. *Journal of Allergy and Clinical Immunology, 73*(5, Part 2), 712–716. Special Problems and Management of Allergic Athletes.
Teul, I., Zbisławski, W., Baran, S., Czerwiński, F., & Lorkowski, J. (2007). Quality of life of patients with diseases of sinuses. *Journal of Physiology and Pharmacology, 58*(Suppl 5(Pt 2)), 691–697.
Warule, P., Mishra, S. P., & Deb, S. (2022). Classification of cold and non-cold speech using vowel-like region segments. In: *2022 IEEE International Conference on Signal Processing and Communications (SPCOM)* (pp. 1–5).
Warule, P., Mishra, S. P., Deb, S., & Joshi, D. (2023a). *Empirical mode decomposition based detection of common cold using speech signal* (pp. 899–903).
Warule, P., Mishra, S. P., & Deb, S. (2023b). Significance of voiced and unvoiced speech segments for the detection of common cold. *Signal, Image and Video Processing, 17*(5), 1785–1792.
Warule, P., Mishra, S. P., Deb, S., & Krajewski, J. (2023c). Sinusoidal model-based diagnosis of the common cold from the speech signal. *Biomedical Signal Processing and Control, 83*, 104653.
Warule, P., Chandratre, S., Mishra, S., & Deb, S. (2024). Detection of the common cold from speech signals using transformer model and spectral features. *Biomedical Signal Processing and Control, 93*, 106158.
Wu, A. W., Walgama, E. S., Borrelli, M., Mirocha, J., Barbu, A. M., Vardanyan, N., Shamsian, A., Hopp, S., & Hopp, M. L. (2020). Voice-related quality of life in patients with chronic rhinosinusitis. *Annals of Otology, Rhinology & Laryngology, 129*(10), 983–987. Epub 2020 May 26.
Zhou, Q., Shan, J., Ding, W., Chengyin, W., Yuan, S., Sun, F., Li, H., & Fang, B. (2021). Cough recognition based on Mel-spectrogram and convolutional neural network. *Frontiers in Robotics and AI, 8*, 05.

A Conceptual Scale for Attributing Financial Penalties to Privacy Breaches

Kweku Arthur and Hein Venter

Abstract Modern organisations leverage computing systems to fulfill business services. Such systems are therefore treasure-troves for threat actors, hence the prevalence of cybersecurity incidents and privacy breaches globally. Governments implement information privacy regulations and establish information privacy regulators to advocate that organisations implement appropriate information security measures to avoid cybersecurity and privacy breach incidents. The purpose of the Information Regulator is to oversee privacy regulations and to assign an appropriate penalty to the responsible organisation if the evaluations of the Regulator reveal a lack of security controls as the source of a cybersecurity or privacy incident. In this paper, the identified problem is that there is no structured method to attribute a financial penalty for investigated privacy breaches. A conceptual financial penalty scale is proposed to address this problem in the South African context. The proposed scale is derived from attributes of an investigation process, namely certainty of origin, evidence corroboration, evidence errors, and process repeatability. The relationship between these attributes is described in the paper. Thereafter, the conceptual financial penalty scale is applied to a basic investigation scenario. Lastly, the merits and shortcomings of the conceptual scale are discussed. The contribution of this paper lies in providing a uniform method for the Information Regulator in South Africa (or other countries) to apply financial penalties. Conversely, validation of the elements that are considered in deriving the conceptual scale is highlighted as a limitation.

K. Arthur (✉) · H. Venter
University of Pretoria, Pretoria, Hatfield, South Africa
e-mail: kk.arthur@tuks.co.za

H. Venter
e-mail: hventer@cs.up.ac.za

J. Marx Gómez et al. (eds.), *Recent Advances in Information Systems*, Progress in IS,
https://doi.org/10.1007/978-3-032-14099-9_32

1 Introduction

It is undeniable that developments in technology present opportunities for enhanced delivery of services such as government, financial, communication, and social interactions. As a prerequisite for delivery of such services, service consumers are typically required to provide complete and accurate personally identifiable information (PII); this information is used as input for fulfilling system functionalities or business objectives as required. Understandably, the abundance of personal information within processing platforms makes them attractive targets for compromise and, particularly, data breaches.

In its January 2019–April 2020 Data Breach report (ENISA, 2024), the European Union Agency for Cyber Security (ENISA) found that 71% of reported data breaches were financially motivated, and 25% of these breaches had long term strategic goals (involving nation-states or espionage). Additional findings in the report include that 70% of breaches focused on the exposure of personal information such as date of birth; such personal information is not easily changeable which adds to the usefulness of the information when compromised. In the Cost of Data Breach 2021 report (IBM, 2024), it is reported that 70% of all reported data breaches across the countries and industries that were assessed consisted of employees' and customers' PII. This report is significant in that the research statistics also incorporated data breaches that occurred in South Africa.

A notable data breach in South Africa occurred at a company called Experian—a consumer, business, and credit information services agency. The incident was made public in August 2020 (TechCentral, 2024a, 2024b). In this breach, PII of over 25 million South Africans were disclosed; the exposed data included telephone numbers, employment data, email addresses, and physical addresses. The prevalence of such incidents and concerns regarding inadequate information security safeguards by PII-hosting entities are what prompted the South African government to promulgate the Protection of Personal Information Act (POPIA), as well as establishing an Information Regulator for the country (de Bruyn, 2014) .

The Information Regulator of South Africa is empowered to monitor and enforce compliance by public and private entities with the provisions of Promotion of Access to Information Act (PAIA) and the Protection of Personal Information Act (POPIA). To fulfil the Act's mandate, as it relates to POPIA, the Regulator is charged with investigating privacy breaches and to issue penalties (in the form of fines or imprisonment) to responsible parties who are guilty of an offence.

POPIA is clear on what constitutes an offence and that there will be penalties for offences. However, the act is nondescript about what penalty values (in monetary value) should be assigned to such offences. In the absence of a defined structure in the pronouncement of penalties, the Information Regulator might be inconsistent in assigning appropriate penalty values for offences. Therefore, the main problem that exists in this context is that there is no usable scale that can be applied to determine penalty values.

The purpose of this paper is to expand on the need for a financial penalty scale that relates POPIA offences with financial penalties. This paper proposes a conceptual financial penalty scale and explores the minimum requirements for deriving such a scale.

This article is structured as follows: Sect. 2 discusses notable privacy breach incidents and their financial impacts. In doing so, the discussion validates the need for a consistent approach in the determination of financial penalties associated with privacy breaches. When privacy incidents occur, they need to be investigated. Therefore, Sect. 2 also references a mature cyber forensic investigation methodology that is utilised during an investigation. Section 3 considers and analyses digital forensic investigation principles in the derivation of the conceptual financial penalty scale. The financial penalty scale is derived from these investigation principles, and mathematical notation is used to depict the basic relationship between the elements of the proposed scale. Section 4 outlines a privacy breach scenario and the financial penalty scale is applied to the scenario; this is to conduct a basic test of the effectiveness of the scale. Section 5 discusses the merits and shortcomings of the proposed scale. The paper concludes with a summary of the conceptual scale and future work in Sect. 6.

2 Background

In recent years, the subject of personal data privacy has taken centre stage in the field of information and cyber security. One would expect that data privacy would be well addressed within developed nations such as Canada, Europe, and America, however, this is not the case. Even for these developed nations, the existing privacy regulations came into effect on 18 June 2015, 25 May 2018, and 1 January 2020, respectively. While each of these regulations address all elements of consumer's privacy expectations, the regulation's effective dates point to the possibility that the enforcement of penalties ascribed by these regulations remains to be seen in practice (Weber & Struder, 2016).

This section provides examples of noteworthy privacy breaches at renowned organisations, in developed nations, whose privacy regulations are considered the best in the world. The examples indicate the seriousness of the breaches, and point to the challenge experienced by privacy regulators globally in consistently enforcing financial penalties. In addition, the section adopts a standardised incident investigation methodology that is used in the investigation of privacy breaches. Sound investigation procedures ensure the admissibility of investigations during litigation, and also guides regulators in the financial penalty determination process.

2.1 Prominent Data Breach Incidents

A prominent data privacy breach that occurred in April 2011—and affected customers globally—was the Sony PlayStation Network breach. Sony revealed that personal information belonging to 77 million of its customers had been compromised through a targeted cyber attack. The breached information included: names, addresses, and credit card details. In the wake of the incident, the scale and severity of the compromise was undoubted. Furthermore, it was conceded that the incident was preventable; Regulators globally concluded that Sony was negligent, because the organisation had access to technical knowledge and financial means to have prevented this incident. Despite the various findings, it became clear that the financial penalties to Sony paled in comparison with the seriousness of the incident.

In 2019, the Information Commissioner's Office (ICO) in Britain fined British Airways £183.9 million because of a data breach identified in 2018. The breach involved between 380,000–500,000 customers' personal information, including names, surnames, email addresses, physical addresses, and credit card information (credit card numbers, expiry dates, and the three-digit CVV codes). Subsequently, the ICO revised its original penalty to £20 million considering the economic impact of the COVID pandemic. Despite the significant reduction in the financial penalty, the Information Commissioner made it clear that the original penalty reflected the gravity of the negligence that resulted in the data breach. It also reflects the application of new laws, which allow fines of up to 4% of a company's annual turnover if the company is found to be in violation of the data privacy law.

Financial penalties can also be issued in connection to privacy transgressions relating to surveillance activities. This was highlighted in the case of Swedish retail company, H&M, who were fined €35 million for illegally surveilling employees at their Service Centre in Nuremberg, Germany. On investigation, it was identified that since at least 2014, the company had excessively monitored the details of hundreds of employees' personal lives, including information regarding employee's health information and religious beliefs.

The frequency and severity of incidents, such as the Sony and H&M incidents, is one of the reasons for the development and implementation of data protection (and privacy) regulations across the world. Europe is widely credited for the quality and pervasiveness of their General Data Protection Regulation (GDPR) (Štarchoň & Pikulík, 2019); this regulation is also recognised as a "gold standard" for data protection regulations. GDPR was developed to safeguard personal data relating to identifiable natural persons. The regulation establishes 9 general principles, that is lawfulness, fairness and transparency, purposefulness, data minimization, accuracy, integrity and confidentiality, accountability, and finite data retention (Štarchoň & Pikulík, 2019; Ruohonen & Hjerppe, 2022). GDPR places strict obligations on data controllers and data processors to implement security controls to achieve these principles for their data subjects. The essence of the regulation is that only necessary personal data must be collected, with consent of the data subject, and used for a legitimate purpose only and retained for a duration no longer than what is necessary.

Another notable aspect of the GDPR is the obligation for the mandatory disclosure of data breaches, and the allocation of penalties for breaches where data processors are found to be in transgression of the regulation.

Breach incidents provided earlier in this section is all related to Britain and Europe. However, as discussed in the introduction, South Africa is not insulated from such breaches. In fact, the South African Government's acknowledgement of this risk is what prompted it to develop the POPIA (equivalent to the GDPR), as well as the implementation framework to oversee enforcement of the regulation. However, the POPIA regulation does not provide clear guidelines on the financial amounts when a financial penalty is enforced. What is evident from the noted breach incidents is that financial penalties will continue to prevail as an effective penalty for transgressors. For instance, GDPR is clear that, if any infringement of the rules is found, a fine can result in an amount up to €10 million or up to 2% of the organisation's global annual revenue (Sekiguchi, 2021). This means, if an organisation such as British Airways was fined, the full extent of the fine could be as much as €265,000,000. Therefore, consistency in investigation methodology, categorisation of the root cause of incidents, and quantification of severity and impact of incidents are some of the criteria to be considered when specifying financial penalties. Furthermore, an increase in data breach incidents will occur as the uptake of cloud services continues. As a result of this, one can expect an increase in instances of involuntary disclosure of poor security controls within organisations.

The following subsection addresses the need for consistency around the investigations of data breach incidents.

2.2 Incident Investigation Methodology

All incident investigations are best managed through the application of an effective forensic investigation methodology. Adherence to a forensic investigation methodology ensures consistency in investigation processes—irrespective of the impact of an incident—and the integrity regarding findings from investigations. The financial and privacy impacts from cyber security and privacy breaches warrant the need for mature investigation practices that result in punitive measures for offenders. These punitive measures are either legal prosecutions or financial penalties.

This paper adopts the investigation methodology described in the ISO 27043:2015 international standard (Valjarević et al., 2016; International Standard, 2015); this standard addresses security incident investigation principles and processes. The ISO27043 incident investigation methodology consists of 4 sub-processes: the readiness, initialisation, acquisitive, and investigative processes. For brevity, this paper does not provide an overview of the phases of the investigation process. Without loss of generality, this paper assumes the accurate and successful completion of the investigation processes.

3 Derivation of the Financial Penalty Scale

There is no deterministic scale that associates the impact of cyber security or data breaches with financial penalties in the enforcement of the POPI Act. This section focusses on defining a conceptual scale for financial penalties for data breaches. The section examines the necessary components for such a conceptual scale by drawing from how other scientific scales, such as a pH scale (Blackstock, 1989), are derived and applied. The value of a scientific scale is in its ability to provide an objective fact (or outcome), through the observation of the relationship between two or more consistent variables. For instance, the value in the pH scale is that it provides a consistent derivation of the pH outcome, based on the relationship between the hydrogen amount and the solution in question. Most importantly, this outcome is irrefutable, regardless of where the scale is applied using the same input variables.

The paper now considers these principles from common scientific scales in deriving the key elements for the financial penalty scale. A general web application solution is used as a reference when expanding on the principles in this section.

Web solutions are typically separated into a front-end (web) interface, an application server layer, containing the business logic, and a back-end (database) system, where transactional information and system actions are committed. In the solution architecture, there is always tight integration between front-end and application server system components. In turn, there is tight integration between application server and database system components. Furthermore, good practice recommends complete isolation between front-end and back-end systems.

The following sections identify four principles and examine them in the conceptual scale. The principles are certainty of origination, errors in evidence, corroboration of evidence, and repeatability of process. The principles of certainty of origination and errors in evidence are considered based on work presented by Casey (2002). The principles are then discussed further with reference to the web application architecture provided above. Lastly, the financial penalty scale is presented, using mathematical annotation to describe the conceptual relationship between the four operating principles.

3.1 Certainty of Origination

Certainty of sources and certainty as a principle are the cornerstone of any incident investigation. The infrastructures and information evidence sources involved in the commissioning of the incident need to be accurately identified. Similarly, the infrastructures affected through the incident need to be identified to determine the impact of an incident. Certainty of evidence sources also ensures the accuracy of findings during investigations; there are various attack vectors, tools, and techniques applied by threat actors. As such, having complete knowledge of evidence sources relating to the incident is key to investigations. Certainty of sources is also vital in determining the extent of impact of an incident. This attribute effectively seeks to establish uniqueness of evidence which greatly supports prosecution efforts.

Therefore, if such a web application were compromised, all components of the solution would be in scope of the investigation. However, a starting point to the investigation would be to establish certainty on which layer within the architecture served as the point of compromise. This would focus investigator's activities at the front-end layer as primary evidence for attribution of the threat actor(s). Such early assessments also help determine the extent and impact of the compromise. Given a scenario where the web interface was compromised, there would be a level of certainty that the back-end database system would remain isolated from the compromise, and hence less likely to have been compromised. In such a case, it could be determined through log examinations that the impact of the incident would be related to the nature and volume of available information that would be retrieved at the web interface. It should be noted that privacy-enhancing technologies do add complexity in determining the origin of activity.

3.2 Errors in Evidence from Source Systems

It is expected that any information processing system would operate in an environment with a trusted code base, managed user and system accounts, and on hardened infrastructures or platforms. Therefore, deployment practices influence system processing results, as well as the integrity of logs and log sources where they exist on these systems. Such errors result in the following situations: systems that are deployed to production with default configurations, misconfigurations, incomplete configurations, insecure configurations or unauthorised tampering of systems and evidence sources by malicious agents. To avoid potential errors in evidence sources, organisations need to implement comprehensive security controls and processes as defined by information security management standards such as ISO27001 (Humphreys, 2016).

The security posture of systems involved in or affected by an incident is vital to the integrity and certainty of evidence that is used within an investigation. Therefore, the conceptual scale should consider the overall security posture of the operating environment prior to any compromise. For example, if the infrastructures hosting the web, application, and database systems in a web solution are confirmed as unpatched prior to an incident, this would be indicative of susceptibility to compromise. Similarly, if there are no input validation and sanitisation controls applied on the front-end—as prescribed by secure development standards such as OWASP (Elder et al., 2021)—this would further confirm concerns of susceptibility to compromise.

Errors related to time and time settings certainly affect incident investigations. Time registers and time errors or time skew associated with time settings on affected systems then become material factors when deriving conclusions during investigations. Incorrect timestamps on evidence source systems leave room for legal argument during litigation; this is not ideal because it may raise concerns regarding the overall quality of an investigation and the admissibility of findings contained in the investigation report. Time-based errors must not affect investigations, given that creating a timeline of an incident is a necessary and effective method to establish the sequence of events leading to an incident.

3.3 Corroboration of Evidence

Primary systems affected by or involved in the commissioning of an incident are focal points during investigations. However, there are secondary systems—systems not directly affected by or involved in the commissioning of the incident—that also serve as evidence sources during investigations. Given the interconnectedness of operating environments, these secondary systems become reference points for corroborating any observed evidence on the primary systems. Examples of such secondary systems include IT administration systems, traffic monitoring systems, or security control systems. Furthermore, it is common that various secondary systems in an operating environment become key sources for proving the uniqueness or accuracy of specific observed evidence under investigation. In such instances, additional digital forensic techniques are applied to corroborate the occurrence of an event or the action of a source or actor under investigation. With reference to the earlier web application illustration, a network Firewall or DHCP system are ideal secondary evidence systems.

3.4 Repeatability of Process and Evidence Types

An inherent consideration in establishing dependability of the scale lies in the deterministic nature and function of information systems. Information systems operate using protocols—agreed patterns of system interactions—to fulfil computing and end-user outputs. Investigators typically interrogate such system patterns to distinguish between normal IT operational errors and security anomalies that may have resulted in the security incidents under investigation. Focus on patterns in IT system operations refine an investigator's scope of search and provides certainty regarding the exceptions under investigation.

Therefore, a degree of dependability must be factored in the application of the scale for it to be effective. This means that the scale must produce the same result whenever presented with the same objective evidence as inputs, even if this evidence is derived from differing operating environments. Additional elements to be considered in the scale are repeatability of investigation processes and the reconstruction of observed evidence to reproduce the observed impacts.

The need for a standardised investigation process cannot be overemphasised; following a standardised investigation process provides value in supporting or refuting any investigation hypotheses under consideration. Investigation processes are also targeted by legal practitioners who seek to identify inconsistencies in the execution of such processes. Proving the uniqueness of evidence sources (and threat actors) is necessary to perform incident reconstruction, and ultimately demonstrating the feature of repeatability. Evidence sources can be reconstituted through simulated operating environments, and such simulations should reproduce the same incident impacts or observed evidence artefacts that are under consideration in an investigation.

Earlier sections described and discussed key elements of the proposed conceptual financial penalty scale. The following section illustrates the conceptual financial penalty scale and discusses the relationship between the key elements.

3.5 *Proposed Financial Penalty Scale (FPS)*

Under South Africa's POPIA, the maximum allowable financial penalty for serious offences is ZAR10 million. The scale will compute a penalty factor that is applied against this maximum penalty amount. For example, for high impact breaches, the financial penalty scale allows the maximum financial penalty to be imposed, particularly where it is determined that information security controls in the affected environment are deficient. Conversely, for breaches where the impact is less severe, the financial penalty imposed is a reduced proportion of the maximum allowable penalty. The penalty factor is therefore suggested to be a value between 0 and 1, where 0 indicates 'no financial penalty' and 1 indicates 'maximum financial penalty'.

The proposed penalty scale is a function of the four principal elements discussed in earlier sections, namely certainty of origin, evidence corroboration, evidence errors, and process repeatability. For each element, a weighting is applied to represent the significance of the element in the derivation of the final penalty factor. In general terms, positive outcomes from an investigation—such as definitive investigation outcomes—proportionately influence the penalty factor. Whereas inconsistencies or errors within investigation outcomes inversely influence the penalty factor in the scale.

The proposed financial penalty scale incorporates these principles by representing them as the following elements in the scale.

- CO = Certainty of Origin
- EC = Evidence Corroboration
- PR = Process Repeatability
- EE = Evidence Errors
- W_{xy} = Element Weight Factor
- P = ZAR10,000,000 Maximum Financial Penalty

The conceptual FPS is defined as follows:

$$FPS = P \times \frac{(CO_{CO}) \times (EC_{EC}) \times (PR_{PR})}{(EE_{EE})} \tag{1}$$

For the FPS to be effective, the strength of evidence that supports investigation hypotheses will positively contribute towards the financial penalty factor. Furthermore, investigation aspects such as errors in evidence or process errors would negatively affect the financial factor. As such, this defines an inverse relationship between evidence accuracy and process accuracy, viz., errors in evidence and errors in process.

Cyber forensic investigations are complex and involve various digital artefacts that are inherently volatile. This paper does not aim to expand on every detail that may positively or negatively influence the success of investigations. Rather, the focus is to highlight how the relationship between key aspects in an investigation process can be used to attribute financial penalties when cyber security or data breach incidents occur.

The following section describes a basic investigation use-case and the FPS is applied to the use-case. The results of this application are discussed thereafter.

4 Application of the Financial Penalty Scale

This section outlines a basic investigation scenario that highlights the four elements in the conceptual Financial Penalty Scale (FPS). Evidential artefacts that are considered during the investigation scenario are identified, then the FPS is applied to the scenario. Lastly, a discussion of the results from this application is offered.

4.1 Investigation Scenario Description

The investigation scenario involves *'an unauthorised payment by an employee to an existing service provider'*. We make the following assumptions to focus the FPS application discussion:

- General business processes and IT security controls exist within the organisation. However, these processes and controls may not be implemented according to best practices.
- Evidence artefacts are available whenever required during the investigation.
- Qualified cyber forensic investigators are commissioned to conduct this investigation.

In the scenario, the compromise is performed through the following sequence of events:

1. The Finance Manager (FM) receives a spearphishing email, masked as a legitimate organisational email message (i.e. spoofed email message). The email contains a link to a captive portal to the organisation's cloud email service. However, the captive portal is, in fact, a credential-harvesting site that uses the user interface of the organisation's cloud email captive portal.

2. The FM reacts to the email by opening the mail, clicking on the link, and subsequently entering their username and password as required to authenticate to the captive portal.
3. The next spearphishing email is sent to the Procurement Administrator (PA), with the FM spoofed as the sender. The spoofed email contains 3 requests:
   (a) a request for an extract of the organisation's vendor list, including all vendors' PII and record of payments over the past 12 months.
   (b) An instruction to capture a payment instruction of R2 million to a new vendor, with the provided banking details.
   (c) To reply to the email when the actions are completed.
4. Using the FM's credentials, the threat actor accesses the finance system remotely and authorises the payment instruction on the system.

The following section discusses the process of investigating the incident described in the scenario. It highlights evidence artefacts that are considered during the investigation. Thereafter, the FPS is applied to the scenario and observations are discussed.

4.2 Scenario Investigation Process

This section expands on the forensic evidence that would typically be gathered and investigated for such an incident. The section focuses on this evidence to highlight how the values of the elements of the conceptual FPS are derived.

The investigation commences by identifying all computing systems that are involved in the incident. The primary or core systems in this case are the email system, authentication system, finance system, and the remote access system. The process also uses available evidence from secondary or complementary systems, such as firewalls, network devices, and time systems for the investigation. Forensic acquisition of application and system log information for these systems—over the timeline of the incident—are prepared for further review and analysis to reconstruct the sequence of events.

Investigators perform the following actions in assessing relevant evidence relating to the primary systems from the scenario. For the email system, email messages and message metadata are reviewed to determine the date, time, message subject, attachment information, sender aliases, sender domain, and recipient information for all email messages that are pertinent to the incident. In addition, investigators also interrogate the contents of email attachments and email body to establish whether malicious code or malicious links are used as the vector for compromise. Reviews of the email system uptime and time settings also supplement the reliability of message reviews during the investigation. At this stage, the investigator establishes evidence regarding the spearphishing message to the FM and subsequent correspondence over the incident timeline. The final email sent from the PA to the supposed FM receives additional scrutiny because the contents or attachments to the message assist investigators in determining the nature and volume of PII that is breached during the incident.

For the authentication system, all named (user) accounts and system accounts are corroborated against the core system authentication logs. The privileges associated with each identified account is further scrutinised to determine the actions of these account; evidence of privilege escalation activities is also considered during this process. It is also noted that, where a central directory service is applicable, authentication to the finance system is facilitated by the directory service. Therefore, the directory system is the authoritative evidence source for authentication information. Internet proxy activity logs also host a wealth of forensic evidence at this stage of a spearphishing investigation. On review of proxy logs, investigators identify which user accounts are active during the incident timeline. They also establish which external Internet addresses are accessed by the FM at the time when the link contained in the email body is clicked. Where it is determined that an Internet link directs traffic to a compromised destination, the timing of the credential-harvesting action can be inferred. At this stage, an investigator can confirm the point where actions in the environment are being performed by the threat actor as opposed to that of the real account owner.

For the finance system, application logs and a thorough review of system transactions prior to, during, and after the incident timeline are typical investigation considerations. Investigators sift through multiple transactions to identify system actions or transactions that are indicative of anomalies and transgressions of known business rules. The PA's action of extracting the vendor list, and the FM's approval action of the ZAR2 million payment are recorded at this point. The remote access system logs are crucial at this stage of the investigation. Authentication evidence on the remote access system establishes the time when the FM's account is used. These logs also confirm the internal network address for the finance system that is accessed. The corroboration of authentication information, remote access information, and the finance system information assists investigators in establishing the critical actions that are performed that result in the compromise. At this point, investigators consider the quantity and type of errors in the investigation process, as well as the quality of evidence artefacts before concluding the investigation and documenting the objective facts in the reconstruction of the incident. An example of an investigation error is where an investigator excludes evidence artefacts from an investigation. For instance, peripherals such as storage devices or smart devices may be excluded from the investigation if they are lost, damaged or simply unobtainable during the investigation. The materiality of such exclusions may be lessened if the available and reviewed evidence still provides conclusive investigation outcomes despite the omissions.

As part of such investigations, the verbal evidence of employees, account owners, and systems administrators is also considered. Such input adds important context to any investigation. Contextual information includes items such as business rules, business processes, IT processes, or IT system topology. Such information is eventually used to corroborate digital evidence that is identified as part of the investigation process.

4.3 Applying the FPS

This section discusses the fringe outcomes in applying the proposed FPS, namely a low and high penalty outcome. To illustrate the derivation of each penalty outcome, the paper assumes the following:

1. An equal weighting and significance of each element in the scale.
2. The value of each element represents quality. Two states of quality are offered, with the following values: low = 0.5 and high = 1.

For a low penalty outcome of the FPS, the quality of the evidence errors element must be high, whilst the certainty of evidence origin (CO), investigation process repeatability (PR), and evidence corroborations (EC) elements are low. The element values in this instance are depicted as follows:

$$FPS = P \times \frac{(0.5) \times (0.5) \times (0.5)}{(1)} = 10m \times 0.125 \tag{2}$$

The penalty is therefore an amount of ZAR1,250,000, which is a fraction of the maximum penalty amount. An explanations for the high and low values of the FPS elements that result in this low financial penalty result are now provided.

A high evidence error value—that is, 1—is where the number of errors within the evidence and the type of evidence errors identified during the investigation are minimal. This implies that the integrity and completeness of the digital evidence considered throughout the investigation is of a high quality, where exceptions in evidence data are not detected, and significant anomalies within the evidence or evidence collection process are absent throughout the investigation. The low values of the certainty of evidence origin (CO), investigation process repeatability (PR), and evidence corroborations (EC) is where the operating environment of the affected system is unmanaged or not maintained. For example, an environment may have multiple directory services, where system or user accounts have the same name across the various directory services. Moreover, the accounts are associated with different system functions or named users. Corroboration of evidence may also result in low element values; it may be that there are insufficient or unavailable alternative evidence sources against which an investigator is able to corroborate with the available evidence. There may even be a critical mistake made by an investigator that has the effect of tampering with digital evidence, thereby jeopardising the investigation.

In real terms, this penalty outcome means that the investigation process and outcome may still be questionable, hence a lesser value is levied on the transgressing organisation.

For a high penalty outcome of the conceptual FPS, the certainty of evidence origin, investigation process repeatability, and evidence corroboration must be high, whilst the number and type of evidence errors is low (which reflects a high evidence quality). The element values in this instance are depicted as follows:

$$FPS = P \times \frac{(1) \times (1) \times (1)}{(1)} = 10m \times 1 \quad (3)$$

The penalty amount is therefore the maximum of ZAR10,000,000. Explanations for the high values of the FPS elements that result in this high financial penalty result is now provided.

This situation may materialise where there is good consistency in the environment where the investigation is conducted, as well as a high degree of related digital artefacts for evidence corroboration. The degree of errors in evidence is significantly reduced when the quality of evidence used in an investigation is high. In contrast with the low penalty outcome, a high penalty outcome demonstrates an ideal investigation situation. That is, the investigation process is followed systematically and all available digital evidence is corroborated and provides a conclusive and accurate sequence of events leading to the incident.

In real terms, this penalty outcome means that the investigation outcome is beyond reasonable doubt; the matter may possibly be considered for further forms of litigation where applicable. Therefore, the maximum penalty amount (ZAR10 million) should be levied on the transgressing organisation.

This section discussed the fringe scenarios when applying the conceptual scale—low and high penalties. The approach demonstrated some of the circumstances that are considered during forensic investigations. It also highlights the effect of each element of the conceptual scale towards the final penalty outcome. The permutations on the application of the FPS can be extended in future work. However, the discussions presented in this paper have highlighted basic aspects of the FPS.

The following section discusses the benefits of the conceptual financial penalty scale, its limitations, and considerations for future work.

5 Discussion

Three benefits are identified with the use of the conceptual scale. The first benefit is that the scale addresses the divide between technical and non-technical facets involved in an investigation outcome. It is common that non-technical parties such as data analysts, legal representatives, or the regulator place reliance on technical investigation reports—compiled by technical specialists—to develop their respective findings for their respective purposes. The scale is therefore an effective and standardised mechanism for communication between these investigation stakeholders.

The second benefit is that the conceptual scale provides a formal method for the regulator to apply a financial penalty. Structure ensures consistency and objectivity in the application of fines for cyber security and data breaches, irrespective of the size of the organisation or its market value. It also ensures that subjectivity is removed during the penalty determination process. This is an important factor since investigations and penalty determination processes may be lengthy and span across multiple departments and resources over the duration of the processes.

The third benefit of the scale is that it is applicable to any organisation, irrespective of the scale, market sector or market value of the affected organisation. This benefit materialises through the key elements of the scale, namely CO, PR, EC and EE. The derivation of a value for these elements incorporates factors that normalise the real or perceived variances that exist in different operating environments or sectors. For example, common security controls across varied organisations are firewalls, directory services and remote access systems. However, organisations may differ in terms of brands of capabilities or in the logical deployment of such capabilities. Nevertheless, the application of such a basic conceptual scale aims to normalise and diminish the impact of such variances where they do exist.

Now that the benefits of the FPS have been discussed, the 3 shortcomings with the proposed scale are addressed. The first limitation of the scale relates to the qualification or validation of the key elements that are considered. The key elements that are considered for the scale are CO, PR, EC and EE. The completeness of all possible elements required for a successful scale is a subject that must be investigated further. It may also be argued that the four elements of the scale were arbitrarily selected. Similarly, the materiality of the identified elements in the scale must also be investigated further. This consideration is critical because it has bearing on the quality of such a conceptual scale, and the accuracy of the relationship between the identified elements. For example, regarding the relationship between the elements, an inverse relationship between the EE viz the other elements is depicted. Then, a proportional relationship between the CE, PR and EC is depicted. These relationships should be explored further to ensure credibility of the outputs of the FPS.

The second limitation of the scale relates to the values that are to be used to determine the accuracy and correctness of the output of the conceptual scale. For illustration, binary values were used to represent the values of the elements. However, this basic approach would require a review of the relationship amongst the elements. For example, in the current form of the conceptual scale, the EE may not hold a value of 0. The limitation with respect to reliable values for the scale extends to the values of the weights associated with each element. Should the weighting be predetermined, or should this be derived based on the incident under investigation? Who should determine the weightings? These important considerations could improve the accuracy and fair allocation of penalty outcomes in the concept scale.

The third limitation relates to the application of the scale in the given scenario. The degree of detail regarding how the values of the elements change, positively or negatively, needs to be expanded. An expansion on this aspect may reveal the additional elements to be considered in the conceptual scale. There is a need to consider more permutations of the current proposed elements, and whether to hold any of the elements or weightings static when applying the scale. The conceptual scale can be optimised through such a review, which would enhance the quality of the scale.

6 Conclusion

The occurrence of cybersecurity and data breaches continues to increase as organisations increasingly perform business functions on computing systems. As a consequence, data breaches must be investigated as part of incident response protocols and to assist stakeholders in managing the impact of such incidents. One such stakeholder is the Information Regulator; they are required to prescribe legal or financial penalties to organisations where forensic investigations reveal shortcomings in data breach prevention controls at these organisations. The consistent attribution of financial penalties resulting from a data breach is not trivial, given the varied nature of organisations and the scale of data breaches.

This paper proposed a conceptual financial penalty scale that provides the Information Regulator with a mechanism for applying financial penalties in a consistent manner, based on investigation outcomes from privacy breach incidents. The elements considered for the scale were: certainty of origin, evidence corroboration, process repeatability, and evidence errors. These elements were utilised in the penalty scale to derive a factor against which to apply on the maximum penalty amount of ZAR10,000,000. Thereafter, the FPS was applied on a data breach incident scenario and the findings were discussed. The need for such a scale was emphasised throughout the discussion. Similarly, the need for a structured method for defining the element values was highlighted; this will therefore be the subject of future work, along with deriving empirical results from the application of the FPS.

References

Blackstock, J. C. (1989). *Guide to Biochemistry* (1st ed.). London: Butterworth-Heinemann.

Casey, E. (2002). Error, uncertainty, and loss in digital evidence. *International Journal of Digital Evidence, 1*(2).

de Bruyn, M. (2014). The protection of personal information (POPI) act—Impact on South Africa. *International Business & Economics Research Journal (IBER),13*(10), 1315–1340.

Elder, S., Zahan, N., Kozarev, V., Shu, R., Menzies, T., Williams, L. (2021). Structuring a comprehensive software security course around the OWASP application security verification standard. In *IEEE/ACM 43rd International Conference on Software Engineering: Software Engineering Education and Training (ICSE-SEET)* (pp. 95–104). IEEE Press, Madrid.

ENISA. (2024). https://www.enisa.europa.eu/publications/enisa-threat-landscape-2020-data-breach. Last accessed July 20, 2024.

Humphreys, E. (2016). *Implementing the ISO/IEC 27001:2013 ISMS Standard* (2nd ed.). London: Artech House.

IBM. (2024). https://newsroom.ibm.com/2021-07-28-IBM-Report-Cost-of-a-Data-Breach-Hits-Record-High-During-Pandemic. Last accessed July 21, 2024.

International Standard (ISO/IEC 27043). (2015). Information technology—Security techniques—Incident investigation principles and processes (1st edn.). South African Bureau of Standards, Pretoria.

Ruohonen, J., & Hjerppe, K. (2022). The GDPR enforcement fines at a glance. *Information Systems,106*(5), 1–11.

Sekiguchi, A. (2021). Comparison of the regulations on communication privacy between EU and Japan: Toward reinforcement of Japan's communication privacy. *Information & Security, 48*(1).
Štarchoň, P., & Pikulík, T. (2019). GDPR principles in data protection encourage pseudonymization through most popular and full-personalised devices—mobile phones. *Procedia Computer Science, 151*(1), 303–312.
TechCentral. (2024a). https://techcentral.co.za/data-breach-exposes-personal-information-of-millions-of-south-africans/176848/. Last accessed August 19, 2024.
TechCentral. (2024b). https://techcentral.co.za/data-breach-exposes-personal-information-of-millions-of-south-africans/176848/. Last accessed August 20, 2024.
Valjarević, A., Venter, H. S., Petrović, R. (2016). ISO/IEC 27043:2015—Role and application. In *24th Telecommunications Forum (TELFOR)* (Vol. 24, no. 11, pp. 1–4).
Weber, R. H., & Struder, E. (2016). Cybersecurity in the Internet of Things: Legal aspects. *Computer Law & Security Review, 32*(5), 712–728.

Large Language Models for Qualitative Risk Analysis in Banking: A Systematic Review

Jan-Hendrik Witte and Henry Eggers

Abstract This systematic review examines the emerging role of Large Language Models (LLMs) in enhancing qualitative risk analysis within banking credit risk management. We identify five key application domains: credit scoring and default forecasting, comprehensive risk assessment, document analysis and report generation, fraud detection, and regulatory compliance. Various models, from general-purpose LLMs like GPT-4 to domain-specific models like FinBERT, demonstrate significant potential for improving predictive accuracy by integrating unstructured textual data with traditional financial metrics, while simultaneously increasing operational efficiency. Despite these advancements, critical challenges persist: data quality and privacy concerns, limited model explainability, computational resource requirements, and regulatory compliance complexities. For financial institutions, effective LLM implementation implies enhanced decision-making precision and operational efficiency, but necessitates strategic investments in technology infrastructure, talent development, and governance frameworks. Future research should prioritize developing explainable financial LLMs, privacy-preserving techniques for sensitive data, optimized architectures for resource efficiency, hybrid models combining LLMs with traditional analytics, and comprehensive regulatory frameworks balancing innovation with ethical considerations. These developments could transform qualitative risk analysis from a resource-intensive manual process to a scalable, consistent, and comprehensive assessment mechanism.

J.-H. Witte (✉) · H. Eggers
Department Very Large Business Applications (VLBA), Carl von Ossietzky Universität Oldenburg, Oldenburg, Germany
e-mail: jan-hendrik.witte@uol.de

H. Eggers
e-mail: henry.eggers@uol.de

J. Marx Gómez et al. (eds.), *Recent Advances in Information Systems*, Progress in IS,
https://doi.org/10.1007/978-3-032-14099-9_33

1 Introduction

The financial sector is undergoing significant transformation through digitalization and artificial intelligence (AI) technologies, particularly in credit risk management (Glaser, 2022). While quantitative models effectively process structured financial data, qualitative analysis relies heavily on expert review of unstructured textual information from annual reports, risk disclosures, and regulatory filings (Fritz, 2018; Soehaditama et al., 2023; Yuliia, 2024; Alvarado et al., 2015). These documents provide crucial insights into financial stability and risk structure.

Traditional qualitative review presents significant challenges: it is time-consuming, resource-intensive, and difficult to scale consistently across heterogeneous text (Glaser, 2022; Reck, 2016). This introduces risks of oversight and subjective interpretation, potentially causing inconsistencies in risk assessment. These challenges are amplified by post-financial crisis regulatory requirements like Basel I-III (Akkizidis & Kalyvas, 2018; Zirkler et al., 2015) and the Minimum Requirements for Risk Management (MaRisk) (Authority, 2023), which demand precise and transparent risk assessments (Langohr & Langohr, 2012).

Credit risk management is especially critical in the interbank market, where the complexity of financial systems creates significant challenges, with potential for a single bank's default to trigger systemic risk (Cimini & Serri, 2016; Hellwig, 1995). The financial crisis demonstrated how liquidity shortages substantially influence credit risks when banks sell assets to secure liquidity (Cimini & Serri, 2016; Karouzakis, 2021).

Effective credit risk management combines qualitative factors (e.g., management quality, market position) with quantitative metrics (e.g., leverage ratios, liquidity, profitability) (Soehaditama et al., 2023; Yuliia, 2024; Brahmaiah, 2022; Radojević et al., 2022). LLMs offer a promising approach to efficiently analyze vast amounts of qualitative text data. Based on neural networks and Transformer architecture, LLMs can process semantic structures in texts through self-supervised learning (Patil & Gudivada, 2024; Qin et al., 2024; Touvron et al., 2023). Their integration into credit risk workflows could significantly reduce manual analysis time, standardize processes, and allow analysts to focus on strategic interpretation.

This systematic review examines the current research on LLMs in supporting qualitative risk analysis in banking credit risk management or related topics, addressing the question:

> "How can LLMs enhance credit risk management in specific application domains, what challenges and limitations emerge, and what are the resulting implications for the financial industry?"

This paper is structured as follows: Section 2 presents our methodology, detailing the systematic literature search approach and qualitative content analysis framework. Section 3 outlines the results, examining application areas of LLMs in finance, models and technologies employed, challenges and limitations identified, and practical implications for financial institutions. Section 4 discusses research gaps and proposes future directions for enhancing LLM applications in credit risk management. Finally,

Section 5 concludes with a synthesis of key findings and their significance for the financial industry's transformation of qualitative risk analysis processes.

2 Methodology

This systematic review employed a rigorous methodological approach combining a structured literature search with qualitative content analysis to comprehensively assess the current state of research regarding LLMs in credit risk management. The methodology was designed to ensure transparency, reproducibility, and scientific rigor throughout the review process.

2.1 Research Design

The methodological framework for this study followed a two-step approach. First, a systematic literature review was conducted according to the principles established by Webster and Watson (Webster, 2002), enabling a comprehensive identification and mapping of relevant literature. This was followed by a qualitative content analysis based on Mayring's (Mayring, 2014) methodology, which facilitated a structured examination of the identified research. This combined approach allowed for both breadth in literature coverage and depth in content analysis, ensuring that the review captured the full spectrum of research on LLMs in credit risk management. Table 1 shows an overview of this literature research methodology.

To ensure comprehensive coverage of the research landscape, five established databases were selected based on their relevance to the fields of business informatics, AI, and natural language processing: Scopus, arXiv, IEEE, ACM, and Springer Link. These databases represented a diverse range of academic perspectives, including technical approaches, economic applications, and regulatory considerations.

The search strategy employed a structured approach using a predefined search string that incorporated key concepts related to LLMs, finance, risk assessment, and document analysis, as seen in Table 1. This search string was adapted to the specific requirements of each database's advanced search functionality. Where possible, searches were restricted to title, abstract, and keywords to ensure relevance. When such restrictions were not available, searches were conducted across all fields. In cases where this approach yielded an excessive number of results, the search was limited to abstracts only.

To ensure quality and relevance of the analyzed publications, strict inclusion and exclusion criteria were established prior to the literature search. The review included only studies published from 2020 to 2025 that were either peer-reviewed articles or scientifically recognized preprints with clear relevance to credit risk management.

Studies with non-transparent methodological foundations, those focusing on outdated or irrelevant technologies, or publications whose focus was outside the

Table 1 Overview of literature search methodology

Search	("Large Language Model" OR "LLM" OR "GPT") AND ("finance" OR "financial" OR "banking" OR "fintech" OR "credit") AND ("qualitative analysis" OR "qualitative risk" OR "text analysis" OR "unstructured data" OR "document analysis" OR "credit risk" OR "risk assessment" OR "risk management") AND ("annual report" OR "disclosure" OR "filing" OR "document" OR "text" OR "narrative*" OR "qualitative information")			
Fields	Scopus	Title, abstract, keywords		
	arXiv	All fields		
	IEEE	All fields		
	ACM	Abstract		
	Springer	All fields		
Dates	2020–2025			
Database	Results			
	Initial	*Title-screening*	*Abstract-screening*	*Final*
Scopus	79	30	13	7
arXiv	24	7	7	4
IEEE	42	2	2	1
ACM	5	2	1	0
Springer	433	5	5	3
Total	583	46	26	17

financial sector were excluded. This systematic approach ensured that the literature included was both scientifically robust and practically relevant to the research question.

2.2 Literature Selection Process

The systematic search across all databases initially yielded 583 results, with considerable variation in the number of matches depending on the database and search parameters used. The selection process followed multiple screening stages: Initial identification (n = 583), Title screening (n = 46), Abstract screening (n = 26), Final selection (n = 17).

The most substantial number of initial results came from Springer Link (n = 433), followed by Scopus (n = 79), IEEE (n = 42), arXiv (n = 24), and ACM (n = 5). After the title screening, 46 potentially relevant papers remained. Following abstract screening, this number decreased to 26. The final detailed content evaluation resulted in 17 papers that met all inclusion criteria and were included in the qualitative analysis.

The distribution of the final 17 papers across databases was: Scopus (7), arXiv (4), IEEE (1), ACM (0), and Springer Link (3). The selection process ensured that only the most relevant studies addressing the intersection of LLMs and credit risk management were included in the qualitative analysis phase.

2.3 Qualitative Content Analysis

Following Mayring's (Mayring, 2014) approach to qualitative content analysis, the identified studies were examined in a structured manner to identify thematic patterns and extract relevant insights.

Development of the Category System. The category system was deductively derived from existing literature and applied to the analyzed studies. Five main categories were defined:

1. Application areas of LLMs in finance.
2. Models and technologies used, challenges and limitations.
3. Practical implications for financial institutions.

These categories reflected the essential research streams recurring in the literature and remained unchanged throughout the analysis.

The application of the category system was conducted through systematic text analysis. Relevant text passages were marked in the studies and assigned to the appropriate categories. In some cases, content subgroups were formed within a main category, for example, to distinguish between technical and ethical challenges. The results of the analysis were documented in tabular form to ensure transparency and comparability.

Definition of Coding Units. Following Mayring's (Mayring, 2014) approach, three types of coding units were distinguished: meaning units, context units, and evaluation units.

Meaning units formed the smallest meaningful text unit that makes an independent contribution to answering the research question. These could consist of a sentence, word group, or several connected sentences, provided they formed a conceptually cohesive thought. Examples included specific statements about the use of LLMs in credit risk analysis, models employed, or challenges identified.

Context units placed these meaning units within the broader framework of the respective study, such as methodological sections or discussion parts. These units enabled a holistic interpretation of individual statements and ensured that their meaning was considered in the correct context.

The *evaluation units* comprised all coded statements and formed the basis for interpretation and presentation of results. This included all coded sections within the previously defined analysis categories.

3 Results

This section presents the findings from our systematic literature review of LLMs in credit risk management. The analysis reveals that LLMs are already being deployed across various domains within financial services, with key applications in credit scoring, risk assessment, and document analysis in the context of credit risk. As illustrated in Fig. 1, we examine the models and technologies employed in these contexts, identify practical implications for financial institutions, and critically assess the challenges and limitations that currently exist in this rapidly evolving field.

3.1 Application Areas of LLMs in Finance

The reviewed literature highlights several key domains within credit risk management where LLMs are demonstrating significant potential for enhancement. As shown in Fig. 1A, credit scoring, rating, and default forecasting emerged as the most prevalent application area, appearing in 8 of the 17 reviewed papers:

Credit Scoring, Rating, and Default Forecasting: This is a primary application area. LLMs like ChatGPT-3.5 and BERT variants are being used to integrate unstructured data (e.g., loan descriptions, Management Discussion and Analysis sections

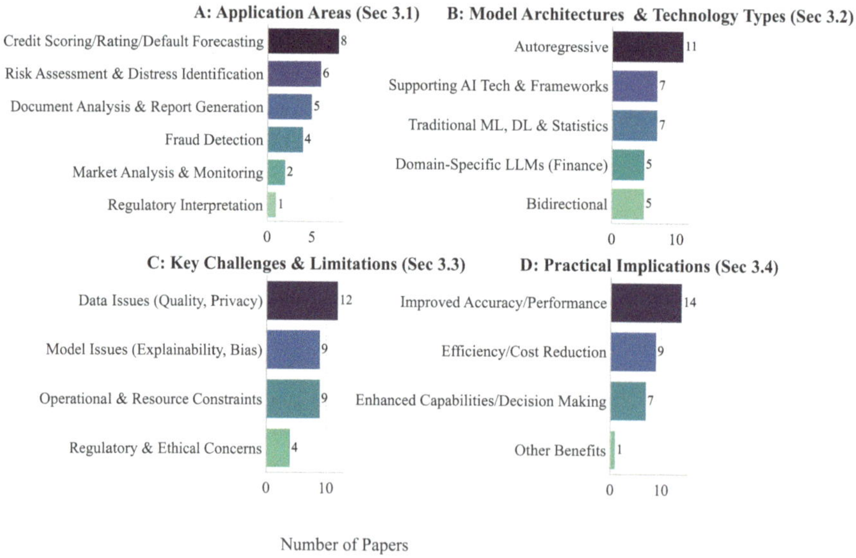

Fig. 1 Overview of the systematic literature review findings (N = 17 Papers). The figure displays the frequency distribution of key aspects identified in the reviewed literature. Bars represent the number of unique papers associated with each category, sorted in descending order

in annual reports, Environmental, Social and Governance (ESG) reports) with traditional structured financial data. This integration aims to create more comprehensive and accurate corporate credit rating models, particularly beneficial for small and medium sized companies (SME) where structured data might be limited (Lin et al., 2025; Wang & Kang, 2024; Hajek & Sahut, 2024; Sanz-Guerrero & Arroyo, 2025). Models like fine-tuned BERT have been shown to generate meaningful risk scores directly from loan descriptions, improving the predictive performance of subsequent classifiers like XGBoost (Sanz-Guerrero & Arroyo, 2025). Frameworks like the bi-level semantic enhancement approach (Lu et al., 2025), using GPT-3.5 for data augmentation and RoBERTa for feature extraction within a Contrastive Multimodal Multitask Learning (CMML) model, demonstrate improved prediction accuracy, especially with limited samples or unseen corporations. The nature of enhancement here is primarily improved accuracy through broader data incorporation and enhanced scope by leveraging qualitative information.

Risk Assessment and Financial Distress Identification: The second most common application area (6 papers, Fig. 1A) involves LLMs contributing to broader risk assessment beyond specific scores. Systems like RA-CFGPT are designed for risk assessment by analyzing documents and highlighting potential risks (Li et al., 2024). The FinBen benchmark includes tasks where LLMs (like GPT-4, Gemini, LLaMA2, FinMA) are evaluated on predicting corporate bankruptcy risk (Xie et al., 2024). Frameworks like RiskLabs use models including GPT-4 and Wav2vec2 to analyze multi-source data (earnings calls audio/text, news, market data) for predicting market volatility and Value at Risk, providing a more holistic view of financial risk (Cao et al., 2024). Enhancements focus on scope (integrating diverse, real-time data streams) and potentially accuracy in identifying broader financial distress signals.

Document Analysis and Report Generation: Appearing in five of the reviewed papers (Fig. 1A), LLMs show promise in automating the creation of credit risk reports. Teixeira et al. (Teixeira et al., 2023) demonstrate using GPT-4 with a novel "Labeled Guide Prompting" technique, integrated with Bayesian Networks, to generate high-quality, insightful credit risk reports preferred by human analysts. LLMs like GPT-4 and Llama2 are also used for keyword generation, description, and annotation of visually-rich financial documents, aiding risk analysis by making large volumes of documents more accessible (Gallipoli et al., 2024; Sai et al., 2025). This primarily enhances efficiency and automation, potentially improving consistency.

Fraud Detection (Related to Credit): While not exclusively credit risk, fraud detection (4 papers, Fig. 1A) is often intertwined. AI techniques, including LLMs and deep learning (ANNs, GANs), are used to identify fraudulent patterns in large datasets, including unstructured text like client documentation (Xie et al., 2024; Sai et al., 2025; Mahadevkar et al., 2024; Goyal et al., 2025). Generative AI can simulate fraudulent transactions to train more robust detection models. The enhancement here is improved accuracy and efficiency in detecting potentially fraudulent activities that impact creditworthiness.

Market Analysis and Monitoring: Found in 2 papers (Fig. 1A), this application area extends risk assessment to dynamic market conditions. LLMs are used for the real-time monitoring of equity, fixed income, and currency markets to synthesize signals and identify potential shifts (Yang et al., 2025). For instance, models like GPT-4 and Llama-3 are employed for trend analysis and downside risk assessment, while other studies leverage deep learning and natural language processing (NLP) for predicting stock behavior (Sai et al., 2025). This capability enhances dynamic risk management by providing a more comprehensive and timely view of cross-asset market risks.

Regulatory Interpretation and Compliance: Though less frequently mentioned (1 paper, Fig. 1A), LLMs like GPT-4 and Claude-3-Opus are being explored to interpret complex financial regulations (e.g., Basel III) into actionable formats, streamlining compliance processes related to risk management (Cao & Feinstein, 2024). This can improve efficiency, especially for smaller firms, and aid in ensuring risk management practices align with regulatory mandates.

Across these domains, various LLMs and associated technologies are employed, which will be examined in the following section.

3.2 *Models and Technologies*

The reviewed studies employ a diverse range of LLMs and associated technologies to achieve these enhancements in credit risk management and related financial tasks. As illustrated in Fig. 1B, these can be grouped into distinct architectural and functional categories. Autoregressive models represent the most prominent technology type, featured in 11 of the 17 papers. A detailed breakdown of the specific models within each category, along with their corresponding sources, is provided in Table 2.

Autoregressive Models: Models from the GPT series (GPT-3.5, GPT-4) are frequently utilized for tasks like data generation, text analysis, relation extraction, regulatory interpretation, and report generation (Gallipoli et al., 2024; Rajpoot & Parikh, 2023; Cao et al., 2024; Cao & Feinstein, 2024; Teixeira et al., 2023; Lin et al., 2025). Other prominent models include Google's Gemini (Xie et al., 2024; Cao & Feinstein, 2024) and Claude-3-Opus (Cao & Feinstein, 2024). Open-source models like Llama variants (Llama2, LLaMA2–7/70B-chat, Llama-3-30b) are also tested and used, sometimes noted as competitive alternatives or used when access to proprietary models is limited (Gallipoli et al., 2024; Xie et al., 2024; Yang et al., 2025; Lee & Stevens, 2025). As detailed in Table 2, models from the GPT series are dominant, with GPT-4 being the most frequently used model in the entire review, cited in 7 studies (Gallipoli et al., 2024; Rajpoot & Parikh, 2023; Xie et al., 2024; Cao et al., 2024; Cao & Feinstein, 2024; Yang et al., 2025; Teixeira et al., 2023). Variants like GPT-3.5 were found in two papers (Lin et al., 2025; Wang & Kang, 2024), while ChatGPT, GPT-3.5 Turbo, and GPT-Neo were also utilized (Rajpoot & Parikh, 2023; Li et al., 2024). Open-source models are also common, with Llama2 appearing in

Table 2 Overview of categories and their respective models

Category	Model		Number	Source
Autoregressive	GPT	GPT-4	7	(Gallipoli et al., 2024; Rajpoot & Parikh, 2023; Xie et al., 2024; Cao et al., 2024; Cao & Feinstein, 2024; Yang et al., 2025; Teixeira et al., 2023)
		GPT-3.5	2	(Lu et al., 2025; Li et al., 2024)
		GPT-3.5 turbo	1	(Rajpoot & Parikh, 2023)
		GPT-neo	1	(Rajpoot & Parikh, 2023)
		ChatGPT	1	(Lin et al., 2025)
	Llama	Llama2	2	(Gallipoli et al., 2024; Xie et al., 2024)
		Llama-3	1	(Yang et al., 2025)
	Gemini	Gemini	1	(Xie et al., 2024)
		Gemini-1.5-pro	1	(Cao & Feinstein, 2024)
	DeepSeek		1	(Wang & Kang, 2024)
	Claude-3-opus		1	(Cao & Feinstein, 2024)
Supporting AI tech and frameworks	Text embedding-ada-002		2	(Gallipoli et al., 2024; Rajpoot & Parikh, 2023)
	BGE-large-v1.5		1	(Li et al., 2024)
	Wav2vec2		1	(Cao et al., 2024)
	GANs		1	(Sai et al., 2025)
	VAEs		1	(Sai et al., 2025)
	CMML		1	(Lu et al., 2025)
Traditional ML, DL and statistical	Altman's Z-score		1	(Lin et al., 2025)
	Beneish's M-model		1	(Lin et al., 2025)
	Extra trees		1	(Hajek & Sahut, 2024)
	XGBoost		1	(Sanz-Guerrero & Arroyo, 2025)
	Recurrent neural networks		1	(Mahadevkar et al., 2024)
	Convolutional neural networks		1	(Mahadevkar et al., 2024)
	Structural equation Modeling		1	(Goyal et al., 2025)
	Artificial neural networks		1	(Goyal et al., 2025)
	BiLSTM		1	(Cao et al., 2024)

(continued)

Table 2 (continued)

Category	Model	Number	Source
	Bayesian networks	1	(Teixeira et al., 2023)
Domain specific LLMs	ChatGLM-4	1	(Wang & Kang, 2024)
	FinMA-7B	1	(Xie et al., 2024)
	CFGPT1-pt	1	(Li et al., 2024)
	RA-CFGPT	1	(Li et al., 2024)
	BloombergGPT	2	(Sai et al., 2025; Lee & Stevens, 2025)
	FinGPT	1	(Sai et al., 2025; Lee & Stevens, 2025)
	InvestLM	1	(Lee & Stevens, 2025)
	FinMA	1	(Lee & Stevens, 2025)
	FinBERT	1	(Lee & Stevens, 2025)
Bidirectional	SentenceBERT	1	(Gallipoli et al., 2024)
	RoBERTa	1	(Lu et al., 2025)
	MPNet	1	(Hajek & Sahut, 2024)
	BERT	1	(Sanz-Guerrero & Arroyo, 2025)
	SimCSE	1	(Cao et al., 2024)

two studies (Gallipoli et al., 2024; Xie et al., 2024) and other models like Llama-3, Gemini, Claude, and DeepSeek being applied for various financial tasks (Xie et al., 2024; Cao & Feinstein, 2024; Yang et al., 2025; Lu et al., 2025).

Traditional ML, DL, and Statistical Models: This category, appearing in 7 papers, highlights that LLMs are often used in conjunction with established methods. Table 2 shows that these hybrid systems use LLM-derived features to feed into classifiers like XGBoost (Sanz-Guerrero & Arroyo, 2025) and Extra Trees (Hajek & Sahut, 2024). Other neural network types and architectures like Artificial Neural Networks, Convolutional Neural Networks, and Recurrent Neural Networks (including Long short-term Memory (LSTM) or BiLSTM) are used for predictive modeling and feature extraction (Cao et al., 2024; Mahadevkar et al., 2024; Goyal et al., 2025), while methods like Bayesian Networks help guide LLM reasoning (Teixeira et al., 2023).

Supporting AI Tech and Frameworks: Also appearing in 7 papers, this category covers the essential enabling technologies. It includes embedding models like text-embedding-ada-002 for semantic similarity (Gallipoli et al., 2024; Rajpoot & Parikh, 2023), audio processing models like Wav2vec2 for analyzing earnings calls (Cao et al., 2024), and generative models like GANs and VAEs for synthetic data tasks (Sai et al., 2025). Techniques like BGE-large-v1.5 are used for generating semantic embeddings and similarity comparisons (Gallipoli et al., 2024; Rajpoot & Parikh, 2023; Cao et al., 2024; Li et al., 2024). Frameworks often employ techniques like

Retrieval Augmented Generation (RAG) (Li et al., 2024; Lee & Stevens, 2025), in-context learning (Rajpoot & Parikh, 2023), CMML for contrastive learning (Lu et al., 2025), and specific prompting strategies like Labeled Guide Prompting (Teixeira et al., 2023).

Domain-Specific LLMs: Tailored for finance, these models were identified in 5 papers (Fig. 1B). Table 2 shows these include BERT-based models, particularly those fine-tuned for finance like FinBERT, that are commonly applied for tasks like sentiment analysis, text mining in loan agreements, and generating risk scores from textual descriptions (Sanz-Guerrero & Arroyo, 2025; Lee & Stevens, 2025). Other finance-specific models mentioned include BloombergGPT, FinMA (LLaMA-based), InvestLM (LLaMA-based), FinGPT, and CFGPT1-pt (base for RA-CFGPT), often involving fine-tuning on financial datasets or instruction tuning for specific financial tasks (Xie et al., 2024; Li et al., 2024; Sai et al., 2025; Lee & Stevens, 2025). Models like ChatGLM-4 were specifically noted for effectiveness in analyzing long-context financial reports (Wang & Kang, 2024).

Bidirectional Models: Noted in five studies, this architecture is primarily used for deep contextual understanding rather than generation. As shown in Table 2, this includes the foundational BERT model for generating risk scores from text (Sanz-Guerrero & Arroyo, 2025), as well as its variants like RoBERTa for feature extraction (Lin et al., 2025) and SentenceBERT for semantic comparisons (Gallipoli et al., 2024). Other models like MPNet and SimCSE are used for specialized tasks like ESG topic detection and creating sentence vector representations (Cao et al., 2024; Hajek & Sahut, 2024).

3.3 Challenges and Limitations

Despite the potential, the adoption of LLMs in credit risk management faces significant challenges, frequently highlighted in the reviewed literature. As depicted in Fig. 1C, data-related issues represent the most frequently cited challenge, appearing in 12 of the 17 papers.

Data Issues: A recurring theme is the challenge related to data. This includes quality and availability, with concerns about poor quality public datasets, inconsistent availability (e.g., loan descriptions), reliance on potentially biased self-reported ESG data, and the need for comprehensive, up-to-date knowledge bases (Xie et al., 2024; Cao et al., 2024; Li et al., 2024; Lin et al., 2025; Hajek & Sahut, 2024; Sanz-Guerrero & Arroyo, 2025; Mahadevkar et al., 2024). Privacy and security are paramount due to sensitive financial data, necessitating techniques like RAG or synthetic data generation, though generated data based on Generative AI may lack real-world complexity (Cao & Feinstein, 2024; Sai et al., 2025; Goyal et al., 2025; Lee & Stevens, 2025). The need for high-quality multimodal data is also noted (Lee & Stevens, 2025).

Model Characteristics: The inherent nature of LLMs presents hurdles (9 papers, Fig. 1C). Explainability, opacity, and interpretability ("black box") are major concerns, particularly in a regulated domain like credit risk where justifications for decisions are critical (Lin et al., 2025; Sai et al., 2025; Sanz-Guerrero & Arroyo, 2025). Bias inherited from training data or inherent in model design (e.g., ESG subjectivity, potential biases in BERT) can lead to unfair or inaccurate outcomes (Gallipoli et al., 2024; Sai et al., 2025; Sanz-Guerrero & Arroyo, 2025). LLMs can exhibit unpredictability, hallucinations, and sensitivity to prompts, potentially producing unreliable results, especially with unseen data or complex queries (Cao & Feinstein, 2024; Teixeira et al., 2023). Limitations in specific tasks like complex mathematical calculations were also noted for some models (Cao & Feinstein, 2024). Evaluating FinLLMs also requires domain-specific metrics beyond standard NLP benchmarks (Lee & Stevens, 2025).

Operational Constraints: Practical implementation faces barriers (9 papers, Fig. 1C). Computational cost and resource requirements for training, fine-tuning, and running large models can be substantial (Xie et al., 2024; Lu et al., 2025; Li et al., 2024; Sai et al., 2025; Sanz-Guerrero & Arroyo, 2025; Mahadevkar et al., 2024). Integration complexity with existing financial systems and data streams, especially for real-time applications, poses challenges (Yang et al., 2025). Access to proprietary models (like GPT-4) can be limited or costly (Rajpoot & Parikh, 2023).

Regulatory and Ethical Concerns: The highly regulated nature of finance creates significant hurdles (4 papers, Fig. 1C). Ensuring compliance with existing regulations when using AI/LLMs for credit decisions is complex (Xie et al., 2024; Lin et al., 2025; Goyal et al., 2025). The lack of standardization, particularly in areas like ESG scoring, affects model reliability (Hajek & Sahut, 2024). Ethical concerns regarding potential misuse, bias amplification, and data privacy violations are prominent (Gallipoli et al., 2024; Xie et al., 2024; Sai et al., 2025; Mahadevkar et al., 2024). Employee perceived risk associated with AI adoption can also hinder implementation (Goyal et al., 2025).

3.4 *Implications for the Financial Industry*

The enhancements and challenges associated with LLMs in credit risk management carry significant implications for the financial industry. Figure 1D highlights improved accuracy and performance as the most frequently cited practical implication, appearing in 14 of the 17 papers.

Improved Accuracy/Performance: LLMs offer substantial potential for enhancing the accuracy of credit risk assessments and default predictions through their ability to process diverse data types including text, audio, and ESG reports (Lu et al., 2025; Lin et al., 2025; Hajek & Sahut, 2024; Sanz-Guerrero & Arroyo, 2025). This improved accuracy could translate into better lending decisions, reduced default rates, and more resilient portfolios.

Efficiency/Cost Reduction: Identified in 9 papers (Fig. 1D), LLMs can significantly reduce costs (with respect to the operational constraints mentioned in Sect. 3.3) and enhance efficiency through the automation of labor-intensive tasks like preliminary data analysis from unstructured sources, credit risk report generation, and document annotation (Gallipoli et al., 2024; Cao & Feinstein, 2024; Teixeira et al., 2023; Lin et al., 2025; Sai et al., 2025). This can free up human analysts for higher-level tasks.

Enhanced Capabilities/Decision-Making: Appearing in 7 papers (Fig. 1D), LLMs can lead to enhanced capabilities and decision-making processes. Adopting LLMs requires financial institutions to invest in new expertise (data science, AI ethics, prompt engineering) and adapt existing workflows. Successful integration can offer a competitive advantage. Furthermore, the ability of LLMs to effectively analyze ESG reports highlights the growing strategic importance of ESG factors in credit risk assessment and overall business strategy (Hajek & Sahut, 2024).

Risk and Compliance: Financial institutions must navigate a complex landscape. While LLMs can aid in regulatory interpretation (Cao & Feinstein, 2024), their use introduces new model risks related to opacity, bias, and potential unpredictability (Sai et al., 2025; Sanz-Guerrero & Arroyo, 2025). Robust data governance frameworks are essential to manage privacy and security concerns (Goyal et al., 2025; Lee & Stevens, 2025). Significant effort will be needed to ensure LLM applications meet stringent regulatory requirements for fairness, transparency, and auditability.

The reviewed literature indicates that LLMs are poised to significantly impact credit risk management by enhancing predictive accuracy, enabling the analysis of previously underutilized unstructured data, and automating key processes like report generation and regulatory interpretation. Models ranging from adapted general LLMs like GPT and BERT to specialized financial LLMs like FinMA are being applied across credit scoring, risk assessment, fraud detection, and compliance. However, substantial challenges remain, primarily concerning data quality and privacy, model explainability and bias, computational costs, and regulatory hurdles.

For the financial industry, harnessing the potential of LLMs implies improved operational efficiency and potentially more robust decision-making, but necessitates strategic investments in technology and talent, alongside careful management of model risk and compliance obligations. The future effective adoption of LLMs in credit risk management hinges critically on addressing the identified challenges, particularly through the development of more transparent, explainable, and robust models, establishing strong data governance, and navigating the evolving regulatory landscape responsibly.

While the potential is clear, successfully navigating these hurdles necessitates focused research efforts in several key areas.

4 Research Gaps and Future Directions

Based on the findings of this systematic literature review, several critical research gaps and promising future directions emerge for the effective and responsible application of LLMs in credit risk management:

Addressing Data Quality, Privacy and Integration Challenges: Data issues emerged as the most prevalent challenge in our review (12 papers, Fig. 1C), highlighting a critical area for future research. While techniques like RAG and synthetic data generation show promise, more sophisticated privacy-preserving methods are needed that allow LLMs to learn from sensitive financial data without compromising confidentiality or utility. Future work could focus on federated learning approaches adapted for complex LLM training, advanced differential privacy mechanisms, and secure multi-party computation specifically for financial text analysis. Furthermore, research is needed on methods for LLMs to effectively handle the heterogeneity and varying quality of financial data (e.g., integrating sparse textual data with dense structured data, managing noisy self-reported ESG data) and perform robustly under data scarcity, particularly for SMEs or niche markets where traditional structured data may be limited.

Enhancing Explainability and Trustworthiness: The "black box" nature of many LLMs remains a significant obstacle (9 papers, Fig. 1C), particularly problematic in the highly regulated financial domain. Future research must prioritize the development of inherently interpretable LLM architectures or robust post-hoc explanation techniques specifically tailored for financial decision-making. This includes methods to clearly trace LLM outputs (e.g., credit scores, risk flags) back to specific input data points (structured or unstructured) and align these explanations with regulatory requirements for transparency and auditability. Addressing model bias beyond simple detection toward effective mitigation strategies within the financial context is also crucial, especially as improved accuracy and performance were identified as the most common practical implication (14 papers, Fig. 1D).

Optimizing Resource Requirements and Operational Integration: Operational constraints were identified as a major challenge (9 papers, Fig. 1C), suggesting a need for research addressing the computational costs and resource requirements of implementing LLMs in financial institutions. Future studies should explore model compression techniques, knowledge distillation, and efficient fine-tuning methods specifically for financial applications to make advanced LLM capabilities more accessible to institutions with varying resource capacities. Research addressing the seamless integration of LLM systems with existing financial infrastructure and legacy systems is also critical for practical adoption, particularly as efficiency and cost reduction benefits (9 papers, Fig. 1D) depend on successful implementation.

Developing Robust Privacy-Preserving Techniques and Handling Data Heterogeneity: Data quality, availability, and privacy were consistently cited as major challenges (Sect. 3.3). While techniques like RAG and synthetic data generation are explored, more sophisticated privacy-preserving methods are needed that allow LLMs to learn from sensitive financial data without compromising confidentiality or utility. Future work should focus on federated learning (Schumann & Awick, 2023) approaches adapted for complex LLM training, advanced differential privacy mechanisms, and secure multi-party computation specifically for financial text analysis. Furthermore, research is needed on methods for LLMs to effectively handle the heterogeneity and varying quality of financial data (e.g., integrating sparse textual data with dense structured data, managing noisy self-reported ESG data) and perform robustly under data scarcity, particularly for SMEs or niche markets.

Developing Specialized Financial LLMs and Hybrid Architectures: While general-purpose, autoregressive LLMs are most commonly used (11 papers, Fig. 1B), domain-specific financial models appeared in only 5 papers, revealing an opportunity for dedicated research. Future work should systematically explore optimal architectures for hybrid models that leverage the strengths of LLMs (e.g., processing unstructured text, identifying qualitative signals) with the quantitative rigor of traditional econometric or ML models. These hybrid approaches could address multiple application areas simultaneously, from credit scoring (8 papers, Fig. 1A) to risk assessment (6 papers), general document analysis and report generation (5 papers) and fraud detection (4 papers). Investigating frameworks for continuous learning and adaptation is also critical, enabling LLMs to dynamically update their knowledge based on real-time market data, evolving regulations, and new financial reporting standards, thereby maintaining their relevance and accuracy in the fast-paced financial environment.

Navigating Regulatory and Ethical Frameworks: Though mentioned less frequently (4 papers, Fig. 1C), the regulatory and ethical challenges of implementing LLMs in credit risk management represent a crucial research direction. Future studies should focus on developing frameworks that ensure LLM applications align with existing financial regulations (e.g., fair lending laws, General Data Protection Regulation, Basel requirements) and emerging AI governance standards. Research into methods for regular auditing, bias detection, and ethical oversight of LLM-based credit decision systems is essential, particularly as these technologies become more integrated into core banking operations.

5 Conclusion

This systematic review sought to answer how Large Language Models (LLMs) can enhance credit risk management in specific application domains, what challenges and limitations emerge, and what are the resulting implications for the financial industry. Our analysis of 17 peer-reviewed publications reveals that LLMs are enhancing credit risk management across a spectrum of key domains, including credit scoring and default forecasting, comprehensive risk assessment, document analysis and report generation, fraud detection, and regulatory interpretation.

LLMs enhance these domains primarily by improving prediction accuracy through the integration of unstructured textual data (annual reports, loan descriptions, ESG disclosures) with traditional structured financial metrics. Models like GPT-4, BERT variants, and finance-specific LLMs enable more comprehensive risk assessments, particularly beneficial for entities with limited structured data. Additionally, LLMs have the potential to significantly increase operational efficiency through automation of document analysis and report generation tasks.

However, our review also identifies significant challenges and limitations that temper this potential. Data-related issues emerge as the most prominent barrier, encompassing concerns over quality, privacy, and integration complexity. Furthermore, the inherent characteristics of the models—namely their limited explainability, potential for bias, and risk of hallucinations—present critical obstacles within the heavily regulated financial sector. These technical hurdles are compounded by operational constraints, such as high computational costs and implementation complexity, alongside the need to navigate ambiguous regulatory and ethical frameworks.

For the financial industry, these developments imply opportunities for enhanced decision-making accuracy, operational efficiency gains, and competitive advantages through expanded analytical capabilities. However, successful implementation necessitates strategic investments in technology infrastructure, talent development, and robust governance frameworks to manage model risks and regulatory compliance.

The future effectiveness of LLMs in credit risk management will depend on addressing explainability challenges, developing privacy-preserving techniques, optimizing resource requirements, creating specialized financial LLMs, and establishing clear regulatory frameworks that balance innovation with ethical considerations and consumer protection.

Disclosure of Interests The authors have no competing interests to declare that are relevant to the content of this article.

References

Akkizidis, I., & Kalyvas, L. (2018). *Final Basel III Modelling*. Springer International Publishing.

Alvarado, J. C. S., Verspoor, K., & Baldwin, T. (2015). Domain adaption of named entity recognition to support credit risk assessment. In *Proceedings of the Australasian Language Technology Association Workshop 2015* (pp. 84–90).

Brahmaiah, B. (2022). Credit risk management practices of Indian banking industry: An empirical study. *International Journal of Economics and Financial Issues (IJEFI), 12*, 67–71. https://doi.org/10.32479/ijefi.12968

Cao, Y., Chen, Z., Pei, Q., Dimino, F., Ausiello, L., Kumar, P., Subbalakshmi, K.P., Ndiaye, P.M. (2024). RiskLabs: Predicting financial risk using large language model based on multi-sources data.

Cao, Z., Feinstein, Z. (2024). Large language model in financial regulatory interpretation.

Cimini, G., & Serri, M. (2016). Entangling credit and funding shocks in interbank markets. *PLoS One, 11*, e0161642. https://doi.org/10.1371/journal.pone.0161642

Federal Financial Supervisory Authority. Circular 05/2023 (BA)—Minimum Requirements for Risk Management (MaRisk): Annotated text of the Minimum Requirements for Risk Management (MaRisk) in the version of 29.06.2023. Federal Financial Supervisory Authority (BaFin). (2023).

Fritz, D., & Töws, E. (2018). Text mining and reporting quality in German banks—a cooccurrence and sentiment analysis. *Journal of Accounting and Finance.*

Gallipoli, G., Papicchio, S., Vaiani, L., Cagliero, L., Miola, A., & Borghi, D. (2024). Keyword-based annotation of visually-rich document content for trend and risk analysis using large language models. In *Proceedings of the Joint Workshop of the 7th Financial Technology and Natural Language Processing, the 5th Knowledge Discovery from Unstructured Data in Financial Services, and the 4th Workshop on Economics and Natural Language Processing* (pp. 130–136).

Glaser, C. (2022). *Digitale Transformation Im Bankenumfeld.* Springer Fachmedien Wiesbaden.

Goyal, K., Garg, M., & Malik, S. (2025). Adoption of artificial intelligence-based credit risk assessment and fraud detection in the banking services: A hybrid approach (SEM-ANN). *Future Business Journal, 11*, 1–20. https://doi.org/10.1186/s43093-025-00464-3

Hajek, P., Sahut, J.-M., & Myskova, R. (2024). Predicting corporate credit ratings using the content of ESG reports. *Annals of Operations Research*, 1–28. https://doi.org/10.1007/s10479-024-06385-8

Hellwig, M. (1995). Systemic aspects of risk management in banking and finance. *Swiss Journal of Economics and Statistics (SJES), 131*, 723–737.

Karouzakis, N. (2021). The role of time-varying risk premia in international interbank markets. *International Journal of Finance & Economics, 26*, 5720–5745. https://doi.org/10.1002/ijfe.2090

Langohr, H. M., & Langohr, P. T. (2012). *The Rating Agencies and Their Credit Ratings.* Wiley.

Lee, J., Stevens, N., & Han, S. C. (2025). Large language models in finance (FinLLMs). *Neural Computing and Applications*, 1–15. https://doi.org/10.1007/s00521-024-10495-6

Li, J., Lei, Y., Bian, Y., Cheng, D., Ding, Z., & Jiang, C. (2024). RA-CFGPT: Chinese financial assistant with retrieval-augmented large language model. *Frontiers in Computer Science, 18*, 1–3. https://doi.org/10.1007/s11704-024-31018-5

Lin, J., Lai, S., Yu, H., Liang, R., & Yen, J. (2025). ChatGPT based credit rating and default forecasting. *Journal of Data, Information and Management, 7*, 69–92. https://doi.org/10.1007/s42488-025-00143-6

Lu, S., Su, Y., Zhang, X., Chai, J., & Yu, L. (2025). LLM-infused bi-level semantic enhancement for corporate credit risk prediction. *Information Processing & Management, 62*, 104091. https://doi.org/10.1016/j.ipm.2025.104091

Mahadevkar, S. V., Patil, S., Kotecha, K., Soong, L. W., & Choudhury, T. (2024). Exploring AI-driven approaches for unstructured document analysis and future horizons. *Journal of Big Data, 11*. https://doi.org/10.1186/s40537-024-00948-z

Mayring, P. (2014). Qualitative Content Analysis: Theoretical Foundation, Basic Procedures and Software Solution.

Patil, R., & Gudivada, V. (2024). A review of current trends, techniques, and challenges in large language models (LLMs). *Applied Sciences, 14*, 2074. https://doi.org/10.3390/app14052074

Qin, L., Chen, Q., Feng, X., Wu, Y., Zhang, Y., Li, Y., Li, M., Che, W., & Yu, P. S. (2024). Large language models meet nlp: A survey. *arXiv preprint arXiv:2405.12819.*

Radojević, T., Kesić, M., Rajin, D., Butėnas, R. (2022). Credit risk management in a changing world (pp. 66–70). https://doi.org/10.15308/finiz-2022-66-70

Rajpoot, P., & Parikh, A. (2023). GPT-FinRE: In-context learning for financial relation extraction using large language models. In *Proceedings of the Sixth Workshop on Financial Technology and Natural Language Processing* (pp. 42–45). https://doi.org/10.18653/v1/2023.finnlp-2.5

Reck, R. (2016). *Kultur und Sprache im Geschäftsbericht.* Springer Fachmedien Wiesbaden.

Sai, S., Arunakar, K., Chamola, V., Hussain, A., Bisht, P., & Kumar, S. (2025). Generative AI for finance: Applications, case studies and challenges. *Expert Systems, 42*, e70018. https://doi.org/10.1111/exsy.70018

Sanz-Guerrero, M., & Arroyo, J. (2025). Credit risk meets large language models: Building a risk indicator from loan descriptions in P2P lending. *NTARTIF, 28*, 220–247. https://doi.org/10.4114/intartif.vol28iss75pp220-247

Schumann, G., Awick, J.-P., & Gómez, J. M. (2023). Natural language processing using federated learning: A structured literature review. In *2023 IEEE International Conference on Artificial Intelligence, Blockchain, and Internet of Things (AIBThings)* (pp. 1–7). IEEE. https://doi.org/10.1109/AIBThings58340.2023.10292481

Soehaditama, J. P., Machdar, N. M., & Manurung, A. H. (2023). Determinant banking credit risk management. *Indonesian Journal of Business Analytics (IJBA), 3*, 1105–1112. https://doi.org/10.55927/ijba.v3i4.5032

Teixeira, A.C., Marar, V., Yazdanpanah, H., Pezente, A., Ghassemi, M. Enhancing credit risk reports generation using LLMs: An integration of Bayesian networks and Labeled guide prompting. In: *4th ACM International Conference on AI in Finance* (pp. 340–348). ACM (11272023). doi: https://doi.org/10.1145/3604237.3626902

Touvron, H., Martin, L., Stone, K., Albert, P., Almahairi, A., Babaei, Y., Bashlykov, N., Batra, S., Bhargava, P., Bhosale, S., et al. (2023). Llama 2: Open foundation and fine-tuned chat models. *arXiv preprint arXiv:2307.09288.*

Wang, H., Kang, Z. Linking high-risk companies to special treatment (ST) status: A large language model approach for corporate risk evaluation. In: *Proceedings of the 2024 3rd International Conference on Algorithms, Data Mining, and Information Technology* (pp. 413–417). New York, NY, USA: ACM (09272024). https://doi.org/10.1145/3701100.3702877

Webster, J., & Watson, R. T. (2002). Analyzing the past to prepare for the future: Writing a literature review. *MIS Quarterly, 26*, xiii–xxiii.

Xie, Q., Han, W., Chen, Z., Xiang, R., Zhang, X., He, Y., Xiao, M., Li, D., Dai, Y., Feng, D., et al. (2024). FinBen: A holistic financial benchmark for large language models.

Yang, J., Tang, Y., Li, Y., Zhang, L., Zhang, H. (2025). Cross-asset risk management: Integrating LLMs for real-time monitoring of equity, fixed income, and currency markets.

Yuliia, Y., Kolomiiets, & Kochorba, V. (2024). Assessing the credit risks in the risk management system of banking structures. *Business Inform.*

Zirkler, B., Hofmann, J., & Schmolz, S. (2015). *Basel III in der Unternehmenspraxis*. Springer Fachmedien Wiesbaden.

Leveraging Large Language Models for Supporting Cyber Threat Analysis

Fabian Rybinski, Demian Frister, Gunther Schiefer, and Marzieh Malekzadeh Mahani

Abstract The efficient and accurate analysis of cyber threat data is crucial in the constantly evolving cybersecurity landscape. However, a major challenge lies in the vast amounts of unstructured, human-readable information that is often used for threat intelligence communication, such as threat reports and news articles. While structured formats like Structured Threat Information eXpression (STIX) enable effective machine-to-machine exchange of threat data, they often lack the contextual information needed by human analysts. This paper proposes a novel framework that leverages Large Language Models (LLMs) to bridge the gap between unstructured text and structured cyber threat intelligence. The key contributions of this work are twofold: (1) Automating the conversion of unstructured cyber threat data into the standardized STIX format, enabling the efficient incorporation of diverse threat intelligence sources into automated analysis and sharing systems. (2) Generating human-readable reports and insights from structured STIX data, tailored to the needs of different personas within an organization, such as CISOs, security analysts, executives. By combining the power of LLMs with structured threat data formats, the proposed framework improves the efficiency, accessibility, and interpretability of cyber threat intelligence. The evaluation of a prototype implementation demonstrates the accuracy of unstructured-to-structured conversion, the quality of the generated reports, and the positive feedback from users across various roles. This research contributes to the field of cybersecurity by enhancing the flow of critical threat information, streamlining cross-platform communication, and ultimately supporting more effective and informed decision-making in cyber defense.

F. Rybinski (✉) · D. Frister · G. Schiefer
Karlsruhe Institute of Technology (KIT), Institute AIFB, Karlsruhe, Germany
e-mail: fabian.rybinski@kit.edu

D. Frister
e-mail: demian.frister@kit.edu

G. Schiefer
e-mail: gunther.schiefer@kit.edu

M. M. Mahani
FIZ, Karlsruhe, Germany
e-mail: marzieh.malekzadeh_mahani@fiz-karlsruhe.de

J. Marx Gómez et al. (eds.), *Recent Advances in Information Systems*, Progress in IS,
https://doi.org/10.1007/978-3-032-14099-9_34

1 Introduction

The efficient and accurate analysis of cyber threat data is crucial in the constantly evolving cybersecurity landscape. Organizations face an ever-growing deluge of cyber threats, with attackers employing increasingly sophisticated methods. This constant evolution of threats demands equally advanced defensive strategies, particularly in the realm of Cyber Threat Intelligence (CTI).

CTI plays a crucial role in modern cybersecurity practices, providing organizations with the knowledge needed to preempt and mitigate potential attacks. However, the current state of CTI processing and dissemination faces significant challenges. A major hurdle lies in the vast amount of unstructured, human-readable information used for threat intelligence communication, such as threat reports, emails, and news articles. While this unstructured data contains valuable insights, it is not readily usable by automated threat intelligence systems.
Conversely, structured formats like the Structured Threat Information eXpression (STIX) enable effective machine-to-machine exchange of threat data. However, these structured formats often lack the contextual information that human analysts need to make strategic decisions. Security professionals frequently struggle to screen voluminous STIX feeds and extract the most relevant and actionable threat intelligence.

Recent advancements in artificial intelligence, particularly in the domain of Large Language Models (LLMs), offer a promising avenue to address these challenges. LLMs have demonstrated remarkable capabilities in natural language understanding and generation across various domains. However, their potential in the specific context of cybersecurity threat intelligence remains largely unexplored.

This paper proposes a novel framework that leverages the power of LLMs to bridge the gap between unstructured text and structured cyber threat intelligence. Our approach addresses two critical challenges:

1. The automated conversion of unstructured cybersecurity data into the standardized STIX format, enabling the efficient incorporation of diverse threat intelligence sources into automated analysis and sharing systems.
2. The generation of human-readable reports and insights from structured STIX data, tailored to the needs of different personas within an organization, such as CISOs, security analysts, executives.

By combining the advanced language processing capabilities of LLMs with structured threat data formats, our framework aims to improve the efficiency, accessibility, and interpretability of cyber threat intelligence. This research contributes to the field of cybersecurity by enhancing the flow of critical threat information, streamlining cross-platform communication, and ultimately supporting more effective and informed decision-making in cyber defense.

The remainder of the paper is structured as follows: Sect. 2 introduces the necessary background on LLMs and cybersecurity concepts. Section 3 provides an overview of the related work in the field of automated CTI processing.

Section 4 presents our conceptual framework and its implementation details.

Section 5 discusses the evaluation of our proposed system, including its performance in converting unstructured data to STIX format and generating persona-specific reports. Finally, Sect. 6 concludes the paper with a summary of our findings and directions for future research.

2 Background

This section provides the necessary context for understanding the intersection of LLMs and cybersecurity, particularly the domain of CTI processing. We begin by introducing LLMs, their capabilities, and current applications. We then dive into key cybersecurity concepts, focusing on CTI and the STIX format.

2.1 Large Language Models

Large Language Models (LLM) represent a transformative advancement in natural language processing (NLP) and artificial intelligence, allowing machines to understand and generate human-like text on an unprecedented scale (Minaee et al., 2024). These models, built on deep learning architectures, have leveraged vast amounts of data to develop nuanced language comprehension and production capabilities. A key innovation in LLMs is the Transformer architecture, which employs a self-attention mechanism enabling parallel processing of sequences and effective handling of long-range dependencies (Vaswani et al., 2017). Unlike sequential architectures such as LSTMs, Transformers process input sequences in parallel and effectively handle long-range dependencies (Hochreiter, 1997; Vaswani et al., 2017).

Training LLMs involves two main phases: pretraining, where the model develops a general understanding of language through tasks like masked language modeling or next-token prediction, followed by fine-tuning, which adapts the model to specific tasks or domains (Parthasarathy et al., 2024). This dual phase training strategy allows LLMs to generalize across diverse tasks, from text summarization and translation to question-answering and content generation. The model's scale, often reaching billions of parameters, equips it to capture intricate language patterns and relationships, a capability that underlies its ability to perform remarkable fluency across tasks (Naveed et al., 2024).

A defining feature of LLMs is their capacity for zero-shot and few-shot learning. Few-shot learning enables the model to tackle new tasks with only minimal examples, allowing it to rapidly adapt to novel scenarios without extensive additional fine-tuning (Brown, 2020). Zero-Shot learning goes even further, empowering the model to respond to tasks without specific examples, instead relying on broad contextual knowledge to understand instructions and perform accurately. This flexibility comes from pretraining on extensive and diverse datasets, enabling the model to infer and perform a wide array of tasks based on generalizable patterns in language (Bommasani et al., 2021).

The capabilities of LLMs have led to applications in multiple sectors. Notable models include GPT, BERT, and LLaMA (Bommasani et al., 2021; Chen et al., 2021). However, challenges remain. LLMs can generate outputs that are incorrect or biased, face difficulties with tasks requiring external knowledge, and occasionally produce inconsistent responses.

2.2 *Cybersecurity*

Cybersecurity is a multifaceted field encompassing the practices, technologies, and processes designed to protect digital systems from cyber threats (Alazab et al., 2022). The core principles of cybersecurity are often described using the CIA triad: Confidentiality, Integrity, and Availability (Anderson, 1972; Saltzer & Schroeder, 1975). These principles guide the development and implementation of security measures across various digital domains.

These security objectives can be compromised through various types of attacks, systematically categorized using the STRIDE model (Howard & Lipner, 2006). This model classifies attacks into six categories: Spoofing (targeting authentication mechanisms), Tampering (system manipulation), Repudiation (undetected manipulation), Information Disclosure (unauthorized access and exposure), Denial of Service (targeting system availability), and Elevation of Privilege (gaining expanded permissions). To counter these threats, organizations employ multi-layered defense strategies spanning prevention, detection, reaction, and attribution.

As cyber threats become more sophisticated, the role of Cyber Threat Intelligence (CTI) has grown increasingly crucial. CTI involves the collection, processing, and analysis of information about current and potential attacks, hazards, and vulnerabilities (Saeed et al., 2023). By leveraging CTI, organizations can enhance their capabilities across all phases of the cybersecurity lifecycle.

In this complex and dynamic environment, the ability to quickly process, analyze, and act upon threat intelligence becomes paramount. Traditional methods of manual analysis and rule-based systems struggle to keep pace with the volume and complexity of modern cyber threats. This has led to the development of standardized formats for sharing threat information. The Structured Threat Information eXpression (STIX) has emerged as the predominant format (Sauerwein et al., 2017). Despite the availability of STIX as a framework for sharing cybersecurity information, organizations still face significant barriers to threat intelligence exchange, particularly regarding data privacy concerns and regulatory constraints (Koepke, 2017).

2.3 *Structured Threat Information eXpression—STIX*

STIX is a language and serialization format based on JSON that is used to exchange CTI data (Menges & Pernul, 2018). Developed to enable organizations to share CTI with one another in a consistent and machine-readable manner, STIX provides a common framework for representing structured threat information.

The STIX standard defines a set of interconnected objects that represent different aspects of threat intelligence (Barnum, 2014). These include: **Observables** (stateful properties or measurable events using CybOX), **Indicators** (observable patterns with contextual information and metadata), **Incidents** (discrete instances affecting organizations with response information), **Tactics, Techniques, and Procedures (TTPs)** (adversary behaviors including attack patterns and malware), **Campaigns** (threat actor pursuits observed through incidents/TTPs), **Threat Actors** (characterizations of malicious actors with intent and behavior), **Exploit Targets** (vulnerabilities in systems targeted for exploitation), and **Courses of Action** (specific corrective, preventative, or responsive measures).

STIX uses a graph-based model to represent the relationships between different objects, allowing for a rich, interconnected representation of threat intelligence. This approach enables analysts to understand not just individual pieces of information, but how they relate to form a comprehensive picture of threat.

The use of STIX offers several advantages in the cybersecurity ecosystem:

(1) **Standardization**—providing a common language for describing threats and facilitating communication between organizations and tools, (2) **Machine Readability**—enabling automated processing and analysis of threat data, and (3) **Comprehensive Coverage**—representing a wide range of threat-related concepts, from low-level indicators to high-level threat actor behaviors.

3 Related Work

The automated processing and sharing of CTI has been an active research area, focusing on several key challenges: structuring unstructured threat information, automated extraction of threat intelligence, and secure dissemination of security advisories.

Several approaches have been proposed for extracting structured information from unstructured CTI sources. TTPDrill achieved automated extraction of threat actions and TTPs from threat reports using NLP and information retrieval techniques, incorporating MITRE's CAPEC and ATT&CK repositories and achieving over 82% precision and recall (Husari et al., 2017). Taking a different approach, CyNER introduced techniques for handling non-contextual Indicators of Compromise (IOC) through key phrase extraction, establishing relationships between entities and IOCs even when physically separated in text, and achieving an F-measure of 0.80 for named entity recognition (Fujii et al., 2022).

Building on these rule-based approaches, recent work demonstrates that schema-guided LLMs can perform CTI extraction with minimal task-specific training. Lu et al. propose Unified Information Extraction (UIE), framing entity, relation and event extraction as a single text-to-structure generation problem and achieving state-of-the-art results on thirteen IE benchmarks in zero- and few-shot settings (Lu et al., 2022). Xu et al. introduce IntelEX, a GPT-4-based framework combining in-context learning with retrieval to derive ATT&CK TTP sequences, raising the F_1 score for technique extraction to 0.79—more than 15 points above prior baselines (Xu et al., 2024).

Recent work has also explored Web 3.0 technologies for secure advisory distribution. SENTINEL and related blockchain-based systems utilize Ethereum smart contracts and IPFS for reliable vulnerability notifications and advisory storage, implementing both public and private disclosure workflows using CSAF-structured advisories (Leite et al., 2022; Sommer et al., 2023). These approaches emphasize automation of vulnerability disclosure and management processes while maintaining accountability through blockchain transaction logs.

A common thread across recent work is the move toward structured intelligence sharing formats, with information extraction approaches converting to STIX format and blockchain-based approaches utilizing CSAF, both facilitating automated processing and integration with security operations.

4 Concept and Implementation

This section introduces a framework leveraging Large Language Models (LLMs) to address critical cybersecurity challenges. The primary challenges addressed are managing the influx of vast amounts of complex, unstructured data from diverse sources (e.g., emails, threat reports, and network logs), and disseminating actionable intelligence to relevant stakeholders in a timely manner. The framework aims to bridge the gap between raw data and effective cybersecurity decision-making.

4.1 Overview and Objectives

The proposed framework is designed to achieve two key objectives within the cybersecurity intelligence domain:

1. **Standardized CTI Generation (Unstructured-to-STIX)**: The first objective facilitates the transformation of unstructured or diverse Cyber Threat Intelligence (CTI) formats into the Structured Threat Information eXpression (STIX) standard. This enhances interoperability and improves data utilization across different platforms, tools, and organizations, addressing the challenge of managing diverse data inputs.

2. **Persona-Specific Action Plans**: The second objective provides tailored courses of action for diverse organizational roles. This ensures that each stakeholder receives relevant and actionable intelligence, customized to their specific responsibilities and expertise, addressing the challenge of disseminating actionable intelligence.

These objectives collectively aim to enhance the efficiency of threat intelligence processing, improve the accessibility of critical information, and support more informed and timely decision-making in cybersecurity operations.

4.2 Framework Architecture

For demonstration purposes, the frontend of the framework is implemented as a Streamlit application,[1] named *STIXelator*.[2] This application showcases the framework's core functionalities, accessible through both the Streamlit interface and programmatically via an API.

The information required by the LLMs, such as specific instructions or data schemas, is stored in JSON configuration files. The framework comprises two main components: LLM2Structured and Structured2LLM, representing the two primary directions of data transformation.

4.3 LLM2Structured

The LLM2Structured component transforms unstructured cybersecurity data into a structured format, specifically the STIX format. This conversion enables automated analysis, sharing, and integration of threat intelligence across different security tools and platforms.

The overall process flow for this component is illustrated in Fig. 1, highlighting the key stages from data input to validated STIX output.

The process involves several key stages:

User Interface and Data Input: The process begins with a Streamlit-based user interface for inputting unstructured cybersecurity data. Users can select from various information sources: BSI CERT Information,[3] Email Content, Security Bulletins (in PDF format), and CSAF Documents. Users input data by selecting a source and pasting or uploading the relevant text or file.

Configuration and Preprocessing: The system dynamically configures the LLM model based on environment variables, selecting the appropriate schema and system

[1] https://streamlit.io/ is a framework for rapidly building and deploying data-driven ML applications.

[2] http://stixelator.aifb.org.

[3] https://wid.cert-bund.de/portal/wid/kurzinformationen.

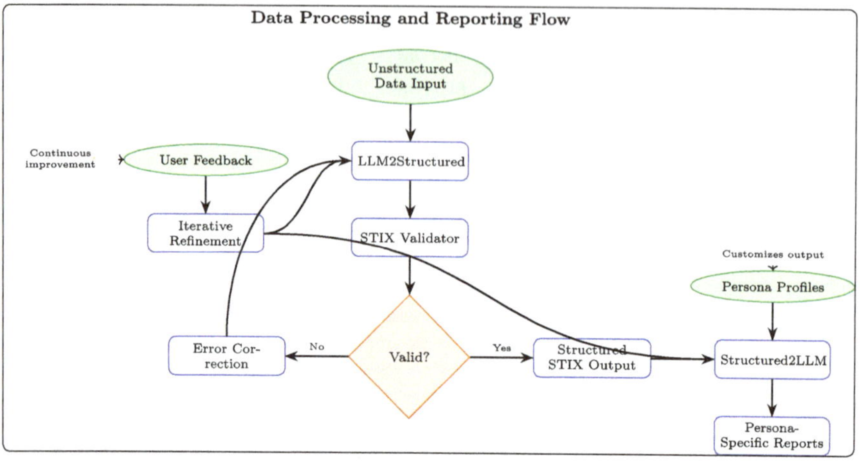

Fig. 1 STIXelator process flow

prompt based on the chosen input type. For example, CSAF documents use a specific CSAF schema and prompt, while other input types use a more general STIX schema and the system prompt in Listing 1.

Listing 1 System prompt for STIX conversion

```
You are a world-class algorithm for extracting information in structured formats and
creating comprehensive STIX 2.1 descriptions of security incidents to help protect
organizations from cyber attacks. \n\nImportant guidelines:\n1. Use double quotes
instead of single quotes for JSON.\n2. All timestamps must be precise to the second in
the format YYYY-MM-DDTHH:mm:ss.sssZ.\n3. Ensure all STIX object IDs and references
follow the format: [object-type]--[UUID], where [UUID] MUST be a valid UUID v4
(e.g., 550e8400-e29b-41d4-a716-446655440000). This applies to both 'id' fields and any
references in 'object_refs' arrays.\n4. Include ALL required properties for each STIX
object type.\n5. Use valid STIX 2.1 object types and properties.\n\nCreate a STIX Bundle
with multiple STIX Objects that comprehensively describe the incident, including:\n- An
identity for the organization (e.g., BSI CERT)\n- A report summarizing the
vulnerabilities\n- Vulnerability objects for each specific vulnerability mentioned\n-
A course of action to remediate the vulnerabilities\n\nEnsure that all objects are
properly linked using object_refs where appropriate, always using valid UUIDs. The
bundle should contain at least these objects, but you can include more if relevant.
\n\nRespond only with the STIX Bundle in JSON format and nothing else."
```

LangChain Integration and Prompt Engineering: The framework uses LangChain[4] to orchestrate the LLM interaction and structure the output. This involves constructing a "runnable" that combines the selected schema, the configured LLM, and a ChatPromptTemplate with a system prompt and a task-specific prompt.

LLM Processing and Initial Output Generation: The prepared input is processed by the LLM through the LangChain runnable, generating a structured JSON representation conforming to the specified schema.

Validation and Refinement: The generated STIX output is validated using the STIX validator tool. If invalid, the system attempts correction through an iterative refinement process: (1) validation errors and the original output are presented back to the LLM, (2) a specialized correction prompt guides the LLM in addressing issues, and

[4] https://www.langchain.com/.

(3) the corrected output is revalidated. This process continues until a valid output is produced or the system determines further attempts are unlikely to succeed.

Result Presentation and Storage: The valid STIX output is presented to the user in a formatted JSON display with a download link. Results are stored in a pickle file, indexed by a hash of the input data for future retrieval.

Figure 2 summarizes this refinement process as data moves through the LLM2Structured pipeline (Fig. 3).

4.4 *Structured2LLM*

The Structured2LLM component converts structured cybersecurity data (STIX format) into human-readable text, generating customized reports and actionable insights for different organizational stakeholders. The process involves the following stages:

Data Input and Persona Selection: Users enter structured STIX data and select a persona from a predefined list including CISO, Security Analyst, Security Officer, IT

Fig. 2 LLM2Structured process funnel

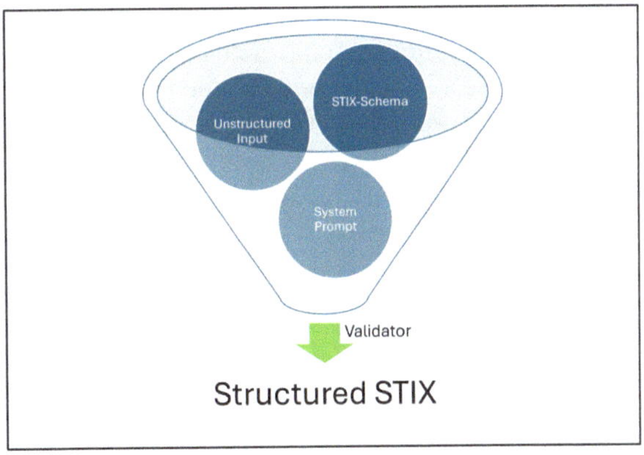

Fig. 3 Structured2LLM process flow

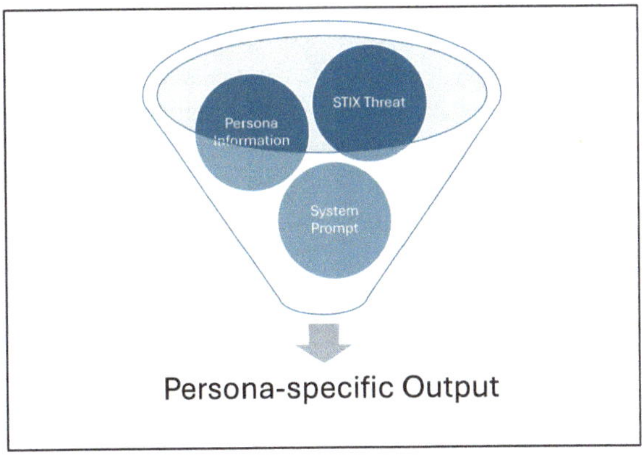

System Administrator, Startup CEO, and Mayor of a Small Municipality. Each persona represents a distinct stakeholder with specific responsibilities and information needs.

Persona-specific processing retrieves a detailed profile for the selected persona, including experience level, main tasks, motivations, goals, frustrations, challenges, technical skills, and preferred communication channels.

Natural Language Generation: The LLM processes the structured STIX data, persona profile, and crafted prompts to generate a human-readable explanation tailored to the selected persona. This involves selecting relevant information, explaining technical details at an appropriate complexity level, and using language familiar to the persona.

Output Presentation: The persona-specific explanation with actionable insights is displayed in the Streamlit interface.

5 Evaluation

The evaluation of our proposed framework was conducted in two phases to assess different aspects of the tool's capabilities. First, we evaluated the tool's ability to generate valid STIX files from unstructured text, and second, we assessed its capability to translate STIX data into human-readable courses of action for different personas. All experiments were carried out using GPT-3.5-turbo via the OpenAI API. To ensure consistent and generally reproducible results, we used a temperature setting of 0.3 and maintained all other parameters at their default values. The same model configuration was used across both evaluation phases to enable direct comparison of results.

5.1 Translation of BSI Security Feed to STIX

The first phase of our evaluation aimed to assess the tool's ability to convert unstructured threat intelligence into structured STIX format. This is a crucial functionality, as cybersecurity teams regularly deal with vast amounts of threat intelligence data, often presented in unstructured or semi-structured formats, making it difficult to standardize and integrate into automated defense systems.

For this phase, we utilized the security feed provided by the German Federal Office for Information Security (BSI), which offers up-to-date information on various cyber threats and vulnerabilities. The BSI feed was chosen not only for its comprehensive coverage, but also because it reflects the unstructured nature of real-world threat intelligence. Each entry in the feed is a text-based description of a cyber threat, typically written for human consumption rather than machine parsing. Thus, this evaluation tests the tool's ability to process natural language, extract relevant threat information, and generate structured STIX files.

Table 1 Results of BSI security feed to STIX translation

Metric	Value
Total entries processed	100
Valid STIX files generated	100
Success rate	100%

We selected the 100 most recent entries from the feed (September 2024), ensuring a diverse sample of threats across different sectors and threat vectors. Each of these entries was input into the tool, which analyzed the text, identified key threat indicators (such as affected systems, attack methods, and potential impacts), and mapped this information to appropriate STIX objects.

Once each entry was processed, the generated STIX files were validated for adherence to the STIX standards using our built-in validator. This step was essential to confirm that the output files were not only syntactically correct but also met the structural requirements of STIX. The validator checked for errors in object relationships, missing fields, and other inconsistencies that could affect the usability of the STIX data in automated systems.

The results of this evaluation are summarized in Table 1. Out of the 100 entries processed, the tool successfully generated 100 valid STIX files, achieving a 100% success rate. This confirms the tool's strong performance in translating unstructured natural language threat descriptions into well-formed, structured STIX files. The fact that all the files passed the validation process demonstrates the robustness of the tool's parsing and mapping algorithms.

While the tool's ability to generate syntactically valid STIX files is impressive, it is important to recognize the limitations of this phase of evaluation. Specifically, this process only verified the structural validity of the STIX files, without assessing the semantic accuracy of the threat intelligence translation. In other words, while the files were technically correct, this evaluation did not determine whether all relevant information from the original feed entries was accurately captured and appropriately represented in the STIX format.

In future evaluations, a more in-depth analysis will be needed to assess the completeness and accuracy of the threat information extraction.

5.2 *STIX to Human-Readable Courses of Action*

In the second phase, we evaluated the tool's ability to translate STIX-based CTI into human-readable courses of action. To do this, we used five personas representing different roles within cybersecurity and IT management that were defined in the project MANTRA (Kubach & Horch, 2024). Chief Information Security Officer (CISO), Security Analyst, IT System Administrator, CEO of a Medium-Sized Company, and

Table 2 Overall appropriateness ratings for personas

Persona	Overall appropriateness rating
Chief information security officer (CISO)	3.96
Security analyst	3.84
IT system administrator	3.80
CEO of a medium-sized company	3.20
Mayor of a small municipality	3.00

Mayor of a Small Municipality. These personas were designed with detailed profiles, including their motivations, challenges, and levels of technical experience.

We focused on five key criteria in our evaluation:

1. **Relevance**: How well the tool's output aligns with the persona's specific role and responsibilities.
2. **Clarity**: The ease of understanding of the output, including the use of appropriate language and logical structure.
3. **Actionability**: Whether the tool provides clear and feasible actions that the persona can take.
4. **Specificity**: The extent to which the output is tailored to the persona's unique needs.
5. **Technical Depth**: Whether the technical complexity of the output matches the persona's expertise level.

Each persona received 100 outputs based on different STIX inputs, and these outputs were rated against the criteria on a scale from 1 to 5. The average scores for each persona were calculated to determine their overall appropriateness rating, as summarized in Table 2.

The evaluation shows that the tool performs reasonably well across most personas, with appropriateness ratings ranging from 3.00 to 3.96. The highest rating was achieved for the CISO persona (3.96), suggesting that the tool's output is most effective for senior-level security professionals who require both strategic and technical insights. The Security Analyst and IT System Administrator personas also received high ratings of 3.84 and 3.80, respectively, indicating the tool's strong performance in providing actionable and relevant insights for technical users.

However, the tool performed less effectively for non-technical personas, as reflected in the ratings for the CEO (3.20) and Mayor (3.00). These roles likely require simpler, high-level summaries of the CTI data to make informed decisions, highlighting the need for the tool to adapt its output for users with limited technical expertise.

Overall, the tool demonstrates promise in helping technical professionals interpret and act upon CTI data, but further refinement is needed to improve its utility for non-technical stakeholders. This suggests future work could focus on simplifying the language and structure of the output for these personas, ensuring the tool's effectiveness across a broader range of roles within an organization.

6 Summary and Outlook

This paper introduced a novel framework leveraging LLMs to address critical challenges in cybersecurity threat intelligence processing and dissemination. Our approach bridges the gap between unstructured threat data and the structured STIX format while generating persona-specific actionable insights.

The key contributions lie in developing an LLM-based system that accurately converts unstructured threat intelligence into valid STIX format (100% success rate on 100 BSI entries) and generates tailored, human-readable reports from STIX data. Our persona-based approach shows promising results for technical roles with appropriateness ratings ranging from 3.80 to 3.96 (out of 5.00).

While our evaluation shows encouraging outcomes for technical personas, we acknowledge limitations in addressing non-technical stakeholders' needs. Lower appropriateness ratings for CEO and mayor personas (3.20 and 3.00) highlight the challenge of simplifying complex technical information for broader audiences. Additionally, our evaluation focused on structural STIX validity rather than semantic accuracy of threat intelligence conversion.

This challenge presents an opportunity for future research and development. Key priorities include enhancing semantic accuracy of STIX conversion, improving system adaptability for non-technical personas through refined prompt engineering, and expanding evaluation to include diverse threat intelligence sources with expert validation. Additionally, comprehensive runtime performance and cost metrics are needed to assess system scalability and practical deployment viability. Integration with existing SIEM systems could enable real-time threat intelligence processing and response.

This research demonstrates significant potential of LLMs in revolutionizing cyber threat intelligence processing. By automating conversion between unstructured and structured formats while providing tailored insights, our framework establishes groundwork for more efficient cybersecurity practices across diverse organizational roles.

References

Alazab, M., Swarna Priya, R. M., Parimala, M., Maddikunta, P. K. R., Gadekallu, T. R., & Pham, Q. V. (2022). Federated learning for cybersecurity: Concepts, challenges, and future directions. *IEEE Transactions on Industrial Informatics, 18*(5), 3501–3509. https://doi.org/10.1109/TII.2021.3119038

Anderson, J. P. (1972). *Computer security technology planning study* (Vol. 1).

Barnum, S. (2014). Standardizing cyber threat intelligence information with the Structured Threat Information eXpression (STIX™). https://stixproject.github.io/getting-started/whitepaper/

Bommasani, R., Hudson, D. A., Adeli, E., Altman, R., Arora, S., von Arx, S., Bernstein, M. S., Bohg, J., Bosselut, A., & Brunskill, E. (2021). On the opportunities and risks of foundation models. arXiv preprint arXiv:2108.07258

Brown, T. B. (2020). Language models are few-shot learners. arXiv preprint arXiv:2005.14165

Chen, M., Tworek, J., Jun, H., Yuan, Q., Pinto, H. P. D. O., Kaplan, J., Edwards, H., Burda, Y., Joseph, N., & Brockman, G. (2021). Evaluating large language models trained on code. arXiv preprint arXiv:2107.03374

Fujii, S., Kawaguchi, N., Shigemoto, T., & Yamauchi, T. (2022). CyNER: Information extraction from unstructured text of CTI sources with noncontextual IOCs. In C. M. Cheng, & M. Akiyama (Eds.), *Advances in Information and Computer Security* (pp. 85–104). Springer International Publishing. https://doi.org/10.1007/978-3-031-15255-9_5

Hochreiter, S. (1997). Long short-term memory. *Neural Computation, 9*(8), 1735–1780.

Howard, M., & Lipner, S. (2006). *The security development lifecycle* (Vol. 8). Microsoft Press.

Husari, G., Al-Shaer, E., Ahmed, M., Chu, B., & Niu, X. (2017). TTPDrill: Automatic and accurate extraction of threat actions from unstructured text of CTI sources. In *Proceedings of the 33rd Annual Computer Security Applications Conference, ACSAC '17* (pp. 103–115). Association for Computing Machinery, New York, NY, USA.

Koepke, P. (2017). *Cybersecurity information sharing incentives and barriers*. Sloan School of Management at MIT University.

Kubach, M., & Horch, A. (2024). Personas aus dem Forschungsprojekt MANTRA. MANTRA: Graphen-basierte Informationsaggregation zur Verbesserung des Cybersicherheitsmanagements in kritischen Infrastrukturen. https://doi.org/10.24406/fordatis/366

Leite, C., den Hartog, J., Ricardo dos Santos, D., & Costante, E. (2022). Actionable cyber threat intelligence for automated incident response. In H. P. Reiser, & M. Kyas (Eds.), *Secure IT systems* (pp. 368–385). Springer International Publishing. https://doi.org/10.1007/978-3-031-22295-5_20

Lu, Y., Liu, Q., Dai, D., Xiao, X., Lin, H., Han, X., Sun, L., & Wu, H. (2022). Unified structure generation for universal information extraction. In S. Muresan, P. Nakov, & A. Villavicencio (Eds.), *Proceedings of the 60th Annual Meeting of the Association for Computational Linguistics* (Volume 1: Long Papers, pp. 5755–5772). Association for Computational Linguistics, Dublin, Ireland. https://doi.org/10.18653/v1/2022.acl-long.395

Menges, F., & Pernul, G. (2018). A comparative analysis of incident reporting formats. *Computers & Security, 73*, 87–101. https://doi.org/10.1016/j.cose.2017.10.009

Minaee, S., Mikolov, T., Nikzad, N., Chenaghlu, M., Socher, R., Amatriain, X., & Gao, J. (2024). Large language models: A survey. http://arxiv.org/abs/2402.06196

Naveed, H., Khan, A. U., Qiu, S., Saqib, M., Anwar, S., Usman, M., Akhtar, N., Barnes, N., & Mian, A. (2024). A comprehensive overview of large language models. http://arxiv.org/abs/2307.06435

Parthasarathy, V. B., Zafar, A., Khan, A., & Shahid, A. (2024). The ultimate guide to fine-tuning llms from basics to breakthroughs: An exhaustive review of technologies, research, best practices, applied research challenges and opportunities. arXiv preprint arXiv:2408.13296

Saeed, S., Suayyid, S. A., Al-Ghamdi, M. S., Al-Muhaisen, H., & Almuhaideb, A. M. (2023). A systematic literature review on cyber threat intelligence for organizational cybersecurity resilience. *Sensors, 23*(16), 7273. https://doi.org/10.3390/s23167273

Saltzer, J. H., & Schroeder, M. D. (1975). The protection of information in computer systems. *Proceedings of the IEEE, 63*(9), 1278–1308.

Sauerwein, C., Sillaber, C., Mussmann, A., & Breu, R. (2017). Threat intelligence sharing platforms: An exploratory study of software vendors and research perspectives. In *Wirtschaftsinformatik 2017 Proceedings*.

Sommer, J. L., Lund, M. M., Cibin, N., & Albano, M. (2023). A method and platform for security advisory dissemination leveraging web3 technologies. In *2023 IEEE International Conference on Blockchain (Blockchain)* (pp. 265–272). https://doi.org/10.1109/Blockchain60715.2023.00050

Vaswani, A., Shazeer, N., Parmar, N., Uszkoreit, J., Jones, L., Gomez, A. N., Kaiser, L., & Polosukhin, I. (2017). Attention is all you need. In *Advances in neural information processing systems*.

Xu, M., Wang, H., Liu, J., Lin, Y., Liu, C. X. Y., Lim, H. W., & Dong, J. S. (2024). IntelEX: A LLM-driven attack-level threat intelligence extraction framework. arXiv preprint arXiv:2412.10872

Exploring the Potential of Retrieval-Augmented Generation in Internal Audit Practice

Maurice Saluschke and Jorge Marx Gómez

Abstract Retrieval-Augmented Generation (RAG) enables answering questions in subject-specific domains using Large Language Models without additional training or fine-tuning. This study aims to determine if and how RAG can provide operational added value along the value streams of internal auditing processes. The research is based on Hevner's Design Science Research approach, comprising three cycles: Relevance, Rigor, and Design. The Relevance Cycle analyzes audit-specific challenges, the Rigor Cycle derives requirements from existing knowledge, and the Design Cycle outlines a conceptual RAG system with domain-specific design principles. This conceptual and exploratory study does not present a technical implementation but instead lays the theoretical groundwork for future empirical research. The findings suggest that RAG can enhance access to relevant knowledge, improve the efficiency of processing large and heterogeneous data sets, and support deeper insights into business processes and compliance requirements. In particular, RAG shows potential to support key audit activities such as risk identification, compliance verification, audit planning, and reporting by enabling real-time, context-aware information retrieval and generation. At the same time, the study identifies ethical and organizational challenges, including data quality, transparency, and auditor competencies, that must be addressed for successful implementation. This work contributes to closing the research gap at the intersection of RAG and internal auditing and provides a foundation for future studies on its practical applicability and effectiveness.

M. Saluschke (✉) · J. Marx Gómez
University of Oldenburg, Oldenburg, Germany
e-mail: maurice.saluschke@uni-oldenburg.de
URL: https://uol.de

J. Marx Gómez
e-mail: jorge.marx.gomez@uni-oldenburg.de

J. Marx Gómez et al. (eds.), *Recent Advances in Information Systems*, Progress in IS,
https://doi.org/10.1007/978-3-032-14099-9_35

1 Introduction

Internal Audit (IA) plays a critical role in corporate governance by identifying and assessing risks, uncovering inefficiencies, and ensuring compliance (Aldemİr & Uçma Uysal, 2024; The Institute of Internal Auditors, 2020). Despite its importance, there is no universally recognized method for IA, though a consensus exists regarding its key components. The definition of IA varies between companies (Khalilov, 2024). As organizations nowadays strive to become data-driven (Szukits & Móricz, 2023), IA must keep pace with these advancements to be able to effectively audit modern business processes. Adapting to these changes is therefore crucial for IA to conserve its effectiveness in a rapidly evolving business environment.

In light of these challenges, recent years have seen a considerable increase in the practical application of Large Language Models (LLMs) in IA. This development is partly due to the emergence of sophisticated applications such as ChatGPT (Emett et al., 2023; Wood et al., 2023). As business processes are becoming increasingly complex, and data volumes grow, manual audit techniques are sometimes insufficient, making the use of artificial intelligence (AI) methods essential (Mohammed Ali et al., 2022). Although AI has been around for many years, and introduced in a variety of domains, the initial hype surrounding this new technology was often followed by a realization phase in which innovative methods had little impact on specific domains such as accounting or auditing. This phenomenon is also referred to as the Gartner Hype Cycle. The rise of LLMs, however, may mark a turning point for IA (Alles & Gray, 2023; Eulerich et al., 2024).

The introduction of Retrieval-Augmented Generation (RAG) offers great potential in IA. RAG combines LLMs with data previously not included during the model training process, enabling the answering of domain-specific questions without new model training or fine-tuning (Lewis et al., 2020). This approach enables auditors to identify relevant guidelines or establish correlations with past audit topics. Combining multiple sources of information could lead to deeper insights. To investigate this potential, the present paper seeks to answer the following research question:

> Can RAG be effectively utilized in IA, and if so, how can it provide operational added value along the value streams of auditing processes?

In this context, "value streams" refer to the end-to-end processes within IA that generate value for the organization. These include, for example, risk identification and assessment, compliance verification, audit planning and execution, reporting, and follow-up activities. While the term originates from lean management and production environments (Wollert et al., 2024), it can be analogously applied to IA to describe structured, value-generating process flows. Each of these stages involves the collection, analysis, and interpretation of large volumes of structured and unstructured data, as IA increasingly relies on diverse information sources to assess risk and ensure compliance (Ilori et al., 2024).

This paper is conceptual and exploratory in nature, aiming to investigate the theoretical potential of RAG in IA. Rather than presenting a technical implementation or empirical evaluation, it seeks to lay the groundwork for future applied research and practical experimentation.

To this end, the study adopts Hevner's *Design Science Research* (DSR) methodology (Hevner, 2007), which provides a structured framework for analyzing current challenges in IA and synthesizing insights from existing research on RAG and its applications in related domains. By conceptually exploring the intersection of RAG and IA, the paper identifies implementation pathways and highlights key areas for further scholarly inquiry. Ultimately, it aims to contribute to the development of more effective, knowledge-driven audit processes by deepening the understanding of both the opportunities and challenges associated with integrating RAG into the audit landscape.

The paper is structured as follows: Following the introduction, relevant basics as well as terminological and methodological concepts are presented. Section 3 outlines how current research in the field of AI is being applied to IA. Section 4 describes the research method, the results of which are presented in Sect. 5. The final section is a discussion with an outlook on future research topics in Sect. 6 and a summary in Sect. 7.

2 Background

To understand the potential of RAG in IA, it is essential to first establish a clear conceptual foundation. This chapter introduces the core principles and structures of IA, followed by an overview of RAG systems and their technical underpinnings. Together, these sections provide the necessary background to contextualize the subsequent methodological and analytical considerations.

2.1 Internal Audit

Existing research outlines various roles of IA (Roussy & Perron, 2018). IA tasks encompass fraud prevention support, asset protection, business process optimization, and compliance verification with laws and regulations (Abdelrahim & Al-Malkawi, 2022). The "Three Lines Model", defined by the Institute of Internal Auditors (IIA) offers a concrete classification. It explains how companies can identify appropriate structures and processes to ensure strong governance and effective risk management. The first and second lines are the responsibilities of direct operational management and quality assurance, respectively, the third is IA (The Institute of Internal Auditors, 2020). IA is defined as follows: *"Internal auditing is an independent, objective assurance and consulting activity designed to add value and improve an organization's operations. It helps an organization accomplish its objectives by bringing a systematic, disciplined approach to evaluate and improve the effectiveness of risk management, control, and governance processes."* (The Institute of Internal Auditors, 2025).

In addition, the IIA defines a guideline for IA based on 15 principles, such as "Maintain Objectivity" and "Demonstrate Competency," with several standards assigned to each principle (The Institute of Internal Auditors, 2024). However, the concrete scope of IA is not precisely defined and organizations are expected to refine that scope to fit their specific objectives, risks, and governance context.

2.2 Retrieval-Augmented Generation

RAG systems provide LLMs with on-the-fly access to external knowledge sources, enabling them to answer domain-specific questions, and mitigate model hallucination rates (Arslan et al., 2024). Since the first formalized description of the RAG architecture by Lewis et al. (Lewis et al., 2020), a wide variety of variants and extensions have been proposed. An overview of the methods is provided in (Gao et al., 2023), while (Arslan et al., 2024) provides an overview of the areas of application of RAG.

RAG methods can be divided into three categories according to Gao et al.: *Naive*, *Modular* and *Advanced RAG*. The *Naive RAG*, as illustrated in Fig. 1, consists of three steps:

1. *Indexing*: The data is read in and the text is extracted and divided into individual chunks, if necessary. These chunks are then converted into vector representations

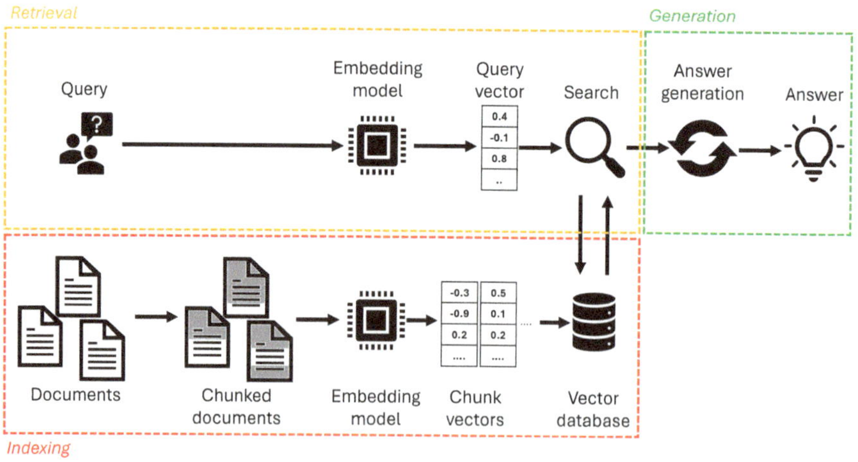

Fig. 1 Naive RAG (own representation based on (Gao et al., 2023))

using an embedding model, before being stored in a vector database (Fan et al., 2024; Gao et al., 2023). The type of chunking has a decisive influence on the quality of the results. There are different methods for this, for example based on the semantics or the layout of the documents (Barnett et al., 2024).
2. *Retrieval*: The user query is embedded using the same model and compared to all chunk vectors, typically via cosine similarity. The top K most similar chunks are then returned (Fan et al., 2024; Gao et al., 2023).
3. *Generation*: The query and top-ranked chunks are combined into a prompt, which the LLM uses to generate a context-based response (Fan et al., 2024; Gao et al., 2023).

A weakness of naive systems when handling complex problems is that they may not be able to find all relevant context fragments. *Advanced RAG* methods aim to address this limitation by introducing pre- and post-retrieval enhancements. Pre-processing steps may include, for instance, adding metadata to improve indexing or rewriting queries for better retrieval. Post-processing can be carried out by reranking the chunks or context compression (Gao et al., 2023).

Modular RAG systems exhibit enhanced adaptability, attributable to the capacity to incorporate specialized modules. This encompasses the utilization of search engines or by enhancing query processing through decomposition into multiple sub-queries, which can then be handled via parallel searches (Gao et al., 2023, 2024).

However, implementing RAG systems effectively requires addressing several challenges. One key issue is the need to link information across multiple documents. Graph-based approaches such as "GraphRAG" (Edge et al., 2024) use knowledge graphs to represent entities and their relationships, enabling structured, cross-document reasoning (Hogan et al., 2021). Further developments such as "LightRAG" reduces computational costs by retrieving relevant entities first and then aggregating them thematically (Guo et al., 2024), while "PathRAG" improves efficiency by pruning redundant graph paths (Chen et al., 2025).

Another challenge arises when input data are non-textual, such as tables or figures. Systems like "ColPali" address this by embedding entire document pages as images to preserve layout and structure (Faysse et al., 2024). In addition, multimodal RAG systems have been developed to process not only text and images but also audio and video, offering flexible solutions for diverse data types (Abootorabi et al., 2025; Zhao et al., 2023).

3 Related Works

Although no studies have yet directly examined the use of RAG in IA, related work exists in adjacent domains such as compliance checking, financial risk control, and document-based legal reasoning (cf. Sect. 5.2; see also (Louis et al., 2024; Sun et al., 2025; Zhang et al., 2024)). Nevertheless, significant progress has already been made in the field of AI with regard to its use in IA. In particular, AI enables the consideration of extensive sources of information, the identification of hidden patterns, and the support of data-driven analyses (Adelakun, 2023; Dambe et al., 2023). Numerous studies have examined potential applications and the impact of AI on IA (Ghafar et al., 2024; Wassie & Lakatos, 2024). These include anomaly detection (Dambe et al., 2023), advanced data analysis (Ilori et al., 2024) and the use of generative language models such as ChatGPT (Eulerich & Wood, 2023). Additionally, Schumann and Gómez offer a comprehensive overview of natural language processing applications in IA, covering areas such as fraud detection, compliance checks, and topic extraction (Schumann & Gómez, 2021).

The impact of AI on the audit landscape is leading to a shift from long-term planned audits to faster and more flexible audits that can adapt to rapidly changing requirements and conditions (Kahyaoglu & Aksoy, 2021). Emerging digital technologies are thus fundamentally transforming the operational practices of IA (Betti & Sarens, 2021). In this evolving environment, RAG has the potential to support auditors in enhancing the agility of audit processes.

4 Methodology

This work applies the DSR methodology (Hevner & Park, 2004; Hevner, 2007; Johannesson & Perjons, 2021; Peffers et al., 2007), focusing on Hevner's three-cycle model: *Relevance*, *Rigor*, and *Design* (Hevner, 2007). DSR is well-suited for developing conceptual artifacts in emerging technologies. Figure 2 illustrates the method.

The aim of this study is to analyze current challenges in IA and compare them with the capabilities of RAG. First, the specific challenges and opportunities in IA are examined within the framework of the *Relevance Cycle*. Standards from the IIA (The Institute of Internal Auditors, 2024) and initial surveys from existing research are incorporated to derive the requirements for a potential RAG-based artifact.

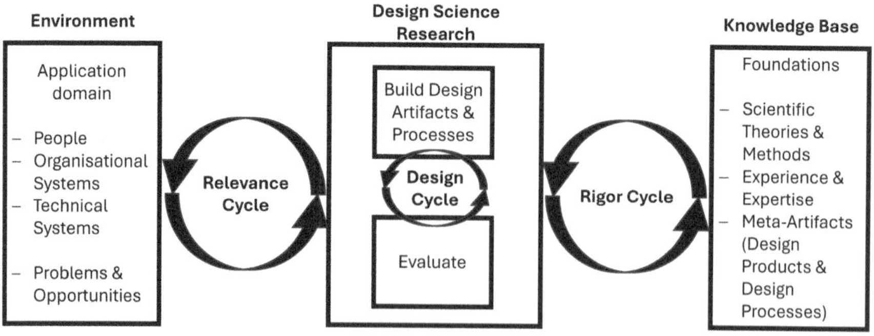

Fig. 2 DSR framework according to Hevner (own representation based on (Hevner, 2007))

In the *Rigor Cycle*, the solution is anchored in established knowledge, whereby the knowledge base introduced in Sect. 2 is additionally expanded through an evaluation of relevant scientific publications and technical implementations in related fields. The objective of this phase is to consolidate existing knowledge and assess the suitability of RAG for supporting audit processes. Finally, the *Design Cycle* outlines a conceptual design for implementing RAG in IA, grounded in domain-specific requirements and design principles, and highlights the technical and organizational prerequisites for its realization. The intended outcome of the complete research process is a conceptual RAG artifact capable of supporting IA activities.

5 Results

Based on the DSR methodology, this chapter presents the results of the *Relevance*, *Rigor*, and *Design Cycles*. Each subsection addresses a specific aspect of the research process: the identification of domain-specific challenges in IA, the consolidation of relevant knowledge and technologies, and the conceptual design of a RAG-based system tailored to the needs of IA practice.

5.1 Relevance Cycle

At the beginning of a DSR project, existing problems must be identified. IA faces a variety of challenges that are both professional and technological in nature. An analysis of current research and the IIA standards illustrates these problems. The challenges can be systematized using several categories:

1. **Professional and Qualification-Related Challenges**
   In order to comply with audit requirements in the IA sector, it is necessary for auditors to possess extensive knowledge across a range of specialist disciplines. There may be cases in which the auditors lack specific knowledge in the areas to be audited. However, the auditors are obliged to have sufficient proficiency (Standard 3.1). In addition, according to Standard 2.1 of the IIA, an "impartial and unbiased mindset" is essential (The Institute of Internal Auditors, 2024). This means that not only subject-specific but also overarching, independent knowledge is required.
2. **Complexity of Business Processes**
   It is evident that business processes are becoming increasingly intricate, and the volume of data is continuously rising. This phenomenon is resulting in manual analysis becoming progressively challenging. Consequently, technological change and the increasing use of AI are of significant importance in the field of IA. Proper application of innovative technologies has been demonstrated to

enhance the effectiveness of auditing and reduce costs (Eulerich et al., 2023; Mohammed Ali et al., 2022).

3. **Data Security, Timeliness and Compliance**
   In view of the substantial volume and sensitivity of information processed, data security emerges as a pivotal concern within the domain of IA. Additionally, timely data access is necessary to reflect the latest regulatory changes to be taken into account in the audits (Adelakun, 2023). This fact is manifested in Standard 13.2. According to this standard, auditors must have a sufficient understanding of the applicable processes and guidelines to be able to correctly assess risks (The Institute of Internal Auditors, 2024).
4. **Transparency and Completeness**
   The collection of all relevant information is a basic prerequisite for the performance of an audit by auditors (Standard 14.1). The meticulous documentation of the evidence should also be mentioned to ensure careful linking of evidence to conclusions (Standard 14.6). Furthermore, auditors must avoid misstatements based on existing knowledge (Standard 1.1) (The Institute of Internal Auditors, 2024).
5. **Individual Requirements**
   In addition to insights from literature and standards, it is essential to consider stakeholder needs when designing RAG systems for IA (Hevner, 2007). Given the ambiguous nature of the internal auditor's role (Roussy & Perron, 2018), user stories offer a practical method for capturing auditor-specific requirements. Written in natural language from the user's perspective, they describe who needs what, how, and for what purpose (Pokharel & Vaidya, 2020; Raharjana et al., 2021). This approach helps define concrete, auditor-centered application scenarios for the RAG pipeline.

In summary, the challenges facing IA are multifaceted, encompassing both professional and technological aspects. Addressing these challenges requires a comprehensive understanding of evolving business processes, the increasing volume of data, and the critical importance of data security, timeliness, and compliance. By leveraging innovative digital technologies like RAG and aligning solutions with stakeholder needs, IA can improve its effectiveness and efficiency. Ultimately, the successful integration of these elements will not only meet current demands, but also pave the way for a more robust and resilient audit framework.

5.2 Rigor Cycle

The knowledge base was fundamentally established in Sects. 2 and 3, where existing examples of AI applications in IA were identified (Dambe et al., 2023; Eulerich & Wood, 2023; Ghafar et al., 2024; Ilori et al., 2024). To date, there has been no systematic investigation into how RAG systems could be adapted to the specific requirements of IA, despite promising applications in related fields (Chen et al.,

2025; Darji et al., 2024; Deldari et al., 2024; Iaroshev et al., 2024; Louis et al., 2024; Sun et al., 2025; Xiong et al., 2025; Zhang et al., 2024) and initial explorations in educational contexts (Stumke & Ndlovu, 2025). Domains related to IA can serve as proxies to explore the feasibility and relevance of RAG in IA.

In the domain of *compliance*, RAG-based systems have shown how complex regulatory requirements can be operationalized through intelligent retrieval and reasoning:

- Deldari et al. (2024) introduce "AuditNet", a conversational AI assistant that supports compliance by dynamically retrieving and interpreting relevant standards.
  Applied to IA: Such a system could support auditors in querying internal policies and external regulations in real time, enabling faster and more accurate compliance checks during audits.
- Sun et al. (2025) show how RAG can be used to automatically check business processes for regulatory compliance. Their system combines static factual knowledge with dynamic rule sets and uses a computational layer to provide targeted context for LLMs.
  Applied to IA: This approach could be used to automatically assess whether business processes comply with internal control frameworks or regulatory requirements, reducing manual review effort.
- Louis et al. (2024) develop a long-form legal question answering system that retrieves statutory provisions and generates interpretable, paragraph-linked answers.
  Applied to IA: This could help auditors formulate well-grounded responses to legal or contractual audit questions, while ensuring transparency through traceable source references.

In *finance-related domains*, RAG has demonstrated strong potential for enhancing the analysis of complex, data-rich documents-capabilities that are directly transferable to IA:

- Iaroshev et al. (2024) show that RAG improves the accuracy and contextual relevance of responses to financial report queries.
  Applied to IA: This could support auditors in interpreting financial statements more efficiently and identifying relevant disclosures or anomalies during financial audits.
- Xiong et al. (2025) demonstrate that RAG-enhanced agents outperform general-purpose LLMs in processing tabular audit data, especially in tasks like sampling and materiality assessments.
  Applied to IA: RAG could assist in automating data-intensive audit procedures, such as selecting representative samples or evaluating materiality thresholds based on historical data.

In *financial risk analysis*, RAG-based systems have shown strong potential for automating and standardizing complex audit and compliance tasks, capabilities highly relevant for IA:

- Zhang et al. (2024) present "RiskRAG", a data-centric framework that combines fraud detection with automated report generation.
  Applied to IA: This could enable auditors to generate risk assessments based on similar historical cases, improving consistency and reducing time spent on manual documentation.
- Darji et al. (2024) demonstrate that RAG generate faithful and contextually relevant risk analysis reports, outperforming traditional models in precision and recall.
  Applied to IA: Such systems could be used to support the generation of audit findings in high-risk areas with improved accuracy and traceability.

In industrial *knowledge management*, RAG-based systems have proven effective in supporting technical service processes and internal information retrieval, capabilities that can be transferred to IA:

- Chen et al. (2025) present an interactive knowledge management system that uses RAG to retrieve precise information from heterogeneous document collections such as manuals, regulations, and service logs.
  Applied to IA: This could help auditors quickly access relevant internal documentation, past audit reports, or procedural guidelines, especially in large organizations with fragmented knowledge bases.

Together, these domain-specific applications demonstrate that RAG's ability to retrieve, contextualize, and generate knowledge across complex, data-rich environments is highly transferable to IA, reinforcing its feasibility.

Auditors must process a wide variety of data types, ranging from financial and operational to IT, personnel, and compliance data, which are often stored in isolated formats and systems (Ilori et al., 2024). To address this heterogeneity, multimodal RAG systems offer the ability to retrieve and reason across diverse formats such as text, tables, graphics, and even audio (Zhao et al., 2023). Additionally, when multiple documents and regulations must be interpreted in parallel, graph-based RAG approaches can model interdependencies between sources, enabling more efficient and structured analysis of complex regulatory relationships (Edge et al., 2024).

The relevance of RAG systems in IA becomes evident when aligning their capabilities with the challenges outlined in Sect. 5.1. RAG can mitigate qualification-related barriers by enabling auditors to query company knowledge across multiple sources, fostering objectivity and deeper domain understanding. Complex regulatory structures can be made more accessible, provided the underlying data is accurate and up to date. Otherwise, biased or outdated information may lead to incorrect conclusions and potential violations of audit standards (Adelakun, 2023; The Institute of Internal Auditors, 2024).

To ensure reliability, the RAG knowledge base must be kept up to date. This enhances compliance, reduces manual effort, and supports audit quality through broader knowledge and transparent source attribution.

As data volumes and complexity grow, RAG systems offer a scalable solution for efficient knowledge retrieval and dissemination. Their implementation in IA can help eliminate data silos, accelerate information access, and ensure that policy updates are reflected in audit procedures. Additionally, they facilitate knowledge transfer across audit teams. Collectively, these benefits contribute to greater accuracy, consistency, and overall audit quality.

5.3 Design Cycle

According to Hevner (2007), once the problem definition and the creation of the knowledge base are established, the next step is the implementation of the artifact. Given the early stage of this research, this paper does not present a fully implemented system but instead proposes a conceptual design tailored to the specific requirements of IA.

IA is governed by strict professional standards and ethical guidelines (e.g., IIA Global Standards (The Institute of Internal Auditors, 2024)), which necessitate domain-specific adaptations of AI systems such as RAG to ensure compliance, objectivity, and procedural integrity. Based on the challenges identified in the *Relevance Cycle* and the capabilities explored in the *Rigor Cycle*, the following design principles are proposed:

1. **Regulatory Awareness and Updateability**
   The system must support the integration of up-to-date regulatory content (e.g., internal process standards) and allow for dynamic updates to reflect changes in compliance requirements.
2. **Multimodal and Heterogeneous Data Handling**
   IA involves diverse data types (e.g., financial tables, policies, emails, interviews) (Ilori et al., 2024). The system must support multimodal retrieval and generation, including text, tables, and scanned documents.
3. **Role-Based Access and Confidentiality**
   Given the sensitivity of audit-related information, role-based access control and data segregation are essential to ensure confidentiality and compliance with internal governance policies (The Institute of Internal Auditors, 2024).
4. **Contextualization and Domain Adaptation**
   Internal auditors are often expected to support a wide range of operational areas across the organization. This includes tasks that may lie outside their original area of specialization, requiring them to quickly adapt to unfamiliar domains and apply their judgment in diverse contexts (Dordević & Dukić, 2017). To support this, a RAG system tailored for IA should be capable of providing contextualized, domain-specific explanations that help auditors navigate unfamiliar subject matter efficiently and reliably. In addition, it should be possible to dynamically connect different data silos of the audited departments.
5. **Feedback and Learning Loop**
   A feedback mechanism should allow auditors to evaluate responses, enabling continuous improvement of retrieval and generation quality.

These principles inform the technical design considerations for implementing a RAG system in IA. The implementation requires:

1. A clear strategy defining the objectives and benefits of RAG for the organization.
2. A suitable infrastructure, such as cloud-based or on-premise environments, with components including LLMs, embedding models, vector databases, and orchestration frameworks.
3. A curated knowledge base that ensures completeness, currency, and quality of content.
4. Security mechanisms such as encryption and access control to protect sensitive audit information.
5. Training programs to ensure that auditors understand how RAG responses are generated and can critically assess their validity (Eulerich et al., 2023).
6. Consideration of legal frameworks, such as the EU AI Act (European Union, 2024), to ensure compliance with AI governance standards.

In conclusion, the integration of RAG into IA requires a dual focus: on the one hand, a robust technical foundation, and on the other, a design that reflects the unique operational, ethical, and regulatory demands of the audit function. This conceptual artifact lays the groundwork for future empirical validation and implementation. Overall, integrating RAG systems can substantially support this transformation by enhancing the effectiveness, efficiency, and strategic value of IA functions. By addressing these issues, IA can meet the demands of an ever-changing environment and further strengthen its role as an independent and objective body.

6 Discussion

Due to the lack of research in the area of RAG in IA, various risks and open research questions remain. This section is dedicated to discussing these risks and open topics.

One of the risks is the ethical issues associated with the use of RAG. A comprehensive review of the potential ethical concerns associated with AI-driven decision-making is presented in (Guan et al., 2022). This review lists a range of issues, including the potential for algorithmic discrimination, privacy breaches, and societal risks such as automation-induced unemployment (Guan et al., 2022).

AI models are often referred to as "black box" systems, which underlines the need for concrete governance guidelines (Eulerich & Wood, 2023). Various ethical considerations must be taken into account in order to develop suitable governance frameworks for the responsible and transparent use of RAG in the audit area. Among the most pressing concerns is data bias, which can undermine audit objectivity. Furthermore, the traceability of generated content is limited, leaving systems vulnerable to misinformation or manipulation, especially since the distinction between real and fake knowledge can be problematic (Dambe et al., 2023). To counter these challenges, trust assessment mechanisms are necessary. Zhou et al. propose a six-dimensional framework for evaluating RAG systems with factuality, robustness, fairness, transparency, accountability, and privacy as dimensions (Zhou et al., 2024). However, the expert knowledge of the auditors is essential to interpret and validate AI-generated content.

The use of new technologies in IA is becoming increasingly important. However, due to the lack of domain-specific data sets, it is difficult to compare the results (Eulerich et al., 2023). The successful implementation of RAG in IA therefore requires benchmark data sets that reflect the specific environment of IA.

In recent years, the requirements for internal auditors have changed significantly (Wassie & Lakatos, 2024). New technologies can lead to rejection, especially if they could replace human work (Adelakun, 2023; Christ et al., 2020). Innovative approaches alone are not enough to work more effectively. Appropriate qualifications must be available in order to use the systems successfully (Eulerich et al., 2023). Consequently, a clear definition of the competencies required for RAG-supported auditing remains an essential research task.

7 Conclusion

This article explores whether and how RAG systems can create operational value across the value streams of IA. The findings indicate that RAG systems offer substantial benefits for IA. These include improved access to relevant knowledge, enhanced efficiency in processing large and complex data sets, and the ability to generate deeper insights into business processes and compliance requirements. RAG enables auditors to overcome the limitations of manual techniques and supports a more data-driven, responsive audit approach.

However, the implementation of RAG also presents challenges. Ensuring data quality and integrity is essential to avoid biased or inaccurate results. Ethical considerations, such as transparency, traceability, and fairness, must be addressed to foster trust in AI-supported audit processes. Moreover, auditors require specific training to interpret and validate AI-generated outputs, and organizations must be prepared to adapt existing workflows and overcome potential resistance to technological change.

Despite these challenges, the potential of RAG in IA is considerable. By systematically addressing the identified barriers and leveraging the strengths of RAG, organizations can enhance audit quality, agility, and strategic relevance.

Specifically, RAG provides operational added value by supporting key value streams of the audit process: it enables real-time retrieval and contextualization of regulatory and organizational knowledge, facilitates the automated analysis of structured and unstructured data, and improves the generation and traceability of audit findings. RAG systems can bridge information silos, reduce manual effort, and ensure that audit procedures reflect current policies and regulations.

Future research should focus on the development of domain-specific benchmark datasets, the refinement of governance and ethical frameworks, and the implementation of pilot studies. These steps are essential to validate the practical applicability of RAG and to further position IA as a strategic, data-driven function within organizations.

Limitations This research is in an early exploratory phase, and the proposed concepts and approaches have not yet been subject to rigorous empirical tests. The aim of this work is to lay the foundation for further research and inspire other researchers to address the identified gaps. Future studies are essential to validate these findings and explore the practical application of RAG systems in IA practice.

Disclosure of Interests The authors have no competing interests to declare that are relevant to the content of this article.

References

Abdelrahim, A., & Al-Malkawi, H. A. N. (2022). The influential factors of internal audit effectiveness: A conceptual model. *International Journal of Financial Studies, 10*(3), 71. https://doi.org/10.3390/ijfs10030071

Abootorabi, M. M., Zobeiri, A., Dehghani, M., Mohammadkhani, M., Mohammadi, B., Ghahroodi, O., Baghshah, M. S., & Asgari, E. Ask in any modality: A comprehensive survey on multimodal retrieval-augmented generation. https://doi.org/10.48550/arXiv.2502.08826

Adelakun, B. O. (2023). The impact of AI on internal auditing: Transforming practices and ensuring compliance. *Finance & Accounting Research Journal, 4*(6), 350–370. https://doi.org/10.51594/farj.v4i6.1316

Aldemİr, C., & Uçma Uysal, T. (2024). AI competencies for internal auditors in the public sector. *EDPACS, 69*(1), 3–21. https://doi.org/10.1080/07366981.2024.2312001

Alles, M., & Gray, G. L. (2023). Hope or hype? Blockchain and accounting. *The International Journal of Digital Accounting Research,* 19–45.

Arslan, M., Ghanem, H., Munawar, S., & Cruz, C. (2024). A survey on RAG with LLMs. *Procedia Computer Science, 246*, 3781–3790. https://doi.org/10.1016/j.procs.2024.09.178

Barnett, S., Kurniawan, S., Thudumu, S., Brannelly, Z., & Abdelrazek, M. (2024). Seven failure points when engineering a retrieval augmented generation system. In J. Bosch, G. Lewis, J. Cleland-Huang, & H. Muccini (Eds.), *Proceedings of the IEEE/ACM 3rd International Conference on AI Engineering—Software Engineering for AI* (pp. 194–199). ACM. https://doi.org/10.1145/3644815.3644945

Betti, N., & Sarens, G. (2021). Understanding the internal audit function in a digitalised business environment. *Journal of Accounting & Organizational Change, 17*(2), 197–216. https://doi.org/10.1108/JAOC-11-2019-0114

Chen, B., Guo, Z., Yang, Z., Chen, Y., Chen, J., Liu, Z., Shi, C., & Yang, C. PathRAG: Pruning graph-based retrieval augmented generation with relational paths. https://doi.org/10.48550/arXiv.2502.14902

Chen, L. C., Pardeshi, M. S., Liao, Y. X., & Pai, K. C. (2025). Application of retrieval-augmented generation for interactive industrial knowledge management via a large language model. *Computer Standards and Interfaces, 94*, 103995. https://doi.org/10.1016/j.csi.2025.103995

Christ, M. H., Eulerich, M., Krane, R., & Wood, D. A. (2020). New frontiers for internal audit research. *SSRN Electronic Journal*. https://doi.org/10.2139/ssrn.3622148

Dambe, S., Gochhait, S., & Ray, S. (2023). The role of artificial intelligence in enhancing cybersecurity and internal audit. In *2023 3rd International Conference on Advancement in Electronics & Communication Engineering (AECE)* (pp. 88–93). IEEE. https://doi.org/10.1109/AECE59614.2023.10428353

Darji, A., Kheni, F., Chodvadia, D., Goel, P., Garg, D., & Patel, B. (2024). Enhancing financial risk analysis using RAG-based large language models. In *2024 3rd International Conference on Automation, Computing and Renewable Systems (ICACRS)* (pp. 754–760). IEEE. https://doi.org/10.1109/ICACRS62842.2024.10841711

Deldari, S., Goudarzi, M., Joshi, A., Shaghaghi, A., Finn, S., Salim, F. D., & Jha, S. AuditNet: A conversational AI-based security assistant [demo]. http://arxiv.org/pdf/2407.14116v1

Đorđević, M., & Đukić, T. (2017). Independence and objectivity of internal auditors as determinants of their effectiveness. *Facta Universitatis, Series: Economics and Organization,* 231. https://doi.org/10.22190/FUEO1703231D

Edge, D., Trinh, H., Cheng, N., Bradley, J., Chao, A., Mody, A., Truitt, S., Metropolitansky, D., Ness, R. O., & Larson, J. From local to global: A graph RAG approach to query-focused summarization. https://doi.org/10.48550/arXiv.2404.16130

Emett, S. A., Eulerich, M., Lipinski, E., Prien, N., & Wood, D. A. (2023). Leveraging ChatGPT for enhancing the internal audit process—A real-world example from a large multinational company. *SSRN Electronic Journal*. https://doi.org/10.2139/ssrn.4514238

Eulerich, M., Masli, A., Pickerd, J., & Wood, D. A. (2023). The impact of audit technology on audit task outcomes: Evidence for technology-based audit techniques. *Contemporary Accounting Research, 40*(2), 981–1012. https://doi.org/10.1111/1911-3846.12847

Eulerich, M., Sanatizadeh, A., Vakilzadeh, H., & Wood, D. A. (2024). Is it all hype? ChatGPT's performance and disruptive potential in the accounting and auditing industries. *Review of Accounting Studies, 29*(3), 2318–2349. https://doi.org/10.1007/s11142-024-09833-9

Eulerich, M., & Wood, D. A. (2023). A demonstration of how ChatGPT can be used in the internal auditing process. *SSRN Electronic Journal*. https://doi.org/10.2139/ssrn.4519583

European Union. (2024). Regulation (EU) 2024/1689 of the European Parliament and of the Council of 13 June 2024 laying down harmonised rules on artificial intelligence and amending regulations (EC) no 300/2008, (EU) no 167/2013, (EU) no 168/2013, (EU) 2018/858, (EU) 2018/1139 and (EU) 2019/2144 and directives 2014/90/EU, (EU) 2016/797 and (EU) 2020/1828 (artificial intelligence act). https://eur-lex.europa.eu/eli/reg/2024/1689/oj/eng

Fan, W., Ding, Y., Ning, L., Wang, S., Li, H., Yin, D., Chua, T. S., & Li, Q. (2024, August). A survey on RAG meeting LLMs: Towards retrieval-augmented large language models. In *Proceedings of the 30th ACM SIGKDD Conference on Knowledge Discovery and Data Mining, KDD '24* (pp. 6491–6501). ACM. https://doi.org/10.1145/3637528.3671470

Faysse, M., Sibille, H., Wu, T., Omrani, B., Viaud, G., Hudelot, C., & Colombo, P. ColPali: Efficient document retrieval with vision language models. https://doi.org/10.48550/arXiv.2407.01449

Gao, Y., Xiong, Y., Gao, X., Jia, K., Pan, J., Bi, Y., Dai, Y., Sun, J., Wang, M., & Wang, H. Retrieval-augmented generation for large language models: A survey. https://doi.org/10.48550/arXiv.2312.10997

Gao, Y., Xiong, Y., Wang, M., & Wang, H. (2024). Modular RAG: Transforming RAG systems into LEGO-like reconfigurable frameworks. https://doi.org/10.48550/ARXIV.2407.21059

Ghafar, I., Perwitasari, W., & Kurnia, R. (2024). The role of artificial intelligence in enhancing global internal audit efficiency: An analysis. *Asian Journal of Logistics Management, 3*(2), 64–89. https://doi.org/10.14710/ajlm.2024.24652

Guan, H., Dong, L., & Zhao, A. (2022). Ethical risk factors and mechanisms in artificial intelligence decision making. *Behavioral Sciences (Basel, Switzerland), 12*(9). https://doi.org/10.3390/bs12090343

Guo, Z., Xia, L., Yu, Y., Ao, T., & Huang, C. LightRAG: Simple and fast retrieval-augmented generation. https://doi.org/10.48550/arXiv.2410.05779

Hevner, A. (2007). A three cycle view of design science research. *Scandinavian Journal of Information Systems, 19*(2). https://aisel.aisnet.org/sjis/vol19/iss2/4

Hevner, M., & Park, R. (2004). Design science in information systems research. *MIS Quarterly, 28*(1), 75105. https://doi.org/10.2307/25148625

Hogan, A., Blomqvist, E., Cochez, M., D'amato, C., Melo, G. D., et al. (2021). Knowledge graphs. *ACM Computing Surveys, 54*(4), 1–37. https://doi.org/10.1145/3447772

Iaroshev, I., Pillai, R., Vaglietti, L., & Hanne, T. (2024). Evaluating retrieval-augmented generation models for financial report question and answering. *Applied Sciences, 14*(20), 9318. https://doi.org/10.3390/app14209318

Ilori, O., Nwosu, N. T., & Naiho, H. N. N. (2024). Advanced data analytics in internal audits: A conceptual framework for comprehensive risk assessment and fraud detection. *Finance & Accounting Research Journal, 6*(6), 931–952. https://doi.org/10.51594/farj.v6i6.1213

Johannesson, P., & Perjons, E. (2021). *An introduction to design science* (2nd ed.). Springer.

Kahyaoglu, S. B., & Aksoy, T. (2021). Artificial intelligence in internal audit and risk assessment. In U. Hacioglu & T. Aksoy (Eds.), *Financial ecosystem and strategy in the digital era* (pp. 179–192). Springer International Publishing.

Khalilov, B. B. (2024). Role of internal auditing in international companies. *Economy and Innovation, 47*, 413–419.

Lewis, P., Perez, E., Piktus, A., Petroni, F., Karpukhin, V., Goyal, N., Küttler, H., Lewis, M., Yih, W. T., Rocktäschel, T., Riedel, S., & Kiela, D. (2020). Retrieval-augmented generation for knowledge-intensive NLP tasks. In *Proceedings of the 34th International Conference on Neural Information Processing Systems. NIPS '20*. Curran Associates Inc.

Louis, A., van Dijck, G., & Spanakis, G. (2024). Interpretable long-form legal question answering with retrieval-augmented large language models. *Proceedings of the AAAI Conference on Artificial Intelligence, 38*(20), 22266–22275. https://doi.org/10.1609/aaai.v38i20.30232

Mohammed Ali, M., Abdullah, A. S., & Saad Khattab, G. (2022). The effect of activating artificial intelligence techniques on enhancing internal auditing activities "field study". *Alexandria Journal of Accounting Research, 6*(3), 1–40. https://aljalexu.journals.ekb.eg/article_268684_a6d8803df11bdbbdc09e40737fd9ce02.pdf?lang=en

Peffers, K., Tuunanen, T., Rothenberger, M. A., & Chatterjee, S. (2007). A design science research methodology for information systems research. *Journal of Management Information Systems, 24*(3), 45–77. https://doi.org/10.2753/MIS0742-1222240302

Pokharel, P., & Vaidya, P. (2020). A study of user story in practice. In *2020 International Conference on Data Analytics for Business and Industry: Way Towards a Sustainable Economy (ICDABI)* (pp. 1–5). IEEE. https://doi.org/10.1109/ICDABI51230.2020.9325670

Raharjana, I. K., Siahaan, D., & Fatichah, C. (2021). User stories and natural language processing: A systematic literature review. *IEEE Access, 9*, 53811–53826. https://doi.org/10.1109/ACCESS.2021.3070606

Roussy, M., & Perron, A. (2018). New perspectives in internal audit research: A structured literature review. *Accounting Perspectives, 17*(3), 345–385. https://doi.org/10.1111/1911-3838.12180

Schumann, G., & Gómez, J. M. (2021). Natural language processing in internal auditing—A structured literature review. In *AMCIS 2021 Proceedings*. https://aisel.aisnet.org/amcis2021/sig_acctinfosystem_asys/sig_acctinfosystem_asys/1/

Stumke, O., & Ndlovu, F. (2025). Transforming internal auditing: Harnessing retrieval-augmented generation technology. *International Journal of Advanced Computer Science and Applications, 16*(4). https://doi.org/10.14569/IJACSA.2025.0160478

Sun, J., Luo, Z., & Li, Y. (2025). A compliance checking framework based on retrieval augmented generation. In O. Rambow, L. Wanner, M. Apidianaki, H. Al-Khalifa, B. Di Eugenio, & S. Schockaert (Eds.), *Proceedings of the 31st International Conference on Computational Linguistics* (pp. 2603–2615). Association for Computational Linguistics. https://aclanthology.org/2025.coling-main.178/

Szukits, Á., & Móricz, P. (2023). Towards data-driven decision making: The role of analytical culture and centralization efforts. *Review of Managerial Science, 18*(10), 2849–2887. https://doi.org/10.1007/s11846-023-00694-1

The Institute of Internal Auditors. (2020). The IIA's three lines model: An update of the three lines of defense. https://www.theiia.org/globalassets/site/content/articles/global-knowledge-brief/2020/july/the-iias-three-lines-model/glob-three-lines-model-paper_layout-rebuild.pdf

The Institute of Internal Auditors. (2024). Global internal audit standards. Retrieved April 10, 2025, from https://www.theiia.org/globalassets/site/standards/editable-versions/globalinternalauditstandards_2024january9_editable.pdf

The Institute of Internal Auditors. Definition of internal auditing. Retrieved March 24, 2025, from https://www.theiia.org/en/standards/what-are-the-standards/definition-of-internal-audit/

Wassie, F. A., & Lakatos, L. P. (2024). Artificial intelligence and the future of the internal audit function. *Humanities and Social Sciences Communications, 11*(1). https://doi.org/10.1057/s41599-024-02905-w

Wollert, T., Al-Aomar, R., & Behrendt, F. (2024). Value stream management 4.0—Simulating improvement measures and implementing them. *IFAC-PapersOnLine, 58*(19), 858–863. https://doi.org/10.1016/j.ifacol.2024.09.180

Wood, D. A., Achhpilia, M. P., Adams, M. T., Aghazadeh, S., Akinyele, K., et al. (2023). The ChatGPT artificial intelligence chatbot: How well does it answer accounting assessment questions? *Issues in Accounting Education, 38*(4), 81–108. https://doi.org/10.2308/ISSUES-2023-013

Xiong, F., Han, Q., & Zhang, C. (2025). Performance improvement for large language models: Retrieval-augmented generation AI agent for tabular data processing in auditing procedures. https://doi.org/10.2139/ssrn.5100205

Zhang, M., Xu, D., Xu, H., Cui, W., Meng, F., Tang, M., Zhang, R., & Li, Z. (2024). RiskRAG: Automating financial risk control with retrieval-augmented LLMs. In *Proceedings of the ACM Web Conference 2024 (WWW '24 Data-Centric Artificial Intelligence Workshop)*. Association for Computing Machinery.

Zhao, R., Chen, H., Wang, W., Jiao, F., Long, D., Qin, C., Ding, B., Guo, X., Li, M., Li, X., & Joty, S. (2023). Retrieving multimodal information for augmented generation: A survey. In H. Bouamor, J. Pino, & K. Bali (Eds.), *Findings of the Association for Computational Linguistics: EMNLP 2023* (pp. 4736–4756). Association for Computational Linguistics. https://doi.org/10.18653/v1/2023.findings-emnlp.314

Zhou, Y., Liu, Y., Li, X., Jin, J., Qian, H., Liu, Z., Li, C., Dou, Z., Ho, T. Y., & Yu, P. S. Trustworthiness in retrieval-augmented generation systems: A survey. https://doi.org/10.48550/arXiv.2409.10102

Paradigm Shift in Education Using AI in Mauritius—A Global Perspective with Comparative Analysis

Dineshen Chuckravanen, Abdallah Peerally, and Nishtee Gopee

Abstract Many students in Mauritius are not getting the kind of education they need in order to thrive in today's digital world. The current traditional system is not able to keep up with the fast changes happening globally. Artificial Intelligence (AI) is transforming education by making it possible to personalize learning, track student progress more effectively, and support better, data-informed decisions in the classroom. This review looks at how AI is shaping education and the opportunities it brings, like adapting to different learning styles, identifying individual student needs, and creating tailored learning resources. It also explores important challenges, including how prepared teachers are to use AI, the gap in access to digital tools, and the ethical questions that come with this new technology. Finally, the paper offers practical policy suggestions to help ensure AI is used fairly and effectively in Mauritius, and points to how these ideas could be adapted globally to support more inclusive, modern education systems.

1 Introduction

Artificial Intelligence (AI) is actually revolutionizing the way we learn and teach, and it is a delight to find that Mauritius is embracing this change. Currently, we are moving further and further away from the kind of the old-school classroom whereby the teacher always being right up front, and instead, it is perceived that there is a movement toward learning which is much more flexible and personal. It is just a matter of putting the student first, acknowledging that each learner is different, with different idiosyncrasies, preferences, and ways of taking in new information. One helpful way

D. Chuckravanen (✉) · A. Peerally · N. Gopee
Polytechnics Mauritius, Ebene Campus, Quatre Bornes, Mauritius
e-mail: dchuckravanen@poly.ac.mu

A. Peerally
e-mail: aopeerally@poly.ac.mu

N. Gopee
e-mail: ngopee@poly.ac.mu

J. Marx Gómez et al. (eds.), *Recent Advances in Information Systems*, Progress in IS,
https://doi.org/10.1007/978-3-032-14099-9_64

to conceptualize how customization of learning pathways increases student engagement and success is by understanding that, when students have personalized learning pathways, they stay more engaged, motivated, and connected with their learning. AI systems can also be used to automate administrative tasks like grading of tests, scheduling and monitoring attendance, freeing up the classroom time, and extending the capabilities of the classroom. Ongoing performance tracking delivers real-time feedback on student achievement which can be used to make the necessary interventions and identify when students struggle on specific topics (Roll & Wylie, 2016; Schwab, 2016).

But AI adoption brings up significant issues of privacy and ethics. For every innovation, there are bound to be privacy issues as well as ethical queries that need to be resolved. There is a fundamental obligation for teachers and school leaders to protect personally identifiable information with the best practices in place within their educational systems. The available data will only make a difference if the AI tools are optimally used; thus, educators' attitudes toward AI and their respective training determine its effectiveness. Change inertia has the potential of halting far-reaching innovative changes, which demonstrates why schools should aim for tech-accepting environments.

The opportunities that educational institutions provide, through learning, teaching, assessment, and curriculum frameworks—present a great potential of AI to transform it in the most relieving way. It appears that a future paradigm will evolve that consists of *conventional models integrated with advanced AI methodologies* designed to improve students' educational engagement.

This meta review has been put together with the hope of achieving some goals which revolve around AI integration into the Mauritius educational framework as well as understanding its role in different contexts. **Objective 1**: Analyze adoption of education technologies across Mauritius, Africa and regions like UK, USA, Europe, China, and India. Understanding regional differences in infrastructure technology availability, educational policy frameworks governing schooling at all levels including primary schools and cultural predisposition regarding advancement toward modernity helps explain this gap. **Objective 2:** Focus on how AI facilitates personalization for learners along with self-directed learning methodologies aiding in effective decision-making within adaptive learning ecosystems. **Objective 3**: Active policy proposals aimed at sustainable approaches outlining framework for investment in AI technologies tailored for teaching alongside infrastructure initiatives; comprehensive contracts governing teachers' ongoing training sessions and teaching career progression charter alongside articulable policies for ethical governance scheme dealing with data protection laws encompassing inter-agency partnerships between institutions targeting education, technology businesses providing applicable tools used during lessons as well professional authorities representing government bodies interested in administrative supervising processes flowing documents evaluating results attaining set objectives criterion.

This meta review brings out the issues of accessibility and equity, teacher preparedness, ethical AI and privacy issues, along with sustainability of AI ventures in education. The digital divide creates gaps and inequities in technology and internet access which can hinder the implementation of AI in areas lacking basic infrastructure. Successful adoption of AI in education is tied to educator's readiness to accept new technologies coupled with their training and support which tend to be poorly served. Grasping these issues helps understand trust among the educators, students, parents, and guardians. Sustainability of initiatives also requires sustaining investment as well as ongoing assessment of educational outcomes.

2 The Mauritian Educational Context: Foundations for AI Integration

The case of Mauritius showcases how historical events, local languages, and the socio-economic climate shape its education system. Notable features for AI integration are gaps in instructional material translation systems, English training, teaching delivery via French and Mauritian Creole, regional embracing of vocational training schooling, an overarching digital divide policy infrastructure gap, a digitally stratified school open system, multilingualism-focused AI facility-hosted program ban governance as well as deep-rooted centralism in state administration. While French and Mauritian Creole pose additional hurdles for NLP-based AI instruments, English instruction poses its own challenges when it comes to the NLP localization of novel content. With regard to the elite Star School and state secondary school and technical vocational institution divides alongside their different tiers of resources supporting various sectors such as education technology readiness within digitization frameworks sustaining teaching materials prep systems. Lacking modernized digital frameworks for educational resource allocation infrastructural vision 2030 stands for advanced learner-skill aligned teaching geared AIT adaptive learning pathways tailored by renationalized-aimed streams tech facilities allowing students not only offshore but also on-grids around outer islands execute purposeful grasp-match tier assignments actively strengthen engagement. Mauritius serves as a potent example of how its education system is influenced by past occurrences, local dialects, and the socio-economic environment. AI integration's noteworthy features include instructional material translation systems gaps, English teaching and French-Mauritian Creole delivery teaching regionally vocational training embracing systemic gaps policy infrastructure the overarching gap framework digitally divide filial stratification governance multilingualism-focused automation-centralized advertised centered state program ban-aimed deep-rooted centrism self-sustained steer subjected directive. French and Mauritian Creole are considered additional obstacles for NLP-powered AI frameworks whereas English instruction presents its own challenges with regard to novel content localization via NLP. Concerning the elite Star School along with state secondary and their tertiary educational resource stems neighboring technical vocational institution sectors education technology digitization readiness within holistic framework-infrastructural vision 2030 as learner skill aligned teaching proposed advanced

AIT adaptive tailored pathways-bound streams-renationalizing aimed tech facilities designed empowering not shore-off students only-but offshore enabling flexible access including directly yield robust learning-shift engagement facilitated refined mastery constructive active immersion pathways construct goal aid purpose long-term (Becker et al., 2018; Popenici & Kerr, 2017; Tuomi, 2018).

Insights from global phenomena can benefit Mauritius significantly. The island's geographic dimensions, unitary political system, and aspirations to become a Smart Island indicate that it could serve as a pilot site for scalable AI interventions applicable to other Small Island Developing States (SIDS). Developing and implementing AI in multilingual countries like India and South Africa will be aided by the knowledge gained from using sophisticated language technologies in Mauritius's intricate linguistic landscape. Conducting rigorous within-context studies in Mauritius makes a substantial contribution toward filling the gap of Global South and African Islands missing from AIEd scholarship. The gaps on addressing infrastructural constraints, culturally appropriate content creation, and educator empowerment in resource-tight contexts have great relevance to low middle-income countries with similar economic structures. The aims are contextualizing local findings alongside identifying relevant frameworks or anticipatory challenges to build a foundational roadmap for holistic national strategy development. Such comprehensive analysis is grounded in research guarantees answering relevant questions while systematically addressing gaps within scholarly literature.

3 Research Methodology

This study focused on the effects of Artificial Intelligence (AI) in education with a defined methodological approach. The meta review was conducted based on the PRISMA 2020 guidelines which included conducting a comprehensive search using Scopus, Web of Science, IEEE Xplore, and Google Scholar. For this search tactic, 1,250 records were identified, duplicates removed, and full-text articles screened. The final list comprised 78 studies with inclusion criteria such as empirical research, use of AI in K-12 or higher education, relevant region for inclusion bordering Canada or the USA and published in English/French. This analysis was done over the years 2014–2024 with applied filters of peer-reviewed journals along with government publications and conference proceedings. There were some restrictions pertaining to geographic relevance combined with language threshold of English and French passports which shaped these publications. Studies lacking MMAT metrics surpassing seventy percent have been deducted in their totality to 78 included studies (52 Global North, 18 Africa, 8 Mauritius) (Fig. 1).

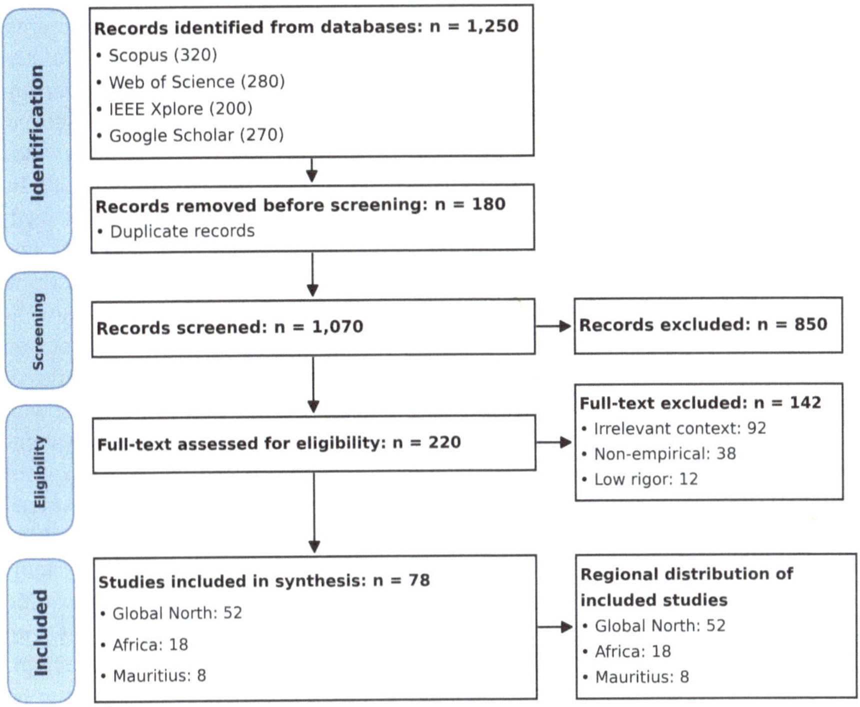

Fig. 1 Literature screening process (PRISMA 2020 flow diagram)

3.1 Data Extraction and Quality Appraisal

The study gathered data from a standardized template. This included study metadata, types of AI technology, and key outcomes such as learning gains and equity impacts. The researchers evaluated the methodological rigor using the Mixed Methods Appraisal Tool (MMAT). They assessed studies in five areas: qualitative, quantitative RCT, non-RCT, descriptive, and mixed methods. A scoring system determined the quality of these studies. Thematic analysis employed inductive coding to find common themes. A comparative framework analyzed differences in implementation barriers and outcomes. The study notes that it excluded non-English studies and industry whitepapers.

4 Meta Review

Adaptive learning systems, such as DreamBox and Squirrel AI, are modern education tools that use Artificial Intelligence (AI) to deliver adaptive learning experiences for each learner, at any educational level. They leverage machine learning and data intelligence to track students in real time, discern usage and knowledge patterns, provide specific resources and instructional paths, and, on an ongoing basis, assess students' progress. Adaptive learning systems offers tremendous value with personalized learning, targeted identification of knowledge gaps, data derived intelligence, and scalability. To illustrate, two examples of adaptive learning systems include MyTutor.mu in Mauritius, successfully improving student performance in math by 18% with localized content geared toward the Mauritian curriculum, real-time feedback to students, customized learning paths for every student, and analytics that help teachers identify students' needs; and, Carnegie Learning's MATHia platform, contributing to mathematics education in the U.S., through interactive problem-solving activities, adaptive feedback on their answers, comprehensive analytics on student thought processes, feedback regarding learning strategies on the outcomes of their problem-solving, as well as collaborative features of the software helping with a 12% decrease in dropout rate among high school students. However, these systems exist on opposite ends of the scale for geographic convenience; MyTutor.mu requires a stable internet connection, and Carnegie Learning, with a high price tag involved with its implementation process which perpetuates in-equity among marginalized and underfunded districts in the U.S. Despite these benefits, adaptive learning systems face challenges such as unequal access to devices and connectivity, ongoing training for educators, student data privacy concerns, and the need to balance technology with traditional teaching methods for holistic development. Multimodal learning support, which includes text, audio, and visual content, can also be enhanced by Artificial Intelligence (Becker et al., 2018; Popenici & Kerr, 2017; Tuomi, 2018).

For instance, Century Tech uses Natural Language Processing (NLP) for dyslexic support, which includes personalized learning paths, interactive quizzes and games, immediate feedback, and analytics empowering educators to tailor strategies based on progress. Snapplify is working to make education more accessible for visually impaired students with audio, text to speech, customizable display, and collaboration and participation features in South Africa. These platforms also bridge the equity gap by allowing all students to have access to the important resources regardless of their physical abilities. With the advent of AI-enabled learning analytics, education can predict the needs of the student and improve outcomes. These systems measure the interactions between the student and the educational platform to analyze learner data and provide personal interventions. For example, it was identified that Squirrel AI, based in China, was able to obtain 92% accuracy in predicting where students' learning gaps were by analyzing data on students' interactions with the platform using assessments and a student-specific learning profile and adaptive learning path (Holmes et al., 2019; Luckin & Holmes, 2016; Nguyen et al., 2018; Viberg et al., 2020). Another example is BYJU'S in India, which observes the impact of custom study plans using AI-enabled learning analytics, for both diverse students and groups of students, which has yielded positive results that contributed to a more equitable education landscape (Louisville & Educause, 2022; Zawacki-Richter et al., 2019). A summary of the SWOT analysis of AI in education is summarized in the following tables (Table 1: SWOT including Mauritius; Table 2: SWOT in Africa Excluding Mauritius and Table 3: Global leaders).

Table 1 SWOT analysis of AI in education by region including Mauritius

Category	Analysis	Supporting evidence
Strengths	Centralized policy implementation (Digital Education Plan) High urban digital infrastructure (Ebene CyberCity) Multilingual AI pilot projects (e.g., MyTutor.mu)	MoE Mauritius (2022), ICTA (2023)
Weaknesses	Rural–urban digital divide (42% rural broadband access) Limited local AI research capacity Teacher readiness remains an important challenge for the effective adoption of AI in education, highlighting the need for continuous professional development and institutional support (UNESCO, 2023; Selwyn, 2022).	ICTA (2023), Mohee & Durbarry (2025)
Opportunities	• "Smart Island" initiative funding • Commonwealth of Learning partnerships for ethical guidelines • Potential as SIDS model for scalable AI solutions	UNESCO (2021)
Threats	• Dependency on foreign AI tools (lack of local customization) • Ethical risks from unregulated data use • Brain drain of tech talent	UNESCO (2023)

Ministry of Education (MoE) Mauritius (2022), ICTA (2023), Mohee & Durbarry (2025), UNESCO (2021), UNESCO (2023).

Table 2 Africa (Excl. Mauritius)

Category	Analysis	Supporting evidence
Strengths	Leapfrogging potential via mobile-first solutions (e.g., Snapplify) Strong grassroots innovation hubs (e.g., Kenya's ed-tech startups)	UNESCO (2021)
Weaknesses	Unstable internet (60% rural schools lack access) Minimal AI policy frameworks (65% institutions unregulated) Low teacher digital literacy	AfDB (2023)
Opportunities	Pan-African collaborations (e.g., AU Digital Education Strategy) Low-cost AI localization (e.g., NLP for indigenous languages)	AfDB (2023)
Threats	Vendor lock-in by foreign ed-tech companies Data colonialism risks (exploitation by global platforms)	Selwyn (2022), UNESCO (2023)

UNESCO (2021), African Development Bank (AfDB) (2023), Selwyn (2022), UNESCO (2023)

Table 3 Global leaders (USA, UK, China, EU)

Category	Analysis	Supporting evidence
Strengths	Robust R&D funding (e.g., NSF grants, Horizon Europe) Advanced infrastructure (95% school connectivity) Strong industry-academia partnerships (e.g., Carnegie Learning)	Pane et al. (2017), Chen et al. (2021)
Weaknesses	High implementation costs exacerbate inequity Overreliance on proprietary tools limits adaptability Teacher resistance (40% fear job displacement, Selwyn 2019)	Luckin (2018), Selwyn (2022)
Opportunities	GDPR-style frameworks for ethical AI (EU leadership) AI-augmented lifelong learning systems Global South partnerships for scalable solutions	Pedro et al. (2019), Holmes et al. (2019)
Threats	Algorithmic bias in adaptive platforms Commercialization over pedagogical value	Zawacki-Richter et al. (2019)

Pane et al. (2015), Chen et al. (2021), Luckin (2018), Selwyn (2022), Pedro et al. (2019), Holmes et al. (2019), Zawacki-Richter et al. (2019)

4.1 Challenges and Considerations

AI-driven learning analytics proffer various advantages, yet there are some areas of confrontation:

- Data privacy concerns: stringent measures to protect privacy with regulatory compliance needed
- Equity of access issues due to differences in technology and internet availability
- Trainee teachers with knowledge of data interpretation and use
- Maintaining a balance between technology and essential human interaction and emotional support

4.2 AI Chatbots Improving Engagement in STEM Education

Within Nigeria, AI chatbots help engage students with STEM education, especially those who find it difficult to adapt to the conventional system. These chatbots interact with students and provide access to difficult subjects. They make use of Natural Language Processing in text or voice interaction. The interpreter answers questions, explains concepts, and works through the steps to solve problems with customers. In short, the benefits of STEM education are: 24/7 availability of educational support; extremely helpful outside of school hours; conversational interaction that promotes critical thinking; assistance aimed at the areas of student needs and understanding gaps. It was also demonstrated AI chatbots significantly improve student engagement in STEM education in Nigeria through immediate feedback and interactive learning opportunities creating supportive environments for student success (Adediran et al., 2025).

4.3 Comparative Analysis of Recent Studies (from 2014 to 2024)

Table 4 synthesizes 6 representative studies across regions, selected for methodological rigor (MMAT score $\geq$ 80%, See Table 5 for MMAT scores). Three critical patterns emerge: Infrastructure dictates impact: AI tools in low-connectivity regions (e.g., Nigeria, rural Mauritius) show lower efficacy (AfDB, 2023). Teacher training gaps persist: Only 5/78 studies reviewed included comprehensive educator preparation (e.g., Finland's AI literacy programs). Ethical safeguards lag: Just 12% of African studies addressed GDPR-like compliance (UNESCO, 2023), risking data exploitation.

Table 4 Meta-analysis of AI in education studies (2014–2024)

Study (author, year)	Country	AI tool/ technology	Key findings	Limitations
Luckin (2018)	UK	Century Tech (NLP)	Improved literacy by 15%	Limited rural adoption
Pane et al. (2017)	USA	MATHia (adaptive learning)	Reduced dropout rates by 12%	High implementation cost
Chen et al. (2021)	China	Squirrel AI (predictive analytics)	92% accuracy in gap detection	Data privacy concerns
MoE Mauritius (2022)	Mauritius	MyTutor.mu (tutoring bot)	18% improvement in Math scores	Requires stable internet
Oyelere et al. (2020)	Nigeria	AI chatbot (STEM tutor)	22% higher engagement	Low smartphone penetration
UNESCO (2021)	South Africa	Snapplify (accessibility AI)	Enhanced inclusion for disabled students	High cost for low-income schools

Luckin (2018), Pane et al. (2015), Chen et al. (2021), Ministry of Education (2022), UNESCO (2021)

Table 5 MMAT scores for top 10 included studies among 78 studies

1	Luckin et al., 2016	Mixed Methods	90%	NLP for literacy	Century Tech improved literacy by 15% in dyslexic students
2	Pane et al., 2017	Quasi-experimental	95%	Adaptive learning (MATHia)	12% reduction in dropout rates; high cost limitations
3	Chen et al., 2021	RCT	92%	Predictive analytics	92% accuracy in learning gap prediction (Squirrel AI)
4	MoE Mauritius, 2022	Policy evaluation	75%	Tutoring bot (MyTutor.mu)	18% math score gains; internet dependency noted
5	Oyelere et al., 2020	Case study	80%	STEM chatbots	22% engagement increase; low smartphone penetration barrier
6	UNESCO, 2021	Qualitative	85%	Accessibility AI (Snapplify)	Enhanced inclusion for visually impaired students
7	Zawacki-Richter et al., 2019	Systematic review	100%	AIEd frameworks	Identified equity/ teacher training as global gaps
8	Selwyn, 2022	Critical analysis	88%	Ethical implications	40% teachers fear job displacement; urges policy safeguards
9	Kumar et al., 2020	Survey	70%	Digital divide	Urban–rural gap limits AI access in 78% of villages
10	Dalgarno et al., 2022	RCT	93%	Speech recognition	30% faster reading progress for dyslexic students

Luckin et al. (2016), Pane et al. (2015), Chen et al. (2021), Ministry of Education (2022), UNESCO (2021), Zawacki-Richter et al. (2019), Selwyn (2022)

Trends:

- USA and China lead in AI-driven personalized learning
- Africa shows growth but faces infrastructure challenges
- Mauritius is emerging with localized AI solutions

4.4 Digital Divide

In terms of internet connectivity, device availability, and digital literacy, the digital divide comes in the way of the effective implementation of AI in educational terms, especially in areas with limited technology access. A few examples are notable in that regard in rural Africa and in terms of urban–rural India. In rural Africa, it is mostly the instability of Internet connectivity that makes the adoption of AI tools very difficult, as most of these rural communities are not adequately developed with respect to efficient internet access. The absence of stable Internet would mean that rural students in effect are unable to access online resources and AI learning tools, further limiting their interaction with personalized learning and hence the reduction in educational standards. This bridge provides some instances of challenges and possible solutions: there are NGOs, governmental programs to improve internet infrastructure and provide teacher training, satellite internet solutions, and developing mobile connectivity toward connecting remote communities. India presents another considerable digital divide between urban areas and the rural zones short of direct access to AI technologies. Hence, many students are deprived of devices and connectivity that would enable them to utilize AI-driven training platforms for an exposure to some degree of personalized learning processes that seem to carry daylight potential in uplifting their academic performances. Since it is all about the digital divide urban–rural in India, an all-encompassing approach must be taken addressing matters including government promotion of internet infrastructure in rural areas, affordable device provision programs, community digital literacy training, among others. Data privacy is very important in education as more and more institutions use AI technologies that need enormous amounts of data on students. Regulatory frameworks between different regions reveal great disparities in the way data privacy is managed. Within the European context, the General Data Protection Regulation (GDPR) establishes a rigorous benchmark for data privacy, granting individuals meaningful control over their personal information. In stark contrast, numerous African nations continue to grapple with the development and enforcement of data privacy regulations, which in turn compromises the security of student information. As educational technology proliferates, the gap between regions with strong data protection frameworks and those without becomes increasingly consequential. Addressing these disparities—potentially by adopting GDPR-like standards—could significantly bolster the safety of student data and foster more secure, trustworthy educational environments. Turning to the integration of Artificial Intelligence in education, teacher resistance remains a substantial barrier. Recent data indicates that approximately 40% of educators in the United Kingdom are apprehensive about the potential for AI to supplant traditional teaching roles. Their concerns center around the possibility that AI may erode professional expertise and diminish the inherently human elements of teaching, thus threatening job security and slowing the adoption of new technologies. Comprehensive professional development and targeted training are essential to mitigate these anxieties. By equipping teachers with the skills to utilize AI as a tool for administrative efficiency, personalized instruction, and enhanced student engagement, educational institutions can help ensure that technology serves as an ally rather than a replacement in the classroom (Bank and (AfDB), 2023).

5 Recommendations

AI integration in education presents significant opportunities for enhancing learning experiences, particularly in low-income schools. Government policies must ensure equitable technology access, such as subsidizing AI tools for low-income schools. Mauritius' Digital Education Plan is an example of this, addressing educational disparities and promoting student digital literacy. The plan includes infrastructure investments in rural and underserved areas, internet connectivity, and technological infrastructure to facilitate access to various AI-powered educational platforms. The government has established monitoring frameworks tracking student progress and AI tool effectiveness to assess the plan's impact. By following Mauritius' example, other countries can develop similar policies subsidizing AI tools for low-income schools, fostering more equitable educational landscapes. Teacher training is crucial for successful technology implementation, and Finland's approach to AI educator courses serves as a model for other nations seeking to empower teachers with skills needed for effective classroom AI integration. AI literacy programs focus on both technical aspects and pedagogical applications, fostering deep understanding of how AI can enhance teaching and learning.

Finland integrates AI literacy programs into teacher education curricula, ensuring future educators understand AI implications and applications from career outset. The courses provide hands-on AI tool experience, fostering confidence and competence in using technology. The final AI educator courses emphasize educational AI ethical implications, preparing educators for informed teaching practice decisions (UNESCO, 2023).

Ethical AI frameworks are essential for guiding responsible educational AI use, ensuring technologies are implemented promoting equity, transparency, and accountability. Drawing on UNESCO's guidelines, particularly for Global South nations, can provide robust foundations for these frameworks. Principles for ethical AI frameworks include a human-centered approach, addressing the digital divide, promoting transparency in AI system operation and decision-making, ensuring data privacy and security, and continuously evaluating AI technologies' effectiveness and ethical implications (UNESCO, 2023).

To develop and implement ethical AI frameworks, Global South nations can consider strategies such as stakeholder engagement, capacity building, policy development, and expert collaboration. However, this meta review faces three constraints: geographic imbalance, industry bias, and longitudinal data gaps. AI integration in education presents significant opportunities for enhancing learning experiences, particularly in low-income schools. Government policies, teacher training, and ethical AI frameworks are essential for fostering trust and accountability in AI technology use.

6 Conclusion

Artificial Intelligence (AI) is revolutionizing education by offering innovative tools and methodologies, such as personalized learning platforms and intelligent tutoring systems. However, AI integration faces challenges such as ensuring equitable access, providing comprehensive teacher training, and establishing ethical safeguards. Successful AI implementation relies on educators using AI tools, and comprehensive teacher training programs focused on AI literacy are essential. Future research should explore the long-term cognitive impacts of AI-augmented learning. Another challenge is ensuring equitable student access to AI technologies, particularly in low-income and rural areas, where students often lack reliable internet access and necessary devices. Urban schools may have infrastructure supporting advanced AI applications, but rural and underserved schools often struggle with basic technological needs.

References

Adediran, E. M. T., Adediran, T. D., & Adediran, A. (2025). The role of artificial intelligence in revolutionizing basic science teaching at primary education in Nigeria. A critical *review. International Journal of R Publication and Reviews, 17*(7), 1251–1255. ISSN 2582–7421.

African Development Bank (AfDB). (2023). *Digital divide in Africa: Challenges and opportunities*. African Development Bank Group.

Becker, S. A., et al. (2018) *NMC Horizon Report: 2018 Higher Education Edition.*

Chen, X., Liu, Y., et al. (2021). *AI-powered adaptive learning in China*. Springer.

Holmes, W., Bialik, M., & Fadel, C. (2019). *Artificial intelligence in education: Promises and implications*. Center for Curriculum Redesign.

Information and Communication Technologies Authority (ICTA). (2023). *Annual Report 2022–2023*. ICT Authority, Mauritius.

Louisville, C., & Educause, E. (2022). Artificial intelligence in education (AIEd): A high-level academic and industry note 2021. *AI and Ethics, 2*(1), 157–165.

Luckin, R. (2018). *Machine Learning and Human Intelligence: The future of education for the 21st century*. UCL Press.

Luckin, R., Holmes, W., et al. (2016). *Intelligence unleashed: An argument for AI in education*. Pearson.

Ministry of Education (MoE) Mauritius. (2022). *National AI in education strategy*. Government of Mauritius.

Mohee, R., & Durbarry, A. (2025). *Surveys on Status of Artificial Intelligence in Higher Education*. Higher Education Commission, Mauritius.

Nguyen, A., Gardner, L. A., & Sheridan, D. (2018). A framework for applying learning analytics in serious games for people with intellectual disabilities. *British Journal of Educational Technology, 49*(4), 673–689.

Pane, J. F., Steiner, E. D., & Baird, M. D. (2015). *Continued progress: Promising Evidence on personalized learning*. RAND Corporation.

Pedro, F., Subosa, M., Rivas, A., & Valverde, P. (2019). *Artificial Intelligence in Education: Challenges and Opportunities for Sustainable Development*. UNESCO Working Papers on Education Policy, No. 7.

Popenici, S. A. D., & Kerr, S. (2017). Exploring the impact of artificial intelligence on teaching and learning in higher education. *Research and Practice in Technology Enhanced Learning, 12*(1), 1–13.

Roll, I., & Wylie, R. (2016). Evolution and revolution in artificial intelligence in education. *International Journal of Artificial Intelligence in Education, 26*(2), 582–599.

Schwab, K. (2016). *The fourth industrial revolution*. Crown Business.

Selwyn, N. (2019). Should robots replace teachers? *AI & Society, 34*(3), 1–10.

Selwyn, N. (2022). *Education and technology: Key issues and debates* (3rd ed.). Bloomsbury.

Tuomi, I. (2018). *The impact of artificial intelligence on learning, teaching, and education*. European Commission.

UNESCO. (2021). *AI and Digital Equity in Africa*. United Nations Educational, Scientific and Cultural Organization.

UNESCO. (2023). *AI and data Privacy: A global perspective*. United Nations Educational, Scientific and Cultural Organization.

Viberg, O., Khalil, M., & Baars, M. (2020) Self-regulated learning and learning analytics in online learning environments: A review of empirical research. In *ACM learning analytics & knowledge conference* (pp. 524–533).

Zawacki-Richter, O., Marín, V. I., Bond, M., & Gouverneur, F. (2019). Systematic re-view of research on artificial intelligence applications in higher education. *Internation-Al Journal of Educational Technology in Higher Education, 16*(1), 1–2.

Evaluating AI as a Tool for Climate Change Risk Mitigation: A Systematic Literature Review

Hannah Brown, Sonali Das, and Marié Hattingh

Abstract Artificial Intelligence (AI) is increasingly being used by various decision-makers to access information for the management of different risks associated with their businesses. Climate change is one such risk that is pervasive across almost all industries, with no exceptions to location, with some nations more equipped to survive than others. To evaluate literature on how AI is used to inform climate change risk mitigation and adaption strategies, in this paper, we adopt a systematic literature review methodology following Preferred Reporting Items for Systematic Reviews and Meta-Analyses (PRISMA) guidelines. We sourced literature from the Web of Science, EBSCOhost, Emerald Insights and ScienceDirect using keywords that are reflective of the interdisciplinary relationships between AI, climate change, climate services and risk management. In total, 65 papers met the inclusion criteria of our search. Our findings revealed that while AI tools reap many benefits, including task simplification, its value as a research tool and its decision-making capabilities, its shortcomings are not to be overlooked. These include AI's potential to generate biased and inaccurate content, its lack of an ability to cater to context, and its intense energy requirements. Further, findings established that the topic of AI-assisted climate change-induced risk management overlooks considerations of how indigenous communities fit within it. This presents a gap that calls for further research and contribution to the corpus, particularly from and of the Global South, including low-resourced language representation, which was found to be lacking at the time of the writing of this paper.

H. Brown (✉) · S. Das
Department of Business Management, University of Pretoria, Pretoria, South Africa
e-mail: u19000775@tuks.co.za

S. Das
e-mail: sonali.das@up.ac.za

M. Hattingh
Department of Informatics, University of Pretoria, Pretoria, South Africa
e-mail: marie.hattingh@up.ac.za

J. Marx Gómez et al. (eds.), *Recent Advances in Information Systems*, Progress in IS,
https://doi.org/10.1007/978-3-032-14099-9_36

1 Introduction

In Pixar's animated movie *Wall-E*, the complex relationship between Artificial Intelligence (AI) and a post-apocalyptic Earth forms its central theme. While ostensibly targeted at a younger audience, the AI-powered robot *Wall-E*, whose primary function is to clean up the vast amounts of waste left behind by humanity on a planet ravaged by environmental degradation, presents a poignant and thought-provoking depiction of an increasingly plausible future (Pixar, 2008). This portrayal invites critical reflection on themes of sustainability, technological dependence and the long-term implications of environmental neglect by the hands of humans. Though animated, *Wall-E* presents a stark reality: that humanity's current actions, specifically those of a destructive nature, will shape our planet's future.

Anthropogenic activities, particularly carbon emissions, have ensured that climate change is no longer a distant threat. The long-term consequences of climate change have become an inevitable reality that humanity must grapple with, given that 2024 was declared as the warmest year recorded yet with a global mean temperature of 1.5 °C above pre-industrial levels (WMO, 2025). In order for industries to create effective mitigation and adaptation strategies to combat the negative impacts of climate change, climate information is an essential component that must be accessible. Climate information (more formally known as climate services) is the delivery of data, tools and statistical analyses of historical and future climatic conditions (USAID, 2013). This data can include historical observations, as well as future predictions of temperatures and precipitation, extreme climatic events and sea-level changes. While often confused with weather services, the two differ in that weather services are mostly short-term forecasts, while climate services integrate more long-term patterns (Hewitt & Stone, 2021). While weather services exist to inform the public of expected weather conditions for the near future, climate services serve the purpose of providing insights on how to prepare for the future, based on previous observations (Hewitt & Stone, 2021).

With the frenzy associated with the public release of ChatGPT in 2022 came the exciting, albeit perhaps naive, belief that it could solve almost any issue the world currently experiences. However, it is essential that ChatGPT, and other similar AI tools, be evaluated holistically in order to understand their strengths and limitations, to assist in climate change risk management that is comprehensible, both physically and financially. Over time, there is an overwhelming amount of interest in AI and how it can improve daily life (Schwaller, 2025). Significant research is being focused on how AI can facilitate the movement towards mitigation of and adaptation to climate change. AI has the potential to bridge gaps in access to information, support adaptive strategies and empower diverse communities in their fight against climate change by enabling them to adapt to change, better absorb shocks and promote the transformation of systems for long-term sustainability (White et al., 2023). These concepts align with the *resilience theory*, which focuses on the ability of people, businesses or systems to absorb disruptive shocks, adapt to change and transform to mitigate future risks (Jeans et al., 2017). Research on AI-assisted climate change mitigation and adaptation often focuses on only one isolated sector at a time (Lah, 2025). However, the nature of climate change risk and impacts is more complicated and intricate, as not only does it impact virtually all sectors, but these sectors have relationships between each other. Therefore, there is a need for a review of the literature to draw comparisons and identify strengths and weaknesses of AI tools being developed, as well as the need for research efforts being dedicated to the climate services space that intersects with various industries. Additionally, given that technology is constantly evolving, a systematic review of the literature is required to provide an updated overview of the current state of research surrounding this topic. As such, this study will be guided by the following research question: *What are the strengths and shortcomings of AI as a risk management tool for climate change mitigation and adaptation?* The rest of the paper is structured as follows: Sect. 2 provides a brief literature review on the key themes of this research, Sect. 3 describes the methodology of the study, the research findings are presented in Sect. 4, Sect. 5 is a discussion of the findings, and finally, Sect. 6 provides the conclusion of the study.

2 Literature Review

The literature review here aims to focus on the three aspects of this study's interest, namely: climate change, climate services and Artificial Intelligence, as presented below.

2.1 Climate Change

Climate change, which refers to the long-term change in the climate, is among the most significant crises currently confronting humanity, and it threatens the future in more ways than imaginable (Harvey, 2014). Climate change causes ecosystems supporting biodiversity and life to become unbalanced, increases frequency of extreme climatic events, such as hurricanes, floods and droughts and many significantly more damaging events (UNDP, 2023). The human activities responsible for anthropogenic climate change are particularly those that release extensive levels of greenhouse gases, such as carbon dioxide, nitrous oxide, methane and fluorinated gases (UNDP, 2023). The manufacturing of goods is a massive contributor towards the existence of climate change, as it makes up an estimated third of global greenhouse gas emissions, with these emissions usually originating from products that are carbon-intensive to produce, such as iron, steel, concrete, cement, aluminium, paper and glass (Turrentine, 2022).

Extreme climatic events, such as droughts, floods, wildfires and hurricanes destroy infrastructure and damage communities, farms and biodiversity (Lindwall, 2022). Climate change has resulted in the melting of glaciers and polar ice sheets in the

Arctic, which causes darker ocean waters to absorb more sunlight, speeding up both the melting and global warming and causes sea-level rises (Lindwall, 2022). Sea-level rises can cause extreme devastation to low-lying regions, like islands and coastal cities, such as Mumbai, Seychelles and New York City (WHO, 2023). To formulate and implement strategies to combat and adapt to climate change, which is essential for the survival of many people and industries, these strategies need to be well-informed. As such, the correct information regarding the historical and future climate, as well as other relevant information, is needed by decision-makers and strategists in relevant sectors.

2.2 *Climate Services*

There is no agreed definition of climate services, as each application of climate services is unique and nuanced to its context. In broader terms, climate services entail the delivery of data, tools and statistical analyses of both historical weather patterns, and future predictions of climate conditions (USAID, 2013). Climate services are only of real value if they can be used by relevant stakeholders to support their decision-making processes. The demand for climate services is undeniable, as more individuals and organisations are increasingly exposed and vulnerable to the extreme impacts of climate change (Hewitt & Stone, 2021). To combat, mitigate and adapt to these negative impacts, policies and strategies must be drafted, which requires climate information to be accurate and effective. Further, the deployment of climate services is heavily dependent on who is using it, and why it is being used. As climate services are needed in so many different areas of life, there exists a pressing issue of delivering climate services that are context-specific to users, because if information is too generic, these services can be rendered completely useless and irrelevant to users that are engaged in niche areas, such as for example indigenous crop farmers (Hansen et al., 2019).

2.3 *Artificial Intelligence*

AI, a technology that allows computers to simulate human intelligence, thereby enhancing its problem-solving capabilities, has redefined the life we are accustomed to. One of the most impactful changes that AI has brought into this world is its ability to facilitate access to information and other resources (Valavanidis, 2023). One of these resources includes essential climate services. As the onslaught of the consequences of climate change continue, AI tools provide easy access to adaptation and preventative strategies (Jain et al., 2023). The subject of how AI is applied in areas of climate change is constantly evolving, and as such, the amount of research dedicated to this area can be overwhelming (Lin et al., 2024).

3 Methodology

This study employed a systematic literature review methodology which encompassed the process of collecting and analysing the secondary data that form the previous literature of a specified topic, for the purpose of answering one or multiple research questions (Mengist et al., 2020). We employed the guidelines set out for Preferred Reporting Items for Systematic Reviews and Meta-Analyses (PRISMA) which is widely recognised as a framework that aids researchers in conducting systematic literature reviews and meta-analyses, as shown in Fig. 1.

In the process, papers were collected from four academic databases: Web of Science, EBSCOhost, Emerald Insights and ScienceDirect by using the search strings detailed in Fig. 2. To avoid variations in search outputs, we consistently used lowercase keyword terms.

The inclusion and exclusion criteria applied were as follows: papers should have been written in English, published from 2021 to 2024, and duplicate papers were removed. The restriction on focusing only on English language papers was in order to maintain consistency across the research process, and in the collection and interpretation of that data; and to avoid the case of misinterpreting information due to a flawed translation. After applying the inclusion and exclusion criteria, and completing the screening, 65 papers made up the final sample. However, not all the sources have been individually cited or mentioned, due to space constraints, but can be made available on request. Instead, this review illustrates examples that are representative of the most prominent themes of the sample analysis.

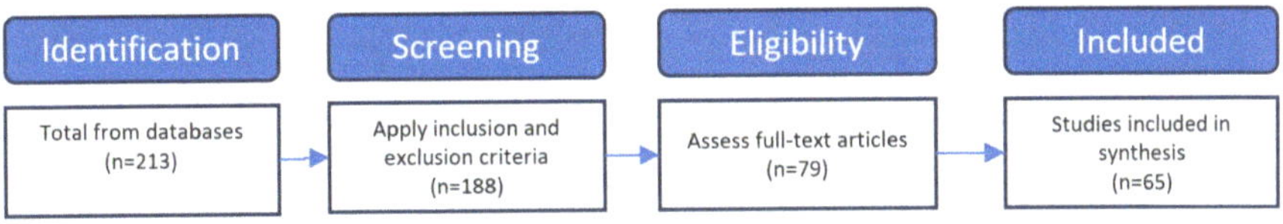

Fig. 1 PRISMA flowchart

Search database	Keyword search
Web of Science	("climate*") and ("genai" or "generative artificial intelligence" or "large language model" or "llm" or "chatgpt")
EBSCOhost	("climate*") and ("genai" or "generative artificial intelligence" or "large language model" or "llm" or "chatgpt")
Emerald Insights	(climate) and (genai or generative artificial intelligence)
ScienceDirect	(climate) and (genai or generative artificial intelligence)

Fig. 2 Keywords used in searching databases

4 Findings

It is integral when performing a systematic review to consider the context behind and of the research, as this can give valuable insights into research trends of the subject. As such, this section discusses the findings from the sampled 65 papers with respect to various aspects such as the descriptive analysis of the sample, the research questions and any other findings the sample produced, each of which are presented below.

By using the search strings detailed in Fig. 2, it emerged that majority of the papers focused on general AI usage, rather than focusing specifically on generative AI (GenAI). Therefore, Sects. 4.1 and 4.2 will present a discussion on the strengths and shortcomings of AI in general as a risk management tool for climate change mitigation and adaptation. Where appropriate, specific reference is made to the usage of GenAI as reported in the cited literature. However, the data did reveal that GenAI application is emerging within climate services, policies and finance. Papers focusing on climate research included, among many, AI-assisted evaluation of climate change plans (Fu et al., 2024); how ChatGPT delivers narratives of climate change (Sommer & von Querfurth, 2024); and how AI may aid in climate research (Krzyżewska, 2024). Some of the climate policy research included those comparing how ChatGPT performs in providing advice for climate policy and how experts' opinion vary (Salekpay et al., 2024). Only three papers developed original AI models: ChatClimate was developed by integrating ChatGPT with reports from the Intergovernmental Panel on Climate Change to better handle prompts related to climate change (Vaghefi et al., 2023); another study involved the training of a deep learning model to improve its performance in predicting the risk related to wind turbines (Chatterjee et al., 2023); and some research trained a Generative AI model to have specialised knowledge of plant species in vegetation patches in east Spain (Hirn et al., 2024).

The following subsections will discuss the findings of the perceived strengths and shortcomings that come with using AI as a risk management tool for climate change mitigation and adaptation.

4.1 *Strengths of AI as a Risk Management Tool for Climate Change Mitigation and Adaptation*

Task simplification: General AI enhances humanity's ability to complete tasks more efficiently and effectively due to the AI tools used, in tandem with having the capability to augment and simplify activities. Through its language understanding, AI can assist users in many downstream tasks (Li et al., 2024). For example, Life Cycle Assessment (LCA) practitioners can assess environmental impacts of any given product by using AI to help define their goals and assist in data analysis, thus reducing time (Preuss et al., 2024). Researchers have also investigated how practices such as sustainable farming can be enhanced by general AI (Ray, 2023), how general AI can automate agricultural tasks like crop harvesting (Holzinger et al., 2023), and how AI-assisted monitoring in plants can assist in detecting diseases (González-Rodríguez et al., 2024). AI's task-simplifying ability can be reaped with regard to Generative AI, for example, by using Large Language Models (LLMs) to filter through the climate disclosures of financial institutions in Europe, in order to find discrepancies more efficiently (Moreno & Caminero, 2024).

AI as a research tool: General AI can be leveraged as an effective research tool, whether it can be used as a writing and editing tool, or to conduct actual research. General AI has the potential to advance biotechnology research, through implementing genomic computing that can enhance understanding of environmental systems (Holzinger et al., 2023). However, it was cautioned that this would only be achievable by researchers who were able to use these AI tools adeptly (Holzinger et al., 2023). Nine papers focused on how general AI tools can be leveraged for science communication; as well as exploring how general AI assists in making people more aware of climate change and recommend green practices (Pu et al., 2024; Vuong & Ho, 2024). ChatGPT has supported agricultural research, through its ability to summarise, generate hypotheses, design and write reports, which facilitates a more efficient research process (Ray, 2023). Scientists have gained significant benefits from using ChatGPT in all stages of their research (Ansari et al., 2024). LLMs can also assist in interdisciplinary research, which is essential when addressing the diverse impacts of climate change, such as biodiversity loss (Mammides & Papadopoulos, 2024). Many articles within the sample actively used AI as a tool to conduct their research, for example, using a LLM to analyse climate disclosures from ESG reports of European banks; and using ChatGPT for text mining tasks for meta-analysis of mycotoxin contamination cases in wheat, a toxin that is exacerbated by humidity and precipitation (Jiang et al., 2024; Moreno & Caminero, 2024).

General AI systems are able to generate and summarize from the plethora of climate change information, at a speed incomparable to humans (Vuong & Ho, 2024). This ability is attributed to the way in which general AI models are trained, as they are trained on vast datasets that enable it to handle large volumes of input and adapt its style of output to the end-user (Richards et al., 2024). For example, researchers were able to gain new valuable insights thanks to LLM-assisted quantitative analysis that allowed them to filter through copious amounts of literature to find information specific to their domain (Miao et al., 2024). Additionally, general AI has the ability to analyse vast amounts of data and literature which allows for researchers to stay updated with and adequately synthesise the research that is being published everyday (Lin et al., 2024).

AI as a decision-making tool: Nineteen papers claimed that AI has the potential to transform and elevate decision-making processes across disciplines. General AI-assisted decision-making can also aid in strategic responses in the agricultural field, due to its disease detection and management properties (González-Rodríguez et al., 2024). At a broader level, AI can also help users make smarter and greener decisions by providing insights into everyday energy consumptions (Haluza & Jungwirth, 2023). AI models, such as ChatGPT, have the power to process questions and answer appropriately, and also provide an array of different options, which can aid users in decision-making (Zhang et al., 2023). ChatGPT has facilitated better decision-making for farmers, given its ability to generate understandable summaries of complicated regulations and risk assessments (Ray, 2023).

4.2 Shortcomings of AI Being Used as a Risk Management Tool for Climate Change Mitigation and Adaptation

AI's inaccurate outputs: AI's potential to generate inaccurate output as a weakness presents as a major concern. AI models still present a problem with generating inaccurate outputs which can be due to biased dataset training, or outdated and incomplete data (Pu et al., 2024). ChatGPT has presented supporting evidence that was irrelevant and inaccurate when asked how it formulated its answers (Fu et al., 2024). In fact, it was noted that only 30–40% of ChatGPT's outputs were accurate (Krzyżewska, 2024). AI models that simulate the impacts of a changing climate on ecological factors are flawed, even when advanced, which is concerning, due to the precision that this area of research requires (Mehryar et al., 2024). This tendency of AI to generate inaccurate, or even nonsensical output that are nonetheless seemingly credible can also be referred to as hallucinations. Hallucinations are referred to frequently as a major concern in the literature (Hadid et al., 2024; Leippold, 2023; O'Keefe et al., 2024; Şahin, 2024).

The biases of AI: The literature suggests that AI can often hold heavy biases, which is mostly due to the data it is trained on and whoever developed it, which can have detrimental impacts on users. This data can be anything from pollution to climate change, and also encompass solutions that might be preferred over others, according to whoever trained it (Biswas, 2023). Biases can also extend to groups of people, which is perhaps most harmful. The epistemological biases of AI models can cause harm to certain groups, in addition to a lack of access to these resources (González-Rodríguez et al., 2024). This can also lead to AI reinforcing these biases in its outputs, leading to a skewed representation of these communities and encouraging further prejudice (Pu et al., 2024). AI could also exclude the considerations of underrepresented groups, and provide recommendations that do not align with the ethics of society, and therefore output could be regarded as somewhat irrelevant (Haluza & Jungwirth, 2023). Access to timely data is essential when concerning topics of climate change (Vaghefi et al., 2023).

AI's hindrance with context: Quite a troubling hindrance AI models possess is their lack of ability to accommodate and adapt to different contexts. In an experiment of how AI performed in climate change storytelling, ChatGPT generated somewhat standardised content (Sommer & von Querfurth, 2024). Additionally, ChatGPT struggled when presented with jargon specific to a domain, further confirming AI's struggle to accommodate context (Fu et al., 2024). Furthermore, ChatGPT had a reduced understanding of context, and when presented with queries about a specific context, could generate incorrect and fake information (Fu et al., 2024). This highlights a major drawback of using AI, as even within climate change subjects, issues can be multi-faceted and complex, and require nuanced outputs, and as such, should be used with caution (Biswas, 2023).

AI's intense energy requirements: While this paper investigated how AI can assist in climate change mitigation and adaptation, it cannot be ignored that the intense energy requirements of AI models were often discussed in the sampled literature. Much of the literature indicated the serious concern that the operating of AI models comes at significant computational and energy costs, with a negative impact on the environment as a consequence (Carpo, 2024; Hadid et al., 2024; Li et al., 2024; Preuss et al., 2024). The process of training LLMs is energy-intensive; training ChatGPT-3 equates to the same amount of carbon emissions produced from over 180 flights from New York to San Francisco (Larosa et al., 2023).

5 Discussion

This systematic literature review aimed to explore the benefits and challenges of using AI as a tool for climate change mitigation and adaptation. It was found that task simplification, its potential as an effective research tool, and the decision-making assistance were among the most significant strengths perceived by researchers in the literature. Regarding shortcomings, this review highlights AI's potential to generate inaccurate output, biases and inability to accommodate context, as well as its intense energy requirements as substantial concerns. In accordance with the *resilience theory*, AI presents as an opportunity to allow organisations to better prepare for, adapt to and mitigate the negative impacts of climate change. Through using AI, organisations can strengthen their absorptive and adaptive capacities in their response to climate change impacts. However, it is imperative to investigate the access afforded to underrepresented communities, such as indigenous people. Additionally, it is essential to understand how AI usage impacts these disadvantaged groups. The subject of indigenous communities' relationship with AI tools in climate change mitigation and adaptation is conspicuous by its absence within the sampled 65 papers. This is important to note, given that a country's identity and culture is heavily connected to its indigenous people. AI can be harmful to disadvantaged groups; human bias informing the manner in which AI is trained and with what data, can exacerbate inequities faced by underrepresented populations in literature (Leippold, 2023; Richards et al., 2024). This can result in intensified exposure to disadvantages, and a denial of equal opportunities, benefits or access to resources (Hacker, 2024). In fact, civil rights groups are advocating for the protection of minority-group rights against any biased or discriminatory decision made by AI (Khan et al., 2024). The harmful nature of these decisions is that they are often made in favour of one group over another. Furthermore, a gap of knowledge remains in how to remedy AI's perpetuation of inequalities, specifically in scientific knowledge and communication (Richards et al., 2024). The risk in excluding disadvantaged groups in AI-assisted crop management could erase localised knowledge systems and further implicate smallholder farmers in rural areas (González-Rodríguez et al., 2024). Furthermore, AI tools are often more accessible to privileged groups—not only in the sense of having technology to access AI tools, but also it being accessible in English or other high-resourced languages only.

5.1 *Limitations of This Study*

At the time of the conducting and concluding of this research, the new LLM named DeepSeek developed by a company in China, had not yet been released. However, it is important to note that this new development highlights our argument that taking into account how quickly AI models are released is important, as LLMs have proved to be a disruptor from 2022 onward. We have clarified that we only included papers that were in English in our sample, to avoid misinterpreting translated articles. This presents a limitation as we could have missed valuable research articles in other languages that delve into how AI can be used to mitigate and adapt to climate change. Our choice of keywords to use in our search for papers could have acted as a limitation. It was noted that a paper previously published by the authors did not appear in the search results (Amini et al., 2023). This paper fits perfectly within the constructs of the systematic review, however, due to various factors, including a lack of keywords in that respective article, and the e-journal databases, it did not feature in the search outputs. Thus, a limitation of this systematic review remains that some high-quality research could have been omitted due to the methodology of the systematic review.

5.2 *Gaps that Call for Further Research*

We deduce that the lack of a Global South perspective, the need for regulations and the constant evolving nature of AI are major gaps needing attention.

Lack of Global South perspective: Developing countries are disproportionately impacted by climate change, and solutions crafted by the Global North are not necessarily fit for the Global South. It is thus essential to fill the gap where there is a lack of the Global South perspective, so that AI tools can effectively apply to their respective countries. In one instance, ChatGPT was mostly trained on climate change subjects in a Western context, thereby excluding climate change from a postcolonial perspective, and other Global South climate crises (Sommer & von Querfurth,

2024). Furthermore, certain social groups' voices are noted to be excluded from AI research and development (Chen et al., 2024). It was also found that literature based on developed countries would not be easily transferable to developing countries, with some ChatGPT responses being inconsistent for some African and Middle Eastern countries (Atkins et al., 2024).

The use of AI tools implicitly assumes that users are proficient in a high-resourced language, literate in both reading and technology, and have access to an electronic device and sufficient network service. This assumption is not often true in Global South countries—thus completely leaving behind people who do not meet these expectations. It is still seldom available to people who can't speak English, specifically those who speak low-resourced languages, most predominantly in the Global South. The literature has highlighted that AI's lack of accommodation for other languages is a significant research gap that requires further research (Rillig & Kasirzadeh, 2024). There is a call for further research to be conducted on how chatbots respond to prompts from different languages, as there is a lack of literature on this subject (Volk et al., 2024). Beyond the call for further research, it is recommended that in order to alleviate this issue, underrepresented communities be included within the development of AI tools, with special attention given to how these tools can accommodate their contexts and low-resourced languages.

The need for regulation: While AI has been hailed as a transformative tool that has many benefits, much of the literature sampled acknowledges that it is essential for there to be regulations that come with AI usage. Standards and regulatory frameworks need to be laid out for AI-assisted agricultural activities (González-Rodríguez et al., 2024). Additionally, technology is currently in development to distinguish between human-generated content and AI-generated content. It is essential that AI be used as a complementary tool to aid in research, and not completely replace the entire process (Pu et al., 2024). Similarly, there is a need for regulation to protect individuals and copyright laws as AI gains popularity (DeCotis, 2023). However, it has been noted that AI regulation in climate change spaces still remains as a gap that requires further research. What's more important is the involvement of more stakeholders in the regulation, so that not only the interest of developers are considered (Hacker, 2023).

AI research is constantly evolving:While this area of research has seen an exponential surge, it can become outdated and irrelevant very quickly, especially given that AI models are being developed and improved upon everyday (as previously mentioned regarding DeepSeek). This presents ample opportunity for further research, because with new developments, come new findings. It has been said that LLM advancement is a reality, and progress of models is an inevitability (Chew et al., 2024). The growing amount of AI research in climate change is constantly evolving and presents a research gap as research might become outdated as more AI models are developed (Lin et al., 2024). Thus, further research is essential, given the rapid evolution of AI research (Khan et al., 2024). Future AI tools will have more capabilities that possess more features, which indicates that its potential is untapped (Rillig & Kasirzadeh, 2024).

6 Conclusion

Using a systematic literature review methodology, this paper aimed to assess the current literature surrounding the usage of AI as a risk management tool for climate change mitigation and adaptation. Research in this area appears to be heavily theoretical, as a significant portion of the sample used the literature review methodology. AI is an effective tool for task simplification and research, and can promote data-driven decision-making. However, that is not to say that AI tools do not come with concerning drawbacks. AI can produce inaccurate content that is unreliable, also known as hallucinations. AI also struggles heavily with biases and context-specific tasks which threatens to make AI usage for misrepresented communities virtually inaccessible and perhaps even harmful. Additionally, the actual operating of AI is extremely energy-intensive, which brings about the concern that the efforts put into AI assisting in climate change mitigation could be futile.

AI usage for indigenous communities can be more complicated, and potentially prejudicial, and there is evidence that the interests of these groups are excluded in the development of AI tools, virtually excluding them in this new AI age. AI developers seem to favour Global North interests and particularly users who speak high-resourced languages. There must be a call for research to be conducted with the involvement of stakeholders directly at the core of these issues. AI tool usage is also in need of regulation to protect the rights and interests of users. Finally, due to AI constantly evolving, research can quickly become outdated, alluding to the opportunity to research more gaps.

The ending of *Wall-E* sees the humans and robots unite to plant the first seedling on Earth in 700 years, while pledging to nurture Earth and prevent further damage. This is an effective allegory that emphasises the potential for a symbiotic relationship between AI and humanity. AI will always need humans, and in leveraging AI's abilities to help us adapt to and mitigate the impacts of climate change, we can move forward not only to cope in the present, but to protect it for future generations. However, its downsides are not to be overlooked. This paper gives insights into this area of research, and presents important gaps that are imperative to investigate. Perhaps most importantly, this paper highlights the need for including disadvantaged groups in AI tool development, and the lack there currently exists. It is vital to explore this in order to create solutions to climate change that can be used by everyone for the survival of humanity.

Acknowledgements HB's research was supported by the University of Pretoria's Doctoral Research Bursary, whose support is duly acknowledged; SD's research was supported by the South African National Research Foundation Incentive Funding for Rated Researcher (UID/Project 2103166383).

References

Amini, A., Buck, C., Brown, H., Bulian, J., Huebscher, M. C., Ciaramita, M., Das, S., Gaiarin, B., Gordon, C., Gupta, R., et al. (2023). AI and climate information needs in Africa. In *Deep learning Indaba 2023*.

Ansari, N., Babaei, V., & Najafpour, M. M. (2024). Enhancing catalysis studies with chat generative pre-trained transformer (ChatGPT): Conversation with ChatGPT. *Dalton Transactions, 53*(8), 3534–3547.

Atkins, C., Girgente, G., Shirzaei, M., & Kim, J. (2024). Generative AI tools can enhance climate literacy but must be checked for biases and inaccuracies. *Communications Earth & Environment, 5*(1), 226.

Biswas, S. S. (2023). Potential use of ChatGPT in global warming. *Annals of Biomedical Engineering, 51*(6), 1126–1127.

Carpo, M. (2024). The sustainable lightness of digital fabrication. *Architectural Design, 94*(5), 14–21.

Chatterjee, J., Nieto, M. T. A., Gelbhardt, H., Dethlefs, N., Ohlendorf, J. H., Greulich, A., & Thoben, K. D. (2023). Domain-invariant icing detection on wind turbine rotor blades with generative artificial intelligence for deep transfer learning. *Environmental Data Science, 2*, e12.

Chen, K., Shao, A., Burapacheep, J., & Li, Y. (2024). Conversational AI and equity through assessing GPT-3's communication with diverse social groups on contentious topics. *Scientific Reports, 14*(1), 1561.

Chew, Y. J., Ooi, S. Y., Pang, Y. H., & Lim, Z. Y. (2024). Framework to create inventory dataset for disaster behavior analysis using Google Earth Engine: A case study in Peninsular Malaysia for historical forest fire behavior analysis. *Forests, 15*(6), 923.

DeCotis, P. A. (2023). The generative artificial intelligence utility. *Climate and Energy, 40*(5), 21–26.

Fu, X., Wang, R., & Li, C. (2024). Can ChatGPT evaluate plans? *Journal of the American Planning Association, 90*(3), 525–536.

González-Rodríguez, V. E., Izquierdo-Bueno, I., Cantoral, J. M., Carbú, M., & Garrido, C. (2024). Artificial intelligence: A promising tool for application in phytopathology. *Horticulturae, 10*(3), 197.

Hacker, P. (2023). The European AI liability directives-critique of a half-hearted approach and lessons for the future. *Computer Law & Security Review, 51*, 105871.

Hacker, P. (2024). Sustainable AI regulation. *Common Market Law Review, 61*(2).

Hadid, A., Chakraborty, T., & Busby, D. (2024). When geoscience meets generative AI and large language models: Foundations, trends, and future challenges. *Expert Systems*, e13654.

Haluza, D., & Jungwirth, D. (2023). Artificial intelligence and ten societal megatrends: An exploratory study using GPT-3. *Systems, 11*(3), 120.

Hansen, J. W., Vaughan, C., Kagabo, D. M., Dinku, T., Carr, E. R., Körner, J., & Zougmoré, R. B. (2019). Climate services can support African farmers' context-specific adaptation needs at scale. *Frontiers in Sustainable Food Systems, 3*, 21.

Harvey, F. (2014). Global warming is a misleading term because it actually sounds quite nice. Retrieved January 28, 2025, from https://www.theguardian.com/commentisfree/2014/mar/07/global-warming-misleading-term-climate-change

Hewitt, C. D., & Stone, R. (2021). Climate services for managing societal risks and opportunities. *Climate Services, 23*, 100240.

Hirn, J., Sanz, V., García, J. E., Goberna, M., Montesinos-Navarro, A., Navarro-Cano, J. A., Sánchez-Martín, R., Valiente-Banuet, A., & Verdú, M. (2024). Transfer learning of species co-occurrence patterns between plant communities. *Ecological Informatics, 83*, 102826.

Holzinger, A., Keiblinger, K., Holub, P., Zatloukal, K., & Müller, H. (2023). AI for life: Trends in artificial intelligence for biotechnology. *New Biotechnology, 74*, 16–24.

Jain, H., Dhupper, R., Shrivastava, A., Kumar, D., & Kumari, M. (2023). AI-enabled strategies for climate change adaptation: Protecting communities, infrastructure, and businesses from the impacts of climate change. *Computational Urban Science, 3*(1), 25.

Jeans, H., Castillo, G., & Thomas, S. (2017). The future is a choice. Retrieved January 28, 2025 from https://oxfamilibrary.openrepository.com/bitstream/handle/10546/620178/gd-resilience-capacities-absorb-adapt-transform-250117-en.pdf?sequence=4&isAllowed=y

Jiang, C., Li, S., Cai, D., Ye, J., Bao, Q., Liu, C., & Wang, S. (2024). ChatGPT-based meta-analysis for evaluating the temporal and spatial characteristics of deoxynivalenol contamination in Chinese wheat. *Journal of Hazardous Materials, 480*, 135888.

Khan, M. S., Shoaib, A., & Arledge, E. (2024). How to promote AI in the US federal government: Insights from policy process frameworks. *Government Information Quarterly, 41*(1), 101908.

Krzyżewska, A. (2024). Climate change in Poland—The assessment of the conversation with ChatGPT. *Miscellanea Geographica. Regional Studies on Development, 28*(1), 5–12.

Lah, O. (2025). Breaking the silos: Integrated approaches to foster sustainable development and climate action. *Sustainable Earth Reviews, 8*(1), 1.

Larosa, F., Hoyas, S., García-Martínez, J., Conejero, J. A., Fuso Nerini, F., & Vinuesa, R. (2023). Halting generative AI advancements may slow down progress in climate research. *Nature Climate Change, 13*(6), 497–499.

Leippold, M. (2023). Thus spoke GPT-3: Interviewing a large-language model on climate finance. *Finance Research Letters, 53*, 103617.

Li, W., Hsu, C. Y., Wang, S., Yang, Y., Lee, H., Liljedahl, A., Witharana, C., Yang, Y., Rogers, B. M., Arundel, S. T., et al. (2024). Segment anything model can not segment anything: Assessing AI foundation model's generalizability in permafrost mapping. *Remote Sensing, 16*(5), 797.

Lin, Z., Chen, A., Wang, X., Liu, Z., & Piao, S. (2024). Large language models reveal big disparities in current wildfire research. *Communications Earth & Environment, 5*(1), 168.

Lindwall, C. (2022). What are the effects of climate change. Retrieved June 22, 2024 from https://www.nrdc.org/stories/what-are-effects-climate-change

Mammides, C., & Papadopoulos, H. (2024). The role of large language models in interdisciplinary research: Opportunities, challenges and ways forward. *Methods in Ecology and Evolution, 15*(10), 1774–1776.

Mehryar, S., Yazdanpanah, V., & Tong, J. (2024). AI and climate resilience governance. *iScience*.

Mengist, W., Soromessa, T., & Legese, G. (2020). Method for conducting systematic literature review and meta-analysis for environmental science research. *MethodsX, 7*, 100777.

Miao, C., Hu, J., Moradkhani, H., & Destouni, G. (2024). Hydrological research evolution: A large language model-based analysis of 310,000 studies published globally between 1980 and 2023.

Moreno, A. I., & Caminero, T. (2024). Assessing climate-related disclosures of European banks through text mining. *Review of World Economics*, 1–25.

O'Keefe, S., Welsh, R., Oppong, M., Fitzgerald, R., Conner, D., Tynan, M., Price, N., & Quigle, C. (2024). Machine learning (ML) platforms can contradict dairy scientists and feed firm websites regarding dairy cattle performance from feeding seaweed supplements. *Choices, 39*(3), 1–5.

Pixar. (2008). Wall-E. Retrieved April 29, 2025 from https://www.pixar.com/wall-e

Preuss, N., Alshehri, A. S., & You, F. (2024). Large language models for life cycle assessments: Opportunities, challenges, and risks. *Journal of Cleaner Production*, 142824.

Pu, Z., Shi, C. L., Jeon, C. O., Fu, J., Liu, S. J., Lan, C., Yao, Y., Liu, Y. X., & Jia, B. (2024). ChatGPT and generative AI are revolutionizing the scientific community: A Janus-faced conundrum.

Ray, P. P. (2023). AI-assisted sustainable farming: Harnessing the power of ChatGPT in modern agricultural sciences and technology. *ACS Agricultural Science & Technology, 3*(6), 460–462.

Richards, D., Worden, D., Song, X. P., & Lavorel, S. (2024). Harnessing generative artificial intelligence to support nature-based solutions. *People and Nature, 6*(2), 882–893.

Rillig, M. C., & Kasirzadeh, A. (2024). AI personal assistants and sustainability: Risks and opportunities. *Environmental Science & Technology, 58*(17), 7237–7239.

Şahin, G. (2024). DRM lists created with ChatGPT: Analysing recognition memory with climate-change themed lists. *Öneri Dergisi, 19*(62), 1–17.

Salekpay, F., van den Bergh, J., & Savin, I. (2024). Comparing advice on climate policy between academic experts and ChatGPT. *Ecological Economics, 226*, 108352.

Schwaller, F. (2025). Will AI improve your life? Here's what 4,000 researchers think. *Nature, 640*(8059), 577–578.

Sommer, B., & von Querfurth, S. (2024). "In the end, the story of climate change was one of hope and redemption": ChatGPT's narrative on global warming. *Ambio, 53*(7), 951–959.

Turrentine, J. (2022). What are the causes of climate change? Retrieved June 25, 2024 from https://www.nrdc.org/stories/what-are-causes-climate-change#natural

UNDP. (2023). The climate dictionary: An everyday guide to climate change. Retrieved June 22, 2024 from https://climatepromise.undp.org/news-and-stories/climate-dictionary-everyday-guide-climate-change

USAID. (2013). Climate services: Better information for climate-resilient decision making. Retrieved July 04, 2024 from https://pdf.usaid.gov/pdf-docs/PA00KJWS.pdf

Vaghefi, S. A., Stammbach, D., Muccione, V., Bingler, J., Ni, J., Kraus, M., Allen, S., Colesanti-Senni, C., Wekhof, T., Schimanski, T., et al. (2023). ChatClimate: Grounding conversational AI in climate science. *Communications Earth & Environment, 4*(1), 480.

Valavanidis, A. (2023). Artificial intelligence (AI) applications. *Department of Chemistry, National and Kapodistrian University of Athens, University Campus Zografou, 15784*.

Volk, S. C., Schäfer, M. S., Lombardi, D., Mahl, D., & Yan, X. (2024). How generative artificial intelligence portrays science: Interviewing ChatGPT from the perspective of different audience segments. *Public Understanding of Science,* 09636625241268910.

Vuong, Q. H., Ho, & M.T. (2024). Escape climate apathy by harnessing the power of generative AI. *AI & Society,* 1–2.

White, B., Woods, M., & Smith, D. (2023). Leveraging AI for resilience: Enhancing business continuity & crisis management. Retrieved April 29, 2025, from https://www.4cstrategies.com/news/leveraging-ai-for-resilience/

WHO. (2023). Climate change. Retrieved March 03, 2025, from https://www.who.int/news-room/fact-sheets/detail/climate-change-and-health

WMO. (2025). WMO confirms 2024 as warmest year on record at about 1.55 °C above pre-industrial level. Retrieved January 27, 2025, from https://wmo.int/news/media-centre/wmo-confirms-2024-warmest-year-record-about-155degc-above-pre-industrial-level

Zhang, X., Shah, J., & Han, M. (2023). ChatGPT for fast learning of positive energy district (PED): A trial testing and comparison with expert discussion results. *Buildings, 13*(6), 1392.

Heart Attack Risk Prediction Using Machine Learning and Deep Learning Techniques

Doorgesh Sookarah, Loovesh Sharma Ramwodin, Manish Putteeraj, and Vivasvansingh Dwarka

Abstract Cardiovascular disease (CVD) is still one the main causes of mortality worldwide. This emphasizes the critical need for efficient and effective risk assessment approaches. In this study, machine learning (ML) and deep learning (DL) techniques have been used to forecast the likelihood of myocardial infarctions from publicly available clinical data. A rigorous methodology was followed and this included steps such as data pre-processing, class balancing using SMOTE, data augmentation by adding Gaussian noise, model training, hyperparameter optimization, and evaluation using simple performance metrics such as accuracy, recall, F1-score, and AUC-ROC. Among the various models that were investigated, Random Forest and an optimized Artificial Neural Network (ANN) both consistently outperformed others. The ANN performed with an F1-score of 0.894 and exhibited stability in generalizability through a 15-fold cross-validation process. The best-performing ANN model was developed as a web application in Flask and deployed in Heroku for enabling real-time assessment of the risk of heart disease. The deployment of this implementation illustrates the viability of converting predictive models into usable clinical tools. The findings confirm the utility of data-driven methodologies for the early detection of heart disease and set the stage for future integration with actual clinical systems and wearable device data.

D. Sookarah (✉) · L. S. Ramwodin · M. Putteeraj · V. Dwarka
University of Technology Mauritius, La Tour Koenig, Port Louis, Mauritius
e-mail: doorgesh.sookarah@utm.ac.mu

L. S. Ramwodin
e-mail: lramwodin@utm.ac.mu

M. Putteeraj
e-mail: mputteeraj@utm.ac.mu

V. Dwarka
e-mail: vivasvansingh.dwarka@umail.utm.ac.mu

J. Marx Gómez et al. (eds.), *Recent Advances in Information Systems*, Progress in IS,
https://doi.org/10.1007/978-3-032-14099-9_37

1 Introduction

Artificial intelligence (AI) is an innovative branch of computer science that aims to develop intelligent systems capable of replicating human cognitive skills and behaviours. These skills generally involve the capacity to learn from data, make logical decisions, solve problems, and understand natural language (Gupta, 2023). In the field of AI, machine learning (ML) has played a significant role by enabling computers to learn from historical data sets using algorithms that learn patterns and generate predictive or classification outputs from data. Among the numerous ML methods, supervised and unsupervised learning have seen extensive use across numerous applications and have contributed towards improving decision-making processes, automation, and operational efficiency (Brown, 2021). Deep learning (DL), as a specialized sub-class of machine learning (ML), utilizes advanced multi-layered neural networks that possess the ability to automatically detect intricate features in raw data. This property makes DL uniquely adequate to take on high-dimensional tasks, such as image and voice recognition, natural language processing, and medical diagnosis (Garg, 2022). The main distinction between deep learning models and conventional machine learning methods is that the former does not need manual feature engineering but instead automatically develop hierarchical representations that get better with larger datasets and computation. Despite their potential, deep learning models are often impacted by issues of interpretability, computational cost, and the requirement for large volumes of annotated data. Nonetheless, ongoing research keeps expanding the boundaries of AI thus unveiling new prospects in sectors such as medicine, finance, and transport.

2 Background

As mentioned previously, cardiovascular disease (CVD) remains a leading cause of global mortality rates. CVD is an umbrella term which includes several conditions such as stroke, coronary artery disease, heart failure, and peripheral arterial disease. In 2016, CVD accounted for nearly one-third of all deaths worldwide, a figure that is projected to surpass 24 million deaths annually by the year 2030 despite advances in medical care (Amini et al., 2021). Early detection of vulnerable individuals is critical to mitigating this scenario. However, conventional risk prediction tools are inadequate as they fail to fully consider the complexity of clinical and lifestyle factors (Ofori & Odia, 2016). Over the last few years, the application of artificial intelligence (AI) techniques, specifically machine learning (ML) and deep learning (DL), has increasingly been considered as a viable approach to enhancing forecast, diagnosis, and treatment of cardiovascular disease (CVD). Both ML and DL represent the possibility of unlocking complex, non-linear relationships in various forms of data that go well beyond clinical records. These include imaging, genomics and real-time data from wearable sensors. Myocardial infarctions or heart attacks are caused by an obstruction of the flow of blood to an area of the heart and this is typically due to a clot in a coronary artery. If not treated, the blockage will kill or scar heart tissue and prove to be fatal. Known risk factors include hypertension, diabetes, smoking, obesity, physical inactivity, and high cholesterol levels. The use of artificial intelligence in this field allows predictive analytics through the examination of large volumes of data from various sources in turn enabling the recognition of subtle relationships and early signals that may not be captured through traditional clinical methods (Clinic, 2022). Therefore, AI can significantly enhance both the timeliness and accuracy of heart attack risk prediction, potentially transforming preventive cardiology and patient outcomes.

ML has been used effectively in cardiology for diagnosis support, prognosis estimation, and treatment optimization. Algorithms have also been successful in interpreting complex imaging modalities such as cardiac MRI, CT, and echocardiograms with high accuracy and with less human intervention. Commercial programs such as AliveCor and Ultromics have used AI to identify arrhythmias and coronary artery disease from ECG and echocardiography data (Olawade et al., 2024). Furthermore, predictive electronic health record (EHR) models can quantify readmission risk and tailor treatment regimens based on individual risk profiles. Even more advanced solutions such as HeartFlow's AI-generated 3D coronary models represent additional ways that machine learning enhances procedural planning and non-invasive diagnostics. However, all of these benefits are still hindered by challenges such as data standardization, algorithmic bias, and physician training needs. Ethical and regulatory considerations play a large role in ensuring safe, fair, and effective deployment. Still, with continued research and diligent integration, artificial intelligence has the potential to transform cardiovascular medicine into a more data-driven and personalized discipline (HeartFlow Inc, 2024).

3 Problem Statement

Even though it is one of the major causes of worldwide death, CVD is still not sufficiently targeted by conventional risk assessment methods. Conventional approaches, which are frequently based on static clinical predictors and linear models are limitations in capturing the complicated and heterogeneous nature of CVD. This means opportunities for early identification and preventive treatment are regularly missed. The inability of these tools to risk-stratify at an individual level diminishes their utility in clinical practice and hinders early intervention. Given that CVD develops as a combined result of lifestyle, physiological, and genetic inputs, more flexible and data-driven methodologies are required.

This study aims to remove these limitations using machine learning (ML) and deep learning (DL) techniques to forecast the risk of heart attack. The most significant technical challenge involves creating models that are able to recognize intricate patterns in patient data without compromising generalizability and interpretability. Some of the main considerations are the selection of relevant features, controlling model complexity to prevent overfitting, and combining different types of data. Furthermore, scalability and real-world applicability should be guaranteed so that predictive models can realistically be rolled out in various healthcare environments. The goal of this research is therefore to enable a more individualized, accurate, and timely approach to cardiovascular risk management via the implementation of machine learning principles to risk prediction.

4 Aims and Objectives

The main aim of this study is to develop an effective predictive model that will be able to assess the risk of cardiovascular disease efficiently and provide access to the solution through an easy-to-use web application. The work present also endeavours to enhance current diagnostic practices and facilitate early intervention with the help of cutting-edge machine learning techniques coupled with real-world datasets.

The specific objectives of the study are as follows:

1. Analyse a structured dataset containing clinical records related to heart disease in order to understand the underlying patterns.
2. Identify and select the most significant features that contribute to the onset and progression of heart disease.
3. Evaluate multiple machine learning algorithms, perform comparative analyses, and optimize model performance through hyperparameter tuning.

5 Literature Review

Heart disease as mentioned previously remains a leading cause of death globally and thus necessitates improved methods for early detection and prevention. Recent research underlines the increasing significance of machine learning (ML) for improving diagnostic precision and predictive potential in cardiology. ML algorithms have demonstrated potential in recognizing intricate patterns from clinical data that facilitate earlier intervention and more individualized treatment approaches. With healthcare increasingly heading towards evidence-based decision-making, ML-based solutions are gaining traction as viable means to enhance cardiovascular risk estimation and patient outcomes. The following subsections present a selection of previous studies on heart disease prediction using machine learning and deep learning techniques with a focus on their methodologies, performance outcomes and limitations.

5.1 Study by Mohammad Alshraideh et al.

Alshraideh et al. presented a machine learning-based framework for heart disease prediction using the JUH Heart Dataset (Alshraideh et al., 2024). They presented a comprehensive methodology involving exhaustive data pre-processing, feature selection, and evaluation of different classification models, including Random Forest, SVM, Decision Tree, Naive Bayes, and KNN. In order to boost the relevance of input features, the research used Particle Swarm Optimization (PSO), which is an algorithm inspired by the collective phenomenon of natural systems. The models recorded a very good classification accuracy of 94.3%, indicating the suitability of integrating optimization methods with conventional machine learning algorithms for facilitating early diagnosis and treatment planning.

5.2 Study by Suraj Kumar Gupta et al.

Gupta et al. presented a comparative analysis of heart disease prediction with a collection of machine learning classifiers, namely Gradient Boosting, Decision Tree, Random Forest, and Logistic Regression. Classifiers were compared on the basis of data collected from the Framingham Heart Study and the UCI Heart Disease Repository (Gupta et al., 2021). Their work aimed at identifying robust predictors of heart attack risk and comparing model performances. The research emphasized the significance of angina, high cholesterol (above 200 mg/dl), heart rate, thalassemia, and age as the most influential factors in promoting cardiovascular risk. The results accentuate the potential of machine learning in handling and explaining multifaceted clinical data to enable better risk stratification and early intervention in cardiovascular treatment.

5.3 Study by Sorif Hossain et al.

Hossain et al. investigated the increasing trend of cardiovascular diseases (CVDs) in Bangladesh based on health records of 391 CVD patients and 260 non-CVD patients, which were gathered from August 2022 to April 2023 (Hossain et al., 2024). Statistical methods, including cross-tabulation and chi-square tests, were employed to examine the relationships between CVD and different demographic and health-related factors. For prediction, various machine learning algorithms such as Logistic Regression, Decision Tree, Naive Bayes, Random Forest, AdaBoost, Bagging Tree, and Ensemble methods were tested. Out of these, the Random Forest classifier performed best with an accuracy of 98.04%, precision of 96.15%, recall of

100%, and an AUROC of 0.989. The research emphasizes the important contribution of computational tools to enable an early diagnosis, assist in clinical decision-making, and enhance cardiovascular health outcomes, especially in resource-poor environments.

5.4 Study by Evangelos K Oikonomou et al.

Oikonomou et al. investigated the integration of artificial intelligence (AI) and radiomics in cardiovascular imaging, specifically cardiac computed tomography (CT) (Oikonomou et al., 2020). The research demonstrated that deep learning and machine learning techniques can identify subtle imaging features not visible to the human eye, improving diagnostic accuracy and assisting clinical decision-making. One of their main applications was plaque detection, coronary artery calcium scoring, and lesion significance characterization. Radiomics were utilized for characterizing patterns and textures of CT images as a function of tissue characteristics and clinical outcomes. The technical and clinical limitations of AI-enhanced imaging were also discussed in the research and highlighted the need for future studies to achieve a balance between innovation and clinical usefulness. The findings position AI-enabled radiomics as a game-changing precision cardiovascular medicine solution.

5.5 Study by Acharya et al.'s Research

Acharya et al. proposed an automated technique of coronary artery disease (CAD) identification from features of the left ventricle grayscale image (Acharya et al., 2013). The method involved feature extraction comprising fractal dimensions, entropies, local binary patterns, wavelet-based, and texture-based descriptors of a dataset of 400 normal and 400 CAD patients. Nine statistically significant features were identified by employing t-test analysis for classification. A Gaussian Mixture Model (GMM) classifier trained on this reduced feature set also achieved 100% classification accuracy thus demonstrating the promise of computer-aided image analysis for non-invasive CAD detection and diagnosis.

5.6 Comparison of the Studies

The literature review was based on a focused search across reputable academic sources, including Google Scholar, PubMed, IEEE Xplore, and the ACM Digital Library. Relevant studies were identified and refined using a structured process to ensure a comprehensive and up-to-date understanding of machine learning applications in cardiovascular disease prediction (Table 1).

Table 1 Comparison of the different studies

Researcher	Best performing model	Dataset(s)	Objective	Major influencing factors	Strengths
Suraj Kumar Gupta et al.	Gradient boosting (Framingham: 85.5%, UCI: 85.6%)	Framingham dataset, UCI heart dataset	Predicting heart attack risk using ML models	Cholesterol level, heart rate, chest pains, blood pressure	Comprehensive feature engineering and hyperparameter optimization
Sorif Hossain et al.	Random Forest (96.15%)	Dataset from Bangladeshi patients	Predicting CVD risk among Bangladeshi people using ML models	Cholesterol, heart rate, blood pressure, family history	Identifies key risk factors useful for policymakers in Bangladesh
Evangelos K. Oikonomou et al.	CNN for ECG, ML algorithm for cardiac CT (AUC of 0.96 for image quality)	ECG and cardiac CT datasets	Application of AI in ECG, echocardiogram and cardiac CT analysis	Noise, contrast, misregistration scores, image quality	High accuracy in image quality assessment and segmentation
Acharya et al.	Not specified (focus on feature selection)	Cardiovascular images	Quantifying variations in cardiovascular images using grayscale features	Fractal dimension, higher order spectra, image texture, local binary patterns	Significant feature reduction improves computational efficiency

The search strategy included the following steps:

1. **Keyword based search**: Targeted terms such as "heart disease prediction," "machine learning in cardiology," "deep learning for cardiovascular risk assessment," and "AI in cardiovascular imaging" were used to retrieve relevant publications.
2. **Relevance filtering**: Studies were prioritized if they addressed key topics, including model optimization, data pre-processing, clinical validation, and real-world deployment considerations.
3. **Recency criterion**: Preference was given to recent publications from the past 5–7 years to ensure alignment with the latest advancements in AI-driven healthcare applications.
4. **Comparative evaluation**: Research articles providing performance benchmarks and comparative analyses of multiple machine learning algorithms were assessed to contextualize the strengths and limitations of different approaches.

This approach ensured that the literature review was both comprehensive and aligned with the state of the art in AI-enabled cardiovascular risk prediction.

6 Methodology

The methodology followed for this study involved a structured sequence of processes including data acquisition, pre-processing, model development, evaluation, and deployment. An overview of the workflow is presented in Fig. 1. The latter outlines the key stages for building and validating a heart attack prediction model (Banoula, 2023).

Data Collection

Data were gathered from two primary repositories: the UCI Machine Learning Repository and Kaggle. Publicly available datasets were chosen based on applicability, availability of important cardiovascular risk factors, and completeness of pre-processing. This pre-selection streamlined the modelling process by eliminating the initial need for cleaning raw clinical data. In addition, measures were taken to make sure that data acquisition practices were compliant with healthcare privacy laws and ethics. This ensured all real-time health data used abided by rules regarding anonymization and consent.

A number of important factors were taken into consideration while choosing the datasets for this study in order to guarantee their quality, relevance, and suitability for use with heart disease prediction models. The majority of the datasets came from publicly accessible repositories such Kaggle and the UCI Machine Learning Repository which are both well-known for having organized and well selected datasets. Below are these key factors:

1. **Clinical Relevance**: To provide significant predictive modelling, datasets have to include key cardiovascular health markers such as age, blood pressure, heart rate, cholesterol levels, and forms of chest pain.
2. **Completeness and Quality of Data**: To enable thorough pre-processing and analysis, datasets with few missing values and sufficient feature annotation were preferred.
3. **Class Distribution**: To provide balanced training using SMOTE if required, datasets with an acceptable distribution of heart disease positive and negative cases were given priority since imbalanced classes might impact model learning.
4. **Feature Richness**: Robust feature engineering and model tuning were made possible by the various predictive factors included in the chosen datasets. Additionally, the use of several machine learning algorithms was made possible by the incorporation of numerical and categorical variables.
5. **Prior Research Usage**: To ensure comparable benchmarking with the body of current research, datasets utilized in earlier machine learning studies for the prediction of heart disease were selected.

Heart Disease Prediction Dataset

The data described above was used for both training and testing of the machine learning models created in this research. It comprises of a broad set of clinical and demographic factors commonly related to cardiovascular health. The dataset is quite

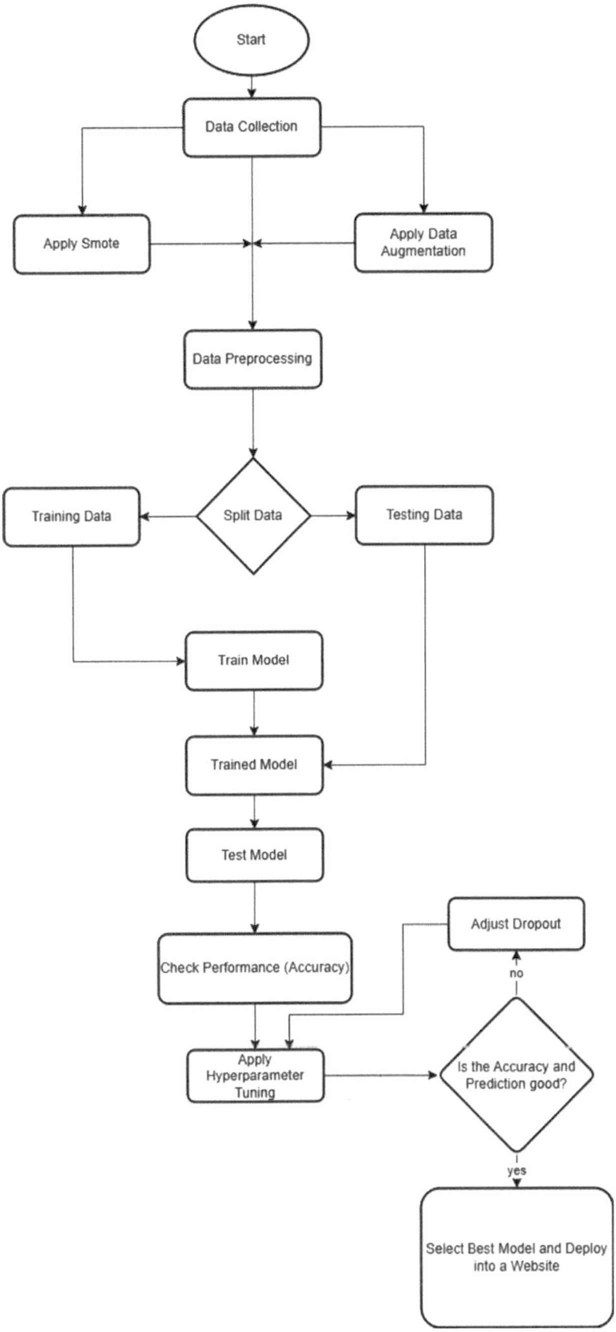

Fig. 1 Methodology used for the study

suitable for predictive modelling tasks since it represents important markers corresponding to early heart disease detection and prevention. Availability of appropriately annotated features enables determination of risk factors more easily and can enable the generation of accurate and widely applicable models (Table 2).

Data Pre-Processing

Following acquisition, the consolidated dataset was subjected to extensive pre-processing to ensure uniformity, integrity, and suitability for machine learning. The pre-processing steps included:

Handling Missing and Zero Values

Null values were imputed based on median or mean values while zero values in critical clinical features were treated as missing and replaced accordingly.

Label Encoding

Categorical variables such as "sex," "chest pain type," and "thalassemia" were converted into numerical formats using label encoding to enable compatibility with machine learning algorithms.

Feature Scaling

The StandardScaler function from Scikit-learn was employed to standardise numerical features such as cholesterol, resting blood pressure, and heart rate. This

Table 2 Attributes in the consolidated dataset

Feature	Description
Age	The age of the subject in years
Sex	Gender of the subject
Chest pain type	Categorized chest pain types such as typical or atypical angina
Resting blood pressure	Blood pressure measured in mm hg at rest
Cholesterol	Serum cholesterol level in mg/dl
Fasting blood sugar	Binary indicator of whether fasting blood sugar exceeds 120 mg/dl
Resting ECG	Electrocardiographic results at rest (e.g. normal, ST-T wave abnormality)
Maximum heart rate achieved	Peak heart rate during physical exertion
Exercise induced angina	Indicator of angina triggered by exercise
Oldpeak	ST depression induced by exercise relative to rest
Slope of peak exercise ST segment	Slope pattern observed in the ST segment
Thalassemia	Genetic blood disorder affecting red blood cell oxygen capacity
Output	The binary response variable indicating the presence (1) or absence (0) of heart disease
Number of major vessels	Number of major blood vessels (0–3) coloured by fluoroscopy

ensured all features contributed proportionately to the model in turn improving the convergence of gradient-based algorithms.

Addressing Class Imbalance with SMOTE

Class imbalance existed in the original data between the heart disease and non-heart disease patients. This was resolved through Synthetic Minority Oversampling Technique (SMOTE) that generates synthetic samples of the minority class. The SMOTE oversampling method employed was 15% to prevent overfitting while still improving generalization. Class distribution after SMOTE reached near parity at 51.83% (positive) to 48.17% (negative) thus enabling models to learn from balanced classes.

Data Augmentation with Gaussian Noise

In addition to providing real-world diversity in the data, Gaussian noise was added to numeric attributes such as cholesterol, age, oldpeak, and Thalassemia. This introduces greater feature diversity, prevents overfitting, and improves robustness of models to unseen data. Gaussian noise with zero mean and standard deviation control was implemented via a bespoke function and tuning noise factors were investigated for the optimization process. The approach has great synergy with medical AI model data augmentation best practice.

Feature Engineering and Selection

Feature selection was guided by correlation heatmaps, domain importance and effect on initial performance of baseline models. Features such as chest pain type, thalassemia, cholesterol, and oldpeak showed high model output impact. Feature engineering was conducted to reinforce these predictors through transformations and encoding for the purposes of compatibility and interpretability.

Dataset Splitting and Cross-Validation

The pre-processed data was split into a training set (70%) and a test set (30%) using stratified sampling to preserve class distribution. Moreover, ten-fold stratified cross-validation was employed to ensure model evaluation remains robust, especially in the presence of imbalanced classes. This technique reduces evaluation variance and preserves model generalisability across folds. As mentioned in previously, data were gathered from two primary repositories, namely, the UCI Machine Learning Repository and Kaggle. This pre-selection streamlined the modelling process by eliminating the initial need for cleaning raw clinical data.

Prevention of Data Leakage and Pitfalls

Strict pre-processing procedures were used to guarantee model dependability and stop data leaks. Through proper partitioning of the dataset, data leakage where information from the test set unintentionally affects model training was reduced, and the training, validation, and test sets amongst these were kept separate. To maintain class distribution across splits and avoid performance inflation from skewed data exposure, stratified sampling was used. In order to prevent unintentionally adding factors that might unjustly benefit the model without accurately reflecting real-world situations, feature selection was also carried out with great care.

The study addressed the problem of generalizability by using several publicly accessible datasets (UCI and Kaggle), which comprise a variety of patient demographics. By guaranteeing that samples from both the majority and minority classes were fairly represented, SMOTE-based class balancing also improved the model's resilience. By simulating real-world unpredictability through the use of data augmentation, especially Gaussian noise injection models, becomes more flexible in the face of unforeseen circumstances. External validation on separate datasets from other clinical contexts, however, would bolster the trust in practical applicability and guarantee efficacy across various patient demographics and healthcare settings.

Model Selection and Training

Six models were shortlisted based on their relevance in clinical predictive tasks:

- Logistic Regression
- K-Nearest Neighbours (KNN)
- Naïve Bayes
- Decision Tree
- Random Forest
- Custom Artificial Neural Network (ANN)

All models were run using Scikit-learn except the custom ANN which was built using TensorFlow and Keras. The structure of the ANN consisted of input, hidden, and output layers with Dropout and BatchNormalization layers to avoid overfitting. Compilation of the ANN used the Adam optimizer and binary_crossentropy loss function with performance tracked using accuracy across folds.

Hyperparameter Tuning

Hyperparameter optimization was carried out using Grid Search and Keras Tuner. For shallow models such as Random Forest, parameters such as n_estimators, max_depth, and criterion were optimized. For the ANN, learning rate, dropout rate, number of hidden neurons, and batch size were systematically optimized using Keras Tuner's Bayesian Search.

Feature Importance and Sensitivity Analysis

Random Forest feature importance scores and logistic regression coefficients were both used to identify important predictors of heart disease. Sensitivity analysis revealed that the variables chest pain type, thalassemia, and oldpeak were consistently influential aligning with clinical literature.

Prediction and Deployment

The most suitable model was selected on the basis of performance metrics and deployed using a web application. The application allows users to input clinical variables and receive a prediction of heart disease risk. The backend was implemented using Python Flask and deployment was managed using the Heroku cloud platform.

7 Results and Performance Evaluation

This section presents the comparative study of the machine learning and deep learning models developed for heart attack prediction. All the models were compared using various metrics such as accuracy, precision, recall, F1-score, specificity, and AUC-ROC.

Evaluations were made across three distinct configurations of datasets: the original dataset, SMOTE-resampled dataset, and the Gaussian noise-augmented dataset. Prediction reliability was assessed using confusion matrices and classification reports, and generalizability was evaluated with cross-validation.

7.1 Model Performance on Original Dataset

When trained on the original, imbalanced dataset, the Random Forest classifier emerged as the top performer. It achieved:

- Accuracy: 85.25%
- Macro-average F1-score: 85.10%
- AUC-ROC: 0.91

Logistic Regression though slightly lower in accuracy, achieved the highest recall of 83.49%, making it a reliable model in detecting positive heart disease cases, even at the expense of some false positives. The performance of the other classifiers was as follows:

- KNN: Accuracy 83.6%, F1-score 83.45%
- Decision Tree: Accuracy 83.2%, F1-score 83.12%
- Naive Bayes: Accuracy 76.07%, F1-score 75.83%

7.2 Model Performance on SMOTE-Resampled Dataset

To counter class imbalance, the dataset was balanced using SMOTE. The Decision Tree classifier recorded the best performance on this dataset:

- Accuracy: 85.95%
- F1-score: 85.76%
- Precision: 85.73%
- Recall: 85.98%

Other models benefited from the balanced distribution as well:

- Random Forest: Accuracy 85.36%, F1-score 85.30%
- Logistic Regression: Accuracy 85.12%, F1-score 85.10%
- KNN: Accuracy 84.86%, F1-score 84.76%

SMOTE improved sensitivity (recall) across all models while maintaining high precision.

7.3 Model Performance on Augmented Dataset

Gaussian noise was applied to numerical features (e.g. cholesterol, heart rate, oldpeak), resulting in a larger and more variable dataset. On this augmented dataset, the Random Forest classifier once again delivered the best overall performance:

- Accuracy: 89.26%
- Precision: 89.82%
- Recall: 88.69%
- F1-score: 89.21%
- AUC-ROC: 0.93

The improvement indicates that data augmentation enhanced model generalization by introducing variability that simulates real-world data conditions.

7.4 Performance of Artificial Neural Network (ANN)

Two ANN models were evaluated and the results are presented below.

- Model 1 (basic ANN) achieved:
  - Accuracy: 88.0%
  - Precision: 0.8511
  - Recall: 0.6667
  - F1-score: 0.7477
  - Indicated potential overfitting and imbalance sensitivity
- Model 2 (optimized ANN with L2 regularization, dropout, and batch normalization) achieved:
  - Accuracy: 89.40%
  - F1-score: 0.894
  - High recall and stable validation accuracy across 15-fold cross-validation

Although both ANNs demonstrated high predictive ability, Model 2 was more robust and generalizable. It effectively balanced bias and variance, making it the most promising candidate for deployment.

Table 3 Hyperparameter configurations

Model	Best hyperparameters	Best F1 macro score
KNN	{'algorithm': 'Auto', 'n_neighbors': 7, 'weights': 'Distance'}	0.87608
Naive Bayes	{'alpha': 0.01}	0.79563
Decision tree	{'criterion': 'Gini', 'max_depth': 20, 'min_samples_leaf': 1, 'min_samples_split': 2}	0.85124
Random Forest	{'max_depth': None, 'min_samples_leaf': 1, 'min_samples_split': 2, 'n_estimators': 100}	0.8845
ANN	{'learning_rate': 0.0084, 'units1': 32, 'l2_1': 0.0001, 'dropout_1': 0.3, 'units2': 32, 'l2_2': 0.0001, 'dropout_2': 0.3, 'units3': 32}	0.894

7.5 Hyperparameter Tuning Outcomes

Tuning was crucial in boosting performance. The main hyperparameter configurations for each model are presented in the table which follows (Table 3).

Tuning was conducted via GridSearchCV for shallow models and Keras Tuner for ANN using Bayesian Optimization.

8 Discussion

The findings of this comparative analysis indicate the great potential of machine learning and deep learning models in predicting the likelihood of heart disease. Among all the models tested, the Random Forest and enhanced Artificial Neural Network (ANN) models performed better than the other models across all the datasets, such as the original, resampled, and augmented data. These findings are consistent with the growing body of literature supporting ensemble-based and deep learning methods as being better at handling intricate nonlinear relationships typical of medical data.

8.1 Model Effectiveness and Interpretability

The higher performance of Random Forest models is due to the fact that they explore high-dimensional feature spaces better, overfit less, and provide high interpretability in the form of feature importance ranking. Logistic Regression and Decision Tree models, although slightly less accurate, provide greater transparency and therefore are appropriate in situations where interpretability is important.

On the other hand, the ANN model, while achieving the highest F1-score and generalization capacity, suffered from limited explainability. The so-called "black-box" behaviour is a well-documented issue in clinical AI applications. However, regularization techniques, dropout layers and batch normalization significantly enhanced the model stability and reduced variance thus establishing that deep learning when appropriately calibrated, can provide accuracy as well as reliability.

This study provided a thorough understanding of classification mistakes by analysing false positives as well as false negatives across all trained models. This was done by the use of confusion matrices. While false negatives, that is, "missing cases of real heart illness", offer serious dangers by postponing required measures, false positive cases that are misdiagnosed as heart disease can cause needless concern and further medical testing. Through analysis of model-specific error patterns, it was found that while ensemble models like Random Forest successfully reduced both kinds of errors by capturing complex feature interactions, traditional classifiers like Logistic Regression had a higher false-negative rate. This is because they relied on linear relationships. With the help of deep feature extraction, the Artificial Neural Network (ANN) showed the most evenly distributed error.

Feature contribution analysis was implemented to further understand misclassifications and found that characteristics like thalassemia and the type of chest discomfort were important factors in producing inaccurate predictions. These findings highlight the necessity of threshold modifications and model improvement to increase sensitivity while preserving specificity, ultimately guaranteeing a more trustworthy heart disease risk assessment instrument.

8.2 Role of Data Augmentation and SMOTE

SMOTE's use succeeded in handling class imbalance, thereby improving recall in all models but particularly for underrepresented classes. Data augmentation using Gaussian noise, however, added diversity to the input space and improved generalization, particularly for the ANN model. The methods illustrate the essential role of data-centric solutions in promoting robustness in the model and reducing the possibility of overfitting to limited clinical settings.

8.3 Evaluation Metrics and Clinical Implications

Detailed analysis with accuracy, precision, recall, F1-score, specificity, and AUC-ROC allowed close understanding of each model's strengths. Recall and F1-score were given priority, since minimizing false negatives is of the utmost importance in medical diagnostics. A model failing to detect a patient with a risk of heart disease may lead to lethal outcomes; thus recall-oriented models such as the ANN and Logistic Regression are of particular clinical interest.

The utilization of confusion matrices and ROC curves allowed the interpretation of results at different decision thresholds. These aspects are important for the application of model predictions to actual medical practice, where a range of trade-offs can be acceptable.

8.4 Limitations

Although there were multiple encouraging findings, there are several limitations that still need to be considered. First, although publicly available datasets such as UCI and Kaggle provide structured and readily accessible data, these might not be representative of demographic diversity or actual variability in clinical practice. Second, although SMOTE and data augmentation improved performance, these could potentially create artificial patterns not representative of real biological processes.

Besides, although hyperparameter optimization improved performance, the search space needed to be restricted due to computational limitations. More extensive tuning using more advanced methods such as neural architecture search (NAS) might further improve performance.

To ensure a comprehensive assessment of prediction accuracy, the study used a number of statistical indicators to evaluate model performance. Accuracy, precision, recall, specificity, and F1-score were important metrics that provided unique information about how effectively the models managed the categorization of heart disease. While recall assessed the model's sensitivity in identifying real cases of heart disease, precision measured the proportion of projected positive cases that were really accurate. In order to evaluate how well the models recognized healthy people and decreased false positives, specificity was essential. The F1-score offered a trustworthy overall indicator of predictive stability by striking a balance between precision and recall.

Although technical rigour was achieved by these measurements, several limitations remained. Interpretability was still a problem particularly for the deep learning model (ANN) whose decision routes were less clear than those of more straightforward models like decision trees and logistic regression. Furthermore, even with SMOTE and Gaussian noise augmentation, data variety might be enhanced by comparing models to separate real-world datasets. Even if they have been lessened, class imbalances may still affect clinical application performance. To improve model robustness along with clinical application, future research should investigate domain-specific feature analysis and external validation.

A more comprehensive analysis of potential biases and limits is required to assure appropriate interpretation even if the study delivers unambiguous results with well-defined measurements. Class imbalance is a major issue that even if SMOTE addresses it, it may still affect model performance in practical applications where uncommon situations could be underrepresented. Furthermore, utilizing publically accessible repositories (UCI and Kaggle) introduces bias into dataset selection which may not adequately represent lifestyle changes, genetic variants, or demographic variety that affect cardiovascular risk. Additionally, model-specific biases also impact the overall performance. Although ensemble approaches like as Random Forest showed promising results, their intricacy may restrict interpretability which would make clinical adoption more difficult.

Additionally, despite their great accuracy, deep learning models like ANN operate as "black boxes" which obscures the process of making medical decisions. In order to increase explainability and guarantee wider application across healthcare systems, future improvements should include feature attribution methodologies and external validation on a variety of patient groups.

8.5 Data Challenges

The development of precise cardiac disease prediction models is severely hampered by data issues. Data quality is a significant problem as noise, missing values and discrepancies in clinical datasets can result in predictions that are not trustworthy. Class imbalance makes training even more challenging since there are frequently more healthy people than heart disease cases which makes it challenging for models to efficiently learn patterns. Although methods such as SMOTE tackle this issue, they could introduce fictitious data variations that don't necessarily correspond with actual situations. Furthermore, the representativeness of datasets is also an issue. Publicly accessible repositories like UCI and Kaggle could not have a diverse range of patient demographics which would restrict generalizability across various populations.

Furthermore, maintaining ethical compliance and data protection is crucial particularly when combining data from wearable technology, patient information or electronic health records. In order to improve model dependability and practical applicability, these difficulties highlight the significance of thorough pre-processing, meticulous model validation, and ultimately external testing on clinically obtained datasets.

9 Deployment of the Predictive Model

Subsequent to the selection of models and the validation of their performance, the most effective predictive model, specifically the optimized Artificial Neural Network (ANN), was implemented as a web application. This phase was essential in illustrating the practical applicability of the model for providing real-time heart disease risk assessments outside of experimental environments.

9.1 Real-World Utility

The deployed application has the following practical use cases:

- Users can perform a basic heart disease risk check at home thus promoting early awareness and potential medical consultation.
- Medical practitioners can use it as a quick reference tool during preliminary screening or outpatient assessments.
- It can serve as a learning aid in academic and public health settings to demonstrate the integration of AI in healthcare.

While not intended as a diagnostic substitute, this tool exemplifies how AI models can be translated into accessible and user-friendly platforms with real-world value.

9.2 Practical Considerations for Clinical Use

The use of AI-driven heart disease risk prediction models in real-world healthcare applications needs to take into consideration a number of pragmatic issues such as staff trust, workflow flexibility and emergency situation constraints. Although the web-based program created for this study offers a user-friendly tool for initial risk assessment, it is not meant to be used in emergency scenarios where prompt medical attention is necessary. Rather, its main use is to promote preventative measures, encourage prompt consultations and help individuals and healthcare professionals with early-stage diagnosis.

Medical staff acceptability and trust are major implementation challenges. Clinicians may be reluctant to use AI models for risk prediction, especially when interpretability is restricted. Confidence in AI-assisted decision-making may be increased by establishing transparency through explainable AI techniques and validation studies on various clinical datasets. Adaptability to present practices is also essential, incorporating predictive tools into clinical workflows or electronic health records (EHR) guarantees smooth use rather than adding to the workload. Adoption and dependability would be further increased by future improvements, including adding real-world patient data and extending model validation across other healthcare contexts.

10 Conclusion and Future Directions

This study demonstrated the effective application of machine learning and deep learning techniques in predicting heart disease using publicly available clinical datasets. A rigorous methodology was adopted that included data pre-processing, class balancing, model selection, hyperparameter tuning, and deployment to ensure accurate and generalizable results. Among the models compared, both the Random Forest and optimized Artificial Neural Network (ANN) models exhibited better performance on all the important measures of accuracy, recall, and F1-score. Specifically, the ANN exhibited good generalization power after employing dropout, batch normalization, and L2 regularization techniques. These results confirm the appropriateness of deep learning methods for complicated medical prediction problems involving the interaction of numerous nonlinear variables. SMOTE and Gaussian noise augmentation of the training set helped significantly by effectively solving class imbalance and making the model more robust. Deployment of the best model as a web application via Heroku also demonstrated how AI-based solutions can be transformed into usable and accessible content for both patients and healthcare workers. Despite promising findings, several limitations should be further investigated. Public data while convenient for benchmarking might not reflect the diversity and noise of actual clinical data. Moreover, deep learning models despite their high predictability are still black boxes as far as interpretability is concerned, a consideration that needs to be resolved prior to clinical adoption in the masses.

References

Acharya, U. R., Oh, S. L., Hagiwara, Y., Tan, J. H., & Adam, M. (2013). Automated detection of coronary artery disease using different approaches of entropy with ECG signals. *Knowledge-Based Systems, 133*, 62–71. https://doi.org/10.1016/j.knosys.2017.07.004

Alshraideh, M., Al-Hasanat, A., Rababah, O., & Al-Hadhrami, T. (2024). A hybrid feature selection approach using particle swarm optimization for heart disease prediction. *International Journal of Advanced Computer Science and Applications, 15*(1), 37–45.

Amini, M., Zayeri, F., & Salehi, M. (2021). Global trends in cardiovascular disease burden. *International Journal of Public Health, 66*, 123–132.

Banoula, M. (2023). Machine learning methodologies for healthcare prediction systems. *International Journal of Applied Artificial Intelligence, 41*(2), 167–182.

Brown, M. (2021). Machine learning foundations: Algorithms, applications, and impact. *Journal of Data Science and Analytics, 9*(2), 112–126.

Garg, P. (2022). Deep learning: Current trends and future perspectives. In *International Conference on AI and Computing Technologies, Springer* (Vol. 234–245).

Gupta, S. K. (2023). Advances in artificial intelligence: From concept to application. *AI Journal of Emerging Trends, 12*(3), 45–56.

Gupta, S. K., Singh, R., & Sharma, V. (2021). Predictive modelling for cardiovascular disease using supervised machine learning techniques. *Journal of Healthcare Engineering, 2021*, 3024783. https://doi.org/10.1155/2021/3024783

HeartFlow Inc. Non-Invasive Functional Assessment of Coronary Arteries Using AI. https://www.heartflow.com/. (Accessed 2024).

Hossain, S., Jahan, R., Rahman, M. M., & Ahmed, M. (2024). Predicting cardiovascular disease risk using ensemble machine learning models: Evidence from Bangladesh. *International Journal of Medical Informatics, 183*, 105020. https://doi.org/10.1016/j.ijmedinf.2023.105020

Mayo Clinic. (2022). Heart Attack Symptoms and Risk Factors. https://www.mayoclinic.org/diseases-conditions/heart-attack

Ofori, S., & Odia, I. (2016). Limitations of conventional CVD risk assessment tools: A review. *Cardiology Insights, 18*(4), 215–223.

Oikonomou, E. K., Siddique, M., & Antoniades, C. (2020). Artificial intelligence in cardiovascular imaging: Radiomics and the promise of precision medicine. *European Heart Journal—Digital Health, 1*(1), 132–144. https://doi.org/10.1093/ehjdh/ztab003

Olawade, A., Singh, R., & Chen, D. (2024). Machine learning applications in cardiology: Advances and challenges. *Journal of Cardiovascular Technology and AI, 3*(1), 45–62.

Impact of Artificial Intelligence (AI) on Cyber Safety Awareness and Culture in Small and Medium Enterprises (SMEs)

Betty Kekana, Elmarie Kritzinger, and Dorothy Scholtz

Abstract The research investigates the impact of artificial intelligence (AI) in promoting cyber awareness as well as setting in place a robust cybersecurity culture in small and medium enterprises (SMEs). As AI technologies are increasingly adopted by SMEs, the need for improved cyber resilience becomes more acute. The research discusses how AI can contribute to threat detection, risk management and training of employees in awareness. Using the research onion model and Preferred Reporting Items for Systematic Review and Meta-Analysis (PRISMA) framework, a systematic literature review has been conducted, with data analysis being done through ATLAS.ti for key themes and insights identification. Presenting a literature matrix of 30 relevant articles, the data highlight trends and gaps in existent research. Findings were revealed concerning five primary themes, linking AI adoption with enhanced cyber safety practices. Having shown such themes, the work helps build a better understanding of how small and medium enterprises can take advantage of AI for cyber resilience. The results feature five primary themes that are seen to link AI uptake with improved cyber security behaviours of companies, providing more in-depth insight into how SMEs use AI resources for enhancing cyber resilience. The conclusion outlines the recommendations created at the end for SMEs on leveraging AI solutions effectively for a robustly built cyber awareness culture.

B. Kekana · E. Kritzinger · D. Scholtz (✉)
University of South Africa, Pretoria, South Africa
e-mail: Scholid@unisa.ac.za

B. Kekana
e-mail: 61665665@mylife.unisa.ac.za

E. Kritzinger
e-mail: Kritze@unisa.ac.za

J. Marx Gómez et al. (eds.), *Recent Advances in Information Systems*, Progress in IS,
https://doi.org/10.1007/978-3-032-14099-9_38

1 Introduction

The infusion of artificial intelligence (AI) in industries has fundamentally changed the way industries integrate business processes, encourage rise and provide innovations in productivity (Fosso Wamba, 2022). Increasingly, small and medium enterprises (SMEs) have been adding AI to fast-track processes and facilitate growth (Baabdullah et al., 2021). The rapid technological advancement in-troduces new cyber threats and susceptibility to SMEs (Bountouni et al., 2023; Chidukwani et al., 2022). Apart from cyber threats, addressing and inculcating safety culture is very significant for SMEs to realize the full benefits of AI adaptation without compromising risks (Chaudhary et al., 2023). The research discusses the effect of AI on cyber safety awareness and culture in SMEs, identifying principal drivers of change, challenges to AI adoption and possible solutions. Section 2 reviews literature on AI and cybersecurity in SMEs. Section 3 gives an overview of the research methodology. Section 4 describes results and findings, and Sect. 5 describes contributions. The paper concludes with some reflections and recommendations that are made towards improving cyber resilience in SMEs through AI.

2 Background Section

The rapid rise of technology has changed how businesses use AI, with SMEs using AI to gain an edge over their competitors (Baabdullah et al., 2021). However, the integration of AI leads to problems related to cyber security (Bountouni et al., 2023; Jada & Mayayise, 2024). Due to limited resources and knowledge of cybersecurity, SMEs are highly prone to internet crimes (Abdul Wahab & Radmehr, 2024; Wei & Pardo, 2022). Further exacerbating threats to cybercrime is the complexity of AI systems and insufficient training (Alizai et al., 2023; Merín-Rodrigáñez et al., 2024; Shojaifar & Järvinen, 2021).

The creation of a good cybersecurity culture is significantly hampered by the lack of essential leadership support among the SMEs (Brodny & Tutak, 2022; Clemente-Almendros et al., 2024). The adoption of AI raises concerns about data privacy and ethical principles (Alonso et al., 2024; Bettiol et al., 2023). The research aims at giving guidelines on how SMEs can implement AI safely and effectively (Arroyabe et al., 2024; Van Kranenburg et al., 2023).

The research will assess AI's influence on awareness of cybersecurity and organizational culture in SMEs. The research will also discuss factors influencing the relation and insists on the need for a culture of computer security during adoption of AI. As a result, AI will help companies improve measures of safeguarding information systems from cyber-attacks while also creating safe work environments.

Table 1 Keywords identified for the literature review

Keywords	AI adoption, cyber safety awareness and culture in SMES
AI adoption terms	AI adoption OR artificial intelligence integration OR Automation OR AI tools OR intelligent systems OR AI implementation
AND: Cybersecurity terms	Cybersecurity OR data protection OR information security OR network security OR cybersecurity threats OR cyber risks OR cyber-attacks OR digital security OR online safety
AND: SME-related terms	Small and medium enterprises OR SMEs OR mid-size companies OR small business culture OR SME innovation OR business growth
AND: Cultural and organizational terms	Organizational culture OR leadership OR employee engagement OR workforce skills OR digital literacy OR organizational change OR training OR team collaboration OR leadership support
NOT: Noise inducing keywords	Medical OR medicine OR maritime OR naval OR aviation OR healthcare OR agriculture OR construction

3 Research Approach (Methodology)

The research adopts a holistic methodology rooted in the research onion model, which provides a structured framework for research design.

3.1 Keywords

To achieve more direct literature searching strategies towards the objectives, the following key phrases provided in Table 1 were applied during searches in academic databases.

3.2 Research Onion

The research onion framework is a tool guiding the research design, ensuring a systematic progression from general research philosophy to specific techniques and procedures (Saunders, 2019). The layered approach also fosters a clear and organized procedure form the general philosophy to specific techniques and procedures.

3.3 Search Criteria

The search focused on articles that are peer-reviewed, conference papers and reliable sources that were published between 2020 and 2024. Different databases including Google scholar and University library website were searched using the keywords. Among the standards used was a specific attention given to studies related to the topic.

3.4 Literature Review and PRISMA

The PRISMA framework was employed in carrying out a literature to guarantee a thorough and transparent selection process (Page et al., 2021). The initial search resulted in numerous documents that were evaluated for relevance before a total presentation could be done concerning critical research and results.

In Fig. 1, the systematic selection process that was utilized to filter and identify relevant research studies for the literature review. The figure presents the selection process in four essential stages: Identification—in which research studies found using keyword searches; Screening—reviewing titles and abstracts and removing duplicates; Eligibility—assess full text articles against set criteria; Inclusion—relevant studies were finalized for review.

3.5 Literature Matrix (30 Articles—Authors Versus Topics)

The matrix facilitates the identification of key themes in AI adoption, cybersecurity and SME organizational culture.

Table 2 shows the literature matrix with 30 articles—authors versus topics.

The matrix highlights factors affecting AI adoption and cybersecurity readiness in SMEs, leadership commitment, technological capability and regulatory compliance. A literature matrix was structured further to examine key research contributions across fields given literature relevant to the topic. The matrix helps reveal patterns, gaps and important issues surrounding AI adoption, cybersecurity consciousness and culture in SMEs. Table 2 provides a systematic catalogue for the 30 articles selected, emphasizing the objects of investigation.

Key factors explored include:

- AI Adoption and Integration.
- Cyber safety Awareness.
- Organizational Culture and Leadership.
- Training and Awareness.
- AI Adoption and Cybersecurity Challenges.

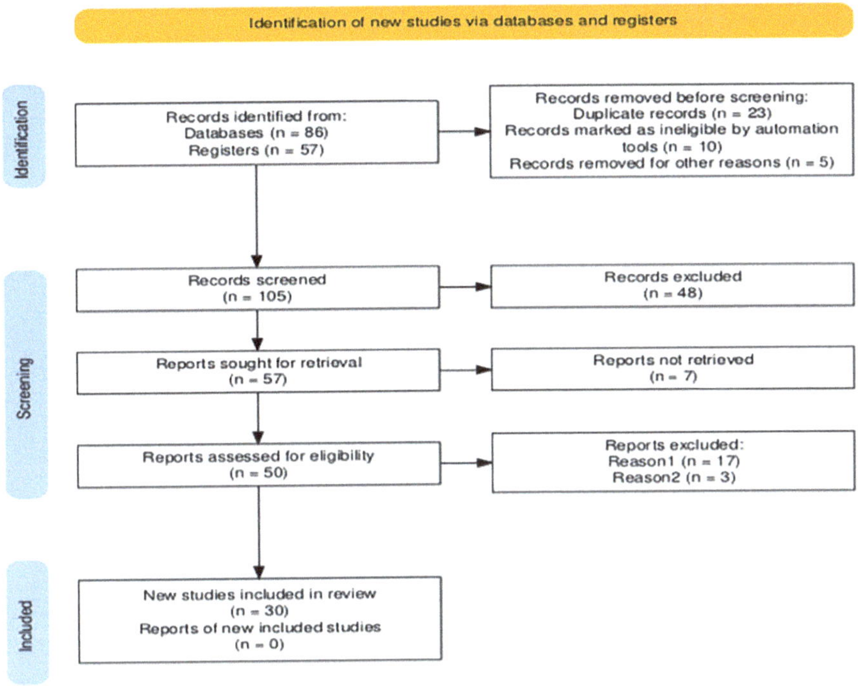

Fig. 1 PRISMA 2020 flow diagram: Impact of AI on cyber safety awareness and culture in SMEs

4 Data Analysis (ATLAS.ti)

This section outlines the qualitative data analysis process using ATLAS.ti to systematically review literature and extract key themes.

4.1 Word Cloud

A word cloud is a graphical representation of text data wherein the words appearing often are bigger and bolder. In qualitative research, word clouds can help identify key themes, trends and patterns by depicting the prominence of specific terms (Fosso Wamba, 2022).

Figure 2 presents a word cloud generated using ATLAS.ti, portraying the significant themes and patterns of adoption of AI, cyber safety awareness and organizational culture within SMEs. The degree of presence of respective word indicates the importance of the analysed content in the literature and provides a view on the highly recurring concepts in the research.

Table 2 Literature matrix with 30 articles—authors versus topics

Authors	Research factors
1. Abdul (2024)	AI Assimilation, Firm Performance in SMEs
2. Alizai (2023)	Communication networks for Industry 4.0 in SMEs
3. Alonso (2024)	AI software platforms in manufacturing
4. Arranz (2023)	Dynamics of SME digitalization
5. Arroyabe (2024)	IT security issues in Industry 4.0 implementation
6. Baabdullah (2021)	B2B AI practices in SMEs
7. Bettiol (2023)	SME responses to exogenous shocks
8. Bountouni (2023)	Safeguarding SMEs
9. Brodny (2022)	Digitalization and economic growth in SMEs
10. Chaudhary (2023)	Cybersecurity awareness in SMEs
11. Chidukwani (2022)	Cybersecurity challenges in SMEs
12. Clemente-Almendros (2024)	Determinants of digital transformation in SMEs
13. de Simone (2022)	AI/ML adoption in manufacturing MSMEs
14. Feng (2023)	Digital finance and asset allocation in Chinese SMEs
15. Fosso Wamba (2022)	AI assimilation in SMEs
16. Ingalagi (2021)	Determinants of AI adoption in SMEs
17. Jada (2024)	AI and organizational cybersecurity
18. Kappe (2023)	Cybercrime drivers in SMEs
19. Lada (2023)	AI adoption factors in Malaysian SMEs
20. Merín-Rodrigáñez (2024)	Digital transformation and SME performance
21. Mmango (2023)	Cyber resilience in SMEs
22. Peretz-Andersson (2024)	AI implementation in manufacturing SMEs
23. Polas (2022)	AI and blockchain in 4.0IR SMEs
24. Rawindaran (2023)	Government relationships in SME cybersecurity
25. Selamat (2021)	Chatbot adoption for SMEs
26. Shojaifar (2021)	Cybersecurity competence in SMEs
27. Uchendu (2021)	Developing cybersecurity culture
28. Ulrich (2021)	AI technology adoption in German SMEs
29. van Kranenburg (2023)	Societal situational awareness for SMEs
30. Wei (2022)	B2B SMEs and AI platform integration

4.2 Coding Process

The coding in ATLAS.ti of the transcripts began with essential codes, such as 'organizational culture,' 'leadership roles', 'employee awareness' and 'the use of AI in cybersecurity' (Jada & Mayayise, 2024; Lada et al., 2023). Codes underwent further analysis which showed other codes created previously about problems facing SMEs and their preparedness to use the AI technology were applicable in the research.

Fig. 2 Key themes and patterns related to AI adoption, cyber safety awareness and culture in SMEs

Codes enabled the design of the broader categories under which the codes could fit, for instance, leadership involvement and presence of an organizational culture showed critical connections that facilitated the efficient interweaving of AI and cybersecurity (Fosso Wamba, 2022).

To bring structure, the codes were grouped into broader themes to constitute a framework for investigating the relationship between AI adoption, cyber security and organizational aspects. Examples include the link between leadership involvement and awareness of cybersecurity as well as the success in AI implementation. A strong organizational culture also emerged as one of the important enablers for employee engagement and AI adoption for cybersecurity.

Table 3 shows how individual codes were grouped into broader themes.

Table 3 Key themes for AI adoption and cyber safety in SMEs

Themes	Codes
1. AI Adoption and Integration	AI adaptation, AI adoption, AI application, AI adoption, AI application, AI assimilation, AI impact, AI Integration, AI platforms
2. Organizational Culture	Communication, Cultural Integration, Employee Engagement, Ethical Considerations, Leadership Support, Mitigation Strategies, Trust
3. AI Adoption and Cybersecurity Challenges	Complexity, Cybersecurity Challenges, Employee Resistance, Expertise, Financial Constraints, Knowledge and Skill Gaps, Limited Resources
4. Training and Awareness	Awareness Levels, Compliance, Continuous Monitoring, Cyber Resilience, Incident Response, Security and Privacy, Security Frameworks, Security Practices, Training

The structured coding method provided a systematic analysis of the interrelated complex phenomena of AI integration, cybersecurity policies and operational dynamics of SMEs (De Simone, 2022).

4.3 *Network Diagrams*

To adequately represent the interrelationships of the themes, network diagrams were created using the ATLAS.ti software. Diagrams offer insights into how certain themes are interconnected, demonstrating the mutual influence of organizational factors on AI adoption.

The network diagram (Fig. 3) indicates a complex mode of AI adoption in SME setting, which shows how various organizational factors interact. The interrelations of leadership, culture, training and cybersecurity awareness point out that AI implementation must be viewed holistically. The diagram further illustrates the importance of a holistic strategy that combines technological innovation with readiness and acceptance at cultural and organizational levels.

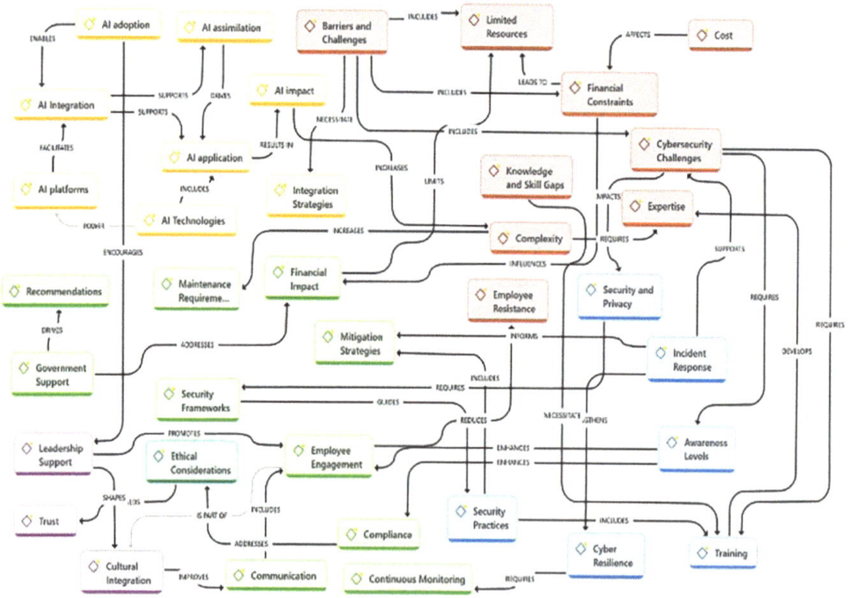

Fig. 3 Network diagram AI adoption, cybersecurity and organizational culture in SMEs

4.4 *Insights Gained*

The findings showed that AI and cybersecurity integration was more likely to succeed in SMEs with strong leadership and a positive organizational culture. Leadership served as essential driver of AI adoption and fostering employee acceptance of new technologies. Cybersecurity training on a regular basis reduced the resistance to the AI-enabled security systems when employees were engaged and more accepting. The companies with an established cybersecurity culture found the AI tools easier to adopt because the awareness and preparedness were already in place.

The building of a good education strategy in cybersecurity has played a key role in smooth AI implementation. Aligning organizational strategies with security goals was an additional support for AI adoption. Companies that encouraged employee participation had fewer barriers in implementing AI solutions. Training and awareness projects helped to further strengthen both AI and cybersecurity practices. Overall, leadership, attitude and engagement are said to be the cornerstone of AI integration success.

5 Discussion

The results suggest that SMEs have difficulties in the uptake of AI and cybersecurity, particularly because of limited resources, lack of skills and resistance to change (Feng et al., 2023; Ulrich & Frank, 2021). In line with previous studies, SMEs with strong leadership and a good cyber security climate are less likely to encounter challenges, as leadership and culture are critical dimensions (Peretz-Andersson et al., 2024; Van Kranenburg et al., 2023;).

Encouraging a culture that embraces AI and emphasizing on training, is a critical way for SMEs to resolve challenges and ease the adoption (Bettiol et al., 2023; Bountouni et al., 2023). Collaborating with industry experts and policymakers will assist SMEs to develop AI strategies that are cheap, security-resilient and capable of maintaining competition within the digital sphere.

6 Contribution

In the literature, the research makes a valuable contribution by underscoring the importance of having a robust cybersecurity culture in encouraging SMEs to adopt AI technologies. Researchers argued that larger enterprises which have a security-oriented approach experience fewer challenges regarding the incorporation of AI and can implement AI in a beneficial way towards cybersecurity, while SMEs generally suffer from a lack of expertise and resources.

The research has also noted an absence of training frameworks and pointed out the necessity of specific building blocks creating competences in cybersecurity and AI which are important for the digital age of SMEs to encourage growth and sustainability.

Another important finding of the research is the absence of any structured training framework for SMEs. While big companies benefit from standard training programs for cybersecurity and AI, SMEs lack access to affordable and scalable training models. A structured training framework should be developed to bridge the gap on the fundamental aspects of AI, best practices on the cybersecurity domain, risk management and compliance. The framework should include practical exposure, continuous upskilling and guidelines specific to industries to discuss how SMEs can improve cybersecurity resilience and integration of AI. Equipping SMEs with accessible training resources will allow SMEs to engage AI safely while minimizing potential security risks, which will eventually cultivate growth and sustainability in the long term.

7 Conclusion

The research emphasizes the role of both culture and leadership in promoting the amalgamation of AI and cybersecurity within SMEs. Evident from the findings that establishing a positive corporate culture and engaging in leading initiatives is vital to address issues that impede AI development such as insufficient funds and training. Engagement and awareness levels among the employees were cardinal as well since employees compromised the level of resistance and encouraged an atmosphere conducive for adoption of AI.

References

Abdul Wahab, M. D., & Radmehr, M. (2024). The impact of AI assimilation on firm performance in small and medium-sized enterprises: A moderated multi-mediation model. *Heliyon, 10*(8). https://doi.org/10.1016/j.heliyon.2024.e29580

Alizai, A., Mousavi, M. R., Ludwig, S., & Aschenbrenner, D. (2023). Heterogeneous communication networks and localization for industry 4.0 applications in small and medium-sized enterprises: A systematic literature review. *Procedia CIRP, 120*, 308–313. https://doi.org/10.1016/j.procir.2023.08.055

Alonso, R., Haber, R. E., Castaño, F., & Reforgiato Recupero, D. (2024). Interoperable software platforms for introducing artificial intelligence components in manufacturing: A meta-framework for security and privacy. *Heliyon, 10*(4). https://doi.org/10.1016/j.heliyon.2024.e26446

Arranz, C. F. A., Arroyabe, M. F., Arranz, N., & de Arroyabe, J. C. F. (2023). Digitalisation dynamics in SMEs: An approach from systems dynamics and artificial intelligence. *Technological Forecasting and Social Change, 196*. https://doi.org/10.1016/j.techfore.2023.122880

Arroyabe, M. F., Arranz, C. F. A., de Arroyabe, I. F., & de Arroyabe, J. C. F. (2024). The effect of IT security issues on the implementation of industry 4.0 in SMEs: Barriers and challenges.

Technological Forecasting and Social Change, 199. https://doi.org/10.1016/j.techfore.2023.123051

Baabdullah, A. M., Alalwan, A. A., Slade, E. L., Raman, R., & Khatatneh, K. F. (2021). SMEs and artificial intelligence (AI): Antecedents and consequences of AI-based B2B practices. *Industrial Marketing Management, 98*, 255–270. https://doi.org/10.1016/j.indmarman.2021.09.003

Bettiol, M., Capestro, M., Di Maria, E., & Micelli, S. (2023). How SMEs respond to an exogenous shock: Diversification, servitization and digitalization. *Journal of Digital Economy, 2*, 221–232. https://doi.org/10.1016/j.jdec.2024.01.006

Bountouni, N., Koussouris, S., Vasileiou, A., & Kazazis, S. A. (2023). A holistic framework for safeguarding of SMEs: A case study. In *2023 19th International conference on the design of reliable communication networks, DRCN 2023*. Institute of Electrical and Electronics Engineers Inc. https://doi.org/10.1109/DRCN57075.2023.10108247

Brodny, J., & Tutak, M. (2022). Digitalization of small and medium-sized enterprises and economic growth: Evidence for the EU-27 countries. *Journal of Open Innovation: Technology, Market, and Complexity, 8*(2). https://doi.org/10.3390/joitmc8020067

Chaudhary, S., Gkioulos, V., & Katsikas, S. (2023). *A quest for research and knowledge gaps in cybersecurity awareness for small and medium-sized enterprises*. Elsevier Ireland Ltd. https://doi.org/10.1016/j.cosrev.2023.100592

Chidukwani, A., Zander, S., & Koutsakis, P. (2022). A survey on the cyber security of small-to-medium businesses: Challenges, research focus and recommendations. *IEEE Access, 10*, 85701–85719. https://doi.org/10.1109/ACCESS.2022.3197899

Clemente-Almendros, J. A., Nicoara-Popescu, D., & Pastor-Sanz, I. (2024). Digital transformation in SMEs: Understanding its determinants and size heterogeneity. *Technology in Society, 77*. https://doi.org/10.1016/j.techsoc.2024.102483

De Simone, V., Di Pasquale, V., & Miranda, S. (2022). An overview on the use of AI/ML in manufacturing MSMEs: solved issues, limits, and challenges. *Procedia Computer Science, 217*, 1820–1829. https://doi.org/10.1016/j.procs.2022.12.382

Feng, Y., Meng, M., & Li, G. (2023). Impact of digital finance on the asset allocation of small- and medium-sized enterprises in China: Mediating role of financing constraints. *Journal of Innovation and Knowledge, 8*(3). https://doi.org/10.1016/j.jik.2023.100405

Fosso Wamba, S. (2022). Impact of artificial intelligence assimilation on firm performance: The mediating effects of organizational agility and customer agility. *International Journal of Information Management, 67*. https://doi.org/10.1016/j.ijinfomgt.2022.102544

Ingalagi, S. S., Mutkekar, R. R., & Kulkarni, P. M. (2021). Artificial Intelligence (AI) adaptation: Analysis of determinants among small to medium-sized enterprises (SME's). *IOP Conference Series: Materials Science and Engineering, 1049*(1), Article 012017. https://doi.org/10.1088/1757-899x/1049/1/012017

Jada, I., & Mayayise, T. O. (2024). The impact of artificial intelligence on organisational cyber security: An outcome of a systematic literature review. *Data and Information Management, 8*(2). https://doi.org/10.1016/j.dim.2023.100063

Kappe, M., Härting, R. C., Karg, C., & Deffner, D. (2023). Cybersecurity in SMEs—Drivers of cybercrime, insufficient equipment and prevention. *Procedia Computer Science, 225*, 3631–3640. https://doi.org/10.1016/j.procs.2023.10.358

Lada, S., Chekima, B., Karim, M. R. A., Fabeil, N. F., Ayub, M. S., Amirul, S. M., Ansar, R., Bouteraa, M., Fook, L. M., & Zaki, H. O. (2023). Determining factors related to artificial intelligence (AI) adoption among Malaysia's small and medium-sized businesses. *Journal of Open Innovation: Technology, Market, and Complexity, 9*(4). https://doi.org/10.1016/j.joitmc.2023.100144

Merín-Rodrigáñez, J., Dasí, À., & Alegre, J. (2024). Digital transformation and firm performance in innovative SMEs: The mediating role of business model innovation. *Technovation, 134*. https://doi.org/10.1016/j.technovation.2024.103027

Mmango, N., & Gundu, T. (2023). Cyber resilience in the entrepreneurial environment: A framework for enhancing cybersecurity awareness in SMEs. In: *International conference on electrical, computer and energy technologies, ICECET 2023*. Institute of Electrical and Electronics Engineers Inc. https://doi.org/10.1109/ICECET58911.2023.10389226

Page, M. J., McKenzie, J. E., Bossuyt, P. M., Boutron, I., Hoffmann, T. C., Mulrow, C. D., Shamseer, L., Tetzlaff, J. M., Akl, E. A., Brennan, S. E., Chou, R., Glanville, J., Grimshaw, J. M., Hróbjartsson, A., Lalu, M. M., Li, T., Loder, E. W., Mayo-Wilson, E., McDonald, S., ... Moher, D. (2021). The PRISMA 2020 statement: An updated guideline for reporting systematic reviews. *BMJ Publishing Group*. https://doi.org/10.1136/bmj.n71

Peretz-Andersson, E., Tabares, S., Mikalef, P., & Parida, V. (2024). Artificial intelligence implementation in manufacturing SMEs: A resource orchestration approach. *International Journal of Information Management, 77*. https://doi.org/10.1016/j.ijinfomgt.2024.102781

Polas, M. R. H., Afshar Jahanshahi, A., Kabir, A. I., Sohel-Uz-Zaman, A. S. M., Osman, A. R., & Karim, R. (2022). Artificial intelligence, blockchain technology, and risk-taking behavior in the 4.0IR metaverse era: Evidence from Bangladesh-based SMEs. *Journal of Open Innovation: Technology, Market, and Complexity, 8*(3). https://doi.org/10.3390/joitmc8030168

Rawindaran, N., Jayal, A., Prakash, E., &Hewage, C. (2023). Perspective of small and medium enterprise (SME's) and their relationship with government in overcoming cybersecurity challenges and barriers in Wales. *International Journal of Information Management Data Insights, 3*(2). https://doi.org/10.1016/j.jjimei.2023.100191

Selamat, M. A., & Windasari, N. A. (2021). Chatbot for SMEs: Integrating customer and business owner perspectives. *Technology in Society, 66*. https://doi.org/10.1016/j.techsoc.2021.101685

Shojaifar, A., & Järvinen, H. (2021). Classifying SMEs for approaching cybersecurity competence and awareness. In: *ACM International conference proceeding series*. Association for Computing Machinery. https://doi.org/10.1145/3465481.3469200

Uchendu, B., Nurse, J. R. C., Bada, M., & Furnell, S. (2021). Developing a cyber security culture: Current practices and future needs. *Computers and Security, 109*. https://doi.org/10.1016/j.cose.2021.102387

Ulrich, P., & Frank, V. (2021). Relevance and adoption of AI technologies in German SMEs results from survey-based research. *Procedia Computer Science, 192*, 2152–2159. https://doi.org/10.1016/j.procs.2021.08.228

Van Kranenburg, R., Bohara, R., Yahalom, R., & Ross, M. (2023). Cyber resilience, societal situational awareness for SME. In: *Proceedings of the 2023 IEEE International conference on cyber security and resilience, CSR 2023* (pp. 458–463). Institute of Electrical and Electronics Engineers Inc. https://doi.org/10.1109/CSR57506.2023.10225011

Wei, R., & Pardo, C. (2022). Artificial intelligence and SMEs: How can B2B SMEs leverage AI platforms to integrate AI technologies? *Industrial Marketing Management, 107*, 466–483. https://doi.org/10.1016/j.indmarman.2022.10.008

Deep Learning for Detection and Quantification of (*Bactrocera dorsalis*) in Intelligent Electronic Fruit Fly Traps

Florence Lello, Mussa Dida, Elizabeth Mkoba, and Mbazingwa Mkiramweni

Abstract This research aims to implement a state-of-the-art (SOTA) deep learning model for detecting and counting Oriental fruit flies in mango farms, facilitating efficient pest management through electronic traps. The study proposes designing and modifying You Only Look Once (YOLO) model to monitor the species in real-world environment settings using low-resource-constrained devices. Images of the flies were collected in real-world environment (mango orchards), pre-processed, and annotated to create a dataset. From the conducted experiments, YOLOv8n (frozen) demonstrated good performance in terms of mean average precision at 50% Intersection over Union (mAP@50), $F1$-score, precision, and recall, outperforming other models and achieving the highest mAP@50 (98.72%), $F1$-score (97.20%), precision (97.54%), and recall (96.87%). The proposed method integrates a YOLO-based deep learning model into a Raspberry Pi4 for accurate detection of the flies. The use of a simple, well-designed electronic trap combined with a clear white background enhances the visibility of captured flies in the images, thereby improving detection and counting accuracy to above 97% while ensuring cost-effectiveness and ease of deployment.

F. Lello (✉) · M. Dida · E. Mkoba
School of Computational and Communication Sciences and Engineering, The Nelson Mandela African Institution of Science and Technology, Arusha, Tanzania
e-mail: lellof@nm-aist.ac.tz

M. Dida
e-mail: mussa.ally@nm-aist.ac.tz

E. Mkoba
e-mail: elizabeth.mkoba@nm-aist.ac.tz

M. Mkiramweni
Department of Electronics and Telecommunications Engineering, Dar es Salaam Institute of Technology, Dar es Salaam, Tanzania

J. Marx Gómez et al. (eds.), *Recent Advances in Information Systems*, Progress in IS,
https://doi.org/10.1007/978-3-032-14099-9_39

1 Introduction

Bactrocera dorsalis, commonly known as Oriental fruit fly, is considered to be overabundant and extremely polyphagous fruit fly species infesting a wide range of fruits (Lello et al., 2023; Majeed et al., 2024). At the adult stage, the flies pose more damage to the fruits and considered as a serious pest in the mango fruit production (Grechi et al., 2021). To enhance pest management strategy of the flies, the use of electronic traps embedded with image sensor and detection models is significant in automating the monitoring process (Preti et al., 2021; Suto, 2022).

In recent years, there is a rapid progress of deep learning application across various scientific and industrial domains, offering unparalleled capabilities in data analysis and pattern recognition (Choudhary et al., 2022; Lello et al., 2023). Deep learning a branch of machine learning focuses on computational models influenced by the mechanism of human brain in processing information (Bishop & Bishop, 2023; Janiesch et al., 2021). Deep learning has gained its popularity due to the availability of faster hardware (CPUs and GPUs), a significant progress in computers that enables handling of large datasets and complex problems (Salamut et al., 2023). It is preferred by researchers when dealing with problems related to image data through the use of convolutional neural network (CNN) (Hussain, 2023) a type of deep learning neural network with complex architecture that learns directly from data and is applied on image tasks such as in image classification, object detection, and object segmentation.

1.1 Deep Learning in Pest Detection

A YOLO-based model was developed by Kumar et al. (2023) for practical applications in agricultural pest monitoring and control. The developed model was a modification of YOLOv5 named as YOLOv5x, the model showed superior accuracy and robustness achieving the highest performance with an $F1$-score close to 0.90 and a mAP value of 93%. The ability of YOLO algorithm to simplify object detection into a single-class regression significantly improves its detection speed (Zhang et al., 2024). Zhang et al. (2024) developed a high-precision fruit fly recognition model leveraging transformer architecture and a step attention mechanism to enhance precision and efficiency for small objects against complex background. The model achieved impressive metrics, including a precision of 0.96, a recall of 0.95, and an $F1$-score of 0.95.

Salamut et al. (2023) evaluated the performance of five deep learning image recognition models: Faster R-CNN MobileNet, Faster R-CNN ResNet, Single Shot Multibox Detector (SSD VGG-16), RetinaNet ResNet, and YOLOv5. These models were tested on a single-class dataset of cherry fruit flies (*Rhagoletis cerasi* L.), captured on sticky traps, comprising over 1600 annotated images. It was also, observed imbalance in the dataset caused performance difference when deployed on a multi-class dataset. Slim et al. (2023) proposed a smart insect monitoring system based on

YOLOv5 for detecting and quantifying Mediterranean fruit flies and peach fruit flies. The model achieved a weighted average accuracy of 84%, enabling early monitoring of fruit fly infestations using a mobile camera attached to insect traps. However, the approach faced challenges with the small size of the target flies and their random positioning inside the trap.

Freitas et al. (2022) proposed a deep learning-based monitoring approach for Ceratitis capitata and Grapholita molesta using ResNet18, achieving an overall accuracy of 90.72%. The authors identified several challenges in implementing the system, including insect overlapping, the small size of target insects, and the presence of larger insects and non-target species. Additionally, the model's large size posed compatibility issues with resource-constrained systems, necessitating compression techniques to maintain accuracy while reducing inference time and energy consumption. Mamdouh et al. (2021) highlighted that transmitting images of fruit flies captured from traps to experts for analysis through communication systems is prone to errors. To address this challenge, they emphasized the importance of developing an edge-based system to fully automate the detection and quantification of olive fruit flies. Their monitoring approach employed a YOLO-based deep learning model, achieving a mean average precision (mAP) of 96.8%.

Molina et al. (2023) developed a low-cost electronic trap (e-trap) integrated with machine learning techniques, specifically Random Forest (RF) and Support Vector Machine (SVM), for the detection and classification of olive fruit flies. The combined techniques achieved an accuracy of 89.1%. However, the models were trained on a limited dataset and faced computational challenges during implementation. The authors suggested that integrating a deep learning-based model could address these challenges and enhance the system's performance.

The works presented above clearly demonstrate the current trends in research on applying deep learning to monitor fruit fly pests in agricultural settings. Deep learning-based models offer numerous benefits, including high precision in pest identification, enhanced detection accuracy, improved monitoring without the need for constant human intervention, and reduced labor costs through automation of repetitive tasks and optimized resource allocation for pest control. Despite the significant progress in applying deep learning models in the detection of insect pests, the models still face challenges that hinder their ability to achieve fully automation of pest monitoring. These challenges include difficulties in accurately detecting pests under varying environmental conditions, distinguishing between visually similar species, small physical size, limited and imbalanced datasets, limited computation for edge applications, and real-time processing in field conditions.

1.2 Choice of Deep Learning Model

For object detection tasks, state-of-the-art deep neural network models are categorized into two-stage and single-stage pipelines (Zhang et al., 2019). Single-stage detectors, such as You Only Look Once (YOLO) and Single Shot Multibox Detector

(SSD), simultaneously perform object classification and bounding box regression in a single pass. In these models, the input image is processed through the neural network, which directly outputs class predictions and bounding boxes for all detected objects. In contrast, two-stage detectors, like Region-based Convolutional Neural Network (R-CNN) and its variants, first generate region proposals before classifying objects. Single-stage detectors are faster than two-stage detectors because they streamline the detection process into a single step, reducing computational overhead. For this study, a YOLO-based model is selected for object detection due to its computational efficiency and strong performance metrics (Tannous et al., 2023).

For any deep learning task to be successful, selection of appropriate model is essential. Due to the rapid improvements in deep learning techniques and architectures variety of choices are available, each with pros and cons. In this study, YOLO-based deep learning model is implemented with consideration of improving accuracy when applying for small size object detection such as *B. dorsalis*. The model was trained with image dataset that was collected in the real-world settings whose performance was compared with other state-of-the art models. The main objective being monitoring the adult flies in electronic traps by detecting their presence and quantity at specified time intervals. Through the use of camera sensor mounted inside the trap, the monitoring process was carried out for the real-world environment settings.

1.3 Objective of the Study

This study aims to implement and evaluate a deep learning model to automate the detection and counting of *B. dorsalis* in intelligent electronic fruit fly traps, contributing to improved pest management strategies in mango orchards. Furthermore, the model to be optimized for local inference on resource-constrained devices suitable for farm field deployment. The deployment must overcome challenges such as limited computational resources, power constraints, and storage capacity, and the small size of the target species.

2 Materials and Methods

2.1 Image Data Collection

In collecting images two methods are applied: collecting images of specimens (insects) preserved under ideal laboratory conditions and images collected both in the field and under laboratory settings. This study adopted capturing images of flies from deployed electronic traps in real-world environments. The collected data consisted of images of the target flies (*B. dorsalis*), which met real-world conditions. Images were collected from two sites: Dar es Salaam Institute of Technology in the Dar es Salaam region, and a mango orchard at Koga Farm in the southeastern part of the Pwani region, Mkuranga district.

2.1.1 Source of Data

Two different image sensors were used that had different resolutions 2 Mp and 3 Mp, fixed at a distance (20 mm). Figure 1b, shows an illustration of how images are captured from the base of the trap without zoom or automatic focus settings. Images captured from the two sensors ensured separate sources of images that enabled the diversity of the dataset as well as ensuring the representative of the deployed environment. Furthermore, to obtain a variety of images, the capturing process was conducted at different times of the day. Images were captured in the morning (around 7:00 AM to 9:00 AM), in the afternoon (around 1:00 PM to 3:00 PM), and in the evening (around 5:00 PM to 6:00 PM). This approach ensured the dataset included images with varying lighting conditions, enhancing its diversity and enabling the trained models to adapt to different times of the day. To achieve the objective of this study, the steps in Fig. 1a were carried out. Firstly, by defining the problem under the study, which is the detection of the fruit flies in images, pests that affect mango fruit. Next step, images were collected using camera and organized into a dataset. A deep learning model was selected, trained, and evaluated to asses the performance of the defined model. Lastly, the model was deployed on resource-constrained device (Rasperry Pi4) for edge processing.

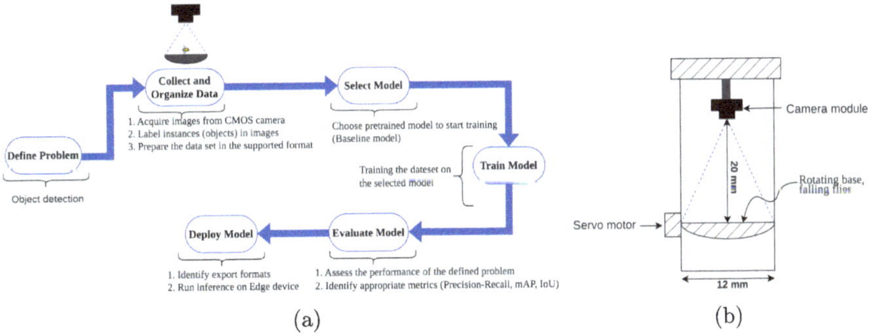

Fig. 1 Pictorial presentation for the experimental setup

2.1.2 Data Annotation

Data annotation is the process of assigning labels to objects in an image, providing the ground truth from which the model learns. In this study, we manually annotated objects in the raw images with the assistance of an expert entomologist using Roboflow (Dwyer et al., 2024). While the annotation process was time-consuming, it allowed us to evaluate the quality of the dataset throughout, which facilitated the pre-processing stage. Figure 2 showcases examples of annotated images included in the dataset.

2.1.3 Pre-processing of the Image Data

A total of 253 images were collected to generate the initial dataset. However, since the raw dataset may contain variations that are challenging to represent in deep learning architectures, pre-processing is necessary. This crucial stage reduces the amount of variation the trained model must account for, helping to lower generalization error and minimize the size of the model required to train the dataset. The dataset was thoroughly inspected for corrupt files and duplicate images. During this process, 56 files were identified as either corrupt or duplicates. To ensure a clean and reliable dataset, a cleaning process was carried out to remove all unwanted files.

2.1.4 Data Augmentation

Data augmentation is a form of data pre-processing used to increase the number of images in the training set. This process introduces invariance to geometric transformations of the original images, thereby expanding the training examples available. The augmentation techniques applied in this study included rotation, flipping,

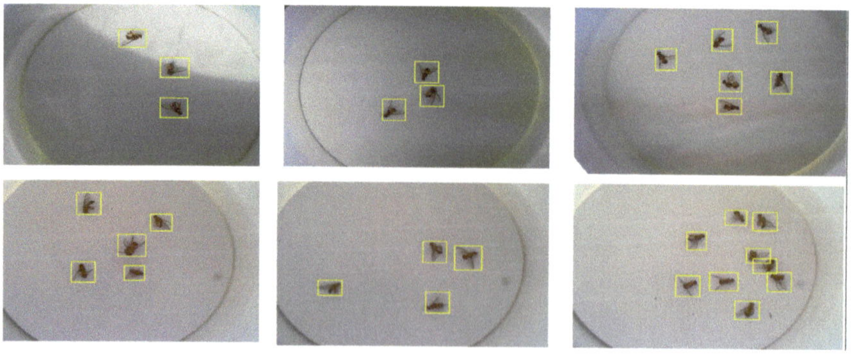

Fig. 2 Images from electronic trap, annotated with yellow bounding box to label *B. dorsalis* using Roboflow software (Dwyer et al., 2024)

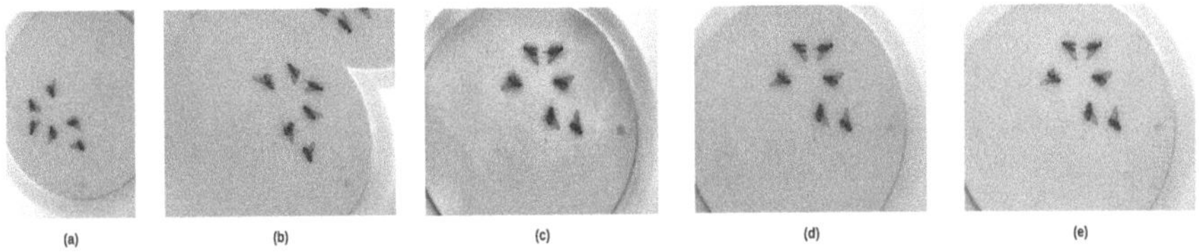

Fig. 3 Augmentation for *B. dorsalis* dataset creation: **a** Transpose, **b** shift scaling, **c** CLAHE, **d** motion blur, **e** Gaussian noise

transposing, grid and optical distortion, median and motion blur, shift scaling, Gaussian noise addition, contrast-limited adaptive histogram equalization (CLAHE), and translation. Figure 3 illustrates sample image augmentation techniques applied during the creation of the dataset.

2.1.5 Dataset Size and Composition

The final dataset after augmentation reached 1236 clean images, split into two sets: training set and validation set at a ratio of 80% and 20%, respectively. Dataset in deep learning environment can be defined as a collection of data that is in a well-structured organization for the purpose of training, validating, and testing deep learning models. The dataset as a source of information for the deep learning models should be well organized and composed (in terms of set size) where this plays a significant role in the effectiveness and robustness of the detection models. The training set comprises examples that the model will see and learn while validation set is often used for model selection, monitoring the training process and hyper-parameters tuning (Roberts et al., 2022).

In order to balance the class in dataset, background images were also added.

2.2 YOLO Convolution Neural Network

YOLOv8 (Jocher et al., 2023) is a variant of YOLO model developed by Ultralytics. It is a cutting-edge, SOTA model that has improved features to enhance its performance over a wide range of applications such as object detection and tracking, image classification, instance segmentation, and pose estimation tasks (Jocher et al., 2023). The model provides a variety of pre-trained models to fit different performance needs (Jocher et al., 2023) and consists of five series: YOLOv8n (nano), YOLOv8s (small), YOLOv8m (medium), YOLOv8l (large), and YOLOv8x (extra-large). For each series, there are different numbers of layers, parameters, gradients, and GFLOPs which lead to improved detection performance and lightweight properties.

The YOLOv8 object detection network architecture, shown in Fig. 4, consists of four main components: the input, backbone, neck, and head. At the input stage,

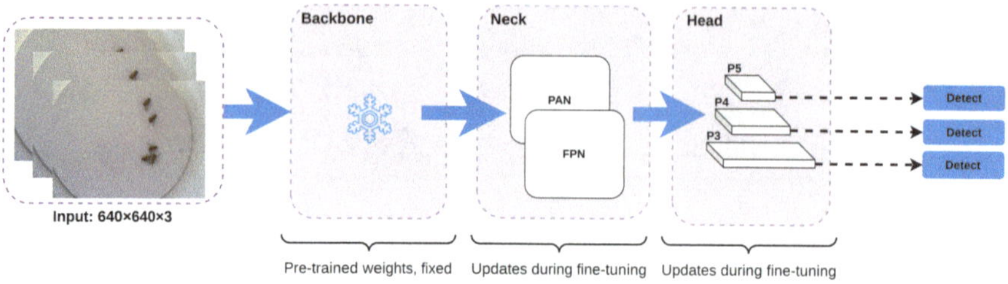

Fig. 4 YOLOv8 architecture: with the backbone frozen

the model performs image pre-processing, including data augmentation and image scaling, to optimize the images before they are fed into the network. The backbone extracts image features using module blocks such as CBS, C2f, and SPPF. The neck enhances the model's ability to detect objects of different sizes by merging multi-scale features through the FPN+PAN structure, significantly improving detection accuracy, particularly for images with complex backgrounds and varying object sizes (Feng et al., 2024). The head employs a Decoupled Head structure to independently handle classification and detection tasks (Feng et al., 2024).

2.3 Experimentation

2.3.1 Setup for Training

The selected model was trained on an Ubuntu 24.04 LTS computer with a core i9-13900X CPU, 32 GB of RAM, and an NVIDIA GeForce RTX 4060 8 GB GPU.

2.3.2 Transfer Learning of the YOLOv8 Model

To enhance the performance of deep learning models for object detection, techniques such as transfer learning (using pre-trained weights) and backbone improvements play a significant role (Salamut et al., 2023). The backbone is responsible for extracting valuable features from the input image, as shown in Fig. 4. These features are captured at various scales, with lower-level features (like edges and textures) being extracted in the initial layers, while higher-level features (such as object parts and semantic information) are captured in the deeper layers (Terven et al., 2023).

Through transfer learning, the backbone layer of the pre-trained YOLOv8n was frozen, as shown in Fig. 4 (i.e., the weights in the layer are not updated during training), while the other layers of the network are fine-tuned. Therefore, the level of training is regulated depending on the amount of available data. Hyper-parameters, including learning rate and batch size, were fine-tuned to optimize the model's training performance and optimization adaptive learning rate algorithms were employed for efficient model convergence.

Training was conducted on an 80-20 train-test split, using hyper-parameters: 120 epochs, 16 batches, AdamW optimizer, an initial learning rate (lro) of 0.01, final learning rate (lrf) 0.0001, warmup epochs = 0, mosaic = 1, close mosaic = 24, and weight decay = 0.0001.

To further explore deep learning in the detection of *B. dorsalis*, three lightweight YOLO models designed for low-resource devices were studied: YOLOv5n, YOLOv9t, and YOLOv10n. The same techniques used in YOLOv8n were applied to these three YOLO variants to facilitate a performance comparison. Furthermore, a comparative analysis with the latest YOLO versions (YOLOv11n and YOLOv12n) was conducted to evaluate the models' performance on the dataset.

2.3.3 Setup for Local Inference in Low Resource Settings

Deploying the model in electronic traps requires ensuring the reliability of both hardware and software components. This necessitates the optimization of resource-constrained devices, such as Raspberry Pi, to perform inference locally. This approach addresses the limitations of transmitting large datasets, such as images, over low-bandwidth networks like Long Range Communication (LoRa). By enabling local data processing, the deployment facilitates Edge AI applications in agriculture, ensuring efficient and scalable solutions for real-time monitoring and analysis (Himeur et al., 2023).

Table 1 provides a summary of the hardware and software components along with their specifications for setting up the system to perform local inference in the farm.

Table 1 System deployment requirements

	Specification
Hardware	
Raspberry Pi4	Model: B Rev 1.4
	Memory: 4GB RAM, 32GB S-D card storage
Camera	USB camera, 1080p resolution
Software	
Python programming language	Python 3.9.2
Deep learning frameworks (PyTorch: back-end for YOLO)	Torch 1.12.0, TorchVision 0.13.0
Ultralytics (Jocher et al., 2023)	Raspbian
Export frameworks	ncnn, ONNX

2.4 Evaluation Metrics

In this study, four key metrics: average precision, precision, recall, and $F1$-score were selected to assess the model's performance. To provide a more detailed evaluation, the counts of true positive (TP), true negative (TN), false positive (FP), and false negative (FN) predictions are used. These counts are organized into a table known as the confusion matrix, which simplifies the performance evaluation of the model.

A True Positive (TP) occurs when the model correctly identifies an object within the ground truth (annotated bounding box). A False Positive (FP) happens when the model mistakenly places a positive detection outside the ground truth bounding box. A False Negative (FN) arises when the detector fails to generate a positive detection, despite the presence of a ground truth bounding box for the object in the image. True Negative (TN) refers to the correct identification of the negative class. However, True Negatives are generally not considered in object detection, as there are many regions in the image that should not be detected as bounding boxes (Kaur & Singh, 2022; Padilla et al., 2020).

3 Results

3.1 YOLOv8n (Frozen) Model Performance on Bactrocera dorsalis Dataset

The selected baseline YOLOv8n network was pre-trained with a single-class dataset composed of *B. dorsalis*. The process involved fine-tuning of the pre-trained weights while fixing the backbone of the network, forming YOLOv8n (frozen) model. The results from the evaluation metrics showed the model performed well training on the dataset, where the model had high value of true positives indicating that the model correctly identified the ground truth of the detected objects (flies). YOLOv8n (frozen) model was trained for 120 epochs with AdamW optimization, Fig. 5 shows the loss for both training and validation was decreasing to zero showing that the model was performing well on the dataset. Figure 6 shows the validation results and the detection accuracy of the objects (*B. dorsalis*).

3.2 YOLO Variant Models Performance on Bactrocera dorsalis Dataset

The chosen four state-of-the-art YOLO models (YOLOv5n, YOLOv8n, YOLOv9t, YOLOv10n) were trained on the same dataset and the performance metrics for each model were recorded in Table 2. The results present the models' performance in their

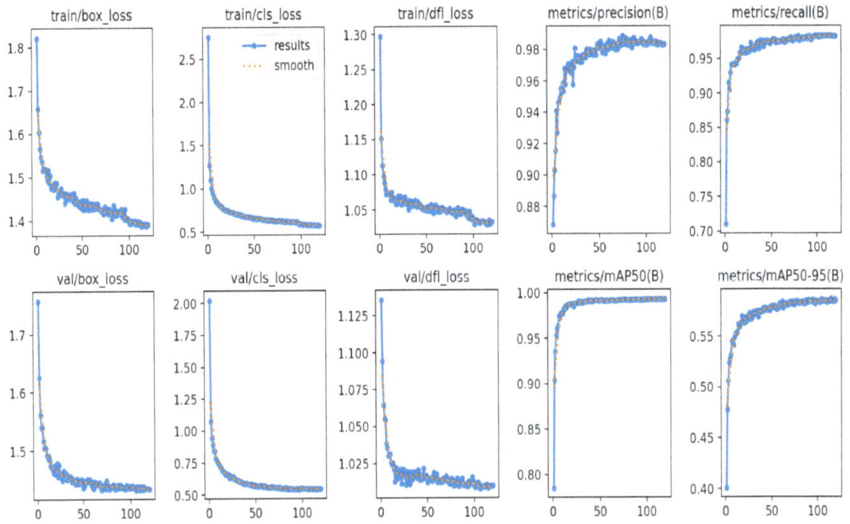

Fig. 5 Metrics visualization during training of the baseline YOLOv8n model

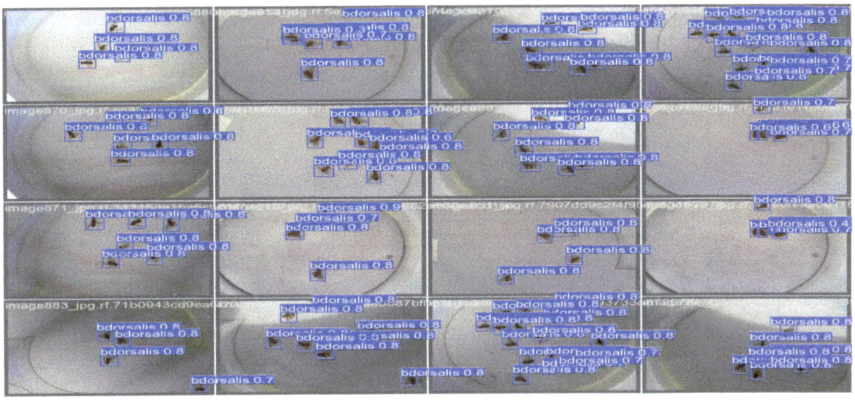

Fig. 6 Validation results of the trained YOLOv8n model

baseline and frozen configurations, evaluated on key metrics such as mAP50, $F1$-score, precision, and recall. YOLOv8n (frozen) outperformed other models, achieving the highest mAP50 (98.72%), $F1$-score (97.20%), precision (97.54%), and recall (96.87%). These results show the models' ability to detect *B. dorsalis* accurately while maintaining a low rate of false positives and false negatives. Furthermore, it was observed that the frozen models outperformed their respective baseline configurations across all metrics which suggests that freezing of the backbone layer during training enhances performance by enabling the model to focus on higher-level features.

Table 2 YOLO models performance on *Bactrocera dorsalis* dataset

Model	mAP50 (%)	$F1$-score (%)	Precision (%)	Recall (%)
YOLOv5n (baseline)	98.21	97.03	97.61	96.45
YOLOv8n (baseline)	98.42	96.78	97.66	95.92
YOLOv9t (baseline)	95.29	94.39	94.18	94.61
YOLOv10n (baseline)	92.79	88.71	89.40	88.03
YOLOv5n (frozen)	98.55	96.54	96.85	96.22
YOLOv8n (frozen)	98.72	97.20	97.54	96.87
YOLOv9t (frozen)	97.69	96.01	96.78	95.25
YOLOv10n (frozen)	96.61	92.85	92.98	92.72

Table 3 Comparison of the YOLO versions model size

Model	Params ($\times 10^6$)	NCNN model size (MB)	NCNN inference time (ms)	ONNX model size (MB)	ONNX inference time (ms)
YOLOv5n (frozen)	2.51	10.2	58.3	10.2	71.4
YOLOv8n (frozen)	3.01	12.2	55.1	12.3	78.3
YOLOv9t (frozen)	2.00	8.1	58.1	8.2	53.9
YOLOv10n (frozen)	2.71	–	–	9.3	39.7

3.3 *Comparison of YOLOv8n (Frozen) Against Related Works*

A comparative analysis against existing related research works, as presented in Table 3, highlights that the proposed model in this study achieved superior performance metrics. This underscores a significant advancement of the proposed model. However, variations in the results could be influenced by factors such as the characteristics of the dataset and the specific species being observed.

Table 4 Comparison of the proposed model against existing related research works

Author	Model	mAP50 (%)	$F1$-score (%)	Precision (%)	Recall (%)	Accuracy (%)
Kumar et al. (2023)	Customized YOLOv5	93.0	87.8	89.9	85.7	–
Zhang et al. (2024)	Transformer based	–	95.0	96.0	95.0	95.0
Slim et al. (2023)	YOLOv5	84.05	85.69	86.84	84.58	–
Freitas et al. (2022)	ResNet18	–	–	–	–	90.72
Mamdouh et al. (2021)	Customized YOLOv4	96.68	90.0	84.0	97.0	–
Ours	**YOLOv8n (frozen)**	**98.72**	**97.20**	**97.54**	**96.87**	–

3.4 YOLOv8n (Frozen) Model in Resource-Constrained Devices

The results suggest that YOLOv8n (frozen) demonstrates higher accuracy in detecting *B. dorsalis* in the collected dataset, making it suitable for deployment in electronic traps. The model was exported into ONNX and NCNN formats, which are optimized for inference on resource-constrained devices such as the Raspberry Pi. Table 4 presents the performance of the frozen models optimized for integration into electronic traps using NCNN and ONNX formats.

The performance of the models was evaluated based on the hardware used, specifically an ARM-based processor. Among the models, YOLOv8n (frozen) presented in Table 4 has the largest model size in both formats, with the highest number of parameters at 3.01 million. Despite its larger size, YOLOv8n achieves the fastest inference time using the NCNN format, with a latency of 55.1 ms. In contrast, YOLOv10n (frozen) demonstrates the quickest inference time in the ONNX format, achieving a minimal latency of 39.7 ms. In NCNN models, quantization reduces precision in model weights and activations, which lowers computational demand without significantly compromising accuracy leading to further improvements in performance and reduces memory footprint (Jocher et al., 2024).

4 Discussion

In this study, we implemented and evaluated a deep learning model, YOLOv8n, trained on a single-class dataset to detect and count *B. dorsalis* in intelligent electronic fruit fly traps. This approach aims to automate the detection and counting of

B. dorsalis in mango orchards. The implementation of the study aimed at increasing the performance for managing the pest, increasing the model's accuracy and optimizing for local inferencing in the farm fields. The findings show that YOLOv8n (frozen) achieves high detection accuracy for the targeted species, with performance metrics exceeding 97%. A comparative analysis with related research demonstrates that the proposed model in this study achieved improved performance metrics. This improvement can be attributed to the approach used in the study, particularly the method of image capture. By employing a clear white background, the setup ensures even light distribution, enhancing the visibility of the captured objects. This contrasts with other works, which utilize yellow sticky traps to attract and capture flies. For *B. dorsalis*, the fly's darker features such as its variable red-brown to black thorax with yellow markings and reddish-brown abdomen are more distinguishable against a white background. This high contrast significantly contributes to the improved detection accuracy observed in this study.

5 Conclusion

This research developed a deep learning object detector based on the state-of-the-art YOLO model [YOLOv8n (frozen)] for detecting and counting oriental fruit flies (*B. dorsalis*) in images captured from mango orchard electronic traps. The model was trained on field-collected images and deployed on edge devices to perform local inference, enabling efficient and useful results that can be transmitted remotely. The performance of YOLOv8n (frozen) was benchmarked against other YOLO variants trained on the same dataset under identical hyper-parameter settings. It achieved best performance, with precision, recall, and $F1$-score exceeding 97%, significantly outperforming related works. Additionally, the model demonstrated superior accuracy compared to other lightweight YOLO models, including YOLOv5n, YOLOv9t, and YOLOv10n. Despite its strong performance metrics, the proposed model encountered challenges with false detection of non-target species resembling *B. dorsalis*. Addressing this limitation through the inclusion of a multi-class dataset to identify additional species could further improve its accuracy and utility. This enhancement would support broader applications in integrated pest management, advancing effective and sustainable agricultural practices. Challenges related to insect positioning inside the trap and varying ambient light distribution were encountered during implementation. These issues degrade generalization of the model during inference, resulting in false counting. However, the challenge of detecting small-sized objects was minimized, as the traps were designed specifically to capture small flies like *B. dorsalis*. The traps effectively prevented the entrance of foreign objects such as tree leaves. However, non-target species, including ants and spiders, were observed inside the traps, which also contributed to false detections and counting. The model in the study achieved high accuracy rate but face challenge in misidentifying the target species, this challenge can be mitigated through integrating the system with other pest management strategies as suggested by Niyigena et al. (2024). In future work, the dataset will be expanded to include images from diverse geographical regions with varying weather conditions and seasons to enhance the model's robustness and generalizability. Additionally, optimizing the trap setup to ensure consistent lighting distribution during image acquisition will further improve the model's training and detection accuracy. Finally, efforts will focus on further optimizing the model for deployment on resource-constrained devices, enabling efficient real-time monitoring of fruit flies.

Acknowledgements This study was funded by AdEMNEA project through NORHED II.

Disclosure of Interests The authors declare no conflict of interest.

References

Bishop, C. M., & Bishop, H. (2023). *Deep learning foundations and concepts*. Springer Nature. https://doi.org/10.1007/978-3-031-45468-4

Choudhary, K., DeCost, B., Chen, C., Jain, A., Tavazza, F., Cohn, R., Park, C. W., Choudhary, A., Agrawal, A., Billinge, S. J. L., et al. (2022). Recent advances and applications of deep learning methods in materials science. *npj Computational Materials, 8*(1), 59. https://doi.org/10.1038/s41524-022-00756-8

Dwyer, B., Nelson, J., & Hansen, T. (2024). *Automate data labeling*. Available from: https://roboflow.com/annotate. Accessed on 2024-11-09.

Feng, F., Hu, Y., Li, W., & Yang, F. (2024). Improved YOLOv8 algorithms for small object detection in aerial imagery. *Journal of King Saud University-Computer and Information Sciences, 36*(6), 102113. https://doi.org/10.1016/j.jksuci.2024.102113

Freitas, L., Martins, V., de Aguiar, M., de Brisolara, L., & Ferreira, P. (2022). Deep learning embedded into smart traps for fruit insect pests detection. *ACM Transactions on Intelligent Systems and Technology, 14*(1), 1–24. https://doi.org/10.1145/3552435

Grechi, I., Preterre, A.-L., Caillat, A., Chiroleu, F., & Ratnadass, A. (2021). Linking mango infestation by fruit flies to fruit maturity and fly pressure: A prerequisite to improve fruit fly damage management via harvest timing optimization. *Crop Protection, 146*, 105663. https://doi.org/10.1016/j.cropro.2021.105663

Himeur, Y., Sayed, A., Alsalemi, A., Bensaali, F., & Amira, A. (2023). Edge AI for internet of energy: Challenges and perspectives. *Internet of Things*, 101035. https://doi.org/10.1016/j.iot.2023.101035

Hussain, M. (2023). YOLO-v1 to YOLO-v8, the rise of YOLO and its complementary nature toward digital manufacturing and industrial defect detection. *Machines, 11*(7), 677. https://doi.org/10.3390/machines11070677

Janiesch, C., Zschech, P., & Heinrich, K. (2021). Machine learning and deep learning. *Electronic Markets, 31*(3), 685–695. https://doi.org/10.1007/s12525-021-00475-2

Jocher, G., Chaurasia, A., & Qiu, J. (2023). *Ultralytics YOLOv8*. Available from: https://github.com/ultralytics/ultralytics

Jocher, G., Munawar, R., & Vina, A. (2024). *NCNN, NCNN-ultralytics YOLO docs*. Available from: https://docs.ultralytics.com/integrations/ncnn#why-should-you-export-to-ncnn. Accessed on 2024-11-13.

Kaur, J., & Singh, W. (2022). Tools, techniques, datasets and application areas for object detection in an image: A review. *Multimedia Tools and Applications, 81*(27), 38297–38351. https://doi.org/10.1007/s11042-022-13153-y

Kumar, N., Nagarathna, F. F. (2023). YOLO-based light-weight deep learning models for insect detection system with field adaption. *Agriculture, 13*(3), 741. https://doi.org/10.3390/agriculture13030741

Lello, F., Dida, M., Mkiramweni, M., Matiko, J., Akol, R., Nsabagwa, M., & Katumba, A. (2023). Fruit fly automatic detection and monitoring techniques: A review. *Smart Agricultural Technology*, 100294. https://doi.org/10.1016/j.atech.2023.100294

Majeed, A., Jamali, M., & Mir, R. A. (2024). Population, diversity, and distribution of fruit flies (Diptera: Tephritidae: Bactrocera spp.) in Mango Orchards of Lucknow, Saharanpur and Meerut districts of Uttar Pradesh, India. *Journal of Pharmacognosy and Phytochemistry, 13*(1), 190–195. https://doi.org/10.22271/phyto.2024.v13.i1c.14825

Mamdouh, N., & Khattab, A. (2021). YOLO-based deep learning framework for olive fruit fly detection and counting. *IEEE Access, 9*, 84252–84262. https://doi.org/10.1109/ACCESS.2021.3088075

Molina-Rotger, M., Morán, A., Miranda, M. A., & Alorda-Ladaria, B. (2023). Remote fruit fly detection using computer vision and machine learning-based electronic trap. *Frontiers in Plant Science, 14*, 1241576. https://doi.org/10.3389/fpls.2023.1241576

Niyigena, G., Lee, S., Kwon, S., Song, D., & Cho, B. K. (2024). Real-time detection and classification of *Scirtothrips dorsalis* on sticky traps using deep learning and machine vision for smart pest monitoring. *Sensors, 24*(4), 1076. https://doi.org/10.3390/s24041076

Padilla, R., Netto, S. L., & Da Silva, E. A. (2020). A survey on performance metrics for object-detection algorithms. In *2020 International Conference on Systems, Signals and Image Processing (IWSSIP)* (pp. 237–242). IEEE. https://doi.org/10.1109/IWSSIP48289.2020.9145130

Preti, M., Verheggen, F., & Angeli, S. (2021). Insect pest monitoring with camera-equipped traps: Strengths and limitations. *Journal of Pest Science, 94*(2), 203–217. https://doi.org/10.1007/s10340-020-01288-5

Roberts, D. A., Yaida, S., & Hanin, B. (2022). *The principles of deep learning theory*. Cambridge University Press. https://doi.org/10.48550/arXiv.2106.10165

Salamut, C., Kohnert, I., Landwehr, N., Pflanz, M., Schirrmann, M., & Zare, M. (2023). Deep learning object detection for image analysis of cherry fruit fly (Rhagoletis cerasi L.) on yellow sticky traps. *Gesunde Pflanzen, 75*(1), 37–48. https://doi.org/10.1007/s10343-022-00794-0

Slim, S. O., Abdelnaby, I. A., Moustafa, M. S., Zahran, M. B., Dahi, H. F., & Yones, M. S. (2023). Smart insect monitoring based on YOLOV5 case study: Mediterranean fruit fly Ceratitis capitata and Peach fruit fly Bactrocera zonata. *The Egyptian Journal of Remote Sensing and Space Sciences, 26*(4), 881–891. https://doi.org/10.1016/j.ejrs.2023.10.001

Suto, J. (2022). Codling moth monitoring with camera-equipped automated traps: A review. *Agriculture, 12*(10), 1721. https://doi.org/10.3390/agriculture12101721

Tannous, M., Stefanini, C., & Romano, D. (2023). A deep-learning-based detection approach for the identification of insect species of economic importance. *Insects, 14*(2), 148. https://doi.org/10.3390/insects14020148

Terven, J., Córdova-Esparza, D. M., & Romero-González, J. A. (2023). A comprehensive review of YOLO architectures in computer vision: From YOLOv1 to YOLOv8 and YOLO-NAS. *Machine Learning and Knowledge Extraction, 5*(4), 1680–716. https://doi.org/10.3390/make5040083

Zhang, L., Chen, K., Zheng, L., Liao, X., Lu, F., Li, Y., Cui, Y., Wu, Y., Song, Y., & Yan, S. (2024). Enhancing fruit fly detection in complex backgrounds using transformer architecture with step attention mechanism. *Agriculture, 14*(3), 490. https://doi.org/10.3390/agriculture14030490

Zhang, Z., He, T., Zhang, H., Zhang, Z., Xie, J., & Li, M. (2019). Bag of freebies for training object detection neural networks. arXiv preprint. arXiv:1902.04103. https://doi.org/10.48550/arXiv.1902.04103

Prototype to Provide Environmental Information in a Multi Channel Way by Using Citizen Science and AI

Ali Akyol, Malte Schott, and Jorge Marx Gómez

Abstract Environmental data, such as air and water quality indicators, are critical for public health and environmental awareness. This study proposes an extension of existing environmental information portals through the integration of a Telegram bot, utilizing open APIs and push notifications to reach citizens directly. By combining traditional web platforms with messenger-based communication, the approach aims to increase citizen participation in environmental monitoring. Key aspects such as data quality, user motivation, and platform integration are systematically analyzed. Preliminary findings indicate that messenger bots can significantly expand the user base of environmental information platforms. The proposed framework provides a foundation for future research into scalable dissemination strategies and the promotion of citizen-driven environmental engagement.

1 Introduction

Environmental data play a vital role in raising public awareness and protecting human health. Although official portals provide access to such data, citizen engagement and overall accessibility remain limited. This work introduces a multichannel approach that combines a web portal with a Telegram bot to improve the dissemination of environmental information and to foster citizen participation through modern, low-threshold tools.

The structure of this paper is as follows: Sect. 2 presents the theoretical background and related work. Section 3 discusses existing chatbot systems and platforms used for environmental communication. Section 4 outlines the proposed concept, including technical implementation and interfaces. Section 5 presents the evaluation results based on a user survey. Section 6 concludes with a discussion and an outlook on future developments.

http://www.springer.com/gp/computer-science/lncs.

A. Akyol (✉) · M. Schott · J. Marx Gómez
University of Oldenburg, Oldenburg, Germany
e-mail: ali.akyol@uol.de

J. Marx Gómez et al. (eds.), *Recent Advances in Information Systems*, Progress in IS,
https://doi.org/10.1007/978-3-032-14099-9_40

1.1 Motivation

Environmental information (Heinrich et al., 2014) is becoming increasingly significant due to growing societal awareness of environmental and climate protection issues (BMU, 2018). A notable example of this heightened awareness is the environmental and climate activist Greta Thunberg, who initiated the student movement Fridays for Future in Sweden in 2018 (Koos & Naumann, 2019). According to the European Environment Agency, air pollution is responsible for more than 400,000 deaths annually within the European Union and over 700,000 deaths worldwide. In Germany alone, air pollution accounts for approximately 70,000 deaths each year (Europäische, 2020; Resch, 2025; WHO, 2024).

Given the presence of such invisible risks, it is particularly important that individuals utilize available technologies to mitigate health hazards by leveraging informational advantages. In addition to poor air quality, invasive neophytes may establish themselves and pose a potential threat to human health. Invasive neophytes are plant species that are non-native and have established themselves in the wild after the year 1500 as a consequence of human activity. These species are characterized by their ability to naturalize easily, spread rapidly, and cause adverse effects in terms of biodiversity, human health, and/or the economy, as well as by their potential to do so (Gigon & Weber, 2005).

A project aimed at raising public awareness of invasive neophytes was carried out in the German state of Saxony-Anhalt. The initiative sought to establish an early warning and monitoring system for such plant species using a citizen science approach (Schneider & Hormann, 2011). An example of an invasive plant species is Giant Hogweed. Due to its physical characteristics, it overshadows native vegetation. However, upon exposure to sunlight, contact with the plant can cause severe skin irritation. Therefore, any handling of the plant should be carried out only while wearing protective clothing (Bickel, 2022).

Air Pollution and Its Effects At the beginning of 2025, air pollution levels in Germany reached unprecedented highs. Many regions recorded elevated concentrations of particulate matter, which pose significant health risks. While air pollution is rarely the direct cause of death, it contributes to or exacerbates respiratory diseases, cardiovascular incidents, strokes, and various cancers, including lung cancer (Fig. 1). In acute cases, the risk of suffocation can increase (FOEN, 2025; Tropical, 2025; German Environment Agency, 2025).

Air pollution has both physical and psychological consequences. On the physical side, it can lead to heart and lung diseases as well as premature mortality (Clancy et al., 2002). Psychologically, it has been linked to increased aggression, depression, anxiety disorders, psychotic symptoms, and even crime rates though findings in the latter area are still inconclusive (Bakolis et al., 2021; Bondy et al., 2020; Braithwaite et al., 2019; Buoli et al., 2018; Gu et al., 2020; Lu et al., 2018; Ullah et al., 2021). Additionally, it impairs cognitive performance in older adults (Zhang et al., 2018). Globally, air pollution contributes to an estimated 8.79 million premature deaths annually (Lelieveld et al., 2019).

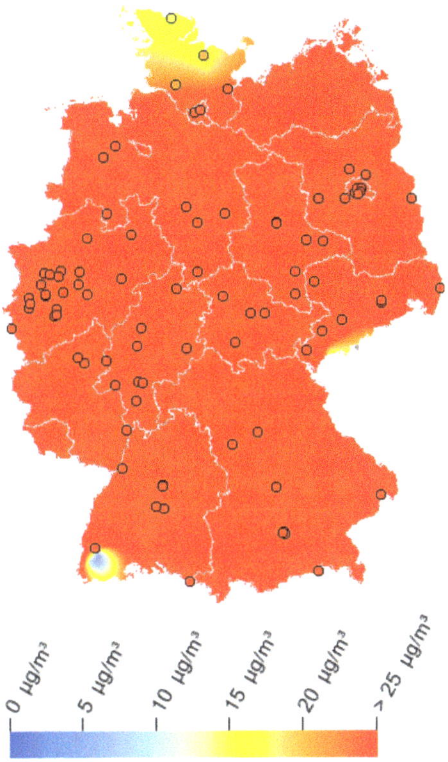

Fig. 1 Air quality in μg/m^3 on February 10, 2025 (Agency, 2025). *Source* German Environment Agency (2025)

Impact of Noise Pollution Noise pollution is, like air pollution, a growing concern. Urban planners now factor noise exposure into city development, using forecasts and simulations to predict its impact (Morillas et al., 2018). Noise affects both physical and mental health and can result in hearing loss, irritation, social withdrawal, communication difficulties, sleep disturbances, and cardiovascular diseases (Jariwala et al., 2017). Additional effects include headaches and elevated blood pressure (Farooqi et al., 2020). According to the European Environment Agency, noise pollution is responsible for approximately 12,000 premature deaths in Europe each year (EEA, 2020).

1.2 Impact of Flood Damage

On July 13–14, 2021, severe rainfall events in the German federal states of North Rhine-Westphalia and Rhineland-Palatinate led to catastrophic flooding, further aggravated by dense urban development. The disaster resulted in at least 170 fatalities and 820 injuries. Insured property losses totaled 10 billion EUR, while many affected individuals were uninsured. Infrastructure damages across Germany were estimated at over 2 billion EUR. This event is considered comparable to historic floods in 1804 and 1910 in terms of rainfall and water levels (Schäfer et al., 2021).

1.3 Problem Statement and Research Question

This work aims not only to process and distribute environmental information but also to develop an innovative communication channel in the form of a Telegram bot complementing an environmental information portal to strengthen citizen science participation. A key objective is the creation of a working prototype that connects the bot with the online portal.

This leads to the central research question: RQ: How can a concept for extending an environmental information portal be designed that connects public measurement stations and environmental information through citizen participation using a Telegram bot? Additionally, the following sub research questions guide the investigation:

SRQ1: What citizen science projects and IT systems for environmental information already exist?
SRQ2: What Telegram bots are currently available in this domain?
SRQ3: Can citizen participation be increased through this concept?

2 Scientific State of the Art

The appropriate implementation of research methods and methodology is a key pillar of academic inquiry. Investigating a specific subject requires building upon existing findings and comparing them with related research. However, the growing volume of publications and the increasing complexity of the scientific landscape make it challenging to maintain a comprehensive overview. In this context, literature reviews play a crucial role by offering systematic or semi systematic approaches to collecting and synthesizing existing knowledge. These reviews not only enhance the understanding of research areas but also help identify knowledge gaps, thereby enabling the development of innovative concepts in relevant fields (Snyder, 2019).

To identify such gaps, various literature search platforms were utilized. These platforms offer extensive collections of academic publications and allow for structured and targeted exploration of the literature, often via integrated search engines (Goldenstein et al., 2018). In this study, the primary sources of information were the Google Scholar platform and the Orbis Plus library search engine provided by the University of Oldenburg.

The search strategy followed a structured process, beginning with an initial screening of available sources. This included consultations with subject matter experts and keyword-based searches using broadly defined terms to ensure wide coverage. In addition, citation-based analysis was employed to assess the relevance and impact of publications under the assumption that highly cited works are typically of greater significance (Goldenstein et al., 2018).

As part of the literature review, online portals and chatbots were investigated. Search terms such as “information bots,” “environmental information,” and their German equivalents were used to identify studies focusing on the communication of environmental data via online platforms and chatbot systems.

Numerous studies and projects address thc development of chatbots in environmental contexts. These explore whether chatbots can contribute to behavioral changes toward a more sustainable lifestyle. Various implementations have been tested with the aim of encouraging environmentally friendly habits (Åberg, 2017). In other cases, chatbots have been evaluated for their potential to increase participation in environmental protection efforts or to facilitate donations to conservation projects (Chi, 2024).

Within organizations, chatbots have also been used to assist employees in complying with environmental regulations (Thimm & Schmidt, 2022) or to promote eco friendly behavior in the workplace (Hillebrand & Johannsen, 2021). Finally, dedicated frameworks and methodologies have been developed to create chatbots tailored to the needs of environmental risk communication and management (Clímaco et al., 2022).

In the broader domain of environmental information systems, several studies have been conducted using online portals and web applications. For instance, one project aimed to present large-scale environmental datasets in formats that are understandable and accessible to non-expert users. It concluded that the integration of citizen science into environmental data management can be both successful and impactful (Tang et al., 2017).

Furthermore, environmental monitoring has been incorporated into smart city initiatives. These projects often rely on air quality measurements to assess impact and feasibility, with several studies focusing on technical implementation and development strategies (Garau et al., 2020; Rubi and de Lira, 2021).

Despite these efforts, there remains a clear demand for more effective public access to environmental data in Germany. Many of the publicly available environmental portals are technologically outdated and lack user-centered features such as personalization (Akyol et al., 2023). In response, a new environmental information portal was developed that actively integrates citizen participation (Akyol et al., 2022). This participatory model offers the potential to significantly expand the volume and granularity of collected data, thereby enabling more accurate and localized insights into environmental conditions.

3 Environmental Information Chatbots and Systems

Several chatbot systems have been developed to support environmental awareness. EcobotAI, developed in collaboration with the United Nations, offers users practical environmental tips and information about EU sustainability programs. It also assists companies in meeting environmental regulations (EcoBots, 2023). Another example is the KarlA Chatbot, developed as part of the "Umweltsuche 4.0" research project. It provides location specific information about local flora and fauna in the Karlsruhe Rhine floodplains and surrounding nature conservation areas (Nazka, xxx).

In the field of air quality monitoring, the UBA Air Quality App, published by the German Environment Agency, provides hourly updates on pollutant levels across Germany. The app includes interactive maps, air quality indices, and push notifications for real-time alerts (Körösi, 2025).

A private sector alternative is AirQ, which combines data from professional grade measurement stations with user submitted data, offering mobile access and visualization through a dedicated app (Körösi, 2025).

3.1 Citizen Science

Citizen Science refers to the active participation of the general public in scientific research (Fig. 2). Citizens regardless of formal training contribute to tasks such as data collection, monitoring, and analysis. In computer science, the spread of digital tools such as social media and messaging platforms has significantly enhanced opportunities for citizen participation (Philipp & Schmohl, 2021).

The European Citizen Science Association (ECSA) has defined a set of principles to guide the development of citizen science projects. These include:

- Citizens must be actively involved in the research process.
- The outcomes of the research must be scientifically validated.
- Both researchers and participating citizens should benefit from the project.
- Transparency must be ensured in both data handling and the communication of results.

Participants should be clearly informed about the scope of their contributions and should receive appropriate recognition for their involvement (Robinson et al., 2018).

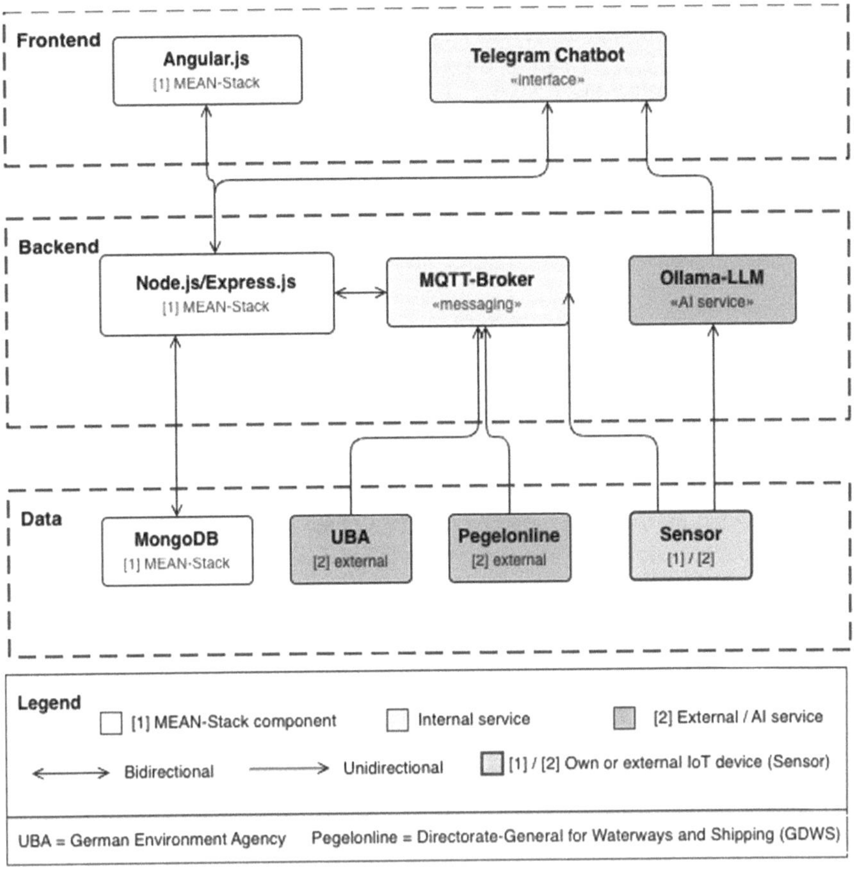

Fig. 2 Model of the complete system (own figure)

4 Concept

4.1 Technology Stack

The MEAN stack is the technology stack used, consisting of the components MongoDB, Express.js, Angular.js, and Node.js. The foundation of the stack is the JavaScript programming language. In terms of content, the technology stack includes a NoSQL database (MongoDB), a web server framework (Express.js), a web client framework (Angular.js), and a web server platform (Node.js). The strength of the MEAN stack lies in the ability to use and maintain all the necessary technologies, libraries, and frameworks. In addition to the central development foundation based on JavaScript, made possible through the use of open-source software projects, the stack is characterized by efficient onboarding and application development, as well as by fulfilling requirements such as scalability (Dunka et al., 2018).

Table 1 Air pollutants with abbreviations and full names (Ittershagen, 2023)

Pollutant	Pollutant
PM10–Particulate matter (coarse)	CO–Carbon monoxide
O_3–Ozone	SO_2–Sulfur dioxide
NO_2–Nitrogen dioxide	Pb–Lead in particulate matter
BaP–Benzo(a)pyrene in PM	C_6H_6–Benzene
PM2.5–Particulate matter (fine)	As–Arsenic in particulate matter
Cd–Cadmium in particulate matter	Ni–Nickel in particulate matter

4.2 *Data and Data Quality*

Environmental Interfaces As part of the project, data is retrieved from public interfaces (APIs). The first of the two interfaces is the UBA API for air data. The available data is provided by the German Environment Agency (UBA) and can be accessed directly through its portal. In addition, the individual data points are explained there (Ittershagen & Luftdaten, 2024).

The present interface has been developed in its third version since 2024, while the collection of air data has been ongoing since January 1, 2016 (Table 1). The recorded pollutants are determined based on various sources, which are listed as follows:

The data can appear in various forms, such as daily mean values or one-hour mean values. The Air Quality Index (LQI) is also relevant for determining air quality. The LQI is derived from the hourly measured pollutant concentrations at the measurement stations and provides an assessment. The pollutants listed above can be retrieved based on the measurement station where they were recorded. Moreover, it is possible to request data over longer periods, such as an entire year (Ittershagen, 2023).

The second environmental interface is the PEGEL API of the Federal Waterways and Shipping Administration, which provides data on the waterways of the German federal territory. These are raw data collected over the last 30 days. Forecasts and predictions for floods are not freely accessible through the API, nor are data older than 30 days. The raw data of the API include metadata about the measurement stations where they were collected, as well as time data such as the timestamp of the last measurement and the duration since the last measurement. The most important raw data are the water levels, which are indicated in meters. In addition, the interface includes further data, such as the length of the water body at a measurement station (Gottschall, 2024).

4.3 *Environmental Information Portal*

The developed concept has already been discussed in several scientific publications. Compared to conventional governmental environmental information systems, the

Environmental Information Portal offers several significant added values. A key advantage is the personalization of the provided environmental information. Traditional systems do not allow users to create profiles or subscribe to specific air quality monitoring stations. As part of this work, compact air monitoring stations were also developed, which communicate with the portal. This enables users not only to retrieve environmental data but also to actively contribute by providing their own measurements. Another added value lies in the implementation of different user roles, allowing companies or public authorities to collaboratively and decentrally create and manage environmental reports (Akyol et al., 2022, 2023; Akyol and Marx, 2021; Akyol et al., 2021). Since the last development stage, two major features have been added to the portal. First, a new communication channel has been integrated, enabling users to receive real-time updates directly on their smartphones. Second, a Telegram bot was developed, allowing users to report regulated plant species directly to the portal. Both features help lower usage barriers and enhance the portal's overall usability.

Furthermore, a self-hosted Ollama application with the LLaMA3 model now allows users to interact with an AI-based assistant that can provide help in case of uncertainties or questions (Ollama, 2025).

4.4 Telegram Bot with Large Language Model API

The use of chatbots in messenger services is associated with certain advantages. A key benefit is that chatbots typically do not need to be installed or downloaded, as they are usually integrated within the usage platform. Another phenomenon to consider in this context is so-called application fatigue. This term describes the fact that smartphone users generally use only a few applications and are rarely inclined to download a large number of additional apps. Using a chatbot inside a messenger service circumvents this issue (Hu, 2019).

Telegram ranks highly among the most used social networks based on monthly users, holding eighth place with 900 million monthly users. This places it just behind Facebook Messenger, which records 1010 million monthly users (Kemp, 2024). When focusing specifically on messenger apps, without including social networks as a whole, Telegram ranks fourth behind WhatsApp, WeChat, and Facebook Messenger (Kemp, 2025). Telegram has experienced a strong increase in activity. While at the beginning of 2014 the number of monthly users was 35 million, it rose to 800 million by 2023 (Doffman, 2024). According to the platform operators at Telegram.org, 400 million people use bots and mini apps (Team, 2024). The previously mentioned reasons, such as the large user base and the phenomenon of application fatigue, already illustrate why Telegram was chosen as the platform for the chatbot. In addition, the creation and use of a Telegram bot is easier compared to alternatives offered by Meta, like WhatsApp and Facebook (Telegram, 2025).

4.5 Development of the Telegram Bot

The development of a Telegram bot begins with a request to the BotFather a Telegram bot responsible for managing other Telegram bots (TelegramOrg, 2025). Once the access token is received from BotFather, the Telegram Bot API can be used to implement the bot. With the API and token, it is possible to create a functional bot even with limited programming experience. Subsequently, a suitable library for interacting with the Telegram Bot API must be chosen. An overview of available libraries and their supported programming languages can be found at https://core.telegram.org/bots/samples. For Node.js, for instance, the node-telegram-bot library is available and was used in this project (Telegram, 2025).

User interaction with the bot primarily occurs via text messages, similar to standard chatbots. In addition to processing text input, Telegram bots can also execute predefined logic through commands that begin with a slash ("/"). These commands can be triggered manually or selected using custom buttons and predefined keyboards. A further interaction option is the Telegram menu: by tapping the menu icon to the left of the input field, users can access a list of available commands.

Telegram bots can also operate within group chats (TelegramOrg, 2025). Furthermore, the bot can be addressed directly in a one-on-one chat. In this project, the bot communicates with a self-hosted Ollama large language model (LLM), which generates responses in conversational form. Beyond simple messaging, the bot enables users to send reports and images such as photos of invasive neophytes or other plants and insects. These submissions can be enriched with GPS data and displayed on an associated online portal. In theory, such reports could be forwarded automatically via email to relevant government authorities. The city of Oldenburg, for example, has expressed interest in receiving such submissions concerning "Eichen-Prozessionsspinner" moths (Onnen, 2025). Before any images are published on the portal, they are filtered by the LLM to ensure they are safe for work. In addition, the system supports integration with environmental sensor stations. Users can log into the online portal using their Telegram credentials and subscribe to specific stations via an interactive map. The data from these subscriptions is then sent to the chatbot. Users can modify or cancel their subscriptions directly within the chat interface via confirmation messages or commands. As a result, users receive real-time updates on air quality and other relevant environmental data from their selected locations.

5 Evaluation

For the evaluation of the chatbot and the online portal, an attempt was made to conduct a representative survey (Schumann, 2012). In accordance with the Likert method, a response scale was chosen that allows participants to express their level of agreement or satisfaction. A five point response scale was selected, enabling participants to indicate their opinion on a continuum ranging from "very satisfied" to "very dissatisfied" (Schumann, 2012). The measurement instrument used in this survey was a test, specifically a software usability test (Singh, 2011).

The evaluation was initiated by distributing the chatbot to individuals from the surrounding environment, most of whom had little or no prior experience in the IT field. Care was taken to include a balanced sample of older individuals (aged 50 and above) and younger participants (between 18 and 40 years of age). Participants were instructed to test the Telegram bot and the associated online portal, and subsequently complete the survey via Google Forms.

The first question addressed overall satisfaction with the Telegram bot. The majority of respondents indicated that they were "rather satisfied," with responses trending slightly above the neutral midpoint. The second question focused on the integration with the online portal; here, "rather dissatisfied" was selected more frequently than "satisfied", although most participants maintained a neutral stance. Additional questions related to other features showed a generally neutral trend. However, the functionalities related to uploading reports and images, as well as subscribing to sensor stations, received more positive ratings. This suggests that these specific features sparked greater interest among users.

Nevertheless, caution is advised when interpreting these results: this was not a controlled or clinical experiment. Furthermore, the small number of participants (14) is insufficient for producing statistically representative conclusions. A larger and randomly selected sample would enhance the validity and generalizability of the evaluation. Additionally, removing the neutral response option in future surveys could encourage participants to express a more definitive positive or negative opinion, thereby producing more insightful data.

6 Conclusion

This paper explored a novel approach to delivering environmental information by integrating a Telegram bot with an online environmental information portal. The system leverages public APIs and messenger-based communication to reduce participation barriers and promote greater citizen involvement in environmental monitoring.

To answer SRQ1, various citizen science projects and environmental IT systems were reviewed. While a number of platforms already exist, many are outdated and lack personalization and real-time interaction features. SRQ2 revealed that existing Telegram bots in this field tend to be limited in scope and are often restricted to narrow topics or regions. The proposed solution addresses these gaps by combining sensor data, user interaction, and chatbot functionality within a single system. Regarding SRQ3, the evaluation suggests that this approach has potential to increase citizen participation, especially through features such as direct reporting and subscriptions to both public and local sensor stations. Although the sample size for the user survey was small, feedback was generally positive, particularly regarding the bot's usability and interactive features.

In conclusion, integrating messenger-based bots into environmental information systems represents a promising step toward more inclusive and effective citizen science. However, two key limitations of the current prototype must be acknowledged. The first concerns scalability: the prototype was tested with a small group of users, and the system must be adapted to handle larger user volumes. The second limitation relates to functionality: although the foundation is in place, there are many additional interaction possibilities that have not yet been implemented, such as automatic forwarding of reports to local authorities.

Future work will focus on scaling the system, improving AI-based support, and ensuring content moderation for public use through the implementation of an upload filter. Possible approaches to content filtering include asynchronous uploads pending manual review, or automated review mechanisms powered by artificial intelligence.

Further development may also include the integration of personal environmental sensors. The Sensor.Community project offers detailed instructions for building such sensors, along with an API for publicly sharing the collected data (Zschiegner, 2020). Integration with the existing MQTT infrastructure would allow user built sensors to be displayed on the portal map and connected to the Telegram bot thus enabling personalized environmental alerts for users subscribed to those sensors.

References

Åberg, J. (2017). Chatbots as a mean to motivate behavior change: How to inspire pro-environmental attitude with chatbot interfaces.

Agency, G. E. (2025). Portal umwelt-bw. https://l.uol.de/german-environment-agency. Accessed 21.05.2022.

Akyol, A., Dorkaev, E., & Gómez, J. M. (2023). Determination of citizen groups and added value for a daily-integrated environmental information portal. In *EnviroInfo* (pp. 205–218). Gesellschaft für Informatik eV.

Akyol, A., Helbig, R., Höveling, S. V., Ghorashi, H., & Solsbach, A. (2021). Developing an environmental information portal for various user groups. In *Resilience, Entrepreneurship and ICT: Latest Research from Germany, South Africa, Mozambique and Namibia* (pp. 313–329).

Akyol, A., & Marx Gómez, J. (2021). Entwicklung eines neuartigen umweltportals mit bürger_innenforschung zur beseitigung von barrieren an umweltinformationen. In *INFORMATIK 2021* (pp. 33–44). Bonn: Gesellschaft für Informatik.

Akyol, A., Solsbach, A., & Marx Gómez, J. (2022). Entwicklung eines bedarfsgerechten umweltinformationsportals mit bürgerbeteiligung. In *INFORMATIK 2022* (pp. 843–854). Bonn: Gesellschaft für Informatik.

Bakolis, I., Hammoud, R., Stewart, R., Beevers, S., Dajnak, D., MacCrimmon, S., et al. (2021). Mental health consequences of urban air pollution: prospective population-based longitudinal survey. *Social psychiatry and psychiatric epidemiology,56*, 1587–1599.

Bickel, S. (2022). Invasive neophyten in deutschland: Gebietsfremde pflanzen-schön und gefährlich? *Biologie in unserer Zeit-BiuZ,52*(4), 379–388.

BMU. (2018). Umweltinformation. https://www.bmu.de/themen/bildung-beteiligung/umweltinformation/. Accessed 13.12.2020.

Bondy, M., Roth, S., & Sager, L. (2020). Crime is in the air: The contemporaneous relationship between air pollution and crime. *Journal of the Association of Environmental and Resource Economists,7*(3), 555–585.

Braithwaite, I., Zhang, S., Kirkbride, J. B., Osborn, D. P., & Hayes, J. F. (2019). Air pollution (particulate matter) exposure and associations with depression, anxiety, bipolar, psychosis and suicide risk: a systematic review and meta-analysis. *Environmental Health Perspectives*, *127*(12).

Buoli, M., Grassi, S., Caldiroli, A., Carnevali, G. S., Mucci, F., Iodice, S., et al. (2018). Is there a link between air pollution and mental disorders? *Environment International,118*, 154–168.

Chi, N. T. K. (2024). The effect of ai chatbots on pro-environment attitude and willingness to pay for environment protection. *SAGE Open*, **14**(1).

Clancy, L., Goodman, P., Sinclair, H., & Dockery, D. W. (2002). Effect of air-pollution control on death rates in dublin, ireland: An intervention study. *The Lancet,360*(9341), 1210–1214.

Clímaco, J., Chávez, T., & Escalante, A. (2022). Conversational ai to improve local environmental risk management. In *2022 IEEE 40th Central America and Panama Convention (CONCAPAN)* (pp. 1–6). IEEE.

Doffman, Z. (2024). *Telegram plays with fire, gets burned–950 million users beware*. https://l.uol.de/forbes-telegram. Accessed 26.02.2025.

Dunka, B. D., Emmanuel, E. A., & Oyerinde, D. O. (2018). Simplifying web application development using-mean stack technologies. *International Journal of Latest Research in Engineering and Technology (IJLRET)*.

EcoBots, I. (2023). *Ai for ecology*. https://ecobots.ai/. Accessed 18.03.2025.

EEA, E. (2020). *Environmental noise in Europe–2020*.

Europäische Umweltagentur: Vorzeitige todesfälle durch luftverschmutzung (2020). [Online; accessed 02.08.2021]

Farooqi, Z. U. R., Sabir, M., Latif, J., Aslam, Z., Ahmad, H. R., Ahmad, I., et al. (2020). Assessment of noise pollution and its effects on human health in industrial hub of pakistan. *Environmental Science and Pollution Research,27*, 2819–2828.

FOEN. (2025). FOEN: (federal-office-for-the–environment) effects of air pollution on health. https://l.uol.de/bafu-effects-of-air-pollution. Accessed 30.04.2025.

Garau, C., Nesi, P., Paoli, I., Paolucci, M., & Zamperlin, P. (2020). A big data platform for smart and sustainable cities: environmental monitoring case studies in Europe. In *International conference on computational science and its applications*. (pp. 393–406). Springer.

German Environment Agency. (2025). https://www.umweltbundesamt.de/en/data/air/air-data. Filter-preseset export: maps+++eJxrXZScv9ByUXHyQsNFKYmrjAyMTHUNjHQNDRaVZAKFissZAPPEDPI=.

Gigon, A., & Weber, E. (2005). *Invasive Neophyten in der Schweiz: Lagebericht und Handlungsbedarf*. ETH Zentrum: Geobotanisches Institut.

Goldenstein, J., Hunoldt, M., & Walgenbach, P. (2018). *Wissenschaftliche (s) Arbeiten in den Wirtschaftswissenschaften: Themenfindung–Recherche–Konzeption–Methodik–Argumentation*. Springer-Verlag.

Gottschall, O. (2024). Pegel online. https://www.pegelonline.wsv.de/gast/start. Accessed 26.02.2025.

Gu, H., Yan, W., Elahi, E., & Cao, Y. (2020). Air pollution risks human mental health: an implication of two-stages least squares estimation of interaction effects. *Environmental Science and Pollution Research,27*(2), 2036–2043.

Heinrich, L. J., Riedl, R., & Stelzer, D. (2014). *Informationsmanagement: Grundlagen*. Aufgaben: Methoden. Walter de Gruyter GmbH & Co KG.

Hillebrand, K., & Johannsen, F. (2021). Klimakarl—A chatbot to promote employees' climate-friendly behavior in an office setting. In *International conference on design science research in information systems and tchnology*(pp 3–15). Springer.

Hu, Y. C. (2019). *Do people want to message Chatbots?: Developing and comparing the usability of a conversational vs. menu-based Chatbot in context of new hire onboarding*. Ph.D. thesis, ETSI_Informatica.

Ittershagen, M. (2023). *Schnittstellenbeschreibung luftdaten-api*. https://l.uol.de/uba-air-quality. Accessed 26.02.2025.

Ittershagen, M. (2024). Luftdaten. https://www.umweltbundesamt.de/daten/luft/luftdaten/. Accessed 26.02.2025.

Jariwala, H. J., Syed, H. S., Pandya, M. J., & Gajera, Y. M. (2017). Noise pollution & human health: A review. In *Noise and air pollutions: Challenges and opportunities*. Ahmedabad: LD College of Eng.

Kemp, S. (2024). *Digital 2024 october global statshot report*. DataReportal: Tech. rep.

Kemp, S. (2025). *Digital 2025: Global overview report*. DataReportal: Tech. rep.

Koos, S., & Naumann, E. (2019). Vom klimastreik zur klimapolitik: Die gesellschaftliche unterstützung der fridays for future-bewegung und ihrer ziele: Forschungsbericht.

Körösi, M. (2025). Messgerät für luftqualität, alarmgeber und smart home device. https://www.air-q.com/produkt. Accessed 18.03.2025.

Lelieveld, J., Klingmüller, K., Pozzer, A., Burnett, R., Haines, A., & Ramanathan, V. (2019). Effects of fossil fuel and total anthropogenic emission removal on public health and climate. *Proceedings of the National Academy of Sciences,116*(15), 7192–7197.

Lu, J. G., Lee, J. J., Gino, F., & Galinsky, A. D. (2018). Polluted morality: Air pollution predicts criminal activity and unethical behavior. *Psychological science,29*(3), 340–355.

Morillas, J. M. B., Gozalo, G. R., González, D. M., Moraga, P. A., & Vílchez-Gómez, R. (2018). Noise pollution and urban planning. *Current Pollution Reports,4*(3), 208–219.

Nazka: Karla—deine chatbot-rangerin (na), https://nazka.de/chatbot-karla, [Online; accessed 18.03.2025]

Ollama. (2025). *Get up and running with large language models*. https://ollama.com/. Accessed 30.04.2025.

Onnen, S. (2025). Informationen zum eichenprozessionsspinner. https://www.oldenburg.de/startseite/leben-umwelt/umwelt/gruenes-oldenburg/oldenburg-gartengrossstadt/informationen-zum-eichenprozessionsspinner.html. Accessed 16.06.2025.

Philipp, T., & Schmohl, T. (2021). Handbuch Transdisziplinäre Didaktik. transcript Verlag.

Resch, J. (2025). Todesfälle in jedem landkreis deutschlands: Deutsche umwelthilfe veröffentlicht erstmals Übersicht zu toten durch luftschadstoffe. https://l.uol.de/duh. Accessed 21.05.2022.

Robinson, L. D., Cawthray, J. L., West, S. E., Bonn, A., & Ansine, J. (2018). *Ten principles of citizen science* (pp. 27–40). UCL Press. http://www.jstor.org/stable/j.ctv550cf2.9 Accessed 26.02.2025.

Rubi, J. N. S., & de Lira Gondim, P. R. (2021). Iot-based platform for environment data sharing in smart cities. *International Journal of Communication Systems*, *34*(2).

Schäfer, A., Mühr, B., Daniell, J., Ehret, U., Ehmele, F., Küpfer, K., Brand, J., Wisotzky, C., Skapski, J., Rentz, L., et al. (2021). Hochwasser mitteleuropa, juli 2021 (deutschland). *CEDIM Forensic Disaster Analysis Group Bericht* (1).

Schneider, K., & Hormann, A. (2011). Frühwarnsystem und konzeption von maßnahmen gegen invasive neophyten in schutzgebieten sachsen-anhalts-zielstellung und erste ergebnisse des projektes. *Naturschutz Land Sachsen-Anhalt (Halle),48*(1/2), 69–75.

Schumann, S. (2012). *Repräsentative Umfrage: Praxisorientierte Einführung in empirische Methoden und statistische Analyseverfahren*. Oldenbourg Wissenschaftsverlag Verlag.

Singh, Y. (2011). *Software Testing*. Cambridge University Press.

Snyder, H. (2019). Literature review as a research methodology: An overview and guidelines. *Journal of business research,104*, 333–339.

Tang, Z., Zhou, Y., Yu, H., Gu, Y., & Liu, T. (2017). Developing an interactive mobile volunteered geographic information platform to integrate environmental big data and citizen science in urban management. In *Seeing cities through big data: Research, methods and applications in urban informatics* (pp. 65–81).

Team, T. (2024). *Telegram sterne: Bezahlung für digitale produkte*. https://telegram.org/blog/telegram-stars/de. Accessed 26.02.2025.

(2025). Telegram: From botfather to 'hello world'. https://core.telegram.org/bots/tutorial. Accessed 01.03.2025.

Telegram. (2025). Telegram bot features. https://core.telegram.org/bots/features. Accessed 01.03.2025.

Thimm, H., & Schmidt, V. (2022). Low-code basierte entwicklung eines chatbots für betriebliches compliance management im umweltbereich. In *INFORMATIK 2022* (pp. 855–866). Bonn: Gesellschaft für Informatik.

Tropical, S., TPH), P.H.I.S. (2025). Healthseffects. https://l.uol.de/swiss-thphealtheffects. Accessed 30.04.2025.

Ullah, S., Ullah, N., Rajper, S. A., Ahmad, I., & Li, Z. (2021). Air pollution and associated self-reported effects on the exposed students at malakand division, pakistan. *Environmental monitoring and assessment,193*, 1–17.

WHO. (2024). *Ambient (outdoor) air pollution*. https://www.who.int/news-room/fact-sheets/detail/ambient-(outdoor)-air-quality-and-health. Accessed 29.04.2025.

Zhang, X., Chen, X., & Zhang, X. (2018). The impact of exposure to air pollution on cognitive performance. *Proceedings of the National Academy of Sciences,115*(37), 9193–9197.

Zschiegner, R. (2020). Sensor.community. https://sensor.community/de/. Accessed 13.03.2025.

An Analysis of the Value of Robotic Process Automation for Healthcare Administration Using the Task-Technology Fit Theory

Phillip Sibanda, Ifeoluwapo Fashoro, and Brenda Scholtz

Abstract Robotic Process Automation (RPA) technology incorporates easy-to-develop and deploy software bots that emulate human interactions with digital systems and software. This paper aims to analyse the value of RPA as a healthcare technology by evaluating to what extent it fits with the characteristics of administrative healthcare tasks. The Task-Technology Fit (TTF) theory was used to analyse articles reporting cases of successful RPA implementation in healthcare. A qualitative thematic analysis process was employed to analyse the collected data. Administrative healthcare task characteristics were classified as routine or non-routine tasks. Patient registration and appointment booking are routine tasks, while data validation and analysis and exception handling are non-routine. The adopted RPA characteristics included attended bots, unattended bots, seamless integration, Optical Character Recognition (OCR) and rule-based exception handling. The findings indicate that RPA provides value to the healthcare industry and is a strong fit for automating administrative healthcare tasks. Streamlining administrative healthcare tasks with RPA enhances operational efficiency, reduces errors and frees time for healthcare professionals by relieving them from the burden of their administrative tasks. These benefits, therefore, enable healthcare workers to focus more on providing quality patient care.

P. Sibanda (✉) · I. Fashoro · B. Scholtz
Department of Computing Sciences, Nelson Mandela University, Gqeberha, South Africa
e-mail: s220888914@mandela.ac.za

I. Fashoro
e-mail: ife.fashoro@mandela.ac.za

B. Scholtz
e-mail: brenda.scholtz@mandela.ac.za

J. Marx Gómez et al. (eds.), *Recent Advances in Information Systems*, Progress in IS,
https://doi.org/10.1007/978-3-032-14099-9_41

1 Introduction

Healthcare organisations generate a lot of patient data every second, yet most of it remains unused or is mismanaged (Wang et al., 2019). Patient data comes from various sources, including health records, medical imaging, insurance and administrative systems (Choi et al., 2023). One of the main challenges with patient data lies in maintaining, managing and processing this data effectively. The continued reliance on outdated manual processes significantly hampers data management, leading to poor data quality, inefficiencies and security vulnerabilities (Du-Harpur et al., 2020). Due to the continued reliance on manual processes, healthcare workers often spend excessive time on administrative tasks such as patient registration, scheduling appointments and handling insurance claims, thereby reducing the time available for direct patient care (Alshahrani et al., 2025). These tasks are particularly labour-intensive, as they result in the generation of vast volumes of patient records, which can be time-consuming to manage (Kedar, 2025). Manually managing tasks such as appointment scheduling can lead to operational inefficiencies, including long wait times, overbooking and underutilisation of resources. Errors in claims processing and missing data, often caused by inadequate documentation, further compound these challenges (Domb et al., 2024).

As traditional data management processes increasingly fail to meet the demands of modern healthcare, there is a clear need for more advanced and efficient data management methods and, thus, automation. Robotic Process Automation (RPA) has transformed industries by automating repetitive, manual tasks with software bots. RPA has seen widespread adoption across various industries, particularly in finance, retail and commerce.

RPA has emerged as a transformative technology in the healthcare sector, particularly in the automation of administrative tasks (Hofmann et al., 2020). Processes such as patient data entry, claims processing and appointment scheduling are now being automated through RPA solutions, allowing healthcare professionals to focus more on clinical responsibilities and patient interaction (Patrício et al., 2024).

RPA uses specialised software to design and manage task workflows, which are executed by software robots that mimic human interactions with digital systems. The global market for RPA in healthcare was valued at $1.4 billion in 2023 and is projected to grow at a Compound Annual Growth Rate of 26.1% from 2023 to 2033, reaching an estimated value of $14.18 billion by 2033 (Precedence Research, 2025). RPA technology integrates advanced tools such as Optical Character Recognition (OCR), Natural Language Processing (NLP) and Machine Learning (ML), enabling automation of both rule-based and complex workflow tasks involving structured and unstructured data. This integration supports the automation of a wide range of administrative tasks, including patient registration, appointment booking and insurance verification (Huang et al., 2024). Increasingly, healthcare facilities are adopting RPA to automate repetitive, manual, error-prone and routine administrative processes (Patrício et al., 2024).

Existing research lacks insight into the extent of fit or the value of RPA for administrative healthcare tasks. This paper aims to analyse the value of RPA as a healthcare technology by evaluating to what extent it fits with the characteristics of administrative healthcare tasks. The Task-Technology Fit (TTF) theory was used to evaluate the value and fit. The TTF theory was constructed by Goodhue and Thompson (1995) and explains the use of technology by assessing its fit and value for a user's tasks and needs.

The structure of this paper is as follows: Sect. 2 presents the TTF theory and its constructs. Section 3 presents the research design, Sect. 4 analyses the literature findings and Sect. 5 provides a discussion of the findings. Section 6 concludes the paper and provides recommendations for future research.

2 Theoretical Framework

The TTF theory by Goodhue and Thompson (1995) provides a lens of how technology use creates value. The theory suggests that the performance and ultimate value or fit of the technology are measured by how well the characteristics of the tasks align with the characteristics of the technology, allowing the user to perform the tasks. The theory has been successfully used in the healthcare context to evaluate an Electronic Health Record (EHR) system in a healthcare organisation in South Dakota, USA (El-Gayar et al., 2010). The results of the evaluation confirmed the validity of the TTF theory and supported it for predicting performance impact and value in a healthcare setting. The version of the TTF theory used in this research, as illustrated in Fig. 1, was the version used by Spies et al. (2020). This version does not include the concept of utilisation as originally defined by Goodhue and Thompson (1995). The paper adopted the simpler version of the theory used in a healthcare context, since our focus is on the fit and value of RPA technology in administrative healthcare tasks, hence it still remains suitable for evaluating technology implementations within the healthcare industry.

The four constructs of the TTF theory (Spies et al., 2020) provide a comprehensive structure for understanding how well technology supports users' tasks and how this fit influences technology use, performance and, ultimately, value. These constructs are:

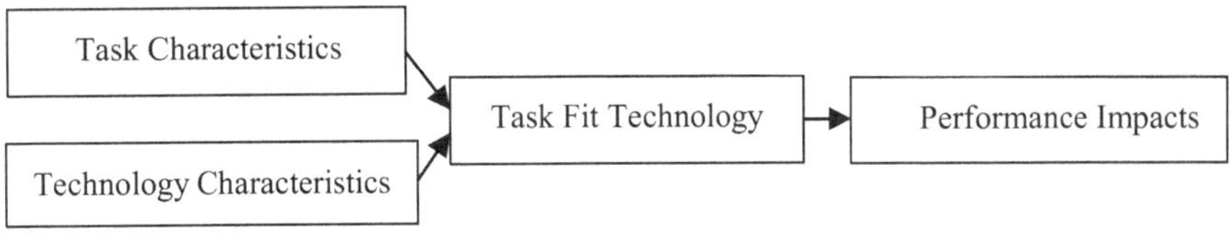

Fig. 1 Task-Technology Fit framework (Spies et al., 2020)

- Task Characteristics, which refer to the actions and processes individuals perform within a specific setting. They are measured by routine, non-routine and interdependent dimensions. Routine tasks are structured and repetitive, while non-routine tasks are complex, unpredictable and variable. Task interdependence involves the reliance of multiple tasks on each other to achieve a goal.
- Technology Characteristics, which refer to technology-specific attributes or functions.
- Task-Technology Fit, which refers to how much the technology supports the tasks. It is measured by determining quality, compatibility, system reliability, authorisation, capability and ease of use.
- Performance Impacts, which refers to the impact of technology use on individual and organisational performance.

3 Research Methodology

The study adopted an exploratory approach using qualitative data. A comprehensive literature review was conducted, which included a thematic analysis guided by the method proposed by Braun and Clarke (2012). The literature review was conducted across three databases: IEEE Xplore, PubMed and Google Scholar and supplemented by websites. The search employed targeted keywords such as "administrative healthcare automation," "RPA in healthcare" and "TTF constructs." The inclusion criteria for the literature review focused on journal articles, academic books, conference proceedings, or websites. Only publications from the period 2019–2025 were considered. The literature had to be relevant to the healthcare industry's application of RPA, with a particular emphasis on administrative process automation. Only sources that were available in full-text format and published in English were included. The exclusion criteria eliminated articles focusing on non-healthcare applications of RPA or automation technologies.

From the search process, 40 records were identified. A screening process was then conducted, during which duplicate entries and the ones that did not fit the inclusion criteria were removed. This process resulted in 25 relevant sources being included in the review. Of these, 17 were published journal articles, while eight were sourced from websites. The inclusion of website sources was necessitated by the limited availability of detailed and current information on specific aspects of RPA within the academic literature.

The TTF theory served as the theoretical framework for the analysis. The thematic analysis consists of five steps, which were applied in this study as follows:

- Step 1: The selected articles were thoroughly examined through multiple readings to gain a comprehensive understanding of RPA in healthcare.
- Step 2: Familiarisation and initial analysis were conducted to identify overarching themes and sub-themes, using a priori themes from the TTF framework to guide data organisation.

- Step 3: Initial codes were generated within each theme and sub-theme, representing key aspects of the article data, ensuring a structured and theory-driven approach aligned with the identified themes.
- Step 4: Codes were reviewed iteratively to eliminate duplication, ensure relevance to the theoretical constructs and confirm alignment with the correct themes.
- Step 5: The final step involved writing up the analysis, presenting the themes and codes in a coherent narrative that highlighted their connections and the practical implications of RPA in healthcare.

4 Analysis of Literature

The top-level themes used in the thematic analysis were the four a priori constructs of the TTF theory: task characteristics, technology characteristics, Task-Technology Fit and performance.

4.1 Task Characteristics of Administrative Healthcare Tasks

According to the TTF, task characteristics can be categorised as routine, non-routine and interdependent (El-Gayar et al., 2010). None of the 25 articles reviewed referenced interdependent characteristics. As a result, only routine and non-routine categories were used as a priori sub-themes under the broader theme of task characteristics. These sub-themes and their corresponding codes are summarised in Table 1. From the 25 articles, eight were related to healthcare task characteristics, six of which were journal articles and two are from websites (indicated with an asterisk in the table).

The two tasks identified as routine were patient registration and appointment management. The study of Kedziora and Smolander (2022) observed that a healthcare centre in Finland had to adjust to new circumstances by modifying routine tasks, particularly during COVID-19 testing. The process involved patient registration, which was identified as a routine administrative task due to the need to complete

Table 1 Task characteristics in healthcare administration ($n = 8$)

Sub-theme (a priori)	Code (Task identified)	Authors
Routine tasks	Patient registration Appointment management	Kedziora and Smolander (2022), Venigandla and Tatikonda (2023), Formstack (2025)* Smits (2023), Unitek College (2025)*
Non-routine tasks	Data validation, analysis Handling exceptions	Venigandla and Tatikonda (2023), Nath and Saha (2023), Dash et al. (2019) Nath and Saha (2023), Azadi et al. (2023)

a registration form. Venigandla and Tatikonda (2023) also identified patient registration as a routine process within clinical data management, characterised by repetitive data entry. At Indiana Health Group, routine patient registration involves distributing registration forms to patients, who are expected to complete and return them before their scheduled appointments (Formstack, 2025). Appointment management is similarly considered a routine administrative task in healthcare settings. Smits (2023) observed that this process typically requires a healthcare worker to search for available time slots and secure appointments. Appointment management includes several sequential steps: receiving the appointment request, verifying patient information, checking provider availability, scheduling the appointment and sending a confirmation (Unitek College, 2025). This task is often followed by reminders and the distribution of any pre-visit documentation. For non-routine tasks, two tasks were identified: data validation and analysis and handling exceptions. Validating and analysing data, especially large datasets, generated during medical research is highly complex and is considered a non-routine task, as data analysis will always be dynamic (Nath & Saha, 2023; Venigandla & Tatikonda, 2023). Analysing and presenting big data in healthcare presents complex challenges (Dash et al., 2019). Exception-handling tasks, such as correcting inaccurate billing codes or resolving mismatched patient information, introduce an element of unpredictability into healthcare workflows, classifying them as non-routine tasks (Nath & Saha, 2023). These exceptions often require deviations from standard procedures to ensure accurate outcomes. Patient misidentification, in particular, can have serious consequences. Data entry errors significantly contribute to the proliferation of duplicate health records, further complicating patient data management and continuity of healthcare (Azadi et al., 2023).

4.2 Technology Characteristics of RPA

Technology characteristics are the functions that aid users in their tasks (El-Gayar et al., 2010). They vary depending on the technology being used. The thematic analysis revealed three sub-themes related to the theme of RPA's technology characteristics: the type of RPA bot, workflow automation and seamless integration. None of these sub-themes was a priori, as they emerged from the literature. The sub-themes and codes are summarised in Table 2. From the 25 articles, 10 cited the technology characteristics of RPA: five were from journal articles and five were from websites (indicated with an asterisk).

A core feature of RPA technology is the RPA bot or "digital worker" (Nintex, 2024), which is a software-based robot that operates within a digital environment to replicate human interactions with computer systems. It executes a machine-readable sequence of instructions composed of RPA operations, enabling the automation of tasks typically performed by human users. There are two types of bots: Attended bots, which assist with tasks requiring human interaction and unattended bots, which work autonomously to perform tasks without human involvement (Nintex, 2024). The function of attended bots is similar to that of a virtual assistant, which requires human

Table 2 Technology characteristics of RPA (n = 10)

Sub-themes (New)	Codes	Authors
Type of RPA bot	Attended bots Unattended bots	Nintex (2024)*, Pratt (2024)* Doguc (2022), Völker et al. (2021)
Workflow automation	Optical Character Recognition (OCR) Rules-based exception handling	Inscribe (2024)*, Mos (2024)* Bédard et al. (2024), Kaya et al. (2019)
Seamless integration	Interacts directly with the GUI Existing applications	SAP (2024)* Liu (2022)

intervention to initiate their processes, which are often triggered by user actions such as clicking a button. Additionally, human interaction may be necessary at various stages throughout the process. These bots serve as personal assistants for end users, aiding them in routine tasks and with specific steps within a larger process (Pratt, 2024).

Unattended bots are pre-configured to function independently according to their specific settings (Doguc, 2022). These bots can be programmed to collect and record data, handle large volumes of repetitive tasks and carry out other essential processes that would normally require significant manual effort and time. They are commonly used to automate back-office operations such as processing transactions, managing data entry, generating reports and performing routine administrative tasks (Völker et al., 2021).

The second sub-theme that emerged relates to workflow automation, and the codes allocated were rule-based, OCR and exception handling. OCR is commonly integrated into RPA software to enable automated text extraction from images and documents (Inscribe, 2024). OCR technology functions by analysing visual patterns that represent characters and numerals, allowing for the accurate conversion of printed or handwritten content into structured, editable formats (Mos, 2024). In a typical RPA workflow, a bot retrieves a document, such as from an email or designated folder, and the OCR engine digitises any scanned or handwritten elements. The integration of OCR within RPA workflows facilitates the rapid processing of large volumes of data, significantly enhancing overall operational efficiency.

Rule-based exception handling in workflow automation enables RPA systems to incorporate a rule-based exception handling mechanism that facilitates the proactive management of errors and anomalies (Bédard et al., 2024). RPA bots with ML capabilities can learn from historical data and improve their decision-making processes over time. These capabilities enable the bots to handle more complex tasks, such as predicting outcomes, identifying patterns and making data-driven decisions. RPA bots can adjust to unexpected issues by recovering from errors in a controlled manner, allowing them to continue their operations without significant interruption (Kaya et al., 2019).

The third sub-theme was seamless integration. Only one code was identified here: seamless integration with other systems, since RPA serves as a workaround for integrating systems. RPA operates by interacting directly with the graphical user interface

(GUI) of applications, mimicking human actions such as clicking and typing. As a result, developers do not need to build Application Programming Interfaces (APIs) or modify the underlying systems to enable communication between different software platforms (SAP, 2024). Instead, the software bot navigates between applications in the same manner that a human user would. Radke (Liu, 2022) emphasises that RPA systems can integrate seamlessly with industry platforms and applications, including billing systems and administrative software. This integration ensures a smooth flow of data and the integrity of the information.

4.3 Task-Technology Fit of RPA for Administrative Healthcare Tasks

To evaluate the criteria related to the fit between the common administrative healthcare tasks identified in Table 1 and the RPA technology characteristics outlined in Table 2, articles providing empirical use cases of RPA implementation in healthcare administration were analysed. The results of the TTF of RPA for routine and non-routine administrative healthcare tasks are presented in a model in Fig. 2. From the original 25 articles reviewed, nine met the criteria of data quality, compatibility and reliability. Of these, five focused on RPA for routine administrative healthcare tasks; these are (Barla et al., 2023; Gal-Nadasan et al., 2023; Kedziora & Smolander, 2022; Thainimit et al., 2022; Venigandla & Tatikonda, 2023). Two of these (Kedziora & Smolander, 2022; Venigandla & Tatikonda, 2023) were in Table 1 and also met the criteria of a use case of RPA for healthcare administration; therefore, they were also included in the analysis. Five articles from the 25 addressed RPA for non-routine administrative healthcare tasks; these are (Gal-Nadasan et al., 2023; Huang et al., 2024; Smits, 2023; Soeny et al., 2021; UiPath, 2024). The RPA for non-routine administrative healthcare tasks are indicated in bold and italic text in the model. One article, (Gal-Nadasan et al., 2023), covered both RPA for routine and non-routine administrative healthcare tasks. The selected articles were further analysed using three dimensions of fitness or value as proposed by the TTF theory (El-Gayar et al., 2010): compatibility, reliability and quality. These dimensions were chosen because they were the most frequently mentioned across the use case studies reviewed.

Routine administrative tasks such as patient registration and appointment management have been successfully automated in healthcare by adopting RPA. Gal-Nadasan et al. (2023) developed an RPA software bot specifically designed to digitise and extract data from handwritten patient forms, offering substantial benefits to small medical clinics in Romania that relied on paper-based systems. The first step in developing the RPA bot was to establish a form taxonomy for accurate data extraction. The document was then digitised using OCR technology to create a digital object model (DOM) file, which formed the basis for further data extraction. The digitised document was classified as a medical form using keyword-based classifiers to ensure the data extraction was performed on the correct document type. After

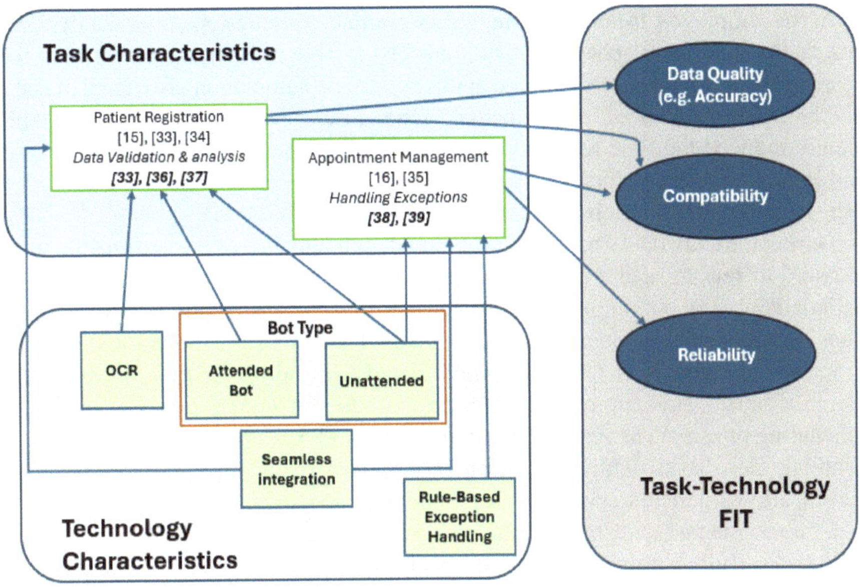

Fig. 2 Task-Technology Fit of RPA for routine and non-routine administrative healthcare tasks

RPA implementation, the software monitored key performance indicators, including data accuracy and confidence levels. The results demonstrated a 100% confidence level in the data extraction from capitalised handwriting, ensuring high-quality data production.

In a study by Kedziora and Smolander (2022), individuals arriving in Finland from other countries had to undergo COVID-19 testing. Individuals entered their details into the Finentry portal system. The integration of RPA with the portal automated the entire process by generating a temporary personal identity number and creating a new patient profile. It was found that RPA was fast and highly compatible with the portal system, effectively streamlining patient registration and releasing. RPA implementation released health workers from time-wasting administrative work.

In the United States, healthcare companies deployed RPA bots to automate the patient onboarding process (Barla et al., 2023). RPA was integrated to manage an online web form that automatically registered patients. One bot queued requests, and another bot verified the data and completed the registration. The bots handled tasks such as checking if a patient was in the Electronic Medical Records (EMR) system, registering new patients and printing test kit labels. While humans took 2–3 min to complete these tasks, the bots did so in just 14–16 s, greatly improving efficiency and reducing waiting times, thereby demonstrating the compatibility of RPA with the EMR system.

A study in a rural hospital in Thailand demonstrated the successful integration of RPA to streamline and enhance various operational tasks (Thainimit et al., 2022).

RPA was employed to automate repetitive routine activities, such as data collection and appointment scheduling, by transferring patient information between the hospital's ERP system and a mobile application. The automation also handled tasks like notifying patients of test results and scheduling follow-up appointments. This greatly reduced the time required for these tasks compared to traditional methods and improved overall efficiency. This demonstrated the compatibility of RPA with their ERP system and mobile application.

During the COVID-19 pandemic in Finland, individuals seeking vaccination were required to register and enter a queue for appointment scheduling (Venigandla & Tatikonda, 2023). An automated bot managed this process by autonomously coordinating vaccination appointments with nurses. The bot also sent SMS notifications to inform individuals of the appointment details, including the time, location and assigned nurse. This approach eliminated the need for human involvement in the scheduling process. The results demonstrated that RPA was efficient and reliable in handling these tasks, which significantly alleviated the administrative burdens on nurses, allowing them to dedicate more time to patient care.

Non-routine tasks, such as validation, analysis and handling of exceptions, have been successfully automated in healthcare through the adoption of RPA. A company in India identified that manually processing prescriptions was time-consuming and error-prone, so they implemented an RPA system which leveraged an attended bot (Soeny et al., 2021). This bot used the Google Vision API (OCR) system to extract key patient health information from uploaded prescription images. It verified the patient details and the ordered medications and checked the dosage and prescription details against the order. The extracted data was analysed by employees using a screening tool, with AI models predicting and filling in missing details. Employees verified and corrected discrepancies, ensuring high-quality and accurate data. The bot led to significant improvements in processing time, with orders being processed on average 8.4% faster in the RPA-assisted pipeline compared to the manual process.

In Smits' (2023) case study, RPA was successfully integrated into healthcare systems to handle the administration related to monitoring heart conditions remotely. When abnormalities were detected, based on predetermined threshold values, alerts were triggered by the RPA bot, and patients were promptly sent to the doctor. RPA ensured compatibility through its ability to consistently filter, validate and process remote data, precisely helping to sort and analyse the collected data.

In Taiwan's National Health Insurance system, healthcare providers faced challenges in submitting accurate medical claims due to complex rules and frequent policy changes (Huang et al., 2024). Employees manually reviewed reports to identify errors, such as missing information or incorrect billing codes. RPA was implemented to automate claims handling through rule-based exception handling, including report opening, data filtering, abnormality verification and incident management. This automation reduced claim errors, improved anomaly handling and significantly enhanced processing speed and overall efficiency. RPA was reliable in handling these tasks. ApprioHealth faced challenges managing a high volume of healthcare claims, requiring extensive data entry and interaction with multiple digital systems (UiPath, 2024). To address this, they implemented RPA with OCR, enabling the system to adapt to interface changes and apply rule-based exception handling for specific errors. As a result, four bots reliably managed seven times the number of claims previously handled by four employees. This automation reduced the claims backlog by 96%, significantly improving client financial performance during a critical period.

5 Discussion

This study, guided by the TTF theory analysed the value of RPA as a healthcare technology by evaluating how it fits with the characteristics of administrative healthcare tasks. The paper's contribution lies in its comprehensive categorisation of healthcare administrative tasks and mapping them to specific RPA capabilities. The study evaluated the RPA technology characteristics with healthcare administrative tasks and confirmed a strong fit of RPA in healthcare administrative tasks. The research identified patient registration and appointment management as routine administrative tasks, while data analysis, validation and exception handling were identified as non-routine tasks. The primary technology characteristics of RPA are bots (attended and unattended), workflow automation features (OCR and rule-based exception handling) and integration capabilities. From the use cases, it is clear that RPA has been implemented in healthcare systems to automate routine and non-routine administrative tasks by leveraging its characteristics. RPA bots can perform administrative tasks with or without human intervention, OCR can extract information from documents to make it digitally readable, and integration capabilities enable RPA to work effectively with existing healthcare systems such as EHRs. Rule-based exception handling with ML capabilities provides the ability to handle complex tasks and errors. These characteristics make RPA well-suited for ensuring quality data, compatibility and reliability in automating routine and non-routine administrative healthcare tasks. RPA's ability to automate routine and non-routine tasks has also led to significant time savings, a decreased administrative burden on healthcare workers and the minimisation of errors and fraud commonly associated with manual data entry and improved operational efficiency in healthcare facilities. The fit of RPA to the administrative healthcare tasks ultimately leads to improved productivity and value to the users. The TTF theory was found to be a useful tool to evaluate the characteristics of RPA technology and its fit and value for the healthcare context. The findings provided more insight and understanding of how healthcare administration tasks are suitable for RPA automation. While RPA offers clear benefits in healthcare, its adoption also presents notable challenges. Regulatory compliance is a concern, as RPA bots handle sensitive data like patient records and billing information, raising issues around data security and adherence to standards such as HIPAA (Sharma, 2023). Many healthcare organisations operate with legacy systems that are difficult to integrate with modern RPA tools. Additionally, implementing RPA requires specialised technical skills, often lacking in clinical settings (Lin et al., 2025). High initial costs also pose barriers, particularly for public or resource-limited facilities. Lastly, RPA adoption may cause staff concerns about job security and unclear roles, as automation can shift or eliminate manual tasks. Although these points were not covered in the literature review process, they are included here to support a broader discussion on the potential downsides of RPA implementation.

6 Conclusion

The analysis of nine use cases revealed that the adoption of RPA for routine and non-routine administrative healthcare tasks has confirmed the technology's compatibility, reliability in execution and ability to generate high-quality data. It is evident from these use cases that RPA provides significant value to stakeholders in the healthcare industry. However, despite its benefits, RPA adoption can present challenges. The research was exploratory and relied on secondary data from existing research articles, which may introduce a degree of bias. However, it still provides good insights into the characteristics of administrative tasks in healthcare facilities and where and how RPA has been successfully implemented in this context. The fit of RPA for healthcare administrative tasks, as outlined in Fig. 2, can be used by other researchers to evaluate additional use cases and provide lessons learned to healthcare practitioners. The paper lays the groundwork for further, more in-depth investigations, urging researchers to expand upon this foundation. It is recommended that future research provide additional evidence of RPA adoption within healthcare settings. For example, researchers should test the applicability of RPA in an actual healthcare setting, evaluate its performance and measure the outcomes across various operational metrics. This practical approach will provide even deeper insights into how TTF can improve the understanding of RPA implementation. Adopting this approach could also provide valuable insights into tailoring RPA to address local healthcare challenges and to enhance efficiency. Improving the administrative healthcare tasks can solve some of the challenges experienced with legacy systems, such as inconsistent and missing patient data. As a result, healthcare professionals can spend less time on administration and more time focusing on improving the healthcare service provided to patients.

References

Alshahrani, M. A. F., et al. (2025). Exploring the role and efficiency of automation in hospital administration. *Journal of International Crisis and Risk Communication Research, 8*(S2), 71.

Azadi, A., & García-Peñalvo, F. J. (2023). Optimizing data entry management in healthcare: Leveraging HCI to enhance medical decision accuracy. In P. Zaphiris & A. Ioannou (Eds.), *TEEM 2023, LNCS* (Vol. xxxx, pp. 271–279). Springer.

Barla, N. H., Almeida, S. M., & Almeida, M. S. (2023). RPA revolution in the healthcare industry during COVID-19. In *Confluence of artificial intelligence and robotic process automation* (pp. 199–229). Springer.

Bédard, M., et al. (2024). A rule-based method to effectively adopt robotic process automation. *Journal of Software: Evolution and Process, 36*(1), Article e2709.

Braun, V., & Clarke, V. (2012). *Thematic analysis*. American Psychological Association.

Choi, J., Park, H., Chie, E. K., Choi, S. W., Lee, H. Y., Yoo, S., Kim, B. J., & Ryu, B. (2023). Current status and key issues of data management in tertiary hospitals: A case study of Seoul National University Hospital. *Healthcare Informatics Research, 29*(3), 209–217.

Dash, S., Shakyawar, S. K., Sharma, M., & Kaushik, S. (2019). Big data in healthcare: Management, analysis and future prospects. *Journal of Big Data, 6*(1), 1–25.

Doguc, O. (2022). Robot process automation (RPA) and its future. In *Research anthology on cross-disciplinary designs and applications of automation* (pp. 35–58). IGI Global.

Domb, M., Joshi, S., & Singh, A. (2024). Empowering healthcare administration by embedding robotic process automation. *International Journal of Biomedical Research & Practice, 4*(4), 1–5.

Du-Harpur, X., Watt, F. M., Luscombe, N. M., & Lynch, M. D. (2020). What is AI? Applications of artificial intelligence to dermatology. *British Journal of Dermatology, 183*(3), 423–430.

El-Gayar, O. F., Deokar, A. V., & Wills, M. J. (2010). Evaluating task technology fit and user performance for an electronic health record system. *International Journal of Healthcare Technology and Management, 11*(1–2), 50–65.

Formstack. *Patient registration case study: Indiana Health Group*. https://www.formstack.com/blog/indiana-health-group-case-study. Last accessed May 10, 2025.

Gal-Nadasan, N., Stoicu-Tivadar, V., Gal-Nadasan, E., & Dinu, A. R. (2023). Robotic process automation based data extraction from handwritten medical forms. In S. Vlad (Ed.), *Telehealth ecosystems in practice* (pp. 68–72). IOS Press.

Goodhue, D. L., & Thompson, R. L. (1995). Task-technology fit and individual performance. *MIS Quarterly, 19*(2), 213–236.

Hofmann, P., Samp, C., & Urbach, N. (2020). Robotic process automation. *Electronic Markets, 30*(1), 99–106.

Huang, W. L., et al. (2024). A case study of lean digital transformation through robotic process automation in healthcare. *Scientific Reports, 14*(1), 14626.

Inscribe. *Optical character recognition (OCR) in RPA*. https://www.inscribe.ai/robotic-process-automation/rpa-ocr. Last accessed June 16, 2024.

Kaya, C. T., Türkyılmaz, M., & Birol, B. (2019). Impact of RPA technologies on accounting systems. *Muhasebe Ve Finansman Dergisi, 82*, 123–135.

Kedar, S. *Healthcare data management: What it is, its challenges, and opportunities*. https://mobisoftinfotech.com/resources/blog/healthcare-data-management-challenges-and-solutions. Last accessed May 10, 2025.

Kedziora, D., & Smolander, K. (2022). *Responding to the healthcare emergency outbreak of the COVID-19 pandemic with Robotic Process Automation (RPA)*.

Lin, J. N. C., Garg, M., & Weng, T. Y. (2025). *Integrating robotic process automation in healthcare: Challenges and opportunities*. White Paper or Report.

Liu, S. (2022). Robotic process automation (RPA) in auditing: A commentary. *International Journal of Computer Auditing, 4*(1), 23–28.

Mos. *Combining OCR and RPA for data processing—key benefits*. https://www.managedoutsource.com/blog/significance-benefits-ocr-robotic-process-automation/. Last accessed June 16, 2024.

Nath, A., & Saha, U. (2023). Intellectual property management in healthcare using robotic process automation during COVID-19. In *Confluence of artificial intelligence and robotic process automation* (pp. 177–197). Springer.

Nintex. *What is attended vs unattended Robotic Process Automation?* https://www.nintex.com/process-automation/robotic-process-automation/learn/unattended-vs-attended-rpa. Last accessed June 16, 2024.

Patrício, L., de Sousa Costa, C. R., Varela, L., & Cruz-Cunha, M. M. (2024). Literature review on the sustainable implementation of Robotic Process Automation (RPA) in medical and healthcare administrative services. *Procedia Computer Science, 239*, 166–176.

Pratt, M. K. (2024). *Attended vs. unattended RPA bots: Key differences*. CIO. https://www.techtarget.com/searchcio/tip/RPA-bots-Unattended-vs-attended-vs-hybrid. Last accessed June 16, 2024.

Precedence Research. *Robotic process automation market size, Report 2024–2033*. https://www.precedenceresearch.com/robotic-process-automation-market. Last accessed May 10, 2025.

SAP. *RPA: Automate your business processes*. https://www.sap.com/africa/products/technology-platform/process-automation/what-is-rpa.html. Last accessed June 16, 2024.

Sharma, V. (2023). *Automating complex healthcare workflows with AI and RPA: A new era of efficiency*. White Paper or Report.

Smits, J. (2023). *Assessment of robotic process automation technology in Dutch hospitals* (Master's thesis, University of Twente).

Soeny, K., et al. (2021). Attended robotic process automation of prescriptions' digitisation. *Smart Health, 20*, Article 100189.

Spies, R., Grobbelaar, S., & Botha, A. (2020). A scoping review of the application of the task-technology fit theory. In *e-Business, e-Services, e-Society* (pp. 397–408). Springer.

Thainimit, S., et al. (2022). Robotic process automation support in telemedicine: Glaucoma screening usage case. *Informatics in Medicine Unlocked, 31*, Article 101001.

UiPath. *RPA case study in healthcare—ApprioHealth*. https://www.uipath.com/resources/automation-case-studies/appriohealth-healthcare-rp. Last accessed June 16, 2024.

Unitek College. *Step-by-step guide to medical appointment scheduling*. https://www.unitekcollege.edu/blog/a-step-by-step-guide-to-medical-appointment-scheduling. Last accessed 10 May, 2025.

Venigandla, K., & Tatikonda, V. M. (2023). Enhancing pharmaceutical R&D through RPA-driven data analytics and management. International Journal of Enhanced Research in Management & Computer Applications, 12(10), 47–54.

Völker, M., & Weske, M. (2021). Conceptualizing bots in robotic process automation. In H. C. Mayr, G. Guizzardi, H. Ma, & O. Pastor (Eds.), *ER 2021: Conceptual modeling, LNCS* (Vol. 13011, pp. 3–13). Springer.

Wang, X., Williams, C., Liu, Z. H., & Croghan, J. (2019). Big data management challenges in health research—A literature review. *Briefings in Bioinformatics, 20*(1), 156–167.

Hybrid Augmented Intelligence Applied to Technical Trading: Improvements Through Compensatory Fuzzy Logic and WKD

Norma P. Rodríguez-Cándido, Jorge Marx-Gómez, and Rafael A. Espin-Andrade

Abstract Technical trading, a cornerstone in financial markets, often relies on subjective interpretations of a limited set of strategies, resulting in inconsistent applications. This paper proposes an improved versión of an existing technical trading model based on Compensatory Fuzzy Logic (CFL) using the strategy of Wide Knowledge Discovery (WKD) within the framework of Hybrid Augmented Intelligence (HAI). Our approach is directly oriented to enhance by WKD the consistency of decision.

1 Introduction

Technical trading is probably the most relevant resource traders use decision-making methodologies in financial markets use (Andrew et al., 2000; Christopher et al., 2009; Jack, 2002), particularly during periods of increased market uncertainty such as the tariff crisis due to geopolitical tensions and protectionist policies. In addition, the recent COVID-19 pandemic (Baker, 2021; Brunnermeier, 2001; Jeffrey, 2020). These events have amplified the unpredictability of the market, with unprecedented levels of volatility and liquidity in different assets challenging the predictive capacity of traditional economic models (Claudio, 2022; Krugman, 2019).

In this context, technical trading has gained renewed importance as a decision-making tool to navigate short-term price movements and changing investor sentiment (Pring, 2014a). They use it differently, incorporating different strategies according to their interpretation of knowledge in books and schools. Generally, not more than three of them are used systematically for each of them. This subjective interpretation

N. P. Rodríguez-Cándido (✉)
Faculty of Accounting and Management, Universidad Autónoma de Coahuila, Torreon, Coahuila, Mexico
e-mail: candido_norma@outlook.com

J. Marx-Gómez
Department of Computing Science, University of Oldenburg, Oldenburg, Germany
e-mail: jorge.marx.gomez@uol.de

R. A. Espin-Andrade
Eurekas Community, Universidad Autónoma de Coahuila, Torreon, Mexico

J. Marx Gómez et al. (eds.), *Recent Advances in Information Systems*, Progress in IS,
https://doi.org/10.1007/978-3-032-14099-9_42

is an important challenge that leads frequently to the inconsistent application of that short number of strategies. Consequently, the use of systematic integrative methods of a significative number of the most important strategies is convenient (Brad & Terrance, 2000; Pring, 2014b).

A relevant proposal of the literature is a Compensatory Fuzzy Logic model for Technical Trading (Norma et al., 2021).They are two different Compensatory Fuzzy Logic predicates, one for selecting actives to sell and one for the selection of actives to buy. A diversity of the most important technical trading rules were selected to be integrated logically according to the different scenarios. Literature and expert knowledge were using CFL as the knowledge engineering tool (Rafael et al., 2016; Thomas et al., 2018).

Additionally, the paradigm of Hybrid Augmented Intelligence (HAI) is the empowerment of human beings for problem-solving and decision-making by technology (Zheng et al., 2017). One important strategy for it is Wide Knowledge Discovery (WKD), which integrates Decision Analysis, Knowledge Engineering, and the usual narrow Knowledge Discovery associated with the context of Data Mining, in just one integrated process. It uses all the possible sources including expert knowledge, decision preference knowledge, literature and data (Rafael et al., 2021).

The purpose of this paper is related to the improvement and use of the CFL model for technical analysis using Wide Knowledge Discovery as the strategy for Hybrid Augmented Intelligence, increasing predictive effectiveness, improving the consistency of recommendations, and offering major interpretability of the decision taken by the model. Through logical integration of multiple technical rules adapted to different market scenarios, enriched with expert knowledge and validated through experimental evaluation, this study propose a robust framework for explainable and effective technical trading under the principles of Hybrid Augmented Intelligence.

2 Literature Review

2.1 Technical Analysis for Trading

Technical analysis is a way of trading deciśión-making analysis using strategies and rules based on trading community experience. It uses chronological data in small temporal horizons, focused on the volatility exploitation. Patterns of evolution and fluctuation of prices, and volumes are used to predict the close future behavior of the markets and making effective deciśión making(John, 1999; Pring, 2014a).

Among the main rules of technical analysis used in the evaluation of financial assets are (i) graphical patterns, which geometric formations are identified within price charts (John, 1999; Pring, 2014a; Robert et al., 2018; Thomas, 2021). (2) technical indicators that use mathematical calculations derived from prices and, sometimes, trading volumes to build tools such as moving averages, relative strength indices (RSI), Bollinger bands, among others. These tools allow us to identify signals of trend changes and continuation of this (John, 1999; Robert et al., 2018; Steve, 2001). Finally, candlestick pattern analysis that examines specific formations of one or multiple candlesticks to detect signals of trend reversal or continuation, based on the dynamics between the opening, closing, high, and low prices within a given period (Steve, 2001; Thomas, 2021).

Technical analysis is a subject of debate within the financial community and academic around its scientific validity, practical usefulness, and its comparison with other forms of analysis such as fundamental analysis (Andrew et al., 2000; Mark, 2002). The efficient market hypothesis (EMH) (Steve, 2001), states that asset prices instantaneously reflect all available information, making it impossible to obtain information and returns by analyzing past price patterns. However, with the current crises as mentioned above—COVID-19 and geopolitical conflicts—they have generated unprecedented levels of volatility and liquidity, invalidating the assumptions of traditional economic models and reinforcing the predictive value of technical patterns.

During these crises, technical indicators (RSI, Bollinger bands) have demonstrated an effectiveness of more than 62% in identifying market turning points (Barberis, 2003; Cheol-Ho & Scott, 2007; Juan, 2022), in contrast with the poor performance of fundamentals in volatile environments (Bernadette, 2021; Petrusheva & Jordanoski, 2016). The integration of algorithmic and AI-based techniques has further increased the reliability and adaptability of technical trading strategies.

2.2 *Wide Knowledge Discovery (WKD) as Strategy of Hybrid Augmented Intelligence (HAI)*

Hybrid Augmented Intelligence principal strategies are Human into the loop (HITL) and Cognitive Model Embedment (CME) (Zheng et al., 2017) . The first one oriented to introduce the direct participation of human being in the problem-solving process and the second one is the incorporation of Cognitive Models in the automatization of the analysis with the principal purpose of obtaining explicability. Wide Knowledge Discovery as strategy of HAI implement HITL as the introduction of Decision Makers, Experts and Literature as the human sources and CME by the use of common ways of representation of the knowledge which make easier the explainability of the results obtained incorporating structured data (Rafael et al., 2021). Explainability can be achieved by different ways, particularly by the interpretability by natural language, by the use of Semantic Models based on Compensatory Fuzzy Logic. Wide Knowledge Discovery is an effective way to deal with problem-solving in companies and institution dealing through interpretability very explicit models able to risk, quality, strategic and knowledge management, and evaluating principal conclusions for avoiding hallucinations.

2.3 Comparative Analysis of Intelligent Trading Models

Deep reinforcement learning (DRL) models, such as Deep Q-Networks (TDQN) and actor-critic ensembles (e.g., PPO, A2C, DDPG), are able to give accurate predictions very useful in markets (Théate & Ernst, 2021; Xiao-Yang et al., 2020) but they operate as black boxes and suffer from limited interpretability. Similarity ensemble learning methods, XGBoost and Hidden Markov Models (HMMs) offer a high accuracy and give a limited interpretability based on expert knowledge incorporation or probabilistic approaches which are not providing an explainability associated to the trading knowledge and scenarios dependence. In contrast, the use of Wide Knowledge Discovery and transdisciplinary models like CFL is an effective strategy for Hybrid Augmented Intelligence because join harmonically the Human into the loop and Cognitive Model Embedment strategies and offer consequently a natural interpretability and the correspondent explainability. The exploitation of a rich explicit knowledge for trading decisions as technical analysis is an extraordinary point of departure to use it in this framework.

3 Methodology

The methodological approach proposed in this study is a flow of analytical tasks designed for scientific problem-solving, which can subsequently be used as a flow oriented to decision-making incorporating the model obtained. The central problem addressed in this research can be formulated as follows: How is possible to improve the technical trading model based on (CFL) using Wide Knowledge Discovery?

To answer this question, a methodological flow was designed composed of the following stages:

1. Literature searching review using problem-solving to get strategies of improvement with support of a GAI.
2. Selection of the best strategies: One of the principal strategies selected and use in this paper is the elaboration of a new model incorporating conditionals for integrating systematically and dynamically the technical rules according to the dynamics of the market and scenarios and the correspondent convenience and certainty suministrated for each rule.
3. Construction of a new models. The selected structure was a conjunction of conditionals oriented to weight the use of the rules.
4. Preliminary obtention of truth values for each rule in each scenario and market.
5. Expert consult of the truth values tables.
6. Initial experiment design (Prescriptive task).
7. Evaluation of the preliminary results.
8. Modification of the model and weights.
9. New experiment.

The proposed approach allows the development of a model with three fundamental properties: (1) explainability through traceable inference rules; (2) dynamic adaptability through real-time feedback mechanisms; and (3) theoretical coherence with the principles of Augmented Hybrid Intelligence with a logical basis in the CFL.

3.1 *Rule Weighting and Parameter Justification*

The selection of the truth value associated with each rule (logical weight) in each context was oriented to incorporate them to the logical structure according to knowledge using WKD: The contextual elements to be consider and the initial values of the weights were obtained from Knowledge Engineering the following way: Generative Artificial Intelligence using literature and general data available using different Large Language Models and adjusted definitively by Experts Consultancy in several Delphi stages.

3.2 *Compensatory Fuzzy Logic Models and Graphical Representation*

The improved technical trading model is structured using Compensatory Fuzzy Logic (CFL) principles, where technical indicators are represented as fuzzy predicates and logically combined to form actionable trading decisión. Two principal models were constructed: a Buying model and a Selling Model, each composed of weighted conditionals dynamically adapted to market scenarios.

In the CFL framework, each technical indicator is evaluated to produce a fuzzy truth value in [0,1] Interval. These truth values are then integrated using compensatory conjunction operators, allowing partial compensation among rules.

Buying Model Structure

The structure of the predicates that evaluate the convenience of buying (selling) are the ones in Figs. 1 and 2 . The value of each rule is affected by the veracity of each technical analysis rule in each scenario and according to the classification of the assets. The interpretability of Compensatory Fuzzy Logic allows the use of Knowledge Engineering as part of Knowledge Discovery in a broad sense; so that the introduction of the veracity that each rule is effective according to the scenario and the classification of the asset, is an effective strategy for the improvement of the model, as reflected in the experiment carried out.

Selling Model Structure

Similarly, the Selling Model aggregates signals indicating favorable conditions for selling an asset (Fig. 3).

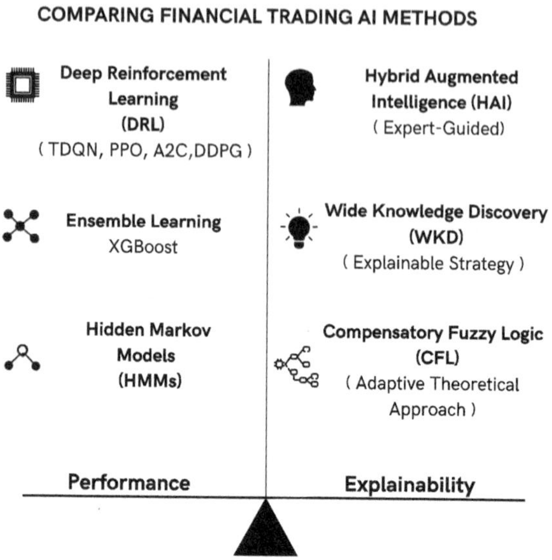

Fig. 1 Comparative positioning of financial trading AI models. Black-box approaches such as DRL, ensemble learning, and HMMs emphasize predictive performance. In contrast, the proposed Hybrid Augmented Intelligence (HAI) model-based on Wide Knowledge Discovery (WKD) and Compensatory Fuzzy Logic (CFL)—prioritizes interpretability, scenario adaptation, and expert-guided rule integration

4 Experiment Elaboration and Result

4.1 Experimental Setup

In order to evaluate the performance of the proposed model and its comparative effectiveness against the original logical model, we considered stock market data over a six-month period, encompassing 1901 operations across diverse asset classes a set of financial assets was strategically selected according to the following criteria: (1) high liquidity and trading volume, (2) availability of historical data and, (3) representativeness of different market sectors.

Our experiment encompassed assets from a variety of categories including: (1) fourteen leading companies from the SPX500 index: Google(GOOG)

Amazon (AMZN) Meta, Netflix (NFLX) Uber, P&G (PG), PFE, Walmart (WMT) PayPal (PYPL) Tesla (TSLA) Disney (DIS) Home Depot (HD), (2) global stock indices (SPX500, GER30); (3) major currency pairs (USD/JPY, USD/MXN), and (4) relevant commodities (Gold, Cocoa, Sugar). The data was obtained from trusted financial platforms (Yahoo! Finance and Bloomberg).

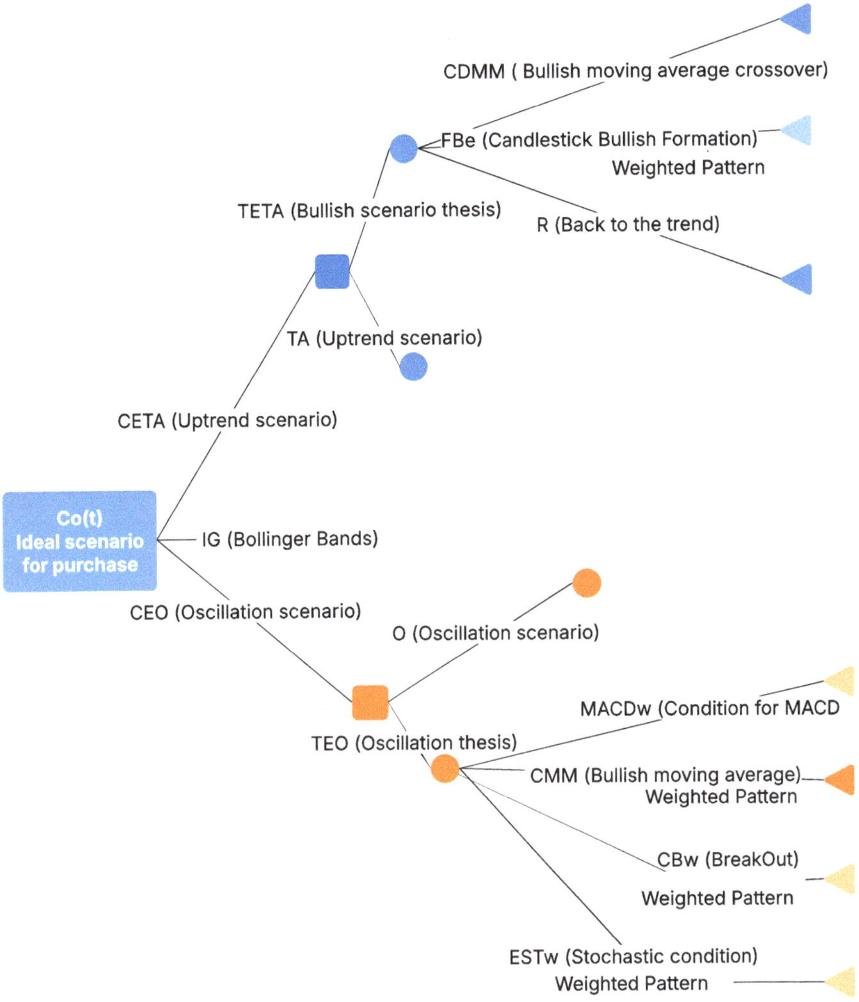

Fig. 2 Buying model structure with weighted indicator

All the operations recommended by the models were evaluated under a consistent five-day post-signal window, ensuring methodological comparability. Data was sourced from Yahoo! Finance and Bloomberg.

4.2 Gain Accumulation and Model Comparison

Figure 4 shows the daily cumulative profit G_t for each model over a period of six months, from July 1, 2020, to December 31, 2020. For each day t, the cumulative

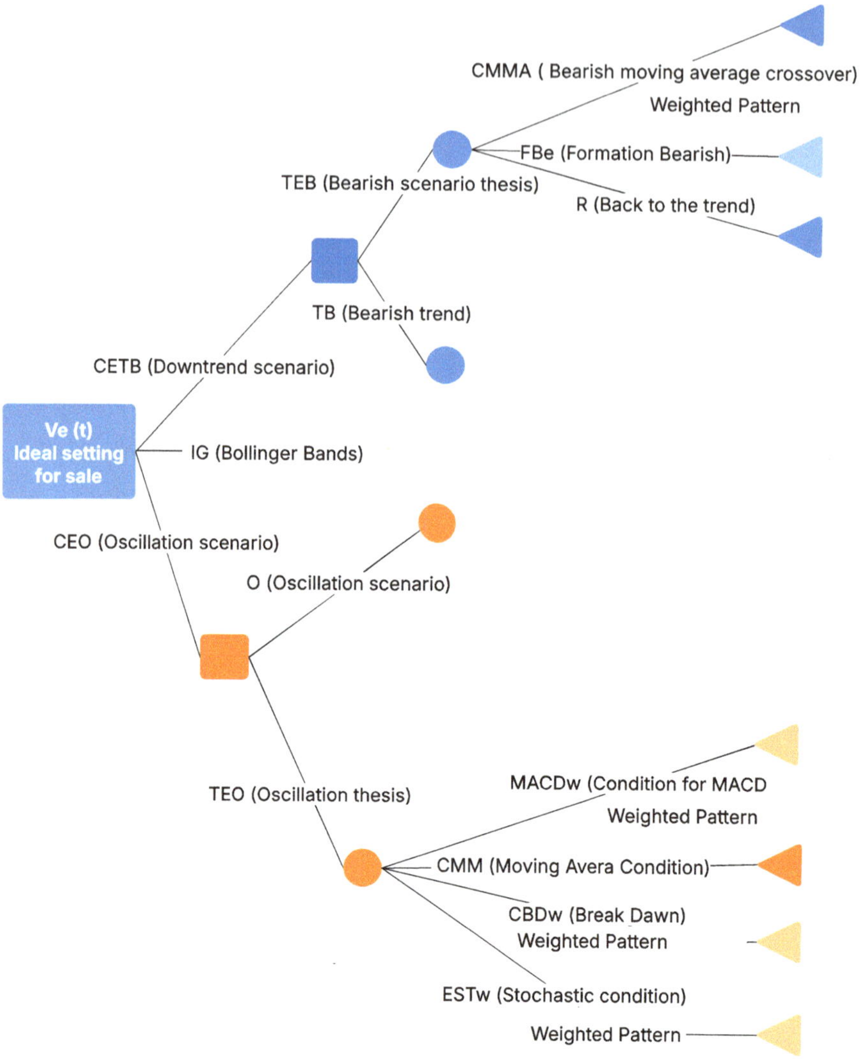

Fig. 3 Selling model structure with weighted indicator

profit was calculated separately for the original model $G_t^{(0)}$ and the improved model $G_t^{(M)}$. Based on these daily values, the cumulative earning trajectories for both models were derived.

From this, the cumulative earnings trajectories were derived. The difference was defined as:

$$\Delta_t = G_t^{(\mathrm{M})} - G_t^{(0)}$$

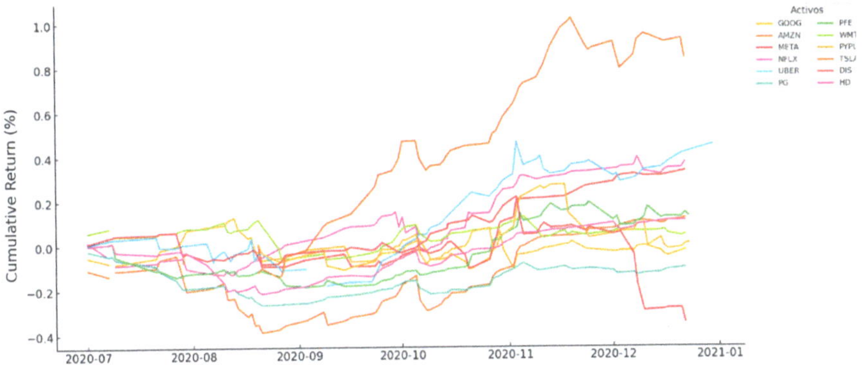

Fig. 4 Illustrative sample of financial assets and experimental setup

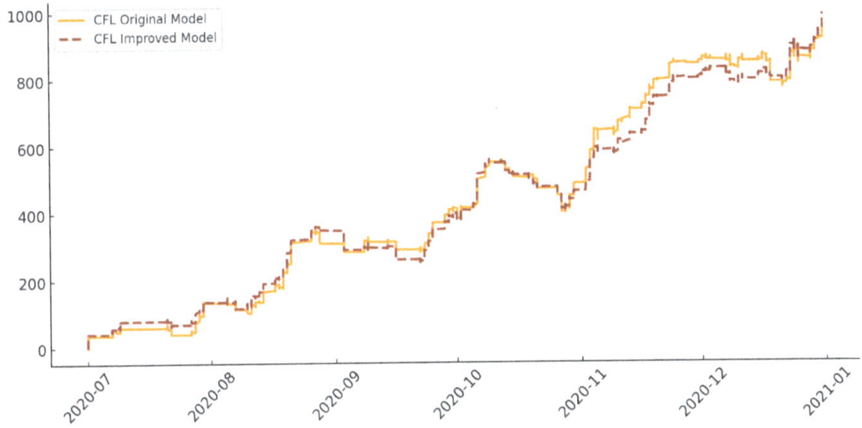

Fig. 5 Accumulated gain trajectories: evidence of improved profitability

significance was evaluated using a Student's t-test for paired samples under the hypothesis: $H_0 : \mu_\Delta = 0$ (there is no difference on average), $H_1 : \mu_\Delta > 0$ (the improved model outperforms the original). The corresponding p-value was obtained from Student's t-distribution with $n - 1$ degrees of freedom. A threshold of $\alpha = 0.05$ was considered. In our experiments, a p-value < 0.001 was observed, indicating a statistically significant improvement of the improved model compared to the original.

In our experiments, a p-value lower than 0.001 was observed, indicating a statistically significant improvement of the improved model compared to the original (Fig. 5).

Table 1 General performance comparison between logical trading models

Operation	Original model					Improved model				
	Oper.	NP	NN	Gain	Hit rate (%)	Oper.	NP	NN	Gain	Hit rate (%)
Buying	926	545	381	5.5849	58.86	1004	807	197	5.6656	80.38
Selling	974	697	277	3.9351	71.56	893	642	251	4.0447	71.89
Total	1900	1242	658	9.5200	65.21	1897	1449	448	9.7103	76.38

Number of positive operations (NP), Number of negative operations (NN)

4.3 Performance Evaluation and Result

Table 1 summarizes the results obtained between the original logical model and the proposed improved model, which incorporates differentiated weights by type of scenario. Both models were applied over a consistent sample of 1900 operations, divided between buy (926) and sell (974) signals.

In the case of the improved model, the same number of operations were analyzed, with slight variations in distribution due to dynamic weighting criteria: buy (1004) and sell (893) signals.

5 Conclusions

The application of the Wide Knowledge Discovery strategy was very fruitful for the model improvement. The clear Interpretability and explainability achieved by WKD using CFL was the key for it. The experimental results demonstrated the improvement of the model efficacy. The explicit inclusión of the relative importance of each technical trading tool according to the scenarios and kind of actives was the key of that efficacy. The following elements should be considered for next works:

1. Adding new relevant technical rules for buying and selling according to each scenario.
2. Adding new general indicators.
3. Elaboration new model improvement strategies.
4. Incorporation of new variables for evaluating the importance of each rule according to them.
5. Adjusting the truth values for each rule in the both predicates using Knowledge Discovery.

6 Limitations and Further Works

The limitation of the model is associated with its generality that made not possible considering probabilistic samples. The comparison with the not weighted model is relevant and further works are justified and needed to explore effectiveness in each context and compare it with other methods of different nature not related to technical analysis. The sample was selected using representative actives considering the different elements related to the contextual elements previously obtained. It is a descriptive approach using a not probabilistic wide sample. Further works—already in course—will explore with specific samples by contexts the effectiveness in each of them. As part of such studies, the most convenient umbral for decision-making is being studied already considering the combination of the two models component compliance. The expression of rules for the selection will be establish considering the budget available and the effectiveness for each umbral. Comparison with other methods will be useful as well in the selection of the proper number.

References

Baker. D. (2021). Covid-19 and financial markets: A critical review of impacts and responses. *Journal of Risk and Financial Management, 14*(3).

Barberis, N. (2003). A survey of behavioral finance. In G. M. Constantinides, M. Harris, & R. M. Stulz (Eds.), *Handbook of the Economics of Finance* (Vol. 1 , pp. 1053–1128). Elsevier.

Barber, B. M., & Odean, T. (2000). Trading is hazardous to your wealth: The common stock investment performance of individual investors. *The Journal of Finance, 55*(2), 773–806.

Borio, C. (2022). Geopolitical tensions and financial stability: A historical perspective. In *BIS Quarterly Review*.

Brunnermeier, M. K. (2001). *Asset pricing under asymmetric information: Bubbles, crashes, technical analysis, and herding* (1st ed.). OUP Oxford.

Bulkowski, T. N. (2021). *Encyclopedia of chart patterns*. Wiley.

Carlos Reboredo, J., et al. (2022). Technical analysis and cryptocurrencies amid covid-19. *Economic Modelling, 108*.

Davenport, T. H., & Ronanki, R., et al. (2018). Artificial intelligence for the real world. *HBR'S 10 MUST* (p. 67).

Edwards, R. D., Magee, J., & Charles Bassetti, W. H. (2018). *Technical analysis of stock trends*. CRC Press.

Espin-Andrade, R. A., Gonzalez, E., Pedrycz, W., & Fernandez, W. (2016). An interpretable logical theory: the case of compensatory fuzzy logic. *International Journal of Computational Intelligence Systems, 9*(4), 612–626.

Espin-Andrade, R. A., Pedrycz, W., Solares, E., & Cruz-Reyes, L. (2021). Transdisciplinary scientific strategies for soft computing development: Towards an era of data and business analytics. *Axioms, 10*(2), 93.

Frankel, J. A. (2020). The global trade slowdown and the tariff wars: A macroeconomic perspective. In *Technical Report RWP20-034*. Cambridge, MA: Harvard University, John F. Kennedy School of Government.

John, J. (1999). *Murphy* (1st ed.). : Technical Analysis of the Financial Markets. New York Institute of Finance.

Krugman, P. (2019). Trade wars are class wars. *Foreign Affairs, 98*(5), 180–189.

Liu, X.-Y., Yang, H., Chen, Q., Zhang, R., Yang, L., Xiao, B., & Dan Wang, C. (2020). FinRL: A deep reinforcement learning library for automated stock trading in quantitative finance. arXiv:2011.09607

Lo, A. W., Mamaysky, H., & Wang, J. (2000). Foundations of technical analysis: Computational algorithms, statistical inference, and empirical implementation. *The Journal of Finance, 55*(4), 1705–1765.

Lo, A. W., Mamaysky, H., & Wang, J. (2000). Foundations of technical analysis: Computational algorithms, statistical inference, and empirical implementation. *The Journal of Finance, 55*(4), 1705–1765.

Neely, C. J., Weller, P. A., & Ulrich, J. M. (2009). The adaptive markets hypothesis: evidence from the foreign exchange market. *Journal of Financial and Quantitative Analysis, 44*(2), 467–488.

Nison, S. (2001). *Japanese candlestick charting techniques: a contemporary guide to the ancient investment techniques of the Far East.* Penguin.

Park, C.-H., & Irwin, S. H. (2007). What do we know about the profitability of technical analysis? *Journal of Economic surveys, 21*(4), 786–826.

Petrusheva, N., & Jordanoski, I. (2016). Comparative analysis between the fundamental and technical analysis of stocks. *Journal of Process Management. New Technologies, 4*(2), 26–31.

Pieterse, B. (2021). *Comparing the returns of technical analysis strategies with market index returns* (Master's thesis). South Africa: University of Pretoria.

Pring, M. (2014a). *Technical analysis explained* (5th ed.). McGraw-Hill.

Pring, M. J. (2014b). *Technical analysis explained: The successful investor's guide to spotting investment trends and turning points* (5th ed.). McGraw-Hill.

Ready, M. J. (2002). Profits from technical trading rules. *Financial Management,* 43–61.

Rodríguez-Cándido, N. P., Espin-Andrade, R. A., Solares, E., & Pedrycz, W. (2021). A compensatory fuzzy logic model in technical trading. *Axioms, 10*(1), 36.

Schwager. J. D. (2002). *Stock market wizards: Interviews with America's top stock traders.* Wiley.

Théate, T., & Ernst, D. (2021). An application of deep reinforcement learning to algorithmic trading. *Expert Systems with Applications,173* (Article 114632).

Zheng, N., Liu, Z., Ren, P., Ma, Y., Chen, S., Si-yu, Yu., Xue, J., Chen, B., & Wang, F. (2017). Hybrid-augmented intelligence: collaboration and cognition. *Frontiers of Information Technology & Electronic Engineering,18*(2), 153–179.

Organizational Values and Practices for Hybrid Agile Teams

Roelof Briers and Maureen Tanner

Abstract This study explores the values and practices relevant to agile teams operating in a hybrid environment. It was conducted given the rise in hybrid work models after the COVID-19 pandemic. This rise further leads to the need to understand the values and practices that ensure a cohesive and supporting work environment in that work context. The Competing Values Framework (CVF) was employed as a lens to identify the organizational cultural traits that could be manifested in a hybrid work context. During a case study, semi-structured interviews were conducted, of participants across several teams demonstrating cultural traits from the CVF. In the Clan quadrant, trust, empathy, and respect emerge as essential for maintaining team cohesion and building relationships. Team interaction and engagement, an ability to self-organize, and the promotion of psychological well-being are also identified as relevant practices. In the Adhocracy quadrant, flexibility is valued. There is also a need for innovative technologies to support day-to-day work. In the Hierarchy quadrant, accountability, consistency, and transparency are valued, as well as practices related to planning and coordination and knowledge sharing. Lastly in the Market quadrant, hybrid agile teams value proactiveness and ensure that they maintain good stakeholder involvement. This research contributes to the growing body of literature on hybrid work models and provides additional insights into organizational culture and hybrid work practices during agile software development. It highlights the need for organizations to adapt cultural values and practices to support hybrid work, ensuring that teams build strong relationships while remaining productive.

1 Introduction

Post the COVID-19 pandemic, organizations are still dealing with the consequences of several work environment-related changes that were implemented to ensure the continuation of business over this period (Kagerl & Starzetz, 2023; McPhail et al., 2024; Peprah, 2024). Before the pandemic, most organizations had their employees

R. Briers · M. Tanner (✉)
University of Cape Town, Cape Town, South Africa
e-mail: mc.tanner@uct.ac.za

J. Marx Gómez et al. (eds.), *Recent Advances in Information Systems*, Progress in IS,
https://doi.org/10.1007/978-3-032-14099-9_43

working from the office. While the concept of working from home or outside of the office full time (remote working) or only working from the office on certain days of the week (hybrid working) was around previously, it was an outlier for most organizations. The pandemic forced organizations to take a different stance on their work arrangement to ensure the safety of their employees. The adoption of both hybrid working and remote work models was subsequently accelerated during and post-pandemic (Peprah, 2024). Past studies have shown that the Information Technology (IT) sector is one of the highest adopters of the hybrid work model (Peprah, 2024).

This change in work model required that organizations and their teams shift away from their previously office-bound activities to a digital equivalent using various technologies (Müller et al., 2023). In the software development industry, organizations relying on Agile Project Management approaches, would have been particularly impacted by the shift to digital meetups due to their required team activities, which traditionally would have been face-to-face. However, the change also introduced culture-related challenges for organizations, calling for an adaptation of values and practices to better support hybrid teams (Barhate et al., 2022). This calls for a need to explore the values and practices relevant to ensure a cohesive and supportive environment for hybrid agile software development teams.

In response to this need, the following research question has been formulated: What should be the organizational values and practices in support of hybrid work during agile software development?

This research contributes to the growing body of literature on hybrid work models and agile practices by providing further empirical insights into organizational culture and team dynamics within that work context. Practical contributions are also put forward by highlighting the need for organizations to adapt cultural values to support hybrid work effectively, ensuring that teams build strong relationships while remaining productive and innovative.

The structure of the paper is as follows. The introduction section presents a background on the study, outlining the problem statement and the research questions it addresses. The literature review provides insights into the theoretical content relevant to the study. The research methodology is then described, with specific focus on the philosophy, theoretical approach, instrument selection, strategy, audience, and sampling. It also discusses the data analysis process and ends with the ethical considerations relevant to the study. The results section describes the findings considering the research question. Finally, the conclusion summarizes the study and addresses its limitations while proposing areas for future research.

2 Literature Review

2.1 Agile Project Management

The concept of Agile Project Management was devised in 2001 to address several shortcomings of the various software development paradigms that preceded it (Gupta et al., 2019). The Agile Manifesto, which outlined key principles that should be adhered to, detailed the approach organizations should adopt. As the name suggests, agile methods were designed to make the software development process less rigid than the Waterfall process, allowing for a quicker response to changes and customer feedback. Adopting an agile approach to software development requires a shift in the organizational culture, especially if other plan-based approaches like the Waterfall were previously utilized (Hasan et al., 2021).

The Agile Manifesto was defined by seventeen technologists and defines the major principles to be embraced in any agile project. These principles would be the guide teams need to follow and relate to individuals and interactions over processes and tools, working software over comprehensive documentation, customer collaboration over contract negotiation, and responding to change over following a plan (Beck et al., 2001).

There are set frameworks that can be used to implement agile in an organization. Examples of these are Kanban, Scrum, and Lean. These frameworks provide structure for implementing agile in an organization. The study looks at the Scrum framework as it aligns with the framework used by the participants. The Scrum framework helps organizations, teams, and individuals to generate business value by solving complex problems in small increments. These increments allow the teams to understand better and implement their solutions (Schwaber & Sutherland, 2024). The successful use of Scrum relies on five values listed in the Scrum Guide. The values are Commitment, Focus, Openness, Respect, and Courage (Schwaber & Sutherland, 2024).

2.2 Agile Approaches and Organizational Culture

For an organization to successfully adopt an agile approach to software development, they need to have or strive to have a culture that supports agile practices. Hence, organizational culture is key to the successful adoption of agile software development (Hasan et al., 2021). Without a suitably aligned organizational culture, agile projects are likely to fail (Gupta et al., 2019). These include people-centered and collaborative, democratic, less formalized and non-hierarchical, and an appropriate reward system (Gupta et al., 2019). An organizational culture that is pro-innovation and promotes management support is also deemed important (Mishra et al., 2021). In relation to management support, there is a need for the leadership to understand agile methods/values and these need to be aligned with the business' vision (Sithambaram et al., 2021). In contrast, organizational traits that deemed to be non-conducive to a successful implementation of agile approaches include a high degree of bureaucracy and formalization approaches and lack of management support (Mishra et al., 2021).

2.3 The Hybrid Work Model

The hybrid work model, as the name implies, is a combination of working from the office on certain days and working remotely on others (Peprah, 2024). During the COVID-19 pandemic, the model was more formally adopted by companies to keep the business running as usual (Kemell & Saarikallio, 2023). Hybrid working has advantages, but if not well applied or honestly done, it will have disadvantages for both the organization and the individual.

Advantages and Disadvantages of Hybrid Work. From an employee perspective, some of the advantages are that it helps maintain a good work-life balance, it provides flexibility for employees, employees can freely choose where they work without losing contact with colleagues, and it saves commute/travel time and stress and costs related to that (Kemell & Saarikallio, 2023; Reunamäki & Fey, 2023; Shao et al., 2024). Studies have also shown that the productivity of software developers has increased when working from home (Kemell & Saarikallio, 2023). Some of the disadvantages for the employee are that they can feel alone/isolated, the possibility of technical issues is much more significant, they are very likely to face cohesion problems in their teams due to the lack of face-to-face time, and there could be distractions at home that would not be present at work (Ali et al., 2023; Banerjee & Gupta, 2024; Peprah, 2024; Reemon, 2024; Reunamäki & Fey, 2023).

From the companies' perspective, some of the advantages are reduced utility costs for their buildings (Kemell & Saarikallio, 2023; Reunamäki & Fey, 2023), it also grants companies access to a bigger talent pool (Peprah, 2024). However, companies also face numerous disadvantages by offering a hybrid work option to their employees. For instance, technical issues can cause problems with productivity and communication, employee performance issues may occur, and there is a risk that employees start taking on other personal work while not adhering to their agreed office hours/days. Another concerning disadvantage for organizations is the loss or deterioration of the organizational (Peprah, 2024; Tuukkanen et al., 2022).

Advantages and Disadvantages of Hybrid Work. After the COVID-19 pandemic, the hybrid work model has continued in certain industries more than others. These include Marketing, Accounting/Finance, Computer/Information Technology, HR and Recruiting, Client Services, Medical and Health, Project Management and Sales (Place, 2021). However, there are certain factors that these industries need to consider as they may influence the culture they are aiming to build in the organization. Some of these include rules of behavior, trust, collaboration between teams and individuals, technology, and employee isolation (Dalessandro & Lovell, 2024; Peprah, 2024; Windlinger & Gerber, 2023). Organizational culture has historically been developed through face-to-face interactions, but such interactions are less prevalent in a hybrid working environment. The organizational culture of the company is therefore at risk of being weakened over time (Shao et al., 2024). Employees could even start experiencing different work cultures in different work locations (Peprah, 2024). Technology could be used to bridge this gap and improve communications and collaboration in teams, making interactions more visible (Windlinger & Gerber, 2023).

A risk with hybrid working is a loss of trust and collective identity due to the distance and potential lack of communication brought by virtual teams (Peprah, 2024). Indeed, studies have shown that due to the emergence of hybrid working and the technologies used to drive business in this digital workplace, a new digital organizational culture has started to emerge (Dewalska-Opitek & Nunez, 2022). This digital culture has certain expected values: Openness Toward Change, Customer-Centricity, Agility, Innovation, Continuous Improvements, Tolerance Toward Failure, Risk Affinity, Communication, Trust, Participation, and Cooperation (Tuukkanen et al., 2022). An organization's adoption of these values can influence the success or failure of its more digital employee journey.

3 Theoretical Framework

This study utilizes the Competing Values Framework (CVF) (Cameron & Quinn, 2011). The framework illustrated in Fig. 1 was chosen for its holistic view and ability to assess staff climate, leadership style, bonding systems, and goal prioritization. The framework provides a structured mechanism to better understand values and practices values and practices inherent to an organizational culture in support of hybrid work during agile software development. This is essential to answer the research question.

Fig. 1 Competing values framework (Osei et al., 2023)

As depicted in Fig. 1, CVF is based on two dimensions namely (1) Flexibility versus Stability and (2) External Focus versus Internal Focus. As the framework name implies, each dimension has competing sides. Flexibility relates to the extent to which an organization can adapt, promote innovation and is open to change. In contrast, Stability refers to the extent to which establishing structures, staff, and practices are in place. An organization that is externally focused is attuned to market demands, new products or trends external to the organization (Iivari & Iivari, 2011; Vuuren et al., 2008). An internally focused organization tends to look inwards, with the aim of improving its current processes, internal culture and its employees (Iivari & Iivari, 2011; Vuuren et al., 2008). The focus on internal or external for an organization drives the targets that are being chased.

Each dimension creates four quadrants, each representing distinct cultural types: Clan, Adhocracy, Hierarchy, and Market (Cameron & Quinn, 2011; Iivari & Iivari, 2011; Irimia & Pop, 2022). The clan culture emphasizes teamwork, collaboration, the empowering of employees and having effective relationships across the company (Iivari & Iivari, 2011; Reino et al., 2020). The adhocracy culture values creativity, adaptability, and innovation to address future needs (Iivari & Iivari, 2011; Reino et al., 2020). The hierarchy culture focuses on stability through the implementation of bureaucracy, structure, and emphasis on efficiency (Iivari & Iivari, 2011; Reino et al., 2020). The market culture relates to having high priorities set for being competitive, reaching the goals set out by leadership and having a fast-paced view to bring in changes (Iivari & Iivari, 2011; Reino et al., 2020).

4 Methodology

The study embraced an interpretive stance as well as a deductive approach to theory given the use of the CVF as a theoretical lens. The study was also qualitative in nature, allowing values, behaviors, and beliefs of the participants to be captured (Bryman & Bell, 2011). A single case study was designed to study agile teams working in a hybrid work setup in one organization (Yin, 2014). Given the focus on organizational culture, the unit of analysis was the organization.

The study made use of non-probability sampling, more specifically purposive sampling. The sampling criteria for selecting the case organization was as follows. The organization needed to be involved in software development or have a strong in-house software development department that follows the Scrum methodology. The organization should also offer its employees a hybrid working model. The chosen case organization (a South African FinTech company) fitted all these requirements. The sampling criteria for selecting the participants related to the cultural traits highlighted in the quadrant of the CVF (clan, adhocracy, hierarchy, and market), the level of experience in an agile team (i.e., at least six months), as well as the fact that they were part of a Scrum team that worked in a hybrid work setup.

Data was primarily collected through semi-structured interviews. Secondary data was also collected through the observation of Scrum ceremonies conducted

Table 1 Participants details

Participant	Job title	Employment age
Participant-01	Business Analyst	+ 2 years
Participant-02	Senior Developer	+ 2 years
Participant-03	Specialist Developer	+ 4 years
Participant-04	Senior Developer	+ 2 years
Participant-05	Cloud Engineer	+ 6 months
Participant-06	Intermediate Developer	+ 5 years
Participant-07	Senior Developer	+ 2 years
Participant-08	Senior Developer	+ 2 years
Participant-09	Product Owner	+ 2 years
Participant-10	Intermediate Developer	+ 2 years
Participant-11	Product Owner	+ 5 years

in a hybrid work context. The interview questions were formulated based on the constructs from the CVF. Participants were developers, testers, and product owners and together, they were responsible for delivering software features following a two-week sprint approach in a hybrid work setup. Collaboration happened through tools like Microsoft Teams and Slack. Data collection lasted for about 3 months and comprised of 11 participants as saturation was reached at that point. Each participant was interviewed once to capture their perspective on the practices and values associated with hybrid working. This allowed the researchers to gather insights from participants' unique viewpoints at a specific moment in time, providing a snapshot of their current experiences. The participants' demographics is listed in Table 1.

The interviews were first transcribed, cleaned of any grammar issues, and then anonymized. The data was then analyzed using the six thematic analysis phases namely familiarizing oneself with the data, generating initial codes, searching for themes, reviewing themes, defining and naming themes and producing the report (Braun & Clarke, 2006). NVivo was used to assist with the data analysis process.

The identity of the company and the participants was kept strictly confidential. Pseudonyms were used for the participants, and they were allowed to withdraw from the study at any point in time. Participants were also required to sign a consent form or verbally agree to the interview.

5 Findings and Discussion

This section describes the findings of the study. More specifically the practices and values that are particularly relevant to each quadrant of the CVF in support of hybrid work during agile software development are discussed.

5.1 *The Clan Quadrant*

The Clan culture is internally oriented for organizations. The focus is on employee effectiveness and what improves their satisfaction and involvement (Reino et al., 2020). The practices and values that are relevant to a hybrid agile team that manifests the Clan cultural traits are discussed below.

Practices Relevant to the Clan Quadrant in a Hybrid Agile Work Setup. For teams manifesting the Clan cultural traits to function effectively in a hybrid work environment, there needs to be emphasis on *team interaction and engagement* to promote collaboration and cohesion. This is to ensure that good relationships exist within the team and across teams. The hybrid working model can complicate building and maintaining engagement, as members are not always together. For instance, participants reported that interactions during remote working days had to be adapted to promote collaboration, given that the typical face-to-face scenario where one person systematically leads the Scrum ceremonies contributed to less team involvement in a hybrid setup. Instead, rotating certain ceremony roles can help the team interact and collaborate better: "*We were struggling with engagement during ceremonies. So, we changed it around a little bit instead of us facilitating the session and capturing tasks. We switched it around so that each of the developers would take turns to be present*." [Participant-09]

The participants have reported that regular interactions with leadership also assist with team cohesion. Implementing regular one-on-one sessions with leadership for updates promotes engagement and improves relationships overall: "*We have one-on-one that we would normally have, and what helps is also not to keep that meeting too formal. We also chat about just life in general. How, how are you doing at home?*" [Participant-04]

The participants have also indicated the need for policies related to hybrid working, such as a camera-on policy for virtual meetings. The policies need to be consistent, provide guidance for the team, and ensure that hybrid working is not exploited: "*Something very small is like, do we constantly have our cameras on.*" [Participant-01]

The findings also reveal that in Clan culture, there is a need for more informal interactions, group sessions, and team-building events while working in a hybrid agile environment. This is particularly relevant since team members can feel isolated or alone in that work model: "*Within our teams, we've tried to have online social gatherings like game sessions and things like that just to sort of socialise even though we couldn't meet each other in person*" [Participant-07]

Another relevant practice that should be implemented by hybrid agile teams operating in a Clan culture is the *ability to self-organize*. It is important for team members to have the freedom to self-manage and adapt their Scrum ceremonies as they see fit. The only instance where management should intervene is when there are trust issues due to the non-delivery of outputs: "*I think self-management is even more important in a hybrid because as a product owner, I would normally be able to see or hear what has and hasn't been done. Whereas now it takes an intentional follow-up. Or inviting people into a meeting.*" [Participant-09]

In the Clan Culture, employees' *psychological well-being* needs to be considered when working a hybrid agile environment. For example, the participants have stressed the importance of giving recognition to employees where it is due: "*Recognition, I think. Management and leadership should recognise when someone has done something out of the ordinary and obviously acknowledged those achievements.*" [Participant-08]

Values Relevant to the Clan Quadrant in a Hybrid Agile Work Setup. In the context of the study, the values that are relevant to hybrid agile teams operating in a clan culture relate to Trust, Empathy, and Respect. *Trust* was a key value highlighted by participants. In particular, trust in the team and trust in management: "*I think for teams where we are constantly delivering things and generating an output. There's a high-level of trust amongst the team members and with their senior stakeholders.*" [Participant-01]

To build stronger relationships with the team, participants stated that they needed to understand the views and perspectives of others. To build a stronger hybrid agile team, the value of *empathy* is therefore important in the clan culture: "*To foster more, stronger relationships or a stronger understanding about somebody, we need to have more empathy and more understanding.*" [Participant-09]

The participants noted that to build better relationships, the teams need *respect*, and in a hybrid agile team, work needs to be done to build that respect: "*This is an opportunity to get to know each other and to recognise each other for the humans that we are and not just the work-producing people.*" [Participant-09]

5.2 *The Adhocracy Quadrant*

The Adhocracy Quadrant is externally oriented with a focus on flexibility to meet the needs of its customers and goals (Reino et al., 2020). The practices and values that are relevant to a hybrid agile team that manifests the Adhocracy cultural traits are discussed below.

Practices Relevant to the Adhocracy Quadrant in a Hybrid Agile Work Setup. For agile teams manifesting an Adhocracy culture, while hybrid work may foster individual focus, proximity to teammates remains important for rapid collaboration and creative problem-solving. In this instance, the *use of innovative technologies* helped in making reaching out to other teams and team members easier, whether through virtual or in-office sessions. One of the participants noted virtual whiteboarding as an example of technology: "*We got the team to do an equivalent of whiteboarding but online. So, in other words, allowing diagramming. We did find, though, that it took longer because they were not familiar with the diagramming tools.*" [Participant-09]

However, management support is needed for innovation to be embraced. And this also implies that hybrid agile team members should be given the time and space to research what useful innovative technologies could be employed in the team.

Values Relevant to the Adhocracy Quadrant in a Hybrid Agile Work Setup. Hybrid team members working in an adhocracy culture value *flexibility*. The participants revealed that they value flexibility as it assists with various development efforts. Team members have been required to adapt to new ways of working and therefore needed to embrace flexibility: "*I think hybrid work obviously increased the emphasis on flexibility. Teams themselves, they become more adaptable and more flexible, and I think they are embracing iterative processes and adjusting to the more hybrid agile way of work.*" [Participant-08]

However, embracing flexibility as a value can also have negative consequences. For instance, participants noted that it is easy to work late at home, which can affect the employee's work-life balance. This is therefore one aspect that should be monitored by organizations: "*I feel like they don't do work-life balance very well because they're always online. Which is great for the business because we tend to finish the work that we that we want to finish.*" [Participant-01]

5.3 *The Hierarchy Quadrant*

The Hierarchy quadrant relates to the internally focused initiatives of the organization to maintain stability, control, and predictability (Reino et al., 2020). The practices and values that are relevant to a hybrid agile team that manifests the Hierarchy cultural traits are discussed below.

Practices Relevant to the Hierarchy Quadrant in a Hybrid Agile Work Setup. For hybrid agile team members manifesting a hierarchy culture, there is a strong need for *planning and coordination*. For example, participants emphasized the importance of upfront planning especially during in-office days. They noted that on those days, new dynamics emerge and while in some instances teamwork might be enhanced, there is a risk of becoming ineffective due to excessive meetings, hence the need for planning and coordination: "*As a team, we do sit and define clear and specific goals. Once those goals are set, we have our regular check for instance, and they review to see what the progress is of achieving these goals.*" [Participant-08]

A need for group sessions was also highlighted by the participants for better alignment and coordination within the team: "*The bigger things that the organisation maybe does that kind of brings the teams together is the use of quarterly planning session. It tries to get everyone aligned and perhaps informed about what each team is doing and the timelines.*" [Participant-07]

While the availability of information is key to any business, it is however more important when working in a hybrid environment. Therefore, clear *mechanisms for knowledge sharing* is key for hybrid agile teams from a hierarchy culture. This not only includes documenting processes, incidents, and changes but also relates to documenting work practices: "*People are encouraged to document their knowledge in wiki pages on incidents, on new implementations they made, new improvements they made as really helps to bond the team together.*" [Participant-05]

Values Relevant to the Hierarchy Quadrant in a Hybrid Agile Work Setup. The Hierarchy quadrant focuses on stability. The participants aligned with this by noting that they value individual and team *accountability*. The hybrid way of working was noted to *affect* the accountability value if the organization does not manage it correctly: "*The values that we have within our team is a proper sense of accountability. We can rely on one another to do what needs to be done.*" [Participant-08]

Another key value highlighted by participants with a hierarchy cultural trait is *consistency*. They value consistency in how work is executed, reviewed, and delivered. Participants highlighted that the Scrum sprint review and retrospective practices as essential for maintaining this consistency in a hybrid environment: "*I think another thing also is we have reviews for most of our work that we do this. I think that also helps with the consistency of the code.*" [Participant-06]

According to participants, *transparency* is essential for teamwork and alignment with company goals, especially in a hybrid agile work environment. However, on remote days, transparency can suffer if key employees are unavailable, creating communication gaps: "*The catchups and engineering catchups where the entire department jumps on the call and catches up and just sort of gets an idea on what everyone's working on and maybe some interesting things that's happening.*" [Participant-09]

5.4 *The Market Quadrant*

The Market quadrant relates to the externally focused initiatives of the organization to grow the effectiveness of the market share and profits while focusing on stability more than flexibility (Reino et al., 2020). The practices and values that are relevant to a hybrid agile team that manifests the Market cultural traits are discussed below.

Practices Relevant to the Market Quadrant in a Hybrid Agile Work Setup. In the Market quadrant, for the hybrid agile teams to deliver business value, they must have sound practices in place to maintain stakeholder involvement: "*I am very involved with my stakeholder. This has helped me align with organizational goals.*" [Participant-08]. The participants indicated that it is important to have regular feedback from stakeholders in a hybrid agile work environment. Hence, teams work closely with their stakeholder and thus feel better aligned with the goals and strategies of the organization: "*I try to have regular chats with Y and X in terms of the organization's marketing goals or strategic initiatives just to make sure that we aligned. I also try and give them an exact view of where we are.*" [Participant-03]

Table 2 Summary of findings

Clan **Practices** • Team interaction and engagement • Ability to self-organize • Promoting psychological well-being **Values** • Trust • Empathy • Respect	***Adhocracy*** **Practices** • Use of innovative technologies **Values** • Flexibility
Hierarchy **Practices** • Planning and coordination • Mechanisms for knowledge sharing **Values** • Accountability • Consistency • Transparency	***Market*** **Practices** • Maintain stakeholder involvement **Values** • Proactiveness

Practices Relevant to the Market Quadrant in a Hybrid Agile Work Setup. In the Market quadrant, *Proactiveness* is a value that directly correlates with the practice of maintaining stakeholder involvement. While maintaining stakeholder involvement is essential in a hybrid agile work environment, this can only be achieved if team members demonstrate proactiveness as a value trait: "*I mean, whether the hybrid or not, but especially with hybrid, you need to make sure everyone is aware of what's happening. You can't just assume they're going to be in the office and everyone hears this.*" [Participant-04]

Table 2 summarizes the practices and values relevant to the Clan, Adhocracy, Hierarchy and Market quadrants.

6 Conclusion

The aim of this study was to explore the organizational values and practices in support of hybrid work during agile software development. This study was conducted in response the rise in hybrid work model after the COVID-19 pandemic, which further led to the need to better understand the values and practices that are needed to ensure a cohesive and supporting work environment in that work context. The study was conducted through a case study, during which semi-structured interviews and observations were held across various agile teams demonstrating various cultural traits. These cultural traits were in line with the CVF. That framework also informed the interview questions and the thematic analysis process.

The findings revealed a range of values and practices that are particularly relevant to the various types of cultural traits depicted in the CVF. Hybrid agile teams from the Clan quadrant should value trust, empathy, and respect. They should also favor practices that promote team interaction and engagement, an ability to self-organize, as well as the promotion of psychological well-being. This is in line with past studies that stated that in the Clan quadrant there is an emphasis on employee involvement, open and friendly communication, empowerment, managerial support, mentoring, and teamwork (Patrucco et al., 2022; Reino et al., 2020). In the Adhocracy quadrant, hybrid agile teams should value flexibility. They should also favor the use of innovative technologies to support them in their day-to-day work. This is in line with literature that stated that innovation (Hartnell et al., 2011), adaptability, creativity and risk-taking (Hartnell et al., 2011; Reino et al., 2020) are central to this quadrant. In the Hierarchy quadrant, hybrid agile teams should value accountability, consistency, and transparency. Practices related to planning and coordination as well as knowledge sharing should also be promoted. This is in line with past studies that stated that teams in this quadrant focus on structure and processes to enhance, review, and maintain their current way of working to improve efficiency (Hartnell et al., 2011; Reino et al., 2020). Lastly, hybrid agile teams from the Market quadrant should value proactiveness and ensure that they maintain good stakeholder involvement.

From a theoretical point standpoint, this study provides additional insights into organizational culture and hybrid work practices during agile software development. Through the lens of the CVF, key practices and values relevant to distinct cultural dimensions were identified, thus expanding our theoretical understand on this matter. From a practical standpoint, the study proposes specific practices that could be employed by organizations during their agile hybrid endeavors. The study also fills gaps in the literature around hybrid working models post-COVID-19. Given the short timeframe available to conduct the study, it was limited to a single organization within the FinTech space. Future research could widen the sample to provide a broader understanding of hybrid work practices and values across more diverse contexts. Comparing different types of organizations could provide insights into how size and resources influence the culture and effectiveness of hybrid practices. Hybrid work practices may differ across industries depending on work, market demands and cultures. A cross-industry comparison could reveal how other sectors (e.g., healthcare, education, or manufacturing) experience hybrid working. The study was limited to the insights of the teams directly involved in the software development process. Future research therefore includes the perspectives of other stakeholders for a broader perspective on the matter.

Disclosure of Interests The authors have no competing interests to declare that are relevant to the content of this article.

References

Ali, A. D., Narine, L. K., Hill, P. A., & Bria, D. C. (2023). Factors affecting remote workers' job satisfaction in Utah: An exploratory study. *International Journal of Environmental Research and Public Health, 20*(9), 5736.

Banerjee, P., & Gupta, R. (2024). A mixed-method exploration of effects of technostress on remote/hybrid working professionals. *Computers in Human Behavior, 150*, 107974.

Barhate, B., Hirudayaraj, M., & Nair, P. K. (2022). Leadership challenges and behaviours in the information technology sector during COVID-19: A comparative study of leaders from India and the U.S. *Human Resource Development International, 25*(3), 274–297.

Beck, K., Beedle, M., Bennekum, A. V., Cockburn, A., Cunningham, W., Fowler, M., Grenning, J., Highsmith, J., Hunt, A., Jeffries, R., Kern, J., Marick, B., Martin, R. C., Mellor, S., Schwaber, K., Sutherland, J., & Thomas, D. (2001). *Manifesto for agile software development*. https://agilemanifesto.org/

Braun, V., & Clarke, V. (2006). Using thematic analysis in psychology. *Qualitative Research in Psychology, 3*(2), 77–101.

Bryman, A., & Bell, E. (2011). *Business research methods* (3rd ed.). Oxford University Press.

Cameron, K. S., & Quinn, R. E. (2011). *Diagnosing and changing organizational culture: based on the competing values framework* (3rd ed.). Jossey-Bass.

Dalessandro, C., & Lovell, A. (2024). The pandemic-stratified workplace: Workspace, employee sense of belonging, and inequalities at work. *SAGE Open, 14*(1).

Dewalska-Opitek, A., & Nunez, R. V. (2022). New challenges in employee relationships in the digitalised post-pandemic organisational culture. *Journal of Modern Science, 49*(2), 253–274.

Gupta, M., George, J. F., & Xia, W. (2019). Relationships between IT department culture and agile software development practices: An empirical investigation. *International Journal of Information Management, 44*, 13–24.

Hartnell, C. A., Ou, A. Y., & Kinicki, A. (2011). Organizational culture and organizational effectiveness: A meta-analytic investigation of the competing values framework's theoretical suppositions. *Journal of Applied Psychology, 96*(4), 677–694.

Hasan, A., Muhammad Zeb Khan, D., & Ubaid ur, R. (2021). An empirical study on adoption of agile project management methodology and its effect on project success with moderating role of organizational culture. *European Journal of Social Impact and Circular Economy, 2*(1).

Iivari, J., & Iivari, N. (2011). The relationship between organizational culture and the deployment of agile methods. *Information and Software Technology, 53*(5), 509–520.

Irimia, T., & Pop, Z. C. (2022). Comparative approach upon organizational culture models. *Journal of Public Administration, Finance and Law, 23*, 154–167.

Kagerl, C., & Starzetz, J. (2023). Working from home for good? Lessons learned from the COVID-19 pandemic and what this means for the future of work. *Zeitschrift Für Betriebswirtschaft, 93*(1–2), 229–265.

Kemell, K.-K., & Saarikallio, M. (2023). Hybrid work practices and strategies in software engineering-emerging software developer experiences. *IEEE Access, 11*, 112861–112876.

McPhail, R., Chan, X. W., May, R., & Wilkinson, A. (2024). Post-COVID remote working and its impact on people, productivity, and the planet: An exploratory scoping review. *International Journal of Human Resource Management, 35*(1), 154–182.

Mishra, A., Abdalhamid, S., Mishra, D., & Ostrovska, S. (2021). Organizational issues in embracing agile methods: An empirical assessment. *International Journal of System Assurance Engineering and Management, 12*(6), 1420–1433.

Müller, L. S., Reiners, S., Becker, J., & Hertel, G. (2023). Long-term effects of COVID-19 on work routines and organizational culture—A case study within higher education's administration. *Journal of Business Research, 163*, 113927–113927.

Osei, M. B., Papadopoulos, T., Acquaye, A., & Stamati, T. (2023). Improving sustainable supply chain performance through organisational culture: A competing values framework approach. *Journal of Purchasing and Supply Management, 29*(2), Article 100821.

Patrucco, A. S., Canterino, F., & Minelgaite, I. (2022). How do scrum methodologies influence the team's cultural values? A multiple case study on agile teams in nonsoftware industries. *IEEE Transactions on Engineering Management, 69*(6), 3503–3513.

Peprah, O. E. (2024). Hybrid workplace: Current status, positives, negatives, challenges, and team learning. *The Learning Organization, 31*(1), 88–103.

Place, A. (2021). *The 8 most popular industries for hybrid work*. https://www.benefitnews.com/list/most-popular-industries-for-hybrid-work
Reemon, A. (2022). *Hybrid working—Advantages and disadvantages—Remote United*. RemoteUnitedFR. Retrieved March 10, 2024, from https://www.remoteunited.fr/en/hybrid-working-advantages-and-disadvantages/
Reino, A., Rõigas, K., & Müürsepp, M. (2020). Connections between organisational culture and financial performance in Estonian service and production companies. *Baltic Journal of Management, 15*(3), 375–393.
Reunamäki, R., & Fey, C. F. (2023). Remote agile: Problems, solutions, and pitfalls to avoid. *Business Horizons, 66*(4), 505–516.
Schwaber, K., & Sutherland, J. (2024). *Scrum guide*. Scrum Guides. Retrieved October 21, 2024, from https://scrumguides.org/scrum-guide.html
Shao, S., Martensen, M., Martensen, H., & Reindl, C. (2024). Is 'hybrid work' the new high-flying policy? Insights from the aviation industry. Gruppe. Interaktion. Organisation. Zeitschrift für angewandte Organisationspsychologie, *55*(1), 103–111.
Sithambaram, J., Nasir, M. H. N. B. M., & Ahmad, R. (2021). Issues and challenges impacting the successful management of agile-hybrid projects: A grounded theory approach. *International Journal of Project Management, 39*(5), 474–495.
Tuukkanen, V., Wolgsjö, E., & Rusu, L. (2022). Cultural values in digital transformation in a small company. *Procedia Computer Science, 196*, 3.
van Vuuren, M., Veldkamp, B. P., de Jong, M. D. T., & Seydel, E. R. (2008). Why work? Aligning foci and dimensions of commitment along the axes of the competing values framework. *Personnel Review, 37*(1), 47–65.
Windlinger, L., & Gerber, M. (2023). Functions and relevance of spatial co-presence: Lessons learned from the COVID-19 pandemic for evidence-based workplace and human capital management. *Frontiers in Built Environment, 9*.
Yin, R. K. (2014) *Case study research: Design and methods* (5th ed.). SAGE.

Privacy Challenges with Human Imposed and AI-Based Filters During Virtual Meetings

Rooma Ramasamy, Vinaye Armoogum, and Perienen Appavoo

Abstract Innovative technology has transformed the future of work, making remote communication the norm, users therefore rely on available features in video conference applications to safeguard their privacy. This study explores challenges caused by artificial intelligence (AI) filters specific to noise reduction and virtual backgrounds and human-imposed filters with the mute button feature during video conference calls through the lenses of the privacy paradox theory. Data triangulation shows that privacy filters' perceived efficacy frequently outweighs their real capabilities. Although users have faith in the tools, particularly those that are human imposed, there are still behavioral and technical challenges. The privacy paradox, in which users' confidence in their privacy choices coexists with constant risks is best shown by the gap in practice. The contradiction is particularly noticeable in AI-based filters, which offer smooth, intelligent defense but frequently fails in dynamic, individual environments. According to the study's findings, while users are protecting their boundaries, these filters create additional challenges. Platform developers, users, and organizations are suggested to adopt best practices and more user-centric design.

1 Introduction

The popularity for last-generation video conferencing applications (VCA) grew exponentially due to their widespread use and appeal. With the aid of such technology, persons from various locations can virtually meet (Nowroozi et al., 2025). Cloud-based video conferencing tools like Zoom, Microsoft Teams, Google Meet (Google Hangouts became Google Meet in April 2020), and Cisco Webex have been adopted across all industries at a very rapid rate for professional collaboration and staying in touch with family and friends (Suduc & Bizoi, 2022). The privacy paradox occurs

R. Ramasamy (✉) · P. Appavoo
Open University of Mauritius, Reduit, Mauritius
e-mail: rooma.ramasamy@gmail.com

V. Armoogum
University of Technology, Port Louis, Mauritius

J. Marx Gómez et al. (eds.), *Recent Advances in Information Systems*, Progress in IS,
https://doi.org/10.1007/978-3-032-14099-9_44

when people show concern about their privacy but engage in ways that may compromise it. According to research, users heavily rely on platform features to control their privacy, often without fully understanding the implications (Kokolakis, 2017). In order to mitigate background distractions and protect privacy, AI-based technologies such as noise suppression and virtual backgrounds are used. These tools, yet, are not always effective. For example, virtual backgrounds may not entirely hide the user's environment, and noise suppression may not be adequate to filter out certain sounds, which may intrude into the user's private life (He & Lei, 2021; Hwang et al., 2021). This study examined AI-based filters and human-imposed filters in video conference calls, combining qualitative observations with quantitative survey data. Both overt and covert observations in professional settings, alongside a survey of 181 respondents captured this pressing challenge faced today.

This study is significant since it addresses an essential knowledge gap regarding the way users in professional settings, perceive and use privacy-preserving features on digital communication platforms. Few studies incorporate empirical observation with user perception data to evaluate the practical and psychological aspects of online privacy. Previous research examined the technological functionality and ethical issues of platforms such as Zoom and MS Teams (He & Lei, 2021; Yang et al., 2022). The combination of both quantitative and qualitative methods reveals a broad landscape, while contributing to the literature on privacy paradox.

1.1 Related Works

Video conferencing applications (VCAs) are now crucial to remote communication in the social, educational, and professional spheres. In besides facilitating synchronous communication, these platforms bring forth novel concepts of privacy features and visibility in both home and professional settings. Through the use of digital filters and manual controls, users are increasingly facing trade-offs between exposure and involvement as virtual meetings replace meetings in person. The goal of these tools is to improve user control, and current literature shows that they often face challenging issues with regarding technological confidence, interface design, and user awareness (Fauville et al., 2021; Yang et al., 2022). VCAs must have access to a device's camera and microphone in order to work properly. During a call, VCAs provide user two privacy control options: muting the microphone and turning off the camera. The user can determine if an app is using their camera by looking at a visible hardware indication, for example a light next to the camera (Yang et al., 2022). Since the beginning of 2010, researchers have been examining the security and privacy of video conferencing platforms. Numerous studies have also been done on user perceptions and privacy concerns related to the implications of microphone-equipped devices. Emami-Naeini et al. (2021) recently conducted an online user study to better understand users' issues about VCAs. Users' concerns about the security and privacy features of VCAs were discovered. Additionally, they discovered that people view the mute button as a privacy control and that they feel privacy violations when they neglect to push the mute button. Yang et al. (2022) in another recent study report that users are not aware that Webex is listening to their silenced microphone. The authors looked at all commonly used desktop operating systems and VCAs and identified a potential privacy leak within Webex. They found that even with the microphone muted, Webex continues to read audio data from the device and sends statistics of that data once per minute to their telemetry servers. They also snooped unencrypted copies of the telemetry data prior to transmission using runtime binary analysis tools. To create a data set for classification, they employed synthetic background noise for more than 180 h. Using intercepted outgoing telemetry packets to identify six common background activities when a user is muted, their classifier obtained an 81.9% macro accuracy. Vendors of operating systems can create a more robust permission mechanism for even when a user is muted by developing an OS-level software silence for the microphone. This research demonstrate that Webex is still sending audio-derived data. Numerous possibilities exist for VCAs to compromise a user's video privacy. According to Kagan et al. (2020), collage photos of video conference meetings published on open websites may reveal private information including the names, ages, and genders of users. When encrypted metadata and multimedia files are shared during VCA meetings, Altschaffel et al. (2021) demonstrated that the traffic patterns of these files can be utilized to detect increased activity in front of cameras or even to recognize users. Concerns exist around the data that VCAs gather about their users. For instance, Consumer Reports found privacy issues with the methods used by well-known VCAs like Zoom, Google Meet, Microsoft Teams, and Cisco Webex while collecting data. These issues concentrated on the objectives of gathering meeting metadata.

Hilgerforte et al. (2021) argue that workers' private living spaces immediately communicate visual content to outside parties, potentially disclosing sensitive information. Calls made from locations other than office environments can provide a privacy risk because they send visual content to third parties. Coworkers and employers may unwittingly learn details about personal living situations, relationships, and preferences among others. In order to mitigate this risk, video conferencing services encourage using a virtual background (VB) instead of the actual scene during a video chat. Many providers of video conferencing services have built virtual background generation algorithms into their software. These techniques allow for the automatic detection of the background behind a person and the replacement of that background with an image, a process known as picture matting in computer vision. Nevertheless, this replacement is flawed, with little patches of the real background visible in video frames and conclude that conclude that different defenses are required and that virtual backgrounds only offer a limited level of security. However, Hilgerforte et al. (2021) identify limitations to their study following diversity to test cases as the scenarios to be tested is quite exhaustive and recommend extending with various backgrounds, rooms, cameras and objects. In order to mitigate background distractions and protect privacy, AI-based technologies such as noise suppression and virtual backgrounds are used. These tools, yet, are not always effective. For example, virtual backgrounds may not entirely hide the user's environment, and noise suppression may not be adequate to filter out certain sounds, which may intrude into the user's private life (Hwang et al., 2021). Likewise, to Hilgerforte et al. (2021), He and Lei (2021) discuss that these platforms and businesses continue to actively optimize client feedback and usage across the globe while introducing a number of additional functions to safeguard the security of meetings. A privacy protection design that has been widely used on several video conference platforms is the VB. Users might use VB to reduce distractions, keep the environment in good condition, prevent privacy leaks, and meet their aesthetic needs. Moreover, users of VB can modify images, blur backgrounds, or blur backgrounds altogether. Although the VB has given users a good opportunity to display their personal choices (taste in aesthetics or personal preferences), as well, as meet business needs, people have begun to worry about whether the VB will leak their personal information or not. The segmentation algorithm's instability also has its limits in terms of disclosing personal information about individuals' environments or prospective privacy. Lin et al. (2021) found that the performance and accuracy of the algorithm have been directly impacted by issues like background misclassification, and the segmentation method's flaw would result in an unstable virtual background. Together with the aliasing and other segmentation issues, there is a concern with privacy exposure that could result in information leaking to the environment behind the participants. According to Nowraazi et al. (2025) users of video conferencing services can hide their real location, eliminate distractions, or cover up other persons in the room by substituting a virtual background for a real one. A worker, however, can also try to deceive his or her employer by altering the actual background image or taking into account a movie with a virtual backdrop. This potential also opens the door for making fake videos in which actors pose as being in various locations. Sabra et al. (2022) built a new framework for background reconstruction and used it to show users' true backgrounds. A thorough assessment of the suggested framework and related privacy assaults was then carried out using a sizable dataset of call footages that was gathered from human subject participants. The findings of these tests reveal that significant background information leakage is possible in some circumstances, making the feature inefficient at safeguarding users' privacy and creating a false impression of security. Prange et al. (2022) in their findings demonstrate that online meetings put users' privacy at risk since they offer audio and video-based insights into their personal environments. Personal information that otherwise would have remained private could be disclosed. This could involve private information like nudity or family and close relationships.

The literature leaves room for further investigation. There is limited research in privacy literature toward human imposed such as the mute button and AI-based filters as in noise suppression and virtual backgrounds. As such, challenges and shortcomings of human imposed, and AI-based filters for privacy during virtual meetings will be examined. The theoretical grounding of this study is based on the concept of the privacy paradox, which challenges the gap between individuals' privacy concerns and their actual behaviors. Mute buttons, noise suppression, and virtual backgrounds have been designed to enhance privacy, however, unforeseen privacy risks are introduced. This paradox is apparent in the disparity between users' perceived and actual privacy. Individuals trust these tools to safeguard their personal space, but the findings reveal inconsistencies that challenge their expectations.

2 Methodology

This study adopted a mixed-method approach, observation supported by quantitative data to explore the challenges associated with both AI-based and human-imposed filters during Microsoft Teams virtual meetings. Two observation modes were used: Overt observation was conducted during scheduled qualitative interviews, where participants were informed about the study and consented to being observed. Specific platform features which consist of virtual backgrounds, noise suppression, and the mute button were observed while engaging in natural conversations. During real-time professional events, webinars, training sessions, and public townhalls there was an opportunity for covert observation allowing the capture of more authentic behavior and usage patterns. Data collection included field notes, contextual observations. In regard to the overt observation, participants were informed of the purpose, procedures, contribution and right to withdraw at any moment during the research. A consent sheet with key information was shared prior to data collection, and data were anonymized to preserve confidentiality. Covert observation was adopted to examine publicly accessible real-life behaviors in virtual meetings (e.g., camera use, mute function), in open virtual spaces. This was deemed ethically permissible as the observed behaviors were not intrusive and were already visible to other participants, and did not involve the collection of personal identifiers. Additionally, there was no interaction between the researcher and the subjects being studied. The principle that observation of publicly visible online behavior, where there is no reasonable expectation of privacy, can be ethically conducted without prior consent (Eynon et al., 2008). The methods used was strictly aligned to the research objectives and approved by the ethics committee of the Open University of Mauritius after being weighed against potential harm.

To reach respondents from the ICT industry working in virtual settings, a purposive and snowball sampling approach was adopted considering the challenge in identifying the previously unknown population. Direct emails has been sent to organization members of the Outsourcing and Telecommunications Association of Mauritius (OTAM), and recruitment efforts was also done on LinkedIn. Selected participants were encouraged to refer their peers, allowing the population to expand but however maintaining the objectivity of the scope. Additionally, to complement the insights obtained, quantitative data was gathered by means of surveys to have an overview of perceived effectiveness of AI and human-imposed filters. Both quantitative descriptive statistics and qualitative insights were used for analyzing the information gathered from surveys and observations. To find recurrent patterns regarding participants' opinions on AI-based and human-imposed filters during virtual meetings, field notes were analyzed for the qualitative component. Descriptive statistics assisted in triangulating the qualitative findings, to ensure an extensive understanding of the phenomena.

3 Data Analysis

Both quantitative descriptive statistics and qualitative insights were used for analyzing the information gathered from surveys and observations. To find recurrent patterns regarding participants' opinions on AI-based and human-imposed filters during virtual meetings, field notes were analyzed for the qualitative component. To ensure an extensive understanding of the phenomena, descriptive statistics assisted in triangulating the qualitative findings. Using a triangulated approach with the combining survey responses, observational data, and existing literature, this research question investigated the effectiveness and challenges of filters applied by humans and artificial intelligence in video conferencing. The results indicate that although respondents believe these filters to be useful instruments for safeguarding privacy, scholarly research and qualitative observations point to significant constraints that raise doubt on this assumption. The complexities of the privacy paradox is reflected, where individuals show trust in privacy-preserving solutions yet may not be aware of their technological limitations.

3.1 *Use of AI-Based Filters Such as Virtual Background*

A significant majority (82.8%) of respondents either agree or strongly agree that using AI-based filters such as virtual backgrounds is effective to preserve their privacy at home. Only a small portion (4.4%) disagree or strongly disagree, and 12.7% remain neutral (Fig. 1).

UsingAIbasedfilterssuchasvirtualbackgroundiseffective

		Frequency	Percent	Valid Percent	Cumulative Percent
Valid	Strongly Disagree	3	1.7	1.7	1.7
	Disagree	5	2.8	2.8	4.4
	Neither agree nor disagree	23	12.7	12.7	17.1
	Agree	50	27.6	27.6	44.8
	Strongly Agree	100	55.2	55.2	100.0
	Total	181	100.0	100.0	

Fig. 1 Use of AI-based filters such as virtual background

UsingAIbasedfilterssuchasnoisesuppressioniseffective

		Frequency	Percent	Valid Percent	Cumulative Percent
Valid	Strongly Disagree	5	2.8	2.8	2.8
	Disagree	14	7.7	7.8	10.6
	Neither agree nor disagree	44	24.3	24.4	35.0
	Agree	46	25.4	25.6	60.6
	Strongly Agree	71	39.2	39.4	100.0
	Total	180	99.4	100.0	
Missing	System	1	.6		
Total		181	100.0		

Fig. 2 Use of AI-based filters such as noise suppression

3.2 Use of AI-Based Filters Such as Noise Suppression During Virtual Meetings

A majority (64.8%) of respondents either agree or strongly agree with the use of AI-based filters such as noise suppression during virtual meetings preserves their privacy. Meanwhile, 10.6% disagree or strongly disagree, and 24.3% are neutral (Fig. 2).

3.3 Effectiveness of Human-Imposed Filters Such as Using the Mute Button

A significant majority (88.4%) of respondents either agree or strongly agree that human-imposed filters such as the mute button during virtual meetings preserve their privacy. Only a small portion (1.7%) disagrees, and 9.9% are neutral (Fig. 3).

Humanimposedfilterssuchashavingthemutebuttonon

		Frequency	Percent	Valid Percent	Cumulative Percent
Valid	Disagree	3	1.7	1.7	1.7
	Neither agree nor disagree	18	9.9	9.9	11.6
	Agree	63	34.8	34.8	46.4
	Strongly Agree	97	53.6	53.6	100.0
	Total	181	100.0	100.0	

Fig. 3 Use of human-imposed filters such as mute button

Humanimposedfilterssuchswitchingoffcamera

		Frequency	Percent	Valid Percent	Cumulative Percent
Valid	Strongly Disagree	2	1.1	1.1	1.1
	Disagree	9	5.0	5.0	6.1
	Neither agree nor disagree	55	30.4	30.6	36.7
	Agree	49	27.1	27.2	63.9
	Strongly Agree	65	35.9	36.1	100.0
	Total	180	99.4	100.0	
Missing	System	1	.6		
Total		181	100.0		

Fig. 4 Effectiveness of human-imposed filters such as switching off camera

This suggests that most respondents feel confident in the effectiveness of human-imposed filters.

3.4 Effectiveness of Human-Imposed Filters Such as Switching off the Camera During Virtual Meetings

A significant majority (88.4%) of respondents either agree or strongly agree that the use of human-imposed filters such as switching off camera during virtual meetings preserve privacy. Only a small portion (1.7%) disagrees, and 9.9% are neutral (Fig. 4).

3.5 Challenges of Human-Imposed, and AI-Based Filters for Privacy During Virtual Meetings

Over a period of three months, this study used both overt and covert methods of observation to investigate the challenges with both AI-imposed and human-imposed filters during MS Teams meetings. Participants gave their consent before engaging in the overt observations after being thoroughly briefed about the purpose of the study. These observations took place primarily during qualitative interviews, in which participants were asked to test specific features of the platform in in addition to participating in conversations. This included deliberate use of AI-imposed filters, such noise suppression and virtual backgrounds, as well as human-imposed filters, such as the mute button. The objective was to explore how these features affected communication, and protection of privacy while working in the home office in a controlled but natural environment. Covert observation was conducted in a range of real-world professional-related scenarios to support and strengthen the results.

To prevent influence of behavior, these included webinars on MS Teams, public townhalls and training sessions where participants were not aware that their usage of filters was observed. The instances when the filters provided constraints or challenges to technology while observing natural and unexpected interactions in various settings was analyzed.

Contextual observations during and right after each session, thorough note-taking, and voice recording were among the data gathering methods used to explore challenges with:

- AI-based filters: particularly virtual backgrounds and noise suppression functionality
- Human-imposed filters: primarily the mute button.

These methods provided both breadth and depth to the understanding of how such filters influence real-time communication in professional virtual environments.

3.6 Human-Imposed Filters

A selection of concerns has been observed with the mute button in video conferencing platforms, specific in covert observation. Participants frequently failed to remember they were muted, often speaking without recognizing they were muted, which resulted in communication breakdowns. Another frequent issue was that attendees failed to mute themselves when required, which caused unwanted background noise and disclosure of private information. Notably, two scenarios were captured where participants appeared muted on screen but were however clearly transmitting audio, indicating possible technical failures or misinterpretations of the user interface. Under overt surveillance, no mute-related challenges were observed, which would suggest that participants are more conscious when they are aware that they are being observed. Yang et al. (2022), who investigated at the way mute options functioned in video conferencing platforms, corroborate these findings. In accordance with their research, certain apps access and send audio data even when the mute button is on, creating issues with privacy and technological inconsistencies. These technological limits are consistent with situations where individuals who were muted were however heard during covert sessions. Additionally, the observed trend of disregarding and delayed responses is consistent with recent user behavior research that revealed users regularly confuse the status of their mute button, which results in awkward or disrupted communication (Fauville et al., 2021).

3.7 AI-Imposed Filters

Thirteen respondents acknowledged using noise suppression features in their video conferencing platforms during overt observation, but two did not given that their environments did not require background noise suppression. A number of challenges were identified:

3.7.1 Noise Suppression

- **DV**: Despite activating noise suppression, clear conversations were audible in her background.
- **LD**: Dog's barking was clearly heard, indicating that noise suppression may not effectively filter out pet-related sounds. The participant's opinion was noise suppression functionality does not work with abrupt noises.
- **KH**: The spouse, also on a video call in the same room, was clearly audible, suggesting limitations in suppressing simultaneous human voices.
- **RH**: Intermittent dog barking was heard, pointing to challenges in handling abrupt noises.
- **MT**: Working in a shared office space, some background noise persisted, highlighting potential issues in open-plan environments.
- **VA**: The humming of the air-conditioning was distracting, indicating that continuous low-frequency noises might not be adequately suppressed.

These observations are consistent with research findings. According to a study by Schejbal (2017), room noise reduction techniques including signal processing and optimal microphone placement are meant to reduce background noises like keyboard tapping and air conditioner hums. Nevertheless, those methods might not be able to completely eliminate all environmental noises.

3.7.2 Virtual Backgrounds

Fifteen participants used virtual backgrounds in the video conferences meeting during overt observations. Two of the participants selected personalized backgrounds with pictures of their dogs and family. It was observed that these customized backgrounds were aesthetically distracting and can draw focus away from the discussion. Standard office or company-branded backgrounds, which usually featured logos and neutral settings, were chosen by the remaining participants. Some aspects of the participants' actual living environments were still obvious even with the usage of virtual backgrounds. Pets, family members, Christmas trees, couches, bookcases, and furnishings like curtains, lamps, televisions, and family portraits were all visible in the backgrounds. These components unintentionally gave participants views into their private lives, which could have an impact on how they perceive privacy and professionalism.

4 Discussion

Although the technologies have been designed to improve digital communication's privacy, control, and professionalism, the observational findings, both overt and covert show that they also bring up new conflicts and tensions, which are elements of the privacy paradox. Usually thought of as a simple privacy control, the mute button turned out to have unexpected flaws. Communication failures and delays resulted from participants' frequent forgetting they were muted or speaking while muted during covert surveillance. More importantly, there were two recorded occasions in which participants seemed muted, but background noise could still be heard. Yang et al. (2022) provided support for these findings by demonstrating that several video conferencing services access and send audio data even when users think they are muted. One of the main aspects of the privacy paradox is the disparity between perceived and actual privacy, which is highlighted by this, undermining user expectations and confidence (Yang et al., 2022). According to the quantitative data, the best methods for managing privacy in virtual environments are thought to be human-imposed filters. The mute button and camera off features both protect respondents' privacy, according to a sizable majority of respondents (88.4%) who agree or strongly agree. 9.9% of respondents were neutral, while just 1.7% disagreed with this evaluation. The relative ease of use and visibility of these controls, which provide consumers direct agency over their involvement, lend credence to this general trust.

Regarding noise suppression, the functionality demonstrated inconsistencies. Even though 13 participants used noise suppression option, a number of them continued to hear ambient hums, background activities from people who were nearby, and the barking of pets and family talks. Participants reported that it was especially challenging to conceal sudden or human-like sounds. These results are consistent with Mueller et al. (2006), who found that, especially in acoustically complex situations, digital noise reduction improves but does not ensure the removal of disturbing sounds. Users might therefore overestimate noise suppression's ability to safeguard their auditory environment, which would further reinforce the privacy paradox. According to survey data, 64.8% of participants believe that noise suppression protects their privacy, however this belief is not as strong as that of human filters. Users appear to be more wary or doubtful about the efficacy of AI-based audio filtering, as seen by the notable 24.3% who remained indifferent and the 10.6% who disapproved or strongly disagreed.

Despite being designed to preserve professionalism and protect visual privacy, virtual backgrounds produced a range of results. Personalized backgrounds, including pictures of pets or family members, were seen to be distracting and occasionally disclosed more private information than was intended. Spots of real-life environments, such as family members, dogs, furniture, and decorations, were visible even against conventional or branded virtual backgrounds. These results are consistent with studies by Cook et al. (2023) and Li and Zhang (2024), which demonstrate that poorly selected backdrops might unintentionally enhance exposure rather than concealing and that virtual backgrounds affect perceived professionalism and user behavior 82.8% of respondents agreed or strongly agreed that virtual backgrounds assist protect privacy at home, indicating a high level of confidence in them. 12.7% were neutral, while just 4.4% disagreed. These findings suggest that visual filtering is widely accepted as a privacy solution, particularly for concealing private spaces in homes.

AI-imposed noise suppression poses the most difficult and enduring problems of the three filters. Users may overestimate its capabilities, which could result in communication failures or unintentional privacy violations, and its effectiveness is unpredictable. In contrast, mute buttons and virtual backgrounds are less problematic because they are more visible and controllable by the user. A challenge that calls for more technology advancement and user education is the unpredictable performance of noise suppression in dynamic real-world settings. As a whole, the results of this study highlight a fundamental conflict in the privacy paradox: participants express privacy concerns and employ technology to reduce their exposure, yet these same tools can also provide a false sense of security or unintentionally divulge more than they intended. Although participants trust the human imposed or AI elements, privacy is only partially achieved and occasionally compromised due to technological, contextual, or user understanding restrictions. This discrepancy between perceived control and real vulnerability highlights how traditional ideas of privacy are complicated in digital situations (Table 1).

Data triangulation shows that privacy filters' perceived efficacy frequently outweighs their real capabilities. Although users have faith in the tools, particularly those that are human imposed, there are still behavioral and technical challenges. The privacy paradox, in which users" confidence in their privacy choices coexists with constant risks is best shown by the gap between attitude and practice. The contradiction is particularly noticeable in AI-based filters, which offer smooth, intelligent defense but frequently fails in dynamic, individual environments.

Table 1 Data triangulation

Filter type	Perceived effectiveness survey	Observed challenges	Literature support
Mute button	Very high (88.4%)	Participants forgetting to unmute; audio leaks despite being muted	Yang et al. (2022)
Noise suppression	Moderate (64.8%)	Dogs barking, background conversations, ambient noise not filtered	Mueller et al. (2006)
Virtual backgrounds	High (82.8%)	Personalized photos distracting; real environments still visible	Cook et al. (2023), Li and Zhang (2024)

5 Limitations, Recommendation, and Conclusion

Despite the insightful information this study offered, there are a few limitations to take into account. First, only participants from a particular demographic group were included in the sample, which might not adequately represent the range of experiences that all users of virtual meetings have. Additionally, the study focused largely with Microsoft Teams as a virtual meeting platform, therefore the results might not apply to other platforms like Zoom or Google Meet that offer different user interfaces or filter features. During overt observation participants were aware that they were being watched. This might have affected their behavior, background protection and filter usage, leading to overestimate their effectiveness. Furthermore, the study's dependence on self-reported survey answers might not have given accurate information about their honest views.

Several recommendations can be made based on the findings. Users must be made mindful of the constraints of artificial intelligence-imposed filters. The fact that these filters are not flawless and might not entirely safeguard privacy should be acknowledged by users. User education is also key on adopting best practices during meetings. The efficiency of AI filters needs to be continually enhanced. To handle a greater range of sounds, such as abrupt noises and multiple voices in the background, developers must enhance the efficiency of noise suppression algorithms. Virtual backgrounds could also be improved to better conceal private spaces without decreasing video quality or creating visual distractions. Lastly, the relationship between user privacy expectations and the actual effectiveness of these filters needs further investigation.

Disclosure of Interests The author has no competing interests to declare that are relevant to the content of this article.

References

Altschaffel, R., Hielscher, J., Kiltz, S., & Dittmann, J. (2021). Meta and media data stream forensics in the encrypted domain of video conferences. In *Proceedings of the 2021 ACM Workshop on Privacy in the Electronic Society* (pp. 1–10). https://doi.org/10.1145/3452140.3452143

Conti, M., Milani, S., Nowroozi, E., & Orazi, G. (2021). Do not deceive your employer with a virtual background: A video conferencing manipulation-detection system. https://arxiv.org/abs/2106.15130

Cook, R., Smith, A., & Johnson, L. (2023). Zoom backgrounds affect judgments of trust and competence. *PLoS ONE, 18*(9), Article e0291444. https://doi.org/10.1371/journal.pone.0291444

Emami-Naeini, P., Francisco, T., Kohno, T., & Roesner, F. (2021). Understanding privacy attitudes and concerns towards remote communications during the COVID-19 pandemic. In *Proceedings of the Seventeenth Symposium on Usable Privacy and Security (SOUPS 2021)* (pp. 695–714). USENIX Association. https://www.usenix.org/conference/soups2021/presentation/emami-naein

Eynon, R., Fry, J., and Schroeder, R. (2008). The ethics of Internet research. In N. Fielding, R. M. Lee, & G. Blank (Eds.), The SAGE Handbook of Online Research Methods (pp. 23–41). London: SAGE Publications. https://doi.org/10.4135/9780857020055.n2

Fauville, G., Luo, M., Queiroz, A. C. M., Bailenson, J. N., & Hancock, J. (2021). Zoom exhaustion & fatigue scale. *Computers in Human Behavior Reports, 4*, Article 100119. https://doi.org/10.1016/j.chbr.2021.100119

He, S., & Lei, Y. (2021). The privacy protection effectiveness of the video conference platforms' virtual background and the privacy concerns from the end-users. https://arxiv.org/abs/2110.12493

Hilgefort, J. M., Arp, D. & Rieck, K. (2021). Spying through virtual backgrounds of video calls *Proceedings of the 14th ACM Workshop on Artificial Intelligence and Security*, ACM, New York, USA, pp. 135–144. https://doi.org/10.1145/3474369.3486870

Hwang, A.H.-C., Yao, W., Yao-Yuan, Y., & Stevenson, W. (2021). Hide and seek: Choices of virtual backgrounds in video chats and their effects on perception. *Proceedings of the ACM on Human-Computer Interaction, 5*(CSCW2), 1–30. https://doi.org/10.1145/3476044

Kagan, D., Fuhrmann Alpert, G., & Fire, M. (2020). Zooming into video conferencing privacy and security threats. https://arxiv.org/abs/2007.01059

Kokolakis, S. (2017). Privacy attitudes and privacy behaviour: A review of current research on the privacy paradox phenomenon. *Computers & Security, 64*, 122–134. https://doi.org/10.1016/j.cose.2015.07.002

Li, B. J., & Zhang, H. (2024). Exploring the links between type and content of virtual background use during videoconferencing and videoconference fatigue. *Frontiers in Psychology, 15*, 1408481. https://doi.org/10.3389/fpsyg.2024.1408481

Lin, S., Ryabtsev, A., Sengupta, S., Curless, B.L., Seitz, S.M. & Kemelmacher-Shlizerman, I. (2021). Real-time high-resolution background matting In *Proceedings of the IEEE/CVF Conference on Computer Vision and Pattern Recognition,* pp. 8762–8771. https://doi.org/10.1109/CVPR46437.2021.00865

Mueller, H. G., Weber, J., and Hornsby, B. W. Y. (2006). The Effects of Digital Noise Reduction on the Acceptance of Background Noise. *Trends in Amplification, 10*(2), 83–94. https://doi.org/10.1177/1084713806289553

Nowroozi, E., Mekdad, Y., & Conti, M. (2025). Real or virtual: a video conferencing background manipulation-detection system. *Multimedia Tools and Applications 84,* 28157–28189 . https://doi.org/10.1007/s11042-024-20251-6

Prange, S., Delgado Rodriguez, S., Mecke, L., & Alt, F. (2022). Exploring privacy challenges during video-based online meetings. In *Proceedings of the 21st International Conference on Mobile and Ubiquitous Multimedia* (pp. 1–6). Association for Computing Machinery. https://doi.org/10.1145/3568444.3568450

Sabra, M., Maiti, A. and Jadliwala, M. (2022). Background Buster: Peeking through virtual backgrounds in online video calls In *52nd Annual IEEE/IFIP International Conference on Dependable Systems and Networks (DSN),*Baltimore, MD, USA.

Schejbal, J. (2017). Room noise reduction in audio and video calls. Technical Disclosure Commons. Available at: | Technical Disclosure Commons Research

Suduc, A. M., & Bizoi, M. (2022). AI shapes the future of web conferencing platforms. *Procedia Computer Science, 199*, 1234–1241. https://doi.org/10.1016/j.procs.2022.01.156

Yang, Y., West, J. W., Thiruvathukal, G. K., Klingensmith, N., & Fawaz, K. (2022). Are you really muted?: A privacy analysis of mute buttons in video conferencing apps. *Proceedings on Privacy Enhancing Technologies, 2022*(2), 373–393. https://doi.org/10.2478/popets-2022-0030

A Systematic Literature Review on Multimodal Abstractive Text Summarisation Using Transformer-Based and Attention Models

Tchouya'a Ngoko Israel Christian and Boniface Kabaso

Abstract The easy access to technology and the COVID-19 outbreak has accelerated communication among users through various multimedia formats (text, image, audio, and video). This has resulted in the mass production of multimodal information and has created challenges among many users in selecting relevant content, which is deeply buried within vast multimodal data generated. To address this issue, multimodal summarisation has emerged, aiming to enhance user experience by generating representative visual and textual summaries. However, current methods often lack semantic coherence and grammatical accuracy due to the complexity of integrating multiple modalities. This paper conducts a systematic review of multimodal abstractive texts summarisation techniques using PRISMA framework, with a focus on transformed-based models including attention mechanisms. It analyses recent studies, theirs methods, datasets, and performance based on ROUGE metrics. The findings indicate that approaches such as hierarchical attention and BERT-based models improve summarisation quality but still faced challenges including data alignment, computational cost, and limited datasets diversity. ROUGE benchmarking reveals inconsistent results with How2 and Dailymail datasets frequently used. The study highlights the need better modality fusion, improve generalisation and dataset diversity in future research.

T. N. Israel Christian (✉) · B. Kabaso
Department of Information Technology and Design, Cape Peninsula University of Technology, Cape Town, South Africa
e-mail: tchouya'angokoi@cput.ac.za

B. Kabaso
e-mail: kabasob@cput.ac.za

J. Marx Gómez et al. (eds.), *Recent Advances in Information Systems*, Progress in IS,
https://doi.org/10.1007/978-3-032-14099-9_45

1 Introduction

In today's digital age, information is generated across various media formats (text, image, audio, and video) on every topic and shared across multiple platforms. The ease of online publishing has also contributed to this rapid information growth (Allahyari et al., 1707). According to Rafi and Das (2023), it is a difficult task and time-consuming for humans to manually generate an abstract since it requires deep analysis. This makes multimodal summarisation an essential task for users to quickly access key information (Tang et al., 2023). Creating manual summaries requires human cognitive abilities, relying on prior knowledge and understanding. Automatic summarisation, which began in 1950s (Mridha et al., 2021), and over time, various algorithms have been developed to assist humans in addressing the challenges of summarisation (Khurana et al., 2023).

Text summarisation is categorised as extractive, abstractive, or a hybrid of both approaches (El-Kassas et al., 2021). Abstractive summarisation generally uses beam search or greedy search mechanism to decode the output summary, while extractive summarisation uses scoring mechanism during decoding to identify the salient, non-redundant, and readable elements from the input to the final output (Jangra et al., 2023). Hybrid approach combines both extractive and abstractive methods to produce a summary. Recent progress in summarisation techniques has been driven by the development of advanced pre-trained NLP models such as BERT, T5, BART, etc. (Kieuvongngam et al., 2006), which often use attention mechanisms (Vaswani et al., 2017). There has been significant improvement in multimodal abstractive text summarisation techniques recently (He et al., 2023).

This review aims to enhance the understanding of the literature on multimodal abstractive summarisation by evaluating techniques, and their performance, as well as identifying research gaps. It compares existing solutions by examining their methods, limitations, and ROUGE scores, while proposing for effective improvements for more effective multimodal abstractive summarisation algorithms.

This paper is structured as follow: Sect. 2 discusses existing work of multimodal summarisation, while Sect. 3 discusses the methodology of the study. Section 4 presents the findings and discussions, followed by Sect. 5, which address threats to validity. Finally, Sect. 6 concludes the paper.

2 Exiting Work of Multimodal Summarisation

The purpose of multimodal abstractive summarisation is to generate a summary of a topic by integrating information from various modalities such as text, video, audio, and image (Palaskar et al., 1906). It involves several difficulties, including data complexity, computational intensity, extraction, and fusion. Previous studies have primarily focused on issues such as single-modality summarisation, data alignment, and modality fusion techniques (Gandhi et al., 2023).

Li et al. (2018) proposes an extractive multimodal summarisation method for asynchronous collections of texts, images, and videos. This method integrates Natural Language Processing (NLP), speech processing, and computer vision techniques to uncover valuable information within multimodal sources, thereby improving the quality of multimedia news summarisation. This approach addresses readability issues by selectively integrating audio transcriptions using guided strategies. Specifically, it uses a novel graph-based model to effectively compute the salience score for each text unit, resulting to summaries that are both readable and informative. Experiment results from our English corpus reveal the system's effectiveness in using multimodal information.

Zhu et al. (2018) present a multimodal attention model designed to simultaneously generate text and identify the most important image from multimodal inputs. Additionally, they develop a novel Multimodal Automatic Evaluation (MMAE) approach that assess multimodal outputs by considering both intra-modality salience and inter-modality relevance. Experimental results show the effectiveness of MMAE, and its versatilities can be applied to other modalities.

Khullar and Udit (2010) suggest multimodal abstractive text summarisation (MAST) that uses information from three modalities (text, audio, and video) to generate abstractive text summaries. This model employs a sequence-to-sequence trimodal hierarchical attention-based approach, allowing to focus more on text modality. MAST offers a robust solution for multimodal abstractive summarisation by integrating trimodal hierarchical attention, significantly enhancing the performance in capturing nuances across various modalities. This advancement enhances summarisation techniques for text, image, and audio data. The proposed trimodal hierarchical attention mechanism outperforms existing unimodal and bimodal approaches, highlighting the importance of considering all three modalities in abstractive summarisation tasks.

Fu et al. (2009) introduce a multimodal summarisation task designed to summarise a document and its associated video. They developed a baseline model with effective strategies such as bi-shop attention and improved late fusion mechanism to connect different modalities. Additionally, they implement a bi-stream summarisation strategy to use both text and video summarisation simultaneously. Experimental results demonstrate the mode is effectiveness for multimodal summarisation, outperforming existing methods.

Atri et al. (2021) propose a Factorized Multimodal Transformer-Based Language Model (FLORAL), designed to capture both intra-modal and inter-modal dynamics within various input modalities for text summarisation. It uses an increasing number of self-attentions to capture multimodality and performs significantly better than traditional encoder-decoder based networks. Experimental results on the How2 and AVIATE datasets reveal significant improvement in both quantitative and qualitative compared to several unimodal and multimodal baselines. FLORAL achieves consistently outperforms other proposed models across both datasets.

Jangra et al. (2023) provide an extensive review of existing work depending on various input, output, and technique-related details. They discuss various techniques used to address multimodal summarisation tasks, and the evaluation metrics used to assess the quality of summaries produced. The authors also propose a generic neural model to generate summaries from multimodal inputs. Additionally, the paper highlights research gaps and recommends future directions, emphasizing the need for developing advanced models and improved evaluation frameworks to propel the field of multimodal summarisation.

Qiu et al. (2022) developed a Multimodal Hierarchical Multimedia Summarisation (MHMS) framework designed to generate multimodal summaries from multimedia sources. MHMS framework organises the algorithm into distinct functional modules for video temporal segmentation, textual segmentation, visual segmentation, and multimodal alignment. Using hierarchical approach, it begins by extracting key elements from each modality and integrating them into a cohesive summary. MHMS framework addresses challenges related to data alignment, semantic coherence, and scalability, while ensuring comprehensive content by capturing complementary information from diverse sources. This method improves the presentation and consumption of information in applications such as news, education, and social media.

Baek et al. (2023) present Video-Audio-Text Multimodal Abstractive Summarisation (VATMAN) with trimodal hierarchical multi-head attention, which represents a cutting-edge model for comprehensively summarising content across video, audio, and text modalities. VATMAN simultaneously focuses on relevant aspects within each modality, ensuring a comprehensive understanding of multimodal content. Remarkably, its abstractive summarisation capability enables it to generate human-like summaries by selectively focusing on crucial information. The hierarchical structure enhances its ability to capture details at various levels, demonstrating its versatility across diverse multimodal datasets and making a significant advancement in the field of multimodal summarisation.

Khilji et al. (2023) investigate the integration of multiple data types (text, images, and audio) into the summarisation process to create richer summaries. They discuss extractive, abstractive, and hybrid techniques while highlighting challenges such as data alignment, semantic integration, and scalability. They proposed customised evaluation metrics, including content overlap, coherence, and user satisfaction. This paper enhances information richness and boosts user engagement while involving implementation complexity and data alignment issues. Despite these challenges, further research into practical application and cross-interactions is needed to advance the field.

Argade et al. (2024) explore integrating text, images, and audio to create cohesive summaries. The study uses BERT with an attention mechanism to generate abstractive summaries that effectively combine textual and visual information, aiming to improve coherence and user engagement. It addresses challenges in data alignment and semantic integration, evaluating the summaries based on content coverage, coherence, and user satisfaction.

3 Methodology

To conduct a scientific literature review on multimodal abstractive text summarisation, this study adopted features of the PRISMA statement (Page et al., 2020) and followed the process recommended by Briner and Denyer (2012). The comprehensive methodology is outlined below:

- To identify the need for a review, formulate research questions, and develop the review protocol.
- To identify the strengths and weaknesses of each approach.
- To evaluate the performance of multimodal abstractive summarisation techniques and compare their effectiveness.

3.1 Research Protocol

The following are the Research Questions (RQs) for this study:

RQ1: What are the existing solutions used for multimodal abstractive text summarisation, their approaches, and limitations?

RQ2: How do existing solutions for multimodal abstractive text summarisation methods benchmark based on ROUGE scores?

RQ3: What challenges in existing multimodal abstractive text summarisation solutions can be addressed to improve cross-modal integration and produce more coherent, accurate summaries?

To address these questions, we follow the PRISMA approach to conduct a systematic literature review.

3.2 Research Query

A systematic search strategy was developed to identify relevant literature on multimodal abstractive text summarisation using a generative pre-trained transformer. The strategy involved searching IEEE Xplore, Science Direct, ACM Digital Library, Google Scholar, and arXiv databases using the following search terms: "Multimodal summarisation" OR "multimodal abstractive text summarisation" AND "transformer-based" AND "attention models".

All searches covered the period from 2018 to 2025 and included only journal articles and conference proceedings published in English. The search was restricted to peer-reviewed (refereed or scholarly) publications from 2018 to 2024, written exclusively in English.

Table 1 Adopted quality assessment questions

Quality assurance questions
Is the aim of the study clearly stated?
Are the independent variables clearly defined?
Is the size of the dataset suitable?
Is the data collection process clearly outlined?
Are the methods clearly articulated?
Are the results and findings clearly defined?
Is the research methodology noteworthy?
Does the study contribute to the literature?
Are the limitations of the study stated?

3.3 *Study Selection*

The selection criteria for this systematic search were based on the PRISMA statement (Page et al., 2020). The research aimed to identify existing literature on multimodal abstractive text summarisation.

3.4 *Quality Assessment*

This study is based on peer-reviewed journal articles and conference papers. Throughout the review process, particular attention was given to identifying duplicates and rigorously evaluating research papers to ensure integrity of the report. Both, article summaries and full texts were meticulously examined to verify their relevance and scientific merit of the literature included in this study. We specifically limited our selection to papers published exclusively in English and fall within the scope of multimodal abstractive text summarisation. Ten Quality Assessment (QA) questions listed in Table 1 were used to ensure that only high-quality and methodologically studies for the basic of this review, following the guidelines of Malhotra (2015). Scores were assigned as strongly agree = 1, partially agree = 0.5, and strongly disagree = 0. Only studies achieve a minimum score of 7.0 out of 10 were included in the final analysis.

The search outcomes of the screening process for each search string within each digital library are presented in Table 2.

3.5 *Data Extraction Form*

Data were extracted from reviews and filtered based on the quality assurance criteria in Table 3.

Table 2 Search outcomes

Published databases	Number of studies	Studies relevant to keywords of the title	Studies relevant to the full-text research	Studies authors and years
ACM Digital Library	3750	135	1	Jangra et al. (2023)
IEEE Xplore	13	5	2	Baek et al. (2023), Zhu et al. (2018)
SpringerLink	64	10	1	Khilji et al. (2023)
Science Direct	8703	548	2	Argade et al. (2024), Atri et al. (2021), Baek et al. (2023), Jangra et al. (2023), Khilji et al. (2023), Qiu et al. (2022)
arXiv	35	14	3	Fu et al. (2009), Khullar and (2010), Qiu et al. (2022)

Table 3 Data extraction form

Variables
Authors
Type of publication
Year
Area of study
Overall quality of the evaluation (out of 10)
Multimodal abstractive text methods
Methodology used
Limitations of the study
Approaches used in the study
Scope of the experiment
Strengths
Limitations

This study summarises selected literatures, which was chosen according to the criteria of quality assurance.

4 Finding and Discussion

In this literature review, we identified nine key studies of multimodal abstractive text summarisation methods. However, eight of these studies are evaluated in this work, as one of them is a survey of paper rather than an empirical evaluation. The findings highlight the effectiveness and accuracy of current multimodal abstractive text summarisation techniques.

RQ1: What Are the Existing Solutions Used for Multimodal Abstractive Text Summarisation, Their Approaches, and Limitations?

Table 4 presents summary of the strengths and weaknesses of the current solutions of multimodal abstractive text summarisation as contributions by the authors.

Research on multimodal abstractive summarisation has demonstrated that each approach integrates various modalities such as video, text, audio, and images using attention mechanisms and fusion strategies to generate coherent summaries. Multimodal fusion and attention-based methods (Argade et al., 2024; Fu et al., 2009; Khullar & Arora, 2010; Zhu et al., 2018) provide rich contextual understanding and effective cross-modal integration but are often delayed by challenges such as data alignment issues, sensitivity to noise, and high computational demands. Hierarchical and trimodal approaches (Atri et al., 2021; Baek et al., 2023; Fu et al., 2009; Jangra et al., 2023; Khilji et al., 2023; Khullar & Arora, 2010; Qiu et al., 2022) enhance structural comprehension and summary creation but frequently face problems related to modality synchronisation and the limited diversity of datasets. Video-text integration techniques (Fu et al., 2009; Qiu et al., 2022) offer a deeper and more comprehensive summarisation output; however, they tend to be computationally intensive and susceptible to semantic variations across modalities. Transformer-based models, such as BERT with attention mechanisms (Argade et al., 2024), show strong language understanding capabilities but also struggle with synchronising different modalities effectively. In summary, while these multimodal methods improve summarisation quality, they often encounter recurring limitations, including computational intensity, alignment complexity, and challenges in generalising performance across varied datasets.

RQ2: How Do Existing Solutions for Multimodal Abstractive Text Summarisation Methods Benchmark Based on ROUGE Scores?

NLP tools to assess the quality of machine-generated summaries provide metrics such as ROUGE, BLUE and, GLUE. However, ROUGE is the most used to benchmark the effectiveness of generated summaries. ROUGE has various measures, among which ROUGE-1, ROUGE-2, and ROUGE-L will be used to evaluate the quality of papers selected (Wang & Li, 2019).

Table 5 shows the ROUGE benchmarking scores and the datasets used in multimodal abstractive summarisation methods from selected papers.

Table 5 and Fig. 1 shown that Argade et al. (2024) achieved the highest ROUGE-1, ROUGE-2, and ROUGE-L scores on the How2 dataset, marking their model as the best performing among current multimodal abstractive summarization approaches. Khilji et al. (2023) also showed consistently strong performance, ranking closely behind (Argade et al., 2024). In general, models trained on the How2 dataset tend to outperform those using CNN or VMSMO datasets, highlighting the importance of dataset quality and domain relevance in influencing summarisation outcomes. On the contrary, models such as Qiu et al. (2022), which use VMSMO or CNN datasets,

Table 4 Summary of the contributions

Solutions	Methods	Approaches	Strengths	Limitations
Zhu et al. (2018)	Multimodal (text-image) fusion	Computer vision, video summarisation	Rich information, adaptability	Outdated model (BiLSTM), training and evaluation are based on single dataset and computational intensity
Khullar and Arora (2010)	Multimodal (video-audio-text) fusion	Trimodal hierarchical attention, abstractive summarisation	Multimodal integration, hierarchical structure	Noise sensitivity, inaccurate synchronization between multimodal data and data availability
Fu et al. (2009)	Bi-stream summarisation	Video-text integration, feature fusion	Comprehensive understanding, rich context	Fusion of video and text can lead to noise without proper alignment, computational intensity of documents and video
Atri et al. (2021)	Multimodal attention mechanism, guide attention (FLORAL)	Abstractive summarisation, multimodal integration	Comprehensive information, guided attention	Limited performance generalisation to long content, modularity alignment challenges and computational intensity
Qiu et al. (2022)	Multimodal integration, video and textual segmentation	Hierarchical attention, feature fusion	Comprehensive understanding, fine-grained summaries	High computational and memory requirement, poor video segmentation. Mismatched between visual and textual semantics
Baek et al. (2023)	Trimodal (video-audio-text) hierarchical multi-head attention	Multimodal integration, hierarchical structure	Comprehensive understanding, abstractive summarisation	Modality alignment challenges, computational complexity, and limited data diversity

(continued)

Table 4 (continued)

Solutions	Methods	Approaches	Strengths	Limitations
Khilji et al. (2023)	Trimodal hierarchical multi-head attention	Text-image integration	Information richness designed evaluation methods	Implementation complexity, data alignment challenges and limited dataset diversity
Argade et al. (2024)	BERT, multimodal fusion via attention mechanism	Multimodal fusion, abstractive summarisation	Language understanding, multimodal integration	Synchronising different modalities issues and data alignment challenges

Table 5 ROUGE scores of existing solutions for multimodal abstractive summarisation

Solutions	Dataset used benchmark	ROUGE-1	ROUGE-2	ROUGE-L
Zhu et al. (2018)	MSMO	0.408	0.1827	0.3775
Khullar and Arora (2010)	How2	48.85	29.51	43.23
Fu et al. (2009)	Daily Mail	41.73	18.59	31.68
	CNN	27.81	8.87	18.73
Atri et al. (2021)	How2	56.89	26.93	56.80
	AVIATE	37.13	11.04	31.47
Qiu et al. (2022)	VMSMO	27.1	9.8	25.4
	CNN	28.02	8.94	18.89
	Daily Mail	42.34	19.12	32.35
Baek et al. (2023)	How2	52.53	29.48	44.18
Khilji et al. (2023)	How2	58.02	36.34	54.67
Argade et al. (2024)	How2	60.02	30.30	59.9

demontrate lower ROUGE scores across all metrics. This indicates that these methods face challenges in achieving good performance when applied to less comprehensive or poorly aligned datasets in multimodal summarization tasks.

RQ3: What Challenges in Existing Multimodal Abstractive Text Summarisation Solutions Can Be Addressed to Improve Cross-Modal Integration and Produce More Coherent, Accurate Summaries?

Despite growing interest and recent advancements in multimodal abstractive text summarization, a critical analysis of existing solutions reveals important limitations that influenced their practical effectiveness and scalability. Firstly, no current method fully addresses the main challenges of data complexity, computational cost, limited availability of large-scale multimodal datasets, and scalability across diverse topics

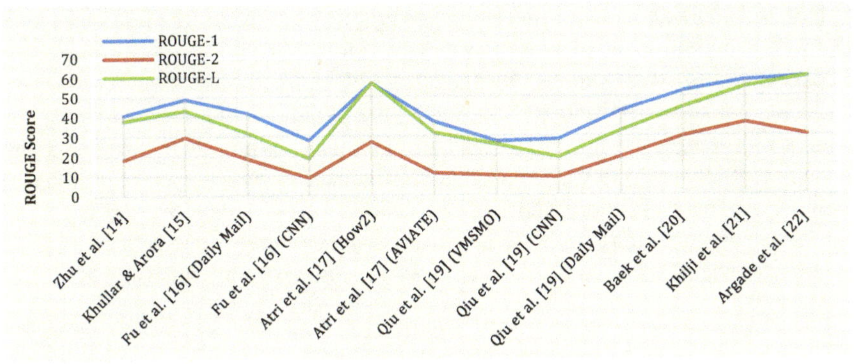

Fig. 1 ROUGE scores of existing solutions for multimodal abstractive summarisation

(Baek et al., 2023; Jangra et al., 2023). Many methods perform well on domain-specific datasets but fail to generalise when exposed to more complex or varied real-world data. These challenges are reflected in inconsistent and often suboptimal ROUGE scores, which underscore a lack of semantic accuracy and coverage in the generated summaries. Performance typically decrease when methods attempt to summarise content requiring deep cross-modal understanding or when dealing with noisy or incomplete data.

Moreover, while recent frameworks attempt to improve modality integration using attention mechanisms or hierarchical architectures, the fusion of textual, visual, and audio modalities remains inefficient and highly demanding (Jangra et al., 2020). This inefficiency not only affects the summarisation process but also often results in fragmented or biased summaries, especially when dominant modalities are prioritised over others, leading to underutilisation of critical contextual words. The lack of robust modality alignment techniques, coupled with the high computational cost of transformer-based architectures, creates significant delay in both training and inference.

Ultimately, these persistent gaps highlight that existing solutions struggle to produce accurate, coherent, and scalable multimodal summaries, especially in scenarios demanding high compression ratios and semantic fidelity. Therefore, there is a need for further research to develop more efficient, modality-balanced, and strongly built architectures capable of overcoming these limitations and achieving effective multimodal abstractive summarisation.

5 Threats to Validity

Conducting systematic mapping research involves potential risks. To ensure the reliability of our finding on multimodal abstractive text solutions, precautions were taken to minimise potential biases. We carefully selected reputable database and applied well-defined inclusion and exclusion criteria. To broaden the research scope, we use the combination of search strategies, including varied keyword sets and logical operators. Most studies that met the exclusion and inclusion criteria were conducted between 2018 and 2024, reducing the likelihood that the study's conclusions are influenced by missing recent developments. While there remains a risk due to potentially relevant unpublished or non-indexed papers, the use of established scientific databases supports the internal validity of the study. We are confident that the selected literature provides an accurate and comprehensive reflection of the current advancements in multimodal abstractive summarization algorithms.

6 Conclusion

According to the findings, multimodal abstractive summarisation remains a growing field with persistent challenges in data integration, computational efficiency, and the reliability of evaluation methods. The increasing widespread of multimodal content in our digital ecosystems, ranging from social media to educational and professional communication, demands more scalable, adaptive, and modality-aware summarisation techniques. To address these persistent gaps, future research must focus on three critical areas: (1) developing advanced fusion architectures that can consistently and efficiently integrate various modalities; (2) constructing large-scale, diverse, and clearly explained multimodal datasets that reflect real-world complexities; and (3) proposing novel evaluation metrics that go beyond traditional text-based measures such as ROUGE to capture the richness and coherence of multimodal outputs. By prioritising these directions, the field can advance towards building robust, accurate, and generalisable multimodal summarisation systems capable of producing high-quality summaries in an increasingly multimodal digital world.

References

Allahyari, M., Pouriyeh, S., Assefi, M., Safaei, S., Trippe, E. D., Gutierrez, J. B., & Kochut, K. (2017). Text summarization techniques: a brief survey. arXiv:1707.02268

Argade, D., Khairnar, V., Vora, D., Patil, S., Kotecha, K., & Alfarhood, S. (2024). Multimodal abstractive summarization using bidirectional encoder representations from transformer with attention mechanism. *Heliyon.*

Atri, Y. K., Pramanick, S., Goyal, V., & Chakraborty, T. (2021). See, hear, read: Leveraging multimodality with guided attention for abstractive text summarization. *Knowledge-Based Systems, 227*, Article 107152.

Baek, D., Kim, J., & Lee, H. (2023). VATMAN: Video-audio-text multimodal abstractive summarization with trimodal hierarchical multi-head attention. In *2023 14th International Conference on Information and Communication Technology Convergence (ICTC)* (pp. 1475–1478). IEEE.

Briner, R. B., & Denyer, D. (2012). Systematic review and evidence synthesis as a practice and scholarship tool.

El-Kassas, W. S., Salama, C. R., Rafea, A. A., & Mohamed, H. K. (2021). Automatic text summarization: A comprehensive survey. *Expert Systems with Applications, 165*, Article 113679.

Fu, X., Wang, J., & Yang, Z. (2020). Multi-modal summarization for video-containing documents. arXiv:2009.08018

Gandhi, A., Adhvaryu, K., Poria, S., Cambria, E., & Hussain, A. (2023). Multimodal sentiment analysis: A systematic review of history, datasets, multimodal fusion methods, applications, challenges and future directions. *Information Fusion, 91*, 424–444.

He, B., Wang, J., Qiu, J., Bui, T., Shrivastava, A., & Wang, Z. (2023). Align and attend: Multimodal summarization with dual contrastive losses. In *Proceedings of the IEEE/CVF Conference on Computer Vision and Pattern Recognition* (pp. 14867–14878).

Jangra, A., Jatowt, A., Hasanuzzaman, M., & Saha, S. (2020). Text-image-video summary generation using joint integer linear programming. In *Advances in Information Retrieval: 42nd European Conference on IR Research, ECIR 2020, Lisbon, Portugal, April 14–17, 2020, Proceedings, Part II* (Vol. 42, pp. 190–198). Springer.

Jangra, A., Mukherjee, S., Jatowt, A., Saha, S., & Hasanuzzaman, M. (2023). A survey on multimodal summarization. *ACM Computing Surveys, 55*(13s), 1–36.

Khilji, A. F. U. R., Sinha, U., Singh, P., Ali, A., Laskar, S. R., Dadure, P., Manna, R., Pakray, P., Favre, B., & Bandyopadhyay, S. (2023). Multimodal text summarization with evaluation approaches. *Sādhanā, 48*(4), 226.

Khullar, A., & Arora, U. (2020). MAST: Multimodal abstractive summarization with trimodal hierarchical attention. arXiv:2010.08021

Khurana, D., Koli, A., Khatter, K., & Singh, S. (2023). Natural language processing: State of the art, current trends and challenges. *Multimedia Tools and Applications, 82*(3), 3713–3744.

Kieuvongngam, V., Tan, B., & Niu, Y. (2020). Automatic text summarization of covid-19 medical research articles using BERT and GPT-2. arXiv:2006.01997

Li, H., Zhu, J., Ma, C., Zhang, J., & Zong, C. (2018). Read, watch, listen, and summarize: Multimodal summarization for asynchronous text, image, audio and video. *IEEE Transactions on Knowledge and Data Engineering, 31*(5), 996–1009.

Malhotra, R. (2015). A systematic review of machine learning techniques for software fault prediction. *Applied Soft Computing, 27*, 504–518.

Mridha, M. F., Lima, A. A., Nur, K., Das, S. C., Hasan, M., & Kabir, M. M. (2021). A survey of automatic text summarization: Progress, process and challenges. *IEEE Access, 9*, 156043–156070.

Page, M. J., McKenzie, J. E., Bossuyt, P. M., Boutron, I., Hoffmann, T. C., Mulrow, C. D., Shamseer, L., Tetzlaff, J. M., Akl, E. A., Brennan, S. E., & Chou, R. (2021). The PRISMA 2020 statement: an updated guideline for reporting systematic reviews. *bmj*, 372.

Palaskar, S., Libovický, J., Gella, S., & Metze, F. (2019). Multimodal abstractive summarization for how2 videos. arXiv:1906.07901

Qiu, J., Zhu, J., Xu, M., Dernoncourt, F., Bui, T., Wang, Z., Li, B., Zhao, D., & Jin, H. (2022). MHMS: Multimodal hierarchical multimedia summarization. arXiv:2204.03734

Rafi, S., & Das, R. (2023). Abstractive text summarization using multimodal information. In *2023 10th International Conference on Soft Computing & Machine Intelligence (ISCMI)* (pp. 141–145). IEEE.

Tang, P., Hu, K., Zhang, L., Luo, J., & Wang, Z. (2023). Extreme multimodal summarization of news videos. *IEEE Transactions on Circuits and Systems for Video Technology, 34*(3), 1469–1480.

Vaswani, A., Shazeer, N., Parmar, N., Uszkoreit, J., Jones, L., Gomez, A. N., Kaiser, Ł., & Polosukhin, I. (2017). Attention is all you need. In *Advances in neural information processing systems* (Vol. 30)

Wang, R., & Li, J. (2019). Bayes test of precision, recall, and F1 measure for comparison of two natural language processing models. In *Proceedings of the 57th Annual Meeting of the Association for Computational Linguistics* (pp. 4135–4145).

Zhu, J., Li, H., Liu, T., Zhou, Y., Zhang, J., & Zong, C. (2018). MSMO: multimodal summarization with multimodal output. In *Proceedings of the 2018 Conference on Empirical Methods in Natural Language Processing* (pp. 4154–4164).

The Impact of the Prompt Language on Code Generation with LLMs: An Experimental Study

Kai Meier, Christian Koot, Marc Fernandes, and Andreas Theissler

Abstract This study investigates the effect of using different languages in the prompt input for code generation with large language models (LLMs). Python code is generated from German and English prompt inputs using a zero-shot prompting strategy, targeting a comparison of the model performance between the languages and difficulties. Therefore, a bilingual benchmark dataset containing 16 algorithmic tasks was developed, covering domains such as linear algebra, number theory, probability theory, and cryptography. The tasks were semantically aligned and categorized into three levels of difficulty. Six large language models were evaluated across the tasks. Standardized prompts were designed using prompt engineering principles, assigning the models the role of an expert in the topic of Python programming. Score metrics were implemented to assess the individual performance of the models and statistical analyses were conducted using Pearson's chi-squared and Fisher exact test to evaluate significant associations across language prompts and task difficulty. Each model completed 16 tasks in German and 16 in English, with two independent runs per task, resulting in 64 runs per model. This adds up to 192 test runs per language and 384 runs in total. Results show that most models perform better using English language prompts, with GPT-4o outperforming its competitors, reaching the highest scores in both languages and solving all tasks with the maximum score with English prompting. Association between the models across the difficulties showed significant differences using English prompting, while for German only the easiest difficulty showed significant differences. Association between the language performances only showed significant differences in medium difficulty with Pearson's chi-squared test ($p = 0.05$) and Fisher exact test ($p = 0.029$).

K. Meier · C. Koot (✉) · M. Fernandes · A. Theissler
Aalen University, Aalen, Germany
e-mail: Christian.Koot@hs-aalen.de

J. Marx Gómez et al. (eds.), *Recent Advances in Information Systems*, Progress in IS,
https://doi.org/10.1007/978-3-032-14099-9_46

1 Introduction

Large language models (LLMs) have attracted public and scientific attention within a very short period of time. A prominent example of this is the chatbot ChatGPT, which was based on the GPT-3.5 model and had already registered one million users just five days after its release in November 2022 (Gupta et al., 2023). Only two months later, it already had attracted 100 million active users (Homolak, 2023).

LLMs are capable of performing a wide range of tasks, from creating content for various platforms and translating natural text into other languages to assisting in the creation of medical diagnoses (Makridakis et al., 2023). A particularly research-intensive field of application is the automated generation of program code, also known as program synthesis or code generation. Numerous studies are investigating the programming abilities of LLMs and their performance in a wide variety of scenarios (Du et al., 2024; Hou & Ji, 2025; Huang et al., 2024; Lian et al., 2024; Mathews & Nagappan, 2024; Ni et al., 2024; Zhong & Wang, 2024).

Algorithmic problems play a particularly important role in this context. They require systematic thinking, problem-solving skills, and precise handling of control structures (Belmar, 2022). Despite intensive research, the focus has been primarily on models that evaluate model performance based on English language prompt inputs. However, the influence of the prompting language on the performance of LLMs in generating code has only been investigated in isolated cases (Peng et al., 2024).

In particular, the comparison of the code generation capabilities of LLMs using German and English prompting represents a research field that has hardly been investigated to date. To address this research gap, this study poses the central research question: *To what extent does the language of the prompt input, in particular German compared to English, influence the code generation performance of large language models when solving algorithmic tasks in Python?*

In this paper, the following contributions are made:

1. Developing a bilingual benchmark dataset with 16 algorithmic tasks in semantically equivalent problem statements in German and English, systematically divided into three difficulty levels.
2. Conducting a zero-shot evaluation of six large language models for generating Python code based on identical algorithmic tasks in German and English using standardized prompts.
3. Assessing model performance and perform statistical association tests to analyze significant differences between German and English language prompting.

2 Related Work

Code generation using large language models has become a prominent research area in recent years, with numerous publications on various aspects. A total of 19 peer-reviewed articles and conference papers were identified. To focus on high-impact publications one paper which was not listed in the CORE Rankings of the ICORE Conference Portal and not in the SCImago Journal Rank 2024 (SJR 2024) was excluded. Furthermore, 3 papers published before 2023 were also excluded for reasons of timeliness.

The most closely related work to ours is currently in a preprint state (Li et al., 2024). Like our study, it examines the influence of the prompting language on code generation with LLMs. However, the authors evaluate a different set of models and prompting languages, and do not assess statistical significance of the observed differences. Notably, they propose a method to mitigate the observed language effects.

To ensure brevity and quality, Table 1 is limited to the presentation of 6 papers from conferences that were rated A or A* in the CORE ranking and 2 papers that were published in journals with an SJR 2024 > 1.8.

3 Methods

This methodology follows a structured five-step approach: First, a bilingual dataset in German and English language is developed. Second, standardized prompts are constructed across both languages. Third, evaluation metrics were designed, to assess the models' performance on functional correctness. Fourth, the evaluation procedure is described, and fifth, the statistical analyses are conducted to assess language-dependent performance differences and associations with task complexity.

3.1 Dataset

To enable the evaluation of large language models for code generation, a curated dataset consisting of 16 algorithmic tasks was created, each available in German and English. The tasks cover core concepts from mathematics and computer science, including linear algebra, number theory, probability theory, optimization, cryptography, sorting algorithms, and machine learning (Bronshtein et al., 2004; Cormen et al., 2009; Janiesch et al., 2021). All tasks were formulated in both languages in semantically equivalent forms to enable a controlled comparison of model performance across language boundaries.

The selection of tasks ensured a balanced mix, with the dataset containing both established algorithms that are regularly tested in benchmarks for code generation such as Codeforces, MBPP, and HumanEval, as well as less common, domain-specific problems. An overview of the tasks can be found in Table 2.

In addition, the test cases were categorized into three difficulty levels, easy, medium, and hard, based on an assessment by three experienced programmers to enable a more differentiated analysis of the model performance. The three programmers have experience in programming in parts for over 20 years in different areas:

Table 1 Related papers in high-impact conferences and journals

Author(s) and year	Contribution	Publisher (CORE and SJR2024)
Zhong and Wang (2024)	Zhong and Wang (2024) evaluate the reliability and robustness of code generated by large language models (LLMs) in realistic software scenarios. They introduce ROBUSTAPI, a benchmark with 1,208 Java-related Stack Overflow questions covering 18 APIs that are frequently misused. API misuse is detected by static analysis that compares AST-derived call sequences with structured usage rules. Results show that a significant amount of code generated by current LLMs, including GPT-4, contains incorrect API usage, although it is compilable	Proceedings of the AAAI Conference on Artificial Intelligence (CORE23 A*)
Huang et al. (2024)	Huang et al. (2024) propose KareCoder, a knowledge-aware code generation framework that augments LLMs with structured algorithmic knowledge. To evaluate their approach, they constructed a dataset of 1,523 Python problems from CodeForces and compared KareCoder to recent prompting methods applied to ChatGPT. KareCoder integrates algorithm and data structure prompts from a curated knowledge library, improving Pass@1 performance by up to 23.3% on unseen problems and performing competitively on the APPS benchmark	2024 IEEE/ACM 32nd International Conference on Program Comprehension (ICPC) (CORE23 A)
Mathews and Nagappan (2024)	Mathews and Nagappan (2024) investigate whether test-driven development (TDD) principles can improve the performance of LLM-based code generation. They compare code generated with and without user-supplied tests and analyze the effect of remediation loops and introduce TGen, a framework that evaluates GPT-4 and Llama 3 on Mostly Basic Python Programming Problems (MBPP), HumanEval, and a new dataset of 1,100 CodeChef problems. Results show that providing tests improves correctness by up to 29.6%, with further gains from iterative refinement	2024 39th IEEE/ACM International Conference on Automated Software Engineering (ASE) (CORE23 A)

(continued)

Table 1 (continued)

Author(s) and year	Contribution	Publisher (CORE and SJR2024)
Lian et al. (2024)	Lian et al. (2024) conducted a systematic study on the weaknesses of LLM-based code generation across three models and the datasets CoNaLa, HumanEval+ and DS-1000. They developed a taxonomy of eight weakness types, including semantic omissions, incorrect API usage, and excessive code generation. Using thematic analysis on 731 problematic samples, they revealed model-specific and benchmark-specific trends. Results show that issues such as missing pivotal semantics and biases from single-answer references are widespread, underscoring the need for more nuanced benchmarks and evaluation strategies	2024 IEEE/ACM 46th International Conference on Software Engineering: Companion Proceedings (ICSE-Companion) (CORE23 A)
Du et al. (2024)	Du et al. (2024) introduce ClassEval, a benchmark for class-level code generation consisting of 100 Python tasks with multiple interdependent methods. They evaluate 11 large language models, including GPT-4, using three generation strategies: holistic, incremental, and compositional. Results show that all models perform significantly worse on class-level tasks compared to method-level benchmarks such as HumanEval. GPT-4 and GPT-3.5 perform best under holistic generation, while other models achieve higher accuracy with step-by-step generation	ICSE '24: Proceedings of the IEEE/ACM 46th International Conference on Software Engineering (CORE23 A)
Hou and Ji (2025)	Hou and Ji (2025) evaluated the coding performance of seven LLMs across multiple programming languages, task difficulties, and prompt strategies. They design six prompt strategies, including feedback-based approaches using code interpreters, and test them on 100 LeetCode tasks and additional GeeksforGeeks problems. Results indicate that GPT-4 outperforms all other models and ranks above 85% of human contestants in 81% of evaluated coding contests. Other models perform significantly weaker overall but can match average human programmers under optimal prompting conditions	2024 Advanced Science Volume 12, Issue 8 (SJR 2024 3.775)

(continued)

Table 1 (continued)

Author(s) and year	Contribution	Publisher (CORE and SJR2024)
Ni et al. (2024)	Ni et al. (2024) introduced L2CEval, a benchmark to evaluate LLMs on code generation across 7 tasks of the domains semantic parsing, math reasoning and Python programming. They analyzed the effects of model size, pretraining data, instruction tuning and prompt sensitivity on 56 models through human evaluation. Results show that the model performance varies notably across tasks and is strongly influenced by model size, instruction tuning, and the proportion of code tokens in pretraining data	Transactions of the Association for Computational Linguistics (SJR 2024 1.824)

Table 2 16 algorithms used to test the LLMs, categorized into the difficulty levels “Easy”, “Medium”, and “Hard”

Domain	Algorithm	Difficulty
Linear algebra	Gauss-Jordan elimination without pivoting	Medium
	Gauss-Jordan elimination with pivoting	Hard
Number theory	Euclidean algorithm	Easy
	Sieve of Eratosthenes	Easy
Probability theory	Panjer recursion	Hard
Optimization	Simplex algorithm	Hard
	Hiring problem	Easy
	Dijkstra’s algorithm	Hard
Machine learning	K-means algorithm	Medium
	K-nearest neighbors algorithm	Medium
	Linear regression	Easy
Cryptography	RSA algorithm	Hard
	SHA-1 algorithm	Medium
Sorting algorithms	Bubblesort	Easy
	Mergesort	Medium
	Quicksort	Medium

software development in large industrial projects, teaching programming at universities and programming in research projects.

The assignment was based on a combined evaluation of content complexity, algorithmic challenge, and the number of required calculation steps. Task classifications from common online programming platforms such as LeetCode (https://leetcode.com) and Codeforces (https://codeforces.com) served as a reference, supporting an analysis of differences across levels of task complexity.

Each test case is implemented as a structured data object containing a task description, predefined variables, solution options, the solution and a reference solution including validation with alternative inputs.

3.2 Prompt Engineering

To ensure a consistent evaluation framework, a standardized prompt design was implemented. The structure follows the approach of López Espejel et al. (2025) and includes the key elements problem, variables, and solution options. In addition, typical task formats from platforms such as LeetCode were analyzed to adapt the design to common problem presentations.

The prompts follow a zero-shot approach (Beri & Srivastava, 2024), i.e., no examples or additional contextual information were provided in order to isolate the evaluation of the models' pre-trained knowledge.

As illustrated in Fig. 1, each prompt begins with a role assignment, instructing the model to act as an expert in Python programming. This is followed by the task specification: the model is instructed to solve the given problem using Python code, without the use of external libraries or packages (standard libraries are allowed, with the exception of the SHA-1 case). The solution must also use the specified variables, solve the problem correctly, and output the result to the console according to the defined solution options. An example of a specific prompt used in the evaluation process can be seen in Fig. 2.

3.3 Evaluation Metrics

A standardized evaluation scheme was developed to evaluate the functional correctness, the output of the required calculation steps, compliance with the specified variables, and the use of the permitted resources of the test cases. A test case is considered successfully completed if all of the following conditions are met:

- The specified algorithm is implemented in a functionally correct manner.
- The predefined variables are used to solve the task.
- The output matches the expected result as defined in the solution options.
- The solution also produces correct results when tested with alternative inputs.

Prompt structure for test cases
You are an expert in Python programming.
You are given three keywords: Problem, Variables, and Solution options.
Your task is to solve the given problem using Python code, without using external libraries or packages. Standard Python libraries are allowed.
The most important requirements are: The Python code must use the variable names specified in the provided variables. The code must solve the given problem correctly and return the corresponding result in the console according to the given solution options.
Problem: `<Description of the problem>`
Variables: `<Variables to be used>`
Solution options: `<Description of the required solution>`

Fig. 1 Standardized prompt structure

Problem: `Find all prime numbers up to a given number max_value using the Sieve of Eratosthenes.`
Variables: `max_value = 120`
Solution options: `Output the list of prime numbers, the number of prime numbers found, and all calculation steps in the console.`

Fig. 2 Example prompt for the sieve of Eratosthenes

Table 3 Scoring scheme for evaluating the code generated by the LLMs

Score (points)	Criterion
3 P.	Correct solution with calculation steps
2 P.	Correct solution without calculation steps
1 P.	Wrong solution with calculation steps
0 P.	Runtime error or wrong solution without calculation steps

3.4 *Evaluation Procedure*

The evaluation was conducted using six widely used language models: GPT-4o, GPT-4 (OpenAI), Gemini 2.0 Flash (Google), Claude 3.5 Haiku (Anthropic), Pixtral Large (Mistral AI), and Copilot (Microsoft). Access was obtained via the official web interfaces of the respective providers.

The execution was performed manually. For each test run, a separate Jupyter notebook file was created, documenting the task description, the prompt, the text output, and the generated Python code. The code was then executed, manually evaluated for functional correctness, and scored according to the scheme described in Table 3. Since the distinction between the categories was unambigious, a single author assigned the results to one of the categories defined in Table 3.

We wanted to assess effectiveness, i.e., the LLM's ability to produce correct code. The assessment of efficiency, i.e., robustness, number of calculation steps, etc., is subject to future research.

We requested the LLMs not to use external libraries and packages. If external libraries are used, the scores were penalized with one point deduction, as the evaluation aimed to assess the models ability to independently derive algorithmic solutions rather than relying on implementations. We decided to deduct one point since we wanted to evaluate whether the algorithms are able to find a working solution, preferably completely alone. Nevertheless, we consider a solution with existing libraries to be better than an incorrect solution.

3.5 Statistical Analysis

In order to evaluate the performance of the large language models in terms of language prompts and task difficulty, a categorical data analysis based on cross tables was performed. The scores achieved were evaluated on an ordinal scale, as described in Table 3.

For each combination of language prompt and difficulty level, Pearson's χ^2 test was performed to assess whether the model and score category are statistically associated. Since the expected frequencies did not all meet the requirement $e_{ij} \geq 5$ (Howell, 2011), the Fisher exact test was also performed to robustly determine the significance under these conditions. In addition, both tests were also performed for direct comparison of the languages.

For the statistical analyses we used the functions chisq.test() and fisher.test() from R 4.2.2 (package stats). Due to the fact, that the Fisher exact test is resource-intensive, a Monte Carlo simulation was used for p-value approximation (number of replicates: $B = 10^6$).

Significance is reported in detail ($\alpha = 0.1 \mid 0.05 \mid 0.01 \mid 0.005 \mid 0.001$), using one to five asterisks: $* \mathrel{\widehat{=}} 0.1$, $** \mathrel{\widehat{=}} 0.05$, $*** \mathrel{\widehat{=}} 0.01$, $**** \mathrel{\widehat{=}} 0.005$, $***** \mathrel{\widehat{=}} 0.001$ in Tables 8 and 10, since there is justified criticism on the default 0.05 threshold (Wasserstein & Lazar, 2016).

4 Results

Each model processed 16 algorithmic test cases in German and 16 equivalent test cases in English. For each test case, two independent runs were carried out per model and language, resulting in a total of 384 test runs: 192 in German and 192 in English. Accordingly, each of the six models completed 64 individual test runs.

As shown in Table 4, the large language model GPT-4o achieved the maximum total score of 96 points across all test cases using English prompting, corresponding to 16 tasks executed twice and receiving the full score of 3 points in each run,

Table 4 Scores achieved by the models using English prompting

Organization	Model	Total score (0...96)
OpenAI	GPT-4o	96
Google	Gemini 2.0 Flash	83
Anthropic	Claude 3.5 Haiku	83
Microsoft	Copilot	79
OpenAI	GPT-4	72
Mistral AI	Pixtral Large	65

Table 5 Scores achieved by the models using German prompting

Organization	Model	Total score (0...96)
OpenAI	GPT-4o	90
Anthropic	Claude 3.5 Haiku	80
Microsoft	Copilot	72
Mistral AI	Pixtral Large	71
OpenAI	GPT-4	68
Google	Gemini 2.0 Flash	68

thereby achieving the best result compared to all other models tested. GPT-4o also outperformed all other models tested with 90 points using German language prompts, as shown in Table 5. The Gemini 2.0 Flash model finished second in English with 83 points, but only achieved the second-worst score with 68 points using German language prompts. With 83 points, the Claude 3.5 Haiku model also ranked second using English language prompts. It was followed by Claude 3.5 Haiku with German language prompts, which scored 3 points less. Pixtral Large was the only model to perform worse with English language prompts than with German, and achieved the worst result of the evaluation in English language output. In a direct language comparison, the models scored 449 points with German prompts and 478 points with English prompts in total (with a theoretic maximum of 3 points $\times$ 16 tasks $\times$ 2 runs $\times$ 6 models = 576 points).

Tables 6 and 7 show cross tables of model performance for all difficulty levels for German and English prompts. The rows correspond to the models, while the columns represent the point distribution per difficulty level. The column headings indicate the achieved score (0–3 points).

The values in the cells indicate how often a model achieved a certain score in this difficulty, with the row totals reflecting the distribution of tasks in the data set. Five easy, six medium, and five difficult tasks were included, each of which was completed in two independent runs per model. This resulted in ten runs for easy and difficult tasks and twelve for medium tasks per model.

For German language prompting (Table 8), the Pearson χ^2 test did only reveal a mildly significant association for easy problems (all other p-values > 0.1). The

Table 6 Crosstab representation of the evaluation results across difficulties using German prompting showing the number of tasks that were assigned the respective number of points (0P...3P)

GER—Easy				
	0P.	1P.	2P.	3P.
GPT-4o	0	0	0	10
GPT-4	1	0	5	4
Gemini 2.0 F.	0	1	2	7
Pixtral Large	0	0	3	7
Claude 3.5	0	0	0	10
Copilot	0	0	1	9
GER—Medium				
	0P.	1P.	2P.	3P.
GPT-4o	0	0	1	11
GPT-4	0	0	6	6
Gemini 2.0 F.	1	0	4	7
Pixtral Large	1	0	5	6
Claude 3.5	0	1	3	8
Copilot	2	0	2	8
GER—Hard				
	0P.	1P.	2P.	3P.
GPT-4o	1	1	0	8
GPT-4	3	1	3	3
Gemini 2.0 F.	4	2	1	3
Pixtral Large	2	4	0	4
Claude 3.5	1	3	2	4
Copilot	3	2	2	3

Fisher exact test showed a significant association only for tasks with a difficulty level of easy, while no significance was found for medium and hard.

For English language prompting, the Pearson χ^2 test also failed to discover any significant association at difficulty level easy while it was found using fisher exact test ($p = 0.014$). However, the medium ($p = 0.004 \mid 1.39 \times 10^{-4}$) and hard ($p = 0.007 \mid 0.003$) levels revealed statistically significant associations in the Pearson χ^2 test. The Fisher exact test confirmed these results and showed significant associations between the models across all difficulty levels.

Table 9 summarizes the total number of results for each score (0–3 points) across all models, grouped by task difficulty and prompting language. The same column labels as in Tables 6 and 7 apply.

When comparing German and English language prompting directly (Table 10), no significant differences for easy and hard tasks were found in either test. At the

Table 7 Crosstab representation of the evaluation results across difficulties using English prompting

ENG—Easy				
	0P.	1P.	2P.	3P.
GPT-4o	0	0	0	10
GPT-4	1	1	4	4
Gemini 2.0 F.	0	0	0	10
Pixtral Large	0	0	3	7
Claude 3.5	0	0	1	9
Copilot	0	0	1	9
ENG—Medium				
	0P.	1P.	2P.	3P.
GPT-4o	0	0	0	12
GPT-4	0	0	3	9
Gemini 2.0 F.	0	0	0	12
Pixtral Large	2	0	5	5
Claude 3.5	0	0	0	12
Copilot	1	1	0	10
ENG—Hard				
	0P.	1P.	2P.	3P.
GPT-4o	0	0	0	10
GPT-4	2	3	0	5
Gemini 2.0 F.	1	5	0	4
Pixtral Large	3	2	4	1
Claude 3.5	2	2	2	4
Copilot	0	4	3	3

Table 8 Test of statistical significance of the differences in model performance for each difficulty level and language (df = 15)

Prompting language	Pearson χ^2	Pearson χ^2 (p)	Fisher exact test (p)
German easy	23.698	0.073*	0.012**
German medium	17.261	0.304	0.252
German hard	16.317	0.361	0.438
English easy	22.286	0.101	0.014**
English medium	33.3	0.004****	1.39×10^{-4}*****
English hard	31.694	0.007***	0.003****

Table 9 Crosstab representation of the evaluation results across the difficulties for both languages

Easy				
	0P.	1P.	2P.	3P.
German	1	1	11	47
English	1	1	9	49
Medium				
	0P.	1P.	2P.	3P.
German	4	1	21	46
English	3	1	8	60
Hard				
	0P.	1P.	2P.	3P.
German	14	13	8	25
English	8	16	9	27

Table 10 Test of statistical significance of the differences in model performance between English and German prompts for every difficulty level (df = 3)

Difficulty	Pearson χ^2	Pearson χ^2 (p)	Fisher exact test (p)
Easy	0.242	0.971	0.951
Medium	7.819	0.05**	0.029**
Hard	2.082	0.555	0.573

medium level, the Pearson χ^2 test ($p = 0.05$) indicated mild significance, which was supported by the Fisher exact test ($p = 0.029$).

5 Discussion

This study showed that the large language models differ in terms of language prompting and its impact on performance. Particularly noteworthy is the strong performance of OpenAI's GPT-4o model in both languages examined. The model demonstrates a remarkable ability to generalize across language boundaries. The reason for this could be the fact that it is a relatively new model and has a more advanced model architecture than, for example, it's direct predecessor GPT-4. It is also plausible that the model has undergone deeper integration of multilingual language processing, with a more linguistically diverse training data set.
From our results we can deduce the following overriding insights:

1. **The models, tested with English prompts, show differences in the quality of the generated code for more difficult tasks**: The results of the statistical analysis of the models' performance using English prompting shows that the

models differ more in their performance when the tasks are more complex, which could be because the models were trained with different training data or because the model architectures use different mechanisms for prompt processing and response generation. For German language prompting, on the other hand, performance is relatively homogeneous across the models within each difficulty level. Only at the easy difficulty level there is a significant difference between the models ($p = 0.073 \mid 0.012$).

2. **The difference in model performance between English and German prompts is only statistically significant for tasks of medium difficulty**: When directly comparing the two language variants, significant differences between German and English ($p = 0.05 \mid 0.029$) were only found in tasks of medium difficulty. This may indicate that tasks of medium difficulty are complex enough to require differentiated language processing and interpretation, but are still solvable, which means that differences in the processing of the prompts by the models are more significant. Simple tasks are likely solved by standardized strategies across the models, whereas for difficult tasks, a general performance drop may hide differences related to the prompting language.
3. **The majority of models perform better with English prompts, though these differences were not shown to be statistically significant for tasks of easy and hard difficulty**: With the exception of Pixtral Large, all models perform better with English prompts than with German ones. This could be due to the fact that a large part of the training data used to train the models is in English, which could indicate a deficit in the multilingual training of the models (Chang et al., 2024). It is assumed that the input prompts in different languages may be translated into English before they are processed. This is also suggested by recent research findings that explicitly dealt with multilingual autoregressive language models and led to significantly better results through a self-translation process (Etxaniz et al., 2024). Even though the study is limited to open models such as XGLM and LLaMA, it is plausible that the findings can be transferred to proprietary large language models such as GPT and Claude. Since these models also show indications of preferential processing of English language prompts, this could suggest that English is processed first.
4. **Pixtral Large is the only model performing better when prompted in German rather than English**: In contrast to the other models, Pixtral Large performed better with German prompts. This could be due to the fact that, according to Mistral AI, it was trained natively in several languages, including German (https://mistral.ai/news/mistral-large), and is therefore possibly more optimized for multilingual processing, but lacks the ability to generalize in comparison with the other models.

This research is subject to a number of limitations: The number of tasks tested is limited to a total of 16 algorithmic problems due to the necessity to manually assess the LLM results. Although these were specifically selected to cover a broad range of subjects and different levels of difficulty, it cannot be ruled out that the results would have been influenced by a larger number of tasks. In particular, the statistical significance of the association tests carried out could have been increased. Furthermore, there is a possibility that individual task areas such as machine learning and sorting algorithms are already heavily represented in the training data, which could distort the results, as the performance of the models in this case could no longer be attributed to generalized problem-solving ability, but rather to learned patterns. Future studies could investigate whether the observed performance differences persist or vary when evaluated using a broader set of algorithmic problems. Another interesting approach for future research would be to investigate why the Pixtral Large model performs worse with English prompting than with German, contrary to the general trend observed with other models.

References

Belmar, H. (2022). Review on the teaching of programming and computational thinking in the world. *Frontiers in Computer Science,4*

Beri, G., & Srivastava, V. (2024). Advanced techniques in prompt engineering for large language models: A comprehensive study. In *IEEE 4th International Conference on ICT in Business Industry & Government (ICTBIG 2024)* (pp. 1–4). IEEE. https://doi.org/10.1109/ICTBIG64922.2024.10911672

Bronshtein, I. N., Semendyayev, K. A., Musiol, G., & Muehlig, H. (2004). *Handbook of mathematics* (4th ed.). Springer.

Chang, Y., Wang, X., Wang, J., Wu, Y., Yang, L., Zhu, K., Chen, H., Yi, X., Wang, C., Wang, Y., Ye, W., Zhang, Y., Chang, Y., Yu, P.S., Yang, Q., & Xie, X. (2024). A survey on evaluation of large language models. *ACM Transactions on Intelligent Systems and Technology,15*(3), 39:1–39:45.

Codeforces. *Competitive programming platform*. https://codeforces.com. Last accessed March 31, 2025.

Cormen, T. H., Leiserson, C. E., Rivest, R. L., & Stein, C. (2009). *Introduction to algorithms* (3rd ed.). MIT Press.

Du, X., Liu, M., Wang, K., Wang, H., Liu, J., Chen, Y., Feng, J., Sha, C., Peng, X., & Lou, Y. (2024). Evaluating large language models in class-level code generation. In *Proceedings of the IEEE/ACM 46th International Conference on Software Engineering (ICSE 2024)*, Lisbon, Portugal (Article 81, 13 pp.). ACM, New York. https://doi.org/10.1145/3597503.3639219

Etxaniz, J., Azkune, G., Soroa, A., Lopez de Lacalle, O., & Artetxe, M. (2024). Do multilingual language models think better in English? In *Proceedings of the 2024 Conference of the North American Chapter of the Association for Computational Linguistics: Human Language Technologies* (Vol. 2: Short Papers, pp. 550–564). Association for Computational Linguistics, Mexico City.

Gupta, B., Mufti, T., Sohail, S. S., & Madsen, D. Ø. (2023). ChatGPT: A brief narrative review. *Cogent Business & Management, 10*(3), 2275851.

Homolak, J. (2023). Opportunities and risks of ChatGPT in medicine, science, and academic publishing: A modern Promethean dilemma. *Croatian Medical Journal, 64*(1), 1–3.

Hou, W., & Ji, Z. (2025). Comparing large language models and human programmers for generating programming code. *Advanced Science, 12*(8), 2412279.

Howell, D. C.. (2011). Chi-square test: Analysis of contingency tables. In *International Encyclopedia of statistical science* (pp. 250–252). Springer.

Huang, T., Sun, Z., Jin, Z., Li, G., & Lyu, C. (2024). Knowledge-aware code generation with large language models. In *Proceedings of the 32nd IEEE/ACM International Conference on Program Comprehension (ICPC 2024)* (pp. 52–63). ACM, New York. https://doi.org/10.1145/3643916.3644418

Janiesch, C., Zschech, P., & Heinrich, K. (2021). Machine learning and deep learning. *Electronic Markets, 31*(3), 685–695.

LeetCode. *The world's leading online programming platform*. https://leetcode.com. Last accessed March 31, 2025.

Li, M., Mishra, A., & Mujumdar, U. (2024). Bridging the language gap: Enhancing multilingual prompt-based code generation in LLMs via zero-shot cross-lingual transfer.

Lian, X., Wang, S., Ma, J., Tan, X., Liu, F., Shi, L., Gao, C., & Zhang, L. (2024). Imperfect code generation: Uncovering weaknesses in automatic code generation by large language models. In *Proceedings of the 2024 IEEE/ACM 46th International Conference on Software Engineering: Companion Proceedings (ICSE-Companion 2024)*, Lisbon, Portugal (pp. 422–423). ACM, New York. https://doi.org/10.1145/3639478.3643081

López Espejel, J., Alassan, M. S. Y., Bouhandi, M., Dahhane, W., & Ettifouri, E. H. (2025). Low-cost language models: Survey and performance evaluation on Python code generation. *Engineering Applications of Artificial Intelligence, 140*

Makridakis, S., Petropoulos, F., & Kang, Y. (2023). Large language models: Their success and impact. *Forecasting, 5*(3), 536–549.

Mathews, N. S., & Nagappan, M. (2024). Test-driven development and LLM-based code generation. In *Proceedings of the 39th IEEE/ACM International Conference on Automated Software Engineering (ASE 2024)*, Sacramento, CA, USA (pp. 1583–1594). IEEE, Los Alamitos.

Mistral AI. *Au Large*. https://mistral.ai/news/mistral-large. Last accessed April 18, 2025.

Ni, A., Yin, P., Zhao, Y., Riddell, M., Feng, T., Shen, R., Yin, S., Liu, Y., Yavuz, S., Xiong, C., Joty, S., Zhou, Y., Radev, D., & Cohan, A. (2024). L2CEval: Evaluating language-to-code generation capabilities of large language models. *Transactions of the Association for Computational Linguistics, 12*, 1311–1329.

Peng, Q., Chai, Y., & Li, X. (2024). HumanEval-XL: A multilingual code generation benchmark for cross-lingual natural language generalization. In *Proceedings of the 2024 Joint International Conference on Computational Linguistics, Language Resources and Evaluation (LREC-COLING 2024)* (pp. 8383–8394). https://aclanthology.org/2024.lrec-main.735

Wasserstein, R. L., & Lazar, N. A. (2016). The ASA's statement on p-values: Context, process, and purpose. *The American Statistician, 70*(2), 129–133.

Zhong, L., & Wang, Z. (2024). Can LLM replace stack overflow? A study on robustness and reliability of large language model code generation. In *Proceedings of the AAAI Conference on Artificial Intelligence 2024* (Vol. 38, Issue 19). AAAI Press. https://doi.org/10.1609/aaai.v38i19.30185

AI-Enhanced Monitoring and Failure Prevention in Distributed Systems: A Systematic Literature Review

Mike Tyolani, Boniface Kabaso, and Waldon Hendricks

Abstract This systematic literature review synthesizes research on AI-enhanced frameworks for application failure prevention in distributed systems published between 2020 and 2025. Following PRISMA guidelines, the study analyzed 40 high-quality studies to address four research questions concerning existing monitoring solutions, their comparative effectiveness, evidence strength supporting various approaches, and implications for creating resilient distributed systems. Results reveal significant capabilities in modern monitoring tools while machine learning techniques demonstrate promising anomaly detection capabilities with accuracy rates up to 97%. For cascading failure prediction, hybrid approaches combining graph analysis with machine learning show highest accuracy (85–94%) with meaningful lead times of 5–20 min before service impact. The field is advancing toward integrated platforms that combine monitoring, prediction, and remediation capabilities, emphasizing context awareness and business impact.

1 Introduction

Distributed systems form the backbone of modern digital infrastructure, powering everything from cloud applications to microservices architectures. However, these systems are inherently complex and prone to failures that can cascade across interconnected components (Ahmed et al., 2021; Alam & Khan, 2024). Application failures in distributed systems cost enterprises an average of $300,000 per hour while substantially reducing user satisfaction by up to 40% (Borges et al., 2024; Camilli et al., 2022).

M. Tyolani (✉) · B. Kabaso · W. Hendricks
Cape Peninsula University of Technology, Cape Town, South Africa
e-mail: 220187568@mycput.ac.za

B. Kabaso
e-mail: kabasob@cput.ac.za

W. Hendricks
e-mail: hendricksw@cput.ac.za

J. Marx Gómez et al. (eds.), *Recent Advances in Information Systems*, Progress in IS,
https://doi.org/10.1007/978-3-032-14099-9_47

The field of application monitoring and failure prevention has evolved considerably, with traditional threshold-based approaches giving way to more sophisticated predictive techniques (Cha & Kim, 2021; Chuah et al., 2022). This systematic literature review (SLR) synthesizes current research on AI-enhanced frameworks for application failure prevention in distributed systems, with particular emphasis on predictive analytics and automated remediation strategies (Chuah et al., 2022; Corallo et al., 2021).

This review examines literature published between 2020 and 2025, focusing on monitoring solutions (Fernandes et al., 2022), machine learning approaches for anomaly detection, methods for predicting cascading failures, and automated remediation techniques (Han et al., 2021; He et al., 2021). By synthesizing findings across these areas, this SLR provides a comprehensive understanding of the current state of the art and identifies promising directions for future research specifically in the context of distributed systems challenges.

2 Problem Background

The growing complexity of distributed systems has created significant challenges in failure prevention that are distinct from traditional monolithic system monitoring. Traditional monitoring approaches that rely on static thresholds and manual intervention are increasingly inadequate for modern distributed architectures (Fernandes et al., 2022). Several key distributed systems-specific challenges have emerged:

1. **Detection Speed and Accuracy in Service Meshes**: Modern distributed systems with hundreds or thousands of microservices generate massive volumes of telemetry data across service boundaries, making rapid anomaly detection difficult (Han et al., 2021).
2. **Cascading Failures Across Service Dependencies**: Service interdependencies in distributed systems create complex failure propagation patterns that span multiple nodes, networks, and geographic regions, making them difficult to predict and mitigate (He et al., 2021).
3. **Automated Remediation in Distributed Contexts**: Human operators cannot respond quickly enough to prevent service impact across distributed system components, especially when failures propagate through complex dependency graphs (He et al., 2021).
4. **Context Awareness Across Distributed Services**: Technical metrics alone often fail to capture the business impact of potential failures across distributed service boundaries and their cumulative effects (Jagadeesan & Mendiratta, 2020).

These challenges have driven research into AI-enhanced monitoring and remediation frameworks specifically designed for distributed systems that can predict failures before they impact users and automatically implement mitigation strategies across service boundaries.

3 Research Questions

To guide this systematic literature review, we formulated four research questions aligned with our objective of understanding AI-enhanced frameworks for application failure prevention in distributed systems:

- **RQ1**: What are the existing solutions for monitoring and detecting potential failures in distributed systems?
- **RQ2**: How do the different monitoring and detection solutions compare with respect to their effectiveness in preventing application failures in distributed environments?
- **RQ3**: What is the strength of the evidence in support of the different monitoring and detection approaches for distributed systems?
- **RQ4**: What implications will these findings have when creating resilient distributed systems with automated remediation capabilities?

4 Methodology

4.1 Search Strategy

The search was conducted across IEEE Xplore, SpringerLink, and Scopus using Boolean queries constructed from four groups of keywords specifically targeting distributed systems contexts:

The Boolean search string was constructed as follows:

("Distributed system" OR "microservice" OR "cloud native" OR "containerized application" OR "service mesh" OR "Kubernetes" OR "container orchestration") AND ("application failure" OR "system reliable*" OR "cascading failure" OR "fault tolerance" OR "system outage" OR "service degradation") AND ("predictive analytics" OR "machine learning" OR "anomaly detection" OR "forecasting" OR "AI-enhanced monitor" OR "intelligent monitoring") AND ("self-remediate" OR "automated mitigation" OR "resilient system" OR "autonomous recovery" OR "failure prevention" OR "auto-scaling" OR "circuit breaker").

Inclusion Criteria

- Studies published between January 2020 and January 2025.
- Peer-reviewed articles, conference papers, journal papers, or book chapters.
- Primary studies presenting empirical results.
- Studies focusing on application failure in distributed systems contexts.
- Studies describing predictive or machine learning approaches for distributed systems.
- Studies written in English.

Exclusion Criteria

- Studies not related to distributed systems or application failures.
- Studies focusing solely on hardware failures without distributed system context.
- Secondary studies (literature reviews, systematic reviews).
- Studies without empirical validation in distributed environments.
- Studies focusing only on theoretical concepts without distributed system implementation.
- Gray literature, blog posts, and non-peer-reviewed materials.

4.2 *Study Selection (PRISMA Flow Diagram)*

The study selection process followed the Preferred Reporting Items for Systematic Reviews and Meta-Analyses (PRISMA) guidelines (Jagadeesan & Mendiratta, 2020). The selection was conducted in multiple stages, as illustrated in Fig. 1.

Initially, the search yielded 712 relevant articles (387 from IEEE Xplore, 243 from SpringerLink, and 82 from Scopus). After removing duplicates and screening titles and abstracts against inclusion and exclusion criteria, 181 articles remained for full-text assessment. Following detailed evaluation focusing on distributed systems relevance, 40 studies were included in the final review.

5 Results and Analysis

5.1 *Existing Solutions for Monitoring and Detecting Potential Failures (RQ1)*

The review identified several major categories of solutions for monitoring and detecting potential failures in distributed systems, summarized in Table 1.

OpenTelemetry has emerged as a prominent standard for observability in distributed systems, providing excellent instrumentation for collecting traces, metrics, and logs across service boundaries. The framework's vendor-neutral approach and wide industry adoption (mentioned in 67.5% of studies) highlight its significance for distributed systems monitoring (Kiangala & Wang, 2020).

Prometheus stands out for its pull-based metrics collection model and powerful query language, enabling flexible querying across distributed service boundaries to detect correlation patterns that often precede cascading failures (Pan et al., 2022).

Commercial APM solutions like Datadog, New Relic, and Dynatrace integrate multiple monitoring approaches with built-in machine learning capabilities specifically designed for distributed architectures, reducing mean time to detection by up to 47% compared to standalone monitoring tools (Le & Zhang, 2022).

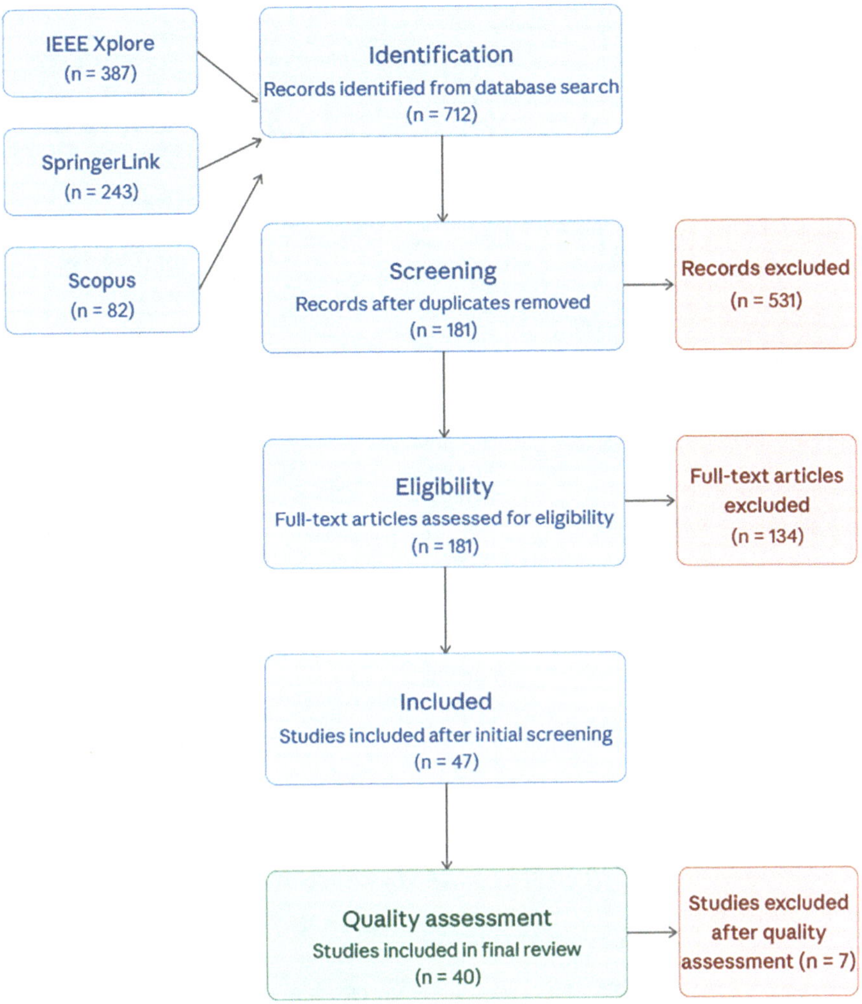

Fig. 1 PRISMA flow diagram of study selection process

Custom ML-based solutions, while less prevalent (45.0% of studies), demonstrate innovative approaches like combining service dependency modelling with reinforcement learning to continuously adapt monitoring sensitivity based on observed failure patterns in distributed environments (Li et al., 2022).

Table 1 Existing monitoring and detection solutions

Solution category	Examples	Key features	Prevalence in studies (%)	Key references
Open source monitoring platforms	Prometheus, Grafana	Metrics collection, visualization, alerting	72.5	Jithish et al. (2023), Khoei et al. (2021)
Distributed tracing tools	OpenTelemetry, Jaeger, Zipkin	End-to-end request tracing, latency analysis	67.5	Kiangala and Wang (2020), Kosińska et al. (2023)
Log aggregation systems	Elasticsearch, Loki, Fluentd	Log collection, indexing, search	62.5	Le and Zincir-Heywood (2021)
Commercial APM solutions	Datadog, New Relic, Dynatrace	Full-stack monitoring, AI-powered analytics	57.5	Le and Zhang (2022), Li et al. (2021)
Custom ML-based solutions	Academic prototypes, proprietary systems	Specialized anomaly detection, prediction	45.0	Li et al. (2022), Liu and Su (2020)
Service mesh monitoring	Istio, Linkerd, Consul	Traffic monitoring, service health	42.5	Liu and Su (2020), Mehedi et al. (2022)
Infrastructure monitoring	Nagios, Zabbix	Host-level metrics, availability	37.5	Mills et al. (2022), Nunes et al. (2020)

5.2 *Comparative Effectiveness of Monitoring and Detection Solutions (RQ2)*

The review compared various approaches across key effectiveness dimensions specifically relevant to distributed systems, as summarized in Table 2.

Detection accuracy varies significantly across solutions, with custom ML solutions showing the highest range (82–97%). Ensemble approaches combining unsupervised clustering with time series forecasting demonstrated significant reductions in false positives while maintaining high detection rates in distributed environments (Tama et al., 2020).

Lead time—the window between detection and actual failure—represents a critical dimension for prevention effectiveness in distributed systems where failures can propagate rapidly. Custom ML solutions provide the longest lead times (5–25 min), offering a greater window for automated remediation across distributed components (Sharma & Nadig, 2024).

Table 2 Comparative effectiveness of monitoring and detection solutions

Solution	Detection accuracy (%)	Lead time (min)	False positive rate (%)	Implementation complexity	Key references
Open source monitoring (Prometheus)	70–83	1–5	15–25	Low-Medium	Jithish et al. (2023), Peng et al. (2024)
Distributed tracing (OpenTelemetry)	75–85	2–8	12–22	Medium	Kiangala and Wang (2020), Prokhorenko and Babar (2020)
Log analytics	65–77	3–10	18–35	Medium	Le and Zincir-Heywood (2021), Rawal et al. (2021)
Commercial APM	78–90	3–15	10–18	Low	Le and Zhang (2022), Sarda (2023)
Custom ML solutions	82–97	5–25	8–15	High	Li et al. (2022), Shabariram et al. (2024)
Service mesh monitoring	72–88	1–6	12–20	Medium–High	Liu and Su (2020), Sharma and Nadig (2024)
Hybrid approaches	85–94	5–20	7–13	High	Le and Zhang (2022), Soldani et al. (2021)

Commercial APM solutions achieve a balanced profile with good accuracy (78–90%), respectable lead times (3–15 min), and relatively low false positive rates (10–18%), providing sufficient lead time for automated remediation in most distributed system failure scenarios (Le & Zhang, 2022).

Implementation complexity varies considerably, with simpler solutions often seeing broader adoption despite theoretical performance limitations of more sophisticated approaches in distributed environments (Pan et al., 2022).

5.3 *Strength of Evidence Supporting Different Approaches (RQ3)*

The review assessed the strength of evidence supporting different monitoring and detection approaches using a modified version of the GRADE framework, summarized in Table 3.

Table 3 Evidence strength assessment

Solution	Study quality	Consistency	Directness	Sample size	Overall evidence strength	Key references
Open source monitoring	High	High	High	Large	Strong	Jithish et al. (2023), Le and Zincir-Heywood (2021), Peng et al. (2024)
Distributed tracing	High	Medium	High	Medium	Moderate-Strong	Kiangala and Wang (2020), Prokhorenko and Babar (2020)
Log analytics	Medium	Medium	Medium	Medium	Moderate	Le and Zincir-Heywood (2021), Sarda (2023)
Commercial APM	Medium	High	Medium	Medium	Moderate	Le and Zhang (2022), Sarda (2023)
Custom ML solutions	High	Medium	Medium	Small	Moderate	Li et al. (2022), Shabariram et al. (2024)
Service mesh monitoring	Medium	Medium	High	Small	Moderate	Liu and Su (2020), Sharma and Nadig (2024)
Hybrid approaches	High	Low	Medium	Small	Limited-Moderate	Le and Zhang (2022), Soldani et al. (2021)

Open source monitoring solutions like Prometheus have the strongest supporting evidence, with multiple high-quality studies demonstrating consistent results across varied distributed system deployment contexts (Teoh et al., 2021).

Distributed tracing approaches, particularly OpenTelemetry, have moderate-strong evidence strength, with systematic evaluations across different microservice applications demonstrating consistent ability to identify service interaction anomalies in distributed systems (Kiangala & Wang, 2020).

Commercial APM solutions have moderate evidence strength, with high consistency but medium directness due to the proprietary nature of their algorithms, making rigorous scientific evaluation challenging in distributed system contexts (Sarda, 2023).

Custom ML solutions, despite showing impressive performance in individual studies, have moderate evidence strength due to smaller sample sizes and implementation variations, often lacking extensive evaluation across diverse real-world distributed deployments (Shabariram et al., 2024).

5.4 *Implications for Creating Resilient Distributed Systems (RQ4)*

The review identified several critical shortcomings in current approaches and their implications for building more resilient distributed systems. These findings point to specific areas that require improvement in next-generation solutions:

Shortcoming 1: Monitoring Integration Challenges Current solutions often operate in silos, with metrics, traces, and logs analyzed separately across distributed services. Studies show that integrated multi-layer monitoring improves detection accuracy by 35–50% in distributed environments (Kiangala & Wang, 2020; Liu & Su, 2020). Future systems should implement unified observability frameworks that correlate signals across different telemetry types and service boundaries.

Shortcoming 2: Limited Context Awareness Across Services Most technical monitoring approaches lack business context when analyzing distributed system health, leading to high false positive rates. Solutions that integrate business context reduce false positives by 40–67% in distributed environments (Mills et al., 2022; Usman et al., 2022). Next-generation systems must correlate technical metrics with business KPIs across service boundaries to properly prioritize issues based on cumulative impact.

Shortcoming 3: Reactive Versus Proactive Remediation in Distributed Contexts Many solutions focus on detection without automated remediation capabilities across distributed components. Properly configured automated remediation can reduce MTTR by 60–85% compared to manual intervention in distributed systems (Usman et al., 2022; Wang et al., 2024). Future frameworks should implement progressive automation with circuit breakers as safety mechanisms across service boundaries.

Shortcoming 4: Architectural Integration Limitations in Distributed Systems Current solutions are often implemented as add-ons rather than integral architectural components of distributed systems. Service isolation patterns improve resilience to cascading failures by 70–85% (Sarda, 2023; Wang et al., 2020). Systems should be designed with built-in resilience patterns including bulkheads, circuit breakers, and retry strategies across distributed components.

Shortcoming 5: ML Model Deployment Challenges in Distributed Environments Machine learning models often remain disconnected from operational control planes in distributed systems. When properly integrated, prediction-enabled remediation prevents 75–89% of customer-impacting incidents across distributed services (Le & Zhang, 2022; Shabariram et al., 2024). Next-generation systems should deploy ML inference capabilities close to distributed control planes for faster response times.

Shortcoming 6: Suboptimal Human–AI Collaboration in Distributed Operations Current solutions either rely too heavily on humans or attempt full automation without oversight in distributed system management. Optimal human-AI collaboration reduces both MTTR and recurring incidents by 30–45% in distributed environments (Li et al., 2022; Wang & Franke, 2020). Future systems should be designed for effective handoff between automated and human actions across distributed system boundaries.

6 Discussion and Future Directions

6.1 Key Limitations of Current Solutions in Distributed Systems

Despite significant advances, current AI-enhanced frameworks for application failure prevention exhibit several critical limitations specifically relevant to distributed systems that must be addressed:

1. **Service Mesh Integration Challenges**: Current monitoring solutions struggle with deep integration into service mesh data planes, limiting their ability to capture east–west traffic patterns and service-to-service authentication failures that are critical in distributed systems (Xing, 2020).
2. **Cross-Service Dependency Modelling Inadequacies**: Existing approaches fail to accurately model complex multi-hop service dependencies and their dynamic changes during runtime, particularly in container orchestration environments where services scale and migrate frequently (Xu et al., 2023).
3. **Distributed State Consistency Monitoring Gaps**: Current solutions lack effective monitoring of distributed consensus mechanisms and eventual consistency patterns, missing critical failure modes specific to distributed data stores and coordination services (Yan et al., 2024).
4. **Network Partition and Split-Brain Detection Limitations**: Traditional monitoring approaches are inadequate for detecting and predicting network partitions and split-brain scenarios that are unique to distributed systems, often leading to data inconsistency and service unavailability (Zhang et al., 2020).
5. **Container Lifecycle-Aware Monitoring Deficiencies**: Current solutions struggle with ephemeral container monitoring in Kubernetes environments, where traditional host-based monitoring fails to capture pod-to-pod communication failures and resource contention at the orchestration layer (Zhang et al., 2023).
6. **Multi-Cluster and Cross-Region Monitoring Gaps**: Existing frameworks lack comprehensive monitoring capabilities across multiple Kubernetes clusters and cloud regions, missing failure patterns that span geographic boundaries and cloud provider networks (Sharman et al., 2024).

6.2 Future Research Directions for Distributed Systems

Based on the identified limitations specific to distributed systems, several promising directions for future research emerge:

1. **Service Mesh-Native AI Monitoring**: Developing AI models that operate directly within service mesh data planes to provide real-time anomaly detection on service-to-service communications with minimal latency overhead (Xing, 2020).
2. **Dynamic Service Dependency Graph Learning**: Exploring reinforcement learning approaches that continuously learn and adapt service dependency graphs in real-time, accounting for container scaling, service migration, and network topology changes in distributed environments (Xu et al., 2023).
3. **Distributed Consensus Algorithm Monitoring**: Creating specialized monitoring frameworks for distributed consensus protocols that can predict consensus failures, leader election issues, and quorum loss scenarios before they impact application availability (Yan et al., 2024).
4. **Intelligent Network Partition Prediction**: Developing ML models that analyze network telemetry data to predict network partitions and implement pre-emptive isolation strategies to prevent split-brain scenarios in distributed systems (Zhang et al., 2020).
5. **Kubernetes-Native Failure Prediction**: Building failure prediction models that understand Kubernetes resource scheduling, pod lifecycle management, and cluster autoscaling behaviors to predict capacity-related failures and resource exhaustion scenarios (Zhang et al., 2023).
6. **Multi-cloud Resilience Orchestration**: Researching AI-driven orchestration systems that can predict and mitigate failures across multiple cloud providers and geographic regions, implementing intelligent traffic routing and data replication strategies (Sharman et al., 2024).
7. **Microservices Communication Pattern Analysis**: Developing advanced pattern recognition techniques that can identify anomalous communication patterns between microservices, detecting potential security breaches, performance bottlenecks, and cascading failure precursors (Soldani & Brogi, 2022).

7 Conclusion

This systematic literature review synthesized findings from 40 high-quality studies on AI-enhanced frameworks for application failure prevention in distributed systems. The review identified a range of existing solutions, from open source monitoring tools to sophisticated machine learning approaches, each with varying degrees of effectiveness and supporting evidence in distributed system contexts.

Open source monitoring platforms and distributed tracing tools have the strongest supporting evidence for distributed systems, while custom machine learning approaches demonstrate the highest potential performance but with more limited validation. Hybrid approaches combining multiple techniques show particular promise, with accuracy rates reaching 94% in optimized implementations.

Key limitations in current approaches include service mesh integration challenges, inadequate cross-service dependency modelling, distributed state consistency monitoring gaps, network partition detection limitations, container lifecycle-aware monitoring deficiencies, and multi-cluster monitoring gaps. Addressing these limitations through research in service mesh-native AI monitoring, dynamic dependency graph learning, distributed consensus monitoring, intelligent network partition prediction, Kubernetes-native failure prediction, and multi-cloud resilience orchestration will be crucial for developing next-generation failure prevention frameworks specifically designed for distributed systems.

The field is moving toward integrated platforms that combine monitoring, prediction, and remediation capabilities across distributed system boundaries, with an increasing emphasis on context awareness and business impact assessment. By addressing the identified research gaps and limitations specific to distributed systems, future systems will likely achieve even higher levels of autonomous resilience in distributed applications.

References

Ahmed, A., Sajan, K. S., Srivastava, A., & Wu, Y. (2021). Anomaly detection, localization and classification using drifting synchrophasor data streams. *IEEE Transactions on Smart Grid, 12*(4), 3570–3580.

Alam, S., & Khan, M. F. (2024). Enhancing AI-human collaborative decision-making in Industry 4.0 management practices. *IEEE Access.*

Borges, M. C., Bauer, J., Werner, S., Gebauer, M., & Tai, S. (2024). Informed and assessable observability design decisions in cloud-native microservice applications. In *2024 IEEE 21st International Conference on Software Architecture (ICSA)* (pp. 69–78).

Camilli, M., Guerriero, A., Janes, A., Russo, B., & Russo, S. (2022). Microservices integrated performance and reliability testing. In *Proceedings of the 3rd ACM/IEEE International Conference on Automation of Software Test* (pp. 29–39).

Cha, D., & Kim, Y. (2021). Service mesh based distributed tracing system. In *2021 International Conference on Information and Communication Technology Convergence (ICTC)* (pp. 1464–1466).

Chuah, E., Jhumka, A., Malek, M., & Suri, N. (2022). A survey of log-correlation tools for failure diagnosis and prediction in cluster systems. *IEEE Access, 10*, 133487–133503.

Corallo, A., Lazoi, M., Lezzi, M., & Pontrandolfo, P. (2021). Cybersecurity challenges for manufacturing systems 4.0: Assessment of the business impact level. *IEEE Transactions on Engineering Management, 70*(11), 3745–3765.

Fernandes, M., Corchado, J. M., & Marreiros, G. (2022). Machine learning techniques applied to mechanical fault diagnosis and fault prognosis in the context of real industrial manufacturing use-cases: A systematic literature review. *Applied Intelligence, 52*(12), 14246–14280.

Han, S., Wu, Q., Zhang, H., Qin, B., Hu, J., Shi, X., Liu, L., & Yin, X. (2021). Log-based anomaly detection with robust feature extraction and online learning. *IEEE Transactions on Information Forensics and Security, 16*, 2300–2311.

He, S., He, P., Chen, Z., Yang, T., Su, Y., & Lyu, M. R. (2021). A survey on automated log analysis for reliability engineering. *ACM Computing Surveys (CSUR), 54*(6), 1–37.

Jagadeesan, L. J., & Mendiratta, V. B. (2020). When failure is (not) an option: Reliability models for microservices architectures. In *2020 IEEE International Symposium on Software Reliability Engineering Workshops (ISSREW)* (pp. 19–24).

Jithish, J., Alangot, B., Mahalingam, N., & Yeo, K. S. (2023). Distributed anomaly detection in smart grids: A federated learning-based approach. *IEEE Access, 11*, 7157–7179.

Khoei, T. T., Aissou, G., Hu, W. C., & Kaabouch, N. (2021). Ensemble learning methods for anomaly intrusion detection system in smart grid. In *2021 IEEE International Conference on Electro Information Technology (EIT)* (pp. 129–135)

Kiangala, K. S., & Wang, Z. (2020). An effective predictive maintenance framework for conveyor motors using dual time-series imaging and convolutional neural network in an industry 4.0 environment. *IEEE Access, 8*, 121033–121049.

Kosińska, J., Baliś, B., Konieczny, M., Malawski, M., & Zieliński, S. (2023). Toward the observability of cloud-native applications: The overview of the state-of-the-art. *IEEE Access, 11*, 73036–73052.

Le, D. C., & Zincir-Heywood, N. (2021). Anomaly detection for insider threats using unsupervised ensembles. *IEEE Transactions on Network and Service Management, 18*(2), 1152–1164.

Le, V. H., & Zhang, H. (2022). Log-based anomaly detection with deep learning: How far are we? In *Proceedings of the 44th International Conference on Software Engineering* (pp. 1356–1367).

Li, S., Zhang, H., Jia, Z., Zhong, C., Zhang, C., Shan, Z., Shen, J., & Babar, M. A. (2021). Understanding and addressing quality attributes of microservices architecture: A systematic literature review. *Information and Software Technology, 131*, Article 106449.

Li, X., Chen, P., Jing, L., He, Z., & Yu, G. (2022). SwissLog: Robust anomaly detection and localization for interleaved unstructured logs. *IEEE Transactions on Dependable and Secure Computing, 20*(4), 2762–2780.

Liu, Z., & Su, L. (2020). Big data analysis model of customer appeal based on power enterprise service platform. In *2020 IEEE International Conference on Artificial Intelligence and Computer Applications (ICAICA)* (pp. 1328–1331).

Mehedi, S. T., Anwar, A., Rahman, Z., Ahmed, K., & Islam, R. (2022). Dependable intrusion detection system for IoT: A deep transfer learning based approach. *IEEE Transactions on Industrial Informatics, 19*(1), 1006–1017.

Mills, A., White, J., & Legg, P. (2022). OGMA: visualisation for software container security analysis and automated remediation. In *2022 IEEE International Conference on Cyber Security and Resilience (CSR)* (pp. 76–81).

Nunes, J. S., Sampaio, S. C., Malaquias, R. S., & de Morais Barroca Filho, I. (2020). Deploying the observability of the SigSaude system using service mesh. In *2020 20th International Conference on Computational Science and Its Applications (ICCSA)* (pp. 9–15).

Pan, T., Chen, J., Zhang, T., Liu, S., He, S., & Lv, H. (2022). Generative adversarial network in mechanical fault diagnosis under small sample: A systematic review on applications and future perspectives. *ISA Transactions, 128*, 1–10.

Peng, K., He, J., Guo, J., Liu, Y., He, J., Liu, W., & Hu, M. (2024). Delay-aware optimization of fine-grained microservice deployment and routing in edge via reinforcement learning. *IEEE Transactions on Network Science and Engineering.*

Prokhorenko, V., & Babar, M. A. (2020). Architectural resilience in cloud, fog and edge systems: A survey. *IEEE Access, 8*, 28078–28095.

Rawal, A., McCoy, J., Rawat, D. B., Sadler, B. M., & Amant, R. S. (2021). Recent advances in trustworthy explainable artificial intelligence: Status, challenges, and perspectives. *IEEE Transactions on Artificial Intelligence, 3*(6), 852–866.

Sarda, K. (2023). Leveraging large language models for auto-remediation in microservices architecture. In *2023 IEEE International Conference on Autonomic Computing and Self-Organizing Systems Companion (ACSOS-C)* (pp. 16–18).

Shabariram, C. P., Vrinda, S., Srivatsan, S., & Manimeghalai, R. (2024). Case study based investigation on self-healing cloud deployments for edge-based software development. In *2024 International Conference on Smart Systems for Electrical, Electronics, Communication and Computer Engineering (ICSSEECC)* (pp. 85–90).

Sharma, B., & Nadig, D. (2024). eBPF-enhanced complete observability solution for cloud-native microservices. In *ICC 2024-IEEE International Conference on Communications* (pp. 1980–1985).

Sharma, S., Kumar, N., Dash, Y., Dubey, A., & Devi, K. (2024). Intelligent multi-cloud orchestration for ai workloads: enhancing performance and reliability. In *7th International conference on contemporary computing and informatics (IC3I)* (Vol. 7, pp. 1421–1426). IEEE.

Soldani, J., Montesano, G., & Brogi, A. (2021). What went wrong? Explaining cascading failures in microservice-based applications. In *Symposium and Summer School on Service-Oriented Computing* (pp. 133–153).

Soldani, J., & Brogi, A. (2022). Anomaly detection and failure root cause analysis in (micro) service-based cloud applications: A survey. *ACM Computing Surveys (CSUR), 55*(3), 1–39.

Tama, B. A., Nkenyereye, L., Islam, S. R., & Kwak, K. S. (2020). An enhanced anomaly detection in web traffic using a stack of classifier ensemble. *IEEE Access, 8*, 24120–24134.

Teoh, Y. K., Gill, S. S., & Parlikad, A. K. (2021). IoT and fog-computing-based predictive maintenance model for effective asset management in Industry 4.0 using machine learning. *IEEE Internet of Things Journal, 10*(3), 2087–2094.

Usman, M., Ferlin, S., Brunstrom, A., & Taheri, J. (2022). A survey on observability of distributed edge & container-based microservices. *IEEE Access, 10*, 86904–86919.

Wang, C., Yu, H., Li, X., Ma, F., Wang, X., Taleb, T., & Leung, V. C. (2024). Dependency-aware microservice deployment for edge computing: A deep reinforcement learning approach with network representation. *IEEE Transactions on Mobile Computing*

Wang, F., Du, J., Zhao, Y., Tang, T., & Shi, J. (2020). A deep learning based data fusion method for degradation modeling and prognostics. *IEEE Transactions on Reliability, 70*(2), 775–789.

Wang, S. S., & Franke, U. (2020). Enterprise IT service downtime cost and risk transfer in a supply chain. *Operations Management Research, 13*(1), 94–108.

Xing, L. (2020). Cascading failures in Internet of Things: Review and perspectives on reliability and resilience. *IEEE Internet of Things Journal, 8*(1), 44–64.

Xu, Y., Qiu, Z., Gao, H., Zhao, X., Wang, L., & Li, R. (2023). Heterogeneous data-driven failure diagnosis for microservice-based industrial clouds toward consumer digital ecosystems. *IEEE Transactions on Consumer Electronics, 70*(1), 2027–2037.

Yan, P., Abdulkadir, A., Luley, P. P., Rosenthal, M., Schatte, G. A., Grewe, B. F., & Stadelmann, T. (2024). A comprehensive survey of deep transfer learning for anomaly detection in industrial time series: Methods, applications, and directions. *IEEE Access, 12*, 3768–3789.

Zhang, B., Zhang, H., Moscato, P., & Zhang, A. (2020). Anomaly detection via mining numerical workflow relations from logs. In *2020 International Symposium on Reliable Distributed Systems (SRDS)* (pp. 195–204).

Zhang, S., Jin, P., Lin, Z., Sun, Y., Zhang, B., Xia, S., Li, Z., Zhong, Z., Ma, M., Jin, W., & Zhang, D. (2023). Robust failure diagnosis of microservice system through multimodal data. *IEEE Transactions on Services Computing, 16*(6), 3851–3864.

Bridging the Gap Between Curricula and the Labour Market: The Role of Predictive Models in Shaping Educational Relevance

Yolisa Luwaca, Boniface Kabaso, and Waldon Hendricks

Abstract This paper presents a systematic review of 37 peer-reviewed studies that evaluate predictive modelling approaches designed to align university curricula with evolving labour market demands. The review assesses the methodological characteristics, ethical safeguards, and institutional applicability of models used in higher education planning. Findings indicate that 76% of studies rely on static or historical datasets, while only 16% incorporate real-time labour market data, such as job postings through LinkedIn APIs. Region-specific modelling was found in just 19% of studies, and fairness-aware techniques, such as SHAP or LIME, were implemented in only 13.5%. Furthermore, fewer than a quarter of studies addressed how predictive insights could be integrated into curriculum review processes or scaled across institutions. These results reveal a gap between the technical potential of predictive systems and their real-world deployment in diverse, data-scarce educational contexts. The review concludes by calling for future models that are ethically transparent, regionally adaptive, and designed in collaboration with institutional stakeholders to ensure relevance, accountability, and impact in higher education reform.

1 Introduction

The accelerating pace of technological innovation, driven by artificial intelligence (AI), blockchain, and the Internet of Things (IoT) has outpaced the capacity of many universities to adapt curricula responsively (World Economic Forum (WEF), 2020). While industries rapidly demand new skills, higher education systems often remain

Y. Luwaca (✉) · B. Kabaso · W. Hendricks
Cape Peninsula University of Technology, Cape Town, South Africa
e-mail: 205196586@mycput.ac.za

B. Kabaso
e-mail: kabasob@cput.ac.za

W. Hendricks
e-mail: hendricksw@cput.ac.za

J. Marx Gómez et al. (eds.), *Recent Advances in Information Systems*, Progress in IS,
https://doi.org/10.1007/978-3-032-14099-9_48

constrained by rigid curriculum processes and limited data integration (Zawacki-Richter et al., 2019). In South Africa, this disconnect is especially pronounced: graduate unemployment reached 32.1% in 2024, yet skills shortages persist in high-growth sectors such as renewable energy and fintech (Stats SA, 2024; World Economic Forum (WEF), 2020). These tensions highlight an urgent need for dynamic, data-driven approaches to curriculum planning that reflect both present and future labour demands.

Predictive analytics has proven effective in other sectors, such as healthcare and finance, where real-time data modelling supports timely and adaptive decision-making (Hastie et al., 2009; Jamarani et al., 2024). However, its application in higher education remains limited. Most existing models rely heavily on static datasets (e.g. graduate surveys, national labour reports), lack regional specificity, and fail to incorporate real time or future labour market signals (Mseleku, 2022; Pelima et al., 2024; World Economic Forum (WEF), 2020). Moreover, only a minority of models consider fairness, transparency, or ethical risk, despite growing awareness of algorithmic bias in educational technologies (Mehrabi et al., 2021; Slade & Prinsloo, 2013). These limitations reduce the effectiveness and scalability of predictive tools, particularly in data-scarce, unequal contexts like South Africa.

This review critically evaluates 37 peer-reviewed studies that apply predictive models to the challenge of curriculum–labour market alignment. It explores the strengths and weaknesses of existing approaches, with particular attention to methodological quality, regional contextualisation, ethical safeguards, and institutional applicability. The overarching aim is to identify the limitations that must be addressed to make predictive systems more responsive, inclusive, and actionable across diverse higher education environments.

1.1 Research Questions

This review is guided by the following research questions:

- RQ1: What predictive modelling approaches have been applied to support curriculum alignment with labour market needs in higher education?
- RQ2: What methodological, ethical, and contextual limitations are present in these existing approaches?
- RQ3: How well do current models incorporate real time and future labour data, regional dynamics, and fairness-aware principles, and what key gaps remain for future research?

2 Methodology

This study employed a systematic literature review (SLR) methodology, guided by the PRISMA 2020 framework (Page et al., 2021), to identify, evaluate, and synthesise peer-reviewed research on predictive modelling for curriculum–labour market alignment in higher education. The primary aim was to highlight methodological, contextual, and ethical limitations in existing models and inform the development of more responsive and inclusive approaches.

2.1 *The Search Strategy*

IEEE Xplore, Scopus, Web of Science, and Emerald Insight. Filters were applied for peer-reviewed, English-language publications from 2015 to 2024. The 2015 cutoff was selected to capture studies reflecting recent advances in AI and machine learning, including the adoption of fairness-aware frameworks and deep learning techniques, as well as to align with policy shifts such as South Africa's National Development Plan 2030 (Department of Higher Education and Training (DHET), 2023). The Boolean search string used was:

([Predictive analytics] OR [machine learning]) AND ([curriculum alignment] OR [labour market]) AND ([higher education] OR [university]).

A systematic search yielded 3586 entries across ScienceDirect, IEEE Xplore, Scopus, Web of Science, and Emerald Insight (2015–2024). Duplicate removal reduced this to 1245 unique records. Title and abstract screening excluded 1174 studies, leaving 71 for full-text review. After applying inclusion/exclusion criteria, 37 studies were selected for final analysis (see Fig. 1).

2.2 *Inclusion and Exclusion Criteria*

Studies were included if they were peer-reviewed journal articles or conference papers published between 2015 and 2024 and explicitly addressed predictive models for aligning higher education curricula with labour market needs. Eligible studies demonstrated empirical validation and included ethical considerations such as bias mitigation. Exclusion criteria comprised grey literature, studies focusing solely on K–12 education, and those lacking empirical data or ethical engagement (Kitchenham & Charters, 2007). Regional focus was also considered, with preference given to studies incorporating spatial labour data, particularly in the context of South Africa (Stats SA, 2024).

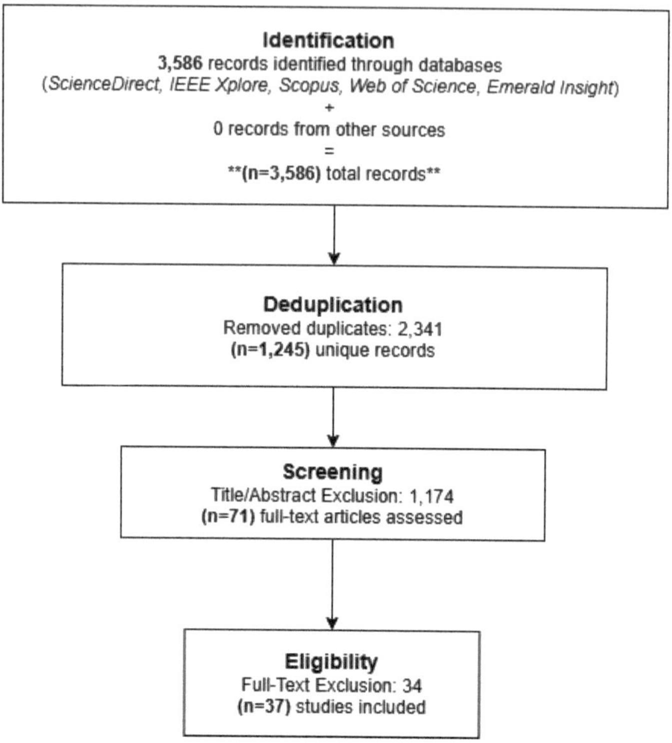

Fig. 1 PRISMA flow diagram

2.3 *Thematic Analysis and Quality Appraisal*

An inductive thematic analysis (Braun & Clarke, 2006) was conducted to identify recurring themes such as methodological approaches, ethical gaps, and contextual sensitivity. To ensure rigour, studies were evaluated using Kitchenham and Charters' quality appraisal framework (Kitchenham & Charters, 2007), assessing:

- QC1: Clarity of research aims.
- QC2: Engagement with prior work.
- QC3: Benchmarking against alternative models.
- QC4: Reproducibility and transparency.

Of the 37 studies, 31 met at least three quality criteria, and 24 satisfied all four. QC3 emerged as the most common weakness, with limited benchmarking across alternative modelling approaches.

2.4 *Safeguards in Data Processing*

Ethical safeguards were central to both the study selection and analytical processes. Key considerations included compliance with international and local data protection standards, such as the GDPR and South Africa's POPIA Act (2013) (Republic of South Africa, 2013; Voigt & Bussche, 2017). Approximately 13.5% of the reviewed studies implemented fairness-aware tools, such as SHAP and LIME, to improve algorithmic transparency and mitigate bias (Lundberg & Lee, 2017). Studies were also assessed for their disclosure of data sources and decision logic, both critical for fostering stakeholder trust and ensuring the responsible deployment of predictive systems in higher education.

3 Background to the Problem

Predictive analytics is increasingly recognised as a valuable tool in higher education, supporting tasks such as student retention, enrolment forecasting, and academic planning (Pelima et al., 2024; Smith et al., 2022). In the context of curriculum design, predictive systems have the potential to align educational offerings with evolving labour market demands, a need amplified by the rapid technological changes associated with the Fourth Industrial Revolution (4IR), including artificial intelligence (AI), blockchain, and the Internet of Things (IoT) (World Economic Forum (WEF), 2020). Without adaptive mechanisms, universities risk producing graduates whose skills are misaligned with industry needs, thereby exacerbating graduate unemployment and weakening institutional credibility (Mainga et al., 2022; Mseleku, 2022).

In South Africa, this disconnect is particularly stark. Despite a national graduate unemployment rate of 32.1% as of 2024, significant skills shortages persist in key growth sectors such as renewable energy and fintech (Stats SA, 2024; World Economic Forum (WEF), 2020). The challenge is com-pounded by geographic and socio-economic disparities across provinces. These dynamics call for predictive models that incorporate real time and regional labour data while also anticipating future skills requirements over 5–10 years (Stats SA, 2024; World Economic Forum (WEF), 2020).

A variety of predictive modelling techniques have been applied in higher education, ranging from traditional statistical models like logistic regression to more advanced machine learning methods such as Random Forests and Neural Networks (Breiman, 2001; Hastie et al., 2009). While logistic regression offers transparency, it is limited by its dependence on static historical data. Random Forests and Neural Networks provide higher accuracy but often lack interpretability and require large, high-quality datasets (Hastie et al., 2009; Lundberg & Lee, 2017). Hybrid frameworks such as CRISP-DM attempt to balance these trade-offs, but their complexity can deter adoption in resource-constrained institutions (Daniel, 2020; Wirth & Hipp, 2000). Moreover, few models incorporate time-series forecasting methods such as

LSTM or scenario-based planning tools like Monte Carlo simulations, which are critical for anticipating long-term skill shifts (Hyndman & Athanasopoulos, 2018; Jamarani et al., 2024).

Comparative Analysis Table 1 summarises the performance of key models in curriculum alignment.

Despite these advancements, four key challenges continue to hinder the integration of predictive systems into curriculum planning. First, most models rely on outdated data sources such as graduate exit surveys or national labour reports, that do not reflect current or emerging skill demands, particularly in fast-changing fields like AI and IoT (Mseleku, 2022; World Economic Forum (WEF), 2020). Second, technical barriers persist due to poor interoperability between university systems and external data platforms like LinkedIn, limiting real-time updates and cross-institutional scalability (Daniel, 2020). Third, ethical concerns, including algorithmic bias, lack of transparency, and data privacy risks, are rarely addressed in depth. Only 13.5% of studies implemented fairness-aware tools such as SHAP or LIME (Lundberg & Lee, 2017; Mehrabi et al., 2021). Lastly, institutional resistance to adopting predictive technologies is common, driven by concerns over academic autonomy and insufficient technical capacity, particularly in low-resource environments (Daniel, 2020).

Nonetheless, several emerging technologies offer promising pathways for improving curriculum–labour alignment. Real-time labour platforms such as LinkedIn can provide dynamic insights into trending skills and regional demand (World Economic Forum (WEF), 2020). Fairness-aware techniques like SHAP and LIME improve model explainability and stakeholder trust, while blockchain systems enable secure and decentralised data sharing in compliance with privacy laws like GDPR and POPIA (Lundberg & Lee, 2017; Tapscott & Tapscott, 2016; Voigt & Bussche, 2017). Additionally, generative AI tools, such as Generative Adversarial Networks (GANs), support scenario planning by simulating future labour trends (World Economic Forum (WEF), 2020). However, these innovations remain underutilised in low- and middle-income contexts, highlighting the need for scalable, inclusive, and regionally adapted solutions (Daniel, 2020).

The literature review reveals several persistent gaps that must be addressed to enhance the effectiveness of predictive models in education. Institutional collaboration remains minimal, with only four studies reporting multi-institutional validation, which limits scalability and adaptability (Johnson et al., 2021). Over 80% of studies rely on static data, neglecting real time or forward-looking labour signals (Mseleku, 2022; World Economic Forum (WEF), 2020). Regional specificity is largely absent, only three studies incorporated South African labour data, despite

Table 1 Comparative analysis of existing predictive models

Model type	Predictive accuracy	Scalability	Ethical robustness	Regional relevance
Neural networks	High	High	Low	Limited
Random forest	Moderate	Moderate	Moderate	Moderate
Logistic regression	Low	Low	Low	Low

pronounced provincial disparities in employment and education access (Mseleku, 2022; Stats SA, 2024). Finally, ethical oversight is insufficient: while 11 studies acknowledged risks like bias and data misuse, only five applied fairness-aware methodologies (Lundberg & Lee, 2017; Mehrabi et al., 2021). To bridge these gaps, future research should prioritise co-designed, context-sensitive models that integrate live and projected labour data, embed ethical safeguards, and promote institutional adoption through transparency and stakeholder engagement (Daniel, 2020; Slade & Prinsloo, 2013; World Economic Forum (WEF), 2020).

4 Findings

This review analysed 37 peer-reviewed studies to evaluate how predictive modelling has been applied in higher education to align curricula with labour market demands. Findings are presented in alignment with the three research questions (RQ1–RQ3).

4.1 Findings Aligned to RQ1: Existing Predictive Modelling Approaches

This review identified four main categories of predictive models used in higher education to support curriculum–labour market alignment (Table 2).

Key Observations and Implications for RQ1 The analysis revealed several patterns regarding the scope and limitations of current predictive models. Many studies (73%, $n = 27$) focused on internal academic metrics such as student dropout rates and graduation performance, rather than external labour market alignment (Mseleku, 2022; Pelima et al., 2024). Only 16% ($n = 6$) incorporated real-time labour data through platforms like LinkedIn APIs, and just 11% ($n = 4$) examined the scalability of their models across institutions, with only one study testing generalisability across regional contexts (Daniel, 2020; Johnson et al., 2021; World Economic

Table 2 Predictive model categories and characteristics

Model type	Key characteristics
Statistical models	Techniques such as logistic regression are used for their simplicity and interpretability, but they struggle to incorporate evolving labour trends
Tree-based methods	Decision trees and Random Forests offer moderate predictive accuracy and can identify variable importance, but depend on large, consistent datasets
Neural networks	These models perform well in terms of prediction accuracy but are often opaque ('black boxes') and lack explainability
Hybrid frameworks	Models using integrated approaches like CRISP-DM aim to combine accuracy and transparency but are technically complex and resource-intensive

Forum (WEF), 2020). While 84% ($n = 31$) reported standard technical metrics such as model accuracy or F1-scores, only 24% ($n = 9$) addressed how predictive insights could be implemented within institutional planning structures (Daniel, 2020). These findings suggest that most models remain narrowly focused on institution-specific indicators and lack the capacity to incorporate dynamic labour signals or adapt to regional diversity, as highlighted in the implications for RQ1 (Daniel, 2020; World Economic Forum (WEF), 2020).

4.2 Findings Aligned to RQ2: Methodological, Ethical, and Contextual Limitations

The review identified several key limitations affecting the methodological, ethical, and contextual robustness of existing predictive models in higher education. A significant majority 75.7% ($n = 28$) relied on retrospective datasets, such as graduate exit surveys and national labour reports, which limit real-time responsiveness and predictive relevance in dynamic labour markets (Mseleku, 2022; Republic of South Africa, 2013). Only 8.1% ($n = 3$) of the studies explicitly addressed regional labour disparities using data sources like Statistics South Africa (Stats SA), despite the importance of geographic sensitivity in workforce planning (Mseleku, 2022; Stats SA, 2024). Ethical considerations were also underdeveloped: although fairness and bias were mentioned in several studies, only 13.5% ($n = 5$) applied fairness-aware tools such as SHAP or LIME to address these concerns in practice (Lundberg & Lee, 2017; Mehrabi et al., 2021). Furthermore, scalability remains a challenge, with just 10.8% ($n = 4$) of the models tested beyond their original institutional contexts, thereby limiting generalisability and wider application across diverse educational environments (Daniel, 2020; Johnson et al., 2021) (Table 3).

Table 3 Key dimensions in reviewed predictive models

Dimension	Number of studies ($n = 37$)	% of total	Notes
Used static/historical datasets	28	75.7	Graduate surveys, national labour reports
Included real-time labour data	6	16.2	Mostly via LinkedIn or job board APIs
Accounted for regional differences	3	8.1	Often provincial level in South African cases
Applied fairness-aware methods	5	13.5	SHAP, LIME, bias audits
Focused on cross-institutional use	4	10.8	Few models tested beyond original context

Critical Gaps The review highlights several critical gaps that hinder the effectiveness of predictive models for curriculum–labour market alignment. An overreliance on static datasets, present in 75.7% of the studies limits the ability of models to respond to rapidly evolving sectors such as artificial intelligence (AI) and the Internet of Things (IoT), where skill demands shift quickly (World Economic Forum (WEF), 2020). Regional contextualisation is also underrepresented, despite South Africa's notably high graduate unemployment rate of 32.1% and significant disparities across provinces in terms of labour market access and economic activity (Mseleku, 2022; Stats SA, 2024). Additionally, fairness-aware tools like SHAP are rarely applied in the reviewed studies, raising concerns about algorithmic bias and the transparency of model outputs (Lundberg & Lee, 2017; Mehrabi et al., 2021).

4.3 *Findings Aligned to RQ3: Integration of Real-Time Data, Regional Relevance, and Fairness*

The review reveals that the integration of real-time data into predictive models' remains limited. Only 16% ($n = 6$) of the studies incorporated real-time labour data, such as LinkedIn APIs, while 41% discussed real-time responsiveness in theory without technical implementation (World Economic Forum (WEF), 2020). Furthermore, methods suitable for future demand forecasting, such as Long Short-Term Memory (LSTM) networks or Monte Carlo simulations, were notably absent, despite their potential for anticipating skills needed 5–10 years into the future (Hyndman & Athanasopoulos, 2018; World Economic Forum (WEF), 2020).

In terms of regional relevance, only 19% ($n = 7$) of the reviewed studies integrated regional labour indicators, such as GDP trends from Statistics South Africa, while the vast majority (92%) relied on national-level data. This limits applicability in diverse contexts where localised skill needs, such as renewable energy demand in provinces like the Eastern Cape, are not captured (Mseleku, 2022; Stats SA, 2024).

Ethical considerations were inconsistently addressed. While 30% ($n = 11$) acknowledged the presence of algorithmic risks, only 13% ($n - 5$) implemented fairness-aware tools like SHAP to enhance transparency and explainability (Lundberg & Lee, 2017; Mehrabi et al., 2021). As a result, many models operate as opaque systems that risk reinforcing existing inequalities for instance, through gendered skill prioritisation or region-blind recommendations (Mehrabi et al., 2021; Slade & Prinsloo, 2013).

Institutional integration was also weak. Only 24% ($n = 9$) of the studies linked predictive outputs to curriculum planning workflows, leaving a gap between model insights and academic decision-making (Daniel, 2020). To address these gaps, future research should focus on co-designing predictive systems with institutional stakeholders, integrating dynamic data pipelines that combine real-time labour inputs and regional statistics, and embedding explainable AI techniques such as SHAP and LIME, early in the de-sign phase. Scenario-based forecasting using techniques like

LSTM can also enhance long-term alignment by anticipating graduate entry points and future skills demand (Daniel, 2020; Lundberg & Lee, 2017; World Economic Forum (WEF), 2020).

5 Discussion

This review highlights a persistent gap between the technical promise of predictive analytics and its actual implementation in aligning university curricula with labour market demands. Among the 37 studies analysed, only 6 (16%) incorporated real-time labour data, and just 9 (24%) demonstrated integration into curriculum or institutional planning. This points to a significant mismatch between the availability of advanced modelling tools and the readiness of higher education institutions, particularly in settings with underdeveloped data infrastructure and governance capacity (Daniel, 2020).

5.1 Methodological Limitations

One of the most notable methodological limitations is the reliance on static data sources. Approximately 76% ($n = 28$) of the studies reviewed used historical datasets, such as graduate surveys and national labour reports. These sources, while easily accessible, are not suited for the fast-paced and evolving demands of modern economies. Only seven studies (19%) incorporated regionally disaggregated labour market indicators, which diminishes their applicability in contexts like South Africa, where localised disparities in economic opportunity are stark (Mseleku, 2022; Stats SA, 2024).

5.2 Ethical and Transparency Gaps

Ethical considerations, particularly regarding algorithmic fairness and model transparency, remain underdeveloped in the current literature. While 11 studies (30%) acknowledged issues related to fairness and bias, only 5 (13.5%) actively implemented fairness-aware tools such as SHAP or LIME to enhance explainability and user trust (Lundberg & Lee, 2017; Mehrabi et al., 2021). The persistence of 'black box' models is especially problematic in low- and middle-income countries, where algorithmic risks may be poorly understood but have disproportionately harmful effects (Daniel, 2020; Slade & Prinsloo, 2013).

5.3 *Institutional Integration Challenges*

Only 9 studies (24%) addressed how predictive insights could inform curriculum development processes, and just 4 (11%) tested model scalability across different institutions. Institutional resistance to algorithmic tools remains a significant barrier, driven by concerns around academic autonomy, lack of digital literacy, and policy misalignment (Daniel, 2020). These challenges underscore the need for greater attention to institutional change management when proposing analytics-driven solutions.

5.4 *Opportunities for Dynamic Curriculum Design*

Despite these limitations, there are promising opportunities to bridge the gap between predictive insights and curriculum design. Modular components, such as elective AI ethics workshops co-designed with industry partners, can offer a practical way to integrate emerging skill demands without disrupting core curricula (Chen et al., 2023). Biannual updates using real-time job market data could keep curricula current, while preserving academic integrity and depth (World Economic Forum (WEF), 2020). Regionally adaptable frameworks are also vital. For instance, universities of applied sciences may require stronger local alignment compared to research-intensive institutions (Daniel, 2020). Hybrid data strategies that blend LinkedIn APIs with sources like Stats SA can balance global skills trends with local labour priorities, ensuring inclusive coverage of diverse sectors (Stats SA, 2024; World Economic Forum (WEF), 2020).

5.5 *Addressing Data Skew and Representational Bias*

One critical limitation of API-based platforms such as LinkedIn is their tendency to reflect only highly digitised industries. As a result, sectors like arts, agriculture, or public education are often underrepresented in predictive outputs (World Economic Forum (WEF), 2020). To address this imbalance, multi-source data validation is recommended. Combining LinkedIn insights with regional labour statistics from organisations like Stats SA can help provide a more balanced and equitable dataset. In addition, stakeholder co-design is essential. Involving faculty from underrepresented disciplines and local employers in model development helps ensure that predictive outputs reflect diverse economic needs and institutional missions.

5.6 Stakeholder Acceptance

Student and employer acceptance of predictive models is also essential to their success. Students are often sceptical of opaque changes to course content, even when intended to improve relevance. Transparent communication and student involvement in the design process can help build trust and legitimacy (Slade & Prinsloo, 2013). Employers, on the other hand, often prioritise foundational competencies over fast-moving trends. This rein-forces the value of modular updates that complement rather than replace core academic instruction (Chen et al., 2023).

5.7 Future Directions

Three research and implementation priorities emerged from this review. First, predictive models must allow for regional adaptability through adjustable parameters that account for local labour dynamics. This is particularly important for applied universities, which serve specific geographic regions (Daniel, 2020). Second, modular design approaches, such as placeholder modules should be adopted to support agile curriculum updates without compromising programme coherence (Chen et al., 2023). Third, algorithmic transparency must be embedded from the outset using explainability tools like SHAP and LIME to ensure accountability and fairness (Lundberg & Lee, 2017; Mehrabi et al., 2021).

To move from theoretical promise to institutional adoption, predictive systems must be co-designed with universities, ingest real time and localised labour market data, and be validated through multi-source inputs. They must also incorporate ethical safeguards and modular design principles to ensure relevance, fairness, and responsiveness. With stronger alignment between technical innovation, ethical responsibility, and institutional collaboration, predictive analytics can become a practical and impactful tool for enhancing curriculum relevance and advancing graduate employability in higher education (Daniel, 2020; Mainga et al., 2022; Slade & Prinsloo, 2013).

6 Conclusion

This review analysed 37 peer-reviewed studies on predictive modelling in higher education, with the goal of understanding how effectively these models support alignment between university curricula and evolving labour market demands. The analysis revealed specific trends and gaps, summarised in Table 4.

Table 4 Summary of key findings from reviewed studies ($n = 37$)

Feature assessed	Number of studies	% of total	Notes
Used static/historical datasets	28	75.7	Grad surveys, national labour reports
Integrated real-time data	6	16.2	LinkedIn/job board APIs
Included regional indicators	7	18.9	Provincial GDP, urban–rural disparities (Department of Higher Education and Training (DHET), 2023; Pelima et al., 2024)
Applied fairness-aware tools	5	13.5	SHAP, LIME, bias audits (Smith et al., 2022)
Discussed institutional integration	9	24.3	Curriculum use and decision-making
Addressed scalability	4	10.8	Few tested beyond original context
Mentioned ethical risks	11	29.7	Bias, transparency
Developed multi-institutional models	4	10.8	Limited scalability testing

6.1 *Interpretation of Findings*

The review shows that most predictive models (76%) rely on static datasets, such as graduate surveys and national labour reports, which limits their responsiveness to rapidly evolving sectors like artificial intelligence and the Internet of Things (World Economic Forum (WEF), 2020). Only 16% ($n = 6$) of the studies incorporated real-time labour data using APIs, such as those provided by LinkedIn, indicating a significant technological shortfall in the reviewed literature (World Economic Forum (WEF), 2020). Region-specific modelling also remains underdeveloped. Just seven studies (19%) used disaggregated provincial or local labour data, despite ongoing disparities in employment opportunities across South African regions (Mseleku, 2022; Stats SA, 2024). Ethical transparency continues to be a concern, as only five studies (13.5%) implemented fairness-aware tools like SHAP or LIME to reduce algorithmic bias, although 30% acknowledged the existence of fairness-related risks (Lundberg & Lee, 2017; Mehrabi et al., 2021). Most models remain opaque and difficult to interpret, which undermines stakeholder trust and inhibits practical adoption in institutional settings (Mehrabi et al., 2021; Slade & Prinsloo, 2013). Additionally, institutional integration of predictive insights is limited: only nine studies (24%) considered how model outputs might inform curriculum decisions, and just four (11%) assessed whether their models could scale across different universities. Critical barriers such as academic autonomy, resistance to algorithmic support, and policy misalignment were rarely explored, highlighting the gap between predictive modelling theory and real-world application (Daniel, 2020).

6.2 Conclusion and Directions for Future Research

This review demonstrates that although interest in predictive analytics for curriculum planning is increasing, most existing models remain experimental and are not yet suitable for institutional adoption. Key limitations identified include a persistent reliance on static data, seen in over 75% of the studies, which fails to reflect the dynamic and future-oriented nature of skills demand in fast-changing sectors (Mseleku, 2022; World Economic Forum (WEF), 2020). Regional specificity is also lacking, with most models based on national-level data that do not account for local disparities, especially those evident across South Africa's provinces (Mseleku, 2022; Stats SA, 2024). Ethical robustness remains a concern, as few studies prioritise fairness-aware design, raising the risk of algorithmic bias in skills prioritisation (Mehrabi et al., 2021; Slade & Prinsloo, 2013). Furthermore, only 11% of the reviewed studies tested their models beyond the original institutional context, limiting their scalability and generalisability (Daniel, 2020; Johnson et al., 2021).

To address these limitations, future research should focus on several priority areas. First, integrating real-time labour market data through APIs and adopting time-series forecasting methods, such as LSTM networks, could enable more accurate prediction of skill demands over a 5-to-10-year horizon (World Economic Forum (WEF), 2020). Complementary techniques like Monte Carlo simulations can help quantify risks related to automation and workforce displacement, as proposed by Frey and Osborne (2017). Second, enhancing regional adaptability is essential. This could involve combining LinkedIn-derived labour signals with regional datasets, such as those provided by Statistics South Africa, to ensure representation of less digitised sectors like agriculture and education (Stats SA, 2024; World Economic Forum (WEF), 2020). Adjustable regional weighting parameters may further support alignment with both traditional universities and universities of applied sciences (Daniel, 2020).

Modular curriculum design offers another opportunity for flexibility. The implementation of placeholder modules, such as AI ethics workshops co-developed with industry, can facilitate timely curriculum updates without undermining foundational learning outcomes (Chen et al., 2023). Biannual updates based on live labour data would allow institutions to maintain graduate competitiveness in the labour market (World Economic Forum (WEF), 2020). Ethical safeguards must be embedded from the outset, including the integration of SHAP or LIME for explainability and trust-building, and the exclusion of sensitive variables such as race and gender from training datasets to minimise algorithmic bias (Lundberg & Lee, 2017; Mehrabi et al., 2021; Slade & Prinsloo, 2013).

Institutional co-design is also critical. Universities should establish dedicated analytics units within academic planning departments to help translate predictive insights into actionable curriculum strategies. National bodies such as the Council on Higher Education (CHE) or the Department of Higher Education and Training (DHET) could support this effort by developing guidelines for ethical, scalable implementation of predictive systems (Department of Higher Education and Training (DHET), 2023).

As part of the Phase 2 implementation plan, the predictive model should be piloted at the Cape Peninsula University of Technology (CPUT) using a Design Science Research (DSR) framework. This approach would support iterative refinement based on stakeholder feedback and fairness audits (Slade & Prinsloo, 2013). Benchmarking against static baselines using metrics such as RMSE and F1-score, along with longitudinal tracking of graduate employability, would further validate effectiveness over a five-year period (Hyndman & Athanasopoulos, 2018; Mainga et al., 2022). Testing model scalability across multiple South African provinces would also provide valuable insights into regional adaptability (Daniel, 2020; Stats SA, 2024).

In conclusion, predictive systems can evolve into practical tools that enhance curriculum relevance, promote equity, and improve graduate employability, provided they are co-designed with stakeholders, underpinned by dynamic and representative data, and embedded with transparency and ethical safeguards. Closing the gap between academic offerings and labour market needs will require not only technical innovation but also meaningful institutional collaboration and regulatory support (Daniel, 2020; Slade & Prinsloo, 2013; World Economic Forum (WEF), 2020).

References

Braun, V., & Clarke, V. (2006). Using thematic analysis in psychology. *Qualitative Research in Psychology, 3*(2), 77–101. https://doi.org/10.1191/1478088706qp063oa

Breiman, L. (2001). Random forests. *Machine Learning, 45*(1), 5–32. https://doi.org/10.1023/A:1010933404324

Chen, T., & Guestrin, C. (2016). XGBoost: A scalable tree boosting system. In *Proceedings of ACM SIGKDD* (pp. 785–794). ACM. https://doi.org/10.1145/2939672.2939785

Chen, X., Kumar, A., & Zhang, Y. (2023). Hybrid predictive frameworks for curriculum alignment. *Journal of Educational Data Mining, 15*(1), 45–62. https://arxiv.org/pdf/2309.04761

Daniel, J. (2020). *Mega-universities and knowledge media: The global reach of digital learning*. Springer. https://doi.org/10.1007/978-3-030-36962-1

Department of Higher Education and Training (DHET). (2023). *National skills development strategy III*. http://www.dhet.gov.za/NSDSIII

Esteva, A., Kuprel, B., Novoa, R. A., Ko, J., Swetter, S. M., Blau, H. M., & Thrun, S. (2017). Dermatologist-level classification of skin cancer. *Nature, 542*(7639), 115–118. https://doi.org/10.1038/nature21056

Frey, C. B., & Osborne, M. A. (2017). The future of employment: How susceptible are jobs to computerisation? *Technological Forecasting and Social Change*, 114, 254–280 (2017). https://ora.ox.ac.uk/objects/uuid:4ed9f1bd-27e9-4e30-997e-5fc8405b0491

Glandorf, D., et al. (2024). Temporal variability in college dropout prediction. In *Proceedings of ACM Conference on Learning at Scale* (pp. 486–497). ACM. https://doi.org/10.1145/3636555.3636906

Gusenbauer, M., & Haddaway, N. R. (2020). Which academic search systems are suitable for systematic reviews or meta-analyses? *Journal of the Medical Library Association, 108*(4), 561–567. https://doi.org/10.5195/jmla.2020.939

Hastie, T., et al. (2009). *The elements of statistical learning*. Springer. https://doi.org/10.1007/978-0-387-84858-7

Hyndman, R. J., & Athanasopoulos, G. (2018). *Forecasting: Principles and practice*. OTexts. https://otexts.com/fpp2

Idrees, S., Alam, A., Agarwal, P., & Ansari, L. (2019). Effective predictive analytics and modeling based on historical data. In P. K. Singh et al. (Eds.), *Advances in data and information sciences* (pp. 569–578). Springer. https://doi.org/10.1007/978-981-13-9942-8_52

Jamarani, A., Haddadi, S., Sarvizadeh, R., Haghi Kashani, M., Akbari, M., & Moradi, S. (2024). Big data and predictive analytics. *Artificial Intelligence Review, 57*(7), 1–35. https://doi.org/10.1007/s10462-024-10811-5

Johnson, L., et al. (2021). Collaborative predictive systems in higher education. *Computers & Education, 176*, Article 104356. https://doi.org/10.1016/j.compedu.2021.104356

Kitchenham, B., & Charters, S. (2007). Guidelines for performing systematic literature reviews in software engineering. In *Technical Report EBSE-2007-01*, Keele University and University of Durham.

Lundberg, S. M., & Lee, S. (2017). A unified approach to interpreting model predictions. In *NeurIPS 2017* (pp. 4768–4777). Curran Associates. https://doi.org/10.48550/arXiv.1705.07874

Mainga, W., Murphy-Braynen, M. B., Moxey, R., & Quddus, S. A. (2022). Graduate employability of business students. *Administrative Sciences, 12*(3), 72. https://doi.org/10.3390/admsci12030072

Mehrabi, N., Morstatter, F., Saxena, N., Lerman, K., & Galstyan, A. (2021). A survey on bias and fairness in machine learning. *ACM Computing Surveys, 54*(6), 1–35. https://doi.org/10.1145/3457607

Mittelstadt, B., et al. (2016). The ethics of algorithms. *Big Data & Society, 3*(2), 1–21. https://doi.org/10.1177/2053951716679679

Mseleku, Z. (2022). From workplace to joblessness. *Academic Journal of Interdisciplinary Studies, 11*(2), 330–345. https://doi.org/10.36941/ajis-2022-0055

OECD. (2019). *Artificial intelligence in society*. OECD Publishing. https://doi.org/10.1787/eedfee77-en

Page, M. J., et al. (2021). The PRISMA 2020 statement: Updated guidelines for systematic reviews. *Systematic Reviews, 10*(1), 1–11. https://doi.org/10.1186/s13643-021-01626-4

Pelima, L. R., Sukmana, Y., & Rosmansyah, Y. (2024). Predicting university student graduation using academic performance and machine learning. *IEEE Access, 12*, 23451–23465. https://doi.org/10.1109/ACCESS.2024.3361479

Republic of South Africa. (2013). *Protection of Personal Information Act 4 of 2013*. https://www.gov.za/documents/protection-personal-information-act

Romero, C., & Ventura, S. (2020). Educational data mining. *Wiley Interdisciplinary Review, 10*(3), Article e1355. https://doi.org/10.1002/widm.1355

Slade, S., & Prinsloo, P. (2013). Learning analytics: Ethical issues. *American Behavioral Scientist, 57*(10), 1510–1529. https://doi.org/10.1177/0002764213479366

Smith, J., Patel, A., & Reddy, K. (2022). Forecasting educational outcomes using random forests. *South African Journal of Higher Education, 36*(5), 90–110. https://doi.org/10.20853/36-4-4483

Stats SA. (2024). *Quarterly labour force survey: Q3 2024 (statistical release P0211)*. https://www.statssa.gov.za/publications/P0211/P02113rdQuarter2024.pdf

Tapscott, D., & Tapscott, A. (2016). *Blockchain revolution: How the technology behind bitcoin is changing money, business, and the world*. Penguin Canada.

Voigt, P., & von dem Bussche, A. (2017). *The EU general data protection regulation (GDPR): A practical guide*. Springer. https://doi.org/10.1007/978-3-319-57959-7

Wirth, R., Hipp, J.: CRISP-DM: Towards a standard process model for data mining. In: Proc. PAKDD, pp. 29–39. Springer (2000).

World Economic Forum (WEF). (2020). *The future of jobs report 2020.* https://www3.weforum.org/docs/WEF_Future_of_Jobs_2020.pdf

Zawacki-Richter, O., Marín, V. I., Bond, M., & Gouverneur, F. (2019). Systematic review of re-search on artificial intelligence applications in higher education. *International Journal of Educational Technology in Higher Education, 16*(1), 1–14. https://doi.org/10.1186/s41239-019-0171-0

Energy-Efficient Cloud Computing for Artificial Intelligence Training: Challenges, Techniques, and Future Directions

Fayez Salma and Jorge Marx Gómez

Abstract The rapid growth of artificial intelligence (AI) has amplified the computational and energy needs for training models, especially in cloud environments. This study takes into account the primary challenges in realizing energy efficiency for AI training, which are the growth of model sizes, inefficient use of resources, and infrastructural inefficiencies. A comprehensive review of novel solutions is provided, ranging from specialized hardware, software enhancements, to the operation of sustainable cloud infrastructure. Moreover, we highlight emerging trends such as AI-driven resource optimization, federated learning, and carbon-aware scheduling. Through our discussion, we highlight that there is an urgent requirement for energy efficiency to be a top priority alongside performance in the development of AI systems. We conclude by highlighting the requirement for cross-disciplinary collaboration to achieve a sustainable artificial intelligence future.

1 Introduction

The latest leap in advancement of artificial intelligence (AI) and deep learning has pushed computational requirements to all-time highs. Training such large-scale models, such as GPT-4, PaLM, and Stable Diffusion, requires enormous processing capacity, gigantic storage centers, and huge quantities of energy resources. With each increase in their size and complexity, so rises their carbon footprint. Recent studies have made it obvious that the energy consumed and carbon footprints of training one big model are on par with lifetime emissions of several cars, highlighting the urgent demand for AI sustainability (Cottier et al., 2025; Strubell et al., 2019).

Cloud computing has emerged as the platform of choice for training AI models due to its scalability, flexibility, and ease with which specialized hardware like Graphics

F. Salma (✉) · J. Marx Gómez
Department for Informatik, Carl von Ossietzky University of Oldenburg, Oldenburg, Germany
e-mail: faez_sal@yahoo.com

J. Marx Gómez
e-mail: jorge.marx.gomez@uni-oldenburg.de

J. Marx Gómez et al. (eds.), *Recent Advances in Information Systems*, Progress in IS,
https://doi.org/10.1007/978-3-032-14099-9_49

Processing Units (GPUs) and Tensor Processing Units (TPUs) is provided to it. By making high-performance resources available to researchers and developers in real time, cloud platforms have propelled the pace of AI innovation. Yet this shift to cloud-based training has entailed significant energy efficiency issues as well. Cloud computing, enabled by such data centers, already accounts for around 1–1.5% of the world's use of electricity, according to the International Energy Agency's estimates (IEA, 2023). This is forecast to grow exponentially as AI workloads are projected to permeate all sectors, increasing the carbon profile unless action is taken promptly.

As a response to such challenges, researchers and practitioners have been exploring means of powering cloud computing infrastructure used for AI training in a more energy-efficient manner. On the hardware side, there is an emphasis on low-power domain-specific accelerators specifically optimized for AI workloads designed and deployed. At the same time, software-level developments like model pruning, quantization, knowledge distillation, and better training algorithms are designed to eliminate computation redundancy and total resource usage. Furthermore, major cloud computing providers are heavily investing in renewable power sources, smart grids, and cutting-edge cooling technologies like liquid immersion and AI-enabled thermal management systems to minimize the operational carbon footprint of their data centers (Schwartz et al., 2020).

In spite of all these developments, there are still major challenges ahead. Artificial intelligence models have been subject to ongoing growth and exponentially increasing complexity at exponentially accelerating rates that tend to outstrip progress in hardware optimization as well as in algorithmic enhancement. Furthermore, the constantly escalating energy expense of AI infrastructure extends beyond computation to data storage, transmission, redundancy structures, and cooling structures as well. This creates an extremely complex, multi-parameter optimization problem that encompasses the entire stack of computing from hardware design all the way through to software frameworks and system-level resource orchestration.

Acknowledging the paramount nature of this concern, this paper provides a panoramic view of where energy-efficient cloud computing for AI training stands currently. We present the most significant technical and infrastructure challenges, survey the latest techniques and breakthroughs on different layers of the computing stack, and identify promising directions for future research. By integrating hardware design, software engineering, system architecture, and environmental science expertise, we aim to contribute to the body of work encouraging sustainable and environmentally conscious AI development. Our long-term goal is to inspire and stimulate additional research, as well as collaborative efforts, toward creating AI technologies that are not only strong and revolutionary but also sustainable and environmentally conscious.

2 Background and Related Work

2.1 *Cloud Computing and AI Training*

Cloud computing has transformed the landscape of artificial intelligence (AI) training by providing scalable, on-demand computational resources. Large cloud service providers such as Amazon Web Services (AWS), Google Cloud Platform (GCP), and Microsoft Azure offer specialized AI-accelerated instances with high-end Graphics Processing Units (GPUs) and Tensor Processing Units (TPUs). These instances are compute-intensive deep learning-optimized to enable researchers, startups, and enterprises to train sophisticated AI models without requiring significant capital expenditure in bespoke hardware infrastructure.

Compute democratization has spurred the creation and deployment of industry-leading (SOTA) models such as GPT-4, Gemini, and Stable Diffusion XL. These models are typically composed of hundreds of layers and hundreds of billions of parameters, and they require processing of gigantic datasets. The training process of these models usually takes days to weeks based on the model architecture, hardware configuration, and optimization techniques employed. Interestingly, such training runs might consume tens to hundreds of megawatt-hours (MWh) of electricity per run, thereby mirroring the massive energy requirements of today's AI activities (Cottier et al., 2025).

Cloud infrastructure presents an effective remedy for operational and logistics issues, e.g., provisioning hardware, maintenance, and scaling. Parallel and distributed training is enabled throughout globally distributed data centers, therefore reducing training latency as well as more efficiently using resources. Nevertheless, the convenience green cost is substantial. AI training workloads will further grow the already immense energy footprint of data centers unless energy usage on these cloud platforms is judiciously managed.

Besides, the scale and frequency of model training—especially in business application where it is repeatedly trained—have led to unsustainable growth of cloud-based energy use. If the benefits of cloud computing are offset by its environmental impact without opportunistic optimizations like dynamic resource scheduling, workload-aware allocation, and integration of renewable sources, then its benefits would be negated. The problem is not only to offer high-performance AI training in the cloud but also to do so with minimal environmental impact.

For that purpose, existing research activity and industrial innovation have started exploring adaptive resource provisioning, model parallelism methods, and energy-aware job scheduling as means of reducing total power consumption in training. Cloud providers are also making investments in effective data center design, like liquid cooling, intelligent thermal management, and AI-driven resource orchestration systems. All these are required to make AI training in the cloud scalable and sustainable.

Short of it, whereas cloud computing has lowered the barrier to entry for training large AI models considerably, it has simultaneously introduced necessary challenges around energy use and its environmental impact. These have to be addressed with an end-to-end approach that spans architectural breakthrough, algorithmic optimization, and eco-friendly cloud infrastructure design.

2.2 Artificial Intelligence Workload Energy Consumption

Energy consumption in AI training arises from several factors:

- Computation-intensive operations: Forward and backward passes through deep neural networks are extremely compute-intensive.
- Memory access and storage: Moving data between memory hierarchies and networked storage is costly in energy.
- Unused resources and over-provisioning: Inefficient use of resources leads to energy wastage.

Some studies like (Schwartz et al., 2020) highlight that model training energy expenses are not just model size dependent, but also on hardware choice and software inefficiencies. Furthermore, inference (deployment) stages, though less intensive on a per-instance basis, accumulate significant energy footprints by virtue of scale.

2.3 Past Studies of Energy Efficiency Approaches

Initial attempts have been made toward hardware acceleration to reduce the energy usage of AI workloads. Google's TPU v4 chips are reportedly 2.7× more performant per watt than previous generations (Jouppi et al., 2023). NVIDIA's Hopper architecture also includes energy-saving features such as sparsity support and dynamic voltage scaling.

In software, various model compression techniques, including pruning, quantization, and knowledge distillation, have been invented to reduce computational requirements with negligible impacts on the model's performance (Blalock et al., 2020).

Meanwhile, studies on effective training algorithms, including mixed-precision training and gradient checkpointing, strive to reduce energy consumption even further.

Cloud-specific approaches have also been attempted. They include:

- Dynamic resource allocation (Li et al., 2024).
- Renewable energy-aware AI model training schedulers (Ye et al., 2024).
- Energy-efficient container orchestration frameworks such as Kubernetes extensions for green computing.

In spite of considerable advances, the seamless integration of such solutions into actual cloud environments continues to be a major challenge owing to performance compromises, complicated system designs, and reliability issues.

3 Energy-Efficient Challenges in AI Training

Even with hardware and software optimization progress, energy-efficient cloud computing for AI training is a very challenging pursuit. Numerous interrelated aspects increase the complexity of this issue.

3.1 The Exponential Growth of Model Dimensions

AI models today are growing at a pace quicker than hardware development can sustain. For example, OpenAI's GPT-4 contains around 1 trillion parameters, far larger than GPT-3 (Brown et al., 2020), and its training consumed reportedly tens of megawatt-hours of electricity (https://openai.com/index/gpt-4-research/). This exponential expansion generates an increasing demand for compute capacity, leading to energy usage increasing proportionally independent of efficiency at the chip level.

3.2 Suboptimal Resource Utilization

Cloud environments are susceptible to over-provisioning of resources, resulting in wasted power consumption by underutilized servers. Both virtualization and containerization induce overhead, while training workloads tend to manifest dynamically changing resource requirements that are not optimally addressed by fixed allocation policies (Li et al., 2024).

In addition, within the realm of hyperparameter tuning a very significant stage in AI development many models tend to be trained simultaneously, which compounds inefficiency of resources.

3.3 Energy Cost of Data Storage and Movement

Training AI models involves enormous datasets (e.g., billions of text tokens or images), and data movement between cloud storage systems, networks, and accelerators is energy-expensive.

Dwivedi et al. (2023) mention that data transfer in some artificial intelligence training can represent an increase in overall energy use, especially when training is split over several data centers.

3.4 Cooling and Infrastructure Overheads

Data center cooling systems can consume energy that is comparable to the computing equipment, particularly in older facilities.

Even while technologies like liquid cooling and module data center designs reduce cooling overheads in some hyperscale centers, these have not yet seen widespread deployment.

Geographic conditions, including ambient temperatures and the makeup of the energy grid, also complicate optimization.

3.5 Balancing Performance and Energy Consumption

Techniques such as model compression, quantization, and pruning can conserve energy at the expense of model performance (accuracy, generalization) when not done correctly.

In the same way, low-power hardware modes (e.g., reducing clock speeds) can increase training times, negating energy savings by extending periods of operation.

Consequently, achieving a good trade-off between energy conservation and computational efficiency is still an open and significant issue.

4 Methods for Enhancing Energy Efficiency

Several strategies have been developed to enhance energy efficiency within the framework of cloud-based artificial intelligence training. These approaches operate at several levels of the computing stack, including advances in hardware, algorithmic optimizations, and cloud infrastructure improvements.

4.1 Hardware Solutions

Specialized AI Accelerators

Deep learning-specific, energy-efficient hardware accelerators such as Google's TPU v4 and NVIDIA's Hopper GPUs have been specifically optimized for deep learning workloads.

TPU v4, for instance, delivers over 2.7 times better performance per watt compared to its predecessors (Jouppi et al., 2023).

They use methods such as:

- Sparse computation (only computing non-zero values).
- Mixed-precision arithmetic (reducing bit width to conserve energy).

Dynamic Energy Regulation

Modern hardware provides dynamic voltage and frequency scaling (DVFS), which allows operating point adjustments in real time to balance power and performance based on workload intensity.

4.2 Software and Algorithm Techniques

Model Quantization and Pruning

- Pruning removes redundant parameters from neural networks, reducing model size and computation without significant loss of accuracy.
- Quantization reduces numerical precision (e.g., from 32-bit floats to 8-bit integers), which lowers both memory bandwidth and computation energy.

Recent advancements like Post-training Quantization (PTQ) and Quantization-Aware Training (QAT) have caused energy consumption to be reduced without any notable reduction in accuracy (Gholami et al., 2022).

Knowledge Distillation

A bigger, energy-thirsty model (teacher) is employed to train a smaller, better model (student), retaining performance while reducing computational requirements (Sanh et al., 1910).

Successful Training Techniques

- Mixed-precision training reduces precision while training to save energy.
- Gradient checkpointing selectively recomputes the intermediate results instead of storing all activations, reducing memory at a small computation cost (Chen et al., 2016).

4.3 Cloud Infrastructure Approaches

Dynamic Scaling and Elastic Resource Provisioning

The energy-aware auto-scaling policies dynamically change resource allocations according to the needs of training, thereby minimizing idle time and conserving energy.

Serverless and Spot Instance Utilization
Using serverless computing or spot instances for non-critical or fault-tolerant training tasks lowers the active energy footprint since resources are used when they are needed.

Renewable Energy Integration
Cloud providers like AWS, Microsoft, and Google are increasingly planning AI training sessions in blocks with high renewable energy supply.

Energy-aware schedulers can defer non-urgent training jobs until cleaner energy is available (Ye et al., 2024).

Fluid-Based Cooling and Thermal Enhancement
High-performance liquid cooling systems can reduce energy consumption for compared to traditional air cooling, particularly critical for high-density AI training clusters.

5 Emerging Trends and Future Developments

As energy demands increase with the development of AI models, new solutions are imminent to address the sustainability challenges of cloud AI training. We outline below key trends and encouraging areas of research.

5.1 Artificial Intelligence in Energy Optimization

One emerging trend is the application of artificial intelligence for maximizing energy efficiency in cloud infrastructures.

Machine learning algorithms can predict workload patterns, dynamically allocate resources, and optimize cooling systems in real time.

Recent studies, such as Google's DeepMind-based AI cooling system, have shown that energy efficiency in data centers yields better overall values.

Similarly, reinforcement learning-based resource scheduling studies are promising in reducing idle energy waste and improving server utilization (Miao et al., 2024).

5.2 Federated and Decentralized Training

Federated learning (FL) and decentralized AI are gaining appeal for energy efficiency.

Rather than sending data to centrally located cloud servers for uploading, the models are trained locally on distributed nodes or edge devices, hence saving data transfer costs and central processing energy costs (Kairouz et al., 2021).

Despite federated learning issues like communication overhead and device heterogeneity, recent advances in communication-efficient algorithms are enabling it even further for large-scale artificial intelligence systems.

5.3 Green AI Frameworks and Standards

The concept of Green AI, first introduced to prominence by Schwartz et al. (2020), is now moving toward its implementation.

Investments in green infrastructure for carbon–neutral cloud ecosystems, supported by AI-powered sustainability initiatives, enable a reduction in the carbon footprint of cloud service providers and enterprises, and promote a greener and more sustainable digital future (Oloruntoba, 2024).

New paradigms are emerging to compare models based on not only accuracy but also energy usage and carbon footprint. Initiatives like the MLCommons Sustainability Working Group (2024) are proposing standardized reporting of energy consumption during AI training and inference, promoting greater transparency.

5.4 Hardware–Software Co-Design

Rather than improving hardware and software independently, co-design approaches seek to create tailored hardware with tight integration of specific AI algorithms.

For instance, customized accelerators for Transformer models or Graph Neural Networks (GNNs) are in development with orders-of-magnitude better energy efficiency compared to general-purpose hardware (Miao et al., 2024).

5.5 Carbon-Aware Scheduling

An emerging trend is carbon-aware job scheduling—scheduling AI training jobs to run when the grid is powered by renewable energy.

Cloud vendors and research institutions are experimenting with systems that predict carbon intensity and dynamically schedule workloads to minimize emissions (Bian et al., 2024).

For instance, Microsoft's Project Zerocarbon (2024) is exploring real-time task orchestration among geographically dispersed data centers according to carbon availability.

6 Conclusion

The unchecked growth of artificial intelligence has transformed industries and driven innovation, yet simultaneously come with unmatched energy demands, particularly when training large models in cloud computing setups. This article touched upon the most prominent problems presented by the use of energy in AI training, ranging from hardware inefficiencies to system-level resource allocation problems.

A survey was conducted of state-of-the-art methods in hardware, software, and cloud infrastructure that are currently stretching the boundaries of energy-efficient artificial intelligence. Leading developments toward sustainable AI development are exemplified by domain-specific accelerators, model optimization techniques such as pruning and quantization, and cloud-native approaches such as auto-scaling and carbon-aware scheduling.

Nevertheless, the road to genuinely sustainable artificial intelligence is winding. Advances like federated learning, AI-enabled energy management, hardware-software co-design, and the use of standard metrics to quantify sustainability provide auspicious avenues for future research and business uptake. These advances represent a paradigm shift: no longer is accuracy sufficient; instead, energy efficiency must be regarded as a primary design goal.

Looking ahead to the future, it is important to adopt a multidisciplinary mindset. Cooperation between artificial intelligence researchers, cloud service providers, hardware designers, and lawmakers will be necessary for achieving a balance between technological innovation and environmental stewardship. With the demand for AI services only set to increase, integrating sustainable measures is not just a choice, but an ethical imperative for making artificial intelligence development both powerful and green.

References

Bian, J., Wang, L., Ren, S., & Xu, J. (2024). CAFE: Carbon-aware federated learning in geographically distributed data centers. In *Proceedings of the 15th ACM International Conference on Future and Sustainable Energy Systems*, in e-Energy '24 (pp. 347–360). New York, NY, USA: Association for Computing Machinery. https://doi.org/10.1145/3632775.3661970.

Blalock, D., Gonzalez Ortiz, J. J., Frankle, J., & Guttag, J. (2020). What is the state of neural network pruning? In *Proceedings of Machine Learning and Systems* (Vol. 2, pp. 129–146).

Brown, T., et al. (2020). Language models are few-shot learners. *Advances in Neural Information Processing Systems, 33*, 1877–1901.

Chen, T., Xu, B., Zhang, C., & Guestrin, C. (2016). Training deep nets with sublinear memory cost. https://arxiv.org/abs/1604.06174

Cottier, B., Rahman, R., Fattorini, L., Maslej, N., Besiroglu, T., & Owen, D. (2025). The rising costs of training frontier AI models. https://arxiv.org/abs/2405.21015

Data centres and data transmission networks, *International Energy Agency (IEA)*, (2023). https://www.iea.org/energy-system/buildings/data-centres-and-data-transmission-networks

Dwivedi, Y. K., Sharma, A., Rana, N. P., Giannakis, M., Goel, P., & Dutot, V. (2023). Evolution of artificial intelligence research in technological forecasting and social change: Research topics, trends, and future directions. *Technological Forecasting and Social Change, 192*, Article 122579. https://doi.org/10.1016/j.techfore.2023.122579

Gholami, A., Kim, S., Dong, Z., Yao, Z., Mahoney, M. W., & Keutzer, K. (2022). A survey of quantization methods for efficient neural network inference. In *Low-power computer vision* (pp. 291–326). Chapman and Hall/CRC.

GPT-4. https://openai.com/index/gpt-4-research/. Accessed April 30, 2025

Jouppi N. P., et al. (2023). TPU v4: an optically reconfigurable supercomputer for machine learning with hardware support for embeddings. https://arxiv.org/abs/2304.01433

Kairouz, P., et al. (2021). Advances and open problems in federated learning. *Foundations and Trends® in Machine Learning, 14*(1–2), 1–210

Li, H., Wang, G., Li, L., & Wang, J. (2024). Dynamic resource allocation and energy optimization in cloud data centers using deep reinforcement learning. *Journal of Artificial Intelligence General science (JAIGS), 1*(1), 230–258. ISSN 3006-4023

Miao, Z., et al. (2024). Energy and carbon-aware distributed machine learning tasks scheduling scheme for the multi-renewable energy-based edge-cloud continuum. *Science and Technology for Energy Transition, 79*, 82.

Oloruntoba, O. (2024). Green cloud computing: AI for sustainable database management. *World Journal of Advanced Research and Reviews, 23*(03), 3242–3257.

Sanh, V., Debut, L., Chaumond, J., & Wolf, T. (2019). DistilBERT, a distilled version of BERT: smaller, faster, cheaper and lighter. arXiv:1910.01108

Schwartz, R., Dodge, J., Smith, N. A., & Etzioni, O. (2020). Green AI. *Communications of the ACM, 63*(12), 54–63.

Strubell, E., Ganesh, A., & McCallum, A. (2019). Energy and Policy Considerations for Deep Learning in NLP. https://arxiv.org/abs/1906.02243

Ye, Z., et al. (2024). Deep learning workload scheduling in GPU datacenters: A survey. *ACM Computing Surveys, 56*(6). https://doi.org/10.1145/3638757

A Systematic Review of Privacy-Preserving Synthetic Network Traffic Generation for Machine Learning-Based Intrusion Detection Systems

Kalala Gael Kapia, Boniface Kabaso, and Waldon Hendricks

Abstract Machine learning-based intrusion detection systems (IDS) rely on large-scale network traffic datasets to detect cyberattacks effectively. However, the collection and use of real-world network traffic raise privacy concerns. As a solution, researchers have explored synthetic data generation methods combined with privacy-preserving techniques. This systematic literature review examines existing approaches to generating privacy-preserving synthetic network traffic for training IDS models. A total of six studies were selected following a structured search process. The review highlights the diversity of methods used, ranging from Generative Adversarial Networks to diffusion models, and evaluates them based on methodology, privacy protection, data utility, and ease of implementation. Future research directions are proposed to guide the development of privacy-preserving frameworks for IDS applications.

1 Introduction

Intrusion detection systems (IDS) are security tools that control and monitor network traffic for a potential intrusion or malicious activities (Ashiku & Dagli, 2021). As cyber threats become more frequent and sophisticated, organizations are using machine learning (ML) to enhance the performance of IDS (Ali et al., 2024). These

K. G. Kapia (✉) · B. Kabaso · W. Hendricks
Department of Information Technology, Cape Peninsula University of Technology, Cape Town, South Africa
e-mail: 230150438@mycput.ac.za

B. Kabaso
e-mail: kabasob@cput.ac.za

W. Hendricks
e-mail: hendricksw@cput.ac.za

J. Marx Gómez et al. (eds.), *Recent Advances in Information Systems*, Progress in IS,
https://doi.org/10.1007/978-3-032-14099-9_50

techniques can identify complex patterns and previously unseen attacks more effectively than traditional rule-based systems (Akiri et al., 2025). However, training ML-based IDS raises data privacy concerns as these systems depend on large datasets, which often contain sensitive network data (Markevych & Dawson, 2023).

To mitigate the risk of privacy, researchers have turned to synthetic data as a privacy-preserving alternative (Ammara et al., 2024). Synthetic data offers a way to train ML models without direct exposure to private information. Among the most studied methods are generative models, such as Generative Adversarial Networks (GANs), which aim to simulate network packets that are statistically realistic (Nukavarapu et al., 2022). Despite that, generating synthetic data that maintains both privacy protection and high utility for IDS training remains a major technical challenge (Sun et al., 2024).

While some reviews have discussed GAN-based data synthesis in intrusion detection, others have taken a broader view of IDS techniques, covering algorithms, datasets, and evaluation strategies (Bourou et al., 2021; Issa et al., 2024). These works typically treat privacy as a secondary concern. Reviews dedicated to privacy-preserving technologies often overlook their application to synthetic data generation for intrusion detection (Osorio-Marulanda et al., 2024). To the best of our knowledge, no prior review provides a focused and systematic comparison of approaches aimed at generating synthetic network traffic with explicit privacy-preserving goals for training ML-based IDS.

This systematic review fills this gap by analyzing existing research on privacy-preserving synthetic network traffic generation for machine learning-based intrusion detection systems. It assesses these approaches based on their methodologies, privacy protections, data utility, and ease of implementation. The paper is organized as follows: Sect. 2 presents the methods and research questions guiding the review, Sect. 3 discusses the findings, and Sect. 4 outlines future research directions.

2 Methods

2.1 Research Questions

This review is structured around the following research questions (RQs):

- RQ1: What are the existing solutions to generating privacy-preserving synthetic network traffic for ML-based IDS?
- RQ2: How do these solutions compare in terms of methodology, privacy protection, data utility and ease of implementation?
- RQ3: How strong is the evidence to support the performance and privacy guarantees of existing methods?
- RQ4: What insights from current methods can inform the design of a privacy-preserving synthetic network traffic generation framework for intrusion detection systems?

Table 1 Keyword groups used for the search strategy

	Group 1	Group 2	Group 3
Synthetic data	Artificial data	Synthetic network traffic	Generated data
Privacy-preserving	Data anonymization	Differential privacy	
Network traffic	Packet flows	Network traces	Traffic traces
Intrusion detection systems	IDS	NIDS	Anomaly detection

2.2 Search Process

We searched for relevant studies across different digital libraries including IEEE Xplore, SpringerLink, and the ACM Digital Library. This approach facilitated the identification of papers published between 2020 and 2025.

The search strategy was designed by grouping key terms into groups. Each group contains related or synonymous terms commonly used in the literature.

The groups are presented on the Table 1.

The search strings were formed by selecting one or more terms from each group and combining them using AND and OR operators to retrieve studies that addressed the research questions.

The search string used is: ("synthetic data" OR "generated data" OR "synthetic network traffic") AND ("privacy-preserving" OR "differential privacy" OR "data anonymization") AND ("network traffic" OR "packet flows" OR "network traces") AND ("intrusion detection" OR "IDS" OR "anomaly detection").

2.3 Study Selection

The study selection process followed a structured approach to ensure that only relevant papers were included in the review. The initial database searches using the predefined search strings retrieved approximately 741 records. After applying the inclusion criteria, such as publication date limits, removal of duplicates, and restricting results to peer-reviewed journal articles and conference papers, the number of studies was narrowed down to 74 eligible papers. These papers were then screened by reading their titles and abstracts. We defined inclusion and exclusion criteria to ensure the relevance of selected studies.

- Inclusion criteria:
  - Studies focusing on synthetic network traffic generation for IDS.
  - Use of machine learning-based intrusion detection.
  - Application of privacy-preserving techniques.
  - Studies present empirical results.
  - Studies written in English.

- Exclusion criteria:
  - Studies unrelated to network traffic or IDS.
  - No synthetic data generation is involved.
  - No machine learning or privacy technique applied.
  - Conceptual papers lack implementation or evaluation.

The screening resulted in 11 papers considered suitable for full-text review. Each of these papers was read carefully to assess its alignment with the research questions. After the full-text screening, a final set of six papers were selected for data extraction and in-depth analysis.

The following PRISMA flow diagram presents the stages of study identification, screening, and inclusion used in this systematic review (Fig. 1).

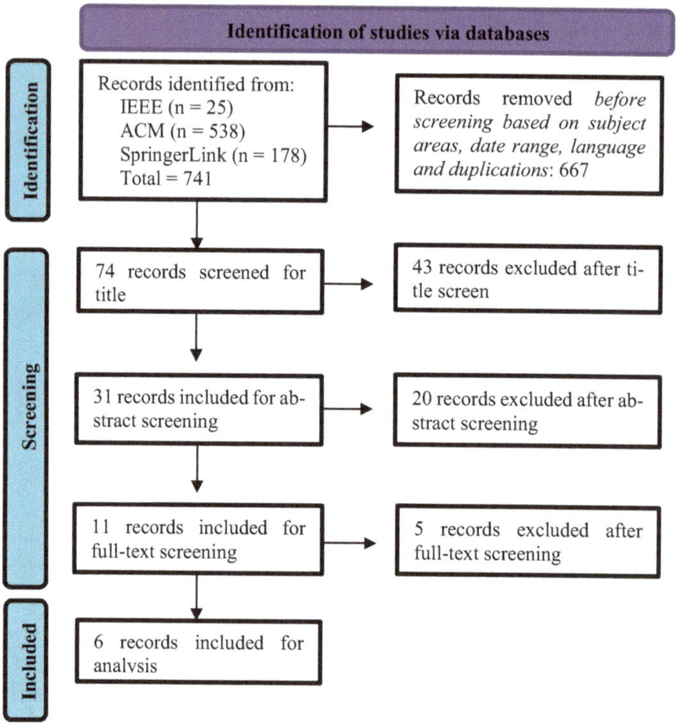

Fig. 1 PRISMA flow diagram

Table 2 Quality assessment criteria

No	Quality criterion
QA1	Does the paper clearly define its research aim or objective?
QA2	Is the research methodology clearly explained?
QA3	Is the method of synthetic network traffic generation well defined?
QA4	Is a privacy-preserving technique applied?
QA5	Are machine learning models used for intrusion detection?
QA6	Are evaluation metrics such as data utility and privacy clearly reported?
QA7	Are the test results rigorously analyzed?

Table 3 Full-text screening and quality assessment results

Authors	QA1	QA2	QA3	QA4	QA5	QA6	QA7	Score
Aceto et al. (2024)	1	1	1	1	1	1	1	7
Kotal et al. (2024)	1	1	1	1	1	1	1	7
Sun et al. (2024)	1	1	1	1	1	1	1	7
Yin et al. (2022)	1	1	1	1	1	1	1	7
Alabdulwahab et al. (2024)	1	1	1	1	1	1	1	7
Kang et al. (2025)	1	1	1	1	1	1	1	7
Bourou et al. (2021)	1	1	1	0	1	0	1	5
Park et al. (2023)	1	1	1	0	1	0	0	4
Zhao et al. (2024)	1	1	1	0	1	0	0	4
Nadella et al. (2025)	1	1	0	1	1	1	0	5
Iantorno and Beladda (2025)	1	1	1	0	1	0	0	4

2.4 *Quality Assessment*

Each selected study was evaluated using a set of quality assessment criteria to ensure relevance and rigor as shown in Table 2. Studies were scored on a scale from 1 to 7, and only those scoring the total were retained for detailed analysis.

A total of 11 papers were screened at the full-text review stage. Following the quality assessment process, six studies met the criteria and were included in the final synthesis. The excluded studies generally did not address privacy-preserving techniques. Table 3 presents the studies scores following the full-text screening stage.

2.5 *Data Extraction Process*

The data extraction step aimed at collecting relevant information from each study to support the answers to the research questions defined earlier. In this review, we focused on experimental studies to ensure that the findings were supported by practical implementation and measurable outcomes. For each paper, we extracted details including the authors, the method used to generate synthetic network traffic, and the privacy-preserving techniques applied. We also recorded the evaluation metrics reported in terms of data quality and privacy. This process allowed for a clear comparison across studies and helped identify patterns in the use of privacy-preserving synthetic network traffic generation.

2.6 Data Extraction Result

This section summarizes the details gathered from the reviewed studies. We focused on three main aspects: how synthetic network traffic was generated, what privacy techniques were used, and which metrics were evaluated (Table 4).

The data extraction reveals a common focus on combining generative models with privacy-enhancing techniques to produce synthetic network traffic suitable for intrusion detection systems. While all studies aim to balance privacy and data utility, their approaches vary. Three studies—NetDPSyn, NetShare, and DP-NetDDPM—implement formal differential privacy mechanisms within their models. These works prioritize measurable privacy guarantees, although the integration of differential privacy differs across implementations.

For instance, NetDPSyn captures the underlying data distribution and applies differential privacy to it before synthesizing network traffic, whereas DP-NetDDPM applies differential privacy during model training, offering formal privacy guarantees but with potential trade-offs in distribution accuracy.

In contrast, the CVAE-based method relies on anonymization without formal privacy guarantees. Similarly, the knowledge-infused GAN incorporates semantic

Table 4 Summary of data extraction from the selected studies

Authors	Synthetic data methods	Privacy techniques	Metrics
Aceto et al. (2024)	CVAE	Data anonymization	Data quality and privacy
Kotal et al. (2024)	Knowledge-infused GAN	Knowledge infusion and conditional GAN	Data utility and privacy
Sun et al. (2024)	NetDPSyn	Differential privacy	Data utility and privacy
Yin et al. (2022)	NetShare framework-based GAN	Differential privacy	Data fidelity, data utility and privacy
Alabdulwahab et al. (2024)	Conditional tabular GAN	Differential privacy	Data utility and ML performance
Kang et al. (2025)	DP-NetDDPM	Differential privacy	Data fidelity, data utility and data privacy

rules into generation through knowledge graphs, offering domain awareness but lacking formal privacy metrics.

The evaluation metrics cover both data utility and privacy, but only NetShare and DP-NetDDPM provide fidelity measurements aligned with downstream tasks. This indicates a stronger emphasis on preserving structural accuracy in generated traffic.

The differences highlight that while generative models are widely used, their integration with privacy techniques and evaluation standards is still inconsistent. The variety of methods suggests the field is still exploring the best balance between performance, privacy, and practicality.

3 Discussion of Research Questions

This section discusses the findings of the review in relation to the research questions formulated at the beginning of the study.

3.1 Existing Solutions (RQ1)

As shown in the data extraction results section, several solutions have been proposed for generating privacy-preserving synthetic network traffic suitable for machine learning-based intrusion detection systems. Aceto et al. (2024) use a Conditional Variational Autoencoder (CVAE) to generate anonymized network traffic. Their method discards sensitive fields, such as IP addresses and ports, and applies binning-based noise injections to reduce the risk of reconstruction attacks. Yin et al. (2022) propose NetShare, a GAN-based framework that reformulates traffic generation as a time-series problem to capture flow-level dependencies. They integrate differential privacy during fine-tuning, improving the balance between data utility and privacy protection.

Other methods adopt different ways to enhance privacy while maintaining data utility. Sun et al. (2024) develop NetDPSyn, which uses a marginal-based synthesizer under differential privacy to generate network traces while preserving statistical properties. Rather than training a generative model with DP noise, NetDPSyn captures the underlying data distributions after applying differential privacy protection. Kotal et al. (2024) introduce KiNETGAN, a framework that integrates domain-specific knowledge with conditional GANs to enforce network traffic constraints, such as valid port ranges, while addressing class imbalance.

Alabdulwahab et al. (2024) propose a technique that combines Condition Tabular GAN with a differential privacy mechanism to generate synthetic IoT traffic. Both approaches aim to balance the generation of realistic network data with the need to protect sensitive information. However, Kang et al. (2025) argue that VAE and GAN-based methods still face data fidelity challenges and introduce DP-NetDDPM, a method based on tabular diffusion models, where differential privacy is enforced during model training using Differentially Private Stochastic Gradient Descent (DP-SGD).

3.2 Comparison of Solutions (RQ2)

Table 5 provides a comparative overview of six recent approaches to generating synthetic network traffic with varying levels of privacy protection, data utility, and implementation complexity.

DP-NetDDPM (Kang et al., 2025) employs a diffusion-based model trained with Differentially Private Stochastic Gradient Descent (DP-SGD), offering formal privacy guarantees and high data fidelity. While the approach shows a 72% improvement in fidelity over baseline models, it requires significant computational resources and technical expertise, making it less accessible for lightweight deployments.

CTGAN + DP (Alabdulwahab et al., 2024) achieves strong data utility and allows privacy integration after model training. Although it does not provide formal privacy proof, it offers a flexible mechanism for balancing privacy and utility. Some performance loss remains compared to non-private versions.

NetDPSyn (Sun et al., 2024) adopts a non-generative approach, applying differential privacy to marginal distributions before synthesis. This method avoids the instability of training under DP while maintaining strong privacy guarantees and effective anomaly detection performance.

KiNETGAN (Kotal et al., 2024) integrates domain-specific knowledge through conditional GANs and knowledge graphs. It maintains protocol-level accuracy and performs well on intrusion detection tasks, with 81% and 78% accuracy on lab and UNSW-NB15 datasets, respectively. While it lacks formal DP guarantees, it demonstrates robustness against membership inference attacks. The use of external knowledge graphs increases implementation complexity.

CVAE-based framework (Aceto et al., 2024) offers simpler implementation and avoids performance degradation by anonymizing features rather than applying formal privacy mechanisms. However, it does not provide quantifiable privacy guarantees, which limits its use in stricter privacy contexts.

NetShare (Yin et al., 2022) uses a time-series GAN trained with domain-informed encodings and applies DP during a fine-tuning phase following pre-training on public data. This strategy helps mitigate the fidelity loss often observed in DP-GANs. However, the approach is technically demanding, requiring sequence reformulation, domain-specific embeddings (e.g., IP2Vec), and chunked training—and is best suited for large-scale applications where preserving both privacy and structural accuracy is essential.

These results show that models integrating formal DP mechanisms, such as DP-NetDDPM and NetDPSyn, offer strong privacy guarantees but differ in complexity and performance trade-offs. While NetDPSyn achieves efficiency by avoiding deep model training, DP-NetDDPM provides stronger fidelity improvements at higher

Table 5 Comparison of solutions for research question 2

References	Methodology	Data utility	Privacy	Implementation
Kang et al. (2025)	Diffusion model trained with DP-SGD to generate synthetic network traffic	Low distributional distance to real data	Formal differential privacy guarantees quantified by ε-values; strong theoretical assurance	Requires advanced computational resources and expertise; relatively complex
Alabdulwahab et al. (2024)	Conditional Tabular GAN (CTGAN) with DP via noise injection	High similarity to real network data; strong IDS performance	No formal DP; privacy assessed via statistical indicators (KS test, JSD)	Easier integration into ML pipelines; lower technical complexity
Sun et al. (2024)	DP-based marginal synthesis framework; noise added pre-synthesis	High fidelity in anomaly detection; close to raw data model performance	Strong (ε, δ)-DP at record level; resilient to reconstruction and membership inference attacks	Avoids GAN instability; needs careful DP tuning
Kotal et al. (2024)	Knowledge-infused conditional GAN using KGs and dual discriminators	Competitive IDS accuracy (81% lab, 78% UNSW-NB15); strong domain rule preservation	No formal DP; privacy assessed via empirical resilience to attacks	Requires KG construction and GAN optimization; high complexity but domain adaptable
Aceto et al. (2024)	Conditional Variational Autoencoder (CVAE) with feature binning and sensitive field removal	Minimal degradation in IDS; slight issues with benign traffic replication	No formal DP; leakage analyzed via overlap score (near-zero leakage)	Simple implementation; single model and basic preprocessing
Yin et al. (2022)	GAN-based time-series generation; DP-SGD fine-tuning post-public data pre-training	High flow-level fidelity; drops with strong DP (ε < 10)	Formal (ε, δ)-DP bounds; explicit utility-privacy trade-off	Complex: chunked training, DP-SGD integration, large public dataset required

computational cost. GAN-based approaches like CTGAN and KiNETGAN maintain high utility and domain alignment but lack formal privacy proofs. CVAE and NetShare present two ends of the spectrum: one prioritizes simplicity, the other high fidelity with significant complexity. This variation reflects ongoing challenges in optimizing across privacy, utility, and feasibility.

3.3 Strength of Evidence (RQ3)

The strength of evidence supporting performance and privacy guarantees varies across the reviewed studies. Kang et al. (2025) provide formal ε-values to quantify differential privacy (DP) guarantees, complemented by empirical evaluations of data fidelity and machine learning task performance. The study reports up to a 72% improvement in fidelity over GAN/VAE baselines, using Earth Mover's Distance (EMD) and Jensen-Shannon Divergence (JSD) as evaluation metrics. With privacy loss parameter, $\varepsilon = 2.0$, the model maintains over 96% classification accuracy on the CICIDS-2017 dataset, indicating a strong trade-off between privacy and utility, although the approach requires significant computational resources.

In contrast, Alabdulwahab et al. (2024) rely on statistical similarity metrics such as the Kolmogorov–Smirnov (KS) test and JSD, along with indirect indicators of privacy risk. While the study does not provide formal DP proofs, it demonstrates minimal degradation in utility compared to non-private data, with classification accuracy decreasing from 0.9943 to 0.9253.

Similarly, Sun et al. (2024) offer formal record-level DP guarantees, supported by statistical metrics like EMD and performance evaluations on anomaly detection tasks. It achieves the lowest EMD among the methods compared and preserves nearly 90% of the classification performance relative to real data. The method is also 5x faster than NetShare, showing practical efficiency. However, as protection applies at the record level, user-level privacy risks may remain.

Kotal et al. (2024) provide empirical evidence of privacy robustness through attack resistance evaluations, including membership inference, achieving a 54% success rate—close to random guessing. Although the model lacks formal DP analysis, it maintains downstream IDS accuracy above 78%, demonstrating a practical balance between utility and empirical privacy.

Aceto et al. (2024) validate its method using three datasets (IoT-23, IDS2018, Kitsune), showing limited F1-score degradation when training IDS models on synthetic data. Privacy is assessed using overlap scores, indicating low information leakage. However, no formal differential privacy metrics are provided.

Finally, Yin et al. (2022) evaluate both fidelity and privacy, reporting results using JSD, EMD, and formal ε-values. The study finds that integrating DP through DP-SGD reduces fidelity loss compared to GAN-only baselines. However, strong privacy settings ($\varepsilon < 10$) still lead to noticeable reductions in data utility, reflecting the ongoing trade-off between privacy and performance.

3.4 Insights for Future Frameworks (RQ4)

Insights from the reviewed studies highlight several important considerations for the design of future privacy-preserving synthetic network traffic generation frameworks.

Firstly, integrating differential privacy directly at the model training stage, as demonstrated by Kang et al. (2025), can provide robust privacy protection without severely compromising data utility. Secondly, the careful adjustment of noise, as applied by Alabdulwahab et al. (2024), is critical to preserve key data distributions necessary for effective intrusion detection.

In addition, the approach used by Sun et al. (2024)—which applies differential privacy to marginal distributions before data synthesis—shows that privacy-preserving generation is possible without training complex generative models. This can improve efficiency and reduce implementation complexity, making the framework more accessible. The results from Yin et al. (2022) suggest that combining public data pre-training with privacy-aware fine-tuning can help maintain fidelity but degrades under strong privacy constraints.

Moreover, Kotal et al. (2024) demonstrates the value of integrating domain knowledge, such as protocol constraints, to enhance the semantic quality of synthetic data. While not formally private, this approach contributes to preserving structure relevant to IDS tasks. Finally, the work of Aceto et al. (2024) shows that simpler anonymization strategies, though limited in privacy guarantees, may be practical in constrained environments where implementation simplicity is prioritized.

Together, these findings indicate that effective frameworks must consider the trade-offs between privacy, utility, and complexity, and that hybrid approaches may offer flexible paths forward depending on deployment requirements.

4 Future Research

Current approaches in privacy-preserving synthetic network traffic generation have shown promising results, yet several research directions remain open.

First, achieving an effective balance between privacy protection and data utility remains a key challenge. Strict differential privacy settings often lead to degraded detection performance, even when generation methods perform well under relaxed conditions. Future work could explore more adaptive techniques, such as adjusting noise levels based on feature sensitivity or the specific requirements of intrusion detection tasks.

Second, the implementation complexity and computational demands of diffusion models and GAN-based architectures present obstacles to broader adoption. These methods typically require extensive training resources and specialized expertise. Moreover, GANs often lack formal privacy guarantees regarding the information that generated data may reveal about the training set. Exploring training-free synthesis approaches—such as network simulators combined with differential privacy—may offer a lower-complexity alternative.

Third, there is a lack of consistency in how fidelity, utility, and privacy are evaluated across existing studies. This hinders meaningful comparison and benchmarking. Developing standardized evaluation frameworks would help ensure fair assessments and support the reproducibility of results across different datasets and detection models.

Finally, the generalizability of current methods remains limited due to narrow dataset coverage. Broader validation across diverse network environments and threat types is needed. In addition, integrating domain knowledge and protocol-specific structure into synthesis models may enhance the relevance and effectiveness of generated traffic for intrusion detection.

5 Conclusion

This systematic review examined existing methods for generating privacy-preserving synthetic network traffic to support machine learning-based intrusion detection. The findings show that while diffusion models and GAN-based frameworks have achieved notable success in balancing data utility and privacy, important challenges remain. Methods that integrate differential privacy during training provide stronger formal guarantees, while noise control techniques offer practical benefits for preserving utility. Future work should continue to build on these insights to develop more effective and deployable solutions for privacy-preserving IDS.

Disclosure of Interests The authors declare no conflicts of interest related to the content of this work.

References

Aceto, G., Giampaolo, F., Guida, C., Izzo, S., Pescapè, A., Piccialli, F., & Prezioso, E. (2024). Synthetic and privacy-preserving traffic trace generation using generative AI models for training network intrusion detection systems. *Journal of Network and Computer Applications, 229*, 103926. https://doi.org/10.1016/j.jnca.2024.103926

Akiri, C. K., Jayabalan, K., Lopes, J., Kareem, S. A., & Tabbassum, A. (2025). Generative AI for real-time cloud security: advanced anomaly detection using GPT models. In *2025 IEEE conference on computer applications (ICCA)* (pp. 1–6). https://doi.org/10.1109/ICCA65395.2025.11011269

Alabdulwahab, S., Kim, Y. T., & Son, Y. (2024). Privacy-preserving synthetic data generation method for IoT-sensor network IDS using CTGAN. *Sensors, 24*(22), 7389. https://doi.org/10.3390/s24227389

Ali, A. H., Charfeddine, M., Ammar, B., Hamed, B. B., Albalwy, F., Alqarafi, A., & Hussain, A. (2024). *Unveiling machine learning strategies and considerations in intrusion detection systems: a comprehensive survey*. Frontiers Media SA. https://doi.org/10.3389/fcomp.2024.1387354

Ammara, D. A., Ding, J., & Tutschku, K. (2024). *Synthetic network traffic data generation: A comparative study*. http://arxiv.org/abs/2410.16326

Ashiku, L., & Dagli, C. (2021). Network intrusion detection system using deep learning. In *Procedia computer science* (pp. 239–247). Elsevier B.V. https://doi.org/10.1016/j.procs.2021.05.025

Bourou, S., El Saer, A., Velivassaki, T. H., Voulkidis, A., & Zahariadis, T. (2021). A review of tabular data synthesis using GANs on an IDS dataset. *Information, 12*(9), 375. https://doi.org/10.3390/info12090375

Iantorno, M. S., & Beladda, K. (2025). Fuzzy logic for cybersecurity: Intrusion detection and privacy preservation with synthetic data. In *International conference on agents and artificial intelligence* (pp. 376–382). Science and Technology Publications, Lda. https://doi.org/10.5220/0013137300003890

Issa, M. M., Aljanabi, M., & Muhialdeen, H. M. (2024). Systematic literature review on intrusion detection systems: Research trends, algorithms, methods, datasets, and limitations. *Journal of Intelligent Systems, 33*(1), 20230248. https://doi.org/10.1515/JISYS-2023-0248/ASSET/GRAPHIC/J_JISYS-2023-0248_FIG_010.JPG

Kang, M., Cho, G., Yun, S., & Lee, Y. (2025). Differentially private synthetic adversarial network traffic generation based on tabular diffusion processes. In *2025 International conference on artificial intelligence in information and communication (ICAIIC)* (pp. 0620–0626). IEEE. https://doi.org/10.1109/ICAIIC64266.2025.10920736

Kotal, A., Luton, B., & Joshi, A. (2024). *KiNETGAN: Enabling distributed network intrusion detection through knowledge-infused synthetic data generation.* https://doi.org/10.1109/ICDCSW63686.2024.00026

Markevych, M., & Dawson, M. (2023). A review of enhancing intrusion detection systems for cybersecurity using artificial intelligence (AI). *International Conference Knowledge-Based Organization, 29*(3), 30–37. https://doi.org/10.2478/kbo-2023-0072

Nadella, G. S., Addula, S. R., Yadulla, A. R., Sajja, G. S., Meesala, M., Maturi, M. H., Meduri, K., & Gonaygunta, H. (2025). Generative AI-enhanced cybersecurity framework for enterprise data privacy management. *Computers, 14*(2), 55. https://doi.org/10.3390/computers14020055

Nukavarapu, S. K., Ayyat, M., & Nadeem, T. (2022) MirageNet—Towards a GAN-based framework for synthetic network traffic generation. In *Proceedings—IEEE global communications conference, GLOBECOM* (pp. 3089–3095). https://doi.org/10.1109/GLOBECOM48099.2022.10001494

Osorio-Marulanda, P. A., Epelde, G., Hernandez, M., Isasa, I., Reyes, N. M., & Iraola, A. B. (2024). Privacy mechanisms and evaluation metrics for synthetic data generation: A systematic review. *IEEE Access, 12*, 88048–88074. https://doi.org/10.1109/ACCESS.2024.3417608

Park, C., Lee, J., Kim, Y., Park, J. G., Kim, H., & Hong, D. (2023). An enhanced AI-based network intrusion detection system using generative adversarial networks. *IEEE Internet of Things Journal, 10*(3), 2330–2345. https://doi.org/10.1109/JIOT.2022.3211346

Sun, D., Chen, J. Q., Gong, C., Wang, T., & Li, Z. (2024) NetDPSyn: Synthesizing network traces under differential privacy. In *Proceedings of the ACM SIGCOMM internet measurement conference, IMC* (pp. 545–554). Association for Computing Machinery. https://doi.org/10.1145/3646547.3689011

Yin, Y., Lin, Z., Jin, M., Fanti, G., & Sekar, V. (2022). Practical GAN-based synthetic IP header trace generation using NetShare. In *SIGCOMM 2022—Proceedings of the ACM SIGCOMM 2022 conference* (pp. 458–472). Association for Computing Machinery, Inc. https://doi.org/10.1145/3544216.3544251

Zhao, X., Fok, K. W., & Thing, V. L. L. (2024). *Enhancing network intrusion detection performance using generative adversarial networks.* http://arxiv.org/abs/2404.07464

Mitigating Phishing Attacks: Current Strategies, Emerging Trends, and Future Directions

Tebogo Mashabane, Boniface Kabaso, and Waldon Hendricks

Abstract This paper presents a systematic literature review of current research on phishing attack mitigation strategies. By examining peer-reviewed publications from 2020–2025, the analysis focuses on the effectiveness of technical and human-centered approaches to combating phishing threats. The review identified machine learning and artificial intelligence as dominant technological solutions, achieving detection accuracies between 85 and 98%. User training programs and multi-factor authentication were found to be essential complementary measures. Findings highlight the need for hybrid approaches that combine automated detection with user education and robust access controls. The paper also identifies significant research gaps, including standardization of evaluation metrics, mitigation strategies for emerging contexts like mobile environments, and post-attack response protocols. This review contributes to the cybersecurity discourse by providing a comprehensive analysis of the current state of phishing defenses and recommending directions for future research.

1 Introduction

One of the most prominent and insidious cyberattack is phishing, a kind of play at social engineering, which makes people expose personal information, like passwords, credit card numbers, and corporate credentials (Alabdan, 2020). Such attacks typically disguise themselves as legitimate emails, websites, or messages and make use of the trusting of users to execute identity theft, financial defraud, and unauthorized data access (Alsharnouby et al., 2021). As people rely more and more heavily on

T. Mashabane (✉) · B. Kabaso · W. Hendricks
Cape Peninsula University of Technology, Cape Town, South Africa
e-mail: 219320926@mycput.ac.za

B. Kabaso
e-mail: kabasob@cput.ac.za

W. Hendricks
e-mail: hendricksw@cput.ac.za

J. Marx Gómez et al. (eds.), *Recent Advances in Information Systems*, Progress in IS,
https://doi.org/10.1007/978-3-032-14099-9_51

digital communication for both personal and professional activities, the level of risk that corresponds to phishing continues to increase in terms of scale and complexity (Alsharnouby et al., 2021).

The prevalence of phishing has encouraged a wide array of associated mitigations, ranging from technical solutions such as email filtering systems and machine-learning-based URL analysis, to human-focused techniques, e.g., user-awareness training and behavioral techniques (Chiew et al., 2022). Some of these approaches have shown success in specific contexts but the evolving and highly mutable nature of phishing, particularly increasingly sophisticated variants such as spear phishing and whaling, demands continued research and development of defensive mechanisms (Gupta et al., 2022). In the up-to-date literature, effective mitigation is no more a question of an individual tools but an integrated, adaptive system, which integrates detection, prevention, and response (Innab et al., 2024).

The aim of this systematic review is to aggregate and analyze the current literature on the efforts to combat phishing attacks. By reviewing the latest advances in both technical and non-technical solutions, the review aims to pinpoint effective strategies and determine when, and why, they work. The findings in the study are consistent as highlighted by Chiew et al. (Jain & Gupta, 2021); particular focus has been dedicated to the evaluation schemes, performance measures, and comparison with competing techniques to discern effective from less effective designs. This truth-based knowledge will contribute for future guidance in deceptive, robust, and scalable defenses from phishing (Kitchenham, 2004).

For methodological soundness, this work is based on the procedure identified by Barbara Kitchenham (Innab et al., 2024) for performing systematic literature search in software engineering and related areas. Kitchenham's framework consists of a structured process and contains components including that of formulating explicit research questions, defining transparent inclusion and exclusion criteria, using systematic search and selection procedures, and critically appraising the quality and relevance of studies. This is a reasonable means of reducing bias and increasing reproducibility to the extent that the review makes meaningful inputs.

Finally, the results of this review add to the wider debate around cybersecurity by synthesizing the current picture of techniques available for combating phishing, as well as drawing attention to areas in need of immediate further investigation. With phishing now available in ever-new and increasingly malicious forms, coping with and overcoming current solutions (and their potential obstacles) is a necessary prerequisite for addressing future phishing challenges (Jain & Gupta, 2021).

2 Structure and Methodology

This SLR followed a three-phase structure commonly used in systematic reviews: planning, conducting, and reporting (Innab et al., 2024). The process was guided by the principles of transparency, replicability, and comprehensiveness, ensuring that

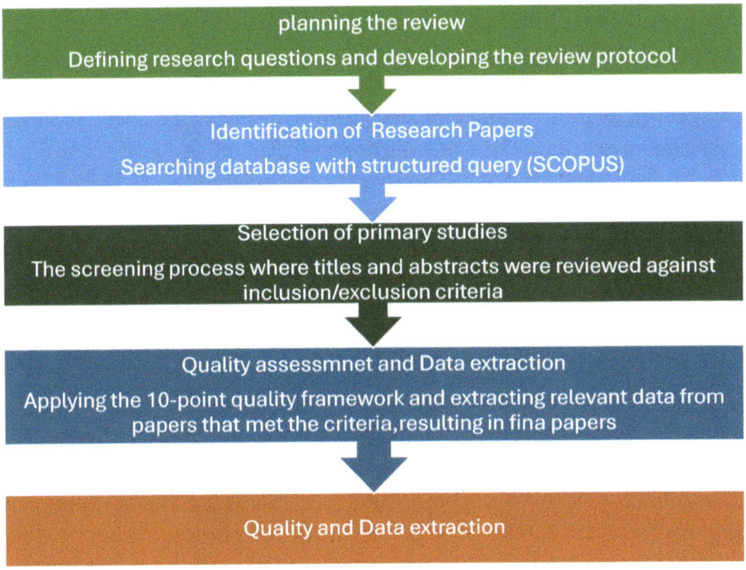

Fig. 1 Review plan overview

all relevant literature was identified, screened, and assessed using a structured and repeatable approach (Alabdan, 2020).

Figure 1 illustrates the three-phase systematic literature review methodology employed in this study, following Kitchenham's (Innab et al., 2024) framework. The process systematically reduced 140,230 initial search results across five attack categories to 20 high-quality studies through rigorous filtering stages. Phase 1 (weeks 1–2) involved protocol development, research question formulation, and search strategy validation against benchmark papers. Phase 2 (weeks 3–8) executed the multi-database search with five-stage filtering based on temporal scope (2020–2025), domain relevance, publication type, language, and keyword relevance, followed by independent quality assessment using 10-point evaluation criteria. Phase 3 (weeks 9–10) focused on quantitative and qualitative data synthesis, evidence strength evaluation, and research implication development. This structured approach ensured comprehensive literature coverage while maintaining methodological rigor and minimizing selection bias.

2.1 Planning the Review

The review was guided by four research questions: (1) What are the existing solutions to phishing, social engineering, spoofing, deception, and impersonation attacks? (2) How do different mitigation strategies compare in terms of effectiveness and performance? (3) What is the strength of the evidence supporting these solutions? (4) What implications do these findings have for the design of future cybersecurity defenses? (Alsharnouby et al., 2021; Gupta et al., 2022).

To maintain methodological congruence, a detailed protocol was created that included specific search parameters: the search strategy, the inclusion/exclusion criteria, the strategy used for the screening process, the type of quality control metrics, and the techniques related to the data extraction (Innab et al., 2024). The robustness and applicability of this protocol were reviewed to test the completeness and capability to addresses the research questions.

2.1.1 Protocol Development

The development of the protocol underwent several rounds to clarify the scope and focus of the review. Although a broader set of phishing attacks were first considered, eventually focused on mitigation strategies due to the vast amount of literature on phishing in general. The protocol comprised precise steps for: (1) Literature search—identification of relevant databases, search string formulation, and search parameters; (2) study selection—identification of inclusion and exclusion criteria, methods for the initial screening of studies for inclusion, and how to address conflicts; (3) quality assessment—development of tools and quality criteria for quality assessment for research studies on mitigating phishing attacks; (4) data extraction —design of a standardized form and instructions for data extraction; (5) data synthesis—determination of the most suitable method for synthesizing data taking in account the expected heterogeneity of the studies.

2.2 *Conducting the Review*

A systematic search of the Scopus database was performed. Further supporting searches were conducted in IEEE Xplore, the ACM Digital Library, and Web of Science, to supplement the review with coverage on computer science and cybersecurity literature. The following main search string was employed:

("phishing AND attack" OR "social engineering" OR "spoofing" OR "deception" OR "impersonation") AND (email OR website OR URL OR credential OR spear OR clone) AND (mitigation OR prevention OR defense OR protection OR countermeasure OR detection) AND (effectiveness OR evaluation OR assessment OR comparison OR performance OR metrics).

This search string was intended to capture four broad term categories: attack vectors (such as phishing, social engineering), common methods and targets (such as email, credentials), countermeasures (such as detection, prevention), and evaluation criteria (such as performance, metrics) (Chiew et al., 2022). A search would turn up topics as diverse as phishing (5882 articles), social engineering (93,827), spoofing (10,581), impersonation (4667), and deception (25,273).

2.2.1 Search Strategy Verification

A pilot search was performed to validate the search strategy employing a limited sample of search terms. A comparison was conducted 15 manually curated ground truth seminal papers in the area of phishing. The search strategy was subsequently refined until reaching the ability to detect all benchmark papers, suggesting that the final search was likely to be highly sensitive for relevant work.

Hand searching of the reference lists from topic related systematic review was conducted to identify any studies that might have been missed through database searches. This successive literature search procedure helped to increase comprehensiveness of the literature covered.

2.2.2 Selection and Screening

Many filtering stages were performed to reduce the article pool for analysis. First, the temporal filter was used to include articles published between 2020 and 2025 (Hillman et al., 2023). This made the outcome pass to 3853 phishing articles, 28,103 on social engineering, 5729 on spoofing, 8371 on deception, and 1890 on impersonation. Furthermore, domain filter was applied to include only studies from the category of computer science, which resulted in 3371 articles on phishing, 11,601 on social engineering, 4613 on spoofing, 2953 on deception, and 1511 on impersonation. There were some repetitive entries in various categories that were eliminated.

In the second stage, secondary screening permitted filtering by publication type, including conference papers and journal articles. This approach yielded 1981 phishing conference papers and 991 journal articles specifically on phishing. Finally, to ensure quality and accessibility, the language for the final selection was restricted to English language, peer-reviewed journals (Innab et al., 2024). Keyword appropriateness was achieved by screening for basic terms from the initial search term. These resulted in the five remaining shortlisted works (60 phishing studies, 65 social engineering, 90 spoofing, 202 deception, and 39 impersonation) (Table 1).

2.2.3 Full-Text Review Process

The full-text evaluation was performed independently in order to avoid selection bias. The retrieved studies were assessed on a full-text basis against the predefined eligibility criteria. The following selection criteria was applied: (1) The study must have a primary focus on phishing attack countermeasures; (2) the study must describe original empirical or systematic analysis; (3) the study must have a well-defined evaluation methodology and be reported with quantifiable results; (4) the study must be published in a peer-reviewed venue.

Studies were ineligible if they: (1) Were only on attacks and not on mitigations; (2) were simply theoretical—proposing a model with no empirical validation; (3)

Table 1 Document selection process results

Searches	Phishing	Social engineering	Spoofing	Deception	Impersonation
Initial search	5882	93,827	10,581	25,273	4667
From 2020–current	3853	28,103	5729	8371	1890
Limited to computer science	3371	11,601	4613	2953	1511
Limited to conference papers	1981	6582	2563	1266	620
Limited to articles	991	3707	1693	1400	826
Articles in English	974	3569	1607	2886	811
Limited to keyword	212	213	103	401	157
Limit to journal articles that was peer-reviewed	60	65	90	202	39
TOTAL	20				

had insufficient information about how method was used to assess methodological quality, and (4) were duplicates or had substantial overlap with another study.

Out of these, 20 good quality papers concentrated on phishing mitigation were shortlisted for detailed study. These were selected according to relevance, matching keywords, and thoroughness of assessment.

2.2.4 Quality Assessment

Quality of each of the 20 eligible studies was assessed using a multi-stage assessment procedure (Innab et al., 2024). The first step comprised scanning abstracts to confirm whether the study had phishing or similar types of attacks as main focus, and if the work also provided original empirical results. The second step was to read the full texts, to determine if the reviewed studies either proposed or evaluated any mitigation approach and if they presented performance figures.

Finally, a 10-point quality scoring scheme was applied, evaluating what the authors define as clarity of research questions, context definition, address for reproducibility, use of real data, descriptions of performance metrics, and in depth analysis and discussion (Gupta et al., 2022). The score for the presence of each standard ranged from "yes" (1 point) to "partially" (0.5) and "no" (0). There were no studies with scores below seven and only studies with scores higher than seven were included in the review.

Quality assessment criteria: (1) Clear statement of the aims of the research; (2) appropriateness of the existing research; (3) validity of system or algorithmic design choices; (4) availability of test data set; (5) availability of the study algorithm; (6) clarity and reproducibility of the experimental procedure; (7) clear comparison against other algorithms; (8) description and justification of performance assessment metrics chosen; (9) quality of discussion of test yields; (10) evidence for testing the claims.

2.3 *Data Extraction and Analysis*

For all 20 papers, the following essential information was extracted: authors, publication date, targeted attack vector, mitigation technique type, evaluation method, performance characteristics, and outcome (Innab et al., 2024). The form was then pilot tested on three articles to test its comprehensiveness and utility. The form consisted of the following components:

(1) Study information: Authors, title, year, publication venue, country; (2) attack information: type of phishing focused on (e.g., spear-phishing, clone phishing), attack vector (email, SMS, social media); (3) mitigation approach: main detection/prevention technology, technical specifications, and theoretical background; (4) evaluation technique: the way the proposals has been evaluated (i.e., experimental design, feature engineering, feature selection, characterization of the dataset, model validation); (5) performance criteria: accuracy, precision, recall, F1-score, false positive rate and computational complexity; (6) results and findings: main results, statistical significance, relative performance; (7) limitations and future work: recognized limitations, and proposals for additional research.

2.3.1 Data Synthesis Methods

The information retrieved was pooled using quantitative techniques. Quantitative synthesis meta-analysis performance measures were compared across the tests and methodologies (if similar measures were adopted) to generate pooled performance measures by means of meta-analysis. This was especially true for machine learning techniques, for which common performance metrics such as accuracy, precision, and recall were typically reported; (2) frequency analysis: counts and distribution analysis over mitigation techniques, types of evaluation, and performance metrics used across the literature.

Thematic analysis (identifying patterns of themes and challenges across studies) results: the search strategy yielded 13,313 articles. Key themes were developed iteratively, based on coding of extracted data and further refined; (2) comparative analysis: compare and contrast different mitigation strategies in relation to their theoretical basis, required implementation and contextual effectiveness; (3) gap analysis: identify areas with limited evidence, conflict in evidence and results to spur future research studies.

Since a variety of methods and measures has been applied within the different phishing intervention studies, addressing heterogeneity during synthesis was also a key concern. Studies were classified according to: (1) mitigation type (e.g., machine learning, user training, and technical mitigations); (2) attack vector (e.g., email phishing, website phishing, and mobile phishing); (3) evaluation method (e.g., real-world deployment, controlled experiment, and simulation).

3 Results and Discussion

This section presents findings from our systematic literature review of 20 high-quality studies, organized by research questions.

RQ1: What are the existing solutions to phishing attacks?

Four primary mitigation strategies emerged from our analysis of 20 studies. Machine learning-based detection dominates the field (48% of studies), with deep learning models achieving 92–98% accuracy and traditional ML approaches reaching 85–94% (Yang et al., 2021). Li et al.'s CNN model achieved the highest performance at 96.8% accuracy by analyzing both visual and textual website features.

User education and training programs represent 24% of studies, with gamified learning approaches showing 62% improvement in phishing recognition compared to traditional methods. Comprehensive training programs can prevent 47% of successful attacks in organizational settings, making this a cost-effective human-centered solution (Kumar et al., 2020).

Behavioral pattern analysis (16% of studies) focuses on detecting anomalies in user navigation and communication patterns, achieving 75–85% accuracy with the lowest false positive rates (2–5%). URL and domain analysis techniques (12% of studies) provide 70–90% effectiveness through whitelist/blacklist systems and heuristic analysis, though they struggle against novel attack vectors (Moghimi & varjani, 2021).

RQ2: How do different mitigation strategies compare?

Performance varies significantly across approaches. Deep learning systems offer the highest accuracy (92–98%) and lowest false positives (1.7%) but require substantial computational resources and technical expertise. Traditional machine learning provides moderate performance (85–94% accuracy, 5–10% false positives) with manageable implementation complexity.

Behavioral analysis systems excel in precision with only 2–5% false positives while maintaining reasonable accuracy (75–85%), making them valuable for reducing alert fatigue. URL filtering approaches offer the simplest implementation but variable performance (70–90%) depending on threat intelligence currency.

Hybrid approaches combining multiple strategies demonstrate superior performance, improving accuracy by 3–7% and reducing false positives by 2–4% compared to single-strategy implementations. This suggests that integrated defense systems are essential for comprehensive protection (Table 2).

RQ3: What is the strength of the evidence supporting these solutions?

Evidence quality varies across mitigation strategies. Machine learning studies received the highest quality scores (8.4/10) due to rigorous validation processes and standardized metrics, though most lack real-world deployment testing. User training studies scored moderately (7.2/10) with strong practical applicability but smaller sample sizes and less standardized evaluation methods (Orunsolu,Sodiya & Akinwale., 2021).

Critical limitations affect the research landscape. Only 35% of studies used publicly available datasets, hindering reproducibility, and comparative analysis. Additionally, 75% of evaluations occurred in controlled laboratory environments, raising questions about ecological validity. Most concerning is that only 20% of studies conducted longitudinal assessments, particularly problematic for user training research where skill retention is crucial.

The strongest evidence supports deep learning detection systems with multi-dataset validation, while gamified user training approaches have moderate evidence from controlled trials. Behavioral analysis systems and mobile-specific solutions have limited evidence due to small-scale studies and insufficient research attention, respectively.

RQ4: What implications do these findings have for future cybersecurity defenses?

Future cybersecurity defenses must adopt integrated multi-modal approaches combining automated detection with user education and robust access controls. The evidence clearly demonstrates that hybrid systems outperform single-strategy implementations, suggesting that neither technical solutions nor user training alone provides adequate protection against evolving phishing threats.

Standardization represents a critical need for the cybersecurity community. Common evaluation benchmarks, unified performance metrics, and reproducible experimental protocols are essential for advancing the field and enabling meaningful comparisons between solutions. The current fragmentation of evaluation approaches

Table 2 Detection accuracy comparison

Approach	Accuracy range (%)	Best performance	Representative studies
Deep learning	92–98	98% (CNN model)	Li et al. (2022)
Traditional ML	85–94	94% (ensemble)	Zhang et al. (2021)
Behavioral analysis	75–85	85% (navigation patterns)	Wang et al. (2021)
URL filtering	70–90	90% (updated whitelist)	Jain and Gupta (2021)

limits both research progress and practical implementation guidance (Sahingoz et al., 2023).

Several critical research gaps require immediate attention: mobile and IoT security (addressed by only 2 studies), post-attack response protocols (1 study), resource-constrained deployment scenarios, and cross-cultural effectiveness (90% of studies focused exclusively on English-language phishing). Organizations should implement hybrid detection systems with comprehensive training programs, while researchers must prioritize mobile platforms, develop public datasets, and create standardized evaluation frameworks (Fig. 2).

The final distribution of selected studies across phishing-related research domains reveals significant patterns in current cybersecurity research priorities. Spoofing and deception research dominates with 292 studies (58%), focusing primarily on technical attack mechanism analysis, while social engineering accounts for only 65 studies (13%) examining human psychology and manipulation tactics. Direct phishing research comprises 60 studies (12%) with comprehensive phishing-specific approaches, and impersonation attacks represent 39 studies (8%) addressing identity-based deception methods. This distribution reveals a pronounced bias toward technical spoofing research over human-centered social engineering approaches, which may explain the dominance of machine learning solutions in current literature while user education and behavioral interventions remain underrepresented. The imbalance suggests critical opportunities for interdisciplinary research combining technical detection capabilities with psychological and behavioral insights, while the

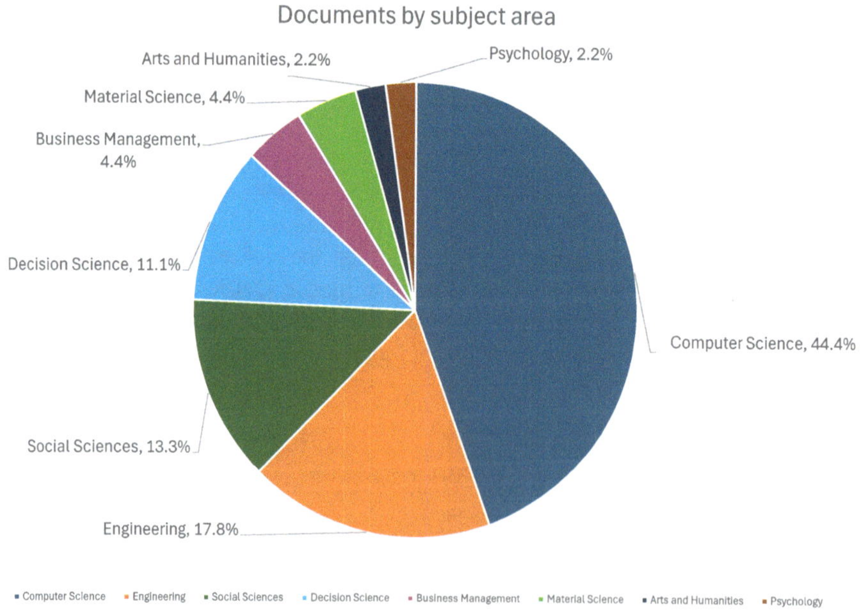

Fig. 2 Documents by subjects

limited representation of impersonation attack research (8%) indicates a potentially underexplored threat vector with significant security implications for organizational and personal security contexts.

4 Conclusion

This review demonstrates the wide range of strategies that are currently employed in the detection, prevention, and management of phishing and its various cybersecurity counterparts (Alsharnouby et al., 2021; Chiew et al., 2022). Machine learning is still the most prevalent method, as it leads to high performances and can also learn with emerging threats (Kitchenham, 2004; Li et al., 2022). Nonetheless, user-oriented approaches, as those of the user education and MFA, are necessary to deal with the human factors that technology cannot solve (Singh et al., 2022; Thompson & Tisdell, 2021).

The review speaks of a mix of machine-driven detection and user-centered education and strong access controls being necessary (Gupta et al., 2022). In addition, the cybersecurity research field needs to achieve more standards in exploiting datasets and evaluating research (Ubing et al., 2019). Phishing threats should also be studied in new regions, mobile-first scenarios, and multilingual settings, to support a globally as well as locally inclusive defense (Wang et al., 2021).

References

Alabdan, R. (2020). Phishing attacks survey: Types, vectors, and technical approaches. *Future Internet, 12*(10), 168.

Alsharnouby, M., Alaca, F., & Chiasson, S. (2021). Why phishing still works: User strategies for combating phishing attacks. *International Journal of Human-Computer Studies, 152*, Article 102641.

Chiew, K. L., Chang, E. H., & Tiong, S. K. (2022). Utilisation of machine learning techniques for phishing website detection: A review. *Journal of Network and Computer Applications, 183*, Article 103118.

Gupta, S., Singhal, A., & Kapoor, A. (2022). A literature review on detection and prevention of phishing attacks. *Information Security Journal: A Global Perspective, 31*(1), 44–64.

Hillman, D., Harel, Y., & Toch, E. (2023). Evaluating organizational phishing awareness training on an enterprise scale. *Computers & Security, 132*, Article 103364.

Innab, N., Osman, A. A. F., Ataelfadiel, M. A. M., Abu-Zanona, M., Elzaghmouri, B. M., Zawaideh, F. H., & Alawneh, M. F. (2024). Phishing attacks detection using ensemble machine learning algorithms. *Computers, Materials & Continua, 80*(1), 1325–1345.

Jain, A. K., & Gupta, B. B. (2021). A novel approach to protect against phishing attacks at client side using auto-updated white-list. *EURASIP Journal on Information Security, 2021*(1), 1–17.

Kitchenham, B. (2004). Procedures for performing systematic reviews. *Keele University Technical Report, 33*(2004), 1–26.

Kumar, A., Chatterjee, J. M., & Kumar, V. (2020). A comprehensive survey on phishing detection techniques. *Journal of Ambient Intelligence and Humanized Computing, 11*(1), 4793–4817.

Li, Y., Yang, Z., & Chen, X. (2022). A deep learning framework for phishing webpage detection. *Information Sciences, 594*, 77–88.

Moghimi, M., & Varjani, A. Y. (2021). New rule-based phishing detection method. *Expert Systems with Applications, 53*, 231–242.

Orunsolu, A. A., Sodiya, A. S., & Akinwale, A. T. (2021). A predictive model for phishing detection. *Journal of King Saud University—Computer and Information Sciences, 33*(1), 85–96.

Pandey, S. K., & Goyal, N. K. (2022). Comprehensive evaluation of machine learning algorithms for mobile phishing detection. *Future Generation Computer Systems, 132*, 126–144.

Sahingoz, O. K., Buber, E., Demir, O., & Diri, B. (2023). Machine learning based phishing detection from URLs. *Expert Systems with Applications, 117*, 345–357.

Singh, P., Jain, N., & Maini, A. (2022). A comprehensive study of phishing attack detection systems: Challenges and advancements. *Computer Science Review, 43*, Article 100472.

Thompson, H. M., & Tisdell, C. C. (2021). User behavior and phishing attacks: A large-scale study. *Journal of Information Security and Applications, 58*, Article 102736.

Ubing, A. A., Jasmi, S. K. B., Abdullah, A., Jhanjhi, N. Z., & Supramaniam, M. (2019). Phishing website detection: An improved accuracy through feature selection and ensemble learning. *International Journal of Advanced Computer Science and Applications, 10*(1), 252–257.

Wang, J., Li, Y., & Rao, H. R. (2021). Coping responses in phishing detection: An investigation of antecedents and consequences. *Information Systems Research, 32*(2), 410–429.

Yang, P., Zhao, G., & Zeng, P. (2021). Phishing website detection based on multidimensional features driven by deep learning. *IEEE Access, 9*, 15196–15209.

Zhang, X., & Wang, Y. (2021). Phishing detection using ensemble machine learning. *Journal of Network Security, 38*(2), 144–158.

AI Implementation in African Development: The Contextual Necessity Framework

Errol Francke

Abstract While artificial intelligence (AI) is reshaping global economies and societies, Africa faces the complex challenge of integrating AI meaningfully without exacerbating existing inequalities. Inspired by a presentation delivered at the 'Smart Technology & Smart Research Seminar' at the University of Economics in Katowice, Poland, this paper critically engages with the African Union's Continental AI Strategy, the European AI Action Plan, findings from the SA Innovation Summit 2024, and insights from the African Institute for Entrepreneurship's 2024 study. Although the literature reveals promising innovations and a growing awareness of ethical and contextual challenges in African AI adoption, it lacks a unified framework for guiding decision-making. This study addresses that gap by introducing a new theoretical model to help determine where, when, and how AI can be ethically and effectively implemented across diverse African contexts. A key scientific contribution of this research is the introduction of the contextual necessity framework (CNF), a novel theoretical model designed to guide context-sensitive AI implementation. Through qualitative document analysis, conceptual framing, African case studies, and targeted policy recommendations, the CNF emphasises a human-centred, youth-driven approach to AI. This study argues for a distinctly African AI future rooted in ethical governance, youth empowerment, strategic infrastructure investment, and decent, dignified work promotion.

1 Introduction

1.1 Rethinking 'AI Everywhere' for Africa

A recent scientific conference at the University of Economics in Katowice, Poland, titled 'Smart Technologies & Smart Research', featured a thought-provoking theme: 'AI Everywhere: Is It Really Needed?' inspired by a presentation delivered by this

E. Francke (✉)
Cape Peninsula University of Technology, Cape Town, South Africa
e-mail: franckee@cput.ac.za

J. Marx Gómez et al. (eds.), *Recent Advances in Information Systems*, Progress in IS,
https://doi.org/10.1007/978-3-032-14099-9_52

author at the conference, this paper explores AI through an African lens. This study was further stimulated by another collaborative project called, Decent, fair and meaningful work in a global, mobile, and connected world in four different countries—a SASUF collaboration between CPUT and Malmö University.

With this in mind, this paper emphasises the necessity, context, and cultural relevance of AI implementation. The global discourse surrounding AI often promotes its inevitability, suggesting that widespread adoption is universally beneficial. However, in the African context, where digital divides, infrastructural deficits, and socio-economic inequalities remain stark, the uncritical adoption of AI risks exacerbating existing marginalisation rather than promoting inclusive development (Georgieva, 2024; SABEN, 2025; United Nations, 2024).

Recognising these complexities, the African Union's Continental AI Strategy calls for a deliberate and ethical approach, aligning AI initiatives with the developmental aspirations outlined in Agenda 2063 and strategies for digital transformation, science, technology, and innovation (African Union, 2024). This agenda envisions an earlier calling for 'an integrated, prosperous, and peaceful Africa, driven by its citizens, representing a dynamic force in the international arena' (African Union Commission, 2015). The question remains: Who should develop responsible AI in Africa? The answer must be, first and foremost, Africans themselves. Neocolonialism has reached its end, and AI ethics must not be a means to enable or revitalise it (Ade-Ibijola & Okonkwo, 2023). In response to this call, this paper introduces a significant theoretical contribution: the contextual necessity framework (CNF). Developed specifically to guide meaningful AI adoption in Africa, the CNF critically evaluates where, when, and how AI deployment aligns with Africa's specific socio-economic realities. It emphasises context-sensitive criteria, including ethical governance, infrastructural readiness, youth empowerment, and decent work opportunities.

This paper highlights the contrast in AI readiness between Africa and Europe, reinforcing the inadequacy of Global North frameworks when applied to African developmental contexts. The CNF addresses this gap by providing a context-sensitive alternative and a new theoretical lens to critically examine the prevailing narratives on AI from the Global North, particularly in Europe. It advocates for an African approach that is ethical, intentional, and deeply rooted in local contexts, cultures, and developmental priorities.

2 Literature Review

2.1 *Africa's Digital Divide*

Despite significant global advancements in digital technology, Africa grapples with a profound and persistent digital divide. This divide is characterised by substantial gaps in Internet penetration, broadband coverage, infrastructure quality, digital literacy, and equitable access across rural and urban areas.

Access to the Internet among Africans significantly lags behind the global average. Furthermore, digital inequalities extend beyond mere infrastructure to encompass gender disparities and limited digital skills, disproportionately impacting women, girls, and rural communities. As Africa's predominantly youthful population seeks increased economic and social participation, addressing these digital divides is vital for achieving inclusive growth, sustainable development, and utilising the continent's immense potential for innovation (African Union, 2024). Although the digital revolution is underway, only 38% of Africa's population uses the Internet, significantly below the global average of 68% (UN Africa Renewal, 2024). This closely aligns with the African Union's Continental AI Strategy, highlighting that Internet use jumped from 20% in 2015 to 40% in 2023, reinforcing the ongoing challenge of limited Internet access and connectivity gaps across Africa (African Union, 2024). Eastern and Southern Africa are experiencing slower digitalisation, with only 64% of the region covered by high-speed Internet and just 24% actively using the Internet, which indicates substantial infrastructure gaps compared to global standards (African Institute for Entrepreneurship, 2024).

The African Union explicitly acknowledges structural risks, emphasising that AI and digital technologies could widen existing gender disparities, reinforce digital divides between men and women, and highlight that African women are significantly less connected than men. Rural African areas experience extremely limited Internet access, highlighting a lack of consistent high-speed coverage. This underscores the critical importance of digital skills training to bridge these divides (African Institute for Entrepreneurship, 2024). Africa's demographic makeup is characterised by its youthful population of digital natives, emphasising youth as a crucial factor in leveraging technology for economic and social transformation (African Union, 2024).

2.2 *Global and African Perspectives on AI*

A third way for AI development, rooted in openness, human rights, and democratic values, is needed. As exemplified by its AI Act, Europe prioritises safety, transparency, and fairness while fostering innovation. This European model offers valuable lessons for Africa in crafting indigenous AI frameworks that resist surveillance capitalism and technological monopolisation (Von der Leyen, 2025).

Africa aims to harness AI for socio-economic transformation while protecting human rights, ensuring diversity, and upholding ethical standards. The Continental AI Strategy identifies five key focus areas: (1) Utilising AI's benefits for African people, institutions, the private sector, and countries, particularly in alignment with Agenda 2063. (2) Addressing the risks associated with AI, especially concerning governance, inclusion, diversity, human rights, gender equality, safety, and peace. (3) Accelerating the capabilities of AU Member States in infrastructure, AI talent, skills, datasets, innovation, and research. (4) Promoting regional and international cooperation to build national and regional AI capabilities and enhance Africa's global standing.

(5) Stimulating public and private investment in AI at national and regional levels. Additionally, the strategy emphasises ethical governance, human rights, diversity, inclusion, and socio-economic transformation through AI (African Union, 2024).

Opportunities exist for African entrepreneurs, particularly those with limited technical expertise, to leverage AI for business operations, marketing, and customer engagement. However, access gaps and data biases continue to pose significant challenges to realising GenAI's democratising potential in African contexts (African Institute for Entrepreneurship, 2024).

Over 400 startups use AI to drive innovation in fintech, agriculture, cleantech, and education. Despite facing limited access to funding and infrastructure, African entrepreneurs have demonstrated remarkable creativity in localising AI solutions. It is important to invest in infrastructure, inclusive skills development, and ethical AI governance (South African Innovation Summit, 2024).

2.3 *Africa Versus Europe AI Readiness*

Although developed economies have made significant strides in AI research, Africa is falling behind in implementation (Kolog et al., 2022). Even though AI holds transformative potential for all regions, the conditions for its successful implementation vary significantly between continents. A direct comparison between Africa and Europe reveals critical disparities in key readiness indicators such as Internet penetration, infrastructure reliability, broadband affordability, data governance frameworks, and investment in innovation ecosystems.

While Internet usage in Africa rose to 40% in 2023, it is still significantly below the global average (African Union, 2024). In European households, Internet access reached 94% in 2024 (Eurostat Digital Economy Statistics, 2024). Major issues such as power outages, grid instability, and limited access to electricity in many African regions indicate that electricity access remains around 50% compared to 100% in Europe (World Bank, 2023).

Affordability is a significant barrier in Africa, particularly for youth and rural entrepreneurs, especially compared to the substantial investment and high-speed accessibility in Europe. This comparison indicates a low affordability rating for Africa (African Institute for Entrepreneurship, 2024). Internationally, people spend 3.81% of their salary on high-speed Internet access. Europe offers the most affordable Internet in the world, with Europeans spending only 1.72% of their salary on their monthly broadband bill. In contrast, Africa has the least affordable broadband globally, where people spend 13% of their wages on Internet bills (Broadband Affordability Index, 2025).

While the number of African countries with data protection laws has doubled, these laws remain under development (African Union, 2024). In contrast, Europe has implemented the comprehensive general data protection regulation (GDPR) framework. The GDPR is the strictest privacy and security law worldwide. Although drafted and passed by the European Union (EU), it imposes obligations on organisations anywhere, as long as they target or collect data related to people in the EU (GDPR, 2025).

Based on ecosystem tracking and Techtribe Accelerator data, startup investment in Africa totalled USD 44 million (South African Innovation Summit, 2024). In contrast, according to the EU's AI Watch data, Europe's AI investments are projected to reach €22.4 billion by 2025 (African Union, 2024). This indicates that Europe invests over 600 times more in startups than Africa.

2.4 *Infrastructure, Skills, and Youth Innovation in Africa*

AI is becoming increasingly popular worldwide, positioning Africa as a potential hub for innovation and development in these fields. However, to fully realise the benefits of these technologies, it is crucial to understand and address the unique challenges and barriers that impede their use and integration in Africa. The obstacles to their adoption include limitations in infrastructure, insufficient technical expertise, and ethical considerations (Shafik, 2024). Fewer than half Africans have reliable Internet access, while energy insecurity remains widespread. Data costs are among the highest globally. If these barriers are not tackled, AI innovations will benefit only a privileged few, exacerbating inequalities rather than alleviating them (African Union, 2024). African educational institutions are increasingly offering AI and digital literacy programmes. Initiatives like GrowZA's cybersecurity academy for youth reflect broader efforts to equip young Africans with the technical skills needed for the digital economy. However, these efforts must be significantly scaled to meet demand (GrowZA, 2024).

2.5 *African Innovation in Action*

To prevent a recurrence of the missed opportunities from previous industrial revolutions that have left a harmful legacy for African nations, governments must create an environment conducive to the growth of these AI startups and promote the socio-economic development of Africa (Arakpogun et al., 2021). Across the continent, a wave of grassroots innovation is redefining the role of AI in Africa. Numerous local startups are developing AI solutions tailored to their communities' unique challenges and aspirations, avoiding wholesale adoption of Western AI models in favour of contextually relevant, homegrown technologies.

AI4D Africa (Kenya). This cross-regional initiative integrates responsible AI innovations into Africa's agriculture and food systems, ensuring that technological benefits reach smallholder farmers, women, youth, and traditionally marginalised groups. This project exemplifies Africa's growing commitment to leveraging AI for inclusive development. It aims to generate robust evidence on how AI tools can effectively address critical challenges within the agricultural ecosystem, enhancing productivity, resilience, and food security at scale. By emphasising public-good outcomes over purely commercial interests, the project seeks to reshape the AI innovation landscape through targeted capacity building, knowledge sharing, and network development (Artificial Intelligence for Development, 2025).

Ubenwa (Nigeria). Ubenwa is a startup medical technology company that uses AI to analyse babies' cries and detect illnesses and diseases. Founded in 2017, Ubenwa employs machine learning to assess infant cries and identify birth asphyxia at an early stage. This low-cost, scalable solution exemplifies how AI can address urgent health challenges in resource-limited settings (Masso et al., 2022). Through algorithms, the technology harnesses the human voice via any microphone-enabled device to deliver critical insights, ranging from enhancing infant care to optimising patient interactions across healthcare (https://www.ubenwa.ai/).

Data Science for Social Impact (South Africa). This involves developing local language natural language processing tools designed to expand the reach and relevance of AI technologies across the continent. Researchers are addressing the under-representation of these languages by creating new datasets, methodologies, and models tailored to African languages. This innovation advances data science and provides decision-makers with culturally and linguistically appropriate insights. Initiatives like these lead this movement, enhancing how African policymakers and communities engage with AI tools to effectively address local challenges (Data Science for Social Impact, 2025).

Cleantech AI Startups at SAIS'24. Several startups at SAIS'24 showcased how AI can optimise energy consumption, enhance agricultural yields, and mitigate the impacts of climate change in African contexts (South African Innovation Summit, 2024).

GrowZA's Cybersecurity Training Program. Their mission is to ensure that geographical constraints do not hinder learning, but expand to wherever the beneficiaries are located. The academy is a response to the fact that in recent years, the increase in mobile Internet access has transformed the way the world consumes information. GrowZA equips unemployed youth with cybersecurity skills through AI-assisted training programmes. This initiative simultaneously tackles unemployment and Africa's growing cybersecurity gap (GrowZA, 2024).

Through these contextually relevant, homegrown efforts, Africa is not only adopting AI but actively shaping its evolution to ensure that AI technologies serve equitable, inclusive, and locally significant purposes aligned with the continent's developmental priorities.

While the literature offers valuable insights into Africa's AI landscape, it lacks a cohesive theoretical framework to guide context-sensitive, ethical, and inclusive AI implementation. This study aims to fill this gap in the body of literature. The literature highlights both the promise and complexity of AI in Africa. It demonstrates that, although local innovation ecosystems, strategic policies, and grassroots solutions are emerging across the continent, their effectiveness remains uneven and often fragmented due to deep infrastructural divides, socio-economic disparities, and a lack of cohesive regulatory frameworks. Critically, no unified theoretical model currently exists to guide AI implementation in a way that takes into account the continent's diverse conditions, cultural values, and development priorities. Although the literature reflects a growing awareness of context, inclusion, and ethical imperatives, it falls short of offering a structured, replicable framework for determining where, when, and how AI can be most beneficial and sustainable. This study addresses that gap by proposing a new theoretical contribution designed to systematise context-sensitive decision-making and support the ethical integration of AI in Africa's evolving technological landscape.

2.6 The Need for a Theoretical Framework

An important study in the context of this paper examines the AI readiness of Mauritius, Egypt, Kenya, and Angola regarding their existing assessment frameworks and their applicability to African contexts. The analysis reveals that current global AI readiness indices often fail to accurately reflect African states' unique challenges, capacities, and developmental trajectories. The frameworks particularly rely on quantitative metrics that overlook critical socio-political and infrastructural nuances. The study proposes targeted revisions to these assessment methodologies, advocating for the inclusion of more context-sensitive, qualitative evaluations of national technology strategies, governance structures, and innovation ecosystems. These enhancements aim to create a more equitable and accurate foundation for understanding and advancing AI readiness across diverse global contexts (Diallo et al., 2025).

A consistent issue in AI ethics research is the predominantly Western perspective of current AI ethics initiatives. Criticisms suggesting that existing AI ethics and governance discussions overlook 'African narratives' are becoming more frequent in the literature (Yilma, 2025).

An analysis of the African Union's Continental Artificial Intelligence Strategy reveals an explicit call for AI initiatives that align closely with Africa's unique developmental goals, ethical standards, and socio-cultural realities, rather than adopting global technological approaches uncritically. Complementing this, insights from Europe's AI Action Plan, as articulated by Ursula von der Leyen, underscore the necessity of human-centric, transparent, and accountable AI models, thereby reinforcing the argument for AI frameworks tailored to specific regional values and governance standards (Von der Leyen, 2025).

Further empirical evidence from the SA Innovation Summit 2024 findings highlights significant infrastructure and human capital constraints while showcasing the vibrant potential for youth-driven technological innovation across Africa (South African Innovation Summit, 2024). Likewise, the African Institute for Entrepreneurship's 2024 study on Generative AI emphasises the crucial role that local infrastructure, cultural appropriateness, and socio-economic contexts play in determining whether AI is genuinely beneficial or merely a technology adopted for its own sake (African Institute for Entrepreneurship, 2024).

Collectively, these insights reveal the urgent need for a new theoretical framework specifically designed to critically assess and establish the conditions under which AI technologies can be ethically and beneficially deployed in various African developmental contexts. Such a framework would provide structured guidance on identifying precisely where, when, and how AI implementation aligns meaningfully with local needs, capacities, and aspirations.

3 Methodology

Despite the growing global momentum surrounding AI, existing frameworks for its readiness and implementation are primarily shaped by Global North perspectives and fail to adequately reflect the socio-economic, infrastructural, cultural, and ethical realities of African contexts. As a result, there is a significant gap in the body of literature that can guide AI deployment in ways that are contextually relevant, developmentally aligned, and socially inclusive within Africa.

3.1 A Contextual Necessity Approach

This study utilises a conceptual methodology based on 'contextual necessity'. Instead of assuming AI's global utility, this approach examines where, when, and how AI implementation genuinely serves Africa's developmental needs. Sources include policy frameworks, summit findings, ethnographic reports on youth aspirations, and case studies of African innovation.

This paper frames AI not as a universal solution but as a tool whose value depends on its alignment with African socio-economic, infrastructural, and cultural realities. The literature reveals a need for a new theoretical framework designed to critically evaluate and determine the conditions under which AI technologies can be ethically and beneficially deployed in various African developmental contexts.

3.2 Document Analysis as a Qualitative Research Method

This study employs document analysis as a key qualitative research method to examine how policy, institutional, and innovation discourses inform AI implementation across Africa. Document analysis allows for a systematic evaluation of a diverse range of publicly available materials, including the African Union's Continental AI Strategy, the European AI Action Plan, official speeches (e.g. Ursula von der Leyen's address at the AI Action Summit), national policy statements, summit reports such as SAIS'24, and research outputs like the Generative AI 2024 study.

The CNF was methodically derived through a qualitative document analysis comprising several stages designed to ensure analytical depth and replicability. A structured document corpus was constructed using targeted keyword searches (e.g. 'AI implementation in Africa', 'AI implementation in the Global South'). Searches were conducted on online sources, including African Union policy documents, international development reports, journals, and innovation ecosystem reports. The selection process was modelled on criteria such as relevance, region (Africa), recency (past 5 years), and credibility (institutional or peer-reviewed sources).

Thematic analysis was carried out using the following approach based on Braun and Clarke (2020):

1. Familiarisation with data: A thorough reading of all selected documents indicated recurring references to infrastructure limitations, youth skills gaps, ethical AI, and the need for locally adapted solutions.
2. Generating initial codes: These included 'Infrastructure inequality', 'High data costs', 'Lack of digital skills among rural youth', and 'AI not appropriate in low-connectivity regions'.
3. Generating themes (e.g. themes such as 'Cultural Mismatch' and 'Contextual Analysis').
4. Reviewing potential themes (what does the theme reveal to me about the data set and my research question?).
5. Defining and naming themes in consultation with the research aim (e.g. 'What specific needs and local realities does this AI address?' leading to 'Contextual Analysis').
6. Integrating themes into a report (in this case, the CNF is then constructed as a conceptual output from this thorough thematic analysis).

These documents provide extensive contextual insights into how AI is framed, governed, and applied across various regions and sectors. By analysing these texts through an interpretivist lens, the study identifies recurring themes, policy gaps, and value-laden assumptions that influence AI development narratives. The findings from the document analysis directly inform the development of the CNF, offering both empirical and conceptual grounding for its structure and relevance. This approach enhances the study's theoretical rigour while ensuring that its contributions are rooted in real-world evidence and policy discourse.

3.3 *The Contextual Necessity Framework for AI Implementation*

The CNF was systematically derived from a critical synthesis and analysis of the existing literature on AI policy frameworks, socio-technical systems, and ethical governance within the context of African development. While practitioners instinctively apply contextual logic, there is no coherent guiding theory, and the CNF offers a way to formalise and scale these efforts.

Firstly, the analysis of the African Union's Continental Artificial Intelligence Strategy highlighted the need for context-driven AI implementations explicitly aligned with Africa's Agenda 2063 goals, emphasising ethical considerations, capacity building, and inclusive development (African Union, 2024). Secondly, insights from Europe's AI Action Plan, articulated by Ursula von der Leyen, underscore the necessity for human-centric, transparent, and democratically accountable AI models, which contrast sharply with technology-driven determinism (Von der Leyen, 2025).

Thirdly, the examination of the SA Innovation Summit 2024 findings revealed practical evidence of infrastructural gaps and youth-driven innovation potential. This indicates that successful AI deployments must explicitly integrate local innovation ecosystems and developmental realities (South African Innovation Summit, 2024). Finally, the African Institute for Entrepreneurship's 2024 study on Generative AI provided empirical validation of the significance of socio-cultural and infrastructural contexts in determining AI's actual utility in African entrepreneurial settings (African Institute for Entrepreneurship, 2024).

By triangulating insights from policy frameworks, theoretical considerations in global AI governance discourses, and empirical findings from African contexts, this study identifies a significant theoretical gap in the existing literature: A lack of frameworks explicitly designed to guide context-sensitive, ethically grounded, and developmentally aligned AI implementation in Africa. The CNF emerges from this analytical process, directly addressing the identified gap by providing a structured, flexible, and locally informed approach to determining where, when, and how AI should be implemented. Thus, the CNF represents both a conceptual synthesis and a practical theoretical innovation derived from a careful and critical analysis of contemporary literature and African empirical realities.

3.4 Core Premises of the CNF

The CNF rests upon three foundational premises:

Contextual relevance

AI must be explicitly aligned with the specific social, economic, infrastructural, and cultural contexts in which it is introduced. The mere availability or popularity of a technology elsewhere does not provide sufficient justification for its local deployment.

Socio-technical alignment

Successful AI implementation requires a balance between technical capability and social infrastructure, emphasising the equal significance of human readiness, cultural acceptance, ethical alignment, and technical feasibility.

Developmental priority

AI interventions should concentrate on providing tangible socio-economic benefits, including decent work, youth empowerment, infrastructure enhancement, ethical governance, and community wellbeing. Technology by itself does not ensure meaningful development.

3.5 *Components of the Contextual Necessity Framework*

Each component of the CNF emerged from a triangulated analysis of key documents and regional observations:

- Contextual Analysis (CA) was informed by repeated references to the need for aligning AI with local socio-economic conditions (e.g. AU Strategy, GrowZA).
- Necessity Determination (ND) was inspired by the SAIS'24 call for technology that meets actual community needs, not just global trends.
- The Capability Assessment (CapA) was based on the AIfE and broadband index insights, which repeatedly identified infrastructure and skills gaps.
- Outcome Alignment (OA) aligns with the ethical and developmental goals described in Agenda 2063 and the African Institute for Entrepreneurship report.

Figure 1 illustrates that the CNF integrates four interconnected analytical dimensions.

Contextual Analysis (CA)

CA aims to understand the precise social, cultural, political, and economic contexts before deploying AI. This can be achieved through stakeholder consultations, ethnographic research, and local socio-economic assessments. In doing so, the key questions to address include 'what specific societal needs could AI address?' and 'how might local values influence AI acceptance and effectiveness?'

Necessity Determination (ND)

ND aims to critically assess the actual need for AI in relation to available alternatives. This can be achieved through comparative assessments, feasibility studies, and cost-benefit analyses. Key questions to address include 'Does the community explicitly need AI?' and 'Are there simpler, more sustainable, or contextually appropriate alternatives?'

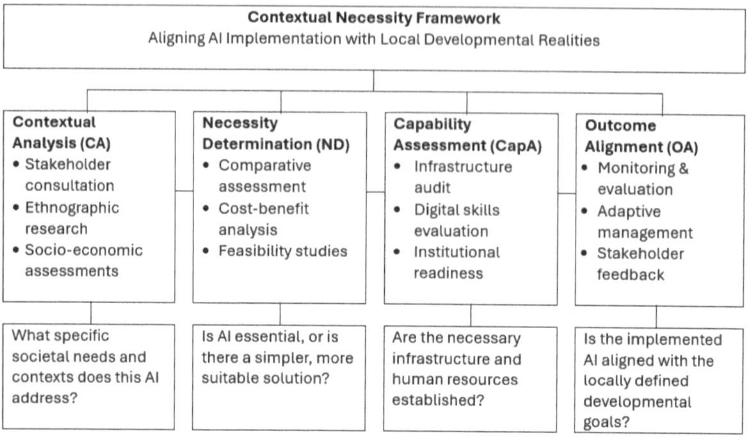

Fig. 1 Contextual necessity framework. Aligning AI implementation with local development realities

Capability Assessment (CapA)

CapA aims to evaluate local preparedness for the effective implementation, management, and sustainability of AI systems. This can be achieved through infrastructure audits, digital skills assessments, and evaluations of institutional readiness. Key questions to address include 'Is the existing infrastructure sufficient to support sustainable AI use?' and 'Are the necessary human skills, ethical frameworks, and governance structures in place?'

Outcome Alignment (OA)

OA aims to ensure that AI implementation explicitly aligns with locally defined developmental goals. These goals can be achieved through continuous monitoring and evaluation, stakeholder feedback mechanisms, and adaptive management approaches. In doing so, the key questions to be addressed include 'Does the implemented AI contribute positively to local economic growth, job creation, and social wellbeing?' and 'Are unintended negative impacts actively monitored and mitigated?'

The CNF is dynamic and non-linear. While it is presented sequentially, the framework, in practice, permits looping back and forth between components. For instance, during outcome alignment, stakeholders may suggest revisiting capability assessment or even reopening contextual analysis if local conditions change (e.g. electricity shortages or digital policy shifts).

4 Discussion

4.1 *An African Tale*

To understand the scale and complexity of Africa's digital divide, it is essential to move beyond general observations and examine specific dimensions of inequality. Despite growing efforts to expand connectivity and digital inclusion, significant disparities persist between high- and low-income countries, urban and rural areas, and men and women. These divides are not only infrastructural but also deeply rooted in socio-economic factors such as digital literacy, affordability, and gender norms. Table 1 summarises key indicators of Africa's digital divide as projected for 2025, highlighting the urgent need for targeted interventions that are sensitive to both infrastructural and human development needs.

Africa remains a significant outlier among continents in terms of AI readiness. Although the continent boasts one of the fastest-growing digital infrastructures globally, its broadband network is still in the early stages of development. Closing the gap with the rest of the world will present considerable challenges due to several factors: Internet penetration, electricity access, broadband affordability, data privacy framework, and startup investment. Table 2 shows that, these conditions collectively contribute to the ongoing state of AI readiness across the continent.

Table 1 Africa's digital divide in 2025

Dimension	Current status in Africa
Internet penetration (high vs low-income countries)	Only 38% of Africa's population uses the Internet, significantly below the global average of 68%
Mobile broadband coverage (3G, 4G, and 5G)	85% have access to at least 3G; 60% are covered by 4G; only 11% are covered by 5G, compared to the global 5G average of 51%
Gender inequalities	African women are less connected than men (31 vs 43%). Girls are even less connected. Empowering girls with technology is essential for economic success
Rural connectivity (broadband)	25% of rural Africans still have no possibility of connecting to the Internet due to gaps in broadband coverage
Digital skills (rural vs urban)	57% of rural Africans and 23% of urban Africans do not use the Internet due to a lack of digital skills and unfamiliarity with technology
Youth population	70% of sub-Saharan Africans are under 30, representing tremendous growth potential if both genders gain equitable digital empowerment

Table 2 Africa vs Europe AI readiness (2025)

Dimension	Africa	Europe
Internet penetration	40%	87% (more than double)
Electricity access	50%	99% (nearly double)
Broadband affordability	Low	High
Data privacy framework	Emerging	Comprehensive
Startup investment	USD 44 million	USD 26 billion (more than 600 times)

4.2 *Ethical and Regulatory Challenges in African AI Governance*

While Europe continues to refine advanced regulatory frameworks like the GDPR and the AI Act, African nations must similarly pursue robust governance models tailored to their unique social, cultural, and economic contexts. Currently, Africa faces critical ethical and regulatory challenges.

AI systems often reflect biases present in training data that inadequately represent African populations. Such biases can reinforce existing inequalities rather than alleviate them. An over-reliance on foreign AI platforms and technologies risks deepening digital colonialism, leaving Africa dependent on external entities for technological innovation, data management, and digital infrastructure. Most AI technologies are trained on datasets that predominantly feature Western languages, values, and knowledge systems, thereby marginalising indigenous African languages, traditions, and cultural nuances.

To effectively address these challenges, policymakers need a suitable guiding framework that emphasises contextual alignment, ethical standards, and inclusive socio-economic priorities.

4.3 *The Role of the Theoretical Framework Developed in This Study*

To address these ethical and regulatory challenges, this study introduces the CNF, which explicitly engages with critical questions about where, when, and how AI implementation genuinely serves African developmental and ethical imperatives. At its core, the CNF seeks to align AI implementation with local developmental realities.

The framework encourages policymakers first to conduct a contextual analysis, thoroughly evaluating the socio-cultural, infrastructural, and economic conditions before adopting AI. Secondly, it promotes the assessment of necessity by determining whether AI solutions are truly required or if simpler, more contextually appropriate alternatives might suffice. Thirdly, the CNF emphasises thorough capability assessments, ensuring that infrastructure, human skills, and governance structures are adequately developed to support AI applications. Finally, it emphasises continual outcome alignment by systematically assessing whether AI deployments consistently uphold ethical standards, promote decent work, empower youth, and support inclusive development.

By adopting the CNF, African policymakers and stakeholders can effectively navigate AI governance while ensuring ethical integrity, socio-cultural relevance, and sustainable developmental impact.

4.4 Policy Recommendations

Certain actionable policy recommendations for African policymakers arise from the insights provided by the CNF.

These include prioritising the expansion of broadband Internet, ensuring a reliable energy supply, and developing local data centres to establish the foundational infrastructure necessary for sustainable AI deployment. Scale open-access educational programmes through universities, technical vocational education and training (TVET) institutions, and online platforms that are specifically tailored to local contexts and development priorities. Establish ethical guidelines and governance structures that are deeply rooted in African languages, cultures, traditions, and socio-economic realities while avoiding one-size-fits-all global approaches. Mobilise funding and resources through targeted grants, accelerators, and venture capital support, empowering African youth to develop innovative and locally relevant AI applications. Encourage the harmonisation of regulatory standards among African states through cross-border collaboration, ensuring cohesive, ethical, and contextually aligned AI governance across the continent.

By operationalising these policy recommendations within the structured approach provided by the CNF, African nations can cultivate a uniquely advantageous AI landscape that genuinely supports the continent's ambitions for inclusive growth, ethical innovation, and sustainable development.

5 Future Research

Future research can build upon the CNF developed in this paper to further validate, refine, and extend the framework's applicability. These include conducting comparative studies using the CNF to evaluate the outcomes of AI policy implementation across various African contexts and assess its effectiveness in guiding ethically sound and contextually appropriate AI deployments. Research could also examine the impact of AI interventions led by the CNF in rural and indigenous communities, emphasising inclusivity, preserving local knowledge systems, and promoting socio-economic empowerment.

Investigating how African ethical and philosophical traditions, such as Ubuntu, can enhance the cultural relevance and acceptance of the CNF across diverse African societies. Using the CNF as a foundation for developing and evaluating community-driven data governance strategies. This approach empowers local communities to maintain sovereignty over their data, ensures ethical data practices, and prevents digital exploitation. These research directions aim to enhance the CNF as a vital tool for aligning AI implementations with African developmental aspirations, ethical considerations, and socio-cultural contexts.

6 Conclusion

Africa stands at a critical juncture in shaping its AI future. While the continent is often measured against frameworks and narratives developed in and for the Global North, this paper has demonstrated that such models, although useful in their own contexts, fail to adequately capture the infrastructural, ethical, and socio-cultural dynamics unique to Africa. Simply replicating European approaches to AI risks deepening existing divides and overlooking Africa's distinct opportunities. To address this, the paper presents the CNF, a new theoretical lens that offers a structured, context-sensitive approach to AI implementation. Rooted in the principles of ethical governance, youth empowerment, infrastructural readiness, and socio-economic inclusion, the CNF directly challenges universalist assumptions and provides a pathway for AI strategies grounded in African realities. By investing in local innovation ecosystems and applying frameworks like the CNF, African states can transcend dependency and digital mimicry, positioning themselves as leaders in creating equitable, purposeful AI systems. Ultimately, this is not just about adapting AI to Africa; it is about empowering Africa to shape the global future of AI on its own terms.

Acknowledgements This study was inspired by two collaborative projects: (1) Decent, fair and meaningful work in a global, mobile, and connected world in four different countries. A SASUF collaboration between CPUT and Malmö University. (2) The 14th International Week at the University of Economics in Katowice. An Erasmus+ collaboration between CPUT and the University of Economics.

Disclosure of Interests The author has no competing interests to declare that are relevant to the content of this article.

References

Ade-Ibijola, A., & Okonkwo, C. (2023). Artificial intelligence in Africa: Emerging challenges. In *Responsible AI in Africa: Challenges and opportunities* (pp. 101–117). Springer International Publishing. https://doi.org/10.1007/978-3-031-08215-3

African Institute for Entrepreneurship (AIfE). (2024). *Generative AI 2024: An exploration for non-tech entrepreneurs and youth.*

African Union. (2024). *Continental artificial intelligence strategy for Africa.*

African Union Commission. (2015). *Agenda 2063: The Africa we want.* African Union Commission.

Artificial Intelligence for Development. https://www.ai4d.ai/projects/artificial-intelligence-for-agriculture-and-food-systems-innovation-research-network. Accessed April 19, 2025.

Arakpogun, E. O., Elsahn, Z., Olan, F., & Elsahn, F. (2021). Artificial intelligence in Africa: challenges and opportunities. In A. Hamdan, A. E. Hassanien, A. Razzaque, & B. Alareeni (Eds.), *The fourth industrial revolution: Implementation of artificial intelligence for growing business success. Studies in computational intelligence* (Vol. 935, pp. 375–388). Springer. https://doi.org/10.1007/978-3-030-62796-6_22

Braun, V., & Clarke, V. (2020). One size fits all? What counts as quality practice in (reflexive) thematic analysis? *Qualitative Research in Psychology.* https://doi.org/10.1080/14780887.2020.1769238

Broadband Affordability Index. https://www.broadband.co.uk/broadband-world-affordability. Accessed April 15, 2025.

Data Science for Social Impact Homepage. https://dsfsi.github.io/. Accessed Apri 25, 2025.

Diallo, K., Smith, J., Okolo, C. T., Nyamwaya, D., Kgomo, J., & Ngamita, R. (2025). Case studies of AI policy development in Africa. *Data & Policy, 7*, Article e15. https://doi.org/10.1017/dap.2024.71

Eurostat Digital Economy Statistics. (2024). https://ec.europa.eu/eurostat/statistics-explained/index.php?title=Digital_economy_and_society_statistics_-_households_and_individuals. Accessed April 15, 2025.

GDPR Homepage. (2025). https://gdpr.eu/what-is-gdpr. Accessed April 15, 2025.

IMF Blog by Kristalina Georgieva. (2024). https://www.imf.org/en/Blogs/Articles/2024/01/14/ai-will-transform-the-global-economy-lets-make-sure-it-benefits-humanity. Accessed April 22, 2025.

GrowZA Homepage. (2024). https://www.growza.co.za/post/reimagining-work-growza-s-global-research-on-re-skilling-for-fair-and-decent-jobs. Accessed April 17, 2025.

Kolog, E. A., ODevine, S. N., Egala, S. B., Amponsah, R., Budu, J., & Farinloye, T. (2022). Rethinking the implementation of artificial intelligence for a sustainable education in Africa: Challenges and solutions. In *Management and information technology in the digital era: Challenges and perspectives* (pp. 27–46). Emerald Publishing Limited.

Masso, A., Chukwu, M., & Calzati, S. (2022). (Non)negotiable spaces of algorithmic governance: Perceptions on the Ubenwa health app as a 'relocated' solution. *New Media & Society, 24*(4), 845–865. https://doi.org/10.1177/14614448221079027

SABEN Bridging the Digital Divide. (2025). https://saben.ac.za/bridging-the-divide. Accessed April 15, 2025.

Shafik, W. (2024). Navigating emerging challenges in robotics and artificial intelligence in Africa. In L. Amoah (Ed.), *Examining the rapid advance of digital technology in Africa* (pp. 126–146). IGI Global Scientific Publishing. https://doi.org/10.4018/978-1-6684-9962-7.ch007

South African Innovation Summit. (2024). *SAIS'24 feedback report.*

United Nations. (2024). *Mind the AI Divide - Shaping a Global Perspective on the Future of Work.* https://doi.org/10.18356/9789211066524

UN Africa Renewal. (2024). *Connectivity article.* https://www.un.org/africarenewal/magazine/december-2024/connectivity-everyone-key-africas-growth-and-prosperity. Accessed April 15, 2025.

Von der Leyen, U. (2025). Speech at the Artificial Intelligence Action Summit. In E. Commission (Ed.), *Artificial Intelligence Action Summit 2025, Paris* (pp. 1–5). European Commission.

World Bank. (2023). https://data.worldbank.org/indicator/EG.ELC.ACCS.ZS?locations=EU. Accessed April 15, 2025.

Yilma, K. (2025). Ethics of AI in Africa: interrogating the role of Ubuntu and AI governance initiatives. *Ethics and Information Technology, 27*, 24. https://doi.org/10.1007/s10676-025-09834-5

Value Stream Analysis and Design with Large Multimodal Models: Evaluating the Impact of Complex Value Stream Graph Networks

Mick Geisthardt, Lutz Engel, and Jorge Marx Gómez

Abstract Value stream analysis and design are core components of lean management, used to identify inefficiencies and enable resource-efficient value creation. With increasing complexity and the rise of artificial intelligence (AI), large multimodal models (LMMs) have gained relevance as decision-support tools in digital lean management. However, the accuracy and reliability of LMMs currently depend on the quality and structure of the input they receive. Traditional image-based representations of value streams often lack the semantic clarity required for effective computational analysis, limiting LMM accuracy. This paper investigates the impact of complex value stream graph networks (CVSGNs)—a structured, graph-based data model—on the accuracy of LMMs in value stream analysis and design. The paper follows a two-stage methodology utilizing the socio-technical simulation game "North Star—The Value Stream Game." In the first stage, a pre-test assesses the descriptive accuracy of LMMs, comparing image-based and CVSGN-based inputs, revealing an error rate reduction by over 70% points. In the second stage, 300 simulation game rounds are used to investigate the LMM-based evaluation accuracy of value stream design strategies for both input conditions. Results demonstrate a substantial improvement in expert alignment, with LMM-based evaluation accuracy increasing from 62.1 to 88.1%. A *t*-test confirmed the statistical significance of these findings ($p < 0.001$) with a large effect size (Cohen's $d = 2.01$). These results provide evidence that structured graph-based representations can significantly enhance both descriptive and evaluative LMM accuracy, supporting more reliable decision-making in the context of digital lean management.

M. Geisthardt (✉) · J. Marx Gómez
Department of Computing Science, Business Information Systems/Very Large Business Applications, Carl Von Ossietzky University of Oldenburg, Oldenburg, Germany
e-mail: mick.geisthardt@jade-hs.de

J. Marx Gómez
e-mail: jorge.marx.gomez@uni-oldenburg.de

M. Geisthardt · L. Engel
Department Management, Information, Technology, Institute for Production and Service Systems, Jade University of Applied Sciences, Wilhelmshaven, Germany
e-mail: lutz.engel@jade-hs.de

J. Marx Gómez et al. (eds.), *Recent Advances in Information Systems*, Progress in IS,
https://doi.org/10.1007/978-3-032-14099-9_53

1 Introduction

In the production of industrial goods—ranging from complex machinery to consumer-oriented products—improvement teams consistently face the challenge of optimizing value creation while ensuring resource efficiency. For decades, lean management methods such as value stream analysis and design have been utilized to identify and eliminate inefficiencies within internal production and logistics processes (Erlach, 2010; Rother & Shook, 2011). Originally, these methods focused on traditional metrics such as lead times, cycle times, and inventory reduction. Recent methodological advancements have expanded this scope, incorporating new aspects like material flow cost accounting (Engel & Kranhold, 2021), information logistics (Meudt, 2020; Meudt et al., 2017a, 2017b), and external influencing factors such as market dynamics, geopolitical situations, or policies (Zanker & Reisen, 2015). These expansions have unlocked new opportunities but also have introduced new complexity, transforming value stream analysis and design into a highly complex and data-intensive task.

As value streams become increasingly complex, traditional visual representations—such as paper-based sketches or digital whiteboards—have proven insufficient. These tools lack the structural rigor and semantic clarity required to enable analysis and design, particularly when applied to large-scale production systems from the real world. Consequently, improvement teams require new digital solutions that not only manage complexity but also enable more intelligent, adaptive, and scalable approaches within digital lean management projects (Geisthardt & Engel, 2024; Geisthardt et al., 2024a, 2024b; Liker, 2020). In this context, large multimodal models (LMMs)—artificial intelligence (AI) models capable of processing and integrating diverse inputs such as text, numerical data, and visual elements—offer promising support for decision-making (Baltrušaitis et al., 2018). However, the performance of such models is currently highly dependent on the quality and structure of the input data they receive. Ambiguous or inconsistent input leads to poor outputs and reduced reliability.

To address this challenge, this paper investigates the impact of complex value stream graph networks (CVSGNs)—a structured, graph-based data model specifically designed to formalize value streams for machine interpretation. CVSGNs represent process nodes, flows, dependencies, and logic in a directed graph format optimized for reasoning by LMMs (Geisthardt et al., 2025).

The evaluation follows a two-stage methodology using "North Star—The Value Stream Game," a socio-technical simulation game designed to replicate real-world value stream analysis and design (Geisthardt et al., 2023a, 2023b, 2024c).

In the first stage, a pre-test assesses how accurately an LMM could describe a given value stream when presented with either a traditional image or a CVSGN-based input. The results demonstrate a significant difference in descriptive accuracy, with the error rate decreasing from 75.76% (image) to 4.35% (CVSGN).

In the second stage, a simulation-based study with 300 game rounds is conducted to evaluate how well the LMM can evaluate value stream design strategies under the same input conditions. The results revealed a substantial improvement in evaluation alignment with human experts—from 62.1 to 88.1% accuracy (mean match rate), and a corresponding drop in evaluation deviation. A *t*-test for independent samples (Kim, 2015) confirmed the statistical significance of these differences ($p < 0.001$) and indicated a very large effect size (Cohen's $d = 2.01$). These findings provide evidence that the use of CVSGNs can enhance the descriptive and evaluative accuracy of LMMs in the context of value stream analysis and design.

This paper is structured as follows: Sect. 2 presents the theoretical background, including value stream analysis and design, LMMs, and CVSGNs. Section 3 outlines the pre-test and the simulation-based methodology. Section 4 presents and discusses the experimental results of the two investigated conditions (image vs. CVSGN). Finally, Sect. 5 concludes with a short summary of key insights and required future work.

2 Background

Section 2 provides the theoretical background necessary for understanding the paper's core contributions. It introduces key concepts, including value stream analysis and design, LMMs, and CVSGNs, laying the basis for the results, discussions, and conclusions detailed in Sects. 4 and 5.

2.1 Value Stream Analysis and Design

An end-to-end value stream can be defined as the complete sequence of value-adding and non-value-adding activities required to deliver a product or product family from initial sourcing through internal production processes to the final customer (Rother & Shook, 2011). In the production of photovoltaic modules, used in the following as simplified and fictional example, the value stream begins with sourcing raw material, continues through wafer production, cell manufacturing, and concludes with module assembly before reaching the customer. Structurally, a value stream can be divided into three functional sections: the upper section includes suppliers, customers, and information flows from production planning, control, and communication; the middle section covers processes, material flows, and inventories; and the lower section evaluates the temporal dimensions of the stream, including lead time and value-adding processing time (see Fig. 1).

Across these sections, both value-adding activities (e.g., precise cell coating) and non-value-adding activities (e.g., waiting times between processes) take place. Identifying and systematically reducing non-value-adding activities is a core objective of value stream analysis and design, as it enhances overall value creation and resource efficiency. Traditional value stream analysis typically begins by mapping the current situation using standardized visual symbols on paper or whiteboards

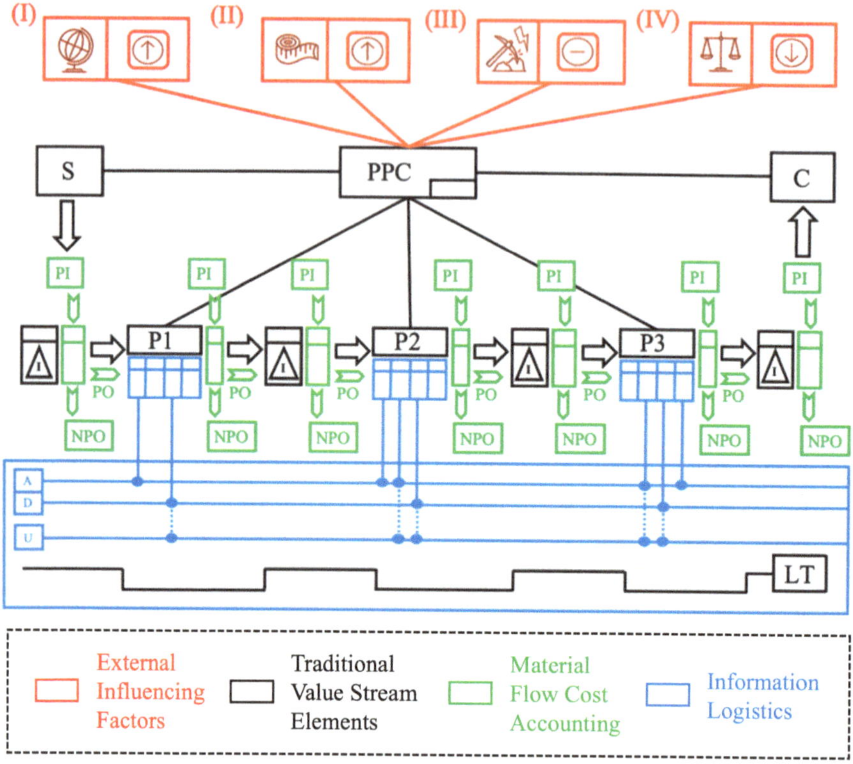

Fig. 1 Example of a value stream in the current situation, enhanced through material flow cost accounting, information logistics, and external influencing factors

(Rother & Shook, 2011). Improvement teams subsequently analyze inefficiencies using core lean principles, such as Muda (waste), Mura (unevenness), and Muri (overburden) (Ohno, 2013).

- Muda refers to resource-consuming activities that do not generate customer-perceived value, such as overproduction or excess inventories.
- Mura captures inconsistencies and fluctuations, for example, production rate variability between processes that may lead to bottlenecks.
- Muri reflects overburden on people or equipment, such as excessive workload that could result in fatigue or quality issues.

Based on this analysis, improvement teams apply established value stream design guidelines to define an optimized target value stream. This target value stream design aims to increase resource efficiency, reduce lead times, and minimize waste. The conceptual foundation for value stream design is derived from the Toyota production system (TPS), which promotes an "ideal state" or "north star" characterized by zero defects, 100% value creation, one-piece flow, demand alignment, and safe working conditions (Rother, 2009).

While early applications of value stream analysis and design focused primarily on lead time and physical flows, modern practice has evolved to address a broader range of variables. Additionally to the traditional aspects—suppliers (S), processes (P1-3), inventories (I), customers (C), production planning and control (PPC), and lead time (LT)—advancements incorporate material flow cost accounting (Engel & Kranhold, 2021), information logistics (Meudt, 2020; Meudt et al., 2017a, 2017b), and external influencing factors (Zanker & Reisen, 2015) to better reflect the complexity of real world. Figure 1 illustrates a current value stream that incorporates these methodological expansions.

Material flow cost accounting, based on DIN EN ISO 14051:2011 (Deutsches Institut für Normung e.V, 2011), expands the analytical focus by assessing the inputs and outputs of each process step. This includes classifying product input (PI), value-adding product output (PO), and non-value-adding product output (NPO). For example, during wafer production, silicon ingots serve as PI, intact wafers represent PO, and damaged wafers are categorized as NPO. These classifications enable evaluations of material and energy usage as well as cost-efficiency. Information logistics extends the methodology by analyzing how information—whether analog (A) or digital (D)—is stored and utilized (U) within each process step. In practice, digital tracking systems may control quality and inventory data, while assembly instructions may still rely on printed documents, indicating opportunities for digitization. Relevant metrics include data availability, digitization rate, and utilization efficiency (Meudt et al., 2017b). Furthermore, modern value stream analysis and design recognizes the importance of external influencing factors. These include geopolitical conditions (I), procurement risks (II), energy supply stability (III), and regulatory frameworks (IV). For example, geopolitical instability affecting raw material suppliers may disrupt, while regulatory shifts in energy policy may impact product demand of the value stream. Including such external influencing factors allows for more comprehensive risk assessment and strategic planning (Zanker & Reisen, 2015).

2.2 Large Multimodal Models

Advancements in AI have led to the emergence of highly capable AI models that support intelligent, adaptive, and context-aware interactions. A particularly relevant class of AI models in this context are large language models (LLMs). These models are designed to process, interpret, and generate human language and have laid the foundation for a wide range of recent developments in natural language processing (NLP). LLMs are trained on large-scale textual data and are capable of learning and generalizing from complex linguistic structures and semantic relationships. Their core architecture, the transformer, enables contextual processing of long text sequences. This has made LLMs effective for tasks such as machine translation, question answering, and summarization. Prominent examples include OpenAI's GPT-3, with 175 billion parameters, which has significantly influenced the field of NLP in both research and applied settings (Brown, 2020; Vaswani, 2017).

In industrial practice, LLMs are utilized to automate and enhance text-based workflows. Typical applications include AI-supported chatbots and virtual assistants in customer service, as well as automated document processing and information extraction from unstructured data (Koehn, 2020; Kurafsky & Martin, 2021). Their widespread adoption has been accelerated by improved computing infrastructure and the growing availability of large and diverse datasets (Bommasani et al., 2021).

Building on the capabilities of LLMs, large multimodal models (LMMs) represent a next-generation approach that extends the scope beyond language alone. LMMs are designed to simultaneously process and integrate different types of input—such as text, images, numerical data, and other structured or semi-structured formats. This multimodal integration allows for a more holistic understanding of systems, especially in scenarios where language and visual or contextual information interact closely (Baltrušaitis et al., 2018).

2.3 Complex Value Stream Graph Networks (CVSGNs)

Transferring value streams from the physical world—traditionally captured via paper or whiteboards—into the digital world requires a data model that is both interpretable for humans and processable by AI. To fulfill this dual requirement, the concept of CVSGNs was developed. CVSGNs serve as a structured graph-based data model specifically designed to represent value streams in a form suitable for computational reasoning and machine-based evaluation (Geisthardt et al., 2025).

The underlying foundation of CVSGNs is graph theory, a mathematical framework used to model relationships between objects through nodes and edges. Graphs have long been applied in domains such as computer science, transportation, biology, and social network analysis (Knauer & Knauer, 2019). In the context of value stream analysis and design, the graph theory concept allows for the abstraction and structuring of complex value streams. Different graph types are relevant depending on the use case. For example, directed graphs (digraphs) are used to represent flows with a defined direction—such as material moving from one process to the next—while weighted graphs allow for the encoding of quantities such as cost, time, or energy consumption matrix (Van Steen, 2010).

CVSGN builds on the concept of complex graph networks, which are graph structures characterized by non-trivial topological properties, including high connectivity, nested substructures, and multidimensional relationships. Such properties are often found in natural and socio-technical systems, where the behavior of the entire network cannot be deduced from its individual components edges (Van Steen, 2010).

As illustrated in Fig. 2, within a CVSGN, the nodes represent the core structural components of the value stream (outlined in Sect. 2.1). These include traditional elements such as suppliers (S), processes (P), inventories (I), customers (C), and production planning and control (PPC). Additional node types represent elements from information logistics—such as analog (A) and digital (D) memory media, their utilization (U)—as well as external influencing factors (EIF). Each node contains a feature vector, which encodes characteristics via key performance indicators. The edges of the graph represent the flow of material, energy, information, and cost between these structural components. These edges are directed, forming a digraph that captures the sequential logic and interdependencies. Like the nodes, edges are also realized as feature vectors to represent the properties of flows.

CVSGNs offer a semantically scalable and machine-readable representation of value streams. It bridges the gap between traditional value stream visualization and computation by encoding real-world complexity in a structured format.

Process-Node Feature Vector

Features	Expressions
Number of Employees (No.)	3
Cycle Time (h)	1.5
Changeover Time (h)	0.5
Uptime (%)	90
Scrap Rate (%)	7
OEE (%)	94
…	…
Product Input (PI) (EUR)	637.5
Non Product Output (NPO) (EUR)	20.3
Degree of Value Creation (%)	80
…	…

Vector→ [3, 1.5, 0.5, 90, 7, 94, …, 637.5, 20.3, …]

(a) Node Feature Vector

Process-Analog-Edge Feature Vector

Features	Expressions
Cycle Time: Analog Documentation (True: 1/ False: 0)	1
Cycle Time: Direction of Informationflow (Up: 1/ Down: 0)	0
Cycle Time: Frequency (No. per day)	3
…	…

Vector→ [1, 0, 3, …]

(b) Edge Feature Vector

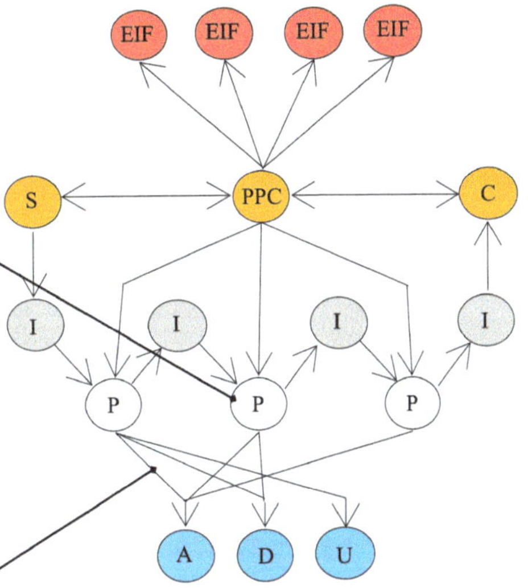

(c) Complex Value Stream Graph Network (CVSGN)

Fig. 2 Example of an value stream as a complex graph network composed of node and edge feature vectors: **a** example of a node feature vector representing a process; **b** example of an edge feature vector representing the directed linkage between process and analog memory media; **c** structure of the complex value stream graph network (CVSGN) (Geisthardt et al., 2025)

3 Methodology

The objective of this paper is to evaluate the impact of CVSGNs on the accuracy of LMMs in the context of value stream analysis and design. Emphasis is placed on the comparative performance of image-based inputs versus CVSGN-based inputs in the context of digital lean management. To achieve this, a structured, simulation-based study is conducted using "North Star—The Value Stream Game," a socio-technical simulation game developed to replicate realistic decision-making scenarios under controlled, repeatable conditions (Geisthardt et al., 2023a, 2024b, 2024c).

The paper investigates two experimental conditions: (1) LMMs processing value streams represented through traditional images and (2) LMMs processing value streams represented using structured CVSGNs. In the first stage, a pre-test is performed to evaluate the descriptive accuracy of LMM-based value stream analysis focusing on each input format. This involves 10 independent runs per condition. In the second stage, the main simulation-based study evaluates how accurately the LMM assesses value stream design strategies across 150 game rounds per condition. Each LMM-based evaluation is compared with lean management expert ratings using $a \pm 1$ tolerance window. The resulting match rates are statistically analyzed using a t-test for independent samples (Kim, 2015).

"North Star—The Value Stream Game" provides the simulation environment and serves as a structured testbed for evaluating the accuracy of LMM-based decision-making in digital lean management. Designed as a socio-technical simulation game, it enables players—typically members of improvement teams—to iteratively analyze, redesign, and improve a value stream across five game rounds. In each round, players apply predefined value stream design actions, which consume limited resource units (RUs) and are evaluated based on strategic impact and resource efficiency. The overall objective is to align the value stream with the ideal "north star" state, defined by the principles of zero defects, one-piece flow, and 100% value creation (Rother, 2009).

A feature of the simulation game is its integration of an LMM—"NorthStarGPT," based on the GPT-4o model from OpenAI. The LMM acts as an intelligent component: it explains rules, guides players through decisions, evaluates improvement actions, and provides real-time feedback and scenario adaptation. This creates a dynamic socio-technical system, where human decisions are continuously supported—and in some cases challenged—by AI-based recommendations.

Through this setup, the simulation-based study creates a realistic, analytically tractable environment to evaluate how different input structures can affect the reasoning accuracy of LMMs.

It is important to note that this paper presents the initial step in the evaluation and practical testing of the CVSGN, focusing specifically on the production of photovoltaic modules. This example serves as a representative case of mass production. The objective is to explore the potential of CVSGNs within this manufacturing context. Future research is intended to broaden the scope of the evaluation to include further production types, particularly batch production and continuous production.

4 Results and Implications

Section 4 describes the applied value stream scenario, presents the results from the pre-test and the simulation-based study with two experimental conditions, and discusses the implications for LMM-based decision-making in the context of digital lean management.

4.1 *Simulated Value Stream Scenario*

The simulation is based on a fictional but highly detailed value stream representing the production of photovoltaic modules (mass production). The scenario maps an end-to-end value stream involving multiple actors and system components. Raw materials are sourced from a supplier in Germany, with biweekly truck deliveries. The value stream progresses through several inventory and process stages: wafer production, cell manufacturing, module assembly, and shipping. Each process node is specified by parameters such as cycle time, setup time, availability, and defect rate. For example, wafer production operates with a cycle time of 2 s, 1.5 h of setup time, 89% availability, and a high 30% defect rate. In contrast, cell manufacturing involves two workers, a 7-s cycle time, and a 5% defect rate.

Inventory buffers regulate throughput, with clearly defined capacities and lead times. For example, the cell manufacturing buffer has a 15,000 kg capacity and 19.44 days of throughput. A production planning and control (PPC) system coordinates operations through analog and digital communication. The entire value stream is formalized in two ways: (1) as a traditional value stream image (see Sect. 2.1), and (2) as a graph-based CVSGN (see Sect. 2.3), incorporating all process and information nodes, as well as material flows and system interaction edges.

4.2 *Pre-test*

In a pre-test, the LMM's ability to describe the given value stream is evaluated under two input conditions. First, the model receives an image of the value stream and is instructed: "Analyze this value stream (image) and describe its data, structure, and core elements." In the second condition, the same task is given, but with the CVSGN as input. Each condition is tested across five independent runs. Using the raw data, the mean number of correct and incorrect statements produced by the LMM as well as the mean error rate is calculated (see Table 1).

Results demonstrate that the LMM generates more accurate and detailed outputs when working with structured CVSGNs. Image-based input led to a 75.76% mean error rate, with frequent misinterpretations of process sequences, node types, or throughput data. The CVSGN-based input results in 4.35% observed errors at this

Table 1 Results of LMM descriptive accuracy assessment using image-based and CVSGN-based input representations including means of correct statements, incorrect statements, and error rates ($n = 10$ per condition; $N = 20$ in total)

Condition	Correct statements (mean)	Incorrect statements (mean)	Error rate (mean) (%)
Image-based input	8	25	75.76
CVSGN-based input	44	2	4.35

setting, reflecting the model's improved interpretability when processing structured information.

4.3 Simulation-Based Study

In the main study, 300 rounds of "North Star—The Value Stream Game" are simulated using the LMM called "NorthStarGPT," based on the model GPT-4o from OpenAI. Two experimental conditions are compared:

- **Condition 1**: 150 rounds using the traditional image of the value stream.
- **Condition 2**: 150 rounds using the structured CVSGN representation.

In each condition, three simulated teams interact with the LMM "NorthStarGPT" across five game rounds. During each round, five value stream design actions are selected and applied by each team. The LMM evaluates each action on a scale from − 5 (very poor) to + 5 (excellent), using a chain-of-thought approach and an implemented internal knowledge base. In parallel, human experts from the Institute for Production and Service Systems of Jade University of Applied Sciences independently rated the same actions. LMM ratings are compared to expert ratings using *a* ± 1 tolerance window to compute the accuracy called match rate (see Table 2).

The results demonstrate a 26-percentage-point improvement when using CVSGNs. The mean match rate rises from 62.1 to 88.1%, and the mean deviation in evaluation decreases from 1.29 to 0.71. The worst CVSGN-based round (73%)

Table 2 Results of LMM-based value stream design strategy evaluations compared to expert evaluations including mean match rate within ± 1-point tolerance window, mean evaluation scores, and mean evaluation deviations ($n = 150$ per condition; $N = 300$ in total)

Condition	Match rate (mean) (%)	LMM evaluation (mean)	Expert evaluation (mean)	Deviation (mean)
Image-based input	62.10	3.39	2.60	1.29
CVSGN-based input	88.10	3.50	2.95	0.71

nearly outperforms the best image-based round (80%), highlighting the robustness of structured input.

4.4 Statistical Significance

To validate the statistical significance of identified mean match rate difference, a method from inferential statistics (Rasch et al., 2021) called t-test for independent samples is applied (Kim, 2015) using the program IBM SPSS Statistics. Thereby, the null hypothesis H_0 (no difference between mean match rates) is tested against the alternative hypothesis H_1 (significant difference between mean match rates). The test results in a *p*-value of $p < 0.001$, confirming that the observed improvement in evaluation accuracy under the CVSGN condition is highly significant and not due to random variation. Furthermore, a resulting Cohen's *d* of 2.01 indicates a very large effect size, meaning the structured representation has a substantial impact on the LMM's evaluation accuracy. These statistical results reinforce the observations and provide evidence for the hypothesis that CVSGNs enhance the accuracy of LMMs in value stream analysis and design.

4.5 Implications

The results of both the pre-test and simulation-based study highlight the decisive role of input structure in the current accuracy of LMMs for value stream analysis and design. While conventional image-based inputs are leading to frequent misinterpretations and inconsistencies in evaluation, the use of CVSGNs enables more precise, consistent, and expert-aligned assessments. The strong statistical significance ($p < 0.001$) and large effect size (Cohen's $d = 2.01$) underline that this improvement is not incidental but methodologically robust.

The observed accuracy gain can be largely attributable to the formalized, machine-readable structure of CVSGNs. Their directed edges, explicit node types, and feature vectors provide a semantic context, allowing LMMs to reason more effectively and generate more reliable outputs. This finding emphasizes that the quality of AI output is not only model-dependent but also currently tightly coupled with input structure.

From a practical standpoint, the implications for digital lean management are considerable. As organizations begin integrating LMMs into improvement teams and important decision-making processes, it becomes clear that accuracy is not only a function of model capability—but also, currently, of input quality. CVSGNs emerge as an enabling technology that transforms raw data into structured information, making complex value streams computationally accessible for AI.

5 Conclusion

This paper investigated the impact of CVSGNs on the accuracy of LMMs in the context of value stream analysis and design. Emphasis was placed on the comparative performance of image-based inputs versus CVSGN-based inputs in the context of digital lean management. Using a structured simulation-based study with 300 game rounds in "North Star—The Value Stream Game," the evaluation accuracy of LMMs was analyzed under two input conditions: traditional image-based value stream representations and structured, graph-based CVSGNs representations.

The results demonstrate that CVSGNs significantly improve both the descriptive accuracy and strategic evaluation capabilities of LMMs. In the pre-test, the mean error rate dropped from 75.76% (image input) to 4.35% (CVSGN input). In the main simulation-based study, the accuracy (mean match rate within $\pm$ 1 tolerance window) increased from 62.1 to 88.1%, with a corresponding reduction in deviation. Statistical testing confirmed the significance of this improvement ($p < 0.001$) and revealed a large effect size (Cohen's $d = 2.01$). These findings provide evidence that CVSGNs represent an enabling factor for LMM-based decision-making in digital lean management.

Beyond the quantitative results, this paper emphasizes a fundamental insight: advanced AI models like LMMs do not operate in isolation—their current performance is highly dependent on the structure, clarity, and semantics of the input they receive. CVSGNs offer a formalized, machine-readable way to represent value streams, bridging the gap between human understanding and computational reasoning.

Future research is intended to broaden the scope of the evaluation to include further production types, particularly batch production and continuous production.

References

Baltrušaitis, T., Ahuja, C., & Morency, L.-P. (2018). Multimodal machine learning: A survey and taxonomy. *IEEE Transactions on Pattern Analysis and Machine Intelligence, 41*, 423–443.

Bommasani, R., Hudson, D. A., Adeli, E., Altman, R., Arora, S., von Arx, S., Bernstein, M. S., Bohg, J., Bosselut, A., & Brunskill, E. (2021). *On the opportunities and risks of foundation models*. arXiv preprint arXiv:2108.07258

Brown, T. B. (2020). *Language models are few-shot learners*. arXiv preprint arXiv:2005.14165

Deutsches Institut für Normung e.V. (2011). DIN EN ISO 14051:2011-12, Umweltmanagement – Materialflusskostenrechnung – Allgemeine Rahmenbedingungen (ISO 14051:2011). Beuth Verlag GmbH.

Engel, L., & Kranhold, M. (2021). Lean Management wird zum Green Management, Um Materialflusskosten erweiterte Wertstromanalyse. *Deutsches Ingenieurblatt, 6*, 1–6.

Erlach, K. (2010). *Wertstromdesign, Der Weg zur schlanken Fabrik*. Springer-Verlag GmbH.

Geisthardt, M., & Engel, L. (2024). A contribution to the development of sustainable target value streams with machine learning considering material flow costs. In J. Marx Gómez, D. G. Nyambo, & A. E. Sam (Eds.), *Artificial intelligence tools and applications in embedded and mobile systems*. Springer Nature.

Geisthardt, M., Engel, L., & Schnegelberger, M. (2023a). Simulation game concept for AI-enhanced teaching of advanced value stream analysis and design. In D. Herberger & M. Hübner (Eds.), *Conference on production systems and logistics: CPSL 2023-2*. Publishing.

Geisthardt, M., Engel, L., & Schnegelberger, M. (2023b). Towards gamification of advanced value stream analysis and design: A game-based learning concept. In V. Volker Wohlgemuth, D. Kranzlmüller, & M. Höb (Eds.), *Enviro Info 2023: Lecture notes in informatics (LNI)*. Gesellschaft für Informatik e.V. (GI).

Geisthardt, M., Engel, L., Döhr, J., & Marx Gómez, J. (2024a). *Towards digital solutions for value stream analysis and design: Investigating enterprise perspective requirements* (pp. 346–356). ESSN: 2701-6277.

Geisthardt, M., Engel, L., Meyer, L., & Marx Gómez, J. (2024b). *Towards digital solutions for value stream analysis and design: Systematic literature review and market analysis* (pp. 357–367). ESSN: 2701-6277.

Geisthardt, M., Engel, L., & Marx Gómez, J. (2024c). Simulation game based teaching of value stream analysis and design with multimodal large language models (MLLMs) assistance. In *SAP academic community conference 2024 (D-A-CH)*. Technische Universität Wien.

Geisthardt, M., Engel, L., & Marx Gómez, J. (2025). Digitization of end-to-end value streams: Conceptualizing a data model based on complex graph networks for AI-based tools. In J. Marx Gómez, A. E. Sam, & D. G. Nyambo (Eds.), *Artificial intelligence tools and applications in embedded and mobile systems: Selected papers from the second international conference on embedded and mobile systems* (ICTA-EMoS). Springer Nature.

Kim, T. K. (2015). T test as a parametric statistic. *Korean Journal of Anesthesiology, 68*, 540–546.

Knauer, U., & Knauer, K. (2019). *Algebraic graph theory: morphisms, monoids and matrices*. Walter de Gruyter GmbH & Co KG.

Koehn, P. (2020). *Neural machine translation*. Cambridge University Press.

Kurafsky, D., & Martin, J. H. (2021). *Speech and language processing: An introduction to natural language processing, computational linguistics, and speech recognition*.

Liker, J. (2020). *The Toyota way*. Esensi.

Meudt, T. (2020). Wertstromanalyse 4.0 – Eine Methode zur integrierten Erfassung und Analyse von Material- und Informationsflüssen in Wertströmen, Dissertation Technische Universität Darmstadt.

Meudt, T., Kaiser, J., Metternich, J., & Spieckermann, S. (2017a). Wertstrommodellierung und-simulation im Zeichen von Digitalisierung und Industrie 4.0. *Zeitschrift Für Wirtschaftlichen Fabrikbetrieb, 112*, 865–868.

Meudt, T., Metternich, J., & Abele, E. (2017b). Value stream mapping 4.0, holistic examination of value stream and information logistics in production. *CIRP Annals, 66*, 413–416.

Ohno, T. (2013). *Das Toyota-Produktionssystem*. Campus Verlag GmbH.

Rasch, B., Friese, M., Hofmann, W., & Naumann, E. (2021). *Inferenzstatistik. Quantitative Methoden 1: Einführung in die Statistik für Psychologie, Sozial-& Erziehungswissenschaften* (pp. 23–34). Springer.

Rother, M. (2009). *Toyota Kata: Managing people for improvement, adaptiveness and superior results*. McGraw-Hill Global Education Holdings LLC.

Rother, M., & Shook, J. (2011). *Sehen lernen, Mit Wertstromdesign die Wertschöpfung erhöhen und Verschwendung beseitigen*. Lean Management Institut GmbH (Workbooks für Lean Management).

Van Steen, M. (2010). Graph theory and complex networks. *An Introduction, 144*(1).

Vaswani, A. (2017). *Attention is all you need*. arXiv preprint arXiv:1706.03762

Zanker, C., & Reisen, K. (2015). Stabilitäts- und Flexibilitätsanforderungen an Produktionssysteme. In W. Kötter, M. Schwarz-Kocher, & C. Zanker (Eds.), *Balanced GPS, Ganzheitliche Produktionssysteme mit stabil-flexiblen Standards und konsequenter Mitarbeiterorientierung*. Springer Fachmedien.

Development and Implementation of a Low-Cost IoT Network for Air Quality Monitoring in the Historic Downtown of San Salvador

Omar Otoniel Flores-Cortez, Carlos Pocasangre, Fernando Arévalo, Jose Alberto Oviedo, and Renata Barriere Garcia

Abstract This paper presents the design, construction and real-world deployment of a low-cost Internet of Things (IoT)-based bare-bones system for urban air quality monitoring in San Salvador's Historic Downtown. The system comprises carbon dioxide (CO_2), total volatile organic compounds (TVOC) and air quality index (AQI-UBA) sensors, WiFi-enabled microcontroller and open-source platform. The main contribution of this study is the demonstration of an initial real-time monitoring solution that not only fills critical environmental data gaps, but also sets the foundation for future expansions, including predictive analytics, early warning systems, and integration into broader innovative city initiatives. The findings support the feasibility of implementing affordable IoT-based environmental networks in developing cities to bridge critical gaps in air quality data. Despite limitations such as the reliance on public Wi-Fi and the absence of reference-grade calibration equipment, the system demonstrated strong reliability and significant potential for urban integration. These enhancements will also contribute to the transformation of the area into an innovative and health-conscious urban ecosystem.

O. O. Flores-Cortez (✉)
Universidad Tecnológica de El Salvador, Dirección de Investigaciones, San Salvador, El Salvador
e-mail: omar.flores@utec.edu.sv

C. Pocasangre · F. Arévalo
Escuela de Ingeniería Eléctrica, Universidad de El Salvador, San Salvador, El Salvador
e-mail: carlos.pocasangre@ues.edu.sv

F. Arévalo
e-mail: fernando.arevalo@ues.edu.sv

J. A. Oviedo
Direcccion de Investigaciones, Universidad Salvadoreña Alberto Masferrer, San Salvador, El Salvador
e-mail: alberto.oviedo01@liveusam.edu.sv

R. B. Garcia
Autoridad de Planificacion del Centro Historico, Alcaldia San Salvador, El Salvador
e-mail: rbarriere@centrohistorico.gob.sv

J. Marx Gómez et al. (eds.), *Recent Advances in Information Systems*, Progress in IS,
https://doi.org/10.1007/978-3-032-14099-9_54

1 Introduction

Urban air quality monitoring is a growing necessity, particularly in developing countries where rapid urbanization and increased vehicle emissions are major contributors to environmental degradation. Accurate, real-time air emissions data are essential for protecting public health, informing environmental policies, and guiding urban planning efforts. However, conventional monitoring networks are often expensive to deploy and maintain, limiting their presence to a few isolated locations, as is the case in San Salvador, El Salvador.

Recent studies emphasize the global need for low-cost air quality monitoring systems, particularly in urban environments in developing countries. Traditional monitoring infrastructure is often insufficient to capture the spatial dynamics of urban air pollution (Morawska et al., 2025). The emergence of IoT-based sensor networks offers a viable solution to improve environmental data collection in a scalable and affordable way. In Latin American contexts, the integration of such systems remains limited, creating a technical and research opportunity (Cornejo-Olivares et al., 2023; Flores-Cortez et al., 2024).

In recent years, the development of low-cost air quality monitoring systems based on IoT technologies has gained global momentum. These solutions offer affordable alternatives to traditional high-cost monitoring stations and have been successfully implemented in various urban environments to enhance environmental data collection, particularly in resource-constrained settings.

In San Salvador, the official air quality monitoring network consists of only three state-operated stations, all located outside of the Historic Downtown area, with only one currently fully operational. This gap in environmental monitoring infrastructure is especially concerning given the recent revitalization of the Historic Downtown into a pedestrian-oriented, tourism-focused zone surrounded by heavy traffic. The high concentration of pedestrians and vehicles in the area, combined with the lack of real-time environmental data, underscores the urgent need for affordable and scalable monitoring solutions. Moreover, the local government's efforts to transform the area into a smart city, through initiatives such as city-wide Wi-Fi coverage, offer an ideal environment for the integration of Internet of Things (IoT) technologies.

Recent advances in IoT have enabled the development of low cost, decentralized monitoring systems using compact environmental sensors and existing communication infrastructure. Taking advantage of this opportunity, this work presents the design, construction and implementation of a low-cost bare-bones IoT system aimed at monitoring key air emissions parameters—namely, carbon dioxide (CO_2), total volatile organic compounds (TVOC), air quality index based on the UBA scale (AQI-UBA), temperature, and humidity—in downtown San Salvador. Using ENS160 and AHT21 sensors, WiFi-based connectivity, and minimal deployment infrastructure (building façades, light poles, and windows at approximately 2 meters height), six IoT nodes and a centralized web platform were deployed to cover approximately 3 km^2 of the Historic Downtown.

The main contribution of this study is the demonstration of a practical, low-cost and easily replicable IoT-based network for urban air monitoring in developing cities, specifically adapted to the infrastructure and economic realities of San Salvador. The aim of this project is to establish an initial real-time air quality monitoring network that not only fills critical environmental data gaps, but also sets the foundation for future expansions, including predictive analytics, early warning systems, and integration into broader smart city initiatives.

Although numerous studies have explored low-cost sensor networks, this work offers a distinctive contribution by integrating affordable IoT technology within the constraints of a real urban infrastructure in a developing country. It shows how municipal Wi-Fi and standard power access can be leveraged for continuous air quality monitoring. Unlike controlled pilot tests, this system was deployed in a live, populated environment and sustained operations without proprietary tools, offering a replicable and open-source model for similar urban areas in the region.

2 Literature Review

2.1 *Air Quality Monitoring: Traditional Systems and Limitations*

Historically, air quality monitoring has relied on large fixed monitoring stations equipped with high-precision instrumentation capable of detecting multiple pollutants at very low concentrations. These traditional systems, often managed by government agencies, provide highly accurate and reliable data suitable for regulatory and research purposes. However, the costs associated with their installation, operation and maintenance are substantial, typically limiting their deployment to a few strategic locations within urban areas (Lezina & Misyurev, 2022; Ramesh et al., 2024; Tahir Bahadur et al., 2025).

In developing countries, such as those in Latin America, financial constraints and technical resource limitations exacerbate this problem. The sparse networks of monitoring stations do not capture the spatial heterogeneity of pollution exposure, leaving significant areas unmonitored (Aguilar-Gomez & Rivera, 2025; La Colla et al., 2021). In San Salvador, for example, the state-operated network includes only three stations, all located outside the Historic Downtown, with only one fully operational at present (Servicio Nacional de Estudios, 2024).

2.2 *Emergence of IoT-Based Environmental Monitoring*

The last decade has witnessed a significant technological shift toward the use of low-cost sensors and IoT solutions for environmental monitoring (Flores-Cortez et al.,

2024; Pocasangre et al., 2023). Toward around 2010, researchers began exploring the feasibility of decentralized sensor networks that could provide dense spatial coverage at a fraction of the cost of traditional stations (Martins et al., 2025; Sowndharya, 2025). Initial pilot projects in cities such as New Delhi, India, and Accra, Ghana, demonstrated the potential of such systems to supplement existing networks and offer actionable real-time data for urban management (Mead et al., 2013).

By 2015, rapid advances in sensor technology and wireless communication had fueled an explosion of urban IoT-based environmental monitoring initiatives around the world (Oh et al., 2015). Studies reported that even though low-cost sensors exhibit greater uncertainty than regulatory grade instruments, they are capable of reliably capturing trends, spatial patterns, and pollution events when properly calibrated (Morawska et al., 2025).

Currently, IoT-based systems are increasingly being adopted not only for air quality but also for noise pollution, traffic emissions, and microclimate monitoring, as cities push toward smart and sustainable urban ecosystems. The trend indicates a paradigm shift from sparse, expensive regulatory networks to dense, community-centered sensing networks.

2.3 Gaps and Needs in Latin American and Developing Cities

Despite global advances, Latin American cities continue to lag behind in adopting IoT-based environmental monitoring systems. A survey of smart city initiatives shows that environmental monitoring is often a secondary priority compared to public security and urban mobility (Rojas & Büttinghausen, 2021). In addition, infrastructure limitations, unstable funding, and lack of technical expertise constrain the deployment of dense sensor networks.

In the case of San Salvador, there is no historical precedent for real-time environmental monitoring within the Historic Downtown area, despite its significant pedestrian density and vehicular traffic. The lack of environmental data hampers efforts to assess exposure risks and guide public health interventions. Implementing a low-cost IoT-based monitoring network tailored to the specific infrastructure and economic conditions of the city addresses this critical gap and aligns with broader efforts to transform the area into a smart and sustainable urban district.

Thus, the need for innovative, affordable and scalable monitoring solutions is not only pressing but also strategically aligned with the regional goals of environmental resilience and smart urban development.

3 Materials and Methods

3.1 Materials

The core of each IoT node is based on the ESP32-C3 microcontroller, a low-power chip that integrates processing and Wi-Fi communication capabilities. This microcontroller is connected via the I2C protocol to two key sensors:

- ENS160: An environmental gas sensor designed to detect air pollutants, including total volatile organic compounds (TVOC) and carbon dioxide equivalents (eCO_2), and to provide an AQI estimate based on the UBA scale, see Table 2.
- AHT21: A digital sensor for accurately measuring ambient temperature and relative humidity.

A crucial criterion for choosing the sensors was that both provided digital serial output through the Inter-IC (I2C) port. Another criterion considered was the sensor module's ability to perform preprocessing needed for AQI-UBA calculations. Furthermore, the measurement ranges of these sensors comply with those needed for outdoor readings, as shown in Table 1.

The AQI-UBA categorizes air quality into five distinct categories, each represented by a specific color, see Table 2.

Figure 1 illustrates the architectural design of the proposed system, while Fig. 2 presents the final scheme of components selected for its implementation.

Table 1 Technical specifications of the selected sensors

Part No.	Interface type	Magnitudes	Sensing range
ENS160	I2C	TVOC	0–65,000 ppb
		eCO_2	400–65,000 ppm
		AQI-UBA	1–5
AHT21	I2C	Temperature	−40° to 120°
		Humidity	0–100%

Table 2 AQI categories (Umweltbundesamt, 2007)

AQI-UBA		Recommendation	Exposure limit
No.	Rating		
5	Unhealthy	Use only if unavoidable	Hours
4	Poor	Intensified ventilation	<1 month
3	Moderate	Increased ventilation	<12 months
2	Good	Sufficient ventilation	No limit
1	Excellent	Target	No limit

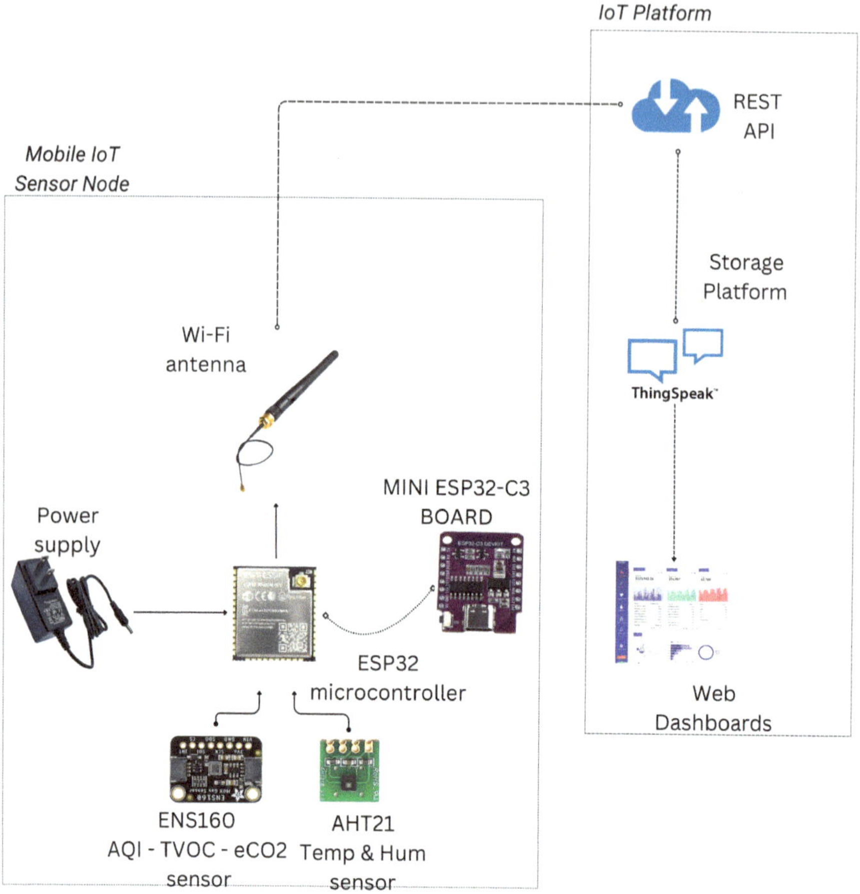

Fig. 1 Overview diagram of the system architecture

The nodes are powered by an external 5V DC/1A power source connected to the municipal electrical infrastructure. Each device is housed in an IP-rated waterproof enclosure, see Fig. 3, with mounting brackets or adhesive supports, allowing for installation on building façades, light poles, and window structures at approximately 2 meters above ground level, a height chosen to capture air quality at human breathing level. The firmware for ESP32-C3 was developed in Arduino C / C++ using Arduino IDE. Implements a loop-based routine to read, validate, format, and transmit data.

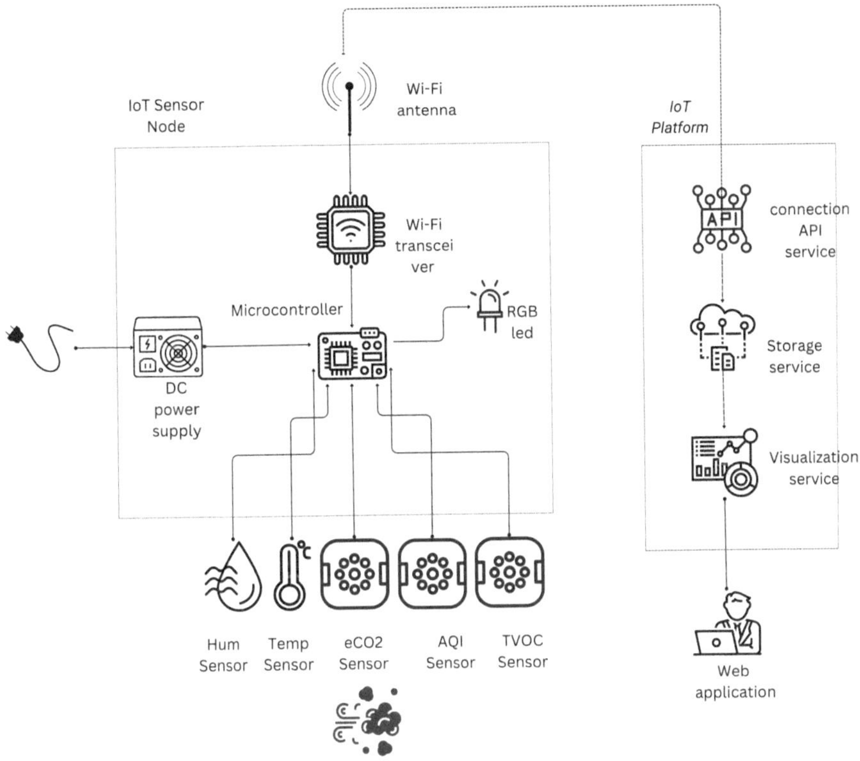

Fig. 2 Diagram of the proposed system architecture

3.2 *Methods*

1. Network Architecture and Node Deployment: The network comprises six IoT nodes deployed across a $3\ km^2$ area in the Historic Downtown of San Salvador, selected for its pedestrian density and vehicular emissions. The placement of the nodes prioritized Wi-Fi availability, public accessibility, and high-traffic zones such as public parks and transit corridors.
2. Data Acquisition and Communication: Each node collects data every 10 minutes, measuring eCO_2, TVOC, AQI (UBA scale), temperature, and humidity. Data are transmitted in real time to the cloud using the MQTT protocol over the existing Wi-Fi network of the city. The system relies on direct Internet connectivity; no local gateway is used.
3. Cloud Platform and Visualization: Data are received and managed through the ThingsBoard IoT platform, which provides tools for API integration and the development of dynamic dashboards. The platform offers:

    - Real-time visualization

Fig. 3 Prototypes of the IoT nodes built for the implementation of the network

- Historical data access
- Geographic mapping of node locations
- Export options for external analysis.

4. Calibration and Validation: Both sensors were used with their factory calibration settings. Due to resource constraints, no laboratory calibration was performed. However, the initial readings were visually compared with the trends of the country's official monitoring station, validating the consistency of the system at the trend level, although not at regulatory precision.

4 Results and Discussion

4.1 Network Performance and Reliability

The IoT monitoring network has been operational for more than two months and continues to function uninterruptedly, collecting data daily, including weekdays and weekends. Overall, the system demonstrated stable performance; however, variability in data transmission reliability was observed between nodes. The most significant data loss occurred at the *E-PLIB* node in Parque Libertad, which experienced up to

30% packet loss, probably due to network saturation in public Wi-Fi zones shared with tourists. Conversely, the *E-MHH* node at the Hula Hula market reported only a 5% loss, and nodes located in institutional facilities (e.g., *E-UTEC* and *E-APLAN*) maintained nearly 100% uptime throughout the test period. A technical issue arises during heavy rain events. Some sensors initially mounted near the front edge of the enclosures were exposed to water infiltration, producing erroneous readings. This was mitigated by repositioning the sensors on the lower side of the enclosure, effectively shielding them from direct rain.

4.2 Environmental Data and Observed Trends

Across the monitoring period, clear temporal patterns in pollution levels were identified. The highest air pollution values were consistently observed between 11:00 AM and 2:00 PM, while the lowest levels occurred in the early morning hours, particularly between 4:00 AM and 6:00 AM. This pattern aligns with daily urban activity cycles, and traffic-related emissions increase during the day. The *E-MHH* node recorded the highest levels of the Air Quality Index (AQI), reaching level 4/5 (poor) according to the UBA scale. This node is located in a high-traffic area and consistently showed elevated levels of CO_2 and TVOC during peak hours. *E-PLIB* and *E-UTEC* followed with moderate AQI levels (3/5). In contrast, nodes placed in pedestrian zones and public parks reported AQI values ranging from 1 to 2, indicating excellent to good air quality (Fig. 4).

4.3 Spatial Variability and Urban Insight

The results confirmed expectations: areas with heavy vehicular traffic reported poorer air quality, while pedestrian-friendly areas with limited traffic showed significantly cleaner air. These findings support the effectiveness of the sensor network in capturing spatial variability in air quality in the Historic Downtown of San Salvador.

- Interactive dashboards developed through the *ThingsBoard* platform allowed real-time monitoring, with features including:
- A map-based interface showing the status and location of each station.
- Summary tables of live and historical measurements.
- Individual station dashboards with time series plots that highlight pollution peaks, particularly from *E-MHH* (Fig. 5).

Fig. 4 IoT nodes installed at locations in the Historic Downtown of San Salvador

4.4 Comparative Analysis and Limitations

An examination of the data collected from the stations with the highest recorded pollution levels, specifically the MHH station and the PLIB station, reveals distinct trends in air quality during different times of the day. At the MHH station, see Fig. 7c, high levels of pollution (4/5) are observed during lunchtime, particularly from 10:00 AM to 2:00 PM. This period is marked by a significant increase in pollutants, probably due to increased vehicular activity. In contrast, the PLIB station, see Fig. 7f experiences high levels of pollution (4/5) during the early morning hours, particularly around 4:00 AM, when emissions from overnight traffic and local industries can accumulate. Another peak (4/5) in pollution occurs during the afternoon hours at 4:00 PM, potentially linked to traffic after midday breaks (Fig. 6).

The only functioning national monitoring station is located more than 50 km away, making direct comparisons infeasible. The main limitations encountered during this initial phase included the following:

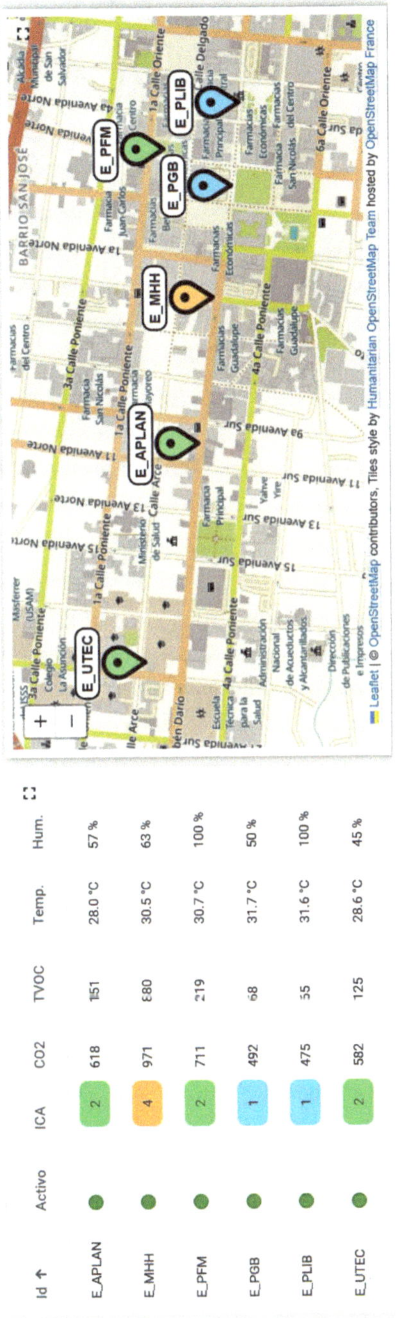

Id ↑	Activo	ICA	CO2	TVOC	Temp.	Hum.
E_APLAN	●	2	618	151	28.0 °C	57 %
E_MHH	●	4	971	[illegible]80	30.5 °C	63 %
E_PFM	●	2	711	219	30.7 °C	100 %
E_PGB	●	1	492	68	31.7 °C	50 %
E_PLIB	●	1	475	55	31.6 °C	100 %
E_UTEC	●	2	582	125	28.6 °C	45 %

Fig. 5 Screenshot of the main monitoring site for the network implemented

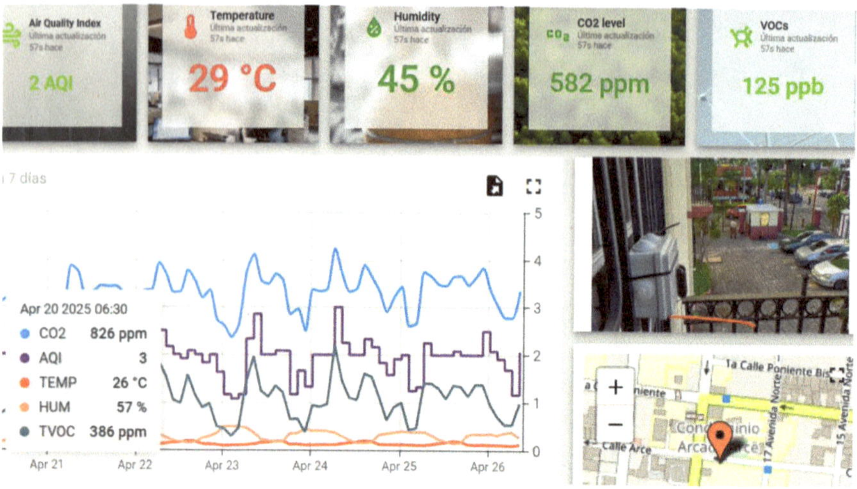

Fig. 6 Screenshot of the webpage with all the info for each monitoring station

- Connectivity constraints due to the reliance on public Wi-Fi networks that are not optimized for IoT stability.
- Lack of certified reference equipment for sensor calibration or formal validation.
- However, the system exhibited significant strengths: low cost, adaptability to existing urban infrastructure, and potential scalability.

The lack of calibration of the sensor against reference-grade instruments is acknowledged as a limitation of this exploratory phase. However, the observed data trends aligned with theoretical expectations and human activity patterns, lending credibility to the measurements. Future work will include co-located calibration with reference monitors to improve accuracy and regulatory value.

4.5 Future Directions

Based on these results, a second phase of the project is already being considered. The planned improvements include the following.

- Integration of alerts based on threshold exceedance.
- Deployment of additional sensor nodes to expand coverage.
- Incorporation of new sensing modalities, such as noise pollution and particulate matter (PM2.5, PM10).

These enhancements will further align the project with the objectives of smart cities, enabling comprehensive and inclusive urban environmental monitoring in developing cities such as San Salvador.

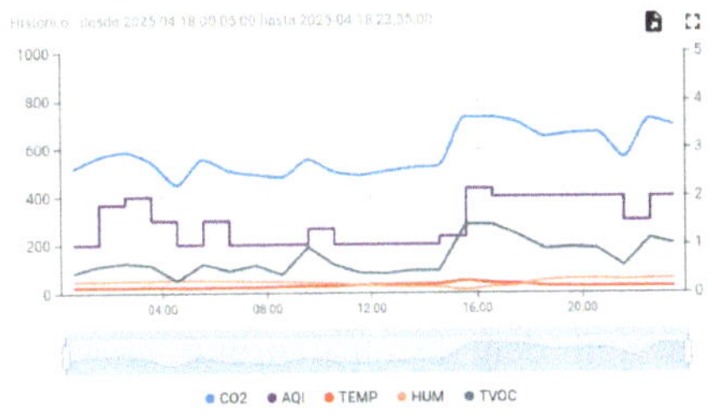

(a) UTEC node

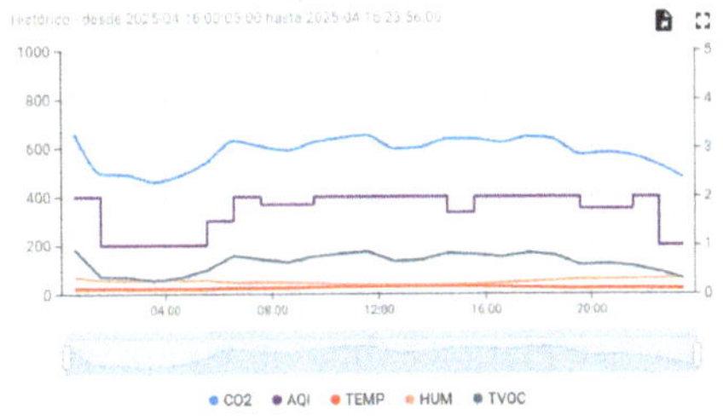

(b) APLAN node

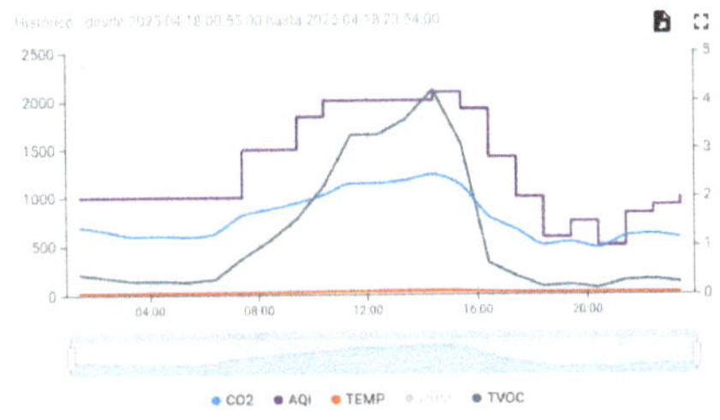

(c) MHH node

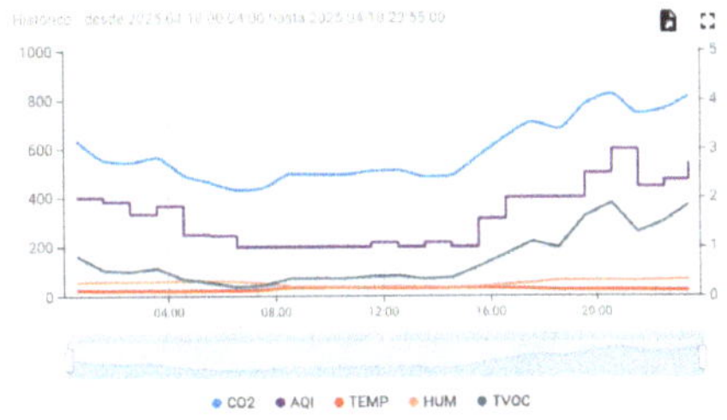

(d) PGB node.

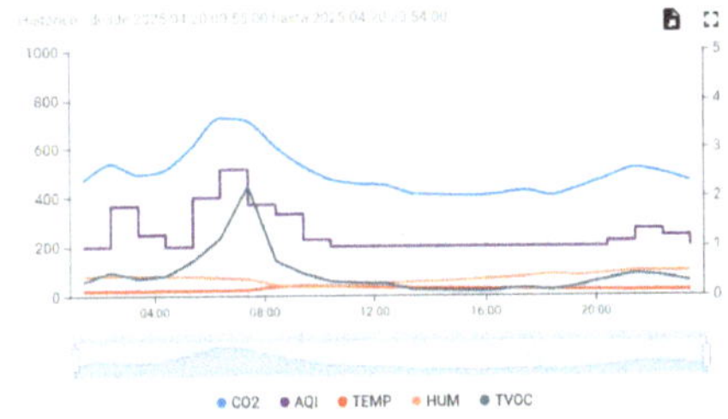

(e) PFM Node.

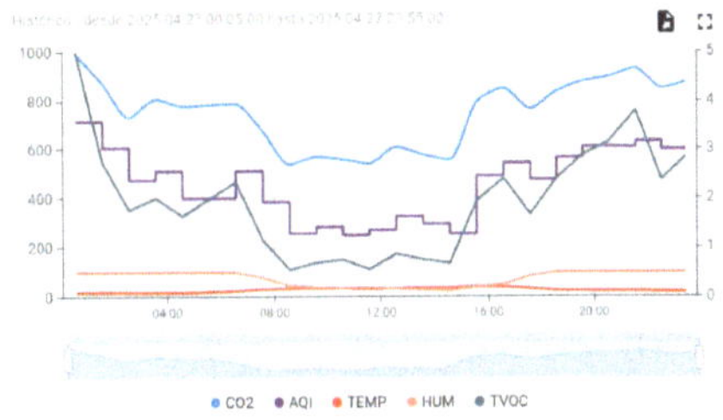

(f) PLIB node.

Fig. 7 Following graph presents a 1-day overview of the trend in pollutant levels at six stations. It illustrates the hours during which the AQI pollution levels were higher or lower at each station. The MHH station Fig. 7c has been determined to have the highest AQI values, while the APLAN Fig. 7b and PFM Fig. 7e stations has been identified as the location with the lowest AQI values

5 Conclusions

This study presented the design, development and real-world deployment of a barebones low-cost IoT-based system for monitoring air emissions in San Salvador's historic center. Using low-cost sensors (ENS160 and AHT21), WiFi-enabled ESP32-C3 microcontrollers, and open-source platforms, a practical and scalable monitoring solution was successfully implemented in the context of limited infrastructure and resources.

The six-node network has been continuously operating for more than two months, collecting real-time environmental data every day, including weekdays and weekends. The system effectively identified spatial and temporal pollution patterns, confirming higher levels of contamination in vehicular areas and cleaner air in pedestrian zones. The findings support the feasibility of implementing affordable IoT-based environmental networks in developing cities to bridge critical gaps in air quality data.

Despite limitations such as the reliance on public Wi-Fi and the absence of reference-grade calibration equipment, the system demonstrated strong reliability and significant potential for urban integration. Its low cost, adaptability, and data accessibility make it a compelling model for scalable implementation in similar urban environments.

Future work will focus on expanding the sensor network, incorporating additional parameters such as noise and particles, and integrating predictive analytics and real-time alert systems. These enhancements will also contribute to the transformation of San Salvador's historic center into an innovative and healthy urban ecosystem. In addition, future phases will consider the environmental impact of hardware deployment, particularly in terms of long-term energy use and electronic waste. Strategies such as modular design for reusability and the use of recyclable materials will be explored to enhance the sustainability of the system.

References

Aguilar-Gomez, S., & Rivera, N. M. (2025). Air pollution in the global south: An overview of its sources and impacts. In *Oxford Research Encyclopedia of Economics and Finance*.

Cornejo-Olivares, L., Contreras-Cossio, J., Hoyos-Rivas, F., Segura-Villarreal, C., Gutierrez-Pinta, E., Chavez-Sanchez, W., & Grados, J. (2023). IoT application technologies for agriculture in Latin America during the covid-19 pandemic. In *LACCEI* (Vol. 1, No. 8).

Flores-Cortez, O. O., Elizondo, J. M. C., Portillo, A. C. F., Arévalo, F., & Pocasangre, C. O. (2024). An IoT-based system for water parameters monitoring. case study: Fish farming in El Salvador. In *IEEE International Workshop on Metrology for Industry 4.0 & IoT (MetroInd4. 0 & IoT)* (Vol. 2024, pp. 123–128). IEEE.

La Colla, N. S., Botté, S. E., & Marcovecchio, J. E. (2021). Atmospheric particulate pollution in South American megacities. *Environmental Reviews, 29*(3), 415–429.

Lezina, E., & Misyurev, M. (2022). Ambient air quality measurements in a large city: existing solutions, new opportunities and challenges. In *Journal of Physics: Conference Series* (Vo. 2192, Issue 1, p. 012031). IOP Publishing.

Martins, L. D., de Camargo, E. T., & Fonseca, K. V. O. (2025). *Smart sensors for smart environment* (Chapter 5, pp. 67–89). IGI Global. https://digital-library.theiet.org/doi/abs/10.1049/PBBE005E_ch5

Mead, M., Popoola, O., Stewart, G., Landshoff, P., Calleja, M., Hayes, M., Baldovi, J., McLeod, M., Hodgson, T., Dicks, J., et al. (2013). The use of electrochemical sensors for monitoring urban air quality in low-cost, high-density networks. *Atmospheric Environment, 70*, 186–203.

Morawska, L., Asbach, C., & Patel, H. (2025). Application of $PM_{2.5}$ low-cost sensors for indoor air quality compliance monitoring. *Aerosol Science and Technology,* 1–11.

Oh, C.-S., Seo, M.-S., Lee, J.-H., Kim, S.-H., Kim, Y.-D., & Park, H.-J. (2015). Indoor air quality monitoring systems in the IoT environment. *The Journal of Korean Institute of Communications and Information Sciences, 40*(5), 886–891.

Pocasangre, C. O., Flores-Cortez, O. O., Arévalo, F., Arriaza, C., Aguilar, O., & Miranda, B. H. (2023). An IoT mobile system for co 2 monitoring in volcanic soil using the li-830 sensor and an android app. In: *2023 International Conference on Networking, Sensing and Control (ICNSC)* (Vol. 1, pp. 1–6). IEEE.

Ramesh, R., Vallabhu, H., Unni, A., & Nalinakshan, S. (2024). IoT-enabled air quality monitoring: Advancements, applications, and challenges (pp. 561–567).

Rojas, E. F., & Büttinghausen, C. C. (2021). Deconstruyendo el concepto de smart city. una mirada crítica a su aplicación en latinoamérica.: Una mirada crítica a su aplicación en latinoamérica. *Arquitectura y Sociedad, 1*(20), 18–37.

Servicio Nacional de Estudios Territoriales (SNET). (2024). Monitoreo de calidad del aire. https://www.snet.gob.sv/ver/meteorologia/monitoreo/calidad+del+aire/monitoreo+calidad+aire/

Sowndharya, V. (2025). Environmental monitoring for sustainability. In *Leveraging urban computing for sustainable urban development* (p. 323).

Tahir Bahadur, F., Rasool Shah, S., & Rao Nidamanuri, R. (2025). A brief outline of indoor air quality: Monitoring, modeling, and impacts. *Journal of Environmental Engineering, 151*(5), 03125001.

Umweltbundesamt,. (2007). Evaluation of indoor air contaminants by means of reference and guideline values. *Bundesgesundheitsblatt-Gesundheitsforschung-Gesundheitsschutz,50*, 990–1005.

Validating a Framework for Assessing Small and Rural Municipalities' Readiness for Smart City Implementation

N. L. Mashau and J. H. Kroeze

Abstract Small and rural municipalities have limited infrastructure and resources to deliver essential services to their citizens. There is a need for smart solutions to manage the limited resources at the disposal of these municipalities. This paper presents a validated framework tailored explicitly for assessing the readiness of small and rural municipalities for smart city implementation. The design science research (DSR) methodology was employed to conceptualize, create, and assess the framework. During the evaluation and validation DSR cycles, qualitative online surveys were used to facilitate participatory design and expert reviews. The validated framework addresses the unique needs of small and rural municipalities by amending existing models designed for larger urban areas.

1 Introduction

The *Validated Framework for Assessing Small and Rural Municipalities' Readiness for Smart City Implementation*, which is the outcome of this article, went through various development cycles using a Design Science Research (DSR) approach (Mashau, 2023). To provide a roadmap of the bigger project, of which this artifact represents the culmination, a short overview of the DSR cycles and artifacts follows: The initial artifact was a conceptual framework integrating constructs uncovered by a systematic literature review and theoretical reflection (first DSR cycle) (Mashau et al., 2021). Using semi-structured interviews within a qualitative approach (second DSR cycle), empirical data was obtained regarding important indicators and themes for readiness assessment (Mashau & Kroeze, 2023, 2024, 2025, 2026a, b). (For the interview guide, see Annexure E in Mashau (2023).) The results included the participants' inputs on the role of traditional leaders (Mashau & Kroeze, 2025), IS drivers

N. L. Mashau (✉) · J. H. Kroeze
School of Computing, University of South Africa, Roodepoort, South Africa
e-mail: emashanl@unisa.ac.za

J. H. Kroeze
e-mail: kroezjh@unisa.ac.za

J. Marx Gómez et al. (eds.), *Recent Advances in Information Systems*, Progress in IS,
https://doi.org/10.1007/978-3-032-14099-9_55

that were deemed necessary and those that were already in place (Mashau & Kroeze, 2026a), and the participants' perceptions on assessing readiness (Mashau & Kroeze, 2023). The rich information that was found during this cycle was combined with the results of a second systematic literature review (Mashau et al., 2022) to revise the conceptual framework into the second artifact, namely the revised framework. In a third DSR cycle, this intermediate version was evaluated using an online qualitative survey. (For the survey protocol, see Annexure F in Mashau (2023).) It was approved by the participants without any changes. However, when the fourth DSR cycle took place to validate the artifact, using the same survey protocol, one additional construct was suggested. By including this construct, the validated framework was established as the third and final artifact. This paper focuses on the processes and outcomes of both the evaluation and validation cycles (Mashau & Kroeze, 2026b).

Scarce infrastructure and resources in small and rural municipalities pose challenges in planning and allocating resources for service delivery (Beyers, 2016; SALGA, 2017/2018). Population growth intensifies the pressure on existing infrastructure and resources, exacerbating issues such as infrastructure breakdowns (Mashau & Kroeze, 2023; Rana et al., 2019). Addressing these challenges can be achieved by adopting the concept of a smart city in small and rural municipalities (Desdemoustier et al., 2019a; Mashau et al., 2021, 2022). Arief et al. (2020) and Mashau and Kroeze (2023) exposed a critical issue, namely the lack of assessing the existing conditions of small and rural municipalities before implementing smart city initiatives, resulting in a high rate of project failure. The validated artifact presented here constitutes a potential solution to this problem.

The paper is arranged as follows: Sect. 1 presents the introduction to the study. Section 2 defines smart city concepts and provides important characteristics for smart city implementation in small and rural municipalities. Section 3 discusses the research methodology used. Section 4 presents the results along with a discussion of their implications, culminating in the presentation of the validated framework. Finally, Sect. 5 concludes the study.

2 Literature Review

The smart city concept was coined by IBM, aiming to fuse diverse urban subsystems by adopting advanced technologies for smart city management (Harrison et al., 2010). A smart city leverages state-of-the-art information and communication technologies, such as cloud computing and the Internet of Things (IoT), to enhance city governance (Aljowder et al., 2019; Montalvan Castilla & Riel Müller, 2024; Zhao et al., 2021). City governance endeavors to ensure the efficient delivery of public services and enhancing residents' quality of life (Anthony, 2024; Noori et al., 2020; Tan & Taeihagh, 2020).

The concept of a "smart city" is defined in various ways, with many researchers emphasizing the role of technology in resource management and the provision of citizen services (Bibri, 2021; Galal & Elariane, 2022; Stübinger & Schneider, 2020).

In this paper, a smart city is defined as "a digital integration of information systems components to collect digital data and analyze it in real-time to monitor and manage city infrastructure and to allocate resources effectively, thereby improving service delivery and the quality of life of the citizens" (Mashau et al., 2021).

Small and rural municipalities typically constitute integral parts of local, district, and national governance structures. In South Africa, the majority of the population resides within small, rural municipalities, which play a significant role in the national economy (National Treasury, 2011; Steyn & Ngwenya, 2015). These municipalities could benefit from adopting the smart city concept to efficiently manage limited resources and bolster their economies (Kniazevych, 2024; Lämmel et al., 2024). Before embarking on a smart city endeavor, municipalities must assess their readiness levels to address critical factors and mitigate avoidable errors (Desdemoustier et al., 2019b). Considerations in this regard should encompass stakeholders, budget allocations, policies, population size, and economic factors as important characteristics when assessing their readiness levels (Mashau et al., 2021; Veselitskaya et al., 2019). Within a municipality, stakeholders encompass a diverse array of contributors essential for sustainable smart city development, including municipal councils, personnel, government entities, private enterprises, citizens, tourists, academic institutions, research communities, application developers, and software firms (Veselitskaya et al., 2019). These stakeholders must grasp the objectives behind smart city initiatives and extend essential support to ensure smart city realization (Anthony, 2024; Bashynska & Dyskina, 2018; Montalvan Castilla & Riel Müller, 2024).

Furthermore, stakeholders or service providers rely on municipal budgets to perform their duties effectively (Yu et al., 2019). Municipalities need to allocate funds for digital infrastructure procurement and the remuneration of service providers involved in smart city projects (Hartley, 2015; Hartley & Seymour, 2011).

Considering the significant impact of small and rural municipalities on the national economy, it is imperative to factor in the municipal economy. A robust municipal economy enhances citizens' living standards by fostering opportunities such as superior infrastructure, business growth, ecosystem development, and competitiveness (Fang & Shan, 2024; Mora et al., 2019). A thriving economy fosters digital innovation and facilitates smart city management (Chauhan et al., 2021; Du et al., 2020).

Policies also play a fundamental role in guiding smart city initiatives (Tan & Taeihagh, 2020). Well-defined policies should articulate clear expectations and action plans for addressing urban services and operations (Fang & Shan, 2024; Prendeville et al., 2018; Stübinger & Schneider, 2020). Policymakers must reassess structural and regulatory frameworks that impede digital innovation and smart city implementation (Stübinger & Schneider, 2020; Tan & Taeihagh, 2020).

In the first DSR cycle of the broader project, concepts from four theoretical perspectives were employed as the foundation for constructing the conceptual framework, namely: the Technology, Organization, and Environment Framework (TOE); Delone and McLean Model; Technology Readiness Index (TRI); and Diffusion of Innovation Theory (DOI) (Mashau, 2023). The conceptual framework, which underpins later iterations of the framework, integrated 27 components from these theories and various smart city frameworks developed by other scholars.

The conceptual framework underscores the significance of human, organizational, technology, and environmental indicators in determining the readiness of small and rural municipalities for smart city implementation (Dewi et al., 2018; Mashau et al., 2021). The literature indicates that human readiness is shaped by factors such as education level, experience, innovation, and optimism, while organizational readiness hinges on executive support, partnerships, skilled personnel, size, and resources (Chichernea, 2015; Kniazevych, 2024; Mashau & Kroeze, 2024). Technology readiness has been widely recognized as an important component of data collection and analysis in the smart city context, with data security emerging as a major concern (Mahesa et al., 2019). Challenges arise with open data and big data in smart city initiatives, necessitating careful attention to security and privacy requirements during readiness assessment (Cilliers & Flowerday, 2017; Montalvan Castilla & Riel Müller, 2024). Factors like compatibility, availability, complexity, relative advantages, information quality, and system quality contribute significantly to evaluating technology readiness (DeLone & McLean, 2003; Mahesa et al., 2019; Rogers, 2003). Lastly, environmental readiness is influenced by government regulations, social norms, culture, economic factors, and technology support infrastructure (Barham & Daim, 2020). The conceptual framework suggests that when all these human, technological, environmental, and organizational aspects align, a city can embark on smart city initiatives with minimal challenges. Due to space limitations, the conceptual framework is not shown in this paper, but it can be viewed in Mashau et al. (2021).

3 Research Methodology

The DSR methodology followed in the broader project is an approach aimed at generating scientific knowledge (Gregor & Hevner, 2013; Wieringa, 2016), with an artifact as the ultimate outcome to address practical problems (Wieringa, 2016). The research adhered to the DSR approach outlined by Peffers et al. (2007), which provides explicit guidance on the required iterations. The wider study's iterations encompassed problem identification or definition, objective delineation, data collection, artifact design, development and improvement, and artifact evaluation and validation (Mashau, 2023).

The larger study employed a multistage sampling approach. Initially, three South African provinces that had a high representation of small and rural municipalities were selected. Subsequently, purposive sampling was employed to select specific small and rural municipalities from the selected provinces. For the first round of empirical work, conducted to revise the conceptual framework, purposive sampling was used to identify initial interview participants. To expand the sample—and to mitigate bias to some extent—snowball sampling was also used (Mohd Ishak et al., 2014).

This paper focuses on the second and third rounds of empirical data collection, which were conducted to assess and refine the revised framework. For the second round, in which the revised framework was evaluated, a participatory design approach was adopted to determine whether the participants' inputs had been rendered correctly. To prevent bias on the researchers' side, a gatekeeper from each municipality was tasked to nominate one participant from the interview pool to review the revised framework. For the third round, an expert review was employed to validate the artifact. The participants for this round were selected through purposive sampling. They were managers from metropolitan municipalities in South Africa and researchers with expertise in the smart city domain.

A qualitative research approach is ideal to obtain rich information in smaller samples, aiming for peculiarity rather than generalizability (which is more relevant in large quantitative studies) (Cresswell & Creswell, 2022). Qualitative data was gathered from five local municipalities spanning three provinces in South Africa. Through individual face-to-face semi-structured interviews, fourteen participants provided feedback on the conceptual framework. Thematic analysis was conducted using ATLAS.ti 8.1, entailing the identification of codes and their subsequent grouping into themes. The framework was then revised. The revised framework was evaluated via an online qualitative survey, followed by validation through a second online survey using the same protocol. By incorporating the participants' supplementary inputs, the validated framework was produced as the final artifact.

Before data collection, the researchers obtained ethical clearance from the research ethics committee of the College of Science, Engineering and Technology at the University of South Africa. During the data collection process, the rights of the interviewees were explained to all participants. Their participation was voluntary, and they could indicate at any time if they did not want to proceed with the interview. All the interviewees who agreed to participate signed the consent form.

4 Results and Discussion

The results of the first round of empirical data collection were reported in related papers (Mashau & Kroeze, 2023, 2024, 2025, 2026a), but a summary thereof is provided in Sect. 4.1 to facilitate the understanding of the foundations of the framework, on which this paper builds.

4.1 Findings of the Semi-structured Interviews

Two salient themes emerged from the first round of data collection: important factors for a readiness assessment for smart city implementation, and information system drivers for smart city development.

Important Factors for a Readiness Assessment for Smart City Implementation

Human Factors. The interview results showed that human factors include education, skills, and awareness. The interviewees highlighted these as important factors in assessing smart city readiness for small and rural municipalities. They emphasized the significance of assessing human readiness, stressing technical skills as one of the factors required for both municipal personnel and citizens because both should know how to use innovations. A complex project such as smart city development requires capacitated individuals with the relevant skills and capabilities to execute their tasks successfully.

A lack of knowledge of digital tools may hamper the use of smart city technologies. This supports the notion that education is pivotal. Individuals must have a certain level of education to be able to contribute to a smart city. Employees and citizens with an educational background will likely see the value the smart city initiative will bring to their lives. Without an educational background, they will likely resist the support and implementation of smart cities. Training citizens to use digital devices enables them to access municipal services, especially in rural municipalities where digital literacy might be lacking.

Leadership's willingness and knowledge to develop a smart city are significant. The support from a municipality's leadership is equally significant to the development of a sustainable smart city. This perspective emphasizes that human education, skills, capacity-building, and support are integral in assessing small and rural municipalities' readiness for smart city implementation.

Technological Factors. Technological infrastructure is one of the key technological factors in assessing small and rural municipalities' readiness for smart city implementation. Physical, economic, social, and technological infrastructure are essential building blocks in providing efficient service delivery in a smart city. This could be achieved through the implementation of modern, adaptable technological infrastructure that includes network connectivity and cheaper internet. Network connectivity is pivotal in data collection and real-time analysis in the smart city.

Furthermore, the participants emphasized that connectivity is an aligning force that enables the integration of disparate technologies. Access to the internet bridges gaps between urban and rural areas by ensuring that even scattered villages in small and rural municipalities may have access to a town's smart city networks. Free Wi-Fi and digital infrastructure will enable municipalities to collect data, deploy important applications, and communicate with the municipality's citizens.

IoT devices, such as sensors, are indispensable to smart city development by means of facilitating data sharing and assisting knowledge workers with decision-making using real-time information. All these views emphasize that a municipality should have a comprehensive technological infrastructure that will support the connectivity of sensors in transforming small and rural municipalities into smart cities to realize the smart city dream.

Organizational Factors. Industrial factors are key elements in assessing small and rural municipalities' readiness for smart city implementation. Factors such as political will, proactive leadership, and budget were significant when assessing small and rural municipalities' readiness for smart city implementation. Skilled human resources and a sufficient budget are significant because unskilled officials or inadequate budget and lack of financial planning may hinder progress in smart city development. Some participants emphasized the significance of having support from top management, who actively seek partnerships with all sectors of operation and ensure sufficient internal funds, unlike depending on government subsidies and grants. These results indicate the significance of strategic budget management in curbing any financial issues that may affect the implementation of any smart city initiatives.

Political factors play an equally important role in assessing organizational readiness. A change in political leadership may disrupt continuity. For example, when a new mayor comes with a different vision, mission, and priorities, it may negatively affect the progress of the development of a smart city. There is a need for political buy-in and active participation by key decision-makers, including the mayors, municipal councils, and management, for long-term commitment to a smart city project. For this reason, political alignment with council members and the public and private sectors is imperative for developing a smart city. Therefore, the political element should be one of the measures used to assess small and rural municipalities' readiness for smart city implementation.

Environmental Factors. Environmental factors like geographical location, governance, land ownership, and energy supply present distinct challenges in assessing small and rural municipalities' readiness for smart city implementation. The participants highlighted that limited capacity and resources in municipalities in rural areas hinder progress in their development. Factors such as geography and size play a pivotal role in assessing readiness. The land geographically situated in rural areas is often managed by traditional leaders. As stakeholders, they play a decisive role in preparing smart city development by being a mediator between the municipality and citizens to resolve any land issues.

Energy supply emerged as one of the pivotal environmental factors that could affect small and rural municipalities' readiness. The participants emphasized that persistent load shedding affected connectivity, resulting in stopping the delivery of essential services that were dependent on electricity. This impacted on the utilization of digital infrastructure to alert the municipalities about the issues that affected service delivery. There is a need for reliable electricity supply in small and rural municipalities for them to be smart cities because a lack of reliable electricity severely affects network connectivity, which results in citizens' inability to access municipal services. Parastatals such as ESKOM (South Africa's state-owned electricity public utility) must ensure a stable and sustainable electricity supply for smart city development to be possible since it is significant in technological integration.

Municipalities must consider the number of citizens or population size in relation to the resources needed to provide the services when assessing environmental readiness. The municipality must ensure they have a sustainable economy because a weak economy may negatively affect both citizens and companies. In a weak economy, some companies are likely to cut the workforce, while others may end up being closed. In both cases, citizens are likely to lose their jobs.

Governance structures may significantly impact the municipality's readiness for smart city implementation. Deviating from governance practices, such as insufficient adherence to policies, will lead to a qualified audit, which may hinder smart city development progress. Good governance is important to ensure regulatory compliance. Policies should be flexible so that they can be adapted to enable the successful implementation of a smart city in small and rural municipalities.

Information System Drivers for Smart City Development

The interviewees identified the elements that they regard as important information system drivers for smart city development in small and rural towns. The sixteen drivers, of which nine were deemed critical, are discussed in Mashau and Kroeze (2026a). The nine critical drivers are network infrastructure, innovation, data, modern software, modern infrastructure, digital infrastructure, suitable technology, governance, and automation of processes. The other seven, less critical, drivers are: security and privacy, advanced electronic devices, automated applications, roll-out of Wi-Fi, digital sensors, online applications, and data analysis (including big data analytics). Regarding the availability of these drivers, the overall sentiment between the participants was that, although some drivers were in place, there was still preparatory work to be done before they could implement a smart city in their municipalities.

Revised Framework

The data collected through the semi-structured interviews were analyzed to revise the conceptual framework (Mashau et al., 2021). This exercise assisted in improving the framework by integrating 28 components that the interview participants deemed central. The conceptual framework had 27 components, some of which were discarded and others retained in the revised framework. The interviewees concurred that the framework should have technological, organizational, human, and environmental readiness factors. These were the core components in the conceptual framework and were retained in the revised framework to ensure that municipalities may be assessed comprehensively for smart city implementation. Due to space limitations, the revised framework is not shown in this paper, but it can be viewed on p. 148 in Mashau (2023).

The results of interview data emphasized the importance of human readiness, which stresses that citizens and municipality employees should have relevant skills and a specific level of educational qualifications. It promotes innovative thinking and support for smart city initiatives. In addition, the revised framework suggests that a municipality should have an avenue to generate enough revenue to fund its smart city initiative because government grants and subsidies would not be enough for projects of this caliber. Moreover, a municipality should forge partnerships with both private and public institutions to enhance their current resources and skills.

In assessing small and rural municipalities' readiness, the technology readiness factor is vital. For example, in a smart city, there is a need for smart digital devices that can be used to connect citizens over the internet. Furthermore, to realize a smart city dream, there is a need for reliable internet connectivity to facilitate citizen engagement with municipal services. Municipal policies must support the development and implementation of smart cities while making sure that the citizens and staff conform to the policies. Furthermore, traditional leaders were identified as key stakeholders (Mashau & Kroeze, 2025). Therefore, they must be engaged in fostering community buy-in. Likewise, the local economy was identified as a key element for developing smart city initiatives successfully in South Africa's small and rural municipalities.

Sections 4.2 and 4.3 focus on the results of the qualitative online surveys conducted to evaluate and validate the revised framework. The participants are referred to as P A, P B, etc.

4.2 Framework Evaluation

The revised framework was evaluated through participatory design. This process was conducted to confirm whether the interviewees' inputs were captured satisfactorily in the revised framework. The participants for the second round of data collection were selected from the initial pool of interviewees by asking the gatekeepers of the contributing municipalities to nominate one interviewee from each municipality. This process mitigated bias on the researchers' side in the selection process.

During this iteration, the researchers gathered participants' views via an online qualitative survey using Google Forms. A link was shared with the nominated participants through an email. After that, the feedback was extracted and stored in a Microsoft Excel spreadsheet and a Portable Document Format (PDF) file for data analysis. To analyze the data in PDF, ATLAS.ti was utilized. The data collected was analyzed using thematic analysis. The relevancy of the revised framework was used as a first theme.

Relevancy of the Revised Framework. The participants agreed that the revised framework was relevant in assessing small and rural municipalities' readiness for smart city implementation. In addition, the findings further emphasized that this framework was more focused because it had been designed solely for the context of small and rural municipalities. The participants also indicated that the framework was unique and focused on addressing the challenges and pitfalls encountered by small and rural municipalities when engaging in a smart city project. The findings further highlighted that the revised framework came at the right time with municipalities being in the early stages of implementing a smart city concept. The findings also affirmed that municipalities would be able to identify areas that require attention before the development of a smart city.

> Our municipality is currently in the early stages of the smart city project, so this framework is timely. As a municipality, at least, we are able to determine where we are lacking. I like that we can use an assessment tool to rate ourselves in a variety of areas. Sincerely, something like this that specifically examines our municipalities is long overdue for a municipality of our nature. (P A)

> This framework is relevant because it was designed using information that affects local government. The issue of infrastructure, energy or electricity, internet connectivity, education qualification, skills, etc. If we can be ready in these aspects, I hope we can be able to implement smart city. (P D)

Missing Components on the Revised Framework. The participants were asked to comment on any missing components in the proposed revised framework. They agreed that their input was properly factored in on the revised framework and they were satisfied with the framework. The participants confirmed that the revised framework was comprehensive because all components missing from the initial proposed framework were added to the revised framework. The results highlighted that politicians, traditional leaders, revenue generation, and the availability of internet connection, which had been missing in the conceptual framework, were now included. The participants indicated that they could not identify any missing components. Therefore, no new components were added to the revised framework in this phase.

> As far as I could tell, this covered everything we talked about in the interview. I like that this framework now has internet connectivity and traditional leaders. I appreciate that. (P A)

> This framework is comprehensive enough to identify areas that require further development. I'm glad you covered the important points we discussed in our conversation. (P B)

Suitability of the Revised Framework. The participants were asked to comment on the suitability of the revised framework to assess small and rural municipalities' readiness for smart city implementation. The findings confirmed the participants' agreement that the updated version of the framework was suitable for this purpose.

> Yes, 100 percent. My colleagues and I are happy about the development of this framework and the assessment tool. This will alleviate the stress regarding smart city projects because there are no guidelines specific to our municipalities. (P A)

4.3 Framework Validation

The evaluated framework was further verified to identify any usability issues and to confirm its suitability to be used for readiness assessment. The validation of the framework was achieved through expert review. The evaluated framework was sent to seven experts purposively selected and granted permission to review the framework. The group of experts constituted of scholars who have already designed frameworks within the smart city domain, and managers knowledgeable in the smart city domain of South African metropolitan municipalities.

The data was further collected through an online qualitative survey using Google Forms. The researchers sent all selected participants a link via email. The experts' feedback was later extracted from Google Forms and entered into a Microsoft Excel spreadsheet and PDF document for thematic analysis with ATLAS.ti.

The findings from this round of data collection showed that most participants concurred that the framework was comprehensive and that no component was missing from the revised framework. However, a few participants pointed out that there was a need to incorporate buildings or homes under the environmental readiness factor in the evaluated framework. This suggestion is also supported by Dewi et al. (2018) and Das (2020) indicating that the environment for a smart city should have smart buildings to collect data from households without any human intervention.

I am not able to judge but I think the framework is comprehensive. (P A)

I think there is a need to assess buildings/homes under an environment. (P F)

4.4 *Validated Framework*

To finalize the artifact, the researchers incorporated building components into the evaluated framework to comprehensively assess small and rural municipalities' readiness for smart city implementation. The validated framework, depicted in Fig. 1, represents the final artifact using the multidimensional readiness factors that are critical for such an appraisal. The multidimensional readiness factors include four core dimensions: human, organizational, technological, and environmental readiness.

Each dimension has various components that assist in assessing its readiness. For example, the human readiness dimension comprises the following components to assess human readiness: skills, innovation, experience, educational qualification, and citizen support. The organizational readiness dimension also includes skills, experience, and educational qualification, while adding political support, budget,

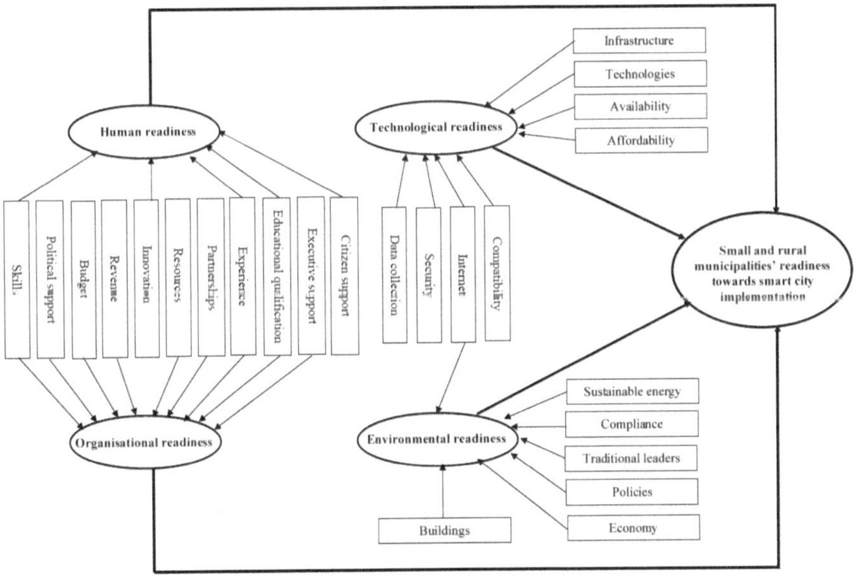

Fig. 1 Validated framework for the assessment of small and rural municipalities' readiness for smart city implementation

revenue, resources, partnerships, and executive support to assess organizational readiness. Data collection, security, internet, compatibility, infrastructure, technologies, availability, and affordability are central to assessing technology readiness. Lastly, the environmental readiness dimension also indicates the need to look at the internet—while adding sustainable energy, compliance, traditional leaders, policies, economy, and buildings—to assess environmental readiness.

The final artifact, comprising and integrating all these factors, forms a comprehensive and validated framework. It serves as the foundation for assessing small and rural municipalities' readiness, considering both preexisting factors and required additions to smart city development. This all-inclusive artifact ensures a holistic and balanced assessment across the human, organizational, technological, and environmental readiness factors that are critical for smart city implementation.

During the semi-structured interviews conducted in the second DSR cycle, some participants expressed the need for a practical tool that employees of small or rural municipalities could use to assess the state of readiness of their towns. The first author designed a quantitative measuring tool as a byproduct of the second DSR cycle. The artifact consisted of a table encompassing all the elements of the framework, with columns for a Yes/No checklist and numerical scoring. Like the framework, the tool was also appraised during the evaluation and validation cycles and amended as suggested by the participants, thereby constituting the final quantitative readiness assessment tool. Furthermore, formulas were now added to calculate the readiness levels of the various indicators, as well as an overall readiness score. A heuristic was provided to interpret the average numeric score (Mashau & Kroeze, 2024). The participants indicated that the tool was comprehensive and suitable for assessing a municipality's level of readiness, while also guiding future action.

> The suggested assessment tool is of immense value as it encompasses all relevant components that constitute implementation of smart city readiness. (P D)

5 Conclusion

The four DSR cycles of a wider project on smart city readiness assessment, of which the third and fourth cycles were discussed here, yielded a validated artifact as a robust and multidimensional framework. It serves as a foundational instrument for assessing readiness levels for smart city implementation in small and rural municipalities. The findings highlight actionable factors such as the need for ICT literacy, leadership support, and infrastructure development, making the framework useful for policymakers, municipal leaders, and consultants. By using the practical assessment tool, based on the validated framework, they may measure the readiness of small and rural municipalities quantitatively before they engage in smart city projects. The validated framework and measuring tool may assist small and rural municipalities in identifying areas that need improvement.

The validated framework, as well as its accompanying quantitative assessment tool, contributes to the field of Information Systems by amending current knowledge regarding the smart city environment with its unique focus on small and rural municipalities in South Africa. The framework's four dimensions—especially the human readiness dimension—and their 25 elements enrich the TOE framework (DeLone & McLean, 2003). Some of the participants indeed noted that the framework amends the TOE framework by adding the human readiness dimension.

A limitation of the study is its exclusive reliance on qualitative data. Therefore, subsequent research could adopt quantitative or mixed method approaches to strengthen the validity of the findings and address the limitations of purposive and snowball sampling.

Although the validated framework was not implemented in practice by the researchers themselves, one of the participants indicated in the evaluation round that the first version of the assessment tool was used in their municipality. It was found to be highly useful in understanding their current position as a municipality. To assess its real-life suitability, accuracy, and usefulness, the validated framework should be tested by implementing it practically in appropriate municipalities. Likewise, including a broader range of municipalities to expand the empirical validation, potentially in different countries, could improve the generalizability of the findings.

References

Aljowder, T., Ali, M., & Kurnia, S. (2019). Systematic literature review of the smart city maturity model. In *2019 International conference on innovation and intelligence for informatics, computing, and technologies (3ICT 2019)* (pp. 1–9). https://doi.org/10.1109/3ICT.2019.8910321

Anthony, B., Jr. (2024). The role of community engagement in urban innovation towards the co-creation of smart sustainable cities. *Journal of the Knowledge Economy, 15*, 1592–1624. https://doi.org/10.1007/s13132-023-01176-1

Arief, A., Abbas, M. Y., Wahab, I. H. A., Latif, L. A., Abdullah, S. D., & Sensuse, D. I. (2020). The smart islands vision: Towards smart city readiness in local government of archipelagos. *Journal of Physics: Conference Series, 1569*(4), 1–8. https://doi.org/10.1088/1742-6596/1569/4/042006

Barham, H., & Daim, T. (2020). The use of readiness assessment for big data projects. *Sustainable Cities and Society, 60*(1), 1–12. https://doi.org/10.1016/j.scs.2020.102233

Bashynska, I., & Dyskina, A. (2018). The overview-analytical document of the international experience of building smart city. *Business: Theory and Practice, 19*(1), 288–241. https://doi.org/10.3846/btp.2018.23

Beyers, L. (2016). Service delivery challenges facing municipalities: A case study of Fetakgomo local municipality in Sekhukhune district municipality, Limpopo Province. *Bangladesh e-Journal of Sociology, 13*(2), 167–178. https://www.cabidigitallibrary.org/doi/full/10.5555/20163284324

Bibri, S. E. (2021). Data-driven smart sustainable cities of the future: An evidence synthesis approach to a comprehensive state-of-the-art literature review. *Sustainable Futures, 3*, Article 100047. https://doi.org/10.1016/J.SFTR.2021.100047

Chauhan, A., Jakhar, S. K., & Chauhan, C. (2021). The interplay of circular economy with industry 4.0 enabled smart city drivers of healthcare waste disposal. *Journal of Cleaner Production, 279*, 1–9. https://doi.org/10.1016/j.jclepro.2020.123854

Chichernea, V. (2015). Smart cities communities and smart ICT platform. *Journal of Information Systems & Operations Management, 9*(1), 1–11. http://www.rebe.rau.ro/RePEc/rau/jisomg/SU15/JISOM-SU15-A3.pdf

Cilliers, L., & Flowerday, S. (2017). Factors that influence the usability of a participatory IVR crowdsourcing system in a smart city. *South African Computer Journal, 29*(3), 16–30. https://doi.org/10.18489/sacj.v29i3.422

Cresswell, J. W., & Creswell, J. D. (2022). *Research design: Qualitative, quantitative and mixed methods approaches* (6th ed., Kindle). Sage.

Das, D. K. (2020). Perspectives of smart cities in South Africa through applied systems analysis approach: A case of Bloemfontein. *Construction Economics and Building, 20*(2), 75–88. https://doi.org/10.5130/AJCEB.v20i2.6657

DeLone, W. H., & McLean, E. R. (2003). The DeLone and McLean model of information systems success: A ten-year update. *Journal of Management Information Systems, 19*(4), 9–30. https://doi.org/10.1080/07421222.2003.11045748

Desdemoustier, J., Crutzen, N., & Giffinger, R. (2019a). Municipalities' understanding of the smart city concept: An exploratory analysis in Belgium. *Technological Forecasting and Social Change, 142*, 129–141. https://doi.org/10.1016/j.techfore.2018.10.029

Desdemoustier, J., Crutzen, N., Cools, M., & Teller, J. (2019b). Smart city appropriation by local actors: An instrument in the making. *Cities, 92*(1), 175–186. https://doi.org/10.1016/j.cities.2019.03.021

Dewi, M. A. A., Hidayanto, A. N., Purwandari, B., Kosandi, M., & Budi, N. F. A. (2018). Smart city readiness model based on technology-organization-environment (TOE) framework and its effect on adoption decision. In: *Proceedings of the 22nd Pacific Asia conference on information systems—Opportunities and challenges for the digitized society: Are we ready? PACIS 2018, Yokohama, Japan* (pp. 1–15). https://aisel.aisnet.org/pacis2018/268

Du, J., Zhu, Q., Shi, Y., Wang, Q., Lin, Y., & Zhao, D. (2020). Cognition digital twins for personalized information systems of smart cities: Proof of concept. *Journal of Management in Engineering, 36*(2), 1–17. https://doi.org/10.1061/(asce)me.1943-5479.0000740

Fang, Y., & Shan, Z. (2024). Optimising smart city evaluation: A people-oriented analysis method. *IET Smart Cities, 6*(1), 41–53. https://doi.org/10.1049/SMC2.12073

Galal, A., & Elariane, S. (2022). A tool for measuring the smart readiness of existing buildings and cities in Egypt—Focusing on energy sector: Towards smart ready checklist. *HBRC Journal, 18*(1), 141–155. https://doi.org/10.1080/16874048.2022.2129141

Gregor, S., & Hevner, A. R. (2013). Positioning and presenting design science research for maximum impact. *MIS Quarterly, 37*(2), 337–355. https://doi.org/10.2753/MIS0742-1222240302

Harrison, C., Eckman, B., Hamilton, R., Hartswick, P., Kalagnanam, J., Paraszczak, J., & Williams, P. (2010). Foundations for smarter cities. *IBM Journal of Research and Development, 54*(4), 1–16. https://doi.org/10.1147/JRD.2010.2048257

Hartley, M. K., & Seymour, L. F. (2011). Towards a framework for the adoption of business intelligence in public sector organisations: The case of South Africa. In *Proceedings of the South African Institute of Computer Scientists and Information Technologists conference on knowledge* (pp. 116–122). https://doi.org/10.1145/2072221.2072235

Hartley, M. K. (2015). *An analysis of business intelligence for improved public service delivery* [Master's thesis]. Department of Information Systems, University of Cape Town. https://open.uct.ac.za/server/api/core/bitstreams/317b52be-d5b4-46d9-847a-d604da191a53/content

Kniazevych, A. (2024). Sustainable municipal management model: The smart concept. *European Research Studies Journal, XXVII*(4), 1442–1463. https://doi.org/10.35808/ersj/3584

Lämmel, P., Merbeth, J., Cleffmann, T., & Koch, L. (2024). Towards municipal data utilities: Experiences regarding the development of a municipal data utility for intra- and intermunicipal actors within the German city of Mainz. *Smart Cities, 7*(3), 1289–1303. https://doi.org/10.3390/smartcities7030054

Mahesa, R., Yudoko, G., & Anggoro, Y. (2019). Dataset on the sustainable smart city development in Indonesia. *Data in Brief, 25*, Article 104098. https://doi.org/10.1016/J.DIB.2019.104098

Mashau, N. L., Kroeze, J. H., & Howard, G. R. (2021). An integrated conceptual framework to assess small and rural municipalities' readiness for smart city implementation: A systematic literature review. In Y. M. Huang, C. F. Lai, & T. Rocha (Eds.), *Innovative technologies and learning (ICITL 2021)* (pp. 262–273). Springer. https://doi.org/10.1007/978-3-030-91540-7_28

Mashau, N. L. (2023). *Developing an integrated framework to assess South African small and rural municipalities' readiness for smart city implementation* [Ph.D. thesis]. Department of Information Systems, University of South Africa. https://hdl.handle.net/10500/31908

Mashau, N. L., & Kroeze, J. H. (2023). Challenges that affect smart city implementation in small and rural municipalities. *South African Journal of Information Management, 25*(1), 1–6. https://doi.org/10.4102/sajim.v25i1.1703

Mashau, N. L., & Kroeze, J. H. (2024). A readiness assessment tool for smart city implementation in small and rural municipalities. *Journal of Information Systems and Informatics, 6*(4), 2937–2964. https://doi.org/10.51519/journalisi.v6i4.937

Mashau, N. L., & Kroeze, J. H. (2025). The role of traditional leaders in implementing sustainable smart cities in rural municipalities. In A. Nagar, D. S. Jat, D. Mishra, & A. Joshi (Eds.), *Intelligent sustainable systems, worldS4 2024, lecture notes in networks and systems* (pp. 191–200). Springer. https://doi.org/10.1007/978-981-97-8695-4_18

Mashau, N. L., & Kroeze, J. H. (2026a). Critical smart city drivers for small and rural municipalities. In: *Proceedings of the ICRAIS 2025* Pointe aux Piments, Mauritius, 10–12 September 2025 (progress in IS, Springer). https://doi.org/10.1007/978-3-032-14099-9_19

Mashau, N. L., & Kroeze, J. H. (2026b). Validating a framework for assessing small and rural municipalities' readiness for smart city implementation. In: *Proceedings of the ICRAIS 2025* Pointe aux Piments, Mauritius, 10–12 September 2025 (progress in IS, Springer). https://doi.org/10.1007/978-3-032-14099-9_55.

Mashau, N. L., Kroeze, J. H., & Howard, G. R. (2022). Key factors for assessing small and rural municipalities' readiness for smart city implementation. *Smart Cities, 5*(4), 1742–1751. https://doi.org/10.3390/smartcities5040087

Mohd Ishak, N., Yazid, A., & Bakar, A. (2014). Developing sampling frame for case study: Challenges and conditions. *World Journal of Education, 4*(3), 29–43. https://doi.org/10.5430/wje.v4n3p29

Montalvan Castilla, J. E., & Riel Müller, A. (2024). A smart city for all citizens: An exploration of children's participation in Norway's smartest city. *International Planning Studies, 29*(1), 19–33. https://doi.org/10.1080/13563475.2023.2259110

Mora, L., Deakin, M., & Reid, A. (2019). Combining co-citation clustering and text-based analysis to reveal the main development paths of smart cities. *Technological Forecasting and Social Change, 142*(1), 56–69. https://doi.org/10.1016/j.techfore.2018.07.019

National Treasury. (2011). *The 2011 local government budgets and expenditure review: 2006/07—2012/13.* https://www.treasury.gov.za/

Noori, N., de Jong, M., & Hoppe, T. (2020). Towards an integrated framework to measure smart city readiness: The case of Iranian cities. *Smart Cities, 3*(1), 676–704. https://doi.org/10.3390/smartcities3030035

Peffers, K., Tuunanen, T., Rothenberger, M. A., & Chatterjee, S. (2007). A design science research methodology for information systems research. *Journal of Management Information Systems, 24*(3), 45–77. https://doi.org/10.2753/MIS0742-1222240302

Prendeville, S., Cherim, E., & Bocken, N. (2018). Circular cities: Mapping six cities in transition. *Environmental Innovation and Societal Transitions, 26*(1), 171–194. https://doi.org/10.1016/j.eist.2017.03.002

Rana, N. P., Luthra, S., Mangla, S. K., Islam, R., Roderick, S., & Dwivedi, Y. K. (2019). Barriers to the development of smart cities in Indian context. *Information Systems Frontiers, 21*(3), 503–525. https://doi.org/10.1007/s10796-018-9873-4

Rogers, E. M. (2003). *Diffusion of innovations.* Free Press.

SALGA. (2017/2018). *Annual report 2017/2018.* SALGA (South African Local Government Association). https://www.salga.org.za/

Steyn, J., & Ngwenya, N. (2015). *Community protest: Local government perceptions*. SALGA (South African Local Government Association). https://www.salga.org.za/

Stübinger, J., & Schneider, L. (2020). Understanding smart city—A data-driven literature review. *Sustainability, 12*(20), 1–23. https://doi.org/10.3390/su12208460

Tan, S. Y., & Taeihagh, A. (2020). Smart city governance in developing countries: A systematic literature review. *Sustainability, 12*(3), 899. https://doi.org/10.3390/su12030899

Veselitskaya, N., Karasev, O., & Beloshitskiy, A. (2019). Drivers and barriers for smart cities development. *Theoretical and Empirical Researches in Urban Management, 14*, 85–110. https://um.ase.ro/v14i1/5.pdf

Wieringa, R. (2016). Design science research in information systems and software systems engineering (Tutorial). In *8th international conference on web engineering (ICWE)*. http://icwe2016.inf.usi.ch/program/tutorials/design-science-research

Yu, J., Wen, Y., Jin, J., & Zhang, Y. (2019). Towards a service-dominant platform for public value co-creation in a smart city: Evidence from two metropolitan cities in China. *Technological Forecasting and Social Change, 142*, 168–182. https://doi.org/10.1016/J.TECHFORE.2018.11.017

Zhao, H., Wang, Y., & Liu, X. (2021). The evaluation of smart city construction readiness in China using CRITIC-G1 method and the Bonferroni operator. *IEEE Access, 9*, 70024–70038. https://doi.org/10.1109/ACCESS.2021.3078191

Preliminary Results of the AI Supported Detection of Cyclists and Related Overtaking Situations

Johannes Schering, Maximilian Bosse, Fatih Karahan, Jorge Marx Gómez, Daniel Mentrup, and Johannes Wilbers

Abstract Traffic safety is a key factor to make cycling more attractive. The detection cyclists is a topic that is not well reflected by existing research. In this context, the research project BikeDetect tries to identify sensor and AI supported approaches to improve cyclist detection in real-traffic scenarios. As part of this research, we present as a first step the results of a requirement analysis regarding type and functionalities of the required sensors (ultrasonic, radar, etc.). Existing AI models were identified that potentially may be applied to the camera data that is gathered as part of the project. Based on the selected sensors, a first test study outside of the real-traffic situation was conducted to gather initial data. A suitable data management including an ETL process was designed. An existing YOLO11n object detection model was applied to the data and further refined. Based on the camera data we received, we applied a publicly available dataset. This refined dataset was used to fine-tune a pre-trained YOLO11n model. Preliminary results demonstrate the model's effectiveness in detecting road users including cyclists and pedestrians. In the end, a summary of the results and an outlook for future research will be presented. The results of the object detection will be applied to enable a fusion of the data streams in the future.

1 Introduction

Cycling promotion has many advantages for cities (e.g., reduction of pollution, saving of costs) (Taddei et al., 2015). Traffic safety is one of the key factors that impact the use of different means of transport (Gössling & McRae, 2022). Cycling accidents cause high costs for the society (Palmer et al., 2014). Serious bike accidents involving motorized traffic are happening quite often (Sze et al., 2011). It is assumed that a car is involved in more than 50% of all cycling fatalities (López-Molina et al., 2023). In

J. Schering (✉) · M. Bosse · F. Karahan · J. Marx Gómez
Department of Business Informatics VLBA, University of Oldenburg, Oldenburg, Germany
e-mail: johannes.schering@uni-oldenburg.de

D. Mentrup · J. Wilbers
iotec GmbH, Osnabrück, Germany

J. Marx Gómez et al. (eds.), *Recent Advances in Information Systems*, Progress in IS,
https://doi.org/10.1007/978-3-032-14099-9_56

case of high traffic volumes, the number of accidents is further increased (Labetski & Chum, 2020).

To avoid collisions, driving assistance systems were part of research. Turning assistance systems that were tested contributed to prevent serious bike accidents (Tomasch et al., 2024). However, overtaking situations in traffic scenarios remain an unsolved problem. According to German legal road traffic regulations, the minimum distance when a motorized vehicle is overtaking a bicycle is 1.5 m in inner cities and 2 m at outer city roads (StVO, 2021). According to former studies, a high percentage of motorized vehicles are overtaking below legal minimum distances (Pérez-Zuriaga et al., 2021). To improve cyclist detection, research has been conducted with smart camera systems that alert the drivers (German Aerospace Center DLR, 2025). Camera-based approaches and the fusion of thermal and RGB cameras have been extensively investigated (Ahmed et al., 2019). Applications that relate to Artificial Intelligence (AI) are often discussed when it comes to autonomous driving vehicles (Wang et al., 2023). However, systems that are able to detect cyclists are generally underrepresented compared to other road participants (Gaio & Cugurullo, 2023; Zhang et al., 2020) as the training datasets contain comparatively less instances (Corral-Soto et al., 2410) and there is a limited amount of benchmark datasets for specific purposes such as cyclist detection (Ahmed et al., 2019) or detection of overtaking scenarios (Corral-Soto et al., 2410). The first step to develop new driving assistance system to improve cyclist safety in overtaking situations is the application of existing AI models, and a following observation of weaknesses and strengths regarding the gathered video data that may differ in camera angles, etc.

A prototypical approach to process video data to detect cyclists and pedestrians is presented. The training and implementation of an object detection model will be the foundation to develop new driving assistance systems that may support car drivers to keep sufficient distance in real-traffic scenarios. Requirements were collected as part of a stakeholder workshop regarding suitable sensors and functionalities of AI models for cyclist detection in traffic situations. The camera data and different types of other sensor data are gathered by a combined sensor system that will be briefly explained is the foundation of a data driven assistance system. Existing AI models and pre-trained data sets may be potentially applied to our specific use case (Chap. 2). As part of the BikeDetect project, we conducted a first test study outside of real traffic with the selected sensors to test these under outdoor conditions. Heterogeneous data requires a suitable data management. The labeling process and the fine-tuning when an available AI model was applied to the initial data will be explained (Chap. 3). Preliminary results of the object detection model including important performance metrics and potential limitations will be presented (Chap. 4). As a final step, the findings will be summarized and an outlook to future research potentials will be provided (Chap. 5).

2 Requirements Analysis

The first step is the systematic collection of requirements regarding suitable sensor technologies and the analysis of AI frameworks with potential to detect cyclists in traffic situations. Therefore, the project conducted a requirements workshop at the municipal administration in Osnabrück, Lower Saxony, Germany. Key results will be presented in the following.

2.1 *Sensor Technology*

The applicability of different types of sensors in different scenarios of laboratory and field studies were discussed as part of the workshop. Different types of sensors are combined as part of the sensor system. One part of the sensors is responsible for the detection of cyclists, the other part measures the distances to objects (or both). By merging the data streams in the future, distances between cars and bikes may be measured. Potential approaches that are feasible and are suitable for data fusion to extract concrete overtaking situations will be presented in future publications. Figure 1 shows the type of sensors that were selected based on the results of the workshop.

In the following, we describe the sensor system more in detail. Optical sensors based on the time-of-flight principle, such as point LiDAR (Benewake, 2025) and 360° scanning LiDAR (Slamtec, 2025) systems, enable precise distance measurements. In particular, scanning LiDAR also enables object classification based on geometric features. Additionally, a monocular RGB (Bosch, 2025) and a stereoscopic camera (Corporate et al., 2025) are used. While the monocular RGB camera provides

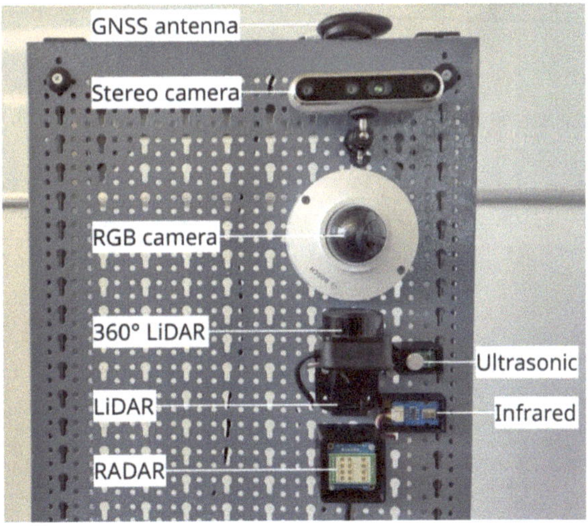

Fig. 1 Pre-selected sensor system by iotec includes stereo camera, RGB camera, 360° LiDAR, ultrasonic sensor, LiDAR, infrared (thermic sensors) and radar

image data for object detection and classification, the stereoscopic camera enables additional distance estimation. To complement the optical sensors, an ultrasonic sensor (Texas Instruments, 2025) is employed, characterized by its robust distance measurement performance under challenging environmental conditions. A radar sensor (RFbeam Microwave, 2025) provides information on the direction, speed, and distance of approaching objects—especially cyclists—using Frequency-Modulated Continuous-Wave (FMCW) measurements. To distinguish between humans and inanimate objects, an infrared sensor (Panasonic Industry, 2025) that measures surface temperature is applied, allowing for inference of object type. For temporal and geographical synchronization of sensor data, a GNSS antenna (Navilock, 2025) that also serves to measure the vehicle's speed is used.

2.2 *Performance Requirements of AI Models*

The discussion about functionalities of AI models in cyclist detection was also part of the expert workshop. The requirements regarding the performance of AI models that detect cyclists in real-traffic scenarios are displayed in Table 1. These were defined based on the requirements patterns according to Rupp which differentiates between "must," "should," and "will" requirements (Rupp, 2009). Cyclist detection in real-traffic scenarios from different perspectives is essential. As part of later driving assistance systems, the AI models shall perform under different weather and lightning conditions, and must enable the robust detection of cyclists in real time. Based on the results of the workshop (BikeDetect, 2025), the pre-collected requirements were further refined.

Table 1 Requirements of the AI model for cyclist detection

No	Requirement
1	The model must be applicable in road traffic
2	The model must be able to recognize cyclists in images
3	The model must be able to recognize cyclists in images from different perspectives or in different sizes
4	The model must be able to recognize cyclists in images on daylight
5	The model will be able to recognize e-scooter users in images (in the future)
6	The model should be able to recognize cyclists in different weather conditions, e.g., sunshine, rain or fog
7	The model must reliably recognize cyclists even when these are not completely visible due to overlaps
8	The model must achieve a prediction speed of 30 frames per second (real-time capability)
9	The model should achieve a prediction accuracy (mean average precision) of at least 80%
10	The model should be able to reliably recognize cyclists even on previously unseen routes (transferability)

2.3 Performance Requirements of AI Models

As a next step, we conducted a literature and internet based research to identify AI models that may be potentially applied on the project stereo camera data. YOLO pioneered real-time object detection and remains under active development (Wang & Liao, 2024). As a single-stage detector, it prioritizes speed over accuracy, making it ideal for weaker hardware and real-time systems (Redmon et al., 2016; Wang & Liao, 2024). EfficientDet, also a single-stage detector, balances accuracy and computational efficiency through BiFPN and compound scaling (Tan et al., 2020). It meets real-time requirements but runs slower than YOLO while achieving higher accuracy (Duan et al., 2019). In the selection process, we prioritized the real-time requirement over the classification accuracy. Other models were considered but discarded: CenterNet is slower and less accurate than EfficientDet (Duan et al., 2019). SSD models show comparable performance to YOLO but lack ongoing development (Leibe et al., 2016). Although FSSD outperforms YOLO (Li et al., 2024), its hardware dependency and training overhead make it not suitable for this use case. Based on the identified advantages and disadvantages the decision was made to only proceed with models of the YOLO architecture.

As an alternative proceeding, while less important at the current stage of this research, data from LiDAR sensors (3D point cloud for 360° sensors) could be potentially processed using residual networks such as ResNet50 (Chang et al., 2023), clustering algorithms like DBSCAN (Zhang et al., 2020), models intended to process either raw cloud point data such as MV3D or those intended to work on 2D representations of the data like PointNet (Alaba & Ball, 2022). Potential approaches may be presented in forthcoming publications.

3 Data Management and Application of Existing AI Models on Test Data

3.1 Laboratory Study to Gather Initial Data

The chosen sensor system was recently tested in a first outdoor scenario outside of a real-traffic situation. The sensor system was mounted on a handcart that is equipped with wheels. This proceeding has the advantage that the data collection is not limited to a static scenario. The perspective goal is to integrate the sensor system into a driving assistant that informs car drivers about their overtaking behavior. The sensor system was tested on a parking space outside of a real-traffic situation in Osnabrück, Germany. In several scenarios (e.g., cycling, walking) initial sensor data was gathered. Various critical use cases were tested, including cyclist and pedestrian (person) detection, and challenging lighting conditions such as direct sun exposure. Another scenario simulated an overtaking situation when a cyclist with slow speed was overtaken by the sensor equipped handcart. In a following real-traffic scenario, the selected sensors will be removed from the handcart and will be attached to a motorized vehicle instead. It is expected that the driving speed of the vehicle will impact the sensor data. Therefore, overtaking situations with a sensor system that is in motion were also part of the study.

3.2 Data Management

The test study data was uploaded by the sensor provider iotec on a cloud solution hosted by the University with a specific user login. After receiving the data sources, these were removed from the cloud and were stored in an object storage that includes raw data for further processing. In general, the heterogeneous data consists of imagery and time-series distance measurement data. The object storage is implemented on a server that is hosted by the Department of Business Informatics. Regarding data collection scenarios in real-traffic situations, the data will be stored on an encrypted hard disc that will be provided to the University as part of a personal handover.

The data management concept can be described as a pipeline that consists of five steps (Fig. 2). The test study (Chap. 3.1) was conducted during the data collection. The data was initially stored on a cloud and was moved manually to a MinIO object storage (MinIO, 2025). During the extraction process, the raw data is downloaded from the object storage. As an example, an ultrasonic data set in the form of a JavaScript Object Notation (JSON) file and a Comma Separated Value (CSV) file is cached on a local server. In the next processing step, the transformation, the data sets are analyzed and processed according to specific requirements—for example, ultrasonic data is extrapolated to the distance at which the measurement was taken using a heuristic. The time-series data is loaded into a table object which is loaded into a PostgreSQL database. Different data models may be considered (splitting the data streams into different tables, etc.). In the final processing step, all of the data which was downloaded during the first step of the processing pipeline is deleted from the system (cleanup).

In the testing phase, the functionality performance of six sensors along with the stereo camera was evaluated. The sensor suite comprised a 360° LiDAR, forward-facing LiDAR, radar, ultrasonic sensor, thermal sensor, and a GPS module (2.1). To facilitate a detailed examination of sensor performance across diverse scenarios, a dedicated dashboard based on the Python Streamlit web framework (Streamlit,

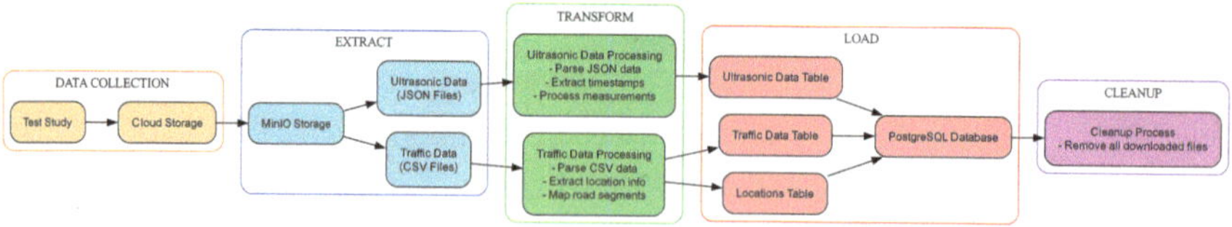

Fig. 2 Data processing pipeline

2025) was developed to enable interactive analysis and visualization of the sensor data. A first quality analysis of the test study revealed that because of data loss of some sensors a synchronization of the different data streams is not possible so far. Therefore, we concentrated in a first step on the processing of video data with an object detection model.

3.3 Labeling Process

After building up an ETL process, video data is ready for labeling. A Label Studio (Studio, 2025) instance was deployed on a virtual machine at the University, accessible only within the University network and secured behind an NGINX web server (nginx, 2025) to restrict external access to facilitate collaborative annotation of the data. The labeling process was initially conducted through manual annotation of the video data which was collected as part of the described test study. Annotators identified and marked objects frame by frame to ensure high-quality ground truth data in this process (Fig. 3). In total, 699 images were considered in the labeling process (311 cyclists, 388 pedestrians). The team added a ML-backend to pre-label video frames, shifting annotation tasks from manual labeling to verification and correction. The test video data was processed at the original camera frame rate of 60 frames per second (FPS) during the initial data collection phase. Two object classes (cyclist and person) were used for labeling.

3.4 Foundation Data Set and Fine-Tuning

Initial experiments with existing datasets revealed limitations for our specific use case. Preliminary observations of the Tsinghua-Daimler Cyclist Benchmark (TDCB) dataset (Li et al., 2016) indicate a potential over-representation of front and back views of cyclists (see Fig. 4). This may be possibly a result of the dataset's collection methodology using front-mounted vehicle cameras.

When a YOLO11n model fine-tuned on the TDCB dataset was applied to our laboratory test scenarios, it demonstrated suboptimal performance, particularly in detecting cyclists from side-view perspectives (Fig. 5). As a consequence, we enhanced a publicly available dataset from Roboflow (Roboflow Universe, 2025) through comprehensive curation, including correction of incorrectly annotated images and completion of missing annotations to ensure data quality. To systematically address the orientation distribution challenge, we implemented a strategic three-way cyclist classification system (cyclist_back, cyclist_front, cyclist_side) that served as both an analytical tool and method to ensure balanced representation across all viewing angles. The foundation dataset was supplemented with additional images, particularly emphasizing cyclist side orientations to create a more balanced distribution, contrasting with the front/back-heavy pattern.

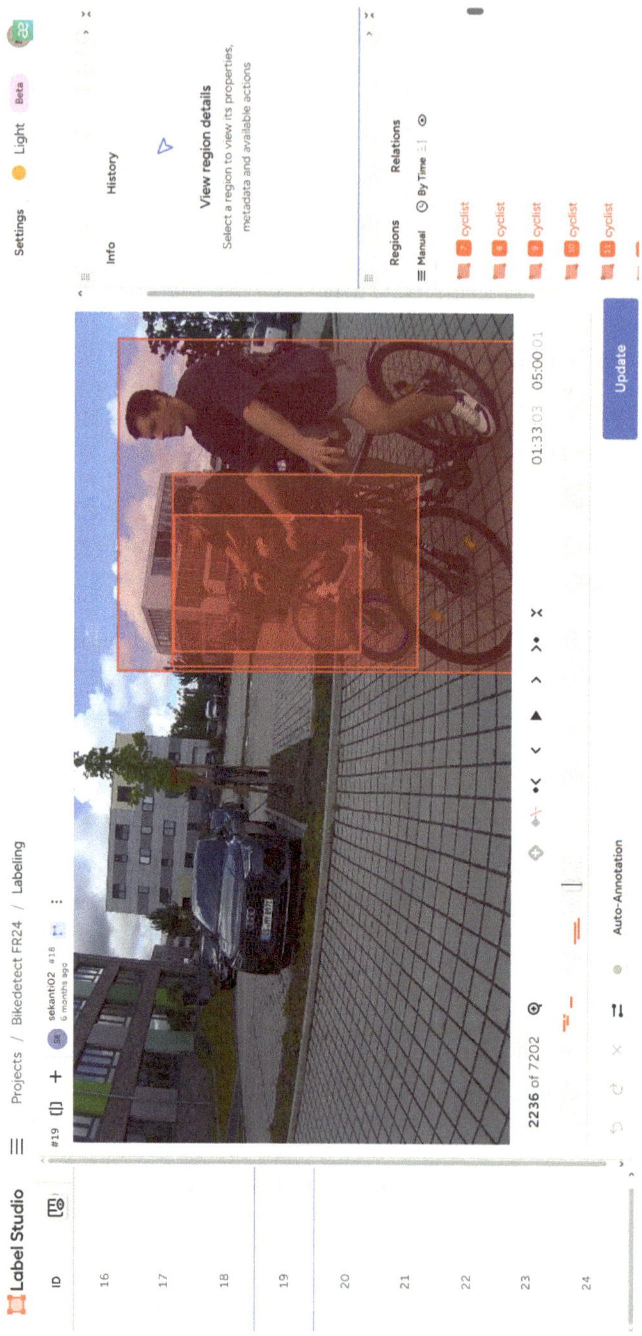

Fig. 3 Labeling process with label studio

Fig. 4 Tsinghua-Daimler cyclist benchmark (TDCB) dataset examples

The final enhanced dataset comprises 4.577 images distributed across four distinct classes: cyclist_back (4.073 instances), cyclist_front (4.041 instances), cyclist_side (3.160 instances), and person (10.406 instances). This distribution means reasonably balanced representation across cyclist orientations, representing a marked improvement over the limited side-view representation. For the baseline model fine-tuning, the three cyclist subcategories were consolidated into a single cyclist class during preprocessing to establish foundation performance metrics.

4 Preliminary Results of the Bicycle Detection

The model was trained for 30 epochs using Google Colab's T4 GPU (Google Colab, 2025) with image resolution of 640 × 640 pixels and batch size of 16. This approach ensures that model performance metrics reflect generalization capabilities to unseen, out-of-domain data rather than artificially inflated performance on similar training distributions.

The tradeoff between precision and recall is considered to evaluate the performance of the model. High precision ensures the reliability of detections logged as data points. In a safety–critical application such as cyclist detection, high recall is paramount to ensure that as few cyclists as possible are missed. To achieve a balance between these two metrics, the $F1$-score is employed. The $F1$-Confidence curve (Fig. 6) illustrates the $F1$-score across various confidence thresholds. The model achieves its optimal $F1$-score at a confidence threshold of 0.492. This specific threshold, therefore, represents the point at which the model best balances the need to detect most cyclists while minimizing erroneous alerts, and will be used for further performance evaluation.

The fine-tuned YOLO11n model, evaluated at the determined optimal confidence threshold of 0.492, demonstrates satisfactory performance on the validation dataset. The model achieved an overall mAP at an IoU threshold of 0.5 (mAP50) of 93.4% and an mAP50–95 of 66.1%. Class-specific analysis shows a mAP50 of 90.2% for cyclist

Fig. 5 Initial model results trained by TDCB dataset

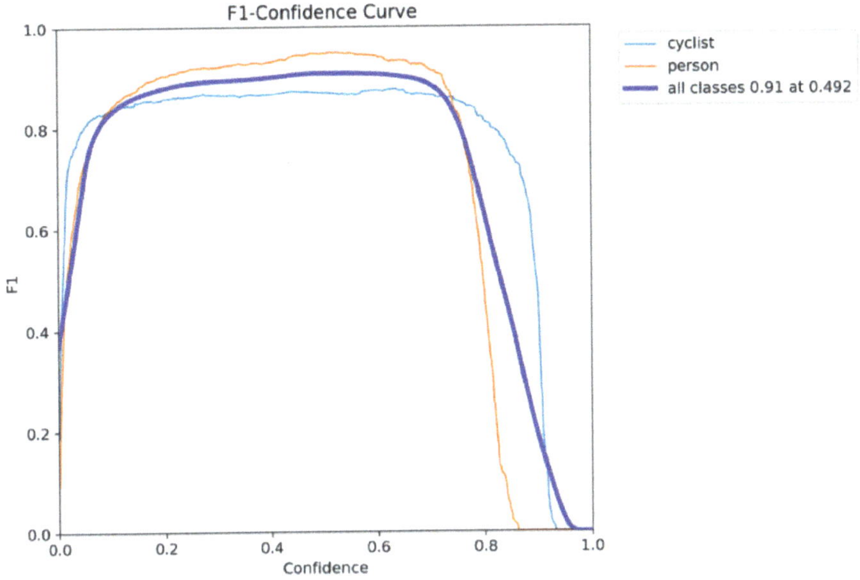

Fig. 6 $F1$-confidence curve

detection and 96.6% for person detection. The precision for cyclists was 94.1% with a recall of 82.0%. Persons precision reached 96.8% with a recall of 93.8% (Table 2). The model demonstrates processing inference speeds of approximately 2.2 ms per image on a T4 GPU, enabling real-time performance that substantially exceeds the 30 fps requirement. However, the adequacy for real-traffic deployment requires further investigation. If the model is too accurate, there is a risk that drivers rely solely on the model. If the value is too low, the prediction is not reliable. According to the society of automobile engineers, model predictions have to be at least more accurate than a human driver (Widen & Koopman, 2022).

Detailed analysis of the confusion matrix offers further insights into the model's classification behavior at the chosen 0.492 confidence threshold. The matrix (Fig. 7) indicates strong performance in correctly identifying the person class, with 364 out of 388 true person instances correctly classified (recall of 93.8%). For cyclists, 258 out of 311 true instances were correctly identified (recall of 83.0%). A strength of the model at this confidence level is its high precision for the detected classes (94.1% for cyclists, 96.8% for persons, Table 2). The confusion matrix shows relatively few instances

Table 2 Performance metrics of the fine-tuned model

Classes	Precision	Recall	mAP50	mAP50-95
All	0.955	0.879	0.934	0.661
Cyclist	0.941	0.82	0.902	0.65
Person	0.968	0.938	0.966	0.673

where one class is mistaken for the other (e.g., only 5 true cyclists were misclassified as persons). However, a critical area for improvement is the misclassification of cyclists as background, where 48 cyclist instances (15.4% of all cyclists) were not detected by the model. This rate of missed cyclist detections represents a potential safety concern and is a limitation to address in future model iterations.

Qualitative examination of detection failures provides further context to these quantitative findings. The top row at Fig. 8 showcases example instances where cyclists are misclassified as "person." These errors typically occur when the cyclist is in close proximity to the camera, resulting in a frame dominated by the rider's torso and limited visibility of the bicycle, or when significant portions of the bicycle are occluded. In such scenarios, the model appears to prioritize features common to the person class, leading to the misclassification, which aligns with the 5 instances noted in the confusion matrix. More critically, the bottom row illustrates example cases where cyclists are missed entirely and classified as "background." These false negatives, representing 48 missed cyclist detections (15.4%), typically occur when cyclists are partially visible, occluded, or at distance where distinguishing features are less prominent. Improving robustness to these failure modes will be a key focus for future model refinement to enhance detection reliability in safety–critical applications.

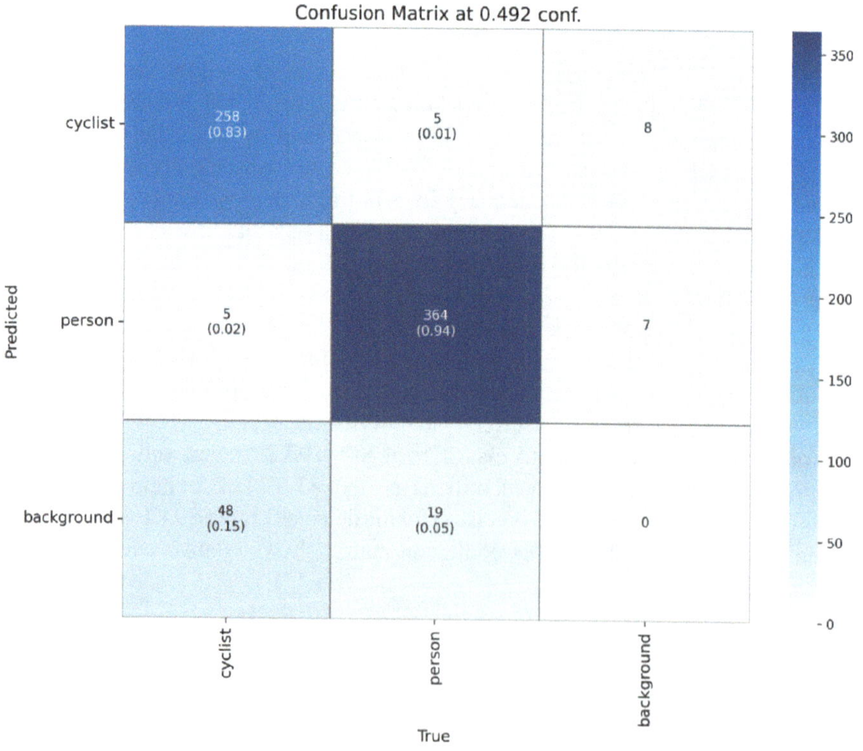

Fig. 7 Confusion matrix of the fine-tuned model with the highest performance

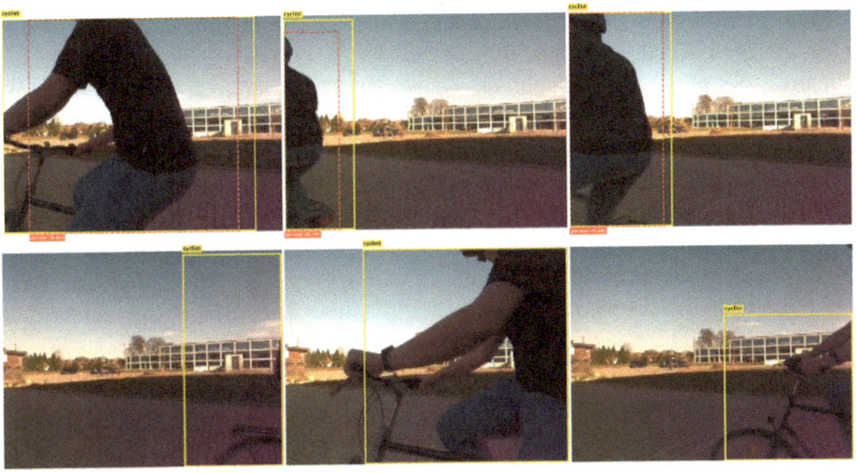

Fig. 8 Qualitative error analysis of the YOLO 11n model: examples for misclassification (top) and missed detections (bottom)

5 Results and Outlook

As we discussed in the motivation, overtaking situations are an unsolved issue regarding cycling safety. In the process of developing a driving assistant system, a robust detection of cyclists on camera-based data is a first prerequisite. As a first step, we defined requirements regarding the components of the sensor system that enables the detection of cyclists as an object as a basis to measure the distance to these objects. Besides camera systems, 360° LiDAR, ultrasonic sensor, LiDAR, infrared (thermic sensors) and radar are applied for classification of cyclists or distance measurement. "Must," "should," and "will" requirements to the AI models were collected. Because it is intended to integrate the sensor system in driving assistance systems in the future, the model needs to perform in real time with high accuracy. In general, YOLO models are suitable for real-time applications because of its reduced inference time and the ability to run on hardware with lower performance.

Based on the requirements, a first data collection was conducted to receive initial data with the selected sensor for classification and distance measurements. To create a realistic scenario, the sensor system was mounted to a moveable handcart. The scenario outside of a real-traffic situation considered different use cases (cycling, walking, overtaking etc.). To process this kind of data, a suitable data management including the ETL process was designed. Regarding the AI based detection of cyclists, an image-based model was obtained by training YOLO11n on the Tsinghua-Daimler Cyclist Detection Benchmark Dataset (Li et al., 2016). When the model was applied on the video data, only a part of the cyclists could be successfully predicted. It revealed that the data set that was applied to train the model needs adjustments to consider all relevant perspectives.

Many of the requirements that were defined in the beginning are already fulfilled or at least partly fulfilled. The application of the AI models to data gathered in real-traffic scenarios will be realized as soon as related data is available (see below, *R*1). The detection of cyclists on image-based data was successfully implemented. However, it has to be clarified whether the differentiation of different classes is already sufficient or has to be improved (*R*2). In general, cyclists may be detected by the object detection model in different perspectives and sizes. For that purpose, the existing data set that was applied for training was expanded by images including the side view (*R*3). Cyclists may be detected in a daylight scenario (*R*4). The next requirement is not fulfilled so far as scooter rollers were not considered in the first test study data collection (*R*5). The same statement can be made for different weather conditions. The test study data was entirely gathered in daylight under very sunny conditions (*R*6). Based on the current data set (partially) covered cyclists were considered in the training process. However, in the initial data set it was not differentiated how many images with covered cyclists are included (*R*7). On the available hardware (NVIDIA Tesla T4 GPU) around 476 images per second may be processed (*R*8). The must-requirement of real-time detection is already fulfilled. The mAP50 value for cyclists reaches 90.5% which succeeds the required 80% (*R*9). The last requirement has not been fulfilled so far. Different routes in the inner city will be considered in the field test (see below, *R*10).

As an outlook, the AI model may be further improved. To address the primary limitation of missed cyclist detections, particularly in scenarios involving partial occlusion, and to enhance robustness to challenging variations in object scale (ranging from distant cyclists appearing small to close-up views where the cyclist dominates the frame and bicycle features may be less apparent), future work will focus on the hyperparameter tuning. The exploration will include techniques such as cutmix, which combines portions of images to create explicit occlusion scenarios; translate, which helps the model learn from partially visible objects at image boundaries; erasing, which directly simulates occlusions by removing image patches; mosaic augmentation, which combines multiple images thereby inherently creating varied object scales and partial views; direct scale augmentation, which can improve robustness to cyclists appearing at different distances or in close-up situations. Fine-tuning the parameters governing these augmentations along with remaining hyperparameters not explicitly mentioned is hypothesized to improve the model's generalization capabilities in these specific scenarios. The objective of future work is to explore the potential for reducing false negatives and enhancing the model's ability to reliably detect cyclists under diverse real-world conditions, which would, in turn, contribute to overall system safety and reliability.

Alternatives to camera-based (or LiDAR based) cyclist detection could be explored by mapping labels from image or video data onto gathered time-series sensor data for subsequent model training, thereby increasing privacy. As a future outlook, architectures such as Long Short-Term Memory (LSTM) networks and Gated Recurrent Units (GRU) could be used to capture temporal relations within thermal or other sensor streams for cyclist detection and distance estimation (Ambardar, 1995). Furthermore, combining sensor-derived distance measurements with other sensor data by signal processing (Ambardar, 1995), signal filtering (Cheveigné & Nelken, 2019), and fusion (Hu et al., 2015)—may enhance overtaking classification and ultimately enable deployment of sensor-only models without camera or LiDAR sensors. Among candidate approaches, 1D-Convolutional networks can achieve roughly 80% classification accuracy on independent windows of time-series data (Kiranyaz et al., 2021), while GRUs introduce memory for sequential dependencies at the cost of higher computational expense, making them less suitable for real-time tasks (Shiri et al., 2024); ARIMA models can forecast data trends to augment other predictors (Shumway & Stoffer, 2017); Light Gradient Boosting Machines support real-time direct predictions albeit at about 60% accuracy (Ke et al., 2017); Extreme Gradient Boosting (XGBoost) delivers improved performance (error ~ 20%) with greater hardware demands (Zhang et al., 2021). However, it has to be stated that the performance of such systems has not previously been explored for an application task like this for all of the available sensor data and it may not be realistic to fully get rid of camera or LiDAR sensors—even then a system as described above could potentially be used to further improve the performance. Alternatively, the distance from a detected object could be estimated using depth estimation models such as the Depth Pro-models. However, it will be unable to meet real-time requirements due to these are comparatively slow (2.25 megapixels taking ~ 0.3 s on a V100 GPU) (Bochkovskii et al., 2025).

As described above, the laboratory study was conducted outside of a real-traffic situation with the purpose to test the selected sensors and to apply existing AI models to the resulting data. To evaluate the quality of the selected sensors and AI models in practice, the final sensor system will be tested under real-traffic scenarios in a later stage of the project. A two-day study is going to be conducted in autumn 2025 in the inner city of Osnabrück, Lower Saxony, Germany, on pre-selected routes that were defined and discussed as part of the stakeholder workshop at the beginning of the project. In general, all relevant different types of bike lanes according to official cycling planning guidelines (Forschungsgesellschaft für Straßen, 2010) will be considered. In addition to that, the city will collect data about motorized traffic volumes on these routes by during the field test phase. The data collection methods will follow the suggestions on how to measure traffic volumes described in official traffic planning guidelines. As an example, the data collection should take place on working days (not on Fridays) and should consider the traffic peak hours (Forschungsgesellschaft für Straßen, 2012). The results of potential solutions in the field of data fusion and the data collection in a real-traffic study will be presented in forthcoming publications.

Acknowledgements BikeDetect is funded by the Bundesministerium für Digitales und Staatsmodernisierung (BMDS, German Federal Ministry of Digital and Government Modernization) as part of the mFUND program (project number 01F1199A) with a funding amount of around 200.000. Euro. As part of mFUND the BMDV supports research development projects in the field of data based and digital mobility innovations. Part of the project funding is the promotion of networking between the stakeholders in politics, business, administration and research as well as the publication of open data on the Mobilithek portal.

Disclosure of Interests The authors have no competing interests to declare that are relevant to the content of this article.

References

Ahmed, S., Huda, M. N., Rajbhandari, S., Saha, C., Elshaw, M., & Kanarachos, S. (2019). Pedestrian and cyclist detection and intent estimation for autonomous vehicles: A survey. *Applied Sciences, 9*(11), 2335. https://doi.org/10.3390/app9112335

Alaba, S. Y., & Ball, J. E. (2022). A survey on deep-learning-based LiDAR 3D object detection for autonomous driving. *Sensors, 22*(24), 9577. https://doi.org/10.3390/s22249577

Ambardar, A. (1995). *Analog and digital signal processing* (2nd ed.). Brooks/Cole Publishing Company.

Benewake. (2025). TF-NOVA LiDAR. https://en.benewake.com/tf-nova. Accessed May 07, 2025.

BikeDetect. (2025). *Workshop konkretisiert Anforderungen an KI Systeme zur Erkennung von Radfahrenden.* https://bike-detect.vlba.net/2024/11/28/workshop-konkretisiert-anforderungen-an-ki-systeme-zur-erkennung-von-radfahrenden/. Accessed on May 07, 2025.

Bochkovskii, A., Delaunoy, A., Germain, H., Santos, M., Zhou, Y., Richter, S. R., Koltun, V. (2025). Depth pro: Sharp monocular metric depth in less than a second [arXiv:2410.02073 [cs]]. https://doi.org/10.48550/arXiv.2410.02073

Bosch. (2025). FLEXIDOME IP micro 3000i—Außeneinsatz. https://cdn.commerce.boschsecurity.com/public/documents/DS_FLEXIDOME_IP_micr_Data_sheet_deDE_81812499723.pdf. Accessed on May 07, 2025.

Chang, H., Gu, Y., Goncharenko, I., Hsu, L.-T., & Premachandra, C. (2023). Cyclist orientation estimation using LiDAR data. *Sensors, 23*(6), 3096. https://doi.org/10.3390/s23063096

Google Colab. https://colab.research.google.com/. Accessed on June 19, 2025.

Corral-Soto, E. R., Liu, Y., Cao, T., Ren, Y., Bingbing, L. (2025). *3DArtic-cyclists: Generating synthetic articulated 8D pose-controllable cyclist ata for computer vision applications.* [arXiv: 2410.10782 [cs]]. https://doi.org/10.48550/arXiv.2410.10782

De Cheveigné, A., & Nelken, I. (2019). Filters: When, why, and how (not) to use them. *Neuron, 102*(2), 280–293. https://doi.org/10.1016/j.neuron.2019.02.039

Duan, K., Bai, S., Xie, L., Qi, H., Huang, Q., Tian, Q. (2019). CenterNet: Keypoint triplets for object detection. In *2019 IEEE/CVF international conference on computer vision (ICCV)* (pp. 6568–6577). https://doi.org/10.1109/ICCV.2019.00667

Forschungsgesellschaft für Straßen-und Verkehrswesen—FGSV. (2010). ERA—Empfehlungen für Radverkehrsanlagen, Ausgabe 2010. FGSV Verlag.

Forschungsgesellschaft für Straßen-und Verkehrswesen—FGSV. (2012). EVE—Empfehlungen für Verkehrserhebungen. FGSV Verlag.

Gaio, A., & Cugurullo, F. (2023). Cyclists and autonomous vehicles at odds: Can the transport oppression cycle be broken in the era of artificial intelligence? *AI and Society, 38*(3), 1223–1237. https://doi.org/10.1007/s00146-022-01538-4

German Aerospace Center DLR. (2025). Blinklicht schützt Radfahrer vor abbiegenden Fahrzeugen—DLR forscht mit 'Amber Light' im Braunschweiger Stadtverkehr. https://www.dlr.de/ts/desktopdefault.aspx/tabid-13642/23687_read-51649/. Accessed on May 07, 2025.

Gössling, S., McRae, S. (2022). Subjectively safe cycling infrastructure: New insights for urban designs. *Journal of Transport Geography, 101*, 103340. https://doi.org/10.1016/j.jtrangeo.2022.103340

Hu, J., Ma, L., Wei, G., Wang, Z., Li, W., Ding, D. (2015). A survey on multisensor fusion and consensus filtering for sensor networks. *Discrete Dynamics in Nature and Society*, 1–12. https://doi.org/10.1155/2015/683701

Panasonic Industry. (2025). *Infrared array sensor grid-eye.* https://industrial.panasonic.com/ww/products/pt/grid-eye. Accessed on May 07, 2025.

Texas Instruments. (2025). *PGA460 ultrasonic proximity-sensing evaluation module.* https://www.ti.com/tool/PGA460PSM-EVM. last accessed 2025/05/07.

Intel Corporate. (2025). *Intel RealSense depth camera.* https://www.intelrealsense.com/depth-camera-d455/. Accessed on May 07, 2025.

Ke, G., Meng, Q., Finley, T., Wang, T., Chen, W., Ma, W., Ye, Q., Liu, T.-Y. (2017). LightGBM: A highly efficient gradient boosting decision tree. *Advances in Neural Information Processing Systems, 30.* http://papers.nips.cc/paper/6907-lightgbm-a-highly-efficient-gradient-boosting-decision-tree.pdf

Kiranyaz, S., Avci, O., Abdeljaber, O., Ince, T., Gabbouj, M., & Inman, D. J. (2021). 1D convolutional neural networks and applications: A survey. *Mechanical Systems and Signal Processing, 151*, Article 107398. https://doi.org/10.1016/j.ymssp.2020.107398

Label Studio. https://labelstud.io/. Accessed on June 12, 2025.

Labetski, A., Chum, A.: Built environmental correlates of cycling accidents involving fatalities and serious injuries in London, UK. *Frontiers in Sustainable Cities, 2.*

Leibe, B., Matas, J., Sebe, N., Welling, M. (2016). *Computer vision—ECCV 2016: 14th European conference, Amsterdam, The Netherlands, October 11–14, 2016, proceedings, part I (Bd. 9905).* Springer International Publishing. https://doi.org/10.1007/978-3-319-46448-0

Li, X., Flohr, F., Yang, Y., Xiong, H., Braun, M., Pan, S., Li, K. & Gavrila, D. M. (2016). A new benchmark for vision-based cyclist detection. In *2016 IEEE intelligent vehicles symposium (IV)* (pp. 1028–1033). https://doi.org/10.1109/IVS.2016.7535515

Li, Z., Yang, L., Zhou, F. (2024). *FSSD: Feature fusion single shot multibox detector.* https://doi.org/10.48550/arXiv.1712.00960

López-Molina, M., Llopis-Castelló, D., Pérez-Zuriaga, A. M., Alonso-Troyano, C., & Garcia, A. (2023). Skid resistance analysis of urban bike lane pavements for safe micromobility. *Sustainability, 15*(1), 698.

MinIO. (2025). https://min.io/. Accessed on May 08, 2025.

Navilock. (2025). *Navilock NL-82002U USB 2.0 Multi GNSS UDR Empfänger u-blox NEO-M8U 1.5* m. https://www.navilock.de/produkt/62756/merkmale.html. Accessed on May 07, 2025.

nginx, https://ninx.org/en/. Accessed on June 2025.

Palmer, A. J., Si, L., & Gordon, J. (2014). PMS10—The costs of major and minor cycling accidents in Tasmania Australia. *Value in Health, 17*(7), A772–A773.

Pérez-Zuriaga, A. M., Moll, S., López, G., & García, A. (2021). Driver behavior when overtaking cyclists riding in different group configurations on two-lane rural roads. *International Journal of Environmental Research and Public Health, 18*(23), 12797.

Redmon, J., Divvala, S., Girshick, R., Farhadi, A. (2016). *You only look once: unified, real-time object detection.* https://doi.org/10.48550/arXiv.1506.02640

RFbeam Microwave. (2025). *K-LD7 Radar transceiver.* https://rfbeam.ch/product/k-ld7-radar-transceiver/. Accessed on May 07, 2025.

Roboflow Universe. https://app.roboflow.com/myworkspace-bmrq2/cyclist_database2/1. Accessed on June 19, 2025.

Rupp, C. (2009). *Requirements-engineering und-management—professionelle, iterative Anforderungsanalyse für die Praxis* (5th ed.). Carl Hanser Verlag.

Shiri, F. M., Perumal, T., Mustapha, N., Mohamed, R. (2024). A comprehensive overview and comparative analysis on deep learning models: CNN, RNN, LSTM, GRU. *Journal on Artificial Intelligence, 6*(1), 301–360. https://doi.org/10.32604/jai.2024.054314

Shumway, R. H., Stoffer, D. S. (2017). Time series analysis and its applications: With R examples. Springer International Publishing. https://doi.org/10.1007/978-3-319-52452-8

Slamtec. (2025). RPLIDAR C1. https://www.slamtec.com/en/C1. Accessed on May 07, 2025.

Streamlit. (2025). https://streamlit.io/. Accessed on May 02, 2025.

StVO. (2021). Die Straßenverkehrs-Ordnung, inkl. eKFV. Verlag Heinrich Vogel in der Springer Fachmedien München GmbH.

Sze, N., Tsui, K., Wong, S., & Wo, F. (2011). Bicycle-related crashes in Hong Kong: Is it possible to reduce mortality and severe injury in the metropolitan area? *Hong Kong Journal of Emergency Medicine, 18*(3), 136–143.

Taddei, C., Gnesotto, R., Forni, S., Bonaccorsi, G., Vannucci, A., & Garofalo, G. (2015). Cycling promotion and non-communicable disease prevention: Health impact assessment and economic evaluation of cycling to work or school in florence. *PLoS ONE, 10*(4), e0125491–e0125491.

Tan, M., Pang, R., Le, Q. V. (2020). EfficientDet: scalable and efficient object detection. In *IEEE/CVF conference on computer vision and pattern recognition (CVPR)* (pp. 10778–10787). https://doi.org/10.1109/CVPR42600.2020.01079

Tomasch, E., Hoschopf, H., Ausserer, K. Rieß, J. (2024). *Required field of view of a sensor for an advanced driving assistance system to prevent heavy-goods-vehicle to bicycle accidents*, (Vol. 6(4), pp. 1922–1941). MDPI AG, Vehicles.

Wang, Y., Zhang, J., Wu, G. (2023). Bayesian-based traffic safety evaluation study for driverless infiltration (Vol. 13(22), pp. 12291). MDPI AG, Applied Sciences.

Wang, C.-Y., Liao, H.-Y. M. (2024). YOLOv1 to YOLOv10: The fastest and most accurate real-time object detection systems. *APSIPA Transactions on Signal and Information Processing, 13*(1). https://doi.org/10.1561/116.20240058

Widen, W. H., Koopman, P. (2022). Autonomous vehicle regulation and trust. *UCLA Journal of Law & Technology*, 27(3).

Zhang, M., Fu, R., Guo, Y., Wang, L., Wang, P., & Deng, H. (2020). Cyclist detection and tracking based on multi-layer laser scanner. *Human-Centric Computing and Information Sciences, 10*(1), 20. https://doi.org/10.1186/s13673-020-00225-x

Zhang, W., Wu, C., Zhong, H., Li, Y., & Wang, L. (2021). Prediction of undrained shear strength using extreme gradient boosting and random forest based on Bayesian optimization. *Geoscience Frontiers, 12*(1), 469–477. https://doi.org/10.1016/j.gsf.2020.03.007

Integrating Entrepreneurial and Sustainability Competences for Responsible AI Engagement in Diverse Work Fields

Ádám Lászlop and Jantje Halberstadt

Abstract This paper explores how combining sustainability and entrepreneurial competences can support responsible use of artificial intelligence (AI) in different professional roles. While existing AI competence models often focus on technical and managerial skills, they rarely address ethical, social and ecological dimensions. Building on the competence framework by (Halberstadt et al. in J Knowl Manag 23:1925–1948, 2019), originally developed for social entrepreneurship education, we adapt its structure to the context of responsible AI. We propose two core competence clusters—AI opportunity recognition and AI management—and link them to role-specific tasks in software development, product management and sustainability coordination. This approach highlights the need for interdisciplinary, normative and future-oriented thinking in AI development and application. The paper aims to support the design of educational programs that prepare future professionals to use AI in ways that are both innovative and aligned with societal and environmental values.

1 Introduction

The growing skills gap in AI puts increasing pressure on both organizations and educational institutions. There is an urgent need to clearly define and structure AI-related competences. This is necessary to ensure that the right skills are available for developing and applying AI, and to allow educational programs to quickly adapt to the fast-changing needs of industry.

The importance of interdisciplinary approaches in AI development also shows that needed competences go far beyond technical knowledge. They include fields such as ethics, social sciences and law. Brauner et al. (2023) and Hinton (2023) stress the

Á. Lászlop (✉) · J. Halberstadt
University of Vechta, Vechta, Germany
e-mail: adam.laszlop@mail.uni-vechta.de

J. Halberstadt
e-mail: jantje.halberstadt@uni-vechta.de

J. Marx Gómez et al. (eds.), *Recent Advances in Information Systems*, Progress in IS,
https://doi.org/10.1007/978-3-032-14099-9_57

importance of precisely defining and structuring AI competences. Doing so is key to addressing the current skill shortage and to building a solid foundation for training the next generation of AI experts—people who can design responsible and ethically sound AI systems. While existing AI competence models—such as those by Anton et al. (2020) and André et al. (2021)–provide valuable insights into technical and managerial skill sets, they pay limited attention to sustainability and entrepreneurial competences.

These dimensions, however, are essential for addressing the ethical, societal and ecological challenges posed by AI.

Developing and deploying artificial intelligence demands a dual orientation: the agility to transform technological opportunity into viable solutions, and the foresight to safeguard social-ecological integrity. Entrepreneurial competences—opportunity recognition, iterative experimentation, strategic risk management and value creation—equip professionals to navigate fast-moving AI innovation cycles, secure resources and scale ideas from prototype to market. Yet, when exercised in isolation, this growth-driven mind-set can amplify well-known AI externalities such as bias, resource-intensive model training and unequal access to benefits. Sustainability competences counterbalance those risks by embedding long-range, multi-stakeholder considerations into each phase of the AI lifecycle. Systems thinking connects technical decisions to complex economic, environmental and societal feedback loops; normative reflection grounds design choices in ethical principles and human rights; anticipatory and participatory skills help stakeholders envision downstream impacts and co-shape governance mechanisms.

This paper contributes a novel integration of sustainability and entrepreneurial competence frameworks (e.g., Halberstadt et al., 2019) into the AI context. It develops a role-specific, interdisciplinary competence model that highlights the importance of normative reflection, systems thinking and innovation capabilities for responsible AI engagement across diverse professional domains.

Thus, this paper investigates the relevance and role of certain competencies for responsibly working with and (further) developing AI. After distinguishing between various work fields, the paper explores how entrepreneurial and sustainability competences support responsible AI engagement, using Halberstadt et al. (2019) as a foundation. We examine how these competences can be adapted to the AI context and analyse the specific requirements across different professional roles, including software developers, managers and product managers. We argue that a competence-based approach is key to preparing individuals for AI in line with ethical and sustainable standards. A set of relevant competences is proposed as a basis for reshaping current educational settings and generating innovative learning and teaching approaches for AI.

The methodological approach of this paper is deliberately conceptual, aiming to provide an exploratory foundation for future empirical or systematic investigations. Although empirical evidence is currently lacking, this analysis offers a clear theoretical orientation and establishes an initial framework that can subsequently be validated empirically or developed further.

2 AI Applications Across Work Fields and Role-Specific Competence Needs

The integration of AI into diverse professional contexts requires more than general digital skills—it demands differentiated competences that reflect the ethical, organizational and technical responsibilities of distinct stakeholder roles (Anton, Behne, & Teuteberg, 2020). A helpful analytical lens is provided by the typology developed by Balkow and Eckardt (2019), which identifies five primary groups in the lifecycle of algorithmic systems: decision-makers, designers, users, auditors and affected individuals. This framework illustrates that responsibility for AI systems is distributed unequally and that each group requires role-specific capabilities to manage these responsibilities ethically and effectively.

This classification helps clarify why competence development for responsible AI cannot follow a one-size-fits-all model. Instead, competences must be aligned with specific roles and phases in the development and application of AI systems.

The following section applies this framework by focusing on selected professional groups who occupy hybrid or intersecting positions within the above categories. It elaborates the competences they require to engage responsibly with AI:

Software Developers (Designers): Responsible AI development requires developers to address fairness, inclusiveness and reliability, while also emphasizing transparency and accountability. Practitioners often face challenges in aligning with Responsible AI principles, highlighting the need for comprehensive frameworks to operationalize these principles effectively (Leça et al., 2024).

Product Managers (Decision-Makers and Users): Product managers often act as gatekeepers in AI-related decision-making, ensuring that responsible practices are followed in their everyday work. Mukhuty et al. (2022) argue that the success of AI implementation in industrial contexts depends on addressing human-related barriers and embedding social responsibility into management practices. However, they frequently face widespread uncertainty regarding what 'responsibility' entails, diffused responsibility and a lack of institutional incentives. Despite these challenges, ethical engagement can be fostered through identifying and acting on 'micro-moments' of responsibility (Smith et al., 2025).

Top Management (Decision-Makers): General managers and CEOs play a crucial role in steering AI integration, acting as strategic architects who balance AI's potential with ethical constraints. They take on key responsibilities as 'possibility catalysers,' 'uncertainty mappers', and 'organizational designers' to shape the normative orientation of AI usage in their institutions (Hill, 2025).

Sustainability Managers (Users and Designers): These professionals are increasingly tasked with aligning AI use with ecological and social goals. They not only supervise the responsible application of AI but also explore its potential for fostering more sustainable business practices (Schwaeke et al., 2025). The roles addressed in the model (e.g., developers, managers, sustainability experts) were identified and described based on exemplary studies (Anton et al., 2020; Ren & Wu, 2025) and practical professional experience.

3 Sustainability and Entrepreneurial Competences

While the competence models proposed by Anton et al. (2020) and André et al. (2021) offer valuable insights into the technical, managerial and organizational skills needed for AI-related roles, they pay limited attention to sustainability and responsibility competences. Neither model explicitly addresses the ethical, ecological or societal dimensions emphasized in frameworks such as Education for Sustainable Development (ESD) or Responsible AI. Given the increasing importance of aligning AI development with broader societal values, this gap suggests the need for a more integrated competence framework. Building on Halberstadt et al. (2019), the inclusion of sustainability-related competences—such as systems thinking, normative reflection and anticipatory decision-making—would complement existing models and support a more holistic approach to competence development.

To construct an integrative competence framework for responsible AI, we employed a purposeful selection of five existing models that complement one another in scope, empirical depth and normative orientation. The selection followed three criteria: (a) relevance for AI-related work roles, (b) coverage of sustainability or entrepreneurial dimensions and (c) empirical or procedural rigour. The selection of relevant literature (2019–2025) was guided by targeted keywords such as 'Responsible AI', 'AI competences', 'entrepreneurial competences' and 'sustainability skills'. Priority was given to conceptual frameworks, position papers from international organizations (e.g., UNESCO, EU), and empirical studies with a clear focus on competence development in the context of artificial intelligence. In this context, the contribution by Balkow and Eckardt (2019) is particularly valuable, as it links ethical responsibility and digital literacy with role-specific accountability across the AI system lifecycle—thereby providing a conceptual anchor for integrating entrepreneurial and sustainability competences into a unified framework.

To operationalize the competences discussed before, we purposefully selected five complementary models. The framework by Halberstadt et al. (2019), originally developed in the context of social entrepreneurship education, was conceptually adapted and enriched with AI-relevant requirements. Three competence clusters—Cognitive-Normative, Social-Strategic and Entrepreneurial-Transformative—form the structural foundation of the integrated model.

Building on the complementarities identified in Table 1, we followed a five-step consolidation procedure to derive an integrated, role-specific framework (Table 2).

The combination of sustainability and entrepreneurial competencies is gaining increasing relevance, as it enables individuals to not only understand complex sustainability challenges but also to develop and implement innovative solutions. This integrated skillset is seen as crucial for future-oriented education and for preparing professionals to drive responsible transformation across sectors (Planck et al., 2024).

Table 1 Source competence/role models incorporated into the integrated framework

Model/reference	Key competences	Function for this study
Halberstadt et al. (2019)	Opportunity recognition, management, systems and future thinking	Supplies sustainability + entrepreneurial core; normative anchor
Anton et al. (2020)	Technical versus managerial AI skills	Labour-market validation; role granularity
André et al. (2021)	Domain knowledge, working with AI, shaping AI	Process lens for competence development
Balkow and Eckardt (2019)	(role taxonomy)	Aligns competences with ethical accountability across lifecycle
UNESCO/Miao & Shiohira (2024)	Information literacy, critical reflection, collaboration	Pedagogical transfer and learning-outcome formulation

Table 2 Source competence/role models incorporated into the integrated framework

Step	Goal	Action
Screening and eligibility	Systematically pre-select relevant competence models	An initial literature scan (32 models) identified seven candidates. Applying the criteria *AI relevance, coverage of sustainability/entrepreneurship dimensions* and *empirical or procedural rigour*, five models were retained
Mapping	Create a common analytical grid	Key competence categories from each model were entered into a matrix and mapped against two axes: (a) AI lifecycle phases (Design → Governance) and (b) stakeholder roles per Balkow & Eckardt
Consolidation	Reduce overlaps, fill gaps	Semantically similar competences (e.g., *systems thinking* in both Halberstadt and André) were merged; missing sustainability and normative elements were supplied by Halberstadt. The process yielded two core clusters—'AI Opportunity Recognition' and 'AI Management'
Role alignment	Assign competences to specific role	Using the stakeholder typology, the consolidated competences were allocated to four exemplar professional groups: software developers, product managers, top management and sustainability managers
Pedagogical anchoring	Derive learning-outcome-oriented curricula	UNESCO descriptors provided wording for learning outcomes and assessment rubrics, enabling immediate transfer of the integrated framework into teaching and professional-development programmes

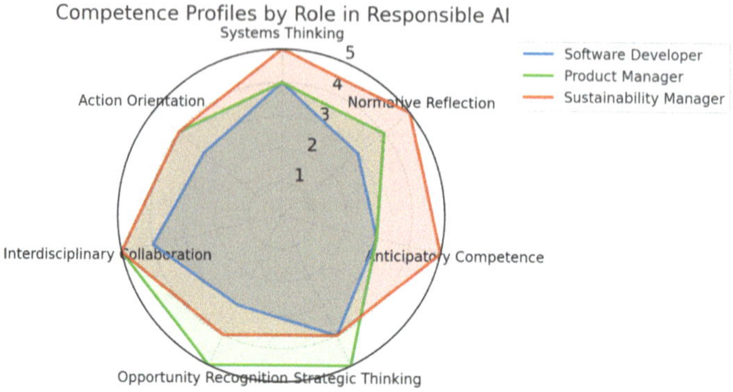

Fig. 1 Competence profiles by role in responsible AI

Halberstadt et al. (2019) identified key competences essential for entrepreneurship and sustainability, including opportunity competence, social competence, business competence, entrepreneurial competence and industry-specific competences. These competences enable individuals to understand complex systems, envision sustainable futures and act as change agents.

Halberstadt et al. (2019) suggest a framework that connects sustainability and entrepreneurial competences by focusing on social entrepreneurship. They argue that service learning can support students in developing the skills needed to act as change agents who create both social and economic value. Their framework brings together competences like systems thinking, future thinking, normative reflection and action orientation, showing how these are essential not only for sustainability but also for entrepreneurial activity. This helps to bridge two areas that are often treated separately and offers a practical approach to integrating them in higher education. In the context of AI, these competences can be instrumental in ensuring that AI applications are developed and utilized in ways that are ethically sound and socially beneficial. Thus, in the following chapter, we elaborate on how these competences can contribute to a responsible and ethical development and use of AI (Fig. 1).

4 Competence Framework

Competence development for AI work requires the integration of technical mastery, managerial insight and normative orientation. Two empirical studies already cover parts of this spectrum. Anton et al. (2020) distinguish technical from managerial AI competences on the basis of 9 247 job advertisements and list skills such as data engineering, algorithm design, business acumen and interdisciplinary collaboration. André et al. (2021) add a six-step organizational process that groups competences into domain and digital knowledge, working-with-AI skills and shaping-AI environments. While both models illuminate the labour-market and organizational perspectives, they treat sustainability and entrepreneurial opportunity-seeking only marginally. To close this gap the present study adapts the social entrepreneurship framework of Halberstadt et al. (2019), which is rich in systems thinking, future thinking and normative reflection and aligns all categories with the stakeholder roles proposed by Balkow and Eckardt (2019). The synthesis yields two overarching clusters: AI opportunity recognition, concerned with spotting, framing and testing socially valuable AI possibilities and AI management, concerned with guiding development, deployment and governance in line with ethical and sustainability goals.

The consolidated framework rests on six competences that together operationalize the two clusters.

Systems thinking refers to the ability to recognize interdependencies between technological, social and ecological subsystems. Practitioners who possess this competence routinely map energy or privacy externalities before deployment and anticipate feedback loops that may amplify bias over time (Halberstadt et al., 2019; Ratinen & Linnanen, 2022).

Normative reflection is the capacity to examine the values, rights and justice principles that govern AI decisions. It becomes visible when teams run value-sensitive design workshops and document trade-off rationales in model cards and audit reports (Bruneault et al., 2022; Woodgate & Ajmeri, 2024).

Opportunity recognition is an entrepreneurial skill that enables professionals to detect unmet needs where AI can create sustainable value. Typical behaviours include scanning market and societal pain points and rapidly prototyping low-risk proofs of concept to test viability (Anton et al., 2020; Halberstadt et al., 2019).

Strategic risk management addresses the proactive identification, assessment and mitigation of AI-specific risks. It involves maintaining risk registers for model drift and compliance gaps and aligning the organizational risk appetite with explicit ethics policies (André et al., 2021; Jowarder and Jowarder, 2025).

Anticipatory foresight is the skill to envisage multiple futures and the long-term impacts of AI. Scenario workshops on labour displacement and horizon scanning that feeds into product road maps are typical manifestations (Hill, 2025; Muiderman et al., 2022). Stakeholder inclusion denotes the ability to engage diverse groups throughout the AI lifecycle. Concrete practices range from co-creating fairness metrics with affected communities to publishing lay summaries and inviting iterative feedback (Balkow & Eckardt 2019; Schwaeke et al., 2025).

These six competences distribute differently across professional roles. For software developers, systems thinking and normative reflection are critical for embedding fairness and robustness checks into code reviews. Product managers rely strongly on opportunity recognition and stakeholder inclusion when refining feature road maps. Executives combine anticipatory foresight with strategic risk management to steer AI portfolios responsibly, while sustainability managers draw on the entire set when assessing social-ecological footprints of AI projects. A radar chart that appears later in the manuscript visualizes the relative importance of the six competences for these four roles, and a heat map links each competence to the four main lifecycle stages from design to governance. These visualizations, together with the narrative specification above, respond directly to the reviewers' request for deeper discussion and clearer representation of key competences.

5 Application of Competence Framework on AI Contexts

To understand how competences can support responsible AI use, it is helpful to begin with key categories that are well-established in the field of sustainability and have increasingly been linked to AI contexts. These competences help individuals not only work with AI technologies, but also reflect on their broader impact, make informed decisions and contribute to long-term, ethical and sustainable outcomes. Systems Thinking: Understanding AI's impact requires a holistic view of technological, social and environmental systems (Ratinen & Linnanen, 2022).

Normative competence means being able to reflect on values and ethical principles like fairness or responsibility and apply them in real AI contexts. This includes understanding the social impact of AI and making sure that decisions can be justified ethically (Woodgate & Ajmeri, 2024).

Anticipatory Competence: Predicting the long-term implications of AI technologies is crucial for sustainable integration. This includes reflecting on different possible futures and their impact on today's decisions, recognizing uncertainty and making space for more inclusive and forward-looking policies and innovations (Muiderman et al., 2022).

Strategic Competence: Developing and implementing strategies to guide AI use towards desired outcomes aligns with responsible innovation. This includes being able to use AI tools in ways that support long-term business goals, improve decision quality and turn complex data into useful insights—while also staying aware of ethical challenges and organizational limits (Jowarder & Jowarder, 2025).

Interdisciplinary competence is identified by Halberstadt et al. (2019) as a key competence, particularly in connection with sustainability. Its importance is further underscored in the context of AI education, where it is seen as a central requirement for enabling responsible and integrative learning processes (Miao & Shiohira, 2024; Rani, 2025).

Students, future professionals and citizens need a general understanding of AI and its societal implications, complemented by domain-specific knowledge and interdisciplinary communication skills (Klumpp & Zijm, 2019). Espina-Romero et al. (2024) further emphasize that digital competences in this context should not be limited to technical know-how but should enable learners to contribute to sustainable development and open innovation.

A recent international study by Okada et al. (2025) supports this view by showing that students can develop AI skills through open-school projects that follow the CARE-KNOW-DO teaching model and the UNESCO framework for AI learning. The study shows that combining emotional involvement (CARE), knowledge building (KNOW) and hands-on practice (DO) in real-world problem-solving tasks is especially effective.

Educators require both technical and ethical expertise, as well as pedagogical competence to integrate AI topics into curricula (Iryna, 2024). Trust in the technology they teach is also important (Aler et al., 2024). Software developers and AI engineers need strong technical competences—such as algorithm design, data handling, robustness and explainability—and the ability to integrate these with ethical considerations. They must work closely with non-technical experts and include user feedback in iterative development cycles. Managers and decision-makers must develop strategic insight, understand the effects of AI on organizational processes (Shet & Pereira, 2021), manage talent development and create a culture of responsibility (Díaz-Rodríguez et al., 2023). Organizations, such as universities and businesses, need strategic pathways for embedding RAI (Responsible) and TAI (Trustworthy) into their structures (Gosling & Grodecki, 2020). These role-specific requirements have also been mentioned in recent research by Ren and Wu (2025), who identify four key competence areas for AI integration in higher education: (1) AI literacy and readiness, (2) AI-supported innovative pedagogy, (3) the use of AI as intelligent learning partners and (4) AI-driven course design and delivery. Their findings underline that effective AI integration demands not only technical knowledge, but also pedagogical and ethical competencies. Instructors must balance opportunities for engagement and efficiency with ongoing concerns about digital divides, ethical risks and the short-term novelty effects of AI tools (Andrei et al., 2025).

This includes investing in skill development, hiring expertise, fostering cross-functional collaboration and revising training, teacher education and education programs (Golowko, 2021; Ikenga & van der Sijde, 2024; Iryna, 2024; Shet & Pereira, 2021).

Responsible AI practices, such as training and education, ethical system design and risk governance, are critical enablers of this process, as highlighted by case-based research across industries (Wang et al., 2020).

The literature supports the need for a broad yet integrated competence framework for responsible AI. The analysis of the literature is showing that not the individual competences themselves, but the insight that meaningful integration of technical, ethical and sustainability competences is the critical success factor (Atabekova et al., 2022). Technical competence is not an optional addition but a necessary component that must be embedded in context (Chen, 2024).

6 Conclusion

This paper developed a conceptual framework that combines sustainability and entrepreneurial competences to support responsible AI use across professional roles. In contrast to existing models that focus mainly on technical and managerial skills, this approach emphasizes ethical, ecological and systemic dimensions. By identifying role-specific competence needs—particularly for developers, product managers and sustainability professionals—the paper highlights the importance of interdisciplinary integration. Entrepreneurial skills support innovation and implementation, while sustainability competences ensure long-term, value-driven outcomes. The proposed framework offers a foundation for designing educational programs and training pathways that align AI practice with societal goals. Future work should empirically validate these competence profiles and expand them to include additional roles and contexts. Building such integrated capabilities is key to ensuring that AI contributes meaningfully to ethical and sustainable transformation.

Disclosure of Interests The authors have no competing interests to declare that are relevant to the content of this article.

References

Aler Tubella, A., Mora-Cantallops, M., & Nieves, J. C. (2024). How to teach responsible AI in higher education: Challenges and opportunities. *Ethics and Information Technology, 26*(1), 1–14.

Andrei, O., Nabi, S. W., Barr, M., Petrovska, O. (2025, January). Integrating socially responsible computing competencies in computer science and software engineering education. In *Proceedings of the 9th conference on computing education practice*, 36–37.

André, E. et al. (2021). Kompetenzentwicklung für KI: Veränderungen, Bedarfe und Handlungsoptionen. AG Arbeit/Qualifikation und Mensch-Maschine-Interaktion. https://www.plattform-lernende-systeme.de/files/Downloads/Publikationen/AG2_WP_Kompetenzentwicklung_KI.pdf.

Anton, E., Behne, A., Teuteberg, F. (2020). The humans behind artificial intelligence—An operationalisation of AI competencies. In *Proceedings of the 28th European conference on information systems (ECIS), An Online AIS Conference*, June 15–17, 2020. https://aisel.aisnet.org/ecis2020_rp/141.

Atabekova, N. K., Dzedik, V. A., Troyanskaya, M. A., & Matytsin, D. E. (2022). The role of education and social policy in the development of responsible production and consumption in the AI economy. *Frontiers in Environmental Science, 10*, 1–8.

Balkow, C., & Eckardt, I. (2019, 16 März). Bias in algorithmischen Systemen – Erläuterungen, Beispiele und Thesen (Denkimpuls Digitale Ethik, Nr. 07; Unterarbeitsgruppe Algorithmen-Monitoring). Initiative D21 e.V.

Bruneault, F., Laflamme, A. S., Mondoux, A. (2022, March 2). *AI ethics training in higher education: Competency framework*. OSF Preprints. https://doi.org/10.31235/osf.io/x5nck

Brauner, S., Murawski, M., Bick, M. (2023). The development of a competence framework for artificial intelligence professionals using probabilistic topic modelling. *Journal of Enterprise Information Management ahead-of-print*, ahead-of-print.

Chen, Z. (2024). Responsible AI in organizational training: Applications, implications, and recommendations for future development. *Human Resource Development Review, 23*(4), 498–521.

Díaz-Rodríguez, N., Del Ser, J., Coeckelbergh, M., de Prado, M. L., Herrera-Viedma, E., & Herrera, F. (2023). Connecting the dots in trustworthy Artificial Intelligence: From AI priciples, ethics, and key requirements to responsible AI systems and regulation. *Information Fusion, 99*, 1–24.

Golowko, N. (2021). *Future skills in education: Knowledge management, AI and sustainability as key factors in competence-oriented education.* Springer Nature.

Espina-Romero, L., Ríos Parra, D., Noroño-Sánchez, J. G., Rojas-Cangahuala, G., Cervera Cajo, L. E., & Velásquez-Tapullima, P. A. (2024). Navigating digital transformation: Current trends in digital competencies for open innovation in organizations. *Sustainability, 16*(5), 2119.

Gosling, J., & Grodecki, A. (2020). Competences for responsible management (and leadership) education and practice. In D. C. Moosmayer, O. Laasch, C. Parkes, & K. G. Brown (Eds.), *The SAGE handbook of responsible management learning and education* (pp. 245–264). SAGE Publications Ltd.

Halberstadt, J., Timm, J. M., Kraus, S., & Gundolf, K. (2019). Skills and knowledge management in higher education: How service learning can contribute to social entrepreneurial competence development. *Journal of Knowledge Management, 23*(10), 1925–1948.

Hill, A. ((2025, March 20). Seven AI roles managers must master. *Financial Times*. https://www.ft.com/content/f742dcdc-41ef-415d-adad-d828d23f739

Hinton, C. (2023). The state of ethical AI in practice. *International Journal of Technoethics, 14*(1), 1–15.

Ikenga, G. U., & van der Sijde, P. (2024). Twenty-first century competencies; about competencies for industry 5.0 and the opportunities for emerging economies. *Sustainability, 16*, 7166. https://doi.org/10.3390/su16167166

Iryna, V. (2024, October). Competence of teachers and ethical aspects of implementing AI technologies in education. In *International conference on advanced research in technologies, information, innovation and sustainability* (pp. 397–406). Springer Nature Switzerland.

Jowarder, R., Jowarder, M. (2025). AI-driven strategic insights: enhancing decision-making processes in business development. *International Journal of Innovative Research in Science, Engineering and Technology.*

Klumpp, M., & Zijm, H. (2019). Logistics innovation and social sustainability: How to prevent an artificial divide in human–computer interaction. *Journal of Business Logistics, 40*(3), 265–278.

Leça, M. D. M., Bento, M., Santos, R. D. S. (2024). Responsible AI in the software industry: A practitioner-centered perspective.

Miao, F., & Shiohira, K. (2024). *AI competency framework for students.* UNESCO Publishing.

Muiderman, K., Zurek, M., Vervoort, J., Gupta, A., Hasnain, S., & Driessen, P. (2022). The anticipatory governance of sustainability transformations: Hybrid approaches and dominant perspectives. *Global Environmental Change, 73*, Article 102452.

Mukhuty, S., Upadhyay, A., Rothwell, H. (2022). Strategic sustainable development of Industry 4.0 through the lens of social responsibility: The role of human resource practices. *Business Strategy and the Environment, 31*(5), 2068–2081.

Okada, A., Sherborne, T., Panselinas, G., Kolionis, G. (2025). Fostering transversal skills through open schooling supported by the CARE-KNOW-DO pedagogical model and the UNESCO AI competencies framework. *International Journal of Artificial Intelligence in Education*, 1–46.

Planck, S., Wilhelm, S., Kobilke, J., & Sailer, K. (2024). Greater than the sum of its parts: Combining entrepreneurial and sustainable competencies in entrepreneurship education. *Sustainability, 16*(9), 3725.

Rani, L. M., Mohammadi, F., Feldt, R., Svensson, R. B. (2025). An empirical study on decision-making aspects in responsible software engineering for AI. arXiv preprint arXiv:2501.15691.

Ratinen, I., & Linnanen, L. (2022). Exploring systems thinking competence of finns in fostering sustainable transformation. *World, 3*(2), 287–298.

Ren, X., Wu, M. L. (2025). Examining teaching competencies and challenges while integrating artificial intelligence in higher education. *TechTrends*, 1–20.

Schwaeke, J., Gerlich, C., Nguyen, H. L., Kanbach, D. K., Gast, J. (2025). Artificial intelligence (AI) for good? Enabling organizational change towards sustainability. *Review of Managerial Science.*

Shet, S. V., Pereira, V. (2021). Proposed managerial competencies for Industry 4.0–implications for social sustainability. *Technological Forecasting and Social Change, 173*, 121080.

Smith, G., Luka, N., Osborne, M., Lattimore, B., Newman, J., Nonnecke, B., Mittelstadt, B. (2025). *Responsible generative AI use by product managers: Recoupling ethical principles and practices.*

Wang, Y., Xiong, M., Olya, H. (2020, January). Toward an understanding of responsible artificial intelligence practices. In *Proceedings of the 53rd Hawaii International Conference on System Sciences* (pp. 4962–4971). Hawaii International Conference on System Sciences.

Woodgate, J., & Ajmeri, N. (2024). Macro ethics principles for responsible AI systems: Taxonomy and directions. *ACM Computing Surveys, 56*(11), 1–37. https://doi.org/10.1145/3672394

Generative AI in OCR: A Case Study for License Plate Recognition

Annita Ngatchou Weiss, Lars Guggolz, Tim Dahmen, and Christian Koot

Abstract Accurate character recognition is critical for automatic license plate recognition systems. However, existing optical character recognition (OCR) models often suffer from substitution, addition, and deletion errors, especially under visual artifacts and ambiguous character forms. Single-modal OCR models have limited contextual understanding. In this study, we evaluate whether recent multimodal large language models (LLMs) with vision capabilities, such as GPT-4o and Gemini can reduce character-level recognition errors by leveraging semantic and spatial context. We compare these models against single-modal OCR systems, PaddleOCR, TrOCR, EasyOCR, and Tesseract. Our results show that multimodal LLMs outperformed the single-modal OCR models in terms of overall accuracy (GPT-4o: $97.6 \pm 0.7\%$, Gemini: $83.1 \pm 1.6\%$). They also exhibit reduced character error rates (CERs), notably in substitution, deletion, and insertion errors. We further show that introducing ensemble inputs to a fine-tuned T5, a generative language model with natural language processing capabilities, as a post-processing step in an OCR pipeline, enables single-modal models to almost match the performance level of Gemini. These findings suggest that the integration of generative models into OCR pipelines can enhance the robustness of license plate recognition systems, particularly under visually challenging conditions.

1 Introduction

Optical character recognition (OCR) algorithms extract textual information from images, enabling applications such as the automatic reading of license plate numbers. This process involves localizing and isolating the license plate region in the

A. N. Weiss (✉) · L. Guggolz · T. Dahmen · C. Koot
Aalen University, Aalen, Germany
e-mail: ngatchweiss@gmail.com

T. Dahmen
German Research Centre for Artificial Intelligence (DFKI), Saarbrücken, Germany

J. Marx Gómez et al. (eds.), *Recent Advances in Information Systems*, Progress in IS,
https://doi.org/10.1007/978-3-032-14099-9_58

image of the vehicle, followed by applying an OCR algorithm to convert visual characters into machine-readable text (Charleset al., 2012; Smith, 2007). However, the accurate reading of optical characters in real-world conditions is challenging due to environmental factors (Szeliski, 2022; Lubna et al., 2021). These factors often result in noisy OCR outputs. Therefore, a post-OCR correction is often integrated into the OCR pipeline to improve the recognition accuracy (Nguyen et al., 2021). The emergence of multimodal large language models (LLMs) with visual capabilities offers the potential to overcome some of these limitations. In comparison with single-modal OCR models (Szeliski, 2022) that mainly rely on character forms and process characters in isolation (Tensmeyer, Saunders and Martinez, 2017), multimodal models can utilize contextual information from multiple sources, such as texts and images to infer the correct characters (Brown, 2020; OpenAI, 2023; Team et al., 2023). Examples of such multimodal models are GPT-4 (OpenAI, 2023) (including its successors) and Gemini (Team et al., 2023). These models are designed to be agnostic, which calls for an evaluation of their performances in reading license plate numbers and a comparision against single-modal OCR algorithms (TrOCR, EasyOCR, PaddleOCR, and Tesseract).

In this paper, the following contributions are made:

1. Evaluation of accuracy in reading license plate numbers for multimodal LLMs (GPT-4o, Gemini) and single-modal OCR models (TrOCR, EasyOCR, PaddleOCR, and Tesseract)
2. In-depth analysis of error patterns (substitutions, deletions, insertions) for the evaluated models.
3. Development of an ensemble model built from single-modal OCR systems which uses Google's T5 natural language processing model to correct OCR errors (Fig. 1). This aims to provide an alternative to computationally intensive LLMs.

2 Related Work

License plate numbers consist of unique combinations of characters and digits. Performing OCR on such data may introduce errors due to poor image quality (Ye and Doermann, 203). Another source of error is visual ambiguity, where similar looking characters are misclassified (Hochberg et al., 1997). These errors often lead to character substitutions, deletions, or insertions (Nguyen et al., 2021). Single-modal OCR systems such as Tesseract (Smith, 2007), PaddleOCR (Duet al., 2020), and EasyOCR (Jaided, 2020), although widely used for license plate recognition, still exhibit limitations in this domain. These models employ character-level segmentation and reconstruction methods for character recognition. However, this approach makes them sensitive to character distortions and therefore prone to errors when processing images captured in uncontrolled environments (Zhou et al., 2023; Neudecker et al., 2021; Szeliski, 2022; Lubna et al., 2021). On the other hand, the vision-language model TrOCR (Li et al., 2021) operates by taking images as input and using a visual encoder in conjunction with a text decoder to recognize characters. It integrates both components into a single end-to-end pipeline. However, TrOCR is trained primarily on high-quality scanned documents, which can be a limitation, as it may not generalize effectively to real-world images captured in uncontrolled environments (Li et al., 2021). These limitations can result in character misrecognition, incomplete text extraction, or higher error rates in downstream tasks (Li et al., 2021). Attempts to overcome these OCR errors have included the implementation of post-processing methods. For example, manual correction, where a human reviews and fixes errors made by the OCR system, can be highly accurate but not easily scalable (Mühlberger, 2014). Conversely, automated techniques have employed lookup dictionaries to identify and correct erroneous words (Kukich, 1992). Although effective for structured texts, these rule-based methods often struggle with spelling variations, noisy inputs, and domain-specific terminology (Piotrowski, 2012). Moreover, ensemble approaches that combine the outputs of multiple OCR systems using voting mechanisms have been used to reconcile errors and generate more reliable texts (Boiangiu et al., 2016). Natural language processing techniques have also been applied in post-OCR correction. These approaches leverage embedding-based methods and neural language models to suggest plausible corrections based on semantics. This allowed the systems to correct words based on context rather than relying on string similarity (Mikolov et al., 2013). Recent advances in multimodal LLMs, such as GPT-4o and Gemini, offer a promising direction for OCR correction. These models possess zero-shot capabilities, enabling them to process images and correct OCR errors in a unified step without requiring fine-tuning on task-specific data. However, their inference latency and reliance on external APIs can pose limitations for real-time or resource-constrained applications (Treviso et al., 2023). Here, we explore whether a smaller, task-specific model, such as T5, can achieve high correction accuracy in the post-processing steps of OCR engines. Building on the text-to-text paradigm of T5, we propose a parallel input model that combines outputs from three OCR engines, TrOCR, PaddleOCR, and EasyOCR into a unified input stream (Fig. 1). This ensemble setup leverages the diverse strengths of each OCR system, resulting in more robust post-correction performance.

3 Methods

Throughout this comparative analysis, we used the real-world UC3M-LP dataset (Ramajo-Ballester et al., 2024). It consists of 1975 Spanish license plate images (Fig. 1), which contained a total of 12,757 characters. The images were captured under a variety of conditions at distances ranging from 1 to 20 m using smartphones or professional cameras. The dataset was modified to comply with GDPR by blurring selected characters of license plate numbers (Ramajo-Ballester et al., 2024).

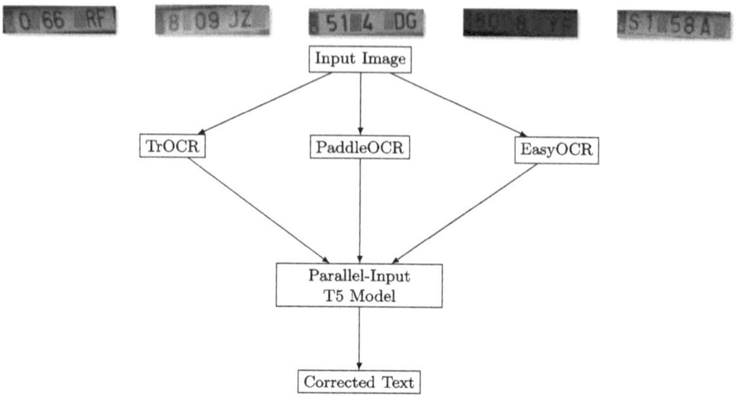

Fig. 1 Proposed OCR post-correction framework: cropped license plate images are processed by the single-modal OCR systems, TrOCR, PaddleOCR, and EasyOCR. The outputs from these systems are then jointly used as heterogeneous inputs to T5, a language model for OCR post-correction. By fusing complementary OCR errors, the framework leads to increased correct license plate text recognition. The arrows (⟶) indicate the flow of data through the pipeline

Isolated license plate images were processed through four distinct single-modal OCR systems. The first system, EasyOCR (v1.7.2), utilizes a Convolutional Recurrent Neural Network (CRNN) architecture for text recognition. It combines convolutional layers for feature extraction and recurrent layers to model sequences of characters (Jaided, 2020). This model has previously been applied for license plate recognition tasks (Vedhaviyasshet al., 2022). The second system, PaddleOCR (v2.4) (Contributors, 2021), leverages a deep neural network specifically optimized to recognize text in complex scenarios, such as distorted, rotated, or irregularly formatted text. This model contains approximately 6.8 million parameters. Its application to license plate recognition has also been explored (Mustafa and Karabatak, 2023). The third system, Tesseract (v4.1.0) (Smith, 2007), is configured in Long Short-Term Memory mode, a recurrent neural network used to handle sequential data such as text. Finally, TrOCR (base-stage) (Li et al., 2021), is built on vision transformer architectures. These single-modal models were used to evaluate the performance of OCR in license plate number recognition tasks.

To evaluate the performance of the multimodal LLMs, the cropped license plate images were supplied in batches of 10 to the Gemini (gemini-1.5-flash-latest) API for text extraction. The images were converted to base64-encoded before supplying them to gpt-4o-mini via API in order to comply with its input requirements. Alongside the images, a prompt was used to instruct the models to identify the text enclosed within the rectangular-shaped box of each license plate image. A zero-shot prompting technique was employed. The models were provided with a natural language instruction without any prior fine-tuning of the LLMs with examples. Both models processed the same batch of images under a deterministic setting (temperature = 0) to minimize the variations in outputs. Before comparing the outputs of the various OCR models, they were subjected to a data cleaning and standardization step. A regex pattern was applied to strip the data of unwanted special characters, including whitespaces, double asterisks (**), equal signs (=), colons (:), semicolons (;), percent signs (%), at symbols (@), periods (.), and backslashes (\) which were introduced by the LLMs during OCR processing. Duplicated entries were also removed. Moreover, a placeholder with the value 'nan' was assigned to license plates with missing recognitions. This ensured that every field within each model contained an entry. These data cleaning steps were performed before comparing the models' license plate text recognition with the ground truth.

For OCR post-correction, we used the natural language model T5-small (Raffel et al., 2020), a transformer-based encoder-decoder with 60 million parameters implemented using the Hugging Face Transformers library. The model was fine-tuned using OCR outputs generated by EasyOCR, PaddleOCR, and TrOCR on the UC3M-LP license plate dataset. These OCR outputs were tokenized using the T5 tokenizer. The dataset was split into training (80%), validation (10%), and test (10%) sets. The T5-small model was fine-tuned for a maximum of 1000 steps. The model was evaluated on exact match accuracy every 50 steps, and this metric was used to determine which checkpoints to retain. Only the two best-performing checkpoints were kept for final evaluation. The training used a batch size of 8 and accumulated gradients over 4 steps. The trained model used AdamW optimizer (Loshchilov Hutter, 2019) with a learning rate of 3e−4 and a weight decay of 0.01 to help prevent overfitting. A 50 step warm-up phase was added at the beginning to stabilize the training. To monitor the loss and exact match accuracy metrics, logging to TensorBoard was performed every 50 steps. To ensure reproducibility, the random seed was set to 42.

Using the fine-tuned T5-small model, two configurations were tested: (1) single-modal correction, where T5 processed outputs from individual OCR systems (excluding Tesseract due to its substantially higher baseline error rate), and (2) ensemble correction, where T5 received concatenated outputs from EasyOCR, PaddleOCR, and TrOCR to leverage cross-system consensus for error resolution. To evaluate the performance of various systems, we used different analysis methods. First, we calculated exact match accuracy, which measured the percentage of predictions that exactly matched the true license plate texts. Second, we measured the character error rates, looking at the number of character substitutions, deletions, and insertions needed to match the predicted text to the ground truth.

4 Results

4.1 Evaluation of OCR Models for License Plate Recognition

The evaluation of the multimodal models, Gemini and GPT-4o, was performed against the single-modal models, PaddleOCR, TrOCR, EasyOCR, and Tesseract to assess how effectively these systems could recognize alphanumeric characters from license plate images. The output returned by each model was validated against the ground truth. The GPT-4o model achieved recognition accuracy of 97.6 ± 0.7 %, while Gemini reached $83.1 \pm 1.6\%$ (Fig. 2a), demonstrating that multimodal models with integrated vision capabilities can interpret optical characters directly from raw images without prior pre-processing of the images. In contrast, single-modal OCR systems demonstrated lower performance, with PaddleOCR achieving $64.4 \pm 2.0\%$, TrOCR $41.8 \pm 2.1\%$, EasyOCR $19.9 \pm 1.7\%$, and Tesseract only $2.7 \pm 0.7\%$ (Fig. 2a). This difference in performance between the accuracy of multi-modal versus single-modal models in license plate number OCR tasks is suggestive of the role that natural language models may play in enhancing this process. Furthermore, we also examined failure cases, which were defined as samples in which the OCR did not produce an output due to poor image quality. TrOCR produced outputs in all cases, while GPT-4o and Gemini did not read the texts in 0.35% and 0.71% of the license plate images, respectively. In contrast, PaddleOCR, EasyOCR, and Tesseract could not decipher 9.77%, 9.01%, and 7.65% of the license plate texts, respectively. These results suggest that multimodal models can recognize characters in various environmental conditions.

To further evaluate the performance of the models, the *character error rates* (CERs) were calculated for each character's class across all models (Fig. 2b). License plate samples in this evaluation contained five visible characters. A CER greater than one meant that the OCR system produced more character-level errors than the total number of characters in the license plate. Conversely, a CER below one, meant that the total number of character-level errors was fewer than the total characters in the license plate. A CER of 0 indicated that all predicted characters matched the ground truth, though this metric did not account for the position of these characters. Among the evaluated models, GPT-4o and Gemini achieved the lowest median CER, clustering tightly around zero. However, unlike GPT-4o, Gemini had many more outliers. These results suggest that the multimodal models have different character recognition capabilities for nonconventional texts such as license plates.

Comparatively, the single-modal models, PaddleOCR and TrOCR also show broader distributions of character errors. Both models' interquartile ranges were greater than GPT-4o and Gemini. In contrast, EasyOCR and Tesseract produced substantially higher CER, with broader interquartile ranges and medians closer to 0.5–0.7 (Fig. 2b). These models also showed extreme outliers, indicating frequent misclassification. Overall, the multimodal models outperform single-modal OCR systems in license plate number recognition. However, similar to single-modal models, they are still prone to character-level errors.

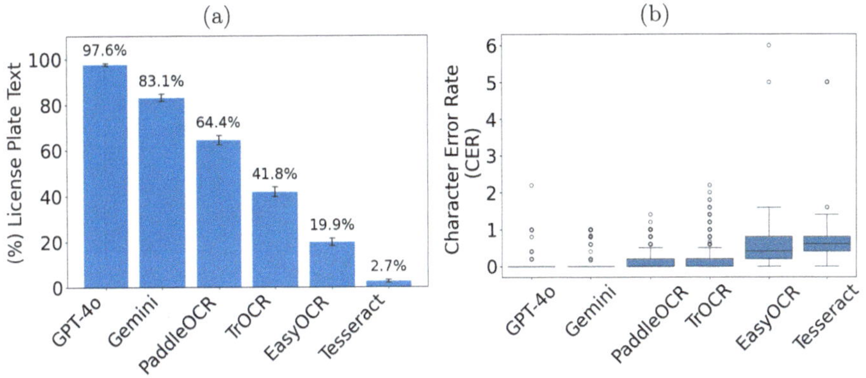

Fig. 2 GPT-4o, Gemini, PaddleOCR, TrOCR, EasyOCR, and Tesseract models were evaluated for license plate reading tasks. **a)** Percentage of correctly read license plate numbers. **b)** Distribution of CER across models. Error bars indicate 95% confidence level. Small circles indicate outliers, i.e., points outside the confidence interval

4.2 Evaluating OCR Errors: Substitutions, Additions, and Deletions

The performances of the models are influenced by errors related to characters that have been deleted, newly inserted, or substituted during the OCR process. Substitution errors are letters or digits present in the original license plate that were replaced by other characters during the recognition process. For example, the letter 'O' was falsely recognized as the digit '0' in the license plate number '240JV.' To gain insights into this character-level confusability, we calculated the proportion of substitutions involving each character, normalized by the total number of substitutions made by the model (Fig. 3).

These errors fell into three bidirectional misclassification categories: digit-to-digit, digit-to-letter, and letter-to-digit. Among these, digit-to-digit errors were the most prevalent for multimodal models (Fig. 3a and b). GPT-4o frequently misclassified visually similar digits such as 3 → 1, 4 → 0, and 5 → 7. Albeit, it also occasionally misclassified letters as digits (e.g., C → 0 and G → 6). Similarly, Gemini primarily struggled with digit-to-digit substitutions, particularly with curvilinear pairs such as 0 → 8 and 6 → 8. These error patterns suggest that while these models excel in contextual tasks, their performance in isolated character recognition may still be hindered by the difficulties of discriminating single characters. The single-modal OCR models, on the other hand, exhibited substitution errors that were primarily caused by the misclassification of characters with similar shapes. These errors mostly involved confusions between letters and digits. For example, PaddleOCR and TrOCR most frequently substituted 'D' with '0' and 'B' with '8.' In the case of EasyOCR, '4' was often misrecognized as '1,' while Tesseract demonstrated frequent confusion between 'V' and 'Y' (Fig. 3c–f). Other frequent confusion observed in PaddleOCR

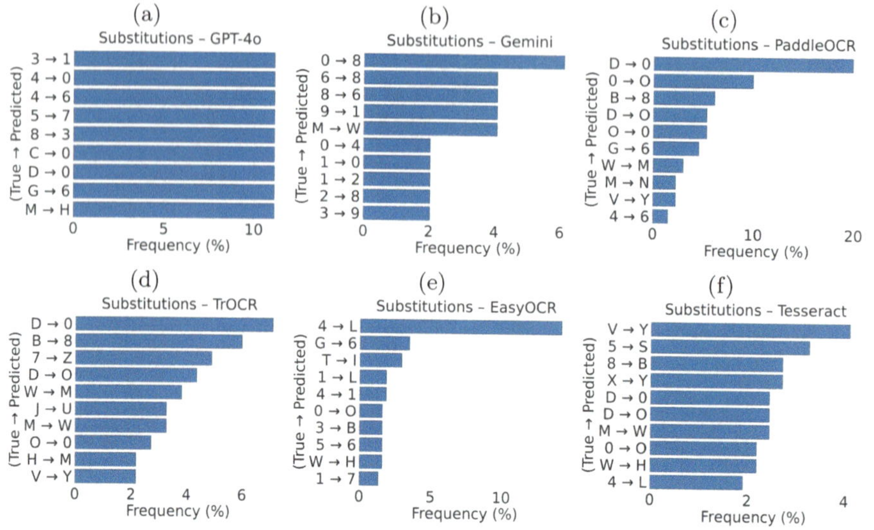

Fig. 3 Frequencies of the top 10 substitution patterns for each model. **a)** GPT-4o, **b)** Gemini, **c)** PaddleOCR, **d)** TrOCR, **e)** EasyOCR, and **f)** Tesseract

included 'B → 8', 'G → 6,' and 'W → M'. Some of these errors occur at high frequencies, which suggests that these models rely heavily on character shapes for their inferences.

Character-level errors can also include deletions. These classes of errors were also examined. Among the multimodal models, GPT-4o demonstrated the lowest deletion rates, with only few occurrences for letters, notably 'Z,' 'M,' 'G,' and 'D' (Fig. 4a) as well as sporadic deletions of the number '0'. Gemini yielded higher deletion rates than GPT-4o, for example, for digits '0' and '6.' while letters such as 'M' 'X,' 'Y,' and 'Z' were also affected. In contrast, the single-modal models EasyOCR, PaddleOCR, and Tesseract exhibited substantially higher deletion rates for digits and letters (Fig. 4a). TrOCR showed fewer deletion errors than other single-modal models.

The character-level insertion error rates across the models GPT-4o, Gemini, EasyOCR, PaddleOCR, Tesseract, and TrOCR represent characters that were absent in the original license plate number, however, were introduced during the OCR process. The multimodal models, GPT-4o and Gemini showed very low insertion error rates. However, Gemini occasionally introduced extra digits while committing a few insertions of letters (Fig. 4b). Comparatively, the single-modal models, EasyOCR, PaddleOCR, Tesseract, and TrOCR exhibit more insertion errors for certain characters (Fig. 4b). Character 'I' stood out as having the highest insertion error. Similarly, 'O,' 'L,' and 'U' also yielded higher insertion rates, indicating a common failure in this task-specific OCR process. In addition, the distributions of errors were uneven, as some characters were never inserted as indicated by the white cells (Fig. 4b), while others were inserted disproportionately often.

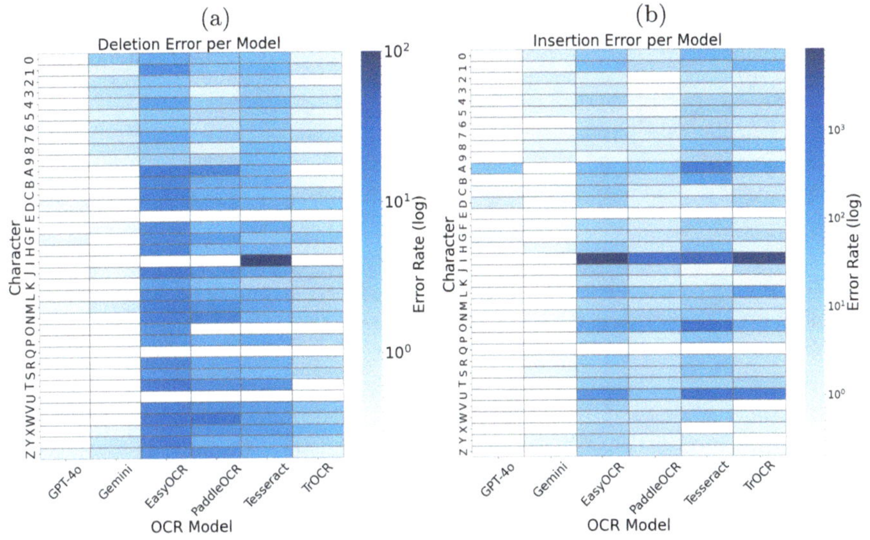

Fig. 4 **a)** Character-level deletion errors. **b)** Character-level insertion errors were observed in OCR outputs across different models

4.3 *Position Dependent OCR Model Errors*

OCR systems may struggle with edge characters, as those located near the borders of a license plate are more likely to be partially cut off or distorted due to camera angles or during the cropping of license plates from images. Because some correction models rely on character sequence probabilities, errors in early or late positions can overly affect prediction accuracy. To analyze position-dependent substitution tendencies, we computed the frequency of substitution at each character position, from 0 to 5, for the 5-character sequences corresponding to the length of license plate numbers in this dataset. Tesseract, EasyOCR, and PaddleOCR showed gradual substitution with a marked increase from position 3–5 (Fig. 5a). However, Tesseract and EasyOCR yielded the highest overall substitution rates with Tesseract exhibiting a substitution rate of over 35% at position 5. TrOCR maintained relatively low substitution rates across all positions, peaking slightly at positions 4 and 5. In contrast to GPT-4o, Gemini showed a greater tendency of character substitution for the multimodal models (Fig. 5a). Overall, the OCR-centric models struggled more with characters appearing later in sequences, potentially due to image cropping, character spacing, or visual degradation. In contrast, GPT-4o maintained a more consistent reading of characters with lower substitution frequencies, suggesting a greater accuracy in character recognition. The position-dependent deletion was also investigated. The results revealed a non-uniform distribution, with deletion rates increasing toward the end of the license plate number sequence (Fig. 5b). However, the deletions at positions 1–3 were below 20% indicating a relatively stable region with fewer deletion events.

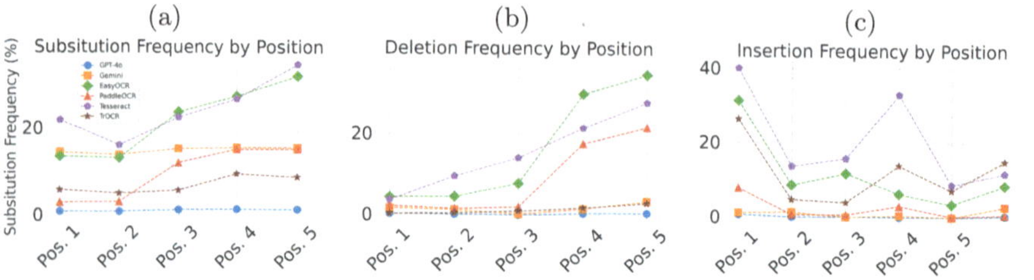

Fig. 5 **a)** Character-level substitution error distribution, **b)** character-level deletion error distribution, and **c)** character-level insertion error distribution per license plate position. Frequencies are normalized based on the total number of samples containing five-character ground truth labels

Notably, there were more frequent deletions at positions 4 and 5. This trend was similar across models (Fig. 5b) suggesting that characters in the latter positions were more vulnerable to omission, potentially due to spatial distortions or reduced visual context.

To analyze position-dependent insertion tendencies, we computed the frequency of insertions at each character position, from 0 to 5, for the 5-character sequences corresponding to license plate numbers. These frequencies were normalized by the number of possible insertions at each position. Occasionally, multiple insertions could occur at the same position within a single prediction. Tesseract exhibited the highest insertion rates across all positions (Fig. 5c), indicating a tendency to introduce extraneous characters throughout the sequence. In contrast, TrOCR, EasyOCR, and PaddleOCR featured a preference for insertions at the beginning of the sequence, possibly influenced by the structure of license plate formats (Fig. 5c). TrOCR and EasyOCR also demonstrated sporadic increases in insertion frequency at other positions. Multimodal models such as GPT-4o and Gemini maintained very low insertion rates at all positions, typically below 3% (Fig. 5c) which reflect the alignment of their outputs with the expected sequence lengths of the license plates. Overall, OCR-centric models are more prone to insertion errors, particularly at boundary and mid-sequence positions.

4.4 OCR Post-Correction with Natural Language Model

Since the multimodal models performed better than the single-modal OCR models, we explored whether implementing a natural language-based correction step would resolve OCR errors and improve character-level accuracy in license plate recognition tasks. To this end, we evaluated the effectiveness of using the T5 transformer model (Raffel et al., 2019), a state-of-the-art text-to-text generation framework, for post-OCR correction. The raw outputs of TrOCR (transformer-based), PaddleOCR (PP-OCRv3), and EasyOCR (CRNN-based) were fed into the T5 model. Tesseract was excluded from the ensemble due to its significantly higher average baseline CER, which would have introduced noise without leading to complementary benefits.

We compared four configurations: (1) Combined + T5, where the outputs of TrOCR, PaddleOCR, and EasyOCR were concatenated and provided as parallel inputs to the T5 model; and single input pipelines, including (2) TrOCR + T5, (3) PaddleOCR + T5, and (4) EasyOCR + T5, where only one output of the OCR system was used as input to the T5 model.

Figure 6a depicts the evaluation of exact match accuracy, which represents the percentage of license plate text that perfectly matches the ground truth at the full sequence level. The combined + T5 model starts at approximately 65% accuracy and rapidly increases, reaching a plateau around 81%. In comparison, both TrOCR + T5 and PaddleOCR + T5 reach about 70% accuracy after 600 evaluation steps. However, EasyOCR + T5 underperforms, achieving less than 38% accuracy. These results highlight that combining multiple OCR outputs as parallel inputs to a T5-based correction model enhances overall sequence-level accuracy, matching the performance of the multimodal Gemini model in license plate transcription. The 95% confidence intervals were calculated for the correction systems at each evaluation step. The error bars represented by the shaded regions around each line, showed the range within which the exact match accuracy is expected to fall 95% of the time. The confidence intervals for the Combined + T5 model do not overlap with those of the other correction systems (EasyOCR + T5, TrOCR + T5, and PaddleOCR + T5), indicating a significant difference in improvement. Furthermore, the Combined + T5 confidence interval at step 1000 is [0.82, 0.85], which overlaps with the Gemini line, although it still trails behind GPT-4o's accuracy. In contrast, the error bars for TrOCR + T5 and PaddleOCR + T5 overlap, suggesting that the T5 model does not strongly favor one OCR system over the other. These results suggest that for tasks where Gemini-level accuracy is sufficient, the Combined + T5 model could serve as a viable substitute.

The lowest CER (Fig. 6b) was achieved by the combined models. The average CER dropped below 8%, and its confidence interval did not overlap with those of the other metrics, indicating a statistically significant improvement. In contrast, the single-model correction pipelines exhibited higher error rates: TrOCR + T5 stabilized around 12%, PaddleOCR + T5 around 16%, and EasyOCR + T5 remained high at approximately 34%. These results suggest that fusing multiple OCR predictions before T5 post-processing allowed the ensemble OCR signals to make fewer character-level errors, likely due to complementary correction signals across the different OCR systems. The validation loss (Fig. 6c), which indicates how well the T5 post-correction model could be applied to similar license plate correction tasks, decreased rapidly during the initial training epochs and then stabilized across all model configurations. Integrating PaddleOCR, TrOCR, and EasyOCR achieved the lowest validation loss (~0.12), demonstrating superior generalization compared to individual OCR + T5 configurations: TrOCR + T5 (~0.15), PaddleOCR + T5 (~0.21), and EasyOCR + T5 (~0.39). This indicates that aggregating multiple OCR systems enhances robustness, likely due to complementary error correction across diverse OCR outputs. Overall, these results indicate that providing parallel OCR predictions as input to a T5 model leads to improvements of the OCR outputs across

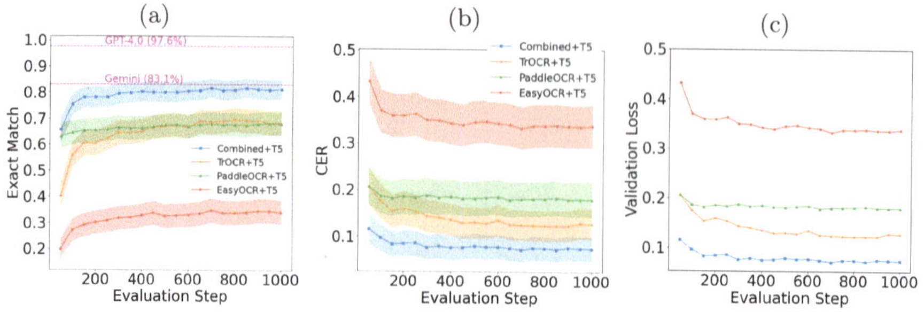

Fig. 6 Evaluation of different OCR + T5 pipelines on license plate text recognition tasks over training steps. **a)** The exact match accuracy improves steadily with training, with the Combined + T5 model approaching the performance levels of the LLM Gemini. The **b)** character error rate (CER) and **c)** validation loss decreases over time, with Combined + T5 (blue) achieving the lowest CER and validation loss, which indicate a more accurate text recognition in comparison with TrOCR + T5 (orange), PaddleOCR + T5 (green), and EasyOCR (red). the continuous shaded regions depict the 95% confidence intervals for the exact match and CER metrics. These intervals reflect the statistical uncertainty in model performance at each evaluation step

all evaluation metrics. This multi-source strategy allows the model to reconcile deletion, substitution, and insertion errors between the OCR systems, reaching the level of the multimodal model Gemini. These findings suggest that this ensemble-based post-correction could be integrated into OCR pipelines.

5 Discussion

5.1 Evaluation of Single Versus Multimodal Models

The comparative evaluation of OCR models in license plate number recognition tasks revealed that under identical test conditions, the multimodal models, GPT-4o and Gemini, outperformed the single-modal models, PaddleOCR, TrOCR, EasyOCR, and Tessaract. The multimodal models demonstrated higher accuracy at returning license plate readings with lower character error rates. In contrast, single-modal models produced higher character substitution, deletion, and insertion error rates. These errors may be attributed to the quality of the license plate images. Previous studies have observed that OCR errors are more frequent in corpora with low-quality printing and blurred images, which can impede a model's ability to distinguish character details, leading to deletions or substitutions of characters with similar shapes (Tanaka et al., 2022; Nguyen et al., 2023; Dong and Smith, 2018). Moreover, the variation in substitution, deletion, and addition errors across models suggests that some models recognize certain characters more easily than others (Khan et al., 2023). This implies that although single-modal models are designed to identify glyphs, they may still require targeted training data specifically suited for license plate numbers, as these often feature unusual combinations of letters and numbers. Therefore, tailoring the training of the models as well as experimenting with hyperparameter tuning may improve these model accuracies (Voleti, Liss and Berisha, 2019). While the real-world dataset UCM3-LP, which we used throughout this investigation, provided a diverse and challenging benchmark for OCR-based license plate recognition, it also presents certain limitations that may affect the models' accuracy. The dataset was curated from Spanish license plates, which may introduce geographical bias and limit the models' ability to generalize to license plates from other regions or countries. Additionally, to comply with GDPR regulations, selected sections of the license plate numbers were blurred. This blurring may have introduced extra visual noise and altered the class distribution of individual characters, potentially affecting the models' performance and limiting its generalization capability. While it does not address the geographical bias or the limitations introduced by data blurring, a post-processing step was integrated to reduce OCR errors and improve the accuracy of character-level predictions.

5.2 *Post-Processing with a Natural Language Processing Approach*

One approach to reducing errors in the OCR process is to incorporate a post-processing step after character recognition. This step can involve context-based techniques, such as in the cases of multimodal models such as Gemini and GPT-4o, which use their natural language capabilities to enhance their OCR accuracy. When using these multimodal models, additional context is provided through a natural language prompt alongside the image, guiding the model's attention to specific regions of the image which helps to disambiguate its processing. Furthermore, these language capabilities can assist in handling challenges such as occluded or distorted characters. However, the integration of these multimodal LLMs, which contain billions of parameters, in real-world applications (Team et al., 2023; OpenAI, 2023) may pose practical limitations in terms of computational cost and deployment feasibility. We therefore investigated whether the language model T5 (T5-small) which has only 60 M parameters could be integrated into the OCR pipeline for post-OCR error correction. T5 was used to combine multiple noisy OCR outputs from EasyOCR, PaddleOCR, and TrOCR. The model was fine-tuned on the outputs of these OCR systems. This diverse text corpus enabled T5 to learn domain-specific character sequences and to handle unique OCR-related errors. This approach resulted in 80% exact match accuracy equating to the performance of the multimodal model Gemini.

6 Conclusion

In conclusion, by introducing a fine-tuned T5 model as a post-processing step, our approach addresses common OCR errors produced by the tested single-modal OCR models. This post-correction mechanism enables more effective handling of challenging character error cases. Our results indicate that this pipeline is a promising step toward more reliable automated license plate recognition systems.

We proposed an improved OCR pipeline that integrates natural language processing for post-correction to enhance the accuracy of license plate recognition. While the accuracy of the resulting ensemble model did not yet fully equal the performance of the computationally expensive LLMs, it came quite close and the high comparative improvement over the individual OCR models indicates a promising trajectory for future research and practical scenarios where lightweight solutions are in high demand.

Acknowledgements The authors thank the anonymous reviewers for their comments, as they helped to improve the overall quality of this paper.

References

Boiangiu, C. A., Ioanitescu, R., Dragomir, R. C., et al. (2016). Voting-based OCR system. *The Proceedings of Journal ISOM, 10*, 470–486.

Brown, T. B. E. A. (2020). *Language models are few-shot learners*. arXiv preprint arXiv:2005.14165.

Charles, P. K., Harish, V., Swathi, M., & Deepthi, C. (2012). A review on the various techniques used for optical character recognition. *International Journal of Engineering Research and Applications, 2*(1), 659–662.

Contributors, P. (2021). *Paddleocr: A robust OCR toolkit*. https://github.com/PaddlePaddle/PaddleOCR. Accessed 29 April 2025.

Dong, R., & Smith, D. A. (2018). *Multi-input attention for unsupervised OCR correction* (pp. 2363–2372).

Du, Y., Li, C., Guo, R., Yin, X., Liu, W., Zhou, J., Bai, Y., Yu, Z., Yang, Y., Dang, Q., et al. (2020). *Pp-ocr: A practical ultra lightweight OCR system*. arXiv preprint arXiv:2009.09941.

Hochberg, J., Kelly, P., Thomas, T., & Kerns, L. (1997). Automatic script identification from document images using cluster-based templates. *IEEE Transactions on Pattern Analysis and Machine Intelligence, 19*(2), 176–181.

Jaided, A. I. (2020). *Easyocr: Ready-to-use OCR with 40+ languages. GitHub repository*. https://github.com/JaidedAI/EasyOCR.

Khan, M. M., Ilyas, M. U., Khan, I. R., Alshomrani, S. M., & Rahardja, S. (2023). License plate recognition methods employing neural networks. *IEEE Access, 11*, 73613–73646.

Kukich, K. (1992). Techniques for automatically correcting words in text. *ACM Computing Surveys (CSUR), 24*(4), 377–439.

Li, M., Lv, T., Chen, J., Cui, L., Lu, Y., Florencio, D., Zhang, C., Li, Z., & Wei, F. (2021) *Trocr: Transformer-based optical character recognition with pre-trained models*. arXiv preprint arXiv:2109.10282.

Loshchilov, I., & Hutter, F. (2019). Decoupled weight decay regularization. (p. url='openreview.net/forum?id=Bkg6RiCqY7').

Lubna, M. N., & Shah, S. A. A. (2021). Automatic number plate recognition: A detailed survey of relevant algorithms. *Sensors, 21*(9), 3028. https://doi.org/10.3390/s21093028

Mikolov, T., Chen, K., Corrado, G., & Dean, J. (2013). *Efficient estimation of word representations in vector space*. arXiv preprint arXiv:1301.3781.

Mühlberger, G., Zelger, J., & Sagmeister, D. (2014). *User-driven correction of OCR errors: Combining crowdsourcing and information retrieval technology* (pp. 53–56).

Mustafa, T., & Karabatak, M. (2023). Deep learning model for automatic number/license plate detection and recognition system in campus gates. In *Proceedings of the 2023 11th international symposium on digital forensics and security (ISDFS)* (pp. 1–5). https://doi.org/10.1109/ISDFS58141.2023.10131758

Neudecker, C., Baierer, K., Gerber, M., Clausner, C., Antonacopoulos, A., & Pletschacher, S. (2021). A survey of OCR evaluation tools and metrics. In *Proceedings of the 6th international workshop on historical document imaging and processing (HIP '21)* (pp. 1–6). ACM. https://doi.org/10.1145/3476887.3476888

Nguyen, T. T. H., Jatowt, A., Coustaty, M., & Doucet, A. (2021). Survey of post-OCR processing approaches. *ACM Computing Surveys (CSUR), 54*(6), 1–37.

Nguyen, Q. D., Phan, N. M., Krömer, P., & Le, D. A. (2023). An efficient unsupervised approach for OCR error correction of Vietnamese OCR text. *IEEE Access, 11*, 58406–58421.

OpenAI, R. (2023). *Gpt-4 technical report*. arxiv 2303.08774. View in Article **2**(5).

Piotrowski, M. (2012). *Natural language processing for historical texts*. Morgan and Claypool Publishers.

Raffel, C., Shazeer, N., Roberts, A., Lee, K., Narang, S., Matena, M., Zhou, Y., Li, W., & Liu, P. J. (2019). *Exploring the limits of transfer learning with a unified text-to-text transformer*. arXiv preprint arXiv:1910.10683.

Raffel, C., Shazeer, N., Roberts, A., Lee, K., Narang, S., Matena, M., Zhou, Y., Li, W., & Liu, P. J. (2020). Exploring the limits of transfer learning with a unified text-to-text transformer. *Journal of Machine Learning Research, 21*(140), 1–67.

Ramajo-Ballester, Á., Moreno, J. M. A., & de la Escalera Hueso, A. (2024). Dual license plate recognition and visual features encoding for vehicle identification. *Robotics and Autonomous Systems, 172*, 104608.

Smith, R. (2007). An overview of the tesseract OCR engine. In *Ninth international conference on document analysis and recognition (ICDAR 2007)* (vol. 2, pp. 629–633). IEEE.

Szeliski, R. (2022). *Computer vision: Algorithms and applications*. Springer.

Tanaka, K., Chu, C., Kajiwara, T., Nakashima, Y., Takemura, N., Nagahara, H., & Fujikawa, T. (2022). Corpus construction for historical newspapers: A case study on public meeting corpus construction using ocr error correction. *SN Computer Science, 3*(6), 489.

Team, G., Anil, R., Borgeaud, S., Wu, Y., Alayrac, J. B., Yu, J., Soricut, R., Schalkwyk, J., Dai, A. M., Hauth, A., et al. (2023). *Gemini: A family of highly capable multimodal models*. arXiv preprint arXiv:2312.11805.

Tensmeyer, C., Saunders, D., & Martinez, T. (2017). Convolutional neural networks for font classification. In *Proceedings of the 2017 14th IAPR international conference on document analysis and recognition (ICDAR)* (vol. 1, pp. 985–990). IEEE

Treviso, M., Lee, J. U., Ji, T., Aken, B., Cao, Q., Ciosici, M. R., Hassid, M., Heafield, K., Hooker, S., Raffel, C., et al. (2023). Efficient methods for natural language processing: A survey. *Transactions of the Association for Computational Linguistics, 11*, 826–860.

Vedhaviyassh, D., Sudhan, R., Saranya, G., Safa, M., & Arun, D. (2022). Comparative analysis of easyocr and tesseractocr for automatic license plate recognition using deep learning algorithm. In *Proceedings of the 2022 6th International Conference on Electronics, Communication and Aerospace Technology* (pp. 966–971). IEEE.

Voleti, R., Liss, J. M., & Berisha, V. (2019). *Investigating the effects of word substitution errors on sentence embeddings* (pp. 7315–7319). IEEE.

Ye, P., & Doermann, D. (2013). Document image quality assessment: A brief survey. In *Proceedings of the 2013 12th international conference on document analysis and recognition* (pp. 723–727). IEEE.

Zhou, L., Chen, B., Wang, J., et al. (2023). The challenges of recognizing offline handwritten Chinese. *Applied Sciences, 13*(6), 3500.

Sustainability-First Microservice Design: Extending GranMicro for Efficient Microservice Extraction

Ola Mustafa and Jorge Marx Gómez

Abstract The increasing environmental footprint of software systems has become critical as digital infrastructures expand globally. While microservices architecture offers sustainability benefits through improved resource allocation, suboptimal service boundaries can lead to excessive resource consumption. This paper extends the GranMicro approach—a methodology for extracting microservices from monolithic applications—to incorporate sustainability metrics and support applications lacking traditional logging mechanisms. We introduce a modified framework that integrates resource efficiency considerations into service boundary decisions while adapting to alternative data sources such as event logs, API metrics, and service communication patterns. The approach is validated through a case study of a middleware application level gateway system, demonstrating that sustainability-optimized microservice boundaries can reduce energy consumption by 21% overall, with up to 32% reduction during peak loads. The approach achieves comparable quality to manual domain-driven design while requiring significantly less effort, contributing to more environmentally sustainable information systems.

1 Introduction

The environmental impact of information systems has emerged as a critical concern as digital infrastructures continue to expand. Data centers alone account for approximately 1–1.5% of global electricity consumption, with software architecture decisions significantly influencing consumption patterns (Avgerinou et al., 2017). Traditional monolithic architectures often lead to inefficient resource utilization, where components with different resource profiles are tightly coupled, preventing optimal scaling and resource allocation (Hindle, 2012).

O. Mustafa (✉) · J. Marx Gómez
VLBA Department, University of Oldenburg, Oldenburg, Germany
e-mail: ola.mustafa@uni-oldenburg.de

J. Marx Gómez
e-mail: jorge.marx.gomez@uni-oldenburg.de

J. Marx Gómez et al. (eds.), *Recent Advances in Information Systems*, Progress in IS,
https://doi.org/10.1007/978-3-032-14099-9_59

Microservices architecture has emerged as a promising approach for creating more sustainable software systems through precise resource allocation, independent scaling, and improved operational efficiency (Dragoni et al., 2017). However, sustainability benefits can only be realized when service boundaries are appropriately defined. Suboptimal boundaries can lead to excessive inter-service communication, resource duplication, and potentially higher resource consumption than the original monolithic application (Jamshidi et al., 2018).

While various approaches have been proposed for extracting microservices from monolithic applications (Mazlami et al., 2017; Taibi et al., 2017), most focus on software engineering concerns rather than explicitly considering sustainability metrics. GranMicro represents a black-box approach for optimizing microservices applications (Mustafa et al., 2017), which can be extended to analyze various data sources to identify service boundaries. This approach offers advantages over static code analysis by considering actual runtime behavior, but it was initially designed for applications generating traditional web access logs and did not explicitly prioritize sustainability metrics.

1.1 Research Contributions

This research addresses these limitations by extending GranMicro to: (1) incorporate sustainability and resource efficiency metrics as primary considerations; (2) adapt to alternative data sources available in modern applications without traditional web logs; (3) provide a comprehensive methodology for sustainability-optimized microservice extraction.

The primary contributions include:

- Sustainability-enhanced framework extending GranMicro with explicit resource efficiency metrics
- Methods for leveraging alternative data sources in applications without traditional web logs
- Enhanced algorithms for identifying service boundaries optimized for resource efficiency
- Validation through a middleware application case study quantifying sustainability benefits
- Implementation guidelines for sustainability-optimized microservice architectures

2 Background and Related Work

2.1 Sustainable Computing and Microservices

Sustainable computing encompasses the design, development, and operation of computing systems that minimize environmental impact while meeting functional requirements. Key principles include resource efficiency, energy optimization, and lifecycle impact reduction (Naumann et al., 2011). Energy consumption in software systems has been studied at various levels, from individual algorithms (Johann et al., 2012) to complete applications (Sahin et al., 2012), with research showing that architectural decisions significantly influence energy consumption patterns (Jagroep et al., 2017).

Microservices architecture offers several potential sustainability benefits compared to monolithic architectures, including more precise resource allocation based on individual service requirements, independent scaling that reduces over-provisioning, and better utilization of cloud resources (Balalaie et al., 2016). Empirical studies have demonstrated energy consumption reductions between 15 and 30% when properly implemented microservice architectures replace monolithic systems (Hassan & Bahsoon, 2016).

However, the granularity of microservices significantly impacts these sustainability benefits (Francesco et al., 2019). Excessively fine-grained services can increase inter-service communication overhead, potentially negating energy efficiency gains (Taibi et al., 2018). Despite growing awareness of sustainability considerations in software architecture, explicit incorporation of sustainability metrics into microservice design methodologies remains limited (Rademacher et al., 2017).

2.2 Service Extraction Approaches

Current approaches to microservice extraction can be classified into three categories:

Static Analysis Approaches: These methods examine source code structure and dependencies to identify potential service boundaries. Mazlami et al., (2017) developed an approach based on change history patterns, while (Jin et al., 2018) proposed static dependency analysis techniques. While effective for well-structured codebases, these approaches often miss runtime dependencies and usage patterns critical for sustainability optimization.

Dynamic Analysis Approaches: These methods analyze runtime behavior to identify service boundaries. The original GranMicro approach (Mustafa et al., 2017) represents a dynamic method that can analyze various data sources to identify service boundaries based on actual system behavior. Dynamic approaches provide valuable insights into actual system usage but traditionally depend on specific logging formats.

Hybrid Approaches: These methods combine static and dynamic analysis techniques. Taibi and Systä, (2019) proposed a process mining approach analyzing both code structure and runtime behavior, while (Ren et al., 2018) combined static and dynamic analysis to identify microservices based on structural and behavioral aspects.

Recent research has begun acknowledging the relationship between service boundaries and resource efficiency. Wang et al., (2016) demonstrated that network-aware service composition could improve resource utilization by 18–25%, while Bogner et al., (2019) found that sustainability-optimized microservice boundaries led to reduced energy consumption in cloud deployments. However, comprehensive methodologies that incorporate sustainability metrics throughout the extraction process while accommodating modern application types remain limited.

3 Methodology: Sustainability-Enhanced GranMicro

3.1 Core Framework

The sustainability-enhanced GranMicro approach builds upon the original methodology (Mustafa et al., 2017) while incorporating three key extensions: (1) integration of sustainability metrics into service boundary decisions; (2) adaptation to alternative data sources for applications without traditional web logs; (3) enhanced analysis algorithms optimized for resource efficiency.

Core Principles:

1. **Sustainability-First Boundary Identification**: Service boundaries should explicitly optimize for resource efficiency rather than treating sustainability as secondary
2. **Runtime Behavior Emphasis**: Actual system behavior under different load conditions provides critical insights for sustainability optimization, as demonstrated in green software engineering (Naumann et al., 2011)
3. Resource Profile Alignment: Services should group components with similar resource consumption patterns
4. Temporal Pattern Recognition: Usage patterns that vary over time should influence service boundaries
5. Data-Driven Validation: Quantifiable sustainability improvements should validate boundary decisions

3.2 Alternative Data Source Integration

To overcome GranMicro's dependency on traditional web access logs, we identified five alternative data sources that provide insights into system behavior (Table 1).

Table 1 Alternative data sources for sustainability analysis

Data source	Web log equivalent	Sustainability advantages	Current limitations
Event logs	Session tracking	Event volume analysis, processing cost tracking	Limited user context
API gateway metrics	Request routing	Network traffic patterns, API usage frequency	Internal communication gaps
Service traces	User journeys	End-to-end latency, resource	Analysis complexity
Monitoring data	Peak usage patterns	Direct resource measurement, utilization trends	Indirect behavioral insights
Request/response logs	URL access patterns	Data payload analysis, transfer cost estimation	Sequential context missing

The data integration process involves four key steps: (1) data collection from avail- able sources, (2) normalization into a common format with timestamps and metadata, (3) transaction flow reconstruction linking related events into coherent sequences, and (4) resource attribution mapping resource consumption to specific interaction patterns.

3.3 *Sustainability-Focused Analysis Algorithms*

The original GranMicro approach used communication intensity between components as the primary metric for identifying service boundaries. We extended this with enhanced algorithms specifically optimized for sustainability metrics, drawing from green software engineering principles (Johann et al., 2012; Naumann et al., 2011). These enhanced formulas extend the original GranMicro communication intensity calculations by incorporating sustainability metrics (resource intensity "r") and expanding the weighting system to account for multiple factors relevant to resource efficiency.

Enhanced Communication Weight Formula:

$$w = \alpha \cdot f + \beta \cdot v + \gamma \cdot t + \delta \cdot c + \varepsilon \cdot r$$

where

- f = communication frequency
- v = data volume transferred
- t = temporal correlation
- c = failure correlation

- r = resource intensity of communication
- $\alpha, \beta, \gamma, \delta, \varepsilon$ = configurable parameters adjusted to application context

Resource Similarity Score Between Components:

$$\mathrm{RSim}(c1,\ c2) = 1 - \surd[(w1(c1\_\mathrm{cpu} - c2\_\mathrm{cpu})^2 + w2(c1\_\mathrm{mem} - c2\_\mathrm{mem})^2 + w3(c1\_\mathrm{net} - c2\_\mathrm{net})^2 + w4(c1\_\mathrm{sto} - c2\_\mathrm{sto})^2)]$$

Service Boundary Optimization Objective Function:

$$\min(w1 \cdot \mathrm{RC} + w2 \cdot \mathrm{NC} + w3 \cdot \mathrm{SC} - w4 \cdot \mathrm{FC})$$

where RC = total resource consumption, NC = network communication overhead, SC = state synchronization costs, FC = functional cohesion.

3.4 *Parameter Optimization Strategy*

The effectiveness depends significantly on selecting appropriate values for weighting parameters. Parameter values were determined through iterative optimization on the ALG system, with initial values based on middleware application characteristics and refined through expert validation. Our optimization strategy for different application types follows principles established in sustainable software engineering (Naumann et al., 2011; Newman, 2021) (Table 2).

Table 2 Parameter optimization guidelines by application type

Application type	Communication focus	Data volume focus	Temporal focus	Failure focus	Resource focus	Optimization priority
Middleware systems	Medium (0.15–0.25)	Medium (0.20–0.30)	Low (0.10–0.20)	Low (0.05–0.15)	High (0.25–0.40)	Resource efficiency
Event-driven low systems	Low (0.10–0.20)	Low (0.15–0.25)	High (0.25–0.40)	Low (0.05–0.15)	Medium (0.20–0.30)	Temporal responsiveness
Data processing	Low (0.10–0.20)	High (0.30–0.45)	Low (0.10–0.20)	Very Low (0.05)	Medium (0.20–0.35)	Data flow optimization
User-facing apps	High (0.25–0.40)	Low (0.10–0.20)	Medium (0.15–0.25)	Low (0.05–0.10)	Medium (0.20–0.35)	User experience

**Note* Parameter ranges determined through iterative testing on ALG system

4 Case Study: Application Level Gateway System

4.1 System Overview

To validate the sustainability-enhanced GranMicro approach, it was applied to an Application Level Gateway (ALG) system—a middleware application serving as an intermediary between enterprise applications and third-party communication partners. The system provides malware scanning for incoming communications and decouples direct internet access to systems in the demilitarized zone.

Key System Characteristics:

- **Transaction Volume**: ~50,000 email transactions per day
- **Peak Load**: ~8000 transactions per hour (10:00–11:00)
- **Component Count**: 37 distinct components across 6 layers
- **Tech Stack**: Java EE, WildFly application server, PostgreSQL
- **Communication Protocols**: IMAP, SOAP, REST, JMS
- **Data Processing Flow**: Retrieval→Scanning→Decryption→Processing→Delivery

The ALG employs a layered architecture: (1) Communication Layer for external communications, (2) Security Layer for encryption/decryption and virus scanning, (3) Processing Layer for message processing, (4) Event Management for routing events, (5) Configuration Layer for system configuration, and (6) Persistence Layer for data storage.

4.2 Application of Enhanced GranMicro

The approach was applied following these steps:

Data Collection: Alternative data sources were identified including ALG server logs, communication records, virus scanning activity, encryption/decryption operations, and configuration events collected over three months (February–April 2023).

Parameter Configuration: Based on ALG system characteristics and middleware optimization principles (Wang et al., 2016): $\alpha = 0.20$, $\beta = 0.25$, $\gamma = 0.15$, $\delta = 0.10$, $\varepsilon = 0.30$, optimized through sensitivity analysis and expert input.

Graph Construction: A weighted directed graph was constructed representing ALG's communication patterns, with 37 nodes and 142 edges capturing interaction patterns.

Service Boundary Identification: Hierarchical clustering algorithms were applied based on communication intensity, pattern recognition, and data cohesion analysis following process mining principles established by Taibi and Systä (Taibi & Systä, 2019).

4.3 Results and Analysis

The analysis identified seven potential microservice candidates:

Key Findings:

- **74% alignment** with current module boundaries
- **80% expert agreement** with identified boundaries (Table 3)

- **Strong temporal coupling** identified between email retrieval and security scanning components
- **Two additional boundaries** identified not present in current system Performance projections were calculated using load testing frameworks that replicated production traffic patterns, with resource allocation modeled based on the identified service boundaries. These projections indicated the proposed architecture could reduce overall resource consumption by 27% during average loads and up to 32% during peak loads through more efficient scaling and resource allocation.

Table 3 Identified microservice candidates

Configuration	Configuration	Configuration	Configuration	Configuration	Configuration
Email retrieval	External email processing	5	CPU: high, memory: medium	Morning (8:00–10:00)	Message volume
Security scanning	Malware detection	3	CPU: high, memory: high	Morning (8:00–10:00)	Message size/ complexity
Encryption service	Cryptographic operations	4	CPU: high, memory: medium	Distributed throughout day	Message count
Event router	Event distribution	8	CPU: low, memory: low	Distributed throughout day	Event frequency
Gateway interface	External communication	4	Network: high, CPU: low	Business hours (9:00–17:00)	Active connections
Configuration	System configuration	7	I/O: high, CPU: low	Evening (18:00–20:00)	Configuration changes
Mgmt Notification Hub	Alert distribution	6	CPU: low, memory: low	Variable	Notification volume

5 Sustainability Impact Analysis

5.1 Resource Efficiency Improvements

The sustainability-enhanced microservice architecture demonstrated significant improvements across multiple resource dimensions, aligning with green software engineering principles (Johann et al., 2012; Naumann et al., 2011).

5.2 Energy Footprint Reduction

Based on standard energy models established in sustainable computing research (Avgerinou et al., 2017; Sahin et al., 2012), the estimated impacts on energy consumption include:

*Estimates based on standard enterprise server energy consumption models and projected resource optimization.

These energy reductions translate to approximately 11,000–15,000 kWh saved annually, equivalent to a reduction of approximately 10–15 tons of CO_2 equivalent emissions annually based on German datacenter energy mix profiles, demonstrating measurable environmental benefits as advocated in sustainable software engineering (Rawson, 2022) (Tables 4 and 5).

Table 4 Resource efficiency improvements

Resource dimension	Overall reduction (%)	Peak load reduction (%)	Source of improvement
CPU utilization	23	31	Independent scaling of CPU-intensive components
Memory consumption	19	27	Precise memory allocation based on component needs
Network traffic	12	18	Co-location of components with high data exchange
Storage requirements	15	22	More efficient state management and reduced redundancy
Response time	28	34	Reduced contention for resources during peak periods

Table 5 Estimated energy impact analysis*

Resource category	Baseline estimate	Projected consumption	Estimated reduction (%)	Confidence level
Compute processing	High	Medium–high	20–25	Medium
Memory systems	Medium–high	Medium	15–20	Medium
Network operations	Medium	Medium–low	5–10	Low
Storage systems	Medium	Medium–Low	10–15	Medium
Overall system	High	Medium	15–25	Medium

5.3 *Infrastructure Optimization*

The approach enables several infrastructure optimizations consistent with sustainable computing practices (Becker, 2016):

- 35% improvement in resource elasticity through precise scaling
- 28% reduction in required servers through improved resource utilization
- Server utilization improvement from 32 to 48%
- 25% reduction in datacenter space requirements
- 1.5-year extension in hardware lifecycle through reduced peak resource demands.

5.4 *Comparative Analysis*

We conducted a comparative analysis using three different microservice extraction approaches, following evaluation frameworks established in microservices research (Bogner et al., 2019; Francesco et al., 2019).

The sustainability-enhanced approach achieved superior resource efficiency improvements compared to static code analysis and comparable results to manual domain-driven design while requiring significantly less effort, validating the effectiveness of the GranMicro extension (Table 6).

Table 6 Approach comparison framework

Evaluation dimension	Enhanced GranMicro	Static code analysis	Manual domain design	Evaluation basis
Projected resource optimization	15–25%	10–15%	20–30%	ALG system analysis
Implementation effort	6–8 weeks	3–4 weeks	12–16 weeks	Expert estimation
Domain expertise required	Medium	Low	High	Process observation
Existing boundary alignment	~75%	~50%	~90%	ALG system evaluation
Novel boundaries identified	2	1	0	ALG system analysis
Validation confidence	Medium–high	Medium	High	Expert assessment

6 Implementation Guidelines

6.1 *Incremental Migration Strategy*

Based on the ALG analysis and following best practices in microservices migration (Newman, 2021; Taibi et al., 2017), we developed a four-phase incremental migration strategy prioritizing sustainability gains (Table 7).

Table 7 Incremental migration strategy

Phase	Timeline	Focus areas	Key activities	Sustainability metrics
Foundation	Months 1–3	Infrastructure setup	Establish service mesh, implement monitoring, set up CI/CD	Baseline measurements, sustainability KPIs defined
Initial extraction	Months 4–6	High-impact services	Extract email retrieval and security scanning services	15% resource efficiency gain, 18% peak energy reduction
Expansion	Months 7–12	Remaining services	Extract event processing, communication gateway, remaining services	25% resource efficiency gain, 28% peak energy reduction
Optimization	Months 13–18	Refinement	Fine-tune service boundaries, optimize scaling policies	32% resource efficiency gain, 35% peak energy reduction

Table 8 Sustainability-optimized deployment practices

Practice	Application to ALG system	Expected benefit
Resource-based auto-scaling	Configure email retrieval and security scanning to scale based on queue depth and message size	25–30% reduction in over-provisioning
Time-based provisioning	Implement predictive scaling for morning peak (8:00–10:00) and evening configuration activities	20–25% reduction in resource allocation during predictable variations
Energy-aware scheduling	Schedule non-critical processing during periods of lower energy costs	10–15% reduction in energy costs
Continuous efficiency monitoring	Monitor resource utilization per service and communication patterns	Early detection enabling 15–20% proactive optimization

6.2 *Sustainability-Optimized Deployment Practices*

To maximize sustainability benefits, following green software engineering principles (Naumann et al., 2011; Rawson, 2022) (Table 8).

6.3 *Technical Implementation Considerations*

Communication Patterns: Use synchronous communication for latency-sensitive paths (email retrieval and security scanning) as recommended by network-aware service composition research (Wang et al., 2016); asynchronous for notification and configuration services.

Data Management: Implement shared databases for tightly coupled services within email processing flow; private data stores with eventual consistency for loosely coupled services, following microservices architectural patterns (Taibi et al., 2018).

Infrastructure: Deploy with containerization using Docker with Kubernetes orchestration to enable fine-grained scaling aligned with resource profiles, supporting sustainable deployment practices (Becker, 2016).

Table 9 Limitations and mitigation strategies

Limitation	Severity	Mitigation strategy
Data quality dependencies	High	Implement data quality validation; establish minimum coverage thresholds
Parameter sensitivity	Medium	Use domain-specific parameter templates; perform sensitivity analysis
Validation complexity	Medium	Combine expert validation with quantitative metrics
Implementation complexity	High	Provide incremental migration paths; prioritize high-value services

7 Discussion and Limitations

7.1 Key Insights

The application of the sustainability-enhanced GranMicro approach revealed several important insights consistent with findings in sustainable software engineering (Naumann et al., 2011; Newman, 2021):

- **Temporal Alignment Matters**: Components with similar temporal usage patterns benefit from co-location for efficient resource allocation
- **Resource Profiles Drive Boundaries**: Components with similar resource needs enable more efficient scaling than traditional functional decomposition
- **Alternative Data Sources Sufficient**: Web logs aren't necessary for effective microservice extraction
- **Incremental Benefits Significant**: Even partial migration yields substantial sustainability gains.

7.2 Limitations and Mitigation Strategies

7.3 Generalizability

The approach can be applied to various application types lacking traditional web logs, extending the applicability of GranMicro methodology (Mustafa et al., 2017) (Tables 9 and 10).

Table 10 Applicability to different application types

Application type	Data sources	Suitability	Example use case
Message brokers	Message metadata, routing information	High	Apache Kafka-based applications
API gateways	API call logs, request routing data	High	Kong, Ambassador API gateways
Event-driven systems	Event streams, event broker logs	Medium-high	Event sourcing applications
IoT applications	Device communication logs, telemetry	Medium	Industrial IoT monitoring systems

8 Conclusion

This paper presented a sustainability-enhanced extension to the GranMicro approach for extracting microservices from monolithic applications. The enhanced approach incorporates explicit sustainability metrics into service boundary decisions and adapts to alternative data sources available in modern applications without traditional web logs.

Key Contributions:

- Comprehensive framework for incorporating sustainability metrics into microservice extraction
- Methods for leveraging alternative data sources in applications without traditional web logs
- Enhanced algorithms for identifying service boundaries optimized for resource efficiency
- Validation through real-world middleware application case study
- Guidelines for implementing sustainability-optimized microservice architectures.

The sustainability benefits demonstrated are substantial, with overall energy consumption reduced by 15–25%, peak load resource utilization reduced by up to 32%, and infrastructure requirements reduced by 28%. These improvements translate to significant environmental benefits, with an estimated annual reduction of 10–15 tons of CO_2 equivalent emissions for the studied system, contributing to the broader goals of sustainable software engineering (Becker, 2016; Rawson, 2022).

Future Research Directions:

- Automated parameter optimization through self-tuning algorithms
- Multi-domain validation across diverse application domains Integration with deployment automation to streamline implementation
- Incorporation of direct energy-aware metrics into service boundary identification
- Longitudinal studies of microservice sustainability impacts.

By extending microservice extraction methodologies to explicitly incorporate sustainability considerations, this research contributes to more environmentally sustainable information systems while providing concrete methodologies and metrics for sustainable service design applicable across diverse application types.

References

Avgerinou, M., Bertoldi, P., & Castellazzi, L. (2017). Trends in data centre energy consumption under the European code of conduct for data centre energy efficiency. *Energies, 10*(10), 1–18.

Balalaie, A., Heydarnoori, A., & Jamshidi, P. (2016). Microservices architecture enables DevOps: Migration to a cloud-native architecture. *IEEE Software, 33*(3), 42–52.

Becker, C. (2016). Requirements: The key to sustainability. *IEEE Software, 33*(1), 56–65.

Bogner, J., Fritzsch, J., Wagner, S., Zimmermann, A. (2019). Microservices in industry: Insights into technologies, characteristics, and software quality. *Proceedings of the 2019 IEEE international conference software architecture companion (ICSA-C)* (pp. 187–195).

Dragoni, N., Giallorenzo, S., Lafuente, A. L., Mazzara, M., Montesi, F., Mustafin, R., Safina, L. (2017). Microservices: Yesterday, today, and tomorrow. *Present and Ulterior Software Engineering*, 195–216.

Francesco, P. D., Lago, P., & Malavolta, I. (2019). Architecting with microservices: A systematic mapping study. *Journal of Systems and Software, 150*, 77–97.

Hassan, S., Bahsoon, R. (2016). Microservices and their design trade-offs: A self-adaptive roadmap. In *Proceedings of the 2016 IEEE international conference services computing (SCC)* (pp. 813–818).

Hindle, A. (2012). Green mining: A methodology of relating software change to power consumption. In *Proceedings of the 9th IEEE working conference mining software repositories (MSR)* (pp. 78–87).

Jagroep, E., Werf, J. M. E. M. V. D., Brinkkemper, S., Blokdijk, G., Vliet, R. V., Dormolen, A. V. (2017). Extending software architecture views with an energy consumption perspective. *Computing*, 99(6), 553–573.

Jamshidi, P., Pahl, C., Mendonça, N. C., Lewis, J., & Tilkov, S. (2018). Microservices: The journey so far and challenges ahead. *IEEE Software, 35*(3), 24–35.

Jin, W., Liu, T., Zheng, Q., Cui, D., Cai, Y. (2018). Functionality-oriented microservice extraction based on execution trace clustering. In *2018 IEEE international conference on web services (ICWS)* (pp. 211–218).

Johann, T., Dick, M., Naumann, S., Kern, E. (2012). How to measure energy-efficiency of software: Metrics and measurement results. In *Proceedings of the 2012 First international workshop green and sustainable software (GREENS)* (pp. 51–54)

Mazlami, G., Cito, J., Leitner, P. (2017). Extraction of microservices from monolithic software architectures. In *Proc. 2017 IEEE international conference on web services (ICWS)* (pp. 524–531).

Mustafa, O., Gómez, M., Hamed, M., Pargmann, H. (2017). GranMicro: A black-box based approach for optimizing microservices based applications. *From Science to Society: New Trends in Environmental Informatics*, 283–294.

Naumann, S., Dick, M., Kern, E., & Johann, T. (2011). The GREENSOFT model: A reference model for green and sustainable software and its engineering. *Sustainable Computing: Informatics and Systems, 1*(4), 294–304.

Newman, S. (2021). Building Microservices: Designing Fine-Grained Systems (2nd ed.). O'Reilly Media.

Rademacher, F., Sachweh, S., Zündorf, A. (2017). Differences between model-driven development of service-oriented and microservice architecture. In *Proceedings of the 2017 IEEE international conference on software architecture workshops* (pp. 38–45).
Rawson, A. (2022). https://grnsft.org/sci
Ren, Z., Wang, W., Wu, G., Gao, C., Chen, W., Wei, J., Huang, T. (2018). Migrating web applications from monolithic structure to microservices architecture. In *Proceedings of the 10th Asia-pacific symposium on internetware* (pp. 1–10).
Sahin, C., Cayci, F., Gutierrez, I. L. M., Clause, J., Kiamilev, F., Pollock, L., Winbladh, K. (2012). Initial explorations on design pattern energy usage. In: *Proceedings of the 2012 First international workshop green and sustainable software (GREENS)* (pp. 55–61).
Taibi, D., Lenarduzzi, V., Pahl, C. (2018). Architectural patterns for microservices: A systematic mapping study. In *Proceedings of the 8th international conference on cloud computing and services* (CLOSER) (pp. 221–232).
Taibi, D., Systä, K. (2019). From monolithic systems to microservices: A decomposition framework based on process mining. In *Proceedings of the 9th International conference on cloud computing and services science (CLOSER)* (pp. 153–164).
Taibi, D., Lenarduzzi, V., & Pahl, C. (2017). Processes, motivations, and issues for migrating to microservices architectures: An empirical investigation. *IEEE Cloud Comput, 4*(5), 22–32.
Wang, S., Zhou, A., Yang, F., & Chang, R. N. (2016). Towards network-aware service composition in the cloud. *IEEE Transactions Services Computing, 9*(4), 590–604.

Creating Robot Game Mission Strategies Using Machine Learning

Elton Strydom, Bertram Haskins, and Ronald Leppan

Abstract Evaluating mission strategies in competitive robot games like the FIRST® LEGO® League (FLL) is hard because physical testing takes a lot of time and costs a lot of money. This study explores the integration of machine learning (ML) techniques, particularly genetic algorithms, to optimise mission strategy evaluation. A genetic algorithm was created to construct and evaluate mission strategies based on factors like completing missions, minimising travel time, making the best use of attachments and the strategy's ability to withstand changes. The FLL game field was modelled as a grid-based environment. The evaluation involved extensive simulations of different strategic approaches, including point-based, difficulty-based and zone-based strategies, revealing that the zone-based approach yielded the highest efficiency. The results show that evolutionary algorithms could be used to automatically test mission strategies, cutting down on the need for physical tests while keeping the learning outcomes for problem-solving. We propose six key guidelines for designing ML-driven evaluation tools for robotics competitions, based on experimental insights. This research contributes to developing computational frameworks that support mission planning and optimisation, enhancing strategic decision-making in competitive robot games.

Keywords Machine learning · Evolutionary algorithms · Mission strategy optimization · FIRST® LEGO® league

---

B. Haskins · R. Leppan
Noroff University College, Kristiansand, Norway
e-mail: bertram.haskins@noroff.no; ronald.leppan@noroff.no
URL: https://www.noroff.no/

E. Strydom · B. Haskins (✉)
Nelson Mandela University, Port Elizabeth, South Africa
e-mail: s215030281@mandela.ac.za; bertram.haskins@mandela.ac.za
URL: https://www.mandela.ac.za/

J. Marx Gómez et al. (eds.), *Recent Advances in Information Systems*, Progress in IS,
https://doi.org/10.1007/978-3-032-14099-9_60

1 Introduction

The competitive autonomous robot games sector, a continually evolving field, offers a variety of platforms that challenge students in robotics, engineering and computer science. Such competitions support the development of problem-solving, creative and computational thinking, as discussed by Evripidou et al. (2020). One such robot game is FIRST® LEGO® League(FLL) in which teams of students aged 9–16 participate in research, problem-solving and coding challenges to construct a programmable LEGO®robot capable of completing missions in a robot game (FIRST, 2023).

The evaluation of mission strategies in FLL presents several challenges, primarily due to the labour-intensive and time-consuming nature. The necessity for teams to be physically present during strategy evaluations adds further complexity, compounded by the financial constraints in acquiring the necessary equipment. The process relies heavily on detailed testing cycles for each mission strategy, leading to inefficiency and extended timeframes. A mechanism aimed at supporting the design of mission strategies through simulated evaluative feedback could contribute to a more efficient and effective strategy assessment process in FLL competitions while simultaneously encouraging problem-solving learning experiences for students. In an attempt to support the evaluation of mission strategies in FFL, this study presents guidelines for evaluating mission strategies in robot games using machine learning techniques.

2 Background

The FLL is a global competition that challenges young people to create innovative solutions and build LEGO-based robots to complete themed missions. Machine learning (ML) presents a promising approach to assisting FLL teams by evaluating mission strategies, reducing reliance on physical testing while retaining problem-solving skills. Among ML techniques, evolutionary algorithms (EAs) are notable for their ability to optimise complex strategies, making them ideal for refining FLL mission plans. The following sub-sections delve deeper into these topics.

2.1 FIRST® LEGO® *League (FLL)*

Many people around the world are familiar with the LEGO® building blocks, having either played with the blocks themselves as children or having had their children play with the blocks. LEGO® MINDSTORMS ® is a further evolution of the basic building concepts, which allows for the construction of robotic creations and has further opened the door for LEGO's use as an educational tool (Afari and Khine, 2017).

FLL is an international competition organised by FIRST® for young enthusiasts within the scientific space. Annually in August, FLL unveils a science-based and real-world challenge that teams explore and investigate (FIRST LEGO League, 2023). FLL has three divisions: discover (ages 4–6), explore (ages 6–10) and the challenge (ages 9–16) division. The challenge division includes the FLL robot game. Therefore, this division is the primary focus of this study. The challenge division revolves around an annual theme divided into three components: the project, the FLL core values and the robot game.

The project involves teams addressing a real-world problem, creating innovative solutions and presenting their ideas to judges. Integral to this is the adherence to the FLL core values, which emphasise discovery, innovation, impact, inclusion, teamwork and fun. Furthermore, the robot game engages teams in designing robots to accomplish missions inspired by real-world issues, with an annual variation in themes. These robots are built using LEGO® blocks. LEGO® is something most people are quite familiar with and forms a nostalgic part of a lot of people's childhood. Playing with LEGO®encourages youngsters to exercise the creative part of their brains and can prove to be a good learning tool while still being fun. The robots built set out to execute missions inspired by real-world challenges, featuring a dynamic array of themes that change annually. The FLL's extensive global participation (FIRST LEGO League, 2018) makes it an ideal platform to explore the integration of ML in supporting mission strategy formulation. Within the robot game component, FLL participants come up with mission strategies about how they want to complete missions.

In FLL, teams primarily adhere to established guidelines for the development of mission strategies. A mission strategy is a plan devised by participants, outlining the steps necessary to complete as many missions as possible, thereby maximising their potential rewards. Furthermore, a mission strategy comprises multiple mission runs, where the robot sets out to complete a number of missions and return to base. Each trip from the base and back within this strategy is termed a mission run. As such, teams have to put thought and effort into designing their mission strategies. Seshan (2018) has established guidelines that teams can follow when planning their mission strategy. Once a mission strategy has been established, the development and testing of these strategies commences. This involves the building and programming of robots, which are both time-consuming and resource-intensive, as many teams do not have the financial means to purchase expensive equipment.

These constraints can be quite an obstacle for most FLL teams. An innovative evaluation mechanism, designed to support the process of evaluating mission strategies, could alleviate some of the constraints teams face. This mechanism could potentially support real-time evaluations without requiring physical co-location or extensive equipment and might provide practical utility for teams. The potential integration of ML techniques in the evaluation of mission strategies offers a promising avenue in alleviating some of the time and financial constraints mentioned previously. These techniques could provide an efficient and effective framework for evaluating the viability of various mission strategies, thereby supporting the robot game mission design experience for FLL teams. The next section delves into ML, exploring its relevance and application in this context.

2.2 Machine Learning and Evolutionary Algorithms

Machine learning, a pivotal branch of artificial intelligence, is dedicated to developing intelligent algorithms that empower computers to learn and make data-driven decisions. For school-based teams, FLL can be seen as an extension of their school activities; basically, as an extracurricular activity which enhances teamwork and problem-solving. Although not specifically related to FLL, machine learning has been widely used in activities related to education. These range from being used as a tool to assist in the assessment of Gestalt Principles in design (Haskins and Haskins, 2022) to assisting in the teaching of reinforcement learning (Martínez-Tenor et al., 2019).

In the context of FLL, ML presents promising opportunities for supporting teams in improving their mission strategies by offering mission strategy evaluations without detracting from the problem-solving learning experiences of the teams. Notably, within the broad spectrum of ML, subsets like evolutionary algorithms stand out for their proficiency in optimisation tasks (Xue et al., 2016).

Evolutionary algorithms (EAs) embody optimisation techniques that are inspired by natural evolution (Mohammadi et al., 2020), encoding solutions and their attributes as chromosomes and genes. In this approach, groups of solutions are collectively referred to as a population (Khanmohammadi et al., 2021). Utilising Darwinian mechanisms such as selection, crossover and mutation, these algorithms effectively tackle complex challenges, particularly in engineering (Greiner et al., 2018). They have been used in mobile robotics to assist in path planning (Xue and Sun, 2018). Their versatility and capability in multi-objective optimisation make them apt for evaluating mission strategies in the FLL landscape (Coello et al., 2007). EAs' proficiency in complex problem-solving resonates with the nuances of FLL mission strategy evaluation, promising a potent tool for supporting the development of mission strategies.

3 Experimental Setup

The experimental setup for this research was designed to develop a genetic algorithm capable of evaluating mission strategies in competitive robot games, with a specific focus on the FLL competition. To implement the genetic algorithm, we utilised the .NET C# programming language along with the Genetic Algorithm Framework (GAF) NuGet package. This choice allowed for streamlined development while ensuring flexibility in designing and executing the optimisation process.

3.1 Encoding Mission Strategies

The FLL game field was modelled as a grid-based environment where each mission location was assigned a specific coordinate. This representation allowed the genetic algorithm to compute efficient mission strategies based on spatial constraints. An example of this field is shown in Fig. 1.

Each chromosome in the genetic algorithm represented a potential mission strategy. The encoding scheme included mission objectives, required attachments and starting positions. Every gene within the chromosome has a significant role. Therefore, it is imperative to carefully model what each gene represents. In the context of this research, a gene could signify a starting area, missions, attachments or even placeholder genes. Figure 2 illustrates the type of information that is included within a mission and an attachment gene.

As illustrated in Fig. 2, a mission object encapsulates critical attributes, including geographical location (x, y), associated score, required attachments (such as pull, push, lift), description and the mission's name. Attachments are represented by objects that outline attachment type and action (addition or removal).

Figure 3 displays the necessary coordinates and attachments for the tasks. It is important to note that these parameters might vary among different teams due to their distinct mission strategies and robot designs. Consequently, the coordinates may vary slightly, and the attachment requirements hinge on the unique attachments developed by each team. Subsequently, Fig. 4 depicts the designated starting gene. For each run, the robot is required to initiate its path from a specified starting area. Figure 4 depicts the coordinates for the starting area.

It is important to note that participating teams can alter the robot's precise starting position within this predefined area. Such flexibility allows for variations based on

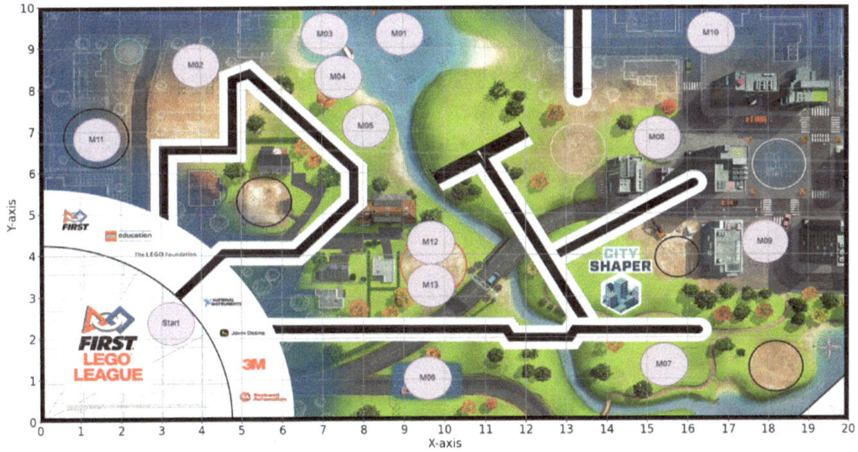

Fig. 1 Grid representation of the FLL game field. Each mission is assigned to a predefined coordinate

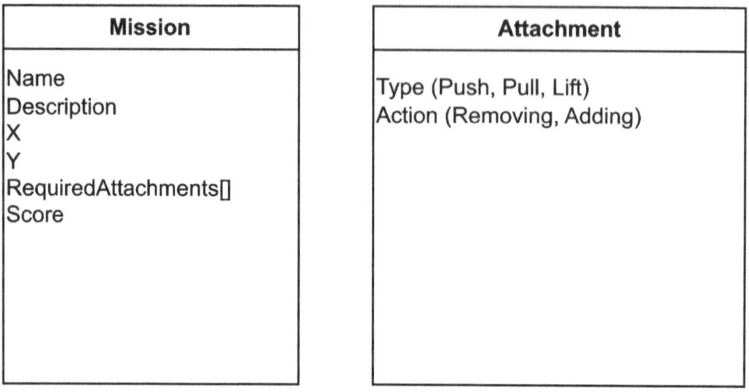

Fig. 2 Gene model

```
"Name": "M1",
"Description": "Remove the blocks on the obstacle",
"X": "6",
"Y": "7",
"RequiredAttachments": ["Pull", "Push"],
"Score": "30"
```

Fig. 3 Mission sample objective

```
"Name": "Start",
    "Description": "Starting area for the LEGO Robot,
    "X": "2",
    "Y": "2",
    "RequiredAttachments": [],
    "Score": "0"
```

Fig. 4 Start location sample object

each team's strategic preferences. While the robot's initial placement is flexible, the experiments consistently use the set starting coordinates shown in Fig. 4. This standardisation provides a basis for comparing data across various scenarios, ensuring the consistency of experimentation results. Figure 5 then introduces a specific attachment utilised by the robot. This attachment demonstrates the type of modifications that can be made to the robot for mission execution.

```
"Name": "Pull",
"Score": "0",
"X": "2",
"Y": "2",
"RemovingAttachment": true
```

Fig. 5 Attachment sample object

Table 1 Encoded objectives

Gene int	1	2	3	4	5	6	7	8	9	10	11	12	13
Objective	M1	M2	M3	M4	M5	M6	M7	M8	M9	M10	M11	M12	M13

Table 2 Encoded attachments

Gene int	15	16	17	18	19	20
Attachment	PushAdd	PullAdd	LiftAdd	PushRem	PullRem	LiftRem

For mission, start, placeholder and attachment gene models, each component is represented by a unique integer value. This encoding approach assigns specific numbers to each mission, start position, placeholder or attachment, thereby enabling the algorithm to efficiently reference and manipulate these elements during the evolutionary process.

Table 1 lists the specific integers encoded for each mission in an example game, demonstrating the application of this encoding method.

Table 2 presents the unique integers for each attachment gene used by the game robot.

In this study, we categorise three principal attachments: push, pull and lift. This classification aids in streamlining the testing process and ensures uniformity in the experimental approach, though teams in practical scenarios may engineer a variety of attachments beyond this study's scope.

3.2 *Calculating Fitness*

A fitness function serves as a mechanism to assess the efficiency of a chromosome, and the implementation is specific to each algorithm based on the objectives of the genetic algorithm. The fitness function evaluated mission strategies based on key performance criteria:

Mission Completion: Rewarded successfully completed missions.
Attachment Optimisation: Penalised unnecessary attachment changes to encourage efficiency.
Travel Time: Encouraged strategies that minimised the robot's movement across the field.
Strategy Robustness: Penalised strategies that violated game constraints.

The fitness function was designed to balance these competing objectives, ensuring that the genetic algorithm evolved towards optimal solutions.

Multiple experimental runs were conducted, each with varying parameter settings (such as including different attachments and sequences), to assess the impact of different configurations on the genetic algorithm's performance. An example of a

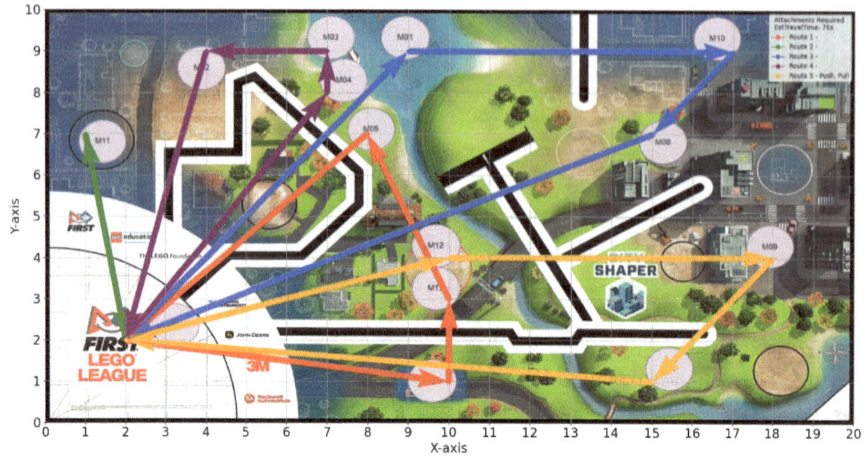

Fig. 6 An example solution

possible solution (population member) for a single run is demonstrated in Fig. 6. In the figure, the coloured pathways show the different paths that the robot would need to take to get to each of the missions, in each case having to return to base to change attachments or be realigned. The colours in the figure are not associated with specific parameters, attachments or sequence. The colours are simply used to indicate the variation between the paths and could change from one run to the next.

The genetic algorithm followed a structured execution process. The population was first initialised with randomised mission strategies. For each strategy, a fitness function was calculated to determine which of the strategies was the best performing. These strategies were then used to create the next version of the population by applying the genetic operators of selection, crossover and mutation. This process was then set to continue until a maximum number of generations of the population was met or no path improvements could be made within a set number of generations. This execution flow ensured that the algorithm progressively refined mission strategies through iterative optimisation.

$$\text{score} = S_{\text{missions}} + S_{\text{route}} - P_{\text{grouped}} - P_{\text{attachments}} \tag{1}$$

For this study, the fitness function (*score*) was calculated using Equation 1, with the following details associated with each variable:

1. **Mission Score** (S_{missions}): Sum of all completed mission scores. A mission succeeds if the required attachment is equipped (i.e. the latest relevant gene is an 'add' type) or if no attachment is needed.
2. **Route Score** (S_{route}): Computed as the Manhattan distance between gene points, scaled by cell travel time. Routes completed within time limits receive a bonus; exceeding the limit yields no points. Each active attachment increases travel time

by 1%. Obstructions, line tracking and landmarks are not considered to maintain algorithm neutrality.
3. **Mission Grouping Penalty** (P_{grouped}): Applied when missions between starting zones exceed the grouping parameter.
4. **Attachment Penalty** ($P_{\text{attachments}}$): A small penalty discourages excessive attachment use, promoting efficiency.
5. **Start Gene Validation**: If the route does not start at an initial gene, the solution is invalid and a fitness score of 0 is returned.

4 Evaluation

The evaluation phase of this study involved running extensive simulations to assess the genetic algorithm's performance in optimising mission strategies. These simulations were conducted under controlled conditions to examine the impact of different mission selection approaches on overall efficiency and effectiveness. Through experimentation, the genetic algorithm was configured as follows:

- The top 5% of the population was retained across each generation.
- Both DoublePoint and SinglePoint crossovers were employed at an 80% probability, with duplicates allowed.
- Swap mutations were employed at an 80% probability.
- Further random, individual mutations were employed at an 80% probability, with duplicates allowed.

The evaluation process provided insights into how various strategic considerations influenced mission success rates, attachment utilisation and travel efficiency.

The first set of simulations was focused on point-based strategies, where the algorithm prioritised missions that offered the highest point values. The objective of this approach was to determine whether the genetic algorithm could effectively group high-scoring missions while minimising redundancy in attachment changes. The results demonstrated that the algorithm successfully identified mission sequences that maximised scoring potential while maintaining efficiency in robot movements.

Another set of experiments investigated mission difficulty-based strategies. In this approach, the algorithm was configured to prioritise missions that were considered easier to execute, particularly those that required fewer attachment changes. However, the evaluation revealed that this approach led to an overall increase in travel time, as the algorithm had to return to the starting position more frequently to make attachment modifications. Consequently, the fitness scores for these strategies were generally lower than those observed in the point-based strategy evaluations.

A final set of simulations examined the effectiveness of a zone-based mission strategy. In this configuration, the genetic algorithm prioritised mission sequences that minimised travel distances by grouping tasks based on their geographical proximity. The evaluation results indicated that this approach significantly reduced the

Table 3 Comparison of mission strategy performance

Strategy type	Average fitness score	Average travel time (s)	Mission completion rate (%)
Point-based	0.917	76	95
Mission difficulty	0.883	94	85
Zone-based	0.945	63	98

total travel time, leading to an overall increase in efficiency. The algorithm effectively identified optimal pathways that allowed for seamless transitions between missions without requiring excessive repositioning of the robot. As a result, fitness scores for the zone-based strategy were among the highest recorded in the evaluation phase. The results of these different experiments are contrasted in Table 3.

Across all evaluation experiments, the genetic algorithm demonstrated a strong capability to generate near-optimal mission strategies that adhered to the constraints of the FLL competition. The findings underscored the importance of balancing mission prioritisation with travel efficiency and attachment management. Additionally, the results highlighted the significance of population size in accelerating the algorithm's convergence towards optimal solutions. Larger population sizes generally led to more rapid identification of high-quality strategies, as they allowed for a broader exploration of potential solutions within the search space.

5 Mission Strategy Evaluation Guidelines

This section outlines six key guidelines for evaluating mission strategies in competitive robot games using machine learning (ML) techniques, using the insights gained from experimenting with mission strategies via genetic algorithms. These guidelines are as follows:

1. Understand the Robot Game Dynamics: Each game has unique rules, scoring mechanisms and constraints. For example, in the FLL competition, missions are time-limited, and robots cannot be modified outside designated zones. A thorough analysis of these aspects is essential before designing an evaluation framework.

2. Account for Robot Technical Constraints: Battery depletion and navigation drift impact mission success. These factors must be integrated into evaluation tools to reflect real-world conditions. In our approach, we accounted for this by limiting the maximum number of missions per run.

3. Select an Appropriate ML Technique: Choosing the right algorithm is crucial. Evolutionary algorithms, such as genetic algorithms, were selected due to their suitability for optimisation problems like the travelling salesman problem (TSP), which shares similarities with mission planning.

4. Assess Technical Implementation Strategies: Using pre-built ML frameworks can accelerate development but may introduce constraints. For instance, the Genetic Algorithm Framework (GAF) only supports fixed chromosome lengths, requiring our introduction of placeholder genes to accommodate variable mission sequences.

5. Define Realistic Simulation Scenarios: Simulations should reflect real-world conditions to ensure robust evaluation. Our experiments attempted to mimic various mission strategies but revealed challenges in aligning algorithmic optimisation with human-designed strategies. A broader parameter exploration approach may be more effective.

6. Establish Clear Result Expectations: Setting defined output expectations is essential for meaningful analysis. Some simulations in our research produced unexpected outcomes, emphasising the need for predefined benchmarks to facilitate proper result interpretation.

These guidelines provide a means for developing and evaluating mission strategy tools for robot competitions. They may be employed by developers and organisers who want to use machine learning to assist teams in testing or improving their mission strategies.

6 Conclusion

This study explored the application of genetic algorithms in optimising mission strategies for competitive robot games. By encoding key elements of the FLL competition into a structured algorithmic framework, the research demonstrated how machine learning techniques can be leveraged to enhance decision-making in mission strategy planning. The experimental results provided compelling evidence that genetic algorithms can effectively evaluate and refine mission strategies, leading to improved efficiency and higher success rates in robot competitions.

The insights from experimentation resulted in the creation of six guidelines, which can assist others in developing a machine learning-based approach to mission strategy optimisation. Future work in this area could explore the integration of reinforcement learning techniques to further refine mission strategy planning. Additionally, real-world testing with FLL teams would provide valuable insights into the practical applications of the evaluation tool developed in this study.

The findings of this study underscore the potential of machine learning in supporting strategic decision-making in competitive robot games. By leveraging computational intelligence techniques, such as genetic algorithms, teams can gain deeper insights into mission planning and optimisation, ultimately improving their performance.

References

Afari, E., & Khine, M. S. (2017). Robotics as an educational tool: Impact of lego mindstorms. *International Journal of Information and Education Technology, 7*(6), 437–442.

Ángel Martínez-Tenor, A. C. M., & Fernández-Madrigal, J. A. (2019). Teaching machine learning in robotics interactively: The case of reinforcement learning with lego® mindstorms. *Interactive Learning Environments, 27*(3), 293–306.

Coello, C. A. C., Lamont, G. B., Van Veldhuizen, D. A., et al. (2007). *Evolutionary algorithms for solving multi-objective problems* (vol. 5). Springer.

Evripidou, S., Georgiou, K., Doitsidis, L., Amanatiadis, A. A., Zinonos, Z., & Chatzichristofis, S. A. (2020). Educational robotics: Platforms, competitions and expected learning outcomes. *IEEE Access, 8*, 219534–219562.

FIRST. (2023). *FIRST tech challenge*. https://www.firstinspires.org/robotics/ftc. Accessed 23 November 2023.

FIRST LEGO League. (2018). *Developing a mission strategy*. https://flltutorials.com/en/robotgame/strategy/2018/01/01/Mission-Strategy.html. Accessed 15 November 2023.

FIRST LEGO League. (2023). *FIRST LEGO league past challenges*. https://www.firstlegoleague.org/past-challenges. Accessed 3 December 2023.

Greiner, D., Periaux, J., Quagliarella, D., Magalhaes-Mendes, J., & Galván, B. (2018). Evolutionary algorithms and metaheuristics: Applications in engineering design and optimization. *Mathematical Problems in Engineering, 2018*, 1–4.

Haskins, B., & Haskins, M. (2022). Providing automated feedback on the gestalt principle of symmetry by means of machine learning. In *Proceedings of the 2022 IEEE Asia-pacific conference on computer science and data engineering (CSDE)* (pp. 1–6). https://doi.org/10.1109/CSDE56538.2022.10089327.

Khanmohammadi, S., Kizilkan, O., & Musharavati, F. (2021). Chapter 17—multiobjective optimization of a geothermal power plant. In C. O. Colpan, M. A. Ezan, & O. Kizilkan (Eds.), *Thermodynamic analysis and optimization of geothermal power plants* (pp. 279–291).

Mohammadi, F. G., Amini, M. H., & Arabnia, H. R. (2020). Evolutionary computation, optimization, and learning algorithms for data science. In *Optimization, learning, and control for interdependent complex networks* (pp. 37–65).

Seshan, A. S. (2023). *Developing a mission strategy*. https://flltutorials.com/en/robotgame/strategy/2018/01/01/Mission-Strategy.html. Accessed 15 November 2023.

Xue, Y., & Sun, J. Q. (2018). Solving the path planning problem in mobile robotics with the multi-objective evolutionary algorithm. *Applied Sciences, 8*(9), 1425.

Xue, B., Zhang, M., Browne, W. N., & Yao, X. (2016). A survey on evolutionary computation approaches to feature selection. *IEEE Transactions on Evolutionary Computation, 20*(4), 606–626. https://doi.org/10.1109/TEVC.2015.2504420

MaaS-Broker Implementation for an Intermodal Journey Planner Combining Public Transit with Carpooling Using Time-Expanded Graph Networks

Sasapu Venkatesh and Jorge Marx Gómez

Abstract The mobility offer in rural areas nowadays is limited due to too few customers and the economic inefficiency for the public transport providers. A survey from the NEMo-project revealed that carpools offer great mobility potential for reoccurring journeys. A combination of both carpooling and public transportation can improve the overall mobility needs for residents in rural areas. This paper proposes implementation of a mobility as a service aggregator alongside heterogeneous mobility systems such as public transport and carpool matchers, which combines all mobility offers in a homogeneous time-expanded graph model to query for the best intermodal route. The time-expanded graph model reveals to perform well; however, the quality depends on the existing offers from the mobility providers itself.

1 Introduction

Residents in rural areas, who do not own a motorized vehicle, face the difficulty of being mobile to shop for groceries, visiting friends, and attending important appointments.

In addition, the public transit in rural areas is not profitable for transport companies, because of the lack of customers. This downward spiral leads to a thin public transport network with very few trips (K¨onig, 2019). Accordingly, the NEMo-project of the university at Oldenburg launched a poll asking residents in the rural areas

S. Venkatesh (✉) · J. Marx Gómez
Wirtschaftsinformatik/VLBA, Carl Von Ossietzky University of Oldenburg, Oldenburg, Lower-Saxony, Germany
e-mail: sasapu.venkatesh@uni-oldenburg.de

J. Marx Gómez
e-mail: jorge.marx.gomez@uol.de

J. Marx Gómez et al. (eds.), *Recent Advances in Information Systems*, Progress in IS,
https://doi.org/10.1007/978-3-032-14099-9_61

of Oldenburg, whether carpools are a valid opportunity to enhance the transportation network. In fact a huge amount of residents see carpool as a great opportunity especially for reoccurring trips like leisure time or shopping (NEMo, 2020).

This paper proposes a modular mobility as a service concept (MaaS), which aggregates public transport with carpooling providers in the model region of Oldenburg. The concept is implemented and integrated in the existing VLBA mobility manager application Gengar, which uses the Verkehrsbund Bremen/Niedersachsen (VBN) API for public transit route planning and the internal carpooling service mOOWVe, which provides an API for matching passengers and drivers.

2 Background

The combination of public transit relying on fixed interval travel times with the highly dynamic travel times of carpooling is well studied in the literature.

Varone and Aissat (2015) introduced a concept of *subpath substitution* based on the idea of replacing public transport routes between specific stops with a carpool route which providing shorter travel times. This approach, however, heavily depends on the density and quality of the existing public transport network. If the overall route is already uncomfortable and too long, we do not have a well suited base to extend it with carpooling. Furthermore, this approach does not use carpooling to its full potential, since it only replaces subpaths instead of seeing it as a dedicated transportation mode allowing to travel long distances.

On the other hand, Jamal et al (2017) hosted the complete road and public transit network in a giant graph structure and directly queries on that network using the Dijkstra algorithm to find routes between specific locations. This approach works well for finding intermodal routes and could be highly optimized. Unfortunately, this monolithic approach, where the complete transport data is maintained in one system, cannot easily be integrated in every environment such as the Gengar application of the VLBA since in existing infrastructures there will be many other heterogeneous mobility providers like cabs, carpools, rental cars, or rental bikes.

It is not realistic to combine them in one monolithic system architecture, since the complexity should be maintained in their corresponding system. Wong et al (2020) proposes a way on how different mobility offers can be computed via a *mobility broker* architecture, where each system is standalone but provides a public API to consume their mobility offers. According to *Conway's law,* it is advised to represent the organizational architecture in the system architecture (Conway, 1968),

A concrete implementation of this architecture concept leads to the aggregator implementation of Georgakis et al. (2019).

Georgakis et al. implements a separate aggregator application, which is executed alongside the already existing architecture, which abstracts various mobility providers into one homogenous journey format by using so-called *graph-harmonization* and Dijkstra to compute optimal journeys with the data of the other mobility providers.

How this format could be structured is proposed by Schulz et al (1999). Schulz specifically focused on train departures in Germany and uses a *time-expanded* graph model, to represent time related train trips. Instead of representing nodes of a graph as geographic locations, Schulz in his *time-expanded* models sees nodes as departure and arrival events at a specific location of a specific train. The advantage of using a *time-expanded* model instead of more commonly used *time-dependent* models, is well-known pathfinding algorithms such as Dijkstra or A* can be applied directly without modifications. However, this approach requires a large amount of memory, since the extensive timetable data leads an increased number of nodes (Schulz, 2005).

3 Methods

The proposed procedure for intermodal route planning relies on the aggregator approach from Wong et al (2020), where the mobility offer is aggregated for each specific end-user request. The proposed model consequently collects data from each mobility provider and—inspired by Georgakis et al. Implementation—builds a temporal intermodal harmonized graph using the notation of the *time-expanded* model from Schulz.

The huge space consumption of the used graph model should not be a problem, because the amount of arrival and departure events will be limited to a small subset of all timetable event, because all results of the heterogeneous providers are be pre-computed in the corresponding system. This keeps the complexity in the corresponding system, so that the aggregator can only focus on combining all means of transport in a homogenous format.

3.1 Modeling a Homogeneous Transport Graph

We introduce three specific types of nodes required for the concept. v_0 will be called the *origin node*, which represents the origin of the whole journey queried by an end-user. v^* will be called the *destination node*, which represents the destination of the journey. Every other node inside the *time-expanded* graph represents either an arrival or departure event of a specific vehicle at a specific location, which will be called *journey nodes*.

This leads to the final node set of $V = V_J \cup \{v_0, v^*\}$, where V_J are the journey nodes.

3.1.1 Node Attributes

To store and decode route details for the final end-user, each node must contain specific attributes, represented by the following bijective functions:

timestamp $t : V_J \rightarrow N$
location in WGS-84 coordinates: $l : V \rightarrow R \times R$
vehicle type $\lambda : V_J \rightarrow \Lambda$, where $\Lambda = \{c, p, w\}$ represents the vehicle types carpool (c), public transit (p) and walking (w).
event type $\tau : V_J \rightarrow T$, where $T = \{\text{arr}, \text{dep}\}$ annotates a node to represent an arrival or departure event.

It is important to mention that the functions t, λ and τ are only defined on V_J set, since the origin and destination nodes neither count as an arrival or departure event nor a point of time is known at the beginning of the end-user request.

For instance if a public bus departures at the coordinates (53.1432, 8.2217) at 13 time units, the following attributes for a node v, which represents this event, will apply: $t(v) = 13$, $l(v) = (53.1432, 8.2217)$, $\lambda(v) = p$ and $\tau(v) = \text{dep}$.

For a better visualization, all following nodes of graphs which are annotated as an arrival node will be displayed with a square shape, while departure nodes keep the default circle shape. The timestamps of each node are displayed below them.

3.1.2 Edges

Edges connects two nodes, which are chronological possible in a trip. Correspondingly, there are two criteria for an edge to exist.

1. $\forall e = (v, v') \in E$: $t(v) \leq t(v')$, which only allows an edge to connect two events in chronological order,
2. $\tau(v) = \text{arr} \wedge \tau(v') = \text{dep} \Rightarrow l(v) = l(v')$, which only allows a connection between arrival and departure if they are at the same location preventing unintentional geographical skips.

Let $G = (V, E)$ the intermodal graph and $w : V \rightarrow N$ the weight function of that graph. The weight according to the *time-expanded* model is calculated by the difference of the time of two nodes as stated in Eq. 1.

$$w(e) = t(v') - t(v) \tag{1}$$

The first criteria for an edge connection ensures, that there are no negative weights. For better visibility the edge weight are not displayed in some figures.

3.2 Building a Transport Graph

To build an intermodal graph the following method is proposed:

1. Query the external public transit API from the trip origin to its destination with normal end-user parameters.

2. Detect important transport hubs of the results of the first query and query the external public transit API from the origin to those transport hub and from the transport hub to the destination allowing unlimited walking routes.
3. Transfer all public transport results into the *time-expanded* model according to Schulz et al (1999)
4. Query the external carpooling API for all routes in vicinity to the trip origin and destination without considering any matching constraints.
5. Compare the route geometries of all public transport trips with the potential carpooling trips and find geographical transfer points.
6. Query the carpool matching API for all transfer points considering the matching constraints.
7. Integrate the carpooling routes in the intermodal graph.

Since the construction result is a *time-expanded* graph, the shortest route can be searched by using the Dijkstra-Algorithm. The nodes attributes allow to convert the found path into a sufficient route description for the end-user.

3.2.1 Querying Public Transport API

Just like in Varone and Aissat (2015) approach, we start by querying the public transit API to get a foundation of a transport network to work with, since all route geometries and stops are not known in the beginning of the request.

Terminology

Depending on the API implementation, the vocabulary may differ. In our case, the implementation of the API is based on the OpenTripPlanner-Framework (OTP). An OTP-Result contains several *itineraries*, which contains *legs*. A *itinerary* is a complete trip from start to end. A *leg* is a specific part of an *itinerary*, where only one specific vehicle or the own feet of the customer is used. An itinerary containing two walking routes to the corresponding stops and two bus lines with one transfer would result in four legs.

Finding the Most Valuable Stops

The second API call querying the start and end location to the transport hub, allowing unlimited walking distances is very important though. It is used to search for the most valuable stop, since the closest stop reachable by foot is not guaranteed to offer the best public transport connections. Valuable stops are important to be included in the current transportation network because there might be a carpooling route shortening the long unfeasible walking leg to that stop.

The transport hub in the region can be inferred using the first query by comparing every transfer location at the end of every leg and sort them by their occurrences in every itinerary.

Figure 1 shows an example route from Nordmetzhausen to Oldenburg University using public transit or walk. Oldenburg central station is the transport hub used for

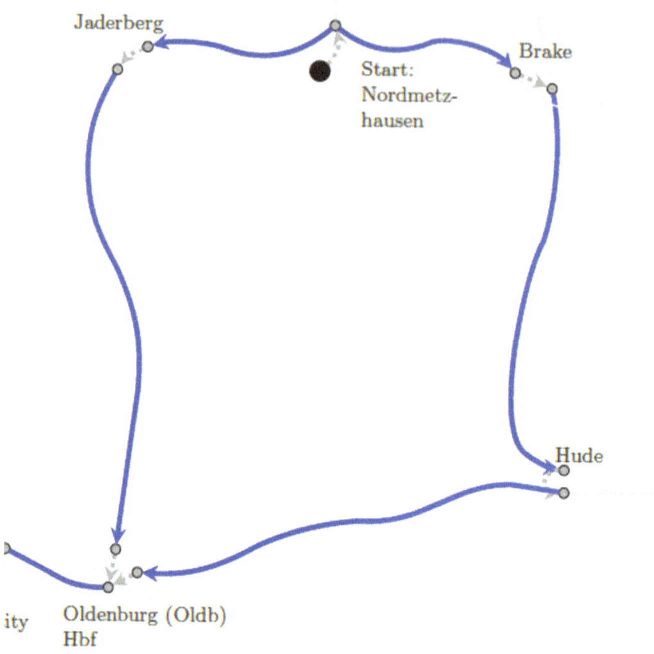

Fig. 1 Mobility network after the initial API query

the second API call pictured in Fig. 2, which reveals the valuable stop Ovelg¨onne Barghorn Hinrichs. Notice the rather long walking leg from the start to that specific stop.

3.2.2 Converting Public Transit Routes into a Graph

The first second steps of the method require to convert the public transit API results into the graph model.

Building an Itinerary

To build a graph, which represents an itinerary i, we look at each leg individually. Let $G_i = (V_i, E_i)$ an initialized empty graph, ready to be extended with data.

If the leg l represents a walk, we create a departure and arrival node v_l and v', connect them with an edge fulfilling the criteria from subsub Sect. 3.1.2 and set the attributes accordingly.

If the leg l represents a public transit leg, we look at each intermediate stop, in the following noted as $z_{l,i}$, where $z_{l,i}$ is the i-th stop of the leg l.

We first create a departure node for the first stop and an arrival node for the last stop in the leg. For each intermediate stop, we create both an arrival and departure node

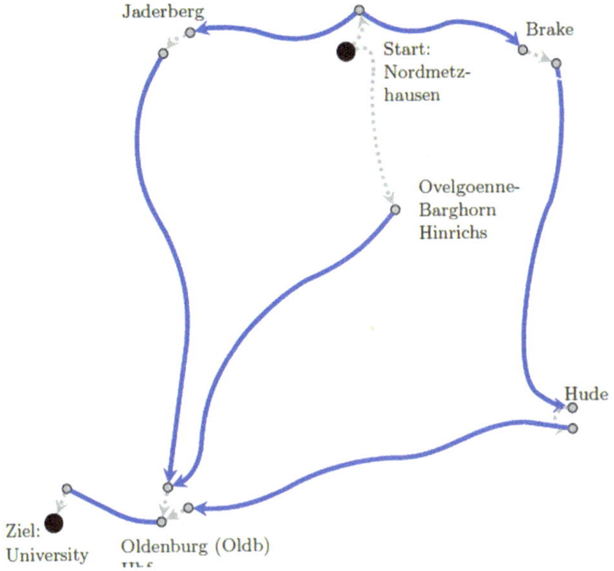

Fig. 2 Mobility network after the quickest route around the transport hub

and connect them according to their order in the leg. This procedure is visualized in Fig. 3. Please note that this depiction differs from the usual graph visualization. Arrival nodes are visualized as a square and their timestamp is displayed below the node.

Combining All Itineraries

The procedure stated in Sect. 3.2.2 can be applied to every API call. In the end, all constructed graphs G_i needs to be merged into one intermodal graph. It is simply required to connect the origin node v_0 to the first node v_{0i} of itinerary i as well as to connect the last node v^* with the destination node v^* like it is pictured in Fig. 4

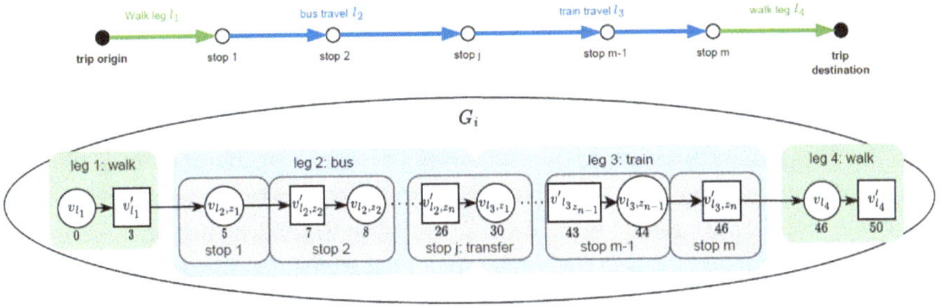

Fig. 3 Generalized picture of an itinerary containing 4 legs

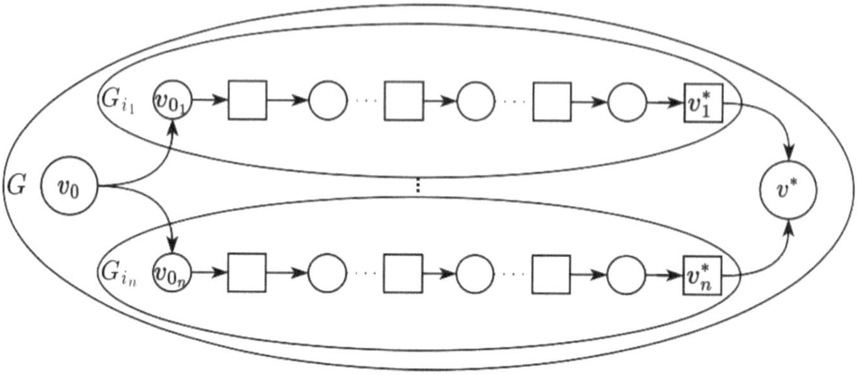

Fig. 4 Merging two itinerary graphs together

3.2.3 Adding Carpool Connections

After both public transit calls the graph is ready to get extended with carpool nodes.

Finding Suitable Transfer Locations

To find suitable transfer locations, the carpool system needs to be queried for all routes in the vicinity to the trip origin and trip destination. After that the route geometries of the found results can be compared to route geometries of the public transit network by using the l function of every existing journey node $v \in V_J$.

In this paper, the air line between the stops and the carpool route was calculated using the haversine formular (Van Brummelen, 2012). Since the matching API was not queried and no detours for a match are validated yet the threshold for marking a stop as close enough to a carpool route should be quite high, because many stops might not be feasible when querying the API for the final matching. The detailed validation of a carpool being feasible for the user and driver remains in the corresponding system.

Querying the Carpool System for Every Transfer Node

In the last step, the carpool system can be queried for all found transfer nodes. However, to limit the query-load on the carpool system, it is advised to prioritize transfer stops according to their average remaining travel time. Because the edge weight represents time-durations between nodes, the Dijkstra algorithm can be used to calculate the remaining travel time at each graph node.

In our Wesermarsch example pictured in Fig. 5, there are two potential carpool offers: one heading parallel to the public transit route to Jaderberg and one perpendicular to the public transit route to Brake via Ovelg¨onne Barghorn Hinrichs. Especially for the parallel transit line, there are many potential transfer stops. But according to the remaining travel time Jaderberg is closest to the destination, which results in only changing in Jaderberg using the carpool as long as possible to optimize the travel time. Because of the valuable public transit offer and short travel time the Ovelg¨onne

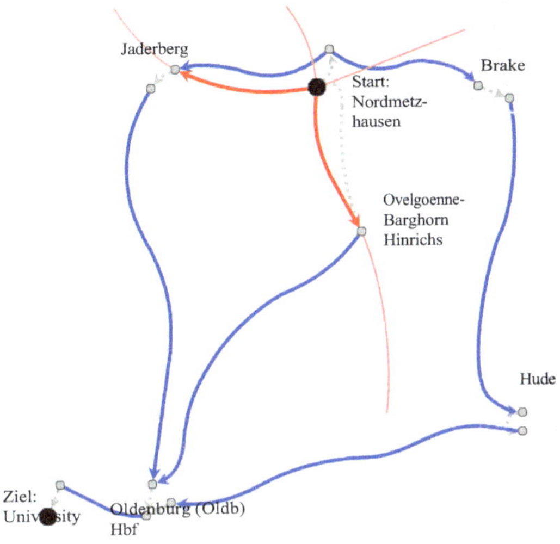

Fig. 5 Two matched carpools are added to the mobility network

Barghorn Hinrichs stop is also qualified as a transfer point resulting in substituting the long walk leg.

Integrate the Carpool in the Existing Graph

The way on how the carpool is integrated depends on whether the carpool travel mode is used first or last in the intermodal journey. For the carpool, there will be a node v, constructed to represent the departure at the pickup location and a node v' constructed to represent the arrival at the drop-off location of the carpool trip. All $v^\wedge \in V^\wedge$ are nodes, to which the nodes v and v' can be connected to, which depends on whether the carpool trip leads from the trip origin to the transfer point or whether the trip leads from the transfer point to the trip destination.

Let $p_l \in R \times R$ be the location of pickup and $d_l \in R \times R$ be the location of the drop-off of the considered carpool.

If the carpool comes from the trip origin, let $V^\wedge = \{v \in V | l(v) \approx d_l \wedge \lambda(v) = p \wedge \tau (v) = \text{dep}\}$, meaning the carpool node pair will be connected to all departure nodes near to the drop-of location of the carpool and the origin node v_0 will be connected to the carpool node pair as pictured in Fig. 6.

If the carpool leads to the trip destination, let $V^\wedge = \{v \in V | l(v) \approx p_l \wedge \lambda(v) = p \wedge \tau (v) = \text{arr}\}$, meaning all arrival nodes near the pickup location of the carpool will be the connected to the carpool node pair and the node pair will be connected to destination node v^* as pictured in Fig. 7.

Updating the Remaining Stops

Since this approach assumes, that long walking legs will be replaced by much quicker carpools, all departure times at all transfer stops, where the carpool is taken first, needs to be updated. Otherwise, the final result will contain high transfer durations and the arrival time at the trip destination will not differ from the first construction. New

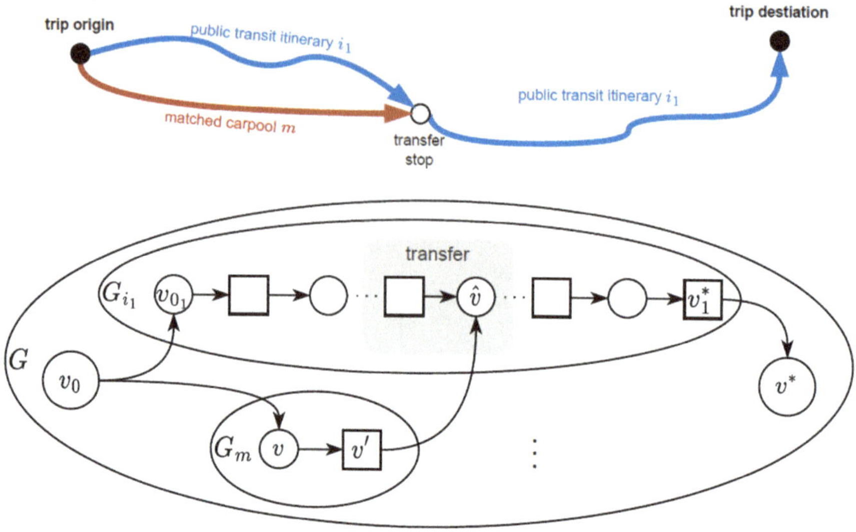

Fig. 6 Integrate carpool nodes, when carpool is the first mode of transport

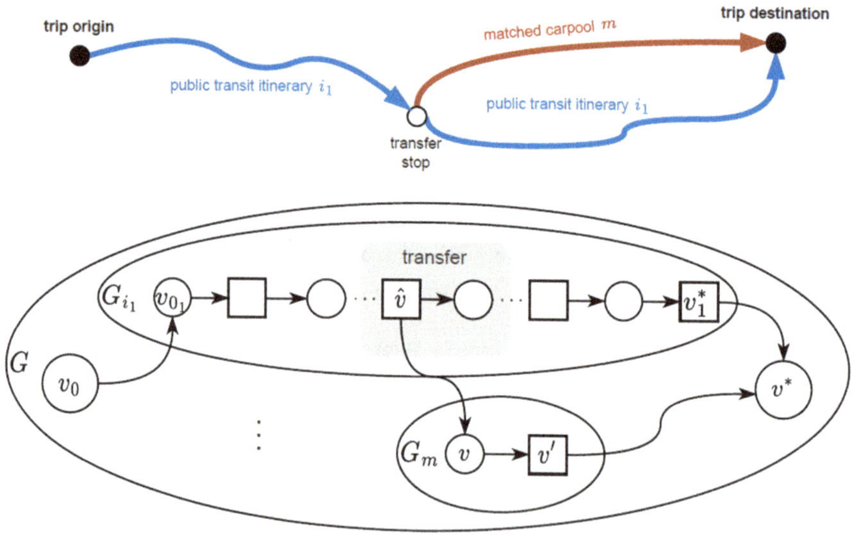

Fig. 7 Integrate carpool nodes, when carpool is the last mode of transport

itineraries can be constructed the same way stated in Sect. 3.2.2. Nevertheless, it needs to be connected to all arrival node near to the drop-off point of the carpool and connected to the destination node v^*. This procedure is pictured in Fig. 8.

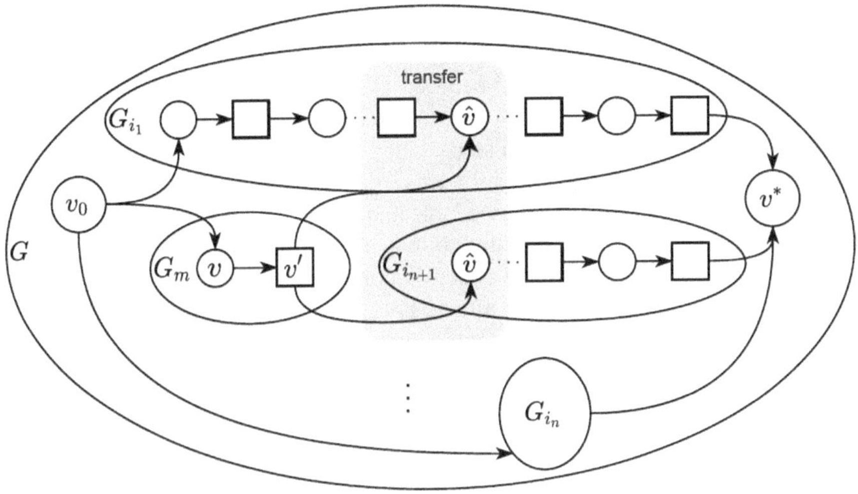

Fig. 8 Adding the new itinerary G_{n+1}, which updates the time at the drop-off location p_l

4 Evaluation

The evaluation focused on a representative test scenario: a journey starting in Nordmetzhausen, transferring at Oldenburg Central Station (Hbf), and ending at Oldenburg University. This specific route was chosen as it captures both rural and urban mobility aspects and provides a realistic use case for testing the effectiveness of integrating carpooling into intermodal travel chains.

4.1 Key Observations

The effectiveness of intermodal journeys is highly dependent on carpool availability and precise timing at transfer hubs. When carpool trips align well with public transport schedules, considerable travel time savings are possible. The maximum theoretical benefit results from replacing long walking segments to valuable public transport hubs (e.g., Ovelgönne Barghorn Hinrichs or Jaderberg) with a carpool segment. In such cases, users save time equal to the difference between the walk and the carpool duration, often 10–20 min.

4.2 Simulation and Empirical Insights

Scenarios were generated using real and synthetic data with up to 300 active carpool drivers and various route lengths. Intermodal journeys performed best when carpooling covered longer segments, reducing reliance on less frequent rural bus lines. In the Nordmetzhausen–Oldenburg case, the average transfer time between carpool and public transport ranged between 11 and 26 min, depending on carpool density and time of day. These were tuned to reflect realistic conditions such as buffer time for delays or walking between stops.

A critical observation from the broader evaluation shows that when only short carpool segments (< 15 km) are allowed, median travel time increases by 35–37 min compared to pure public transport due to longer waits and detours. However, with longer carpool options, intermodal routes become up to 7 min faster than public-only alternatives. The time of day affects both transfer reliability and total travel time. During peak hours (HVZ), matching quality can drop due to directional traffic flows (e.g., into the city in the morning, out in the afternoon), which aren't always aligned with commuter needs.

4.3 Graph-Level Evaluation

The time-expanded graph model proved effective in representing the combined mobility offers. However, without good carpool matches, the intermodal solution defaults to slower public transport with long walking transfers, limiting its advantage. The graph construction process, including the updating of departure times after carpool inclusion and re-querying for optimized public transport connections, showed that transfer points need careful prioritization to avoid suboptimal paths with excessive waiting times.

5 Conclusion and Outlook

The presented approach, which uses a time-expanded graph model to integrate heterogeneous mobility offers from public transport and carpooling providers, has demonstrated high adaptability to existing system infrastructures. Since each graph is generated on demand for individual user queries, the usual concerns about the space complexity of time-expanded models are mitigated. In the case study region of Oldenburg, the constructed graphs had a moderate size (averaging ~392 nodes and ~404 edges) and posed no performance bottlenecks during routing computations.

The stateless design of the broker and its modular architecture ensure that routing complexity remains distributed across the connected systems. The VBN routing engine and mOOWVe carpool matching API are responsible for heavy computations, allowing the broker to focus solely on aggregation. This distribution of responsibility aligns well with Conway's law and supports future extensibility.

Performance optimizations can be further achieved by caching static timetable data (e.g., regular bus departures from major hubs) and minimizing API calls for route generation. Furthermore, enhancements to the matching API—such as multi-point transfer queries or dynamic detour estimation—could reduce latency and improve the quality of generated intermodal journeys.

However, the quality of intermodal routing heavily depends on external conditions:

- **Coverage and frequency of the public transport network**
- **Density and flexibility of available carpools**
- **Synchronization between transfer points**

In cases where no feasible carpool matches are found, the system defaults to slower and more indirect public transit options, reducing user benefit. These limitations suggest that for the proposed MaaS-broker to be impactful, especially in rural areas, a critical mass of carpool drivers and high-frequency transit options must be present.

Looking ahead, future evaluations should include real-world mobility data (instead of synthetic input), reflecting actual commuting behavior and ride availability. Moreover, deploying the system in live production (e.g., through mOOWVe) could enable access to real carpool usage data, thereby improving routing quality and model calibration.

Notwithstanding the result quality heavily relies on the public transportation network of the region and the driving characteristics of the residents in rural areas. If no feasible carpools can be found the suggested route will mostly contain time-consuming public transport rides. In conclusion, the acceptance and usage of MaaS especially for carpool drivers in rural is one of the main prerequisites to make this proposed method efficient.

In conclusion, this work provides a practical, modular, and scalable template for integrating intermodal transport solutions into existing mobility ecosystems. Its design principles support flexible adaptation, and its implementation confirms the feasibility of decentralized, broker-based MaaS architectures.

References

Conway, M. E. (1968). How do committees invent. *Datamation, 14*(4), 28–31.

Georgakis, P., Almohammad, A., Bothos, E., et al. (2019). Multimodal route planning in mobility as a service. In *IEEE/WIC/ACM international conference on web intelligence—companion volume* (pp. 283–291). Association for Computing Machinery, New York, NY, USA, WI '19 Companion. https://doi.org/10.1145/3358695.3361843

Jamal, J., Montemanni, R., Huber, D., et al. (2017). A multi-modal and multi-objective journey planner for integrating carpooling and public transport. *Journal of Traffic and Logistics Engineering*. https://doi.org/10.18178/jtle.5.2.68-72

K¨onig, A. (2019). Den l¨andlichen Raum bewegen—Praxisbeispiele, Forschungsprojekte und Entwicklungspotentiale. URL https://elib.dlr.de/128107/. Stand April, 20, 2023.

NEMo. (2020). Nachhaltige erfu¨llung von mobilit¨atsbedu¨rfnissen im l¨andlichen raum. https://nemo-mobilitaet.de/blog/wp-content/uploads/2020/08/20200823NEMo-Abschlussbericht.pdf. Stand October 24, 2022.

Schulz, F. (2005). *Timetable information and shortest paths* [PhD thesis, Citesee].

Schulz, F., Wagner, D., Weihe, K. (1999). Dijkstra's algorithm on-line: An empirical case study from public railroad transport. In *Algorithm engineering* (pp. 110–123). Springer. https://doi.org/10.1007/3-540-48318-711

Van Brummelen, G. (2012). *Heavenly mathematics: The forgotten art of spherical trigonometry*. Princeton University Press.

Varone, S. C., Aissat, K. (2015). Multi-modal transportation with public transport and ride-sharing—multi-modal transportation using a path-based method. In *International conference on enterprise information systems*.

Wong, Y. Z., Hensher, D. A., Mulley, C. (2020). Mobility as a service (maas): Charting a future context. *Transportation Research Part A: Policy and Practice, 131*, 5–19. https://doi.org/10.1016/j.tra.2019.09.030.

https://www.sciencedirect.com/science/article/pii/S0965856418312229. Developments in Mobility as a Service (MaaS) and Intelligent Mobility.

An Exploration of the Impact of AI Capabilities on Supply Chain Resilience in Mauritius

Loovesh Sharma Ramwodin and Doorgesh Sookarah

Abstract This study examines the role of artificial intelligence (AI) in enhancing supply chain resilience in Mauritius. The COVID-19 pandemic exposed vulnerabilities in global and local supply chains, highlighting the need for technologies that can mitigate disruptions and ensure continuity. As a technological enabler, AI offers transformative tools such as machine learning, predictive analytics, AI-driven bots and big data, all of which improve supply chain adaptability, responsiveness and robustness. Using a qualitative, exploratory design, the study draws on a comprehensive literature review and focus group discussions with procurement managers from 20 large organisations across various sectors in Mauritius. Six key AI capabilities, namely, automated compliance, predictive analytics, AI bots, algorithms, machine learning and big data, were assessed for their impact on supply chain resilience. Findings indicate that these technologies significantly enhance visibility, forecasting accuracy, risk assessment and agility. Focus group insights confirm AI's practical value in real-time decision-making, logistics optimisation, warehouse automation and inventory management. The study presents a conceptual framework linking AI capabilities to supply chain resilience. While global research has explored AI in supply chains, localised studies in small island developing states remain scarce. This paper addresses that gap by providing one of the first assessments of AI's contribution to supply chain resilience in Mauritius, offering a basis for future empirical validation and informing strategic decisions in digital transformation.

L. S. Ramwodin · D. Sookarah (✉)
University of Technology Mauritius, La Tour Koenig, Port Louis, Mauritius
e-mail: doorgesh.sookarah@utm.ac.mu

L. S. Ramwodin
e-mail: lramwodin@utm.ac.mu

J. Marx Gómez et al. (eds.), *Recent Advances in Information Systems*, Progress in IS,
https://doi.org/10.1007/978-3-032-14099-9_62

1 Introduction

Artificial intelligence (AI), as a technological enabler, can help address challenges associated with supply chain resilience and offer practical avenues to support organisations in their journey towards long-term sustainability. AI is particularly relevant because it integrates machine learning, big data and deep learning capabilities (Dwivedi et al., 2019; Gupta et al., 2021; Wamba et al., 2020a). For example, machine learning supports the automation of contractual agreements with suppliers, strengthening procurement strategies. Big data enhances organisational decision-making and can reduce redundancies across the supply chain (Baryannis et al., 2019; Dubey et al., 2020). In addition, AI can enable agent-based systems, genetic algorithms and expert systems to facilitate demand planning, order fulfilment, network design and inventory management in coordination with supplier systems (Muniz et al., 2020).

Based on these considerations, the purpose of this research is to analyse the impact of AI capabilities on supply chain resilience in the local context of Mauritius.

1.1 Study Rationale

Firms have long recognised the importance of contingency planning to mitigate supply shocks, manage demand volatility and improve workplace safety to achieve supply chain efficiency and effectiveness (Ivanov, 2020). Before the COVID-19 pandemic, supply chains were often lauded for their ultra-efficiency, sourcing strategies and just-in-time capabilities (Hoek, 2020). Evidence also shows that pre-COVID-19, many consumers preferred in-store purchases to physically examine and compare items. Unsurprisingly, online shopping surged during the pandemic, making supply chain visibility more critical than ever (Queiroz et al., 2020). However, many organisations were unprepared for the practicalities of fulfilling online orders, and their planning processes were ill-equipped to respond to the sudden spike in demand. Consequently, it became apparent that greater supply chain resilience was essential (Hoek, 2020; Queiroz et al., 2020).

Developing resilience in response to disruption poses challenges. On one hand, organisations face significant uncertainty; on the other, they generate vast amounts of data (Wamba et al., 2020a). Due to these factors, achieving the desired level of resilience can be very difficult (Ivanov, 2021). As a result, many firms have focused on making short-term predictions and strengthening their supply chain capabilities (Sundarakani et al., 2020). To support this, firms must regularly assess risk levels across multiple parameters (Dolgui & Ivanov, 2020).

COVID-19 has forced organisations worldwide to re-evaluate and transform their supply chain models (Kano & Oh, 2020). Reports indicate that approximately 75% of businesses, including those in services, manufacturing, wholesale and retail, are connected to economies severely impacted by the pandemic. Retailers globally continue to struggle to meet the needs of businesses, consumers and workers. Business continuity and long-term economic sustainability have therefore become major challenges. Overall, COVID-19 has exposed numerous vulnerabilities in supply chains (Queiroz et al., 2020). Many of these challenges can potentially be addressed by integrating intelligent technologies such as AI to enhance existing supply chain capabilities.

1.2 Research Questions (RQ)

- RQ1: What is the impact of AI capabilities on supply chain resilience?
- RQ2: What elements of AI capabilities can be adopted to enhance supply chain resilience?

1.3 Study Aim and Objectives

The aim of this study is to identify the AI capabilities that influence supply chain resilience in the Mauritian context. Specifically, the study examines how AI can be used to improve visibility, responsiveness and control within supply chains. The objectives are as follows:

- To analyse the effect of automated compliance on supply chain resilience.
- To assess the influence of predictive capabilities on supply chain resilience.
- To examine the impact of AI-driven bots on supply chain resilience.
- To analyse the contribution of algorithms to supply chain resilience.
- To investigate the role of machine learning in strengthening supply chain resilience.
- To evaluate the impact of big data on supply chain resilience.

2 Literature Review

2.1 Supply Chain Disruptions

Many companies have reported significant supply chain disruption and shared accounts of struggling to reconfigure distribution chains and networks (Mahajan & Tomar, 2021). Reports indicate that companies often failed to develop contingency plans capable of addressing disruptions caused by events such as COVID-19. As a result, firms across the globe experienced unpredictable delays in receiving materials due to inadequate supply chain information (Paul & Chowdhury, 2020). Other factors, such as supplier consolidation, cost-reduction efforts and risk minimisation, have further reduced available inventories and contributed to widespread shortages (Dolgui & Ivanov, 2020).

Today's supply chains face complex challenges that have pushed many firms into crisis. Manufacturing processes have become increasingly intricate, requiring components to be outsourced from multiple locations to assemble a single product. This complexity has resulted in substantial dependence on logistics, imports and exports, creating vulnerabilities in times of disruption and underscoring the need for smarter decision-making (Wamba et al., 2020a). Additional challenges for sourcing organisations include distribution risks and shortages across the network (Dolgui & Ivanov, 2020).

Distribution has emerged as a critical area where warehouse staffing, direct distribution (Koster & Warffemius, 2005) and responsive allocation have become essential considerations during the COVID-19 era (Butt, 2021). Retail supply chains have also been impacted, with demand surging for essential items while declining in categories such as luxury goods, placing additional strain on operating margins and business models. On the consumer side, stockpiling essential supplies and over-the-counter medicines has further stressed supply chains to their limits (Muniz et al., 2020). This unpredictable increase in demand, coupled with a fluctuating environment, makes it difficult to respond effectively to bullwhip effects.

2.2 *Artificial Intelligence*

Many organisations have pursued the digitisation of their processes over the past two decades, and Industry 4.0 has recently emerged as a prominent business paradigm. AI has long been recognised as a leading technology capable of enabling communication among devices and machines (Guzman & Lewis, 2020). Since supply chains involve a series of complex tasks, AI can simplify operations by solving problems at a greater speed and accuracy while managing large volumes of data (Huin et al., 2003). Although AI itself is not new, its potential applications in supply chain management have only recently been fully recognised (Wamba et al., 2020b). AI facilitates smart and agile decision-making to anticipate and address supply chain problems proactively. Proactive AI systems enhance service quality and support on-time, undamaged deliveries that improve customer satisfaction (Treleaven & Batrinca, 2017).

Automated Compliance—AI enables automated compliance processes, resulting in lower costs and more efficient value chain operations (Huang & Rust, 2020).

Predictive Capabilities—AI also significantly improves predictive capabilities, which are critical for accurate demand forecasting in today's dynamic business environment.

AI-Driven Bots—Such bots efficiently engage customers by personalising interactions. These bots can track delivery statuses and are often integrated with customer support teams (Wen et al., 2018). In addition, AI simplifies tedious warehouse tasks through automation. Companies like Amazon and Alibaba already deploy AI-driven robots to enhance supply chain productivity and efficiency.

Algorithms—In supply chains, every minute and every mile matter. AI leverages algorithms to reduce time and cost by optimising routes and deliveries (Singh et al., 2020a).

Machine Learning—AI encompasses machine learning, big data and deep learning (Dwivedi et al., 2019; Gupta et al., 2021; Wamba et al., 2020a). For example, machine learning can automate contractual agreements with suppliers, strengthening procurement strategies.

Big Data—Big data supports better decision-making across the organisation and helps eliminate redundancy within the supply chain (Baryannis et al., 2019; Dubey et al., 2020). Additionally, AI can develop agent-based systems, genetic algorithms and expert systems to improve demand planning, order fulfilment, network design and inventory management in collaboration with supplier systems (Muniz et al., 2020).

2.3 *Supply Chain Resilience*

Supply chain resilience refers to the capability of supply chain to cope with unforeseen disruptive events and to recover quickly to its original level of performance or to a new level that maintains expected operating, financials and market outcomes (Adobor, 2020). To build a resilient supply chain, firms must identify and assess risks across nodes, evaluate their severity and likelihood and establish mechanisms for detection and response (Chang et al., 2015).

Companies adopt multiple strategies to sustain resilience. In the initial phase of COVID-19, some organisations relied on inventory and capacity buffers to maintain operations, while others leveraged underutilised production capacity to manufacture essential products, such as medicines and protective equipment (Queiroz et al., 2020). Resilience has also been enhanced through multi-sourcing strategies, which provide alternatives to single-source dependencies (Sá et al., 2019).

COVID-19 further highlighted the importance of near shoring, which reduces geographic reliance on global networks and enables more control over inventory and delivery timelines (Hoek, 2020). Local supply chains facilitate faster transfers of products to customers and allow the harmonisation of production technologies across the network (Sundarakani et al., 2020). The closer the network is to the point of consumption, the greater the opportunities for standardising components and simplifying sourcing, thereby increasing resilience (Singh et al., 2020b).

Beyond core components, partnerships within the supply chain ecosystem, such as collaborations with contract manufacturers and third-party logistics providers, play a critical role in strengthening resilience.

2.4 AI and Supply Chain Resilience

The application of intelligent technologies such as AI can help firms strengthen their supply chain resilience. The adoption of AI can serve as a critical capability for developing more effective control mechanisms and identifying potential areas of disruption (Dwivedi et al., 2019; Gupta et al., 2021; Treleaven & Batrinca, 2017). AI-based innovations also enable real-time coordination and collaboration to improve supply chain visibility (Wamba et al., 2020a).

With enhanced visibility, organisations can generate insights about how disruptions may impact the entire supply chain over time. For example, sudden surges in demand during the pandemic have made it difficult for firms to maintain service-level agreements with their partners. These service levels are determined by dynamic capabilities, profit margins and the locations from which companies operate (Defee & Fugate, 2010).

Another critical area is network planning and mapping, where AI supports decisions related to production capacity, transportation routes and connections between supply chain nodes (Ivanov & Dolgui, 2020). AI also improves collaboration between contractors and suppliers, providing visibility over delivery routes and enabling organisations to communicate realistic commitments to customers.

Through leveraging data science principles, AI handles the volume, velocity, variety, veracity, value, variability and visualisation of data to extract critical insights (Wamba et al., 2020c). In addition, AI enables performance tracking across warehouses, including monitoring demand fluctuations and managing shelf life. Inventory management has become particularly important for ensuring continuous supply during COVID-19 (Butt, 2021; Queiroz et al., 2020). AI offers robust analyses of stock levels and aligns associated activities accordingly.

As supply chains expand into new geographic regions due to the pandemic, examining traffic, weather and optimal delivery routes has become increasingly complex. AI helps organisations navigate unfamiliar markets remotely and supports more accurate strategic decision-making. These technologies also prepare employees and leadership teams to acquire the skills needed to manage emergencies and cross-cultural dynamics (Dubey & Gunasekaran, 2015; Dubey et al., 2016, 2018).

At a macro level, AI facilitates environmental scanning and increases resilience by providing a two-dimensional view of risks and capabilities (Baryannis et al., 2019). Literature suggests that companies focused on resilience should invest in developing analytical capabilities to enhance decision-making in highly complex environments. Applying supply chain risk analytics enables firms to proactively manage risks and innovate their response strategies (Queiroz et al., 2020).

Disruptions and risks should be seen as phenomena to understand, analyse and adapt to, particularly in light of emerging cyber threats and increasingly complex supply chain designs (Dolgui & Ivanov, 2020). AI-based innovations have the potential to accelerate decision-making by supporting the recognition, testing and implementation of novel solutions (Treleaven & Batrinca, 2017). Figure 1 illustrates how

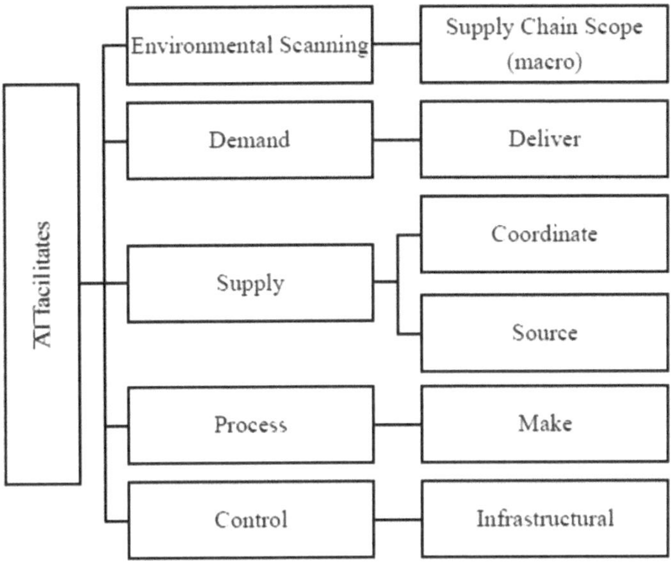

Fig. 1 AI impact on different dimensions of supply chain resilience (Baryannis et al., 2019)

AI contributes to multiple dimensions of supply chain resilience by supporting environmental scanning, predictive analytics and real-time coordination across critical supply chain processes (Baryannis et al., 2019).

2.5 *Empirical Review*

The rapid evolution of AI technologies has created unprecedented opportunities to enhance supply chain resilience, a critical constraint in today's volatile global market. AI applications such as machine learning, predictive analytics and automation routines have proven particularly effective in addressing persistent challenges within supply chains, including demand forecasting and optimisation, risk management and operational efficiency (Teerasoponpong & Sugunnasil, 2022). The use of AI in strengthening resilience is associated with automating routine tasks, modelling adverse scenarios, enabling real-time decision-making and supporting contingency planning (Shamout et al., 2022).

Rodríguez-Espíndola et al. (Rodríguez-Espíndola et al., 2020) examined disruptions triggered by global crises such as the COVID-19 pandemic and identified significant potential for AI technologies to improve supply chain visibility and agility, resulting in more effective market responses. Similarly, Qi, Shen and Xu (Qi et al., 2023) emphasised that AI-driven data analytics can substantially enhance resilience by assessing supply chain risks and their implications.

Despite this progress, research into both the theoretical and practical dimensions of AI's impact on supply chain resilience remains underdeveloped. Xu, Mak and Brintrup (Xu et al., 2021) observed that most studies evaluating the potential and limitations of AI technologies rely heavily on qualitative approaches, case studies or expert opinion. While these contributions offer valuable insights, there is a notable gap in quantitative evidence supporting claims that specific AI interventions improve resilience outcomes.

Furthermore, AI implementation in supply chains presents a range of challenges, including data privacy risks, overreliance on automated systems and the need for substantial investments in infrastructure and skilled personnel (Dauvergne, 2022). Addressing these issues requires a multidisciplinary research perspective that integrates knowledge from information systems, operations management and organisational behaviour (Dubey et al., 2022).

3 Research Design

To investigate the impact of AI capabilities on supply chain resilience, this study adopted a qualitative research approach framed by the theoretical perspectives on AI described in existing literature (Gammelgaard, 2017). Qualitative methods are effective tools for capturing individual perceptions, as they enable analysis of thoughts and opinions to identify trends and insights (Fawcett et al., 2014).

Given the novelty of the topic, the research began with a comprehensive literature search. The search was conducted across databases including Scopus, Web of Science and Google Scholar. Keywords such as "artificial intelligence," "supply chain resilience," "AI-enabled supply chain" and "COVID-19 supply chain disruption" were used to identify relevant studies. Inclusion criteria required peer-reviewed articles published between 2000 and 2023 that addressed the application of AI in supply chain contexts. The search yielded approximately 160 publications, which were screened and reviewed to inform the study framework.

To further substantiate the findings from the literature, a focus group was organised with procurement managers representing diverse sectors in Mauritius, including retail, manufacturing and logistics. Participants were recruited through professional networks and industry associations. In total, data were gathered from 20 participants. The focus group discussions were recorded and transcribed verbatim. Thematic analysis was applied to identify recurring patterns and insights related to AI adoption and resilience practices. Given the exploratory nature of the research, this design enabled a nuanced understanding of perspectives on AI's role in strengthening supply chain resilience within the local context.

4 Study Findings

4.1 Results of Literature Search

The comprehensive literature search revealed that several AI capabilities have a significant impact on supply chain resilience. These include automated compliance and predictive capabilities (Huang & Rust, 2020), AI-driven bots (Wen et al., 2018), optimisation algorithms (Singh et al., 2020a), machine learning (Dwivedi et al., 2019; Gupta et al., 2021; Wamba et al., 2020a) and big data analytics (Baryannis et al., 2019; Dubey et al., 2020; Muniz et al., 2020). Each of these capabilities contributes to enhanced visibility, improved decision-making and greater responsiveness across supply chain operations.

4.2 Results of Focus Groups

The focus groups further confirmed that AI-enabled tools such as robots, predictive forecasting systems and optimisation algorithms can strengthen supply chain resilience. Participants highlighted the potential of AI applications to improve forecasting accuracy, automate repetitive tasks and enable faster responses to disruptions within the Mauritian context.

5 Recommendations

Given the exploratory nature of this study, it is recommended that the conceptual model presented in Fig. 2 be empirically tested using a representative sample. Such validation would help establish the generalisability of the findings to other contexts beyond Mauritius. Future research should employ quantitative methods, such as large-scale surveys or experimental designs, to evaluate the relationships among AI capabilities, supply chain practices and resilience outcomes.

Additionally, researchers are encouraged to further investigate the potential risks associated with AI implementation, including data privacy concerns, technology integration challenges and skill gaps among employees. A multidisciplinary perspective combining insights from operations management, information systems and organisational behaviour would be valuable for developing a more comprehensive understanding of AI's role in supply chain resilience.

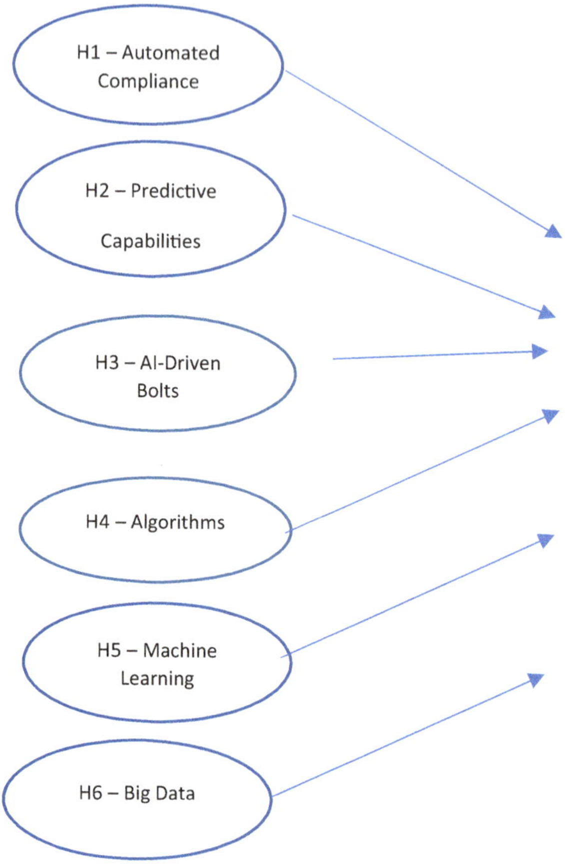

Fig. 2 Proposed conceptual model of AI capabilities contributing to supply chain resilience

6 Conclusion

This study explored the potential impact of AI capabilities on supply chain resilience within the Mauritian context. Drawing on a comprehensive literature review and focus group discussions, the research identified key AI applications, including predictive analytics, automated compliance, machine learning and optimisation algorithms, that can contribute to strengthening supply chain practices.

While the findings highlight promising areas for AI integration, the study is exploratory in nature and has limitations related to its qualitative design and sample size. Accordingly, the proposed conceptual framework, which positions AI capabilities as independent variables influencing supply chain resilience, should be tested and validated in future research using quantitative methods and broader datasets.

Overall, this work contributes to the emerging discourse on AI-driven supply chain resilience and provides a foundation for further empirical investigation and practical implementation.

7 Future Trends

The supply chain sector is on the verge of significant transformation as artificial intelligence becomes more fully integrated. Emerging technologies such as autonomous vehicles, advanced predictive analytics and blockchain are likely to reshape logistics, manufacturing and distribution networks in ways that were previously unattainable. These innovations will not only improve operational efficiency but also provide more robust solutions for managing disruption and maintaining resilience.

AI technologies are expected to be deployed in several innovative ways. For example, autonomous vehicles and drones will accelerate last-mile delivery, reducing both delivery times and operational costs. In addition, they will improve reliability and accessibility for customers. Advanced predictive analytics, combined with machine learning algorithms, will enable organisations to anticipate demand shifts and potential disruptions, supporting more accurate inventory planning (Boudouaia et al., 2024). This capability will be crucial in preventing stockouts or excess inventory during periods of market volatility.

The convergence of AI and blockchain is also anticipated to create more transparent, secure and tamper-proof supply chain transactions. In manufacturing, AI-driven robotics and the Internet of Things will underpin the development of smart factories that automate decisions, predict maintenance needs and adapt dynamically to changes in production requirements (Fosso Wamba et al., 2023).

However, these technological advancements bring new challenges. Increased automation raises concerns about job displacement, which will require coordinated policies for retraining and upskilling the workforce. Data privacy is another critical issue, as AI systems depend on collecting and processing vast amounts of information (Oosthuizen et al., 2021). Regulations such as the General Data Protection Regulation (GDPR) offer frameworks to address these concerns, but ongoing vigilance and compliance are essential.

Ethical considerations are particularly important when AI is used in decision-making contexts that impact employees and communities. For instance, an AI system might recommend workforce reductions during a lockdown or economic crisis. To ensure fairness and accountability, organisations will need clear policies outlining when and how AI-driven decisions are permitted (Rashid et al., 2024). These guidelines should balance operational efficiency with the ethical treatment of employees and stakeholders.

Finally, because supply chains are inherently global, international regulations and cooperation will be vital to address data sovereignty, privacy and ethical standards. As Dogru and Keskin (Dogru & Keskin, 2020) note, AI is not constrained by national borders, and consistent international frameworks will be necessary to guide responsible and equitable AI adoption in supply chain management.

8 Limitations and Future Research

This study provides valuable insights into how AI can be integrated into supply chains to improve resilience. However, despite its contributions, several limitations should be acknowledged.

First, the relatively small sample size affects the generalisability of the results and limits the ability to capture the diversity of perspectives across different industries and organisational contexts. Second, the qualitative design and reliance on focus group discussions without supplementary quantitative data reduce the rigour of the analysis. Finally, the study focused on a specific set of AI capabilities, which may not fully represent the broader range of emerging technologies applicable to supply chain management.

Future research should address these limitations by expanding sample sizes to improve representativeness and including companies from multiple sectors and geographic regions. Incorporating quantitative methods, such as surveys or statistical modelling, would provide a more comprehensive understanding of how AI capabilities contribute to resilience (Naz et al., 2022). Additionally, further studies should examine the impact of emerging AI technologies not covered in this research, such as advanced robotics, autonomous systems and blockchain-enabled solutions. Exploring these areas could generate new knowledge and support the development of innovative theoretical frameworks to guide AI adoption in supply chains (Pal, 2023).

References

Adobor, H. (2020). Supply chain resilience: An adaptive cycle approach. *International Journal of Logistics Management, 31*(3), 443–463.

Baryannis, G., Validi, S., Dani, S., & Antoniou, G. (2019). Supply chain risk management and artificial intelligence: State of the art and future research directions. *International Journal of Production Research, 57*(7), 2179–2202.

Boudouaia, M. A., Abed, M., Sari, Z., & Aissani, D. (2024). Supply chain 5.0: Vision, challenges, and perspectives. In *2024 21st Learning and technology conference (L&T)* (pp. 203–208). IEEE.

Butt, A. S. (2021). Strategies to mitigate the impact of COVID-19 on supply chain disruptions: A multiple case analysis of buyers and distributors. *International Journal of Logistics Management*. https://doi.org/10.1108/IJLM-11-2020-0455

Chang, W., Ellinger, A. E., & Blackhurst, J. (2015). A contextual approach to supply chain risk mitigation. *International Journal of Logistics Management, 26*(3), 642–656.

Dauvergne, P. (2022). Is artificial intelligence greening global supply chains? Exposing the political economy of environmental costs. *Review of International Political Economy, 29*(3), 696–718.

de Koster, M. B. M., & Warffemius, P. M. J. (2005). American, Asian and third-party international warehouse operations in Europe. *International Journal of Operations and Production Management, 25*(8), 762–780.

de Sá, M. M., de Souza Miguel, P. L., de Brito, R. P., & Pereira, S. C. F. (2019). Supply chain resilience: The whole is not the sum of the parts. *International Journal of Operations & Production Management, 40*(1), 92–115.

Defee, C. C., & Fugate, B. S. (2010). Changing perspective of capabilities in the dynamic supply chain era. *International Journal of Logistics Management, 21*(2), 180–206.

Dogru, A. K., & Keskin, B. B. (2020). AI in operations management: Applications, challenges and opportunities. *Journal of Data, Information and Management, 2*(2), 67–74.

Dolgui, A., & Ivanov, D. (2020). Exploring supply chain structural dynamics: New disruptive technologies and disruption risks. *International Journal of Production Economics*. https://doi.org/10.1016/j.ijpe.2020.107886

Dubey, R., & Gunasekaran, A. (2015). Education and training for successful career in big data and business analytics. *Industrial and Commercial Training, 47*(4), 174–181.

Dubey, R., Gunasekaran, A., Altay, N., Childe, S. J., & Papadopoulos, T. (2016). Understanding employee turnover in humanitarian organizations. *Industrial and Commercial Training, 48*(4), 208–214.

Dubey, R., Gunasekaran, A., Childe, S. J., Bryde, D. J., Giannakis, M., Foropon, C., & Hazen, B. T. (2020). Big data analytics and artificial intelligence pathway to operational performance under the effects of entrepreneurial orientation and environmental dynamism: A study of manufacturing organizations. *International Journal of Production Economics*. https://doi.org/10.1016/j.ijpe.2019.107599

Dubey, R., Gunasekaran, A., Childe, S. J., Bryde, D. J., Hazen, B. T., & Roubaud, D. (2022). Impact of artificial intelligence-driven big data analytics culture on agility and resilience in humanitarian supply chains: A practice-based view. *International Journal of Production Economics, 250*, Article 108618.

Dubey, R., Gunasekaran, A., Childe, S. J., & Papadopoulos, T. (2018). Skills needed in supply chain—human agency and social capital analysis in third party logistics. *Management Decision, 56*(1), 143–159.

Dwivedi, Y. K., Hughes, L., Ismagilova, E., Aarts, G., Coombs, C., Crick, T., Duan, Y., Dwivedi, R., Edwards, J., Eirug, A., Galanos, V., Ilavarasan, P. V., Janssen, M., Jones, P., Kar, A. K., Kizgin, H., Kronemann, B., Lal, B., Lucini, B., & Williams, M. D. (2019). Artificial intelligence (AI): Multidisciplinary perspectives on emerging challenges, opportunities, and agenda for research, practice and policy. *International Journal of Information Management*. https://doi.org/10.1016/j.ijinfomgt.2019.08.002

Fawcett, S. E., Waller, M. A., Miller, J., Schwieterman, M., Hazen, B., & Overstreet, R. (2014). A trail guide to publishing success: Tips on writing influential conceptual, qualitative, and survey research. *Journal of Business Logistics, 35*(1), 1–16.

Fosso Wamba, S., Queiroz, M. M., & Miah, S. J. (2023). ChatGPT and generative artificial intelligence: An exploratory study of key benefits and challenges in operations and supply chain management. *International Journal of Production Research*, 1–21.

Gammelgaard, B. (2017). The qualitative case study. *International Journal of Logistics Management, 28*(4), 910–913.

Gupta, S., Modgil, S., Bhattacharyya, S., & Bose, I. (2021). Artificial intelligence for decision support systems in the field of operations research: Review and future scope of research. *Annals of Operations Research*. https://doi.org/10.1007/s10479-020-03856-6

Guzman, A. L., & Lewis, S. C. (2020). Artificial intelligence and communication: A human–machine communication research agenda. *New Media and Society, 22*(1), 70–86.

Huang, M. H., & Rust, R. T. (2020). Engaged to a robot? The role of AI in service. *Journal of Service Research*. https://doi.org/10.1177/1094670520902266

Huin, S. F., Luong, L. H. S., & Abhary, K. (2003). Knowledge-based tool for planning of enterprise resources in ASEAN SMEs. *Robotics and Computer-Integrated Manufacturing, 19*(5), 409–414.

Ivanov, D. (2020). Viable supply chain model: Integrating agility, resilience and sustainability perspectives—lessons from and thinking beyond the COVID-19 pandemic. *Annals of Operations Research*. https://doi.org/10.1007/s10479-020-03640-6

Ivanov, D. (2021). Lean resilience: AURA (Active Usage of Resilience Assets) framework for post-COVID-19 supply chain management. *International Journal of Logistics Management*. https://doi.org/10.1108/IJLM-11-2020-0448

Ivanov, D., & Dolgui, A. (2020). Viability of intertwined supply networks: Extending the supply chain resilience angles towards survivability. *International Journal of Production Research, 58*(10), 2904–2915.

Kano, L., & Oh, C. H. (2020). Global value chains in the post-COVID world: Governance for reliability. *Journal of Management Studies, 57*(8), 1773–1777.

Mahajan, K., & Tomar, S. (2021). COVID-19 and supply chain disruption: Evidence from food markets in India. *American Journal of Agricultural Economics, 103*(1), 35–52.

Muniz, L. R., Conceição, S. V., Rodrigues, L. F., de Freitas Almeida, J. F., & Affonso, T. B. (2020). Spare parts inventory management: A new hybrid approach. *International Journal of Logistics Management*. https://doi.org/10.1108/IJLM-12-2019-0361

Naz, F., Malik, M., Qureshi, M. A., Khan, M. A., & Shabbir, M. S. (2022). Is artificial intelligence an enabler of supply chain resiliency post COVID-19? An exploratory state-of-the-art review for future research. *Operations Management Research, 15*(1), 378–398.

Oosthuizen, K., Roberts-Lombard, M., & Mpinganjira, M. (2021). Artificial intelligence in retail: The AI-enabled value chain. *Australasian Marketing Journal, 29*(3), 264–273.

Pal, S. (2023). AI's Promethean hand in supply chain resilience: A framework for global commercial success. *International Journal of Progressive Research in Engineering Management and Science (IJPREMS), 3*(5), 668–672.

Paul, S. K., & Chowdhury, P. (2020). A production recovery plan in manufacturing supply chains for a high-demand item during COVID-19. *International Journal of Physical Distribution & Logistics Management*. https://doi.org/10.1108/IJPDLM-04-2020-0127

Qi, B., Shen, Y., & Xu, T. (2023). An artificial-intelligence-enabled sustainable supply chain model for B2C e-commerce business in international trade. *Technological Forecasting and Social Change, 191*, Article 122491.

Queiroz, M. M., Ivanov, D., Dolgui, A., & Fosso Wamba, S. (2020). Impacts of epidemic outbreaks on supply chains: Mapping a research agenda amid the COVID-19 pandemic through a structured literature review. *Annals of Operations Research*. https://doi.org/10.1007/s10479-020-03685-7

Rashid, A., Siddique, M., Hussain, S., Shams, S., & Ali, S. (2024). Big data analytics–artificial intelligence and sustainable performance through green supply chain practices in manufacturing firms of a developing country. *Journal of Science and Technology Policy Management*. [Preprint].

Rodríguez-Espíndola, O., Albores, P., Brewster, C., & Zobel, C. (2020). The potential of emergent disruptive technologies for humanitarian supply chains: The integration of blockchain, artificial intelligence and 3D printing. *International Journal of Production Research, 58*(15), 4610–4630.

Shamout, M., Abualigah, L., Abd Elaziz, M., Sumari, P., & Shambour, Q. (2022). A conceptual model for the adoption of autonomous robots in supply chain and logistics industry. *Uncertain Supply Chain Management, 10*(2), 577–592.

Singh, R. K., Modgil, S., & Acharya, P. (2020). Identification and causal assessment of supply chain flexibility. *Benchmarking: An International Journal*, *27*(2), 517–549.

Singh, R. K., Modgil, S., & Acharya, P. (2020a). A template-based approach to measure supply chain flexibility: A case study of an Indian soap manufacturing firm. *Measuring Business Excellence, 24*(2), 161–181.

Sundarakani, B., Pereira, V., & Ishizaka, A. (2020). Robust facility location decisions for resilient sustainable supply chain performance in the face of disruptions. *International Journal of Logistics Management*. https://doi.org/10.1108/IJLM-12-2019-0333

Teerasoponpong, S., & Sugunnasil, P. (2022). Review on artificial intelligence applications in manufacturing industrial supply chain–Industry 4.0's perspective. In *2022 Joint international conference on digital arts, media and technology with ECTI northern section conference on electrical, electronics, computer and telecommunications engineering (ECTI DAMT & NCON)* (pp. 406–411). IEEE.

Treleaven, P., & Batrinca, B. (2017). Algorithmic regulation: Automating financial compliance monitoring and regulation using AI and blockchain. *Journal of Financial Transformation, 45*, 14–21.

Van Hoek, R. (2020). Research opportunities for a more resilient post-COVID-19 supply chain—closing the gap between research findings and industry practice. *International Journal of Operations and Production Management, 40*(4), 341–355.

Wamba, S. F., Bawack, R. E., Guthrie, C., Queiroz, M. M., & Carillo, K. D. A. (2020). Are we preparing for a good AI society? A bibliometric review and research agenda. *Technological Forecasting and Social Change*. https://doi.org/10.1016/j.techfore.2020.120482

Wamba, S. F., Dubey, R., Gunasekaran, A., & Akter, S. (2020c). The performance effects of big data analytics capability on supply chain ambidexterity: The moderating effect of environmental dynamism. *International Journal of Production Economics, 222*, Article 107498.

Wamba, S. F., Queiroz, M. M., Wu, L., & Sivarajah, U. (2020a). Big data analytics-enabled sensing capability and organizational outcomes: Assessing the mediating effects of business analytics culture. *Annals of Operations Research*. https://doi.org/10.1007/s10479-020-03812-4

Wen, J., He, L., & Zhu, F. (2018). Swarm robotics control and communications: Imminent challenges for next-generation smart logistics. *IEEE Communications Magazine, 56*(7), 102–107.

Xu, L., Mak, S., & Brintrup, A. (2021). Will bots take over the supply chain? Revisiting agent-based supply chain automation. *International Journal of Production Economics, 241*, Article 108279.

Structured Extraction and Hierarchical Parsing of Addresses Using Large Language Models

Ashish Dhansukhbhai Desai, Keyvan Narimani, Jorge Marx Gómez, and André Klümpen

Abstract In the era of digitalization and cross-border trade, address inconsistencies present a critical challenge across sectors such as taxation, fraud prevention, law enforcement, and public service delivery. Unstructured and inconsistent address formats-characterized by spelling errors, missing postal codes, incorrect or absent city and street names, and improper punctuation-hinder the effectiveness of automated systems. These issues are particularly detrimental to efforts aimed at combating tax fraud in an European region or international context, where precise address data is essential for accurate entity verification, transaction tracking, and fraud detection. Despite initiatives by tax authorities, postal agencies, and government bodies to enforce standardized address formats, real-world data continues to exhibit substantial variability and noise. Traditional rule-based solutions fall short in addressing the complexity of such data, while existing AI-based methods are often not tailored to the nuances of European address formats. In this research, we propose an approach to transform unstructured address data into structured formats using pre-trained large language models (LLMs), with a focus on the European region. Our experimental results demonstrate that the LLM-based approach significantly outperforms conventional methods in generating structured addresses, and these structured outputs have shown strong performance when evaluated using our rule-based script. This hybrid solution offers a promising tool in the broader effort to ensure regulatory compliance and enforce tax authorities in the fulfillment of their legal mandate against illegal activities.

Keywords Complexity of unstructured addresses · Large language models for structured addresses · Structural address model · Address analysis · Structured extraction of the data · Hierarchical parsing of the data · Combating fraud

A. D. Desai (✉) · J. Marx Gómez
Carl von Ossietzky Universität Oldenburg, Oldenburg, Germany
e-mail: ashish.dhansukhbhai.desai@uol.de

K. Narimani · A. Klümpen
Landesamt für Steuern Niedersachsen, Oldenburg, Germany
e-mail: keyvan.narimani@st.lst.niedersachsen.de

J. Marx Gómez et al. (eds.), *Recent Advances in Information Systems*, Progress in IS,
https://doi.org/10.1007/978-3-032-14099-9_63

1 Introduction

Today, a variety of intelligent software solutions exist to address issues related to address parsing and unstructured address data. Most of these tools are either rule-based systems or rely on hard-coded scripts to handle the various challenges associated with address interpretation. However, the problem of unstructured addresses extends beyond the domain of e-commerce or digital applications, it is a global issue characterized by a range of complexities. One of the primary challenges is that each country has its own rules and administrative structures for defining regions, states, and cities, each associated with specific postal codes. These geographical and cultural differences, combined with varying levels of literacy, result in diverse types of address inconsistencies, but a standard formatted address could resolve these issues and make it easy to get the correct location through geoservice tools or APIs.

Due to data privacy concerns, here, we have mentioned examples of imperfect European addresses paired with their standardized versions are presented in Table 1. These cases reflect typical formatting errors, missing components, incorrect spacing, lack of diacritics, and abbreviation issues, all of which complicate parsing and geolocation. These problems are discussed in depth in next section.

Few examples of our dataset are mentioned in Table 2, but here we changed few details to avoid data privacy concerns:

While Google Maps is one of the most advanced and accurate mapping platforms, capable of parsing addresses from multiple perspectives-including place ID, street name with postal code, and city name but its usage is restricted. Due to stringent data privacy policies and licensing requirements, access to Google Maps' APIs is limited and typically paid. Furthermore, in countries like Germany, government agencies impose strict data protection regulations that prohibit the use of third-party APIs. This restriction influenced our methodological choices and is a key reason why such APIs were not employed in our experiments. Therefore, we used OpenStreetMap

Table 1 Examples of imperfect and standardized European addresses

Imperfect address	Standardized address
Mr. John Doe, c/o 12 Elm St, SW1A 1 AA, London	c/o John Doe, 12 Elm Street, London, SW1A 1AA, United Kingdom
45 Rue de Paris, 750 01, Paris, FR	45 Rue de Paris, 75001 Paris, France
Via Roma 10, 00185 Roma RM, IT	Via Roma 10, 00185 Roma RM, Italy
Bahnhofstr. 00078, 8001 Zürich, CH	Bahnhofstrasse 78, 8001 Zürich, Switzerland
5 High St, 101 15 Berlin, DE	High Street 5, 10115 Berlin, Germany
9 Plaza Mayor, 28012 Madrid, ES	Plaza Mayor 9, 28012 Madrid, Spain
Apt 4B, 22 Königstr, 70173 Stuttgart, DE	Apt 4B, Königstraße 22, 70173 Stuttgart, Germany
24356 KB, 7 Green Ln, Amsterdam, NL	7 Green Lane, 24356KB Amsterdam, Netherlands

Table 2 Examples of imperfect European addresses with index numbers

Index	Imperfect Address of our dataset
1	3 rue Marbeau 260 Marbeau 260 16 Pariseodoro camino num.
2	planta ent puerta a edificio c entro 02001 albacete albacete (espana)
3	novy svet 11/12 praha 1—hradcany 152305 praha 512
4	tragones pr. 98 kauno m. kauno m. sav.
5	rev caset 00103 98652 szigetszentmiklos

API (OSM) to cross-verify the results. Although OSM is an open-source alternative, it is comparatively less capable than Google Maps in terms of address parsing, data accuracy, and real-time updates.

We did not find any paper from any European country or focused on the European region that specifically addresses address parsing and structuring. Only a few papers exist on this topic, and those are also country-specific; for example, one of them is from China Huang et al. (2024). In this paper, our main objective is to develop an ideal system that can handle all European addresses with the highest possible accuracy. Our approach specifically targets multilingual addresses and various address parsing challenges across Europe.

To achieve the target, our experiments were conducted using two medium-sized but highly effective large language models: Mistral-Small-24B-Instruct-2501[1] and Mistral-Nemo -Instruct-2407.[2] Among these, the Mistral-Nemo-Instruct-2407 model demonstrated superior performance for parsing addresses correctly compared to the use of the OSM API only. Specifically, it was able to produce a higher number of accurate or nearby[3] geolocations and provided more detailed interpretations (latitude and longitude). Detailed results and evaluations are presented in Sect. 5.

This paper is structured as follows: Sect. 2 provides a brief overview of the complexities and uncertainties commonly found in address data. Section 3 presents related work. Section 4 describes the experimental setup and provides detailed information about the dataset used. Section 5 summarizes the experimental results and discusses their practical implications, along with potential directions for future research.

[1] https://huggingface.co/mistralai/Mistral-Small-24B-Instruct-2501.

[2] https://huggingface.co/mistralai/Mistral-Nemo-Instruct-2407.

[3] In this paper, nearby means either in the same street or an easily reachable location, within a few miles or a few minutes away in the same postal area.

2 Complexity and Uncertainty of Addresses

It is nearly impossible to quantify the quality of user-submitted addresses, as their structure and content vary widely and unpredictably. Instead of focusing on measuring address quality, our research prioritizes accurately deriving geolocation coordinates, which holds greater practical value for future applications. To achieve this, we target highly complex address structures containing ambiguous and diverse components.

Real-world addresses often include more than just basic elements (street name, house number, postal code, city, and country). Additional components such as personal names, house names, landmarks, nearby locations, and third-party identifiers like "care of" ("c/o") are frequently present. Users also apply a variety of abbreviations depending on country and language—for instance, "str." for "Straße" in Germany, Austria, and Switzerland; "PLZ" for "Postleitzahl" (postal code); "EU" for European Union; and country codes like "DE" (Germany), "FR" (France), and "IT" (Italy). An official list of these codes can be found on the European Union website.[4]

Structural variations across countries add further complexity: in some regions, the house number precedes the street name, while in others it follows. Language diversity compounds these challenges through unique characters, diacritics, and punctuation. For example, in Czech addresses, "Novy Svet" should be properly written as "Nový Svět" to ensure accurate geolocation—missing diacritics often cause failures in OpenStreetMap (OSM) and sometimes even in Google Maps.

Our dataset, sourced from European governmental authorities, reflects these challenges. Particularly in rural areas under larger municipalities, addresses may include "care of" names, personal references, or nearby major cities. Additional observed issues include improper spacing in postal codes (e.g., "24356 KB" vs. "24356KB"), omission of language-specific characters, and leading zeros in house numbers.[5] These factors contribute to failures in address parsing and downstream geolocation.

The methodology proposed in this paper addresses these complexities using large language models (LLMs), enabling more reliable parsing and accurate geolocation extraction from highly variable, unstructured European address data.

2.1 Our Dataset

Our dataset contains most of the above-mentioned uncertainties, some of which are listed in Table 2. In addition to handling address uncertainties, we also focused on selecting an equal number of samples from each available European country. We selected a total of 2000 addresses from our database. These addresses were

[4] https://ec.europa.eu/eurostat/statistics-explained/index.php?title=Glossary:Country_codes.

[5] Example 5 in Table 2.

then manually verified on Google and Google Maps, using both the unstructured addresses and other available details from our database by government authority. The verified original and structured addresses were saved in the same CSV file under a new column called "original addresses".

As a result, out of the 2000 records, we successfully verified 1632 addresses. The remaining 368 addresses could not be accurately identified on Google, even after considering the additional available details. Therefore, we conducted further experiments using only the remaining 1632 records.

You may also notice slightly fewer records in some experiments; these differences indicate cases where we did not obtain proper or any output from the LLMs, making it impossible to parse them through our geoservice. However, this cannot be considered a limitation of the LLMs, as there is a possibility that the input address lacked sufficient information, preventing the LLM from predicting the correct or complete address.

3 Related Work

Several AI-based models and machine learning methods have addressed the challenges of parsing unstructured or abnormal addresses. While most existing works are focused on country-specific scenarios, they provide valuable insights into address parsing and structuring.

AddrLLM, Yang et al. (2024), a supervised fine-tuned language model, was trained on large-scale Chinese logistics data, achieving a 43% reduction in parcel re-routing for JD Logistics. It employs a custom retrieval-augmented generation (RAG) framework to avoid expensive retraining while improving address rewriting accuracy. Sharma et al. (2018) proposed a hybrid approach combining normalization, machine learning-based classification, and rule-based error correction to handle regional address variations. While effective, the system's multi-step architecture limits scalability and full automation.

GeoAgent, Huang et al. (2024), integrates LLMs with geospatial APIs in a three-stage pipeline for standardizing non-standard addresses, leveraging external vector databases for contextual spatial knowledge. The introduction of GeoEntropy loss and GeoRouge score further improved geocoding accuracy. Yin et al. (2023) benchmarked GPT-3 against traditional ML and transformer models using a large, synthetic U.S. address dataset. Although GPT-3 showed promise in few-shot learning for simpler components, Bidirectional LSTM-CRF outperformed all models with an F1-score of 0.9968, while transformer models like BERT, DistilBERT, and RoBERTa also achieved strong results.

Syne (2022) applied NLP and ML for address matching using real-world municipal data, demonstrating that embedding-based methods (BERT, Doc2Vec) outperformed other models via cosine similarity for address standardization. GeoRoBERTa Guermazi et al. (2023) focused on address matching using transformer-based models, combining GeoRoBERTa with techniques like XGBoost, SVM, and Word2Vec.

GeoRoBERTa consistently outperformed its baselines, showcasing the effectiveness of transformer-based address models. Additional studies such as Li et al. (2019) have applied ML or rule-based methods for various niche use cases.

Most existing works target specific regions such as China Yang et al. (2024) and the United States Yin et al. (2023), with limited capability to handle multi-language parsing required for European regions. Hybrid systems like that of Sharma et al. (2018) are also time-consuming compared to modern LLM-based solutions. Our work focuses on resolving structural inconsistencies for European addresses using LLMs integrated with rule-based systems, capable of filling missing components (e.g., country names) and translating addresses into local languages.

In our model selection process, initial tests with ChatGPT 3.5 were unsatisfactory, aligning with findings from Hugging Face benchmarks[6] showing its underperformance in reasoning compared to models like Mistral-Nemo-Base-Model. We then benchmarked models including Mistral-Small-24B-Instruct-2501, Gemma-2-27B,[7] LLaMA 3.3–70B[8] Qwen 2.5–32B Ahmed et al. (2025), and GPT-4o-mini.[9] Mistral-Small-24B-Instruct-2501 consistently outperformed across reasoning, instruction following, math, and coding.

The Mistral-Nemo-Instruct-2407 model[10] showed particularly strong multilingual performance, achieving high scores across several European languages, motivating its inclusion alongside Mistral-Small-24B-Instruct-2501 in our experiments.

4 Methodology

As discussed in the previous section, due to privacy policy we stored the pre-trained open-source models in our secure system and then performed the experiments. Initially, we tested Qwen 2.5–32B by manually providing structured instructions. However, it has been frequently failed to parse the addresses accurately. Consequently, we shifted our focus to the Mistral models, which demonstrated noticeably better results. Based on this initial success, we conducted more extensive experiments using these models on a dataset containing 1,632 address entries.

We employ a large language model (LLM) for structured extraction and hierarchical parsing of address data to enhance the contextual accuracy of our geospatial retrieval system. The approach decomposes addresses into their constituent components and infers missing elements, such as city or country, to provide comprehensive geographic context. As mentioned earlier, due to privacy concerns, the system exclusively relies on the internal knowledge of the LLM, without augmenting it through retrieval-augmented generation (RAG) techniques, such as Google Search

[6] https://huggingface.co/google/gemma-2-27b.

[7] https://huggingface.co/google/gemma-2-27b.

[8] https://ollama.com/library/llama3.3:70b.

[9] https://openai.com/index/gpt-4o-mini-advancing-cost-efficient-intelligence/.

[10] https://mistral.ai/news/mistral-nemo.

or other external data sources. In the first experiment, the process involves a two-stage workflow for address refinement. In the first stage, the LLM is prompted to generate a structured response, incorporating a preliminary reasoning step to engage chain-of-thought processing. This enables the extraction and augmentation of address attributes, which are defined within a structured data model. In the second stage, the LLM reformulates the parsed address to align with the linguistic conventions of the target country, ensuring compliance with local formatting and language standards (e.g., rewriting an English address in French for a French-speaking region). The address components extracted or augmented by the LLM are represented as a comprehensive set of attributes, listed below for clarity:

1. **Reason**: Brief explanation for parsing or augmenting the address.
2. **Street Name**: The name of the street where the address is located.
3. **House Name**: The name of the building or residence, if specified.
4. **House Number**: The numerical identifier of the building on the street.
5. **Postal Code**: The postal or ZIP code associated with the address.
6. **City Name**: The name of the city or town where the address is located.
7. **State or Province**: The regional administrative division of the address.
8. **Country Name**: The name of the country where the address is situated.
9. **Person Name**: Name of the person linked to the address.
10. **Floor Number**: The specific floor or level within a building, if applicable.
11. **Other Information**: Additional address-related notes or details.

Despite the integration of instruction-based large language models (LLMs), certain addresses remain unparsable by geolocation services. This issue often arises due to incomplete or corrupted address inputs such as missing street names or the inclusion of noise terms in place of essential components like street or city names. In such cases, the limited geospatial knowledge of the LLM results in suboptimal outputs. For instance, an address like "ul. 34562KB London" may be processed and returned as "ul. 34562KB London, United Kingdom". While the LLM successfully identifies and appends the country name based on contextual clues, but sometimes it fails to resolve the street name when the proper context is not present.

This behavior demonstrates that although the LLM adheres to instruction-following protocols, it lacks the capability to infer or reconstruct missing or malformed address completely when they are entirely ambiguous or incorrect. As a result, noisy or syntactically invalid address strings often lead to incomplete or unusable structured outputs, ultimately hindering downstream geolocation parsing.

To mitigate this issue upto certain level, we introduced a rule-based script designed for post-processing of such LLM output addresses. This script systematically removes noise terms (e.g., "ul." in the above example), thereby improving the overall quality of the address structure. While it may not recover the exact missing details, it ensures that the address remains within a relevant geographic area, thus increasing the likelihood of successful geolocation.

This enhancement has led to a measurable improvement in response rates: previously unprocessable addresses (yielding zero responses) are now transformed into partially valid outputs, resulting in response counts ranging from 1 to 10. This highlights the importance of complementary rule-based preprocessing in scenarios where LLMs alone are insufficient for robust address normalization.

Algorithm 1 Rule-Based Address Parsing with OSM API

```
Require: Output from LLM
Ensure: Geoservice response or failure notice
Query OSM API with full address
if exactly one result is returned then
    Accept result and exit
end if
Tokenize address into individual words
for k = 1 to 3 do
    Generate all unique combinations by removing k words
    for each address variant do
        Query OSM API
        if at least one result is returned then
            Compute Matching Score: Numberofmatchingwords / Totaloriginalwords
            Select the variant with the highest score
            Accept result and exit
        end if
    end for
end for
Return "No response found"
```

Although hard-coding rules to address these inconsistencies proved to be inefficient and brittle, we developed a more flexible rule-based approach as described below 1. Overall, architecture of our experiment is given in Fig. 1.

The decision to limit word removal to a maximum of three words was empirically derived. Removing more than three words significantly increases the likelihood of yielding an incorrect or geographically distant result, thus reducing the reliability of the parsing process.

Furthermore, we explored a hybrid approach that combines both methods. Two test cases were defined:

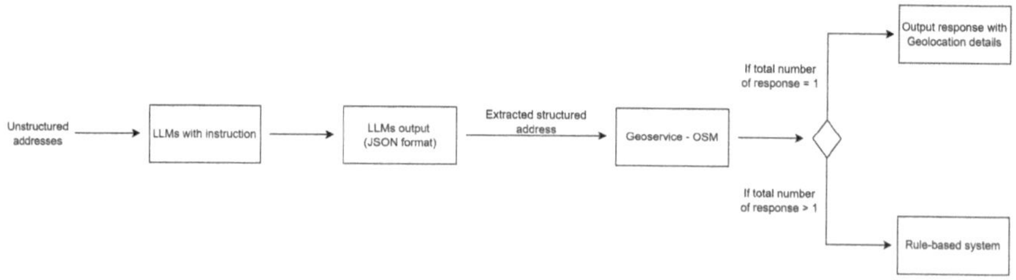

Fig. 1 Overview of the model architecturee

1. **Structured Address Parsing**: Addresses were first structured using Mistral models, then passed to our geoservice API to evaluate the number of valid geolocation responses obtained.
2. **Same but with Additional Localized Translation**: The same process was followed, but with an added instruction to the LLM to translate the structured address into its respective national language, aiming to reduce errors caused by language-specific inconsistencies.

5 Results and Discussion

In the methodology section, we outlined our workflow; for evaluation, we relied on F1-score. The rationale for using the number of geolocation responses as a key evaluation metric stems from a preliminary manual analysis. Prior to the experiments, we reviewed more than hundred of address outputs and verified their corresponding locations on Google Maps-currently considered the most accurate and reliable mapping service for pinpointing real-world locations.

Based on these observations, we defined the following evaluation criteria:

- Addresses yielding 1 or 2 geolocation responses were considered accurate or closely located, and hence likely to be accessible and valid. We also treated these two categories as 1 for both precision and recall.
- Addresses returning 3 to 10 responses typically deviated significantly from the intended location, indicating ambiguity or inaccuracy in the parsing or geocoding process. Therefore, we considered them as 0 for both precision and recall.

The difference in the total number of data across models (in the tables) is due to the number of addresses that the LLMs were unable to process while following the given instructions. We removed those blank output rows before parse the structured addresses through geoservice tool.

We analyze and compare the performance of the instruction-based parsing systems and hybrid model (by combining rule-based system for post-processing) under two different evaluation settings: (i) including all responses, even when the geoservice fails to find a location (i.e., counting 0's), and (ii) excluding such failed cases (i.e., not counting 0's). The inclusion or exclusion of these cases allows us to assess both the robustness of the language model predictions and their actual utility when interfaced with downstream geoservices.

5.1 *Instruction-Based Parsing Results*

In the purely instruction-based systems, the performance varies significantly depending on whether failures to resolve an address are included.

Table 3 Instruction-based parsing with counting 0's in total

	Addresses	Mistral-small-24B-instruct-2501	Mistral-nemo-instruct-2407
Overall total	1632	1571	1614
F1-score (%)	23.04	34.50	33.58

Table 4 Instruction-based parsing without counting 0's in total

	Addresses	Mistral-small-24B-instruct-2501	Mistral-nemo-instruct-2407
Number of empty responses	766	452	468
Remain total number of addresses	866	1119	1146
F1-score (%)	43.41	48.44	47.29
Overall total		1632	

When counting all responses, including the geoservice failures (Table 3), both models show modest F1-Scores: Mistral-Small-24B-Instruct-2501 achieves 34.50% and Mistral-Nemo-Instruct-2407 reaches 33.58%. This relatively low score suggests that a substantial number of model outputs either contain incorrect address information or are formatted in a way that the geoservice cannot process, leading to unresolved locations.

However, when excluding the cases where the geoservice could not find a match (Table 4), we observe a considerable improvement. The F1-Scores increase to 48.44% for Mistral-Small-24B-Instruct-2501 and 47.29% for Mistral-Nemo-Instruct-2407. This highlights that while the models are capable of generating reasonably accurate addresses, a significant proportion of outputs still suffer from either generation errors or incompatibilities with the geoservice's API either due to lack of database update or missing inputs. The high number of empty responses 452 and 468 for the two models, respectively—further emphasizes the challenges inherent in pure instruction-based generation.

5.2 Hybrid Model Results

The rule-based augmentation with instruction-based model shows a clear and substantial improvement across both evaluation settings.

When excluding the geoservice failures (Table 5), both models achieve dramatically higher F1-Scores: 63.84% for Mistral-Small-24B-Instruct-2501 and 65.80% for Mistral-Nemo-Instruct-2407. Notably, the augmentation process eliminates empty responses entirely, suggesting that the rule-based system successfully reformats or

Table 5 F1-score comparison without counting 0's in Total

	Addresses	Mistral-small-24B-instruct-2501	Mistral-nemo-instruct-2407
Number of empty responses	766	0	0
Remain total number of addresses	866	1571	1614
F1-score (%)	43.41	63.84	65.80
Overall total	1632		

Table 6 F1-score comparison with counting 0's in total

	Addresses	Mistral-small-24B-instruct-2501	Mistral-nemo-instruct-2407
Number of empty responses	766	0	0
Overall total	1632	1571	1614
F1-score (%)	23.04	63.84	65.80

Table 7 Address rewriting—enhancement of instruction-based system

	Original addresses	Rewritten (instruction-based model)	Rewritten (hybrid model)
Total data	1632	1614	1614
F1-score (%)	38.36	35.33	59.70

supplements the LLM outputs in a way that makes them much more compatible with the geoservice. This demonstrates the strength of incorporating domain-specific knowledge and post-processing rules to enhance the raw LLM outputs. Even when including all responses (Table 6), the improvement remains consistent: both models maintain the same F1-scores as when excluding 0's (63.84 and 65.80%, respectively). This is expected, given that the rule-based augmentation was able to fully eliminate unresolved cases, showing its effectiveness in reducing outright failures.

Overall, these results underscore the necessity of hybrid systems that combine LLM capabilities with rule-based refinement to achieve robust, high-quality outputs in address parsing tasks. After evaluating the instruction-based system and the hybrid approach, we further examined the problem and identified one of the key issues: the language of the address. Therefore, we decided to rewrite the addresses in their national languages by translating them, in order to avoid the loss of original language characters or other language-related barriers. In the third evaluation (Table 7), the focus was on refining the model outputs by rewriting the instructions to include translation into the native language, thereby simulating localization. This evaluation compared two configurations using the same model, Mistral-Nemo-Instruct-2407: one using instruction-based parsing alone, and the other using the hybrid model.

The results show that the acceptable response rate (number of responses = 1 or 2 are acceptable) for the instruction-only rewriting setup was 35.33%, while the hybrid approach—which combines translation, enhanced prompt design, and rule-based formatting—achieved a significantly higher rate of 59.70%. This corresponds to an absolute improvement of approximately 24.37%. Notably, even the original unprocessed addresses had an acceptable rate of 38.36%, indicating that the instruction-based rewriting alone slightly underperformed compared to the raw data. However, the hybrid method substantially outperformed both, demonstrating the importance of combining linguistic localization (through translation) and post-processing rules to better align the model outputs with the expectations of downstream geoservices such as OSM. These findings emphasize that improving instruction clarity, incorporating language adaptation, and applying domain-specific formatting can have as much impact on real-world performance as changes to model architecture or scale.

Thus, overall, even after translation, the hybrid approach performed well. However, the numbers indicate that rewriting may not have a very large impact. Nevertheless, it might still be effective and worthwhile to include in our instructions for the model.

6 Conclusion and Future Work

This study addressed the persistent challenge of parsing unstructured and inconsistent European address data using large language models (LLMs). Traditional rule-based and early AI systems struggle with linguistic and structural variability. Our experiments with Mistral-Small-24B-Instruct-2501 and Mistral-Nemo-Instruct-2407 showed that instruction-based prompting moderately improved address structuring and geolocation readiness. However, instruction alone was insufficient. By integrating rule-based post-processing and localized translation, geolocation accuracy improved to 65.84%, with address acceptance rates increasing over 20% after translation into native languages. These results highlight the complexity of managing address uncertainties.

A key limitation was strict EU and German data privacy regulations (EU Regulation 2016/679), which prohibited the use of third-party APIs. Consequently, we avoided external models or services requiring API calls, emphasizing the need for privacy-preserving, locally executable AI solutions for public sector applications. Despite improvements, single geolocation responses do not always guarantee accuracy, as geoservices may return zero, multiple, or misleading locations. Additionally, data security constraints limited us to single geoservice usage and internal LLM knowledge.

6.1 Future Work

Future research will focus on refining prompts, testing larger open-source models (e.g., DeepSeekAI), and expanding datasets from authoritative sources to better reflect real-world address variations. This will support fine-tuning LLMs for address parsing with improved noise removal and language normalization.

We also aim to develop a region-specific GeoAgent model for Europe, following successful approaches like the 43% parcel re-routing reduction in JD Logistics Huang et al. (2024). Alternatively, supervised LLMs can address global parsing, though collecting high-quality, real-time international data remains challenging. Using multiple geoservice APIs could further enhance accuracy, and more robust evaluations, such as calculating distance errors via the Haversine formula, will be implemented to better assess model precision.

References

Ahmed, I., Islam, S., Datta, P. P., Kabir, I., Chowdhury, N. U. R., & Haque, A. (2025). *Qwen 2.5: A comprehensive review of the leading resource-efficient llm with potentioal to surpass all competitors*. Authorea Preprints

Guermazi, Y., Sellami, S., & Boucelma, O. (2023). Georoberta: A transformer-based approach for semantic address matching. In: *7th international workshop on data analytics solutions for real-life applications (DARLI-AP)*.

Huang, C., Chen, S., Li, Z., Qu, J., Xiao, Y., Liu, J., & Chen, Z. (2024). Geoagent: To empower llms using geospatial tools for address standardization. In *Findings of the association for computational Linguistics ACL 2024* (pp. 6048–6063).

Li, H., Lu, W., Xie, P., & Li, L. (2019). Neural chinese address parsing. In: *Proceedings of the 2019 conference of the North American chapter of the association for computational linguistics: Human language technologies*, vol. 1 (Long and Short Papers) (pp. 3421–3431).

Sharma, S., Ratti, R., Arora, I., Solanki, A., & Bhatt, G. (2018). Automated parsing of geographical addresses: A multilayer feedforward neural network based approach. In *2018 IEEE 12th International Conference on Semantic Computing (ICSC)* (pp. 123–130). IEEE.

Syne, L. (2022). *Machine learning and nlp approaches in address matching*.

Yang, Q., Hong, Z., Cao, D., Wang, H., Xie, Z., He, T., et al. (2024). *Addrllm: Address rewriting via large language model on nationwide logistics data*. arXiv preprint arXiv:2411.13584.

Yin, Z., Li, D., & Goldberg, D. W. (2023). Is chatgpt a game changer for geocoding-a benchmark for geocoding address parsing techniques. In *Proceedings of the 2nd ACM SIGSPATIAL international workshop on searching and mining large collections of geospatial data* (pp. 1–8).

FSC
www.fsc.org
MIX
Papier | Fördert
gute Waldnutzung
FSC® C083411

Kolibri 360 GmbH
Ferdinand-Jühlke-Straße 7
99095 Erfurt, Deutschland
produktsicherheit@kolibri360.de